LADY MARGARET HALL LIBRARY
1975
OXFORD

302209046Q

# Igneous Petrology

McGRAW-HILL INTERNATIONAL SERIES IN THE EARTH AND PLANETARY SCIENCES



AGER Principles of Paleoecology
BERNER Principles of Chemical Sedimentology
BROECKER AND OVERSBY Chemical Equilibria in the Earth
CARMICHAEL, TURNER, AND VERHOOGEN Igneous Petrology
DE SITTER Structural Geology
DOMENICO Concepts and Models in Groundwater Hydrology
EWING, JARDETZKY, AND PRESS Elastic Waves in Layered Media
GRANT AND WEST Interpretation Theory in Applied Geophysics
GRIM Applied Clay Mineralogy
GRIM Clay Mineralogy
HOWELL Introduction to Geophysics
HYNDMAN Petrology of Igneous and Metamorphic Rocks
JACOBS, RUSSELL, AND WILSON Physics and Geology
KRAUSKOPF Introduction to Geochemistry
KRUMBEIN AND GRAYBILL An Introduction to Statistical Models in Geology
LEGGET Geology and Engineering
MENARD Marine Geology of the Pacific
MILLER Photogeology
OFFICER Introduction to the Theory of Sound Transmission
RAMSAY Folding and Fracturing of Rocks
ROBERTSON The Nature of the Solid Earth
SHROCK AND TWENHOFEL Principles of Invertebrate Paleontology
STANTON Ore Petrology
TOLSTOY Wave Propagation
TURNER Metamorphic Petrology
TURNER AND VERHOOGEN Igneous and Metamorphic Petrology
TURNER AND WEISS Structural Analysis of Metamorphic Tectonites

# Igneous Petrology

IAN S. E. CARMICHAEL    FRANCIS J. TURNER
JOHN VERHOOGEN

*Department of Geology and Geophysics*
*University of California, Berkeley*

**McGraw-Hill Book Company**

*New York   St. Louis   San Francisco   Düsseldorf   Johannesburg   Kuala Lumpur*
*London   Mexico   Montreal   New Delhi   Panama*
*Paris   São Paulo   Singapore   Sydney   Tokyo   Toronto*

**Library of Congress Cataloging in Publication Data**

Carmichael, Ian S E
Igneous petrology.

(McGraw-Hill international series in the earth and planetary sciences)
"Replaces and greatly amplifies much of the content of the first fifteen chapters of . . . Igneous and metamorphic petrology [by F. J. Turner and J. Verhoogen] (McGraw-Hill, 1960)"
1. Rocks, Igneous. I. Turner, Francis J., joint author. II. Verhoogen, John, date, joint author. III. Title.
QE461.C37 552'.1 74-5093
ISBN 0-07-009987-1

**Igneous Petrology**



234567890 MAMM 7987654

*This book was set in Times Roman. The editors were Robert H. Summersgill and M. E. Margolies; the designer was Janet Durey Bollow; the production supervisor was Sam Ratkewitch. The drawings were done by Innographics.*
*The Maple Press Company was printer and binder.*

# Contents

# Preface

This book replaces and greatly amplifies much of the content of the first fifteen chapters of our *Igneous and Metamorphic Petrology* (McGraw-Hill, 1960). That text has been completely rewritten and entirely new illustrations, more than twice the original number, have been provided. The new book seeks to portray the greatly expanded knowledge of the chemistry, mutual relations, and geologic setting of igneous rocks and to present the changing emphasis of research over the past two decades. Like its predecessor it is intended as a general text for advanced students, research workers, and teachers in geology. It presumes general familiarity with igneous petrography and with appropriate aspects of background thermodynamics as set out in Chapter 2 of our earlier book. The book falls into two sections, each of which could serve as the basis for a one-semester course in petrology, the one emphasizing geochemistry, the other stressing the mineralogical diversity of igneous rocks against a global tectonic background.

The first seven chapters cover in a variety of ways the nature of magma and the crystallization processes that ultimately yield igneous rocks. Igneous petrology has long since reached a stage where the application of thermodynamics is profitable if not essential. Thus classification of igneous rocks is treated in terms of activities of principal chemical components. These are used also to estimate temperatures and pressures of igneous mineral assemblages and to evaluate behavior of the volatile components. Other topics treated at length are properties of liquids and phenomena of diffusion and crystallization—fields commonly woefully neglected—and the use of mineral geothermometers and geobarometers. Reversing a conventional approach, we treat evolutionary trends of crystallizing magmas primarily in terms of the now voluminous data of mineral chemistry. Conclusions so reached are tested in the light of experimental data on silicate systems. It is no longer sufficient to interpret magmatic behavior entirely or even principally in terms of simple systems of few components. If there is a general theme that runs through the first part of our book, it is the compelling need and growing opportunity to strengthen petrology as a science through rigorous test and prediction rooted in thermodynamics.

Overtones from the same theme should be perceptible through the second half, which presents the descriptive side of igneous petrology—rock associations, each conforming only in a broad sense to pattern, viewed in the setting of modern tectonics. Even more than in our earlier book, we seek to display the almost infinite variety of igneous rock series by treating individual case histories and referring to well-documented accounts of others. Inevitably this approach reveals the waning power of once-popular universal petrologic patterns (e.g., magma types) and of all-embracing, experimentally founded rival models of petrogenesis, each still fervently espoused by its advocates.

High-pressure experiment and the growing impact of geophysics on petrology, with the stimulus of plate tectonics never far removed, have combined to lead petrologic thinking more and more to problems (e.g., relating to site and source materials) of magma formation. Isotopic and trace-element chemistry enter significantly into modern discussion of such topics. The natural conclusion to the first section of the book (Chapter 7) is a review of constraints placed by geophysics and terrestrial chemistry on possible sources and situations for generation of magmas in the mantle and the crust. The second section returns full cycle to the same topic; this time we examine it in the light of other constraints imposed by the magmas themselves.

To limit total coverage to reasonable brevity, we have compromised in several ways. Some descriptive aspects still adequately covered in *Igneous and Metamorphic Petrology* are treated here in condensed form, with references to specific passages in the earlier book. Other important topics (e.g., pegmatites, magmatic aspects of ore deposition, and lunar petrology) are largely beyond the field of our competence and so have been omitted altogether. For literature prior to 1958 the reader is referred to appropriate sections of the earlier book. Here we have concentrated on newer works published in English in readily accessible journals. Unavoidably this policy fails to do justice to work of equal or greater merit published in less accessible journals or in languages which unfortunately are unfamiliar to most English-speaking readers.

It is a pleasure, in conclusion, to acknowledge gratefully special assistance from a few of the many whose ideas and help have contributed to what we have written: Professor Bert Nordlie provided unpublished data on magmatic gas of Kilauea; Dr. S. Moorbath and Dr. K. O'Nions have been free with advice and information in the fields of isotope and rare-earth chemistry; a group of young colleagues in Berkeley in the course of seminars critically and helpfully reviewed early drafts of the manuscript; among past and present graduate students whose ideas likewise have helped to shape our book we mention especially Professor J. W. Nicholls; and for sympathetic patience in the laborious task of typing the manuscript we owe much to Fanchon Lewis and Dione Carmichael.

IAN S. E. CARMICHAEL
FRANCIS J. TURNER
JOHN VERHOOGEN

# List of Symbols

## Frequently Used Symbols

FUNDAMENTAL CONSTANTS

| | |
|---|---|
| $R$ | gas constant 1.9872 cal mole$^{-1}$ deg$^{-1}$ |
| $h$ | Planck's constant $6.62 \times 10^{-27}$ erg s$^{-1}$ |
| $k$ | Boltzmann's constant $1.3805 \times 10^{-16}$ erg deg$^{-1}$ |
| $N_o$ | Avogadro's number $6.026 \times 10^{23}$ mole$^{-1}$ |

PHYSICAL UNITS[1]

| | |
|---|---|
| cm | centimeter |
| m | meter |
| km | kilometer |
| g | gram |
| kg | kilogram |
| kb | kilobar |
| s | second |
| min | minute |
| m.y. | $10^6$ years |
| b.y. | $10^9$ years |
| cal | calorie (= 4.184 joules = 41.84 cm$^3$ bar) |
| mole$^{-1}$, deg$^{-1}$ | per mole, per degree |
| ppm | parts per million (by weight) |

THERMODYNAMIC QUANTITIES AND PROPERTIES

| | |
|---|---|
| $a$ | activity |
| $C_p$ | heat capacity at constant pressure (cal mole$^{-1}$ deg$^{-1}$) |
| $C_v$ | heat capacity at constant volume (cal mole$^{-1}$ deg$^{-1}$) |
| $c$ | concentration (number cm$^{-3}$) |
| $E$ | internal energy (cal mole$^{-1}$) |
| $f$ | fugacity |
| $G$ | Gibbs free energy (cal mole$^{-1}$ deg$^{-1}$) |
| $H$ | enthalpy (cal mole$^{-1}$ deg$^{-1}$) |
| $K$ | equilibrium constant |
| $P$ | pressure (bar) |
| $p$ | partial pressure of a gas component (bar) |
| $S$ | entropy (cal mole$^{-1}$ deg$^{-1}$) |
| $T$ | absolute temperature (degrees Kelvin) |
| $V$ | volume (cm$^3$; or cal bar$^{-1}$) |
| $X$, $Y$ | molar fractions |
| $\alpha$ | coefficient of thermal expansion (deg$^{-1}$) |
| $\beta$ | compressibility (bar$^{-1}$) |
| $\gamma$ | activity coefficient or fugacity coefficient |
| $\delta$ | isotope fractionation factor (for example, $\delta_{O^{18}/O^{16}}$) |
| $\Delta$ | finite change in any quantity; for example, $\Delta V$ is the change in volume that occurs in a reaction or transformation |
| $\eta$ | viscosity (poise: g cm$^{-1}$ s$^{-1}$) |
| $\mu$ | chemical potential (cal mole$^{-1}$) |
| $\rho$ | density (g cm$^{-3}$) |
| $\tau$ | surface tension or surface energy (dyne cm$^{-1}$ or erg cm$^{-2}$). |

### Subscripts and Superscripts

| | |
|---|---|
| $\alpha$ | as a superscript, designates a phase. |
| $i$ | as a subscript, designates a component; thus $\mu_i^\alpha$ is the chemical potential of component $i$ in phase $\alpha$; $p_i$ = partial pressure of $i$ in gas. |
| $P$, $T$ | as subscripts, indicate fixed, constant pressure or temperature; thus $V_{P,T}$ is the volume at pressure $P$ and temperature $T$; $\left(\frac{\partial V}{\partial P}\right)_T$ is the partial derivative of volume with respect to pressure at constant temperature. The notation $V_{1;298^\circ}$ means volume at 1 bar and 298.15°K. |
| $E$ | as a subscript, indicates equilibrium; thus $T_E$ is the equilibrium temperature for a reaction. |
| $m$ | as a subscript, indicates melting temperature; $\Delta S_m$ is the entropy of fusion; $\Delta V_m$ is the change in volume on melting. |
| $r$ | as a subscript, indicates reaction; $\Delta S_r$ = reaction entropy. |

| | |
|---|---|
| $^\circ$ | as a superscript, indicates standard state; thus $\Delta G^\circ$ is the free-energy change of a reaction when all reactants and products are in their standard states. |
| $^-$ | overbar indicates a partial molar quantity; thus $\bar{S}_i^\alpha$ is the partial molar entropy of component $i$ in phase $\alpha$. |
| liq, sol, soln, ss | as superscripts mean, respectively, in a liquid phase, in a solid phase, in solution, and in solid solution. Phase names as superscripts relate the quantity in question to that phase; for example, $a_{S_2}^{\text{pyrite}}$ is activity of sulfur in pyrite. |
| $\Delta J^\circ$ | change in the quantity $J$ (for example, $H$, $S$, $G$) for 1 mole in the standard states. |
| $\Delta J_f^\circ$ | change in $J$ accompanying formation of 1 mole from elements, all in their standard state. |
| $\Delta J_{\text{mix}}$ | change in $J$ upon mixing of 1 mole of solution. |
| $\Delta J_{\text{mix}}^{\text{ideal}}$ | change in $J$ upon mixing 1 mole of an ideal solution. |
| $\Delta J_{\text{mix}}^{\text{xs}}$ | excess change in $J$ upon mixing 1 mole of solution. |
| $\Delta J_m$ | change in $J$ upon melting. |
| $\Delta J_{\text{vap}}$ | change in $J$ upon vaporization. |

## LOGARITHMS

| | |
|---|---|
| log | common logarithm (base 10) |
| ln | natural logarithm (base 2.71828); $\ln x = 2.303 \log x$ |
| $e$ | exponential (2.71828) |

[1] *Note on Système international d'unités (SI units)*

The basic SI units are the meter (m), kilogram (kg), second (s), kelvin (K), and mole (mol). Some of the derived units with their names are given below:

Energy, the joule, J; kg $\text{m}^2\ \text{s}^{-2}$

Force, the newton, N; kg m $\text{s}^{-2} = \text{Jm}^{-1}$

Pressure, the pascal, Pa; kg $\text{m}^{-1}\ \text{s}^{-2} = \text{Nm}^{-2}$

Almost all tables of constants and thermodynamic data of use to geologists are given in cgs units rather than in the new kg m s (SI) units. We have continued to use the cgs system in this book, especially since tables of thermodynamic data are given in calories and kilocalories rather than in J or kJ, where by definition 1 thermochemical calorie = 4.184 J. As an example of the SI system, 1 bar = 101.972 KPa; and the heat capacity of one kg formula weight would be given in units of J $\text{kmol}^{-1}\ \text{K}^{-1}$ rather than cal $\text{mole}^{-1}\ \text{deg}^{-1}$ as in this book.

# Igneous Petrology

# Magma and Igneous Rocks

# 1

## Magma: Nature, Cooling Behavior, Properties, and Chemistry

### NATURE AND CONSTITUTION

The most abundant material erupted from volcanoes is some kind of silicate melt. At the source of eruption the melt commonly encloses suspended crystals, and there may be significant quantities of gas trapped as bubbles. Largely molten heterogeneous material of this kind is called magma. It consolidates on the surface as lava flows or as extrusions of one type or another; or it may be ejected explosively into the atmosphere to fall as fragments of crystals or glass that then become the main component of pyroclastic rocks. In the depths of the earth beneath centers of volcanism there exist, presumably, large bodies of magma that may freeze completely without being erupted at the surface. There are other magmas of totally different composition—carbonate, sulfide, or iron oxide. These are by and large extremely rare, so that little confusion arises if magma is taken to refer to a melt of silicate composition, the appropriate adjective being added for the three rare types.

Magmas erupted on the surface of the earth are hot and often dangerously explosive and mobile, so geologists have generally been unable to sample the material. But over the years, volcanologists have succeeded in measuring lava temperatures, estimating lava viscosity, and recording the changing composition of magma as the eruptive cycle of a volcano progresses. Studies of this type, particularly those of the Hawaiian Volcano Observatory, have produced new information on the dilation of a volcano before eruption, on magma rheology, and on the thermal regime of a body of lava such as a flow or a lake. These Hawaiian data are necessarily limited in that they concern but one of many kinds of magma and one of many eruptive patterns during a minute fraction

($10^{-4}$ to $10^{-5}$) of the life cycle of a volcano. But it is this type of information that provides much of the background for this broad discussion of magma, and we reserve for later chapters a more detailed treatment of such topics as the volatile components of a magma or the properties of silicate liquids.

The Hawaiian studies on basaltic magma itself are also the natural outcome of more than 70 years' experimental investigation of silicate systems and metallurgical slags. Of the early earth scientists who saw in these systems the means to unravel the complexities of crystallization of igneous rocks (summarized in Vogt, 1921–1923), J. H. L. Vogt was one of the most influential and successful, and many of his ideas were incorporated in the classic *Natural History of Igneous Rocks* (Harker, 1909, pp. 174–179). But the most illuminating insight into the origin of igneous rocks was that of N. L. Bowen, who blended decades of experimental investigation at the Geophysical Laboratory of the Carnegie Institute of Washington with data on the rocks themselves; his *Evolution of the Igneous Rocks* (1928) still remains the most influential exposition of igneous petrogenesis.

In the last century knowledge of the properties and composition of magmas was gleaned almost entirely from their consolidation products, the igneous rocks. And today it is still the rocks themselves, especially lavas, that provide the most complete information on the composition and diversity of magmas. Most igneous rocks are composed almost entirely of minerals with various constituent metals and semimetals (Si, Al, Fe, Mg, Ca, Na) coordinated with oxygen. So it has become universal practice to report the analyses of igneous rocks in terms of their oxide components. However, it is most unlikely that these reported components (for example, $SiO_2$, $Al_2O_3$, CaO, $Na_2O$) actually represent the configurations or species which exist at high temperature in the liquid or melt phase. In this book, we shall be principally concerned with the simple components in igneous rocks, since it takes techniques other than those of the analytical chemist to identify or predict the species which exist throughout the cooling history of a magma. Of the components in igneous rocks, silica ($SiO_2$) predominates, forming between 35 and 75 percent by weight over the whole range of magmas found as lavas. It is this preponderance of silica, and the occurrence of silicates, that gave rise to an erroneous concept of silica acidity and basicity. This nevertheless has persisted as the basis of a broad and useful classification of magmas and igneous rocks: thus acid, intermediate, and basic refer to magmas of relatively high (75 to 66 percent), intermediate (66 to 52 percent), and low (52 to 45 percent) silica content.

The physical state (or species) of the various oxide components in any magma is not known, although certain deductions may be made from the properties of analogous synthetic silicate systems. Magmas of high silica percentage are often extremely viscous; they tend to congeal to glass and form large, dome-shaped extrusions on the earth's surface. Acid lava (rhyolite), when crystalline, is composed of quartz and alkali feldspars, minerals with unusually low entropies of melting or, in other words, minerals whose framework structures tend to persist in the liquid state. Basic magma has more $Na^+$, $Ca^{++}$, $Mg^{++}$, and $Fe^{++}$ relative to Si + Al than acid magma, so that the Si–O–Al three-dimensional network "molecules" are smaller and the viscosity is correspondingly lower than in acid magmas. This explains why basaltic magma typically forms more extensive flows and is far less frequently found quenched to a glass than acid magma.

Many igneous rocks, especially those erupted on the earth's surface, show by their vesicularity that they solidified from magmas saturated with a gas phase. The generation, composition, and cooling history of the magmatic gas phase is dealt with in Chap. 6; however, it is reasonable to suggest that at no very great depth beneath a volcano the gas phase was an integral part (dissolved) of the silicate melt, which only became saturated as it moved into the low-pressure region near the earth's surface. Judging from the composition of volcanic gases collected from active vents, or extracted in vacuo from volcanic rocks, the predominant component is invariably water. Since this water may well be of atmospheric or groundwater origin, it has become customary to list separately the "atmospheric" gases ($H_2O$, $N_2$, $O_2$, and A) from the "active" gases (for example, B, CO, $CO_2$, $CH_4$, $NH_3$, $H_2$, HCl, HF, $H_2S$, S, $SO_2$, and $SO_3$) (White and Waring, 1963), among which $CO_2$ predominates but which in total may amount to less than 1 percent of the whole.

Magma as seen on the earth's surface encompasses a considerable range in composition and almost invariably contains one or more solid phases (feldspar, pyroxene, etc.) enclosed in a silicate melt whose properties (viscosity, temperature) are related to its composition; these will further change if a gas phase is generated as the lava erupts, flows, and cools on the earth's surface.

## COOLING BEHAVIOR

A magma, cooling through a given temperature interval, is affected by physical and chemical reactions which, according to the moderation theorem of thermodynamics (for example, Turner and Verhoogen, 1960, p. 41), must run in such a direction as to be exothermic: for example, condensation of gas, crystallization of a liquid, chemical reactions giving off heat. Some of these reactions may, at first sight, seem anomalous; for instance vaporization, an endothermic process, can take place during cooling if it is accompanied by simultaneous crystallization which releases enough heat. Even a simple fluid with the composition of albite plus water, cooling under constant external pressure, may pass through a complicated sequence of condensation, crystallization, boiling, resorption of early formed crystals, recondensation, and so forth (cf. Turner and Verhoogen, 1960, chap. 14).

If the magma may be considered as a closed system, that is, if it does not exchange matter with its surroundings and the initial amount of each component is known, then according to Duhem's theorem the composition and relative amounts of the various phases present at any temperature can be found from the bulk composition (initial mass of components) of the system and the value of one other variable, intensive or extensive (for example, the total volume of the system). We should thus expect magmas of different composition to differ in their sequence of crystallization, even if the physical conditions are identical. Conversely, the same magma evolving under different physical conditions must necessarily react differently. The sequence of events in a magma cooling under constant external pressure is not the same as in a magma cooling at constant volume. Since the physical and mechanical conditions which prevail in magmatic surroundings are not clearly understood and probably are highly variable, it is difficult to predict from theoretical or experimental grounds any generally applicable course of behavior of magmas undergoing crystallization.

There is even greater latitude of behavior if the magma does not behave as a closed system—and there is little doubt that, indeed, most magmatic systems are not closed. In such a magma, since the number of components is very large, the number of intensive variables required to define the state of the system in terms of the phase rule is also very large, except perhaps at the latest stages of crystallization when the number of phases is also large and the variance is small. Thus our knowledge of the physical conditions prevailing in a magma is usually insufficient to allow us to make any precise prediction of its behavior. Failure to realize that a magma is in general a multivariant system has at times led to much discussion, which, because it refers to hypothetical conditions not even approximately described, is of no petrological significance. There is generally no way of saying that "the magma now will do this" or "the magma now must do that"; all we can say, at most, is that the magma might, or again might not, behave in a specified manner. We find ourselves in a situation all too familiar to the geologist who attempts to use thermodynamic or other experimentally derived data: the system is so complex and the variables so numerous that the number of possible models of behavior is embarrassingly large. In fact, the only concrete evidence of how a given magma has behaved is usually that afforded by the chemistry, mineralogy, and fabric of the resulting rock, if indeed these features are amenable to accurate interpretation.

The problem is further complicated by possible uncertainty as to whether a magma was in internal equilibrium and also in equilibrium with its surroundings. It may even be suggested that, in some respects, the cooling of a magma is not a reversible process. Fractional melting of a rock does not reproduce, in reversed order, the successive phenomena of fractional crystallization of the corresponding magma (Presnall, 1969). Irreversibility also could arise, for instance, if the magma were not in thermal equilibrium, that is, if the temperature at any given moment were not uniform throughout the melt. The fact that many magmas, intruded at different times and at different places, seem to have behaved in the same general way is no proof that their evolution is an equilibrium process, since irreversible processes are also known to obey certain laws. For example, they tend to run in such a manner that the rate of production of entropy is minimized, and they may obey a generalized form of the moderation theorem.

Difficult as the prediction of the behavior of a magma may be, we can nevertheless make a general distinction between magmas which cool at depth and magmas, such as lava flows, which cool on the surface. This difference is due, in part, to changes in equilibrium dependent on the difference in pressure in these two contrasted environments, and in part to differences in the kinetics of the cooling process. At the surface, cooling is comparatively rapid, with the result that crystallization need not occur at all, the magma congealing instead to a glassy, metastable state. Or where crystallization does occur, certain reactions may fail to run to completion; olivine, for instance, may be transformed only partially into pyroxene, the mineral phase stable at lower temperature in the presence of excess silica. It is to be noted that the rate of cooling depends not only on depth but even more on the dimensions (surface-area/volume) of the intrusive body and the outward temperature gradient. From the heat-conduction law, cooling rates are proportional to the square of some typical length—for example, the thickness of a sheet or the radius of a cylindrical body of magma. In one of Jaeger's (1957*a*, 1957*b*, 1968) cooling models, a horizontal intrusive sheet of basaltic magma intruded at 1050°C under a cool rock cover 200 m thick (initially at 25 to 50°C) would crystallize completely at 850°C in 700 years. A similar sheet 700 m thick under a cool cover 350 m thick would take 9000 years. A vertical sheet of granitic magma 2000 m thick intruded at 800°C into warm country rock (100°C) would crystallize to the center (at 600°C) only after 64,000 years. In general the rate of cooling seems much the same whether the body cools at a depth of 100 or 1000 m. Typical features of rapidly cooled masses may thus be found in thin, tabular bodies intruded at considerable depths but may be absent from thick bodies injected near the surface.

An important difference between magmas which cool at or near the surface and magmas which cool at great depths is correlated with the behavior of the volatile components, mainly water. The solubility of water in silicate melts appears to increase, within certain limits, with increasing pressure, the molar volume of water vapor being much larger at low pressure than the partial molar volume of water in the melt. Magmas which reach the surface may thus lose the greater part of their water content and perhaps their Cl, F, and other volatile or light components. Because of their low molecular weight, the loss of volatiles may have a considerable effect in raising the crystallization temperatures of the silicate phases, since the chemical potential of a component is related to its molecular concentration or mole fraction.[1] Small amounts of water, $CO_2$, or HF therefore markedly affect the chemical potentials of other components of the melt; and this significantly changes the melting points of the various silicates

[1] Five percent, by weight, of $H_2O$ in a liquid of $NaAlSi_3O_8$ composition corresponds to 43.4 mole percent $H_2O$.

involved. Rapid release of pressure and attendant reduction of water content of the melt may thus affect crystallization much as does rapid chilling.

## PHYSICAL PROPERTIES

Certain aspects of magmatic behavior, flow, extrusion, intrusion, crystal settling, and fractionation depend on physical properties such as density and viscosity of the silicate melt, both these in turn being temperature-dependent. Thus of the physical properties of magmas, we shall discuss only the temperature and density here, reserving until later chapters such topics as viscosity, surface tension, and diffusion rates, which are more conveniently considered in a general discussion on the properties of silicate liquids.

TEMPERATURES To measure the temperature of a flowing, incandescent lava may be a hazardous task. It can safely be done from a distance by means of an optical (glowing filament) pyrometer, but the readings require corrections that are not easily made. Almost all lavas contain suspended crystals in varying amounts, so that the temperature of a flowing lava does not generally correspond to its liquidus temperature (that temperature above which no solid phase is stable). Lavas are also most unlikely to flow at temperatures below the solidus (that temperature below which the liquid phase is unstable). Thus laboratory estimates of liquidus and solidus temperatures of lavas (minus their volatile component) provide maximum and minimum limits for the existence of a silicate liquid. For lavas with few suspended crystals (phenocrysts), the liquidus temperature may be within a few degrees of the temperature of eruption. Total absence of phenocrysts, suggesting that the lava was superheated above the liquidus temperature at the time of eruption, is rare. One recorded case is a peralkaline obsidian of Chabbi volcano, Ethiopia (Macdonald and Gibson, 1969).

Acid lavas of high viscosity flow slowly, and eruptions are often accompanied by large amounts of air-fall material, so that few if any field measurements of their temperatures have been made. It is also difficult or impossible to determine the liquidus or solidus temperature of an acid lava in the laboratory because of its high viscosity. We must resort therefore to indirect methods using mineral geothermometers, with which the partitioning of an element or an isotope (oxygen) between two minerals in equilibrium depends on temperature. If the two minerals used to estimate the liquidus temperature of an acid lava (obsidian) are the only two crystalline phases, or alternatively if the total amount of phenocrysts is less than, say, 5 percent, then the equilibration temperature of the mineral pair may closely approximate the liquidus temperature. Mineral geothermometers cannot be used, in general, to estimate the solidus temperature.

A large amount of reliable temperature data for the basic magma of the Hawaiian volcanoes Kilauea and Mauna Loa has been summarized by Macdonald (1963). He concludes that Hawaiian basaltic magmas approach the surface between 1050 and 1200°C, and that they may continue to flow below 800°C. Besides these estimates we now have thermocouple measurements on two Kilauea lava lakes, Makaopuhi and Alae, that are still largely liquid (1968–1970) beneath a solid crust. T. L. Wright et al. (1968) found that the maximum temperature in the lake today is 1160°C, but extrapolating to the composition of the early pumice, the initial magma temperature was close to 1200°C. Comparison of thermocouple determinations in this lake with measurements by the filament pyrometer shows the latter to be consistently 35 to 75°C too low. In the Alae lava lake, Peck et al. (1966) found that at 980°C the lava contains only 8 percent glass and that at temperatures below this the glass persists metastably. The solidus temperature, or the lower limit of flow, is therefore likely to be above Macdonald's estimate of less than 800°C.

Temperatures of various types of lava are summarized in Table 1-1. The Hawaiian values are derived from thermocouple and laboratory melting data and are true lava temperatures; the Parícutin and African values were obtained by optical pyrometer. The remaining lavas are predominantly glassy, and among their few phenocrysts are magnetite-ilmenite solid solutions, the composition of which depends on temperature (Buddington and Lindsley, 1964). These mineral temperatures represent quenched equilibria and may approximate the respective liquidus temperatures; they may also approximate the extrusion temperature of the lavas.

Although the temperatures of intrusive magmas cooling in the earth's crust are very difficult to estimate, the extrusion temperatures of lavas of comparable composition given in Table 1-1 suggest an upper limit. This seems not unreasonable as a first estimate, unless temperature, particularly in the case of acid magma, increases as the magma rises to the surface, possibly as a result of viscous heating, that is, production of heat by internal friction associated with viscous flow (H. R. Shaw, 1969).

**Table 1-1** Extrusion temperature estimates of various lava types

| Locality | Lava type | Temperature | Source |
|---|---|---|---|
| Kilauea, Hawaii | Tholeiitic basalt | 1150–1225°C | T. L. Wright et al. (1968) |
| Parícutin, Mexico | Basaltic andesite | 1020–1110°C | Zies (1946) |
| Nyiragongo, Congo | Nephelinite | 980°C | Sahama and Meyer (1958) |
| Nyamuragira, Congo | Leucite basalt | 1095°C | Verhoogen (1948) |
| Taupo, New Zealand | Pyroxene rhyolite: pumice flows | 860–890°C | Ewart et al. (1971) |
| | Amphibole rhyolites: lavas, ignimbrites, pumice flows | 735–780°C | Ewart et al. (1971) |
| Mono Craters, California | Rhyolite lavas | 790–820°C | Carmichael (1967*a*) |
| Iceland | Rhyodacite obsidians | 900–925°C | Carmichael (1967*a*) |
| New Britain, Southwest Pacific | Andesite pumice | 940–990°C | Heming and Carmichael (1973); Lowder (1970) |
| | Dacite lava, pumice | 925°C | Heming and Carmichael (1973); Lowder (1970) |
| | Rhyodacite pumice | 880°C | Heming and Carmichael (1973); Lowder (1970) |

The melting (solidus) temperatures of all igneous rocks in contact with water vapor decrease as the pressure of water increases, simply because the solubility of water increases with pressure. But if, with increasing water pressure, the partial molar volume of water in the melt phase becomes greater than the molar volume of water in the gas phase, then the melting curve will pass through a minimum. In general it appears unlikely that this will occur at pressures equivalent to those of the continental crust; but at higher pressures (20 to 25 kb) the liquidus, and presumably the solidus, of tholeiitic basalts have a well-defined minimum (I. A. Nicholls and Ringwood, 1973). At water pressures between 4 and 7 kb, equivalent to depths of between 12 and 21 km in the crust, the melting curve of acid magma has been suggested to have a weak minimum (Kadik and Khitarov, 1969). All in all, the temperatures of intracrustal magmas are likely to be lower than those on the surface, especially if they are saturated with water. The probable range is 700 to 1200°C; the lower values are for water-saturated granitic, the higher for intermediate and basic magmas.

DENSITY We have to make a distinction here between the density of a magma, composed of a variable proportion of solid phases together with a liquid and perhaps a gas, and the density of a liquid having an identical composition to that of the magma. Most density data have been obtained either on silicate liquids above their crystallization temperatures or on silicate liquids quenched to a glass; in general the density of a silicate liquid or of a glass is related to its composition, particularly its water content. Values measured at room temperature for natural anhydrous glasses (obsidians) range from about 2.4 g $cm^{-3}$ for acid liquids to about 2.9 g $cm^{-3}$ for basic glasses. Lower values ($\sim$10 percent) seem appropriate for corresponding melts at 900 to 1200°C; and at extrusion temperatures the density of Hawaiian basaltic lavas containing a small number of crystals has been estimated as 2.73 g $cm^{-3}$ (Macdonald, 1963). Measurements on artificially prepared basaltic liquids at 1250°C gave values of 2.63 and 2.79 g $cm^{-3}$. Bottinga and Weill (1970) have calculated the partial molar volumes of the rock-forming oxide components in silicate liquids as a function of temperature. This important study allows the density of any magmatic liquid to be calculated and, most importantly, illustrates the very large effect of small amounts (weight percentage) of water in decreasing the density of a liquid. These authors have been able to show that, in magmas becoming increasingly iron-rich as crystallization proceeds, the density of the liquid residue increases, whereas if there is no iron enrichment in the liquid, its density decreases.

Most early crystallizing minerals, especially pyroxenes and olivines, are notably denser than any silicate melt from which they separate. Therefore, provided that the size and shape of crystals and the viscosity of the melt permit, these crystals tend to sink under gravity. The accumulation of minerals under gravity is generally accepted to produce the layering in numerous basic layered intrusions. However, the density contrast between plagioclase feldspar (2.7 g $cm^{-3}$) and an enclosing anhydrous iron-rich basic liquid appears to be small, if not reversed; the petrographic and field evidence of many layered intrusions indicates that plagioclase sinks more slowly than accompanying ferromagnesian minerals. Leucite (2.47 g $cm^{-3}$) and possibly the other feldspathoids are the only minerals which demonstrably rise and float in their enclosing magma (Sahama, 1960, pp. 150–152).

## CHEMICAL COMPOSITION

The chemical composition of magma is inferred from that of its surface consolidation products, the volcanic rocks. By far the most abundant element is oxygen: atomic abundance 58 to 65 percent; by weight, 45 to more than 50 percent; by volume, uniformly close to 94 percent. Thus the composition of a volcanic rock, our best clue to that of corresponding magma, can be expressed in terms of a number of oxides. The results of conventional bulk chemical analysis of any igneous rock are in fact presented as oxide percentages by weight. Most abundant is $SiO_2$, typically 35 to 75 percent. Next in abundance is $Al_2O_3$: 12 to 18 percent in most rocks, with a maximum rarely exceeding 20 percent in the intermediate $SiO_2$ range; lower values are typical of olivine-rich volcanic rocks and peralkaline rhyolites. FeO (and $Fe_2O_3$), MgO, and CaO, which together total 20 to 30 percent in rocks low in $SiO_2$, all decrease as the silica content rises. In highly siliceous volcanic rocks (rhyolites) they may total less than 5 percent. $Na_2O$ and $K_2O$ typically are 2.5 to 4 percent and 0.5 to 5 percent respectively; both increase with $SiO_2$. Much higher values—up to 8 percent $Na_2O$ and 6 percent (very rarely 10 percent) $K_2O$—are found only in highly alkaline rocks of intermediate silica content. Some idea of the general compositional range of magmas, minus their volatile contents, can be gained from widely cited "average" values for common rocks compiled by Nockolds (1954) (for example, Table 1-2).

**Table 1-2** Average composition (oxides, wt %) of five classes of volcanic rocks*

| CONSTITUENT | RHYOLITE | DACITE | ANDESITE | BASALT | PHONOLITE |
|---|---|---|---|---|---|
| $SiO_2$ | 73.66 | 63.58 | 54.20 | 50.83 | 56.90 |
| $TiO_2$ | 0.22 | 0.64 | 1.31 | 2.03 | 0.59 |
| $Al_2O_3$ | 13.45 | 16.67 | 17.17 | 14.07 | 20.17 |
| $Fe_2O_3$ | 1.25 | 2.24 | 3.48 | 2.88 | 2.26 |
| FeO | 0.75 | 3.00 | 5.49 | 9.05 | 1.85 |
| MnO | 0.03 | 0.11 | 0.15 | 0.18 | 0.19 |
| MgO | 0.32 | 2.12 | 4.36 | 6.34 | 0.58 |
| CaO | 1.13 | 5.53 | 7.92 | 10.42 | 1.88 |
| $Na_2O$ | 2.99 | 3.98 | 3.67 | 2.23 | 8.72 |
| $K_2O$ | 5.35 | 1.40 | 1.11 | 0.82 | 5.42 |
| $P_2O_5$ | 0.07 | 0.17 | 0.28 | 0.23 | 0.17 |
| $H_2O$ | 0.78 | 0.56 | 0.86 | 0.91 | 0.96 |
| Total | 100.0 | 100.0 | 100.0 | 100.0 | 100.0† |

* From Nockolds (1954).
† Total includes 0.23% Cl and 0.13% $SO_3$.

There are common plutonic rocks, too, whose gross chemical composition is close to many of the averages listed in Table 1-2. Some plutonic classes, such as peridotites and anorthosites, lack precise volcanic equivalents; others, such as muscovite granites, are far more voluminous than their rare volcanic equivalents, while plutonic equivalents of the highly potassic lavas of the Leucite Hills, Wyoming, have a completely different mineral makeup.

With the advent of chemical analysis by x-ray fluorescence spectrography, and neutron-activation techniques we now have extensive information on the relative quantities of a host of trace elements in igneous rocks—Li, V, Cr, Co, Ni, Cu, Zn, Rb, Sr, Y, Zr, Nb, Ba, Ta, Pb, Th, U, and others. These are expressed in parts per million (ppm) by weight. Often derived from these values are ratios generally thought to be petrogenically significant: K/Rb, K/Ba, Rb/Sr, Nb/Ta, Th/U, and Cd/Zn. Table 1-3 illustrates the chemical variety encountered in associated lavas of widely different composition in a single volcanic field.

Of increasing significance with respect to the origin and history of magmas are isotopic abundances of certain elements. For the radioactive elements, these abundances are reported as a ratio of the daughter isotope to a nonradiogenic isotope; thus we report $Sr^{87}/Sr^{86}$, $Pb^{206}/Pb^{204}$, and $Pb^{207}/Pb^{204}$, where $Sr^{86}$ and $Pb^{204}$ are the stable isotopes. Of the nonradiogenic isotopes, significant variation has been found in $O^{18}/O^{16}$, $S^{34}/S^{32}$, and perhaps also the silicon and hydrogen isotopes, all of which hold great promise in deciphering the later magmatic, or cooling, history of igneous rocks; the oxygen isotopes are a particularly powerful tool in this regard.

## Igneous Rocks: Occurrence, Mineralogy, and Texture

### VOLCANIC AND PLUTONIC ROCKS

Magmas erupted as lava from volcanic vents solidify as volcanic rocks whose igneous origin is beyond doubt. Within the crust, and especially in the depths below volcanic regions, there must exist temporary reservoirs of magma; and presumably some such bodies solidify completely without reaching the surface. Igneous rocks of this nature have been described from many dissected volcanic provinces, such as the island of Mull in western Scotland or the Pacific island of Tahiti, and, as might be expected, are mineralogically and structurally similar, though usually not identical, to volcanic rocks of proved igneous origin. Starting, therefore, with volcanic rocks, it is possible to draw up mineralogical and structural criteria by means of which many igneous rocks may be recognized as such. Differences in one or more of the many variable factors influencing magmatic evolution in the plutonic (deep-seated) as contrasted with the volcanic (surface) environment can be expected to be reflected in corresponding differences in rock fabric and to a lesser extent in mineral assemblages. It is therefore customary to distinguish between *volcanic* and *plutonic* igneous rocks. Some writers recognize an intermediate (*hypabyssal*) class to include rocks that

**Table 1-3** Chemical composition of three representative lavas from the Talasea volcanic field, New Britain*

| | | BASALT | | | ANDESITE | | | RHYOLITE | | |
|---|---|---|---|---|---|---|---|---|---|---|
| Element | Oxide | Oxides, wt % | Elements, wt % | Atomic % | Oxides, wt % | Elements, wt % | Atomic % | Oxides, wt % | Elements, wt % | Atomic % |
| O | | | 45.00† | 60.97 | | 47.00† | 61.83 | | 48.85† | 63.19 |
| Si | $SiO_2$ | 51.57 | 24.10 | 18.86 | 58.93 | 27.54 | 21.21 | 72.19 | 33.74 | 25.28 |
| Ti | $TiO_2$ | 0.80 | 0.48 | 0.22 | 0.62 | 0.37 | 0.17 | 0.33 | 0.20 | 0.09 |
| Al | $Al_2O_3$ | 15.91 | 8.42 | 6.86 | 16.87 | 8.93 | 7.16 | 12.62 | 6.68 | 5.21 |
| Fe | $Fe_2O_3$ | 2.74 | 7.39 | 2.91 | 2.71 | 5.41 | 2.09 | 3.14 | 3.07 | 1.16 |
| | FeO | 7.04 | | | 4.51 | | | 1.12 | | |
| | MnO | 0.17 | | | 0.16 | | | 0.05 | | |
| Mg | MgO | 6.73 | 4.06 | 3.67 | 2.92 | 1.76 | 1.57 | 0.58 | 0.35 | 0.30 |
| Ca | CaO | 11.74 | 8.39 | 4.60 | 6.87 | 4.91 | 2.65 | 2.07 | 1.48 | 0.78 |
| Na | $Na_2O$ | 2.41 | 1.79 | 1.71 | 3.68 | 2.73 | 2.57 | 3.45 | 2.56 | 2.34 |
| K | $K_2O$ | 0.44 | 0.37 | 0.21 | 1.63 | 1.35 | 0.75 | 3.70 | 3.07 | 1.65 |
| | $P_2O_5$ | 0.11 | | | 0.20 | | | 0.02 | | |
| | $H_2O^+$ | 0.35 | | | 0.86 | | | 0.56 | | |
| | $H_2O^-$ | 0.10 | | | 0.15 | | | 0.24 | | |
| Total | | 100.11 | | 100.01 | 100.11 | | 100.00 | 100.07 | | 100.00 |

* From Lowder and Carmichael (1970).
† Oxygen calculated by difference.

| Trace Elements | Concentration, ppm (by wt) | Concentration, ppm (by wt) | Concentration, ppm (by wt) |
|---|---|---|---|
| Nb | ND‡ | 5 | 10 |
| Zr | 30 | 70 | 160 |
| Y | 5 | 15 | 20 |
| Sr | 355 | 525 | 245 |
| Rb | 5 | 20 | 60 |
| Th | 15 | ND | 15 |
| Pb | 1 | 1 | 5 |
| Ga | 15 | 10 | 10 |
| Zn | 80 | 75 | 35 |
| Cu | 175 | 80 | 25 |
| Ni | 100 | 30 | 30 |
| V | 275 | 200 | 30 |
| Cr | 125 | ND | 5 |
| Ba | 150 | 320 | 575 |
| Ce | 35 | 30 | 35 |
| Nd | 15 | 20 | 15 |
| Pr | 10 | 10 | 5 |
| La | 15 | 15 | 20 |
| | Ratios by wt | Ratios by wt | Ratios by wt |
| K/Rb | 740 | 675 | 512 |
| K/Ba | 25 | 42 | 53 |
| Rb/Sr | 0.014 | 0.038 | 0.245 |
| $Sr^{87}/Sr^{86}$ | 0.7035 | 0.7035 | 0.7035 |

‡ ND, not detected.

have crystallized at moderate depth. The need for such a class is doubtful. Many of the "hypabyssal" rocks are identical, or nearly so, with surface lavas, and it would seem more satisfactory to broaden the volcanic class to include surface lavas and near-surface intrusive rocks of similar structure and mineralogy. This is the course adopted in this book. The term *plutonic* is reserved for those igneous rocks that have crystallized at sufficient depth or from bodies of magma massive enough to impart distinctive characters to be discussed later.

## MODE OF OCCURRENCE

Typical volcanic rocks occur as flows extruded on the earth's surface. Individual flows range from a few centimeters to perhaps 300 m in thickness but seldom attain a length greater than 200 km. In volcanic complexes resulting from repeated fissure eruptions, areas of several hundred thousand square kilometers may be covered by flows aggregating 1000 m or more in total thickness. Characteristic features commonly displayed by lava flows are a scoriaceous, blocky, or ropy upper surface; a baked, oxidized dust or ash layer beneath the under surface; and columnar jointing with a regular trend normal to cooling surfaces, or platy jointing parallel either to cooling surfaces or to the plane of flow, especially near the surface. Near-surface intrusive volcanic rocks may occur in the form of more or less cylindrical vertical necks, representing magma that has solidified in conduits of eroded volcanoes, or as tabular sheets. The latter are known respectively as dikes or sills, according to whether they trend across or parallel to the dominant structure (for example, bedding) of the invaded rocks. Fragmented volcanic debris is the principal component of pyroclastic rocks formed by accumulation of material explosively ejected from some volcanoes. The individual fragments have the mineralogical and textural characteristics of volcanic rocks.

Bodies of plutonic rock vary greatly in form and extent. The smallest are dikes and veins a few centimeters wide; the larger masses outcrop continuously over areas measured in thousands of square kilometers. It is beyond the scope of this book to discuss the detailed forms of plutonic intrusion, but some of the commoner types, classified conventionally according to form and relation to invaded rocks, are noted briefly.

*Sills* Tabular bodies concordant with the major structure, for example, bedding or foliation, of the invaded rocks. Large sills hundreds of meters thick and extending laterally for many kilometers are usually of basic composition.

*Laccoliths* Sheetlike bodies with a flat base and domed roof above which the invaded strata have been arched concordantly at the time of intrusion. Most laccoliths are composed of acid rather than basic rocks.

*Phacoliths* Curved, lensoid masses injected along and concordant with the arches and troughs of folded strata.

*Lopoliths* Roughly sheetlike or lensoid bodies with upper and lower surfaces that are concave upwards, the general configuration being connected with sagging of the floor rocks under the load of the thickening intrusions. Some of the largest known plutonic intrusions (composed mainly of basic rocks), once considered to be lopoliths, are now known to narrow downward, with a funnel-shaped cross section that in no way implies sagging of the floor.

*Dikes* Tabular, often vertical or steeply dipping sheets that cut across the trend of structure (for example, bedding) of the invaded rocks.

*Ring dikes* Steeply dipping dikes of arcuate or annular outcrop, formed by uprise of magma along major steeply conical or cylindrical fractures bounding central collapsed blocks. Ring-dike complexes, some many kilometers wide, represent eroded root regions of volcanic centers.

*Batholiths* Large, intrusive bodies with steeply dipping walls and lacking any visible floor. Batholiths are typically composed of acid rocks (granite, granodiorite, and related rocks). The largest of these bodies outcrop continuously over many thousands of square kilometers.

*Stocks (bosses)* Similar in form and composition to batholiths but are smaller in size (with outcrops less than about 100 $km^2$).

*Plutons* A term which embraces all intrusive bodies of igneous rock. It is a convenient term when the intrusion conforms to none of the preceding definitions or when its geometric shape is unknown. Many batholiths, such as the Sierra Nevada batholith of California, are composite intrusions composed of numerous separate plutons individually emplaced at intervals collectively spanning many millions of years.

## CONSTITUENT MINERALS

The principal mineral constituents (crystalline phases) of igneous rocks are a restricted number of silicates and aluminosilicates of Ca, Mg, Fe, Na, and K, oxides of Fe and Ti, and in siliceous rocks some form of crystalline silica. These also accommodate most of the minor elements which substitute for major elements of comparable charge or ionic radius—$Cr^{3+}$ for $Fe^{3+}$, $Ni^{++}$ for $Mg^{++}$, $Rb^{+}$ and $Ba^{++}$ for $K^{+}$, $Sr^{++}$ for $Ca^{++}$, $Li^{+}$ for $Mg^{++}$, and so on. Some minor elements, however, give rise to special accessory minerals such as zircon (Zr), apatite (P, F, Cl), pyrrhotite (S), and hauyne (Cl, $SO_3$).

MINERALS OF VOLCANIC ROCKS Peculiar to volcanic rocks, and reflecting the tendency for rapid cooling and quenching of magma under surface conditions, is the crystallization or preservation of metastable phases. Foremost among these, especially in the more siliceous rocks, is glass—a metastable liquid undercooled to the point where extreme viscosity has slowed crystallization to a halt. Some minerals, metastable over the cooling range of the solidified rock, were previously stable at liquidus temperatures. Such are disordered volcanic feldspars quenched to low temperatures at which a more ordered condition is stable. Other minerals were metastable from the moment of crystallization. Most of these illustrate the operation of Ostwald's step rule: a significantly undercooled system such as glass often precipitates more stable but nevertheless metastable crystalline phases by steps that involve minimal entropy change. Thus the high-entropy forms of silica—cristobalite and tridymite—may crystallize from glass or gas at temperatures within the stability range of quartz ($<870°C$).

Possibly the most revealing example of mineral metastability in basaltic magma is the prehistoric Makaopuhi lava lake of Kilauea (Evans and Moore, 1968). Equilibrium in the various mineral phases, pyroxene, olivine, and plagioclase was approached only at depths in excess of 50 m, where the magma may have taken almost 30 years to cool from 1200°C to below the solidus (980°C). Although the pyroxenes in the top 5 m or so of this lava lake show a considerable compositional range and are clearly metastable, pyroxenes identical to slowly cooled equilibrium assemblages have been found in thin (6 to 10 m), but notably iron-rich, basaltic lavas (Carmichael, 1967*b*). Obviously both cooling rate and fluidity contribute to the generation or preservation of metastable mineral compositions.

Neglecting accessory and rare minerals, the main constituent crystalline minerals of volcanic rocks are as follows.

*Crystalline silica, $SiO_2$*

1. Quartz is common.
2. Tridymite and cristobalite are more restricted but are common in the groundmass of rhyolites. Phenocrysts of tridymite, inverted to quartz, occur in granophyres. At about 800°C, cristobalite has been deposited as a lining to vesicles in the Alae basaltic lava lake, Hawaii (Peck et al., 1966), and it is also found in the vesicles of many lavas.

*Feldspars* Feldspars may be described in terms of three principal end members: potassium feldspar, $KAlSi_3O_8$; sodium feldspar or albite, $NaAlSi_3O_8$; calcium feldspar or anorthite, $CaAl_2Si_2O_8$. At magmatic temperatures there is complete substitution of $Na^+$ for $K^+$ in the alkali-feldspar series, and of $Ca^{++}Al^{3+}$ for $Na^+Si^{4+}$ in the plagioclase series. At lower temperatures isomorphous substitution is restricted, especially in the alkali feldspars, and unmixing may occur, giving submicroscopic intergrowths of two phases. Those optically homogeneous feldspars that can be identified as intergrowths of two phases only by the use of x-ray techniques are termed *cryptoperthites*. High- and low-temperature polymorphs of all three end members and of the two solid-solution series, alkali feldspar and plagioclase, are known. The high-temperature forms are distinguished by a disordered (high-entropy) distribution of Al among equivalent four-coordinated lattice sites.

There are two series of feldspars in volcanic rocks:

1. Sanidine-anorthoclase series: alkali feldspars ranging from monoclinic potassic types (sanidines with small 2V) to triclinic sodic types (anorthoclases with 2V 40 to 50°); intermediate varieties are sanidine cryptoperthites (monoclinic) and anorthoclase cryptoperthites (triclinic); the optic axial plane is normal to {010} throughout the series. High-temperature sanidine, with axial plane parallel to {010}, forms from sanidine or orthoclase on heating above 1000°C. Comparable natural volcanic sanidines are rare. The sodic phase in volcanic cryptoperthites is high-temperature albite. Solid solution between $KAlSi_3O_8$ and $CaAl_2Si_2O_8$ is very restricted, so that alkali feldspars with a medium to high K content typically contain less than 3 percent anorthite. Slight substitution of $Fe^{3+}$ for $Al^{3+}$ is general ($Fe/Al < 0.01$); but in rocks with exceptionally low Al/K the content of $KFeSi_3O_8$ may be very much higher (10 to 18 percent). Trace elements substituting for $K^+$ are $Ba^{++}$, $Sr^{++}$, and, in smaller amounts, $Rb^+$.

2. Plagioclase series: triclinic sodium-calcium feldspars ranging continuously from albite to anorthite. Optical and x-ray data show that the volcanic plagioclases do not form a high-temperature series comparable with plagioclases prepared by dry synthesis in the laboratory. Rather they represent various states of transition from high- to low-temperature forms. Oligoclase and andesine of trachytic lavas may contain as much as 10 to 20 percent $KAlSi_3O_8$ (these potassic plagioclases grade into calcic anorthoclases having a similar significant content of $CaAl_2Si_2O_8$). Plagioclase in the medium to calcic range usually contains no more than 2 to 5 percent $KAlSi_3O_8$. Trace elements other than $Fe^{3+}$ (for Al) and Sr (for Ca) are negligible. The ratio Rb/Sr consistently is very low, so that the initial isotopic composition of Sr ($Sr^{87}/Sr^{86}$) in plagioclase may survive virtually unchanged for geologically long periods.

*Pyroxenes* As a first approximation common volcanic pyroxenes can be described in terms of four end members: diopside, $CaMgSi_2O_6$; hedenbergite, $CaFeSi_2O_6$; enstatite, $MgSiO_3$; ferrosilite, $FeSiO_3$. There is also a sodic series in which an essential role is played by acmite, $NaFeSi_2O_6$. In the monoclinic pyroxenes small but significant amounts of $Al^{3+}$ substitute for $Si^{4+}$ in tetrahedral sites; and $Al^{3+}$, $Cr^{3+}$, $Fe^{3+}$, $Ti^{3+}$, and $Na^+$ together take the place of some 10 percent of the six-coordinated ions $Ca^{++}$, $Mg^{++}$, $Fe^{++}$. There are five series of volcanic pyroxenes:

1. Augite: somewhat aluminous monoclinic pyroxenes approximating Ca(Mg, Fe)$Si_2O_6$ (2V mostly 45 to 60°, with lower values in titaniferous augites). Common augites of basalts and andesites are diopsidic types with Fe/(Mg + Fe) = 0.2 to 0.4; but the augites found in some siliceous volcanics are ferroaugites with Fe/(Mg + Fe) = 0.5 to 0.7 (Carmichael, 1960). Volcanic augites contain (Mg + Fe) above the amount required by the ideal formula Ca(Mg, Fe)$Si_2O_6$. Titanium, which amounts to less than 1 percent $TiO_2$ by weight in most augites, is an essential constituent ($TiO_2$ = 3 to 5 percent) of aluminous augites typical of many alkaline basalts.

2. Orthorhombic pyroxene: often loosely called hypersthene, approximates (Mg, Fe)$SiO_3$ and contains only minor Ca ($CaSiO_3$ < 5 percent). Typical volcanic hypersthenes are relatively magnesian—Fe/(Mg + Fe) = 0.2 to 0.5.

3. Pigeonite: monoclinic pyroxene (Mg, Fe)$SiO_3$, close to hypersthene in composition but with significantly higher Ca ($CaSiO_3$ slightly less than 10 percent) (2V 0 to 25°, axial plane normal to {010}). Under conditions of slow cooling, as in diabase, pigeonite may invert to hypersthene with exsolution of augite.

4. Subcalcic augite: spans the compositional interval between augite and pigeonite (2V 10 to 30°, axial plane parallel to {010}). Coexistence of augite and pigeonite, or inverted pigeonite, in diabases and gabbros demonstrates an immiscibility gap in the series Ca(Mg, Fe)$Si_2O_6$–(Mg, Fe)$SiO_3$. In volcanic rocks, the width of this gap is approximately 30 atom percent Ca for magnesian pyroxenes and perhaps 20 atom percent for more iron-rich pyroxenes. Coexisting groundmass pyroxene compositions (considered to be in equilibrium) from various volcanic rocks illustrating this immiscibility gap are:

| AUGITE | PIGEONITE | |
|---|---|---|
| $Ca_{40}Mg_{50}Fe_{10}$ | $Ca_7Mg_{70}Fe_{23}$ | Hawaiian basalt (Evans and Moore, 1968, p. 95) |
| $Ca_{38}Mg_{46}Fe_{16}$ | $Ca_9Mg_{54}Fe_{37}$ | Icelandic basalts and basaltic andesites (Carmichael, 1967*b*, p. 1827) |
| $Ca_{31}Mg_{38}Fe_{31}$ | $Ca_{12}Mg_{41}Fe_{47}$ | |
| $Ca_{34}Mg_{42}Fe_{24}$ | $Ca_8Mg_{62}Fe_{30}$ | Japanese andesite, Hakone (Nakamura and Kushiro, 1970, p. 268) |

Subcalcic augite found in the groundmass of many basalts and andesites spans this gap in the Fe/(Mg + Fe) range 0.4 to 0.7. It is regarded as a metastable quench product (Muir and Tilley, 1964*b*; Evans and Moore, 1968, fig. 8).

5. Aegirine: $NaFe^{3+}Si_2O_6$; the monoclinic solid-solution series aegirine-

augite approximately $NaFe^{3+}Si_2O_6$–$Ca(Mg, Fe)Si_2O_6$, with variable, in some cases very low, aluminum.

In addition to Al, Cr, $Fe^{3+}$, Ti, and Na in augite and pigeonite, and Ca in orthopyroxenes, members of the pyroxene group tolerate minor substitution of a number of other elements in sixfold coordination: Mn in ferroaugites (MnO $\approx$ 1 percent: Carmichael, 1960) and in relatively iron-rich orthopyroxene, and Ni and V in magnesian augite and orthopyroxene.

*Olivines, $(Mg, Fe)_2SiO_4$* End members are forsterite (Fo), $Mg_2SiO_4$; and fayalite (Fa), $Fe_2SiO_4$. Common olivine of basaltic lavas is magnesian, mostly in the range $Fo_{80}Fa_{20}$ to $Fo_{65}Fa_{35}$. Iron-rich olivine—hortonolite and fayalite ($Fe_{50}Fa_{50}$ to $Fo_0Fa_{100}$)—is confined to the more siliceous lavas: trachytes and rhyolites. Basaltic olivine habitually contains a few tenths of 1 percent of each of the oxides CaO, NiO, and MnO. Particularly important is Ni, because the principal host of this element among common volcanic minerals is magnesian olivine. The Ni content rises and the Mn content falls as Mg/Fe increases (cf. Moore and Evans, 1967). Iron-rich olivine may attain 2 to 3 percent MnO and up to 1 percent CaO.

*Amphiboles*

1. Most amphiboles in volcanic rocks are common hornblende or are members of the pargasite-hastingsite series $NaCa_2(Mg, Fe)_4Al(Al_2Si_6O_{22})(OH)_2$, with limited substitution of K for Na, F for (OH), and $Fe^{3+}$ for Al. Ti can be very variable and all gradations to kaersutite (4 to 10 percent $TiO_2$) are found, sometimes within a single crystal (Carmichael, 1967*a*, p. 57). Little is known of the trace-element pattern of volcanic hornblendes. Oxyhornblende is an optically distinctive type (deep red-brown) with unusually high $Fe^{3+}/(Fe^{++} + Fe^{3+})$ found in some lavas.

2. Cummingtonite, $(Mg, Fe)_7Si_8O_{22}(OH)_2$, is becoming increasingly recognized as a phenocryst in acid volcanics. It is sometimes associated with hornblende, although Ewart et al. (1971) consider them mutually exclusive in New Zealand pumice ash. Most cummingtonites have Mg/(Fe + Mg) = 0.7 to 0.5, and similar ratios are found in the associated hornblendes.

3. Riebeckite, $Na_2Fe_3^{++}Fe_2^{3+}Si_8O_{22}(OH)_2$, is the most widely distributed alkali amphibole and can occur in alkali rhyolites, trachytes, and phonolites. Al, Zr, Ti, Mn, Ca, and K may substitute to some degree for atoms in sixfold coordination.

4. Richterite, another aluminum-poor alkali amphibole, has the generalized formula $Na_2CaMg_5Si_8O_{22}(OH)_2$. Iron-rich representatives are found as phenocrysts in alkali rhyolites; magnophorite, a magnesian potassium-rich variety with notable Ti, Ba, and Sr, is found in uniquely potassic lavas of Leucite Hills, Wyoming, and of Western Australia.

*Micas* Volcanic micas are usually magnesian biotites approximating phlogopite $K(Mg, Fe)_3AlSi_3O_{10}(OH)_2$, with $Fe^{3+}$ replacing $Al^{3+}$ and some additional $Al^{3+}$ in place of $Mg^{++}$ and $Si^{4+}$. They can contain up to 9 percent $TiO_2$ (3 to 5 percent is very common), giving the mineral a distinctive red-brown pleochroism. Minor substitutions are Na, Ba (up to 3 weight percent), and Rb for K; Ni, Cr (in phlogopite), and Mn in octahedral sites. The incorporation of F rather than (OH) stabilizes biotite at high temperatures and low volatile pressures, and fluorophlogopite is commonly found in the groundmass of alkaline basalts (basanites and mugearites) and in highly potassic lavas and dike rocks (lamprophyres). Oxidation of biotite (often F-poor) phenocrysts to aggregates of iron oxides and pyroxenes is common in andesitic and to a lesser degree in rhyolitic lavas (cf. Larsen et al., 1937, p. 900).

*Feldspathoids* The term *feldspathoid* conveniently covers a number of structurally unrelated sodic and potassic aluminosilicates in which the atomic ratio (Na + K)/Si exceeds the alkali-feldspar value, 0.33. They are confined to rocks with high (Na + K)/Si and Al/Si ratios. Feldspathoids thus crystallize only from alkaline melts having low silica activity; and this as we shall see later has important consequences regarding the ultimate distribution and concentration of certain ions—$Cl^-$, $(SO_4)^{--}$, $(CO_3)^{--}$, $Zr^{4+}$, $Ti^{4+}$, and others—in individual mineral phases and in the igneous rocks themselves.

1. Leucite, $KAlSi_2O_6$, is a mineral restricted, because of instability at high pressure, to volcanic rocks. In contrast to alkali feldspars, mutual substitution of alkali ions in leucite is very limited. Even where leucite and nepheline coexist the atomic ratio Na/(K + Na) in leucite does not exceed 0.13 to 0.15. Leucite can accommodate small amounts of $Fe^{3+}$ for $Al^{3+}$; Ba is insignificant or undetectable. Untwinned varieties are sometimes nonstoichiometric.

2. Nepheline is represented by the formula $Na_3K(Al_4Si_4O_{16})$. Substitution between Na and K is limited by difference in coordination: Na in eightfold, K in ninefold coordination. Some sodic volcanic nephelines, quenched from high temperatures, have a small excess of Si above the ideal Si/Al ratio of 1:1; in other words there is limited solid solution of alkali feldspar in nephelines. Some quenched phenocrysts in highly potassic East African lavas show considerable solid solution between nepheline and kalsilite, $KAlSiO_4$. Microscopically these have unmixed to nepheline perthites, one phase of which approximates ideal nepheline and the other kalsilite with minor dissolved nepheline—Na/(K + Na) = 0.01 to 0.1. Kalsilite as a discrete phase is also found in these lavas.

3. The sodalite group includes sodalite, $Na_8Al_6Si_6O_{24}Cl_2$; nosean, $Na_8Al_6Si_6O_{24}SO_4$; and hauyne, similar to nosean but with some $CaSO_4$ substituting for NaCl.

4. Various sodic zeolites include analcite, $NaAlSi_2O_6H_2O$, and natrolite, $Na_2Al_2Si_3O_{10}2H_2O$. It is doubtful whether any of these crystallize in the magmatic crystallization interval.

*Iron titanium oxides* Iron titanium oxides are of particular interest because they provide both a geothermometer and the principal record of paleointensity and paleodirection of the earth's magnetic field.

1. Ulvospinel, $Fe_2TiO_4$, and magnetite, $Fe_3O_4$, form a complete cubic solid-solution series ($\beta$ series) at magmatic temperatures. Partial substitution is found of Mg, Mn, Zn for $Fe^{++}$ and Al, V, and Cr for $Fe^{3+}$. Secondary oxidation of a $\beta$ phase typically generates ilmenite lamellae in a magnetite host. A $\beta$ phase is almost ubiquitous in volcanic rocks and generally contains between 75 and 15 percent ulvospinel. Oxidation can produce nonstoichiometric phases and maghemite (cubic $\gamma$-$Fe_2O_3$).

2. Ilmenite, $FeTiO_3$, and hematite, $\alpha$-$Fe_2O_3$, form a partial solid-solution series ($\alpha$ series) at magmatic temperatures. Most volcanic ilmenites are one-phase and fall in the range 3 to 25 percent hematite. There is some substitution of Mg and Mn, and restricted amounts of Al and V. Solid solutions within a restricted part of the middle composition range have the capacity of magnetic self-reversal.

3. Pseudobrookite, $Fe_2O_3 \cdot TiO_2$, and ferropseudobrookite, $FeO \cdot 2TiO_2$, form an orthorhombic solid-solution series ($\omega$ series) above 600°C. An $\omega$ phase often results from the high-temperature oxidation of magnetite-ilmenite intergrowths (Watkins and Haggerty, 1967); rarely pseudobrookite occurs as a primary phase in alkali-rich basic rocks. Little is known of the minor element substitutions, but Mg is common.

4. Rutile, $TiO_2$, is rare in volcanic rocks and results from secondary oxidation of ilmenite-hematite solid solutions.

*Apatite* $Ca_5(PO_4)_3(OH, F)$ This is a ubiquitous accessory mineral in volcanic rocks. It is the principal repository of P, the rare earths, and often Sr, F, and OH); in the absence of a sulfide mineral, S as sulfate is concentrated in apatite.

MINERALS OF PLUTONIC ROCKS Compared with surface conditions the higher pressures, somewhat lower crystallization temperatures, and slow cooling rate of the plutonic environment leave a distinctive imprint on mineralogical detail of plutonic rocks. Glass fails to survive, hydrous minerals are more varied and more in evidence, solid solutions partially unmix as they cool, and metastable phases are generally lacking. Particular features of plutonic mineralogy are discussed in the following paragraphs.

*Feldspars* Alkali feldspars fall into two series: microcline albite and orthoclase albite. The sodic member of each is low-temperature albite, which seems to be stable below 700°C. At the potassic end microcline, which is triclinic, is stable at temperatures below a value variously estimated at between 500 and 700°C. Orthoclase (monoclinic) is structurally intermediate between microcline and its high-temperature polymorph sanidine; it is distinguished optically from the latter by its higher 2V (40 to 80°). In both microcline and orthoclase, substitution of $Na^+$ for $K^+$ is limited, so that optically recognizable perthites (microperthites) are characteristic of the middle ranges of both alkali-feldspar series. The sodic phase of these perthites is low-temperature albite. Plutonic alkali feldspars, like their volcanic counterparts, contain small amounts of $Ca^{++}$ and $Fe^{3+}$: $Fe/Al < 0.01$ and $Ca/(K + Na) = 0.01$ to 0.02.

In plagioclase of plutonic rocks, distribution of Al and Si atoms among tetrahedral lattice sites approaches, to varying degrees, a completely ordered pattern. In any particular plutonic environment, for example, in the large stratiform basic intrusions, the degree of ordering, as indicated by optical and x-ray parameters, tends to be constant. The structure and cell dimensions of low-temperature plagioclase vary with composition. This leads to recognizable unmixing effects at the sodic end of the series: oligoclase of plutonic rocks is unmixed on a submicroscopic scale to an intergrowth (peristerite) of nearly pure albite and calcic oligoclase. Substitution of minor elements follows the same pattern as in volcanic plagioclase.

*Pyroxenes* In plutonic rocks pigeonite is typically absent, having inverted to orthorhombic pyroxene. Subcalcic augite is unknown. Unmixing phenomena are widespread: orthopyroxene contains exsolved diopsidic lamellae parallel to {100}; in augite exsolution lamellae presumed to be $(Mg, Fe)SiO_3$ separate paralle to {001} or {100}. Distribution of Al, $Fe^{3+}$, Ti, Na, and Mn is much the same in plutonic as in volcanic pyroxenes. Common in magnesian plutonic augite is Cr ($Cr_2O_3$ up to 1 percent), Ni, and V (each up to a few hundred ppm).

*Amphiboles* Common hornblende is more widely distributed in plutonic than in volcanic rocks because its stability range and chemical variety are extended as the activity of $H_2O$ rises. Two varieties virtually confined to alkaline rocks are barkevikite and arfvedsonite. Barkevikite is a brown ferrous hornblende low in Al, with high $Fe^{++}/Fe^{3+}$ and $Fe^{++}/Mg^{++}$; it lacks the high Ti that marks the ferrous kaersutites. Arfvedsonite is a highly sodic amphibole approaching $Na_3(Mg, Fe^{++})_4(Fe^{3+}, Al)Si_8O_{22}(OH)_2$, with appreciable K and Ca replacing Na.

*Micas* Micas, like hornblendes (and for the same reason), are more varied and more widely distributed in plutonic than in volcanic rocks.

1. Muscovite, $KAl_2(AlSi_3O_{10})(OH)_2$, with appreciable Na and Li substituting

for K, is confined to aluminous members of the granite family, especially granite pegmatites.

2. Biotite is represented by the formula $K(Fe, Mg)_3(AlSi_3O_{10})(OH)_2$; $Fe^{3+}$ and $Al^{3+}$ substituting for $Mg^{++}$ are compensated for by additional $Al^{3+}$ in the tetrahedral sites. Typically $Fe^{++}/Mg^{++} > 1$, and $Fe^{++}/Fe^{3+}$ is high; minor elements include F, Ti, Ba, Na, Mn, and traces of Rb. The latter is significant in radiometric dating of biotites because the ratio Rb/Sr is universally high—between 2 and 1000.
3. Lithium micas are widely distributed in syenite and granite pegmatites. The commonest are the lepidolites, with $Li^+$ substituting for six-coordinated $Al^{3+}$ in a series between ideal muscovite $KAl_2(AlSi_3O_{10})(OH)_2$ and an ideal lithium end member $KLi_2Al(Si_4O_{10})(OH)_2$. Lepidolites contain significant F, Rb, Cs, and Mn ($Rb_2O$ and $Cs_2O$ may reach 2 to 3 percent by weight). There are also lithium biotites in which $Li^+$ substitutes appreciably for $Mg^{++}$.

*Feldspathoids* Nepheline is by far the most abundant plutonic feldspathoid. It closely approaches the ideal composition $Na_3K(Al_4Si_4O_{16})$. Of the sodalite group only sodalite itself is important. Cancrinite, $Na_6Ca(Al_6Si_6O_{24})(CO_3)$, with minor substitution of $(SO_4)^{--}$ for $(CO_3)^{--}$, is not uncommon. Leucite is unknown in deep-seated plutonic rocks.

*Iron titanium oxides* Because of the ease with which iron titanium oxides respond to a changing environment, they rarely retain their high-temperature composition. Low-temperature oxidation of both $\alpha$ and $\beta$ phases is common, particularly in granites, and may result in intergrowths or assemblages of rutile, two $\alpha$ phases, a Ti-poor magnetite, or maghemite.

## MODE AND NORM

The mineral composition of an igneous rock, expressed as weight or volume percentages of microscopically identified minerals, is the *mode* (modal composition) of the rock.

The *norm*, as the name implies, was designed to normalize the chemical composition of any igneous rock so that it could be compared with other igneous rocks. In the calculation of the norm, the various oxide components of the rock, determined by chemical analysis, are combined in a sequence of steps to form a series of normative mineral components. This sequence of oxide combinations was formulated so that one procedure would allow the calculated normative minerals to correspond broadly with those present modally in any igneous rock which crystallized slowly at low pressure. Thus normative olivine (*ol*), nepheline (*ne*), and leucite (*lc*) are incompatible with normative quartz (*Q*), in accordance with the paragenesis of these minerals; similarly normative hypersthene (*hy*) and nepheline (*ne*) are mutually exclusive. In North America and Britain the weight norm devised by Cross, Iddings, Pirsson, and Washington (CIPW) is in common use,[1] whereas in continental Europe the Niggli molecular norm is favored.

In the CIPW norm, complex hydrous modal minerals, such as amphibole or mica, are represented by simpler normative components; thus modal muscovite, $KAl_2AlSi_3O_{10}(OH)_2$, is represented by $KAlSi_3O_8$ (*or*), $Al_2O_3$ (*C*), and $H_2O$. For many plutonic rocks of simple mineralogy (no hydrous phases apart from apatite), there is often a very good correlation between the mode and the norm, but at the other extreme, for the glassy volcanic rocks, correlation is obviously poor. Two examples are given in Table 1-4.

## TEXTURE

The mutual relationships in space of the various components of a rock (crystals, parts of crystals, multigranular aggregates, or microscopically irresolvable

**Table 1-4** Chemical composition (oxides, wt %), CIPW norm, and modal composition (wt %) of two diabases from Palisades sill, New Jersey (K. R. Walker, 1969*b*)

| | CHEMICAL COMPOSITION | | NORM | | | MODE | | |
|---|---|---|---|---|---|---|---|---|
| | 1 | 2 | | 1 | 2 | | 1 | 2 |
| $SiO_2$ | 51.81 | 47.41 | | | | | | |
| $TiO_2$ | 1.16 | 0.89 | Quartz, *Q* | 1.61 | — | | | |
| $Al_2O_3$ | 13.36 | 8.66 | Orthoclase, *or* | 3.28 | 2.39 | Quartz and | | |
| $Fe_2O_3$ | 2.24 | 2.81 | Albite, *ab* | 16.66 | 11.53 | micropegmatite | 1.5 | — |
| FeO | 8.44 | 11.15 | Anorthite, *an* | 25.82 | 16.21 | Plagioclase | 40.3 | 25.2 |
| MnO | 0.17 | 0.20 | Diopside, *di* | 23.00 | 13.14 | Augite | 46.9 | 16.9 |
| MgO | 9.25 | 19.29 | Hypersthene, *hy* | 24.02 | 25.91 | Bronzite | 4.3 | 23.6 |
| CaO | 11.12 | 6.76 | Olivine, *ol* | — | 23.77 | Olivine | — | 24 |
| $Na_2O$ | 1.98 | 1.35 | Magnetite, *mt* | 3.25 | 4.18 | Opaque oxides | 4.0 | 3.0 |
| $K_2O$ | 0.59 | 0.43 | Ilmenite, *il* | 2.22 | 1.65 | Biotite | 0.5 | 3.0 |
| $P_2O_5$ | 0.14 | 0.10 | | | | Sphene | — | Tr. |
| $H_2O^+$ | 0.25 | 1.45 | Apatite, *ap* | 0.34 | 0.34 | Apatite | Tr. | Tr. |
| $H_2O^-$ | 0.23 | 0.11 | Water, $H_2O$ | 0.48 | 1.56 | Alteration products | 2 | 5 |
| Total | 100.74 | 100.61 | | 100.68 | 100.68 | | 99.5 | 100.7 |

*Explanation of column headings*

1 Diabase 10 m above base (Walker's specimen 865–60)

2 Olivine diabase 20 m above base (Walker's specimen 824–60)

[1] For details of procedure see such standard works as Iddings (1909, pp. 348–393), Johannsen (1939), Barth (1952, pp. 76–82).

groundmass materials) constitute what has variously been termed the *fabric*, *structure*, or *texture* of the rock. *Fabric* should be interpreted as a whole; and fine distinction between such terms as *texture* and *structure* seems unwarranted, especially since there is still no uniformity as to such usage. In this book all three terms are considered as synonymous.

VOLCANIC TEXTURES Collectively volcanic textures reflect the interplay of two rate processes: nucleation of crystals and crystal growth. Both rates increase from zero at the equilibrium crystallization temperature, reach a maximum, and then decline as the temperature falls (Chap. 4). Where crystal growth is diffusion-controlled, as is likely in magmas where the crystals have a different composition from the enclosing liquid, the rate of crystal growth is dominated by the rate at which material can diffuse through the liquid to become part of the growing crystal. Diffusion rates are inversely proportional to temperature and to viscosity, so that if temperature falls relatively quickly and the liquid residue becomes increasingly siliceous, the viscosity of this liquid may rise dramatically and the crystal growth rate may go to zero. In general we should expect comparatively few crystal nuclei with relatively high growth rates in slowly cooled basic magma, in contrast to many nuclei with small growth rates in quickly cooled basic magma. In acid magmas, the viscosity of the liquid may be so high or increase so much with a small drop in temperature that both nucleation and crystal growth are inhibited—the liquid becomes a glass. However, it is rare to find an obsidian devoid of minute crystals or microlites.

One of the commonest and most characteristic textures of volcanic rocks is *porphyritic* structure, in which large crystals (phenocrysts) of one or more minerals are set in a finely crystalline or glassy groundmass. In many rocks minerals which occur as phenocrysts are also present in this fine-grained base. The sharp break in grain size between phenocrysts and groundmass is correlated with some corresponding change in conditions prevailing during freezing of the magma. Such a break occurs where slow cooling of magma deep within the crust has given way to rapid cooling following uprise of the magma and extrusion at the surface or injection into cooler rocks of the upper crust. That phenocrysts of volcanic rocks do in general belong to the earlier stages of magmatic crystallization is confirmed by prevalence of high-temperature minerals among them. Laboratory experiments have shown, for example, that at low pressure olivine crystallizes early from melts which later yield magnesium pyroxene and that anorthite is the high-temperature member of the plagioclase series. Olivine and anorthite-rich plagioclase are common phenocryst minerals in rocks containing pyroxene and more sodic plagioclase as the chief constituents of the groundmass. The occurrence of leucite phenocrysts in a nepheline-sanidine groundmass has similar significance. But to assume an early origin for all large crystals of igneous rocks still is not justified.

Another very general assumption in interpreting fabric of volcanic rocks is that idiomorphic crystals with sharply defined outlines finished crystallizing earlier than allotriomorphic grains of another mineral with irregular boundaries. This is borne out by the general tendency for phenocrysts to have idiomorphic outlines and for almost invariable idiomorphism of crystals of any mineral against enclosing glass. But there are certain exceptions to this rule. Olivine phenocrysts in basalt are commonly rounded in outline, while the quartz phenocrysts in acid volcanics of the quartz-porphyry and rhyolite groups are in many cases not only rounded but deeply embayed. Here the phenocrysts seem to have crystallized early and then, under changing conditions, to have reacted with or become partially redissolved in the magma, so that an initially idiomorphic form has been partially or completely destroyed by corrosion.

In many thick basaltic lavas and in basic sills, the central portion has a characteristic texture which has been called *ophitic*, *diabasic*, or *doleritic*. In these rocks, randomly oriented laths of plagioclase are enclosed by large, irregularly bounded crystals of augite, often of a titaniferous variety. The origin of this texture has puzzled many petrologists, but as we shall see later in Chap. 4, the nucleation and growth rates of a compound are also related to its entropy of melting. Since the melting entropies of diopside and anorthite are different (Table 4-2), and since the nucleation and growth rates can be shown to be proportional and inversely proportional respectively to the entropy of melting, the development of ophitic texture is an understandable result of an intermediate cooling rate.

Glass is a common and characteristic constituent of volcanic rocks. In basic rocks, it is typically present in subordinate amounts in the groundmass or it is altogether lacking. But in acid lavas it can be the most abundant constituent, and some rocks of the rhyolite and trachyte families (obsidians and pitchstones) are made up almost entirely of glass within which swarms of embryonic crystals (crystallites) may be seen beneath the microscope. Even basic magmas may solidify as glass if chilling is unusually rapid, as when flowing lavas encounter sea water, or when magma at the margins of a dike or sill is locally chilled by sudden contact with cooler rocks. Beneath the microscope or under a hand lens many glassy rocks show minute, curved, sometimes partially concentric cracks due to expansion induced by hydration. This is termed *perlitic* structure. It is highly developed in perlites, which are rhyolitic glasses rich in water incorporated at low temperature (submagmatic). In other glassy rocks, especially in members of the rhyolite family, needlelike crystals of feldspar, accompanied by cristobalite, have developed radially from scattered nuclei to build up spherical bodies (*spherulites*) ranging from a fraction of a millimeter to several centimeters in diameter. Under the microscope spherulites can in many instances be shown to have grown after flow of the magma had ceased, that is, after it had reached an essentially glassy condition. Since glass at ordinary temperatures is a metastable material, it is liable to crystallize (devitrify) with passage of time,[1] especially if subjected to raised temperatures resulting from burial. Hence glassy rocks are

[1] Glassy rocks $4 \times 10^9$ years old are found on the moon, which suggests that migration of water rather than time alone is the controlling factor in devitrification of old terrestrial glasses.

rare, though not unknown, in Paleozoic as compared to Tertiary formations. In their place are uniformly fine-grained (aphanitic or cryptocrystalline) volcanic rocks, within which persistent perlitic cracks may survive as evidence of a former glassy state.

Many volcanic rocks are riddled with small spheroidal or tubular cavities attributed to escaping bubbles of water vapor and other gases. This *vesicular* structure results from the generation or exsolution of a volatile phase; sometimes this is loosely called boiling. The most striking example of vesiculation is *pumice*, the glassy froth formed from rapidly extruded magma, frequently of rhyolitic composition.

Many volcanic rocks, especially those that have crystallized from viscous acid magmas, show a tendency for parallel alignment of various constituent fabric elements. This alignment is usually correlated with flow of the partly crystalline magma (for example, R. H. Clark, 1952; Spry, 1953). Well-known instances are subparallel alignment of feldspar laths in the groundmass of lavas (*trachytic* texture of dacites, trachytes, and phonolites; *pilotaxitic* texture of basalts); parallel orientation of alternately crystalline and glassy bands in finely banded rhyolites; and parallel alignment of tubular vesicles in pumice. The direction of flow may be either parallel or at right angles to the direction of linear parallelism of prismatic or lathy crystals. It can be established with certainty in oriented sections in which rotation of phenocrysts or microcorrugation of flow laminae can be detected. In such rocks the flow direction is normal to the axis of rotation or of corrugation.

PLUTONIC TEXTURES The most characteristic features of the plutonic fabric are its holocrystalline state (glass being absent) and relatively coarse equigranular texture. In basic rocks the structure is typically *gabbroid* (allotriomorphic granular) in that completely allotriomorphic grains dominate the fabric. In most acid and intermediate plutonic rocks there is a distinct tendency on the part of some essential minerals (notably micas, hornblende, feldspars) to develop as grains with subidiomorphic outlines. The resultant structure is *granitoid* (hypidiomorphic granular). It is now realized that some details of the granitoid fabric are due to postmagmatic processes of diffusion and unmixing in a completely solid rock (cf. Tuttle, 1952). Albitic rims on crystals of potash feldspar at contacts with plagioclase in granites seem to have originated in this way.

A special feature of some acid plutonic rocks is *graphic* or *micrographic* structure. Large grains of feldspar, usually orthoclase or microcline, individually enclose many small, imperfectly developed bodies of quartz which maintain constant crystallographic orientation throughout a large part, or all, of a given host crystal. Some such intergrowths possibly represent products of eutectic crystallization of the two minerals concerned, since the structure is closely similar to that of a metallic eutectic. But it is also probable that many intergrowths, especially in pegmatites, are the result of partial replacement of one mineral by the other. In some granitic rocks potash feldspar is replaced marginally by an intergrowth of vermicular quartz in plagioclase. This is called *myrmekite*. Unmixing structures are almost universal in crystals of alkali feldspars, pyroxenes, and ilmenite.

Some degree of banded (gneissic) structure and parallel alignment of tabular or prismatic crystals or elongated mineral clots is common in plutonic rocks. The structure may be primary, the result of flow of the partly crystalline magma during intrusion or convection or spasmodic turbulence affecting the magma after intrusion; or it may result from the sorting of crystals of different sizes and densities as they settled under gravity in a still body of liquid magma. Alternatively, it may be a secondary (metamorphic) structure imposed during deformation of the completely crystalline rock. The origin and significance of gneissic structure in any given instance may be difficult or impossible to determine with certainty. Moreover, there are rocks of metasomatic (metamorphic) origin whose mineralogy and structure are similar to those of truly igneous rocks and whose banded fabric is inherited from a foliated or bedded structure in the parent rock.

Open cavities are much rarer in plutonic than in volcanic rocks. It may be that the water content of the magma is steadily depleted by crystallization of hydroxyl-bearing minerals (mica, hornblendes) under plutonic conditions, so that generation of a gas phase is restricted to a very late stage when crystallization of magma is too advanced to permit vesiculation. Some plutonic rocks, especially granites, have sparsely scattered, angular cavities into which idiomorphic or faceted crystals extend. The fabric of such rocks is termed *miarolitic*. It is likely that miarolitic cavities represent late emission of gas in small quantities toward the end of crystallization.

## APPROACH TO CLASSIFICATION

So far we have generalized regarding essential characteristics of igneous rocks—composition, mineralogy, texture—that permit much to be inferred regarding the nature and history of the parent magmas. It should already be obvious that there is almost infinite variety in the way in which such genetically determined characteristics are combined in rocks of different kinds, such as basalt, rhyolite, granite. To systematize this very extensive information, to bring out regular patterns where such exist, and to provide a descriptive language of petrology, we must set up some form of rock classification. This has been a formidable task for petrologists; and the student today must understand something of the way this problem has been faced and how classification has developed to meet new concepts and new data in a rapidly evolving science. To this general topic we devote the ensuing chapter.

# Classification and Variety of Igneous Rocks

# 2

## Development of Classificatory Concepts

During the latter half of the last century, expanding use of the petrographic microscope revealed the great mineralogical and textural variety displayed by igneous rocks. Traditional classification developed mainly to systematize rapidly growing descriptive information of this kind. Thus in the classic compilations of Zirkel and of Rosenbusch we find elaborate classifications based essentially on mineralogical and textural criteria. It is in this pioneer work that current usage of rock nomenclature and classification is rooted.

It came to be realized that both the mineralogy and texture of an igneous rock are determined by its chemistry and cooling history, and that rock chemistry in turn must reveal something of still earlier processes concerned in generation of magma and its subsequent evolution. Moreover, petrologists of the last century, familiar with problems of taxonomy in paleontology, sought petrographic criteria that might be related to the origin of an igneous rock, the subject of petrogenesis. The difficulty was to decide which criteria were the most significant petrogenically, an uncertainty still with us today.

As more became known about the mineralogy of igneous rocks, chemically experienced petrologists such as H. S. Washington began to concentrate on rock chemistry—for many years through routine analysis of a dozen oxide components. Chemical data accrued so rapidly early in this century that purely chemical classifications of rocks were proposed. One of these, proposed by W. Cross, J. P. Iddings, L. V. Pirsson, and H. S. Washington, was based on a formal calculation of the oxide components of the rock into a series of standard mineral components called the CIPW norm; the proportions or combinations of these normative components were then used to pigeonhole the rock into the CIPW quantitative rock classification. Today the CIPW norm is still widely used, but the rather cumbersome CIPW classification of rock analyses has fallen into disuse.

The relative merits of mineralogical as opposed to chemical classifications of igneous rocks were the subject of wide discussion during the first decades of this century (for example, Harker, 1909, pp. 360–377; Shand, 1927; Bowen, 1928, pp. 321–322). Genetic implications were increasingly emphasized. But as late as 50 years ago the nomenclature of igneous petrology was in a state of chaos. On the one hand was overattention to mineralogical trivia and excessive proliferation of rock names based on mineralogical and textural distinctions of dubious significance. But it was an emphasis not without value, for there is no doubt that rocks and minerals were the better described, a tradition unfashionable today. On the other hand chemical classifications were devised and elaborated to provide for all conceivable combinations of rock-forming minerals, common or rare, and in any proportions. Overelaboration and failure to appreciate the essential mineralogical and chemical characteristics of rocks were sterilizing petrology—the inescapable outcome of classification pursued for its own sake.

S. J. Shand saw the situation clearly. Recognizing the stifling influence of tradition and historic accident on rock nomenclature, he wrote (1927, p. 122): "Some of our most used names, like granite, porphyry, obsidian and basalt, are of great antiquity; others including gabbro, diabase, diorite, trachyte and phonolite date from the beginning of the nineteenth century, when geology was still undecided whether such rocks were igneous, or were chemical precipitates from the primeval ocean." And he proposed to remedy the situation (1927, p. 101):

> In attempting to set up a classification to take the place of the disorderly collection of names that does duty as a classification now, what is required of us is that we should pay less attention than we have done in the past to the individual peculiarities of rocks, and more to the laws that govern their formation. The laws are those of physics and chemistry. The problem before us, then, is to elicit the physico-chemical significance of the various geographical, geological, textural and mineralogical characters that eruptive rocks present, and to express them in a classification.

To this purpose Shand developed his concept of silica and alumina saturation to which we shall turn shortly. On this idea, and the new significance that it gave to an outmoded terminology, we lean heavily in this book.

## Mutual Gradation between Rock Types

Variety among igneous rocks is great, but not infinite. Perhaps their most characteristic feature is that their variety tends to be gradational, whether the criteria in question are chemical, mineralogical, or textural. One example is the Quaternary volcanic province of western Nicaragua described by McBirney and Williams (1965, table 4), in which we find lavas with 48 to 50 percent $SiO_2$ that every petrologist would call basalt and others with 55 to 58 percent $SiO_2$ that are typical andesites. What do we call still other lavas with 52 percent $SiO_2$? Some would say basalt, others basaltic andesite; but the terminology is not of prime significance. What is important is to recognize the complete chemical and mineralogical gradation that exists between the basalts and the andesites of this part of Nicaragua.

In any discussion of the character of chemical or mineralogical transition in a series of associated rocks, it is often convenient to refer collectively to rocks at either end or in the intermediate range of the series. To fill this need, some rather loose terminology has grown up, referring respectively to silica content and to relative proportions of dark- and light-colored minerals. These properties are mutually dependent, since light-colored minerals, with certain exceptions such as nepheline and calcic plagioclase, are more siliceous than dark.

## GRADATION IN SILICA CONTENT

To denote silica content we use the terms *acid* and *basic* in the sense already defined for magmas. In this book we set no precise numerical limits to this usage. But to gain some idea of its implications the reader may wish to note some limits commonly employed elsewhere (for example, Williams et al., 1955, p. 27):

*Acid* $SiO_2 > 66$ percent
Example: granites (average 72 percent); granodiorites (67 percent)

*Intermediate* $SiO_2$ 52 to 66 percent
Example: andesites (average 57 percent); trachytes (62 percent)

*Basic* $SiO_2$ 45 to 52 percent
Example: basalts (average 48 to 51 percent)

*Ultrabasic* $SiO_2 < 45$ percent
Example: peridotites (average 41 to 42 percent); nephelinites (40 percent)

Usage is loose enough for most rocks to be referred to an appropriate category simply on the basis of the relative proportions of their minerals.

## GRADATION IN COLOR

With respect to color, rock-forming minerals fall into two contrasted groups:

1. *Felsic* minerals, with low density and light color (contributing to the white, pale gray, and pink tones in rock color): quartz, feldspars, feldspathoids
2. *Mafic* minerals, denser and dark in color (contributing to the green, brown, and black color tones): pyroxenes, amphiboles, olivines, and biotite

Igneous rocks likewise can be classed as felsic or mafic, depending on the preponderant mineral group. There is also a rough correspondence with the respective acid and basic divisions defined by silica content. *Ultramafic* is a term widely used for plutonic rocks completely lacking felsic minerals, since these form a natural group whose identity is sharper than that of the ultrabasic group in which, together with unrelated nepheline-olivine–bearing volcanics, they are artificially grouped by silica content. More or less synonymous with felsic and mafic are the color-index terms *leucocratic* and *melanocratic* respectively.

## Saturation Concept

The two most abundant components of almost all igneous rocks are $SiO_2$ and $Al_2O_3$. Intuitively, perhaps, it would seem likely that a parameter based on these components could provide a significant criterion for rock classification. Zirkel had already grasped something of this in his emphasis on the mutual incompatibility of quartz and the silica-deficient silicates olivine and feldspathoids in framing his classification. Shand, with his balanced background in the chemistry and mineralogy of igneous rocks, went further. He proposed a classification in terms of the parallel concepts of silica and alumina saturation.

Igneous minerals can be divided into two groups: those which are compatible with either quartz or tridymite and those which are never associated with a primary silica mineral. Representatives of the two groups of minerals, the saturated and the unsaturated, are:

| SATURATED | UNSATURATED |
|---|---|
| All feldspars | Leucite |
| Pyroxenes (Al- and Ti-poor) | Nepheline |
| Amphiboles | Sodalite, hauyne, nosean |
| Micas | Magnesian olivine |
| Fayalitic olivine | Melanite garnet |
| Spessartine-almandine | Perovskite |
| Sphene | Melilite |
| Zircon | Corundum |
| Tourmaline | Augite (Al- and Ti-rich) |
| Topaz | |
| Magnetite | |
| Ilmenite | |
| Apatite | |

Notice that a mineral does not have to contain silica to be compatible with quartz or another silica mineral. From the association of these two groups of minerals, Shand divided igneous rocks into three classes:

1. *Oversaturated rocks*, which contain a primary silica mineral
2. *Saturated rocks*, which contain neither quartz nor an unsaturated mineral
3. *Unsaturated rocks*, which contain unsaturated minerals

The parallel concept of alumina saturation stems from the 1:1 alkali/aluminum ratio of feldspars and feldspathoids; any excess or deficiency of Al (on a molecular basis) with respect to the alkalis in a rock is reflected in the mineralogy. This leads to four more classes of rocks, each independent of silica saturation:

1. *Peraluminous rocks*, in which the molecular proportion of $Al_2O_3$ exceeds ($CaO + Na_2O + K_2O$); corundum (*C*) appears in the norm. Characteristic minerals are muscovite, topaz, tourmaline, spessartine-almandine, corundum, andalusite, and sillimanite.
2. *Metaluminous rocks*, in which $Al_2O_3 < (CaO + Na_2O + K_2O)$ but $Al_2O_3 > (Na_2O + K_2O)$; anorthite (*an*) is prominent in the norm. Al-bearing dark minerals such as biotite, hornblende, and melilite are typical.
3. *Subaluminous rocks*, in which $Al_2O_3 \simeq (Na_2O + K_2O)$; normative *an* is small. The only minerals with essential $Al_2O_3$ are feldspars and feldspathoids.
4. *Peralkaline rocks*, in which $Al_2O_3 < (Na_2O + K_2O)$ and, much more rarely, $Al_2O_3 < K_2O$; acmite (*ac*), sodium silicate (*ns*), and rarely potassium silicate (*ks*) appear in the norm. Alkali ferromagnesian minerals such as aegirine, riebeckite, richterite, and cossyrite are ubiquitous.

Today little use is made of the distinction between metaluminous and subaluminous rocks. Shand's classification is anchored in mineralogy, particularly that of the slowly cooled rocks, where metastability of composition or phase is unlikely. But it can also be applied to fine-grained or glassy volcanic rocks if a chemical analysis is available to provide calculated normative mineral assemblages.

At the end of this chapter we show that Shand's saturation concept can be treated in terms of the chemical potentials or activities of $SiO_2$ and $Al_2O_3$. But we defer discussion of this aspect until we have considered the commonly used rock names and the rocks that they denote. The reader will therefore be more aware of the variety of rocks and magmas, and of the mineralogical and chemical criteria used to characterize them when these are later interwoven with the thermodynamic treatment.

## Classification

For the purposes of this book we have adopted a rock classification with generally accepted terminology based mainly on readily observed mineralogical criteria. It is only to make discussion intelligible and as free as possible from ambiguity that a systematic classification is necessary in a book dealing with petrology rather than descriptive petrography. But the traditions of petrography persist to blur or confuse much petrological discussion, for it is a subject littered with unnecessary rock names, often of only local significance and typically celebrating only minor distinctions. Rock names are usually completely uninformative about the features of the rock they are supposed to denote, and a particular criterion tends to be forgotten or perhaps worse, changed within a few years. Even today, there would be little consensus among earth scientists regarding the features a rock has to display before it deserves the name *andesite*, some stressing its geologic environment, others its chemical composition, and others its mineralogy. We have therefore illustrated each common rock type with a representative chemical analysis, together with a calculated CIPW norm. We start with families of volcanic rocks, since it is these that most closely reflect the compositions of magmas. Many families of plutonic rocks are broadly analogous in composition to corresponding volcanic groups—gabbro to basalt, syenite to trachyte, and so on. But many plutonic rocks during protracted cooling have become modified in composition by interaction with their surrounding rock-water envelope, which oxygen-isotope studies have been particularly successful in revealing. To this exchange of water and oxygen must doubtless be added other elements, alkalis, silicon, and heavy metals, all in varying degrees, but which in sum make a plutonic rock unrepresentative of the magma which cooled to produce it. For evidence of the composition of magma, volcanic rocks seem better suited.

### FAMILIES OF VOLCANIC ROCKS

*The basalt family* in the broad sense of the term includes basic volcanic rocks with calcic plagioclase ($An_{>50}$), augite, and in many cases olivine as essential constituents. The family is divided into two groups by the presence or absence of a calcium-poor pyroxene (pigeonite or orthopyroxene). In the CIPW norm, this division approximately corresponds to the presence or absence of substantial *hy*, the "critical phase of silica undersaturation" of Yoder and Tilley (1962). Representative analyses of the following basalt types are given in Table 2-1:

1. *Tholeiitic basalts.* Olivine is present in *olivine-tholeiites* and in substantial amount in tholeiitic *picrites* but otherwise is often absent. Augite and pigeonite and/or orthopyroxene occur, but subcalcic augite is also found. Sometimes these basalts are iron-rich with large amounts of iron titanium oxide, interstitial siliceous glass, or rare quartz. All tholeiitic basalts have high normative *hy*. Olivine tholeiites have abundant modal olivine, and *ol* in

smaller amount in the norm. Normative *Q* is typical of tholeiitic basalts poor in or lacking modal olivine. Oceanic varieties have high $TiO_2$ (>2.5 percent); continental examples have less (~1 percent). *Basaltic andesite* is a variety transitional to andesite.

2. *High-$Al_2O_3$ basalt.* This basalt is typically nonporphyritic with $Al_2O_3$ > 17

**Table 2-1** Representative analyses and CIPW norms of basalt types

| | 1 | 2 | 3 | 4 | 5 | 6 | 7 |
|---|---|---|---|---|---|---|---|
| $SiO_2$ | 47.01 | 51.57 | 50.83 | 48.27 | 52.28 | 46.53 | 43.52 |
| $TiO_2$ | 3.20 | 0.80 | 3.44 | 0.89 | 0.94 | 2.28 | 2.45 |
| $Al_2O_3$ | 15.57 | 15.91 | 12.67 | 18.28 | 15.75 | 14.31 | 15.76 |
| $Fe_2O_3$ | 2.32 | 2.74 | 3.10 | 1.04 | 3.28 | 3.16 | 2.82 |
| FeO | 11.57 | 7.04 | 11.39 | 8.31 | 4.88 | 9.81 | 7.14 |
| MnO | 0.20 | 0.17 | 0.25 | 0.17 | 0.16 | 0.18 | 0.16 |
| MgO | 5.25 | 6.73 | 4.19 | 8.96 | 4.76 | 9.54 | 9.57 |
| CaO | 9.77 | 11.74 | 8.18 | 11.32 | 8.30 | 10.32 | 12.28 |
| $Na_2O$ | 3.00 | 2.41 | 3.24 | 2.80 | 3.44 | 2.85 | 3.02 |
| $K_2O$ | 0.31 | 0.44 | 0.87 | 0.14 | 2.08 | 0.84 | 1.43 |
| $P_2O_5$ | 0.32 | 0.11 | 0.75 | 0.07 | 0.36 | 0.28 | 0.41 |
| Rest* | 1.64 | 0.45 | 0.94 | 0.22 | 3.57 | 0.14 | 1.16 |
| Total | 100.16 | 100.11 | 99.85 | 100.47 | 99.80 | 100.24 | 99.72 |
| *Q* | — | 2.26 | 4.44 | — | 3.98 | — | — |
| *or* | 1.87 | 2.60 | 5.56 | 0.56 | 12.29 | 5.28 | 8.45 |
| *ab* | 25.39 | 20.39 | 27.25 | 23.58 | 29.11 | 20.04 | 4.48 |
| *an* | 28.10 | 31.29 | 17.51 | 36.97 | 21.39 | 23.63 | 25.22 |
| *ne* | — | — | — | — | — | 2.20 | 11.42 |
| *di* | 15.17 | 21.32 | 15.75 | 15.23 | 6.41 | 20.89 | 25.93 |
| *hy* | 11.92 | 16.05 | 16.04 | — | 13.76 | — | — |
| *ol* | 5.93 | — | — | 20.55 | — | 18.48 | 13.25 |
| *mt* | 3.36 | 3.97 | 4.41 | 1.39 | 4.76 | 4.53 | 4.09 |
| *il* | 6.08 | 1.52 | 6.54 | 1.67 | 1.79 | 4.41 | 4.65 |
| *ap* | 0.76 | 0.26 | 1.68 | 0.17 | 0.85 | 0.67 | 0.97 |

*Explanation of column headings*

1 Olivine tholeiite, Albemarle Island, Galápagos (McBirney and Williams, 1969, p. 121, no. 63)
2 Tholeiite, Talasea, New Britain (Lowder and Carmichael, 1970, p. 27, no. 311)
3 Tholeiitic lava, Thingmuli, Iceland (Carmichael, 1964*b*, p. 439, no. 10)
4 High-alumina basalt, Medicine Lake Highlands, California (Yoder and Tilley, 1962, p. 363, no. 16)
5 Hypersthene shoshonite lava, Yellowstone Park, Wyoming (J. Nicholls and Carmichael, 1969*a*, p. 60, no. 118A)
6 Alkali olivine basalt, prehistoric flow, Hualalai, Hawaii (Yoder and Tilley, 1962, p. 362, no. 20)
7 Basanite flow, Korath Range, Lake Rudolph, Kenya (F. H. Brown and Carmichael, 1969, p. 251, no. k8)

* Includes $H_2O$, $CO_2$, and so on.

percent and $TiO_2 \cong 1$ percent. Olivine and subcalcic augite are common. Often it is associated with andesites on the continents. It is a widespread product of submarine volcanism in ocean basins ($<2$ percent $TiO_2$ and $\sim 0.2$ percent $K_2O$).

3. *Shoshonite.* This consists of plagioclase phenocrysts (jacketed by sanidine), olivine, augite, and calcium-poor pyroxene. High $K_2O/Na_2O$ ratios. Apparently confined to continents.

4. *Alkali olivine basalts.* This is a group of basalts all containing abundant olivine and a single pyroxene—augite with substantial Al and Ti. Accessory kaersutite or phlogopite is present in some rocks. Minor normative *ne* and abundant *ol* are characteristic. *Basanite* has higher normative *ne*, small amounts of modal nepheline ($\pm$leucite), and abundant olivine; *tephrite* is olivine-free basanite. Magnetite solid solutions in this group contain substantial Mg and Al. Varieties with leucite in place of nepheline are denoted by the leucite prefix. *Trachybasalt* is olivine basalt with high ($Na_2O + K_2O$), sometimes with barkevikite or kaersutite (current usage covers a wide range of rocks).

The *andesite-rhyolite family* includes intermediate to acid volcanic rocks. Some can be very rich in iron and peralkaline. Representative analyses and CIPW norms are given in Table 2-2.

1. *Andesite, porphyrite.* These rocks consist of andesine or labradorite and some combination of augite, orthopyroxene, hornblende. They may have bytownite phenocrysts, and some basic types contain olivine. This group is characteristic of island arcs and continental margins. *Icelandite* is an iron-rich, aluminum-poor variety restricted to oceanic islands.

2. *Dacite, rhyodacite.* These are acid volcanic rocks with andesine, biotite, hornblende and/or pyroxenes, and in some cases quartz and sanidine. Glassy types can be identified only by chemical analysis.

3. *Rhyolite, quartz porphyry.* These are acid volcanic rocks often with quartz, oligoclase, sanidine, and minor ferromagnesian minerals (biotite, hypersthene, hornblende, ferroaugite, and fayalitic olivine). Glassy types (*obsidians* or *pitchstones*) can be identified only by chemical analysis. Sodic (peralkaline) rhyolites (for example, *pantellerites, comendites*) contain anorthoclase and sodic ferromagnesian minerals such as aegirine, riebeckite, cossyrite.

4. *Latite.* This type is transitional between andesites and trachytes; plagioclase (andesine or labradorite) and sanidine are both abundant; interstitial quartz is present in some; and there is a high $K_2O/Na_2O$ ratio. *Trachyandesites* are similar but may contain normative *ne*.

**Table 2-2** Representative analyses and CIPW norms of the andesite-dacite-rhyolite family

| | 1 | 2 | 3 | 4 | 5 | 6 |
|---|---|---|---|---|---|---|
| $SiO_2$ | 58.57 | 62.74 | 67.73 | 72.19 | 76.21 | 70.13 |
| $TiO_2$ | 0.64 | 0.56 | 0.50 | 0.33 | 0.07 | 0.30 |
| $Al_2O_3$ | 19.87 | 16.53 | 15.44 | 12.62 | 12.58 | 7.97 |
| $Fe_2O_3$ | 3.20 | 1.71 | 0.69 | 3.14 | 0.30 | 2.77 |
| FeO | 2.73 | 2.14 | 2.40 | 1.12 | 0.73 | 5.27 |
| MnO | 0.15 | 0.07 | 0.06 | 0.05 | 0.04 | 0.26 |
| MgO | 1.74 | 3.24 | 1.30 | 0.58 | 0.03 | 0.07 |
| CaO | 7.51 | 6.20 | 3.35 | 2.07 | 0.61 | 0.55 |
| $Na_2O$ | 4.25 | 4.08 | 3.85 | 3.45 | 4.05 | 7.46 |
| $K_2O$ | 0.74 | 1.18 | 3.25 | 3.70 | 4.72 | 4.24 |
| $P_2O_5$ | 0.10 | 0.16 | 0.15 | 0.02 | 0.01 | 0.04 |
| Rest | 0.63 | 1.31 | 1.15 | 0.80 | 0.52 | 1.19 |
| Total | 100.13 | 99.92 | 99.87 | 100.07 | 99.87 | 100.25 |
| *Q* | 12.67 | 17.43 | 22.98 | 33.18 | 33.06 | 28.75 |
| *or* | 4.37 | 6.97 | 18.90 | 21.86 | 27.80 | 25.06 |
| *ab* | 35.96 | 34.52 | 32.49 | 29.19 | 34.06 | 17.39 |
| *an* | 32.96 | 23.30 | 15.29 | 8.02 | 2.50 | — |
| *di* | 2.93 | 5.09 | — | 1.64 | 0.30 | 2.19 |
| *hy* | 4.52 | 7.38 | 6.82 | 0.68 | 0.92 | 8.68 |
| *ac, ns* | — | — | — | — | — | 15.90 |
| *mt* | 4.64 | 2.48 | 0.93 | 2.82 | 0.46 | — |
| *il* | 1.21 | 1.06 | 0.91 | 0.63 | 0.15 | 0.57 |
| *ap* | 0.24 | 0.38 | 0.34 | 0.15 | — | 0.09 |

*Explanation of column headings*

1 Andesite flow, Mt. Misery, St. Kitts, West Indies (P. E. Baker, 1968*b*, table 6, no. 15)
2 Andesite, Mt. Shasta, California (A. L. Smith and Carmichael, 1968, table 2, no. 72)
3 Pyroxene dacite, Medicine Lake Highlands (Carmichael, 1967*b*, table 5, no. 13)
4 Rhyodacite, Talasea, New Britain (Lowder and Carmichael, 1970, table 2, no. 279B)
5 Rhyolitic obsidian, Mono Craters, California (Carmichael, 1967*a*, table 5, no. 18)
6 Pantellerite obsidian, Lake Naivasha, Kenya (Nicholls and Carmichael, 1969*a*, table 1, no. 121R)

The *trachybasalt-trachyandesite family* includes basic to intermediate volcanic rocks with augite and frequently olivine; it is more feldspathic (high $Na_2O + K_2O$) than basalts. Normative *ne* is more common than normative *Q*. Representative analyses are given in Table 2-3.

1. *Trachybasalt, trachyandesite.* Olivine, augite with notable (Al + Ti), and labradorite-andesine are present in some, along with amphibole and/or phlogopite. Accessory nepheline, less commonly leucite, is sporadic.

2. *Hawaiite* has andesine, anorthoclase, olivine, augite, and interstitial phlogopite. *Mugearite* has oligoclase but is otherwise similar. *Tristanite* and *benmoreite* are closer to trachyte, the former having high $K_2O/Na_2O$.

**Table 2-3** Representative analyses and CIPW norms of trachybasalt and trachyte-phonolite families

| | 1 | 2 | 3 | 4 | 5 | 6 |
|---|---|---|---|---|---|---|
| $SiO_2$ | 47.55 | 49.52 | 46.66 | 53.22 | 57.73 | 54.55 |
| $TiO_2$ | 2.71 | 3.18 | 2.44 | 1.81 | 1.18 | 1.60 |
| $Al_2O_3$ | 16.38 | 17.72 | 16.01 | 17.72 | 17.05 | 19.09 |
| $Fe_2O_3$ | 2.77 | 2.55 | 3.52 | 2.58 | 2.55 | 1.75 |
| FeO | 7.84 | 5.66 | 8.35 | 6.14 | 4.35 | 3.78 |
| MnO | 0.20 | 0.18 | 0.20 | 0.21 | 0.28 | 0.16 |
| MgO | 6.40 | 3.42 | 4.76 | 2.79 | 1.11 | 1.76 |
| CaO | 8.41 | 7.58 | 8.96 | 5.39 | 3.10 | 4.07 |
| $Na_2O$ | 4.46 | 4.94 | 4.56 | 6.00 | 6.81 | 9.06 |
| $K_2O$ | 2.11 | 3.88 | 1.86 | 2.28 | 4.27 | 3.64 |
| $P_2O_5$ | 0.72 | 1.09 | 0.74 | 1.08 | 0.34 | 0.20 |
| Rest | 0.79 | 0.44 | 2.41 | 0.94 | 1.19 | 1.03 |
| Total | 100.34 | 100.16 | 100.47 | 100.16 | 99.96 | 100.69 |
| *Q* | — | — | — | — | — | — |
| *or* | 12.47 | 22.93 | 10.99 | 13.34 | 25.23 | 21.51 |
| *ab* | 22.86 | 22.48 | 25.76 | 47.16 | 47.57 | 32.36 |
| *an* | 18.44 | 14.72 | 17.72 | 14.73 | 3.70 | 0.68 |
| *ne* | 8.06 | 10.46 | 6.95 | 1.99 | 5.09 | 24.00 |
| *di* | 14.48 | 12.68 | 14.40 | 3.68 | 7.95 | 14.90 |
| *ol* | 12.27 | 4.17 | 9.92 | 8.55 | 2.39 | 0.17 |
| *mt* | 4.02 | 3.70 | 5.10 | 3.71 | 3.70 | 2.54 |
| *il* | 5.15 | 6.04 | 4.63 | 3.50 | 2.24 | 3.04 |
| *ap* | 1.71 | 2.51 | 1.62 | 2.69 | 0.81 | 0.47 |

*Explanation of column headings*

1 Trachybasalt lava, Cima Dome, Mojave Desert, California (A. L. Smith and Carmichael, 1969, p. 918, no. 252)
2 Leucite-bearing trachybasalt, Tristan da Cunha (P. E. Baker et al., 1964, table 6, no. 125)
3 Hawaiite, East Otago, New Zealand (Coombs and Wilkinson, 1969, table 6, no. 4)
4 Mugearite, Molokai, Hawaii (Macdonald, 1968, table 2, no. 8)
5 Trachyte lava, Mt. Suswa, Kenya (Nash et al., 1969, table 10, no. W134)
6 Phonolite lava, Viejo caldera, Las Canadas, Tenerife (Ridley, 1970, table 4, no. 34)

The *trachyte-phonolite family* includes volcanic rocks of intermediate silica content, high in ($Na_2O$ + $K_2O$) and low in CaO. Representative analyses are given in Table 2-3 [col. (5) and (6)].

1. *Trachyte*. Trachytes are approximately silica-saturated alkaline lavas with dominant sanidine or anorthoclase accompanied by minor pyroxene, biotite, or hornblende. Small amounts of either quartz or feldspathoid may occur. *Porphyry* is a porphyritic variety. *Kenyte* is a variety transitional to phonolite with anorthoclase phenocrysts and interstitial nepheline. Trachytes with Al < (Na + K) have riebeckite, aegirine, or cossyrite.

2. *Phonolite.* This type includes undersaturated alkaline lavas with abundant feldspathoid (nepheline, sodalite, nosean, or more rarely leucite), sanidine or anorthoclase, aegirine or aegirine augite. Sometimes there is melanite and/or phlogopite. *Tinguaite* is a porphyritic variety.

The *lamprophyre family* includes ultrabasic and basic dike rocks rich in alkali and in iron and magnesium. Typically they are porphyritic with ferromagnesian minerals in two generations: biotite and/or barkevikite, with augite, olivine, and rarely melilite as possible additional minerals. Feldspars are plagioclase or orthoclase, usually confined to groundmass and commonly deuterically altered. Some lamprophyres are chemically equivalent to lavas of the nephelinite family. For average chemical compositions of lamprophyre types, see Metais and Chayes (1963).

The *nephelinite family* includes mineralogically varied ultrabasic and basic volcanic rocks with a high content of alkali. Augite, olivine, accessory phlogopite, and possibly perovskite ($CaTiO_3$) are present. Feldspar is usually absent, but feldspathoids are typical.

1. *Nephelinite.* Nepheline; augite often rich in Ti, Al, and Na; olivine; and possibly accessory amounts of leucite and feldspar. *Etindite* also has perovskite and hauyne.
2. *Leucitite.* Leucite, olivine, augite, with a small amount of nepheline. *Madupite* has abundant phlogopite, augite, and perovskite in a leucite groundmass.
3. *Melilite basalt.* Melilite, olivine, augite, with or without nepheline, but rarely leucite.
4. *Limburgite.* Olivine and augite in an ultrabasic, alkaline, glassy groundmass.
5. *Ugandite*, *katungite*, and *mafurite* contain kalsilite ($KAlSiO_4$) and small amounts of nepheline or leucite, together with augite, olivine, and phlogopite.

## FAMILIES OF PLUTONIC ROCKS[1]

The *gabbro family* (cf. basalts) covers a wide range of basic plutonic rocks consisting mainly of plagioclase (labradorite or bytownite), augite, and in most rocks olivine.

1. *Olivine gabbro.* Plagioclase, augite, olivine, and minor hypersthene; *olivine norite* with olivine and hypersthene but augite subordinate or absent.

[1] Approximately equivalent volcanic families are noted in parentheses.

2. *Gabbro.* Plagioclase and augite, rarely hornblende; *norite* with plagioclase, hypersthene, subordinate augite, rarely biotite. Minor interstitial quartz in some rocks.
3. *Anorthosite.* Plagioclase with only minor pyroxene and/or olivine.
4. *Pyroxenite.* Enstatite or augite or both.
5. *Troctolite.* Calcic plagioclase and olivine.

The *peridotite family* (no volcanic equivalent) includes ultrabasic plutonic rocks consisting mainly of olivine; feldspar is typically absent, but some types have minor calcic plagioclase.

1. *Dunite.* Olivine, minor enstatite, and chrome spinel.
2. *Harzburgite.* Olivine, enstatite, minor chrome spinel.
3. *Lherzolite.* Olivine, enstatite, diopside; minor chrome spinel or pyrope garnet.

The *alkali-gabbro family* (cf. alkali olivine-basalt group) includes the basic plutonic rocks, whose higher alkali content leads to the appearance of such minerals as biotite, barkevikite, orthoclase, anorthoclase, nepheline, or analcite in association with calcic plagioclase, augite, and olivine.

1. *Teschenite.* Analcite-bearing olivine gabbro, usually with barkevikite and titaniferous augite.
2. *Theralite, essexite.* Olivine gabbros containing orthoclase (or anorthoclase), nepheline, brown hornblende, and minor biotite.
3. *Picrite* (transitional to peridotites). Olivine, titaniferous augite, minor plagioclase, and analcite.
4. *Shonkinite.* Augite, olivine, biotite, potash feldspar; sometimes leucite or pseudoleucite. Occasionally has primary pseudobrookite.
5. *Kentallenite* (transitional to monzonite). Plagioclase, potash feldspar, augite, olivine, biotite.

The *diorite family* (cf. andesites) includes intermediate plutonic rocks consisting essentially of andesine, hornblende, and/or pyroxene. *Ferrodiorites* (the nearest volcanic equivalent is *icelandite*) have augite, orthopyroxene, and/or iron-rich olivine and are transitional to gabbro. *Quartz diorite* (*tonalite*) has quartz, biotite, hornblende, and andesine, and it is transitional to the granodiorite family (the volcanic equivalent is *dacite*).

The *granodiorite-granite family* (equivalent to dacites, rhyodacites, rhyolites) includes acid oversaturated rocks consisting mainly of quartz, potassium feldspar (usually perthitic), and sodic plagioclase, with hornblende and/or biotite.

1. *Granodiorite* (equivalent to *rhyodacite*). Andesine-oligoclase dominant, potash feldspar subordinate. Biotite and/or hornblende.
2. *Quartz monzonite, adamellite*. Potash feldspar and oligoclase-andesine subequal. Biotite and/or hornblende.
3. *Granite*. Potash feldspar dominant, oligoclase generally subordinate; biotite alone or with hornblende or muscovite.
4. *Soda granite*. Albite or albite-oligoclase dominant, with small amounts of aegirine or sodic amphibole. *Granophyre* has a characteristic intergrowth of anorthoclase or sanidine and quartz, together with accessory pyroxene, iron-rich olivine, or biotite.
5. *Granite vein rocks. Pegmatite*, coarse-grained and mineralogically complex. *Aplite*, fine-grained white rock consisting of quartz, albite, potash feldspar, and muscovite; sometimes accessory almandine.

The *syenite family* (cf. trachyte-phonolite family) in the broad sense of the term includes saturated or undersaturated plutonic rocks of intermediate silica content, high in alkali.

1. *Syenite*. Saturated rock with potash soda feldspar, subordinate albite-oligoclase, and one or more mafic minerals (hornblende and/or biotite; less commonly, augite). Minor amounts of either quartz or feldspathoid mark respective transitions to granites and to nepheline syenites.
2. *Monzonite* (volcanic equivalent is *latite*). Like syenite, but with subequal amounts of potash feldspar and oligoclase-andesine. Feldspathoidal monzonites are the plutonic equivalent of *tristanites*.
3. *Nepheline-syenite group*. Undersaturated feldspathoidal rocks. A sodic feldspathoid (nepheline, sodalite, analcite, cancrinite) and potash soda feldspar are typically accompanied by one or more alkaline mafic minerals (biotite, alkali hornblende, aegirine).

## Rock Associations, Magma Types, and Magma Series

### ROCK SUITES, KINDREDS

Harker (1909, pp. 90–93) recognized distinctive mineralogical and chemical differences between the Cenozoic volcanic rocks occurring round the respective borders of the Pacific and the Atlantic Oceans. The former, typified by the basalts, andesites, and rhyolites of the Andes and the western United States, are calc-alkalic as later defined by Peacock (1931); the latter, including olivine basalts, nephelinites, basanites, and phonolites of the Canaries, Azores, and other Atlantic islands, are alkalic. Accordingly, Harker set up two contrasted branches (*suites*) of igneous rocks; Pacific (calc-alkalic) and Atlantic (alkalic). They were defined in terms of mineralogical criteria, and the fundamental difference between them was attributed to magmatic evolution along divergent lines controlled in some way by differentiation in radically different tectonic environments. As time went on other suites were proposed: spilitic, Arctic, Mediterranean. Although the nomenclature was predominantly geographic, the diagnostic criteria were mineralogical and chemical. Thus the distinctive characteristic of Mediterranean rocks—such as the lavas of Vesuvius—is high $K_2O$ with the resultant appearance of potassium feldspar and leucite in typical rocks.

By the time Tyrrell's *Principles of Petrology* appeared in 1926, increasingly obvious diversity among more limited igneous provinces and irregularity in their geographic distribution rendered the concept of broad geographically defined rock suites inadequate. "Atlantic" rocks appeared in mid-Pacific islands and in New Zealand along the southwest Pacific margin. Lavas erupted from Tertiary times onward in the north Atlantic province of Iceland prove to be consistently "Pacific" in chemistry and mineralogy. It remained clear, nevertheless, that igneous provinces, whatever their geographic situation, tend to conform to recognizable patterns. Such patterns of rock association Tyrrell termed *rock kindreds*. One such is typified by alkali olivine basalts and associated trachytes and phonolites. Another, conspicuous along the Pacific margins, is characterized by andesites, dacites, and rhyolites. The concept of the rock kindred has proved extremely useful in systematizing data and clarifying ideas on igneous provinces, and also in furnishing a comprehensive basis for current models of petrogenesis (for example, Turner and Verhoogen, 1960).

Allied to the concept of rock suites is that of the *petrographic province*, which is any region within which igneous rocks are related in space and in time. By virtue of this association, there is a presumption that common genetic factors generate the assemblage, hence the modified term *petrogenic province*. There is no clear-cut limit to a province; it is much like a thermodynamic system in that it depends upon the point of view of the observer. Thus the Cenozoic volcanic province of the western United States includes the Pliocene-Recent andesite province of the High Cascades. A more localized example is the late Cenozoic province of East Otago, New Zealand, active over 2.5 million years and composed of alkali olivine basalts, nephelinites, mugearites, and trachytes. Several cycles of eruption, each embracing the full range of rock types, are displayed in an area of only 500 km$^2$ around the city of Dunedin. Among these volcanics several chemical *trends* or *lineages* have been recognized (Coombs and Wilkinson, 1969).

In any petrogenic province, where the various rock types may have been derived by a particular process from a common parent, certain attributes of that parent may be displayed in all the derivatives. One example could be high Zr, Nb, Mo, and Zn, or perhaps high $K_2O/Na_2O$, or in terms of isotopes high $Sr^{87}/Sr^{86}$ ratios throughout the whole rock suite. This common genetic factor is called *consanguinity*.

## VOLCANIC AND PLUTONIC ASSOCIATIONS

After a lifetime study of igneous rocks, Daly (1933, p. 41) concluded: "The igneous rocks of the globe belong chiefly to two types: granite and basalt. . . . To declare the meaning of the fact that one of these dominant types [granite] is intrusive and the other [basalt] extrusive, is to go a long way towards outlining petrogenesis in general." With this quotation as a text, Kennedy and Anderson (1938, pp. 24–33) proposed a twofold grouping of igneous rock associations.

*Volcanic associations* were to include . . ."not only the superficial lava flows and vent intrusions but, in addition, all intrusive masses which are genetically related to a cycle of volcanic activity and originate in the same magmatic source."

*Plutonic associations* were to "comprise the great subjacent stocks and batholiths together with the diverse minor intrusions of such abyssal masses." These were considered to be restricted to orogenic belts and in fact to include all plutonic intrusions geographically associated with orogeny.

A *diatremic association* has been suggested for those igneous rocks "injected by gas fluidization rather than as magmatic liquids" (Harris et al., 1970, p. 197). An example of this association is kimberlite and the brecciated minettes (lamprophyres) of some volcanic necks.

Individually, most recognized patterns (kindreds) of igneous rock association fit one or the other of these categories. Some indeed, such as the associations respectively dominated by peridotites and anorthosites, are plutonic and lack effusive equivalents. But there are significant anomalies. To place the great layered basic intrusions with the volcanic group, as was done by Kennedy, brings out their chemical similarity to basalts, but it can also obscure the significance of their plutonic environment and evolution controlled by differentiation under deep-seated conditions. Again the grouping tends to distort such issues as the possible relation between the plutonic granodiorite-granite association and the chemically similar volcanic equivalents, and the role of acid lavas in orogenic provinces.

## MAGMA TYPES AND MAGMA SERIES

In 1924, the memoir on the Scottish isle of Mull was published (E. B. Bailey et al., 1924) and became one of the pillars of igneous geology. In this volcanic complex of cone sheets, lavas, ring dikes, and intrusions, there are many rock types, coarse- to fine-grained, gabbro to granophyre; but in terms of chemical composition their variety is far more restricted. Thus gabbro, dolerite (diabase), and basalt are different rock types. But in chemical composition they are so similar that they were considered collectively to represent a single magma type. Two main basaltic magma types were recognized in Mull: the nonporphyritic central type, in allusion to the fine-grained lavas found in the caldera; and the plateau magma type, named for the extensive plateau lavas on which the central volcanoes had been built. Several other magma types were recognized in Mull. These include an intermediate magma type represented by lavas and dike rocks of intermediate silica content and an acid magma type composed of felsites (rhyolites) and granophyres. The Mull authors believed that these four magma types were genetically related and together formed the *Mull normal magma series*, with the early and voluminous plateau magma type the probable parent. The concept of a magma series, or genetically related magma types, is used widely today. It must be remembered, however, that whereas a magma type can be defined objectively in terms of chemical criteria, genetic implications essential to the definition of any magma series are necessarily subjective. Over the past 50 years ideas have changed, and there has been little agreement as to whether or not the rocks of Mull represent several magma series.

Daly and Bowen had long recognized and expounded—though by no means on parallel lines—the unique parental role of basaltic magmas in what came to be called magma series. Surveying this phenomenon on a global scale, Kennedy (1933) found what he considered analogs of the two basaltic magma types of Mull, both worldwide in their distribution. He thus proposed two *worldwide* basaltic magma types, defined on a mineralogical-chemical basis:

1. *Tholeiitic magma type* (nonporphyritic central type of Mull): characterized in part by pigeonite (enstatite augite), augite, a siliceous residue that is often glassy, and rarely olivine.
2. *Olivine-basalt magma type* (plateau type of Mull): contains olivine, only one pyroxene (an augite, typically titaniferous), and has an interstitial residuum of alkali feldspar, nepheline, or zeolites.

In association with each of these two worldwide magma types, there is a variety of igneous rocks, or magma types, which under the influence of Bowen (for example, 1928) were presumed to be derivatives by some process of crystal fractionation. Thus the tholeiitic magma type was conceived to be the parent of the *tholeiitic magma series*, with rhyolite or granophyre as the typical end product. On the other hand, the *olivine-basalt magma series* was generally conceded to be a progression of magma types ending in liquids of trachytic or phonolitic composition.

Tilley (1950) recognized that tholeiitic lavas were not confined to the continental environment, as had been thought by Kennedy, since they formed a substantial part of the island of Hawaii. He also introduced the modified name *alkali olivine-basalt magma series*, the prefix denoting higher alkali content ($Na_2O + K_2O$) with respect to the associated tholeiitic series of comparable silica contents. The prefix also serves to emphasize that although olivine is found in both of Kennedy's magma types, only in the tholeiitic is it in reaction relationship with melt.

Although these two magma series dominate the earth's volcanic history, transitional basaltic magma types are found. High-alumina basalts may be such a type. The case for other magma series is not yet conclusive. One proposed, the *shoshonite magma series* (Joplin, 1968), since renamed *association*, characteristically has high $K_2O/Na_2O$. Its members are widespread but confined to the continents. It is conceived to have both a silica-saturated and an unsaturated branch, with leucite as the typical feldspathoid. To summarize the present uncertain status of magma series, which soon may also encompass a nephelinitic series, we quote from a recent summary on magma types (Turner, 1970, pp. 341–343):

> Over the past twenty years many students of volcanic rocks have found the framework presented by the two classic magma-types too narrow to accommodate all basaltic magmas. New magma-types have been proposed; and as basaltic magma-types have proliferated, questions inevitably arise as to their genesis, evolution and possible mutual relations. As early as 1950 Tilley was inquiring as to the status of high-alumina basalts of northern California. Kuno (1960) endorsed the concept of a high-alumina basalt magma-type and explored its character and distribution in Japan and elsewhere. Is there only one high-alumina type? And what if any are its relationships to tholeiitic and alkali-basalt magmas? And what of nephelinites and basanites which, though commonly associated with more "normal" alkali olivine basalts are chemically much further from these than are alkali basalts of Hawaii from Kilauean olivine tholeiites. A recent addition is the oceanic tholeiite magma-type of Engel, Engel and Havens (1965) with its worldwide distribution across the ocean basins and distinctive chemical character—very low $K_2O$ ($<0.2$ percent), Ti, Rb, Sr, U, Zr and uniquely high Na/K. At an opposite extreme are the oceanic potassic basalts of Tristan da Cunha—is this another magma-type?
>
> Seen in the light of these remarks the basaltic magma-types cover a broad spectrum. Probably more than one type is primitive. Others may be derivative. On a more refined but still legible scale each has its unique character: the magma-types are not just two or three but are infinite in number.

With this in mind, it is still convenient to use the terms tholeiite, olivine tholeiite and alkali basalt to express in a general way an important chemical criterion—degree of silica saturation—of some naturally associated basalts or of a particular basaltic magma. But to do so does not imply identity of origin for any two types so designated. The basalts of Mauna Kea in Hawaii and those of the Mid-Atlantic ridge can both be called olivine tholeiite. But they are not chemically identical in other respects; and each set has been uniquely generated.

A further note of warning on these generalizations: as our analytical capacity to resolve increasingly small quantities with precision increases, it has been found that one volcano within a petrographic province may repeatedly erupt lavas that in some respects are chemically unique compared with products of nearby volcanoes. Again, one volcano may produce lavas which show distinctive and consistent gradational features (trends or lineages); a particularly good example of this is Kilauea on the island of Hawaii, each of whose historic summit eruptions has produced lava series that are distinct chemically, not only from lavas of other episodes of eruptions but also from lavas erupted synchronously from the flanks of the same volcano (T. L. Wright and Fiske, 1971). To be of value, the broader generalizations of magma series and magma types must neither obscure nor be undermined by the distinctive features of a single eruption or a single volcano, whose duration and volume may be but a small part of the province of which it forms part.

## GENETIC IMPLICATIONS OF MAGMA SERIES

THE CONCEPT OF PARENTAL QUALITY: CLARIFICATION OF TERMS Implicit in the concept of magma series is that magmas of different composition have independent derivation from a common crystalline source or through continuous evolution along lines of magmatic descent. The notion that a particular magma plays some kind of fundamental role in a series, or even among magma series in general, has been conveyed by partially synonymous usage, with a variety of emphases, of the terms *primitive*, *primary*, and *parental*.

To denote loosely some combination of original and parental qualities inferred from copious effusion and early appearance in a magmatic cycle, some writers have called certain basaltic magmas "primitive" (for example, Powers, 1955, pp. 95, 96; Macdonald, 1968, p. 518). This is the usage adopted in Verhoogen et al. (1970, p. 302) with emphasis on inferred direct derivation by melting of deep source rocks. The same term, however, has come to be used to denote inferred direct relationship between magma and *primordial* mantle material. The chemical nature of the latter is a matter for speculation; and equally speculative must be any set of chemical criteria set up arbitrarily as an index of magmatic "primitiveness." The same criteria in fact have been interpreted in precisely opposite ways. Engel et al. (1965) recognize a highly distinctive minor-element pattern in basaltic lavas on the deep flanks of the ocean ridges (Chap. 8). By chemical analogy with certain stony meteorites they called these basalts "primitive." Other writers believe that this identical pattern implies strong departure from primordial chemistry—by depletion in many of the elements in question. The term *primitive* has come to be meaningless except with reference to any postulated model. It will be dropped from further use in this book.

*Primary* magmas have been defined more objectively to cover magmas of uniform composition rising in great volume from deep sources, with no evidence of derivation from other magma of the same igneous cycle. Basaltic magmas especially have been so designated (Bowen, 1947, p. 266; Barth, 1952, p. 179). So too have granitic (Daly, 1933, p. 214) and ultramafic magmas (Holmes, 1932; H. H. Hess, 1938)—the latter solely because of inferred mantle origin and chemical impossibility of derivation from other silicate magmas. Thus *primary* applied to magmas came to have a strictly negative genetic connotation: lack of magmatic parentage. Elsewhere we have attempted to attach a more positive meaning to the term by invoking the obvious complementary implication of the original definition: primary magmas were redefined to include direct products of melting of source rocks of any kind (Turner and Verhoogen, 1960, pp. 431–434). This conceptual usage has been followed a good deal in current discussion of the relative roles of melting and later crystallization in magmatic evolution (for example, O'Hara, 1965; D. B. Clarke, 1970). Apart from ambiguity arising from this dual usage the issue has now become further clouded with growing realization of another complicating factor. Isotopic evidence suggests that magmas issuing at intervals, even from the same volcano, may have been generated by independent episodes of remelting of solid source rocks that themselves crystallized significantly earlier from the same or mutually related magmas of a previous cycle. To minimize further confusion in this book we avoid reference to primary magma. Where the topic demands it, especially when discussing the nature of magmatic sources in Chap. 13, we shall refer to melts generated from solid sources, and since unmodified, as *primary melts*.

More amenable to productive treatment, and free from ambiguous usage of terms, is the parental relation that exists between some magmas and their derivatives in individual magma series. The relation can be clearly stated in terms open to testing by prediction and evaluation in the light of chemical and mineralogical investigation and of field observation. Bowen's great petrogenic synthesis illustrates the immense value of this concept; his central theme is the parental role of basaltic with respect to more felsic derivative magmas (Bowen, 1928, pp. 21, 63, 319–320; Bowen, 1947, pp. 266, 271–274).

PARENTAL AND DERIVATIVE MAGMAS In many petrogenic provinces one magma appears to qualify uniquely for the parental role. It is chemically capable of giving rise, by accepted processes operating as it cools, to chemically different magmas represented by other rocks of the province. Of these processes fractional crystallization, partly because of Bowen's compelling advocacy, has received special emphasis. Another is partial fusion as developed by Bowen to account for generation of parental magmas in the first place. Where recycling of magmas plays a part in petrogenesis it may be difficult to distinguish mutual relationship of magmas through selective fusion of common source materials from the imprint of fractional crystallization in a simple magmatic cycle.

One requirement of *parental magma* is that under the particular conditions of differentiation, it must have the highest liquidus temperature, and thus by cooling and crystal settling give rise to derivative magmas. Usually a high liquidus temperature is correlated with abundance of those components with a high melting temperature, namely calcic plagioclase, magnesian olivines, and pyroxenes. Melting studies at 1 bar show this correlation to be very close, but at high pressures acid magma may have greater thermal stability than intermediate magma. Additional evidence for the identification of a parental magma comes from abundance patterns of those trace elements which also are affected by differentiation.

Geologists have often required a parental magma to be the most abundant in a petrogenic province; although why volcanoes should necessarily erupt lava in direct proportion to the volume of a progressively changing liquid in the underlying magma column has never been made clear. However, it is singularly unsatisfying, and even unconvincing, if after lengthy analysis of both minerals and rocks the deduced parental magma is represented in the most insignificant rock type. The same problem arises in volcanic provinces where rhyolitic lava and ash have been erupted in such large quantity that they vie with the associated basalts or andesites for an independent role. Are there two parental magmas? Despite an evident capacity of the basic magma to yield the acidic derivative, petrologists are influenced by the absence of intermediate compositions. Thus the identification of a parental magma depends not only on its chemical properties, but also on the nature and volume of the associated lavas.

## Chemical Variation Diagrams

Chemical variation within the rocks of one magma series, or among the rocks of different petrogenic provinces, can be conveniently illustrated by means of *variation diagrams.* In these, analytical data such as the oxide components are plotted in various ways to illustrate whatever aspect of variation the investigator wishes to emphasize. In the past, these diagrams have had two general uses: one descriptive to show and compare the general aspects of chemical variation in and between series; the other to suggest, from the shape of the curves and their relative trends, some model combination of factors by which they might have been generated. By and large, descriptive use has been rather unrevealing, except as a means by which the student can grasp something of the general chemistry of igneous rock series. Use as a basis for testing petrogenic models is still valuable, though now tending to be replaced by computer methods that can handle a greater number of chemical components.

The Harker diagram (Fig. 2-1), a rectangular plot of the weight percentages of each principal oxide against the common ordinate $SiO_2$, has been in use for over 60 years. Underlying its genetic interpretation is a general presumption that in the course of evolution in a consanguineous magma series there is continuous increase in $SiO_2$ content; but demonstrably this is not always true. It has also been generally inferred, particularly with respect to basalt-andesite-rhyolite magma series, that the principal factor controlling the configurations of the

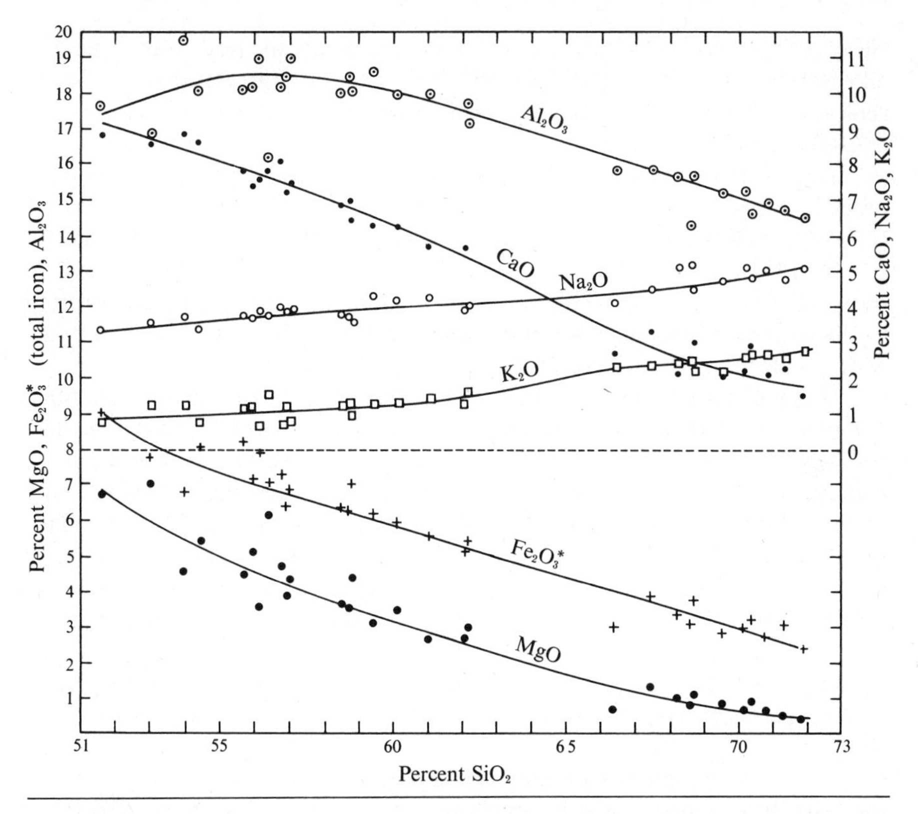


**Fig. 2-1** Variation diagram (Harker type) for volcanic rocks of Crater Lake, Cascades volcanic province, northwestern United States. (After Williams, 1942.)

oxide curves is some process of fractional crystallization (Chap. 3).[1] On this assumption local, and especially terminal, scattering of plotted points is thought to show that the rocks in question were never entirely liquid but were products of crystal accumulation. With this in mind, following the advice of Bowen (1928, p. 117–122), petrologists tend to exclude from the variation diagram analyses of porphyritic or coarse-grained rocks, since these may not necessarily represent points on a continuous *liquid line of descent*. Glassy or fine-grained lavas on the other hand may define such a line; and a diagram restricted to analyses of such rocks is often used to show what materials must be subtracted and what added to generate the curves in question.

One commonly used description of rock suites is derived solely from a Harker variation diagram. This is Peacock's (1931) alkali-lime index, given by the weight percentage of $SiO_2$ at which the weight percentages of CaO and ($Na_2O$ + $K_2O$) are equal. Peacock arbitrarily distinguished four chemical

[1] Chayes (1964*c*, p. 235) considers this assumption statistically unsound.

classes of igneous rocks on the basis of this index: alkalic (index less than 51), alkali-calcic (51 to 56), calc-alkalic (56 to 61), and calcic (greater than 61). Many tholeiitic suites, whether on the continental margins or in the oceanic islands, are calc-alkalic by this criterion.

Larsen (1938) found that a better correlation of chemistry with the presumed order of magmatic evolution was given by plotting oxide percentages against $[(1/3SiO_2 + K_2O) - (CaO + MgO + FeO)]$. Nockolds and Allen (1953, p. 116; 1954) modified this factor to demonstrate any absolute enrichment in iron and principally to show the variation of minor and trace elements in relation to the major components; the modified factor is $[(1/3Si + K) - (Ca + Mg)]$ with metals on a weight percentage basis (Fig. 3-1).

Triangular diagrams are used to plot the mutual relations of any three components $A$, $B$, and $C$, with $(A + B + C)$ recalculated to 100. The most widely used is undoubtedly $(Na_2O + K_2O)$, $(FeO + Fe_2O_3)$, and MgO, or the metal equivalents, and is a modification of the type introduced by Wager and Deer (1939) to show the extreme relative enrichment in iron of the Skaergaard intrusion (Fig. 2-2). However, it is difficult to correlate any break or change in trend of the curves in this diagram with the onset or cessation of crystallization of a particular phase. Another triangular diagram in widespread use is a plot of the three normative components, *Q*, *ab*, and *or*, which in acid rocks may amount to 90 percent or more of the rock and so may be convincingly related to the experimental results on the synthetic system $SiO_2$–$NaAlSi_3O_8$–$KAlSi_3O_8$–$(H_2O)$.

As crystallization of a silicate magma proceeds, the residual liquid becomes increasingly enriched in the low-melting components, or, in other words, the feldspars and the ferromagnesian minerals become progressively enriched in (Na + K) and (Fe + Mn) respectively. Simpson (1954) proposed a *felsic index* $[(Na_2O + K_2O) \times 100]/(Na_2O + K_2O + CaO)$ to be plotted against Wager and Deer's *mafic index* $[(FeO + Fe_2O_3) \times 100]/(MgO + FeO + Fe_2O_3)$. Wager (1956) used a modified version of these indices, namely the molecular normative ratio 100 $ab/(ab + an)$ and the atomic ratio $100(Fe'' + Mn)/(Fe'' + Mn + Mg)$; this has in its turn been modified to $100(ab + or)/(ab + or + an)$ and $100(Fe'' + Fe''' + Mn)/(Fe'' + Fe''' + Mn + Mg)$. A plot of this type for a suite of lavas from an Icelandic volcano showed an abrupt change in trend where titaniferous magnetite became an early crystallizing, and thus fractionated, phase (Carmichael, 1964*b*).

However, the most generally used index of magmatic evolution is the differentiation index (DI) proposed by Thornton and Tuttle (1960); this is simply the weight percentage of the (CIPW) normative salic components quartz (Q), albite (*ab*), orthoclase (*or*), nepheline (*ne*), leucite (*lc*), and kalsilite (*kp*). Some of these components are incompatible (for example, quartz and feldspathoid; albite and kalsilite), so that the value of the DI is independent of silica saturation. This index also has a satisfying thermodynamic basis in that these normative components correspond to minerals with low entropies of melting,

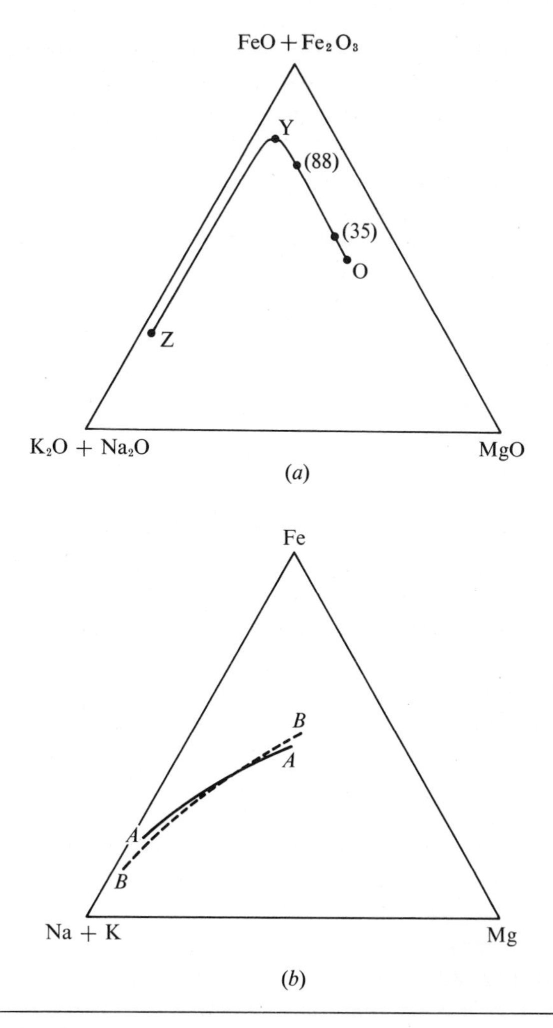


**Fig. 2-2** Variation diagrams for alkali, iron, and magnesium. (a) Weight percentages of ($Na_2O$ + $K_2O$), (FeO + $Fe_2O_3$), and MgO computed for successive liquid fractions, Skaergaard intrusion, Greenland (Fig. 9-13), O, marginal gabbro (assumed parental magma); Y–Z late granophyric liquid fractions ($<1$ percent original volume of magma); numbers in parentheses give estimated percentage solidification at each point. (b) Weight percentages of (Na + K), Fe, and Mg in lavas of Crater Lake and Mt. Shasta (*A*) and of Mt. Lassen (*B*), northwestern United States. (After Nockolds and Allen, 1953.)

which as we shall see demands their concentration in low-temperature liquids.

The most detailed use of a variation diagram in detecting the varied influences that could control an evolving magma type was made by Wilcox (1954) on the lavas of Parícutin, Mexico. Here a combination of crystal separation and incorporation of siliceous crustal material (found as xenoliths) was shown to be necessary to reproduce the observed lava variation. Lately, the task of trying to match an observed liquid line of descent from a possible parent in terms of the separation of the observed (and analyzed) phenocryst assemblage has been done for 12 or more components by matrix algebra and a computer (Bryan et al., 1969).

Sometimes it appears that the case for crystal fractionation as the dominant, if not sole control, of the diversity of igneous rocks has been accepted too readily and often on insubstantial evidence. Certainly, from now on, it will have to be established *numerically* that a close fit can be obtained between observed and calculated liquid compositions; and even then crystal separation will have been shown only to be a viable mechanism and not necessarily the only one capable of producing the observed result.

## Thermodynamic Approach to Classification

Fortified, or perhaps overwhelmed, by the preceding discussion of rock names and the compositions of the rocks or magmas they denote, the reader may perhaps wonder if there is not a better or more rational way of treating the diversity of rocks. We offer one here, very much as an initial approach, our hesitation kindled not so much by the principle of the method but on the necessarily oversimplified treatment of the properties of solid solutions which it requires. Also, surprisingly little is known of the composition of many of the essential minerals in the whole spectrum of igneous rocks, despite an evident consensus to the contrary.

As a starting point, we return to Shand's concept of silica saturation, which can be represented thermodynamically by establishing a measure of the "effective or active concentration" of silica, or of any other component, in a silicate melt. This is the concept of activity, or of the more fundamental function, the chemical potential. If, for example, forsterite ($Mg_2SiO_4$) and clinoenstatite ($MgSiO_3$) coexist in equilibrium with a liquid, then the activity of silica is fixed if the temperature and pressure are fixed. Any addition of liquid silica to the liquid would only produce more pyroxene at the expense of olivine until the olivine was used up. We therefore have a "silica buffer," the precise value of which can be calculated at any temperature and 1 bar from the reaction

$$\underset{\text{forsterite}}{Mg_2SiO_4} + \underset{\text{glass}}{SiO_2} = \underset{\text{clinoenstatite}}{2MgSiO_3} \tag{2-1}$$

The equilibrium constant for this reaction can be written as

$$K = \frac{(a_{MgSiO_3}^{pyrox})^2}{a_{Mg_2SiO_4}^{ol}\, a_{SiO_2}^{liq}} \tag{2-2}$$

where $a$ refers to the activity of the subscript component in the superscript phase.

In the presence of the pure solids $Mg_2SiO_4$ and $MgSiO_3$, which therefore have unit activity, Eq. (2-2) reduces to

$$K = \frac{1}{a_{SiO_2}^{liq}} \tag{2-3}$$

If Eq. (2-3) is substituted into the well-known relationship

$$\Delta G = \Delta G_r^\circ + RT \ln K$$

we have at equilibrium ($\Delta G = 0$)

$$\log a_{SiO_2}^{liq} = \frac{\Delta G_r^\circ}{2.303RT}$$

where $\Delta G_r^\circ$ is the standard Gibbs free-energy change of the reaction and $T$ is the temperature in degrees Kelvin. Activity has no units but is a ratio (at constant temperature) of the fugacity of $SiO_2$ in the chosen state, here taken as the liquid phase of a lava, compared to a standard state, which in this example we have stipulated to be silica glass. In other words, the activity of silica measures the difference between the chemical potential of silica in the liquid lava phase and that in the standard state, silica glass. In nature, neither olivines nor pyroxenes are pure magnesian end members, and in each Fe/Mg increases as crystallization proceeds. This will affect the calculated values of silica activity, but for the moment we shall ignore all substitutions of this type and consider only reactions with minerals in their standard states, that is, with unit activity.

The basalt family of igneous rocks has been subdivided into two groups by the presence or absence in the groundmass of a calcium-poor pyroxene, which can be either pigeonite or orthopyroxene. This calcium-poor pyroxene is a mineralogical expression of a reaction relationship between olivine and melt, and we have followed Tilley (1950) in using this as the principal characteristic of the tholeiitic magma type. Another group of basalts, the alkali olivine basalts, contains only one pyroxene (an augite with substantial amounts of Al and Ti) together with olivine. Reaction (2-1) therefore represents a silica-activity boundary between tholeiitic basalts, with or without olivine, and the alkali olivine basalts which do not display this reaction relation. The calculated variation of silica activity ($a_{SiO_2}^{liq}$) with temperature for reaction (2-1) at 1 bar is shown in Fig. 2-3 (for thermodynamic data, see Table 3-3).

Associates of tholeiitic basalts are commonly rhyolites, often with quartz or tridymite in the groundmass. In a liquid in equilibrium with quartz crystals, $a_{SiO_2}^{liq}$ can be calculated from the reaction

$$\underset{\text{glass}}{SiO_2} = \underset{\text{quartz}}{SiO_2} \tag{2-4}$$

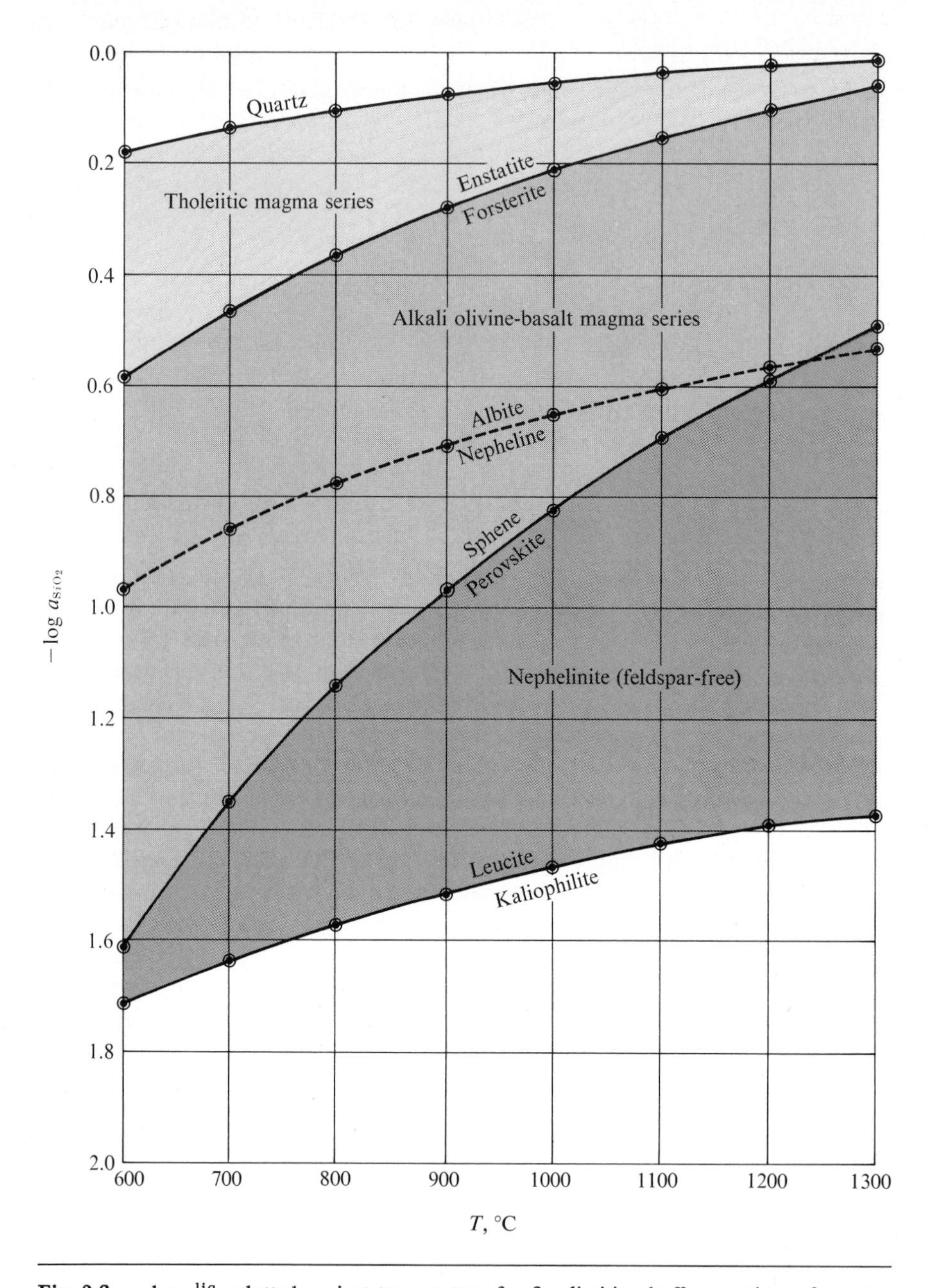


**Fig. 2-3** $-\log a_{SiO_2}^{liq}$ plotted against temperature for five limiting buffer reactions: from top downwards the curves correspond respectively to Eqs. (2-4), (2-1), (2-7) (dashed), (2-5), (2-6). Inferred fields are shown for three principal series of basic magmas.

and the variation of log $a^{\text{liq}}_{SiO_2}$ with temperature, in equilibrium with pure quartz, is shown in Fig. 2-3. Because of the small difference in $\Delta G^\circ_f$ of quartz and tridymite, the glass-tridymite curve is almost identical with the quartz curve. Tholeiitic basalts and their associated basaltic andesites, andesites, dacites, and rhyolites are, in terms of $a^{\text{liq}}_{SiO_2}$, bounded by the two reactions (2-1) and (2-4); and if we assume a temperature span from 750 to 1225°C for this magma series, we can define an area in terms of temperature and log $a^{\text{liq}}_{SiO_2}$ to which the tholeiitic magma series is confined (Fig. 2-3).

The alkali olivine-basalt magma series includes trachytes and phonolites, but it is not common for this series to include rock or magma types in which feldspar is absent. We therefore have to find a reaction that separates those magmas or rocks which contain feldspar and a feldspathoid from those which contain only a feldspathoid (± melilite). Since perovskite has never been reported to coexist with feldspar in volcanic rocks, we can use reaction (2-5) to separate the feldspar-bearing rocks from those in which feldspar is absent and which we shall conveniently label *nephelinites*.

$$\underset{\text{perovskite}}{CaTiO_3} + \underset{\text{glass}}{SiO_2} = \underset{\text{sphene}}{CaTiSiO_5} \qquad (2\text{-}5)$$

The calculated variation of log $a^{\text{liq}}_{SiO_2}$ in equilibrium with pure sphene and perovskite has been plotted in Fig. 2-3; it defines the lower limit to the alkali olivine-basalt magma series, whose upper limit is of course reaction (2-1). At lower activities of silica than those defined by the sphene-perovskite reaction is the whole range of mafic feldspathoidal lavas. These rocks, which we have loosely called nephelinites, have their lower limit of silica activity defined by the reaction

$$\underset{\text{kaliophilite}}{KAlSiO_4} + \underset{\text{glass}}{SiO_2} = \underset{\text{leucite}}{KAlSi_2O_6} \qquad (2\text{-}6)$$

Kaliophilite (kalsilite) and leucite are a common groundmass assemblage in the potassic lavas of Uganda.

As can be seen in Fig. 2-3, we can separate the major magma types and their congeners into three broad groups in terms of silica activity; in each case the reactions are simple representations of natural groundmass assemblages. Other reactions which define silica activity come readily to mind, particularly those involving a feldspar and a feldspathoid, such as

$$\tfrac{1}{2}\underset{\text{nepheline}}{NaAlSiO_4} + \underset{\text{glass}}{SiO_2} = \tfrac{1}{2}\underset{\text{albite}}{NaAlSi_3O_8} \qquad (2\text{-}7)$$

which would subdivide further the alkali olivine-basalt magma series. Further reactions involving melilite, zircon, and wollastonite have been discussed by Carmichael et al. (1970) and Nicholls et al. (1971); the variation of $\Delta G^\circ_r/2.303RT$ with temperature for these reactions is shown in Table 3-3.

Although Shand's silica-saturation concept can be represented in thermodynamic terms and even precise values calculated if both the temperature and the composition of the solid solutions forming the buffer assemblages are known, this is not so readily done for $Al_2O_3$. Thus rather than attempt to treat the activity of $Al_2O_3$ in a liquid, we use another component, $Na_2O$, which can be related to the composition of the plagioclase in the following way:

$$\underset{\text{liquid}}{Na_2O} + \underset{\text{clinoenstatite}}{MgSiO_3} + \underset{\text{anorthite}}{6CaAl_2Si_2O_8} = \underset{\text{albite}}{2NaAlSi_3O_8}$$

$$+ \underset{\text{diopside}}{CaMgSi_2O_6} + \underset{\substack{\text{Ca-Tschermak's}\\ \text{molecule}}}{5CaAl_2SiO_6} \tag{2-8}$$

and with all the solid components in their standard states, namely pure solids of unit activity,

$$\Delta G = \Delta G_r^\circ + RT \ln \frac{1}{a_{Na_2O}^{liq}} \tag{2-9}$$

and at equilibrium ($\Delta G = 0$)

$$\log a_{Na_2O}^{liq} = \frac{\Delta G_r^\circ}{2.303RT} \tag{2-10}$$

where the activity of $Na_2O$ in the liquid is the ratio of the fugacity of $Na_2O$ in the lava's liquid phase compared to that of the standard state of pure $Na_2O$ liquid.

In reality, the solid components of Eq. (2-8) will form solid solutions, and the activity of each component will thereby be reduced relative to the standard state. The equilibrium constant for Eq. (2-8) is written as

$$K = \frac{(a_{NaAlSi_3O_8}^{plag})^2 \, a_{CaMgSi_2O_6}^{pyrox} (a_{CaAl_2SiO_6}^{pyrox})^5}{a_{Na_2O}^{liq} a_{MgSiO_3}^{pyrox} (a_{CaAl_2Si_2O_8}^{plag})^6} \tag{2-11}$$

Before we can substitute typical groundmass mineral compositions into Eq. (2-11), we have to consider the relationship between activity $a$ and mole fraction $X$ of a component in a solid solution. This is considered in some detail in Chap. 4, so it suffices to state here that an assumption of ideal mixing ($\Delta H_{mix} = 0$) does not necessarily require that $a = X$. For the high-temperature plagioclase feldspars we shall nevertheless assume as an approximation that the mole fraction of albite, $X_{NaAlSi_3O_8}^{plag}$, can be substituted for $a_{NaAlSi_3O_8}^{plag}$ and similarly for anorthite. But for an augitic pyroxene the situation is rather perplexing. In an equilibrium assemblage of augite and orthopyroxene, the requirement of equilibrium is that the activity of enstatite, $a_{MgSiO_3}^{pyrox}$ be the same in both phases;

there is no doubt that $X^{augite}_{MgSiO_3} \neq a^{augite}_{MgSiO_3}$, and unfortunately even the calculation of the mole fraction of $MgSiO_3$ in an augite cannot be performed unambiguously. The same is also true of the calculation of $X_{CaAl_2SiO_6}$ for an analyzed augite, and the relationship between $X_{CaAl_2SiO_6}$ and $a_{CaAl_2SiO_6}$ is a complex one. Verhoogen (1962) has suggested that $a_{CaAl_2SiO_6} = X^2(2 - X)/4$, where $X$ is the mole fraction of $CaAl_2SiO_6$. Nicholls and Carmichael (1972), involved with the same problem, assumed a disordered standard state for synthetic $CaAl_2SiO_6$ for which $\Delta G^\circ_f$ is known, so that the 4 in the denominator is replaced by 1 (complete Al/Si disorder in the tetrahedral sites).

But with some hesitation, we ignore what we cannot treat and assume for all minerals that $X = a$ for each component, thus ignoring the entropy-of-random-mixing term for the pyroxene. This will make any calculated value of log $a^{liq}_{Na_2O}$ highly uncertain. But it is the overall pattern which we wish to illustrate here, and if we do the same for all, then the relative positions of each magma type may be sustained when additional information is available on the mixing properties of the augitic pyroxenes. Our goal is not to calculate a realistic value for log $a_{Na_2O}$ but rather to show how variation in this parameter may be used to classify rocks and magmas.

With values for the three pyroxene components, we can calculate log $a^{liq}_{Na_2O}$ (using tables of free-energy data) for a range of groundmass compositions with coexisting plagioclase at various temperatures. Analyses of groundmass augitic pyroxenes taken from the literature gave the average values shown in Table 2-4,[1] where $X$ refers to mole fraction. Substituting the relevant pyroxene mole fractions we obtain a set of values for log $a^{liq}_{Na_2O}$ for different plagioclases in equilibrium with corresponding liquids. For zoned pyroxene and feldspar, several values could be calculated corresponding to the cores, middle zones, and outer rims of each type of crystal.

In Fig. 2-4, common rock types are located on a rectangular grid whose coordinates are normative plagioclase composition and $a^{liq}_{SiO_2}$. Since $a^{liq}_{Na_2O}$ is

**Table 2-4** Mole fractions of three components in average groundmass augitic pyroxenes

| | $X_{CaAl_2SiO_6}$ | $X_{CaMgSi_2O_6}$ | $X_{MgSiO_3}$ |
|---|---|---|---|
| Tholeiitic basalts | 0.02 | 0.47 | 0.24 |
| Andesites | 0.02 | 0.47 | 0.25 |
| Alkali olivine basalts, | | | |
| basanites, tephrites | 0.06 | 0.52 | 0.11 |
| Mugearites | 0.06 | 0.46 | 0.07 |
| Trachytes (phonolitic) | 0.02 | 0.39 | 0.10 |
| Rhyolites (metaluminous) | 0.01 | 0.27 | 0.14 |

[1] The values of these three components derived from a pyroxene analysis depend on the order of combination of the oxides; our method was developed so as to include electron microprobe analyses which do not have values for $Fe_2O_3$.

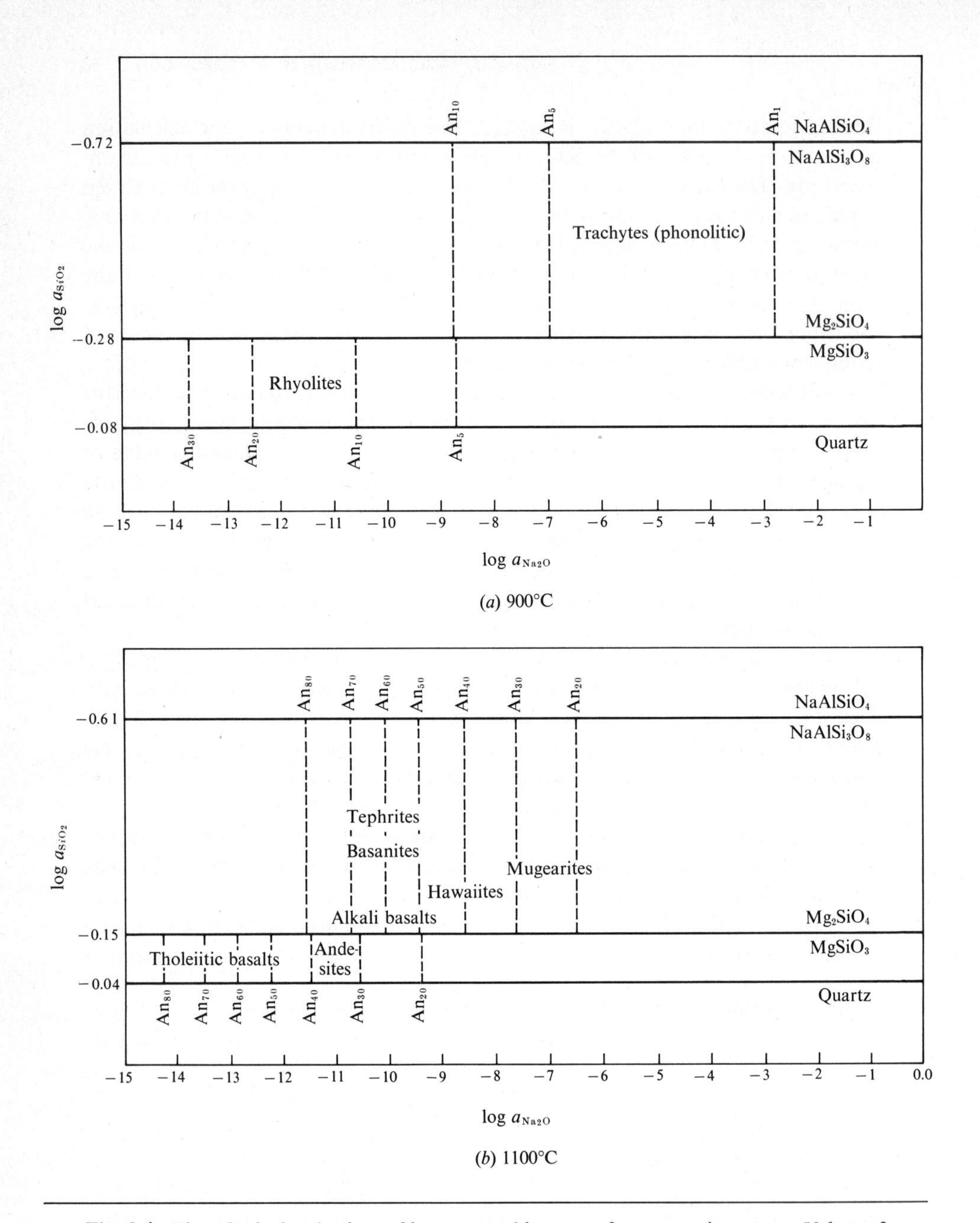


**Fig. 2-4** Plot of calculated values of log $a_{SiO_2}$ and log $a_{Na_2O}$ for magmatic systems. Values of log $a_{SiO_2}$ at 900 and 1100°C are those calculated for Eqs. (2-4), (2-1), and (2-7) as in Fig. 2-3. Values of log $a_{Na_2O}$ were calculated from compositions of coexisting plagioclase and pyroxene (data from Table 2-4).

directly related to plagioclase composition, Fig. 2-4 is really an activity-activity diagram of a type familiar to low-temperature petrologists. It is designed to show the stability fields of common rocks in terms of the logarithms of the activities of silica and soda. Thus at 1100°C the field of tholeiitic basalts is bounded by $a_{SiO_2}^{liq}$ activities for reactions (2-1) and (2-4) and $a_{Na_2O}^{liq}$ activities compatible with crystallization of labradorite, $An_{50}$ to $An_{70}$. Andesites at this same temperature have higher $a_{Na_2O}^{liq}$ than tholeiites because their feldspar is more sodic, although their average pyroxene is much the same (Table 2-4). Alkali olivine basalts, basanites, and tephrites plot between the $a_{SiO_2}^{liq}$ boundaries of reactions (2-1) and (2-5); but their different pyroxene composition displaces equivalent plagioclase compositions to higher values of $a_{Na_2O}^{liq}$ in comparison to tholeiitic basalts. In the upper part of Fig. 2-4 the calculated curves for plagioclase rhyolites and phonolitic trachytes show that a rhyolite and a tholeiite can have comparable values for $a_{Na_2O}^{liq}$; obviously the effect of temperature almost exactly matches, in an opposite sense, the effect of different plagioclase and pyroxene compositions. These two plots in Fig. 2-4 are really two isothermal planes in a $T$–$a_{SiO_2}^{liq}$–$a_{Na_2O}^{liq}$ system, calculated at a pressure of 1 bar, appropriate to the groundmass mineral assemblages of lavas.

At some future date, data will become available on the mixing relations of the solid solutions participating in these reactions and will allow meaningful values to be calculated for the soda activity; mineral geothermometers will also eventually place temperature ranges on all igneous assemblages. But in the meantime, we can generalize the type of data shown in Fig. 2-4 and so point the way to a thermodynamic classification of volcanic rocks.

In Fig. 2-5, the variation of silica activity with both temperature and changing composition of the buffer solid solutions is indicated by representing the reaction in question as a zone. The distance between zones has been greatly exaggerated, since the increase in $a_{SiO_2}^{liq}$ from a tholeiite at 1150°C to a quartz-bearing rhyolite at 850°C is small (Carmichael et al., 1970). On the other axis we have plotted the anorthite component of the plagioclase feldspar, using either the average modal feldspar or the normative composition, either of which is related to $a_{NaO_2}^{liq}$ and also to temperature through the plagioclase geothermometer (Kudo and Weill, 1970). Some of the commonly used names of volcanic rock types have been plotted in Fig. 2-5, but the precise location is really immaterial because rocks show all gradations from one type into another.

The two trends of rock names in Fig. 2-5, one toward high silica activity and the other to low silica activity, essentially depict the tholeiitic magma series and the alkali olivine-basalt magma series. In the former we have shown two further trends. The iron-rich trend includes icelandites (as the iron-rich equivalent of andesites), dacites, and rhyolites, both of which have ferroaugite and iron-rich olivines. The iron-poor trend, typically found in the continental margins, often has andesite as the most voluminous rock type, while the associated dacites and rhyolites have magnesian pyroxenes. The alkali olivine-basalt magma series can have representatives which are relatively potassic, whereas

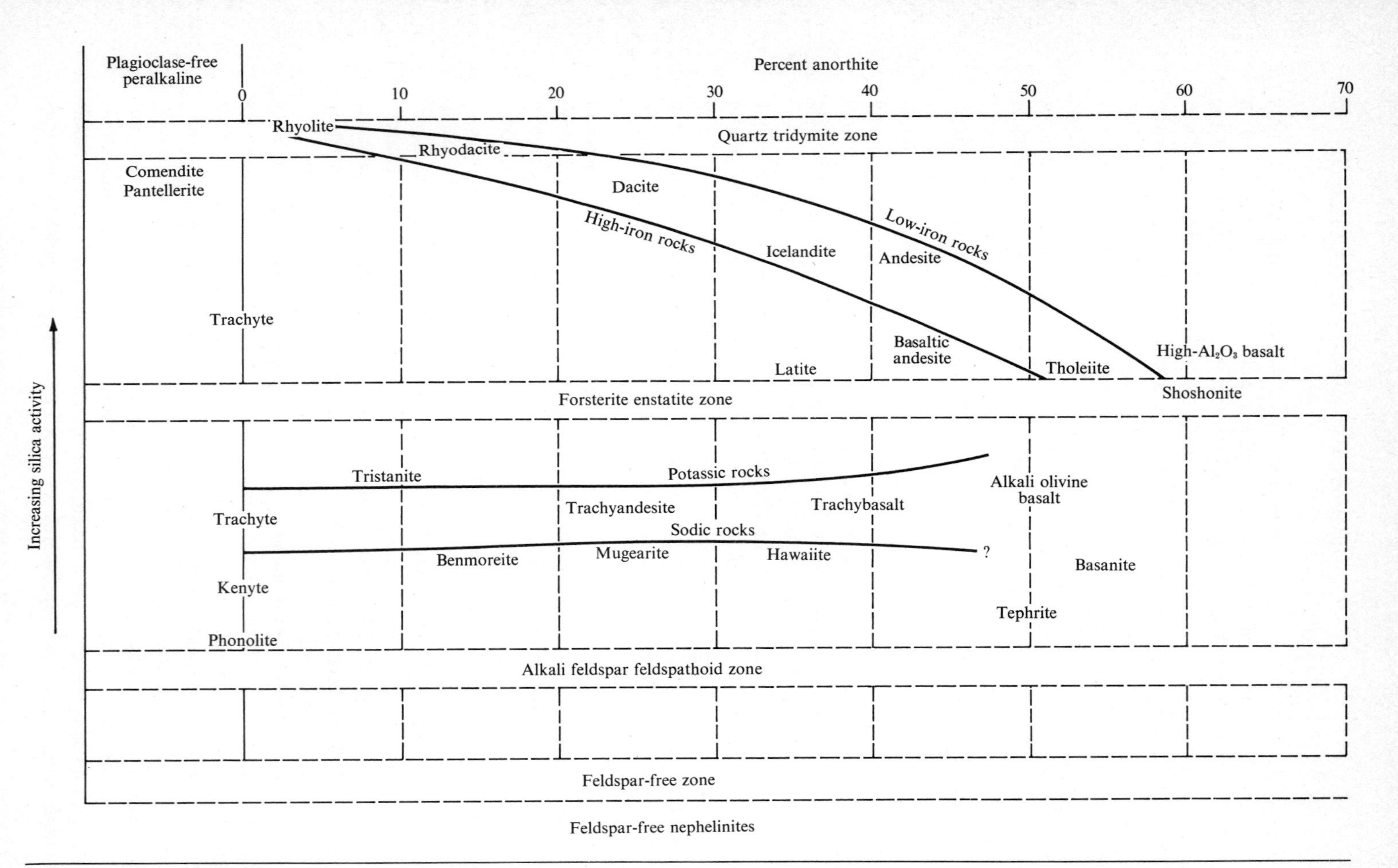


**Fig. 2-5** Schematic tabulation of volcanic rocks in terms of normative plagioclase composition, 100 *an*/(*ab* + *an*), and silica activity $a_{SiO_2}$, based on Fig. 2-4.

others (for example, the lavas of Hawaii) are relatively sodic; these two tendencies have been shown in Fig. 2-5. Within the earth's igneous economy, the great bulk of volcanic rocks are those shown in Fig. 2-5. This is convincing testimony to the dominant influence of the low entropies of melting of silica and alkali feldspar, whereby those components become concentrated in the residual liquids of fractional crystallization.

# Factors in Magmatic Evolution

# 3

## Introduction

The genesis of igneous rocks can be considered in three successive steps: those factors which determine how and where a liquid is generated in the mantle or lower crust; those conditions which influence the liquid's passage toward the surface; and lastly those processes near the surface which may complete the variation caused by the generation and migration of the liquid and so account for the considerable but finite diversity of igneous rocks. The first stage is largely concerned with heat and how it is localized to form a partial melt from either an ultramafic mantle or siliceous crustal rocks. The extent of partial melting, pressure, temperature, composition, and particularly the chemical potentials of the volatile components of the source material all influence the composition of this partial melt. The site of magma generation, and therefore the composition of the derived liquids, is controlled by whatever controls the local temperature and the melting temperature. As explained in Chap. 7, melting may occur in a mantle upwelling wherever the temperature of the rising mass exceeds the local melting temperature, which in turn is determined by pressure and composition (including volatiles). In the second stage, the upward migration of the magma traverses pressure-dependent mineral stability zones, so that if the magma is held at some level with an opportunity to cool, settling of the early formed crystals, whose compositions are also pressure-controlled, will change the liquid's composition. In the third stage, in the top 20 to 30 km or perhaps down to 60 km in the continental regions, the intruding magma, if not born of crustal rocks, comes into contact with them and their fluids. Any interaction here further changes the composition of the liquid, so that by a variety of processes magmas corresponding to the whole range of igneous rocks are developed.

Concisely stated, igneous petrogenesis is concerned with the generation of a magmatic liquid, whether it be silicate, carbonate, or sulfide, followed by the cooling path taken to eliminate this liquid. And usually the only evidence that a geologist has to decipher this complex series of events is the end product, an igneous rock and perhaps its envelope.

Petrologists are generally agreed that magma which has moved into the earth's crust does not have the full range of composition of the volcanic rocks emitted on the surface. In this chapter we look at those processes which modify in one way or another a magma's composition, particularly in the upper reaches of the earth where pressures are low and the opportunity for interaction with sedimentary or metamorphic crustal material is high.

## Evolutionary Mechanisms

Magmatic evolution, which may give rise to directly related rock series, involves the interplay of a number of possible mechanisms customarily grouped under the general headings of differentiation, assimilation, and magmatic mixing. These we shall now briefly review. Later they will be examined in detail in the light of experimental data on silicate-melt equilibria. How they operate in nature will become clearer when we turn to case histories of petrogenesis in the descriptive section of the book.

Concepts of magmatic evolution still in vogue today have their roots in American literature of the latter part of the last century. Iddings recognized the common parentage of mutually associated rocks of different composition and it was he who called them consanguineous. Becker (1897*a*) wrote essays on differentiation (then seen as splitting of liquid magma into chemically contrasted fractions) and on fractional crystallization (today in its broader current usage considered the principal mechanism of differentiation). The first comprehensive coverage of the whole field—one never excelled for clarity of exposition—is Harker's *Natural History of Igneous Rocks* (1909). Differentiation and assimilation came to be considered rival mechanisms of petrogenesis—witness the debate between two masters of the day, Daly (1918) and Bowen (1919). The controversy has subsided. And now magmatic evolution is seen as a composite process, the interplay of several mechanisms uniquely blended in each individual episode of magmatic activity.

### MAGMATIC DIFFERENTIATION

Magmatic differentiation covers all processes by which a magma, initially homogeneous on a large scale, ultimately gives rise to igneous rock masses of contrasted composition. Of all the possible operations which could conceivably change the composition of a cooling magma, and thus by separation of the liquid account for the diversity of igneous rocks, differentiation by crystallization is one of the most easily appreciated. Since the principal rock-forming minerals are solid solutions, the initial composition of each mineral is, in general, enriched in the high-temperature components in relation to the liquid from which they precipitate. Therefore the minerals which are mutually associated during crystallization are those which precipitate over the same range of temperature (for example, forsteritic olivine and diopside, or fayalitic olivine and ferroaugite; forsteritic olivine and labradorite rather than oligoclase; oligoclase and sanidine; and sanidine and fayalite). There is an equally strong tendency for antipathy between minerals of widely different crystallization range (for example, diopside and sanidine, muscovite and labradorite).

Obviously as soon as crystallization starts, particularly in zones of low pressure, the composition of the residual liquid differs both from the initial liquid and from the first batch of crystals; in short, the liquid is differentiated. And the more thorough the withdrawal of early formed crystals, the longer is the crystallization path of this liquid in terms of both temperature and composition.

FRACTIONAL CRYSTALLIZATION Fractional crystallization is any mechanical process by which early formed crystals are prevented from equilibrating with the liquid from which they grew. Thus a basaltic magma could give rise to a succession of residual liquids, each of progressively changing composition; and for each liquid fraction there would be a complementary crystal fraction, represented either as an accumulation of crystals or as zones of changing composition within each mineral. Each fraction, liquid or solid, could become a separate rock of unique composition. The broad chemical pattern of fractional crystallization is for successive residual liquids to become enriched in silica and the alkalis, and in iron relative to magnesium. Since the solids are generally solutions, they tend to be enriched in their respective high-temperature components, namely magnesium, calcium, and aluminum compared to later crystal or liquid fractions.

*Gravity fractionation* Sinking of crystals of heavy minerals in a less dense liquid (gravity differentiation) is effective in the earlier stages of evolution when the liquid phase is still dominant and has not yet become too viscous (or too dense) to prevent settling of the crystals. A vivid example of the sinking of heavy crystals was shown in the 1840 eruption of Mauna Loa, Hawaii, where the summit lava was olivine-poor but the simultaneous lava of the lower-flank eruption was rich in dense olivine phenocrysts.

Layers rich in olivine, iron ore, or pyroxene in the large basic intrusives illustrate the effectiveness of crystal settling from slowly moving bottom currents of liquid. Indeed the rock types in which the effect of some form of crystal settling has been observed include gabbros (almost all), syenites, nepheline syenites, and granodiorites (Wager and Brown, 1967). Even in granites, accumulation of sinking minerals to form well-marked layers has been observed despite the high viscosity of the magma; calculations by Shaw (1965) have shown that crystals (2 mm diameter) of feldspars and ferromagnesian minerals would sink between 100 m and 1 km in 1000 years, a short interval in the cooling of a large granitic pluton. In basic magma, one restriction on crystal settling may be the increasing density of the iron-enriched liquid; Bottinga and Weill (1970) have calculated that the density contrast between plagioclase and liquid in the Skaergaard intrusion may have become small or even reversed in the later cooling history.

Floating of light crystals has been far less commonly observed, and leucite is the prime example. Sahama (1960) has observed giant leucites floating on the surface of a lava lake, and the myriads of small leucite crystals concentrated along the roof of the Shonkin Sag "headed" dike of Montana (Buie, 1941) could have floated up in the same way.

*Flowage differentiation* Model experiments on a fluid flowing with suspended particles show that the particles migrate toward the higher-velocity (axial) region and, by analogy, indicate that a flowing magma could redistribute and concentrate suspended crystals (Bhattacharji and Smith, 1964). Provided that the flow is laminar, a state which depends on velocity, viscosity, and the shape and size of the conduit, the suspended crystals migrate to the axial zone of the flowing magma. This mechanism is considered to account for the distribution of olivine phenocrysts in the thin picritic dikes of Skye (Gibb, 1968). It also seems probable that axial concentration of olivine persists as the liquid-crystal system is intruded as sills, whereupon the olivine distribution could again be modified by gravity differentiation (Bhattacharji, 1967; Simkin, 1967). Other possible examples of flow differentiation of suspended crystals coupled with a later gravitational effect are found in thin (less than 2 m) basic intrusive sheets in Iceland (Blake, 1968*a*) and in basaltic lavas (Fuller, 1939); on a more massive scale, well-developed mineral layering in a granodiorite is considered to result from flow redistribution in a magma which initially contained 50 to 60 percent by volume of crystals (Wilshire, 1969).

Obviously flow differentiation could affect the distribution of crystals in magma as it moves from its source region through the upper mantle toward the surface; if differentiation is thorough, a liquid initially saturated with olivine could develop marginal zones purged of crystals.

*Late processes: "filter pressing" and dilatation* When crystallization of a magma under plutonic conditions is far advanced, the crystals build a continuous mesh in the pores of which is the residual liquid. If the mass is now squeezed by movements in the wall rock, the residual liquid may be pressed out to form a separate body of differentiated magma. Alternatively if the crystal mesh is rifted by tensional forces, the residual liquid tends to fill the voids so formed according to the principle of dilatancy (Emmons, 1940). This process is termed *autointrusion.* One of many examples attributed to it is an irregularly defined body of syenite, 20 m in thickness, in a differentiated basic sill in the Shiant Isles, Scotland (F. Walker, 1930). Emmons drew attention to the possible effects of fracturing, in quartzites and similar brittle wall rocks, in causing outward migration of residual magmas into the voids so formed. He cited the case of differentiated red granophyres in the roof regions of North American gabbro intrusions, where these invade quartzites and similar brittle rocks.

*Chemical models* So influential are the detailed studies of the great layered intrusives in igneous petrology, of which the most glittering example is the Skaergaard intrusion, that in the mind's eye of many petrologists a column of magma underlying any volcano may be a layered intrusion in the making. The increasingly precise chemical data now available point more and more toward the uniqueness of each volcano and its products; these same chemical data on the lavas and their crystals provide the most critical evidence for or against crystal fractionation as the dominant genetic process. It is no longer sufficient to show that a series of lava analyses falls on, or defines, a smooth curve in one or more variation diagrams and therefore to claim that, ipso facto, crystal fractionation is the controlling factor. Nowadays (1970s) with the electronprobe and the computer, it is necessary to show that the liquid line of descent can be reproduced solely by fractionation of the early formed (and analyzed) crystals from the supposed parental liquid, and so for each successive liquid to the next derivative down the whole liquid line. In each case a materials-balance equation is set up (within limits arbitrarily set by the observer) which matches, to the observed difference in composition between liquid *A* and its supposed derivative *B*, a calculated combination of the analyzed phenocrysts (pyroxene, olivine, etc.) found in *A*.

The materials balance can be further refined, and an undesirable gulf thereby bridged between trace-element geochemists and petrologists, by using certain trace elements, since phenocrystic magnetite, olivine, and plagioclase incorporate V, Ni, and Sr, respectively. The variation of these telltale elements in the liquid line of descent may place additional restrictions on the composition of a fractionated crystalline assemblage. Using this type of data, S. R. Taylor et al. (1969*a*, p. 13) have shown that the derivation of many andesites from typical basaltic parents cannot have involved removal of magnetite in significant amounts (1 to 2 percent) in the fractionated crystalline assemblage; similarly, the higher concentration of Sr in the andesites of Mt. Shasta, California, compared with that in the associated basalts, eliminates plagioclase as a possible early-crystallizing phase in any model of derivation from basaltic parent magma (Peterman et al., 1970*a*). If the liquid line can be reproduced by calculation, then at least crystal fractionation is a *possible* process; but there is of course no proof that it did in fact generate the particular lava suite. If the liquid line of descent cannot be reproduced, and in the authors' experience this is more likely in alkali olivine-basalt suites than in tholeiitic, then either fractional crystallization was not the only mechanism or the minerals found as phenocrysts were not the only ones involved in the generation of the suite. In such cases, crystal fractionation may have occurred in a pressure regime where the high-pressure phases were stable and the low-pressure phases were unstable. Crystals or xenoliths indicative of crystallization in a high-pressure regime are often found in alkali basalts and their derivatives (cf. Binns et al., 1970).

LIQUID IMMISCIBILITY Mixtures of $SiO_2$ with divalent cations form immiscible liquids at high silica concentrations, but the compositional span of the immiscibility gap is constricted as the radius of the ion increases. Thus at 1695°C the gap is approximately 40 mole percent $SiO_2$ in the system $MgO$–$SiO_2$, 30 mole percent for CaO, and 20 mole percent for SrO; for the larger cations Pb and Ba, there is no liquid immiscibility (Kracek, 1933). Mixtures of single-charged ions (Li, Na, Rb, K) with $SiO_2$ do not exhibit liquid immiscibility, nor do compositions with $Al_2O_3$. One exception to this generalization has been found in the system $KAlSiO_4$–$Fe_2SiO_4$–$SiO_2$ (Roedder, 1951); here the immiscible region is not close to the $SiO_2$ apex.

From time to time liquid immiscibility has attracted the attention of petrologists as a possible explanation of the common association of two contrasted rock types without, or with only rare, intermediate members; thus gabbro-granophyre or basalt-rhyolite are the classic examples, and just recently a gabbro-syenite immiscible relationship has been advocated. Is there any indication that a cooling silicate magma can unmix to give basalt on the one hand and rhyolite or perhaps trachyte on the other (Holgate, 1954; W. Hamilton, 1965, p. 35)? Petrographic evidence coupled with experiments on synthetic silicate systems have been so convincingly interwoven by Bowen (1928, pp. 7–19) that most petrologists are inclined to dismiss the hypothesis without further ado. Two of Bowen's arguments highlight his objections: firstly, continued cooling of a pair of immiscible liquids in the laboratory generates droplets of one conjugate liquid within the other, and this continues unless further cooling closes the miscibility gap (as in the celebrated nicotine-water system). Secondly, if two immiscible liquids are in equilibrium, any crystalline phase in equilibrium with one must be so with the other. Petrographic evidence to support the continued formation of immiscible droplets (as seen in sulfide-silicate liquid pairs) which would now appear as circumscribed crystalline assemblages enclosed one by the other is lacking, as also are crystals whose *compositions* are common to both rhyolite and basalt.

Although the possibility of silicate liquid immiscibility in magmas seems generally unlikely because the two liquids would have quite different polymeric structures and thus be of contrasted composition, petrologists have continued to search for the elusive evidence of immiscibility in rocks. Tomkeieff (1942) suggested that the large amounts of silica and chlorophaeite filling amygdules and veins of basalts in Scotland represented a liquid fraction, rich in iron and silica, which separated from basaltic magma under the influence of high concentrations of water. However, chlorophaeite, unlike virtually every other mineral in a basalt lava, is remarkably homogeneous in composition (Carmichael, 1967*a*), and sometimes its distribution can be related to the zeolite zones which are demonstrably younger than the lavas in which they are found.

There remain rather puzzling igneous small-scale structures called *ocelli*, which are subcircular (subspherical) areas filled with minerals ($\pm$ zeolites) often similar to those found in the enclosing rock but in different proportions. They are typically found in small intrusions of lamprophyre, especially camptonites (Ramsay, 1955), and also in the cognate xenoliths of extremely potassic lavas, where each ocellus is surrounded by tangential phlogopite crystals (Carmichael, 1967*b*, fig. 6) reminiscent of crystals surrounding an ascending bubble in an industrial flotation process. Perhaps the ocelli result from the crystallization of a vapor phase, rather than a second liquid, which at high confining pressures can contain a large amount of dissolved silicate material and be of high density. Philpotts (1972) has described ocelli which "neck" due to the diapiric rise of the less dense fluid or liquid forming the ocelli; although this is most convincing evidence of the coexistence of two silicate liquids, the case for their mutual immiscibility is less certain.

Clear examples of liquid immiscibility are occasionally found, however, and magnetite-apatite rocks, which can even occur as lavas, have been shown to be immiscible at high temperature with liquids of dioritic composition, their common associate in dikes (Philpotts, 1967). Separation of a sulfide liquid in a Kilauea basalt has been observed by Skinner and Peck (1969), and doubtless the coagulation and sinking of sulfide liquid droplets (containing Cu, Fe, Ni, S) accounts satisfactorily for the sulfide minerals in the lower zones of many layered intrusions.

Liquid immiscibility on a small scale may be a much more widespread phenomenon than is generally acknowledged, particularly in acid rocks, where the evidence of fluid inclusions, so often overlooked, seems conclusive. In quartz and feldspar of the ejected granitic blocks of Ascension Island, Roedder and Coombs (1967) found three principal types of fluid inclusions, which represent two liquids (one silicate and one NaCl) and a hydrous fluid. Since oligoclase was one of the earlier minerals to precipitate, the immiscible NaCl liquid and aqueous fluid are likely to have been present throughout much of the granitic crystallization interval. The limited solubility of molten NaCl in a liquid of $NaAlSi_3O_8$ composition saturated with water has been shown experimentally (Koster van Groos and Wyllie, 1969), so that it seems likely that small drops of NaCl liquid form in many magmas rich in silica and feldspar.

Whether large bodies of magma can ever form as a result of liquid immiscibility is still an open, and disputed, question; but experimental results on the system $Na_2CO_3$–$NaAlSi_3O_8$ (Koster van Groos and Wyllie, 1966) suggest that a sodium-carbonate liquid could separate from a silicate liquid. Thus an origin of the sodium-carbonate lava of Oldoinyo Lengai volcano, Tanzania (Dawson, 1966) by some such unmixing process seems entirely plausible.

## MIXING OF MAGMAS

As early as 1851 Bunsen suggested that mixing of two contrasted parent magmas—the one basaltic, the other rhyolitic—would account for the range of variation observed in lavas of the basalt-andesite-rhyolite series in Iceland and elsewhere. This theory was later elaborated by Durocher, who explained all igneous rocks as hybrid[1] derivatives of two primary worldwide magmas (Harker, 1909, pp. 333–336). Growing petrographic knowledge soon demonstrated the total inadequacy of this mechanism to explain the wide range of known igneous mineral assemblages. The chemical and mineralogical variation encountered even in a single common association of rocks (such as nephelinite, melilite basalt, basanite, mugearite, trachyte, phonolite) is far too complex, when considered in detail, to be compatible with the simple mixing of any two end members. Mingling of magmas is no longer considered to be a principal factor in evolution of such continuous magma series.

This is not to rule out completely the possible role of magmatic mixing. It has been invoked to explain what seem to be anomalous, rather complex, mineral associations, especially among phenocrysts of volcanic rocks (for example, Larsen et al., 1938, pp. 255–257, 429; Turner and Verhoogen, 1960, pp. 171, 276). Cited in support of mixing is petrographically obvious evidence of disequilibrium, for example, widespread association of plagioclase phenocrysts of markedly different composition in andesites and dacites. Other explanations admittedly are possible and indeed in some recent accounts are given preference. Failure to establish equilibrium is perhaps likely, for example, during assimilative reaction between basic magma and "granitic" wall rock, or when crystals sink down a vertical composition gradient in a differentiating magma body (for example, Larsen and Cross, 1956, p. 280).

The evidence of magma mixing near the surface suggests that it must operate at depth to be effective. One well-documented example is in eastern Iceland (Gibson and Walker, 1963), where basaltic and acid magmas, liquid at the same time, have been erupted simultaneously in varying proportions to form composite lavas fed by composite dikes. Liquid basalt here never formed a homogeneous solution with acid magma, since even small blebs of basalt have a thin, glassy skin. Rocks containing basaltic xenoliths, obviously once liquid, result from this intermingling of two magmas. However, at depth in the Tertiary magmatic hearths beneath the island of Skye, a liquid of ferrodiorite composition has mixed completely with acid magma to produce the marscoite suite (Wager et al., 1965). The only mineralogical evidence of mixing in the resultant rocks is anomalously calcic plagioclase and quartz surrounded by pyroxene—just the evidence used to postulate mixing in the San Juan province (Larsen et al., 1938).

[1] By general usage *hybrid* and *contaminated* are terms applied to igneous magmas (or resultant rocks) modified respectively by igneous and by sedimentary materials.

## ASSIMILATION

Intrusive magma is seldom in a state of chemical equilibrium with the rocks into which it is injected, although it may be in equilibrium with one or more constituent minerals of the invaded rock. Reaction between magma and wall rock must, therefore, be a normal accompaniment to igneous intrusion. Thereby the magma (in most cases a silicate melt with suspended crystals of one or more solid phases) becomes modified by incorporation of material originally present in the wall rock. This broad process of modification is termed *assimilation*.[1]

The following general principles, first clearly stated by Bowen, determine the mechanism by which assimilation can proceed in any given instance:

1. It is unlikely (as recognized long ago by Becker, 1897*a*, p. 37) that large masses of magma in the deep crust are ever substantially superheated above liquidus temperatures; most lavas contain phenocrysts, so that an excess of temperature above a magma's liquidus temperature is unlikely to be common. Thus fusion of country rock, though under some conditions possible, cannot significantly increase the volume of the melt phase. Magma, with a specific heat of perhaps 0.25 cal $g^{-1}$ $deg^{-1}$, cooling through a small interval of superheat, could liquefy only minor amounts of rock (heat of fusion 75 cal $g^{-1}$). The heat of fusion or of any other reaction between magma and country rock must be supplied by precipitation from the magma of a thermally equivalent quantity of crystalline material. This, of course, will consist of mineral phases in equilibrium with the magma.

2. A magma may partially or completely melt a mineral aggregate with some lower temperature range of fusion. The melt phase thereby becomes modified, though there is no significant change in its total volume. Later differentiation of the modified melt may inaugurate new lines of liquid descent; or the ultimate result may simply be to increase the quantity of a late liquid fraction whose composition is much the same as if no reaction had occurred. Thus basaltic magma may assimilate granite by fusion into the melt phase whose volume remains unchanged; but subsequent differentiation will yield a correspondingly increased quantity of the normal acid (granitic) residual liquid.

3. A magma may react with but never melt crystalline phases whose melting range is above that of the magma. The tendency is for such phases to be converted to minerals with which the melt is already in equilibrium, by some process of chemical exchange with the melt. Calcic plagioclase, brought into contact with rhyolite magma from which oligoclase is separating, itself

[1] This usage corresponds with that adopted by S. J. Shand (1943, pp. 90–94). Some petrologists restricted the term to the process of simple solution or melting of solid rock by invading magma (cf. Daly, 1933, p. 288; Grout, 1941, p. 1533).

becomes converted to oligoclase. The composition of the melt phase thereby becomes changed by depletion in Na and Si relative to Ca and Al.

Assimilation may thus be pictured as a complex process of reciprocal reaction between magma and invaded rock. Certain minerals present in the wall rock may become partially or completely melted and in this way incorporated into the liquid fraction of the magma. Others are changed by a process of ionic exchange (reaction) into those crystalline phases with which the liquid was already saturated. Yet other minerals, by chance compatible with the invading magma, persist unchanged as they break out from the transformed partially melted matrix that surrounded them and drift away into the reacting magma. The end product is a modified, partially crystalline magma. In many cases the proportion of liquid in this diminishes as reaction proceeds. When with continued cooling the mass becomes completely crystalline, it has given rise to a hybrid or contaminated igneous rock which was at no time entirely liquid and which is made up of material contributed partly by the original magma and partly by the wall rocks. It may then be impossible to recognize any sharp boundary between intrusive and invaded rocks. Toward the intrusive contact the metamorphosed wall rock becomes increasingly affected by metasomatism as a result of chemical exchange with the adjacent magma, until it acquires a composition close to, or identical with, that of the contaminated igneous rock into which it merges.

## Trace Elements in Magmatic Evolution

Eight elements predominate in the earth's crust; these are oxygen, silicon, aluminum, iron, magnesium, calcium, sodium, and potassium, and they constitute about 98 percent by weight of the crust. Conventionally it is these elements which are the *major* elements of igneous rocks, the remainder being the minor or trace elements. The next three most abundant elements of the crust (excluding H) are Ti, P, and Mn; they are customarily called *minor* elements, and the remainder are called *trace* elements. If the abundance in the whole earth, rather than just the crust, were considered, then Ni, Cr, Co, and S would become major elements, and K and Ti would be trace elements.

The studies of Goldschmidt (1937, 1954) have had a lasting and powerful influence on our understanding of the way in which trace elements distribute themselves in minerals. These empirical "rules" can be briefly stated:

1. If two ions have the same radius and charge, they will enter into solid solution in a given mineral in amounts proportional to their abundances. A trace element is "camouflaged" by a major element (for example, Rb is camouflaged by K in alkali feldspar).

2. If two ions have similar radii and the same charge, the smaller ion will be preferentially incorporated by the solid solution. The chemical bond is

weaker for the larger ion, which is evident in a lower melting point. An example is enrichment of olivine in $Mg^{++}$ (0.65Å) in preference to $Fe^{++}$ (0.76Å).

3. If two ions have similar radii and different charges, the ion with the higher charge will be preferentially incorporated by the solid solution. If the trace element has a higher charge than a major element, it is "captured" by the major element and becomes concentrated in the early crystallizing fractions of the solid solution. If the trace element has a lower charge than the major element, it is "admitted" by the major element and tends to be concentrated in the late-crystallizing solid solutions.

These "rules" set the scene for geochemical research into the distribution and abundance of elements for almost 20 years, and as more exceptions were found, modification and additions to the rules were made, two of the best critiques being by D. M. Shaw (1953) and Ringwood (1955*a*, 1955*b*).

Ringwood considered the electronegativity of an ion in influencing its incorporation by a solid, using electronegativity as an indication of bond strength. Thus if two ions have similar radii and charge, the one with the lower electronegativity (stronger bond) will be preferentially incorporated in the solid solution. But anomalies or inconsistencies persisted, and in a series of papers Burns and Fyfe (1964, 1966, 1967) emphasized repeatedly that any rules governing the distribution of trace elements in solid solutions could not ignore the properties of the phase, often liquid, from which they were withdrawn; their 1967 review put to rest many an old geochemical chestnut.

The generality of the "rules" of trace-element distribution foundered on the first set of transition elements, Sc to Zn, each of which has an incomplete *d* electron shell in one of the common oxidation states. When a free transition-metal cation is influenced by the electrical field of a neighboring anion, the transition-metal ion loses its spherical symmetry, and the energy involved depends upon its coordination or the number of anions surrounding it. This energy is called *crystal-field stabilization energy* and is thoroughly described in a short but elegant book by Burns (1970).

Using the crystal-field stabilization energy of the first set of transition elements in octahedral or tetrahedral sites in spinels (normal and inverse), a term called *octahedral site-preference energy* can be derived, which is a partial measure of the affinity of a transition element for an octahedral site in a crystal, its normal (sixfold) coordination in silicate minerals. Burns (1970) predicts the following order of uptake of the transition elements from a cooling silicate magma, based on the octahedral site-preference energy displayed in spinels:

$$\text{Ni} > \text{Cu} > \text{Co} > \text{Fe} > \text{Mn} \geq \text{Ca, Zn} \qquad \text{for } M^{++} \text{ ions}$$

$$\text{Cr} > \text{Co} > \text{V} > \text{Ti} > \text{Fe} \geq \text{Sc, Ga} \qquad \text{for } M^{3+} \text{ ions}$$

This sequence is matched by the order of depletion of elements in the residual liquids of the Skaergaard intrusion (Curtis, 1964), with the possible exception of Cu. The rocks of this intrusion, and the deduced composition of the successive residual liquids, are perhaps the most thoroughly and completely analyzed in all petrology (one exception being the lunar rocks), largely by the work of Vincent and his coworkers, who have provided trace-element data of the highest quality; this rather scattered information has been brought together by Wager and Brown (1967).

One example of the variation of the transition elements in lavas, comparable in many ways to the calculated liquid compositions of the Skaergaard intrusion, is a Tertiary lava series from eastern Iceland (Thingmuli) which ranges in composition from olivine tholeiite to rhyolite or granophyre (Carmichael, 1964*b*). The concentrations of the transition elements for this series are shown in Fig. 3-1; the abscissa is a modified Larsen factor, conventionally taken as an index of magmatic evolution. Seven of the nine elements develop a maximum, sometimes poorly developed, after which the abundance progressively diminishes in successive lava types. The curve of Fe ($Fe^{++}$ + $Fe^{3+}$) reaches a maximum in the tholeiitic basalts, and it is in these lavas that iron titanium oxides crystallize early and would therefore be available for crystal fractionation. The approximate positions of the maxima have been shown and their order, in terms of the Larsen index, corresponds to the sequence in which $M^{++}$ and $M^{3+}$ ions are incorporated into the early crystallizing solid assemblage. It is this solid assemblage which is believed to control the Thingmuli liquid line of descent. The order of element depletion, with $Fe^{++} + Fe^{3+}$ combined as [Fe], is

$$\mathrm{Ni > Cu > Co > [Fe] > Mn > Zn} \qquad \text{for } M^{++} \text{ ions}$$

$$\mathrm{Cr > V > [Fe]} \qquad \text{for } M^{3+} \text{ ions}$$

which is exactly the order predicted by the octahedral site-preference energy for spinels.

Elements which, in addition to Si, are tetrahedrally coordinated with oxygen in silicate liquids form three-dimensional networks and are often called the network-forming elements. Many of these are excluded from tetrahedral coordination in silicate minerals (Ringwood, 1955*b*) and become concentrated in low-temperature residual liquids. Thus such elements as Ga, Ge, Sn, Zr, Hf, P, As, Mo, Nb, and Ta are typically concentrated in later liquid fractions or alternatively form independent crystal phases. The variation of several of these elements in the Thingmuli lava series is shown in Fig. 3-2. Both P and Zr show maxima, presumably because apatite and zircon successively became stable phases on the liquidus surface and were therefore available for inclusion in a fractionated crystalline assemblage. A similar P maximum occurs in the residual liquids of the Skaergaard intrusion, when apatite becomes one of the accumulating crystal phases. The remaining elements—As, Mo, Sn, Ga, and Ge

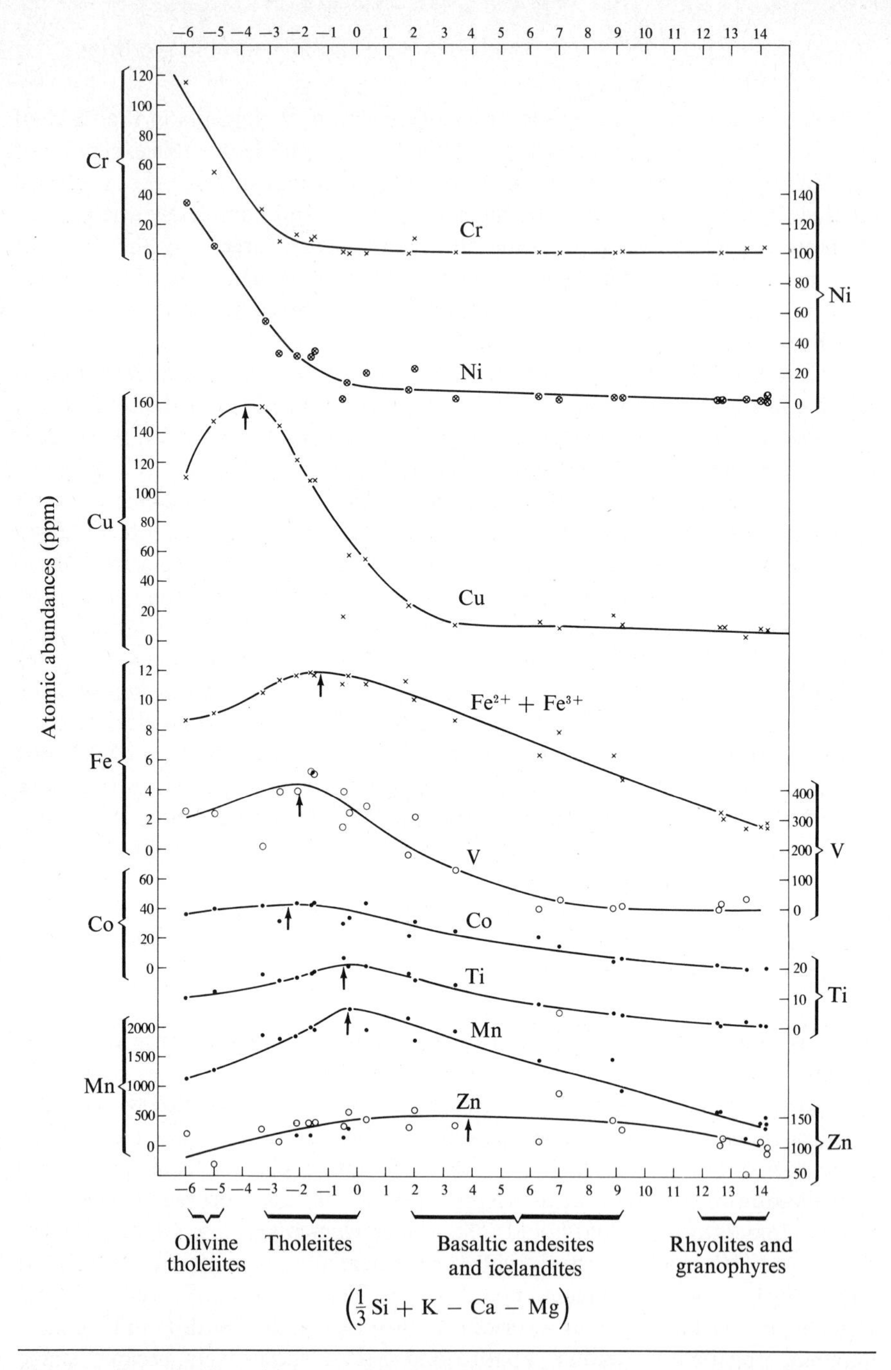


**Fig. 3-1** Concentrations (ppm) of transition elements in volcanic rocks of the Thingmuli series plotted against a modified Larsen factor ($\frac{1}{3}$Si + K − Ca − Mg). The approximate position of the maximum on each curve is marked with an arrow.

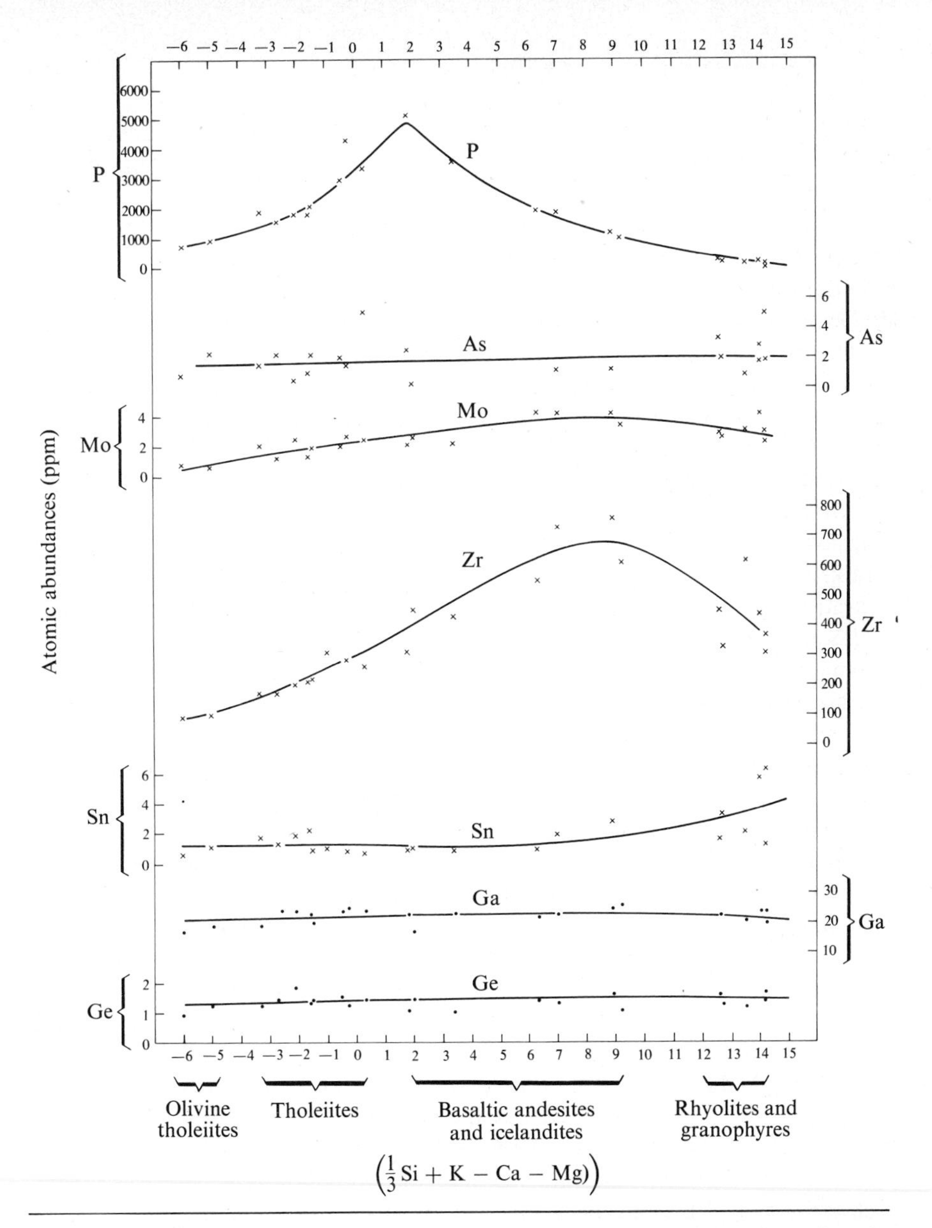


**Fig. 3-2** Trace-element concentrations (ppm) in volcanic rocks of the Thingmuli series plotted against a modified Larsen factor ($\frac{1}{3}$Si + K − Ca − Mg). Note maxima on P and Zr curves.

(Fig. 3-2)—all show a small but progressive increase in lower-temperature, more acid liquids.

The work of Nockolds and Allen (1953, 1954, 1956) on countless samples from many petrogenic provinces shows a general pattern of distribution of trace elements in magmatic evolution. Low-temperature, salic magmas are enriched in Li, Na, K, Rb, Cs, Ba, Pb, Tl, U, Th, Y, and the rare earths in addition to the elements noted above. They are depleted, relative to their basaltic associates, in the first transition metals and Sr.

Trace-element geochemists* have sought to use variation in the ratios of similar elements such as K/Rb, K/Cs, Zr/Hf, Zn/Cd, Nb/Ta, and Mo/W as evidence for unusual or protracted liquid lines of descent. But despite numerous studies there has been little success in correlating K/Rb of a lava series with an oceanic or continental environment, or with magma conceived to be derivative or parental. In most basic magma, K/Rb varies between 1200 and 400, much of the variation being in Rb rather than K, whereas trachytes and rhyolites commonly have values between 600 and 200. One example will suffice to illustrate that K/Rb can remain persistently invariable in rock types of contrasted origin.

In Table 3-1 the average concentration of trace elements in four types of acid lava is set down. One of these is a pantellerite, the most extreme type of peralkaline rhyolite, which has, in contrast to the other rhyolites, high concentrations of Nb, Ta, Mo, Zn, Zr, and Cd, and low Ba and Sr, a pattern common to all peralkaline lavas—phonolite, trachyte, and rhyolite alike. A less extreme peralkaline rhyolite is a comendite, which has the same general pattern of trace elements but in lesser amounts. Many petrologists believe that these two acid lava types result from a protracted liquid line of descent which would presumably generate both the systematically high and the low concentrations of the trace metals shown in Table 3-1. The third type of rhyolite is the average of those from Thingmuli volcano in eastern Iceland, which are probable derivatives of a parental magma of olivine-tholeiite composition. These have less of the distinctive pantellerite trace elements and more Sr and Ba. Lastly there is the average rhyolite from the Taupo volcanic zone of New Zealand, which Ewart and Stipp (1968) believe to result from the partial fusion of Mesozoic eugeosynclinal sediments. But despite all the vagaries of their diverse origin, all four acid lava types have virtually identical values of K/Rb; and Zn/Cd and Zr/Hf show little significant variation.

Although there can be a considerable range of K/Rb, even in rhyolites, its significance is uncertain. The basaltic lavas of the ocean floors often have high values, and many ratios do show a significant variation within a petrogenic province or even within the lavas of one volcano; but any widespread use of such ratios as K/Rb as a general index of magmatic evolution or origin appears unwarranted.

So far we have been concerned solely with the distribution of trace elements during cooling of a magma, whereas much present-day research uses the abundance and distribution of trace elements in basaltic lava as an indication of the conditions and composition of their sources in the mantle. Some elements,

* See, for example, Gunn (1965), Erlank and Hofmeyr (1966). Lessing et al. (1963), and Butler and Smith (1962) for data on K/Rb and K/Cs; also Butler and Thompson (1965, 1967) for data on Zr/Hf and Zn/Cd.

**Table 3-1** Selected abundances of trace elements in acid lavas (values in ppm)*

| | PANTELLERITES OF PANTELLERIA | COMENDITES OF NEW ZEALAND | RHYOLITES OF THINGMULI, ICELAND | RHYOLITES OF TAUPO, NEW ZEALAND |
|---|---|---|---|---|
| Ga | — | 36 | 22.5 | 16 |
| Nb | 320 | 72 | 25 | 20 |
| Ta | 17.8 | 7.4 | — | — |
| Mo | 14.9 | 18.3 | 4.1 | 2 |
| Sn | 0.8 | 6.5 | 1.8 | 3 |
| Ni | 1.2 | — | 0.9 | — |
| Co | 0.6 | — | 1.0 | — |
| Cu | 3.6 | 4.2 | 8.6 | 6 |
| Zn | 440 | 290 | 95 | — |
| Zr | 1800 | 1250 | 385 | 160 |
| Hf | — | 19.5 | — | 4 |
| Y | 145 | 157 | 70 | 40 |
| Cd | 0.52 | 0.45 | 0.20 | — |
| Sr | $<5$ | 2 | 90 | 125 |
| Pb | 15 | 31 | 8.6 | 30 |
| Ba | $<10$ | 21 | 1000 | 870 |
| Rb | 175 | 143 | 130 | 108 |
| Th | — | 20.2 | — | 11.3 |
| U | — | 6.0 | — | 2.53 |
| K/Rb | 230 | 239 | 235 | 250 |
| Nb/Ta | 18 | 10 | — | — |
| Zn/Cd | 850 | 645 | 625 | — |
| Zr/Hf | — | 64 | — | 45 |

* Data taken from Ewart et al. (1968*a*, 1968*b*), Nicholls and Carmichael (1969*b*), Butler and Thompson (1967), Butler and Smith (1962). The data on the Thingmuli lavas (unpublished) were obtained by x-ray fluorescence, colorimetry, and polarography.

such as K, Rb, U, Th, and Zr, can be accommodated only with difficulty in the pyroxene, olivine, garnet, or spinel that are likely to be stable in the source region of a basaltic magma. It is these elements, dispersed among the solid phases of which they are not stoichiometric components, that are believed to become relatively concentrated in a liquid fraction formed by fusion of the source rock. The melting behavior of any solid containing a dispersed element $i$ can be represented by a diagram of the type shown in Fig. 3-3; the distribution of the element $i$ between liquid and solid is given by $K_i^{\text{liq/sol}} = X_i^{\text{liq}}/X_i^{\text{sol}}$. For an element such as Ni which remains preferentially concentrated in the residual crystalline phases (olivine, pyroxene) $K_i^{\text{liq/sol}} < 1$ (Fig. 3-3*a*). For elements like K and Rb which preferentially enter the first liquid fraction, $K_i^{\text{liq/sol}} > 1$ (Fig. 3-3*b*). In general the distribution coefficient of any element is a function of pressure, temperature, and, with the exception of ideal solutions, composition.

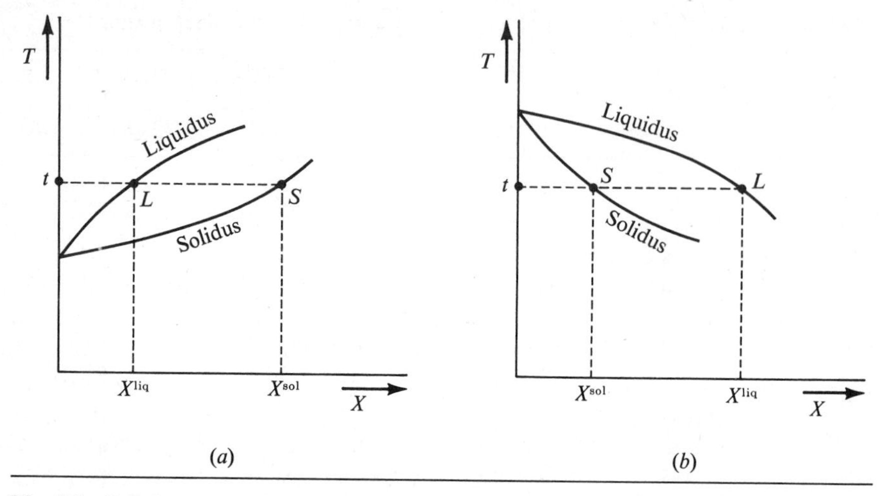


**Fig. 3-3** Relative concentrations of dispersed components $i$ in associated solid ($S$) and liquid ($L$) phases as a function of temperature. $X^{sol}$ and $X^{liq}$ are respective concentrations in solid and liquid at temperature $t$.
$(K_i^{liq/sol})_t = X^{liq}/X^{sol}$.

Consider the fusion of a solid phase in which the initial concentration of $B$ atoms per unit volume is $c_B^{\circ\,sol}$, and assume that the liquid and solid maintain equilibrium (for example, no diffusion gradients). Let the concentration of $B$ atoms in the liquid be $c_B^{liq}$ when a fraction $F$ of the liquid has formed. Since the liquid is in equilibrium with the solid, the concentration of $B$ atoms in the solid is given by

$$c_B^{sol} = \frac{c_B^{liq}}{K_B^{liq/sol}}$$

where $K_B^{liq/sol}$ is the distribution coefficient of $B$ between liquid and solid.

The total amount of $B$ remaining in the solid is

$$V(1 - F)\frac{c_B^{liq}}{K_B^{liq/sol}}$$

where $V$ is the volume of the whole assemblage (liquid + solid). If a further small volume fraction $VdF$ of liquid is formed, then the concentration of $B$ atoms in this volume is given by

$$c_B^{liq} = -\frac{1}{V}\frac{d[V(1 - F)c_B^{liq}/K_B^{liq/sol}]}{dF}$$

The first liquid formed has a concentration of $B$ given by $c_B^{sol} \cdot K_B^{liq/sol}$, so that the equation may be integrated between the limits $c_B^{liq} = c_B^{\circ\, sol}$, $c_B^{liq} = c_B^{liq}$, and $F = 0$, $F = F$—which gives

$$c_B^{liq} = K_B^{liq/sol}\, c_B^{\circ\, sol}(1 - F)^{(K_B^{liq/sol} - 1)}$$

This equation can only be an approximation since it assumes that $K_B^{liq/sol}$ is a constant independent of temperature and composition.

Equations of this type have been used by Gast (1968) and D. M. Shaw (1970)—as discussed more fully in Chap. 13—to test hypotheses of whether or not various basalt magma types (and particularly their concentration of trace and minor components) could be generated by partial fusion, to a varying extent, of uniform mantle material (represented in the equation by $c_B^{\circ\, sol}$). If the fraction of melting is small, then as a first approximation a small increase in the liquid fraction could be considered as an isothermal process; however, in the general petrological application of this equation to the fusion of mantle material there is little experimental evidence to support this assumption.

## Effect of Temperature and Pressure on Mineral Equilibria

Perhaps the most important factor in understanding magmatic evolution is a knowledge of the temperature and pressure (or depth) of the magma's source region or indeed of any region where the magma's residence time allowed its composition or crystallization path to become substantially modified. In its passage toward the earth's surface, a magma presumably moves from a higher to a lower pressure region, ultimately being erupted on the earth's surface or contained as an intrusion high in the crust. The only inherited evidence of this pressure-temperature gradient from source to surface may be the phenocrysts, xenocrysts, or ultramafic nodules contained by many lavas.

In this section on geothermometry and geobarometry we discuss some of the various techniques that can be used to estimate temperature and pressure of a mineralogical assemblage. The emphasis is therefore rather different from that in Chap. 1 where temperature measurements on flowing lavas were recorded. Here we consider the information that can be gleaned from the coexistence of several minerals, assumed in the absence of contrary evidence to be in equilibrium, regarding the temperature and the pressure at which they equilibrated. Although the estimates of pressure and temperature at which, for example, a particular lava equilibrated with its contained ultramafic nodules does not explicitly indicate those of the source region, it is not unreasonable to consider them at least as lower limits, since some nodules may contain minerals stable only at high pressures and temperatures.

Particular emphasis is put on a thermodynamic approach in what follows, for in principle this allows the reader to calculate a geothermometer or geobarometer for whatever mineral assemblage he happens to be studying, provided of course that thermodynamic data are available. Sometimes a simple thermodynamic model suffices in the absence of more complete data, and one example is given of this. This treatment is not intended to belittle those experimental studies on which so much of our temperature information for igneous rocks is based, but unless the reader has access to often expensive and specialized facilities, such methods are beyond his reach. Simple thermodynamic calculations, performed in a few hours at most, can often constrict the range of temperature or pressure to an extent that one of several alternative hypotheses can be eliminated.

## GEOTHERMOMETRY

The distribution of an element between two phases in equilibrium depends upon temperature and pressure. Therefore to be of value in petrology a mineral geothermometer should be independent of pressure, or put in thermodynamic terms, $(\partial \Delta G_r^\circ / \partial P)_T = \Delta V^\circ \simeq 0$, where $\Delta V^\circ$ is the sum of the molar volumes of the products less that of the reactants all in their standard states. A reaction with $\Delta V^\circ = 1 \text{ cm}^3 \text{ mole}^{-1}$ contributes 239 cal to $\Delta G_r^\circ$ at 10 kb, which is well within the uncertainty of $\Delta G_r^\circ$ for most reactions. There are many potential geothermometers in which the effect of pressure can be safely ignored, but few have been calibrated by experiment and fewer have been calculated because of the paucity of thermodynamic data.

Because of the large amount of published data on the distribution of elements between naturally coexisting minerals, it is appropriate to start with a geothermometer of this type.

SOLID-SOLID EXCHANGE REACTIONS A typical reaction, substantially the Ni geothermometer of Hakli and Wright (1967), is

$$\underset{\text{diopside}}{CaMgSi_2O_6} + \tfrac{1}{2}\underset{\text{Ni olivine}}{Ni_2SiO_4} = \underset{\text{Ni diopside}}{CaNiSi_2O_6} + \tfrac{1}{2}\underset{\text{forsterite}}{Mg_2SiO_4} \tag{3-1}$$

so that at equilibrium ($\Delta G = 0$) we can write for the solid solutions

$$\mu^{\text{pyrox}}_{CaNiSi_2O_6} + \tfrac{1}{2}\mu^{\text{ol}}_{Mg_2SiO_4} - \mu^{\text{pyrox}}_{CaMgSi_2O_6} - \tfrac{1}{2}\mu^{\text{ol}}_{Ni_2SiO_4} = 0 \tag{3-2}$$

where $\mu$ is the chemical potential of the subscript component in the superscript phase. But for each component[1] we have

$$\mu^{\text{pyrox}}_{CaNiSi_2O_6} = \mu^{\circ}_{CaNiSi_2O_6} + RT \ln a^{\text{pyrox}}_{CaNiSi_2O_6}$$

$$\mu^{\text{ol}}_{Mg_2SiO_4} = \mu^{\circ}_{Mg_2SiO_4} + RT \ln a^{\text{ol}}_{Mg_2SiO_4}$$

$$\mu^{\text{pyrox}}_{CaMgSi_2O_6} = \mu^{\circ}_{CaMgSi_2O_6} + RT \ln a^{\text{pyrox}}_{CaMgSi_2O_6}$$

$$\mu^{\text{ol}}_{Ni_2SiO_4} = \mu^{\circ}_{Ni_2SiO_4} + RT \ln a^{\text{ol}}_{Ni_2SiO_4}$$

[1] For a discussion of components, see Chap. 4.

If these four equations are substituted into Eq. (3-2), we find with rearrangement

$$\mu^{\circ}_{CaNiSi_2O_6} + \tfrac{1}{2}\mu^{\circ}_{Mg_2SiO_4} - \mu^{\circ}_{CaMgSi_2O_6} - \tfrac{1}{2}\mu^{\circ}_{Ni_2SiO_4}$$

$$= RT \ln \frac{a^{pyrox}_{CaMgSi_2O_6}(a^{ol}_{Ni_2SiO_4})^{1/2}}{a^{pyrox}_{CaNiSi_2O_6}(a^{ol}_{Mg_2SiO_4})^{1/2}} \tag{3-3}$$

The left-hand side of Eq. (3-3) is equal to the Gibbs free-energy change of the reaction for all the components in their standard states, namely pure solids at the temperature of interest and 1 bar; it is therefore constant at any temperature and is represented by $\Delta G^{\circ}_r$. Accordingly, the quantity in the brackets on the right-hand side of Eq. (3-3) is also a constant: it is the reciprocal of the equilibrium constant $K$. Equation (3-3) is an example of the important thermodynamic relationship

$$\Delta G = \Delta G^{\circ}_r + RT \ln K$$

which at equilibrium, $\Delta G = 0$, gives

$$\Delta G^{\circ}_r = -RT \ln K$$

This relationship, in one form or another, is the basis of almost all geothermometers and geobarometers.

Since the concentration of Ni in both olivine and pyroxene is usually small, a few thousand parts per million at most, we can assume that the activities of $CaMgSi_2O_6$ and $Mg_2SiO_4$ in Eq. (3-3) are unity.[1] Thus Eq. (3-3) reduces to

$$\Delta G^{\circ}_r = RT \ln \frac{(a^{ol}_{Ni_2SiO_4})^{1/2}}{a^{pyrox}_{CaNiSi_2O_6}} \tag{3-4}$$

For dilute solutions, such as Ni in olivine and pyroxene are assumed to be, the activity of the Ni component is related to its mole fraction by Henry's law, and we have

$$a^{ol}_{Ni_2SiO_4} = k' X^{ol}_{Ni_2SiO_4}$$

and

$$a^{pyrox}_{CaNiSi_2O_6} = k'' X^{pyrox}_{CaNiSi_2O_6}$$

[1] This does not imply that the activity of either component in a natural augite or olivine is unity; if these were known then they should be substituted into Eq. (3-3).

where $k'$ and $k''$ are the Henry's law constants of proportionality. Substituting these two equations into Eq. (3-4) with rearrangement gives

$$\log \frac{(X^{\text{ol}}_{\text{Ni}_2\text{SiO}_4})^{1/2}}{X^{\text{pyrox}}_{\text{CaNiSi}_2\text{O}_6}} = \frac{\Delta G^\circ_r}{2.303RT} + \log \frac{k''}{(k')^{1/2}} \tag{3-5}$$

Since $\Delta G^\circ_r/2.303RT$ is constant at a fixed temperature, as also is $\log k''/(k')^{1/2}$, we may write for the general case of a trace component $i$ distributed between two solid solutions $\alpha$ and $\beta$ with the stoichiometric coefficients $a$ and $b$, respectively

$$\log \frac{(X^\alpha_i)^a}{(X^\beta_i)^b} = \frac{\Delta G^\circ_r}{2.303RT} - \log \frac{(k^\alpha_i)^a}{(k^\beta_i)^b} = \log K_D \tag{3-6}$$

which is a form of Nernst's distribution law. $K_D$, frequently called the distribution coefficient, should not be confused with the true equilibrium constant $K$. By comparing Eq. (3-6) with Eq. (3-3), it may be seen that only for ideal dilute solutions, in which the activities of the solvents ($Mg_2SiO_4$ and $CaMgSi_2O_6$) are taken as unity, does $K_D$ become equal to the equilibrium constant $K$. In an ideal dilute solution, the activity of component $i$ is equal to the mole fraction of $i$, in which case the Henry's law constants are unity.

The Nernst distribution law [Eq. (3-6)] in principle applies to all trace and those major components which mix ideally and is the basis of many postulated geothermometers, such as Barth's (1962) feldspar geothermometer. The law has also been widely used in the analysis of metamorphic mineral assemblages (Kretz, 1959). The major problem with its general application to igneous rocks lies in commonly unknown values for $\Delta G^\circ_r/2.303RT$ at any temperature, in the unknown values of the Henry's law constants, or in the uncertainty of the assumption of ideal mixing. The Henry's law constants are often expressed as activity coefficients[1] and are necessarily so outside the dilute region, where the component $i$ is no longer present in trace amounts.

The stoichiometric coefficients are often neglected, so that it is always advisable to write out the postulated exchange reaction, formulate the equilibrium constant, and reduce this, if valid, to a form of the Nernst law. There always remains the difficulty of defining the mole of any component. One example arises from Eq. (3-1). Instead of writing $\frac{1}{2}Mg_2SiO_4$ and $\frac{1}{2}Ni_2SiO_4$ for the two olivine components, we could also write $MgSi_{0.5}O_2$ and $NiSi_{0.5}O_2$ (Darken and Schwerdtfeger, 1966); the standard free-energy change of reaction (3-1) would remain the same, but the stoichiometric coefficients for olivine in

[1] The value of the activity coefficient in the Henry's law region depends upon the choice of the standard state. It is unity if the standard state is a hypothetical standard state arrived at by extrapolating the Henry's law region to zero solvent; it is not unity if the standard state is pure solute, which is the common standard state for solid solutions.

the equilibrium constant [Eq. (3-3)] would now be unity rather than $\frac{1}{2}$. The "correct" way, or the formulation that makes the mole fraction of a component most closely correspond to its activity, can be established only by comparing thermodynamic calculation with experimental result.

The equilibrium constant $K$, and presumably $K_D$, change with temperature because $(\partial \Delta G_r^\circ / \partial T)_P = -\Delta S$, although $\Delta S$ may be small ($\sim$0.5 eu) for many solid-solid exchange reactions. If $\Delta S$ is small, or in other words if there are equal amounts of similar phases present both as reactants and as products, then $\Delta G_r^\circ$ will not change greatly with temperature, and it is permissible to use the standard free-energy change at 298°K for all temperatures. Often, however, even $\Delta G_r^\circ$ at 298°K is unknown, as are the Henry's law constants (if applicable), so that to be of use most geothermometers of the type represented by Eq. (3-1) have to be calibrated.[1]

This Hakli and Wright (1967) have done by analyzing (for Ni) coexisting pyroxene and olivine sampled at known temperatures from a cooling Hawaiian tholeiitic lava lake. Since the concentration of Ni is small in all phases, they have reported the distribution coefficient $K_D$ in terms of concentration (parts per million) rather than mole fractions. However, it is unlikely that this calibration is appropriate for all augite-olivine assemblages, since the activity of the $CaMgSi_2O_6$ component in pyroxene appears to vary with the composition of the pyroxene, especially if it contains titanium and aluminum in greater amounts than in the Hawaiian pyroxenes.

*The orthopyroxene-clinopyroxene geothermometer* Calcium-rich pyroxene and magnesian orthopyroxene are immiscible over a wide range of temperature and composition, and since the immiscibility region for the magnesian end members is comparatively unaffected by pressure (Davis and Boyd, 1966), the composition of coexisting pyroxenes in natural assemblages has often been used as a geothermometer (Boyd, 1969).

Because the experimental equilibria were based on pyroxenes of comparatively simple composition, the application of these experimental results to more complex natural minerals is hazardous, although increasingly less so as the natural minerals approach the synthetic in composition.

The reaction investigated is essentially

$$\underset{\text{enstatite-diopside (opx) ss}}{Mg_2Si_2O_6} = \underset{\text{diopside-enstatite (cpx) ss}}{Mg_2Si_2O_6}$$

By making some simplifying assumptions about the mixing properties of orthopyroxene and augite, Wood and Banno (in press) were able to approximate

[1] A detailed thermodynamic formulation of the factors which affect the distribution of trace elements between crystals and magma is given by Banno and Matsui (1973). For use as a geothermometer, experimental calibration will usually be required; an example is the distribution of Eu between plagioclase and liquid (Weill and Drake, 1973).

the activities of $Mg_2Si_2O_6$ in both. They showed that temperature ($T$, °K) was related to the composition of coexisting pyroxenes by

$$T = \frac{-10202}{\ln\left(\dfrac{X^{cpx}_{Mg_2Si_2O_6}}{X^{opx}_{Mg_2Si_2O_6}}\right) - 7.65\,X^{opx}_{Fe} + 3.88(X^{opx}_{Fe})^2 - 4.6}$$

where the numerator may be identified with $-\Delta H_r^\circ$ and the constant (4.6) in the denominator with $\Delta S_r^\circ$ for the pyroxene reaction. Their results reproduced almost all the experimental data on coexisting pyroxenes to within 60°C. One example of the application of this geothermometer is to the orthopyroxene-augite pairs of the Bushveld layered intrusion (Atkins, 1969) which show decreasing equilibration temperatures with increasing iron content. A maximum of 1150°C to a minimum of about 1000°C for the stratigraphically highest pair in the layered series was obtained, which agrees with the temperature of 1000°C for the orthopyroxene-clinopyroxene inversion of a Bushveld pigeonite (G. M. Brown, 1967) obtained from rather higher in the layered succession.

*Example of the estimation of free-energy data for a solid-solid exchange reaction* Since apatite and phlogopite (as a component in biotite) are found together in a wide variety of volcanic rocks and both show a range of fluorine-hydroxyl substitution, a geothermometer based on $F^-$-$(OH)^-$ exchange between these minerals would be of some value. The exchange reaction can be represented by

$$\underset{\text{fluorphlogopite}}{[KMg_3AlSi_3O_{10}]_{1/2}F} + \underset{\text{hydroxy apatite}}{Ca_5(PO_4)_3OH}$$

$$= \underset{\text{fluorapatite}}{Ca_5(PO_4)_3F} + \underset{\text{hydroxy phlogopite}}{[KMg_3AlSi_3O_{10}]_{1/2}OH} \qquad (3\text{-}7)$$

$$\Delta V_{1;298^\circ} = 0.3\ \text{cm}^3 \qquad \Delta S_{1;298^\circ} = 0.2\ \text{cal deg}^{-1}\ \text{mole anion}^{-1}$$

in which the mineral formulas are written to include only one $F^-$ or $(OH)^-$, rather than two or four $F^-$, which would imply that the $F^-$ group (two or four) would exchange as a single unit with $(OH)^-$. The reaction, as written, also simplifies the equilibrium constant since the stoichiometric coefficients are unity. At equilibrium, Eq. (3-7) can be described as

$$\frac{-\Delta G_r^\circ}{2.303RT} = \log\frac{a^{phlog}_{[KMg_3AlSi_3O_{10}]_{1/2}OH}}{a^{phlog}_{[KMg_3AlSi_3O_{10}]_{1/2}F}} + \log\frac{a^{ap}_{Ca_5(PO_4)_3F}}{a^{ap}_{Ca_5(PO_4)_3OH}} \qquad (3\text{-}8)$$

and if ideal mixing is assumed, we can replace the activities by the corresponding mole fractions. If $X$ and $Y$ are respective mole fractions of hydroxy phlogopite and fluorapatite in coexisting phlogopite and apatite, then

$$\frac{-\Delta G_r^\circ}{2.303RT} = \log\frac{X}{1-X} + \log\frac{Y}{1-Y} \qquad (3\text{-}9)$$

If $\Delta G_r^\circ$ is known, then by substitution of a series of values for $X$ ($0 < X < 1$), the corresponding values of $Y$ can be calculated for each temperature. Unfortunately, $\Delta G_r^\circ$ is not known, so Stormer and Carmichael (1971*b*) circumvented this deficiency by considering analogous compounds for which free-energy data are available.

In the mica structure, $F^-$ or $(OH)^-$ is associated with the octahedral cations $Mg^{++}$ and $Al^{3+}$, and the structure of hydroxy phlogopite can be represented by a central layer of brucite $[Mg_3(OH)_6]$ with four of the six $(OH)^-$ ions replaced by oxygen of the tetrahedral layer. In apatite, $F^-$ or $(OH)^-$ is surrounded by three calcium atoms, which suggests that Eq. (3-7) could be modeled by

$$Mg_{1/2}F + Ca_{1/2}(OH) = Mg_{1/2}(OH) + Ca_{1/2}F \tag{3-10}$$

$$\Delta G_{10}^\circ = -3.4 \text{ kcal} \qquad \Delta V_{1;298^\circ} = 1.75 \text{ cm}^3$$

$$\Delta S_{1;298^\circ} = 1.03 \text{ cal deg}^{-1} \text{ mole}^{-1}$$

where $\Delta G_{10}^\circ$ signifies $\Delta(G_r^\circ)_{1;298^\circ}$. However, there is one complication: namely, that in this model reaction $Mg(OH)_2$ and $MgF_2$ do not have the same structure, nor do $Ca(OH)_2$ and $CaF_2$. Thus there is an increment in the standard free-energy change of Eq. (3-10) which is not present in Eq. (3-7). This is reflected by the larger value of $\Delta S$ in Eq. (3-10) in comparison to Eq. (3-7). Although this contribution to the standard free-energy change of the reaction is not likely to be large, it should be established that this is the case. The method chosen by Stormer and Carmichael (1971*b*) was to consider the reaction at 1 bar, 298°K:

$$\underset{\text{solid}}{Mg_{1/2}F} + \underset{\text{gas}}{H_2O} = \underset{\text{solid}}{Mg_{1/2}OH} + \underset{\text{gas}}{HF} \qquad \Delta G_{11}^\circ = +17.6 \text{ kcal} \tag{3-11}$$

which can be written as two separate reactions: the structural transformation

$$\underset{\text{sellaite structure}}{Mg_{1/2}F} \rightarrow \underset{\text{brucite structure}}{Mg_{1/2}F} \qquad \Delta G_{12}^\circ \tag{3-12}$$

and the exchange reaction

$$\underset{\text{brucite structure}}{Mg_{1/2}F} + \underset{\text{gas}}{H_2O} \rightarrow \underset{\text{brucite}}{Mg_{1/2}OH} + \underset{\text{gas}}{HF} \qquad \Delta G_{13}^\circ \tag{3-13}$$

so that

$$\Delta G_{11}^\circ = \Delta G_{12}^\circ + \Delta G_{13}^\circ$$

If we let Eq. (3-11) model the exchange reaction

$$\underset{\text{fluorphlogopite}}{[KMg_3AlSi_3O_{10}]_{1/2}F} + \underset{\text{gas}}{H_2O} = \underset{\text{hydroxy phlogopite}}{[KMg_3AlSi_3O_{10}]_{1/2}OH} + \underset{\text{gas}}{HF} \tag{3-14}$$

then on the assumption that both the gas and the phlogopite components mix ideally, we can write for Eq. (3-14)

$$\frac{-\Delta G_r^\circ}{2.303RT} = \log \frac{X_{\mathrm{HF}}^{\mathrm{gas}}}{X_{\mathrm{H_2O}}^{\mathrm{gas}}} + \log \frac{X_{\mathrm{OH\text{-}phlog}}^{\mathrm{phlog}}}{X_{\mathrm{F\text{-}phlog}}^{\mathrm{phlog}}} \tag{3-15}$$

and substituting $\Delta G_{11}^\circ$ [Eq. (3-11)] into Eq. (3-15), we can calculate the mole fraction of $X_{\mathrm{HF}}^{\mathrm{gas}} = (1 - X_{\mathrm{H_2O}}^{\mathrm{gas}})$ in equilibrium with any value of $X_{\mathrm{F\text{-}phlog}}^{\mathrm{phlog}} = (1 - X_{\mathrm{OH\text{-}phlog}}^{\mathrm{phlog}})$. The curve so calculated (Fig. 3-4*a*) is in good agreement with an experimental curve showing the distribution of fluorine between phlogopite and $H_2O$–HF gas (Munoz and Eugster, 1969), which suggests that $\Delta G_{12}$, the structural transformation increment, is small in comparison to $\Delta G_{13}$, the exchange energy. Ludington (1973) has subsequently determined the equilibrium constant for Eq. (3-14) and found it to be +15.5 kcal, within 12 percent of the value calculated from Eq. (3-11).

Returning to the apatite-phlogopite geothermometer represented in Eq. (3-7), we may use Eq. (3-10) to model this with some certainty since in this case the structural transition contributions of the Mg and Ca compounds tend to cancel, and in any case they are likely to be small as we have seen above. Since $\Delta S_{1;298^\circ}$ for Eq. (3-10) is not large, there is only a small change of $\Delta G_{10}^\circ$ with temperature, and we can use the value of $\Delta G_{10}^\circ$ at 298°C, −3.4 kcal, at reasonable geological temperatures (<1200°C), so that if this value is substituted into Eq. (3-9), values of *X* corresponding to chosen values of *Y* are obtained.

The calculated curves are shown in Fig. 3-4*b*, together with the F/OH ratios of coexisting apatites and phlogopites in volcanic rocks. Obviously some pairs record reasonable magmatic temperatures, whereas others, despite the large F content of the apatite, show anomalously low temperatures because of the low F content of their coexisting micas. This has been interpreted as independent evidence for postmagmatic exchange of $F^-$ for $(OH)^-$, revealed by the distribution of oxygen isotopes (Epstein and Taylor, 1967). The $F^-$–$(OH)^-$ distribution of an apatite-phlogopite pair can also be used as an indicator of low-temperature exchange with an aqueous fluid, to which a mica is more susceptible than apatite. The reader should be aware that equilibrium may not be achieved in many igneous rocks, and that assumption of equilibrium underlies all calculations of this type.

In rocks which when liquid contained both HF and $H_2O$, the very large free-energy change of Eq. (3-11) is reflected in the strong preferential enrichment of phlogopite in F (Fig. 3-4). For phlogopite and apatite to be in exchange equilibrium with fluorine and hydroxyl at magmatic temperatures, we should expect F to occupy approximately 85 percent of the (OH) sites in apatite and between 60 and 75 percent in the coexisting biotite (Fig. 3-4*b*).

This type of simple thermodynamic approach can also be used to model the distribution of F between coexisting biotite and muscovite, the former being the more enriched in F. Analogous reasoning suggests that Cl should be

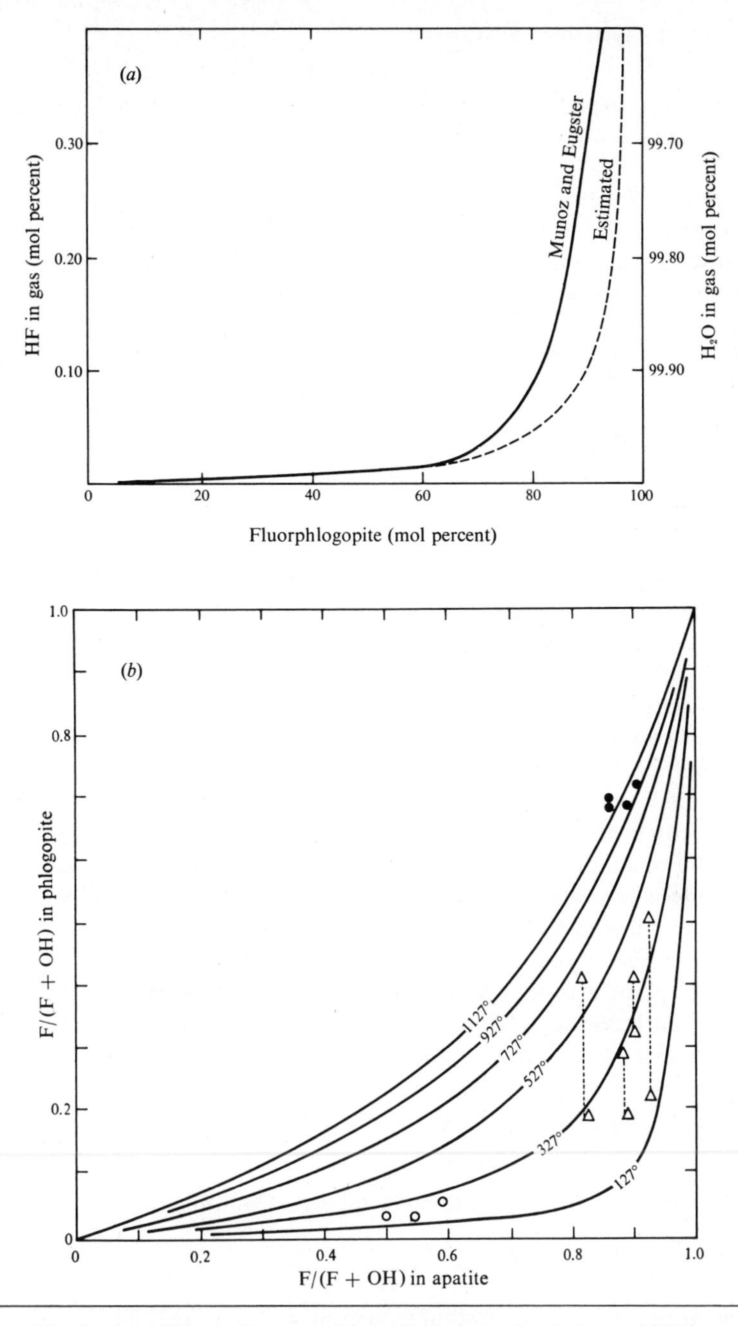


**Fig. 3-4** Distribution of fluorine between phlogopite and associated phases. (*a*) Phlogopite and gas at 700°C and 2 kb (after Munoz and Eugster, 1969); dashed line is computed curve for 298°K and 1 bar (after Stormer and Carmichael, 1971*b*). (*b*) Phlogopite and apatite, calculated as a function of temperature at $P = 1$ bar (after Stormer and Carmichael, 1971*b*). Data for volcanic rocks: solid dots, leucite trachytes, Roman province; triangles, ultrapotassic rocks, Leucite Hills, Wyoming; open circles, dacites, Mt. Lassen, California.

preferentially incorporated by apatite in comparison with phlogopite, in accord with observations on volcanic assemblages (Stormer and Carmichael, 1971*b*).

SOLID-GAS GEOTHERMOMETERS Solid-gas geothermometers, one of which has become widely and successfully applied to igneous and particularly volcanic rocks, involve the concept of a buffer reaction, which we only briefly touched on in Chap. 2. This is illustrated by the following reaction:

$$\underset{\text{magnetite}}{4Fe_3O_4} + \underset{\text{gas}}{O_2} = \underset{\text{hematite}}{6Fe_2O_3} \tag{3-16}$$

for which the equilibrium constant is

$$K_{P,T} = \frac{(a^{\text{hem}}_{Fe_2O_3})^6}{(a^{\text{mag}}_{Fe_3O_4})^4 f_{O_2}}$$

so that at any constant temperature and pressure the fugacity[1] of oxygen $f_{O_2}$ is fixed by the two pure solids (unit activity) in equilibrium. If oxygen is added, then to restore equilibrium magnetite is transformed to hematite and the fugacity of oxygen is reduced to its equilibrium value at that temperature. Thus an assemblage of pure hematite and pure magnetite constitutes an oxygen *buffer*, which continues as such only as long as both solid phases coexist.

In contrast to a buffer assemblage, there are minerals or compounds of variable composition such as pyrrhotite $Fe_{(1-X)}S$ and wustite $Fe_{(1-X)}O$ which change in composition (at constant temperature) in response to any change in the fugacity of sulfur or oxygen, respectively. These minerals are said to *define* the fugacity of the volatile component, for if the temperature can be estimated by some other method, then the composition of each mineral will allow a value of $f_{S_2}$ or $f_{O_2}$ to be derived at that temperature.

At this point it is appropriate to digress upon the effect of total pressure on an oxygen buffer, or how hydrostatic pressure changes the fugacity of the volatile component. For a gas we have, by definition

$$\left(\frac{\partial \mu}{\partial P}\right)_T = RT\left(\frac{\partial \ln f}{\partial P}\right)_T = \bar{V} \tag{3-17}$$

where $\bar{V}$ is the partial molar volume of the gas. From the relationship

$$d\Delta G = \Delta V dP - \Delta S dT$$

we have, at equilibrium at constant temperature $T$, for the solid-gas assemblage

$$d(\Delta G)_T = \Delta V^{\text{sol}}\, dP - \bar{V}^{\text{gas}}\, dP = \Delta V^{\text{sol}}\, dP - RT\left(\frac{\partial \ln f}{\partial P}\right)_T dP \tag{3-18}$$

[1] See Chap. 6 for a discussion of fugacity.

If it is assumed that $\Delta V^{sol}$ is independent of pressure, and noting that $(d\Delta G)_T = 0$, then Eq. (3-18) can be integrated between the limits of $P$ bar and 1 bar to give, with rearrangement,

$$\ln \frac{(f_{O_2})_P}{(f_{O_2})_1} = \frac{\Delta V^{sol}(P - 1)}{RT} \tag{3-19}$$

For Eq. (3-16) $(\Delta V^{sol})_{298°} = 3.55 \text{ cm}^3 \text{ mole}^{-1}$ oxygen. The effect of increasing pressure to 5000 bar at 1000°C can be calculated using Eq. (3-19), and keeping consistent units,* we find

$$\log (f_{O_2})_{5000} - (\log f_{O_2})_1 = \frac{3.55 \times 5000}{2.303 \times 1.987 \times 1273 \times 41.84} = 0.073$$

An increase from 1 to 5000 bar pressure on the solid assemblage at 1273°K increases the logarithm of the equilibrium oxygen fugacity by 0.073.

It is important to stress that there is nothing in the usage of the term *fugacity*, or in the thermodynamic properties thereof, that necessitates a gas phase. The requirement of equilibrium is that the chemical potential of any component must be the same in all phases—gas, liquid, or solid—and one way of expressing the chemical potential is by means of the fugacity.

*The iron-titanium-oxide geothermometer* Although less than 10 years in use, the iron-titanium-oxide geothermometer so carefully calibrated by Lindsley (Buddington and Lindsley, 1964) is already classic. It can be represented by the equation

$$\underset{\substack{\text{ulvospinel-magnetite ss} \\ (\beta \text{ phase})}}{X\,Fe_2TiO_4 + (1 - X)Fe_3O_4} + \tfrac{1}{4}O_2 = \underset{\substack{\text{ilmenite-hematite ss} \\ (\alpha \text{ phase})}}{X\,FeTiO_3 + (\tfrac{3}{2} - X)Fe_2O_3} \tag{3-20}$$

in which both the reactants and the products are solid solutions, the former being the cubic, ulvospinel-magnetite solid-solution series (the $\beta$ series), whereas the products are the hexagonal ilmenite-hematite series (the $\alpha$ series).

The equilibrium constant can be written as

$$K_{P,T} = \frac{(a_{FeTiO_3}^{\alpha\text{ phase}})^X (a_{Fe_2O_3}^{\alpha\text{ phase}})^{3/2 - X}}{(a_{Fe_2TiO_4}^{\beta\text{ phase}})^X (a_{Fe_3O_4}^{\beta\text{ phase}})^{1 - X} (f_{O_2})^{1/4}}$$

and from the previous discussion it is clear that the oxygen fugacity is defined if the activities of all the solids are fixed; and the compositions of the two solid solutions, being temperature-dependent, are uniquely determined by both

* 41.84 cm³ bar = 1 cal, or 1 cm³ = 0.0239 cal bar⁻¹.

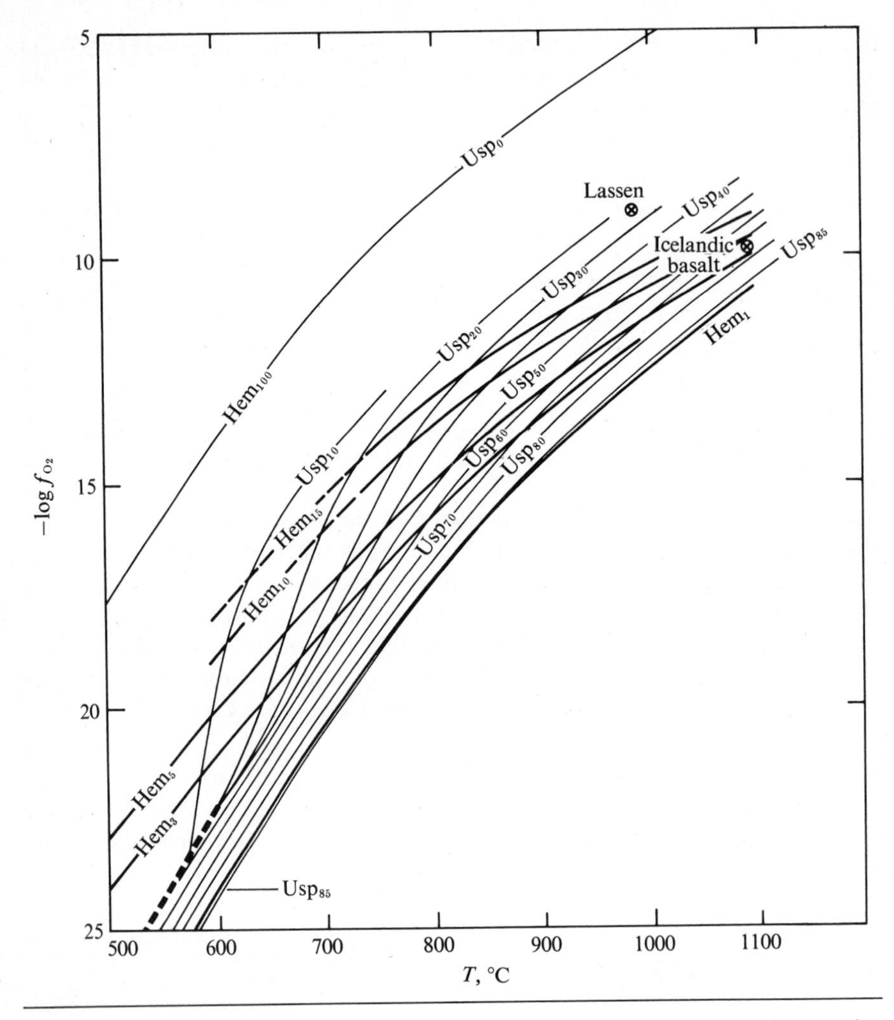


**Fig. 3-5** Compositions, in mole percent, of coexisting ilmenite-hematite and magnetite-ulvospinel solid solutions as a function of temperature and oxygen fugacity shown as $-\log f_{O_2}$. (After Buddington and Lindsley, 1964.)

temperature and $f_{O_2}$. The calibration curves published by Buddington and Lindsley (1964) are shown in Fig. 3-5; if the composition of a magnetite solid solution is known in terms of the mole fraction of $Fe_2TiO_4$, and the coexisting ilmenite solid solution in terms of the mole fraction of $Fe_2O_3$, then the temperature and the logarithm of $f_{O_2}$ of their equilibration can be obtained.

Naturally occurring iron titanium oxides may contain several other components (Table 3-2), such as $MgO$, $MnO$, $ZnO$, $Al_2O_3$, $V_2O_3$, and $Cr_2O_3$, but in general they do not amount to a large proportion of the whole unless the mineral is rich in $Cr_2O_3$. There is some uncertainty about how these additional components affect the curves of Fig. 3-5; Speidel (1970) has shown that MgO tends to raise $f_{O_2}$ at a fixed temperature, but the effect of the other components is not known. A further difficulty arises from the method of recalculating iron-titanium-oxide analyses, which today are almost always obtained by the electron microprobe and which therefore do not separately report FeO and $Fe_2O_3$. Several alternative procedures for the recalculation of these analyses into the necessary components have been published (Anderson, 1968; Carmichael, 1967*b*).

In Table 3-2, two pairs of iron-titanium-oxide analyses are shown as obtained by the electron microprobe; one is from an Icelandic tholeiitic basalt and the other from a dacitic lava from Lassen Park, California. The calculated amounts of $Fe_2O_3$ are shown for each analysis and, for the magnetite solid solutions, the amount of the ulvospinel component. The two pairs of analyses are plotted in Fig. 3-5, and the derived values of $T$ and $f_{O_2}$ are set down in Table 3-2 (the Lassen dacite assemblage requires a judicious interpolation of the curves).

**Table 3-2** Analyses of two pairs of coexisting one-phase iron titanium oxides*

| | ICELANDIC THOLEIITIC BASALT | | LASSEN DACITE | |
|---|---|---|---|---|
| | MAGNETITE ss | ILMENITE ss | MAGNETITE ss | ILMENITE ss |
| $SiO_2$ | 0.23 | 0.17 | 0.05 | 0.06 |
| $TiO_2$ | 27.1 | 49.5 | 7.33 | 37.6 |
| $Al_2O_3$ | 1.22 | 0.13 | 1.65 | 0.28 |
| $Cr_2O_3$ | 0.03 | — | 0.09 | 0.03 |
| $V_2O_3$ | 0.71 | 0.10 | 0.81 | 0.20 |
| FeO | 67.0 | 47.8 | 82.3 | 56.4 |
| MnO | 0.83 | 0.49 | 0.49 | 0.43 |
| MgO | 0.19 | 1.09 | 1.60 | 2.12 |
| CaO | 0.19 | 0.16 | 0.01 | 0.02 |
| ZnO | 0.12 | — | 0.10 | — |
| Total | 97.6 | 99.4 | 94.4 | 97.1 |
| *Calculated* | | | | |
| $Fe_2O_3$ | 13.4 | 6.4 | 52.5 | 29.7 |
| FeO | 55.0 | 42.1 | 35.0 | 29.6 |
| Total | 99.0 | 100.1 | 99.6 | 100.0 |
| Mol % ulvospinel | 77.3 | | 20.9 | |
| Mol % $Fe_2O_3$ | | 6.3 | | 28.7 |
| $T$; $\log f_{O_2}$ | 1090°C; −10.0 | | 980°C; −9.1 | |

* Data from Carmichael (1967*a*,*b*).

Buddington and Lindsley (1964) suggest an error of $\pm 30°C$ for temperature and $\pm 1$ log unit for oxygen fugacity, and presumably this increases for natural minerals with their more complex composition. The only comparable data from other methods agree reasonably well with the iron-titanium-oxide temperatures. Thus an Icelandic basalt (Table 3-2, left) has an experimentally determined liquidus temperature of 1113°C (Tilley et al., 1968) compared with 1090°C for the oxides; a group of New Zealand rhyolites has an average oxide temperature of about 760°C (735 to 780°C) compared with an average oxygen-isotope temperature of near 780°C (695 to 860°C) although, at this temperature the oxygen-isotope geothermometer is not precise (Ewart et al., 1971).[1] Lastly, the data of Grommé et al. (1969) obtained on a cooling lava lake show that the iron-titanium-oxide temperatures may be higher ($\sim 100°C$) than the measured liquid temperatures of the liquid which encloses them.

The iron-titanium-oxide geothermometer can be widely used because iron titanium oxides are almost ubiquitous. In plutonic rocks, however, they are prone to late or postmagmatic alteration, and in alkali-rich lavas, such as basanites and nephelinites, ilmenite is typically absent.

*The pyrrhotite-pyrite geothermometer* The pyrrhotite-pyrite geothermometer, calibrated by Toulmin and Barton (1964), is restricted to temperatures below 742°C, the upper stability limit of pyrite. It can be represented by the reaction

$$FeS + \tfrac{1}{2}S_2 = FeS_2 \qquad (3\text{-}21)$$

and the equilibrium constant at any temperature and pressure is written as

$$K_{T,P} = \frac{a_{FeS_2}^{pyrite}}{a_{FeS}^{pyrrh}\,(f_{S_2})^{1/2}}$$

where the activity of FeS in pyrrhotite $Fe_{(1-X)}S$ is referred to troilite (FeS) as the standard state. The experimental calibration included a novel method for measuring the fugacity of sulfur. From this, the activity of FeS in pyrrhotite was determined and related to the composition of pyrrhotite, which is expressed in terms of FeS. Surprisingly, considering the ease with which the composition of pyrrhotite can be determined, little use has been made of this geothermometer, especially in acid plutonic rocks, where its rather low upper temperature limit is no handicap. However, the composition of pyrrhotite by itself defines $f_{S_2}$, so that if the temperature is known, then a value for sulfur fugacity can be obtained. Several results of this type for volcanic rocks are discussed in Chap. 6.

A PLAGIOCLASE GEOTHERMOMETER Plagioclase phenocrysts are so commonly enclosed by a glassy or finely crystalline matrix in volcanic rocks that pe-

[1] Satisfactory agreement with isotopic temperatures for a variety of lava types has also been recorded by Anderson et al. (1971).

trologists for many years have tried to find a relationship between the composition of the liquid and that of the plagioclase crystals, based essentially on Bowen's (1913) investigation of the plagioclase system. The only successful attempt at formulating a thermometer on these lines is that of Kudo and Weill (1970), who considered the exchange reaction

$$\underset{\text{plagioclase}}{(NaSiO_{2.5})AlSi_2O_{5.5}} + \underset{\text{liquid}}{(CaAlO_{2.5})AlSi_2O_{5.5}}$$

$$= \underset{\text{liquid}}{(NaSiO_{2.5})AlSi_2O_{5.5}} + \underset{\text{plagioclase}}{(CaAlO_{2.5})AlSi_2O_{5.5}} \tag{3-22}$$

or equivalently in terms of the oxide components:

$$\tfrac{1}{2}\mu_{Na_2O}^{liq} + \mu_{SiO_2}^{liq} + \mu_{CaAl_2Si_2O_8}^{plag} - \mu_{NaAlSi_3O_8}^{plag} - \mu_{CaO}^{liq} - \tfrac{1}{2}\mu_{Al_2O_3}^{liq} = 0 \tag{3-22$a$}$$

But, introducing molar activity coefficients $\gamma$

$$\mu_{NaAlSi_3O_8}^{plag} = \mu_{NaAlSi_3O_8}^{\circ} + RT(\ln X_{NaAlSi_3O_8}^{plag} + \ln \gamma_{NaAlSi_3O_8}^{plag})$$

$$\mu_{CaAl_2Si_2O_8}^{plag} = \mu_{CaAl_2Si_2O_8}^{\circ} + RT(\ln X_{CaAl_2Si_2O_8}^{plag} + \ln \gamma_{CaAl_2Si_2O_8}^{plag})$$

$$\mu_{CaO}^{liq} = \mu_{CaO}^{\circ} + RT(\ln X_{CaO}^{liq} + \ln \gamma_{CaO}^{liq})$$

$$\mu_{Na_2O}^{liq} = \mu_{Na_2O}^{\circ} + RT(\ln X_{Na_2O}^{liq} + \ln \gamma_{Na_2O}^{liq})$$

$$\mu_{Al_2O_3}^{liq} = \mu_{Al_2O_3}^{\circ} + RT(\ln X_{Al_2O_3}^{liq} + \ln \gamma_{Al_2O_3}^{liq})$$

$$\mu_{SiO_2}^{liq} = \mu_{SiO_2}^{\circ} + RT(\ln X_{SiO_2}^{liq} + \ln \gamma_{SiO_2}^{liq})$$

If we assume that high-temperature plagioclase solid solutions mix ideally as written in Eq. (3-22) ($\gamma = 1$),[1] then by substituting this array of equations into Eq. (3-22*a*) we obtain

$$\frac{\mu_{NaAlSi_3O_8}^{\circ} + \mu_{CaO}^{\circ} + \tfrac{1}{2}\mu_{Al_2O_3}^{\circ} - \mu_{CaAl_2Si_2O_8}^{\circ} - \mu_{SiO_2}^{\circ} - \tfrac{1}{2}\mu_{Na_2O}^{\circ}}{2.303RT}$$

$$= \log \frac{X_{CaAl_2Si_2O_8}^{plag}}{X_{NaAlSi_3O_8}^{plag}} + \log \frac{(X_{Na_2O})^{1/2} X_{SiO_2}}{X_{CaO}(X_{Al_2O_3})^{1/2}} + \log \frac{(\gamma_{Na_2O})^{1/2}\gamma_{SiO_2}}{\gamma_{CaO}(\gamma_{Al_2O_3})^{1/2}} \tag{3-23}$$

[1] By considering experimental results on the nonideal mixing of plagioclase feldspar, Mathez (1973) was able to refine this geothermometer and improve the results for lavas with calcic plagioclase.

where $X$ and $\gamma$ refer to the mole fractions and activity coefficients respectively of the oxide components in the liquid. The left-hand side of the equation is the negative standard free-energy change $-\Delta G_r^\circ$ with all the components in their standard states (namely, pure liquids or solids at the temperature of interest and 1 bar). Although thermodynamic data exist to calculate $-\Delta G_r^\circ$, Kudo and Weill took another approach which, put in the form they used, requires that Eq. (3-23) be simplified. If we let $\sigma = X_{Ab}/X_{An}$ and $\lambda = (X_{Na_2O}X_{SiO_2})/(X_{CaO}X_{Al_2O_3})$, where the two exponents ($\frac{1}{2}$) have been dropped (for the moment), then Eq. (3-23) can be written

$$\frac{-\Delta G_r^\circ}{2.303RT} = \log\frac{\lambda}{\sigma} + \log\frac{(\gamma_{NaO_2})^{1/2}\gamma_{SiO_2}}{\gamma_{CaO}(\gamma_{Al_2O_3})^{1/2}} \tag{3-24}$$

Kudo and Weill then postulated that silicate liquids behave as regular solutions or that for any component $i$, $\log\gamma_i = C\phi/T$, where $C$ is a constant (incorporating the dropped exponents in the mole fraction term) and $\phi$ is a function of composition.[1] Thus for any two plagioclase-liquid pairs (1 and 2) at the same temperature, we may write

$$\left(\frac{-\Delta G_r^\circ}{2.303RT}\right)_1 = \log\frac{\lambda_1}{\sigma_1} + \frac{C_1\phi_1}{T}$$

and for the second pair

$$\left(\frac{-\Delta G_r^\circ}{2.303RT}\right)_2 = \log\frac{\lambda_2}{\sigma_2} + \frac{C_2\phi_2}{T}$$

Since both pairs are at the same temperature, the left-hand sides of both equations are equal and we find that

$$\log\frac{\lambda_1}{\sigma_1} + \frac{C_1\phi_1}{T} = \log\frac{\lambda_2}{\sigma_2} + \frac{C_2\phi_2}{T}$$

and assuming that $C_1 = C_2 = C'$, then

$$C'\left(\frac{\phi_1 - \phi_2}{T}\right) = \log\frac{\lambda_2\sigma_1}{\sigma_2\lambda_1}$$

or

$$C' = \frac{T}{\phi_1 - \phi_2}\log\frac{\lambda_2\sigma_1}{\sigma_2\lambda_1} \tag{3-25}$$

[1] This specific form of regular solution behavior assumes that all the coefficients of the power series expansion of the mole fractions representing ln $\gamma_i$ are equal; this is not usual for regular solutions.

Kudo and Weill further assumed that the function of composition $\phi$ can be expressed in the terms of the mole fractions of the oxide components in the liquid ($X_{CaO} + X_{Al_2O_3} - X_{Na_2O} - X_{SiO_2}$), so that for any two plagioclase-liquid pairs, $C'$ can be calculated. By taking sets of plagioclase-liquid pairs at various temperatures, an average value of $C'$ can be calculated by least-squares regression. One other experimental variable is the pressure of water $p_{H_2O}$, which lowers the equilibrium temperatures, so that additional sets of values for $C'$ can be calculated for 0.5, 1.0, and 5.0 kb of water pressure. Kudo and Weill report their results for the curve appropriate to 0.5 kb water vapor pressure as

$$2.303 \log \frac{\lambda}{\sigma} + 1.29 \times 10^4 \frac{\phi}{T} = 11.05 \times 10^{-3} T - 17.86$$

in which one solution for $T$ can be discarded as physically unreasonable. The mean error, or the difference between the calculated and observed temperature for 40 plagioclase-liquid pairs, is $-1°$, with a standard deviation of $\pm 34°$; and on the assumption that the experimental temperatures are correct this could also be expressed as a deviation in the mole fraction of $CaAl_2Si_2O_8$ in the measured plagioclase composition of $\pm 0.037$.

This geothermometer has been used to compare the respective temperatures determined from plagioclase and the iron titanium oxides for a group of obsidians with phenocrysts of plagioclase. Because zoning of composition is commonly found in plagioclase and is typically absent in the iron titanium oxides, the response time to changing temperature of the plagioclase thermometer is likely to be much longer than that of the iron titanium oxides. Comparative values are shown in Fig. 3-6. Since the feldspars are zoned, temperatures appropriate to the compositions of both the cores and the margins of the crystals have been calculated. In Fig. 3-6, the correlation between the iron-titanium-oxide temperatures and the plagioclase outer-zone temperatures is excellent either at 0.5 kb $p_{H_2O}$ or with no water. This suggests that all these rocks crystallized at $p_{H_2O}$ lower than about 1 kb. It would take less than 2 kb $p_{H_2O}$ to bring the calculated plagioclase core temperatures into agreement with the iron-titanium-oxide temperatures, so that limits of $p_{H_2O}$ can be derived by combined use of the two geothermometers.

OTHER MINERAL GEOTHERMOMETERS Since there is no single geothermometer that can be applied to the whole range of igneous rocks, we note a few below which, from time to time, have been used to extract estimates of temperature from mineral assemblages.[1]

At high temperatures, sodic nepheline can take considerable amounts of

[1] Others that have been proposed are excluded as insensitive at igneous temperatures—notably those based on the temperature dependence of the partitioning of cations between nonequivalent lattice sites. These tend to reequilibriate with cooling.

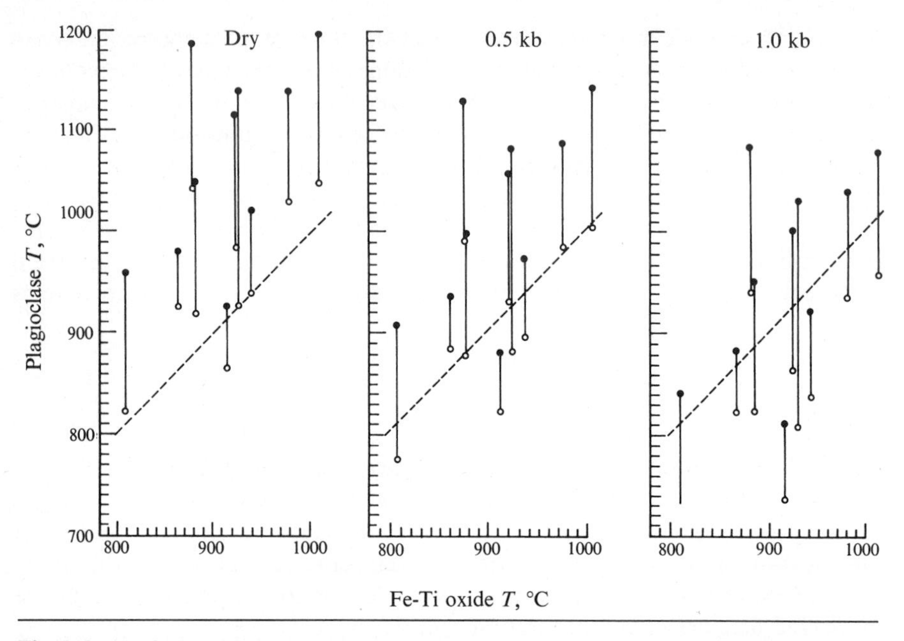


**Fig. 3-6** Plagioclase geothermometer temperatures for the cores (filled circles) and rims (open circles) of plagioclase phenocrysts plotted against the coexisting iron-titanium-oxide equilibration temperatures. The plagioclase results have been calculated assuming three sets of conditions: dry, 0.5, and 1.0 kb water pressure, respectively. The dashed line indicates the locus of correspondence between the two methods of temperature estimation.

alkali feldspar into solid solution, and the extent of this has been calculated by D. L. Hamilton (1961). It is more convenient to represent this nepheline solid solution in terms of "excess silica," rather than alkali feldspar, since the more potassic nephelines of igneous rocks are found either with kalsilite or leucite rather than feldspar; the nepheline–excess silica geothermometer has accordingly been calibrated for equilibria involving both feldspar and kalsilite.

Ever since Tuttle (1949) showed that the temperature of the $\alpha$-$\beta$ quartz inversion varied (inversely) with the temperature of growth, the possibility of using quartz solid solutions as a geothermometer seemed good. Recently, Dennen et al. (1970) have calibrated the $Al_2O_3$ content of quartz in equilibrium with either feldspar or another aluminous phase. The low content of Al (100 to 300 ppm equivalent to 100 to 1000°C) demands unusually clean mineral separations and certainty that the determined Al values are independent of aluminous minerals (for example, dawsonite) found in the fluid inclusions of quartz.

The liquidus temperatures at 1 bar for a wide variety of basaltic and other lavas have been determined by Yoder, Tilley, and Schairer (for example, 1967, 1968). For the lavas of each volcanic province, volcano, or even one eruption, the liquidus temperatures tend to vary linearly with the ratio MgO/(MgO +

$FeO + Fe_2O_3$) (in weight percent), although the slope may differ for lavas accumulative in olivine or other phases. With judicious interpolation, the liquidus temperature of any lava of comparable type at 1 bar can be estimated, although care has to be taken to ensure that the experimental liquidus phases correspond to those found in the lava.

OXYGEN-ISOTOPE GEOTHERMOMETRY The distribution of a stable or non-radioactive isotope between two phases is temperature-dependent. Consider for the moment two different crystalline phases, such as quartz and magnetite. At any given temperature, the free energies of $SiO_2^{18}$ and $SiO_2^{16}$ are slightly different, as are those for $Fe_3O_4^{18}$ and $Fe_3O_4^{16}$. Thus the free-energy of the quartz or magnetite containing both $O^{18}$ and $O^{16}$ depends on the ratio $O^{18}/O^{16}$. This dependence is not the same for quartz and magnetite, because the vibration frequencies of oxygen are different in the two crystals. Thus transfer of $O^{18}$ out of one phase may lower its free energy by an amount greater than the increase in free energy of the phase into which the $O^{18}$ is transferred. When the two phases come to equilibrium, or the free energy of the assemblage is minimum, the $O^{18}/O^{16}$ ratios of the two phases are usually different and temperature-dependent. No theory has yet been developed to predict this effect quantitatively.

Generally the heavy isotope is concentrated in the solid phase which is the more tightly bound, or the one which has the lowest heat capacity ($C_V$) expressed on a gram-atom basis. Corundum has a lower heat capacity than quartz, which in turn is lower than magnetite, so that we should expect that in any pair of these minerals corundum would be enriched in the heavy isotope $O^{18}$ and magnetite in the lighter $O^{16}$.

The experimentally determined composition of oxygen is expressed by

$$\delta_{O^{18}/O^{16}} = \left[\frac{(O^{18}/O^{16})_{\text{sample}}}{(O^{18}/O^{16})_{\text{standard}}} - 1\right] 1000 \tag{3-26}$$

so that $\delta_{O^{18}/O^{16}}$ is the per mil deviation of the sample relative to a standard (standard mean ocean water, SMOW). A value of $\delta_{O^{18}/O^{16}} = +20$ per mil, for example, means that the sample is 2 percent richer (heavier) in $O^{18}$ than the standard. Because the isotopic fractionation between solids and gases tends to be large and experimentally convenient, the temperature calibrations of minerals are usually determined with respect to water vapor; the results are then transposed to the form shown in Fig. 3-7. The quartz-magnetite oxygen fractionation is the largest among common minerals found in igneous rocks, and so provides the most sensitive geothermometer.

The oxygen-isotope geothermometer is virtually insensitive to pressure because the volume terms are insignificant. The fractionation becomes less with increasing temperature, so that its upper practical limit is less than about 1200°C. Isotopic fractionation in general disappears at high temperatures where

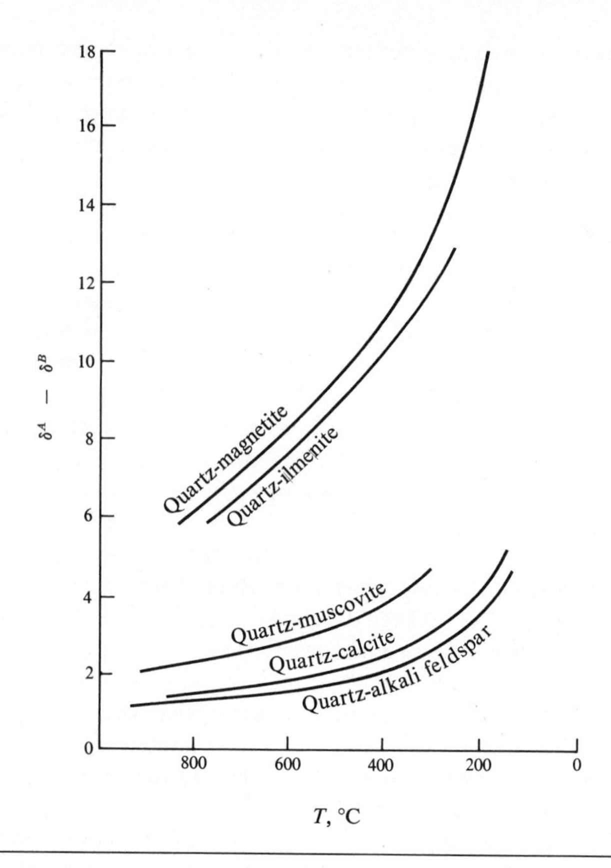


**Fig. 3-7** Experimentally determined oxygen-isotope fractionation $[(\delta_{O^{18}/O^{16}})^A - (\delta_{O^{18}/O^{16}})^B]$ between coexisting phases $A$ (quartz) and $B$. (After H. P. Taylor, 1967*a*, p. 125.)

the energy of an oscillator becomes simply $kT$, regardless of the mass of the vibrating atom or of the strength of the bond it forms with neighboring atoms.

Typical oxygen-isotope temperatures of acid igneous rocks (Taylor, 1967*a*) are: granites, 730 to 785°C (cf. Table 1-1); and granitic pegmatites and aplites, 580 to 730°C. Taylor and his coworkers have also taken their geothermometer out into the rock envelope of granitic plutons to give an invaluable record of the thermal effect of a cooling mass of magma (for example, Shieh and Taylor, 1969).

*Nonradiogenic isotopes in magmatic evolution* The isotopic composition of oxygen, and other nonradiogenic elements (notably sulfur, silicon, and hydrogen) in igneous rocks can be significant in other respects on which we shall digress at this point. If the oxygen-isotope analyses of various lava types reported by H. P. Taylor (1968) represent the original isotopic composition of the liquids, then most magmas—basaltic, nephelinitic, andesitic, and trachytic—

have rather similar $\delta_{O^{18}/O^{16}}$ ratios of about +5.5 to +7.0 per mil. Acid *rocks* may depart significantly from this range, since some plutonic granites have values in excess of +20 or alternatively much lower values (<5.2 per mil). Fresh rhyolitic obsidians, however, often have $O^{18}/O^{16}$ values within the basalt-andesite range, and it is likely that the oxygen of much acid magma does not differ from that of more basic magmas.

Taylor has interpreted the wide isotopic range of acid rocks as being the result of (1) interaction during cooling with meteoric water causing low $O^{18}/O^{16}$ values (the Skaergaard and Scottish Tertiary granophyres are examples) and (2) derivation of acid magma from $O^{18}$-rich metasediments by assimilation or partial fusion, resulting in granites with unusually high $O^{18}/O^{16}$ values.

In the previous discussion of isotopic geothermometry, it was noted that isotopic fractionation between two phases increases with falling temperature and that the heavy isotope tends to concentrate in the solid phase which has the lowest heat capacity ($C_v$). Fractionation of quartz, the common mineral most enriched in $O^{18}$, could deplete a magma in $O^{18}$, whereas fractionation of magnetite would enrich the magma in $O^{18}$. Feldspar, depending upon its composition, may either enrich or deplete the magma in $O^{18}$ (H. P. Taylor, 1968). To predict the effect of crystal fractionation on the oxygen-isotope composition of a magma is complex, if not unrewarding, since it depends upon the amounts and species of the solid phases and particularly on the temperature at which fractionation occurs. Hydrous magmas crystallizing at low temperatures would be more susceptible to change in their isotopic composition, and an alternative explanation (by Taylor) for the origin of plutonic granites rich in $O^{18}$ is that they were derived by low-temperature fractionation of a more basic parent. For acid rocks with $O^{18}/O^{16}$ within the range of basalts and andesites, the oxygen isotopes strongly suggest a high-temperature origin, presumably by some form of differentiation of a hot basic liquid.

The isotopic composition of sulfur ($S^{34}/S^{32}$) has also been found to vary within basalts of various types. Sulfur in these rocks occurs as sulfide ($S^{--}$) or sulfate ($SO_4^{--}$), the former being predominantly found in pyrrhotite and the latter in apatite or, in the more silica-deficient alkali basalts and nephelinites, as nosean or hauyne. The isotopic ratios of sulfide and sulfate sulfur measured by Schneider (1970) show a considerable range in basaltic rocks, but nevertheless some generalizations are possible. Tholeiitic lavas are lightest with their total sulfur having $\delta_{S^{34}/S^{32}} = -0.3$ per mil; olivine basalts are heavier with $\delta_{S^{34}/S^{32}} = +1.3$ per mil; and feldspathoidal lavas and nephelinites heavier still with +3.1 per mil; sulfate sulfur becomes heavier in the same order.

Since the following reaction runs to the right at 1000 to 1200°C,

$$H_2S^{34} + S^{32}O_2 = H_2S^{32} + S^{34}O_2$$

Schneider suggests that lavas degassing $SO_2$, the most abundant sulfur gas in tholeiitic lavas (Chap. 6), would preferentially lose their heavy sulfur. Degassing of sulfur by cooling basaltic lavas has been noted by J. G. Moore and Fabbi (1971), and it is this type of mechanism which Schneider believes to account for the sulfur isotopic variations between tholeiitic and alkali olivine-basalt magma.

The systematic examination of lava types, corresponding to the whole range of natural liquids, for variation in nonradiogenic isotopic abundances offers a fruitful but as yet untouched (except for oxygen) field for research.

PROBLEM OF PLUTONIC GEOTHERMOMETRY Despite the considerable number of igneous geothermometers, not all of which have been mentioned here, there is a surprising lack of temperature information on the cooling paths of the common plutonic rocks. No doubt this is in part due to the extensive re-equilibration of many large intrusions with their rock-water envelope as temperature falls. It is apparent that, with one or two notable exceptions, detailed data on the cooling path of acid plutons, each of which is likely to be as unique as the composition of its magma, are to all intents and purposes completely lacking.

## GEOBAROMETRY

The vertical stress induced by gravity in a layer of rock is proportional to its density and increases linearly with depth. The horizontal component of stress induced by gravity depends on the vertical component and on the elastic constants of the rock. When the difference between vertical and horizontal stresses exceeds the strength of the rock, so that it begins to flow, the two components become equal and the pressure is said to be hydrostatic. This condition presumbably is reached in the earth's crust at a depth of a few kilometers; below this depth, the hydrostatic or lithostatic pressure $P$ increases with depth $h$ as

$$\frac{dP}{dh} = g\rho \tag{3-27}$$

where $\rho$ is density. Since the average density of the crust is about 2.8 to 2.9 g cm$^{-3}$ and that of the uppermost mantle about 3.3 g cm$^{-3}$, pressure increases with depth at an average rate of roughly 0.3 kb km$^{-1}$.

As long as a rock retains some strength, the pressure in a cavity may be smaller or larger than the lithostatic pressure, and a pore fluid filling a cavity (for example, groundwater or magmatic gas) may be at a pressure $P_{\text{fluid}} \neq P_{\text{litho}}$. $P_{\text{fluid}}$ may be controlled by the density of the fluid, as in a system of connected pores filled with fluid and extending to the earth's surface, or it may be controlled by the physicochemical parameters of the magma. If a body of magma is enclosed in wall rock permeable to the fluid phase (water, $CO_2$, etc.) but not to the silicate melt, the melt will be subjected to a lithostatic pressure corresponding to the depth of the magma chamber, which may differ from

$P_{fluid}$. When two or more phases in a system at equilibrium are at different pressures, osmotic conditions are said to prevail, and these may be important. For instance, as we shall see, increasing the pressure on the melt can increase or decrease the concentration of water dissolved in it, depending on whether the fluid pressure is equal to or less than the pressure on the melt. Since, in general, $P_{fluid}$ is not known, it is commonly assumed that at depths greater than a few kilometers $P_{litho} \equiv P_{fluid}$.

The effect of pressure on the free energy of a reaction at constant temperature is given by

$$\left(\frac{\partial \Delta G}{\partial P}\right)_T = \Delta V_T \tag{3-28}$$

Integration between 1 bar and $P$ bar gives

$$\Delta G_T = \Delta G_{1,T} + \int_1^P \Delta V_T \, dP \tag{3-29}$$

where $\Delta G_{1,T}$ is the free energy at 1 bar and temperature $T$. The integral cannot be evaluated unless the variation of $\Delta V$ with pressure is known. Perforce, $\Delta V_T$ is often assumed to be independent of pressure. In this latter case, Eq. (3-29) can be integrated to give

$$\Delta G_T = \Delta G_{1,T} + \Delta V_T(P - 1)$$

If $\Delta V_T$ is zero, then pressure does not affect the equilibrium; if $\Delta V_T$ is negative, then pressure favors the reaction as $\Delta G_{P,T}$ becomes more negative. If $P$ is large, then $(P - 1)$ can be taken as equal to $P$.

In reference works molar volumes for solids and liquids are tabulated for the standard conditions of one atmosphere (1 atm = 1.013 bar) and 298°K, so that if we wish to calculate $\Delta V$ for any reaction involving solids and liquids at any temperature and pressure above the standard conditions, we have to take account of the effect of both $P$ and $T$ on the molar volumes of the products and of reactants. For gaseous components, the molar volume at any pressure and temperature can be expressed in terms of partial pressure for a mixture of ideal gases or in terms of fugacity for a real gas.

The effect of temperature on the molar volume of a mineral can be calculated if the thermal expansion of the mineral is known. The isobaric coefficient of thermal expansion is defined as

$$\alpha = \frac{1}{V}\left(\frac{\partial V}{\partial T}\right)_P \tag{3-30}$$

where $\alpha$ is generally a positive number having the dimensions of $deg^{-1}$, and $V$ is the molar volume of the mineral at the temperature at which $\alpha$ is measured. If we ignore the variation of $\alpha$ with temperature, noting that $\alpha$ is small, then Eq. (3-30) can be rearranged to give for any small change in temperature

$$V_{1\ \text{bar}} = V_{1;298^\circ}(1 + \alpha\Delta T) \tag{3-31}$$

where $\Delta T = T - 298$.

Although the data on the thermal expansions of rock-forming minerals tabulated by Skinner (1966) show that $\alpha$ does vary with temperature, for almost all petrological calculations $\alpha$ can be assumed to be constant and an appropriate value chosen from Skinner's tables (Skinner, 1966, table 6-8).

The effect of pressure on the molar volume of a mineral, liquid, or glass can be calculated if the compressibility $\beta$ is known. The isothermal coefficient of compressibility is defined as

$$\beta = -\frac{1}{V}\left(\frac{\partial V}{\partial P}\right)_T \tag{3-32}$$

Since $\partial V/\partial P$ is intrinsically negative, $\beta$ is a positive number having the dimensions $bar^{-1}$. $V$ is the molar volume at the pressure at which $\beta$ is measured. If we assume that $\beta$ is independent of pressure, then for any small change Eq. (3-32) can be rewritten as

$$V_T = V_{1,T}(1 - \beta_T\Delta P) \tag{3-33}$$

where $\Delta P = P - 1$ and $\beta_T$ is the isothermal compressibility at temperature $T$.

In general, both $\alpha$ and $\beta$ depend on both $P$ and $T$; note from Eq. (3-30) and (3-32) that $(\partial\alpha/\partial P)_T = -(\partial\beta/\partial T)_P$. The compressibility $\beta$ commonly increases with increasing temperature and decreases with increasing pressure; substances become less compressible as they are more compressed.[1] The effect of pressure on the compressibility of minerals and rocks is of great importance in interpreting seismic data on the constitution of the deep mantle and core, where the pressure is very large, but this effect may be neglected as a first approximation in petrological problems at pressures less than 20 or 30 kb. The temperature dependence of $\alpha$ is very large near the absolute zero of temperature, where $\alpha$ itself goes to zero, but becomes rather small at high temperatures; thus both $\alpha$ and $\beta$ are regarded as constant in the temperature-pressure

[1] One exception to this generalization is $SiO_2$ glass, which becomes more compressible with increasing pressure up to 30 kbar.

range considered here, and the volume at temperature $T$ [Eq. (3-31)] and pressure $P$ [Eq. (3-33)] is related to the volume at 298°K, 1 bar as

$$V = V_{1;298^\circ}(1 + \alpha\Delta T)(1 - \beta P) \tag{3-34}$$

where $\Delta T = T - 298$ and $P \gg 1$.

Consider the reaction

$$\underset{\text{forsterite}}{Mg_2SiO_4} + \underset{\text{glass}}{SiO_2} = \underset{\text{enstatite}}{2MgSiO_3} \tag{3-35}$$

$$43.79 \qquad 27.27 \qquad 2(31.47) \qquad V_{1;298^\circ}\ \text{cm}^3\ \text{mole}^{-1}$$

The molar volumes given for the three participating phases of Eq. (3-35), show that the reaction volume $\Delta V_{1;298^\circ} = -8.12$ cm$^3$. By including the coefficients of thermal expansion and compressibility, we can compare $\Delta V$ calculated at, say, 1227°C and various pressures with the value at 298°C and 1 bar. The pertinent data, taken from Skinner (1966) and Birch (1966), are set down below. Note, however, that $\alpha$ and $\beta$ change with temperature and pressure, respectively, and that the value for $\beta$ is applicable only to pressures of less than 15 kb; assuming $\alpha$ and $\beta$ to be constant does not introduce significant errors at moderate temperatures (<1400°C) and pressures (<50 kb).

| | $\alpha$(deg$^{-1}$) | $\beta$(bar$^{-1}$) |
|---|---|---|
| Forsterite | $44 \times 10^{-6}$ (at 800°C) | $0.79 \times 10^{-6}$ |
| $SiO_2$ glass | $1.62 \times 10^{-6}$ (0–1000°C) | $2.45 \times 10^{-6}$ (at 390°C) |
| Enstatite | $33 \times 10^{-6}$ (at 800°C) | $1.01 \times 10^{-6}$ |

These data substituted into Eq. (3-34) give the following values for the reaction volume at 1227°C (1500°K):

| PRESSURE | 5.0 kb | 10.0 kb | 20.0 kb | 30.0 kb | 40.0 kb |
|---|---|---|---|---|---|
| $\Delta V_{P,1500^\circ}$, cm$^3$ | −7.80 | −7.61 | −7.24 | −6.83 | −6.49 |
| $\Delta V_{1;298^\circ} - P\Delta V_{P,1500^\circ}$, cal | 38 | 122 | 421 | 925 | 1558 |

The effect of including $\alpha$ and $\beta$, the coefficients of thermal expansion and compressibility, on the standard free energy of the reaction can be illustrated by tabulating (above) $\Delta V_{1;298^\circ} - P\Delta V_{P,1500^\circ}$ in calories and comparing these values to $(\Delta G_r^\circ)_{1500^\circ}$, which is −798 cal. Obviously for pressures in excess of 20 kb the difference between $\Delta V_{1;298^\circ}$ and $\Delta V_{P,1500^\circ}$ has a substantial effect on the standard free-energy change of the reaction. A general rule of thumb for solid-solid and solid-liquid reactions is to include $\alpha$ and $\beta$ in the calculations of

$\Delta V$ for pressures above 20 to 30 kb unless the reaction has an unusually small free-energy change. Often for a reaction of petrological interest, $\alpha$ and $\beta$ are unknown for one or more of the components, so that we have to be content with using $\Delta V_{1;298°}$ at all pressures and temperatures. If a gas is present as a component in a reaction, the values for the fugacity [Eq. (3-18)] must be included in the calculations.

The calculation of an equilibrium curve for the coexistence of two solids over a considerable range of pressure and temperature can be illustrated by the graphite-diamond equilibrium. The reaction is formally written

$$\underset{\text{graphite}}{\mathrm{C}} = \underset{\text{diamond}}{\mathrm{C}} \tag{3-36}$$

At equilibrium $\Delta G = 0$ and from Eq. (3-29)

$$\int_1^P \Delta V_T \, dP = -\Delta G_{1,T}$$

The equation for the variation of $\Delta V$ with pressure and temperature is given by Eq. (3-34) and takes the form

$$\Delta V = \Delta V_{1;298°} + (V_d\alpha_d - V_g\alpha_g)\Delta T$$
$$- (V_d\beta_d - V_g\beta_g)P - (V_d\alpha_d\beta_d - V_g\alpha_g\beta_g)P\Delta T \tag{3-37}$$

where the subscripts $d$ and $g$ refer to diamond and graphite respectively, and $\Delta T = T - 298$. Substitution of the data below,

| | $S_{1;298°}$ | $V_{1;298°}$ | $\alpha(\text{deg}^{-1})$ | $\beta(\text{bar}^{-1})$ | $(\Delta G_f^\circ)_{1;298°}$ |
|---|---|---|---|---|---|
| Diamond | 0.566 | 3.417 | $10.5 \times 10^{-6}$ | $2.0 \times 10^{-7}$ | 693 |
| Graphite | 1.372 | 5.298 | $28.5 \times 10^{-6}$ | $3.0 \times 10^{-6}$ | 0 |

changed to consistent units (cal bar$^{-1}$), and integration of Eq. (3-37) gives, for equilibrium at any $P$ and $T$,

$$-\Delta G_{1,I} = -(0.04496 + 2.75 \times 10^{-6}\Delta T)P$$
$$+ (0.1818 \times 10^{-6} + 5.327 \times 10^{-12}\Delta T)P^2$$

If values of $\Delta G_{1,T}$ $(= \Delta G_r^\circ)$ at different temperatures are substituted, the quadratic equation can be solved and the physically unreasonable value for $P$ rejected. We have plotted the curve calculated to 2000°K in Fig. 3-8, having extrapolated the free-energy data of diamond (Robie and Waldbaum, 1968) to 2000°K. We

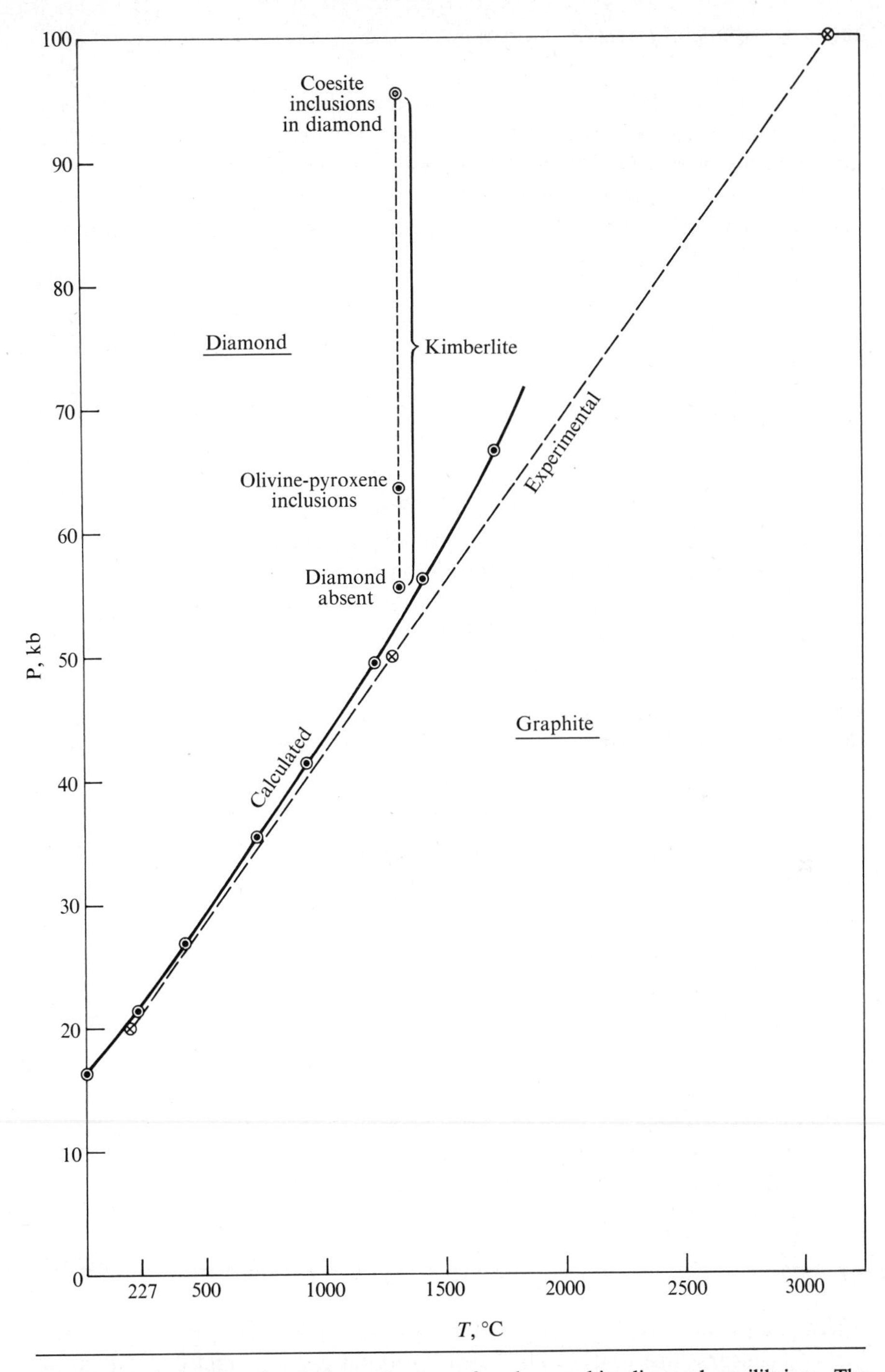


**Fig. 3-8** Calculated and experimental curves for the graphite-diamond equilibrium. The calculated pressures (at 1327°C) are for the equilibration of kimberlite with zircon baddeleyite (diamond absent) and with olivine-pyroxene and coesite inclusions in diamond.

shall use this calculated curve, or the corresponding experimental curve (Fig. 3-8), when discussing the calculated pressures at which certain magmas associated with diamond equilibrated with spinel or garnet peridotite.

The slope of an equilibrium curve such as the graphite-diamond curve is given by the Clausius-Clapeyron equation; this is obtained from the relationship

$$d\Delta G = \Delta V dP - \Delta S dT \tag{3-38}$$

so that at equilibrium $d\Delta G = 0$ and

$$\frac{dP}{dT} = \frac{\Delta S}{\Delta V} = \frac{\Delta H}{T\Delta V} \tag{3-39}$$

If the slope of the equilibrium curve is required at $P$ and $T$, then the values for the reaction entropy and volume at 298°K and 1 bar have to be corrected to these conditions. The effect of pressure and temperature on $\Delta V$ has already been illustrated; the effect of temperature on $\Delta S$ is given by

$$\Delta S_{1\ \mathrm{bar}} = \Delta S_{1;298^\circ} + \int_{298^\circ}^{T} \frac{\Delta C_p}{T}\, dT \tag{3-40}$$

where $\Delta C_p$ is the difference between the heat capacity (at constant pressure) of the products and that of the reactants. The effect of pressure on reaction entropy is found by integrating the relevant Maxwell relation

$$\left(\frac{\partial S}{\partial P}\right)_T = -\left(\frac{\partial V}{\partial T}\right)_P = -\alpha V \quad \text{from Eq. (3-30)} \tag{3-41}$$

Thus the entropy change at pressure $P$ is related to the reaction entropy at 1 bar (at constant temperature) as

$$\Delta S_T = \Delta S_{1,T} - \int_1^P (V_p\alpha_p - V_r\alpha_r)\, dP \tag{3-42}$$

where the subscripts $p$ and $r$ refer to products and reactants, respectively.

For the example of graphite and diamond, the calculated equilibrium pressure at 298°K is 16.8 kb. The slope of the curve at 298° is given by combining Eqs. (3-42) and (3-37)

$$\left(\frac{dP}{dT}\right)_{16,800;\ 298^\circ} = \frac{\Delta S_{1;298^\circ} - \int_1^{16,800} (V_d\alpha_d - V_g\alpha_g)\, dP}{\Delta V_{16,800;\ 298^\circ}} \tag{3-43}$$

which gives a value of 0.0195 $\mathrm{kb}^{-1}\ \mathrm{deg}^{-1}$, which, as can be expected, is lower than the linear slope (0.0273 $\mathrm{kb}^{-1}\ \mathrm{deg}^{-1}$) found experimentally above 40 kb (S. P. Clark, 1966). The graphite-diamond curve illustrates one of the difficulties in using minerals as geobarometers. If both diamond and graphite are adjudged to have coexisted in equilibrium in a rock and the temperature is known, then a precise value for pressure can be obtained. If graphite is absent and the temperature is unknown, the occurrence of diamond provides only a minimum value for pressure. It is rare to find in igneous rocks coexisting minerals which define a pressure-temperature curve of the type shown in Fig. 3-8, so that in general only maximum or minimum estimates of pressure are obtainable, usually at an unknown temperature. By considering several mineral reactions, it is sometimes possible to constrain further the estimates of pressure and temperature. We return later to a procedure used to calculate a precise value of pressure for a point, or segment, of a *PT* path traversed by a particular magma.

The experimental work on mineral stability upon which most estimates of pressure for igneous assemblages are based can be divided into two types. In one $H_2O$ acts as an inert pressure medium and is not incorporated by the solids, except perhaps at extreme pressures (Sclar, 1970). In the other $H_2O$ participates in the reaction, either by being incorporated into a hydrous mineral or by being dissolved in the liquid phase. For the most part igneous rocks lack mineral assemblages capable of defining pressure with any precision.[1] We cite below two solid-solid reactions, one applicable to acid magma and the other to basic or ultrabasic mineral assemblages.

THE COEXISTENCE OF QUARTZ AND TRIDYMITE Tridymite, although commonly found in the vesicles of lavas, also occurs in some granophyres, where it is now pseudomorphed by quartz. Examples are found among the granophyres of the Scottish Tertiary province and also in the upper granophyres of the Skaergaard intrusion, where the mineral relationships indicate that the cooling liquid intersected the quartz-tridymite equilibrium curve (D. H. Lindsley et al., 1969).

The quartz-tridymite[2] equilibrium curve (Fig. 3-9) has been determined by Tuttle and Bowen (1958). If the inversion curve of another mineral in the Skaergaard granophyres intersected the quartz-tridymite curve, then both temperature and pressure would be obtainable for the Skaergaard granophyric crystallization path. The mineral used by Lindsley et al. was hedenbergite, which inverts to a wollastonite solid solution at high temperature. The inversion interval of the naturally occurring mineral is shown in Fig. 3-9. Since the natural

[1] *Precision* is used here and throughout this book in the same way as in analytical chemistry, so that a set of values is considered precise if their standard deviation is small. The mean value may be close to, or far from, the "true" value, which is a measure of accuracy. In general, few methods of estimating pressure are precise; most are of unknown accuracy.

[2] There is some dispute about whether tridymite can exist as a stable phase in the $SiO_2$ system. Some argue that water and alkalis are essential components required for tridymite to crystallize; others maintain that the additional components only act as a catalyst (cf. Rockett and Foster, 1967). To minimize this uncertainty, Lindsley synthesized tridymite with a variety of additional components from a natural assemblage but could detect no departure from the curve in Fig. 3-9.

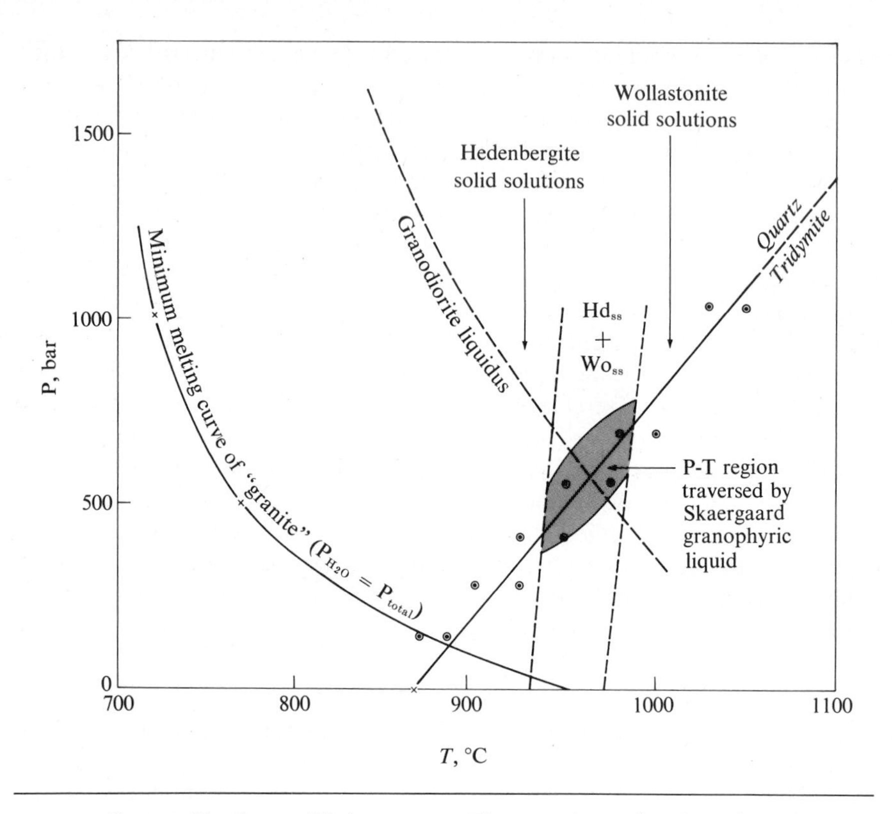


**Fig. 3-9** Quartz-tridymite equilibrium curve. The experimentally determined inversion interval (between dashed lines) of a Skaergaard hedenbergite (inverted wollastonite) is from Lindsley et al. (1969); the estimated *PT* zone through which the later Skaergaard liquid cooled is shown. The granite solidus and granodiorite liquidus (both water-saturated) are from Tuttle and Bowen (1958) and from Burnham (1967), respectively.

hedenbergite is believed to be the product of the inversion of ferrowollastonite, both reaction curves are bracketed in the natural assemblage, so that the pressure-temperature segment traversed by the cooling granophyre spans 950 to 980°C and 600 ± 100 bar. This experimental investigation of a natural igneous assemblage has provided one of the first precise estimates of pressure in a cooling magma.

Because of the high temperature of the quartz-tridymite inversion at 1 bar (867°C), the presence of tridymite as a stable phase in acid rocks can only rarely be used to give a minimum value for pressure, since the minimum melting curve of granite cuts the tridymite-quartz curve at very low pressure (Fig. 3-9). G. M. Brown (1963) has investigated the melting relationships of tridymite granophyres and delineated a possible cooling path for them based on the assumption that $p_{H_2O} = P_{total}$.

PLAGIOCLASE, SPINEL, AND GARNET IN BASIC AND ULTRABASIC ROCKS Experimental investigation of the reaction

$$\underset{\text{forsterite}}{Mg_2SiO_4} + \underset{\text{anorthite}}{CaAl_2Si_2O_8} = \underset{\text{garnet}}{CaMg_2Al_2Si_3O_{12}} \tag{3-44}$$

has shown that it proceeds with an intermediate step in which spinel is stable, and since the components form solid solutions there is an appreciable interval over which the products and reactants can coexist. Kushiro and Yoder (1966) represent the steps of the reaction in this way:

$$\underset{\text{anorthite}}{2CaAl_2Si_2O_8} + \underset{\text{olivine}}{2Mg_2SiO_4}$$

$$= \underset{\text{diopside ss}}{CaMgSi_2O_6 \cdot xCaAl_2SiO_6} + \underset{\text{orthopyroxene ss}}{2MgSiO_3 \cdot xMgAl_2SiO_6}$$

$$+ (1 - x)\underset{\text{anorthite}}{CaAl_2Si_2O_8} + (1 - x)\underset{\text{spinel}}{MgAl_2O_4} \qquad (0 < x < 1)$$

The observed slope of the equilibrium curve, $dT/dP$, is 200° kb$^{-1}$. The second reaction step is

$$\underset{\text{diopside ss}}{CaMgSi_2O_6 \cdot xCaAl_2SiO_6} + \underset{\text{orthopyroxene ss}}{2MgSiO_3 \cdot xMgAl_2SiO_6}$$

$$+ (1 - x)\underset{\text{anorthite}}{CaAl_2Si_2O_8} + (1 - x)\underset{\text{spinel}}{MgAl_2O_4} = \underset{\text{garnet}}{2CaMg_2Al_2Si_3O_{12}} \tag{3-45}$$

with an observed slope of 60° kb$^{-1}$.

Using this simple example to model more complex natural assemblages, we may use the occurrence of plagioclase, spinel, or garnet—together with olivine—to indicate facies of increasing pressure for basic and ultrabasic rocks. Because the natural components all form solid solutions, we cannot directly use the experimental results to estimate pressure, even if spinel and plagioclase, or spinel and garnet, coexist, as they may. By taking another approach based on the activities of $SiO_2$ and $Al_2O_3$ in natural assemblages, we can use this simple experimental representation of a facies sequence to calculate estimates of pressure and temperature.

EARLY CRYSTALLIZING PHASES IN ACID MAGMA Many rhyolitic lavas contain phenocrysts of quartz, fayalitic olivine or orthopyroxene, and iron titanium oxides. We may represent the olivine-bearing assemblages by the reaction

$$\tfrac{2}{3}\underset{\text{magnetite}}{Fe_3O_4} + \underset{\text{quartz}}{SiO_2} = \underset{\text{fayalite}}{Fe_2SiO_4} + \tfrac{1}{3}\underset{\text{gas}}{O_2} \tag{3-46}$$

and in terms of chemical potentials Eq. (3-46) becomes

$$\tfrac{2}{3}\mu_{Fe_3O_4} + \mu_{SiO_2} = \mu_{Fe_2SiO_4} + \tfrac{1}{3}\mu_{O_2} \qquad (3\text{-}46a)$$

so that by substitution, and at equilibrium at 1 bar,

$$0 = \mu^\circ_{Fe_2SiO_4} + \tfrac{1}{3}\mu^\circ_{O_2} - \mu^\circ_{SiO_2} - \tfrac{2}{3}\mu^\circ_{Fe_3O_4} + RT\ln\frac{(f_{O_2})^{1/3}a^{ol}_{Fe_2SiO_4}}{a^{qu}_{SiO_2}(a^{titanomag}_{Fe_3O_4})^{2/3}} \qquad (3\text{-}46b)$$

Since the sum of the standard chemical potentials ($\mu^\circ$) is equal to the standard free-energy charge of the reaction ($\Delta G^\circ_r$), Eq. (3-46*b*) becomes

$$0 = \Delta G^\circ_r + RT\ln\left[\frac{(f_{O_2})^{1/3}a^{ol}_{Fe_2SiO_4}}{a^{qu}_{SiO_2}(a^{titanomag}_{Fe_3O_4})^{2/3}}\right]_{1\text{ bar}} \qquad (3\text{-}46c)$$

Because the phenocrysts did not precipitate at 1 bar but at some unknown pressure $P$, we have to include the effect of pressure on the standard free-energy change of this reaction. This is given by Eq. (3-29), so that if $\Delta V^\circ_r$ of the reaction is independent of pressure, then

$$0 = \Delta G^\circ_r + \Delta V^\circ_r(P-1) + RT\ln\left[\frac{(f_{O_2})^{1/3}a^{ol}_{Fe_2SiO_4}}{a^{qu}_{SiO_2}(a^{titanomag}_{Fe_3O_4})^{2/3}}\right]_{P,T} \qquad (3\text{-}46d)$$

where the activity terms are now those at $P$ and $T$, and not at 1 bar and $T$ as in Eq. (3-46*c*). If Eq. (3-46*d*) is rearranged as

$$0 = \frac{\Delta G^\circ_r}{2.303RT} + \frac{\Delta V^\circ_r}{2.303RT}(P-1) + \log\frac{(f_{O_2})^{1/3}a^{ol}_{Fe_2SiO_4}}{a^{qu}_{SiO_2}(a^{titanomag}_{Fe_3O_4})^{2/3}} \qquad (3\text{-}46e)$$

Then the variation of $\Delta G^\circ_r/2.303RT$ with temperature can be expressed in the form of an equation $(A/T) + B = \Delta H^\circ_r/2.303R - \Delta S^\circ_r/2.303R$ and is obtained by least-squares regression of $\Delta G^\circ_r/2.303RT$ over a range of temperatures. The volume change of the reaction $\Delta V^\circ_r$ is assumed to be constant and to have the value of $(\Delta V_r)_{1;298^\circ}$; $\Delta V^\circ_r/2.303R$ can therefore be represented by a constant $C$. In Table 3-3, values of $\Delta G^\circ_r/2.303RT$ and $\Delta V^\circ_r/2.303RT$ are given for several reactions where $SiO_2$ is a reactant, but the standard state of $SiO_2$ has been chosen as silica glass rather than as quartz [cf. Eq. (3-46)]. The choice of a standard state is arbitrary, and the reason for this standard state will become evident later in this chapter. Obviously by adding the two reactions

$$\underset{\text{magnetite}}{\tfrac{2}{3}Fe_3O_4} + \underset{\text{glass}}{SiO_2} = \underset{\text{fayalite}}{Fe_2SiO_4} + \underset{\text{gas}}{\tfrac{1}{3}O_2} \qquad (3\text{-}47)$$

$$\underset{\text{quartz}}{SiO_2} = \underset{\text{glass}}{SiO_2}$$

**Table 3-3** Values for $\Delta G_r^\circ/2.303RT = (A/T) + B$ and $\Delta V_r^\circ/2.303RT = C/T$ for the listed reactions*

| EQ. NO. | | $A/T$ | $B$ | $C/T + D$ |
|---|---|---|---|---|
| 2-5 | $\alpha CaTiO_3 + SiO_2$ (glass) $= CaTiSiO_5$ | $-2203/T$ | $+0.905$ | $-0.0274/T$ |
| 2-6 | $KAlSiO_4 + SiO_2$ (glass) $= KAlSi_2O_6$ | $-685/T$ | $-0.928$ | $+0.0064/T$ |
| 2-7 | $\frac{1}{2}NaAlSiO_4$ ($\beta$) $+ SiO_2$ (glass) $= \frac{1}{2}NaAlSi_3O_8$ ($T < 907°C$) | $-871/T$ | $+0.031$ | $-0.0216/T$ |
| | $\frac{1}{2}NaAlSiO_4$ ($\gamma$) $+ SiO_2$ (glass) $= \frac{1}{2}NaAlSi_3O_8$ ($T > 907°C$) | $-855/T$ | $+0.012$ | $-0.0216/T$ |
| 3-47 | $\frac{2}{3}Fe_3O_4 + SiO_2$ (glass) $= Fe_2SiO_4 + \frac{1}{3}O_2$ | $8270/T$ | $-2.81$ | $-0.05517/T$ |
| | $SiO_2$ (glass) $= SiO_2$ ($\beta$ quartz) | $-309/T$ | $+0.183$ | $-0.02393/T$ |
| | $\frac{1}{3}Fe_3O_4 + SiO_2$ (glass) $= FeSiO_3 + \frac{1}{6}O_2$ | $4467/T$ | $-1.63$ | $-0.04787/T$ |
| 3-35 | $Mg_2SiO_4 + SiO_2$ (glass) $= 2MgSiO_3$ | $-1034/T$ | $+0.597$ | $-0.04241/T$ |
| 3-58 | $CaMgSi_2O_6 + \frac{1}{2}SiO_2$ (glass) $+ Al_2O_3$ (liquid) $= CaAl_2Si_2O_8 + \frac{1}{2}Mg_2SiO_4$ | $-6776/T$ | $+0.859$ | $+0.2157/T$ $-0.72339 \times 10^{-4}$ |
| 3-62 | $Mg_2SiO_4 + Al_2O_3$ (liquid) $= MgSiO_3 + MgAl_2O_4$ | $-8131/T$ | $+2.632$ | $+0.1344/T$ $-0.72339 \times 10^{-4}$ |
| | $(\bar{V}_{SiO_2} - \hat{V}^\circ_{SiO_2})/2.303RT$ | | | $-0.00470/T$ $+1.344 \times 10^{-6}$ |
| | $(\bar{V}_{Al_2O_3} - \hat{V}^\circ_{Al_2O_3})/2.303RT$ | | | $+0.18061/T$ $-0.67002 \times 10^{-4}$ |

* Data from J. Nicholls et al. (1971); J. Nicholls and Carmichael (1972).

we obtain $\Delta G_r^\circ/2.303RT$ and $\Delta V_r^\circ/2.303RT$ for Eq. (3-46*e*). From Table 3-3 we find

$$\frac{8270}{T} - 2.81 - \frac{0.05517}{T}(P - 1) + \frac{309}{T} - 0.183 + \frac{0.02393}{T}(P - 1)$$

$$= \frac{8579}{T} - 2.993 - \frac{0.03124}{T}(P - 1) \tag{3-46f}$$

remembering to change the sign of the glass-quartz reaction (Table 3-3) to conform with the reaction written in the opposite sense above.

We use as a specific example (Nicholls et al., 1971) a rhyolite from Mono Craters, California, which contains phenocrysts of quartz, fayalitic olivine, magnetite solid solution, and ilmenite solid solution that amount in total to 5.2 percent (volume). It is therefore reasonable to assume that the temperature (and oxygen fugacity) of equilibration of the iron titanium phenocrysts applied to the whole phenocryst assemblage. The data for this rhyolite are

$$T = 1083°K \qquad \log f_{O_2} = -14.7 \qquad X^{\text{titanomag}}_{Fe_3O_4} = 0.564$$

$$X^{\text{ol}}_{Fe_2SiO_4} = 0.893$$

If the olivine solid-solution series is taken as an ideal solid solution, then $a^{ol}_{Fe_2SiO_4} = (X^{ol}_{Fe_2SiO_4})^2$. And since the titanomagnetite solid-solution series is approximately ideal over a restricted composition-temperature range (Heming and Carmichael, 1973), then $a^{titanomag}_{Fe_3O_4} = X^{titanomag}_{Fe_3O_4}$. Lastly quartz is assumed to be pure and to have unit activity. By substituting Eq. (3-46*f*) into Eq. (3-46*e*) and rearranging we obtain

$$P = \frac{8579 - 2.993T + T(\frac{1}{3}\log f_{O_2} - \log a_{Fe_2SiO_4} - \frac{2}{3}\log a_{Fe_3O_4})}{0.03124}$$

so that by substitution of the appropriate values for the activities

$$P = 3.33 \text{ kb}$$

which is equivalent to a depth of about 10 km.

As acid magma migrates to the surface, the quartz crystals may no longer remain in equilibrium, which may account for the embayed or corroded outlines commonly found in quartz phenocrysts of acid lavas. Many rhyolites contain orthopyroxene phenocrysts in place of an iron-rich olivine, and an analogous reaction to Eq. (3-46) can be used to calculate pressure for these (Table 3-3). But granites do not commonly contain olivine, and many do not have orthopyroxene either, although they may contain augite. For a granite which contains sphene, magnetite, ilmenite, biotite, and augite, the following reaction, independent of water fugacity, is representative;

$$\underset{\text{magnetite}}{\tfrac{2}{3}Fe_3O_4} + \underset{\text{sphene}}{CaTiSiO_5} + \underset{\text{quartz}}{SiO_2} + \underset{\text{phlogopite}}{\tfrac{1}{3}KMg_3AlSi_3O_{10}(OH)_2}$$

$$= \underset{\text{annite}}{\tfrac{1}{3}KFe_3AlSi_3O_{10}(OH)_2} + \underset{\text{ilmenite}}{FeTiO_3} + \underset{\text{diopside}}{CaMgSi_2O_6} \qquad (3\text{-}48)$$

and therefore

$$\Delta G = 0 = \frac{\Delta G_r^\circ}{2.303RT} + \frac{\Delta V_r^\circ}{2.303RT}(P - 1)$$

$$+ \log \frac{[a^{biot}_{KFe_3AlSi_3O_{10}(OH)_2}]^{1/3} a^{ilm}_{FeTiO_3} a^{augite}_{CaMgSi_2O_6}}{[a^{biot}_{KMg_3AlSi_3O_{10}(OH)_2}]^{1/3} a^{qu}_{SiO_2} a^{sphene}_{CaTiSiO_5} (a^{mag}_{Fe_3O_4})^{2/3}} \qquad (3\text{-}48a)$$

Free-energy data are known for all these components, so that $\Delta G_r^\circ/2.303RT$ can be obtained and hence a value of $P$ at the equilibration temperature of the iron titanium oxides. However, in most granites titanomagnetite has recrystallized because of reequilibration during the cooling, so that in general the iron titanium oxides cannot be used as a geothermometer. The titanium content of biotite, nevertheless, has been shown to be a sensitive geothermometer in quartz-bearing rocks (Wright, in press).

For granites without pyroxene, the occurrence of hornblende, biotite, sphene, and iron titanium oxides can be used to obtain the equilibration pressure at any known temperature or over any estimated temperature range, if the following reaction is recast into the form of Eq. (3-48*a*):

$$\underset{\text{sphene}}{6CaTiSiO_5} + \underset{\text{magnetite}}{4Fe_3O_4} + \underset{\text{quartz}}{18SiO_2} + \underset{\text{phlogopite}}{5KMg_3AlSi_3O_{10}(OH)_2}$$

$$= \underset{\text{annite}}{2KFe_3AlSi_3O_{10}(OH)_2} + \underset{\text{potassium feldspar}}{3KAlSi_3O_8} + \underset{\text{ilmenite}}{6FeTiO_3}$$

$$+ \underset{\text{tremolite}}{3Ca_2Mg_5Si_8O_{22}(OH)_2} + 2O_2 \tag{3-49}$$

This balances the hydroxyl groups so that the reaction does not depend on obtaining a value for the fugacity of water. It is necessary, however, to estimate the activity of $Ca_2Mg_5Si_8O_{22}(OH)_2$ in the hornblende; this is given by the probability of the $Ca_2Mg_5Si_8O_{22}(OH)_2$ configuration (Chap. 4) occurring in the hornblende, which in the simple case is assumed to be fluorine-free. Amphiboles contain four pseudo-octahedral cation sites (called $m_1$, $m_2$, $m_3$, and $m_4$) together with a vacant site which in hornblende is partly filled but which in tremolite is unfilled. The large cations like Ca occupy the large $m_4$ site, and since the metal-oxygen distance and the site distortion of the $m_1$ and $m_3$ sites are similar, their cation occupancies can be regarded as identical. As a first approximation we may write

$$a^{\text{hornblende}}_{Ca_2Mg_5Si_8O_{22}(OH)_2} = (1 - X^{\text{vacant}}_{\text{Na,K}})(X^{m_4}_{\text{Ca}})^2(X^{m_1,m_3}_{\text{Mg}})^3(X^{m_4}_{\text{Mg}})^2$$

where $X$ represents the fraction of sites filled by Ca, Mg, etc. in the superscript sites, the underlying assumption being that the cations on each site mix ideally.

The first of the granite equations has been found by one of the authors to yield not unreasonable results, but since the calculation of the amphibole formula from the analyzed oxide components of hornblende is not unambiguous, the use of Eq. (3-49) can lead to variable results.

### LIQUID COMPONENTS AND THE CALCULATION OF PRESSURE AND TEMPERATURE

We have seen in Chap. 2 that equilibrium coexistence of such minerals as leucite and sanidine, olivine and orthopyroxene, sphene and perovskite, or nepheline and albite will define the activity of silica, and the value of this can be calculated at any temperature from free-energy data. Lavas are quenched on the surface of the earth at 1 bar, and if the temperature of quenching is known, or can be reasonably estimated, then by using the appropriate analyzed groundmass mineral assemblage the activity of silica in the lava can be calculated. To take an example: in a lava with a groundmass of leucite and sanidine, such as a wyomingite, the groundmass assemblage is represented by the reaction

$$\underset{\text{leucite}}{KAlSi_2O_6} + \underset{\text{glass}}{SiO_2} = \underset{\text{sanidine}}{KAlSi_3O_8} \tag{3-50}$$

and at 1 bar pressure we can write

$$\log a_{SiO_2}^{liq} = \frac{\Delta G_r^\circ}{2.303RT} + \log a_{KAlSi_3O_8}^{san} - \log a_{KAlSi_2O_6}^{leuc} \tag{3-50a}$$

In the following discussion, we in effect reverse the *PT* path that a lava took from source to surface, and in so doing we assume that the concentrations of all the components which determined the mineralogical composition of the lava as it crystallized on the surface remain unchanged in the inverse path. By using the requirement of equilibrium, we may calculate the pressure and temperature at which a lava equilibrated with a stipulated mantle assemblage, an enclosed ultramafic nodule, or the assemblage of phenocrysts or megacrysts contained in the lava.

In a silicate liquid, the chemical potential of silica is given by

$$\mu_{SiO_2}^{liq} = \mu_{SiO_2}^{\circ} + RT \ln a_{SiO_2}^{liq} \tag{3-51}$$

where $\mu_{SiO_2}^{\circ}$ is the chemical potential of silica in the chosen standard state, namely silica glass. If Eq. (3-51) is differentiated with respect to pressure at constant temperature and composition, rearranged, and the appropriate substitutions made, then we obtain

$$\left(\frac{\partial \log a_{SiO_2}^{liq}}{\partial P}\right)_T = \frac{\bar{V}_{SiO_2} - \hat{V}_{SiO_2}^{\circ}}{2.303RT} \tag{3-52}$$

where $\bar{V}_{SiO_2}$ is the partial molar volume of silica in the silicate liquid and $\hat{V}_{SiO_2}^{\circ}$ is the molar volume of silica in the standard state. Values of $\bar{V}_{SiO_2}$ as a function of temperature have been calculated by Bottinga and Weill (1970) and are constant over a wide composition range. Since the thermal expansion of pure silica glass is trivial, we may take $\hat{V}_{SiO_2}^{\circ}$ to be independent of temperature. If $\bar{V}_{SiO_2} - \hat{V}_{SiO_2}^{\circ}$ is assumed to be independent of pressure, then Eq. (3-52) can be integrated to give

$$\log (a_{SiO_2}^{liq})_T = \log (a_{SiO_2}^{liq})_{1,T} + \frac{\bar{V}_{SiO_2} - \hat{V}_{SiO_2}^{\circ}}{2.303RT}(P - 1) \tag{3-53}$$

and the value of $(\bar{V}_{SiO_2} - \hat{V}_{SiO_2}^{\circ})/2.303RT$ is given in the form $C/T + D$ in Table 3-3.

The effect of temperature on the activity of silica is given by differentiating Eq. (3-51) with respect to temperature to obtain

$$\left(\frac{\partial \log a_{SiO_2}^{liq}}{\partial T}\right)_P = -\frac{\overline{H}_{SiO_2} + \hat{H}_{SiO_2}^{\circ}}{2.303RT^2} \tag{3-54}$$

but so far as we are aware there are no data on the partial molar enthalpies of $SiO_2$ or any other component in a silicate liquid. This problem can be circumvented if it is postulated that lavas behave as regular solutions. J. Nicholls and Carmichael (1972) showed that this assumption is valid for liquids saturated with corundum in the $Na_2O$–$Al_2O_3$–$SiO_2$ system, and since we intend to use the regular solution formulation over a moderate temperature range at constant composition, it may not be unreasonable. One of the properties of a regular solution (Chap. 4) is that the logarithm of the activity coefficient of any component, say, $SiO_2$, is given by

$$\log \gamma_{SiO_2}^{liq} = \log a_{SiO_2}^{liq} - \log X_{SiO_2}^{liq} = \frac{\phi_{SiO_2}}{T} \tag{3-55}$$

where $\phi_{SiO_2}$ is a composite function dependent upon composition but independent of temperature. Rearranging Eq. (3-55) gives

$$\log (a_{SiO_2}^{liq})_P = \frac{\phi_{SiO_2}}{T} + \log X_{SiO_2}^{liq} \tag{3-56}$$

so that for a liquid of constant composition if the activity of silica is known at any temperature, and also the mole fraction of silica, then $\phi_{SiO_2}$ can be calculated, which in turn will allow $\log a_{SiO_2}^{liq}$ to be calculated at any other temperature. The presence in lavas of groundmass mineral assemblages of diverse composition often allows the activity of several other components in the lava to be defined; we use liquid $Al_2O_3$ here, but depending on the composition of the lava and on the purpose of the calculations, it has proved fruitful to use liquid $K_2O$, $CaMgSi_2O_6$, and $TiO_2$, all of which are readily defined by the mineral assemblages of many lava types. For a lava containing a groundmass assemblage of augite (cpx), plagioclase, and orthopyroxene (opx), we may write

$$\underset{\text{diopside}}{CaMgSi_2O_6} + \underset{\text{glass}}{SiO_2} + \underset{\text{liquid}}{Al_2O_3} = \underset{\text{anorthite}}{CaAl_2Si_2O_8} + \underset{\text{enstatite}}{MgSiO_3} \tag{3-57}$$

and

$$\log a_{Al_2O_3}^{liq} = \frac{\Delta G_r^{\circ}}{2.303RT} + \log \frac{a_{MgSiO_3}^{opx} a_{CaAl_2Si_2O_8}^{plag}}{a_{CaMgSi_2O_6}^{cpx} a_{SiO_2}^{liq}} \tag{3-57a}$$

The analogous reaction for olivine in place of enstatite is

$$\underset{\text{diopside}}{CaMgSi_2O_6} + \underset{\text{glass}}{\tfrac{1}{2}SiO_2} + \underset{\text{liquid}}{Al_2O_3} = \underset{\text{anorthite}}{CaAl_2Si_2O_8} + \underset{\text{forsterite}}{\tfrac{1}{2}Mg_2SiO_4} \tag{3-58}$$

and for lavas without feldspar, which are loosely called nephelinites,

$$\underset{\text{diopside}}{CaMgSi_2O_6} + \underset{\text{liquid}}{Al_2O_3} = \underset{\substack{\text{Ca-Tschermak's}\\ \text{pyroxene}}}{CaAl_2SiO_6} + \underset{\text{glass}}{\tfrac{1}{2}SiO_2} + \underset{\text{forsterite}}{\tfrac{1}{2}Mg_2SiO_4} \tag{3-59}$$

All three of these equations (Table 3-3) allow a value for $\log a^{\text{liq}}_{Al_2O_3}$ to be obtained from a groundmass assemblage of a lava, quenched at 1 bar and at a temperature which is conveniently given by the iron-titanium-oxide thermometer; thus $\phi_{Al_2O_3}$ can be calculated [Eq. (3-55)]. From the foregoing discussion on silica activity, we may use one of a number of reactions (Table 3-3) appropriate to all natural lava compositions to obtain a value for silica activity, also at 1 bar and the quench temperature.

The variation, as a function of pressure and temperature, of the activity of $SiO_2$ in the liquid is given by substituting Eq. (3-53) into Eq. (3-56) to give

$$\log (a^{\text{liq}}_{SiO_2})_{P,T} = \log X^{\text{liq}}_{SiO_2} + \frac{\phi_{SiO_2}}{T} + \frac{\bar{V}_{SiO_2} - \hat{V}^{\circ}_{SiO_2}}{2.303RT}(P - 1) \tag{3-60}$$

where $P$ and $T$ are the unknown values we wish to calculate.

An exactly analogous equation for $Al_2O_3$ is

$$\log (a^{\text{liq}}_{Al_2O_3})_{P,T} = \log X^{\text{liq}}_{Al_2O_3} + \frac{\phi_{Al_2O_3}}{T} + \frac{\bar{V}_{Al_2O_3} - \hat{V}^{\circ}_{Al_2O_3}}{2.303RT}(P - 1) \tag{3-61}$$

Equations (3-60) and (3-61) give the variation of $\log a^{\text{liq}}_{SiO_2}$ and $\log a^{\text{liq}}_{Al_2O_3}$ as a function of pressure and temperature in the *absence* of a solid buffer assemblage.

We can now use whatever mineral assemblage is appropriate—be it an idealized mantle, an enclosed xenolith, or a phenocryst assemblage—to equate with Eqs. (3-60) and (3-61), the requirement of equilibrium being that the chemical potentials of $SiO_2$ and $Al_2O_3$ should be the same in the liquid lava at $P$ and $T$ as those defined by the crystalline assemblage. Since we have chosen the same standard states for both the liquid and the crystalline assemblage, namely $SiO_2$ glass and $Al_2O_3$ liquid, the activities also have to be equal at equilibrium.

One example of a crystalline assemblage would be a peridotite, which as shown by the experimental results touched on earlier [Eqs. (3-44) and (3-45)], could have as an aluminous phase either plagioclase, spinel, or garnet as pressure increases. For a spinel peridotite, we can represent the assemblage of orthopyroxene, olivine, and spinel by

$$\underset{\text{forsterite}}{Mg_2SiO_4} + \underset{\text{liquid}}{Al_2O_3} = \underset{\text{enstatite}}{MgSiO_3} + \underset{\text{pleonaste}}{MgAl_2O_4} \tag{3-62}$$

and using the same assumptions as those for Eq. (46*e*) we obtain

$$\log a_{Al_2O_3}^{liq} = \frac{\Delta\hat{G}_r^\circ}{2.303RT} + \frac{\Delta\hat{V}_r^\circ(P-1)}{2.303RT} + \log \frac{a_{MgAl_2O_4}^{spinel} a_{MgSiO_3}^{opx}}{a_{Mg_2SiO_4}^{ol}} \tag{3-62a}$$

and because the coexistence of olivine and orthopyroxene in equilibrium also defines silica activity, from Eq. (3-35) we have

$$\log a_{SiO_2}^{liq} = \frac{\Delta\hat{G}_r^\circ}{2.303RT} + \frac{\Delta\hat{V}_r^\circ(P-1)}{2.303RT} + \log \frac{(a_{MgSiO_3}^{opx})^2}{a_{Mg_2SiO_4}^{ol}}$$

If values appropriate to the composition of olivine, orthopyroxene, and spinel in a typical spinel peridotite are substituted for the activity terms in these equations, then by collectively representing them together with Eqs. (3-60) and (3-61) in terms of the constants given in Table 3-3, we obtain for $SiO_2$ in the liquid:

$$\log a_{SiO_2}^{liq} = \log X_{SiO_2}^{liq} + \frac{\phi_{SiO_2}}{T} + \left(1.344 \times 10^{-6} - \frac{0.00470}{T}\right)(P-1) \tag{3-63}$$

For $SiO_2$ defined by the solid assemblage:

$$\log a_{SiO_2}^{liq} = \frac{A_{SiO_2}}{T} + B_{SiO_2} + \frac{C_{SiO_2}}{T}(P-1) + M_{SiO_2} \tag{3-64}$$

where

$$M_{SiO_2} = \log \frac{(a_{MgSiO_3}^{opx})^2}{a_{Mg_2SiO_4}^{ol}}$$

For $Al_2O_3$ in the liquid:

$$\log (a_{Al_2O_3}^{liq})_{P,T} = \log X_{Al_2O_3}^{liq} + \frac{\phi_{Al_2O_3}}{T} + \left(\frac{0.18061}{T} - 0.67002 \times 10^{-4}\right)(P-1) \tag{3-65}$$

For $Al_2O_3$ defined by the solid assemblage:

$$\log a^{\text{liq}}_{Al_2O_3} = \frac{A_{Al_2O_3}}{T} + B_{Al_2O_3} + \left(\frac{C_{Al_2O_3}}{T} - 0.72339 \times 10^{-4}\right)(P - 1) + L_{Al_2O_3} \tag{3-66}$$

where

$$L_{Al_2O_3} = \log \frac{a^{\text{spinel}}_{MgAl_2O_4} a^{\text{opx}}_{MgSiO_3}}{a^{\text{ol}}_{Mg_2SiO_4}}$$

And for each pair of equations:

$$\frac{T(\log X^{\text{liq}}_{SiO_2} - B_{SiO_2} - M_{SiO_2}) + (\phi_{SiO_2} - A_{SiO_2})}{(C_{SiO_2} + 0.0047) - T(1.344 \times 10^{-6})} = P - 1 \tag{3-67}$$

$$\frac{T(\log X^{\text{liq}}_{Al_2O_3} - B_{Al_2O_3} - L_{Al_2O_3}) + (\phi_{Al_2O_3} - A_{Al_2O_3})}{(C_{Al_2O_3} - 0.18061) - T(5.337 \times 10^{-6})} = P - 1 \tag{3-67a}$$

And by rearrangement:

$$\begin{aligned} T^2[&(1.344 \times 10^{-6})(\log X^{\text{liq}}_{Al_2O_3} - B_{Al_2O_3} - L_{Al_2O_3}) \\ &- (5.337 \times 10^{-6})(\log X^{\text{liq}}_{SiO_2} - B_{SiO_2} - M_{SiO_2})] \\ &+ T[(\log X^{\text{liq}}_{SiO_2} - B_{SiO_2} - M_{SiO_2})(C_{Al_2O_3} - 0.18061) \\ &+ (1.344 \times 10^{-6})(\phi_{Al_2O_3} - A_{Al_2O_3}) \\ &- (5.337 \times 10^{-6})(\phi_{SiO_2} - A_{SiO_2}) \\ &- (C_{SiO_2} + 0.0047)(\log X^{\text{liq}}_{Al_2O_3} - B_{Al_2O_3} - L_{Al_2O_3})] \\ &+ (C_{Al_2O_3} - 0.18061)(\phi_{SiO_2} - A_{SiO_2}) \\ &- (C_{SiO_2} + 0.0047)(\phi_{Al_2O_3} - A_{Al_2O_3}) = 0 \end{aligned} \tag{3-68}$$

and the physically plausible result for $T$ can be substituted in Eq. (3-67) to obtain $P$.

As an example, assume that a basaltic lava was quenched on the earth's surface to an assemblage of olivine, diopsidic augite, plagioclase, and iron titanium oxides. This lava also contains xenoliths of lherzolite, and the problem is to calculate at what temperature and pressure the lava and its xenoliths could have been in equilibrium.

For the xenoliths and the groundmass of the lava we have the following mineral data together with the equilibration temperature and oxygen fugacity of the iron titanium oxides; the mole fractions of $SiO_2$ and $Al_2O_3$ in the lava are also given:

| | $T$ | $\log f_{O_2}$ | $X_{Fe_3O_4}^{titanomag}$ | $X_{CaMgSi_2O_6}^{cpx}$ | $X_{Mg_2SiO_4}^{ol}$ | $X_{CaAl_2Si_2O_8}^{plag}$ | $X_{MgSiO_3}^{opx}$ | $X_{MgAl_2O_4}^{spinel}$ | $X_{SiO_2}^{lava}$ | $X_{Al_2O_3}^{lava}$ |
|---|---|---|---|---|---|---|---|---|---|---|
| Basaltic lava | 1253 | −11.7 | 0.323 | 0.493 | 0.680 | 0.466 | — | — | 0.520 | 0.100 |
| Xenoliths | — | — | — | — | 0.899 | — | 0.850 | 0.650 | — | — |

For the lava we have, from Eq. (3-47) in Table 3-3 and assuming ideal mixing of the solid solutions,

$$\log (a_{SiO_2}^{liq})_{1;1253^\circ} = \frac{8270}{T} - 2.81 + \log \frac{a_{Fe_2SiO_4}^{ol}(f_{O_2})^{1/3}}{(a_{Fe_3O_4}^{titanomag})^{2/3}}$$

$$= -0.723$$

and from Eq. (3-55) we have

$$\phi_{SiO_2} = T(\log a_{SiO_2}^{liq} - \log X_{SiO_2}^{liq})$$

$$= -611.9$$

Similarly the activity of $Al_2O_3$ in the lava is given by Eq. (3-58), Table 3-3:

$$\log (a_{Al_2O_3}^{liq})_{1;1253^\circ} = \frac{-6776}{T} + 0.859 + \log \frac{(a_{Mg_2SiO_4}^{ol})^{1/2} a_{CaAl_2Si_2O_8}^{plag}}{a_{CaMgSi_2O_6}^{cpx}(a_{SiO_2}^{liq})^{1/2}}$$

$$= -4.3545$$

and from Eq. (3-55) we obtain

$$\phi_{Al_2O_3} = -4203.2$$

From Table 3-3 we find that the activities of $SiO_2$ and $Al_2O_3$, defined by the assemblage of olivine, orthopyroxene, and spinel, are given by Eqs. (3-35) and

(3-62); accordingly we can tabulate the values of the pertinent constants to be substituted into the quadratic equation of $T$.

$$A_{SiO_2} = -1034; \quad B_{SiO_2} = 0.597; \quad C_{SiO_2} = -0.04241;$$

$$M_{SiO_2} = -0.0487; \quad A_{Al_2O_3} = -8131; \quad B_{Al_2O_3} = 2.632;$$

$$C_{Al_2O_3} = 0.1344; \quad L_{Al_2O_3} = -0.1652$$

and by substitution of these values—together with $\phi_{Al_2O_3}$, $\phi_{SiO_2}$ and $X_{Al_2O_3}$, $X_{SiO_2}$—values for the equilibration temperature and pressure can be obtained. The values are 1173°C and 19.9 kb.

An analogous procedure for a garnet peridotite would use the equation

$$\underset{\text{forsterite}}{Mg_2SiO_4} + \underset{\text{enstatite}}{MgSiO_3} + \underset{\text{glass}}{SiO_2} + \underset{\text{liquid}}{Al_2O_3} = \underset{\text{pyrope}}{Mg_3Al_2Si_3O_{12}} \tag{3-69}$$

which in turn could be combined with Eqs. (3-60), (3-61), and (3-35) to derive a value for $P$ and $T$ at which the liquid lava would be in equilibrium with a garnet peridotite, with appropriate values substituted for the composition of the solid components.

Because there is considerable uncertainty in the free energy of formation ($\Delta G_f^\circ$) of pyrope garnet, it has been found convenient to use liquid $CaMgSi_2O_6$ in place of liquid $Al_2O_3$, since all model mantle compositions contain a diopsidic augite and virtually every basic lava type also contains a calcium-rich pyroxene. Using $CaMgSi_2O_6$ and $SiO_2$ the calculated equilibration pressures and temperatures of basic lavas with spinel lherzolite are closely concordant with those using $Al_2O_3$ and $SiO_2$, indicating that the thermodynamic data are internally consistent (Carmichael and Wood, in press).

The technique of calculating the pressure and temperature at which the activities of any two components in a lava equal those defined by a solid assemblage can be used in many different ways. For example, Bacon and Carmichael (1973) calculated the pressures and temperatures at which an ascending basaltic lava equilibrated with its ultramafic xenoliths, its megacrysts, and its phenocrysts, so that three stages in the $PT$ path of ascent could be established.

This same approach has been used (J. Nicholls et al., 1971) to calculate the pressure at which a kimberlite at an assumed temperature of 1327°C (Boyd, 1969) equilibrated with the olivine and orthopyroxene inclusions found in diamonds (Meyer and Boyd, 1969; Meyer, 1970) and also with the coesite inclusions of diamonds (Meyer, 1968); the calculated values (at 1327°C) are 63.5 and 95.6 kb, respectively (Fig. 3-8). Since kimberlite also contains zircons with an outer rim of baddelyite (Nixon et al., 1963), a pressure can be calculated at which the kimberlite liquid equilibrated with zircon and baddelyite. This calculated pressure of 55.7 kb also falls in the diamond stability field (Fig. 3-8),

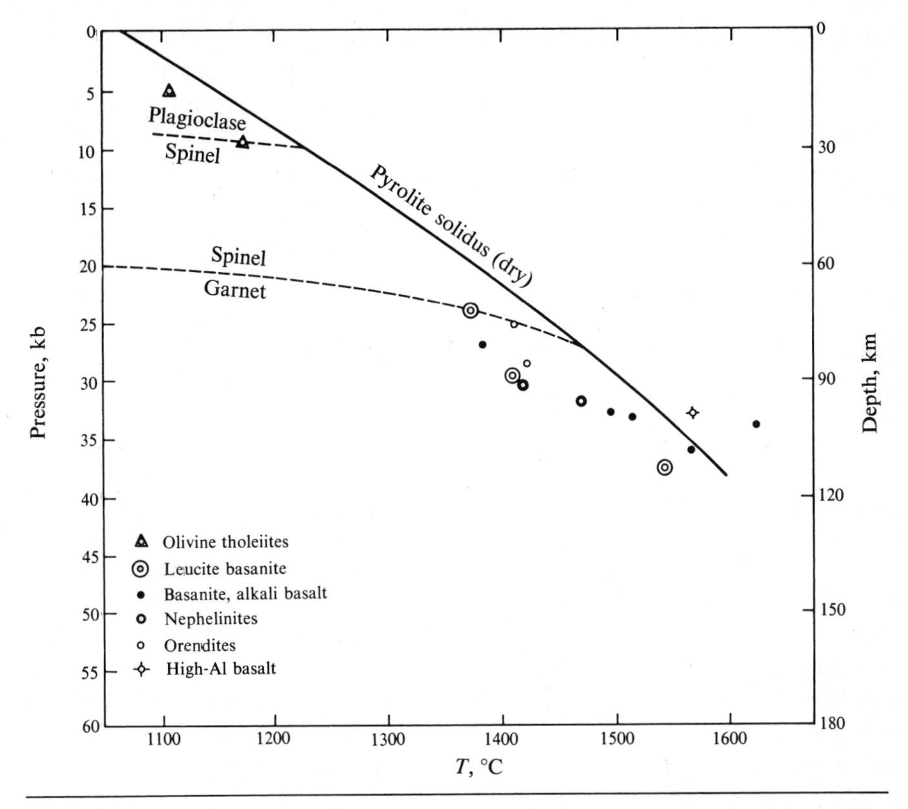


**Fig. 3-10** Calculated equilibration pressures and temperatures of various lava types with a model mantle of olivine ($Fo_{90}$), orthopyroxene ($En_{90}$), and diopsidic augite. (After Carmichael and Wood, 1974.)

but since neither zircon nor baddeleyite has been reported as inclusions in diamond, presumably the discrepancy of 5 or 6 kb in Fig. 3-8 arises from both errors in the free-energy data and particularly in the necessary assumption that $\Delta V_{1;298^\circ}$ can be used to these extreme pressures.

Although the petrological implications of the equilibration of various lava types with a model mantle composition are not appropriately discussed here, it is worthwhile to highlight some of the results which have been obtained using this method. In Fig. 3-10 the calculated temperatures and pressures are plotted for the equilibration of various lava types with a model mantle of olivine, orthopyroxene, and diopsidic augite. The points all fall close to the dry solidus of pyrolite, suggesting that these various lava types were not saturated with water at the stage of their generation. Of the various lava types, the olivine tholeiites equilibrate with the model mantle in the *PT* region in which plagioclase is stable, whereas the alkali lava types fall in the garnet field; however, since Cr-rich spinel is stable to extreme pressures, being found as inclusions in diamond, it is certain that Cr spinel can extend to higher pressures and temperatures than those shown in Fig. 3-10, and the calculated results should not be construed as indicating that these lava types were necessarily generated by partial fusion of garnet lherzolite rather than Cr-rich spinel lherzolite.

Although the procedures outlined above allow the pressures and temperatures of equilibration to be calculated for virtually all known lava types with a mantle of any specified composition, it cannot be rigorously established that each lava was generated by partial fusion under these calculated conditions until it can be shown that the chemical potentials of *all* the components in the lava equate with the postulated source material. This is a very difficult task, since it may prove to be impossible to define the chemical potential, or activity, of many components either in the lava or in the model mantle. However, the results on two components are a necessary step in considering many components, and it may soon be possible to include the effect of a changing concentration of water on the conditions of the mantle-lava equilibration.

GEOBAROMETRY AND WATER In reactions in which it participates as a component, water can have a dramatic effect on the displacement of equilibrium, so that care has to be taken in using a mineral assemblage, or a reaction involving water as a component, as a basis for estimating pressure and temperature of a natural assemblage. Water is such a convenient experimental pressure medium, and so hastens equilibrium, that petrologists could be lulled into accepting it in the same preeminent role in natural silicate equilibria. That this is unlikely in general is indicated by the best estimate of Kilaeua magmatic gas (Chap. 6), which contains only about one-third water. Again as a generalization, the assumption that $P_{total} = p_{H_2O}$ is quite unwarranted unless calculations can be made on the equilibration of phenocrysts at depth to show that this is so (Wood and Carmichael, 1973).

The most familiar example of the effect of water is the melting reaction of silicates, which, because water dissolves in the liquid, decreases the melting temperature as the concentration of water increases. However, the first increments of water are the most effective in depressing the melting, or liquidus, temperature, and a typical curve steepens at higher values of $p_{H_2O}$ (Fig. 3-9). This steepening is the result of the large compressibility of steam, so that the partial molar volume of steam in the gas phase (assumed to be pure $H_2O$) $\hat{V}^{\circ}_{H_2O}$ decreases with pressure faster than the partial molar volume of steam $\bar{V}_{H_2O}$ in the silicate liquid. This is exemplified in the results of Burnham and Davis (1971), who have measured $\bar{V}_{H_2O}$ in liquids of $NaAlSi_3O_8$ composition over a range of pressure and temperature; values of $\bar{V}_{H_2O}$ at two different pressures and temperatures are compared to the corresponding values for pure steam (Burnham et al., 1969) below:

| | $\bar{V}_{H_2O}$ in $NaAlSi_3O_8$ liquid | $\hat{V}^{\circ}_{H_2O}$ (steam) |
|---|---|---|
| 900°C, 1 kb | 24.0 cm$^3$ mole$^{-1}$ | 92 cm$^3$ mole$^{-1}$ |
| 700°C, 10 kb | 13.3 cm$^3$ mole$^{-1}$ | 19.4 cm$^3$ mole$^{-1}$ |

It seems likely that at high pressures $\hat{V}^{\circ}_{H_2O}$ becomes less than $\overline{V}_{H_2O}$, so that the water-saturated liquidus of various lava types will show a temperature minimum, such as that found in the region of 20 to 25 kbar for a liquid of tholeiitic basalt composition (Nicholls and Ringwood, 1973). Expressed in another way, the isothermal solubility of water in silicate liquids will pass through a maximum with increasing pressure, and it has been suggested that this solubility maximum stabilizes the low velocity zone in the upper mantle (F. J. Spera, in press).

Silicate liquids which are unsaturated with water also have a positive melting curve ($P$ increases with $T$). Under conditions of limited availability of water, the first melting fraction may take place along the water-saturated curve, with negative slope. However, as the amount of liquid increases with insufficient water to saturate it, the melting curve becomes positive (Burnham, 1967, fig. 2-9). By and large the melting reaction is of little use as a geobarometer-geothermometer because of the uncertainty of the role and amount of water.

The stability of hydrous minerals at high pressures and temperatures provides a clue to the way in which water could exist in the earth's mantle. Again because of the large compressibility of supercritical water, the breakdown reaction of a hydrous mineral, the dehydration reaction, can have a reaction volume $\Delta V_r$ which changes from positive to negative, so that the equilibrium curve will have a maximum. Fyfe (1970*b*) has suggested that the maximum thermal stability, where $\Delta V_r$ tends to zero, of hydrous amphiboles is approximately 1000°C and of hydrous phlogopitic mica somewhat greater. This has been confirmed by Lambert and Wyllie (1970), who have melted a variety of igneous rocks and showed that the stability curve of hornblende is initially positive ($dT/dP$), reaches a maximum temperature close to 1000°C, and then at 18 or so kilobars $p_{H_2O}$, bends sharply back to low temperatures; thus hornblende becomes unstable at pressures above 25 kb at 600°C. There seems to be no doubt that hydrated minerals are restricted to the upper (low-pressure) parts of the mantle—especially if $p_{H_2O} < P_{total}$, since hydrous mineral breakdown temperatures will then be lowered. At greater depths or higher pressures in the mantle, hydroxylated pyroxenes seem possible; Sclar (1970) was able to produce $Mg_2SiH_4O_6$ at high pressure, resulting from the substitution of $(H_4O_4)^{4-}$ for $(SiO_4)^{4-}$.

Some recent experiments on the stability of a hydrous phase in equilibrium with a liquid have been interpreted to show that reducing the fugacity of $H_2O$ increases the stability of a hydrous mineral. This is in apparent contradiction to the thermodynamic requirements of equilibrium, and as an example we may take the breakdown reaction of anthophyllite:

$$\underset{\text{anthophyllite}}{Mg_7Si_8O_{22}(OH)_2} = \underset{\text{enstatite}}{7MgSiO_3} + \underset{\text{quartz}}{SiO_2} + \underset{\text{steam}}{H_2O}$$

If all the solid phases are pure and thus have unit activity, and if $\Delta V_r^{\circ}$ is assumed to be independent of pressure, then we may write for equilibrium at pressure and temperature

$$\Delta G = 0 = \Delta G_r^\circ + \Delta V_r^\circ(P - 1) + RT \ln f_{H_2O}$$

and

$$= \Delta H_r^\circ - T\Delta S_r^\circ + \Delta V_r^\circ(P - 1) + RT \ln f_{H_2O}$$

so that

$$T_E = \frac{\Delta H_r^\circ + \Delta V_r^\circ(P - 1)}{\Delta S_r^\circ - R \ln f_{H_2O}} \tag{3-70}$$

where $\Delta H_r^\circ$ and $\Delta S_r^\circ$ are both positive for the dehydration reaction as written. If the fugacity of water[1] is reduced below the equilibrium value of $p_{H_2O} = P_{total}$, then the size of the denominator will increase and $T_E$ will be reduced; in short, temperature falls at constant $P_{total}$. Experiments (for example, Holloway and Burnham, 1972) on the stability of amphibole in the presence of a liquid have indicated that where $p_{H_2O} = P_{total}$, the stability or breakdown temperature of amphibole is lower than where $p_{H_2O} < P_{total}$, caused by the dilution of the $H_2O$ component by $CO_2$ or an inert gas. This is apparently in contradiction to the dictates of thermodynamics. To return to our example of anthophyllite: in equilibrium with a liquid, the activity of the solid products would no longer be unity since otherwise the liquid would be saturated with quartz and enstatite (in which case $f_{H_2O}$ would be buffered by the equilibrium coexistence of reactants and products). If quartz and enstatite are not present in the liquid, their activity (compared to solid standard states) must be less than unity, so that Eq. (3-70) becomes

$$T_E = \frac{\Delta H_r^\circ + P\Delta V_r^\circ}{\Delta S_r^\circ - R \ln f_{H_2O} - 7R \ln a_{MgSiO_3}^{liq} - R \ln a_{SiO_2}^{liq}} \tag{3-71}$$

It is obvious that reducing the activity of both $MgSiO_3$ and $SiO_2$ reduces $T_E$. It is well known that $CO_2$ is comparatively insoluble in silicate liquids, and these apparently discrepant experimental results can be reconciled with the thermodynamic constraints if it is assumed that the activities of $MgSiO_3$ and $SiO_2$ are lower in a liquid with a lot of water in solution than they are in a liquid with little water and trivial amounts of $CO_2$. It should be reemphasized that the only way the thermal stability of $Mg_7Si_8O_{22}(OH)_2$ can be increased above that of the standard state (pure solids) conditions is by changing the composition of the amphibole and thus reducing the activity of $Mg_7Si_8O_{22}(OH)_2$.

[1] Note that $f_{H_2O} = \gamma_{H_2O} p_{H_2O}$.

One considerable uncertainty in using a hydrous phase as a geobarometer-geothermometer is the role of fluorine, which if incorporated in place of the hydroxyl group, as in phlogopite, greatly increases the thermal stability in comparison to the hydrous analog. Little is known of the fluorine contents of hydrous minerals in fresh igneous rocks, or of the thermal stability of hydrous minerals with partial substitution of $F^-$ for $(OH)^-$. To make matters worse, evidence for the initial presence of F can disappear, since F in mica is readily displaced by water at low temperatures. Thus magma, cooling and reacting with its rock-water envelope, could develop a hydrous mineral assemblage whose thermal stability has no relation to the earlier history of the rock. If the hydrous mineral contains pristine water and little or no F, it may have acted as a water buffer at high temperature, so that with the appropriate free-energy data, values for the fugacity of water are obtainable. In short, since any assumptions of $p_{H_2O} \leq P_{total}$ are largely guesswork, and there is the additional uncertainty of the role of fluorine, the occurrence of a hydrous mineral offers little secure evidence for use as geobarometer-geothermometer.

## Magmatic Evolution: Petrology Today

It seems that a pervasive element in petrologic investigation and growth of ideas is a quest for regularity of behavior, for conformity to a recognizable recurrent pattern. That regularity exists cannot be denied. Patterns revealed by research have been sustained, usually in modified form, in the light of new discovery and increasingly abundant and more sophisticated data. Yet everywhere appear anomalies or departures from strict order that testify to the unique character of every province, every igneous episode, every upsurge of magma from the depths. To distinguish two contrasted basaltic magma types was valid and stimulating. But to force every basaltic magma into one or the other category is to deny a unique element of its genesis. Kilauean basalts by any definition are tholeiitic, and yet the summit lavas are chemically quite distinct from those of the flanks. Again, one of the more surprising examples of magmatic uniqueness has been found in the Quaternary obsidians of northern California and adjoining regions. Despite their overall similarity in composition, the concentration or relative proportions of their trace elements allows each locality to be "fingerprinted," and thus the distribution of Indian obsidian artifacts can be traced to their respective sources.

As in other sciences, thinking and investigation in igneous petrology tend to follow changing fashion. Today we are less concerned than were our predecessors with rock nomenclature and classification. Some of the old controversies that flourished in the scientific climate of half a century ago have subsided—usually by reconciliation of ideas once thought to be radically opposed: differentiation versus assimilation in magmatic evolution; granitization (metasomatism) versus magmatic crystallization in the "granite controversy." The significance of other problems seems to have waned: for example, the genesis of alkaline rocks presupposing that they represent an abnormal evolutionary line. Some appear, if this is possible, to be largely solved: the mechanics of differentiation; the courses of assimilative reaction between common magmas and common rocks; petrogenesis of layered basic intrusions. Yet other classic problems survive as controversial material in today's petrology: petrogenesis of ultramafic intrusions; origin of Precambrian anorthosites; relations of parental magmas to structural and tectonic environment; identity and possible relations of magma types. Outstanding today, as they were yesterday, are two major questions. What are the source materials of magmas and how are magmas generated from them? How can we account for the large but finite variety of known igneous rocks?

The distinctive flavor of modern igneous petrology has two aspects. First there is a more rigorous approach, a sharper quantification in the light of thermodynamic prediction and experimental geochemistry. Second is the necessity to modify older models of petrogenesis and develop new ones in harmony with new information in fields newly explored under the impetus of new techniques and abundantly funded programs in geophysics, geochemistry, oceanography, and (now in its infancy) lunar petrology. Radiometric dating has greatly sharpened appreciation of the role in petrology of that peculiarly geological quantity, time. Geophysical models and chemical data on distribution of isotopes and minor elements in igneous rocks have brought within the field of legitimate speculation the basic problem of where, how, and from what magmas originate. Some of today's fashionable topics relate to quantitatively insignificant rocks of exotic composition, such as kimberlites and carbonatites. Others spring from exploration of major regions hitherto virtually unknown—the ocean floors, and now the moon. Others, such as relation between igneous phenomena and ore genesis, are old problems enlivened by an influx of novel information.

The problems of today's petrology, like those of yesterday, emerge from the cumulative data of case histories as these are scrutinized in the light of experimental and theoretical geochemistry against a background of acceptable geophysical models. Before proceeding to case histories we review some of this quantitative background.

# Liquids and Solids: An Introduction to Their Properties

## Scope of Treatment

This chapter is divided into two parts. The first deals with the liquid state and such matters as crystal nucleation and growth, whereas the second considers solid-liquid equilibrium and particularly the properties of solutions.

Liquids, especially those of silicate composition, are the essence of igneous petrology, which, succinctly put, can be conceived as the generation of the liquid state followed later by its elimination. Our knowledge of natural silicate liquids, or magmas, is largely confined to their compositional variety; and the effort involved in gathering this type of information vastly exceeds the measurements of the physical or thermodynamic properties of magma. Magma is not the easiest material to study, being hot and often explosive, so that we must draw from the related fields of materials science and glass technology for data to illustrate the diverse properties of silicate liquids. Possibly the reader will hesitate to accept the results or implications taken from apparently simple two- or three-component systems, often rich in alkali, as being applicable or analogous to the far more complex natural equivalents. Perhaps if nothing else, this approach will highlight the paucity of data on magma and provoke the reader to borrow from the materials scientists their variety of experimental techniques and to obtain data which would then make most of what follows redundant. This note of exhortation is not intended to minimize the efforts of those scientists who have measured natural liquid properties or have developed a theoretical approach. To those geologists of the past, such as Vogt, Morey, and Bowen, are to be added less than a dozen today whose work provides almost all the substance of our knowledge of the properties of *natural* silicate liquids.

This chapter does not treat all the varied properties of liquids, so that, depending on the reader's interest, he may find much omitted which he considers of importance; thus there is only a brief discussion of the structure of liquids. We have set out to emphasize those properties, such as viscosity and crystal growth rate, whose impact on the crystallization history of a silicate liquid is a matter of common observation.

## The Liquid State: Thermal and Kinematic Properties

In a liquid, the forces of cohesion between the molecules, atoms, or ions are sufficiently strong to perpetuate a condensed state, but they are not strong enough to prevent translational motion of the individual atoms or molecules. It is these moving particles which introduce disorder in a liquid in comparison to the solid, but at least near the melting point their number is small and a regularity of structure persists which is not completely broken down until the liquid becomes a gas. However, there is not a continuous gradation of properties between the liquid and crystalline state at the melting point. Silicate minerals fuse with a small increase in volume (5 to 15 percent), which indicates that the arrangement of the molecules in the liquid, near or at the melting point, must be quite similar to that of the solid. Another indication that the cohesive forces between the molecules or atoms decrease only slightly at the melting point is that the molar heat of fusion $\Delta H_m$ is always very much smaller than the molar heat of vaporization $\Delta H_{vap}$; the values for $SiO_2$ are $\Delta H_m = 1.95$ kcal mole$^{-1}$ and $\Delta H_{vap} = 123.5$ kcal mole$^{-1}$. Furthermore, the specific heat, or the heat capacity, is usually affected only slightly by fusion, the values above the melting temperature usually being higher than those below it. Several examples are shown in Table 4-1.

**Table 4-1** Heat capacities of solids and liquids at their melting temperatures

| | $T_m$, °K | $C_p^{sol}$, cal mole$^{-1}$ deg$^{-1}$ | $C_p^{liq}$, cal mole$^{-1}$ deg$^{-1}$ |
|---|---|---|---|
| $SiO_2$ | 1996 | 18.47 | 21.66 |
| $Na_2Si_2O_5$ | 1147 | 62.91 | 62.35 |
| $CaTiSiO_5$ | 1670 | 51.29 | 66.80 |
| $KMg_3AlSi_3O_{10}(F)_2$ | 1670 | 128.75 | 164.00 |

At high temperatures, near the critical temperature, the properties of a liquid become increasingly like those of a gas, so that over the liquid stability range the liquid's properties vary from the nearly perfect order of a crystal to the nearly perfect randomness of a gas; it is this variation which has so far inhibited any general theoretical treatment of the liquid state. Some theories have treated a liquid as a solid with long-range disorder and short-range order, since the increase in volume on melting of many silicates indicates an increase of between 2 and 5 percent in intermolecular spacing; in other theories a liquid has been considered as an extremely imperfect gas.

Both these approaches suggest a progressive change in crystal-like properties to gaslike properties, and it is perhaps difficult therefore to see why solids have sharp melting points. The melting of a solid is probably the most familiar of all reversible reactions, and we can write at equilibrium ($\Delta G = 0$) for the reaction crystal $\rightleftarrows$ liquid

$$T(S^{\text{liq}} - S^{\text{sol}}) = H^{\text{liq}} - H^{\text{sol}} \tag{4-1}$$

or

$$T\Delta S_m = \Delta H_m$$

where $\Delta S_m$ is the change in entropy at the melting temperature and $\Delta H_m$ is the corresponding enthalpy change. At the melting temperature, the entropy term, representing the greater randomness of the liquid, equals the enthalpy term, and the crystal melts. The sharp transition is due to the fact that crystals can accept a certain amount of short-range disorder, and indeed their free energy is decreased by small numbers of vacancies. However, if disorder becomes extensive, then the long-range regularity of the crystalline state is destroyed,[1] and the "irregularity rapidly spreads throughout the entire specimen; disorder in a crystal is contagious" (Moore, 1965, p. 704).

In general the varied theories of liquid structure, particularly well summarized by Christian (1965), tend to treat the solid-liquid transformation as quasi-continuous. The semiregularity of liquid structure has been inferred from x-ray diffraction, which indicates the distance between neighboring atoms and the average number of neighbors. Although in silicate liquids the distance between neighboring atoms tends to increase on melting, in alkali halide liquids the distance between the nearest neighbors decreases while that of the next-nearest neighbors increases in comparison to the solid. The coordination number of atoms in a liquid is usually between 8 and 11, compared to 12 in a close-packed solid. Bernal (1964) has given a geometric picture of a simple liquid near its freezing point. The atoms constitute a close-packed, disordered array in which individuals are coordinated, on average, by fewer neighbors than in the solid. His predicted radial distribution function, or the probability of finding the center of another atom at the end of a vector of length $r$ drawn from the origin, agrees well with that measured for liquids.

Eyring and his coworkers have provided, under the name of the significant-structure theory of liquids, an empirical model of a liquid which leads, in many instances, to good quantitative predictions of its properties. The theory assumes that when melting occurs a certain number of particles leave their solid lattice positions to move elsewhere; the solid accordingly expands by the so-called

[1] Superheating of a solid above its melting temperature is impossible unless it is a perfect crystal.

excess volume $V - V_S = \Delta V_m$.† The presence of vacancies or holes at the now unoccupied lattice sites confers new properties on adjacent particles, which now have additional sites available to them. This increases the degeneracy, and therefore also the entropy, of the system. A particle jumping into a hole exchanges vibrational for translational motion and acquires, in effect, gaslike properties; the number of such particles is, however, restricted by the fact that, in order to be able to jump into a hole, a particle must have an excess energy which is taken to be proportional to the energy of sublimation (or evaporation) of the solid. The liquid is then conceived as consisting of a mixture of solidlike particles, vibrating much as they do in the solid, and gaslike particles, the number of which depends on the excess volume of melting and on the sublimation energy. The entropy of melting arises in fact from the increased positional degeneracy (also proportional to $\Delta V$) and from the additional translational freedom. The heat capacity at constant volume of a liquid would, to a first approximation, be of the form

$$C_v^{\text{liq}} = \frac{V_S}{V} C_v^{\text{sol}} + \frac{V - V_S}{V} C_v^{\text{gas}}$$

where $C_v^{\text{sol}}$ and $C_v^{\text{gas}}$ are, respectively, the heat capacities of solid and gas at the melting temperature. This relation is only approximate insofar as it neglects possible additional degrees of freedom in the liquid, such as rotation of groups of particles or molecules and possible electronic effects.

## LIQUIDS AND GLASSES

Some liquids when rapidly cooled below their equilibrium melting point $T_m$ may fail to crystallize and instead form "amorphous" solids which lack the regular crystallographic arrangement of particles characteristic of the crystalline state. Such amorphous solids are commonly called *supercooled liquids* or *glasses*, although the two states are not identical.

The transition crystal $\rightleftarrows$ liquid at $T_m$ entails a discontinuous jump in some of the first-order thermodynamic functions, enthalpy, entropy, or volume; it is therefore a "first-order transition." However, when a glass-forming liquid is cooled metastably below $T_m$ no discontinuity (at $T_m$) is observed in the first-order thermodynamic properties, nor in the heat capacity or the coefficient of thermal expansion, both of which are second-order thermodynamic functions; the "transition" liquid $\rightleftarrows$ supercooled liquid is continuous in terms of these. However, continued cooling of the supercooled liquid causes a jump in the heat capacity and in the coefficient of thermal expansion, and the temperature of this discontinuity is called the glass transformation temperature $T_g$. The transition supercooled liquid $\rightleftarrows$ glass is accordingly a second-order transformation since the first-order thermodynamic functions are continuous at $T_g$.

† The symbolic notation in this paragraph is Eyring's.

Above $T_g$, supercooled liquid is stable relative to glass; below $T_g$ the reverse is true; both are metastable compared to a crystalline phase. The glass transformation temperature has also been defined empirically in terms of viscosity, which will increase as the supercooled liquid is cooled. The temperature at which the viscosity exceeds $10^{13}$ poises has been widely used in glass technology to define $T_g$.

In Fig. 4-1, we have plotted data to illustrate the relationship between liquid, supercooled liquid, and glass. The heat contents ($H_T - H_{298}$) of glass and liquid for three compositions—Pyrex, $SiO_2$, and $Na_2Si_2O_5$—are shown in Fig. 4-1*a* and are most complete for $Na_2Si_2O_5$, a compound for which there are considerable experimental data for a wide range of properties. Above the melting temperature, the $H_T - H_{298}$ curve for $Na_2Si_2O_5$ is straight, and the intersection at $T_m$ with the gently curved $H_T - H_{298}$ glass curve is, as we have noted above, a characteristic property of a second-order transition. The extrapolated, or metastable, $H_T - H_{298}$ liquid curve intersects the glass curve again at a much lower temperature (~660°K, Fig. 4-1), and it is this low-temperature intersection which J. S. Haggerty et al. (1968) have used to define $T_g$, the glass transformation temperature; this intersection is within 20° of the temperature defined by the viscosity exceeding $10^{13}$ poises (Fig. 4-1). The data for $SiO_2$ are also shown, but the glass transformation temperature defined by the viscosity is only approximate. However, for second-order transformations, it is the change in heat capacity, equivalent to the slope of the $H_T - H_{298}$ curves, which clearly depicts the glass–supercooled-liquid transformation.

The heat capacity of a solid tells us a great deal about the thermal motions of the atoms or molecules. Measurements of heat capacity can be made either at constant volume $C_v$ or at constant pressure $C_p$; the latter is the most frequently determined parameter. Although the difference between the two is readily appreciated for a gas, for a solid $C_v$ can be visualized as the amount of heat required to increase the thermal agitation of the atoms, whereas $C_p$ includes this amount of heat $C_v$ plus the amount required to expand the solid in response to the thermal agitation. In Fig. 4-1*b* the heat capacity values for Pyrex, $Na_2Si_2O_5$, and $SiO_2$ glasses and liquids are plotted. J. S. Haggerty et al. (1968) noted that the heat capacity of Pyrex glass approached a value of $3R$ at a temperature close to $T_g$, the glass transformation temperature defined by viscosity measurements. To interpret this observation, we recall that, as shown theoretically by Einstein, the vibrational energy of an atom rises from zero at 0°K to a maximum value $kT$ (where $k = R/N_0$, $N_0$ being Avogadro's number) for each degree of vibrational freedom. Thus for an atom to vibrate in three mutually perpendicular directions, the vibrational energy per mole at sufficiently high temperature is $E = 3RT$, and the corresponding heat capacity at constant volume

$$C_v = \left(\frac{\partial E}{\partial T}\right)_V = 3R \tag{4-2}$$

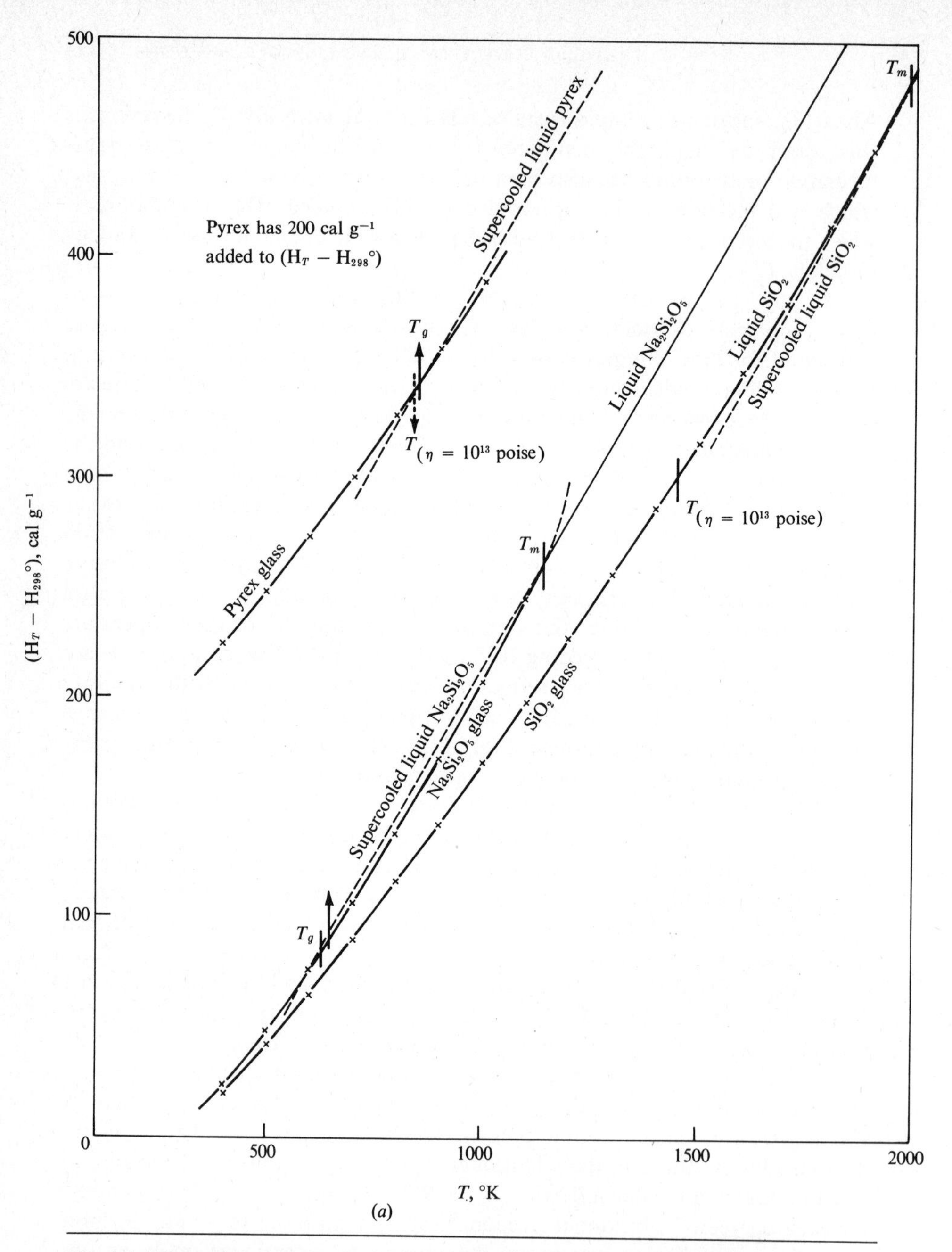


**Fig. 4-1** Thermal data for glasses and supercooled liquids. (*a*) Heat contents $(H_T - H_{298}°)$ of supercooled liquids and glasses (note that Pyrex has been displaced for clarity); $T_m$ is temperature of melting, $T_g$ is the glass transformation temperature, and $T_\eta$ is the temperature at which the viscosity of the glass is $10^{13}$ poises. (*b*) Heat capacities for glass and liquid [$3R$ values for $C_p$ (per g atom) are also shown]; note the difference in heat capacity between glass and supercooled liquid at $T_g$.

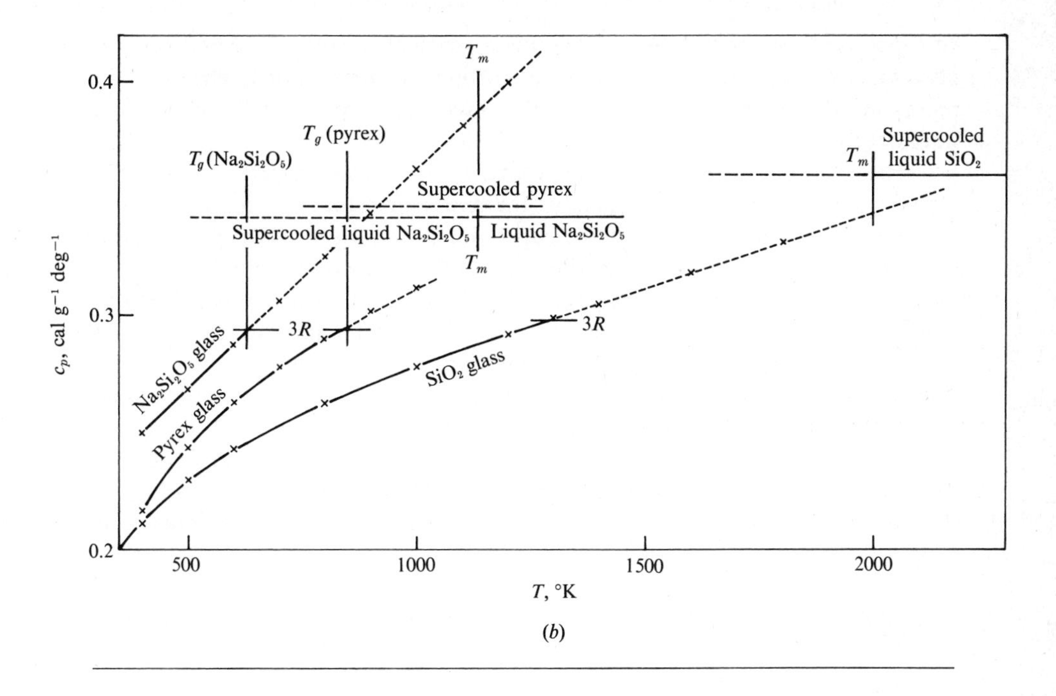


(*b*)

Measurements of heat capacity are usually made at constant pressure to give $C_p = (\partial H/\partial T)_P$. The relation between $C_p$ and $C_v$ is

$$C_p - C_v = \frac{TV\alpha^2}{\beta} \tag{4-3}$$

where $V$ is the molar volume, $\alpha$ is the coefficient of thermal expansion, and $\beta$ is the coefficient of isothermal compressibility. The difference between the two is quite small, and it amounts to 0.003 cal g$^{-1}$ deg$^{-1}$ for $Na_2Si_2O_5$ at $T_g$ = 650°K. For Pyrex glass, J. S. Haggerty et al. (1968) find the difference to be less than their experimental error in measuring $C_p$. Neglecting this small difference, one can say that changes in $C_p$ such as occur in first- and second-order transformation points reflect acquisition, or loss, of modes of thermal motion. At the melting point, for example, some of the vibrational degrees of freedom of the crystal may be replaced by translational degrees of freedom in the liquid. Similarly the glass transformation temperature $T_g$ may be interpreted as the temperature at which appreciable translational mobility sets in.

With this last comment in mind, it has become useful to define an "ideal glass" state (Angell, 1968) which for kinetic reasons is unattainable. If internal equilibrium in the liquid state is conceived to involve an exchange between vibrational and configurational energy states, then in principle a configurational ground state should be accessible; the ground state has zero configurational entropy,[1] and the temperature where this is reached is called $T_0$. Since viscous flow or diffusion can take place only through configurational changes, $T_0$ represents the temperature at which the viscosity of an "ideal" supercooled liquid becomes infinite and the rate of diffusion is zero.

The glass transformation temperature $T_g$ is defined operationally by the discontinuities in the heat capacity, thermal expansion, or compressibility functions or by the viscosity of the supercooled liquid exceeding $10^{13}$ poises. These parameters vary with the cooling rate of the supercooled liquid and attain a limiting value as cooling of longer and longer duration is carried out. There is therefore a limiting value of $T_g$, namely $T_0$, which is the temperature of a true thermodynamic second-order equilibrium transformation but which is nevertheless metastable relative to the crystalline state. Thus $T_0$ is the thermodynamic low-temperature limit to the liquid state (supercooled).

Several attempts have been made to find a relationship between $T_g$ and $T_m$ for glass-forming compounds. Simha and Boyer (1962), using organic polymers as examples, suggested that $T_g$ and the volume thermal expansions of liquid and glass were related so that

$$(\alpha^{\text{liq}} - \alpha^{\text{glass}})T_g = k' \tag{4-4}$$

Empirically it has been shown that

$$\alpha^{\text{liq}}T_m = k'' \tag{4-5}$$

where $k''$ is a constant independent of the crystalline species, so that in combination, Eqs. (4-4) and (4-5) indicate that $T_g/T_m \approx k$. Sakka and Mackenzie (1971) found that $k$ has a value close to 2/3 for many inorganic glasses, irrespective of whether $T_m$ represents the melting temperature of a compound, or the liquidus temperature of an incongruently melting compound. Using the data of Arndt and Haberle (1973) on feldspar liquids and glasses, the values of $k'$ [Eq. (4-4)] and the ratio $T_g/T_m$ are given in Table 4-2.

**Table 4-2** Glass-transition temperatures of feldspar liquids

| | $NaAlSi_3O_8$ | $KAlSi_3O_8$ | $CaAl_2Si_2O_8$ |
|---|---|---|---|
| $\alpha^{\text{liq}}$ | $45 \times 10^{-6}$ | $42 \times 10^{-6}$ | $84 \times 10^{-6}$ |
| $\alpha^{\text{glass}}$ | 22.2 | 18.3 | 14.7 |
| $T_g$, °C | 763 | 905 | 813 |
| $T_m$, °C | 1118 | 1520 | 1552 |
| $T_g/T_m$ | 0.68 | 0.60 | 0.52 |
| $k'$ (Eq. 4-4) | $17.4 \times 10^{-3}$ | $21.4 \times 10^{-3}$ | $56.3 \times 10^{-3}$ |

[1] For a discussion of configurational heat capacity, see Prigogine and Defay (1954, pp. 293–299).

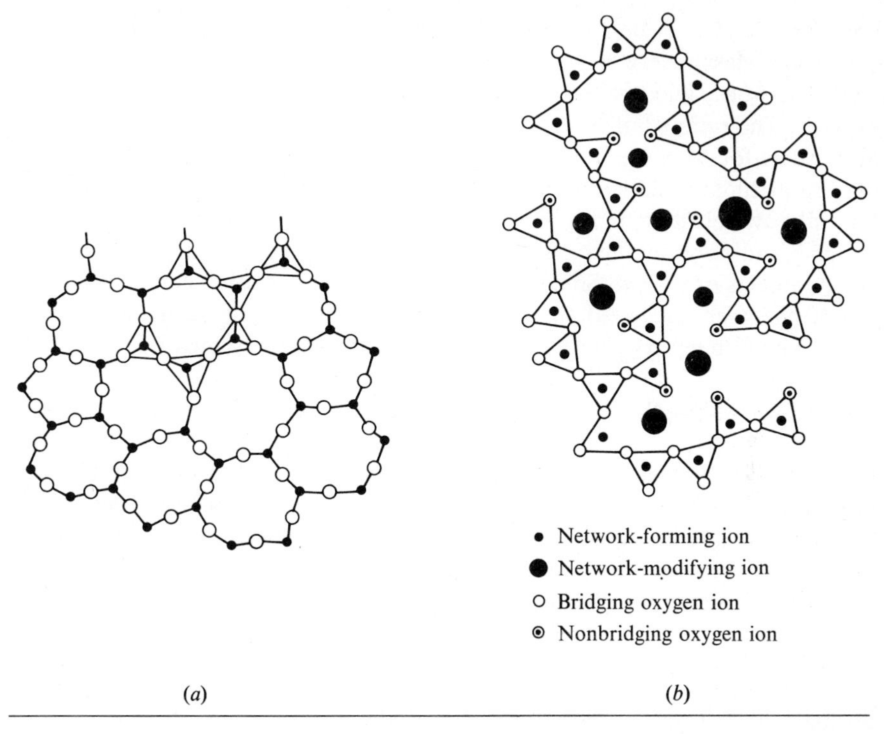


**Fig. 4-2** Structure of (*a*) pure silica glass and (*b*) a silica-rich glass with additional component ions. (After Lee, 1964.)

A rough estimate for $T_g$ for a rhyolitic obsidian can be made from the relationship $T_g/T_m = 2/3$. A liquid of this composition has a liquidus temperature of about 1050°C (Burnham, 1967) so that $T_g \approx 700°C$, which is not far removed from the temperature (735°C) at which almost anhydrous (<0.1 percent $H_2O$) obsidian has a viscosity of $10^{13}$ poises (Friedman et al., 1963). Shaw (1968) found an indication of a glass transformation in obsidian at about 700°C by the change in slope of the diffusion coefficient for water.

Not all liquids can be readily quenched to form glasses. Glass, in fact, generally occurs only in substances with molecules of complex shapes or whose molecules are linked in the liquid state to form long chains and networks. In glass-forming oxides (such as $SiO_2$), the three-dimensional O—Si—O networks characteristic of the crystalline form persist to a large extent in the liquid above $T_m$; this liquid correspondingly has a large viscosity, and the entropy of melting is relatively small. A two-dimensional picture (after Zachariasen) of the arrangement Si—O—Si linkages in pure silica glass is shown in Fig. 4-2*a*; in Fig. 4-2*b* the network has been modified by ions such as $Ca^{++}$ or $Na^+$ in the interstitial positions. Charge neutrality is maintained by the nonbridging oxygen ions in the vicinity of the network-modifying ion. Small ions such as $Ge^{4+}$, $B^{3+}$, and $Al^{3+}$ can replace $Si^{4+}$ as network formers. Hydroxyl will, as we shall see later, break the Si—O—Si bonds rather as a network-modifying cation. In an ideal pure silica glass, every oxygen will be linked as ≡Si—O—Si≡ forming the $SiO_4^{4+}$ network, and every bridging oxygen will be coordinated by six other oxygens. The ratio $\mathscr{R}$ of oxygen to network-forming cations (Si, Al) is two for $SiO_2$, three for $MgSiO_3$ and four for $Mg_2SiO_4$. In natural liquids, the value of $\mathscr{R}$ is approximately 2.01 for rhyolites and 2.33 for basalts (Lacy, 1955); and it is this value which determines, in an anhydrous liquid, the ideal proportion of bridging and nonbridging oxygens. Thus a nonbridging oxygen will have three other contacts of its own $SiO_4^{4+}$ group and a number of other contacts as well.

It has been suggested that there is an equilibrium between oxygen ions $O^{--}$, nonbridging oxygen $O^-$, and bridging oxygen represented by

$$O^{--} + \equiv Si—O—Si \equiv \rightleftarrows 2(\equiv Si—O^-) \tag{4-6}$$

Douglas et al (1965) have used this approach in studying the redox equilibria in glasses, while P. C. Hess (1971) uses it in conjunction with a simple ionic model to obtain the distribution of silica polymers in two-component silica melts.

Calculations of the polymerization, that is, the formation of "multi-molecular" discrete ion groups such as

$$(Si_3O_9)^{6-} + 3SiO_2 = (Si_6O_{15})^{6-} \tag{4-7}$$

in binary silicate systems have been made by P. C. Hess (1971) and by Flood and Knapp (1968). Most theoretical studies so far have been related to binary systems, which are of course a long way from the situation in a natural silicate liquid; it is a field which deserves more attention than we have given it here.

## REACTION AND TRANSPORT RATES IN LIQUIDS

Rates of diffusion, viscous flow, nucleation, and chemical reaction, indeed of all the transport phenomena of liquids, are found to increase exponentially with temperature. Arrhenius first suggested this relationship, which can generally be expressed as

$$Y = Ae^{-E_y/RT} \tag{4-8}$$

or

$$\ln Y = \ln A - \frac{E_y}{RT} \tag{4-8a}$$

or in differential form

$$\frac{d \ln Y}{dT} = \frac{E_y}{RT^2} \tag{4-8b}$$

where $Y$ represents the viscosity $\eta$, rate of chemical reaction $K$, and so forth; $E_y$ is a constant to be considered as an "activation" energy for the process (the factor $R$, the gas constant, is introduced to make the exponential factor dimensionless); and $A$ is a "frequency factor" with the dimension of reciprocal time. Both $A$ and $E$ can be determined by plotting the natural logarithm of a measured reaction rate against $1/T$. The plot is, according to Eq. (4-8$a$), a straight line with slope $-E_y/R$. Its intercept (at $1/T = 0$, or $T = \infty$) is $\ln A$. Not infrequently, $A$ or $E$ or both are found to depend on temperature or on other variables such as pressure $P$.

The exponential term $e^{-E_y/RT}$ in Eq. (4-8) is of the form of the Boltzmann factor $e^{-\varepsilon/kT}$ which measures the fraction of all particles in a system in equilibrium, which have energy $\varepsilon$ in excess of the ground state (lowest level) (Verhoogen et al., 1970, p. 239). The activation energy $E_y$ in Eq. (4-8) can then be understood to mean that the rate of a reaction is proportional to the number of particles that have a certain minimum amount of energy required for them to be able to participate in the reaction.

Eyring has generalized the Arrhenius equation (4-8) by introducing the concept of an activated complex, which is a necessary intermediate state in a reaction. This state has a higher energy than the initial state and thus forms an energy barrier. Consider, for instance, the reaction by which reactants $X$ and $Y$ form a product; we write this reaction as

$$X + Y \rightarrow (XY)^* \rightarrow \text{product}$$

where $(XY)^*$ represents the activated complex. The rate of reaction depends on the concentration of the activated complex, which Eyring assumes to be determined by an equilibrium constant $K^*$.

$$K^* = \frac{[XY]^*}{[X][Y]} \tag{4-9}$$

where square brackets represent concentration; the equilibrium constant, in turn, is determined by the usual thermodynamic relation

$$-RT \ln K^* = \Delta G^{\circ *} \qquad \text{or} \qquad K^* = e^{-\Delta G^{\circ *}/RT} \tag{4-10}$$

where $\Delta G^{\circ *}$ is the free energy of formation of the activated complex from the reactants (all assumed to be in their standard states). The reaction rate $K$ can then be written as

$$K = K^* C \tag{4-11}$$

where $C$ is a frequency factor which turns out, from quantum mechanical considerations, to be

$$C = \frac{kT}{h} \tag{4-12}$$

where $k$ is Boltzmann's constant and $h$ is Planck's constant. Since

$$\Delta G^{\circ *} = \Delta H^{\circ *} - T\Delta S^{\circ *} \tag{4-13}$$

where $\Delta H^{\circ *}$ and $\Delta S^{\circ *}$ are the enthalpy and entropy of activation respectively, we may obtain

$$K = \frac{kT}{h} e^{\Delta S^{\circ *}/R} e^{-\Delta H^{\circ *}/RT} \tag{4-14}$$

which, by comparison with Eq. (4-8), shows that in Eq. (4-8) the frequency factor $A$ must be identified as $(kT/h)e^{\Delta S^{\circ *}/R}$, whereas $E$ stands for $\Delta H^{\circ *}$. In general, $\Delta H^{\circ *} = \Delta E^{\circ *} + (P\Delta V^{\circ *})$, where $\Delta V^{\circ *}$ is the "activation volume," that is, the difference between the volumes of the activated complex and reactants. In certain reactions $P$ has to be taken as the internal pressure[1] rather than the external pressure.

Needless to say, the precise identification of the activated complex and its standard free energy of formation is not always easy or even feasible. In diffusion processes in a crystal, the activated complex presumably represents a diffusing

[1] A perfectly general thermodynamic equation of state which applies to all substances is given by

$$\left(\frac{\partial E}{\partial V}\right)_T = T\left(\frac{\partial P}{\partial T}\right)_V - P \tag{4-15}$$

and for an ideal gas $(\partial P/\partial T)_V = R/V$, so that $(\partial E/\partial V)_T = 0$. For nonideal gases and liquids, $(\partial E/\partial V)_T \neq 0$ and is called the internal pressure $P_{\text{int}}$; the magnitude of this reflects the cohesive forces, or the interaction, between the molecules. At small confining pressures, $P_{\text{int}}$ can be very large for some liquids and indicates that the energy required to push back the surroundings is small or negligible compared to the energy required to pull the molecules or atoms apart. For many liquids, $P_{\text{int}}$ is of the order of $10^4$ bars (water is 20,000 bars at 25°C and 1 bar) and can be calculated from the relationship

$$P_{\text{int}} = T\frac{\alpha}{\beta} \tag{4-16}$$

which would give a value for $Na_2Si_2O_5$ liquid at 1400°C of approximately $3.6 \times 10^3$ bars.

The internal pressure is obviously related to surface tension, and in the theories of transport in silicate liquids the $P_{\text{int}}\Delta V^*$, or the activation volume term, can be a significant fraction of the activation energy, as was found in the electrical conductivity of $Na_2O$–$SiO_2$ and $K_2O$–$SiO_2$ liquids (Tickle, 1967).

atom which has left its initial lattice position and is squeezing its way between adjacent atoms to reach its final lattice position. We note from Eq. (4-14) that, other things being equal, a reaction proceeds faster when $\Delta S^{\circ *}$ is positive than when it is negative. The former case corresponds, for instance, to acquisition of more degrees of freedom (for example, translational). If $\Delta S^{\circ *}$ has the same sign as $\Delta S_r^{\circ}$, the entropy change of the reaction itself, then it can be seen from Eq. (4-14) that the rate will be faster in the direction in which entropy increases, as in melting or vaporization, rather than in the opposite direction (crystallization or condensation). Rates of crystallization and melting of $Na_2Si_2O_5$ and of cristobalite are illustrated in Fig. 4-10.

## DIFFUSION IN LIQUIDS

Diffusion is a spontaneous process that tends to maintain a constant concentration of any component everywhere within a phase. Since the chemical potential of any component of a homogeneous phase at constant $T$ and $P$ increases with its concentration, and since matter tends spontaneously to flow from a high potential to a lower one, equilibrium may be reached within a phase only when the chemical potential of each component is uniform throughout this phase. The force $F_{i,x}$ acting on any particle $i$ in the direction of $x$ is

$$F_{i,x} = -\frac{1}{N_o}\frac{\partial \mu_i}{\partial x} \tag{4-17}$$

where $N_o$ is Avogadro's number, introduced here to reduce the force per mole $-\partial\mu_i/\partial x$ to its value per particle. Now let us suppose a frictional resistance to the movement of particle $i$ which is proportional to its velocity; then in a steady state the velocity $v_{i,x}$ will be such that this frictional resistance is equal to $F_{i,x}$. Let $B_i$ be the "coefficient of mobility" of the particle $i$, that is, the velocity it assumes under unit force. Then

$$v_{i,x} = B_i F_{i,x} = \frac{B_i}{N_o}\frac{\partial \mu_i}{\partial x} \tag{4-18}$$

and the rate of flow of $i$—that is, the number $s_{i,x}$ of moles of $i$ flowing in unit time across a unit surface normal to $x$—will be

$$s_{i,x} = c_i v_{i,x} = -\frac{c_i B_i}{N_o}\frac{\partial \mu_i}{\partial x} \tag{4-19}$$

$c_i$ being the number of moles of $i$ per unit volume.

For an ideal solution

$$\mu_i = \mu^{\circ} + RT \ln c_i \tag{4-20}$$

Thus

$$\frac{\partial \mu_i}{\partial x} = \frac{RT\, \partial \ln c_i}{\partial x} = \frac{RT\, \partial c_i}{c_i\, \partial x} \tag{4-21}$$

and therefore

$$s_{i,x} = -D_i \frac{\partial c_i}{\partial x} \tag{4-22}$$

where $D_i = RT\, B_i/N_o$ and is the *diffusion coefficient* of *i*. Equation (4-22) expresses Fick's law. It states that the rate of flow in any direction per unit surface and per unit time is proportional to the concentration gradient in that direction. The diffusion coefficient has dimensions of $cm^2\ s^{-1}$. The rate of flow $s_{i,x}$ must of course be expressed in terms of the same units as $c$; if $c$ is given in mole $cm^{-3}$, $s_{i,x}$ refers to the number of moles crossing a unit surface in unit time. If there is no mass transfer,[1] $D_i$ is called the self-diffusion coefficient of species $i$; when mass transfer is involved, $D_i = \tilde{D}_i$, an interdiffusivity which can be expressed in terms of the diffusivities (diffusion coefficients) of the individual species.

Often in liquid silicate systems, $D_i$ is independent of concentration, as in the ideal case it should be. The temperature dependence of $D_i$ is given by an equation of the familiar Arrhenius form:

$$D_i = A_D\, e^{-E_D/RT} \tag{4-23}$$

where $A_D$ is a constant and $E_D$ is called the activation energy of diffusion for a specified component. In some liquid silicate systems, the value of $E_D$ corresponds to $E_\eta$, the activation energy for viscous flow, but in many cases this is not so.

The diffusion coefficients of many elements have been measured in molten silicate glasses of varied composition, and two examples will suffice here. The diffusion rate for He is almost 20 times greater than that for Ne (Frischat and Oel, 1967), as would be expected from the size of the atoms (Ne 1.1Å; He 0.8Å), but each requires virtually the same energy for the diffusional step. In liquid $Na_2Si_2O_5$, Borom and Pask (1968) have measured the diffusion coefficients of $Na^+$, $Fe^{++}$, $Fe^{3+}$, and $Si^{4+}$. Although the activation energies for the four were very similar (30.4 kcal $mole^{-1}$) and smaller than $E_\eta$, there was a great difference in the rates (at 900 to 1100°C), which correspond to

$$Na^+ \gg Fe^{++} > Fe^{3+} \gg Si^{4+}$$

with $Fe^{++}$ an order of magnitude greater than $Fe^{3+}$.

[1] Typically where molecules of particles are diffusing through similar molecules or particles; one example is the diffusion of $D_2O$ in $H_2O$.

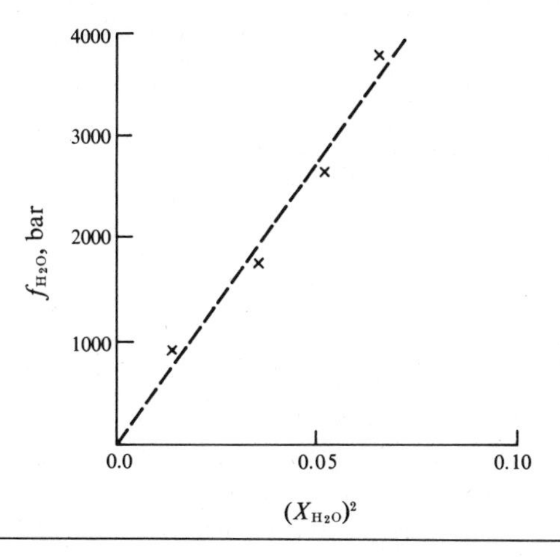


**Fig. 4-3** Fugacity of water plotted against the square of the mole fraction of water in a granite pegmatite liquid at 900°C. (Data from Burnham and Jahns, 1962.)

The relative rates of diffusion in natural solids and liquids are exemplified in the often extraordinarily complex zoning of plagioclase and titanium-bearing augitic pyroxene. Bottinga, Kudo, and Weill (1966) have shown that the growth of plagioclase phenocrysts is controlled by such varied diffusion rates, and they have used the determined concentration gradients adjacent to the crystal's surface as a measure of the relative diffusion rates of the plagioclase components. Evans and Moore (1968) found in their studies of the mineralogy of a cored section through the Makaopuhi lava lake on Kilauea that slow cooling, even below the solidus temperature, tends to obliterate the extreme compositional zoning found in the quickly cooled crystals.

However, in silicate liquids it is the diffusion of water that is perhaps of the greatest interest and perplexity. It is unlikely that water passes into solution in a silicate liquid simply as molecules of $H_2O$, for Burnham and Jahns (1962) and Hamilton et al. (1964) noticed that the mole fraction of water in both granitic and basaltic liquids is proportional to $(p_{H_2O})^{1/2}$. This relationship is shown in Fig. 4-3 for a granitic liquid (pegmatite) at 900°C, at pressures ($p_{H_2O}$) below 4000 bars. There is a convincing straight-line relationship between the fugacity of water $f_{H_2O}$ and the square of its mole fraction (Fig. 4-3). This squared dependence suggests a reaction of the type

$$H_2O + O^{--} = 2(OH)^- \tag{4-24}$$

which in terms of a silicate network glass structure can also be represented as

$$H_2O + \equiv Si—O—Si\equiv \rightleftarrows 2(\equiv Si—OH) \tag{4-25}$$

where $\equiv$Si signifies the other three bonds to oxygen. The concentration of (OH) in molten silica glasses is also proportional to $(p_{H_2})^{1/2}$ (Bell et al., 1962), so that there is an analogous hydrogen reaction which can be written as

$$\underset{\text{dissolved}}{\tfrac{1}{2}H_2} + \underset{\text{silica}}{—O—} \rightleftarrows \underset{\text{hydroxl}}{-OH} \tag{4-26}$$

or as

$$Si^{4+}2O^{--} + \tfrac{1}{2}H_2 = Si^{4+}O^{--}OH^{-} \tag{4-27}$$

where —O— represents a bridging oxygen connected to two silicon atoms. In simple terms, both $H_2O$ and $H_2$ react with Si—O—Si linkages in the glass or liquid, and by replacing the bridging oxygen with 2OH, will, if carried far enough, diminish the viscosity of the liquid.

Because the rate at which $H_2O$ and $H_2$ penetrate silica glass is dependent on the concentration of OH in the glass, the measured diffusivities $D'$ are not diffusion coefficients in the strict sense. However, values have been obtained on silica glass with a previously established equilibrium concentration of hydroxyl and also with the diffusing $H_2O$ and $H_2$ producing hydroxyl. Typical values (Bell et al., 1962) for the latter condition for silica glass at 1050°C are

$$D'_{H_2} = 2.35 \times 10^{-6} \text{ cm}^2 \text{ s}^{-1}$$

$$D'_{H_2O} = 9.51 \times 10^{-10} \text{ cm}^2 \text{ s}^{-1}$$

and for oxygen in silica glass (Roberts and Roberts, 1966) at the same temperature

$$D'_{O_2} = 1 \times 10^{-12} \text{ cm}^2 \text{ s}^{-1}$$

For silica glass with an equilibrium concentration of OH under the experimental conditions, the diffusion of $H_2O$ is faster because of the compositional dependence of diffusivity; a typical result for a similar glass composition at 1050°C (Burn and Roberts, 1970) is

$$D'_{H_2O} = 11 \times 10^{-10} \text{ cm}^2 \text{ s}^{-1}$$

It can be seen that $H_2$ diffuses well over a thousand times faster than $H_2O$ and almost a million times faster than $O_2$. Burn and Roberts (1970) also found a break in the temperature dependence of $D'_{H_2O}$ at a temperature close to the glass transformation temperature.

H. R. Shaw (1966) has reported that the temperature dependence of the diffusion of $H_2O$ in obsidian is similar to that for silica glass but that the rate is about 100 times greater; Shaw (1968) also found a change in slope at 700°C of $D_{H_2O}$ plotted against $1/T$, which could represent a glass transition in obsidian.

Since the two reactions (4-25) and (4-26) are reversible, they give a clue to the component which permits an oxygen buffer to control the chemical potential of oxygen throughout a mass of silicate liquid. If scattered crystals of fayalite, magnetite, and quartz are present in a silicate liquid, then at equilibrium the fugacity of oxygen will be fixed at any temperature and pressure. Obviously, because the diffusion rate of $O_2$ is so small, considerable fluctuations of the chemical potential, or concentration of oxygen, could occur, and equilibrium would be slowly achieved. $H_2$ diffuses much more rapidly than either $H_2O$ or $O_2$, so that it seems probable that the rate-controlling reaction for the attainment of equilibrium, namely, the constancy of the chemical potential of $O_2$, will be $H_2$ diffusion in the silicate liquid. Obviously $H_2$ will fix $f_{O_2}$ or $\mu_{O_2}$ at any $T$ and $P$ through the dissociation reaction of water, which in turn will be coupled with the reversible reactions of the type (4-25) and (4-26) in a silicate liquid.

## VISCOSITY OF SILICATE LIQUIDS

One of the characteristic properties of silicate liquids, particularly those rich in $SiO_2$, is that they are readily supercooled to become glasses, and as we have noted above, they do so at high values of viscosity, an indication of strong molecular or particle interaction. Viscosity is the response of a liquid to stress, which can be a shear stress or a compressive stress; thus a liquid has a shear and volume viscosity, although only the former is considered here.

The shear viscosity $\eta$ of a liquid is the ratio of a shear stress $\sigma$ to the corresponding shear strain rate $de/dt$

$$\eta = \frac{\sigma}{de/dt} \tag{4-28}$$

Equivalently, the strain rate can be replaced by a velocity gradient as shown in Fig. 4-4, which illustrates shear flow between parallel plates. The velocity gradient $dv/dz$ also indicates the strain rate, since the strain is measured by the angle $\alpha$. If $\eta$ is independent of $\sigma$, the viscosity is said to be Newtonian. If the shear stress $\sigma$ is measured in dynes $cm^{-2}$ and the strain rate in $s^{-1}$, then the unit of viscosity, called a poise, is dyne s $cm^{-2}$ (equivalent to g $s^{-1}$ $cm^{-1}$).

The variation of shear viscosity with temperature for many liquids is given by the Arrhenius equation

$$\eta = A_\eta \, e^{-E_\eta/RT} \tag{4-29}$$

where $E_\eta$ is the activation energy per mole. In Eyring's theory of liquids, the activation energy for viscous flow is assumed to be proportional to the energy of sublimation and comparable to the energy required for a particle to jump into an adjacent hole. To show how $E_\eta$ may be calculated, consider the experimental results of H. R. Shaw (1963) on a rhyolitic liquid containing 6.2 percent water

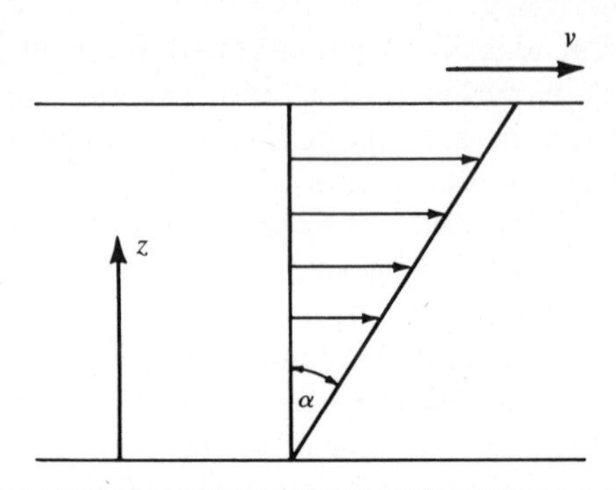


**Fig. 4-4** Shear flow between parallel plates.

at 2000 bars confining pressure. His results are plotted in Fig. 4-5*a*, and one set of data (at 2000 bars; 6.2 percent $H_2O$) is tabulated below.

| $T$, °C | log $\eta$ | $1/T$, °K |
|---|---|---|
| 700 | 6.52 | 0.001028 |
| 750 | 6.05 | 0.000978 |
| 800 | 5.57 | 0.000932 |
| 850 | 5.25 | 0.000890 |

The slope at each point in Fig. 4-5, multiplied by $2.303R$, will give a value of $E_\eta$. The slope between 700 and 750°C is

$$\frac{\Delta \log \eta}{\Delta 1/T} = \frac{6.52 - 6.05}{0.001028 - 0.000978} = 9400 \tag{4-30}$$

$$E_\eta = 2.303 \times 1.987 \times 9400 = 43.0 \text{ kcal mole}^{-1} \tag{4-31}$$

The average slope of the four data points gives a value for $E_\eta$ of 42.1 kcal mole$^{-1}$. Substitution of this value of $E_\eta$ will give the value of $A_\eta$, the pre-exponential constant in Eq. (4-29).

The effect of water on the viscosity of a rhyolitic obsidian is also shown in Fig. 4-5. At constant temperature, an increase in water markedly decreases the viscosity and also decreases the activation energy for viscous flow, $E_\eta$. The value of $E_\eta$ for an obsidian with less than 0.1 percent water is 122 kcal mole$^{-1}$—which is close to that for pure (anhydrous) $SiO_2$ liquid, 123 kcal mole$^{-1}$. Although continued addition of water decreases viscosity, $E_\eta$ falls to a limiting value near 43 kcal mole$^{-1}$; this can be seen in Fig. 4-5, where the slopes of the viscosity curves for obsidian with 4.3 and 6.2 percent water are similar. The effect of increasing water concentration on reducing the viscosity of a rhyolitic obsidian is shown in Fig. 4-6; the area bounded by the granite solidus and the granodiorite liquidus presumably represents the viscosity-temperature region of acid magma in the absence of crystals and bubbles of gas.

The viscosity of a silicate liquid depends on its composition: it is a matter of common record that basaltic magma[1] flows more readily than acid magma. By

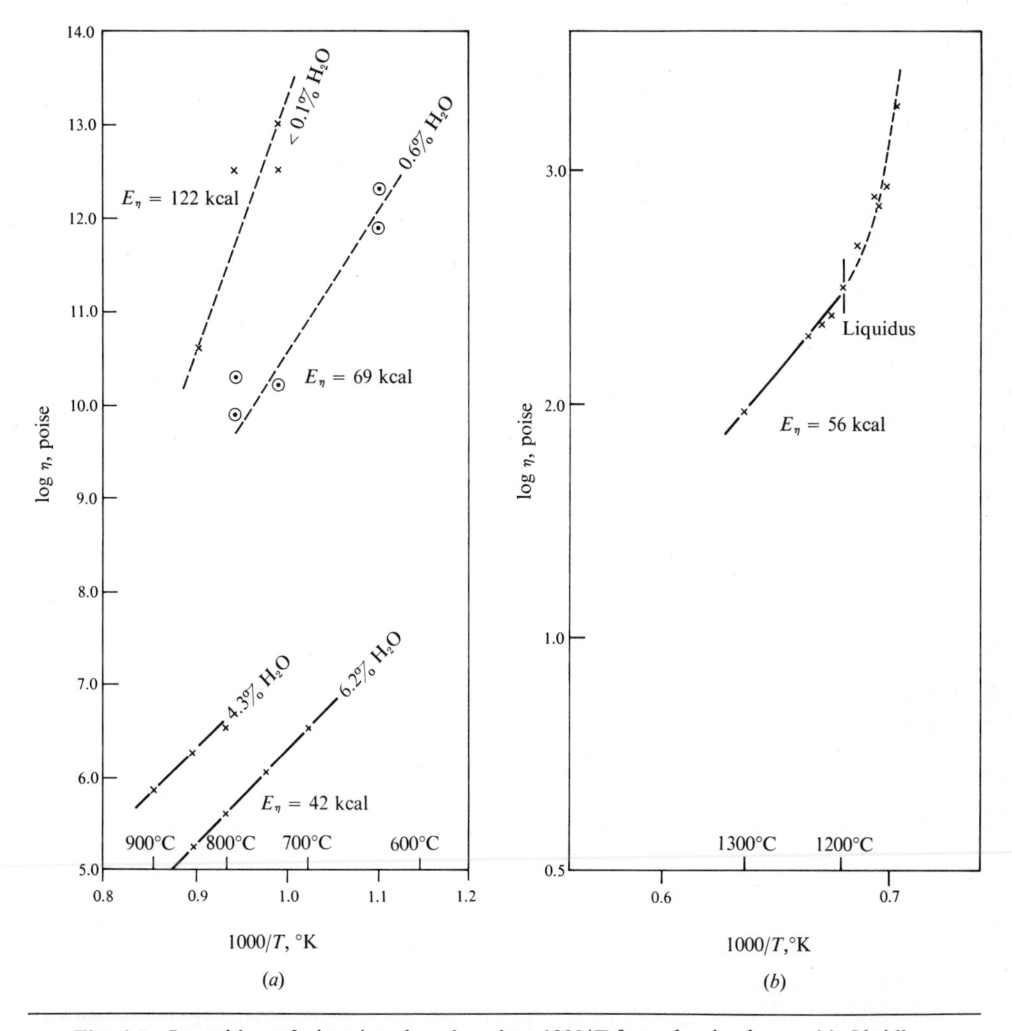


**Fig. 4-5** Logarithm of viscosity plotted against $1000/T$ for volcanic glasses. (*a*) Obsidians with varied amounts of water (labeled); values of $E_\eta$, the activation energy for viscous flow, are also shown (data from Friedman et al., 1963; H. R. Shaw, 1963). (*b*) Values for basaltic glass and liquid (data from H. R. Shaw, 1969).

[1] Experimental data for basaltic liquid and glass are given in Fig. 4-5*b*.

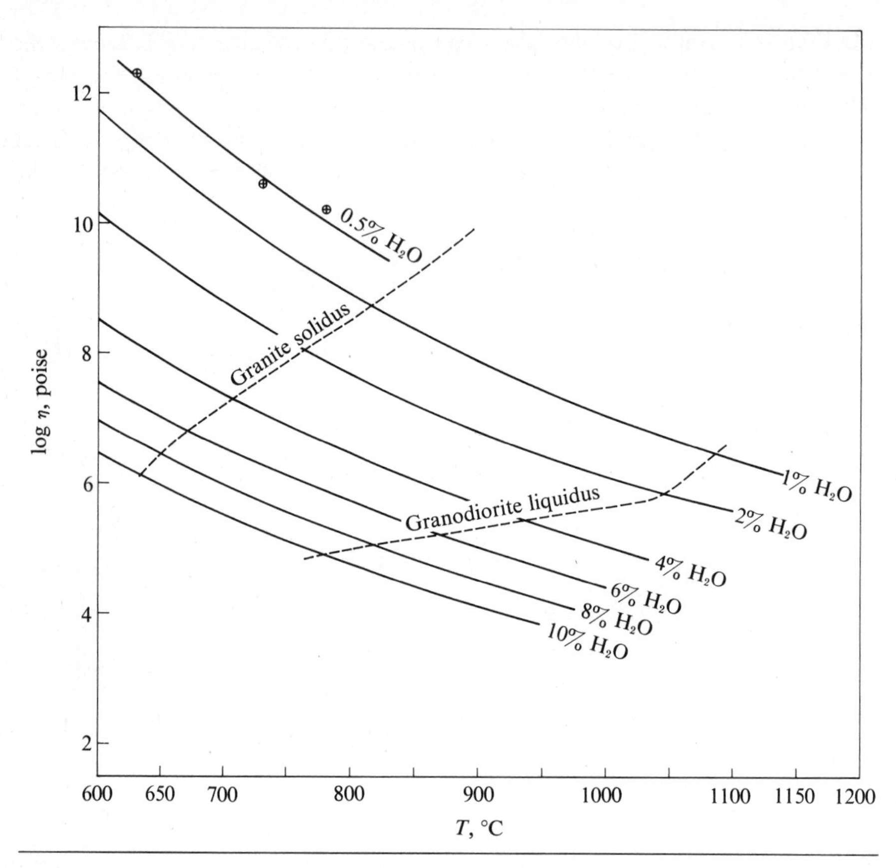


**Fig. 4-6** Generalized curves after H. R. Shaw (1965) showing the viscosity of water-saturated acid magma (obsidian) as a function of temperature. The concentration of water, in wt %, is shown. Three values for obsidian with 0.5% $H_2O$ are taken from Friedman et al. (1963). Burnham (1967) gives a similar plot; and the position of the granite solidus and granodiorite liquidus are taken from his paper.

using viscosity data on synthetic silicate liquids, Bottinga and Weill (1972) were able to calculate the contributions (positive or negative) made by suitably chosen components to the viscosity of a liquid. The oxide components they used included the following: $KAlO_2$, $NaAlO_2$, $CaAl_2O_4$, $SiO_2$, $TiO_2$, FeO, MgO, CaO, $Na_2O$, and $K_2O$. By expressing the chemical composition of any lava in terms of these components, the viscosity contribution for each component can be summed and the viscosity of the liquid lava calculated. H. R. Shaw (1972), by taking a slightly different approach based essentially on the contribution of individual components to $E_\eta$, was able to reduce to four the number of compositional coefficients required to calculate the viscosity of a silicate liquid. He also calculated an expression for the contribution of dissolved $H_2O$. The calculated (anhydrous) viscosities at 50° intervals for six contrasted lava types are shown in Fig. 4-7, and in each case the calculated curves are extrapolated to a plausible liquidus temperature. Clearly at constant temperature viscosity increases with $SiO_2$, and the slope $E_\eta$ of the viscosity curves decreases as the viscosity falls. A ugandite lava is among the most silica-poor of all natural silicate liquids and accordingly has the lowest calculated viscosity of any terrestrial lava known to us.

Macedo and Litovitz (1965) developed an equation for the temperature dependence of viscosity which includes an expression for the free volume of the

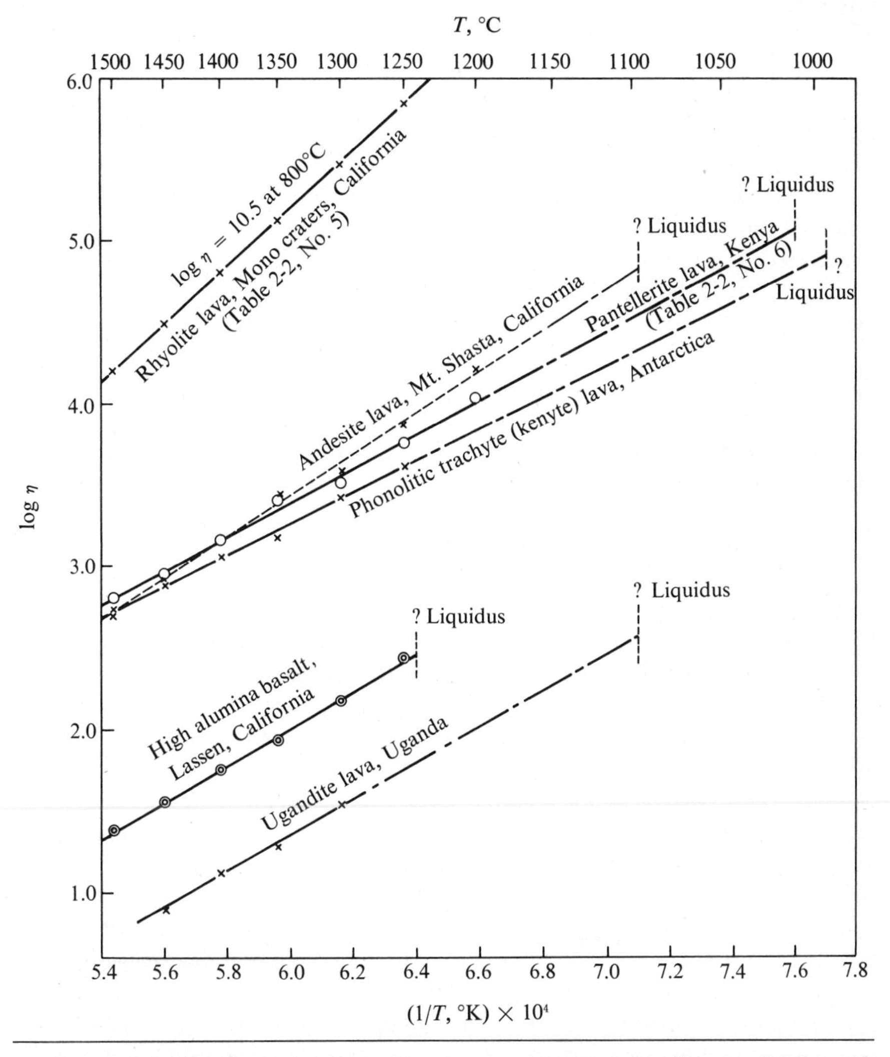


**Fig. 4-7** Logarithms of calculated viscosities of various lava types plotted against $(1/T) \times 10^4$. Possible liquidus temperatures for each lava are also shown. Calculations are based on the data of Bottinga and Weill (1972).

liquid. They postulated that a molecule must attain sufficient energy to jump into a vacant site and that a vacant site is available into which the particle can jump. This second consideration is part of the free-volume approach to the structure of liquids (Eyring and Ree, 1961), since the decrease in density of a solid on melting can be construed as increasing the free volume. We should expect that the free volume will be squeezed out of the liquid as the liquid is compressed, and for many liquids the compressibility shows a change in slope once this (~3 to 5 percent) has been achieved. The equation of Macedo and Litovitz is

$$\eta = A\, e^{E_\eta/RT + \gamma V_o/V_f} \tag{4-32}$$

where $V_o$ approximates the close-packed molecular volume; $V_f = V - V_o$, where $V$ can be related to the increase in volume due to thermal expansion of the liquid and $\gamma$ is a factor to allow for the overlap of free volume. The application of this equation to the non-Arrhenius behavior of the viscosity of $B_2O_3$ established its validity. The viscosity of liquids will show Arrhenius behavior [cf. Eq. (4-8)] if the liquid or glass has a small thermal expansion or, in other words, if the free volume is independent of temperature (for example, $SiO_2$ glass); alternatively if $V_f$ is proportional to temperature, then $\log \eta$ will also show a straight-line relationship with $1/T$.

The effect of pressure on the viscosity of liquids has been treated by Andrade (1934), and the Macedo-Litovitz equation may be modified to include the effect of pressure. An equation of the type

$$\eta = A\, e^{(E_\eta^* + PV_\eta^*)/RT} \tag{4-33}$$

has been suggested, where $V_\eta^*$ is the activation volume for viscous flow. However, $V_\eta^*$ is difficult to determine and is probably related to the free volume $V_f$. Experimentally the pressure dependence of viscosity of hydrous rhyolitic liquids has been within the limits of measurement up to 7.5 kb (Burnham, 1967; Shaw, 1963), and similar results have been obtained for hydrous basaltic liquids (Scarfe, 1973).

The contributions of Lacy (1965, 1967, 1968), one of the few geologists to become concerned with the structures of silicate glasses and liquids, represent a valuable statistical approach to viscosity. Lacy has calculated the probability of finding discrete ion groups for various values of $\mathscr{R}$ (the ratio of oxygen to network-forming cations), and he has shown also that the logarithm of the fluidity ($1/\eta$) is related to the probability of finding a nonbridging oxygen in any randomly chosen $SiO_4$ group. From this he models the flow mechanism by gliding, or juxtaposition, of the nonbridging oxygens. The temperature dependence of the fluidity is attributed to an increase in the coordination of each nonbridging oxygen ($O^-$), reaching a limiting value at a temperature where each $O^-$ touches three oxygens of an external $SiO_4$ group.

Although silicate liquids above their liquidus temperatures behave as Newtonian fluids, magmas charged with crystals or gas bubbles do depart significantly from the Newtonian model. In some fluids, flow is initiated only after a critical shear stress is exceeded (yield value), but thereafter the shear stress has a constant ratio to the strain rate—these fluids are called Bingham bodies.

The flow properties of Hawaiian tholeiitic basalt magma have been studied by H. R. Shaw et al. (1968) and H. R. Shaw (1969). Above the liquidus temperature, the magma behaves as a Newtonian fluid; the temperature dependence of viscosity is shown in Fig. 4-5*b*. Note that the activation energy $E_\eta$ is less than for rhyolitic obsidian with small amounts of water but greater than for obsidian with large amounts of water. Below the liquidus temperature, a mixture of nonvesicular basaltic liquid and up to 20 percent of crystals by volume still behaves as a Newtonian liquid (within the low shear rates of the experiments), but either with gas bubbles (less than 10 percent) or with an increased crystal content, the viscosity of the mixture increases along the steeply ascending curve of Fig. 4-5 rather as for a Bingham body. Measurements of viscosity with more than 30 percent of crystals were not possible. Below the magma's liquidus temperature, the measured viscosities are *apparent* viscosities, or the ratio of *total* shear stress to strain rate; an apparent viscosity is identical with the Newtonian viscosity for a Newtonian liquid, but for a magma system behaving like a Bingham body, the values of the apparent viscosity depend on the shear rate, or rate of flow (H. R. Shaw et al., 1968).

## Rates of Crystallization and Nucleation

It has been over 60 years since Harker (1909) discussed the experimental evidence for "the power of spontaneous crystallization," or crystal nucleation, and used the results of Tamman and Doelter to illustrate his discussion. In the years since, there have been few experiments on crystal nucleation in liquids of geological interest or indeed on the rate of crystal growth. Winkler's paper (1947) is a landmark in this respect; however, we can with some liberty use results on simple silicate systems to illustrate the type of data that are needed and their implications for magmas. First we shall look at some relationships and results concerning crystal growth, since, superficially at least, they seem more generally related to common petrographic observation.

### THE SOLID-LIQUID INTERFACE

The surfaces of a perfect crystal can be divided into closely packed and stepped planes. In the primitive cubic system, only {100} faces are closely packed if only the nearest neighbor's interactions are considered, and all other faces will be made up of {100} steps. In the face-centered cubic system, only the {111} and {100} planes are closely packed if only the nearest neighbor's bonds are considered, and all other faces will be made up of {111} or {100} steps. The step is typically only one atom thick.

At and above room temperature, both the steps and the planes have some disorder; thus the steps may be irregular in plan, with indents or projections known as Frenkel kinks. These kinks increase in number with temperature, and it is at the crotch of a kink that atoms are added in crystal growth, because of the greater number of unsatisfied bonds.

The solid-liquid interface need not be sharp on an atomic scale but rather a diffuse region with some of the order of the solid; the surface of the solid is also likely to be greatly disordered.

There are three ways in which a crystal boundary can migrate normal to itself. (1) If atoms can cross the boundary region and attach themselves simultaneously and independently at all points on the crystal interface. This process of growth is continuous, and a diffuse interface on which atoms can attach themselves at any site has a high degree of "roughness." (2) If the interface is stepped on an atomic scale, and atoms can be attached only at these steps. The interface therefore grows laterally, each surface being covered by the lateral advance of a step, so that the interface moves forward by a distance equal to the thickness of the step. This kind of growth surface is considered "smooth," since atoms can be added only at specific sites. (3) If a dislocation appears on a crystal face, a step is produced and the crystal grows upward in a spiral generated about the point where the dislocation emerges. Screw-dislocation growth of this kind is self-perpetuating and will prevent two-dimensional, or lateral step, growth.

K. A. Jackson (1967) has used the concept of interface "roughness" to predict morphology of crystals growing from liquids of their own composition. He identifies the entropy change as the factor controlling whether or not a crystal face will grow; the entropy change for an atom in a liquid transferring itself to a "rough" surface will be small, whereas the entropy change for an atom attaching itself to a smooth surface will be large. A quantitative relationship thus derived by Jackson shows that for values of $\Delta S_m < 2R$ the crystal interface will grow isotropically, or at the same rate in all directions, whereas for large values of $\Delta S_m$ the crystals will be faceted and the growth anisotropic. Experiments have shown that the morphologies are not greatly dependent on the growth rate or on the undercooling, which is the difference between the growth temperature and the equilibrium temperature.

These results on simple one-component systems are not directly applicable to multicomponent liquids that differ in composition from the growing crystal. Here the diffusion of atoms in the liquid may be the rate-controlling process. However, most petrographers would acknowledge that sphene, ilmenite, and phlogopite are usually euhedral in igneous rocks, but cristobalite rarely is. To an experimental petrologist who has crystallized liquids whose compositions are close to rhyolite, the absence of facets on the quartz grown in the quartz stability field is commonplace. These admittedly crude generalizations are in broad agreement with the entropies of fusion shown in Table 4-3.

## GROWTH OF CRYSTALS IN A LIQUID

In a one-component system such as a metal or perhaps $SiO_2$, the rate of crystal growth is limited by the rate at which atoms or particles can attach themselves to the crystal interface or by the rate at which the latent heat of crystallization can be removed from the region of crystal growth. This rate of heat removal may be the limiting factor. However, if the composition of the growing crystal and the liquid are different, as in most natural silicate liquids, the crystal will grow only if the appropriate atoms arrive at the interface. In this case the growth rate is controlled by the rate of diffusion of the appropriate atoms in the liquid. This type of crystal growth we shall consider later.

For crystal growth limited by removal of heat from the interface, two possibilities exist: the heat may be removed either through the crystal or through the liquid. If the heat is removed through the crystal,[1] then the liquid will be hotter than the crystal interface. Any growth projection of the crystal into this hotter liquid will cause it to melt, and so the projection will cease to grow. Planar or slightly curved interfaces are therefore stable.

If heat is lost through the liquid, then at some distance from the interface this is cooler and so provides a favorable site for enlargement of small projections of the interface. A gently curved or planar face is therefore unstable, and dendritic growth is characteristic. However, the common occurrence of dendritic crystals in the residual glass of basaltic lavas is, as we shall see later, more likely to result from diffusion-controlled growth.

The rate at which a crystal grows from a liquid of its own composition has been treated by K. A. Jackson (1967). The simplest case is continuous growth, namely, attachment of the atoms at any surface site, giving a rough surface. Assuming that each atom at the crystal-liquid interface moves independently, then the individual atomic processes of fusion and crystallization are simple activated processes.[2] The net rate for growth that can take place at any surface site is given by

$$\varepsilon = \varepsilon_A - \varepsilon_L \tag{4-34}$$

$$= A_{\varepsilon_A} e^{-q_c/kT} - A_{\varepsilon_L} e^{-q_m/kT} \tag{4-35}$$

where $\varepsilon_A$ and $\varepsilon_L$ are the rates at which atoms respectively arrive at and leave the crystal, and $q_c$ and $q_m$ are the corresponding activation energies. Then we have

$$q_m - q_c = \Delta h_m \tag{4-36}$$

[1] An example of this could be the border group of the Skaergaard intrusion, where crystals appear to have grown perpendicular to the intrusive contact.

[2] Special symbols are used in this section. Energy quantities $g$, $h$, $s$ (per atom) correspond to $G$, $H$, $S$; $m$ denotes fusion, $c$ crystallization, and $s$ surface.

the latent heat of fusion per atom. For equilibrium

$$\varepsilon_A = \varepsilon_L \tag{4-37}$$

or

$$\frac{A_{\varepsilon A}}{A_{\varepsilon L}} = e^{-\Delta h_m/kT_E} \tag{4-38}$$

where $T_E$ is the equilibrium temperature. By substitution of Eqs. (4-36) and (4-38), Eq. (4-35) gives

$$\varepsilon = A_{\varepsilon_A} e^{-q_c/kT}(1 - e^{\Delta h_m/kT_E - \Delta h_m/kT}) \tag{4-39}$$

If $\Delta T = T_E - T$, then

$$\varepsilon = A_{\varepsilon_A} e^{-q_c/kT}(1 - e^{-\Delta h_m \Delta T/kT_E T}) \tag{4-40}$$

Equation (4-40) can be put on a molar basis by multiplying the exponential terms by $N_o$, Avogadro's number; and, recalling that the entropy of fusion is

$$\Delta S_m = \frac{\Delta H_m}{T_m} \tag{4-41}$$

where $\Delta H_m$ is the molar heat of fusion, we obtain[1]

$$\varepsilon = A_{\varepsilon_A} e^{-Q_c/RT}(1 - e^{-\Delta S_m \Delta T/RT}) \tag{4-42}$$

Since the atom movements at the crystal interface are similar to those elsewhere in the liquid, $Q_c$ should be approximately the activation energy for diffusion $E_D$ in the liquid [Eq. (4-23)]. Thus the first term on the right-hand side, $A_{\varepsilon_A} e^{-Q_c/RT}$, can be related to a diffusion coefficient $D'$.

$$D' \approx A_{\varepsilon_A} e^{-Q_c/RT} \tag{4-43}$$

Experiments on many silicate liquid systems have shown that $D'$ can be approximately represented by the Stokes-Einstein hydrodynamic equation

$$D' = \frac{mRT}{\eta} \tag{4-44}$$

where $m$ is a complicated factor which depends on particle size and jump

[1] $Q_c$ is the *molar* equivalent of $q_c$, Eq. (4-40).

distance. On the basis of Eq. (4-43), we may substitute Eq. (4-44) into Eq. (4-42) to obtain

$$\varepsilon = \frac{m'RT}{\eta}(1 - e^{-\Delta S_m \Delta T/RT}) \tag{4-45}$$

where $m'$ now includes a factor for the fraction of sites at the crystal interface where molecules can be preferentially added.

If $\Delta S_m \Delta T$ is small, Eq. (4-45) can be reduced to

$$\varepsilon \approx \frac{m'}{\eta} \Delta S_m \Delta T \tag{4-46}$$

or expressed in another way, the product of the crystal growth rate and the viscosity is proportional to the undercooling $\Delta T$ and the entropy of fusion.

If $Q_c$ in Eq. (4-42) is identified as the activation energy for crystal growth $E_\varepsilon$, then on the assumption that $\Delta S_m \Delta T$ is small, Eq. (4-42) can be rewritten as

$$\varepsilon = A_\varepsilon e^{-E_\varepsilon/RT} \Delta S_m \Delta T/RT \tag{4-47}$$

The crystal growth rate for surface nucleation therefore varies with the entropy of fusion and with $\Delta T$, the undercooling of the liquid, which is often called the motivating potential. This is resisted by the rate at which atoms or molecules move through the liquid to the crystal interface; the growth rate is thus inversely proportional to the viscosity $\eta$.

K. A. Jackson (1967) has also developed an equation for the growth rate for a crystal growing by spiral steps, or by screw dislocation. This is proportional to $\Delta S_m$ and to $(\Delta T)^2$.

Much of what has been stated above is idealized, with a fair number of simplifying assumptions introduced to show the types of parameters which influence the growth rate of crystals. For the reader anxious to delve more deeply into this complex topic, we recommend K. A. Jackson (1967) and Christian (1965) as starting points.

GROWTH AND MELTING OF $Na_2Si_2O_5$ AND $SiO_2$ CRYSTALS As examples of the rates of growth and melting, we have used the data of Meiling and Uhlmann (1967) and Wagstaff (1968, 1969) on the rates for nucleated, or seeded, liquids having the composition $Na_2Si_2O_5$ and $SiO_2$,† respectively.

† Winkler's (1947) data on the growth and nucleation of nepheline are perhaps closer to a natural crystal growth system, but they lack the related and supporting data available for $SiO_2$ and $Na_2Si_2O_5$.

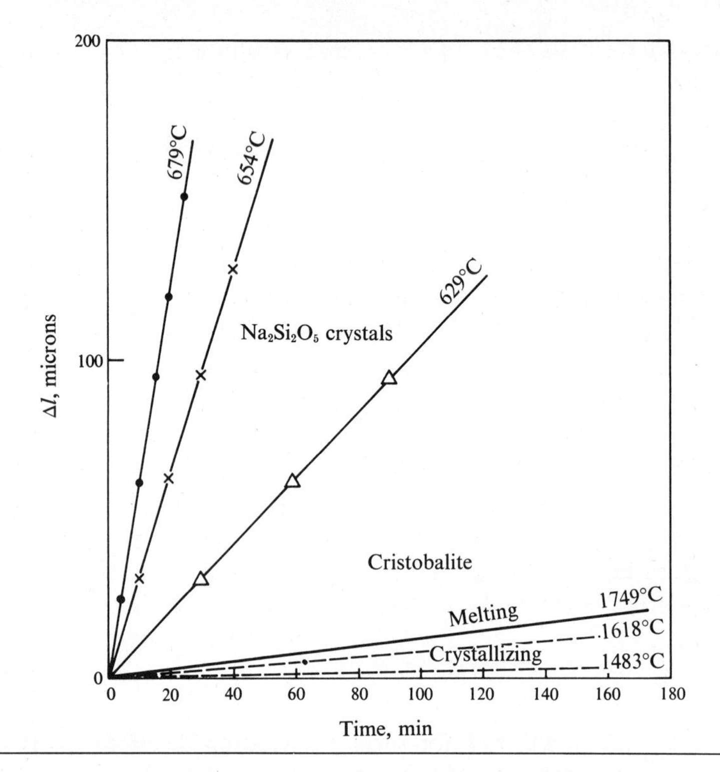


**Fig. 4-8** Rates of growth of crystals of $Na_2Si_2O_5$ at 629° to 679°C and of cristobalite at 1483° and 1618°C. $\Delta l$ is cumulative change in length. Melting of cristobalite at 1749°C is also shown. (Data from Meiling and Uhlmann, 1967; Wagstaff, 1968, 1969.)

The rate of growth of both $Na_2Si_2O_5$ and $SiO_2$ crystals is linear with time (Fig. 4-8); crystals grow faster at higher than at lower temperatures; melting is also linear for cristobalite. From the slow rate of growth of cristobalite (at high temperature) compared with that of $Na_2Si_2O_5$ (at lower temperatures) it is obvious that both viscosity, $\eta^{SiO_2}$ and $\eta^{Na_2Si_2O_5}$, and entropy of fusion, $\Delta S_m^{SiO_2}$ and $\Delta S_m^{Na_2Si_2O_5}$, are influential.

The motivating potential for crystal growth has been considered as the difference $\Delta T$ between the crystal growth temperature $T$ and the equilibrium or melting temperature $T_m$. It can be seen in Fig. 4-9 that the maximum growth rate of both $Na_2Si_2O_5$ and cristobalite occurs just a few degrees below $T_m$, and thereafter falls as $\Delta T$ increases. This type of growth-rate curve has long been familiar in the geological literature (for example, Harker, 1909), but what is perhaps less well known is that the rate of crystal growth for a certain under-cooling $\Delta T^-$ is not matched by the rate of melting for the same amount of overheating $\Delta T^+$. Both $Na_2Si_2O_5$ and cristobalite melt at faster rates for a given $\Delta T^+$ than those at which they crystallize for the same value of $\Delta T^-$ (Fig. 4-10). This may be an example of $\Delta S^{\circ*}$, the entropy of activation, having the same sign as the entropy of melting (entropy or randomness increase in melting), so that the reaction runs faster toward the high-entropy side.

For materials having entropies of fusion less than $2R$, it was already suggested above that crystals would grow by a continuous mechanism and would not be faceted; such is the case with cristobalite. For this type of crystal growth, Eq. (4-46) shows that $\varepsilon\eta/\Delta T$ plotted against $\Delta T$ should give a horizontal line, and within the limits of error this is what Wagstaff found.

Where $\Delta S_m > 2R$, the crystals are likely to be faceted, with growth limited by nucleation; for such a growth regime $\varepsilon\eta/\Delta T$ plotted against $\Delta T$ is a line of positive slope. In Fig. 4-11, the data for $Na_2Si_2O_5$ show just this, and, as expected, the plot of $\varepsilon\eta/(\Delta T)^2$ against $\Delta T$ is horizontal. This illustrates that the growth mechanism of $Na_2Si_2O_5$ crystals is not the same as that of cristobalite, and the observed anisotropic growth of $Na_2Si_2O_5$ is characteristic of a substance with a large $\Delta S_m$ (=7.4 eu). However, the melting curve of $Na_2Si_2O_5$ has a slope different from the crystallization curve (Fig. 4-11) and suggests that, in melting either atoms are removed from sites different from those which are used in growing or $\Delta S^{\circ *}$ is influencing the process.

From Eq. (4-47) it can be seen that a plot of log $\varepsilon/\Delta T$ against $1/T$ would give a slope corresponding to an activation energy $E_\varepsilon$ for crystal growth. The

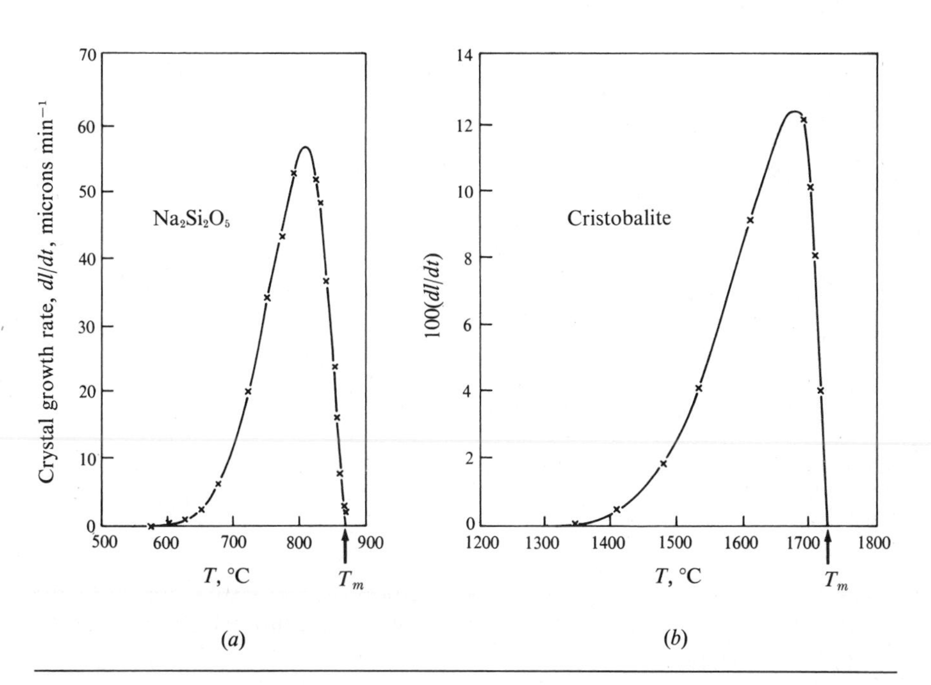


**Fig. 4-9** Growth rate $dl/dt$, where $l$ is length in microns and $t$ is in minutes plotted against temperature $T$; melting temperature is $T_m$. (*a*) $Na_2Si_2O_5$. (*b*) Cristobalite; note the difference in rate scale. (Data from same sources as Fig. 4-8.)

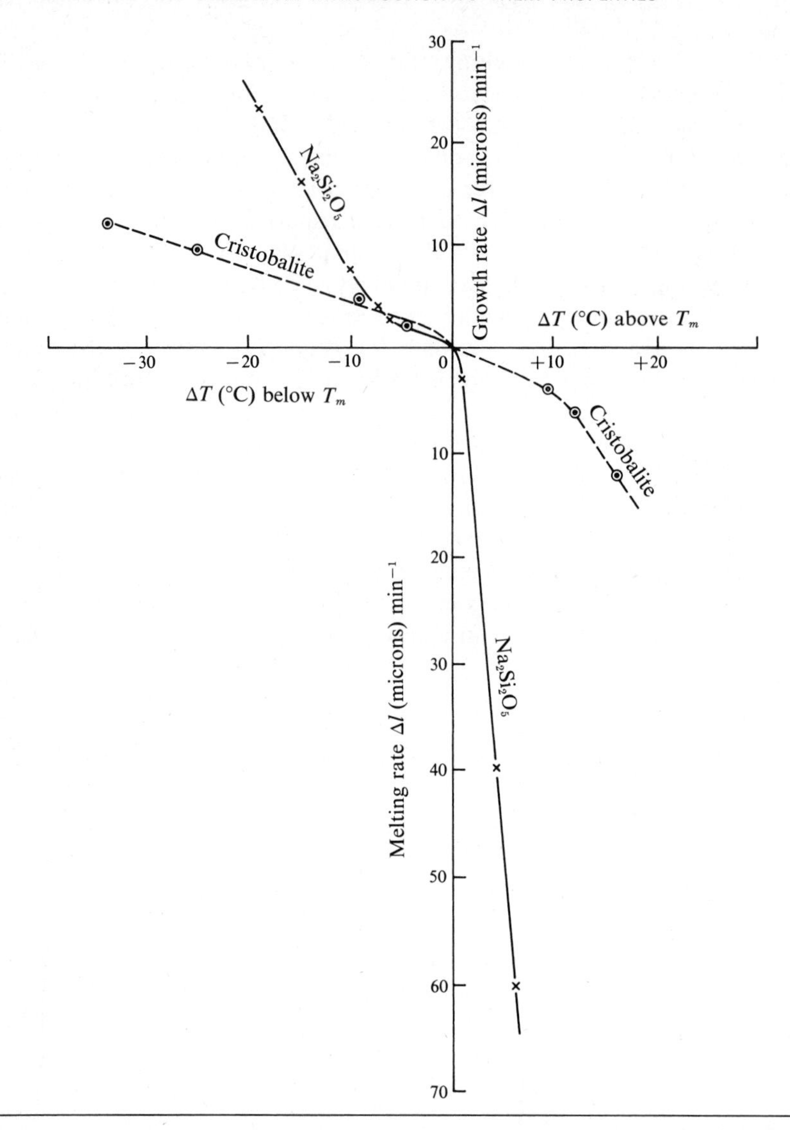


**Fig. 4-10** Growth and melting rates (in microns per minute) for crystals of $Na_2Si_2O_5$ and cristobalite plotted against $\Delta T = T - T_m$. (Data from Meiling and Uhlmann, 1967; Wagstaff, 1968, 1969.)

data points for both melting and crystallization of cristobalite, shown in Fig. 4-12, can be well represented by the line which passes through them. However, this line is the equation for

$$\eta = A\, e^{-E_\eta/RT}$$

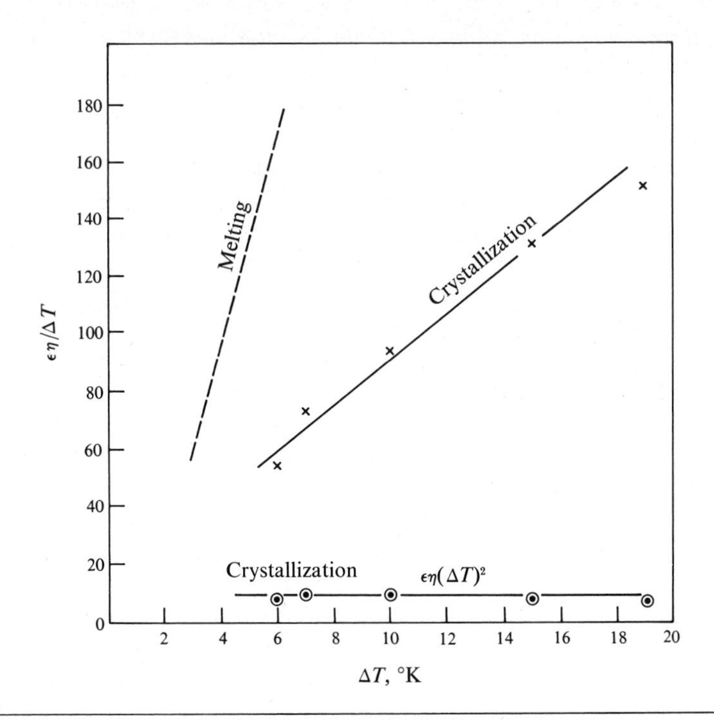


**Fig. 4-11** Product of growth rate $\varepsilon$ and viscosity $\eta$ (in poises) for $Na_2Si_2O_5$ divided by $\Delta T$, plotted against $\Delta T$, the undercooling below the equilibrium temperature $T_m$. Note that the curve for $\varepsilon\eta/(\Delta T)^2$ is horizontal. Melting curve ($\Delta T$ opposite in sign) is shown dashed. (Data from Meiling and Uhlmann, 1967.)

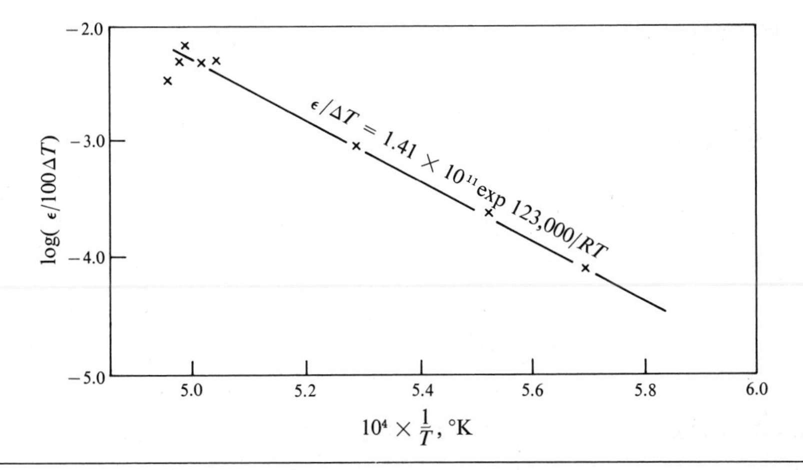


**Fig. 4-12** Plot of log ($\varepsilon/100\Delta T$) against $10^4 \times 1/T$ for melting of cristobalite; $\varepsilon$ is in microns $\text{min}^{-1}$, $\Delta T$ is undercooling below $T_m$. Line has the slope of $E_\eta$, the activation energy of viscous flow. (Data from Wagstaff, 1968, 1969.)

the viscosity of silica glass, with a suitable correction made to $A$. It seems clear that the activation energy for shear viscosity $E_\eta$ in $SiO_2$ liquids is identical to the activation energy for crystal growth, namely, 123 kcal mole $^{-1}$.

Up to this point we have considered only the growth rate of crystals growing from a liquid of their own composition. In magmas the composition of crystals is invariably different from the liquid in which they grow. Growth is therefore controlled largely by diffusion of substance. This complex subject (Christian, 1965), is now considered rather briefly.

DIFFUSION-CONTROLLED GROWTH The rate at which a crystal interface advances into a liquid of *constant* composition is independent of interface position and hence of time. This situation does not apply when the crystal has a composition different from that of the liquid. If, for example, the crystal is richer in $B$ than is the liquid, then the liquid will be depleted in $B$ in a region around the interface. Continued growth requires diffusion of substance, and as the crystal increases in size, the effective distance over which diffusion takes place will also increase. When the interface formed, it is likely that processes near the interface controlled the rate of growth, but eventually volume diffusion will become the limiting factor if the diffusing particles are readily attached to the crystal interface. The crystal will grow as fast as the diffusion rate will allow it.

Let us assume that we have spherical particles of a solid in equilibrium with a liquid and that the solid is enriched in $B$ atoms compared to the liquid. The concentration of $B$ atoms in the liquid will be less than in the solid; in other words $c_B^{\text{liq}} < c_B^{\text{sol}}$. Where $r$ is the radius of the particles, we have [Eq. (4-22)], for the diffusion of a number $n_B$ of $B$ atoms across unit area in a time $\delta t$, given a diffusion coefficient $D_B$,

$$D_B \left(\frac{\partial c}{\partial r}\right)_r \delta t = n_B \tag{4-48}$$

so that the concentration of $B$ atoms in the solid $c_B^{\text{sol}}$ is a function of $r$ and $t$. If the interface advances a distance $\delta r$ in time $\delta t$,

$$(c_B^{\text{sol}} - c_B^{\text{liq}})\,\delta r = n_B \tag{4-49}$$

These $B$ atoms can come only from diffusion, so that Eq. (4-48) can be equated with Eq. (4-49) to give the growth rate $\delta r/\delta t = \varepsilon$:

$$\varepsilon = \frac{\delta r}{\delta t} = \frac{D_B}{c_B^{\text{sol}} - c_B^{\text{liq}}} \left(\frac{\partial c}{\partial r}\right)_r \tag{4-50}$$

For many purposes $(\partial c/\partial r)_{r=r'}$ can be approximated by $\Delta c/y_B$, where $\Delta c_B$ is the concentration of $B$ in the liquid at the interface minus the concentration of $B$ in the liquid remote from the crystal interface; $y_D$ is the effective diffusion distance. Equation (4-50) can therefore be rewritten as

$$\varepsilon \approx \frac{D_B}{c_B^{\text{sol}} - c_B^{\text{liq}}} \frac{\Delta c_B}{y_D} \tag{4-51}$$

The effective diffusion distance $y_D$ must continually increase as the spherical crystal grows and so the growth rate decreases.

The occurrence of dendritic crystals in nature may be a good example of diffusion-controlled growth. If it is assumed that the growth rate of a cube-form crystal is determined by diffusion of atoms in the liquid, then the eight corners can draw atoms from a much larger volume of liquid than any other position on the cube. The centers of the edges are less favored but can draw from a larger volume than the centers of the faces, so that the corners will grow most rapidly. Crystals of hopper form or dendritic shape will be produced. The axes of dendritic growth are usually symmetrical to the most closely packed planes, and in the face-centered and body-centered cubic classes the dendritic growth axis is the zone axis [100].

## NUCLEATION

The fundamental reason for the appearance of nuclei of a new phase in a homogeneous phase is the existence of fluctuations, that is, local deviations from the normal state. If the initial phase is stable, these transient deviations disappear spontaneously in a very short time; but if the phase is unstable, they may start to grow, forming embryos of the new phase. If the initial phase is metastable, only fluctuations of finite magnitude will grow. These large fluctuations are very infrequent, their frequency decreasing exponentially as their magnitude increases.

Nucleation that occurs completely at random throughout a system is called *homogeneous* nucleation. This requires that the parent phase, solid, liquid, or gas, is chemically, structurally, and energetically identical throughout. In practice, however, crystal imperfections, impurities, or the rough walls of the containing vessel create sites where nucleation occurs; nucleation at such energetically preferred sites is called *heterogeneous* nucleation.

The formation of nuclei in a liquid can be represented by the reaction

$$\text{Liquid} \rightarrow \text{solid} + \text{solid interface}$$

where the energy associated with the interface is small or trivial for large crystals but large for the formation of nuclei. This energy, associated with the formation of the interface, is related to surface tension.

SURFACE TENSION The cohesive forces in a liquid, related to the internal pressure [Eq. (4-16)], are familiarly illustrated by the tendency of water to form droplets. It is these forces that resist the tendency of the liquid's surface to extend. If the surface tension is $\sigma$, then the work $dw$ done on the surface in extending its area by $dA_s$ is

$$dw = \sigma \, dA_s \tag{4-52}$$

The surface tension also contributes to the free energy of the drop, since $\sigma$ is also defined as

$$\sigma = \left(\frac{\partial G}{\partial A_s}\right)_{T,P,n_i} \tag{4-53}$$

From this, it follows that the total derivative of the free energy accompanying the formation of a drop is given by

$$dG = V\,dP - S\,dT + \sigma\,dA + \Sigma\mu_i\,dn_i \tag{4-54}$$

so that at constant $P$, $T$, and number of particles $n_i$ in the drop ($dn_i = 0$)

$$dG = \sigma\,dA_s \tag{4-53$a$}$$

which can be integrated to give

$$\sigma = \frac{G}{A_s}$$

if $\sigma$ is independent of the area $A_s$.

The use of the term *surface tension* is clear when applied to liquids, since it represents the tendency of the liquid to achieve a minimum surface area. For solids, the term *surface free energy* is used instead since, unlike liquids, their surface area cannot be changed reversibly. The units of surface tension are dyne $cm^{-1}$, whereas the surface free energy of a solid is reported in units of erg $cm^{-2}$ (1 erg = 1 dyne cm).

Reiss and Mayer (1961) have had some success in calculating $\sigma$ from an equation of state assuming a hard-sphere model of the liquid; this has yet to be applied to liquids approaching in composition those of natural silicate liquids. The variation of surface tension (air interface) with temperature for basalt, andesite, and rhyolite liquids (above their liquidus temperatures) has been measured by McBirney and Murase (1971), and the values for these rock types are shown in Fig. 4-13.

The pressure $P_{int}$ inside a drop of liquid is higher than the external pressure $P_{ext}$, as can be shown by the following argument. If a small volume $dV$ of liquid

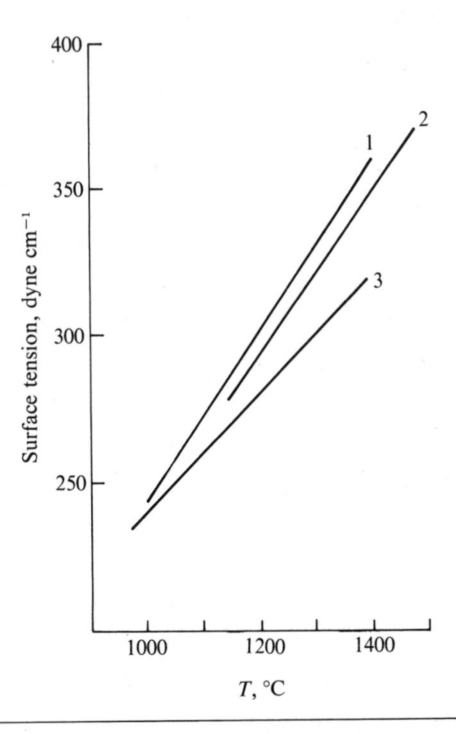


**Fig. 4-13** Variation of surface tension with temperature for various rocks and glasses above their liquidus temperatures in air. 1, Columbia River basalt; 2, Mount Hood andesite; 3, Newberry caldera rhyolite obsidian. (Data from McBirney and Murase, 1971.)

is transferred from the bulk state of liquid to the inside of a drop, the work required is

$$(P_{\text{int}} - P_{\text{ext}})\, dV$$

which must be equal to the work required to extend the surface of the drop [Eq. (4-52)].

We have therefore

$$(P_{\text{int}} - P_{\text{ext}})\, dV = \sigma\, dA_s$$

The volume of a sphere is given by

$$V = \tfrac{4}{3}\pi r^3$$

and

$$dV = 4\pi r^2\, dr$$

and the surface area of a sphere is

$$A_s = 4\pi r^2$$

and

$$dA_s = 8\pi r\, dr$$

so that by substitution we obtain

$$(P_{\text{int}} - P_{\text{ext}})4\pi r^2\, dr = \sigma 8\pi r\, dr$$

or

$$P_{\text{int}} = \frac{2\sigma}{r} + P_{\text{ext}} \tag{4-55}$$

At 1200°C and 1 bar the surface tension of basalt is 350 dynes $\text{cm}^{-1}$ (Fig. 4-13), so that the pressure inside a liquid drop of basalt of radius $10^{-3}$ cm (cf. Pelée's hair[1]) is given by

$$P_{\text{int}} = \frac{2 \times 350}{10^{-3} \times 10^6} + 1 = 1.7 \text{ bars}$$

The surface free energy of crystals can be measured by determining the heat capacity of very large crystals and very small crystals; the difference is due to the contribution of the surface.[2] For minute crystals, the edges also make a contribution to the surface free energy. Swalin (1962, pp. 192–197) gives several examples of the use of the quasi-chemical approach to calculate the surface free energy and records satisfying agreement with measured values.

NUCLEATION OF A SOLID IN A LIQUID OF ITS OWN COMPOSITION A new phase never forms exactly at the temperature and pressure at which it is in equilibrium with the parent phase; there must be a certain amount of overstepping into the metastable or unstable region of the parent phase. The reason for this is that the affinity of the transformation at the equilibrium point is zero, the definition of equilibrium. But if the affinity is zero, so necessarily is the velocity of the transformation.

"No experimental technique has yet been devised that permits direct

[1] If a strand of Pelée's hair is considered as a cylinder, then $r_1 = 10^{-3}$ cm and $r_2 = \infty$; therefore $P_{\text{int}} = 1.35$ bars.

[2] The surface free energy of crystals can be obtained from the work required to cleave a crystal (Brace and Walsh, 1962) or by determining the heat.

observation of the formation of a nucleus. The major obstacle to progress... is that of measuring the rate of formation of nucleii when there is no means of observing them" (Burke, 1965, p. 98). The nucleation rate $\varepsilon$ is defined as the number of nucleii of the phase being produced in unit time per unit volume of the parent phase.

In a liquid of one component, consider the number of atoms in a spherical crystallite as $n$, those in the liquid as $N$, and the free energy per atom or particle in the liquid and solid as $g^{\text{liq}}$ and $g^{\text{sol}}$, respectively. The surface free energy of the crystallite is more convenient to cope with if we assume the shape of the crystallite to be spherical; if it is not, then the following equations can be modified to take account of the edge effects on the surface free energy. The free-energy change of the liquid system accompanying the generation of crystallites is given by

$$\Delta G = \frac{4\pi r^3}{3v}(g^{\text{sol}} - g^{\text{liq}}) + 4\pi r^2\sigma \tag{4-56}$$

where $r$ is the radius of the crystallite, $v$ is the volume of the particle or atom in the solid, and $\sigma$ is the surface free energy of the spherical nucleus (ignoring strain effects). If we consider the number of atoms or particles in the crystallite, rather than its radius, we obtain

$$\Delta G = n(g^{\text{sol}} - g^{\text{liq}}) + \lambda\sigma \tag{4-57}$$

where $\lambda = 4\pi r^2$, the surface area of the spherical crystallite containing $n$ atoms or particles. Recalling that the surface area of a particle is proportional to the $\frac{2}{3}$ power of its volume, we can rewrite Eq. (4-57) to give

$$\Delta G = n(g^{\text{sol}} - g^{\text{liq}}) + \phi n^{2/3}\sigma \tag{4-58}$$

where $\phi$ is a shape factor; $\phi = \lambda/n^{2/3}$ and for a perfect sphere $\phi = 4\pi r^2 n^{2/3} = (36\pi)^{1/3}v^{2/3}$.

If the liquid is supersaturated, then $g^{\text{sol}} - g^{\text{liq}}$ is negative and the first term in Eq. (4-56) will be negative; the second term in Eq. (4-56) is always positive since the surface free energy is always positive. Because these two terms are proportional to $r^3$ and $r^2$, respectively, the curve of $\Delta G$ against $r$, or $n$, will first increase and then decrease. The maximum on the curve is given by $\partial\Delta G/\partial r = 0$, which gives for the critical radius of the spherical crystallites $r_c$

$$r_c = -\frac{2\sigma v}{g^{\text{sol}} - g^{\text{liq}}} \tag{4-59}$$

A crystallite of radius $r_c$ is in metastable equilibrium with the liquid, and crystallites with $r < r_c$ tend to dissolve since an increase in size increases $\Delta G$, whereas crystallites with $r > r_c$ tend to grow since an increase in radius decreases $\Delta G$; these latter crystallites are usually called nuclei, whereas the former are embryos. If Eq. (4-59) is substituted into Eq. (4-56), we obtain the maximum value of the free-energy increase $\Delta G_c$:

$$\Delta G_c = \frac{4\pi r_c^2 \sigma}{3}$$

Thus the free-energy increase is equal to one-third of the surface free energy of the crystallite of critical size. The value of $\Delta G_c$ will be reduced if there are impurities or if there are rough walls to the container.

Using Eq. (4-56), it is possible to calculate the radius of the crystallites as a function of the free energy or of the undercooling below the equilibrium temperature $T_E - T$, represented by $\Delta T$. The calculated results for crystallites of forsterite forming in $Mg_2SiO_4$ liquid are shown in Fig. 4-14, assuming that the crystallites are spherical. The surface free energy of the solid relative to the liquid has been taken as 1000 ergs $cm^{-2}$, which is perhaps rather large.[1] It is clear from Fig. 4-14 that all embryos are unstable at the equilibrium temperature ($\Delta T = 0$) and that the critical radius of the crystallites becomes progressively smaller as $\Delta T$ increases or $(g^{sol} - g^{liq})$ becomes more negative.

Turning now to the generation rate of crystal nuclei, the free energy of an embryo crystal is given by Eq. (4-58), which is

$$\Delta G = n(g^{sol} - g^{liq}) + \phi n^{2/3} \sigma$$

From the condition that $\partial \Delta G / \partial n = 0$, we obtain the critical number $n_c$ of atoms or particles in the crystallites of critical radius $r_c$:

$$n_c = \left[ \frac{-2\phi\sigma}{3(g^{sol} - g^{liq})} \right]^3$$

Substitution of this equation into Eq. (4-58) gives

$$\Delta G_c = \frac{4\phi^3 \sigma^3}{27(g^{sol} - g^{liq})^2} \tag{4-60}$$

A rate equation relating the nucleation rate per unit volume can be obtained by making some simplified assumptions, and Turnbull and Fisher (1949) have

[1] This value, taken for illustrative purposes only, corresponds to the value of the surface energy for MgO for a solid-air interface. Measurements of alkali silicate free-energies for solid-glass interfaces are given by Matusita and Tashiro (*J. Noncrystalline Solids*, 1973).

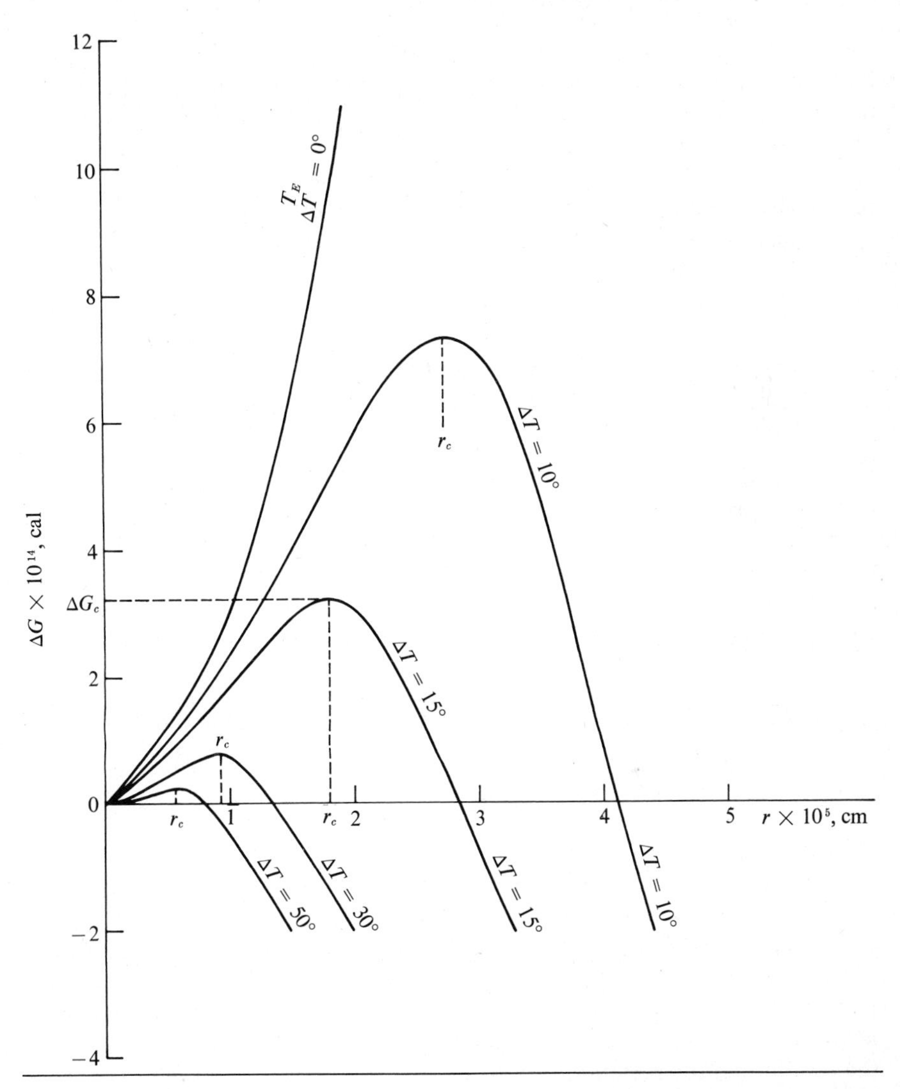


**Fig. 4-14** Free energy of formation of a spherical crystallite of forsterite in $Mg_2SiO_4$ liquid, as a function of radius $r$ for a series of temperatures $\Delta T$ below the equilibrium (melting) temperature $T_E$; $\Delta T = (T_E - T)$.

given an equation for the number of crystal nuclei $\zeta$ generated in a unit volume per second as

$$\frac{NkT}{h} e^{-(\Delta G_c - \Delta g^*)/kT} = \zeta \tag{4-61}$$

where $N$ is the number of atoms per unit volume, $\Delta G_c$ is the maximum free energy corresponding to the critical size of the crystal nuclei, and $\Delta g^*$ is the free energy per atom for transfer of an atom across the liquid-solid interface; in metal systems $\Delta g^*$ is often close to $E_\eta/N_o$, the activation energy of viscous flow.

Let the undercooling below the equilibrium temperature, $T_E - T$, be represented by $\Delta T$. At the melting temperature, $g^{\text{liq}} = g^{\text{sol}}$, the free energies (per atom) in the liquid and solid are equal, and

$$g^{\text{liq}} - g^{\text{sol}} = 0 = \Delta h_m - T_E \Delta s_m \tag{4-62}$$

where $\Delta h_m$ and $\Delta s_m$ are the enthalpy and entropy differences between an atom in the liquid and one in the solid at the equilibrium temperature. If the effect of a small change in temperature on $\Delta h_m$ and $\Delta s_m$ is neglected, we obtain from Eq. (4-62)

$$\Delta h_m = T_E \Delta s_m$$

Hence

$$g^{\text{liq}} - g^{\text{sol}} = \Delta s_m(T_E - T) = \Delta s_m \Delta T \tag{4-63}$$

and we can substitute Eqs. (4-63) and (4-60) into Eq. (4-61) and obtain

$$\zeta = N \frac{kT}{h} e^{-4\phi^3\sigma^3/27k(\Delta s_m)^2T(\Delta T)^2} e^{-(\Delta g^*/kT)} \tag{4-64}$$

Using data for $Na_2Si_2O_5$ taken from Meiling and Uhlmann (1967)—$T_E = 1147°\text{K}$; $\sigma = 100$ ergs cm$^{-2}$; $\Delta s_m = 7.411/N_o$; $V = 400 \times 10^{-24}$ cm$^3$; $\Delta g^* = E_\eta = 39{,}800$ cal mole$^{-1}$; and $\phi = (36\pi)^{1/3}V^{2/3}$—the nucleation rate $\zeta$ of $Na_2Si_2O_5$ crystals in $Na_2Si_2O_5$ liquid was calculated and plotted in Fig. 4-15. Note that the peak of the calculated nucleation curve is approximately 25° below the peak of the measured crystal growth curve (Fig. 4-9) and is 84° below the equilibrium temperature.

For nepheline, Winkler (1947) has shown that the nucleation rate curve peaks about 40° below the crystal growth curve and 60° below the equilibrium temperature; H. R. Shaw (1965) has discussed some intriguing ramifications of these. It is profitable, despite the apprehension the reader may feel at looking at Eq. (4-64), to appreciate the important parameters, namely, that the logarithm of the nucleation rate is inversely proportional to $(\Delta S_m)^2$ and to $(\Delta T)^2$, the undercooling. This must be contrasted to the crystal growth rate, which is directly proportional to $\Delta S_m$ and to either $\Delta T$ or $(\Delta T)^2$.

Before discussing the combined effects of nucleation and crystal growth rates on the development of minerals in a cooling natural silicate liquid, it is necessary to look at the nucleation rate of a solid which changes in composition

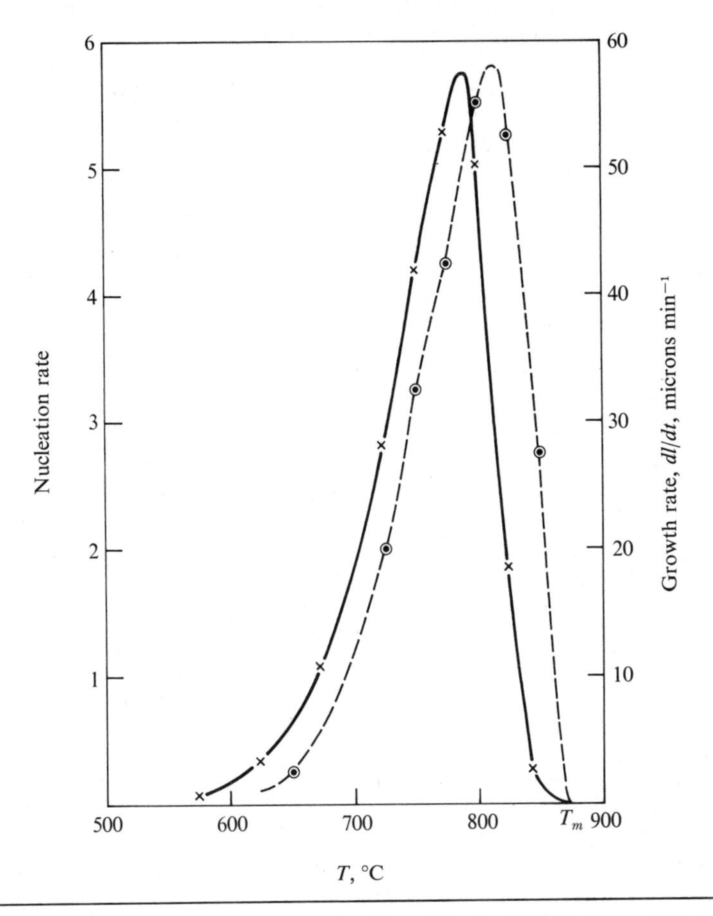


**Fig. 4-15** Calculated nucleation rate ($\zeta \times 10^{-26}$, $cm^{-3}s^{-1}$) of $Na_2Si_2O_5$ nuclei in $Na_2Si_2O_5$ liquid plotted against temperature. Dashed curve shows the experimentally measured crystal growth rate of $Na_2Si_2O_5$ (Fig. 4-9).

or one in which $g^{sol} - g^{liq}$ in Eq. (4-63) changes with the continued nucleation of a phase differing in composition from the liquid.

NUCLEATION OF A SOLID SOLUTION IN A LIQUID Consider a single solid phase (solid solution) precipitating from a liquid of unrestricted range in composition. At a temperature where a solid of a certain composition is stable in equilibrium with liquid, the free-energy composition curves have the form of the curves shown in Fig. 4-16. A liquid initially of composition $X$† has a free energy per atom represented by $Q$, whereas if this cools so that $X^{sol}$ is in equilibrium with $X^{liq}$, the free energy per atom in the system falls to $P$. The free-energy

† In the ensuing discussion $X$ denotes the molar fraction of some given component in an initial liquid or in coexisting superscript phases.

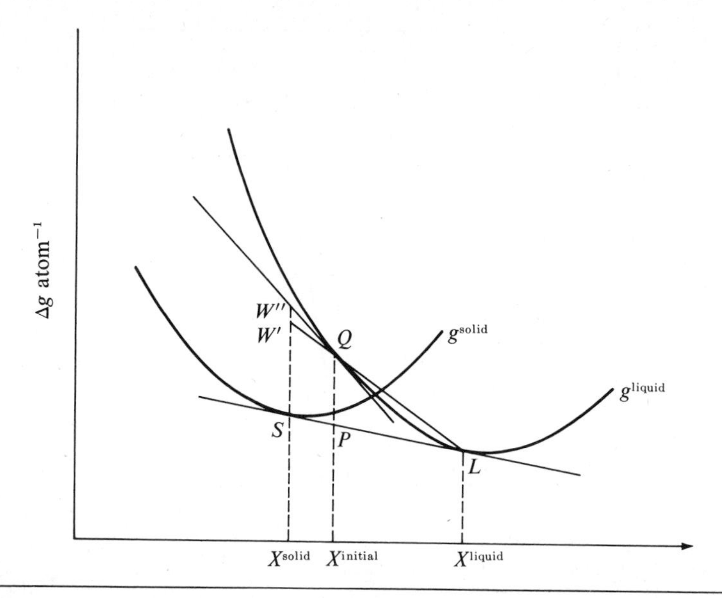


**Fig. 4-16** Variation of free energy per atom ($\Delta g$ atom$^{-1}$) as a function of composition for a liquid solution in equilibrium with a solid solution at constant temperature. $SL$ is tangent to the liquid curve at $L$ and to the solid curve at $S$. $W''Q$ is tangent to the liquid curve at $Q$. $X$ is molar fraction of some selected component, decreasing from left to right.

change accompanying the precipitation of the solid solution $X^{\text{sol}}$ is therefore $-QP$ per atom in the system. Since the two phases, liquid and solid, are in the proportions of $LP$ and $PS$, respectively, the free-energy change per atom in the solid is $-W'S$. This is the net driving force for the crystallization reaction, but it is not the driving force for the nucleation of the solid.

When a few solid nuclei have formed, the composition of the liquid is essentially unchanged; the change in free energy per atom is given by $-W''S$, where $W''Q$ is the tangent to the liquid free-energy curve at $Q$.

To obtain the rate at which nuclei of a solid solution are produced, we have to replace $g^{\text{sol}} - g^{\text{liq}}$ in Eq. (4-58) by an equivalent term which is dependent upon the initial liquid composition $X$. Thus if $\Delta g = g^{\text{sol}} - g^{\text{liq}}$ is the free-energy change per particle in the nucleus of composition $X^{\text{sol}}$ precipitating from a liquid of composition $X^{\text{liq}}$, then

$$\Delta g(X^{\text{liq}}, X^{\text{sol}}) = g^{\text{sol}}(X^{\text{sol}}) - g^{\text{liq}}(X^{\text{liq}}) - (X^{\text{sol}} - X^{\text{liq}})\left(\frac{\partial g^{\text{liq}}}{\partial X}\right)_X \tag{4-65}$$

and the solid nucleus of $n$ atoms having the composition $X^{\text{sol}}$ will have the following free-energy change [cf. Eq. (4-58)]:

$$\Delta G = n\Delta g(X^{\text{liq}}, X^{\text{sol}}) + \sigma(X^{\text{sol}}, X^{\text{liq}})n \tag{4-66}$$

where $\sigma$, the surface free energy of the solid, is also a function of composition of the solid and liquid. If it is assumed that the solid nucleus has the equilibrium composition, then the variation of $\sigma$ with $X^{sol}$ can be neglected, and we obtain from Eq. (4-60)

$$\Delta Gc = \frac{4\phi^3\sigma^3}{27\Delta g(X^{liq}, X^{sol})^2} \tag{4-67}$$

We can therefore proceed to Eq. (4-64) to obtain a nucleation rate, keeping in mind that $\Delta g$ continuously varies with both temperature and composition. To the best of our knowledge, there are no data for silicate systems that allow the free-energy–composition curves of Fig. 4-16 to be constructed. It is therefore not possible to calculate the nucleation rate of a solid solution at this stage.

## NUCLEATION AND CRYSTAL GROWTH IN A COOLING MAGMA

From the discussion of Fig. 4-14, it is important to realize that the precise liquidus temperature of a magma cannot be derived from a geothermometer involving a solid phase; a degree of undercooling is always necessary to nucleate a solid phase. Obviously both melting a crystalline assemblage and crystallizing a liquid of the same composition are necessary to define the equilibrium liquidus temperature.

It is now worthwhile to look at a geological example, noting that the crystal nucleation rate is inversely proportional to $(\Delta S_m)^2$ and $(\Delta T)^2$, whereas the crystal growth rate is proportional to $\Delta S_m$ and either $\Delta T$ or $(\Delta T)^2$, the former for continuous growth and the latter for screw-dislocation growth.

The entropies of fusion of many common rock-forming minerals are given in Table 4-3; however, the use of the pure mineral components ignores the entropy-of-mixing term arising from the mixing of these components within the magma. It is inconceivable that this mixing term could be ideal for all the components in Table 4-3, but since there are few data to show the variation of the mixing entropy of these components with a silicate liquid (magma), we shall, with some reservation, ignore it for our purposes here.

**Table 4-3** Entropy of fusion $\Delta S_m$ for some rock-forming minerals. $\Delta S_m = \Delta H_m/T_m$; values in entropy units per atom in gram-formula-weight

| | | | |
|---|---|---|---|
| Cristobalite | 0.32 | Fluorphlogopite | 2.21 |
| Albite | 0.75 | Sphene | 2.22 |
| Sanidine | 0.77 | Magnetite | 2.52 |
| Anorthite | 1.22 | Ilmenite | 2.64 |
| Diopside | 1.88 | Fluorite | 1.4 |
| Forsterite | 1.94 | Halite | 3.13 |
| Enstatite | 2.03 | Sylvite | 3.01 |
| Fayalite | 2.11 | $H_2O$ | 1.75 |

If the nucleation rate curves peak at lower temperatures than the crystal growth-rate curves (Fig. 4-15; Winkler, 1947) for all the minerals in Table 4-3, then we should expect quickly cooled liquids to generate more nuclei than slowly cooled liquids. Our discussion so far would lead us to expect that a mineral with a low entropy of fusion, for example a plagioclase, would generate far more nuclei than olivine or pyroxene, but the growth rate of the feldspar would lag behind the two ferromagnesian minerals. Wager (1961) has given some measurements on the number of crystals of various types in a cubic centimeter of the chilled margin of the Skaergaard intrusion, and also in a coarse-grained equivalent rock 30 m away. An extract of his data (number of crystals per cubic centimeter) is given here:

| | PLAGIOCLASE | PYROXENE | OLIVINE | IRON ORE |
|---|---|---|---|---|
| Skaergaard contact | 4600 | 38 | 1900 | 13 |
| Coarse rock 30 m away | 2700 | 16 | 75 | 22(?) |

In a qualitative way, the predictions using Fig. 4-15 and Table 4-3 are substantiated by the Skaergaard intrusion, namely, that in the concentration of nuclei in the quickly cooled margin, the predominance of plagioclase is in accord with the fusion entropy relationships. But in the same liquid cooling slowly, olivine grew much faster and incorporated its scattered nuclei, eventually ophitically enclosing much of the plagioclase. The rate equations coupled with the data in Table 4-3 provide a reasonable qualitative explanation for ophitic texture, with the abundant initial plagioclase nuclei later to grow relatively slowly, being enveloped in faster-growing pyroxene. The puzzle seems to be why more lavas are not ophitic, and the most obvious explanation seems to be that as this texture is commonly found in the interior of flows, the cooling rate allows the crystals to approach their peak growth rates (Fig. 4-15), whereas in the more quickly cooled margins there was insufficient time at the peak growth rate and a larger number of nuclei. Iron titanium oxides, as well as olivine, often partially or completely enclose plagioclase crystals in ophitic or subophitic arrangement, which is again in qualitative agreement with the data of Table 4-3. The entropy of fusion, which is so influential in controlling the composition of a eutectic, is also obviously a quantity of considerable and visible importance in influencing the texture of igneous rocks. This qualitative approach ignores the effect of water; in high concentration it doubtless affects the crystal growth rate and nucleation rate as it markedly decreases both the viscosity and $E_\eta$, the activation energy for viscous flow.

One other geological observation illustrates the difficulty of nucleation: the frequent occurrence of lines of gas bubbles or vesicles along shear planes in obsidian lavas. It seems obvious that the necessary energy for bubble nucleation is gathered not from the latent heat of crystallization, since there are no crystals, but from "viscous heating" (H. R. Shaw, 1969) or the energy released by flow of the obsidian lava under the constant stress of the gravitational field.

## Chemical Conditions Governing Solid-Liquid Equilibrium

### SYSTEMS OF ONE COMPONENT

The temperature at which a solid is in equilibrium with a liquid of its own composition is generally known as the melting point $T_m$ of the substance. At this point, the chemical potential of the component is the same in both the solid and the liquid phases. If the temperature is increased beyond $T_m$, the chemical potential of the liquid decreases faster than that of the solid since the entropy of the liquid is greater than that of the solid. At all temperatures above the melting point at constant pressure, the liquid phase is therefore the stable one. Similarly, if the pressure is increased, the chemical potential of each phase increases proportionally to its molar volume, and since the volume of the liquid is generally greater than that of the solid, the chemical potential of the liquid in most cases increases faster and the solid becomes the stable phase. If temperature and pressure are increased simultaneously, equilibrium between the liquid and solid can be maintained only if the effect of temperature is offset exactly by the effect of pressure. This is expressed by Eq. (3-39) and is known as the Clausius-Clapeyron equation.

Generally, but not always, the heat of melting $\Delta H_m$ and the change in volume attendant in melting $\Delta V_m$ have the same sign, so that the melting temperature increases with pressure. Since both $\Delta H_m$ and $\Delta V_m$ depend upon temperature and pressure, it is conceivable that a point on the melting curve might be reached where $dT/dP$ becomes zero and then negative. This would mean that there would be a maximum melting point corresponding to a definite value of pressure beyond which the melting point would decrease with further increasing pressure. It has been suggested also that $\Delta H_m$ and $\Delta V_m$ might simultaneously become zero at a "critical" point at which the melting curve would stop and beyond which the distinction between liquid and solid states would vanish. Although no evidence for the existence of a critical melting point has yet come to light, maxima in the melting curves have been found in several metals; they appear to be generally associated with phase changes. For ionic substances, the melting point continues to increase up to the highest pressures reached experimentally, while the slope of the melting curve decreases slowly. The temperatures and heats of melting under normal conditions for a large number of minerals may be found in compilations by S. P. Clark (1966), Kelley (1960), Robie and Waldbaum (1968), and Stull and Prophet (1971).

There are many compounds, including a number of common minerals, which when heated to a certain temperature (incongruent melting point) decompose to give two phases, one of which is a liquid. That is, the liquid which forms at this temperature does not have the composition of the original substance, and a new solid phase is thus formed at the same time as the liquid. For instance, protoenstatite ($MgSiO_3$) when heated to 1557°C forms a liquid richer in $SiO_2$ than protoenstatite and a solid (forsterite, $Mg_2SiO_4$) correspondingly richer in MgO. Such behavior is known as *incongruent melting*. Strictly speaking, substances which melt incongruently do not form one-component systems.

## TWO-COMPONENT SYSTEMS

Study of equilibria in two-phase, two-component systems covers a wide range of phenomena known as *solubility* or *melting*. The difference between the two terms is one of usage, not of principle. When we study equilibrium between solid NaCl and an aqueous salt solution, we refer to the solubility of salt; when solid albite is in equilibrium at high temperature with a solution of albite in liquid diopside, we refer to the *melting* of albite. Similarly, when we refer to solutions we make no essential distinction between *solvent* and *solute*, although the most abundant of the two components is conventionally assumed to be the solvent.

The solubility of a solid in a given solvent is the concentration of this substance in a saturated solution, that is, in a solution which may exist in equilibrium with an excess of the solid. Equilibrium between a solid and a solution is reached when the chemical potential of the substance is the same in both phases. It is important to consider how this could occur. To do so, consider first the equilibrium between the solid and a liquid of its own composition. At the melting point $T_m$, the two chemical potentials are equal, and we can write for the component $A$, be it $CaMgSi_2O_6$, $CaAl_2Si_2O_8$, or any other substance,

$$\mu_A^{\text{sol}} = \mu_A^{\text{liq}} \tag{4-68}$$

But

$$\mu_A^{\text{liq}} = \mu_A^{\circ\,\text{liq}} + RT \ln a_A^{\text{liq}} \tag{4-69}$$

and

$$\mu_A^{\text{sol}} = \mu_A^{\circ\,\text{sol}} + RT \ln a_A^{\text{sol}} \tag{4-70}$$

where $\mu_A^{\circ\,\text{liq}}$ and $\mu_A^{\circ\,\text{sol}}$ are the chemical potentials of component $A$ in the respective standard states, namely, pure liquid $A$ and pure solid $A$ at 1 bar. Note that equilibrium demands that the chemical potentials be equal, and not the activities; the activities will be equal at equilibrium only if the same standard state is taken for both products and reactants.

Combining Eq. (4-69) with Eq. (4-70) gives

$$\log a_A^{\text{liq}} - \log a_A^{\text{sol}} = \frac{\mu_A^{\circ\,\text{sol}} - \mu_A^{\circ\,\text{liq}}}{2.303RT} \tag{4-71}$$

If Eq. (4-71) is differentiated with respect to temperature at constant pressure, we obtain

$$2.303R\left[\frac{\partial(\log a_A^{\text{liq}} - \log a_A^{\text{sol}})}{\partial T}\right]_P = \left(\frac{\partial \mu_A^{\circ\,\text{sol}}/T}{\partial T}\right)_P - \left(\frac{\partial \mu_A^{\circ\,\text{liq}}/T}{\partial T}\right)_P \tag{4-72}$$

But

$$\left(\frac{\partial \mu_A^{\circ}/T}{\partial T}\right)_{P,B} = \frac{-\hat{H}_A^{\circ}}{T^2} \tag{4-73}$$

where $\hat{H}_A^\circ$ is the molar enthalpy of $A$ in the standard state, either pure solid $A$ or pure liquid $A$, at 1 bar pressure and the temperature of interest.

Substituting Eq. (4-73) into Eq. (4-72) we get

$$2.303R\left[\frac{\partial(\log a_A^{\text{liq}} - \log a_A^{\text{sol}})}{\partial T}\right]_P = \frac{\hat{H}_A^{\circ\,\text{liq}} - \hat{H}_A^{\circ\,\text{sol}}}{T^2} \tag{4-74}$$

But $\hat{H}_A^{\circ\,\text{liq}} - \hat{H}_A^{\circ\,\text{sol}} = \Delta H_m$, the heat of melting of pure $A$. Before Eq. (4-74) can be integrated, we have to know how $\Delta H_m$ varies with temperature at constant pressure. This is given by

$$(\Delta H_m)_T = (\Delta H_m)_{T_m} + \int_{T_m}^{T} \Delta C_p^\circ \, dT \tag{4-75}$$

where $T$ is some temperature below $T_m$ and $\Delta C_p^\circ$ is the difference between the heat capacities of the products and reactants again in their standard states.

The variation of heat capacity at constant pressure with temperature, °K, for a substance is usually given in the empirical form (for example, Kelley, 1960):

$$C_p = a + bT - \frac{c}{T^2} \tag{4-76}$$

where $a$, $b$, and $c$ are constants.

If we let the difference between the heat capacities of the pure products and reactants be

$$\Delta C_p^\circ = A + BT - \frac{C}{T^2} \tag{4-77}$$

then we can substitute Eq. (4-77) into Eq. (4-75) and integrate it between the limits $T_m$ and $T$. Thus

$$(\Delta H_m)_T = (\Delta H_m)_{T_m} + A(T - T_m) + \frac{B}{2}(T^2 - T_m^2) + C\left(\frac{1}{T_m} - \frac{1}{T}\right) \tag{4-78}$$

If Eq. (4-78) is substituted into Eq. (4-74) and again integrated between $T$ and $T_m$, we obtain

$$\log\frac{(a_A^{\text{liq}})_T}{(a_A^{\text{liq}})_{T_m}} - \log\frac{(a_A^{\text{sol}})_T}{(a_A^{\text{sol}})_{T_m}}$$
$$= \frac{1}{2.303R}\left[\left\{\left(A + \frac{B}{2}T_m\right)T_m - (\Delta H_m)_{T_m} - \frac{C}{T_m}\right\}\left(\frac{1}{T} - \frac{1}{T_m}\right) + \left\{\frac{B}{2}(T - T_m) + \frac{C}{2}\left(\frac{1}{T^2} - \frac{1}{T_m^2}\right) + A\log\frac{T}{T_m}\right\}\right] \tag{4-79}$$

At the melting temperature $T_m$, the activities of both the solid and the liquid are unity. This equation, sometimes called the cryoscopic equation, can be used to calculate the liquidus or solubility curve of a solid containing $A$, in a liquid whose composition is changing because of the increasing amount of another component. If the solid containing $A$ is pure $A$, so that it has unit activity, then the left-hand side of Eq. (4-79) reduces to log $(a_A^{\text{liq}})_T$. If the solution of $A$ in the liquid is ideal, we can write, because $\gamma_A^{\text{liq}} = 1$,

$$\log a_A^{\text{liq}} = \log X_A^{\text{liq}} \tag{4-80}$$

so that the right-hand of Eq. (4-79) is equal to Eq. (4-80). We can therefore plot $X_A^{\text{liq}}$ against $T$, and the curve so obtained is the solubility or melting curve showing, as a function of temperature, the change in mole fraction of $A$ in an ideal liquid solution saturated with pure solid $A$. Obviously for $T = T_m$, $X_A^{\text{liq}} = 1$, so that the curve starts from a point representing the melting point of pure $A$.

The reader should be aware of possible confusion in evaluating Eq. (4-77). We have shown the difference between the heat capacity of a glass and that of a liquid of identical composition. The transformation we are concerned with here [Eq. (4-68)] is a melting reaction, so that the heat capacity of the liquid state rather than the corresponding glass should be used in Eq. (4-77).

It is often assumed that $\Delta C_p^\circ$, the difference between the heat capacities of the products (liquid) and the reactants (solid), is zero or close to it. The worth of this assumption is illustrated in Table 4-4 for four common rock-forming minerals. Obviously the assumption is not entirely satisfactory; it would be more correct to assume that $\Delta C_p^\circ$ is constant, so that only the first two terms in Eq. (4-78) would remain. However, if $\Delta C_p^\circ = 0$ or can be safely assumed to be so, then the integration of Eq. (4-74) is simple, and between the limits $T_m$ and $T$ we get

**Table 4-4** Difference between the heat capacities of various solids and their liquids ($\Delta C_p^\circ$) at the melting temperature $T_m$ and $T_m - 100°$

| | $T_m$, °K | $\Delta C_p^\circ$ at $T_m$, cal mole$^{-1}$ deg$^{-1}$ | $\Delta C_p^\circ$ at $T_m - 100°$, cal mole$^{-1}$ deg$^{-1}$ |
|---|---|---|---|
| $CaTiSiO_5$ | 1670 | +15.5 | +16.1 |
| $Fe_2SiO_4$ | 1490 | + 7.34 | + 8.33 |
| $KMg_3AlSi_3O_{10}F_2$ | 1670 | +35.2 | +37.1 |
| $FeTiO_3$ | 1640 | +12.8 | +13.2 |

$$\log a_A^{\text{liq}} - \log a_A^{\text{sol}} = \frac{(\Delta H_m)_{T_m}}{2.303R}\left(\frac{1}{T_m} - \frac{1}{T}\right) \tag{4-81}$$

Since melting or crystallization is a reversible reaction, the entropy of fusion, $\Delta S_m$, is given by

$$\Delta S_m = \frac{\Delta H_m}{T_m} \tag{4-82}$$

which substituted into Eq. (4-81) gives

$$\log a_A^{\text{liq}} - \log a_A^{\text{sol}} = \frac{-\Delta S_m}{2.303RT}(T_m - T) \tag{4-83}$$

Making the same assumptions about ideality as before, we can write for a liquid saturated with pure solid $A$

$$\log X_A^{\text{liq}} = \frac{-\Delta S_m}{2.303RT}(T_m - T) \tag{4-84}$$

It is important to realize that the solubility of $A$ in an ideal solution is independent of the nature of the other components in the liquid.

Inspection of Eq. (4-84) shows that if the entropy of fusion of a compound is small, then $X_A^{\text{liq}}$, or the fraction of $A$ in the liquid, will be relatively large for any given value of $T_m - T$; conversely if $\Delta S_m$ is large, then the fraction of $A$ will decrease very rapidly with a small drop in temperature. The general effect of melting temperature $T_m$ and the entropy of fusion $\Delta S_m$ on the shape of liquidus or solubility curves is shown in Fig. 4-17.

The calculated liquidus curve for $SiO_2$, using Eq. (4-84), is shown in Fig. 4-19, and for $T_m - T$ equal to 500°C, the fraction of $SiO_2$ in the liquid decreases only to 0.85 despite this very large drop in temperature. The very low entropy of fusion of cristobalite also indicates that much of the structural order of the solid persists in the liquid state. Many inorganic substances have entropies of fusion close to 2 cal deg$^{-1}$ g atom$^{-1}$, which as a generalization forms the basis of Richard's rule. Common rock-forming minerals often have much lower values of $\Delta S_m$, as the data given in Table 4-3 illustrate. Since minerals with low entropies of fusion will form eutectics, or low-melting compositions, in which they are the predominant components, it is evident that residual liquids resulting from continued fractional crystallization of a basaltic parent will be rich in alkali feldspar and silica.

The magnitude of the entropy of fusion of a component also affects the shape of the liquidus curve. Assume that the solid component is pure solid $A$ melting at $T_m$ and that $\Delta C_P^\circ$ is zero; then by rearranging Eq. (4-81) we find that

$$T = \frac{\Delta H_m}{R} \frac{1}{(\Delta H_m/RT_m) - \ln X^{\text{liq}}} \tag{4-85}$$

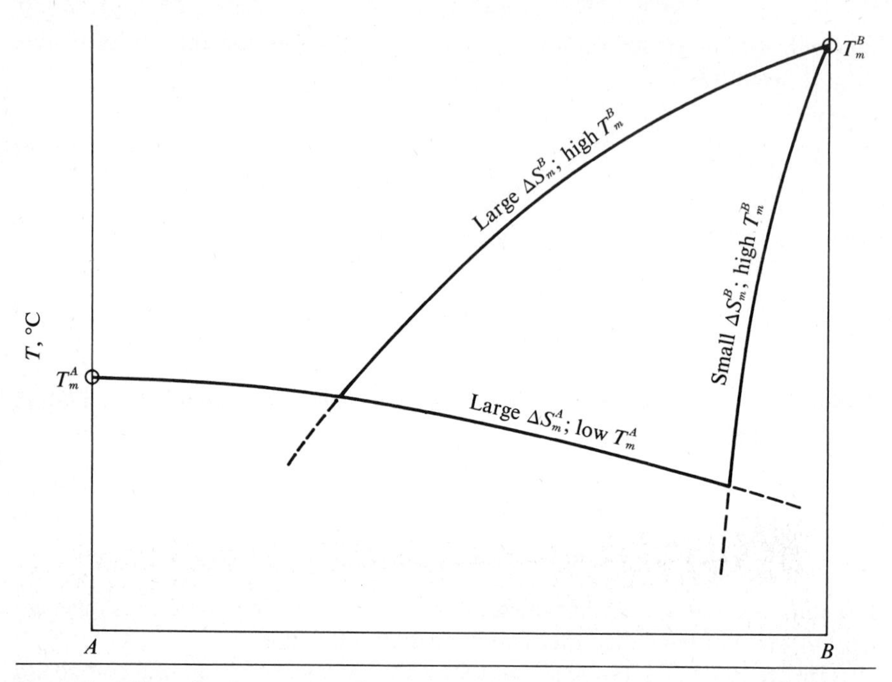


**Fig. 4-17** Effect of a large or small entropy of melting $\Delta S_m$ and a high or low melting temperature $T_m$ of compounds $A$ and $B$ on the general shape of a liquidus or solubility curve. Component $B$ is shown as having alternatively high or low $\Delta S_m$.

and

$$\frac{\partial T}{\partial X_A^{\text{liq}}} = \frac{\Delta H_m}{R} \frac{1}{X_A^{\text{liq}}[(\Delta H_m/RT_m) - \ln X_A^{\text{liq}}]^2} \tag{4-86}$$

so that

$$\frac{\partial^2 T}{(\partial X_A^{\text{liq}})^2} = \frac{\Delta H_m}{R}\left[\frac{1}{(X_A^{\text{liq}})^2[(\Delta H_m/RT_m) - \ln X_A^{\text{liq}}]^2}\left(\frac{2}{(\Delta H_m/RT_m) - \ln X_A^{\text{liq}}} - 1\right)\right] \tag{4-87}$$

The sign of the second derivative, which gives the curvature of the liquidus, depends upon

$$\frac{\Delta H_m}{RT_m} - \ln X_A^{\text{liq}} \gtrless 2$$

For liquids whose composition is nearly pure $A$, $X_A^{\text{liq}} \rightarrow 1$ so that $\ln X_A^{\text{liq}} \rightarrow 0$; by making use of Eq. (4-82) the second derivative will be positive if

$$\frac{\Delta H_m}{RT_m} = \frac{\Delta S_m}{R} < 2$$

or $\Delta S_m < 4$ cal mole$^{-1}$ deg$^{-1}$, and the second derivative will be negative if $\Delta S_m > 2R$.

These two cases are shown in Fig. 4-18, in which it can be seen that substances with low entropies of fusion (Table 4-3) have inflection points in their liquidus curves, whereas those with large entropies of melting do not. Experimental determination of the cristobalite liquidus shows an inflection point in the systems $Na_2O$–$SiO_2$ and $K_2O$–$SiO_2$, which of course must be present in all systems with $SiO_2$ unless liquid immiscibility intervenes.

Bowen (1928, p. 177) suggested that the deviation between the calculated [Eq. (4-84)] and the experimental liquidus curves for the system $CaAl_2Si_2O_8$–$CaMgSi_2O_6$ could arise from a heat-of-mixing term ($\Delta H_{\text{mix}} \neq 0$) since the "solubility" is not ideal. The calculated liquidus curves and their experimental equivalents for the two systems $CaAl_2Si_2O_8$–$SiO_2$ and $CaAl_2Si_2O_8$–$CaMgSi_2O_6$ are shown in Figs. 4-19 and 4-20, respectively. There is only a small deviation between the calculated and the observed liquidus curves for $CaMgSi_2O_6$ (Fig. 4-20), which suggests that the liquid is, to all intents and purposes, ideal. But for $CaAl_2Si_2O_8$ there is a large discrepancy, with the experimental curve

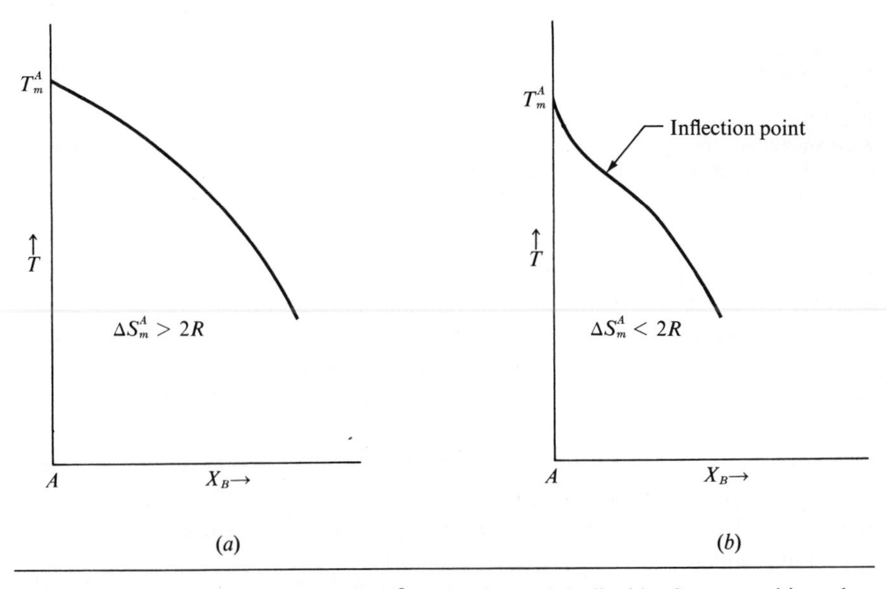


**Fig. 4-18** Effect of the magnitude of $\Delta S_m^a$ on the shape of the liquidus for compositions close to pure $A$ in a two-component system $A$–$B$.

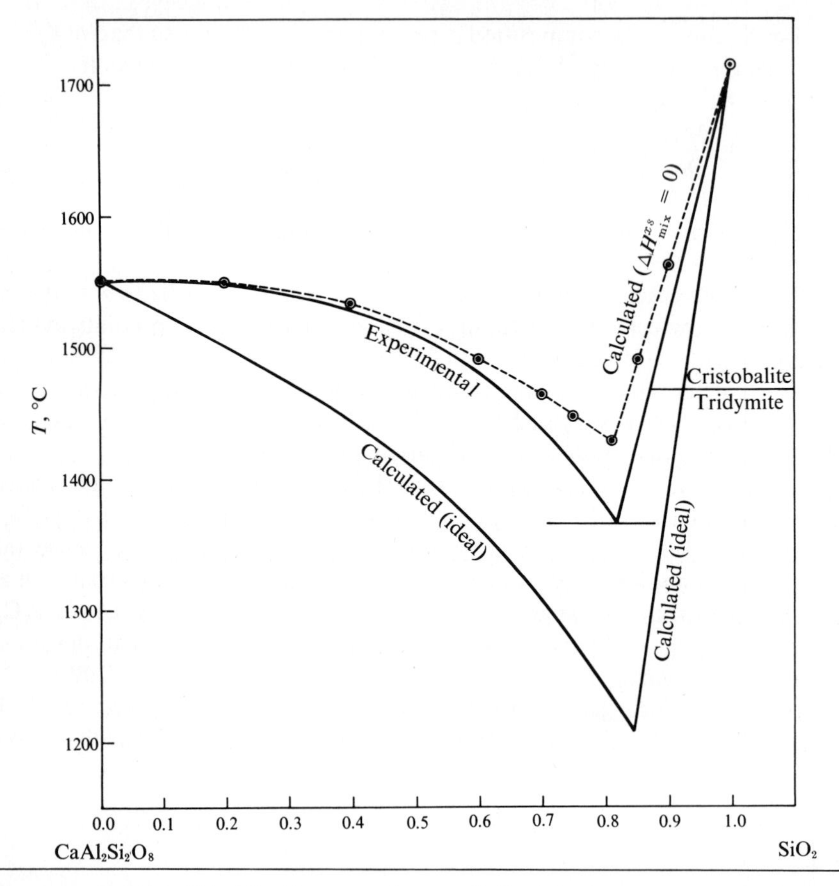


**Fig. 4-19** Calculated ideal liquidus curves (lower solid lines) for the system $CaAl_2Si_2O_8$–$SiO_2$. Also shown are the experimental liquidus curves (upper solid lines) and calculated liquidus curves (dashed) assuming the type of polymerization suggested by Flood and Knapp (1968), with $\Delta H^{xs}_{mix} = 0$.

falling below the calculated or ideal curve. And yet in the system $CaAl_2Si_2O_8$–$SiO_2$ (Fig. 4-19), the experimental liquidus is well above the ideal curve; obviously the solubility of the $CaAl_2Si_2O_8$ component in the liquid is influenced by the nature of the other components. In Fig. 4-20, we speak of the experimental anorthite curve having negative deviation, whereas in Fig. 4-19 the experimental curve has positive deviation.

How could this deviation arise, and how can it be used to obtain some idea of the properties of these liquids? In the case of $CaAl_2Si_2O_8$, as for any component $A$, the chemical potential in a liquid at temperature $T$ is given by Eq. (4-69), which can be rewritten in the form

$$\mu_A^{liq} = \mu_A^{\circ liq} + RT \ln X_A^{liq} \gamma_A^{liq} \tag{4-88}$$

But for an ideal solution of $A$, we have

$$\mu_A^{\text{ideal liq}} = \mu_A^{\circ\,\text{liq}} + RT \ln X_A^{\text{liq}} \tag{4-89}$$

and if Eq. (4-89) is subtracted from Eq. (4-88), we get the difference between the chemical potential of $A$ in any solution and in an ideal solution:

$$\mu_A^{\text{liq}} - \mu_A^{\text{ideal liq}} = \mu_A^{\text{xs}} = RT \ln \gamma_A^{\text{liq}} \tag{4-90}$$

where $\mu_A^{\text{xs}}$ is called the excess chemical potential of $A$ in the liquid, or the excess partial molar free energy of $A$.

If $\gamma_A^{\text{liq}}$ is less than unity, then $\mu_A^{\text{xs}}$ is negative as in the example of $CaAl_2Si_2O_8$ in Fig. 4-20; if $\gamma_A^{\text{liq}}$ is greater than unity, then $\mu_A^{\text{xs}}$ is positive (Fig. 4-19). Ideal behavior ($\gamma_A^{\text{liq}} = 1$) is unlikely to occur in solutions which contain molecules of contrasted types, where the forces between unlike molecules are different from those between like molecules. As a generalization, to be considered later in more detail, a positive value of $\mu_A^{\text{xs}}$ or $\gamma_A > 1$ is a reflection of a repulsive interaction between unlike molecules in solution, which at lower temperatures can generate an immiscibility region. Negative $\mu_A^{\text{xs}}$ or $\gamma_A < 1$ indicates an attractive interaction, which may result in the formation of an intermediate compound.

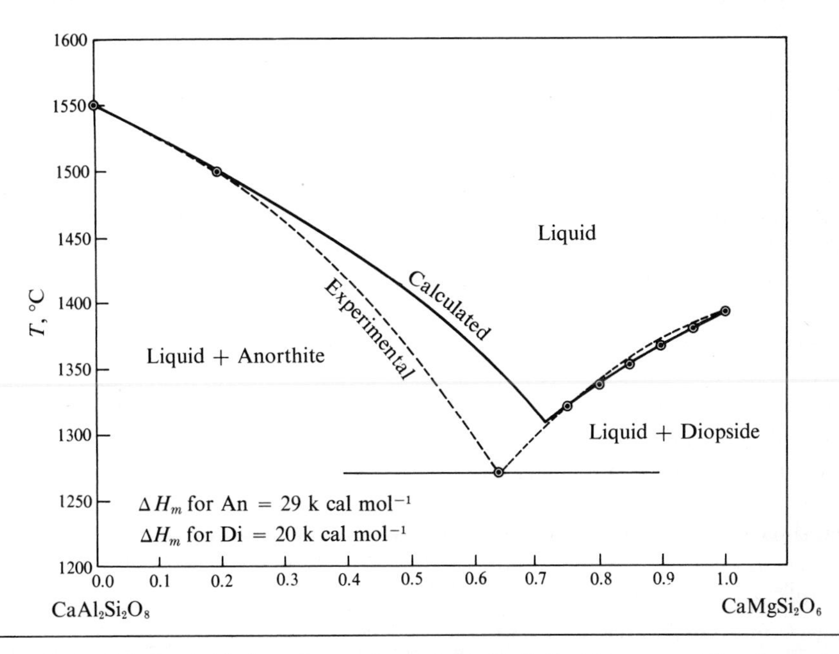


**Fig. 4-20** Calculated ideal and experimental (dashed) liquidus curves for the system $CaAl_2Si_2O_8$–$CaMgSi_2O_6$.

Another approach to understanding the cause of the deviation between the ideal and actual liquidus curves in the $CaAl_2Si_2O_8$–$SiO_2$ system has been taken by Flood and Knapp (1968). They have suggested that silicate liquids could be considered, from a statistical point of view, as a mixture of randomly distributed groups of atoms, for example $AlO_4$ and $SiO_4$ tetrahedra, with cations accompanying the $AlO_4$ groups to maintain electrical charge on the tetrahedral group. It has been assumed that the bonding energies of these groups are relatively small, so that their activity is principally determined by the mixing entropy of the groups; in other words, $\Delta H_{mix}$ is assumed to be zero.

The relative partial molar entropy of component $A$ in an ideal liquid is given by

$$\Delta \bar{S}_A^{liq} = \bar{S}_A^{liq} - \hat{S}_A^{\circ\, liq} = -R \ln X_A^{liq} \tag{4-91}$$

where $\bar{S}_A^{liq}$ is the partial molar entropy of $A$ in the liquid, $\hat{S}_A^{\circ\, liq}$ is the molar entropy of $A$ in a liquid of pure $A$, and $X_A^{liq}$ is the mole fraction of $A$ in the liquid. In the example of $CaAl_2Si_2O_8$, Flood and Knapp have suggested that liquids rich in the $CaAl_2Si_2O_8$ component in the system $CaAl_2Si_2O_8$–$SiO_2$ contain polymeric ions, and that the dissociation of $CaAl_2Si_2O_8$ in the liquid can be represented by

$$CaAl_2Si_2O_8 \rightarrow \tfrac{1}{2}[Ca_24AlO_23SiO_2] + \tfrac{1}{2}SiO_2 \tag{4-92}$$

so that

$$\Delta S_{CaAl_2Si_2O_8}^{liq} = -R \ln [K(X_{group})^{1/2}(X_{SiO_2})^{1/2}] \tag{4-93}$$

where $X_{group}$ is the mole fraction of $[Ca_24AlO_23SiO_2]$ and $X_{SiO_2}$ is the mole fraction of $SiO_2$, less the amount contained in the polymeric group. $K$ is a constant calculated from a liquid of pure $CaAl_2Si_2O_8$ composition and equals 2. If ideality is assumed, that is, if $\Delta \bar{H}_{CaAl_2Si_2O_8}^{liq}$ is ignored in the relationship

$$RT \ln a_{CaAl_2Si_2O_8}^{liq} = \Delta \bar{H}_{CaAl_2Si_2O_8}^{liq} - T\Delta \bar{S}_{CaAl_2Si_2O_8}^{liq} \tag{4-94}$$

then we get from Eq. (4-93)

$$RT \ln a_{CaAl_2Si_2O_8}^{liq} \approx RT \ln [K(X_{group})^{1/2}(X_{SiO_2})^{1/2}] \tag{4-95}$$

and thus

$$a_{CaAl_2Si_2O_8}^{liq} \approx K(X_{group})^{1/2}(X_{SiO_2})^{1/2} \tag{4-96}$$

The calculated liquidus curve for this polymeric arrangement, which in Flood and Knapp's paper (1968) coincided with the experimental $CaAl_2Si_2O_8$

liquidus, falls above the experimental curve in Fig. 4-19, presumably because we used a slightly different value for the heat of melting of anorthite. For the silica liquidus, Flood and Knapp used the analogous relationship

$$\Delta\bar{S}^{\text{liq}}_{\text{SiO}_2} \approx -R \ln X^{\text{liq}}_{\text{SiO}_2} \tag{4-97}$$

but

$$X^{\text{liq}}_{\text{SiO}_2} = \frac{n_{\text{SiO}_2}}{n_{\text{SiO}_2} + n_{\text{Al}_4\text{O}_8}} \tag{4-98}$$

where $n_{SiO_2}$ and $n_{Al_4O_8}$ are the number of moles of $SiO_2$ and $Al_4O_8$ in any given liquid saturated with cristobalite in the system $CaAl_2Si_2O_8$–$SiO_2$. Again our calculated curve (Fig. 4-19) falls above the experimental cristobalite curve, which, if nothing else, highlights the importance of having well-established values for heats of melting $\Delta H_m$. By choosing different polymeric arrangements it is possible, no matter which heat of fusion is used, to calculate a liquidus curve which coincides closely with the experimental liquidus curves in Fig. 4-19. By analogy with this and other nonideal systems such as $NaAlSi_3O_8$–$SiO_2$, it seems likely that natural liquids such as rhyolites whose composition can closely approach these simple silica-rich systems also contain polymeric ions.

In the discussion so far we have seen that the liquidus of a simple binary system can be calculated if the heats of melting and the melting temperatures of the two components are known. Although it is strictly incorrect to ignore the heat-capacity terms for the melting reaction [Eq. (4-84)], their inclusion is often not significant [Eq. (4-79)] for the experimental errors or uncertainties in values $\Delta H_m$ usually outweigh the heat-capacity terms in the calculation of the liquidus or solubility curves. The deviation of the experimental liquidus curve from the ideal or calculated curve reflects an interaction between the molecules, atoms, or particles in the liquid. This interaction can be considered to contribute to a heat-of-mixing term ($\Delta H_{\text{mix}} \neq 0$) which can be either positive or negative; Flood and Knapp in another approach ignored $\Delta H_{\text{mix}}$ and treated the same liquids as a random mixture of polymeric groups, and their calculated liquidus curves agreed well with the experimental curves for the same components. This contrast in emphasis, which is shown here schematically, high-

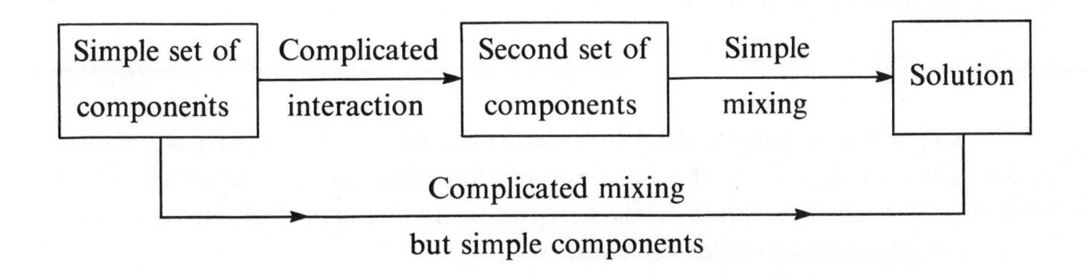

lights alternative methods that have been used in treating the mixing properties of solutions. So it seems worthwhile to delve more deeply into the properties of binary solutions, particularly with regard to the enthalpy and entropy of mixing and how they affect the behavior of solutions.

## Properties of Solutions

### MIXING IN BINARY SOLUTIONS

IDEAL BINARY SYSTEM Although the treatment which follows is largely concerned with solid solutions, it applies equally to liquids; however, it is the paucity of data on the mixing properties of silicate liquids that eliminates the liquid state from providing examples which are more readily found in naturally represented solids.

In an ideal solution, the chemical potential of any component $A$ is given by

$$\mu_A = \mu_A^\circ + RT \ln X_A \tag{4-89}$$

where $\mu_A^\circ$ is the chemical potential of $A$ in pure $A$ and $X_A$ is the mole fraction of $A$. It is obvious that $\mu_A$ is reduced as the mole fraction of $A$ is decreased. At constant composition the chemical potential of $A$ also changes in response to a change of pressure or temperature, and this change is given by

$$\mu_A = \bar{V}_A \, dP - \bar{S}_A \, dT \tag{4-99}$$

so that at constant temperature ($dT = 0$) and constant composition ($dX_A = 0$)

$$\left(\frac{\partial \mu_A}{\partial P}\right)_{T,X_A} = \bar{V}_A \tag{4-100}$$

where $\bar{V}_A$ is the partial molar volume of $A$ in the solution (cf. Fig. 4-21).

If Eq. (4-89) is differentiated at constant temperature and composition ($d \ln X_A = 0$),

$$\left(\frac{\partial \mu_A}{\partial P}\right)_{T,X_A} - \left(\frac{\partial \mu_A^\circ}{\partial P}\right)_T = 0 \tag{4-101}$$

and by substituting Eq. (4-100)

$$\bar{V}_A = \hat{V}_A^\circ \tag{4-102}$$

or the partial molar volume of $A$ is equal to the molar volume of pure $A$, $\hat{V}_A^\circ$, at all values of $X_A$. Since the behavior of the other component in the ideal solution is governed by Eq. (4-104), there can be no change in volume on mixing the pure components to form the solution.

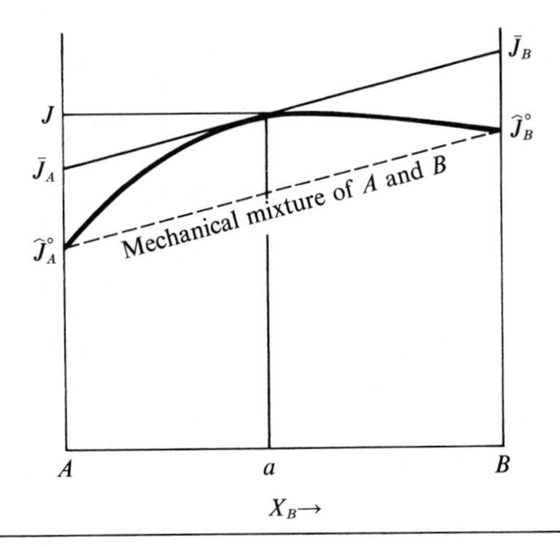


**Fig. 4-21** Molar property $J$ of an ideal binary solution, $A$–$B$. The composition $a$ has partial molar quantities given by the intercepts of the tangent $\bar{J}_A$ and $\bar{J}_B$. $\hat{J}_A^\circ$ and $\hat{J}_B^\circ$ are the molar properties of pure $A$ and pure $B$, respectively. $J$ is the property of solution $a$.

By differentiating Eq. (4-99) at constant pressure ($dP = 0$) and constant composition ($dX_A = 0$) and substituting in Eq. (4-89), we find that

$$\left(\frac{\partial \mu_A}{\partial T}\right)_{P,X_A} = -\bar{S}_A = -\hat{S}_A^\circ + R \ln X_A \tag{4-103}$$

where $\bar{S}_A$ is the partial molar entropy of $A$ in the solution and $\hat{S}_A^\circ$ is the molar entropy of $A$ in a solution of pure $A$. Since

$$\mu_A^{\text{soln}} - \mu_A^\circ = \bar{H}_A - \hat{H}_A^\circ - T(\bar{S}_A - \hat{S}_A^\circ) \tag{4-104}$$

by substituting Eqs. (4-89) and (4-103) we get

$$\bar{H}_A - \hat{H}_A^\circ - T(-R \ln X_A) = RT \ln X_A$$

or

$$\bar{H}_A = \hat{H}_A^\circ \tag{4-105}$$

Therefore the properties of an ideal binary solution, liquid, solid, or gas (Fig. 4-21), are given by

$$V^{\text{ideal soln}} = X_A \hat{V}_A^\circ + X_B \hat{V}_B^\circ \tag{4-106}$$

Similarly

$$H^{\text{ideal soln}} = X_A \hat{H}_A^\circ + X_B \hat{H}_B^\circ \tag{4-107}$$

and necessarily

$$E^{\text{ideal soln}} = X_A\hat{E}_A^\circ + X_B\hat{E}_B^\circ \tag{4-108}$$

or, put in words, every solution in an ideal system has a volume per mole $V$, an internal energy $E$, and an enthalpy $H$, which are all linear functions of composition.

ENTROPY OF IDEAL MIXING, $\Delta S_{\text{mix}}^{\text{ideal}}$ Mixing is a spontaneous process involving the intermingling of particles in gases, liquids, and solids. There are two principal aspects of the mixing process: (1) the spreading of particles over positions in space and (2) the sharing of the available energy between the particles. Complete mixing increases the configurational randomness of the system but could also decrease the thermal randomness; equilibrium is achieved when the overall randomness is maximum. For an ideal solution we may calculate the entropy of mixing, using a statistical approach.

Assume that there are $N_A$ particles of $A$ and $N_B$ particles of $B$, and that they are completely interchangeable without affecting the internal energy ($\Delta E_{\text{mix}} = 0$). There are $N_A + N_B$ available sites, so that the first particle is placed on any one of the $N_A + N_B$ sites, the second on any of the $N_A + N_B - 1$ sites, the third on $N_A + N_B - 2$ empty sites, and so forth. Thus there are $(N_A + N_B)!$ possible arrangements; however, none of the $N_A$ atoms is distinguishable from any other, nor are the $N_B$ atoms or particles mutually distinguishable; so that the total number of possible arrangements must be divided by $N_A!$ the number of possible interchanges between the $A$ particles and by $N_B!$ the interchanges between the $B$ particles. Thus the number of distinguishable arrangements is

$$W = \frac{(N_A + N_B)!}{N_A!N_B!} \tag{4-109}$$

The number of accessible states of a system is represented by $W$, which, in other words, is the number of different microscopic states that constitute a given macroscopic state. $W$ is related to entropy through the Boltzmann relationship

$$S = k \ln W$$

or through

$$S_2 - S_1 = k \ln \frac{W_2}{W_1} \tag{4-110}$$

where subscript 2 represents the final state and subscript 1 the initial state. In the initial state of pure components of $N_A$ $A$ particles and $N_B$ $B$ particles, the number of distinguishable arrangements in each is given by

$$W_{\text{initial } A} = \frac{N_A!}{N_A!} = 1$$

and

$$W_{\text{initial } B} = \frac{N_B!}{N_B!} = 1$$

so that by substitution of Eq. (4-109) in Eq. (4-110) we obtain

$$S_2 - S_1 = \Delta S_{\text{mix}}^{\text{ideal}} = k \ln \frac{(N_A + N_B)!}{N_A! N_B!}$$

$$= k[\ln (N_A + N_B)! - \ln N_A! - \ln N_B!]$$

and by Stirling's approximation for large numbers ($\ln N! = N \ln N - N$) we find that

$$\Delta S_{\text{mix}}^{\text{ideal}} = -k \left( N_A \ln \frac{N_A}{N_A + N_B} + N_B \ln \frac{N_B}{N_A + N_B} \right)$$

Since $N_A/(N_A + N_B) = X_A$, the atom or particle fraction of $A$, and similarly for $B$, this becomes

$$\Delta S_{\text{mix}}^{\text{ideal}} = -k(N_A \ln X_A + N_B \ln X_B) \tag{4-111}$$

which is a positive quantity.

If Eq. (4-111) is multiplied through by $N_o$, Avogadro's number, to put it on the basis of moles of particles, and remembering that $R/N_o = k$, we find that

$$\Delta S_{\text{mix}}^{\text{ideal}} = -R(n_A \ln X_A + n_B \ln X_B) \tag{4-112}$$

where $n_A$ and $n_B$ are the numbers of moles of $A$ and $B$, respectively.

Since the entropy of mixing is not zero and for a binary solution is symmetrical about the midpoint $X_A = X_B$, with a value of 1.38 entropy units, the free energy of mixing is given by

$$\Delta G_{\text{mix}}^{\text{ideal}} = -T\Delta S_{\text{mix}}^{\text{ideal}} \tag{4-113}$$

so that by substitution of Eq. (4-112)

$$\Delta G_{\text{mix}}^{\text{ideal}} = RT(n_A \ln X_A + n_B \ln X_B) \qquad (4\text{-}114)$$

or on the basis of 1 mole of binary solution, obtained by dividing through by $n_A + n_B$,

$$\Delta G_{\text{mix}}^{\text{ideal}} = RT(X_A \ln X_A + X_B \ln X_B) \qquad (4\text{-}115)$$

The total free energy of an ideal two-component solution is given by

$$G^{\text{ideal soln}} = n_A \mu_A^\circ + n_B \mu_B^\circ + \Delta G_{\text{mix}}^{\text{ideal}} \qquad (4\text{-}116)$$

and by substitution of Eq. (4-114)

$$G^{\text{ideal soln}} = n_A \mu_A^\circ + n_B \mu_B^\circ + RT(n_A \ln X_A + N_B \ln X_B) \qquad (4\text{-}117)$$

Since

$$\mu_A = \left(\frac{\partial G^{\text{soln}}}{\partial n_A}\right)_{P,T,n_B}$$

by definition, we may recover

$$\mu_A = \mu_A^\circ + RT \ln X_A$$

by differentiating Eq. (4-117) with respect to $n_A$.

Equation (4-115) represents the driving force to form a solution rather than a mechanical mixture of two end members with the same structure; this driving force increases with temperature ($\Delta G_{\text{mix}}^{\text{ideal}}$ becomes more negative) and is the reason for the more extensive solid solution found in high-temperature igneous minerals in comparison to the same minerals equilibrating in a lower-temperature environment. The curves of $\Delta G_{\text{mix}}^{\text{ideal}}$ and $\Delta S_{\text{mix}}^{\text{ideal}}$ are shown in Fig. 4-22; and as temperature is increased, the free-energy curve becomes depressed to more negative values and the solutions become more stable in comparison to a mechanical mixture of the two end members.

Both liquids and solids in the system $CaAl_2Si_2O_8$–$NaAlSi_3O_8$ (Bowen, 1913) are ideal—since the calculated liquidus and solidus curves correspond with the experimentally determined liquidus and solidus (Bowen, 1928). The olivine solid-solution series ($Mg_2SiO_4$–$Fe_2SiO_4$) was for many years considered to be ideal, but recent research has shown that there are two nonequivalent lattice sites, one of which prefers iron and the other magnesium; this solid-solution series is therefore not strictly ideal since there is an energy term, or site-preference energy, involved in the substitution of an iron atom for a mag-

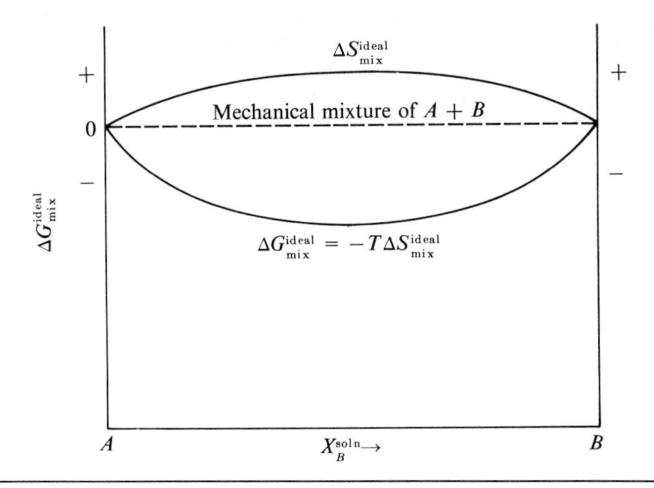


**Fig. 4-22** Hypothetical diagram showing the entropy and free energy of mixing at some constant temperature $T$ for an ideal solution with the reference state taken as a mechanical mixture of the pure components $A$ and $B$.

nesium atom [cf. R. W. Williams (1972), Nafziger (1973)]; this term is unlikely to exist in olivine liquids, which can be taken as ideal.

For an ideal solid solution in equilibrium with an ideal liquid solution, Eq. (4-81) is rewritten as

$$\log \frac{X_A^{liq}}{X_A^{sol}} = \frac{(\Delta H_m)_A}{2.303R}\left[\frac{1}{(T_m)_A} - \frac{1}{T}\right] \tag{4-118}$$

for the $A$ component and as

$$\log \frac{1 - X_A^{liq}}{1 - X_A^{sol}} = \frac{(\Delta H_m)_B}{2.303R}\left[\frac{1}{(T_m)_B} - \frac{1}{T}\right] \tag{4-119}$$

for the $B$ component, the mole fraction of which is equal to $1 - X_A$.

NONIDEAL BINARY SOLUTIONS Few solid solutions are ideal except at extreme dilution, so that it is often convenient to show the extent of the deviation of the solution from the corresponding quantities of an ideal solution; this is done by means of the *excess* thermodynamic functions. The excess molar volume of mixing $\Delta V^{xs}_{mix}$, for example, is equivalent to the actual volume per mole less that which would result from ideal mixing of the end members; this is written as

$$\Delta V^{xs}_{mix} = V^{soln} - V^{ideal}_{mix} \tag{4-120}$$

which from Eq. (4-106)

$$= V^{soln} - (X_A \hat{V}^\circ_A + X_B \hat{V}^\circ_B) \tag{4-121}$$

Other equations involving excess thermodynamic quantities are

$$\Delta H_{\text{mix}}^{\text{xs}} = H^{\text{soln}} - H_{\text{mix}}^{\text{ideal}} \tag{4-122}$$

which from Eq. (4-107)

$$= H^{\text{soln}} - (X_A \hat{H}_A^\circ + X_B \hat{H}_B^\circ) \tag{4-123}$$

Similarly

$$\Delta E_{\text{mix}}^{\text{xs}} = E^{\text{soln}} - E_{\text{mix}}^{\text{ideal}} \tag{4-124}$$

which from Eq. (4-108)

$$= E^{\text{soln}} - (X_A \hat{E}_A^\circ + X_B \hat{E}_B^\circ) \tag{4-125}$$

Since $\Delta V_{\text{mix}}^{\text{xs}}$, $\Delta E_{\text{mix}}^{\text{xs}}$, and $\Delta H_{\text{mix}}^{\text{xs}} = 0$ for an ideal solution, any value for the volume, internal energy, and enthalpy terms on mixing is an excess, or a deviation from ideality. The excess entropy of mixing relative to a mechanical mixture of pure components can be written as

$$\Delta S_{\text{mix}}^{\text{xs}} = \Delta S_{\text{mix}}^{\text{soln}} - \Delta S_{\text{mix}}^{\text{ideal}} \tag{4-126}$$

It follows from the excess equations noted above that the excess free energy of mixing for 1 mole of solution is given by

$$\Delta G_{\text{mix}}^{\text{xs}} = \Delta H_{\text{mix}}^{\text{xs}} - T\Delta S_{\text{mix}}^{\text{xs}} \tag{4-127}$$

and by combining Eqs. (4-126) and (4-127) we obtain for a nonideal binary solution

$$\Delta G_{\text{mix}}^{\text{soln}} = \Delta G_{\text{mix}}^{\text{ideal}} + \Delta G_{\text{mix}}^{\text{xs}} \tag{4-128}$$

$$= \Delta H_{\text{mix}}^{\text{xs}} - T(\Delta S_{\text{mix}}^{\text{xs}} + \Delta S_{\text{mix}}^{\text{ideal}}) \tag{4-129}$$

$$= \Delta H_{\text{mix}}^{\text{xs}} - T\Delta S_{\text{mix}}^{\text{soln}} \tag{4-130}$$

$\Delta H_{\text{mix}}^{\text{xs}}$ can be either positive or negative; the former, as will be shown later, indicates a *repulsive* interaction between $A$ and $B$ atoms, molecules, or ions. This often results in an immiscible region, which, however, is constricted or eliminated at higher temperatures as the term $T\Delta S_{\text{mix}}^{\text{soln}}$ overwhelms the positive $\Delta H_{\text{mix}}^{\text{xs}}$ term; thus the immiscible region in solids or liquids cannot extend without limit to high temperatures. A negative value for $\Delta H_{\text{mix}}^{\text{xs}}$ indicates an *attractive* interaction between unlike atoms, molecules, or ions, which can result in the formation of an intermediate compound at low temperatures.

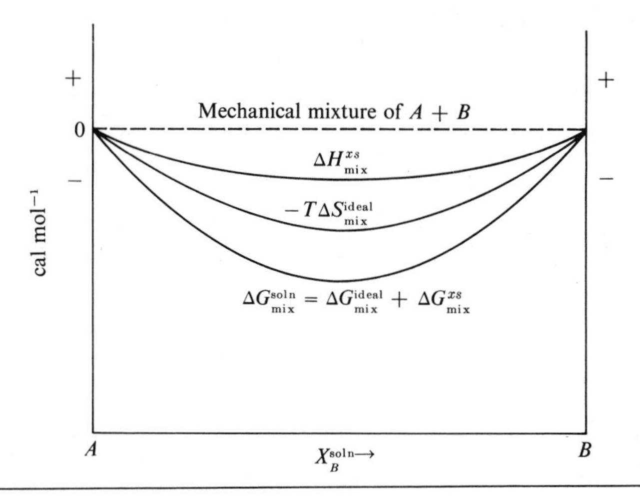


**Fig. 4-23** Hypothetical diagram showing the respective curves of $\Delta H^{xs}_{mix}$, $-T\Delta S^{ideal}_{mix}$, and $\Delta G^{soln}_{mix}$; for regular (symmetrical) solution *A–B* with negative $\Delta H^{xs}_{mix}$.

A solid solution of *A* and *B* with an attractive interaction between the *A* and *B* atoms or a negative $\Delta H^{xs}_{mix}$ is illustrated in Fig. 4-23. Because $\Delta H^{xs}_{mix}$ has the same sign as the sum of $-(T\Delta S^{ideal}_{mix} + T\Delta S^{xs}_{mix})$, solutions in a system of this type are more stable relative to the pure end-member components than is an ideal solution (Fig. 4-22).

A repulsive interaction or a positive $\Delta H^{xs}_{mix}$ results in a free-energy diagram more complex than Fig. 4-23, and an immiscible region is found at low temperatures ($T_c$ in Fig. 4-24*c*). At a higher temperature $T_2 > T_c$ (Fig. 24*a*) $\Delta G^{soln}_{mix}$ is slightly asymmetric; but at any temperature $T_1$ below $T_c$, the curve of $\Delta G^{soln}_{mix}$, which is the sum of the $\Delta H^{xs}_{mix}$ and $-T\Delta S^{soln}_{mix}$ curves, develops two minima near *L* and *M* (Fig. 4-24*b*). The lowest possible value of $\Delta G^{soln}_{mix}$ that can be attained by any bulk composition lying between *L* and *M* is given by the tangent to $\Delta G^{soln}_{mix}$, which, it is important to note, need not intersect the curve of $\Delta G^{soln}_{mix}$ at the two minima (Fig. 4-25). Thus the free energies of *L* and *M* individually may differ considerably from the free energies of the two compositions represented by the two minima on the $\Delta G^{soln}_{mix}$ curve, but in sum they are the lowest possible and hence the most stable assemblage. Thus all compositions between *L* and *M* will, for equilibrium to be achieved, unmix into the two compositions *L* and *M*, and we have a solvus, or immiscible region, as illustrated in Fig. 4-24*c*. As the temperature is increased, the two minima *L* and *M* approach one another, and at the critical temperature $T_c$, corresponding with the crest of the solvus, they will coincide.

It is evident from Fig. 4-24*b* that all compositions lying between pure *A* and *L* and between pure *B* and *M* have lower free energies as solid solutions than as corresponding mechanical mixtures *L-A* and *M-B* respectively. As the temperature is lowered below $T_2$, the two terminal solid solutions $A_{ss}$ and $B_{ss}$ (Fig. 4-24*c*) become closer in composition to pure *A* and pure *B*, respectively.

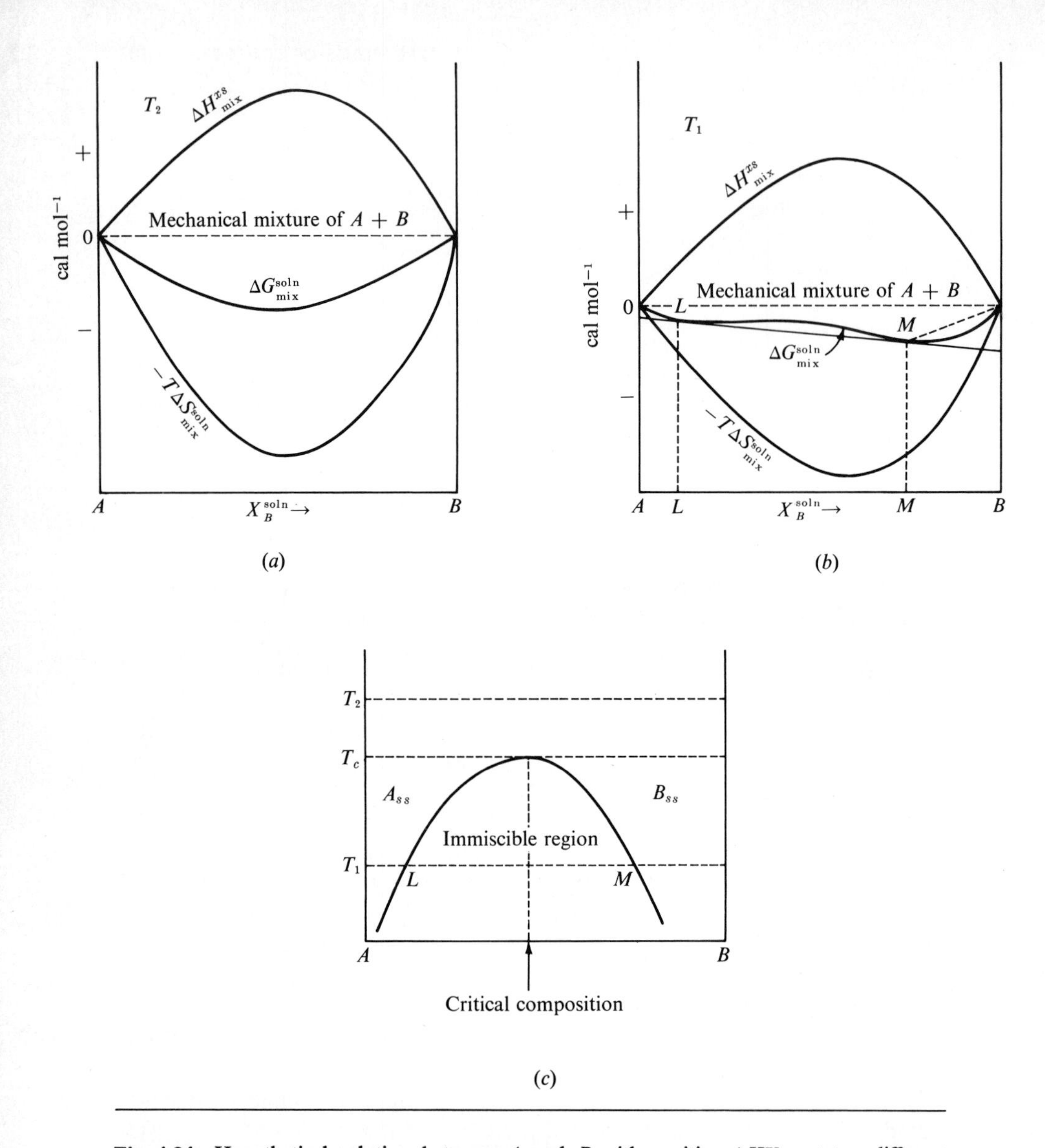


**Fig. 4-24** Hypothetical solution between $A$ and $B$ with positive $\Delta H^{xs}_{mix}$ at two different temperatures. (*a*) Free-energy-of-mixing curve at temperature $T_2$. (*b*) The same solution at temperature $T_1$ ($< T_c$) showing two minima in $\Delta G^{soln}_{mix}$ curve, which do not correspond to the two points of tangency $L$ and $M$. For compositions between $A$ and $L$, and $M$ and $B$, the stable phases are $A$ solid solutions and $B$ solid solutions, respectively. Tangent to the $\Delta G^{soln}_{mix}$ curve gives the compositions of the two phases forming the stable assemblage for any bulk composition between $L$ and $M$. (*c*) Temperature-composition diagram showing the relationship of (*a*) and (*b*) to the immiscible region, or solvus. $T_c$ is the critical temperature, and the critical composition is shown by the arrow.

This introduces the problem of the thermodynamic stability of pure phases, particularly at the high temperatures of igneous rocks. However, rather than treat the properties of a nonideal solution as the general case, it is simpler, for illustration, to consider the particular case of a *regular* solution. This has a $\Delta H_{\text{mix}}^{\text{xs}}$ term but an entropy of mixing which is that of an ideal solution. Thus we may write

$$\Delta G_{\text{mix}}^{\text{reg soln}} = \Delta H_{\text{mix}}^{\text{xs}} - T\Delta S_{\text{mix}}^{\text{ideal}} \tag{4-131}$$

The use of this type of solution does not restrict the following conclusions which, in a different formulation, can be shown to apply to all types of solutions. Let

$$\Delta H_{\text{mix}}^{\text{xs}} = X_A(\bar{H}_A^\circ - \hat{H}_A^\circ) + X_B(\bar{H}_B - \hat{H}_B^\circ) \tag{4-132}$$

where $\bar{H}_A$ and $\bar{H}_B$ are the partial molar enthalpies of $A$ and $B$ in the binary solution and $\hat{H}_A^\circ$ and $\hat{H}_B^\circ$ are the enthalpies of pure $A$ and pure $B$. Again $\Delta\bar{H}_A = \bar{H}_A - \hat{H}_A^\circ$ and $\Delta\bar{H}_B = \bar{H}_B - \hat{H}_B^\circ$; then by substitution Eq. (4-132) becomes

$$\Delta H_{\text{mix}}^{\text{xs}} = X_A\Delta\bar{H}_A + X_B\Delta\bar{H}_B \tag{4-133}$$

By substituting Eqs. (4-113), (4-115), and (4-133) into Eq. (4-131) we find that

$$\Delta G_{\text{mix}}^{\text{reg soln}} = X_A\Delta\bar{H}_A + X_B\Delta\bar{H}_B + RT(X_A \ln X_A + X_B \ln X_B) \tag{4-134}$$

Noting that $X_A = 1 - X_B$, Eq. (4-134) may be differentiated with respect to $X_B$ at constant temperature to give

$$\left(\frac{\partial\Delta G_{\text{mix}}^{\text{reg soln}}}{\partial X_B}\right)_{P,T} = \Delta\bar{H}_B - \Delta\bar{H}_A + RT\ln\frac{X_B}{1 - X_B} \tag{4-135}$$

If the solid solution of $A$ is nearly pure $A$, so that $1 - X_B \approx 1$, $\Delta\bar{H}_A$ can be assumed to be zero; thus Eq. (4-135) becomes

$$\left(\frac{\partial\Delta G_{\text{mix}}^{\text{reg soln}}}{\partial X_B}\right)_{P,T} \approx \Delta\bar{H}_B + RT\ln X_B \tag{4-136}$$

For a dilute solution, $\Delta\bar{H}_B$ is independent of composition, and regardless of its sign, as $X_B \to 0$, $RT \ln X_B \to -\infty$. Thus

$$\left(\frac{\partial\Delta G_{\text{mix}}^{\text{reg soln}}}{\partial X_B}\right)_{P,T}$$

becomes more negative at sufficiently small values of $X_B$. This means that the addition of small amounts of any solute will lower the free energy of the component $A$ and thus pure compounds are thermodynamically unstable. Geologists are possibly less aware than they might be that because of Eq. (4-136) rock-forming minerals are unlikely to be stoichiometric, although any deviation from stoichiometry may require more sensitive analytic techniques than those used up to now. Such minerals as feldspars and pyroxenes are likely to show some solid solution with, for example, silica or with whatever additional associated components may be present. Thus it is not surprising that Carman and Tuttle (1967) found that alkali feldspars from rhyolites contained a stoichiometric excess of silica in contrast to those from phonolites, a relatively silica-poor rock.

Because of Eq. (4-136), all curves of $\Delta G_{\text{mix}}^{\text{soln}}$ as a function of composition, as in Fig. 4-24*b*, must initially become negative as small, or perhaps minute, amounts of solute are added. Thereafter, because of a large positive $\Delta H_{\text{mix}}^{\text{xs}}$ term, the curve of $\Delta G_{\text{mix}}^{\text{soln}}$ may also become positive, especially at low temperatures, where $T\Delta S_{\text{mix}}^{\text{soln}}$ is dominated by $\Delta H_{\text{mix}}^{\text{xs}}$.

Up to this point we have rather cursorily considered nonideal binary solutions, particularly with respect to the effect of a positive or negative $\Delta H_{\text{mix}}^{\text{xs}}$ term. However, it may help the reader to appreciate more about the behavior of solutions if the $\Delta H_{\text{mix}}^{\text{xs}}$ term is examined in more detail or from another point of view.

## QUASI-CHEMICAL APPROACH TO SOLUTIONS

In any solution, liquid or solid, there is an interaction between different types of particles or atoms. There is a strain-energy (mechanical) effect, a valence (coulombic) effect, and a chemical effect, represented by the bonds between atoms; all these effects are interrelated. However, if the first two contributions to the energetics of the nonelectrolyte liquid or solid are ignored and only the chemical effect is considered (thus quasi-chemical approach), then the energetics of a solution can be considered in terms of breaking or forming bonds between atoms. In its simplest form this approach, summarized by Guggenheim (1952), ignores any influence of atomic size over several atomic distances and is concerned only with the chemical bonds between nearest atomic neighbors. The theory also requires that there be no volume change on mixing, a point which is emphasized again below.

In a solid solution of $A$ and $B$ atoms, there are $N_A$ atoms of $A$ and $N_B$ atoms of $B$; there are three types of bonds: those between the like atoms, namely$A$—$A$ and $B$—$B$, and those between the unlike atoms, $A$—$B$. Only the nearest neighbors are considered here. With each bond there is an associated interaction-energy term which we shall designate as $\varepsilon_{AA}$, $\varepsilon_{BB}$, and $\varepsilon_{AB}$. The crystal contains $N_A$ atoms of $A$, each of which has $Z$ bonds, $Z$ being the coordination number, but each $A$—$A$ bond has $2A$ atoms of type $A$; therefore if $Y_{AA}$ is the number of $A$—$A$ bonds, then the number of $A$ atoms in $A$—$A$ bonds is $2Y_{AA}/Z$. Each $A$—$B$ bond has only one $A$ atom, so that the number of $A$ atoms associated with $A$—$B$ bonds is $Y_{AB}/Z$.

Therefore the total number of $A$ atoms, $N_A$, is given by

$$N_A = \frac{2Y_{AA}}{Z} + \frac{Y_{AB}}{Z} \quad \text{or} \quad Y_{AA} = \frac{ZN_A - Y_{AB}}{2} \tag{4-137}$$

and similarly

$$N_B = \frac{2Y_{BB}}{Z} + \frac{Y_{AB}}{Z} \quad \text{or} \quad Y_{AB} = \frac{ZN_B - Y_{AB}}{2} \tag{4-138}$$

The total molar internal energy of the solution $E^{\text{soln}}$ is given by the number of each bond type multiplied by the interaction energy associated with the particular bond. It follows then that

$$E^{\text{soln}} = Y_{AA}\varepsilon_{AA} + Y_{BB}\varepsilon_{BB} + Y_{AB}\varepsilon_{AB} \tag{4-139}$$

which from Eqs. (4-137) and (4-138) gives

$$E^{\text{soln}} = \frac{ZN_A - Y_{AB}}{2}\varepsilon_{AA} + \frac{ZN_B - Y_{AB}}{2}\varepsilon_{BB} + Y_{AB}\varepsilon_{AB} \tag{4-140}$$

$$E^{\text{soln}} = \tfrac{1}{2}ZN_A\varepsilon_{AA} + \tfrac{1}{2}ZN_B\varepsilon_{BB} + Y_{AB}(\varepsilon_{AB} - \tfrac{1}{2}\varepsilon_{AA} - \tfrac{1}{2}\varepsilon_{BB}) \tag{4-141}$$

We are particularly interested in the change in energy which accompanies the formation of a solution from the pure components. Thus the internal energy difference of the reaction $N_A$ atoms of $A$ + $N_B$ atoms of $B$ = $N_A + N_B$ in solution is $\Delta E^{\text{xs}}_{\text{mix}}$ and is given (cf. Eq. (4-124)] by

$$\Delta E^{\text{xs}}_{\text{mix}} = E^{\text{soln}} - \hat{E}^{\circ\,\text{pure components}} \tag{4-142}$$

Equation (4-141) gives the internal energy of the $AB$ solution, from which we have to subtract the internal energy of the pure components $A$ and $B$ so as to get $\Delta E^{\text{xs}}_{\text{mix}}$. If both $A$ and $B$ atoms in the pure liquids or solids have the same coordination $Z$ as in the solution, then in pure $A$, $N_A$ atoms have $ZN_A/2$ bonds of $A$—$A$ type and similarly $N_B$ atoms have $ZN_B/2$ bonds of $B$—$B$ type.

The energies will be $\frac{1}{2}ZN_A\varepsilon_{AA}$ and $\frac{1}{2}ZN_B\varepsilon_{BB}$, which, if subtracted from Eq. (4-141), gives for $\Delta E^{\text{xs}}_{\text{mix}}$

$$\Delta E^{\text{xs}}_{\text{mix}} = Y_{AB}[\varepsilon_{AB} - \tfrac{1}{2}(\varepsilon_{AA} + \varepsilon_{BB})] \tag{4-143}$$

In an ideal solution $\Delta E^{\text{xs}}_{\text{mix}} = 0$, so that

$$\varepsilon_{AB} = \tfrac{1}{2}(\varepsilon_{AA} + \varepsilon_{BB}) \tag{4-144}$$

This could be coincidentally true, however, and an ideal solution is more strictly defined as one where $\varepsilon_{AA} = \varepsilon_{BB}$. If there is no volume change on mixing, then the internal-energy term $E$ may be identified with the enthalpy $H$.

REGULAR SOLUTIONS In regular solutions the intermolecular or interatomic forces are no longer equal, and moreover the atoms or molecules may also be more unequal in size. However, these differences are sufficiently small that the entropy of mixing has the ideal value. Although the term *regular solution* has been extended to solutions which do not show ideal mixing (for example, Thompson, 1967), we shall retain the more restricted definition.

In one mole of a solution $X_A = N_A/N_o$ and $X_B = N_B/N_o$, where $N_o$ is Avogadro's number. The probability that a particular $A$ atom will be on a given site is $X_A$, and the probability that a $B$ atom will be on the neighboring site is $X_B$. The probability that $A$ and $B$ atoms will be both simultaneously on these sites is $X_A X_B$; the probability that a $B$ atom will be on the first site (rather than an $A$ atom) is given by $X_B$ and that the nearest neighbor will be an $A$ atom by $X_A$, so that the total probability is $2X_A X_B$.

The total number of bonds in a crystal is $\frac{1}{2}ZN_o$, so that the total number of $A$—$B$ bonds, $Y_{AB}$, is given by the total number of bonds multiplied by the probability of an $A$—$B$ bond. Therefore

$$
\begin{aligned}
Y_{AB} &= 2X_A X_B \tfrac{1}{2} Z N_o \\
&= X_A X_B Z N_o
\end{aligned}
\tag{4-145}
$$

Substituting into Eq. (4-143) gives

$$\Delta E^{\text{xs}}_{\text{mix}} = X_A X_B Z N_o [\varepsilon_{AB} - \tfrac{1}{2}(\varepsilon_{AA} + \varepsilon_{BB})] \tag{4-146}$$

so that if

$$\omega = [\varepsilon_{AB} - \tfrac{1}{2}(\varepsilon_{AA} + \varepsilon_{BB})] Z N_o \tag{4-147}$$

$$\Delta E^{\text{xs}}_{\text{mix}} = X_A X_B\, \omega \tag{4-148}$$

and the energy of mixing is a symmetrical function about the midpoint composition.* $\omega$ is called the interaction function or parameter and can be positive or negative. If the volume change on mixing is zero or can be neglected, then $\Delta E^{\text{xs}}_{\text{mix}}$ can be identified with $\Delta H^{\text{xs}}_{\text{mix}}$ and therefore

$$\Delta H^{\text{xs}}_{\text{mix}} = X_A X_B\, \omega \tag{4-149}$$

If the interaction between unlike atoms is attractive, then $\Delta E^{\text{xs}}_{\text{mix}}$ is negative[1]

* Note the symmetry of Fig. 4-23.

[1] Attractive forces lead to negative energies.

and $\varepsilon_{AB}$ is less than the average of $\varepsilon_{AA}$ and $\varepsilon_{BB}$. A negative $\Delta E_{\text{mix}}^{\text{xs}}$ often results in long-range order or in the formation of an intermediate compound. Conversely, in solutions with a positive $\Delta E_{\text{mix}}^{\text{xs}}$, or ones in which the interaction between like atoms is stronger than between unlike atoms, there is a tendency to separate into $A$-rich and $B$-rich phases; this tendency is increasingly restricted with increasing temperature because of the $T\Delta S_{\text{mix}}^{\text{soln}}$ term.

The activity coefficient $\gamma_A$ in a regular solution can be related to the interaction parameter $\omega$. From Eq. (4-123) we have

$$\Delta H_{\text{mix}}^{\text{xs}} = H^{\text{soln}} - (X_A\hat{H}_A^\circ + X_B\hat{H}_B^\circ) \tag{4-150}$$

and by substituting Eq. (4-132) we obtain

$$H^{\text{soln}} = X_A\bar{H}_A + X_B\bar{H}_B \tag{4-151}$$

so that

$$dH^{\text{soln}} = \bar{H}_A\,dX_A + \bar{H}_B\,dX_B \tag{4-152}$$

and because $X_A + X_B = 1$, then $dX_A = -dX_B$ and $dH^{\text{soln}} = (\bar{H}_B - \bar{H}_A)\,dX_B$ or

$$\bar{H}_B = \bar{H}_A + \frac{dH^{\text{soln}}}{dX_B} \tag{4-153}$$

By substituting Eq. (4-153) rearranged for $\bar{H}_A$, we find that

$$\bar{H}_B = H^{\text{soln}} + (1 - X_B)\frac{dH^{\text{soln}}}{dX_B} \tag{4-154}$$

and similarly for $\bar{H}_A$

$$\bar{H}_A = H^{\text{soln}} - X_B\frac{dH^{\text{soln}}}{dX_B} \tag{4-155}$$

Substituting Eq. (4-150) into Eq. (4-155), we obtain

$$\bar{H}_A = (1 - X_B)\hat{H}_A^\circ + X_B\hat{H}_B^\circ + \Delta H_{\text{mix}}^{\text{xs}} - X_B\frac{d[(1 - X_B)\hat{H}_A^\circ + X_B\hat{H}_B^\circ + \Delta H_{\text{mix}}^{\text{xs}}]}{dX_B} \tag{4-156}$$

After substituting Eq. (4-149) for $\Delta H^{xs}_{mix}$ and differentiating, we get

$$\bar{H}_A - \hat{H}^\circ_A = (1 - X_A)^2 \omega \tag{4-157}$$

From Eq. (4-104) we have

$$\mu^{soln}_A - \mu^\circ_A = \bar{H}_A - \hat{H}^\circ_A - T(\bar{S}_A - \hat{S}^\circ_A) = RT \ln X_A + RT \ln \gamma_A \tag{4-158}$$

and substituting Eqs. (4-157) and (4-103) we arrive at

$$\ln \gamma_A = \frac{(1 - X_A)^2 \omega}{RT} \tag{4-159}$$

Thus $\gamma_A$, the activity coefficient of $A$ in a regular solution, is directly related to the interaction function $\omega$, and when $\omega$ is positive, or like atoms are more attracted than unlike, log $\gamma_A$ is positive and $\gamma_A > 1$. Conversely, if unlike atoms are more attracted than like, or $\omega$ is negative, then log $\gamma_A$ is negative and $\gamma_A < 1$.

Regular solution behavior is possibly more common in liquids than in solids, and since the partial molar volumes of the various oxide components in a wide range of silicate liquids are not disparate from the molar volumes of the pure oxides (Bottinga and Weill, 1970), it is a plausible hypothesis that silicate magmas may behave as regular solutions. Thus Kudo and Weill (1970) used a function based on Eq. (4-159) for a variety of silicate liquids in their development of a plagioclase thermometer; to the extent that their calculated and observed temperatures agreed, the postulate, or the evidence that rhyolitic and basaltic liquids behave as regular solutions, seems reasonable at this point. Shaw (1964) also used a regular solution approach with a negative $\Delta H^{xs}_{mix}$ to model liquid-silicate–water solutions.

Since the mixing properties for almost all the common igneous minerals are unknown, we usually have to assume ideal mixing ($\Delta H^{xs}_{mix} = 0$) of their various components in any thermodynamic treatment of mineral assemblages. This is not entirely satisfactory, but by using Eq. (4-159) it is possible to determine the sign of log $\gamma_A$ since in all solid-solution series which unmix at low temperatures—such as the alkali feldspars, ilmenite hematite, ulvospinel magnetite, and certain pyroxene compositions—log $\gamma_A$ is positive.

THE SOLVUS AND THE SPINODAL The boundary of the immiscible region, which in solids is called a solvus, can also be examined by using the quasi-chemical approach. At any temperature below that of the crest of the solvus, which is called the critical temperature $T_c$ (Fig. 4-24*c*), the free-energy-of-mixing curve for an asymmetrical solid solution could look like the curve in Fig. 4-25. It is

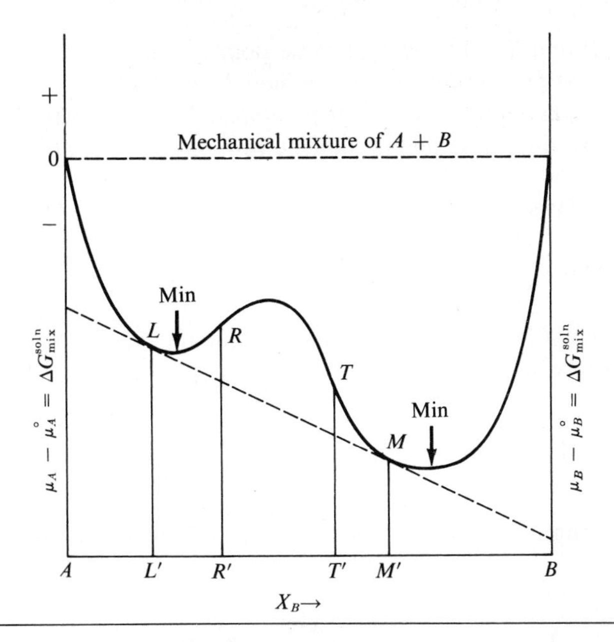


**Fig. 4-25** Hypothetical curve of $G_{mix}^{soln}$ for a strongly asymmetric solution. $L$ and $M$ are the two points of tangency, and the two minima on $\Delta G_{mix}^{soln}$ clearly depart from these. The two points of inflection, or spinodes, are at $R$ and $T$. Extrapolation of the tangent to the ordinates gives values for $\mu_A - \mu_A^\circ$ and $\mu_B - \mu_B^\circ$ in the solid phases $L$ and $M$.

important to remember that the points of tangency at $L$ and $M$, defining the stable assemblage for bulk compositions lying between $L$ and $M$, do not coincide with the two minima on the free-energy curve, since the requirement for equilibrium is that the chemical potential of each component should be the same in each phase and not that each phase should have the lowest free energy.

The chemical potential of $B$ in the coexisting solid solutions $L$ and $M$ is given by the intersection of the tangent with the free-energy ordinate at $X_B = 1$. From Eq. (4-70) we can write

$$\mu_B^{sol\,L} = \mu_B^{sol\,M} = \mu_B^\circ + RT \ln a_B^{sol\,L,M}$$

where $\mu_B^\circ$ is the chemical potential of $B$ in pure solid $B$ at the same temperature and pressure, and is of course zero on the $\Delta G_{mix}^{soln}$ ordinate. By analogous extrapolation of the tangent to $X_A = 1$, we can calculate the chemical potential, or the activity, of $A$ in the two solids $L$ and $M$.

The locus of the two points $L$ and $M$ with respect to temperature (at constant pressure) defines the solvus of the system $A$–$B$. As the temperature is raised, $L$ and $M$ come closer together until at the critical temperature $T_c$ they coincide. For a regular (symmetrical) solution $X_A = X_B$ at $T_c$.

In the free-energy curve in Fig. 4-25 there are two points of inflection, or spinodes, at $R$ and $T$. The locus of these points with respect to temperature, again at constant pressure, defines the spinodal which influences the nucleation of a phase during cooling. At a higher temperature than that represented in Fig 4-25, the two inflection points coincide with one another at $T_c$; accordingly $\partial^2 \Delta G_{\text{mix}}^{\text{soln}}/(\partial X_B)^2 = 0$ at $T_c$.

For a regular solution we can write [Eqs. (4-149) and (4-115)]

$$\Delta G_{\text{mix}}^{\text{reg soln}} = X_A X_B \, \omega + RT(X_A \ln X_A + X_B \ln X_B) \tag{4-160}$$

and noting that $X_A = 1 - X_B$, we can differentiate Eq. (4-160) to obtain the following for a regular solution:[1]

$$\left(\frac{\partial \Delta G_{\text{mix}}^{\text{reg soln}}}{\partial X_B}\right)_{T_c,P} = (1 - 2X_B)\omega + RT \ln \frac{X_B}{1 - X_B} \tag{4-161}$$

and by differentiating Eq. (4-161) we obtain the following for the two inflection points:

$$\left(\frac{\partial^2 \Delta G_{\text{mix}}^{\text{reg soln}}}{(\partial X_B)^2}\right)_{T_c,P} = -2\omega + \frac{RT}{X_B(1 - X_B)} = 0 \tag{4-162}$$

which by rearrangement gives

$$T = \frac{2X_B(1 - X_B)\omega}{R} \tag{4-163}$$

so that as $T$ is positive, $\omega$ will also be positive; for a regular solution $T$ will have a maximum value at $X_A = X_B = 0.5$. It is this maximum value which is the critical temperature $T_c$, so that by substitution of $X_B = 0.5$ into Eq. (4-163) we find that

$$T_c = \frac{\omega}{2R} \tag{4-164}$$

By substituting for $\omega$, we may obtain values for $\gamma_A$ by using Eq. (4-159). From Eq. (4-147) we recall that

$$\omega = ZN_o[\varepsilon_{AB} - \tfrac{1}{2}(\varepsilon_{AA} + \varepsilon_{BB})]$$

so that the greater the repulsive interaction between unlike atoms, the higher the critical temperature.

[1] For nonregular (asymmetric) solutions, $\partial \Delta G_{\text{mix}}^{\text{soln}}/\partial X_B \neq 0$ at the critical temperature, but the second and third derivatives do equal zero at $T_c$.

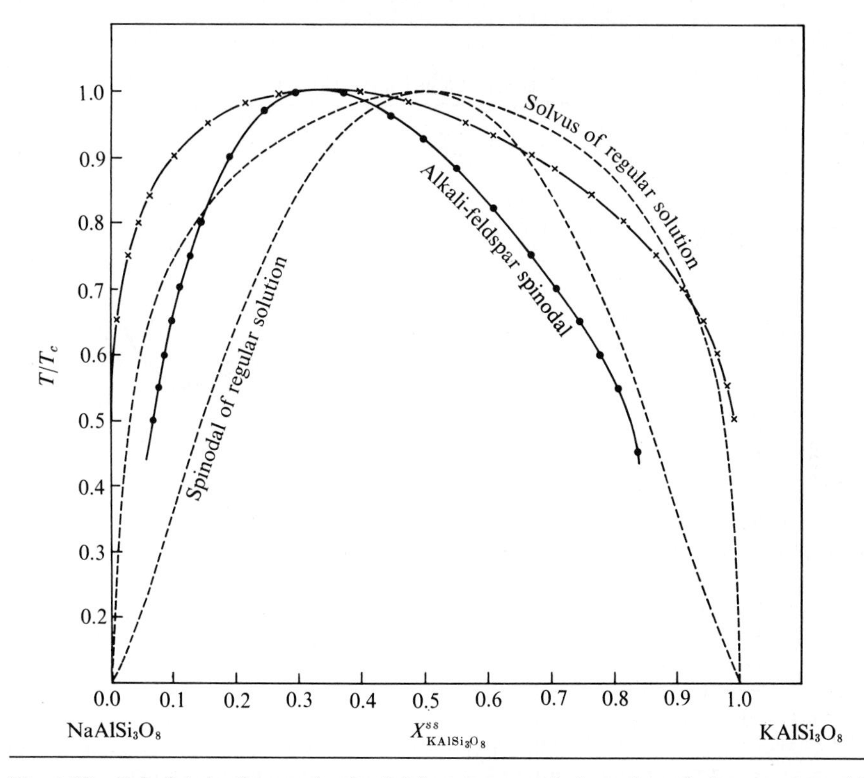


**Fig. 4-26** Calculated solvus and spinodal for a symmetrical regular solution together with the corresponding curves (Waldbaum and Thompson, 1969) for the high-temperature alkali feldspars at 1 bar.

By substituting $\omega = 2RT_c$ [Eq. (4-164)] into Eq. (4-161) we obtain

$$\frac{T}{T_c} = \frac{2(1 - X_B)}{\ln(1 - X_B) - \ln X_B} \tag{4-165}$$

for the equation of the solvus, and by making the same substitution into Eq. (4-163) we get for the spinodal

$$\frac{T}{T_c} = +X_B(1 - X_B) \tag{4-166}$$

The calculated solvus and spinodal for a regular solution are plotted in Fig. 4-26 together with the corresponding curves for the familiar asymmetrical alkali-feldspar system (Waldbaum and Thompson, 1969). Although we have focused attention on a regular (symmetric) solution because the equations are simpler to deal with, unmixing is not restricted to this type of solution behavior but can occur in any nonideal solution with a positive $\Delta H^{xs}_{mix}$.

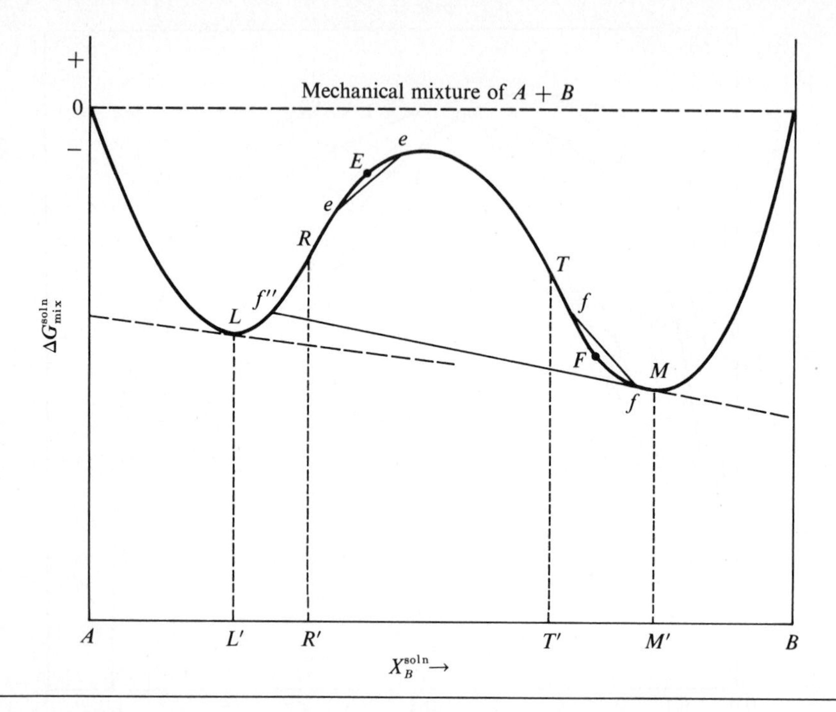


**Fig. 4-27** Diagram to illustrate change in $\Delta G^{soln}_{mix}$ with growth of compositional fluctuations. A crystal of composition $E$ would become more stable if it decomposed to $e$ and $e$ at constant temperature; a crystal $F$ would become less stable if it contained fluctuations of $f$ and $f$—only by nucleating $f''$ can it minimize its free energy. $R$ and $T$ are the spinodes and $L$ and $M$ are the points of tangency on the $\Delta G^{soln}_{mix}$ curve. Reference state taken as a mechanical mixture of pure components $A$ and $B$.

Any homogeneous phase, such as a crystal, contains random compositional fluctuations on an atomic scale, the growth of which results in exsolution of one phase in the host of another. The free-energy changes which result from these fluctuations are illustrated by the solvus and spinodal, although the rate at which a fluctuation can grow is presumably limited by diffusion. Fluctuation of two types can occur; the first is small in volume or extent but large in degree in the sense that a large compositional difference exists between the embryonic phase and the host. The second type of fluctuation is large in extent, but the nuclei have only a small, but variable, difference in composition to the host; these fluctuations tend to have diffuse boundaries unlike the first type of fluctuation.

Referring to Fig. 4-27: in any crystal, such as $E$, whose composition lies between the spinodes $R$ and $T$, a compositional fluctuation can grow (at constant temperature) because even a small change in composition lowers the free energy of the crystal, as shown by the line $e$–$e$; this is called spinodal decomposition or exsolution. Contrast this to a crystal with the composition $F$ outside the spinodal but within the solvus ($L$–$R$ and $T$–$M$), where an increase in the difference between the composition of the fluctuation and the host raises the free energy, as shown by the line $f$–$f$. Only by the nucleation of another phase such as $f''$ could the free energy of the crystal $F$ be reduced at constant temperature. This ignores of course any contribution by the interface to the free energy.

Crystals whose composition lies between $A$ and $L'$ or between $M'$ and $B$ are *stable* at the temperature represented in Fig. 4-27; crystals whose composition lies between $L'$ and $R'$ or between $T'$ and $M'$ are *metastable*; and those whose composition lies between the spinodes $R'$ and $T'$ are *unstable*. Decomposition, or the growth of nuclei in an unstable crystal, involves an increase in interface free energy but a decrease in free energy in terms of composition. A reaction of this type tends to be less sluggish than those of similar type in metastable crystals where the growth of nuclei involves an increase in free energy due to the interface term and to the composition of the growing phase. Spinodal decomposition, or exsolution, has been found by Owen and McConnell (1971) in alkali feldspar, and the nature of the compositional variation in nepheline found by Brown (1970) suggests spinodal decomposition. The whole subject of exsolution has been well summarized by Yund and McCallister (1970), who take particular account of interface effects.

Many silicate phase diagrams exhibit the effects of a positive $\Delta H^{xs}_{mix}$, and solidus-liquidus relationships with a minimum such as the alkali feldspars or the melilite series are both good examples of $(\Delta H^{xs}_{mix})^{sol}$ of the solid being more positive than $(\Delta H^{xs}_{mix})^{liq}$ of the liquid. All systems with a positive $(\Delta H^{xs}_{mix})^{sol}$ should show a miscibility gap, but unless $(\Delta H^{xs}_{mix})^{sol}$ is large, $T_c$ will be so low that the kinetics of the unmixing will be prohibitively slow; thus a solvus, so far unreported, is to be expected in the melilite ($Ca_2Al_2SiO_7$–$Ca_2MgSi_2O_7$) series. As $(\Delta H^{xs}_{mix})^{sol}$ becomes more positive, the critical temperature of the solvus is raised, and as $(\Delta H^{xs}_{mix})^{liq}$ becomes more positive, the temperature of the liquidus minimum is lowered, so that systems with large positive values of $\Delta H^{xs}_{mix}$ form eutectics as a result of the solvus intersecting the solidus-liquidus curves. Muir (1954) has suggested that in the crystallization path of pyroxenes in basaltic magma, the pyroxene solidus ceases to intersect the pyroxene solvus as the pyroxenes become iron-rich, and a minimum relationship is generated, so allowing only one pyroxene, an augite, to crystallize thereafter. This set of relationships implies that $(\Delta H^{xs}_{mix})^{sol}$ of magnesian pyroxenes is greater than that for the iron-rich pyroxenes, and since the solvus-solidus intersection depends on $(\Delta H^{xs}_{mix})^{liq}$ we should expect it to vary in temperature or composition from one magma type to another and especially from basalt to rhyolite.

With this short treatment of some factors that affect $\Delta H^{xs}_{mix}$, which has excluded any consideration of long- and short-range order, we turn to two petrologically important and familiar examples of what has been touched on so far. It is worth noting that $\Delta H^{xs}_{mix}$ is dominated at low pressures by the internal energy term $\Delta E^{xs}_{mix}$ and that $\Delta V^{xs}_{mix}$ makes a significant contribution only at high pressures.

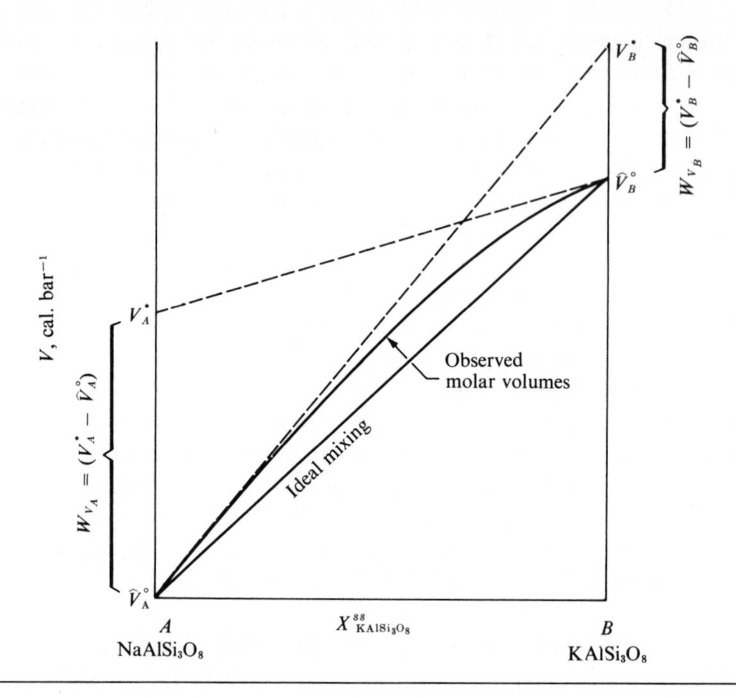


**Fig. 4-28** Molar volumes of a solid solution of components $A$ and $B$ showing departure from ideality. Dashed lines denote limiting slopes of the molar volume curve and correspond to the partial molar volumes at infinite dilution $V_A^*$ and $V_B^*$. (After Waldbaum and Thompson, 1968.)

## THE HIGH-TEMPERATURE ALKALI FELDSPARS: AN EXAMPLE OF POSITIVE $\Delta H_{\text{mix}}^{\text{xs}}$ AND COMPARABLE CRYSTAL STRUCTURES

Up to this point, we have been concerned mainly with the behavior of a binary symmetrical solution, the relationship between two components with similar structures and similar volumes. But many common rock-forming minerals are not so well behaved; as examples of two such systems we look at the high-temperature alkali feldspars, a landmark study by Waldbaum and Thompson, and by way of contrast, the system Fe–$S_2$, which illustrates many aspects of a binary system not found in the alkali feldspars.

As it is the volume properties of a solution which many readers will find most easy to visualize, we can use these to illustrate the form of the thermodynamic excess functions discussed above. Let the volume $V^{\text{soln}}$ per mole of a binary solid-solution series be given in terms of a power series function of the solute $X_B$ as follows:

$$V^{\text{soln}} = a + bX_B + cX_B^2 + dX_B^3 + \cdots \tag{4-167}$$

where $a$, $b$, $c$, and $d$ are constants derived by least-squares analysis of the volume data; a curve of this type is shown in Fig. 4-28. The experimental data do not justify inclusion of the fourth order-term. Although this type of power series, called a Maclaurin series, is commonly used to fit sets of experimental data such as the free energy of a solution, great care has to be taken to ensure that the derivatives of the chosen series agree with experimental measurement of the relevant properties (E. J. Green, 1970).

When the $B$ component is absent ($X_B = 0$), we have from Eq. (4-167)

$$V^{\text{soln}} = a = \hat{V}_A^\circ \tag{4-168}$$

where $\hat{V}_A^\circ$ is the molar volume of pure $A$.

When $X_B = 1$, we find

$$V^{\text{soln}} = a + b + c + d = \hat{V}_B^\circ \tag{4-169}$$

where $\hat{V}_B^\circ$ is the molar volume of pure $B$.

The slope of the tangent to the volume curve at $X_B = 0$ (Fig. 4-28) is given by differentiating Eq. (4-167):

$$\left(\frac{\partial V^{\text{soln}}}{\partial X_B}\right)_{P,T(X_B=0)} = b \tag{4-170}$$

which is the partial molar volume of $B$ in $A$ at infinite dilution. This tangent intercepts the $B$ ordinate (Fig. 4-28) at $V_B^*$, the value of which is given by

$$V_B^* = \hat{V}_A^\circ + 1\left(\frac{\partial V^{\text{soln}}}{\partial X_B}\right)_{P,T(X_B=0)} \tag{4-171}$$

or from Eq. (4-170)

$$V_B^* - \hat{V}_A^\circ = b \tag{4-172}$$

Similarly (Fig. 4-28)

$$V_A^* = \hat{V}_B^\circ - 1\left(\frac{\partial V^{\text{soln}}}{\partial X_B}\right)_{P,T(X_B=1)} \tag{4-173}$$

but since

$$\left(\frac{\partial V^{\text{soln}}}{\partial X_B}\right)_{P,T(X_B=1)} = b + 2c + 3d \tag{4-174}$$

then

$$-(V_A^* - \hat{V}_B^\circ) = b + 2c + 3d \tag{4-175}$$

From Eqs. (4-168), (4-169), (4-172), and (4-175) we get

$$a = \hat{V}_A^\circ \qquad b = V_B^* - \hat{V}_A^\circ \qquad c = (V_A^* - V_A^\circ) - 2(V_B^* - V_B^\circ)$$

$$d = (V_B^* - \hat{V}_B^\circ) - (V_A^* - \hat{V}_A^\circ)$$

which can be substituted back into Eq. (4-167) to give

$$V^{\text{soln}} = \hat{V}_A^\circ + (V_B^* - \hat{V}_A^\circ)X_B + [(V_A^* - \hat{V}_A^\circ) - 2(V_B^* - \hat{V}_B^\circ)]X_B^2 + [(V_B^* - \hat{V}_B^\circ) - (V_A^* - \hat{V}_A^\circ)]X_B^3 \tag{4-176}$$

Inspection of Fig. 4-28 shows that if $V_A^* = \hat{V}_A^\circ$ and $V_*^B = \hat{V}_B^\circ$, the solution is ideal and $c = d = 0$ [Eq. (4-167)], and Eq. (4-176) reduces to

$$V^{\text{soln}} = X_A\hat{V}_A^\circ + X_B\hat{V}_B^\circ \tag{4-177}$$

As shown in Fig. 4-28, let

$$W_{V_A} = V_A^* - \hat{V}_A^\circ \tag{4-178}$$

and necessarily

$$W_{V_A} = -(c + 2d) \tag{4-179}$$

and

$$W_{V_B} = V_B^* - V_B^\circ \tag{4-180}$$

and

$$W_{V_B} = -(c + d) \tag{4-181}$$

Then if

$$W_{V_A} = W_{V_B} = W_V \tag{4-182}$$

the solution is symmetrical (Fig. 4-28) and Eq. (4-176) becomes

$$V^{\text{soln}} = X_A\hat{V}_A^\circ + X_B\hat{V}_B^\circ + W_V X_A X_B \tag{4-183}$$

If, however, $W_{V_A} \neq W_{V_B}$, we have an asymmetric solution the volume of which is given by

$$V^{\text{soln}} = X_A\hat{V}_A^\circ + X_B\hat{V}_B^\circ + W_{V_B}X_B X_A^2 + W_{V_A}X_A X_B^2 \tag{4-184}$$

From Eq. (4-121), the excess volume of mixing $\Delta V_{\text{mix}}^{\text{xs}}$ is given by

$$\Delta V_{\text{mix}}^{\text{xs}} = V^{\text{soln}} - (X_A \hat{V}_A^\circ + X_B \hat{V}_B^\circ)$$

so that by subtracting Eq. (4-121) from Eq. (4-183) we find for a symmetrical solution

$$\Delta V_{\text{mix}}^{\text{xs}} = W_V X_A X_B \tag{4-185}$$

and for an asymmetrical solution [Eq. (4-121)]

$$\Delta V_{\text{mix}}^{\text{xs}} = W_{V_B} X_B X_A^2 + W_{V_A} X_A X_B^2 \tag{4-186}$$

We have seen that the quasi-chemical approach for a symmetrical solution led to Eq. (4-148):

$$\Delta E_{\text{mix}}^{\text{xs}} = X_A X_B \, \omega$$

and we may write for 1 mole of a symmetrical solution

$$\Delta E_{\text{mix}}^{\text{xs}} = W_E X_A X_B \tag{4-187}$$

where $W_E = \omega$, the interaction parameter*

$$\Delta V_{\text{mix}}^{\text{xs}} = W_V X_A X_B \tag{4-188}$$

$$\Delta S_{\text{mix}}^{\text{xs}} = W_S X_A X_B \tag{4-189}$$

$$\Delta G_{\text{mix}}^{\text{xs}} = W_G X_A X_B \tag{4-190}$$

and it follows that

$$W_G = W_E + PW_V - TW_S \tag{4-191}$$

or in the case of an asymmetric solution, where for example $W_{V_A} \neq W_{V_B}$, we can write

$$W_{G_A} = W_{E_A} + PW_{V_A} - TW_{S_A} \tag{4-192}$$

and the $W$'s can be treated just like any standard thermodynamic function. J. B. Thompson (1967) and Waldbaum have derived these and other formulas

* $W_{E_A}$, $W_{G_A}$, and $W_{S_A}$ are related to $E_A$, $G_A$, and $S_A$, respectively, in the same way that $W_{V_A}$ is to $V_A$ [Eqs. (4-178) and (4-182)].

together with values for the excess functions of the high-temperature alkali feldspars. The experimental data required for their evaluation are measurements of molar volumes or unit-cell size, conjugate compositions in the immiscible region (solvus), and free energies of formation. Waldbaum and Thompson (1969) conclude their study with an equation of state for the high-temperature alkali feldspars which we quote here:

$$\Delta E^{xs}_{mix} = 6326.7 X_{Ab} X^2_{Or} + 7671.8 X_{Or} X^2_{Ab} \text{ cal mole}^{-1} \quad (4\text{-}193)$$

$$\Delta V^{xs}_{mix} = 0.0925 X_{Ab} X^2_{Or} + 0.1121 X_{Or} X^2_{Ab} \text{ cal bar}^{-1} \text{ mole}^{-1} \quad (4\text{-}194)$$

$$\Delta S^{xs}_{mix} + \Delta S^{ideal}_{mix} = 4.6321 X_{Ab} X^2_{Or} + 3.8565 X_{Or} X^2_{Ab} - R(X_{Ab} \ln X_{Ab} + X_{Or} \ln X_{Or}) \text{ cal deg}^{-1} \text{ mole}^{-1} \quad (4\text{-}195)$$

$$\Delta S^{ideal}_{mix} = -R(X_{Ab} \ln X_{Ab} + X_{Or} \ln X_{Or}) \text{ cal deg}^{-1} \text{ mole}^{-1} \quad (4\text{-}196)$$

which can be combined to give

$$\Delta G^{xs}_{mix} = G^{soln} - \Delta G^{ideal}_{mix} = (6326.7 + 0.0925P - 4.6321T) X_{Ab} X^2_{Or} + (7671.8 + 0.1121P - 3.8565T) X_{Or} X^2_{Ab} \quad (4\text{-}197)$$

where $T$ is temperature (°K), $P$ is pressure (bars), and $X_{Ab}$ and $X_{Or}$ are the mole fractions of the components $NaAlSi_3O_8$ and $KAlSi_3O_8$, respectively. The numerical coefficients in this group of equations are called Margules parameters and have been calculated and smoothed from experimental data spanning the $PT$ range of 2 to 10 kb and 500 to 700°C. Thus extrapolation far outside this $PT$ range introduces errors, although they may be small.

Using the values of Eqs. (4-193) to (4-197), we have plotted (Fig. 4-29) the mixing properties of the alkali feldspars at 10 kb and 800°C, a temperature above the critical temperature at this pressure; thus the curve of $\Delta G^{soln}_{mix}$ does not have two minima (cf. Fig. 4-25) or two inflection points. The sum of the $\Delta E^{xs}_{mix}$ and $P\Delta V^{xs}_{mix}$ curves would correspond to $\Delta H^{xs}_{mix}$, which is obviously strongly positive at these temperatures and pressures. The $\Delta G^{soln}_{mix}$ curve, or the molar free energy of mixing, is the sum of the $\Delta E^{xs}_{mix}$, $P\Delta V^{xs}_{mix}$, and $-T\Delta S^{soln}_{mix}$ curves, which is markedly asymmetric despite the apparent symmetry of its component curves. The curve for $\Delta G^{xs}_{mix}$ has not been plotted, but it would be given by the sum of $\Delta E^{xs}_{mix}$, and $P\Delta V^{xs}_{mix}$, $- T(\Delta S^{soln}_{mix} - \Delta S^{ideal}_{mix})$ and, as seen in Fig. 4-29, would be slightly positive at this temperature and pressure. This type of free-energy diagram is not generally available for many other solid solutions of geologic interest, but it seems that the near future will bring results of the thermo-

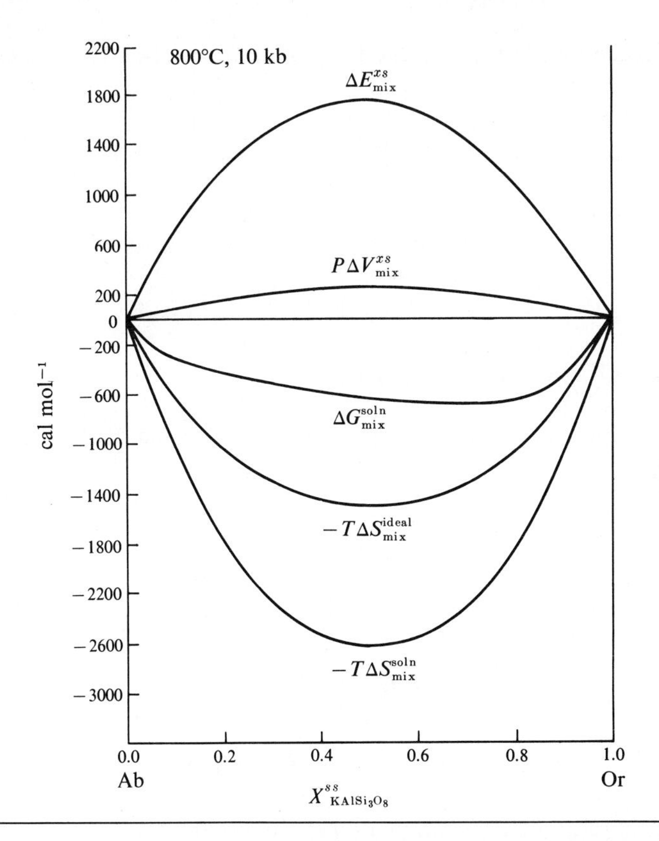


**Fig. 4-29** Mixing properties of high-temperature alkali feldspars at 800°C and 10 kb. Note the asymmetry of the $\Delta G^{soln}_{mix}$ curve despite the apparent symmetry of $\Delta E^{xs}_{mix}$, $P\Delta V^{xs}_{mix}$, and $T\Delta S^{soln}_{mix}$ curves. The curve for $\Delta G^{xs}_{mix}$ (not shown) would correspond to the sum of the curves $\Delta E^{xs}_{mix}$ and $P\Delta V^{xs}_{mix}$, minus the distance separating the two $T\Delta S$ curves, and would be slightly positive. Reference state is mechanical mixture of pure end members at 800°C and 10 kb. (After Waldbaum and Thompson, 1969.)

dynamic analysis of a number of two- and perhaps three-component systems over a range of pressures and temperatures; perhaps before this book is in press, data for the muscovite-paragonite (Eugster et al., 1972), ilmenite-hematite, and diopside-enstatite systems will be completed. But for geologists or petrologists there is another, perhaps more immediate use to which a study of this type can be put; the alkali feldspars are found in many igneous rocks, and often their paragenesis can be used to derive information about the composition of the aqueous solution or perhaps about the fugacity or activity of another component. It is therefore necessary to know the relationship between the activity of a component and its mole fraction. This information can readily be extracted for the two components of the alkali-feldspar series.

The free energy per mole of a binary solution is taken from Eq. (4-117) which gives

$$G^{\text{soln}} = X_A\mu_A^\circ + X_B\mu_B^\circ + RT(X_A \ln X_A + X_B \ln X_B) + \Delta G_{\text{mix}}^{\text{xs}} \tag{4-198}$$

so that substituting Eq. (4-190) for an asymmetrical solution we find

$$G^{\text{soln}} = X_A\mu_A^\circ + X_B\mu_B^\circ + RT(X_A \ln X_A + X_B \ln X_B) + W_{G_A}X_AX_B^2 + W_{G_B}X_BX_A^2 \tag{4-199}$$

The chemical potential of $A$ in this binary solution is given by [cf. Eq. (4-154)]

$$\mu_A = G^{\text{soln}} - X_B\left(\frac{\partial G^{\text{soln}}}{\partial X_B}\right)_{P,T} \tag{4-200}$$

so that by differentiating Eq. (4-200) we get

$$\left(\frac{\partial G^{\text{soln}}}{\partial X_B}\right)_{P,T} = (\mu_B^\circ - \mu_A^\circ) + RT \ln \frac{X_B}{X_A} + W_{G_A}X_B(3X_A - 1) - W_{G_B}X_A(3X_B - 1) \tag{4-201}$$

and substituting Eqs. (4-201) and (4-199) in Eq. (4-200) gives

$$\mu_A = \mu_A^\circ + RT \ln X_A + X_B^2[W_{G_A} + 2(W_{G_B} - W_{G_A})X_A] \tag{4-202}$$

But since, from Eq. (4-70)

$$\mu_A = \mu_A^\circ + RT \ln a_A$$

then

$$\log a_A = \log X_A + \frac{X_B^2[W_{G_A} + 2(W_{G_B} - W_{G_A})X_A]}{2.303RT} \tag{4-203}$$

and by using the appropriate values of $W_{G_A}$ and $W_{G_B}$ we can calculate the activities of the $NaAlSi_3O_8$ and $KAlSi_3O_8$ components in high-temperature alkali feldspars. The activity curves for temperatures up to 1200°C and at 1000 bars pressure are shown in Fig. 4-30. These curves allow the activity of either component to be determined at 1000 bars and at any temperature if the composition of the high-temperature alkali feldspar is known.

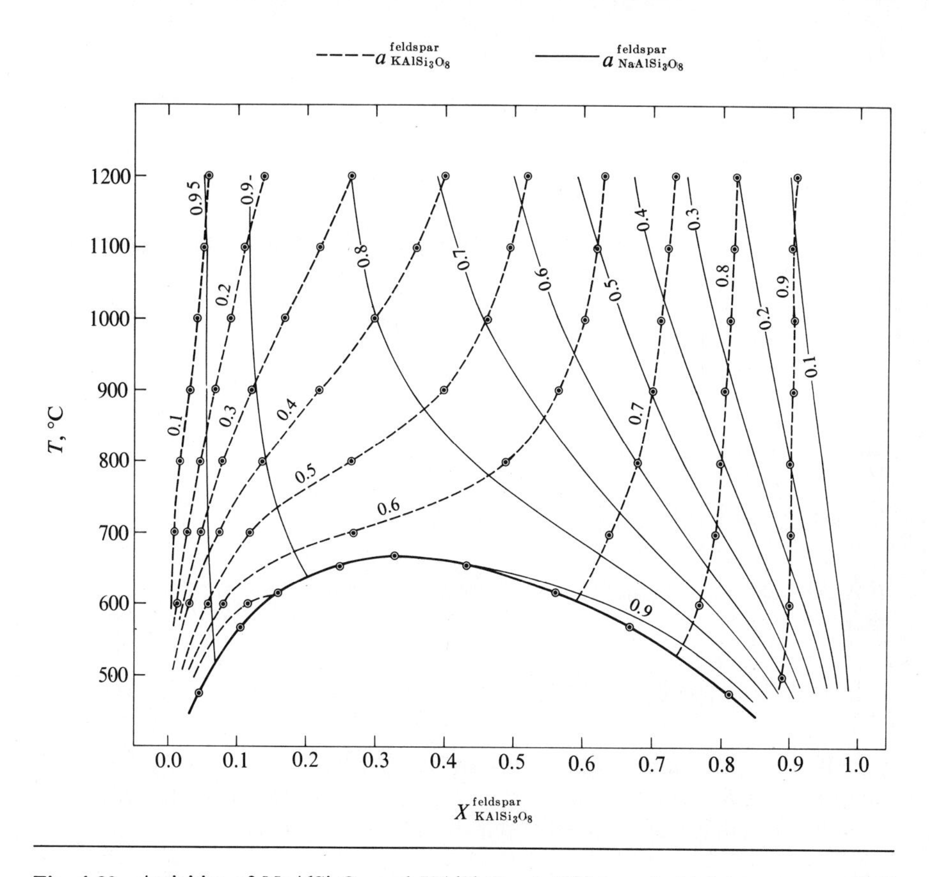


**Fig. 4-30** Activities of $NaAlSi_3O_8$ and $KAlSi_3O_8$ at 1000 bars in high-temperature alkali feldspars plotted as a function of mole fractions of the two components.

The discussion so far has stressed the mixing properties of the high-temperature alkali feldspars, and the entropy and enthalpy changes accompanying this mixing. But if a low-temperature feldspar is heated and changes to a high-temperature modification there will be a contribution to the entropy caused by the increasing randomness of the Al and Si atoms in the tetrahedral sites; this is called configurational entropy[1] and it may be calculated by using Eq. (4-109). In one cell of high-temperature $KAlSi_3O_8$ there will be one Al and three Si atoms distributed randomly over four tetrahedral lattice sites. Thus

$$W_{mix} = \frac{4!}{3!1!} = 4$$

[1] For a discussion of configurational entropy in melilites and spinels see Waldbaum (1973).

and for 1 mole

$$\Delta S_{\text{mix}}^{\text{ideal}} = N_o k \ln 4$$

$$= 1.39R$$

The crystallographic unit cell of sanidine contains four formula units ($Z = 4$), so that now we have

$$W_{\text{mix}} = \frac{16!}{12!4!} = 1820$$

and for 1 mole

$$\Delta S_{\text{mix}}^{\text{ideal}} = 1.88R$$

If $N_o$Al and $3N_o$Si are assumed to be randomly distributed over $4N_o$ lattice sites, then for 1 mole

$$\Delta S_{\text{mix}}^{\text{ideal}} = 2.25R$$

If these $N_o$Al and $3N_o$Si atoms are randomly distributed over $4N_o$ lattice sites, there is a finite probability that configurations other than $KAlSi_3O_8$ will occur. The chances of all possible Al–Si configurations are

$$\begin{aligned}
KAl_4O_8 &= 1 \text{ in } 256\\
KSi_4O_8 &= 81 \text{ in } 256\\
KAlSi_3O_8 &= 108 \text{ in } 256\\
KAl_2Si_2O_8 &= 54 \text{ in } 256\\
KAl_3SiO_8 &= 12 \text{ in } 256
\end{aligned}$$

in which it is reassuring to find $KAlSi_3O_8$ as the most probable configuration. It is usually assumed in tables of thermodynamic data that high-temperature solids are disordered over 1 mole of cells, so that the entropy of sanidine is greater than that of microcline (ordered Al–Si distribution) by $2.25R$. But this is not necessarily so at the temperature of interest, since disordering over a mole of cells entails the presence of cells carrying a net electrical charge ($-3$ for $KAl_4O_8$, $+1$ for $KSi_4O_8$, and so on and occurring with the frequencies just stated; of 256 cells only 108 would be electrically neutral. The charge unbalance entails in turn an electric (coulomb) potential energy $\Delta E_e$. Thus complete disordering over a mole of cells could occur only at a temperature $T$ such that $T(2.25 - 1.39)R \geq \Delta E_e$.

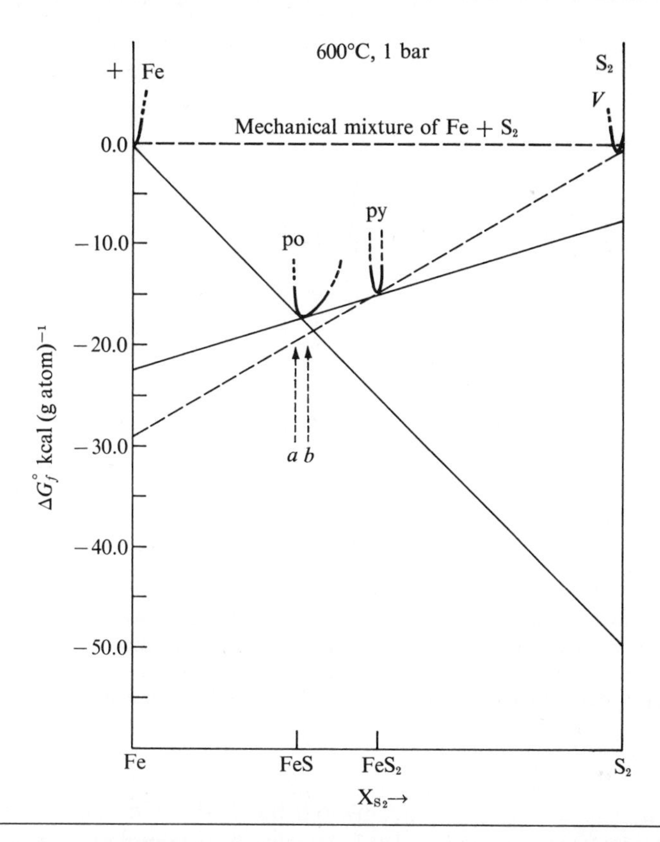


**Fig. 4-31** Free-energy–composition diagram for the system Fe–$S_2$ at 600°C and 1 bar. The free-energy curves bounding the single-phase fields for pyrite (*py*) and pyrrhotite (*po*) are hypothetical except that the one for pyrrhotite must be tangent at the two points vertically above the arrows *a* and *b*. The tangent above *a* corresponds to the stoichiometric composition FeS. The intersections of the three tangents with the ordinates have been taken from Toulmin and Barton (1964). Reference states are pure metallic iron and ideal $S_2$ gas at 600°C and 1 bar.

## THE SYSTEM Fe–$S_2$: AN EXAMPLE OF NEGATIVE $\Delta H^{xs}_{mix}$ AND DIFFERENT CRYSTAL STRUCTURES

In this system which has as end members elements in their standard states, there are two intermediate compounds: pyrrhotite, which has limited solid solution represented by the formula $Fe_{1-x}S$; and pyrite, $FeS_2$. Using the experimental results of Toulmin and Barton (1964), a free-energy diagram has been constructed at 600°C and 1 bar (Fig. 4-31). Single-phase fields are enclosed by curved lines showing the variation of free energy with composition; thus any phase with a range of composition such as pyrrhotite has a broad, U-shaped curve, while those such as metallic iron, $S_2$ gas, and pyrite, which have a restricted compositional range, have steep-sided and narrow, V-shaped free-energy troughs. The reader should note that the free energies of formation

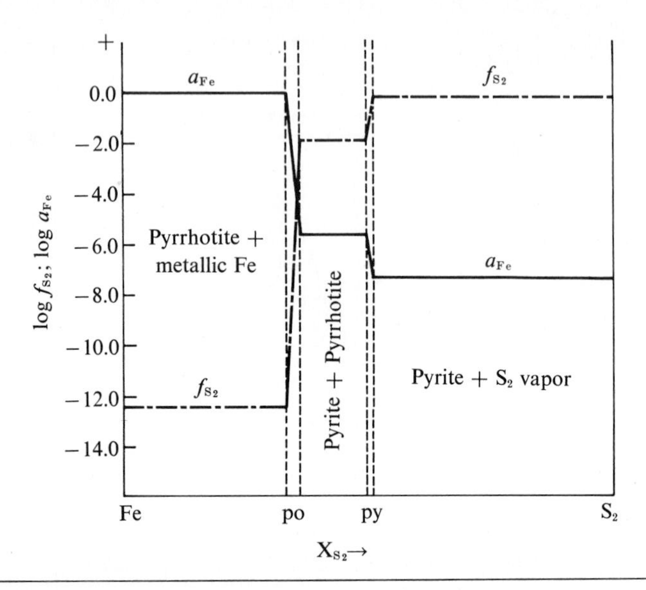


**Fig. 4-32** Activity diagram for the system Fe–$S_2$ at 600°C and 1 bar. Dashed lines represent the range of composition of pyrrhotite (*po*) and pyrite (*py*), the latter being exaggerated. The stable phases in the two-phase fields are shown, together with the curves for log $f_{s_2}$ (dashed line) and log $a_{Fe}$ (solid line). Reference state is pure metallic iron and ideal $S_2$ gas at 600°C and 1 bar.

($\Delta G_f^\circ$) of pyrrhotite and pyrite are plotted on the basis of a gram-atom rather than the more familiar gram-formula-weight. It is apparent in Fig. 4-31 that as a general case the more negative the free energy of the intermediate phase, the more restricted is the solution in the end-member phases.

The tangent joining pyrite to pyrrhotite (Fig. 4-31) impinges on the pyrrhotite free-energy curve at *b*, which is a composition different from the pyrrhotite in equilibrium with metallic iron; this latter composition is given by the point of tangency at *a* and corresponds to stoichiometric FeS. The intercepts of the three tangents (Fig. 4-31) with the compositional ordinates, Fe = 1 and $S_2$ = 1, give values for the activities or fugacities of Fe and $S_2$, referred to the standard state of pure metallic iron and an ideal $S_2$ gas, respectively (at the temperature of interest). In the absence of metallic iron as a phase, the activity of Fe, $a_{Fe}$, has been calculated from a relationship of the type

$$\underset{\text{solid}}{Fe} + \underset{\text{gas}}{S_2} = \underset{\text{pyrite}}{FeS_2} \tag{4-204}$$

since values of $f_{S_2}$ have been experimentally determined (Toulmin and Barton, 1964).

The data for the activities are shown in Fig. 4-32, where it is obvious that in the two-phase regions the activities of both Fe and $S_2$ are fixed. In the one-phase regions, however, there can be large changes in the activity of either component. Thus $a_{Fe}$ in pyrrhotite coexisting with metallic iron is unity, and log $a_{Fe}$ is zero; but log $a_{Fe}$ in pyrrhotite coexisting with pyrite is much lower, namely $-5.6$. In the range of composition of pyrrhotite, represented by two dashed vertical lines corresponding to the points of tangency *a* and *b* in Fig. 4-31, both $a_{Fe}$ and $f_{S_2}$ change by several orders of magnitude. Even in pyrite, which shows little if any departure from stoichiometric $FeS_2$, the activities of Fe and $S_2$ change in the one-phase pyrite field, whose width has been exaggerated in Fig. 4-32. This raises an important and often neglected question of the departure of many common minerals from their stoichiometric composition, a question that will now be briefly discussed.

## Miscellaneous Petrologic Considerations

### A COMMENT ON MINERAL STOICHIOMETRY

The change in activity of a component across a one-phase field illustrates an important point often overlooked by petrologists, since this change in activity can occur only if the phase in question changes composition. Thus minerals such as stoichiometric $FeS_2$ must vary in composition depending upon their coexisting phases, although the departure from stoichiometry may take analytic methods more sensitive than those now available to establish this. Many other sulfides whose analyses show little if any departure from stoichiometry are stable over a range of sulfur activities and behave similarly (Sato, 1966). This departure from a fixed composition is also found in silicates, one example being albite, which at $1065 \pm 5°C$ in the system $NaAlSiO_4$–$SiO_2$ (Schairer and Bowen, 1956) can coexist with tridymite ($a_{SiO_2} \sim 1$) or nepheline ($a_{SiO_2} \sim 0.15$). This variation of silica activity across the one-phase albite field necessarily requires some variation in composition; Carman and Tuttle (1967) have been able to demonstrate this experimentally.

Pure substances in a strict sense are thermodynamically unstable, and the deviation of igneous minerals from an assumed stoichiometric composition, although seemingly increasing the complexity of deciphering the cooling history of igneous rocks, offers enormous potential value in tracing the effects of components in low concentration or ones which have long since vanished.

### INCONGRUENT MELTING

Incongruent melting, as mentioned earlier, occurs when a solid 1 transforms at a definite temperature into a solid 2 and a liquid 3. This solid 2 melts gradually as the temperature is raised above the incongruent melting point, so that there is a certain range of temperature in which a solid and a liquid phase coexist. Conversely, when a liquid of composition 1 is cooled from a high temperature, the first crystals to form consist of 2. These crystals separate gradually over a certain range of temperature in which the composition of the residual liquid must, of course, also change. Finally, a unique temperature is reached; this is the incongruent melting point of 1, and at this temperature the liquid 3 reacts with solid 2, which is made over into solid 1. There are many petrologically important examples of incongruent melting. For instance, potassium feldspar ($KAlSi_3O_8$) melts incongruently at 1150°C to form leucite ($KAlSi_2O_6$) and a liquid richer in $SiO_2$ than orthoclase; complete melting of the leucite so formed occurs only at about 1533°C. Similarly, mullite ($Al_6Si_2O_{13}$) melts incongruently at 1810°C to form corundum ($Al_2O_3$) and liquid; monticellite ($CaMgSiO_4$) forms periclase (MgO) and liquid at 1503°C; hematite ($Fe_2O_3$) forms from acmite ($NaFeSi_2O_6$) at 990°C; and as mentioned before, protoenstatite ($MgSiO_3$) melts incongruently at 1557°C to form forsterite ($Mg_2SiO_4$) and liquid. In this latter case, the temperature interval in which forsterite exists in equilibrium with the liquid is small, since the system becomes wholly liquid below 1600°C.

Pressure can affect incongruent melting behavior either way. At high pressure, both enstatite and sanidine melt congruently; but anorthite and albite melt incongruently, the former to corundum and liquid and the latter to jadeite and liquid.

## MULTICOMPONENT SYSTEMS

The only difference between two-component and multicomponent systems is that the latter are much more difficult to study experimentally and to represent graphically; otherwise the principles involved are exactly the same. Broadly speaking, the order of crystallization from an ideal multicomponent solution is in the sequence of decreasing effect of temperature on solubility; any component whose solubility remains constant or increases as the temperature is lowered through a certain range does not precipitate at all in this range. It should be noted that this has nothing to do with the magnitude of the solubility; component 1 may be far more soluble than component 2 and yet precipitate before 2 when the solution is cooled through a certain range. If the solution is not ideal, the chemical potential of any component is affected by changes in the molar fractions of the other components, so that the solubility of 1 might either increase or decrease with decreasing temperature simply because some other component crystallized in this range.

Processes of crystallization in magmas and hydrothermal solutions are further complicated by the fact that changes in temperature are usually associated with changes in pressure, as when a magma rises to the surface. The order of crystallization is then found to depend on the relative changes in pressure and temperature, and a solution or magma of given composition may show different sequences of crystallization under different conditions of cooling and expansion. The effect of change in pressure is particularly marked when the magma is associated with a gas phase of large molar volume, since relatively small changes in pressure notably affect the chemical potentials of the components of the gas phase.

## PHASE RULE

The object of the phase rule is to state how many things we must know about a system in order to predict all its other properties and characteristics. The answer depends in the first place on what properties and characteristics we want to predict. In some cases we may wish to know only the values of the intensive variables (for example, temperature, pressure) which characterize the system, while in other cases we may wish also to know something about the extensive variables, such as the mass and volume of each phase. We must state also what factors we are allowed to neglect. For instance, it turns out in many cases of practical importance that the chemical potentials depend on, and therefore the conditions for equilibrium involve, only the pressure, temperature, and concentration of the various constituents in the various phases. In other cases, however, such factors as the curvature of the interfaces separating various phases, or position in a gravitational field or in an electrical field, become important in determining the states of equilibrium of a system and must therefore be taken into account as additional variables.

If we now consider more particularly the usual case in which the chemical potentials depend only on pressure, temperature, and molar fractions, and if we do not care for any information concerning the extensive or "capacity" characteristics of the system (mass, volume, etc.), the well-known answer to our question is that, in a system of $c$ components distributed in $\psi$ phases, all intensive variables—$P$, $T$, and molar fractions of all $c$ components in all $\psi$ phases—may be determined whenever the values of $(c + 2 - \psi)$ of these variables are given. This quantity $(c + 2 - \psi)$ is the variance $\omega$ of the system. A system with a negative variance cannot be in equilibrium; for variance 0 there is only one pressure and one temperature at which the system may consist of the given number of phases, the composition of each of these being determined by these particular values of $P$ and $T$. For variance 1, we may assign arbitrarily the pressure or the temperature or the molar fraction of any one component in any one phase and so fix unique values of the other variables such that the system will be in equilibrium with the number of phases indicated; and so on.

This phase rule is obtained simply by counting the number of unknowns and the number of relations to determine them. Let us consider, for instance, a system of $c$ components in $\psi$ phases. Equilibrium requires the chemical potential of each component to be the same in all phases in which it is present, so that we have $c(\psi - 1)$ equations stating that the system is in equilibrium. Now the chemical potentials depend on pressure, temperature, and the molar fraction of each component in every phase; there are thus altogether $2 + c\psi$ variables $P$, $T$, $X_I^1, \ldots, X_c^\psi$. Not all these molar fractions are mutually independent, of course, since we know that for each phase $\Sigma_i X_i = 1$; thus we have $\psi$ additional relations of this type. All told, there are $c(\psi - 1) + \psi$ relations between $c\psi + 2$ variables. Elementary algebra tells us that in such a case if we assign arbitrary values to $\omega = c\psi + 2 - [c(\psi - 1) + \psi] = c + 2 - \psi$ variables, we shall be able to determine all the other variables from these arbitrary values.

The relation

$$\omega = c + 2 - \psi \tag{4-205}$$

expresses the Gibbs phase rule, which says that the state of any system is defined, as far as the intensive variables $P$, $T$, $X_1^1, \ldots, N$ are concerned, by $\omega$ of these quantities.

If any component $i$ is not present in any phase $\gamma$, there is no equilibrium condition involving $i$ and therefore we have one equation less; but we have the additional relation $X_i^\gamma = 0$, and the variance therefore remains unchanged.

Attention should be called again to the significance of the number 2 that appears in the phase rule. This follows from the assumption that the intensive variables required to describe the state of the system consist only of concentrations, one pressure, and one temperature. This assumption may not necessarily be valid; for instance, systems may be mechanically constructed so that two or more parts of the system are at different pressures. This increases the number of variables and also the variance. If calcium carbonate is in equilibrium with CaO and $CO_2$, all phases being at the same pressure, the system has two components and three phases and is therefore univariant, meaning that for equilibrium temperature automatically determines the pressure; but if the gas phase $CO_2$ and solid phases $CaCO_3$ and CaO are at different pressures, the variance is 2, meaning that we may choose both the temperature and the pressure acting on the solid phases and still be able to find a gas pressure at which the system will be in equilibrium.

In some cases the meaning of *component* is clear, but this is not always so. If a system contains water, and water only, there is only one component, which is water. But if the system consists of a solution of water and albite ($NaAlSi_3O_8$) it may be asked whether the components are $H_2O$–Na–Al–Si–O, or $H_2O$–$Na_2O$–$Al_2O_3$–$SiO_2$, or some other combinations of molecules much more complex than any of the preceding ones. Clearly, the choice of components should be such that the chemical composition of each and every phase of the system can be stated in terms of these components. This could be accomplished by taking as components all the $s$ chemical species present in the system; but this number $s$ is not necessarily equal to the number $c$ that enters Eq. (4-205), since some of these species may be capable of reacting and each such reaction introduces a new equilibrium condition that reduces the variance by 1. For instance, in a gas phase in which CO, $CO_2$, and $O_2$ are present, $s = 3$; but there is a possible reaction

$$CO + \tfrac{1}{2}O_2 = CO_2$$

which introduces an equilibrium condition in which the sum of the free energies of the reactants equals the sum of the free energies of the product. Since the variance is defined as the difference between the number of variables and the number of conditions that these variables must satisfy, the phase rule becomes

$$\omega = s - r + 2 - \psi$$

where $r$ is the number of distinct chemical reactions that can occur between the species or the phases. The number of independent components is then $c = s - r$. For instance, in a system consisting of brucite $Mg(OH)_2$, periclase MgO, and water, $s = \psi = 3$, but $r = 1$ because of the reaction

$$MgO + H_2O = Mg(OH)_2$$

and $c = 2$. This is also the number of components we should have obtained directly by noting that the composition of each and every phase can be described in terms of the two components MgO and $H_2O$.

No ambiguity regarding variance can arise from the choice of components if one is careful to remember the definition of variance: *the number of variables minus the number of relations between them.* For example, a system consisting of albite and water could be alternatively described in terms of the two components $H_2O$ and $NaAlSi_3O_8$ or in terms of the four components $H_2O$–$Al_2O_3$–$Na_2O$–$SiO_2$. The latter choice introduces additional compositional variables which are, however, automatically determined by the stoichiometric formula of albite. The obvious choice is to pick the smallest possible number of components, namely, two in the case just mentioned. Two components, however, would be inadequate if the solid phase were to dissolve nonstoichiometrically in water, since the composition of the solution could not be expressed in terms of water and $NaAlSi_3O_8$ only.

In a system of given composition the number of chemical species may change with changing physical conditions, for example, temperature. For instance, at high temperature water partially dissociates

$$H_2O = H_2 + \tfrac{1}{2}O_2$$

There would thus be three species instead of only one, and a single chemical reaction between them: $c = s - r = 2$. Actually, this does not change the variance if neither hydrogen nor oxygen reacts with any of the other phases present, since we have the additional relation that the molar fraction of hydrogen is twice that of oxygen. If some phases are susceptible to oxidation, oxygen must be regarded as an additional component; but now there is yet another relation—the oxidation reaction—and the variance remains unchanged.

In most cases, petrologists find it convenient to choose simple oxides (such as $SiO_2$, CaO, $Al_2O_3$) as components. They also commonly use mineral names (for example, albite) to designate components. This usage is of course incorrect, because *albite* refers exclusively to a solid phase with distinct physical and crystallographic properties which are not present when the substance is, say, dissolved in a melt. Because it is inconvenient to talk of "the component $NaAlSi_3O_8$," mineral names are commonly used in a loose sense.

In general the phase rule limits both variance and the number of phases in a system in equilibrium. In spite of the large number of constituent elements in any igneous rocks, the number of associated minerals (phases) is remarkably small. Moreover among themselves the phases show strikingly consistent chemical relationships. It is these generalizations that make rock classification possible and geochemically significant. They suggest moreover that common igneous rocks closely approximate divariant equilibrium—an assumption that underlies our whole approach to formulating and unravelling the problems of petrogenesis. Finally we note that although the phase rule is concerned with equilibrium conditions, it cannot discriminate between metastable and stable equilibrium.

## NONEQUILIBRIUM CONDITIONS

A characteristic feature of melting or crystallization phenomena in systems of more than one component is that they do not occur entirely at one temperature. Rather they extend over a range of temperatures which, in the case of silicate melts, may cover several hundred degrees. The expression *melting point of a rock* is therefore misleading; instead we should refer to its *melting range.* Melting and crystallization phenomena in multicomponent systems must therefore be thought of as continuous processes involving gradual changes in the composition of the liquid and also, in many cases, changes in the composition of the solid phases which have already crystallized (for example, incongruent melting, solid solutions). Such changes do not occur instantaneously; it takes time to supply the required components by diffusion to the faces of a growing crystal. A close approach to equilibrium can therefore be maintained only if the rate of cooling is carefully adjusted to the rate at which the various processes involved in crystallization can occur. But the rate of cooling is determined by physical factors which depend mainly on the environment and have very little to do with the properties of the magma itself. We must remember, furthermore, that the rate of most reactions decreases exponentially with decreasing temperature, so that if a reaction has not had time to run to completion at high temperature, there is little likelihood that it will be able to do so at a lower temperature. Hence the conditions of disequilibrium remain "frozen" in the rock. A well-known example of this is, of course, the glassy groundmass of many volcanic rocks. It is not only a matter of rate of cooling versus rate of crystallization or of reaction which produces such effects, since a reaction may be prevented from running to completion merely by separating the reactants at an appropriate time. For instance, if early formed crystals settle by gravity and accumulate in some portion of the magma, they may become unable to react with the bulk of the remaining liquid. Or the liquid itself may be squeezed out by mechanical action (filter-press action) and become unable to react with the previously formed crystals. Petrographic examples of such arrested reactions are common: olivine cores surrounded and partially replaced by orthopyroxene rims, augite rimmed with hornblende, zoned crystals of plagioclase, etc. The interpretation in cases such as the first mentioned is that the reaction

$$\text{Olivine} + \text{silicate liquid} \rightarrow \text{orthopyroxene}$$

failed to run to completion at the incongruent melting point either because of rapid cooling or because of lack of a sufficient amount of liquid in contact with the olivine crystals.

A noteworthy feature of such fractional crystallization is that it tends to increase the diversity of products that can be obtained under different conditions from a given magma; it also tends to extend the range of temperature over which crystallization takes place. We later show in specific instances how fractional crystallization may lead to an assemblage of phases (for example, olivine and quartz) which is metastable at ordinary temperature with respect to some reaction (for example, olivine + quartz → orthopyroxene) that would normally have occurred if static conditions had been maintained. Igneous rocks are therefore by no means always stable (at high temperatures) assemblages of minerals. The final products of crystallization that can be obtained from a given magma depend to a large extent on the kinetics of the processes involved and not only, as often stated, on the conditions for thermodynamic equilibrium determined from laboratory experiments. Experimental investigation remains nevertheless an essential step in the understanding of processes that are operative in magmas, and it is to this study that we turn next.

# Crystallization Paths of Igneous Minerals at Low Pressures

# 5

## Introduction

Microscopic study of igneous rocks, with particular reference to textural relations of constituent minerals, has yielded much information on the order in which different mineral phases precipitate during cooling of magmas. On the basis of petrographic experience, and especially the criterion of relative idiomorphism, Rosenbusch (1882) drew up a series of empirical rules which he considered to be generally applicable—admittedly with many exceptions—to crystallizing magmas. For example, the iron oxides and minor constituents of a magma tend to crystallize early as idiomorphic crystals (iron ores, zircon, apatite, sphene, etc.), whereas the main constituents tend to crystallize in order of increasing silica content, so that the residual liquid is at all stages more siliceous than the essential minerals which have already crystallized.

So numerous are the exceptions to these "rules" that some writers consider them of little value (Bowen, 1915; Shand, 1943), and yet, as we have seen in Chap. 4, there is a broad relationship between the entropy of fusion of a mineral and its morphology, nucleation rate, and growth rate. However, such is the complexity and diversity of igneous rocks that it would be surprising if any simple set of rules, or any single thermodynamic property, could be used as a general explanation for the sequence of crystallization and habit of minerals in magmas.

With the advent of the electron microprobe, immensely detailed studies of mineral composition have become commonplace and have demonstrated crystallization paths of solids, stable or metastable, even in rocks of small grain size, hitherto intractable. Possibly no development since the first investigation of rocks in thin section has been so fruitful or so successful in contributing to our knowledge of the crystallization history of igneous rocks.

This great volume of mineral data, taken from all kinds and types of igneous rocks, augers well for a comprehensive understanding of their origin; subsequent thermodynamic treatment of analyzed mineral assemblages will allow estimates of pressure, temperature, and the chemical potentials of many, and eventually all, of the common components present in igneous rocks. The treatment hereafter seeks to display the great variety of magmas or igneous rocks, regardless of geological setting, in readily comprehensible systems, but inevitably with some distortion of which the reader should be wary. We hope that the clarity of the general relationships so depicted outweighs any apprehension the reader may feel at seeing a complex natural system reduced to a manageable and portrayable number of components. To those accustomed to or trained in the detailed analysis of phase diagrams, our later approach may lack rigor; but any deficiency in this regard, we submit, is balanced by the desirability of the reader, particularly a student, getting a feel for igneous rocks.

Any introduction to the crystallization behavior of igneous rocks must pay deserved tribute to the six decades or so of precise studies in silicate systems by the Geophysical Laboratory of the Carnegie Institution of Washington. This type of experimental investigation, associated for all time with N. L. Bowen and J. F. Schairer, has in the last 20 years spread to universities the world over, largely by those associated at one time or another with the Geophysical Laboratory. Much of the early work has been summarized and applied to igneous petrogenesis by Bowen in "The Evolution of the Igneous Rocks" (1928). The geometrical treatment necessary to trace the compositional path of a cooling liquid in a multicomponent system has been well described by Bowen (1928), Osborn and Schairer (1941), and Levin et al. (1964). For systems containing two oxidation states of iron, Muan and Osborn (1956) and Muan (1958) provide explicit details on the additional complexities that arise when the total composition of the liquid plus the solids is not constant—in other words, when oxygen is added to or subtracted from the condensed phases as crystallization proceeds.

We shall not attempt to summarize this extensively recorded work here; rather our intention is to bring together a little of the experimental work on silicate systems, and in conjunction with data on minerals in igneous rocks, interweave a generalized picture of the crystallization path of some common minerals formed in igneous rocks. Much information of the compositional path that a mineral traces as it crystallizes can be revealed only by detailed study with the electron microprobe, whereas, by and large, experimental investigation of silicate systems has been concerned with the pressure-temperature regime of minerals and crystallization paths predicted therefrom. In the examples that follow, the deductions made from rocks represent a large proportion of the sum of our knowledge of crystallization; in other cases, here omitted, the results of experiments either have no natural counterpart or petrologists have been slow to find, or look for, the natural equivalent. We have found it convenient to limit the discussion of igneous rocks in this chapter to those that crystallized in a low-pressure environment. Quite arbitrarily we have taken 10 kb as the upper limit of pressure, roughly equivalent to that at the base of the continental crust. This is the pressure regime in which the vast majority of exposed igneous rocks crystallized and for which most experimental data are available. Greater pressures, 20 to 100 kb, are appropriate to mineral stability in the earth's mantle and therefore to the sporadic samples in the geologic record which exhibit internal equilibration at great depths.

It is important to emphasize at the outset some of the uncertainties inherent in the approach taken here. There is a danger in using data from volcanic rocks to portray the crystallization path of a particular mineral group: that rapid cooling may have produced metastable compositions. However, comparison with slowly cooled plutonic rocks sometimes resolves this uncertainty. Again, unless the composition of phenocrysts exhibits clear evidence of considerable depth, there is little indication available to a petrologist of the extent to which pressure has affected the composition of the phenocryst phases. Many phenocrysts are generally accounted to be intratelluric, having crystallized during the magma's ascent to the earth's surface, so that there is considerable latitude for variation in the effect of pressure. The use of mineral data from plutonic rocks does not always circumvent these uncertainties, since in many slowly cooled rocks, particularly those of granitic composition, extensive subsolidus recrystallization often obliterates the original composition of the crystals which might have been in equilibrium with liquid. The analogous uncertainties in experimental work arise from the generally unknown detailed composition of either the liquid or the crystals, and particularly in the unknown effect of water. It is experimentally convenient to saturate a synthetic silicate liquid with water, since this greatly accelerates reaction; but to what extent this saturation affects the equilibrium composition of both the liquid and the solid is frequently unknown. This uncertainty is particularly troublesome in relating experimental results to their natural equivalents, because natural liquids may be undersaturated with water for a considerable part of their crystallization interval. So, with some hesitation about isolating these varied effects, we shall attempt to interweave data of varied types into a consistent but generalized picture.

## The Feldspar System

The major mineral component in the overwhelming majority of igneous rocks which have crystallized on or near the surface is feldspar, whose three essential components are $KAlSi_3O_8$, $NaAlSi_3O_8$, and $CaAl_2Si_2O_8$. And yet experimental determination of the mutual relationship between these three components over a range of temperatures and pressures has so far proved difficult. The feldspars are a good example of how the results of detailed examination of a wide variety of igneous rocks, taken in conjunction with experimental data, allow the broad features of the ternary feldspar system to be delineated. Because the feldspars exhibit extraordinarily complex relationships of unmixing and ordering at lower temperatures, much of which has become a field of inquiry all its own, we shall first consider the crystallization of feldspar as exemplified in the groundmass of volcanic rocks where the pressure is constant and where the feldspar will usually be the high-temperature, disordered modification. We shall also ignore until later the incongruent melting relationship of $KAlSi_3O_8$; thus the field of leucite, its complementary melting product, is not considered in the next series of figures. Some of the data we use are new, and these are combined with extensive investigations by a number of authors on the crystallization of feldspar both in volcanic rocks and in synthetic systems (for example, Carmichael, 1963, 1965; James and Hamilton, 1969; Tuttle and Bowen, 1958).

Using the experimental data of Bowen and Tuttle (1950) on the alkali-feldspar ($KAlSi_3O_8$–$NaAlSi_3O_8$) system, of Bowen (1913) on the plagioclase-feldspar ($NaAlSi_3O_8$–$CaAl_2Si_2O_8$) system, together with those of Schairer and Bowen (1947) and Franco and Schairer (1951), a schematic diagram has been drawn to illustrate the ternary feldspar system at high temperature and low pressure. In Fig. 5-1 three surfaces are depicted within the triangular temperature-composition prism; the liquidus, the solidus, and the solvus. The liquidus surface is represented by curved parallel lines dipping down from the melting temperatures of $NaAlSi_3O_8$, $KAlSi_3O_8$, and $CaAl_2Si_2O_8$ to intersect in the liquidus field boundary *EF*. This liquidus field boundary extends over most of the area of the triangle and separates the field in which plagioclase is the early crystallizing phase from that in which alkali feldspar is the first phase to crystallize. The solidus surface (stippled) connects with the liquidus at four points: the melting temperatures of $NaAlSi_3O_8$, $KAlSi_3O_8$ (neglecting the incongruent melting to leucite), and $CaAl_2Si_2O_8$, and the minimum melting composition *M* in the alkali-feldspar front face. The solvus is a dome-shaped surface truncated by two vertical planes: the $CaAl_2Si_2O_8$–$KAlSi_3O_8$ join and the alkali-feldspar solvus *BCD*, where *C* is the critical composition corresponding to the crest of the solvus. The top of the solvus dome intersects the solidus along *HP′PSS′GEH*.

The liquidus field boundary *EF* intersects the underlying solvus dome only at *E* (the $KAlSi_3O_8$–$CaAl_2Si_2O_8$ eutectic) and thereafter runs with falling temperature toward the alkali-feldspar minimum *M*, which, however, it does not reach, fading out just before at *F*. For liquids saturated with water at 4 kb or more confining pressure, the liquidus field boundary extends to the alkali-feldspar join where it intersects *C*, the critical composition. This liquidus field boundary represents compositions of liquids in equilibrium with two feldspars—a plagioclase (or anorthoclase) and a sanidine. Two hypothetical three-phase triangles, at two different temperatures, are shown in Fig. 5-1; these triangles are parallel to the base of the prism and are labeled *L′P′S′* and *LPS*, where *L* represents the liquid, *P* the plagioclase feldspar, and *S* the sanidine. The co-existing feldspars lie on the intersection of the feldspar solidus (stippled) with the underlying solvus.

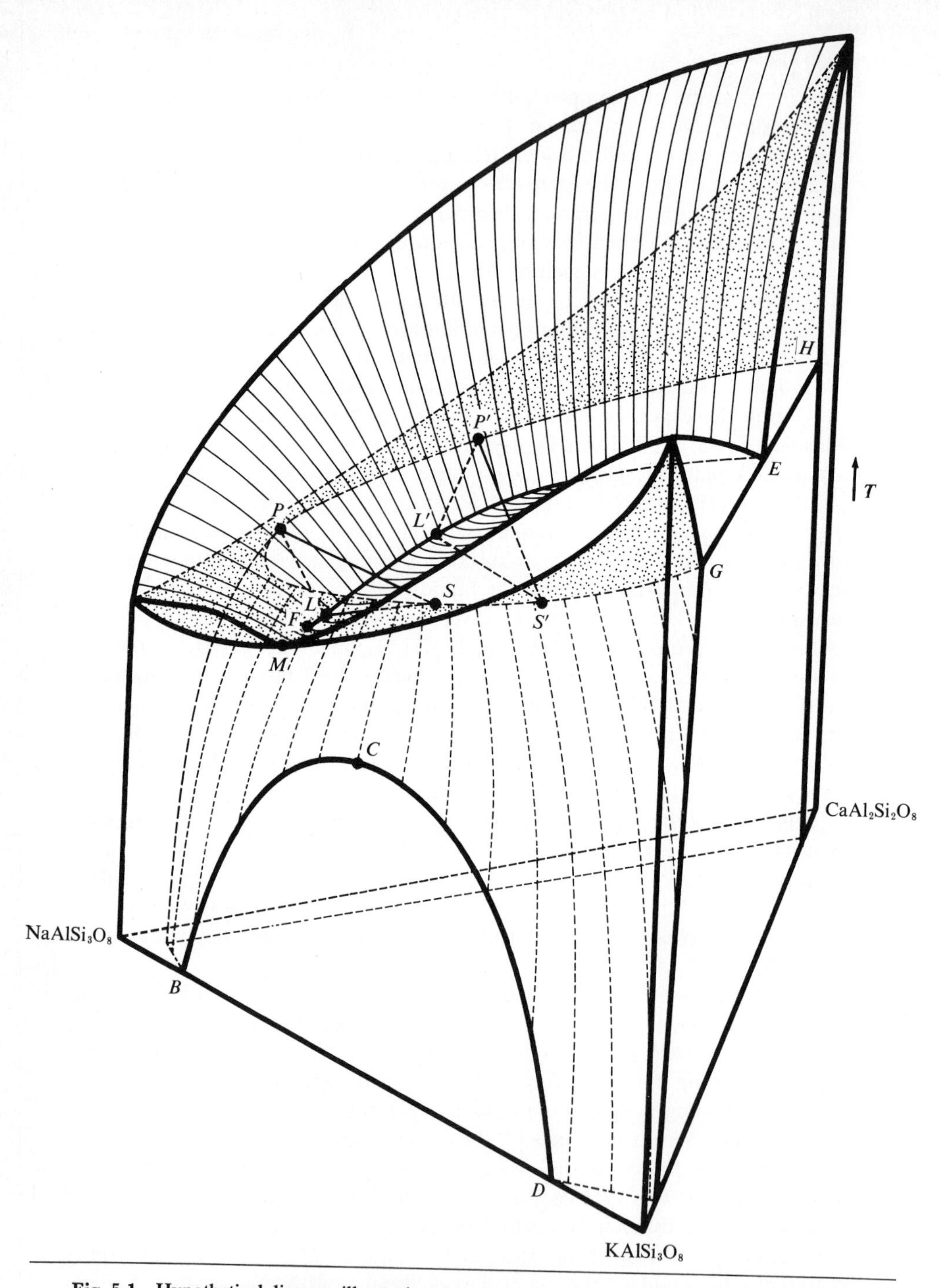


**Fig. 5-1** Hypothetical diagram illustrating the relationship between $CaAl_2Si_2O_8$, $NaAlSi_3O_8$, and $KAlSi_3O_8$ at low water-vapor pressures. Incongruent melting of sanidine has been neglected.

## SOLID SOLUTION IN TERNARY FELDSPARS

From the general relationships shown in Fig. 5-1 it is to be expected that plagioclase in equilibrium with sanidine will contain more $KAlSi_3O_8$ component than plagioclase alone; in the same way sanidine coexisting with plagioclase will contain more $CaAl_2Si_2O_8$ component than a sanidine of otherwise similar composition crystallizing by itself. Presumably the most extensive ternary solid solution occurs at high temperatures and low pressures; liquids saturated with extensive amounts of water crystallize at lower temperatures than anhydrous liquids, and as the effect of pressure is to raise the feldspar solvus (Fig. 5-1), the effect of water is to reduce the extent of ternary solid solution.

Accordingly we have used the analyzed groundmass feldspars in two shoshonitic lavas and a nepheline trachyte to illustrate the range of compositions of both plagioclase and sanidine when they coprecipitate (Fig. 5-2*a* and 5-2*b*). In both examples the plotted compositions represent the composition of the feldspar at various points along several traverses from the cores of the crystals to their margins. Thus the arrows represent the direction in which the crystals change in composition from core to margin, and they illustrate the paths traced out on the solidus by plagioclase and sanidine. These are fractionation paths.

It is reasonable to assume that each plagioclase-sanidine pair would, with slightly different crystallization conditions, eventually have outer rims of the same composition; in other words they would both zone toward a common composition which must be between the most sodic composition of each; this common composition, which is unknown, is represented by point *C* in Fig. 5-2. For these two rock types, the solidus paths of the plagioclase and the sanidine represent a segment of the solidus-solvus intersection shown in Fig. 5-1.

Where plagioclase crystallizes without sanidine in the groundmass of basalts and basaltic andesites, which span the temperature range of the two shoshonites, the $KAlSi_3O_8$ content should be lower than in the plagioclase of two-feldspar assemblages. A typical composite solidus path for a group of olivine tholeiites, basaltic andesites, and icelandites (Carmichael, 1967*a*) is distinctly less rich in $KAlSi_3O_8$ than the plagioclase found in shoshonites (Fig. 5-4). It is of interest that, in a prehistoric tholeiitic lava lake of Kilauea, the solidus paths of plagioclase show an increase in $KAlSi_3O_8$ with progressively slower cooling, perhaps because of an increasing activity of water (Evans and Moore, 1968).

Because the composition of a natural liquid will influence the composition of the precipitating alkali feldspar (cf. Fig. 5-2*a* and *b*), it is not possible to construct the trace of the limit of ternary solid solution which will apply to all natural occurrences of plagioclase and sanidine. However, feldspars do not exhibit unrestricted ternary solid solution, and the extent of this solid solution in natural feldspars has been discussed by Vogt (1926, pp. 48–52) and by Tuttle and Bowen (1958, p. 132). The limit for natural feldspars, taken from J. V. Smith and MacKenzie (1958), is shown in Fig. 5-12: it is taken *generally* to correspond with *HP′PSS′G* in Fig. 5-1 projected onto the base, and it therefore separates the one-feldspar region from the field of two feldspars.

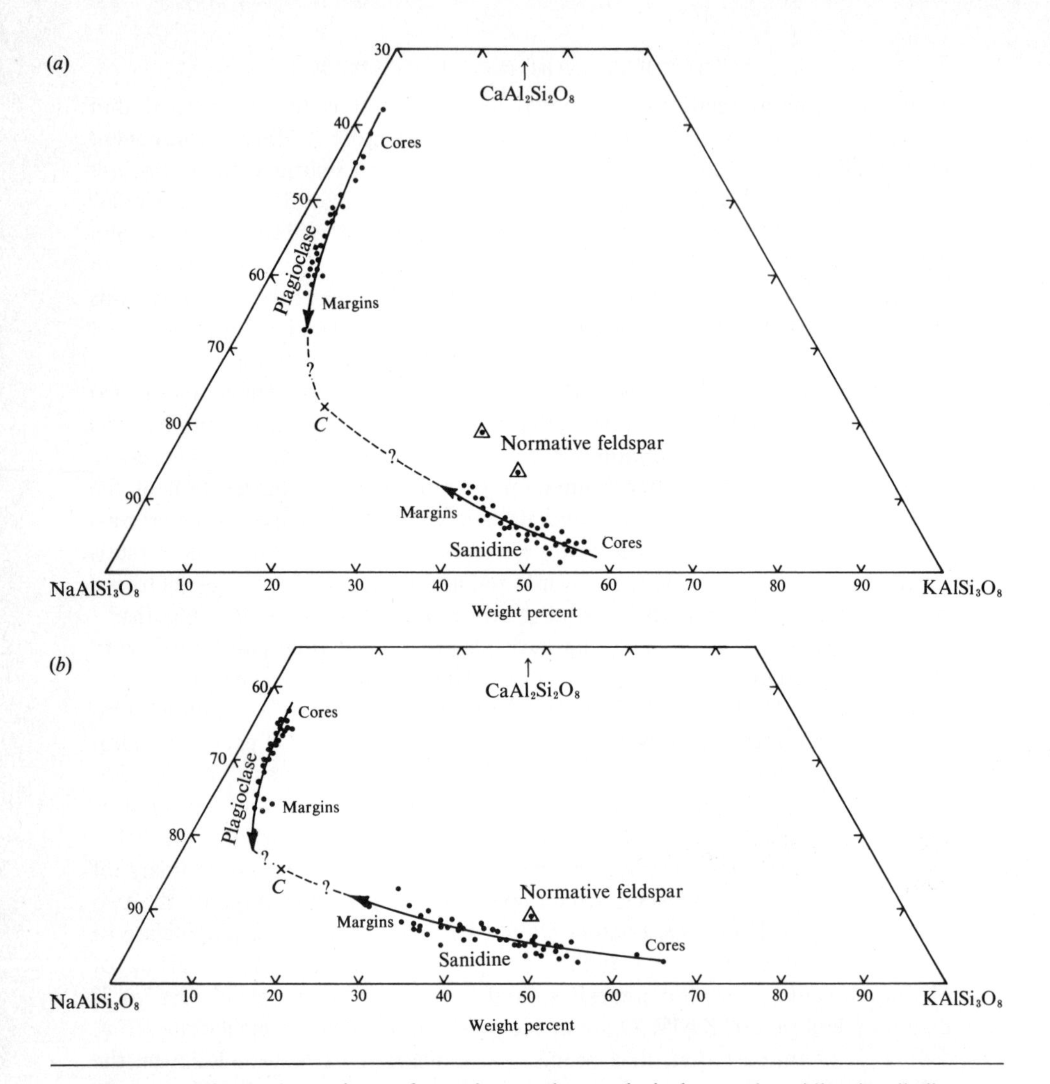


**Fig. 5-2** Microprobe analyses of zoned groundmass plagioclase and sanidine in alkaline lavas. Arrows show trends of zoning from cores to margins, with *C* as possible ultimate unattained composition of both phases. Triangles indicate CIPW normative compositions of rocks in question. (*a*) Two shoshonites (data from J. Nicholls and Carmichael, 1969*a*). (*b*) A nepheline trachyte; note more extensive zoning of sanidine and more sodic plagioclase compared with (*a*) (unpublished data).

It has been stated above that at high temperatures the liquidus boundary curve in the ternary feldspar system fades out before reaching the alkali-feldspar join (Fig. 5-1); this is the necessary consequence of the alkali-feldspar system having a liquidus minimum rather than a eutectic. The possible phase relationships, from a theoretical standpoint, which could result from the termination of the liquidus field boundary have been completely reviewed by Stewart and Roseboom (1962). The evidence of the rocks themselves does not reduce Stewart and Roseboom's examples to any unique solution. However, in volcanic rocks it is consistent with the following explanation, taken largely from Tuttle and Bowen (1958, pp. 131–135).

The data shown in Fig. 5-2 substantiate Tuttle and Bowen's premise that, in assemblages of two feldspars plus liquid, each feldspar will become more sodic as crystallization proceeds until, in certain cases, a common composition is reached (not in rhyolites, however). This common composition is represented by *C* in Fig. 5-2. While two feldspars coexist in equilibrium, the liquid, in terms of its feldspathic components, will be confined to a boundary curve such as the line *EF* in Fig. 5-3; but at the temperature at which both feldspars attain the same composition *C*, the boundary curve will terminate (Fig. 5-3). The evidence from volcanic rocks suggests that this boundary curve, whose precise position in Fig. 5-3 will vary with the bulk composition of the liquid (or broadly speaking rhyolitic, basaltic, or phonolitic), terminates very close to the alkali-feldspar sideline. The close approach is indicated by the existence of plagioclase and sanidine phenocrysts, presumably in equilibrium, in lavas with only small amounts of calcium ($\sim$0.5 percent) or of normative *an*.

This boundary curve has another unusual property in addition to fading out before reaching the sideline; toward its termination it becomes a resorption rather than a coprecipitation curve (Tuttle and Bowen, 1958, pp. 133–134). The exact point at which coprecipitation gives way to resorption as the liquid nears *F* depends upon the composition of the liquid (the feldspathic components thereof). Rahman and MacKenzie (1969) have given a geometrical method for determining this point for any set of feldspar phenocrysts if the composition of the liquid is also known. This predicts that in certain Italian trachytes, the plagioclase phenocrysts should be resorbed when their composition is about $An_{35-40}$, under condition of equilibrium cooling. This prediction is in accord with the embayed or corroded margins of the plagioclase phenocrysts found in these lavas.

The orientation of tie lines joining coexisting plagioclase and sanidine depends upon the composition of the liquid in which they formed, the temperature, and the pressure. In general the plagioclase phenocrysts of trachytes are richer in anorthite, and the sanidine is richer in albite, than the corresponding phenocrysts in rhyolites; in phonolitic liquids the situation is uncertain because few of these precipitate two coexisting feldspars. An example of this general divergence is shown in Fig. 5-4, in which the coexisting phenocrysts of Italian trachytes (with only small amounts of normative *Q* or *ne*; Rahman and

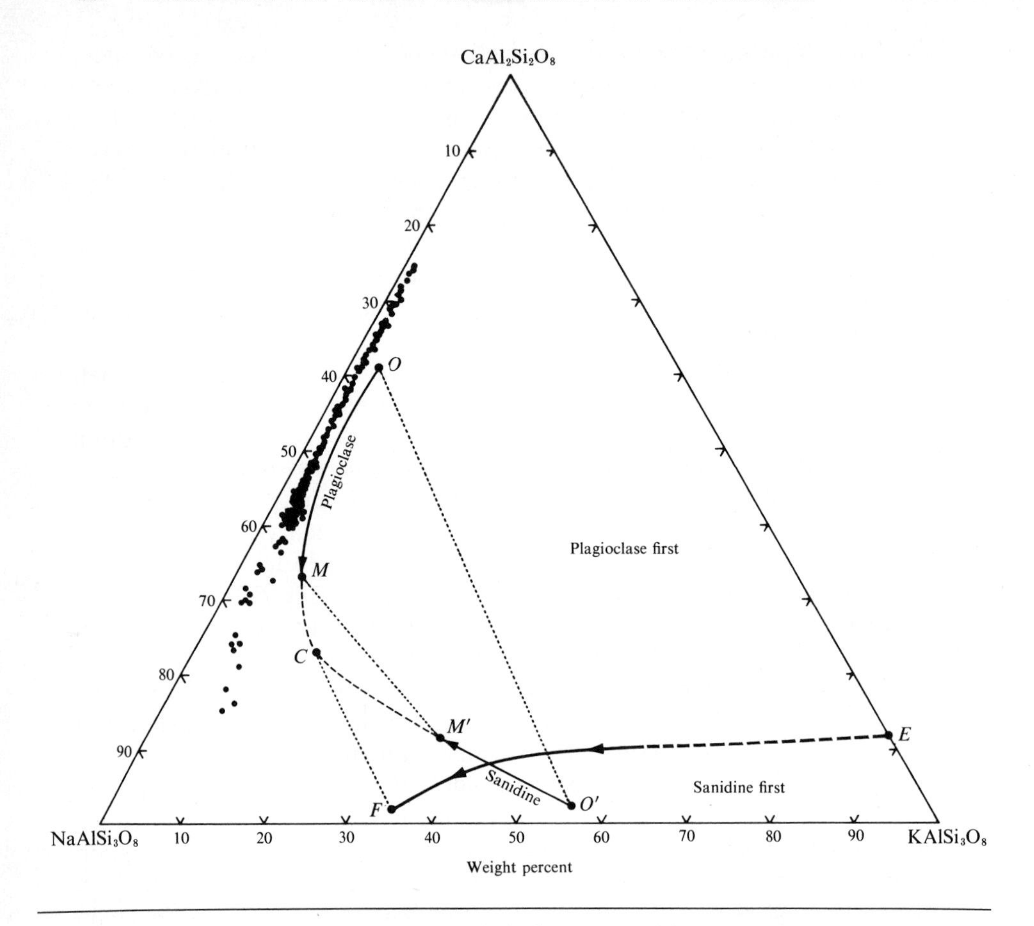


**Fig. 5-3** Coexisting zoned groundmass plagioclase and sanidine of shoshonite lavas (Fig. 5-2*a*) trending toward a possible common composition *C*. Ties connect compositions of coexisting phases—cores *O*,*O′*, marginal zones *M*,*M′*. Curve *EF* (of Fig. 5-1) represents compositions of liquids with which two feldspars coexist in equilibrium (neglecting incongruent melting of $KAlSi_3O_8$). Plagioclase crystallizes first from liquids above, sanidine from liquids below, *EF*. Dots near the left-hand margin are compositions of zoned plagioclases from a suite of less potassic lavas—olivine tholeiites, tholeiites, and icelandites.

MacKenzie, 1969) are shown in comparison with the corresponding data for a suite of Quaternary quartz-bearing obsidians from California. This variation in composition between the two sets of feldspar phenocrysts arises predominantly from a difference in composition (expressed in terms of the feldspathic components) between average trachyte and rhyolite; it can also arise on occasion from the phase relationships which affect quartz-bearing liquids. In these, enrichment of early crystallizing alkali feldspar in $NaAlSi_3O_8$ is restricted to a greater degree than in the corresponding feldspar in trachytic liquids.

In those lavas which precipitate plagioclase and sanidine in the groundmass, such as shoshonites (Fig. 5-3), it is interesting to note that the orientations of the two tie lines joining the most extreme compositions of plagioclase to coexisting sanidine are nearly parallel to those of the phenocrysts in the Italian trachytes (shown in Fig. 5-4); this suggests that partial equilibrium was maintained in the shoshonite lavas as these groundmass feldspars traced out fractionation paths on the solidus surface.

**Fig. 5-4** Compositions of coexisting phenocryst feldspars in Italian trachytes (data from Rahman and Mackenzie, 1969) and in Californian rhyolite obsidians (unpublished data). Note that the lines for trachytes are subparallel to those for shoshonite of Fig. 5-3.

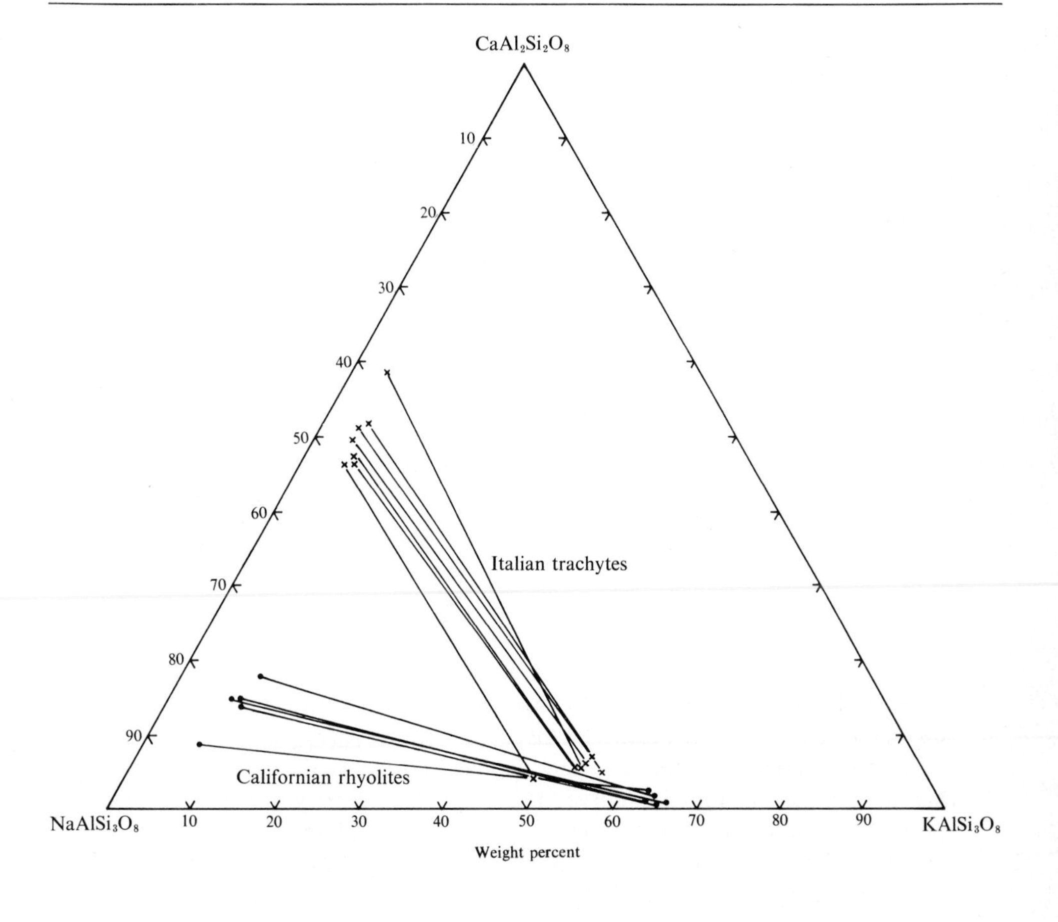

## The Salic Tetrahedron: The System $CaAl_2Si_2O_8$–$NaAlSiO_4$–$KAlSiO_4$–$SiO_2$

It is appropriate next to examine the more important aspects of the crystallization of rhyolites, trachytes, and phonolites in terms of both their feldspars and the other accompanying salic minerals. For this we look at a tetrahedron whose apices are $CaAl_2Si_2O_8$, $NaAlSiO_4$, $KAlSiO_4$, and $SiO_2$; this is conveniently named the salic tetrahedron (to borrow a term from the authors of the CIPW norm), which encompasses the feldspathoids, the feldspars, and quartz or tridymite.

This system, as yet undetermined experimentally, is so complex, if the evidence of the natural counterparts is anything to go by, that it is separated here into two parts: the silica-rich side and the silica-poor side. Thus the rhyolitic and quartz-trachyte liquids are, for this immediate purpose, separated from those of phonolite, nepheline-trachyte, and nephelinite composition.

### THE RHYOLITE TETRAHEDRON: THE SYSTEM $CaAl_2Si_2O_8$–$NaAlSi_3O_8$–$KAlSi_3O_8$–$SiO_2$

Before considering the crystallization of the main types of acid liquids in terms of these four components, it is necessary to look at some highlights of the subsystem $NaAlSi_3O_8$–$KAlSi_3O_8$–$SiO_2$, which for liquids saturated with water at various confining pressures forms the subject of an important memoir by Tuttle and Bowen (1958). The phase relationships for water-saturated liquids at 1000 bars confining pressure are shown in Fig. 5-5. The liquidus surface dips steeply down from the $SiO_2$ apex to intersect the alkali-feldspar liquidus surface at the field boundary *ws*; this boundary separates the field in which quartz is the initial solid phase from that in which a solid solution of $NaAlSi_3O_8$–$KAlSi_3O_8$ composition crystallizes first. The minimum temperature on the boundary curve is labeled *m*, and the composition of the alkali feldspar in equilibrium with quartz and liquid at *m* is given by *m′*. A liquid with the composition of the minimum strictly has a vanishingly small temperature interval in passing from the all-liquid to the all-solid state; it is frequently called an isobaric minimum or, loosely, a ternary minimum. Note that *m′* does not coincide with *M*, the minimum composition in the binary $NaAlSi_3O_8$–$KAlSi_3O_8$ system.

The fractionation path of any liquid which starts to crystallize in the feldspar field can be predicted precisely from the experimental data on equilibrium relationships between solid and liquid (Tuttle and Bowen, 1958, fig. 30). Such a path is called a fractionation curve, but it has no physical reality on the liquidus surface; with one exception, all fractionation curves (of which there are an infinite number) originate at either the $NaAlSi_3O_8$ or the $KAlSi_3O_8$ apex. One, the unique fractionation curve, originates at the binary minimum *M* and intersects the liquidus boundary curve at *q*, slightly displaced from the ternary isobaric minimum *m* (Fig. 5-5). The unique fractionation curve cannot be crossed by any fractionating liquid, and thus its presence can be used to

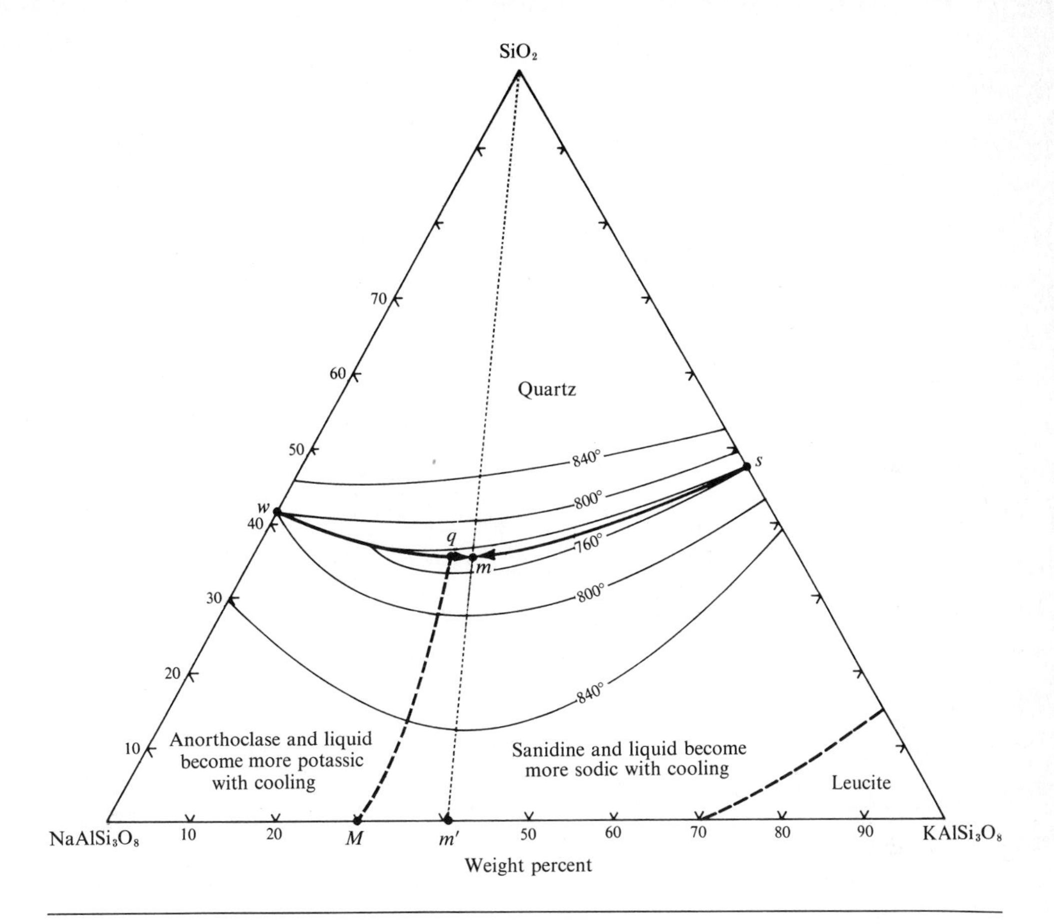


**Fig. 5-5** The liquidus surface for water-saturated liquids at 1000 bars confining pressure in the system $NaAlSi_3O_8$–$KAlSi_3O_8$–$SiO_2$; the boundary curve *wqms* separates the quartz from the feldspar field, and *m* is the isobaric minimum. Isotherms show the configurations of the liquidus surface, which is divided into two areas by the unique fractionation curve *qM* separating liquids which become more sodic from those which become more potassic as cooling proceeds. (After Tuttle and Bowen, 1958.)

distinguish two types of liquids. All liquids which commence their crystallization in the area $NaAlSi_3O_8$–*w*–*q*–*M* become more potassic as crystallization proceeds, as do their crystallizing feldspars. All liquids in the area $KAlSi_3O_8$–*m′*–*m*–*s* are opposite in sense and become more sodic as they cool, as do their feldspars. For those liquid compositions which fall in the area *M*–*q*–*m*–*m′*, the crystallizing feldspar first becomes more sodic, and then, particularly along the segment of the boundary curve *q*–*m*, the trend is reversed and both liquid and feldspar become more potassic. The only well-documented example of this reversal,

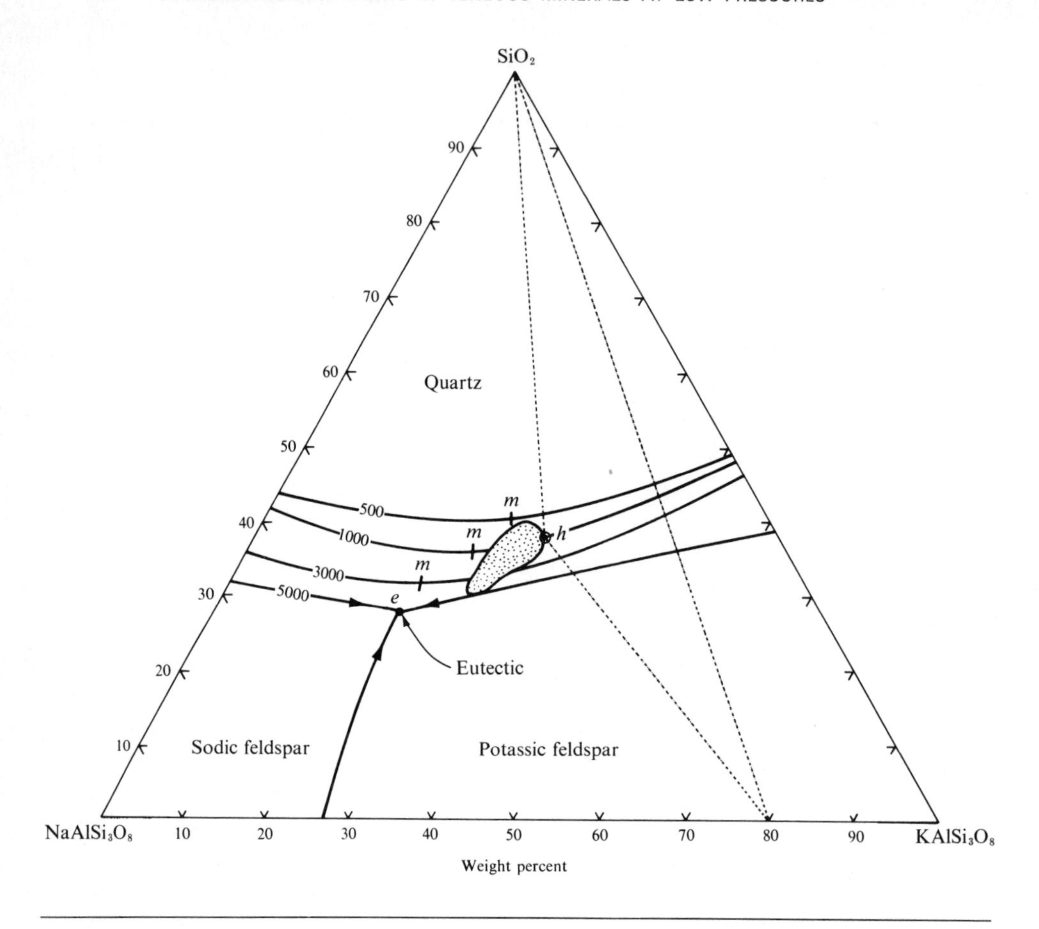


**Fig. 5-6** Curves for water-saturated liquids in equilibrium with quartz and alkali feldspar at indicated confining pressures (500, 1000, 3000, and 5000 bars). Isobaric minima are labeled *m* except at 5000 bars, where a ternary eutectic *e* is generated by intersection of the alkali-feldspar solvus with the liquidus surface (Fig. 5-1). Plots of normative *Q*, *or*, and *ab* of analyzed granites concentrate in the stippled area. The dotted three-phase triangle illustrates the initial phases to precipitate from liquid *h* at 1000 bars. (Data from Tuttle and Bowen, 1958.)

however, is not in a rhyolite but in the feldspar crystals of a phonolite where much the same phase relations prevail (Nash et al., 1969).

The anhydrous composition of *m*, the isobaric minimum, changes as the water content of the liquid changes, and the positions of the minimum at various confining pressures are shown in Fig. 5-6. At pressures above 4 kb, approximately equivalent to 9 weight percent $H_2O$, the temperature of the liquidus surface has been sufficiently depressed to intersect the alkali-feldspar solvus, and a boundary curve and eutectic are generated (Fig. 5-6). The influence of water in depressing the liquidus surface in temperature is augmented by its concomitant effect in raising the temperature of the alkali-feldspar solvus.

Tuttle and Bowen (1958) showed that a high proportion of the published analyses of rhyolites and granites, when plotted in terms of their CIPW components *or*, *ab*, and *Q*, concentrate close to the experimental minima (Fig. 5-6). That this is a potent argument for the magmatic origin of granite is not our concern here, but by comparing the natural with the experimental, the restriction of the composition of alkali feldspar in equilibrium with liquid and quartz can be readily seen. For the synthetic composition labeled *h* in Fig. 5-6, which in terms of these three components is slightly more potassic than the concentration of granitic analyses, the composition of the first feldspar to precipitate in equilibrium with quartz and liquid (at 1000 bars $p_{H_2O}$) is $Or_{80}$, as shown by the three-phase triangle in Fig. 5-6. It is not surprising therefore that phenocrysts of alkali feldspar more potassic than about $Or_{75}$ have not been found in rhyolites, and their presence in granites betokens subsolidus recrystallization rather than precipitation from a silicate liquid. Highly potassic sanidines, around $Or_{90}$, similar in composition to the orthoclase "phenocrysts" of many granites, have yet to be found in rhyolites and among volcanic rocks are found only in rare potassic lavas such as orendites.

Many rhyolites which at first sight could be considered as the natural equivalents of the system portrayed in Fig. 5-6 contain plagioclase, with or without sanidine, as an early crystallizing phase. The addition of $CaAl_2Si_2O_8$ as a fourth component requires combining the relationships depicted in Figs. 5-3 and 5-5; thus a boundary curve such as *EF* in Fig. 5-3 will extend into the four-component tetrahedron as a surface, the two-feldspar surface *FEGH* in Fig. 5-7. Liquids confined to this surface will be in equilibrium with plagioclase and sanidine, and, at the intersection of the two-feldspar surface with the surface of the $SiO_2$ volume (*HG* in Fig. 5-7), liquids will be in equilibrium with quartz, plagioclase, and sanidine. For liquids containing small amounts of water, the two-feldspar surface will fade out before reaching the base of the tetrahedron for the reasons discussed on page 225; it will also continue to be a resorption surface in the region close to its termination. This surface for liquids saturated with water at 1000 bars confining pressure has been located by James and Hamilton (1969).

We may now consider the crystallization of three types of silica-rich liquids in terms of the four components shown in Fig. 5-7; each of the three has a comparable natural analog from which many properties used to construct Fig. 5-7 have been derived.

ACID LIQUIDS WITH ONE FELDSPAR When a liquid is cooled to the temperature of the liquidus, a plagioclase feldspar will start to crystallize on the solidus, the relationships corresponding approximately to *P*1–*L*1 (Fig. 5-7). With further cooling, the plagioclase will react with the liquid and, moving down the solidus, will become more sodic; the liquid will react continuously

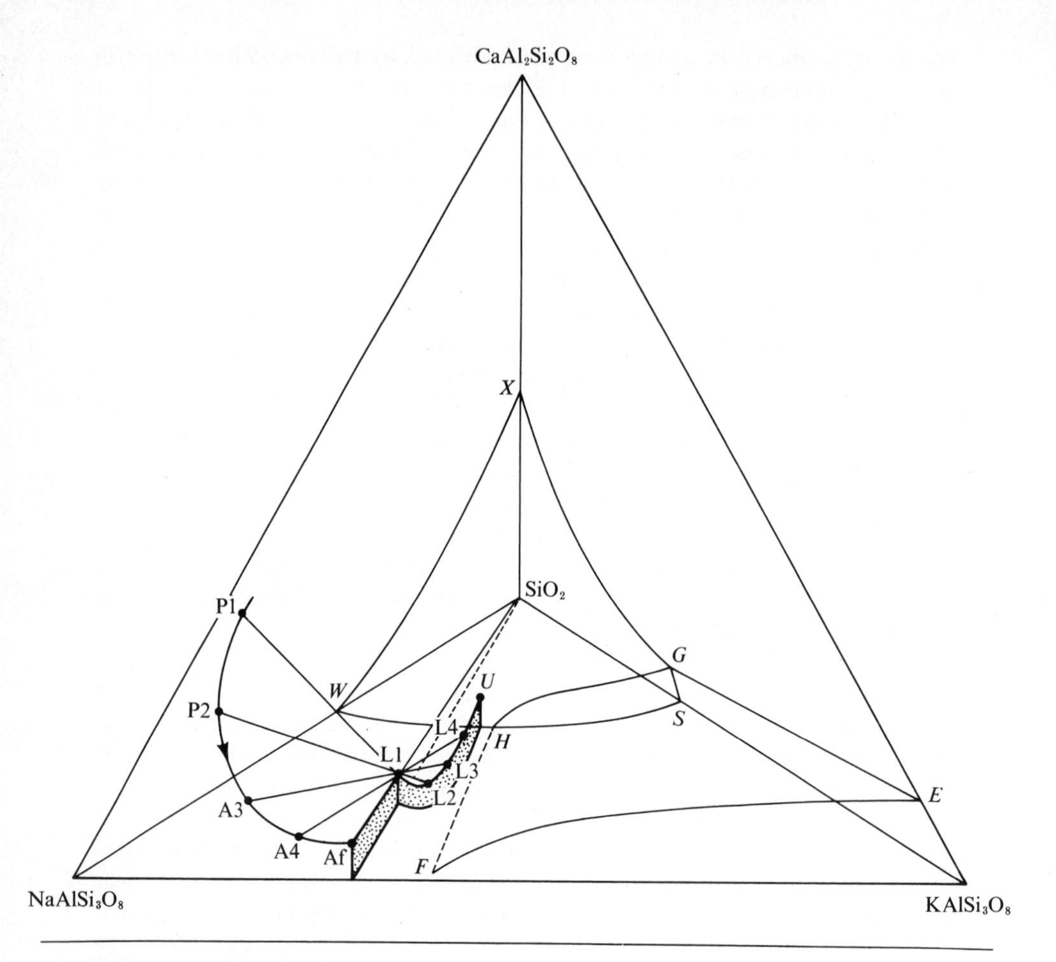


**Fig. 5-7** Possible relationships in the system $CaAl_2Si_2O_8$–$NaAlSi_3O_8$–$KAlSi_3O_8$–$SiO_2$. *WS* represents the quartz-alkali-feldspar boundary curve (cf. Fig. 5-5); *FEGH* is the two-feldspar surface extending into the tetrahedron from the boundary curve *EF* (Fig. 5-3). On the left it curves down to terminate along *FH* before reaching the base of the tetrahedron. *GH* is the intersection of the two-feldspar surface with the bounding surface of the quartz field. Compositions of feldspars *P1–A4–Af* lie in the front face; *P1–A4* are joined by ties to their respective liquids *L1–L4*. (After Carmichael, 1963.)

with the phenocrysts to maintain equilibrium with falling temperature, and its composition will change along a curved path toward the quartz-feldspar surface, becoming at the same time impoverished in $CaAl_2Si_2O_8$ and enriched in silica (Fig. 5-7, $L1 \rightarrow L2$).

If at any time the reaction between the plagioclase and the liquid is impeded, the plagioclase phenocrysts will become zoned; and if the liquid is chilled, the plagioclase will be enclosed by a groundmass of alkali feldspar and quartz.

According to the degree of chilling, a porphyritic obsidian, rhyolite, or granophyre may result; this texture of plagioclase phenocrysts enclosed by an intergrowth of alkali feldspar and quartz is typical of many granophyres.

With continued cooling under equilibrium conditions, the plagioclase will move down the solidus surface and become first more sodic and then potassic, and the feldspar will lose the distinguishing features of a plagioclase to become an anorthoclase (Fig. 5-7, *A*3–*L*3). The liquid will continue to be enriched in silica, but the rate at which calcium is removed from the liquid will decrease, whereas the rate at which potassium is removed from the liquid will increase as the phenocrysts change in composition from plagioclase to anorthoclase. This variable rate of depletion of calcium and potassium from the liquid, together with its enrichment in silica, will cause the liquid to change composition along a curved path toward the quartz-feldspar surface. Because of the higher calcium content of the phenocrysts in relation to the liquid, the liquid will tend to move toward the base of the tetrahedron (Fig. 5-7), near which it will eventually intersect the quartz-feldspar surface (Fig. 5-7, *U*). Here quartz will begin to precipitate, and the alkali-feldspar phenocrysts will continue to react with the liquid to become more potassic, the liquid moving toward the ternary minimum. Crystallization will cease when a straight line drawn from the silica apex to the composition of the precipitating feldspar *Af* intersects the initial composition of the liquid.

Fractionation of an acid liquid forces the liquid more rapidly toward the base of the tetrahedron (Fig. 5-7) and thence to the ternary minimum. Fractionation may be achieved by zoning of the feldspar phenocrysts, giving crystals with cores of plagioclase zoned continuously to anorthoclase.

The natural equivalents of this type of acid liquid are the rhyolites, obsidians, and granophyres found in association with the Tertiary volcano Thingmuli of eastern Iceland (Carmichael, 1963); in terms of the four components represented in Fig. 5-7, they could be derivatives of a basaltic parent.[1]

ANORTHITE-RICH ACID LIQUIDS WITH TWO FELDSPARS Crystallization of these liquids is illustrated schematically in Fig. 5-8; here the initial composition *L*1 lies well above the two-feldspar surface and on the feldspar side of the feldspar-quartz surface. On cooling, the first crystalline phase to appear is a plagioclase feldspar *P*1, in equilibrium with the liquid *L*1. As the temperature falls, the liquid will be enriched in silica and move downward on a curved course to the two-feldspar surface, the plagioclase reacting with the liquid to become more sodic. When the liquid reaches the two-feldspar surface, the plagioclase (now *P*2) is joined by an alkali feldspar *A*2, and the path changes sharply in direction. It now moves along the two-feldspar surface toward the quartz-feldspar surface,

[1] Such a relationship is supported by the similarity of $Sr^{87}/Sr^{86}$ ratios in these rhyolites and basalts.

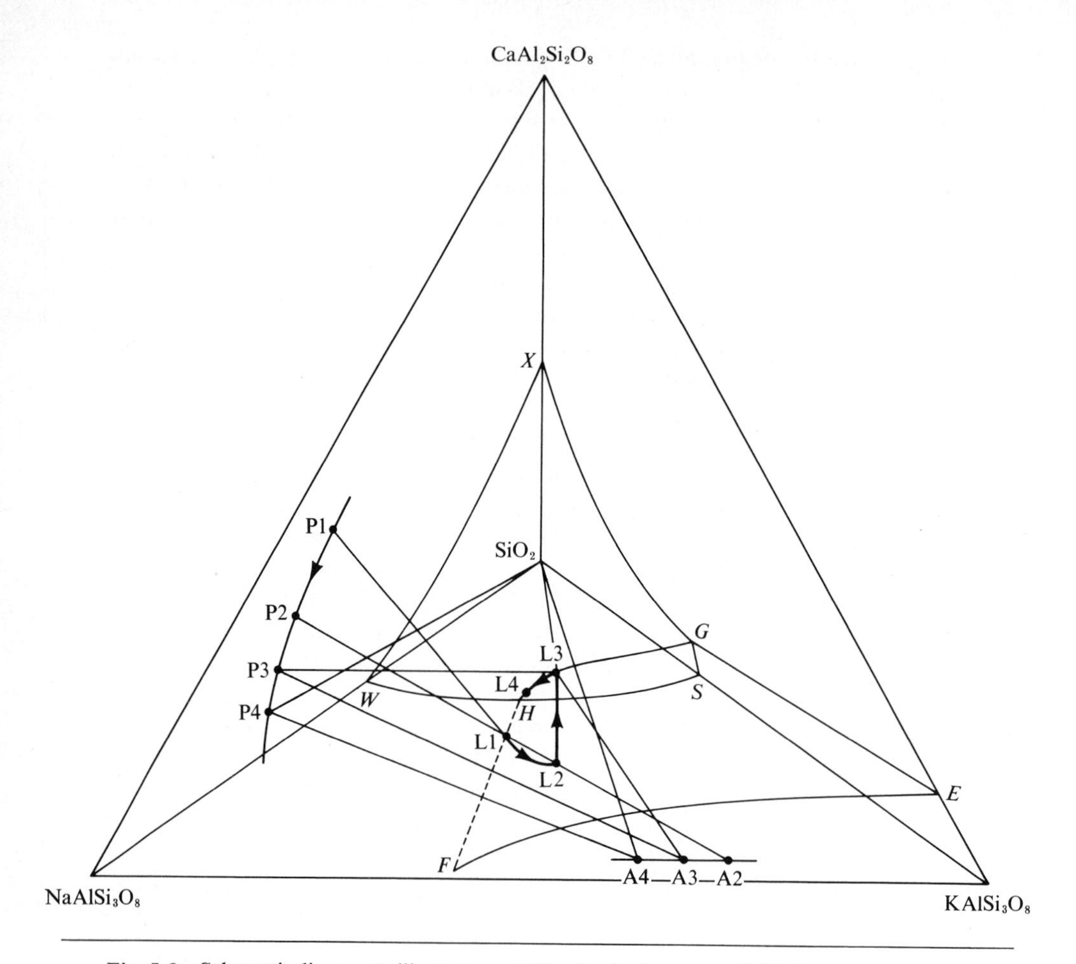


**Fig. 5-8** Schematic diagram to illustrate crystallization in the system $CaAl_2Si_2O_8$–$NaAlSi_3O_8$–$KAlSi_3O_8$–$SiO_2$. The plagioclase feldspars *P*1–*P*4 and alkali feldspars *A*2–*A*4 are plotted on the front face of the tetrahedron, whereas the liquids *L*1–*L*4 fall inside the tetrahedron.

while the plagioclase and sanidine phenocrysts become more sodic by continuous reaction. When the liquid has the composition *L*3, quartz starts to precipitate, by which time the plagioclase has the composition *P*3 and the alkali feldspar *A*3. Further cooling of the liquid changes its composition from *L*3 to *L*4 by the continuous separation of quartz, plagioclase, and sanidine until at *L*4 it is eliminated, since the plane *P*4–*A*4–$SiO_2$ now intersects the initial composition *L*1. The final crystalline product is composed of plagioclase *P*4, alkali feldspar *A*4, and quartz. The compositions of *P*4 and *A*4 are in some measure pressure-dependent, and the trend of the coexisting feldspar tie lines illustrated in Fig. 5-4 will depend on the water-vapor pressure obtaining during crystallization (Yoder et al., 1957).

ANORTHITE-POOR ACID LIQUIDS WITH TWO FELDSPARS These liquids lie under the two-feldspar surface *FEGH*, and thus sanidine is the first feldspar to precipitate. It is these liquids which are predominantly affected by the resorption of plagioclase. If the initial composition contains more $CaAl_2Si_2O_8$ than the sanidine can hold in solid solution, then plagioclase will be present throughout the crystallization interval; but if the amount of $CaAl_2Si_2O_8$ is less than the sanidine can contain as it becomes progressively more sodic, then the resorption will be complete if equilibrium is maintained throughout the crystallization interval.

The added complications of the partial or complete resorption of early crystallizing plagioclase, which depends upon the initial composition of the liquid, do not markedly change the crystallization path in comparison with $CaAl_2Si_2O_8$–rich liquids (Fig. 5-8), except that sanidine rather than plagioclase is the first feldspar to precipitate.

All liquids, on moving down the two-feldspar surface (Fig. 5-8, *FEGH*), will begin to resorb the previously precipitated plagioclase, and the extent of resorption will depend, under conditions of equilibrium, upon the composition of the initial liquid. If this can be expressed only in terms of quartz, alkali feldspar, and plagioclase, then some will remain and the liquid will not leave the two-feldspar surface; but if the initial composition is such that it can be completely represented by alkali feldspar and quartz, then the plagioclase will be completely resorbed and the liquid will then move off the surface *FEGH* toward the base and thence toward the ternary minimum.

Fractional crystallization of liquids similar in composition to those discussed above will ensure that they reach the base of the tetrahedron and thereafter the ternary minimum. Crystallization of this type is most easily achieved by zoning of the feldspar phenocrysts, so that plagioclase zoned continuously to alkali feldspar (anorthoclase) will be found with sanidine zoned toward more sodic compositions.

It must now be clear that the crystallization of natural acid liquids which contain even small amounts of $CaAl_2Si_2O_8$ cannot be completely described solely in relation to the system $NaAlSi_3O_8$–$KAlSi_3O_8$–$SiO_2$. The amount of calcium in a potassic acid liquid determines only whether plagioclase is precipitated before or after sanidine; if it is small, and thus the composition of the initial liquid lies beneath the two-feldspar boundary surface, then the early precipitation of sanidine causes the composition of the liquid to move toward the $NaAlSi_3O_8$–$SiO_2$ join (Fig. 5-8) so that it will later intersect the falling two-feldspar surface. We conclude that natural acid liquids containing calcium (later to form the $CaAl_2Si_2O_8$ component of plagioclase) will, at some stage in their cooling histories, yield a plagioclase, and if the liquid is sufficiently potassic, the plagioclase will be joined by a sanidine; whether or not sanidine accompanies the plagioclase depends solely on the $K_2O/Na_2O$ ratio of the initial liquid.

Rhyolites with phenocrysts of quartz, plagioclase, and sanidine are commonly found in ignimbrite flows in the Basin-Range province of the western United States; many Quaternary rhyolitic lavas in California also display this mineralogy. The granophyres associated with the Tasmanian diabases (McDougall, 1962) are also of this mineralogical type, and here the evidence of the strontium isotopes suggests widespread modification of the parental magma with crustal material.

The corresponding crystallization paths for slowly cooled acid magma are even more difficult to decipher. The presence of large amounts of water saturating the liquid requires that the two-feldspar surface *FEGH* intersect the base of the tetrahedron in Fig. 5-8. Fractional crystallization of the liquid may be unsuccessful in purging it of plagioclase, and there is a possibility that a biotite or muscovite may affect the precipitation of sanidine, since both micas contain the components of alkali feldspar. Lastly, continued slow cooling will allow extensive recrystallization of the alkali feldspar as it exsolves an albite component on intersecting the alkali-feldspar solvus. Two generations of plagioclase are possible: one, the original plagioclase crystallizing from the liquid; the other, the exsolution product of the recrystallizing alkali feldspars. There is much need for the investigation of the cooling path of high-level granite plutons whose original mineralogical relationships have not been changed by reaction during cooling with the rock-water envelope.

PERALKALINE RHYOLITES: FELDSPAR-LIQUID RELATIONSHIPS A rather rare group of rhyolites, the comendites and the more extreme pantellerites, has in the last dozen years been the subject of much debate,[1] particularly with respect to their feldspars and the relationship of these to the enclosing glassy groundmass. In one way or another the controversy arose from the alumina undersaturation (molecular $Na_2O + K_2O > Al_2O_3$) of these rocks, which removes them from the rhyolite tetrahedron and which is reflected mineralogically in the presence of sodic pyroxenes, rarely sodic amphibole, and commonly aenigmatite. The beautiful but extremely rare mineral tuhualite has been found only in the New Zealand comendites.

A typical comendite differs from a pantellerite in its relative paucity of the mafic components (Table 5-1), less extreme enrichment in such trace elements as Nb, Ta, Zr, Mo, Zn, and Cd (which characterize all peralkaline volcanics), and smaller alumina undersaturation. However, we shall use only data on pantellerites to illustrate some of the unusual features of peralkaline rhyolites. In

[1] Contributors to or participants in this debate include: Carmichael (1962), Chayes and Zies (1962), Carmichael and MacKenzie (1963), Chayes (1964*b*), Bailey and Schairer (1964), R. N. Thompson and MacKenzie (1967), Nicholls and Carmichael (1969*b*), D. K. Bailey and Macdonald (1969), Gibson (1972).

**Table 5-1** Representative compositions of pantellerite, comendite, Icelandic rhyolite, and Californian rhyolite*

| | 1 | 2 | 3 | 4 | | 1 | 2 | 3 | 4 |
|---|---|---|---|---|---|---|---|---|---|
| $SiO_2$ | 69.81 | 73.06 | 74.96 | 75.04 | Nb | 320 | 72 | 26 | 20 |
| $TiO_2$ | 0.45 | 0.23 | 0.23 | 0.07 | Ta | 17.8 | 7.4 | — | — |
| $Al_2O_3$ | 8.59 | 9.76 | 12.55 | 12.29 | Mo | 14.9 | 18.3 | 2.5 | — |
| $Fe_2O_3$ | 2.28 | 2.74 | 1.72 | 0.33 | Sn | 0.8 | 6.5 | 2.6 | — |
| FeO | 5.76 | 2.70 | 0.71 | 0.71 | Ni | 1.2 | — | 0.7 | <5 |
| MnO | 0.28 | 0.13 | 0.04 | 0.05 | Co | 0.6 | — | 1.3 | 10 |
| MgO | 0.10 | 0.10 | 0.02 | 0.04 | Cu | 3.6 | 4.2 | 9 | <5 |
| CaO | 0.42 | 0.32 | 0.90 | 0.58 | Zn | 440 | 290 | 125 | 38 |
| $Na_2O$ | 6.46 | 5.64 | 4.41 | 4.03 | Zr | 1800 | 1250 | 400 | 110 |
| $K_2O$ | 4.49 | 4.34 | 3.65 | 4.66 | Hf | — | 19.5 | — | — |
| $P_2O_5$ | 0.13 | 0.02 | 0.04 | 0.01 | Y | 145 | 157 | 45 | 28 |
| $H_2O^+$ | 0.14 | 0.46 | 0.65 | 1.81 | Cd | 0.52 | 0.45 | 0.20 | — |
| $H_2O^-$ | 0.05 | 0.26 | 0.34 | 0.20 | Sr | <5 | 2 | 120 | 5 |
| Cl | 0.76 | 0.12 | — | — | Pb | 15 | 31 | 9 | 40 |
| | | | | | Rb | 175 | 143 | 120 | 190 |
| Total | 99.72 | 99.88 | 100.22 | 99.82 | K/Rb | 213 | 252 | 252 | 204 |

* Data taken from Carmichael (1962, 1964*b*, 1967*a*); Ewart et al. (1968*b*); Nicholls and Carmichael (1969*b*).

*Explanation of column headings*
1 Pantellerite
2 Comendite
3 Icelandic rhyolite
4 Californian rhyolite

many of these lavas, the often abundant anorthoclase phenocrysts are surrounded by glass, which most petrologists would consider to represent a liquid quenched in equilibrium with the feldspar phenocrysts.

It has been shown that, during crystallization of most liquids in the system $NaAlSi_3O_8$–$KAlSi_3O_8$–$SiO_2$, both liquid and feldspars become more sodic, or alternatively both become more potassic. That this tendency persists in natural acid liquids can be seen in Fig. 5-9*a*, where each plotted point shows the $Na_2O/(Na_2O + K_2O)$ ratio of the alkali-feldspar phenocrysts (abscissa) and the obsidian in which it is found (ordinate). The Icelandic trend is defined by three obsidians with anorthoclase phenocrysts, and the Californian trend, of Quaternary obsidians, by three rhyolites with phenocrysts of sanidine. Both trends have a positive slope, the former showing enrichment in potassium and the latter in sodium, as crystallization progresses.

Comparable data for a suite of pantellerites are more complete than for either of these groups, and for each rock two points may be plotted: the first is defined by $Na_2O/(Na_2O + K_2O)$ of the anorthoclase phenocrysts in conjunction with the rock in which they are found, whereas the second point is defined by the alkali composition of alkali-feldspar microlites in combination

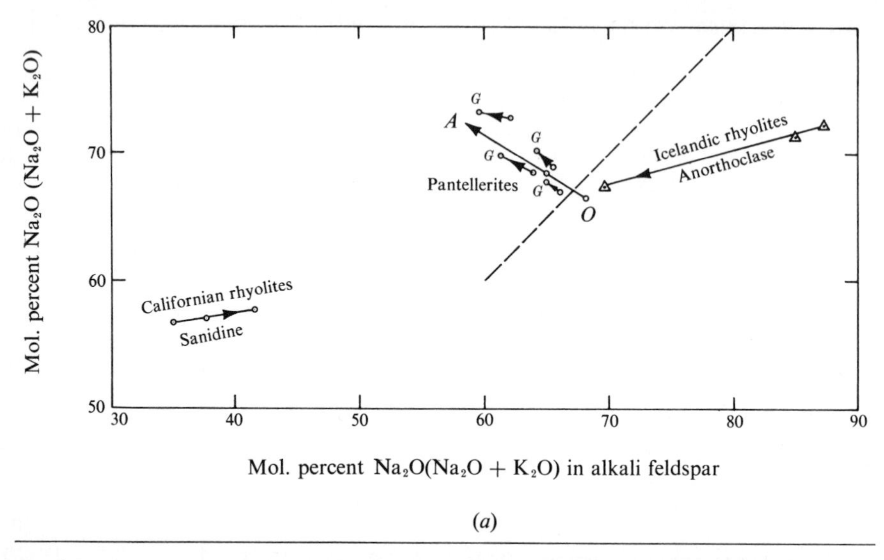


**Fig. 5-9** (*a*) Plot of $Na_2O/(Na_2O + K_2O)$ of feldspar phenocrysts against the same ratio in their associated obsidians or residual glasses. For the pantellerites, *G* refers to the pair (residual glass) – (microlitic feldspar), which is joined to a circle representing the associated rock-phenocryst pair; the arrows represent direction of crystallization of each pair. The long arrow *OA* represents the direction of increasing peralkalinity. Note that the slope of the pantellerite trend is negative, whereas both the Icelandic and Californian rhyolites have a positive trend but are opposite in sense, the former becoming more potassic and the latter increasingly sodic. Dashed line represents the locus for identical alkali ratios in obsidian and in feldspar. (*b*) An identical plot to (*a*) except that the ordinate is now the ratio of the CIPW normative feldspar, $ab/(ab + or)$. Note that there is a general tendency for the most potassic feldspar in the pantellerites to be associated with the most *or*-rich obsidian or glass: here the trend is positive in conformity with the Californian and Icelandic metaluminous rhyolites.

with that of the residual glass which encloses them. Each pantellerite has a crystallization path of negative slope, and the whole group of pantellerites defines a trend which intuitively seems anomalous. This trend is that the most potassic feldspar is found in the most sodic pantellerite, or conversely that the most sodic feldspar precipitates from the most potassic liquid—cooling allows the feldspar to become more potassic while the liquid becomes more sodic. This hallmark of pantellerites contrasts directly with the trends of the metaluminous (nonperalkaline) rhyolites of Iceland and California.

However, because the pantellerites contain more $Na_2O + K_2O$ than $Al_2O_3$, it is not possible for the totality of alkali to become incorporated in feldspar at some later stage in the crystallization path. Perhaps this striking pantellerite trend of negative slope (Fig. 5-9*a*) is related to alumina undersaturation. One way to test this hypothesis would be to calculate the CIPW normative feldspar in these rocks and residual glasses; but this procedure requires a decision as to how much $K_2O$ to allot to $Al_2O_3$ to form normative *or*. Should it be all, or should it be in proportion to the $K_2O/Na_2O$ ratio in the rock or residual glass or perhaps to some other arbitrarily determined factor? The

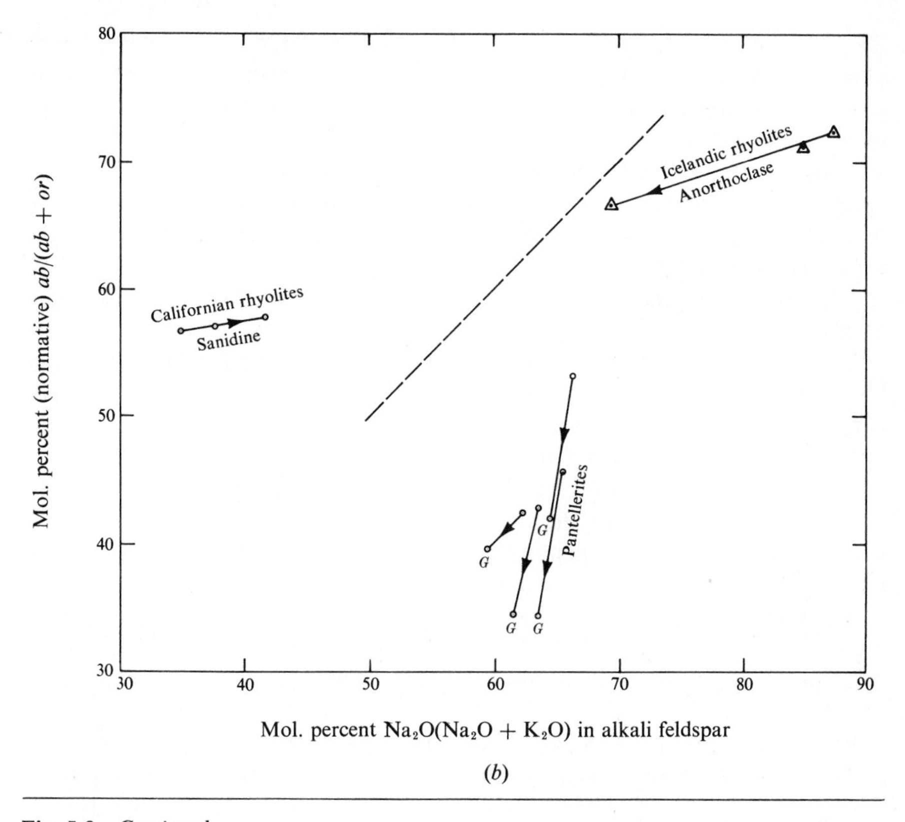


**Fig. 5-9** *Continued.*

authors of the CIPW norm suggest that it should be all, since few if any ferromagnesian minerals containing $K_2O$, but no $Al_2O_3$, are found in igneous rocks. In other words any calculated excess of alkali in the computation of the normative feldspar would be $Na_2O$, in general accord with the frequent occurrence of sodic ferromagnesian minerals. But this is a purely arbitrary procedure which, if it is to be used as an illustration of a genetic property of these acid liquids, should have a more secure foundation.

Fortunately thermodynamics has something to say on this matter. In the following exchange reactions involving sanidine and high albite

$$2NaAlSi_3O_8 + K_2O_{(c)} = 2KAlSi_3O_8 + Na_2O_{(c)}$$

$$(\Delta G_r^\circ)_{1000} = -30.3 \text{ kcal}$$

$$2NaAlSi_3O_8 + 2KOH_{(l)} = 2KAlSi_3O_8 + 2NaOH_{(l)}$$

$$(\Delta G_r^\circ)_{1000} = -16.0 \text{ kcal}$$

$$2NaAlSi_3O_8 + K_2CO_{3(c)} = 2KAlSi_3O_8 + Na_2CO_{3(l)}$$

$$(\Delta G_r^\circ)_{1000} = -15.3 \text{ kcal}$$

the standard free-energy changes at 1000°K are all large and negative, which indicates that in all cases potassium is preferentially combined with aluminum to form feldspar. Thompson and MacKenzie (1967) essentially established the same preferential incorporation of potassium in feldspar, where the additional component was alkali metasilicate. Here then is a thermodynamic basis for the petrographic observation that potassium is found only in a ferromagnesian mineral without aluminum or, alternatively, that in the computation of the normative feldspathic components in the residual liquids of pantellerites, $K_2O$ should first be combined with $Al_2O_3$, and to the remaining $Al_2O_3$, $Na_2O$ should be allotted to form albite. Hence in these rocks the excess alkali over that required to combine with alumina to form feldspar will be $Na_2O$. This rather long digression can be summarized thus: in all peralkaline lavas the ratio $K_2O/Na_2O$ and the calculated normative feldspar ratio *or/ab* cannot be identical; but in all other types of rhyolitic liquids, these two ratios are identical.

If, instead of the ratio $Na_2O/(K_2O + Na_2O)$ for the peralkaline rocks and their individual glasses, the computed CIPW normative feldspars are used, we find the very steep but positive trend shown in the lower part of Fig. 5-9*b*. In this we see clearly the association of sodic feldspar with a rock or glass rich in *ab*, or alternatively that the most potassic feldspars are found in liquids which are rich in normative *or*. The CIPW procedure for peralkaline rhyolites has the merit of making their trends conform to those of metaluminous rhyolites which are of course unchanged by the CIPW normative computation. By considering only the $Na_2O/(Na_2O + K_2O)$ ratios of the pantelleritic rocks and residual glasses, rather than the normative components, it would be difficult to predict that the precipitating feldspar would become more potassic; this of course is the path that would be predicted by considering the normative components.

It is difficult to depict the phase relationship of peralkaline rhyolites in terms of the limited number of components which are adequate for the non-peralkaline rhyolites. Their peralkalinity obviously affects the crystallization path of their feldspars, and several attempts have been made to portray this without distortion (D. K. Bailey and Macdonald, 1969; Thompson and MacKenzie, 1967); the characteristic loss of alkali and chloride during crystallization contributes to the difficulty of representing this diagramatically. To a petrologist, one salient feature of pantellerites is the very restricted range of composition of both the feldspar phenocrysts and the coexisting feldspar microlites in contradistinction to all other rhyolites, which suggests that their crystallization path has alkali feldspar of virtually fixed composition as one termination. Only in these lavas does anorthoclase behave almost as a compound of fixed composition.

Both comendites and pantellerites characteristically have high concentrations of certain trace elements in comparison with Icelandic or Californian rhyolites; representative data are set out in Table 5-1. It can be seen that peralkaline acid lavas have relatively high Nb, Ta, Mo, Zn, Zr, Y, and Cd, which many petrologists (for example, Ewart et al., 1968*b*) believe to result from a long liquid line of descent from a basaltic parent; and yet such element pairs as K/Rb (often an indicator of differentiation) are quite undisturbed by this postulated history (Table 5-1).

Peralkaline rhyolites are found in both oceanic and continental environments (D. K. Bailey and Macdonald, 1970). The best-known occurrences are the Mediterranean island of Pantelleria (Washington, 1913, 1914), the Ethiopian rift valley (Gibson, 1971), and the Kenya rift valley near Lake Naivasha. Plutonic rocks of broadly equivalent composition are found on the Atlantic island of Ascension as xenoliths in the ejecta (Roedder and Coombs, 1967), in the ring complexes at the Oslo graben (Permian), in the White Mountain province of New Hampshire (Jurassic), and in the younger (Jurassic) granitic ring complexes of Nigeria (Jacobson et al., 1958).

## THE PHONOLITE PENTAHEDRON: THE SYSTEM $CaAl_2Si_2O_8$–$NaAlSi_3O_8$–$NaAlSiO_4$–$KAlSi_3O_8$–$KAlSiO_4$

Although igneous rocks which contain feldspar, together with nepheline or leucite or both, represent only a small or even insignificant part of the igneous economy, they also account for an amazingly large proportion of the plethora of petrographic titles. The salic constituents of these rocks[1] can be represented in the phonolite pentahedron, which is the feldspathoidal side of the salic tetrahedron. This pentahedron (Fig. 5-10) is bounded by the now-familiar plane $CaAl_2Si_2O_8$–$NaAlSi_3O_8$–$KAlSi_3O_8$, which intersects the leucite volume and which contains the two-feldspar boundary curve. Some common rock types we use to illustrate crystallization in this system, which is undetermined experimentally, are also shown in Fig. 5-10.

THE SYSTEM $NaAlSiO_4$–$KAlSiO_4$–$SiO_2$ The phase relationships for compositions without $CaAl_2Si_2O_8$ as a component, corresponding to the base of the pentahedron, have been studied by Schairer (1957), who determined the anhydrous system, by D. L. Hamilton and MacKenzie (1965), and by Fudali (1963), who investigated the behavior of liquids saturated with water at 1000 bars confining pressure. Their results are shown in Fig. 5-11, in which it can be seen that the effect of pressure is to constrict the field of leucite, one of the few minerals whose melting curve has a negative slope. At water vapor pressures above 2500 bars, $KAlSi_3O_8$ no longer melts to leucite plus liquid.

The boundary curve separating the alkali-feldspar field from that of nepheline shifts only slightly under the influence of water at 1000 bars, and it contains a minimum $m$ which is analogous in properties to that found in the rhyolite system. Analyses of phonolites and nepheline syenites, when plotted in terms of their salic normative constituents, concentrate close to, but on the potassic side of, the isobaric minimum at 1000 bars (D. L. Hamilton and MacKenzie, 1965, figs. 4 and 5). At the point where the nepheline alkali-feldspar field boundary intersects the leucite field boundary, there is a ternary

[1] This ignores for the moment such minerals as hauyne, nosean, and sodalite, which are common in feldspathoidal assemblages.

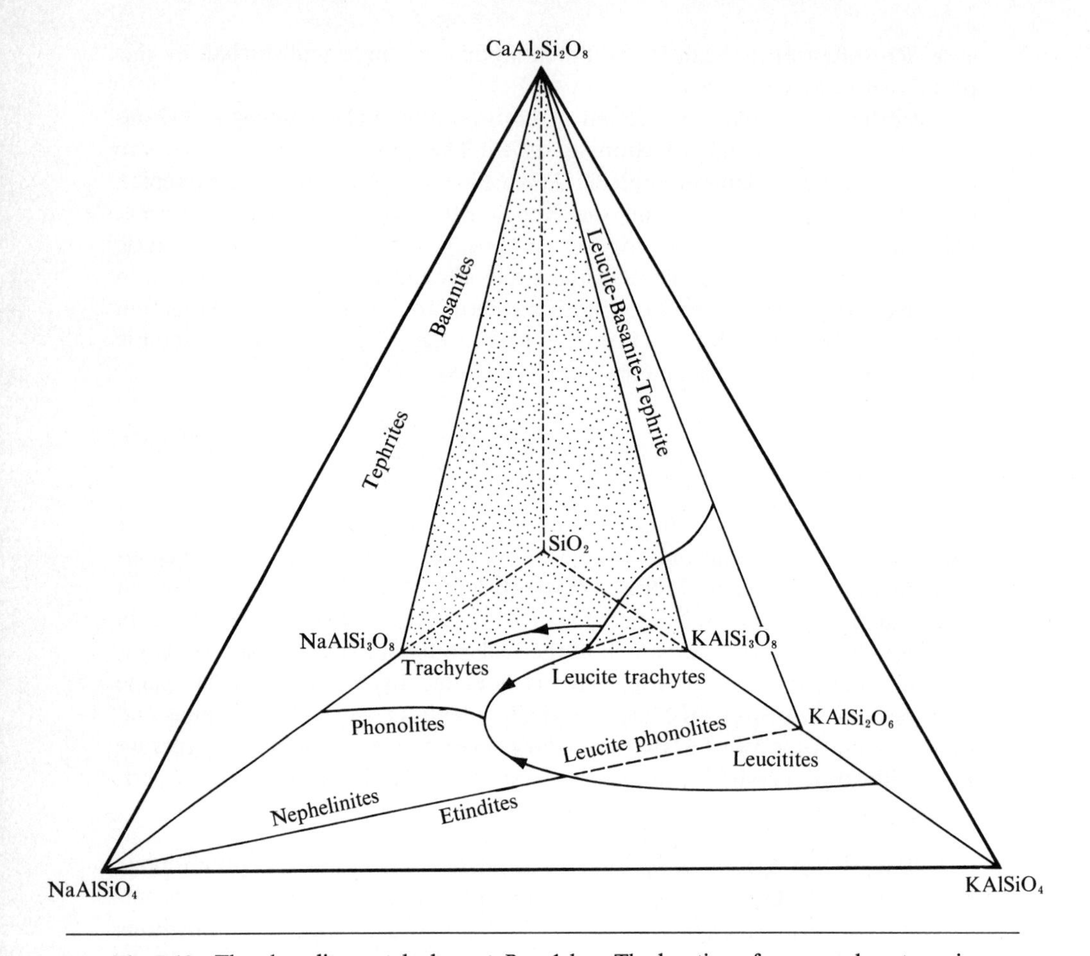


**Fig. 5-10** The phonolite pentahedron at $P = 1$ bar. The location of common lava types in terms of feldspar composition and feldspathoid minerals is shown schematically. The boundary curve of liquid, plagioclase, and sanidine (Fig. 5-1) is shown in the stippled ternary feldspar plane and is truncated on the right by the leucite boundary. The phase relationships in the base of the pentahedron are shown in Fig. 5-11.

or pseudoternary reaction point labeled $R'$ and $R$. At this point, leucite is resorbed by liquid in its passage to the ternary or isobaric minimum, which is the eventual compositional goal of all liquids fractionating in this system.

Two other features found experimentally are of interest in this system. Fudali (1963) found substantial substitution of $NaAlSi_2O_6$ for $KAlSi_2O_6$, and the extent of this at 1000 bars is shown by the line extending from $KAlSi_2O_6$ (leucite). The pseudoleucites of plutonic rocks, which are intergrowths of sanidine and nepheline resulting from subliquidus recrystallization, sometimes contain extensive amounts of $NaAlSi_2O_6$ (Fudali, 1963) and correspond in extent to that found experimentally; but as yet only minor sodium substitution has been recorded in the leucite phenocrysts of volcanic rocks (Fig. 5-13). The other point of interest concerns the variation in the composition of nepheline as a function of temperature. In Fig. 5-11 the extent of solid solution in nepheline, which is often colloquially spoken of as "excess silica" in nepheline, progressively decreases with temperature. Herein nepheline thus offers a potential geothermometer (D. L. Hamilton, 1961). Other complexities are too intricate to be treated fairly here; they involve the properties of the leucite alkali-feldspar field boundary and the resorption of leucite by liquid, which Fudali (1963) has explained in detail; another is the location of the fractionation curves in this system, which both D. L. Hamilton and MacKenzie (1965) and Fudali (1963) consider at some length.

There are nepheline trachytes for which chemical data can be compared with Hamilton and MacKenzie's experimental results. In Fig. 5-12 the compositions $1F$ and $2F$ of anorthoclase phenocrysts in two kenytes (nepheline trachytes) have been plotted in terms of the three feldspar components and projected to the alkali-feldspar join. The glassy groundmass $1G$ of the first, which encloses sporadic microphenocrysts of nepheline, represents a residual liquid on a natural feldspar-nepheline boundary. Coincidentally this is rather close to the experimental boundary at 1000 bars $p_{H_2O}$. The groundmass of the second rock is a fine intergrowth of alkali feldspar and nepheline (whose composition indicates crystallization above 775°C); the tie between the groundmass phases gives a composition $2G$ close to 1G. Coincidentally again the rock composition $2R$ is very close to an experimental three-phase boundary (Fig. 5-12) whose configuration, however, departs from the natural tie. Another three-phase boundary has a feldspar termination close to $1F$, but it is generated by a liquid whose composition is very different from that of the kenyte. We

**Fig. 5-11** Phase relationships between nepheline, alkali feldspar, and leucite at 1 bar (solid lines) and at 1000 bars water-vapor pressure (dashed lines); $m$ is the isobaric minimum at 1000 bars. Also shown are extent of sodium substitution in leucite at 1000 bars and variation of "excess silica" in nepheline at 1000 bars as a function of temperature (bottom left).

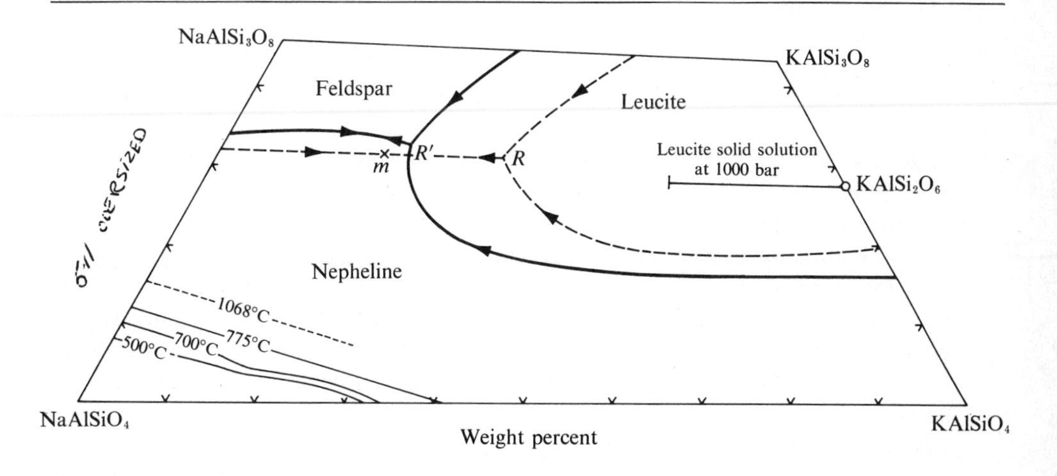

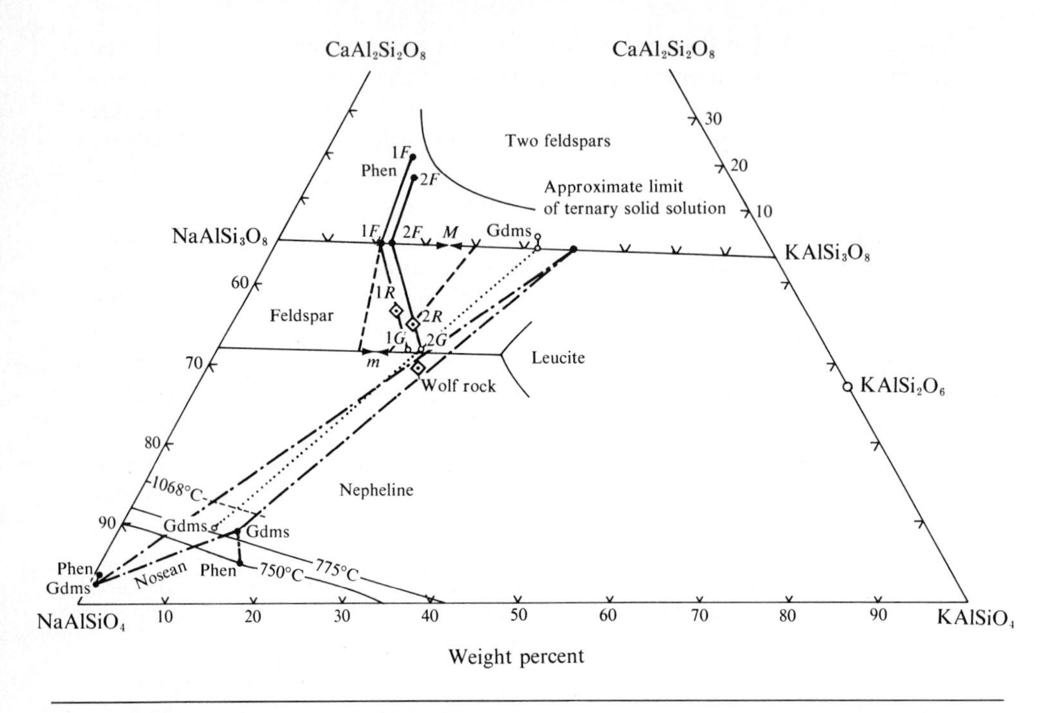


**Fig. 5-12** Part of the phonolite pentahedron with the ternary feldspar plane laid flat (rotated into the plane $KAlSiO_4$–$NaAlSiO_4$–$SiO_2$). Experimental boundaries are taken from Fig. 5-11. Feldspar compositions (1*F*, 2*F*) plotted in ternary feldspar plane and projected onto the alkali-feldspar join. Compositions of rocks are shown as squares. Mineral compositions (and glass) are plotted as open or filled circles with coexisting phases joined by dashed or dotted ties. Two lines (one passing through 2*R* and the other terminating at 1*F*) are experimentally determined three-phase boundaries originating at the alkali-feldspar join.

conclude that additional components in the trachytes, which render them somewhat peralkaline, markedly divert the paths of feldspar crystallization from those of experimental analogs lacking such components.

Of the phonolites that commence their crystallization in the nepheline field, we may take as an example the nosean phonolite of Wolf Rock, Cornwall (Tilley, 1959). Sanidine, nosean, and nepheline all occur both as phenocrysts and as groundmass constituents, and their analyses have been plotted in Fig. 5-12. The triangle whose apices are the compositions of the three groundmass phases includes the composition of the rock, as it should, and the feldspar-nepheline tie line is broadly parallel to those found experimentally. The composition of the nepheline phenocrysts indicates that they crystallized at a lower temperature than did the groundmass nepheline. Although this is possible, it seems unlikely; and hidden by this mean composition is a systematic range and variation which F. H. Brown (1970) attributes to unmixing of alkali feldspar from a nepheline host.

LEUCITE PHONOLITES AND RELATED LAVAS In phonolite lavas more potassic than the two kenytes just discussed, common assemblages are leucite, plagioclase, sanidine, and either nepheline or one of the sodalite group. Such rock types, leucite trachytes, and leucite phonolites are common in the Roman volcanic province, and the mineralogical data (unpublished) on three representatives are shown in Fig. 5-13. In this group, phenocrysts of plagioclase, leucite, and in one case hauyne are found, and their compositions are plotted in Fig. 5-13; note the restricted amount of sodium substitution in the leucite in

**Fig. 5-13** Part of the phonolite pentahedron (with the ternary feldspar plane rotated as in Fig. 5-12). Compositions of plagioclase phenocrysts and of groundmass sanidine and plagioclase are represented by dots, with arrows showing direction of zoning from cores to margins. Composition of coexisting nepheline, or hauyne, and leucite is also shown, with arrows again showing outward direction of zoning. Normative salic constituents of the three lavas are represented by squares (unpublished data).

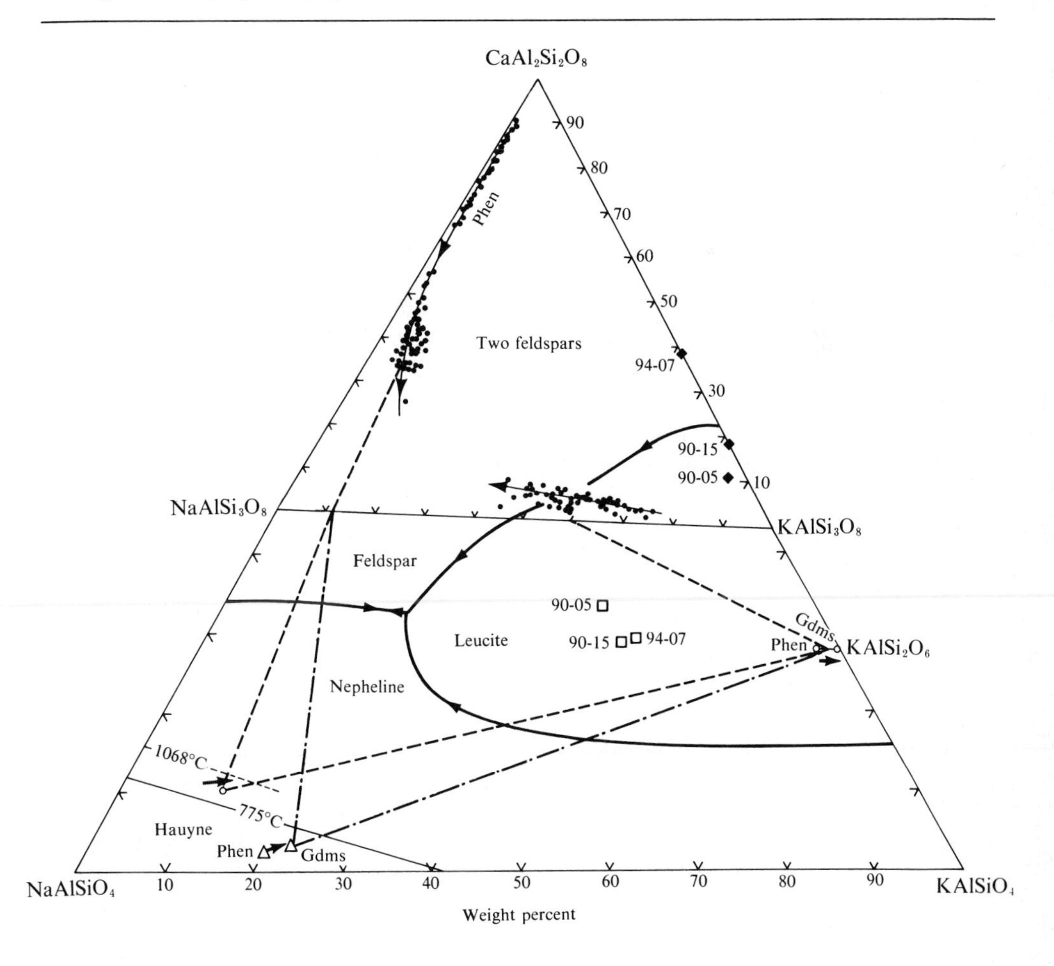

contrast to that found experimentally (Fig. 5-11). In the groundmass of each rock plagioclase and sanidine coexist, both becoming more sodic toward their crystal margins. The feldspathoidal minerals, in contrast to the feldspars, become more potassic toward the crystal margins, and the groundmass phases (leucite and either nepheline or hauyne) are markedly more potassic than their corresponding phenocryst phases. It is this type of mineralogical data, coupled with the data on the rocks themselves (that is, normative salic composition) that allows the construction of a schematic diagram of the type shown in Fig. 5-14.

In Fig. 5-14, the surface *ABEF* is the projection of the two-feldspar boundary of Fig. 5-7 into the phonolite pentahedron. Presumably it acts as a resorption surface toward its lower termination, which must occur slightly above the base of the pentahedron at low pressures. Crystalline phases in equilibrium with liquids on this curve are as follows: along *AB*, nepheline, plagioclase, and sanidine; along *BF*, leucite, plagioclase, and sanidine; along *BC*, leucite, nepheline, and sanidine; at *B*, all four solid phases. The lateral extent of the two-feldspar surface *ABEF* toward the sodic side of the pentahedron is not well known, but nepheline trachytes containing phenocrysts of plagioclase and sanidine can be used to locate its position (Carmichael, 1965).

The surface of the leucite volume has an unusual shape because two slices have been made through it; the first is the ternary feldspar plane, and the second is an arbitrary plane parallel to the join $NaAlSi_2O_6$–$KAlSi_2O_6$ to aid the perspective view.

Assume a liquid composition above the two-feldspar surface and above the surface of the leucite volume, namely in the plagioclase volume. On cooling such a liquid, the first phase to appear is plagioclase *P*1 in equilibrium with liquid *L*1. As the temperature falls, the liquid will be enriched in potassium and will move downward on a curved path, continuously reacting with the plagioclase, which becomes more sodic. When the liquid reaches the surface of the leucite volume, leucite joins plagioclase (now *P*2) and its path changes sharply in direction. It moves down the leucite surface, plagioclase becoming more sodic and leucite slightly more potassic (Fig. 5-13), until the liquid intersects the curve *BF*. At this temperature, plagioclase (now *P*3) and leucite are joined by sanidine *S*, and the liquid *L*3 is at the apex of a tetrahedron. With continued cooling under equilibrium conditions, each feldspar becomes more sodic as the liquid moves toward *B*, at which point its composition remains unchanged while all four solid phases crystallize together. Crystallization may cease before the liquid reaches *B*, and it will do so at the temperature at which the compositions of the coexisting plagioclase, sanidine, and leucite become coplanar with the initial liquid composition.

In many lavas, nepheline or leucite is found with sodalite ($3NaAlSiO_4 \cdot NaCl$) or nosean ($6NaAlSiO_4 \cdot Na_2SO_4$). Thermodynamic evidence (Stormer and Carmichael, 1971*a*), coupled with that of rocks, suggests that both these minerals will precipitate at higher silica activities than are attained along the curve *HMC* and may be expected to precede nepheline in the cooling path of

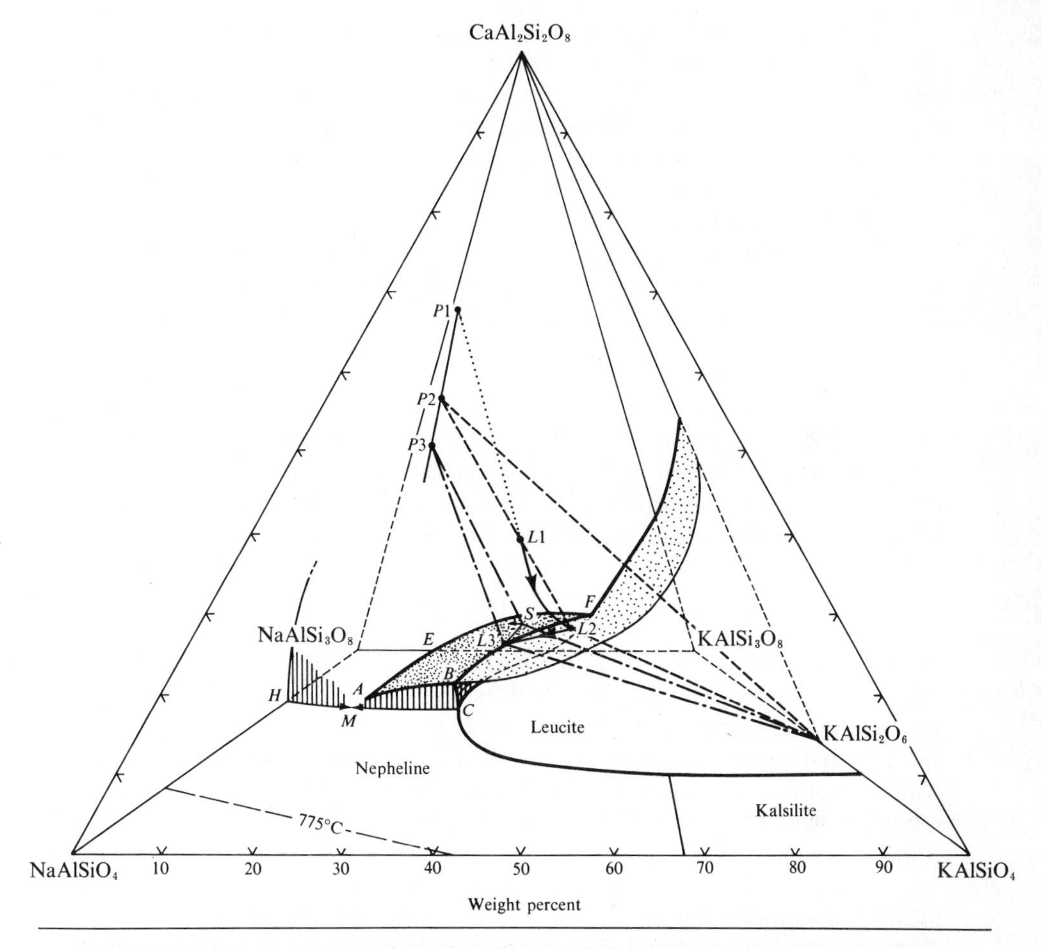


**Fig. 5-14** Schematic diagram of the phonolite pentahedron (with the ternary feldspar plane rotated as in Fig. 5-12) at a few hundred bars pressure. *ABEF* is the two-feldspar surface, which terminates along *AF* before reaching the base of the pentahedron; it intersects the surface of the nepheline volume (shown partly cut away) along *AB* and the leucite volume along the three-phase boundary *BF*. The surface of the leucite volume is truncated at the back by the ternary feldspar plane and in front by a section cut parallel to $KAlSi_2O_6$–$NaAlSi_2O_6$. The composition of *L*1 lies above the two-feldspar plane and in front of the ternary feldspar plane; *P*1, *P*2, *P*3, and *S* all lie in the ternary feldspar plane.

any trachyte. In many lavas, such as the trachytes of Mt. Suswa, Kenya (Nash et al., 1969), this is in fact the case. In more potassic phonolites (which contain leucite), one of the trio, sodalite, nosean, or hauyne, usually the last-named, may be expected to precipitate somewhere along the curve *BF*, but necessarily before the liquid reaches *B*. In the leucite basanites of Vesuvius, the presence of sodalite in the groundmass without sanidine suggests that its precipitation may prevent the residual liquid from reaching the two-feldspar surface *ABFE*.

In feldspathoidal lavas that contain feldspar, the feldspars become more sodic, and the feldspathoids more potassic, as crystallization proceeds. With up to four salic phases, all solid solutions and all changing composition to a greater or lesser extent, it is difficult to represent clearly a liquid's cooling path, especially if one of the feldspathoids is a member of the sodalite group. The generalized picture shown in Fig. 5-14, if considered against a background of this complexity, will suffice.

In conclusion, it is appropriate to emphasize that, in liquids with low silica activities, feldspar of any type is unstable. Since in igneous rocks perovskite is never associated with feldspar, the upper limit of silica activity is given by

$$\underset{\text{perovskite}}{CaTiO_3} + \underset{\text{glass}}{SiO_2} = \underset{\text{sphene}}{CaTiSiO_5}$$

Accordingly, lava types which contain kalsilite and nepheline, or leucite and kalsilite, or all three, will not precipitate feldspar, and the equivalent of the $CaAl_2Si_2O_8$ component will be represented mineralogically either as pyroxene—$CaAl_2SiO_6$ component—or by melilite.

## THE "PLAGIOCLASE EFFECT" IN SALIC LAVAS

Bowen (1945) has shown that in synthetic mixtures of $NaAlSi_3O_8$ with a calcium-bearing second component such as $CaSiO_3$, the feldspar which initially precipitates is a calcium-bearing plagioclase. This preferential incorporation of calcium, and necessarily $Al_2O_3$, into sodic feldspar crystallizing from a liquid with no normative $CaAl_2Si_2O_8$ component has been called the "plagioclase effect." It has been found to occur in all synthetic systems where $NaAlSi_3O_8$ is one component and a calcium-bearing component is the other.

This preferential entry of CaO and $Al_2O_3$ into feldspar generates the peralkaline condition—namely, a molecular excess of ($Na_2O + K_2O$) over $Al_2O_3$—in the residual liquid. This can also be demonstrated in many nepheline trachytes, and one of the kenytes discussed previously provides a good natural example. In Fig. 5-15, all compositions lying on the alkali-feldspar join contain components with a 1:1 alkali/alumina ratio. All compositions above this plane contain more alumina than this; or, in other words, they contain the component $CaAl_2Si_2O_8$. All compositions which fall beneath the alkali-feldspar join are peralkaline, since $(Na_2O + K_2O) > Al_2O_3$. Lines of equal $Na_2O/K_2O$ ratio radiate from the $Al_2O_3$ apex and show the relative enrichment of alkali in coexisting phases.

The composition of the kenyte initial liquid $1R$ falls above the alkali-feldspar join, and precipitation of anorthoclase phenocrysts $1F$ drives the residual liquid through the 1:1 alkali-alumina join into the peralkaline sector. A similar tendency is shown by a trachyte from Mt. Suswa, Kenya (Nash et al., 1969), except that in this lava the initial liquid was already peralkaline, so that the precipitation of the anorthoclase phenocrysts containing $CaAl_2Si_2O_8$

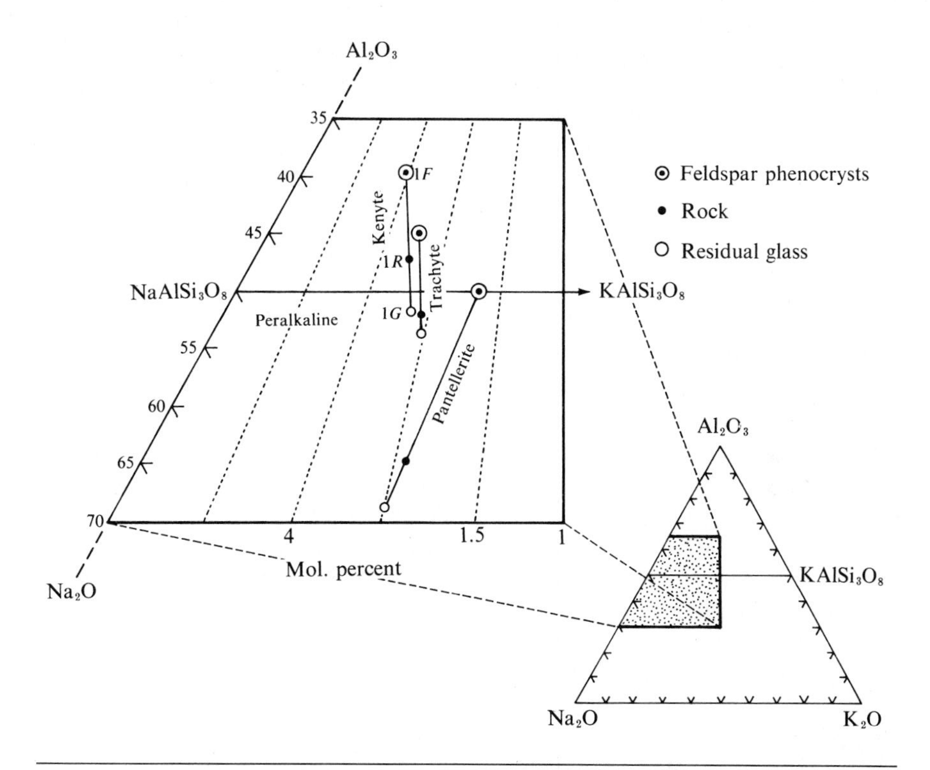


**Fig. 5-15** The molecular proportion of $Na_2O$, $K_2O$, and $Al_2O_3$ of the feldspar phenocrysts, rocks, and residual glasses of a kenyte (Fig. 5-12), a nepheline trachyte, and a pantellerite. Peralkaline compositions lie below the $NaAlSi_3O_8$–$KAlSi_3O_8$ join. Ratio of $K_2O/Na_2O$ is constant on each dotted line. Circled points, phenocrysts; solid dots, rocks; circles, glass.

enhanced this tendency. An additional example is a pantelleritic obsidian, whose molecular excess of ($Na_2O$ + $K_2O$) over $Al_2O_3$ is far more extreme. Here precipitation of anorthoclase phenocrysts almost devoid of $CaAl_2Si_2O_8$ also increases the peralkalinity of the residual liquid. All three trends are shown in Fig. 5-15.

One contrast between the quartz-bearing pantellerite and the nepheline trachyte is significant. Peralkaline nepheline trachytes not only precipitate anorthoclase feldspars containing substantial amounts of $CaAl_2Si_2O_8$, but the feldspar crystallization path often shows extensive enrichment in $KAlSi_3O_8$; exceptionally it may reverse itself so that a feldspar zoned toward $NaAlSi_3O_8$ has an outer rim changing composition toward $KAlSi_3O_8$ (Nash et al., 1969, fig. 9). In contrast, the anorthoclase precipitating from a silica-rich peralkaline liquid contains trivial amounts of $CaAl_2Si_2O_8$ ($<0.3$ percent) and shows extremely restricted zoning. It is this contrast in the behavior of the feldspars that has led Nash et al. (1969) to postulate fractionation curves of complex shape for liquids corresponding to the Mt. Suswa trachytes.

The generation of a peralkaline liquid residue is frequently found in a wide variety of rock types, being manifested by the green sodic rims of titanaugite or by the occurrence of aenigmatite. On a large scale, perhaps the alkaline border facies of a Skye granite intruding dolomite (Tilley, 1949) is the natural equivalent of the experimental generation of a peralkaline liquid when small amounts of calcium carbonate are added to $NaAlSi_3O_8$ liquid (Watkinson and Wyllie, 1969).

## Salic Minerals and Phlogopite in Igneous Rocks

Biotites of volcanic rocks tend to be more magnesian than those found in the corresponding plutonic rock, and the contrast between those of rhyolites and granites is particularly clear. Many volcanic specimens fall within the phlogopite composition range, and they may in addition contain a substantial amount of $Fe_2O_3$, up to 10 percent $TiO_2$, 5 percent BaO, and lesser amounts of $Cr_2O_3$ ($<1$ percent) and NiO ($<0.2$ percent). The presence of a phlogopite mica in the groundmass of such lavas as mugearites, basanites, and leucite tephrites testifies to its thermal stability at pressures approaching 1 bar. This immediately suggests that the phlogopite in these lavas is not solely a hydroxyl variety, since Wones (1967) has shown that hydroxyl phlogopite breaks down near 700°C at 1 bar; the incorporation of F markedly increases the thermal stability of all hydroxyl minerals, and fluorphlogopite is found to melt congruently at 1397°C at 1 bar.

A commonly studied breakdown reaction of phlogopite is

$$\underset{\text{phlogopite}}{2KMg_3AlSi_3O_{10}(OH)_2} = \underset{\text{kalsilite}}{KAlSiO_4} + \underset{\text{leucite}}{KAlSi_2O_6} + \underset{\text{forsterite}}{3Mg_2SiO_4} + \underset{\text{steam}}{2H_2O}$$

illustrating the relationship of salic minerals to phlogopite, which forms the subject of the following discussion. Almost all experimental work on phlogopite involves the hydroxyl variety, but because most pristine phlogopites in volcanic rocks contain substantial amounts of fluorine, we shall assume that the effect of increasing the activity of fluorine is the same as for water so far as stabilizing mica is concerned. The experimental technique of saturating liquids with water at various confining pressures, and thus increasing its concentration or activity, can be regarded as having the same effect as increasing the concentration of fluorine.

The system $KAlSiO_4$–$Mg_2SiO_4$–$SiO_2$, a suitable starting point, is shown in Fig. 5-16. This system contains two eutectics, one involving kalsilite, forsterite, leucite, and liquid, and the other enstatite, sanidine, tridymite, and liquid; the field of sanidine is minute. There is a thermal divide on the $KAlSi_2O_6$–$Mg_2SiO_4$ join which liquids cannot cross, so that no single cooling liquid has the capacity to reach both eutectic compositions simply by varying the conditions of crystallization. Natural anhydrous equivalents of this ternary system are not known to us; those that most closely correspond contain notable amounts of phlogopite, whose ideal anhydrous composition is plotted in Fig. 5-16.

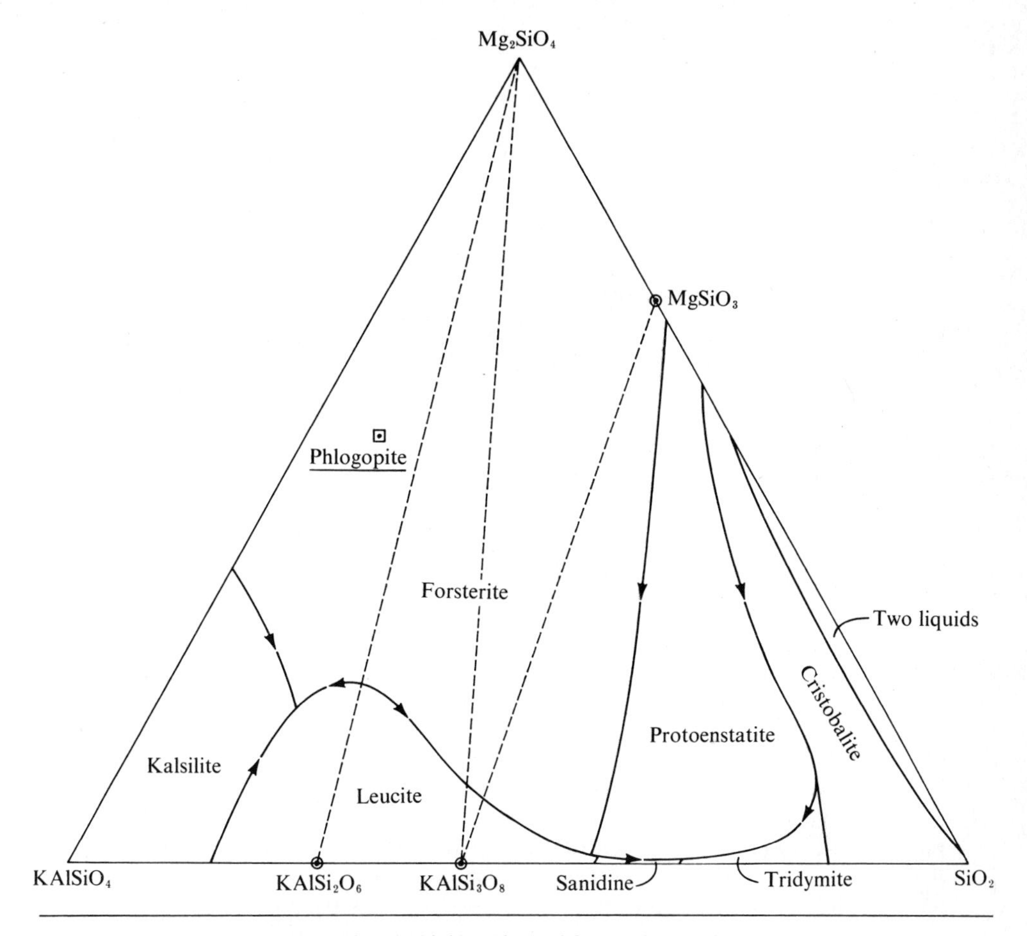


**Fig. 5-16** The system $Mg_2SiO_4$–$KAlSiO_4$–$SiO_2$. (After Luth, 1967.)

With water as an additional component in the system $KAlSiO_4$–$Mg_2SiO_4$–$SiO_2$, the temperature of the water-saturated liquidus surface becomes depressed to intersect the phlogopite volume. The general relationships found by Luth (1967) at 1000 and 2000 bars confining pressure are shown in Fig. 5-17, in which it can be seen that phlogopite first appears on the liquidus surface at silica-rich compositions and then extends, with increasing water concentration, to penetrate the original thermal divide, which it thereby nullifies. There will be a certain concentration of water, or confining pressure, at which phlogopite is stable on both sides of the anhydrous thermal divide without penetrating it. There will also be a unique set of conditions—an invariant point at which kalsilite, leucite, forsterite, phlogopite, and liquid all coexist—that will correspond to the temperature at which the water-saturated liquidus just intersects the top of the phlogopite volume; there is another corresponding invariant point involving leucite, sanidine, phlogopite, forsterite, and liquid. The phlogopite field in Fig. 5-17, at full extent, effectively separates the salic phases from olivine; it is

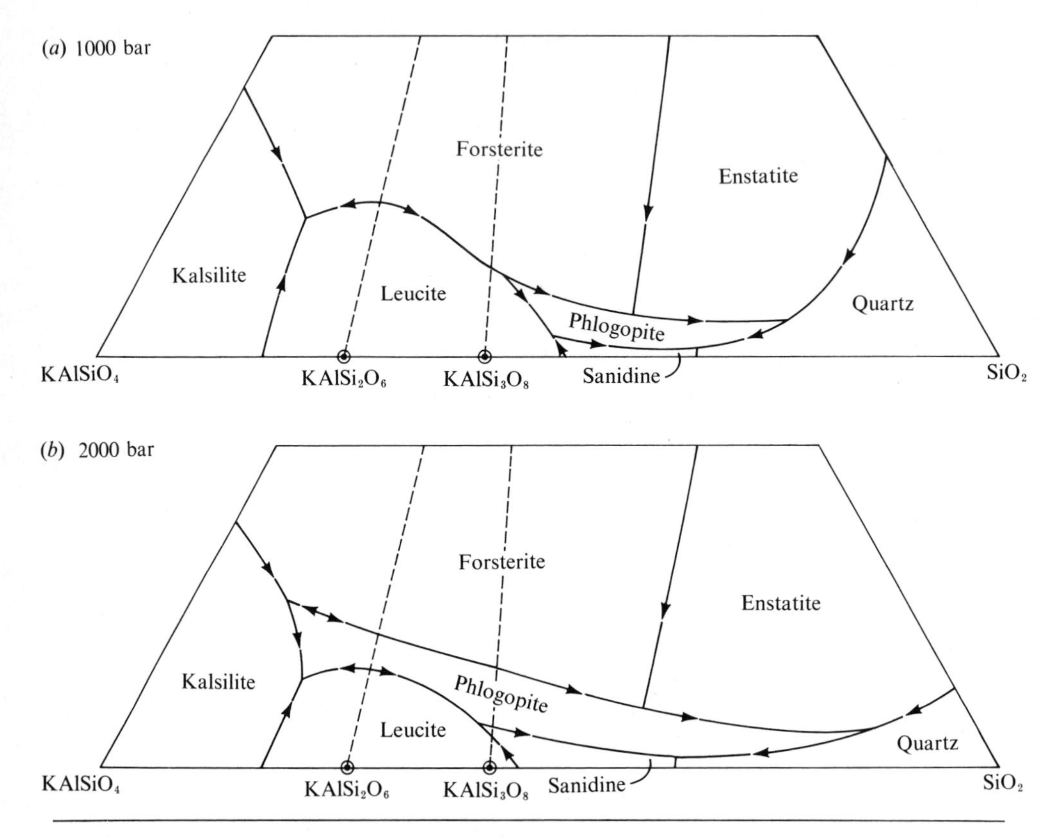


**Fig. 5-17** The system $Mg_2SiO_2$–$KAlSiO_4$–$SiO_2$–$H_2O$ with water-saturated liquids: (*a*) at 1000 bars and (*b*) at 2000 bars confining pressure (after Luth, 1967). Note the enlarged phlogopite field and reduced leucite field with increasing water concentration.

a "roof" through which cooling liquids, with insufficient ($H_2O$ + F) to convert olivine to phlogopite, are unable to penetrate.

Natural equivalents of the hydrous system $KAlSiO_4$–$Mg_2SiO_4$–$SiO_2$ are very rare, but if we consider $NaAlSiO_4$ and $NaAlSi_3O_8$ as additional components, then a considerable variety of feldspathoidal, but mafic, lavas have mineral assemblages represented in terms of this pentahedron. As examples, we look at three lava suites: those from the Leucite Hills, Wyoming (Carmichael, 1967*c*), those from the Toro-Ankole province of Uganda (F. H. Brown, 1971), and lastly those from the Korath Range of southern Ethiopia (F. H. Brown and Carmichael, 1969).

## LEUCITE HILLS, WYOMING

Among the lavas and plugs of this area, three major rock types occur: wyomingite, which is composed of phlogopite phenocrysts in a groundmass of leucite, diopside, apatite, and sometimes glass; orendite, which again has phlogopite phenocrysts (but in a groundmass of leucite and sanidine) together with diopside and apatite; and madupite, with diopside phenocrysts enclosed by poikilitic crystals of phlogopite set in a groundmass of diopside, leucite, and apatite. Olivine phenocrysts with reaction rims of phlogopite occur in many specimens.

Accessory minerals in the groundmass of the wyomingites and orendites are magnophorite (which is a potassic amphibole), priderite ($K_2Ti_8O_{16}$), wadeite ($K_4Zr_2Si_6O_{18}$), and occasional chrome-spinel microphenocrysts with reaction rims of phlogopite; iron titanium oxides are otherwise absent. In the madupites, perovskite and titanomagnetite are ubiquitous accessories together with wadeite. These lavas are the most potassic in the igneous record, and they have no recorded plutonic equivalent. Representative analyses of each rock type are given in Table 5-2. All have low $Na_2O$ compared to $K_2O$; in the CIPW norm, there is the unusual and rare component *ks*, since, on a molecular basis, $K_2O > Al_2O_3$ and is expressed mineralogically by priderite, wadeite, and iron sanidine or iron leucite.

**Table 5-2** Representative analyses and CIPW norms of wyomingite, orendite, and madupite lavas from the Leucite Hills, Wyoming*

| | 1 | 2 | 3 | 4 | | 1 | 2 | 3 | 4 |
|---|---|---|---|---|---|---|---|---|---|
| $SiO_2$ | 50.23 | 55.43 | 55.14 | 43.56 | *Q* | — | 6.03 | 1.88 | — |
| $TiO_2$ | 2.30 | 2.64 | 2.58 | 2.31 | *Z* | 0.37 | 0.42 | 0.40 | 0.40 |
| $ZrO_2$ | 0.25 | 0.28 | 0.27 | 0.27 | *or* | 47.76 | 53.12 | 56.51 | 1.57 |
| $Al_2O_3$ | 10.15 | 9.73 | 10.35 | 7.85 | *lc* | 6.00 | — | — | 32.09 |
| $Cr_2O_3$ | 0.06 | 0.02 | 0.04 | 0.04 | *ne* | — | — | — | 0.19 |
| $Fe_2O_3$ | 3.65 | 2.12 | 3.27 | 5.57 | | | | | |
| FeO | 1.21 | 1.48 | 0.62 | 0.85 | *th* | 0.62 | 0.82 | 0.71 | 0.92 |
| MnO | 0.09 | 0.08 | 0.06 | 0.15 | *ac* | 7.60 | 4.35 | 6.71 | 2.21 |
| MgO | 7.48 | 6.11 | 6.41 | 11.03 | *ks* | 1.81 | 6.01 | 3.62 | — |
| CaO | 6.12 | 2.69 | 3.45 | 11.89 | *wo* | 7.27 | 0.88 | 1.33 | 19.72 |
| SrO | 0.32 | 0.27 | 0.26 | 0.40 | *en* | 6.29 | 15.21 | 15.96 | 17.04 |
| BaO | 0.61 | 0.64 | 0.52 | 0.66 | *fo* | 8.65 | — | — | 7.31 |
| $Na_2O$ | 1.29 | 0.94 | 1.21 | 0.74 | | | | | |
| $K_2O$ | 10.48 | 12.66 | 11.77 | 7.19 | *cm* | 0.09 | 0.03 | 0.06 | 0.06 |
| $P_2O_5$ | 1.81 | 1.52 | 1.40 | 1.50 | *il* | 2.69 | 3.28 | 1.40 | 2.08 |
| $H_2O^+$ | 2.34 | 2.07 | 1.23 | 2.89 | *hm* | 1.02 | 0.62 | 0.95 | 4.81 |
| $H_2O^-$ | 1.09 | 0.61 | 0.61 | 2.09 | *tn* | — | 2.24 | 4.52 | — |
| $SO_3$ | 0.35 | 0.46 | 0.40 | 0.52 | *pf* | 1.51 | — | — | 2.07 |
| Total | 99.83 | 99.75 | 99.59 | 99.51 | *ap* | 4.29 | 3.60 | 3.32 | 3.55 |

* Data taken from Carmichael (1967*c*).

*Explanation of column headings*

1 Wyomingite
2 Wyomingite
3 Orendite
4 Madupite

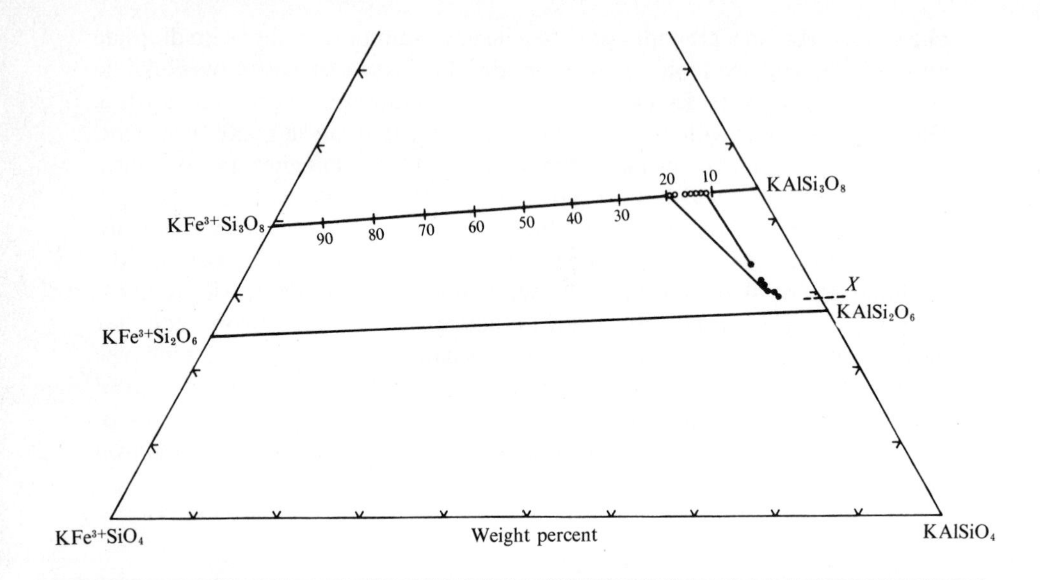


**Fig. 5-18** Composition of iron-rich sanidine (open circles) and leucite (solid dots) in orendite lavas with coexisting phases joined. Note the large amount of "excess silica" in the leucites compared to that at *X* found experimentally at 1000 bars water-vapor pressure. (After Carmichael, 1967*c*.)

The compositions of both leucite and sanidine are unusual, since the former is nonstoichiometric with a substantial amount of "excess silica." The experimental limit of excess silica (or solid solubility of $KAlSi_3O_8$) in $KAlSi_2O_6$ at 1000 bars (Fudali, 1963) is shown in Fig. 5-18 and is very much less than that found in these natural groundmass leucities. Both leucite and sanidine contain large amounts of ferric iron, and sanidine with up to 20 percent $KFe^{3+}Si_3O_8$ is common (Fig. 5-18). The amount of $Na_2O$ substituting in both minerals is minimal, since the feldspars contain less than 2 percent $NaAlSi_3O_8$.

If the relevant normative components[1] of the Leucite Hills lavas are plotted in the system $KAlSiO_4$–$Mg_2SiO_4$–$SiO_2$ (Fig. 5-19), the wyomingites and orendites parallel the phlogopite-olivine field boundary. This is in broad agreement with the occurrence of sporadic olivine phenocrysts enclosed by reaction rims of phlogopite. On ascending from the depths at which the phenocryst assemblages formed, the concentration of ($H_2O$ + F) in the liquid would decrease, the field of phlogopite would contract, and the field of leucite would expand, so that on the earth's surface these lavas would be in the leucite field. The resorbed margins of the phenocrysts of phlogopite, and its absence as a

[1] Normative *ac* is taken as a salic component in these rocks, since its presence is a direct result of modal iron sanidine and iron leucite.

groundmass constituent, show that the concentration of ($H_2O$ + F) on the surface was not large enough to stabilize this mica. This is not always the case in lavas, as the next examples show.

## TORO-ANKOLE REGION, UGANDA

This part of the Western Rift Valley of East Africa lies within the potassic volcanic province made classic by Holmes (1942, 1952; Holmes and Harwood, 1937; Combe and Holmes, 1945). The lavas are silica-poor mafic rocks and have recently been investigated by F. H. Brown (1971), from whose account most of the data on which the following discussion is based have been drawn. The lavas

**Fig. 5-19** Relevant normative constituents of the madupite, wyomingite, and orendite lavas of Leucite Hills, Wyoming, plotted in the system $Mg_2SiO_4$–$KAlSiO_4$–$SiO_2$ (cf. Fig. 5-17). (After Carmichael, 1967*c*.)

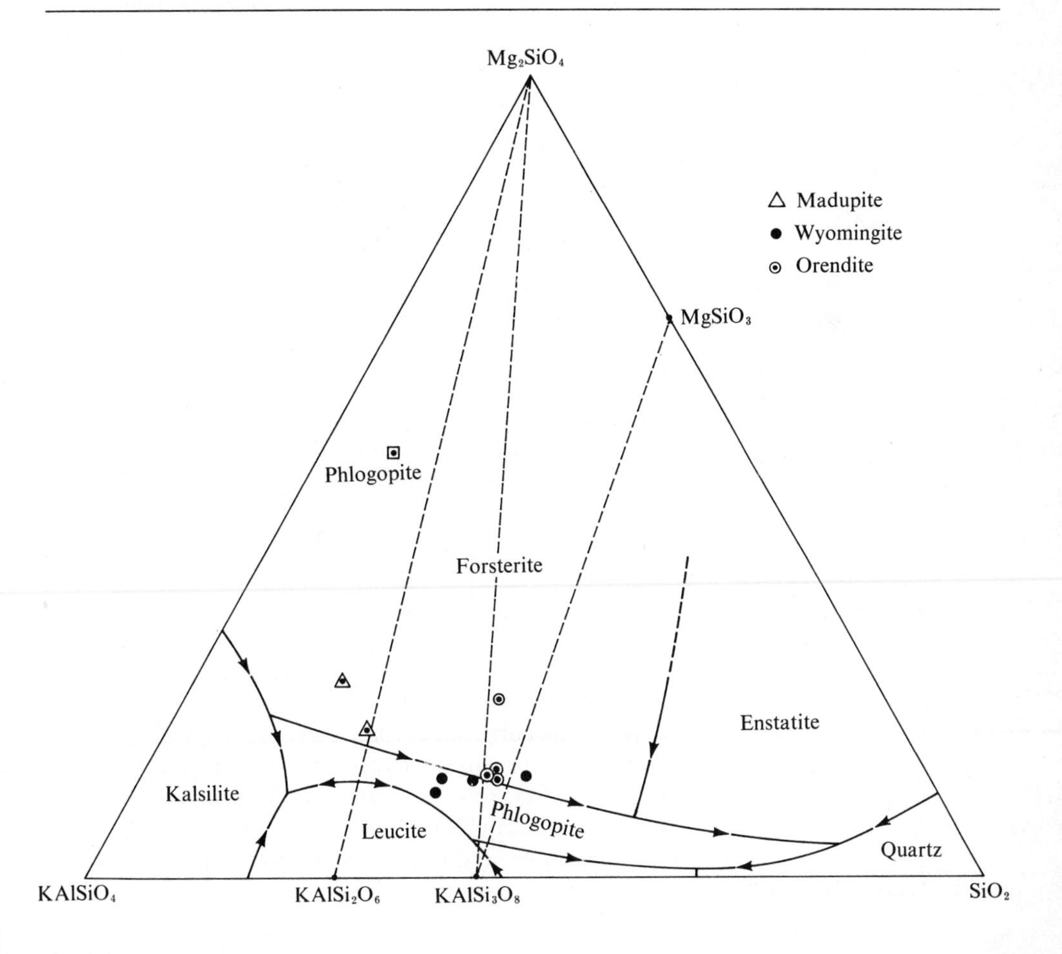

**Table 5-3** Representative analyses and norms of mafic lavas from the Toro-Ankole region containing kalsilite, leucite, and nepheline in the groundmass*

| | 1 | 2 | 3 | | 1 | 2 | 3 |
|---|---|---|---|---|---|---|---|
| $SiO_2$ | 40.52 | 36.71 | 39.18 | | | | |
| $TiO_2$ | 5.28 | 5.54 | 6.29 | *lc* | 9.91 | 6.64 | 16.19 |
| $Al_2O_3$ | 8.17 | 9.30 | 9.86 | *ne* | 11.46 | 7.67 | 12.39 |
| $Fe_2O_3$ | 6.33 | 9.44 | 8.03 | *kp* | 5.41 | 15.50 | 5.06 |
| FeO | 6.29 | 4.24 | 5.13 | *ac* | 0.01 | 5.42 | 1.24 |
| MnO | 0.21 | 0.26 | 0.23 | *wo* | 25.17 | 22.52 | 24.32 |
| MgO | 11.57 | 6.34 | 6.63 | *en* | 21.76 | 15.83 | 16.51 |
| CaO | 12.55 | 14.08 | 13.31 | *fo* | 4.94 | — | — |
| SrO | 0.22 | 0.34 | 0.25 | *mt* | 5.66 | — | — |
| BaO | 0.19 | 0.25 | 0.23 | *il* | 10.03 | 9.51 | 11.33 |
| $Na_2O$ | 2.50 | 2.40 | 2.87 | *hm* | 2.43 | 7.57 | 7.60 |
| $K_2O$ | 3.75 | 6.05 | 5.00 | *pf* | — | 0.91 | 0.55 |
| $P_2O_5$ | 0.44 | 1.11 | 0.85 | *ap* | 1.04 | 2.63 | 2.01 |
| $H_2O^+$ | 1.31 | 1.82 | 1.08 | *cc* | — | 2.96 | 0.80 |
| $H_2O^-$ | 0.40 | 0.42 | 0.55 | | | | |
| $CO_2$ | — | 1.30 | 0.35 | | | | |
| Total | 99.73 | 99.60 | 99.84 | | | | |

* Data taken from F. H. Brown (1971).

*Explanation of column headings*
1–3 Potash ankaratrites

contain no feldspar, but they do include one or more of the feldspathoids nepheline, kalsilite, and leucite, in addition to olivine, diopside, perovskite, magnetite, and phlogopite. They are varieties of potassium-rich nephelinites, and the three examples whose analyses are given in Table 5-3 are potash ankaratrites in the nomenclature of Holmes. All three examples contain kalsilite, leucite, and nepheline in the groundmass, although only one contains groundmass phlogopite (Table 5-3, no. 1); this same sample has olivine phenocrysts rimmed by phlogopite; the other two are devoid of olivine and phlogopite.

In Fig. 5-20, the average compositions of the groundmass salic phases are plotted; kalsilite shows a range from $Kp_{98}$ to $Kp_{94}$, leucite is virtually pure $KAlSi_2O_6$, and nepheline shows some small variation. All three lavas have normative salic components which are contained by the triangle connecting the analyzed minerals. This suggests that a field boundary, where liquid coexists with nepheline and kalsilite, is present, and its projected location is represented by the dashed curve with arrows. This curve should not be confused with that found by Schairer (1957); it arises solely because lavas crystallize at temperatures below the crest of the $NaAlSiO_4$–$KAlSiO_4$ solvus ($\sim$1100°C) (Tuttle and Smith, 1958). On this boundary, liquids will be in equilibrium with pyroxene, perovskite, etc., in addition to kalsilite and nepheline, and it is the projection of this which is shown in Fig. 5-20.

By combining Luth's work on the system $KAlSiO_4$–$Mg_2SiO_4$–$SiO_2$–$H_2O$ with Carman's (1969) on the analogous system $NaAlSiO_4$–$Mg_2SiO_4$–$SiO_2$–$H_2O$ together with the relationships shown in Fig. 5-11, a hypothetical model of the phase volumes in the tetrahedron $NaAlSiO_4$–$KAlSiO_4$–$Mg_2SiO_4$–$SiO_2$ can be constructed. In this schematic representation, it has been assumed that sodium phlogopite and potassium phlogopite form a continuous solid-solution series at high temperature, analogous to the muscovite-paragonite series. It is also assumed that the concentration of ($H_2O$ + F) in the liquid is sufficient to stabilize phlogopite but not sufficient for the phlogopite volume to extend over the top of the leucite volume. The volume of glaucophane has been ignored because this phase is not found in volcanic rocks.

Numbered curves and points *ABC* in Fig. 5-21 represent liquids in equilibrium with the following crystalline phases:

Curve 1: leucite, forsterite, kalsilite
Curves 2, 8: nepheline, forsterite, and leucite or nepheline
Point *B*: leucite, nepheline, kalsilite, forsterite
Curve 7: leucite, nepheline, kalsilite
Curve 3: forsterite, phlogopite, leucite
Curve 4: phlogopite, leucite, sanidine
Curve 6: forsterite, phlogopite, nepheline
Point *A*: nepheline, leucite, forsterite, phlogopite
Curve 5: nepheline, phlogopite, leucite (which becomes resorbed toward *C*)
Point *C*: nepheline, phlogopite, leucite, sanidine

**Fig. 5-20** Average compositions (solid circles) of leucite, nepheline, and kalsilite coexisting in groundmass of Uganda potash ankaratrites. Normative salic components of the three lavas are shown at points 1, 2, 3 (Table 5-3). Dashed line with arrows is projected position of the boundary where kalsilite and nepheline coprecipitate in lavas.

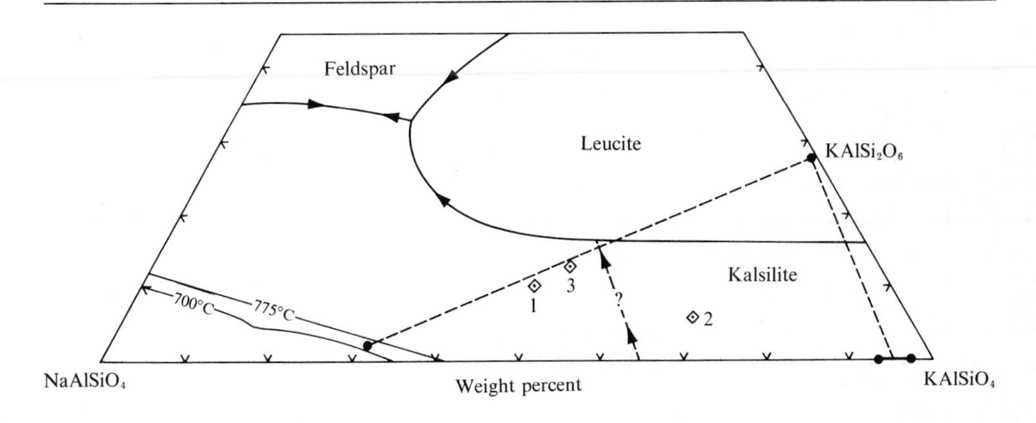

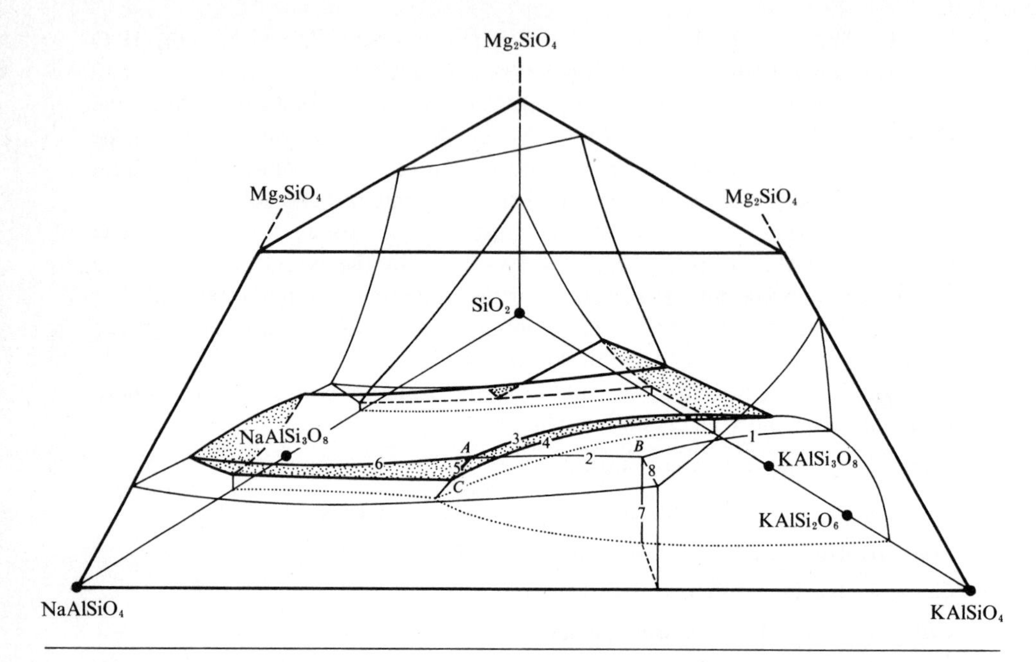


**Fig. 5-21** Hypothetical phase relations in the system $NaAlSiO_4$–$KAlSiO_4$–$Mg_2SiO_4$–$SiO_2$–$H_2O$ assuming complete solid solution between sodium and potassium phlogopite. Shaded surfaces are nepheline-phlogopite, leucite-phlogopite, enstatite-phlogopite, and the phlogopite fields in the ternary systems $KAlSiO_4$–$MgSiO_4$–$SiO_2$ and $NaAlSiO_4$–$Mg_2SiO_4$–$SiO_2$. (After F. H. Brown and Carmichael, 1969.)

A liquid cannot leave *C* to move toward the base of the tetrahedron until phlogopite has been eliminated; nor can it move away from *C*, saturated with phlogopite, while leucite still remains.

Among the whole range of the Toro-Ankole lavas, F. H. Brown (1971) reports the following phlogopite-free groundmass assemblage:

Olivine-kalsilite-leucite (cf. curve 1 in Fig. 5-21)

Olivine-kalsilite-nepheline (cf. curve 8 in Fig. 5-21)

Kalsilite-nepheline-leucite (cf. curve 7 in Fig. 5-21)

Relationships appropriate to higher concentrations of ($H_2O$ + F), exemplified by lavas of the same kind but with phlogopite in the groundmass, are shown in Fig. 5-22. Numbered curves and lettered points show liquids in equilibrium with the following crystalline phases:

Curve 9: kalsilite, forsterite, phlogopite

Point *C*: nepheline, kalsilite, forsterite, phlogopite

Curves 11, 12: leucite, phlogopite, and nepheline or kalsilite

Point *B*: nepheline, kalsilite, leucite, phlogopite

Within a certain range ($H_2O$ + F), curve 10 could coincide with 11, and curve 9 with 12, in which case *B* would also coincide with *C*.

The following groundmass assemblages have been recognized in the Toro-Ankole lavas:

Olivine-phlogopite-kalsilite (cf. curve 9 in Fig. 5-22)

Olivine-phlogopite-kalsilite-leucite (cf. point *C* coinciding with point *B*)

Phlogopite-kalsilite-nepheline-leucite (cf. point *B*)

Olivine-phlogopite-leucite-nepheline (cf. curve 10 coinciding with curve 11)

If the presence of diopsidic augite and the various other ferromagnesian minerals is neglected, it can be shown that the diverse salic assemblages found with phlogopite and olivine in the groundmass of these Uganda lavas can be simply represented in terms of an expanding phlogopite volume in Fig. 5-21. This expansion is presumably caused in nature by fluctuating concentrations of ($H_2O$ + F) coupled with consumption of the salic components as phlogopite precipitates; the occurrence of occasional olivine and phlogopite phenocrysts in some variants is also easily represented.

**Fig. 5-22** Hypothetical diagram similar to Fig. 5-21 except that the concentration of $H_2O$ + F is sufficiently great for the phlogopite volume to cover the leucite volume.

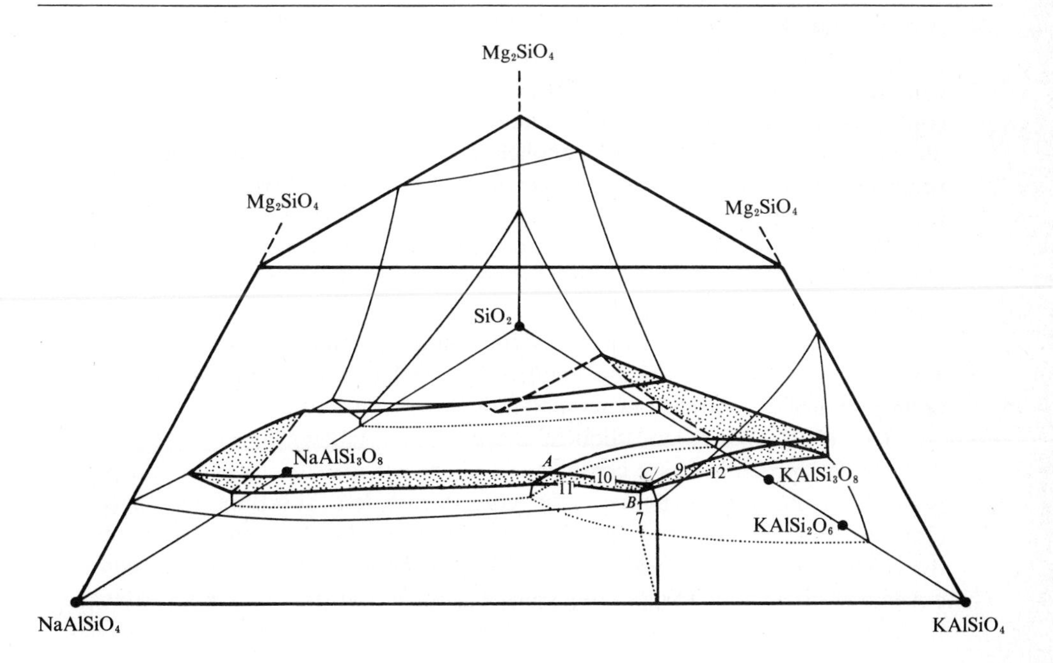

**Table 5-4** Analyses and CIPW norms of two basanites from the Korath Range, Ethiopia, with olivine, phlogopite, leucite, nepheline, and plagioclase in the groundmass*

| | K8 | K10 | | K8 | K10 |
|---|---|---|---|---|---|
| $SiO_2$ | 43.52 | 44.54 | *or* | 8.45 | 10.75 |
| $TiO_2$ | 2.45 | 2.16 | *ab* | 4.48 | 8.40 |
| $Al_2O_3$ | 15.76 | 15.50 | *an* | 25.22 | 18.60 |
| $Fe_2O_3$ | 2.82 | 4.63 | *ne* | 11.42 | 14.15 |
| FeO | 7.14 | 4.98 | *wo* | 13.55 | 13.54 |
| MnO | 0.16 | 0.19 | *en* | 9.56 | 10.93 |
| MgO | 9.57 | 9.12 | *fs* | 2.82 | 1.01 |
| CaO | 12.28 | 11.14 | *fo* | 10.00 | 8.26 |
| $Na_2O$ | 3.02 | 4.08 | *fa* | 3.25 | 0.84 |
| $K_2O$ | 1.43 | 1.82 | *mt* | 4.09 | 6.71 |
| $P_2O_5$ | 0.41 | 0.61 | *il* | 4.65 | 4.10 |
| $H_2O^+$ | 0.83 | 0.77 | *ap* | 0.97 | 1.44 |
| $H_2O^-$ | 0.24 | 0.15 | *cc* | 0.20 | 0.09 |
| $CO_2$ | 0.09 | 0.04 | Rest | 1.07 | 0.92 |
| Total | 99.72 | 99.73 | | | |

* Data taken from F. H. Brown and Carmichael (1969).

## KORATH RANGE, SOUTHERN ETHIOPIA

To show how the phlogopite "roof" in Fig. 5-22 can effectively resist penetration by residual liquids and thus prevent precipitation of alkali feldspar, we turn to two basanitic lavas from the Korath Range in southern Ethiopia. This small group of volcanoes is built up of basanite and tephrite lavas, the latter often containing large phenocrysts of Ti-poor kaersutite. The basanites typically contain phenocrysts of olivine and titanaugite set in a groundmass of plagioclase, olivine, phlogopite, leucite, and nepheline, and it is this assemblage which provides the basis for the ensuing discussion. The analyses of two representative basanites are shown in Table 5-4, together with their norms, in which, despite the presence of modal leucite, this component is absent (this is a common feature of basic rocks and need not concern us here). The analyses of the salic minerals of these two lavas are plotted in Fig. 5-23, in which it may be seen that alkali feldspar is absent, even as outer rims on the plagioclase crystals. Leucite apparently has an excess silica content, and nepheline has a composition which perhaps indicates a rather low temperature.

The initial salic composition of the two lavas (shown in Fig. 5-23) can perhaps be represented by a combination of the calculated normative feldspar (*an–ab–or*) and the normative salic constituents less anorthite (*ne–kp–Q*). However, because of the extensive solid solution of $Al_2O_3$ in the titanaugite, the calculated normative feldspar is unlikely to be representative of the components available to a precipitating feldspar; this is also true, to a lesser extent,

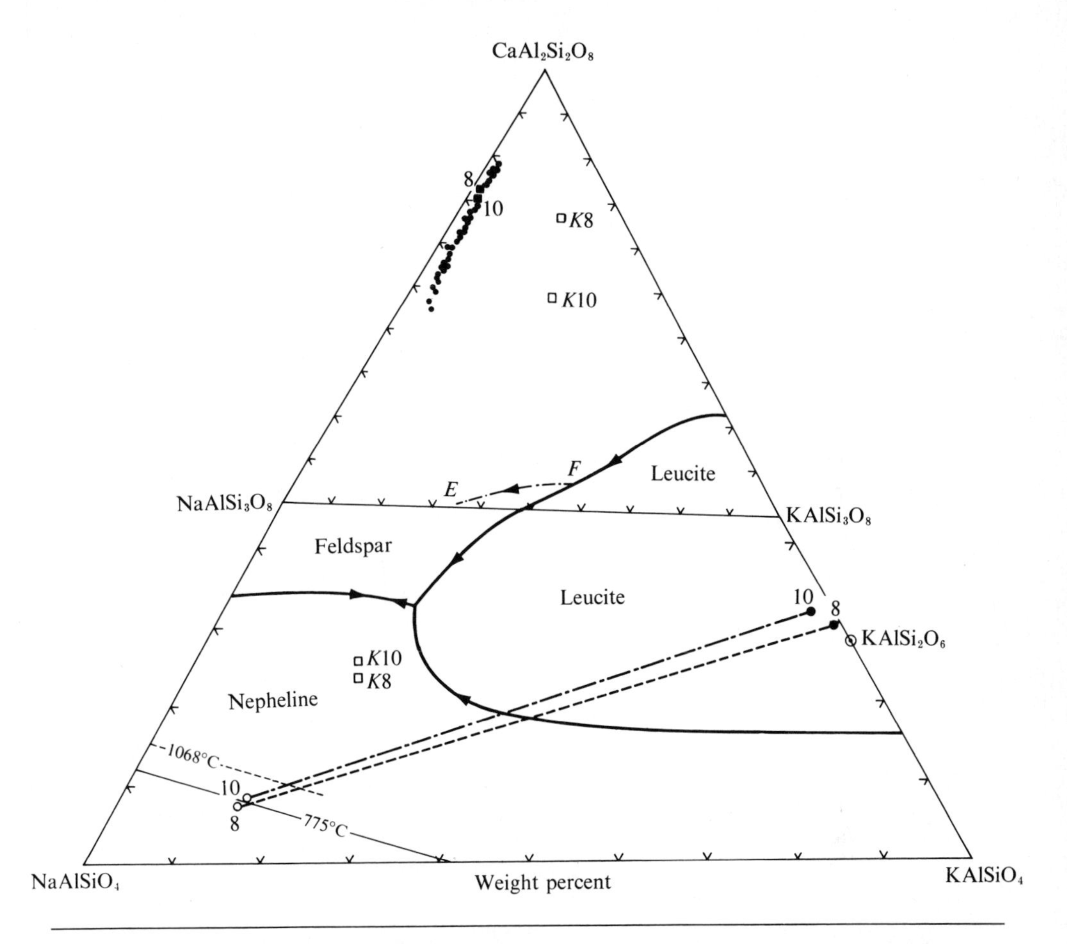


**Fig. 5-23** Two leucite basanites (Table 5-4) plotted on part of the phonolite pentahedron (with the ternary feldspar plane rotated as in Fig. 5-12). Composition of zoned groundmass plagioclase is shown as dots near the apex; compositions of leucite and nepheline are shown respectively as filled and open circles. Normative salic constituents of the two lavas are represented by open squares, average modal plagioclase compositions by solid squares. (After F. H. Brown and Carmichael, 1969.)

of the components *ne* and *kp* and *Q*, because of the presence of $Na_2O$ in the titanaugite. Viewed in the background of our earlier discussion of the phonolite pentahedron, it would be anticipated that alkali feldspar would precipitate at some stage in the crystallization paths of both basanites.

If attention is now focused on the $Mg_2SiO_4$ component, rather than on $CaAl_2Si_2O_8$, we find that the normative composition of these lavas appropriately falls in the olivine field of Fig. 5-24, and that their residual liquids, in terms of these three components, could be prevented from reaching the sanidine field if phlogopite was stable. That this is likely to persist into the sodic system is

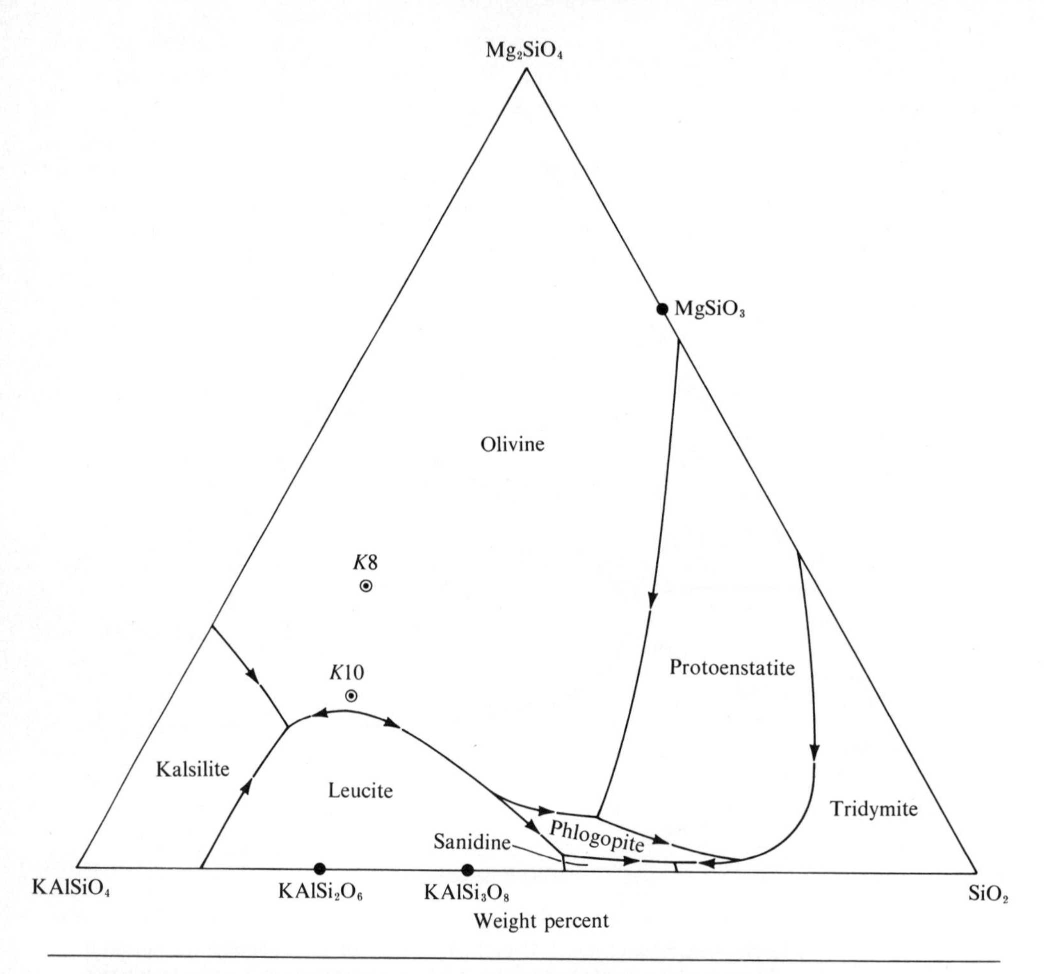


**Fig. 5-24** Normative components of two leucite basanites (Table 5-4) plotted in the system $KAlSiO_4$–$Mg_2SiO_4$–$SiO_2$–$H_2O$ (cf. Fig. 5-17).

substantiated by both basanites, in which the absence of alkali feldspar indicates that the residual liquids were in equilibrium with olivine, phlogopite, nepheline, and leucite. These liquids, now represented by the groundmass, are at nature's equivalent of point *A* in Fig. 5-21.

## Biotite Solid Solutions in Igneous Rocks

Up to this point we have considered only the reaction relationship of olivine to liquid with phlogopite as product. In some lavas this went to completion and the residual liquid successfully penetrated the phlogopite roof (or alternatively the roof contracted), while in others this proved to be impossible. In essence it would seem that success or failure for a lava to precipitate phlogopite is very much bound up with the concentration of F in the liquid. Because both olivine and phlogopite are solid solutions, then for that stage of the crystallization interval in which both coexist in equilibrium, both could change toward more iron-rich compositions. Many zoned phlogopite phenocrysts which show an increase in iron toward the crystal margins also become enriched in $TiO_2$ and BaO while becoming depleted in $Al_2O_3$, $Cr_2O_3$, MgO, and NiO.

In the Shonkin Sag laccolith of Montana, the original magma crystallized to an assemblage of olivine, biotite, sanidine, leucite (now pseudoleucite), titanomagnetite, and calcium-rich pyroxene zoned to acmite. The chilled margin grades upward into a coarser-grained facies which changes in composition to become syenitic, and except in the last stages of crystallization, olivine and biotite coexisted throughout. Nash and Wilkinson (1970) have shown that both olivine and biotite become more iron-rich toward the crystal margins. The maximum compositional range for olivine is small, $Fa_{22}$ to $Fa_{40}$; but the associated biotite, which often mantles the olivine grains, shows a more extreme range, and in one rock a biotite with a core of $annite_{24}$ ranges to $annite_{96}$ at the margins. In the late-stage syenite, in which olivine is absent, the biotite is pure annite but for substantial amounts of manganese.

Nash and Wilkinson chose to make the assumption that in each rock the cores of the olivine crystals were in equilibrium with the cores of biotite, and that the compositions of the crystal margins were also determined by equilibrium. On this basis they calculated two temperatures for each rock and were therefore able to determine the crystallization limits of the intrusion.

Although they formulated their approach in a rather different way, in substance they used the known standard free-energy change for each of the following reactions, substituting in the logarithmic terms the analyzed mineral compositions:

$$\underset{\text{fayalite}}{3Fe_2SiO_4} + \underset{\text{leucite}}{3KAlSi_2O_6} + O_2 = \underset{\text{magnetite}}{2Fe_3O_4} + \underset{\text{sanidine}}{3KAlSi_3O_8} \tag{5-1}$$

so that

$$\log f_{O_2} = \frac{\Delta G_r^\circ}{2.303RT} + \log \frac{(a_{Fe_3O_4}^{titanomag})^2 (a_{KAlSi_3O_8}^{san})^3}{(a_{Fe_2SiO_4}^{ol})^3 (a_{KAlSi_2O_6}^{leuc})^3}$$

and[1]

$$\underset{\text{annite}}{2KFe_3AlSi_3O_{10}(OH)_2} + O_2 = \underset{\text{sanidine}}{2KAlSi_3O_8} + \underset{\text{magnetite}}{2Fe_3O_4} + \underset{\text{steam}}{2H_2O} \tag{5-2}$$

[1] The standard free-energy change for Eq. (5-2) was taken from Wones and Eugster (1965) as modified by Wones (1972). These two writers have formulated the dependence of the composition of biotite on oxygen fugacity, which is of course implicit in Eq. (5-2).

so that

$$\log f_{O_2} = \frac{\Delta G_r^\circ}{2.303RT} + \log \frac{(a_{KAlSi_3O_8}^{san})^2 (a_{Fe_3O_4}^{titanomag})^2 (f_{H_2O})^2}{(a_{KFe_3AlSi_3O_{10}(OH)_2}^{biot})^2}$$

By relating $p_{H_2O}$ to the lithostatic load of the overlying sedimentary column, they were able to plot $\log f_{O_2}$ against $T$ for both reactions. At equilibrium, the right-hand side of Eq. (5-1) must equal that of Eq. (5-2), so that a graphic solution for each olivine-biotite-sanidine-leucite-titanomagnetite assemblage was obtained.

## Muscovite in Granitic Rocks

According to Williams et al. (1955, p. 137), "most granites[1] are peraluminous—that is they contain alumina in molecular excess over the sum of potash, soda and lime." The most common mineralogical expression of this is muscovite, which in many granites is found as coarse interstitial grains, usually accounted to be of primary origin. Muscovite is absent in volcanic rocks, which suggests that the pressure-temperature-composition conditions necessary for its precipitation are outside the range of volcanic rocks.

Two breakdown reactions of muscovite are of significance to its occurrence in granitic rocks:

$$KAl_2AlSi_3O_{10}(OH)_2 = KAlSi_3O_8 + Al_2O_3 + H_2O$$

$$KAl_2AlSi_3O_{10}(OH)_2 + SiO_2 = KAlSi_3O_8 + Al_2SiO_5 + H_2O$$

in which we leave the standard states unlabeled so that the right-hand sides of both equations can be seen as components in a granitic liquid. Both dehydration reactions involving sanidine and corundum, or quartz, sanidine, and andalusite/sillimanite, have been evaluated by Turner (1968), and the *PT* curves shown in Fig. 5-25 are taken from Evans (1965); also shown in Fig. 5-25 is the solidus curve for granite.

The intersection of the muscovite breakdown curve with the granite solidus at about 2 kb $p_{H_2O}$ would suggest that muscovite would be stable in granitic liquid above this pressure. Note that if the granitic liquid contains quartz or has a high silica activity (as it must to be a granite), then the intersection is displaced to about 3.5 kb.

Confusion can arise from a diagram such as Fig. 5-25, because it contains nothing explicit about the role of composition in the precipitation of muscovite in granitic magma. The granite whose melting curve is shown in Fig. 5-25 does not have the appropriate *initial* composition since it is not peraluminous, and therefore it will never enter the muscovite field (Lambert et al., 1969, fig. 5).

[1] Granites in the strict sense, with alkali feldspar $>2/3$ total feldspar.

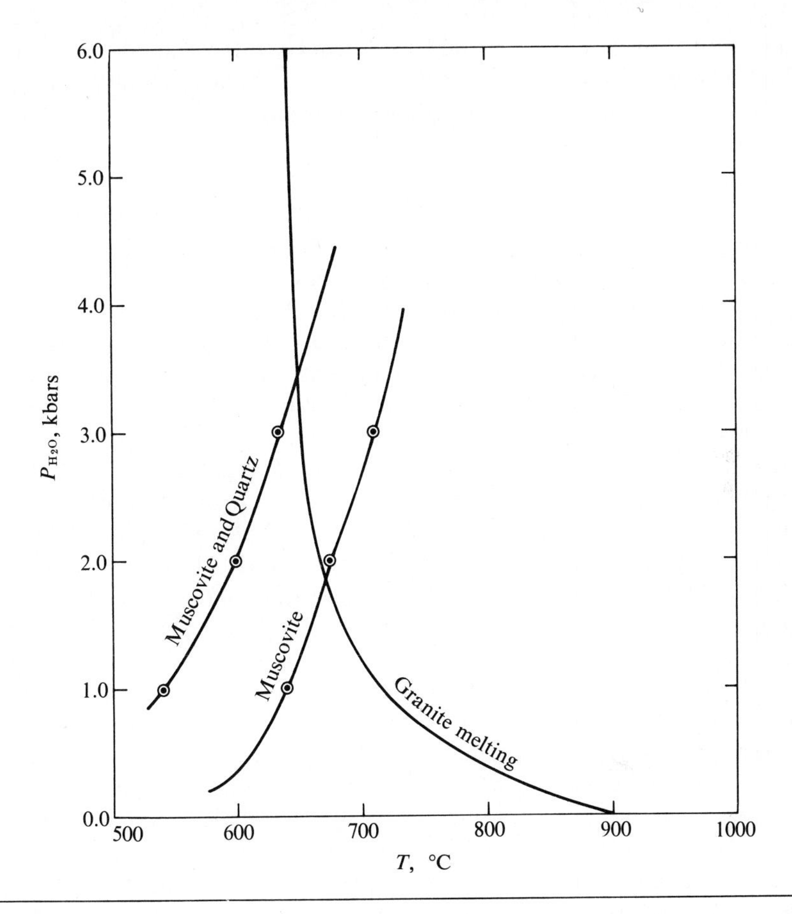


**Fig. 5-25** Stability of muscovite and of muscovite plus quartz as a function of water-vapor pressure. Melting curve of granite is also shown. (Data from Tuttle and Bowen, 1958; Evans, 1965.)

There are in general two ways that a metaluminous granite magma could become peraluminous: one is by fractionation of a phase containing alkalis in excess of $Al_2O_3$ (which, so far as the writers are aware, is unknown) and the other is by modification of the liquid's composition by an aqueous vapor phase. Burnham (1967) has shown that the solubility of the granitic components in the aqueous vapor phase increases with pressure and that the alkalis are preferentially withdrawn, thus tending to make the liquid peraluminous. Perhaps, by interaction of the vapor phase with a metaluminous granitic liquid, muscovite could become stable in granitic liquids at pressures near 4 kb (Fig. 5-25); but this liquid cannot have the same composition as the starting material.

Luth et al. (1964) consider that muscovite granites result from the purging effect of a high-pressure aqueous phase removing alkali from the granitic liquid; their contention is strongly supported by the rarity of andalusite rhyolites, which would seem to be the anhydrous equivalents of muscovite granites. Indeed if peraluminous acid magmas were commonplace, andalusite or garnet granites and rhyolites would not perhaps be so rare.[1]

## Pyroxenes

### THE SYSTEM $Mg_2SiO_4$–$CaMgSi_2O_6$–$SiO_2$

Most of the common rock-forming pyroxenes are solid solutions made up essentially of the five components $MgSiO_3$, $FeSiO_3$, $CaSiO_3$, $CaAl_2SiO_6$, and $NaFe^{3+}Si_2O_6$; small amounts of $TiO_2$, $Cr_2O_3$, MnO, and $Fe_2O_3$ (in low-$Na_2O$ pyroxenes) are usually present. However, the crystallization behavior of most igneous pyroxenes can be portrayed in terms of the three components $MgSiO_3$, $FeSiO_3$, and $CaSiO_3$ (cf. Fig. 5-27); some pyroxenes, especially those of silica-poor rocks, contain substantial amounts of $Al_2O_3$ and $TiO_2$. We are, as with the feldspars, concerned with pyroxene crystallization paths as they can be deduced both from natural assemblages and from experimental systems. A suitable starting point which highlights the essential features of two great groups of magma, the tholeiitic and the alkali olivine basalt, is the system $Mg_2SiO_4$–$CaMgSi_2O_6$–$SiO_2$ (Kushiro, 1972*b*). The general relationships are shown in Fig. 5-26, and two properties of this iron-free system have analogies to the crystallization of pyroxenes in magmas. Firstly, there is the incongruent melting behavior of $MgSiO_3$* or, alternatively, the reaction relationship of olivine and liquid to produce calcium-poor pyroxene. Thus along the boundary $MgSiO_3$–*B* (Fig. 5-26) the liquid precipitates calcium-poor pyroxene while at the same time resorbing olivine. Any liquid whose composition falls in the area $Mg_2SiO_4$–*A*–*B*–$MgSiO_3$ will display this reaction relationship and in a simple way may be taken to represent the tholeiitic magma type.

Secondly, for liquids whose compositions fall in the area *D*–$Mg_2SiO_4$–*A*, there is no reaction relationship of olivine to liquid, and when the liquid reaches the boundary curve *AD*, a diopside solid solution precipitates which becomes progressively richer in $CaMgSi_2O_6$ as the liquid moves toward *D*. To the crystallization of olivine and diopside solid solution along *AD*, *A* being a thermal maximum on the boundary curve, we may liken the alkali olivine basalts and their congeners where olivine and augite are present throughout most, if not all, the crystallization interval.

Because there is no continuous solid-solution series between $CaMgSi_2O_6$

[1] From time to time discussion has arisen on the significance of peraluminous rhyolites in the igneous record with respect to the more abundant peraluminous granites. But here any argument can become circuitous, because several writers (for example, Ewart, 1966; Ewart et al., 1971) have used the presence of normative corundum in rhyolites to postulate that they are the fusion products of metasediments containing muscovite.

* The incongruent melting of $MgSiO_3$ is eliminated at pressures greater than 7 kb but it persists in the presence of water-saturated liquids to over 20 kb $p_{H_2O}$.

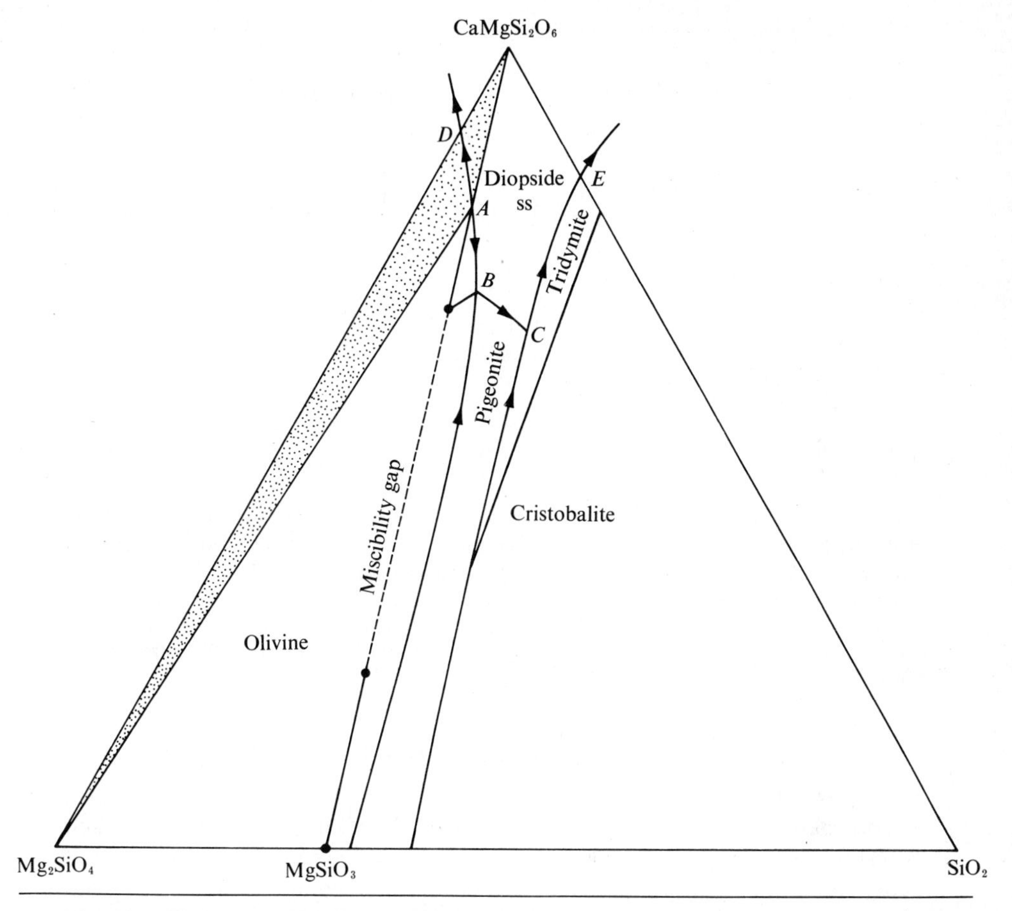


**Fig. 5-26** The system $Mg_2SiO_4$–$CaMgSi_2O_6$–$SiO_2$ at 1 bar. Liquids in the stippled area precipitate olivine and calcium-rich pyroxene along *AD*, the pyroxene becoming richer in the $CaMgSi_2O_6$ component. (After Kushiro, 1972.)

and $MgSiO_3$ at liquidus temperatures, or indeed at magmatic temperatures, there will be a range of liquid compositions in equilibrium with a diopside solid solution and a calcium-poor pyroxene. A liquid at *B*, a ternary invariant point, is in equilibrium with olivine, a calcium-rich pyroxene, and a calcium-poor pyroxene. Liquids along the boundary curve *BC* are in equilibrium with two pyroxenes, again a calcium-rich and a calcium-poor variety; at *C*, tridymite becomes stable on the liquidus.

Any *fractionating* liquid which precipitates olivine in the area $Mg_2SiO_4$–*A*–*B*–$MgSiO_3$ may reach *B* and then pass along the boundary curve to *C*. Thus a liquid at *B* can be seen as a simplified *Q*-normative olivine tholeiite, while liquids along *BC* can be taken to represent olivine-free tholeiites, basaltic andesites, and some andesites.

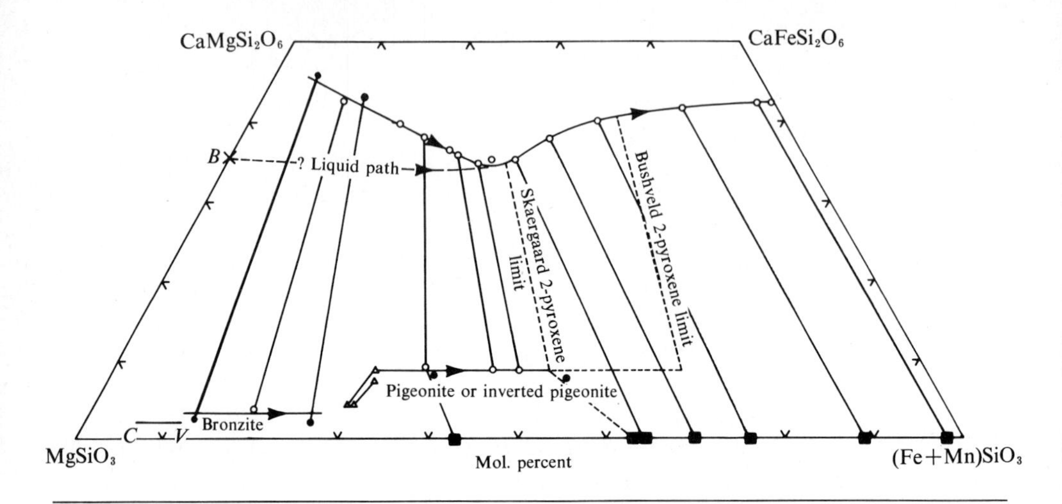


**Fig. 5-27** Composition of pyroxenes in rocks of the Skaergaard intrusion (open circles) and of the Bushveld intrusion (filled circles). Pyroxenes become more iron-rich as crystallization proceeds, as shown by arrows on pyroxene trends. Composition of coexisting iron-rich olivines is shown as solid squares on the base of the trapezium. Also shown are compositions of coexisting pigeonite and orthopyroxene in a tholeiitic andesite (open triangles) and an unusual clinoenstatite *CV* from Cape Vogel, Papua. (Data from G. M. Brown, 1957; Brown and Vincent, 1963; Atkins, 1969.)

## PYROXENES IN SLOWLY COOLED THOLEIITIC INTRUSIONS

The most thoroughly studied pyroxene assemblage of fractionated basaltic magma is that from the Skaergaard intrusion[1] (G. M. Brown, 1957; Brown and Vincent, 1963), which we may use as the measure of the equilibrium variation of pyroxenes in a basic liquid continuously changing in composition. Throughout the 2500 m of the layered series, the accumulating pyroxene crystals show a progressive change in composition which is displayed in part in Fig. 5-27. Using these data and those from the South African Bushveld intrusion (Atkins, 1969), which can be used as a substitute for the hidden lower part of the Skaergaard intrusion, the generalized crystallization paths of pyroxenes in slowly cooled tholeiitic magma can be portrayed (Fig. 5-27).

The calcium-rich pyroxene trend starts close to the $CaMgSi_2O_6$ corner (Fig. 5-27) and shows initial depletion in $CaSiO_3$ (and Cr) as the pyroxenes become more iron-rich, the overall trend being concave to the $CaMgSi_2O_6$–$CaFeSi_2O_6$ join. After reaching a low point of about 35 percent $CaSiO_3$, increasing iron enrichment is accompanied by an increase in $CaSiO_3$, and the trend flattens as it moves toward the iron-rich side of the trapezium. Near the top of the Skaergaard intrusion, the iron-rich hedenbergites show an unusual mosaic texture which Wager and Deer (1939) interpreted as representing the inversion product of an iron-wollastonite solid solution; synthetic hedenbergite

[1] Described in detail in Chapter 9.

($CaFeSi_2O_6$) inverts at 965°C to an iron wollastonite (Bowen et al., 1933). This is one of few known natural occurrences of an inverted iron wollastonite, and, in conjunction with tridymite, it has been used to obtain an estimate of total pressure in this intrusion.

The coexisting calcium-poor pyroxenes exhibit more diversity. The lower part of the Skaergaard intrusion, which although hidden is believed to be represented as a condensed sequence in the marginal rocks, contains orthopyroxene, as does the Bushveld intrusion. These pyroxenes commonly fall in the composition range $Fs_{13}$ to $Fs_{28}$ (Fig. 5-27) and characteristically contain between 3 and 4 percent $CaSiO_3$ (now partially exsolved as augite lamellae). In the upper, exposed, layered series, the orthopyroxene contains more numerous augite lamellae, and the composition of these pyroxenes now has about 9 percent $CaSiO_3$. Because homogeneous one-phase monoclinic pyroxenes identical in composition to these orthopyroxenes plus their lamellae are found in volcanic rocks (pigeonites), these Skaergaard pyroxenes are interpreted as inverted pigeonites. Their inversion involves a change to orthorhombic symmetry and exsolution of $CaSiO_3$ as augite lamellae; but relict monoclinic structures—{001} lamellae and {100} twins—still persist.

In a few volcanic rocks, the groundmass contains both orthopyroxene and pigeonite, and the analyses of Nakamura and Kushiro (1970) are plotted in Fig. 5-27 to illustrate a possible equilibrium assemblage intersecting a natural orthorhombic-monoclinic inversion curve. This volcanic pair illustrates the increase of $CaSiO_3$ in pyroxene as the magma crystallization curve crosses from the orthorhombic pyroxene stability field into the monoclinic.

Pigeonites do not show unrestricted iron enrichment, and pyroxenes more iron-rich than $Fs_{70}$ are not found in igneous rocks. Pigeonite may cease to crystallize long before this, and there is a marked difference between the composition of the last pigeonite to precipitate in the Skaergaard compared with that in the Bushveld intrusion (Fig. 5-27). The point at which pigeonite disappears is called the limit of the two-pyroxene field. At the other extreme, calcium-poor pyroxenes (orthopyroxenes) more magnesian than about $Fs_{13}$ are not known, although monoclinic enstatites (possibly an inversion product of protoenstatite) have been found in a rare terrestrial rock (Dallwitz et al., 1966); these contain even less $CaSiO_3$ than bronzites (Fig. 5-27).

The crystallization paths of the calcium-poor and calcium-rich pyroxenes seem best represented by the liquidus of the magma intersecting a pyroxene solvus dome, one of whose limbs is broadly parallel, but concave, to the $CaMgSi_2O_6$–$CaFeSi_2O_6$ join and the other to the $MgSiO_3$–$FeSiO_3$ join. This being the case, the augite trend in Fig. 5-27 corresponds to the solidus-solvus intersection, as does the calcium-poor pyroxene trend, except that this will cross the orthorhombic-monoclinic inversion curve near $En_{65}$.

The relationships at the limit of the two-pyroxene field are not known, one suggestion for the cessation of pigeonite precipitation being a reaction relationship of liquid and pigeonite (or orthopyroxene) to produce iron-rich olivine. Pyroxenes close to $FeSiO_3$ have no stability field at low pressures (J. Lindsley

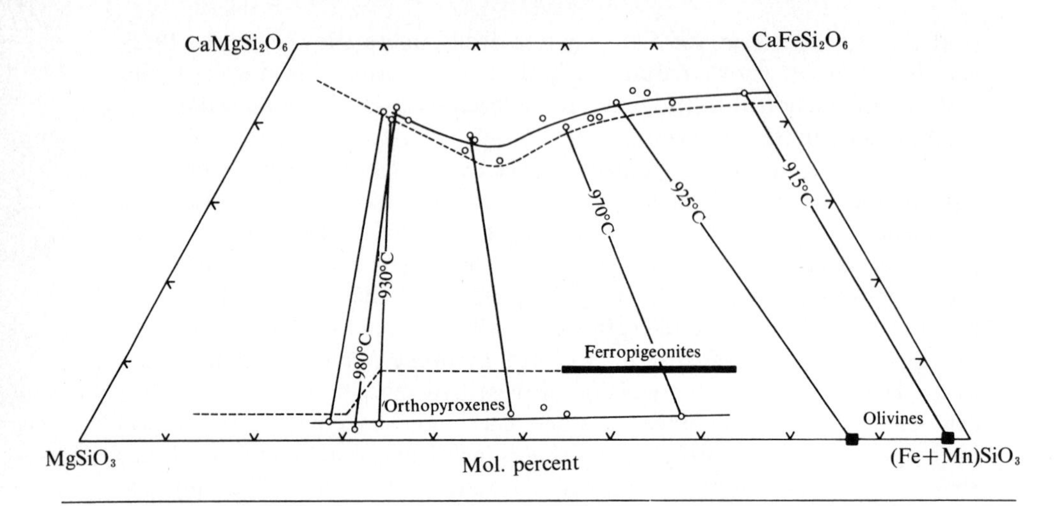


**Fig. 5-28** Pyroxenes and olivines found as phenocrysts in acid lavas; temperatures (°C) of equilibration of coexisting iron titanium oxides are shown on the tie lines (after Carmichael, 1967*b*). The Skaergaard pyroxene trends are dashed (cf. Fig. 5-27). The compositional range of ferropigeonites in certain Hebridean acid rocks is also shown as the solid horizontal bar (Emeleus et al., 1971).

and Munoz, 1969). The evidence of the layered intrusions, however, shows that there is no correspondence between the disappearance of pigeonite as a crystallizing phase and the precipitation of olivine, since the two overlap. Accordingly Muir (1954) has suggested that the two-pyroxene limit may be due to the migration of the liquidus to the calcium-rich limb of the solvus, eventually decoupling itself to become a liquidus minimum. If this is so, there is no satisfactory explanation of why this decoupling should occur at such different compositions in the Skaergaard as compared with the Bushveld intrusion (Fig. 5-27). On the assumption that Muir's hypothesis is correct, a possible liquid path (in projection) is shown in Fig. 5-27; the point *B* corresponds to *B* in Fig. 5-26. It is at the limit of the two pyroxene field that the liquid will lie close to the augite solidus-solvus intersection.

## PYROXENE PHENOCRYSTS IN RHYOLITIC ROCKS

Pyroxene phenocrysts of acid lavas cover almost the whole composition range of pyroxenes from slowly cooled tholeiitic intrusions (Fig. 5-28). However, the augite phenocrysts lie on a trend slightly displaced toward the $CaMgSi_2O_6$–$CaFeSi_2O_6$ join, and the coexisting calcium-poor pyroxenes are orthopyroxene rather than pigeonite. These orthopyroxenes contain a little less $CaSiO_3$ than those of basic intrusions, and all magnesian pyroxene phenocrysts of acid lavas may be distinguished from their counterparts of tholeiitic intrusions by significantly higher MnO.

Since the pyroxene phenocrysts in acid lavas coexist with microphenocrysts of iron titanium oxides, we may use the temperatures of their equilibration as an indication of the temperature at which the pyroxene phenocrysts precipitated. These temperatures, shown in Fig. 5-28, indicate that for the composition range represented by the augite-orthopyroxene pairs, the orthorhombic-monoclinic inversion curve of the calcium-poor pyroxenes must be at temperatures greater than 980°C. Not all acid rocks contain orthopyroxene, however, and in some, though at an unknown temperature, there is ferropigeonite (Emeleus et al., 1971).

## PYROXENE ASSEMBLAGES IN THOLEIITIC BASALTS

If volcanic rocks crystallize at higher temperatures than plutonic rocks, then we should expect a rather narrower miscibility gap between volcanic augites and calcium-poor pyroxenes. Sometimes this is found to be the case, as in a tholeiitic suite of lavas from Iceland (Fig. 5-29). However, metastable pyroxene phases are frequently found in volcanic rocks, particularly the more magnesian varieties, and these can range across the miscibility gap (Fig. 5-30). Metastable, or quench, pyroxenes (subcalcic augites) have been described by Muir and Tilley (1964*a*) and by D. Smith and Lindsley (1971), who show that they are confined to the chilled flow margins. Evans and Moore (1968) record that, in a prehistoric tholeiitic lava lake of Hawaii, the slower the cooling, the closer the pyroxene trend approached that of the tholeiitic intrusions.

By and large, the calcium-rich pyroxenes of tholeiitic rocks, from plutonic

**Fig. 5-29** Composition of groundmass pyroxenes in a suite of olivine tholeiites, tholeiites, and icelandites from Iceland. Skaergaard pyroxene trends shown as dashed lines. (Data from Carmichael, 1967*a*.)

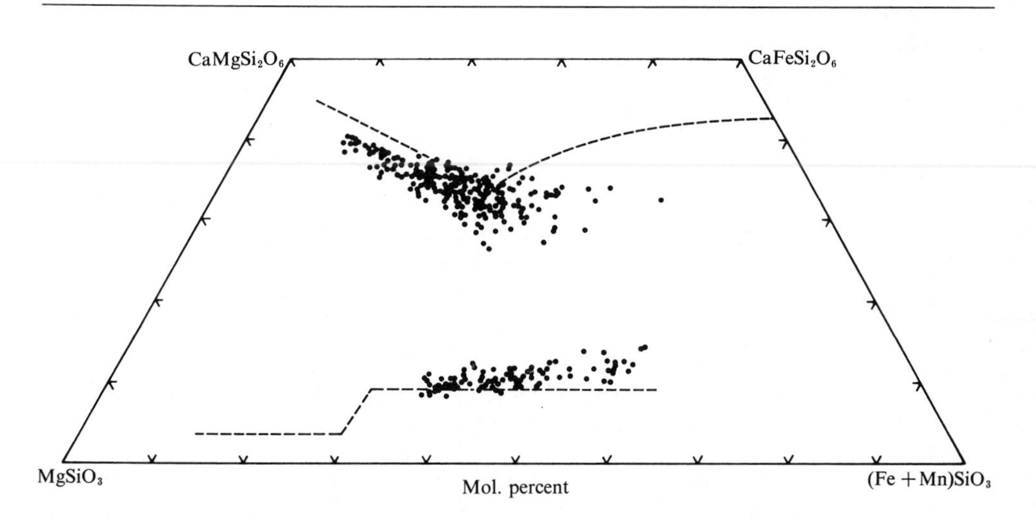

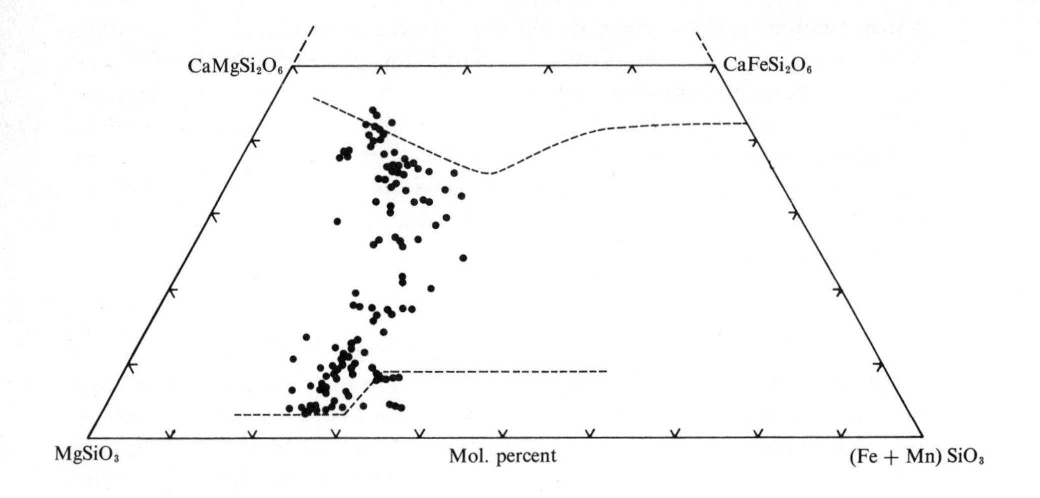


**Fig. 5-30** Compositions extending across the pyroxene miscibility gap, found in the groundmass pyroxenes of basic lavas from New Britain. Skaergaard pyroxene trends are shown dashed. (Data from Lowder, 1970.)

and volcanic environments alike, contain less than 2.5 percent $Al_2O_3$, which decreases with increasing iron; $TiO_2$ is commonly in the range 0.9 to 1.1 percent, and $Na_2O$ is usually less than 0.4 percent. In combination these compositional ranges set the tholeiitic augites apart from those found in the alkali olivine basalts and their congeners.

## PYROXENES IN ALKALI-BASALT MAGMA

The crystallization path of pyroxenes in alkali-basalt magma, which are usually accompanied by olivine, may be compared with the system $Mg_2SiO_4$–$CaMgSi_2O_6$–$SiO_2$ (Fig. 5-26), where the pyroxene becomes richer in $CaMgSi_2O_6$ as the liquid changes in composition along *AD*, olivine also being present. Slowly cooled intrusions of alkali-basalt magma do not display the complex pyroxene relationships that are commonplace in the tholeiitic intrusions. One layered pluton, the Kap Edvard Holm intrusion of eastern Greenland, has been studied in some detail, particularly with respect to the pyroxene path of crystallization (Deer and Abbott, 1965; Elsdon, 1971), which is shown in Fig. 5-31.

In lavas of more diverse composition, covering a considerable range of silica activity, the calcium-rich pyroxenes match their plutonic equivalents or show the same general feature of limited iron enrichment, and they contain substantial amounts of $Al_2O_3$, usually taken to be a solid solution of $CaAl_2SiO_6$ (Ca-Tschermak's molecule). Some representative pyroxene compositions from basanites, etindites, and leucite phonolites are plotted in Fig. 5-31. These analyses show that, in terms of the three components $CaSiO_3$, $MgSiO_3$, and $FeSiO_3$, they straddle the $CaMgSi_2O_6$–$CaFeSi_2O_6$ join. The crystallization paths, represented by arrows, are broadly in accord with our analogy to the system $Mg_2SiO_4$–$CaMgSi_2O_6$–$SiO_2$, except that the natural pyroxenes contain substantial amounts of $Al_2O_3$ and $TiO_2$, the significance of which is taken up below.

## THE ROLE OF $Al_2O_3$ AND $TiO_2$ IN CALCIUM-RICH PYROXENES

Kushiro (1960) and Le Bas (1962) have shown that aluminum increases systematically in calcium-rich pyroxenes in relation to decreasing silica concentration of the host magma. This correlation was examined by Verhoogen (1962), who concluded that silica activity would directly affect the aluminum, and hence titanium, content of a pyroxene. Furthermore, he showed, despite common practice in the calculation of pyroxene formulas, that Ti is unlikely to be found in *Z* sites (Si), Mg being much more favored, and also that the incorporation of Ti into the pyroxene structure would be accompanied by Al in *Z* sites. Yagi and Onuma (1967) postulated a hypothetical pyroxene component, $CaTiAl_2O_6$, essentially based on this reasoning, and they determined its solubility in $CaMgSi_2O_6$; they found the equivalent of 4 percent $TiO_2$ at 1 bar.

**Fig. 5-31** Composition of calcium-rich groundmass pyroxenes found in various silica-poor lava types; arrows show direction of zoning from core to margin. Augites from part of the slowly cooled Kap Edvard Holm intrusion are also shown. Skaergaard calcium-rich pyroxene trend is dashed.

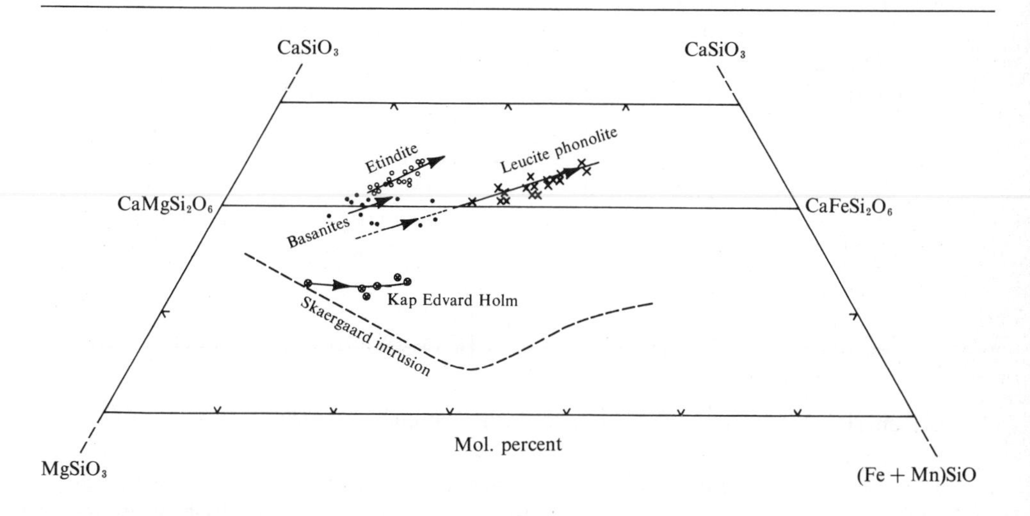

**Table 5-5** Representative compositions of calcium-rich pyroxenes (groundmass) in lavas of varying silica activity, decreasing from left to right

| | | OLIVINE THOLEIITE | LEUCITE BASANITE | LEUCITE PHONOLITE | ETINDITE | MELILITE BASALT |
|---|---|---|---|---|---|---|
| $SiO_2$ | | 52.4 | 46.6 | 40.3 | 43.8 | 53.3 |
| $TiO_2$ | | 0.97 | 1.60 | 2.43 | 4.39 | 0.82 |
| $Al_2O_3$ | | 1.77 | 8.6 | 11.3 | 8.2 | 0.58 |
| FeO | | 9.8 | 7.9 | 16.6 | 9.5 | 5.8 |
| MnO | | 0.19 | 0.22 | 0.56 | 0.24 | 0.28 |
| MgO | | 16.2 | 12.0 | 5.3 | 10.4 | 14.7 |
| CaO | | 18.4 | 22.8 | 23.0 | 22.7 | 24.2 |
| $Na_2O$ | | 0.32 | 0.29 | 0.54 | 0.77 | 0.36 |
| $Z$ | Si | 1.942 | 1.746 | 1.594 | 1.670 | 1.973 |
| | $Al^{iv}$ | 0.058 | 0.254 | 0.406 | 0.330 | 0.025 |
| $XY$ | $Al^{vi}$ | 0.019 | 0.126 | 0.121 | 0.038 | — |
| | Ti | 0.027 | 0.045 | 0.072 | 0.126 | 0.023 |
| | Fe | 0.304 | 0.247 | 0.549 | 0.303 | 0.179 |
| | Mn | 0.006 | 0.007 | 0.018 | 0.008 | 0.009 |
| | Mg | 0.895 | 0.667 | 0.312 | 0.591 | 0.811 |
| | Ca | 0.731 | 0.915 | 0.975 | 0.927 | 0.960 |
| | Na | 0.023 | 0.021 | 0.041 | 0.057 | 0.026 |
| %Al in $Z$ | | 2.9 | 12.7 | 20.3 | 15.5 | 1.2 |

The following reaction illustrates the role of $Al_2O_3$ in pyroxene:

$$\underset{\text{pyroxene component}}{CaAl_2SiO_6} + \underset{\text{glass}}{SiO_2} = \underset{\text{anorthite}}{CaAl_2Si_2O_8} \quad (5\text{-}3)$$

This reaction demonstrates that low silica activity will favor the incorporation of Al in $Z$ sites (contrast jadeite, $NaAlSi_2O_6$). Accordingly in Table 5-5 we show five representative analyses of augites from lavas of widely different silica activities, all of which crystallized at or near 1 bar. In augite from an olivine tholeiite, only 2 percent Al is found in the $Z$ group, whereas up to 20 percent can exist in augite from lavas of lower silica activity.

In lavas of low silica activity where feldspar is unstable, melilite is often found, but in this event the coexisting augite may contain only small amounts of $Al_2O_3$ (Table 5-5, last column). This may be represented by the following reaction:

$$\underset{\text{gehlenite}}{\tfrac{2}{3}CaAl_2SiO_7} + \underset{\text{forsterite}}{\tfrac{1}{3}Mg_2SiO_4} + \underset{\text{glass}}{SiO_2} = \underbrace{\tfrac{2}{3}CaMgSi_2O_6 + \tfrac{2}{3}CaAl_2SiO_6}_{\text{diopside solid solution}} \quad (5\text{-}4)$$

which shows that low silica activity will diminish the $Al_2O_3$ content of the pyroxene.

One frustrating characteristic of many alkali basalts and their relatives is the absence of ilmenite, since the presence of titanomagnetite alone prohibits the use of the iron-titanium-oxide geothermometer. We may represent this by the reaction

$$\underset{\text{ilmenite}}{2FeTiO_3} + \underset{\text{anorthite}}{2CaAl_2Si_2O_8} = \underset{\text{titanpyroxene}}{2CaTiAl_2O_6} + \underset{\text{fayalite}}{Fe_2SiO_4} + \underset{\text{glass}}{3SiO_2} \quad (5\text{-}5)$$

in which low silica activity will drive the reaction to the right; with elimination of ilmenite, an assemblage of plagioclase, titanaugite, olivine, and a silicate liquid would result.

## SODIC PYROXENES

Acmite ($NaFe^{3+}Si_2O_6$) is stable over a range of oxygen fugacities (Bailey, 1969), and it can exist where magnetite and quartz would be reduced to fayalite. However, highly sodic pyroxenes are rarely, if ever, found with olivine in igneous assemblages. There is now substantial evidence to show that there is a continuous solid solution between pyroxenes along the $CaMgSi_2O_6$–$CaFeSi_2O_6$ join and acmite (Nash and Wilkinson, 1970; Sutherland, 1967; Yagi, 1953, 1966). Unlike the crystallization paths of the calcium-rich and calcium-poor pyroxenes, which tend to be comparable in all rocks in which they occur, crystallization paths of calcium-rich pyroxenes becoming enriched in acmite are varied. One such example is the Shonkin Sag laccolith (Nash and Wilkinson, 1970), in which the pyroxene in almost every rock defines a different path toward $NaFeSi_2O_6$ (Fig. 5-32). Yagi (1966) has suggested that this variation is due to

**Fig. 5-32** Crystallization trends of sodium enrichment in calcium-rich pyroxenes from different parts of the Shonkin Sag laccolith. (Data from Nash and Wilkinson, 1970.)

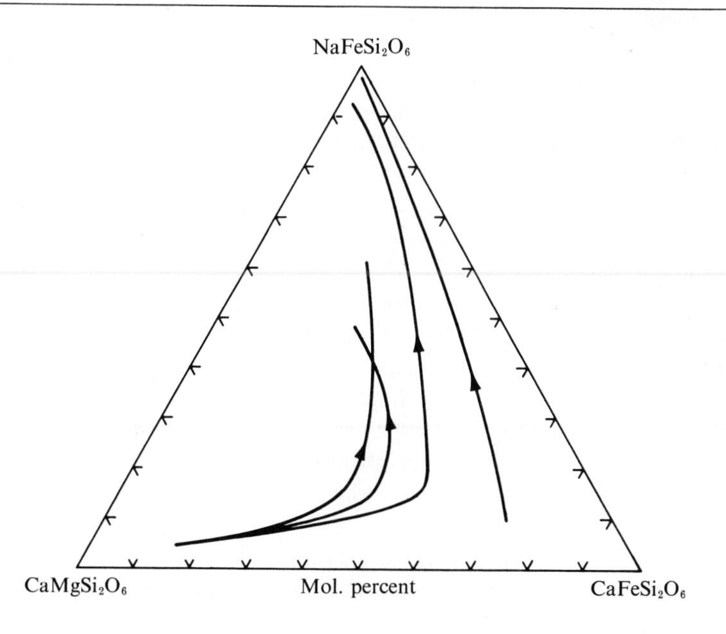

differences in oxygen fugacity, but the occurrence of all but the most sodic pyroxenes with olivine and titanomagnetite in the Shonkin Sag intrusion suggests that oxygen fugacity (if equilibrium was maintained) decreased progressively and systematically. Nevertheless, because acmite (or pyroxenes close in composition to it) is not found with olivine and magnetite, it would seem that its precipitation does not require an increase in oxygen fugacity, but rather that it would fall less rapidly with temperature than would be the case if olivine and titanomagnetite were present.

## Amphiboles

General statements about the precipitation and composition of amphibole in cooling magma are virtually impossible to make. Not only do the igneous amphiboles show a wide range of composition throughout at least five solid-solution series, but even single crystals (phenocrysts) may show extreme compositional variation within a few microns. The more commonly found varieties are:

1. Pargasite-hastingsite (hornblende): $NaCa_2(Mg,Fe)_4(Fe^{3+},Al)Al_2Si_6(OH)_2$ often with substantial amounts of $TiO_2$ (then called kaersutite)
2. Riebeckite: $Na_2(Fe^{++}, Mg)_3Fe_2^{3+}Si_8O_{22}(OH)_2$
3. Arfvedsonite: $Na_3(Fe^{++}, Mg)_4(Fe^{3+}, Al)Si_8O_{22}(OH)_2$
4. Richterite: $Na_2Ca(Mg, Fe^{++})_5Si_8O_{22}(OH)_2$
5. Cummingtonite: $(Mg,Fe^{++})_7Si_8O_{22}(OH)_2$

A member of the pargasite-hastingsite series is the most common amphibole in volcanic rocks, in which it characteristically shows resorption or a reaction rim (sometimes) made up of titanomagnetite, pyroxene, and glass. This texture is interpreted to result from the instability of amphibole in a low-pressure (or higher-temperature) environment. There is little clear evidence for any consistent correlation between the composition of the amphibole and that of the lava, although Jakes and White (1972) suggest that hornblendes in island-arc andesites may be distinguished from those of continental andesites by their higher Al in fourfold coordination. Volcanic hornblendes typically have more $TiO_2$ than their plutonic equivalents, which frequently coexist with sphene. Amphiboles of kaersutite type are found in a wide range of magma types (for example, rhyolites, basanites, tephrites) and have also been found in ultramafic nodules in many alkaline lavas (for example, Tristan da Cunha; P. E. Baker et al., 1964). In plutonic rocks, hastingsite is common in granites, syenites, and nepheline syenites.

The alkali amphiboles riebeckite, arfvedsonite, and richterite are found in peralkaline lavas; they are the only amphiboles in the groundmass of lavas and

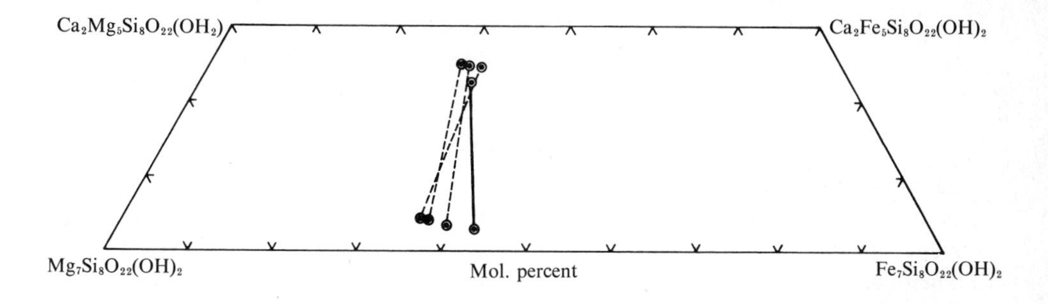


**Fig. 5-33** Composition of coexisting cummingtonite and hornblende occurring as phenocrysts in acid lavas. The equilibrium assemblage is shown by a full line. (After Klein, 1968.)

are therefore stable at magmatic temperatures at about 1 bar pressure. Examples are fluorine-bearing potassium richterite (magnophorite) in the orendite lavas of the Leucite Hills, Wyoming, and arfvedsonite in crystalline comendites of Mayor Island, New Zealand.

It is the occurrence of cummingtonite and hornblende phenocrysts in rhyolites that provides the closest analogy to the pyroxenes. Klein (1968) has analyzed four coexisting amphibole pairs with the results shown in Fig. 5-33. Of these pairs, only one, in his opinion, represents an equilibrium assemblage.

The temperature of the breakdown of any hydrous amphibole increases with pressure, but not without limit. Since one of the breakdown products is water, there will be a *PT* maximum on the amphibole stability (dehydration) curve, as increasing pressure reduces the volume and entropy of water, and $dP/dT$ will become infinite at this maximum; at pressures above this the temperature of the dehydration will decrease. Thus it is theoretically possible to dehydrate an amphibole (or any hydrous mineral) by pressure alone at room temperature. Experimental data on the breakdown of hydrous synthetic amphiboles have been summarized by W. G. Ernst (1968). These results show that no hydrous amphibole could be stable at 1 bar at lava temperatures, so that their occurrence, noted above, is due to fluorine, which amounts to about $\frac{2}{3}$ of the hydroxyl structural group.

Pargasite [$NaCa_2Mg_4AlAl_2Si_6O_{22}(OH)_2$] is the most refractory of the magnesian amphiboles, and in all cases the ferrous analog breaks down at lower temperatures and pressures than the magnesian end member; the stability range of the ferrous amphiboles depends upon the additional variable of oxygen fugacity. Amphiboles such as riebeckite-arfvedsonite which contain essential $Fe^{3+}$ are stable at lower oxygen fugacities than those where magnetite and quartz are reduced to fayalite. Although amphiboles in igneous rocks are not commonly found with olivine, kaersutite has been recorded with fayalite, titanomagnetite, and quartz in rhyolitic lavas (Carmichael, 1967*b*); here oxygen fugacity approximates closely the synthetic fayalite-magnetite-quartz oxygen buffer. Generally, the oxygen fugacity of phenocryst assemblages containing amphibole are several log units higher than this; data are available for cummingtonite (Ewart et al., 1971) and hornblende (Lowder, 1970) in acid lavas.

## Iron Titanium Oxides

Although iron titanium oxides are present in almost all igneous rocks, they are often found in only rather small amounts, which has frequently relegated their study to the petrographic observation that "accessory opaque oxides occur." Their great importance to paleomagnetism has, however, necessitated their detailed investigation with the result that geophysicists may often know far more about their mineralogy and paragenesis than petrologists do. Moreover, experimental gas-liquid-solid equilibria and their application to basaltic genesis have indicated that the role of the iron titanium oxides in controlling or influencing a basaltic liquid's fractionation path may be of great significance.

All the naturally occurring iron titanium oxides are prone to oxidation during cooling, and they are probably the most sensitive mineral indicators of the pristine nature of a rock; the rather complex oxide mineral assemblages that can result from this subsequent oxidation have been thoroughly studied by S. E. Haggerty (1971) and Watkins and Haggerty (1967). This aspect of their mineralogy will be neglected here, since we are principally concerned with the composition and species of iron titanium oxides which precipitate from a silicate liquid and remain as single-phase crystals with no sign of exsolution.

At high temperatures, there is continuous solid solution in each of three two-component series and restricted solid solution between the different series (Fig. 5-34). The composition bands along the three solid-solution series will become narrower with falling temperature, again because of the entropy-of-mixing term. The recalculation procedures of analyses obtained on the electron-microprobe, where the two oxidation states of iron cannot be determined, usually assume that the composition of an iron titanium oxide is restricted to the appropriate binary join.

ILMENITE-HEMATITE SERIES Most analyzed ilmenites contain less than 15 percent $Fe_2O_3$ in solid solution and, apart from small variations in MgO and MnO, change little in composition as crystallization proceeds. Ilmenites from the layered series of the Skaergaard intrusion (Vincent and Phillips, 1954) show progressive depletion in MgO, $Al_2O_3$, and $V_2O_3$ (with increasing height in the intrusion), a trend almost identical with that found in the olivine-tholeiite-rhyolite series of Thingmuli (Carmichael, 1967*a*). Magnesian ilmenites have not been found in volcanic rocks, but they are characteristically associated with kimberlites; manganiferous ilmenites are sometimes found in peralkaline acid lavas. Possibly because of the type of reaction discussed on page 275, lavas with low silica activities rarely contain ilmenite.

ULVOSPINEL-MAGNETITE SERIES Throughout an evolutionary magma sequence, this series shows some depletion in the $Fe_2TiO_4$ component in response to falling

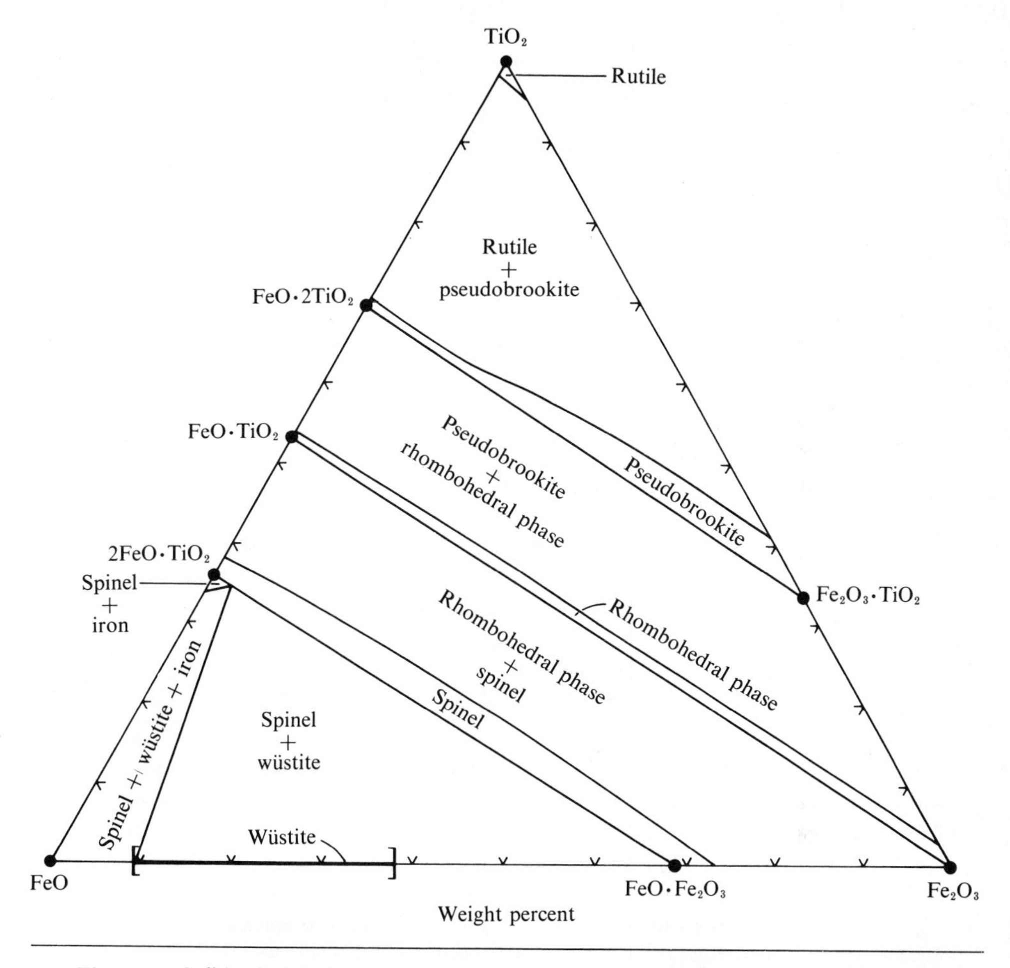


**Fig. 5-34** Solid solutions in the system $FeO$–$Fe_2O_3$–$TiO_2$ at 1300°C: pseudobrookite series; rhombohedral (ilmenite-hematite) series; spinel (ulvospinel-magnetite) series. (*After R. W. Taylor*, 1964.)

temperature. Titanomagnetite also becomes depleted in $Cr_2O_3$, $Al_2O_3$, $V_2O_3$, and MgO in more fractionated liquids, with simultaneous slight enrichment in ZnO and MnO. When ilmenite and titanomagnetite coprecipitate, the cubic phase is relatively enriched in the three-valent oxides, and the rhombohedral phase in the two-valent minor oxides.

The range of composition in terms of FeO, $Fe_2O_3$, and $TiO_2$ of one-phase coexisting titanomagnetite and ilmenite in basalts, andesites, dacites, and rhyolites is shown in Fig. 5-35. There is some indication that the titanomagnetites of basaltic lavas of the continental margins and island arcs are lower in $Fe_2TiO_4$ than those of oceanic tholeiites; the former typically contain between 30 and 45 percent ulvospinel, the latter between 60 and 80 percent.

Titanomagnetites in basic lavas of low silica activity frequently contain

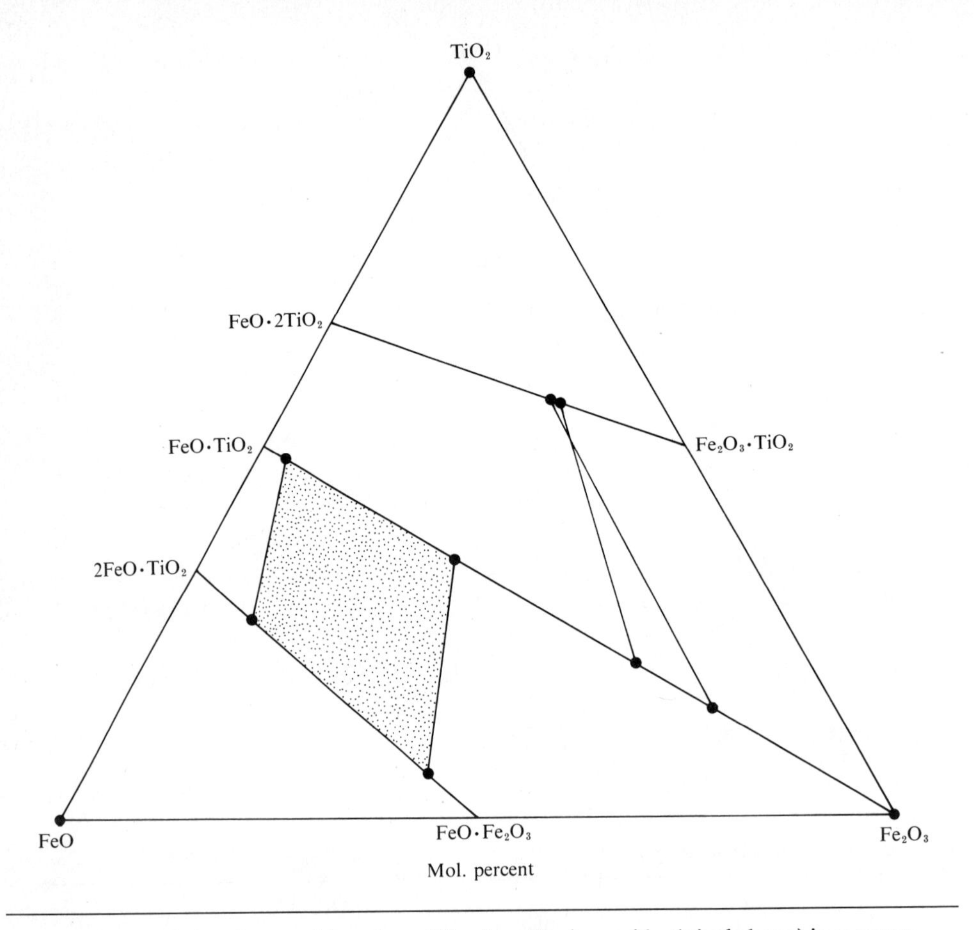


**Fig. 5-35** Limits of composition of coexisting iron-titanium oxides (stippled area) in common magma types. Tie lines (right) show two natural pseudobrookite-titanohematite assemblages.

more $Al_2O_3$ and MgO than those of tholeiitic lavas; typical values would be 3 to 5 percent $Al_2O_3$ and 1 to 3 percent MgO, compared to 1 to 2 percent $Al_2O_3$ and 0.5 to 1.5 percent MgO for tholeiitic lavas. It appears then that the $Al_2O_3$ content of titanomagnetite is governed by a reaction of the type

$$\underset{\text{pleonaste}}{MgAl_2O_4} + \underset{\text{glass}}{SiO_2} = \underset{\substack{\text{pyroxene}\\\text{component}}}{MgAl_2SiO_6} \tag{5-6}$$

where $MgAl_2SiO_6$ is in solid solution in pyroxene, so that in liquids of low silica activity the left-hand side of the reaction is favored.

PSEUDOBROOKITE SERIES Primary pseudobrookite, often with substantial amounts of MgO (5 to 6 percent), is rare and occurs in the groundmass of some shonkinites and alkali gabbros together with titanohematite (Carmichael and Nicholls, 1967); the compositions of coexisting phases are shown in Fig. 5-35. In both rock types there are phenocrysts of a rare and unusual yellow pyroxene rich in ferric iron. Because pseudobrookite is unstable below 600°C, its presence in an iron-titanium assemblage resulting from subsequent oxidation is used to "date" the time of this oxidation, since it is unlikely that a lava would be heated to 600°C many years after cooling without betraying this in the silicate minerals; oxidation is therefore likely to have occurred during subsolidus cooling of the lava.

CHROMITE SERIES In many magnesian basalts, small, discrete grains of chromite are found, and small inclusions of chromite in olivine phenocrysts are common. One of the more detailed studies of chromite in basaltic liquid is that of Evans and Wright (1972), who showed that in Kilauea magma the chromite becomes progressively poorer in $Cr_2O_3$, MgO, and $Al_2O_3$ and enriched in $Fe_2TiO_4$ with cooling; they also demonstrated a close correlation between the magnesian contents of chromite and of the enclosing olivine. Since chromite and olivine continue to equilibrate below solidus temperatures, some care has to be taken before accepting the compositions of coexisting chromite and olivine in slowly cooled intrusions as representative of high-temperature equilibration.

## IRON TITANIUM OXIDES IN FRACTIONAL CRYSTALLIZATION

The role of oxygen as a component in controlling the fractionation paths of a liquid in the system $MgO$–$FeO$–$Fe_2O_3$–$SiO_2$ (Muan and Osborn, 1956) was extended by Osborn (1959) to basaltic liquids. Two contrasted fractionation paths of cooling liquid may exist: in one the liquid becomes rich in iron and depleted in silica; in the other the liquid becomes silica-rich without concomitant iron enrichment. Presnall (1966) showed that the addition of diopside does not nullify these contrasted fractionation paths, and he also stressed that both cooling liquids precipitated magnetite. Thus for a liquid to become iron-rich, the amount of fractionated magnetite must be small compared to the ferromagnesian silicates, whereas to become silica-rich, the amount of magnetite would be much larger in the fractionated solid assemblage.

Osborn (1959) saw in these experimental results an ingenious way to explain the contrast between the basaltic and andesitic lavas of the continental margins, and the iron-rich trend, which although found in some basaltic provinces, is best displayed in the residual liquids of the Skaergaard intrusion. The lavas of the Cascade province of the western United States, for example, show little or no iron enrichment, presumably because of fractionation of magnetite crystals. Petrographic examination of many of these lavas shows that titanomagnetite is absent as an early crystallizing phase, for it is found only as granules in the glassy residuum. If these lavas have been generated by a process of fractional crystallization involving titanomagnetite, then they have become completely purged of the oxide phase necessary for their generation. So it behooves one to look at the concentration of FeO, $Fe_2O_3$, and $TiO_2$ that a basaltic liquid requires before titanomagnetite and ilmenite precipitate. From analyses of lavas with one-phase iron titanium oxides, it appears that between 2.0 and 3.5 percent $Fe_2O_3$ and between 0.9 and 1.2 percent $TiO_2$ are required if both titanomagnetite and ilmenite are to precipitate as part of the early crystallizing assemblage. These values tend to be higher than those found in many fresh Cascade basaltic lavas, so it is unlikely that these have been purged of their iron-titanium-oxide phenocrysts. Osborn's (1962) hypothesis is not easy to substantiate on the evidence of the very lavas from which he drew support for one of the more stimulating petrological ideas in the past 30 years.

## FERRIC-FERROUS EQUILIBRIUM IN LIQUIDS

The ratio $Fe_2O_3/FeO$ in a basaltic liquid presumably determines whether or not titanomagnetite precipitates during the crystallization interval. Its value in silicate liquids (at constant pressure) will depend on temperature, oxygen fugacity, and the composition of the liquid. Paul and Douglas (1965*a*) have shown that, in alkali-silicate glasses equilibrated at constant temperature and oxygen fugacity, the ferric-ferrous ratio is influenced by the molar concentration of alkali and the specific alkali ion present, potassium being more "oxidizing" than sodium (Fig. 5-36). Similar results have been obtained for Cr (Nath and Douglas, 1965) and Ce (Paul and Douglas, 1965*b*) in alkali-borate glasses (Paul and Lahiri, 1966). Paul and Lahiri have also shown the temperature dependence of the $Mn^{++}$–$Mn^{3+}$ equilibrium in alkali-borate glasses, a factor that becomes increasingly effective with increase in alkali (Fig. 5-36).

The ferric-ferrous equilibrium in alkali-silicate glasses has been represented by Paul and Douglas (1965*a*) by the ionic equation

$$Fe^{++} + \tfrac{1}{4}O_2 = Fe^{3+} + \tfrac{1}{2}O^{--} \tag{5-7}$$

and hence

$$K_{P,T} = \frac{[Fe^{3+}][O^{--}]^{1/2}}{[Fe^{++}][O_2]^{1/4}}$$

where [ ] represents activity. Thus, the ferric-ferrous equilibrium in an alkali-silicate or alkali-borate liquid is a function of the activity (fugacity) of oxygen, the activity of the oxygen ion, and a reaction constant. The implications of the data obtained by Douglas and his coworkers have been essentially confirmed for natural basaltic and andesitic liquids by the experimental results of Fudali (1965), who equilibrated a variety of rock types at temperatures above or close to the liquidus, so that their ferric-ferrous ratios are undisturbed by crystal-liquid equilibrium. In Fig. 5-37, the $Fe^{3+}/Fe^{++}$ ratios of five natural liquids

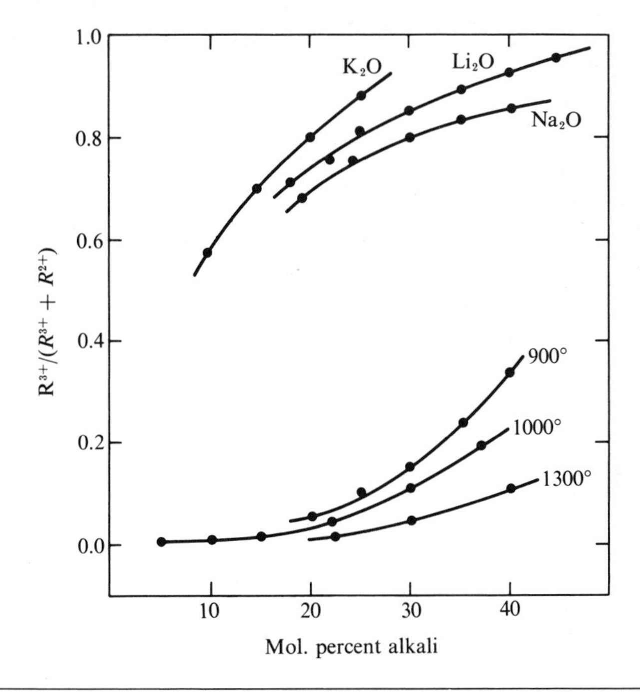


**Fig. 5-36** Upper curves: variation of $Fe^{3+}$ and $Fe^{++}$ with increase in alkali for silicate glasses equilibrated in air at 1400°C (after Paul and Douglas, 1965*a*). Lower curves: variation of $Mn^{3+}$ and $Mn^{++}$ with increase in $Na_2O$ in sodium-borate glasses equilibrated in air at the three temperatures (°C) indicated (after Paul and Lahiri, 1966).

equilibrated at 1200°C have been plotted—first against mole percentages of $Na_2O$ and $K_2O$, and second against molecular percentages of the three normative feldspar components. For both oxygen fugacities ($10^{-8.0}$ and $10^{-6.0}$ bar) at 1200°C there is a tendency for the proportions of $Fe^{3+}$ in these liquids (rocks) to increase with increase in alkali or feldspathic components.

Although the researchers in glass equilibria have found the concept of $O^{--}$ activity to be of value, it is more difficult for us to appreciate $O^{--}$ activity because it is not easy to define a standard state independent of other components. The results in Fig. 5-37 can therefore be represented by

$$2FeO + \tfrac{1}{2}O_2 = Fe_2O_3\dagger$$

and

$$K_{P,T} = \frac{a_{Fe_2O_3}}{(a_{FeO})^2 f_{O_2}} = \frac{X_{Fe_2O_3}\,\gamma_{Fe_2O_3}}{(X_{FeO})^2 f_{O_2} (\gamma^2_{FeO})^2} \qquad (5\text{-}8)$$

† This equation could be written $FeO + \tfrac{1}{4}O_2 = FeO_{1.5}$. This may have the effect of making the activity coefficients closer to unity, and in this sense only, it is more "correct."

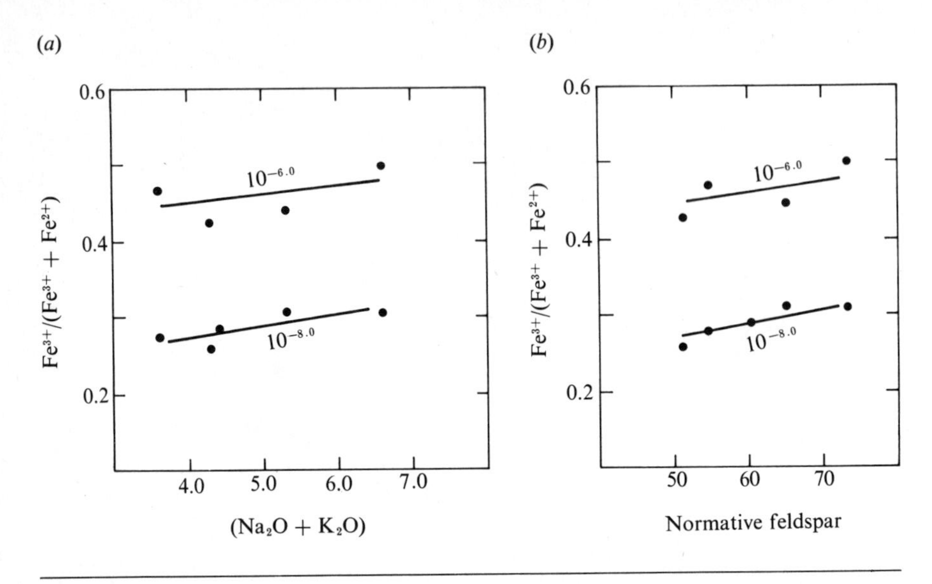


**Fig. 5-37** Variation of $Fe^{3+}$ and $Fe^{++}$ at $f_{O_2} = 10^{-8.0}$ and $10^{-6.0}$ bar in analyzed basaltic and andesitic liquids at 1200°C (at or below liquidus temperatures) plotted against (*a*) mole percent of $(Na_2O + K_2O)$ and (*b*) molecular percent of normative feldspar. (Data from Fudali, 1965.)

where $X_{Fe_2O_3}$ and $X_{FeO}$ refer to the mole fraction of the components in the liquid, and $\gamma$ is the corresponding activity coefficient; this equation takes account of the dependence of the ferric-ferrous equilibrium on composition (Fig. 5-37) by means of the activity coefficients. Thus basalts rich in alkali, such as basanites, crystallizing at exactly the same temperature and $f_{O_2}$ as an alkali-poor tholeiitic basalt will have a higher $Fe^3/Fe^{++}$ ratio and therefore different activity coefficients; thus it is generally impossible to deduce from the $Fe^{3+}/Fe^{++}$ ratios of various basaltic lava types any reliable indication of whether $f_{O_2}$ was either relatively high or relatively low. Moreover, temperature also has an effect on $Fe^{3+}/Fe^{++}$ in basaltic liquids as shown by the results of G. C. Kennedy (1948), who equilibrated a basalt liquid at temperatures above the liquidus in air (Fig. 5-38). There are, however, no data on a range of basaltic liquids to indicate whether the influence of temperature on the $Fe^{3+}/Fe^{++}$ equilibrium is more marked in alkali-rich liquids, as would be suggested by the results of Paul and Lahiri (Fig. 5-36).

In the cooling and crystallization of a natural silicate liquid, there are at least three factors other than liquid-crystal equilibrium that will tend to change the $Fe^{3+}/Fe^{++}$ ratio in the residual liquid. This ratio will tend to increase with alkali or with the normative feldspathic or feldspathoidal components, and this tendency has been called the *alkali-ferric-iron effect* (Carmichael and Nicholls, 1967). This effect will be enhanced by falling temperature, to a degree that becomes progressively more effective as the alkali content of the liquid increases (Fig. 5-36). Finally, the contribution due to variation of oxygen fugacity is difficult to evaluate; if the oxygen fugacity decreases, which would seem to be the average or normal trend, then presumably there will be a tendency for the $Fe^{3+}/Fe^{++}$ ratio in the residual liquid to decrease.

The combined effect of increasing alkali content, temperature, and oxygen fugacity on the ferric-ferrous ratio will perhaps be shown for the whole spectrum of natural liquids if it is assumed that this ratio in rocks has not been markedly changed during and particularly after their crystallization. In Fig. 5-39 the ferric-ferrous ratios of the average volcanic rock analyses compiled by Nockolds (1954) have been plotted against their respective alkali contents. There is considerable scatter, no doubt due in part to subsequent oxidation of the iron titanium oxides, but nevertheless a general tendency for the ratio to rise with increase in total alkali; residual alkali-rich liquids would therefore seem, in general, to become progressively more "oxidized" as fractionation proceeds.

## ABSENCE OF IRON TITANIUM OXIDES

Absence of iron titanium oxides in some cooling silicate liquids can arise from reactions which eliminate either or both from the crystallization sequence. A typical metaluminous Californian rhyolitic obsidian contains about 0.3 percent $Fe_2O_3$ and 0.7 percent $TiO_2$ (Table 5-1), and yet microphenocrysts of titanomagnetite and ilmenite are sparse but ubiquitous. In contrast, pantelleritic (and comenditic) obsidians containing 2.3 percent $Fe_2O_3$, 5.8 percent FeO, and 0.4 percent $TiO_2$ almost never contain an iron titanium oxide. When they do, it is ilmenite rather than titanomagnetite; but ilmenite does not occur as a microphenocryst if aenigmatite is present. J. Nicholls and Carmichael (1969*b*) have

**Fig. 5-38** Variation of $Fe^{3+}$ and $Fe^{++}$ for a basaltic liquid equilibrated in air at various temperatures above that of the silicate liquidus. (Data from G. C. Kennedy, 1948; *NM5*.)

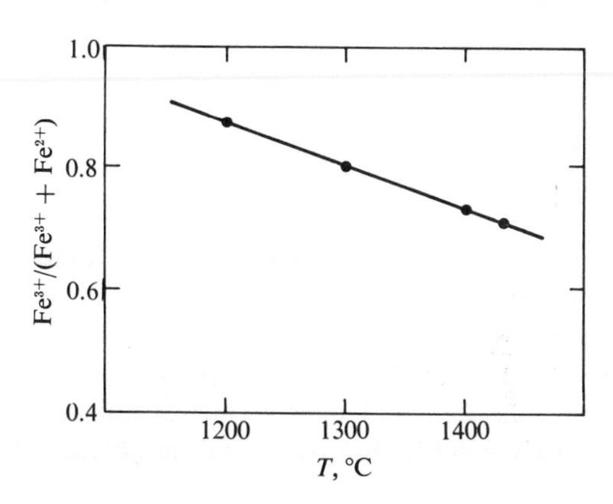

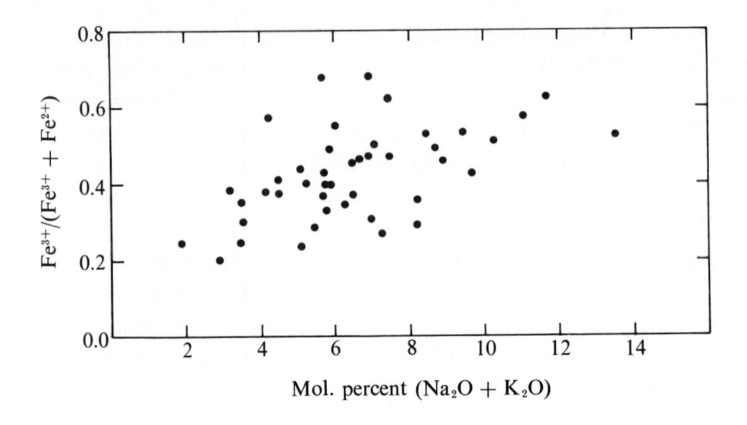


**Fig. 5-39** Values of $Fe^{3+}/(Fe^{3+} + Fe^{++})$ and of molecular percentage of $(Na_2O + K_2O)$ for average compositions of volcanic rocks. (Data from Nockolds, 1954.)

suggested that this antipathetic relationship could be represented by the reaction

$$\underset{\text{ilmenite}}{FeTiO_3} + \underset{\text{acmite}}{4NaFeSi_2O_6} = \underset{\text{aenigmatite}}{Na_2Fe_5TiSi_6O_{20}} + \underset{\text{liquid}}{Na_2Si_2O_5} + \underset{\text{gas}}{O_2} \tag{5-9}$$

and that the absence of titanomagnetite could be represented by

$$\underset{\text{magnetite}}{4Fe_3O_4} + \underset{\text{liquid}}{6Na_2Si_2O_5} + \underset{\text{quartz}}{12SiO_2} + O_2 = \underset{\text{acmite}}{12NaFeSi_2O_6} \tag{5-10}$$

The products of the two reactions taken in conjunction represent the common occurrence of aenigmatite (taken as a pure phase), a sodic pyroxene, and quartz in a sodic peralkaline liquid. If Eqs. (5-9) and (5-10) are recast to

$$\log f_{O_2} = \frac{\Delta G_r^\circ}{2.303RT} - \log a_{Na_2Si_2O_5}^{liq} + 4 \log a_{NaFeSi_2O_6}^{pyrox} + \log_{FeTiO_3}^{\alpha\ phase}$$

and

$$\log f_{O_2} = \frac{\Delta G_r^\circ}{2.303RT} - 4 \log a_{Fe_3O_4}^{titanomag}$$

$$- 6 \log a_{Na_2Si_2O_5}^{liq} + 12 \log a_{NaFeSi_2O_6}^{pyrox}$$

then the variation of $\log f_{O_2}$ with temperature for each reaction can be plotted with all the components in their standard states (unit activity). This is done in Fig. 5-40, where it can be seen that the two curves intersect and bound a field in $T$-$f_{O_2}$ space in which iron titanium oxides will be absent ("no-oxide" field); in this field aenigmatite and acmite will coexist. Unfortunately the position of the two curves is subject to considerable error (±5 log units) which arises from the free-energy estimates of aenigmatite and acmite made by Nicholls and Carmichael. One of the possible lower terminations of the "no-oxide" field appropriate to peralkaline rhyolites is the field of riebeckite-arfvedsonite solid solutions (Ernst, 1968); this will doubtless be extended to higher temperatures than those shown by the substitution of F for (OH).

**Fig. 5-40** Fugacity of oxygen plotted against temperature (after J. Nicholls and Carmichael, 1969*b*), showing the "no-oxide" field (stippled). Abbreviations: Hm, hematite; Mt, magnetite; Ac, acmite; Q, quartz; Fa, fayalite; Nds, sodium disilicate; Aen, aenigmatite; Im, ilmenite. The dashed curve is the locus of the intersections of the two curves Ac/Mt–Nds–Q and Ac–Im/Aen–Nds for varying activities of Nds (labeled 0.1 to 1) at unit activity of acmite, magnetite, and ilmenite. The intersection of the two curves at $X$ represents activity of acmite 0.5, with unit activity of all other components. Solid circles 1 and 2 are two pantellerite obsidians; aenigmatite is present in 1, but not in 2.

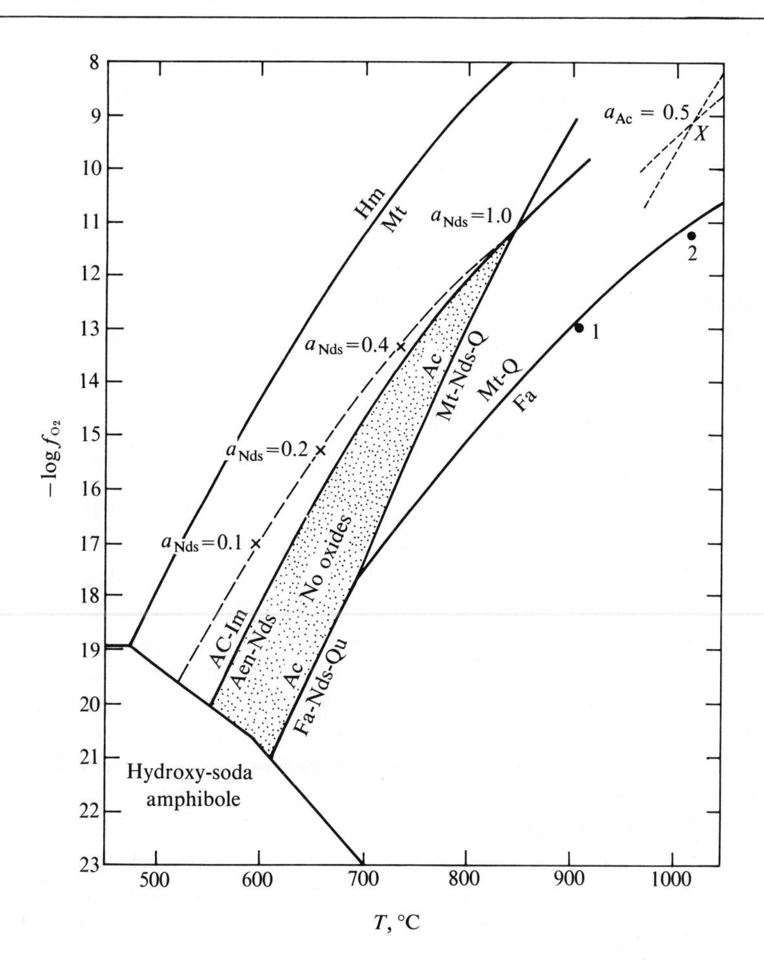

If the pyroxene in Eq. (5-10) is not pure acmite, but rather a solid solution with an acmite component, then the intersection of the two reaction curves will shift to higher temperatures; an example is shown in Fig. 5-40, where the activity of acmite is taken as 0.5. Conversely, reduced activities of $Na_2Si_2O_5$ will require intersections at lower and lower temperatures (Fig. 5-40); the activity of $Na_2Si_2O_5$ can be taken as a measure of the peralkalinity $[(Na_2O + K_2O) > Al_2O_3]$ of a peralkaline lava, which increases in their residual liquids (Fig. 5-9). A peralkaline rhyolite containing phenocrysts of fayalite, quartz, and perhaps ilmenite (Fig. 5-40, no. 1) will have an oxygen fugacity close to the synthetic FMQ buffer curve, and as it cools it will enter the "no-oxide" region in which aenigmatite and a sodic pyroxene precipitate at the expense of the iron titanium oxides. Many peralkaline rhyolites will obviously start their crystallization in the "no-oxide" field; whether or not they do so will depend on temperature, oxygen fugacity, and peralkalinity (activity of $Na_2Si_2O_5$).

## Interpretation of Data on Hydrous Systems

We conclude this chapter with a note of warning about pressure and the interpretation of cooling or fusion paths of hydrous liquids. So long as water does not enter into reaction with the equilibrium phases, as in the reaction

$$\underset{\text{forsterite}}{Mg_2SiO_4} + \underset{\text{quartz}}{SiO_2} = \underset{\text{enstatite}}{2MgSiO_3} \qquad (5\text{-}11)$$

it matters not at all whether the experiments are performed dry or with water as a pressure medium; the results at, say, 5 kb will be the same. But should water dissolve in one of the phases, as in the reaction

$$\underset{\text{forsterite}}{Mg_2SiO_4} + \underset{\text{liquid}}{SiO_2} = \underset{\text{enstatite}}{2MgSiO_3} \qquad (5\text{-}12)$$

then the equilibrium with liquid saturated by water at 5 kb confining pressure will be quite different from that at 5 kb without water. Furthermore, in liquids which contain water but are unsaturated by it, the equilibrium will not necessarily correspond to liquids saturated with water at the same pressure, as Luth (1969) takes pains to illustrate. This is a most fruitful field for analysis: the effect of pressure on liquid-solid equilibrium containing water as a component but without a vapor phase. This after all is undoubtedly the common situation in magmas, except perhaps those close to or on the earth's surface. The studies of Luth (1969), Robertson and Wyllie (1971), and Kushiro (1972*a*) are a small but essential start in modeling in detail the crystallization or fusion paths of hydrous liquids in equilibrium with solid solutions.

# Magmatic Gases and Volatile Components

## Introduction

Many igneous rocks have a texture which is reasonably ascribed to the existence of a gas phase, the most familiar examples being vesicular lava and pumice. But magmas that cool at depth may also generate a gas phase and submarine lava with vesicles is found up to depths of 4000 m or more (J. G. Moore, 1965). Granites and granophyres of fair-sized bodies have drusy or miarolitic cavities into which project beautifully faceted crystals of quartz and feldspar, and the occurrence of coarse-grained pegmatites enclosed in more equigranular granite is considered to represent the coeval crystallization of an exsolved aqueous fluid (dense gas) and silicate magma (Jahns and Burnham, 1969). But it is the eruption of a volcano with an ejected gas cloud billowing to 5000 m or more that is the most spectacular evidence of magmatic gas; indeed many volcanologists see gas as the cause of eruption itself and certainly as the propellant for extensive and catastrophic ash showers.

The components of a magma which under favorable conditions enter the gas phase are generally and conveniently referred to as volatile components. They are of course also present in the liquid phase, and although typically in much smaller concentration than in the gas, this need not invariably be so. Although there is generally no way of telling what the relative proportions of the volatile components may have been, the concept of mineral-gas buffers sometimes allows estimates of fugacities of certain components to be calculated if equilibrium can be assumed. Generally the volatile components make up only a small percentage by weight of the common igneous rocks. However, the abundance of the most common component, water, is often suspect in view of the evidence of extensive oxygen and hydrogen isotopic exchange which occurs at low temperatures and, in the case of many obsidians, with extensive hydration. Thus reliable values for the pristine water content of various types of plutonic rocks are almost impossible to give; values for certain volcanic rocks are given later (Table 6-5).

Because of the difficulty of circumscribing a magmatic gas with any set of properties that geologists could either agree on or determine, the gas is usually conceived to be dominantly hydrous, and historically it has pretty well "achieved" whatever a petrologist required it to do. There is, however, good reason to suppose that relatively small quantities of volatile components may appreciably modify the behavior of silicate melts. Because their molecular weights are low compared with those of most silicates or metallic oxides, their mole fractions may be large in spite of their small percentages by weight. The mole fraction of water in a water-albite melt containing 6 percent $H_2O$ by weight is 0.483, and that of water-saturated forsterite liquid at 20 kb (Kushiro, 1969) is 0.65. Thus the effect of a small amount (by weight) of water on the chemical potential of albite should be comparable with that of a much larger amount of some silicate of high molecular weight. Moreover, the molar volume of a volatile component, particularly when in the form of a gas phase at moderate pressure, is large, and the effect of pressure changes on equilibrium should thus be greater in a system containing minor volatiles than in the corresponding "dry" system.

In this chapter we briefly look into the composition of volcanic gases and how they may be related to the mineral composition of the main types of igneous rock. However, in order to do this, we delve a little into thermodynamics, particularly with regard to the concept of fugacity and activity, so that by using the principle of solid-gas buffers (Eugster and Wones, 1962) and assuming equilibrium the fugacities of certain components may be calculated.

## Properties of a Gas

The distinction between a gas and a liquid under ordinary conditions is quite easy to make, since the liquid is many times denser than the gas. For instance, water vapor at 1 bar and 520°C is in the tenuous state that we would call a gas, but at 10,000 bars the same water vapor has a density of 1.0 g $cm^{-3}$. This density is closer to that of a liquid than a gas, so that we might suspect that the properties of this dense gas are much like those of a liquid. In geologic environments, the density of a gas phase may vary between that of a gas at very low pressure and that of a highly compressed fluid; and the change between the two extremes may be brought about continuously.

A perfect gas by definition obeys the gas law

$$\frac{RT}{P} = V \tag{6-1}$$

where $V$ is the volume of 1 mole of gas. The fugacity coefficient $\gamma$ of a gas essentially measures its departure from perfect behavior; a perfect gas therefore has $\gamma = 1$. All gases become perfect or ideal at sufficiently low pressures and high temperatures. For example, water vapor behaves as an ideal gas above 1200°C at pressures less than 100 bars.

All the thermodynamic properties of a perfect gas (free energy, entropy, etc.) are determined by pressure and temperature, but for nonideal gases it is necessary to know also the value of the fugacity coefficient, which itself depends on temperature and pressure in a manner that cannot be predicted from thermodynamic theory only. In gaseous mixtures of several components, the fugacity coefficient of each depends also on the relative proportions of the other components. This dependence of the fugacity coefficient on composition is slight in gases at low pressure, but under high pressure, where the density of the gas is large and where the molecules of the various components are relatively close together, intermolecular forces come into play and affect the fugacity coefficients. This would mean, for example, that the thermodynamic properties of water vapor would not be exactly the same in a dense gas consisting of water and $CO_2$ as they would be in one consisting of water and HCl. Very little experimental work has been done so far in this connection, so that it is difficult to predict the behavior of dense gaseous mixtures within given ranges of temperature and pressure.

## LIQUID-GAS EQUILIBRIUM: CRITICAL POINT

Because a system of one component in two phases is univariant, a pure liquid at any given pressure may be in equilibrium with its pure vapor only at one temperature. The relation between this temperature and the corresponding pressure is given by the Clausius-Clapeyron relation:

$$\frac{dT}{dP} = \frac{T\Delta V_{\text{vap}}}{\Delta H_{\text{vap}}} \tag{6-2}$$

where $\Delta H$ is the heat of vaporization and $\Delta V$ is the change in volume when 1 mole passes from the liquid to the gas state.

In contradistinction to the behavior of a solid in equilibrium with its own liquid (melting), it is found that both $\Delta H$ and $\Delta V$ in Eq. (6-2) simultaneously tend to zero when temperature and pressure are raised. When this happens, the liquid "boils" without absorption of heat or change in volume; $dT/dP$ becomes indeterminate, and the boiling-point curve ceases. The two phases, gas and liquid, become indistinguishable. The pressure $P_c$ and the temperature $T_c$ at which this happens are respectively the critical pressure and the critical temperature. Values of $P_c$ and $T_c$ for certain gases are listed in Table 6-1.

**Table 6-1** Critical temperatures and pressures and molar volumes of various gases

| GAS | $T_c$,°C | $P_c$, bars | MOLAR VOLUME AT CRITICAL POINT, $cm^3\ mole^{-1}$ | DENSITY AT CRITICAL POINT, $g\ cm^{-3}$ |
|---|---|---|---|---|
| $CO_2$ | 31.1 | 73.8 | 94 | 0.468 |
| HCl | 51.4 | 82.6 | 87 | 0.419 |
| $SO_3$ | 218.3 | 84.9 | 126 | 0.635 |
| $SO_2$ | 157.6 | 78.8 | 122 | 0.525 |
| $H_2S$ | 100.4 | 90.0 | 98 | 0.348 |
| $H_2O$ | 374.2 | 221.1 | 56 | 0.322 |

At any pressure or temperature greater than $P_c$ or $T_c$, respectively, the substance may be brought continuously, without any change of phase, from a state in which it behaves as a perfect gas to a state in which its density is comparable with that of an ordinary liquid and its activity coefficient is very different from 1. In such a state a "fluid" presumably resembles an ordinary liquid in many of its properties, for example, viscosity and solvent power, but one would nevertheless hesitate to call it a "liquid," since it may be brought continuously into a state which would generally be described as gaseous. It is for this reason that we shall usually refer to supercritical phases as gases, but we must remember that if the pressure is sufficiently high and the temperature sufficiently low these "gases" may behave in many respects like ordinary liquids.

The critical pressure of water is only 218.3 atmospheres or 221 bars, a value which may be attained in the earth at a depth of less than 1 km. Pure water in the crust at a depth greater than 1 km is therefore normally in the supercritical state, regardless of its temperature.

## FUGACITY

We have mentioned the fugacity coefficient as a measure of the departure of a real gas from ideal gas behavior; the quantity $p_i\gamma_i = f_i$ is the fugacity of a gas whose pressure is $p_i$, where $i$ refers to a particular component in the gas phase. The fugacity function is not restricted to gases, but is another way of expressing the chemical potential $\mu_i$ or the partial molar free energy $\overline{G}_i$ of any component $i$ in a solid, liquid, or gas.

The chemical potential $\mu$ can be visualized as a measure of escaping tendency, since matter flows from regions of high potential to those of low. A familiar example is that of a gas contained in two isolated chambers connected by a porous plug; the gas at the higher pressure flows into the lower-pressure chamber, the system coming to equilibrium when the pressures on either side of the plug equalize. Because solids and liquids have finite, but often very low, vapor pressures, we can use escaping tendency for these states as well. However, we cannot compare the escaping tendencies, or chemical potentials, of different substances since there is no way to determine absolute values of $\mu$. It was to circumvent this problem of absolute values and also the nonideality of many vapors and gases that the concept of fugacity was proposed. The fugacity $f$ of a pure gas is defined by the relationship

$$d\mu = d\hat{G} = RTd \ln f \tag{6-3}$$

together with the restriction that as the pressure in the gas approaches zero, the ratio of the fugacity to the pressure $f/P$ approaches unity. In tenuous gases at very low pressure, there is no interaction between molecules, and the gas behaves as a perfect gas; the fugacity then becomes equal to the pressure.

For any isothermal process, the change in chemical potential or partial molar free energy of a pure gas is given by

$$\mu_2 - \mu_1 = \hat{G}_2 - \hat{G}_1 = RT \ln \frac{f_2}{f_1} \tag{6-4}$$

For any set of phases at equilibrium (liquid, solid, or gas), the chemical potential of any component must be the same in all; it follows then that at equilibrium the fugacity of $i$ must also be the same in all phases.

The vapor pressures of many saltlike solids are so low that they are difficult to measure, but because we are often interested in *changes* in fugacity, we can circumvent this difficulty by using the change in *relative fugacity* or in a fugacity ratio.

The relative fugacity, or activity $a$, is defined as the ratio of the fugacity $f$ in any state (for example, liquid or gas solution) to the fugacity $f^\circ$ in a standard state at the same temperature. Thus

$$a = \frac{f}{f^\circ} \tag{6-5}$$

By convention the standard state of a gas is unit fugacity at the temperature of interest. Thus

$$\frac{f}{f^\circ} = \frac{f}{1} = a \tag{6-6}$$

It can be seen that although fugacity has the same units as pressure, the activity $a$ is a ratio without units, although for a gas it is numerically equal to the fugacity. By identifying $\mu^\circ$ as the chemical potential of a gas in the standard state of unit fugacity $f^\circ$, we obtain, by substituting Eq. (6-6) into Eq. (6-4),

$$\mu - \mu^\circ = RT \ln \frac{f}{f^\circ} = RT \ln f \tag{6-7}$$

At sufficiently low pressure all gases become perfect and the fugacity becomes equal to the pressure, or the fugacity coefficient $\gamma$ becomes unity. Thus because

$$f_i = p_i \gamma_i \tag{6-8}$$

hence

$$f_i \to p_i \quad \text{as } P \to 0$$

where $p_i$ is the partial pressure of $i$. But

$$p_i = PX_i \tag{6-9}$$

where $P$ is the total confining pressure and $X_i$ is the mole fraction of $i$. Therefore for any gas, by substituting Eqs. (6-8) and (6-9) into Eq. (6-7),

$$\mu_i = \mu_i^\circ + RT \ln PX_i \gamma_i \tag{6-10}$$

where $\mu_i^0$ is dependent only on temperature. Although the fugacity coefficient $\gamma$ becomes unity at sufficiently low pressure, it is dependent on pressure, temperature, *and* composition of the gas if the pressure $P$ is large.

The reader should be careful to distinguish between $\mu^\circ$ for a gas which is independent of pressure [Eq. (6-10)] and $\mu^\circ$ for a solid or liquid which is *not* independent of pressure. The distinction between $\mu^\circ$ for a gas and for the two condensed states is solely a matter of convenience, in the choice of standard states. For a solid and liquid we write by convention

$$a_i = X_i \gamma_i \tag{6-11}$$

where $X_i$ is the mole fraction and $\gamma_i$ is the activity coefficient of component $i$. Substitution of Eq. (6-11) into Eq. (6-7) gives

$$\mu_i = \mu_i^\circ + RT \ln X_i \gamma_i \tag{6-12}$$

where $\mu_i^\circ$ is now a function of temperature *and* pressure.

Although the concept of fugacity was conceived with gases in mind, it is applicable to liquids and solids, and its power in dealing with the volatile components of igneous rocks lies in the fact that the fugacity function does not require any conditional statement about the existence of a gas phase. Although fugacity is measured in units of pressure,[1] it does not carry with it the con-

[1] Strictly it is dimensionless.

notation of a gas phase in the same way as partial pressure. Volatile components are soluble in both liquids and solids to some extent, but it is only when the solubility limit has been exceeded that an equilibrium gas phase is generated. The great power of the fugacity function is that if we can determine the fugacity of a volatile component, or indeed the chemical potential of any component, then this will be its value in all phases regardless of their number (liquid, solid, or gas), provided that they are in equilibrium with one another.

## A MAGMATIC GAS PHASE

The factors which control or influence whether or not a magma becomes saturated with a volatile component have been examined from a completely general point of view by Verhoogen (1949), whose formulation is reproduced below. If the reader wishes to think of a particular component, rather than $i$, there will be little distortion of reality if he identifies this with water, even though it appears that much of this component in volcanic gas may be meteoric rather than juvenile.

The most obvious requirement for generation of a gas phase is that the sum of the partial pressures of all the components is equal to or greater than the confining pressure $P$ on the magma. The problem, then, is to calculate the partial pressure of component $i$. Let $X_i^{\text{liq}}$, $\gamma_i^{\text{liq}}$, and $\overline{V}_i^{\text{liq}}$ be respectively the mole fraction, activity coefficient, and partial molar volume of $i$ in the silicate liquid (magma); let $p_i^\circ$ be the vapor pressure of pure liquid or solid $i$ at temperature $T$ of the magma. This means that at $T$ and $p_i^\circ$ the pure solid or liquid is in equilibrium with its pure vapor, which, if $p_i^\circ$ is small, will be a perfect gas.

When $i$ passes from pure solid or liquid $i$ at $T$, $p_i^\circ$, to its actual state in the melt ($T$, $P$, $X_i^{\text{liq}}$), its chemical potential [Eq. (6-12)] changes by

$$\mu_{T,P,X_i^{\text{liq}}} - \mu_{T,p_i^\circ} = \int_{T,p_i^\circ}^{T,P,X^{\text{liq}}} \overline{V}_i \, dP + RT \int_1^{X_i^{\text{liq}}\gamma_i^{\text{liq}}} d\ln X_i^{\text{liq}}\gamma_i^{\text{liq}} \tag{6-13}$$

where $X_i^{\text{liq}}\gamma_i^{\text{liq}} = a_i^{\text{liq}}$, the activity of $i$ in the magma.

The fugacity of $i$ in the gas phase in equilibrium with the magma must change by an equal amount, which from Eq. (6-7) is

$$\Delta\mu_i = RT \int_{p_i^\circ}^{PX_i^{\text{gas}}\gamma_i^{\text{gas}}} d\ln f_i \tag{6-14}$$

Since $PX_i^{\text{gas}} = p_i$, the partial pressure of $i$ in the gas phase, then by equating Eq. (6-13) with Eq. (6-14) we get

$$RT \ln \frac{p_i\gamma_i^{\text{gas}}}{p_i^\circ} = |\overline{V}_i|(P - p_i^\circ) + RT \ln X_i^{\text{liq}}\gamma_i^{\text{liq}} \tag{6-15}$$

where $|\bar{V}_i|$ is the mean value of $\bar{V}_i$ between the limits of integration. Rearranging,

$$\log p_i = \frac{|\bar{V}_i|(P - p_i^\circ)}{2.303RT} + \log \frac{X_i^{\text{liq}}\gamma_i^{\text{liq}}p_i^\circ}{\gamma_i^{\text{gas}}} \tag{6-16}$$

or

$$X_i^{\text{gas}} = \frac{p_i^\circ X_i^{\text{liq}}\gamma_i^{\text{liq}}}{\gamma_i^{\text{gas}}P} \exp \frac{|\bar{V}_i|(P - p_i^\circ)}{2.303RT} \tag{6-17}$$

so that the mole fraction of $i$ in the gas, $X_i^{\text{gas}}$, is influenced not only by $P$, $T$, and the vapor pressure of the pure liquid or solid, $p_i^\circ$, but also by the partial molar volume of $i$ in the magma ($\bar{V}_i$), together with the activity of $i$ in the magma ($X_i^{\text{liq}}\gamma_i^{\text{liq}}$) and the fugacity coefficient of $i$ in the gas ($\gamma_i^{\text{gas}}$). Thus two substances with the same vapor pressure $p_i^\circ$, but with different partial molar volumes, may be present in the gas phase in different proportions.

For a gas phase to exist, the sum of the partial pressures of all the components must equal $P$, the confining pressure. Thus from Eq. (6-16)

$$\sum_i \frac{\gamma_i^{\text{liq}}}{\gamma_i^{\text{gas}}} X_i^{\text{liq}} p_i^\circ \exp \frac{|\bar{V}_i|(P - p_i^\circ)}{2.303RT} \geq P \tag{6-18}$$

where the summation is over all the components in the magma. If a component, such as $H_2O$ or $SO_2$, has a critical temperature $T_c$ below that of the magma (Table 6-1), then $p_i^\circ$ will have no physical meaning. This can be allowed for by taking a temperature $t$ less than $T_c$ where the vapor pressure is $p_i^{\circ\prime}$ and then by adding the change in chemical potential for the temperature interval $T_c - t$ to both Eqs. (6-13) and (6-14). This will involve the partial molar entropy $\bar{S}_i$ in both the gas and the magma, the former usually being larger, so that the generation of a gas phase for a component like water will depend on a $T\Delta\bar{S}$ term, where $\Delta\bar{S} = \bar{S}_i^{\text{gas}} - \bar{S}_i^{\text{liq}}$.

Although it is evident that the partial molar volume, which can be positive, negative, or zero, of any component in the magma will have an influence on the generation of a gas, the only data on partial molar volumes at the time of writing are those for water in $NaAlSi_3O_8$ liquids (Burnham and Davis, 1971) and for several rock-forming oxides in silicate liquid (Bottinga and Weill, 1970). It is therefore difficult to predict whether or not a particular body of cooling magma will become saturated at the time of emplacement. Only volcanoes provide the opportunity to sample magmatic gas directly, although the fluid inclusions trapped in minerals are of great additional value as a qualitative guide, especially now that there seems to be some experimental evidence that inclusions do not leak.

## Volcanic Gases

The composition of volcanic gases, whether they are collected at the source of eruption or from a fumarole on a tongue of lava or by heating quenched volcanic glass in a vacuum, shows an extreme amount of variation. It is therefore difficult to make any general statement about their composition and particularly to relate this either to the chemistry of the magma or to the minerals contained therein. Following Matsuo (1960), we shall attempt to relate the composition of magmatic gas to the calculated fugacities of various volatile components defined by mineral assemblages. This is a complex topic, not so much because of the necessary assumptions of equilibrium but because of the many factors that can change the composition of a volcanic gas.

It seems reasonable to postulate that magmas of different composition have volatile components in different proportions. These will change during cooling of a gas as it equilibrates with the magma. If the gas separates from the magma and interacts with wall rock on its way to the surface, then not only will the proportions of the constituents change, but material may become added to or lost from the gas. Because most volcanic eruptions show a declining explosivity as the eruption proceeds, it is likely that the gas emitted during the initial stages of the eruption is significantly different from that given off as the magma chamber becomes depleted. Certainly the composition of the lava itself often shows a progressive change. Gas collected at the throat of the volcano may also differ from that collected from a fumarole on top of a lava flow some distance from the site of eruption. Over and beyond all the variation related to magma, temperature, and wall-rock interaction is that resulting from an unknown amount of atmospheric or meteoric contamination. Many gas analyses show substantial amounts of $N_2$, Ar, and $O_2$, which are usually regarded as telltale indices of the extent of atmospheric contamination. Couple all these factors with the difficulty of gas analysis, particularly before the days of gas chromatographs, and a little of the hazard of interpreting gas analyses becomes evident.

D. E. White and Waring (1963) have made a compilation of the analyses of volcanic gases, sublimates, and the gases driven from igneous rocks by heating. In order to isolate possible effects of contamination, they reported separately the amounts of $H_2O$, Ar, $N_2$, and $O_2$ in each gas analysis, and the remaining constituents, called the "active" gases, were recalculated to 100 percent. One series of analyses, collected from an extrusive dome of hypersthene dacite in the Japanese volcano Usu, is a superb example of the craft of an analytical chemist; selected examples of these analyses for a range of collection temperatures are given in Table 6-2. Water is the most abundant component; of the "active" gases, $CO_2$ predominates, and $CH_4$, in very small amounts, is the only other carbon gas. $SO_2$ is the dominant sulfur gas at higher temperatures, but at lower temperatures $H_2S$ becomes more abundant; HCl is invariably more abundant than HF, and the ratio HF/HCl falls with temperature. Substantial amounts of $H_2$ are found together with small quantities of $NH_3$; B, P, and As were also

**Table 6-2** Analyses of gases collected from fumaroles at Showa-shinzan, Usu volcano, Japan Dome of hypersthene dacite extruded in 1944–1945*

| FUMAROLE No. | $T$, °C | $CO_2$ | $CH_4$ | $NH_3$ | $H_2$ | HCl | HF | $H_2S$ | $SO_2$ | $P_4O_{10}$† | TOTAL ACTIVE GASES, % | $H_2O$, % |
|---|---|---|---|---|---|---|---|---|---|---|---|---|
| A–1(9081) | 750 | 4700 | 5.8 | 4.3 | 1808 | 389 | 199 | 7.2 | 120 | 0.0016 | 0.723 | 99.25 |
| A–6a | 700 | 3620 | 8.3 | 0.4 | 1450 | 510 | 209 | 37 | 89 | 0.0019 | 0.592 | 99.39 |
| C–2 | 645 | 3660 | 8.0 | 0.6 | 1210 | 490 | 200 | 30 | 91 | 0.0012 | 0.569 | 99.41 |
| B–4a | 464 | 7825 | 12.0 | 8.6 | 440 | 130 | 76 | 9.2 | 10 | 0.0019 | 0.859 | 99.10 |
| A–4a | 460 | 3550 | 3.2 | 1.1 | 811 | 569 | 306 | 15 | 10 | 0.0022 | 0.537 | 99.24 |
| C–4 | 430 | 3230 | 6.2 | — | 1400 | 669 | 230 | 45 | 89 | 0.0053 | 0.567 | 99.41 |
| B–1 | 328 | 8485 | 14.2 | 0.7 | 660 | 140 | 62 | 100 | 13 | 0.0030 | 0.948 | 99.00 |
| B–4b | 300 | 7820 | 12.8 | 0.2 | 380 | 150 | 45 | 87 | 36 | 0.0011 | 0.853 | 99.07 |
| A–6b | 203 | 3900 | Nil | 0.5 | Nil | 748 | 83 | 52 | 70 | 0.0008 | 0.486 | 99.1 |
| C–3(9063) | 194 | 1970 | 4.1 | 0.26 | 351 | 120 | 11 | 110 | 13 | — | 0.258 | 99.72 |

Components of low volatility in fumarole A–1(9081) but collected almost 5 years before the samples above. Values in ppm by weight

| | | | |
|---|---|---|---|
| $SiO_2$ | 253 | Cu | 0.03 |
| Al | 15 | Zn | 0.5 |
| Fe | 1.3 | As | 0.7 |
| Ca | 4.6 | Ag | 0.003 |
| Mg | 32 | Sn | 0.03 |
| Na | 22 | Pb | 0.03 |
| K | 15 | | |

* Results given in volume percent ($\times 10^4$) except for water. Data taken from D. E. White and Waring (1963).

† Reported as $P$ in mg per 1000 liters; recalculated to $P_4O_{10}$ gas in volume percent ($\times 10^4$).

found in minute quantities in the condensate of the gas samples. A more complete analysis of the condensate shows elements that are abundant in the magma (Si, Al, Fe, Mg, Ca, K, and Na) together with traces of the rarer elements Cu, Zn, Pb, and others.

Despite the extreme compositional variations in analyzed volcanic gases, certain generalizations are possible. Qualitatively there is considerable uniformity, with water the most abundant component, often comprising 90 percent or more of the gas. The carbon gases are next, particularly $CO_2$, although CO is the predominant carbon gas at the Zavaritskii crater of Kliuchevskii volcano, Kamchatka. Of the sulfur gases, $SO_2$ is typically more abundant than $H_2S$ at high temperatures, with the reverse relationship at low temperatures, but there are many exceptions to this. HCl is the dominant halogen gas, with subordinate amounts of HF. $H_2$, though very variable, is a substantial component of many active gases, particularly at Kliuchevskii volcano.

In fumarole sublimates, NaCl, KCl, and $NH_4Cl$ are conspicuous, together with fluorides and sulfates of Na, K, Ca, Mg, and Fe. An intriguing aspect of fumarole chemistry has been found in several Central American volcanoes by Stoiber and Rose (1970). Fumaroles near the main eruptive vents are sulfate-rich, while those on the lava flows are Cl-rich; and a decrease in the Cl/sulfate ratio in the vent fumaroles indicates an impending volcanic eruption.

If the composition of a volcanic gas is recalculated in terms of the more abundant elements, for example as the ratio H:O:C:S, then it is possible to calculate the equilibrium partial pressures of the various gaseous species for different temperatures. This requires solution of several simultaneous equations of the type

$$2H_2 + O_2 = 2H_2O$$

$$H_2 + CO_2 = CO + H_2O$$

$$4H_2 + CO_2 = CH_4 + 2H_2O$$

$$4H_2 + 2SO_2 = S_2 + 4H_2O$$

$$2H_2 + S_2 = 2H_2S$$

for which the values of the equilibrium constants can be calculated from free-energy data. Solutions for two or three equations in turn have been given by Ellis (1957) and Heald (1968), who were able to show that magmatic gases often closely approach equilibrium. In more detailed calculations, Heald et al. (1963) used the element ratio of a gas collected at Kilauea, Hawaii (H:O:C:S equal to 275.5:142.2:2.680:1.000) to compute the partial pressures of all possible species over a range of temperature and pressure.

The most thorough investigation of the equilibrium composition of volcanic gas is that made by Nordlie (1971) using the Kilauea gas analyses reported by Shepherd (1938). Nordlie identified the various contaminating gases, such as $H_2O$, Ar, $N_2$, $SO_2$, and $O_2$ in the gas samples, by using a computer to calculate the proportions of the various species with increasing amounts of contamination. He showed (1971, fig. 6) that if $H_2O$ was added to one gas, all the other 13 analyses fell on the "water-addition curve," a remarkable coincidence if due to chance. Nordlie's conclusion, after extensive investigation, was that the 14 analyzed Kilauea gases were all derived from a single magmatic gas; the composition of the least contaminated of these is given in Table 6-3. The dominant species in this best estimate of magmatic gas are $H_2O$, $CO_2$, and $SO_2$ in approximately equal proportions, and the calculated partial pressures of the various species as a function of temperature (at 1 atmosphere total pressure) are shown in Fig. 6-1. The Kilauea gas composition is considered to be that of the gas released from the erupting lava. To what extent this gas has been modified during the ascent of the magma to the surface is unknown, but "an optimistic

**Table 6-3** Logarithm of the partial pressures of various species in Nordlie's best estimate of Kilauea magmatic gas; total pressure is 1 atmosphere

| | 800°C | 900°C | 1000°C | 1100°C | 1200°C |
|---|---|---|---|---|---|
| CO | −3.19 | −2.77 | −2.44 | −2.19 | −2.02 |
| $CO_2$ | −0.44 | −0.44 | −0.44 | −0.45 | −0.45 |
| $H_2$ | −3.28 | −3.00 | −2.79 | −2.63 | −2.53 |
| $H_2O$ | −0.57 | −0.57 | −0.57 | −0.57 | −0.57 |
| $S_2$ | −2.63 | −2.65 | −2.74 | −2.91 | −3.24 |
| $SO_2$ | −0.44 | −0.44 | −0.44 | −0.44 | −0.44 |
| $SO_3$ | −7.00 | −6.65 | −6.33 | −6.03 | −5.72 |
| $O_2$ | −12.94 | −11.43 | −10.12 | −8.95 | −7.84 |
| $Cl_2$ | −13.63 | −13.12 | −12.68 | −12.27 | −11.87 |
| HCl | −3.53 | −3.53 | −3.53 | −3.53 | −3.52 |
| $H_2S$ | −2.77 | −2.87 | −3.01 | −3.21 | −3.53 |
| $CH_4$ | −9.68 | −9.36 | −9.14 | −9.03 | −9.07 |
| COS | −9.16 | −9.14 | −9.20 | −9.36 | −9.70 |
| $CS_2$ | −8.54 | −8.48 | −8.52 | −8.72 | −9.18 |

viewpoint would suggest that the derived magmatic composition is actually close to the juvenile gas composition present at depth. However, changes brought about by increased pressure, varying solubilities in the magma, contamination at depth, et cetera, must be evaluated before this becomes a viable conclusion" (Nordlie, 1971, p. 460). Decreasing pressure, such as would occur in ascending magma, tends to drive reactions toward the side with the larger number of moles of gas. Heald et al. (1963), using a different estimate of Kilauea gas composition, showed that (at 1227°C) $SO_2$ was replaced by $H_2S$ as the dominant sulfur gas at high pressures.

There is often some discrepancy between the calculated equilibrium temperature of the analyzed gas and the temperature of collection. This discrepancy could result from a failure to quench the high-temperature equilibrium or from atmospheric contamination of the magmatic gas at the collection site, which would disturb the equilibrium composition. Nordlie's best estimate of the composition of magmatic gas, corrected for contamination, has a composition that indicates equilibration at approximately 1150°C. Heald et al. (1963) found that equilibrium temperatures were between 100 and 300° lower than the collection temperatures; but Matsuo (1960) found the reverse relationship at Usu volcano, where it appears that the equilibrium compositions of the Showa-shinzan gases are higher than their collection temperatures—as if high-temperature equilibrium has been quenched. Mueller (1970) has also postulated similar high-temperature equilibration with respect to HF/HCl, the gas composition remaining unchanged after separating from the magma at liquidus temperatures.

We shall use the calculated equilibrium partial pressures of the Kilauea gas components (Fig. 6-1), which at low pressure (1 atmosphere) and high

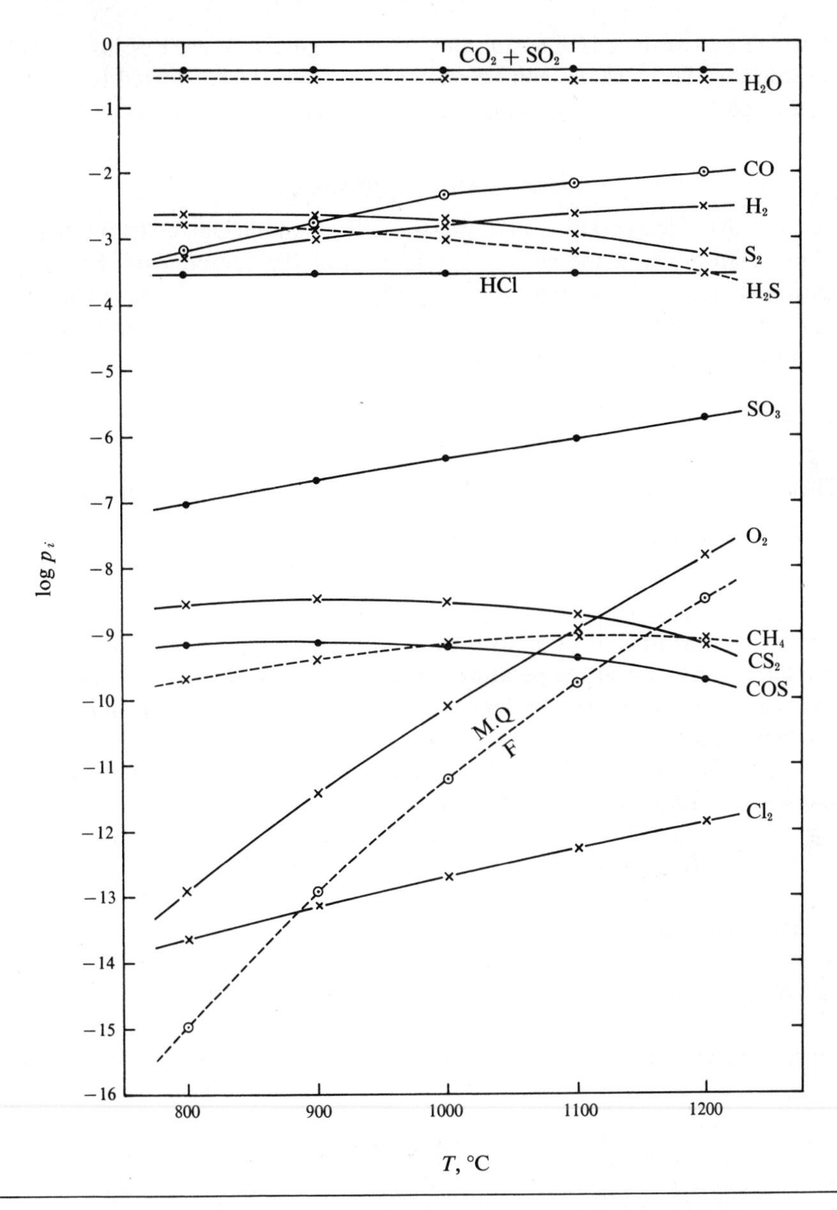


**Fig. 6-1** Calculated change in equilibrium partial pressure with temperature for the more abundant species of a Kilauea gas sample (data from Nordlie, 1971). The fugacity of oxygen ($f_{O_2}$) defined by the fayalite-magnetite-quartz buffer is shown as a dashed line. Total pressure is 1 atmosphere.

temperature can be assumed to equal their respective fugacities, as typical of basaltic gas. These values can be related in turn to the calculated fugacities of various components using free-energy data for solid-gas buffers.

## Solid-Gas Equilibrium

At any given temperature and pressure, the equilibrium constant fixes the relationship between the chemical potentials of the components. Equilibrium constants are usually written in terms of fugacities for the volatile components and activities for the solid and liquid components. As an example, take the reaction

$$\underset{\text{fayalite}}{3Fe_2SiO_4} + \underset{\text{gas}}{O_2} = \underset{\text{magnetite}}{2Fe_3O_4} + \underset{\text{quartz}}{3SiO_2} \tag{6-19}$$

Then the equilibrium constant $K$ at fixed temperature and pressure is equal to

$$K_{P,T} = \frac{(a_{Fe_3O_4}^{\text{titanomag}})^2 (a_{SiO_2}^{\text{quartz}})^3}{(a_{Fe_2SiO_4}^{\text{ol}})^3 f_{O_2}} \tag{6-20}$$

Obviously if crystals of pure fayalite, magnetite, and quartz are in equilibrium at a fixed pressure and temperature, then the fugacity of oxygen $f_{O_2}$ is constant and we have an oxygen buffer—commonly abbreviated to FMQ (fayalite-magnetite-quartz).

The value of $K$ in Eq. (6-20) may be experimentally determined (Wones and Gilbert, 1969) or it may be calculated from the relationship

$$\Delta G = \Delta G_r^\circ + RT \ln K \tag{6-21}$$

At equilibrium $\Delta G = 0$, so that

$$\Delta G_r^\circ = -RT \ln K \tag{6-21a}$$

By rearrangement, Eqs. (6-20) and (6-21*a*) can be written as

$$\log f_{O_2} = \frac{\Delta G_r^\circ}{2.303RT} + \log \frac{(a_{Fe_3O_4}^{\text{titanomag}})^2 (a_{SiO_2}^{\text{quartz}})^3}{(a_{Fe_2SiO_4}^{\text{ol}})^3} \tag{6-22}$$

where $\Delta G_r^\circ$ is the standard free energies of formation ($\Delta G_f^\circ$) of the products less those of the reactants. Since the solid phases are all in their standard states, namely, pure solids at the temperature of interest and 1 bar, their activities are unity, and the logarithmic term on the right of the equation is zero.

The use of solid-phase oxygen buffers in experimental geochemistry was pioneered by Eugster (Eugster and Wones, 1962), and a summary of the range of gas composition (up to four components) over which solid buffers can be used has been given by Eugster and Skippen (1967). Coexisting solid phases at equilibrium can fix the chemical potential of a volatile component in nature just as in a laboratory experiment. Because the assemblage of fayalite, magnetite, and quartz is not uncommon in rhyolitic obsidians, it is convenient to plot the values of $f_{O_2}$ for Eq. (6-22) over a range of temperature as a common standard of reference (as in some of the diagrams in this chapter). In most igneous rocks, however, neither olivine nor magnetite is found as a pure solid, so that the activities of the $Fe_2SiO_4$ and $Fe_3O_4$ components (in a *stable* phase) are smaller than those of the pure phases. In addition, quartz is not present in most igneous rocks, and to take account of this the more convenient standard state of $SiO_2$ glass is substituted for quartz (Chap. 2). Although it is experimentally convenient to have a gas phase when using the oxygen buffer, the chemical potential of oxygen, $\mu_{O_2}$, is fixed by the equilibrium assemblage of fayalite, magnetite, and quartz irrespective of the existence of a gas or liquid phase.

The effect of pressure on the equilibrium fugacity of oxygen has been treated in Chap. 3, as has also the distinction between an assemblage which defines rather than buffers the fugacity of a volatile component. Any assemblage of solid phases whose compositions are fixed in experimental reactions is a buffer, but where the solid phases have variable composition as in nature (that is, in solid solutions), the assemblage does not buffer, but defines, the fugacity of the volatile component.

It seems to be a matter of considerable importance to determine whether or not a particular igneous mineral assemblage buffered the chemical potential of a volatile or mobile component, or, alternatively, equilibrated with this component whose potential was externally controlled. We thus find the same perplexing problems of internal and external buffers in igneous rocks that are familiar to the metamorphic petrologist. This point is taken up again below, particularly with regard to the chemical potential of oxygen, which can greatly influence the crystallization path of a basaltic magma.

## Contributing Components (S, P, $CO_2$, Cl, F, $H_2O$, $O_2$)

### SULFUR IN MAGMAS AND IGNEOUS ROCKS

There is little information on the solubility of sulfur in magma; Skinner and Peck (1968) found 380 ppm by weight in a quenched tholeiitic glass at 1065°C. One of the writers determined the sulfur content of residual glass in rhyolitic obsidians which contained rare crystals of pyrrhotite—150 to 85 ppm for a temperature range of 925 to 850°C. The solubility of sulfur in silicate liquids falls with temperature. It can be reduced by adding silica, an industrial process which Vogt (1921) used as an analogy for the formation of immiscible sulfide liquids in nature. In general the solubility of sulfur in silicate liquids is controlled essentially by the activity of the FeO component; addition of silica or any other oxide, or an increase in the oxidation state (increased oxygen content), will depress the solubility of sulfur in a silicate liquid (Maclean, 1969).

Although several surveys of sulfide minerals in igneous rocks have been made (for example, Newhouse, 1936), there is an acute need for more detailed information on these minerals in volcanic rocks or in slowly cooled differentiated intrusions of diverse composition. Investigations by Desborough (1967) and by Wager et al. (1957) on sulfide minerals in fractionated basaltic magma are but a beginning. As a broad generalization, pyrrhotite ($Fe_{1-x}S$) appears to be the dominant primary sulfide of igneous rocks, but it is often accompanied by varying amounts of chalcopyrite ($CuFeS_2$), pentlandite [$(Fe,Ni)_9S_8$], or sphalerite (ZnS).

In the Alae tholeiitic lava lake of Hawaii, two sulfide phases are found; one is a copper-rich pyrrhotite solid solution and the other was an immiscible sulfide liquid which crystallized to a mixture of chalcopyrite, pyrrhotite, and magnetite (Skinner and Peck, 1969). Comparable assemblages with rather more copper but less oxygen have been found in tholeiitic basalts from Kilauea (Desborough et al., 1968). It does not seen unreasonable to conclude that a pyrrhotite solid solution, perhaps with notable amounts of Cu and Ni, is the predominant phase crystallizing in basaltic magma; certainly the evidence of the large economic sulfide masses associated with basic magma, as at Sudbury, Canada (Naldrett and Kullerud, 1967), does not controvert this.

One of the most fruitful experimental studies in mineral equilibria is that of Toulmin and Barton (1964) on a part of the Fe–S system. They were able to relate sulfur fugacity $f_{S_2}$ to the composition of pyrrhotite, expressed as a mole fraction of troilite (FeS), over a range of temperature. Thus the composition of a pyrrhotite can be used to define $f_{S_2}$ if the equilibration temperature is known. If pyrrhotite is accompanied by pyrite in equilibrium, the composition of pyrrhotite can be used to obtain both $f_{S_2}$ and temperature up to a maximum of 742°C, the incongruent melting temperature of pyrite. However, there are virtually no published data for granites or rhyolites which one could apply to this geothermometer, presumably because sulfide minerals are prone to react or reequilibrate at low temperatures; for example, pyrrhotite with substantial amounts of Cu will unmix to cubanite and chalcopyrite solid solutions (Yund and Kullerud, 1966).

The Rabaul caldera of eastern New Britain has been draped by a series of recent pumice flows ranging from andesite to rhyolite, and in each flow are small crystals of pyrrhotite enclosed by titanomagnetite (Heming and Carmichael, 1973). The analyses of these pyrrhotites, recalculated in terms of mole fraction FeS, are plotted in Fig. 6-2, which is taken from Toulmin and Barton (1964). Using an estimate of temperature derived from the coexisting iron titanium oxides, values of $\log f_{S_2}$ can be obtained. In all these pumice flows, the pyrrhotite precipitated before the iron titanium oxides, so that there is some uncertainty about using the iron-titanium-oxide equilibration temperature; but despite this, there is a general increase in $\log f_{S_2}$ from approximately $-2.0$ to $+0.1$ with temperature.

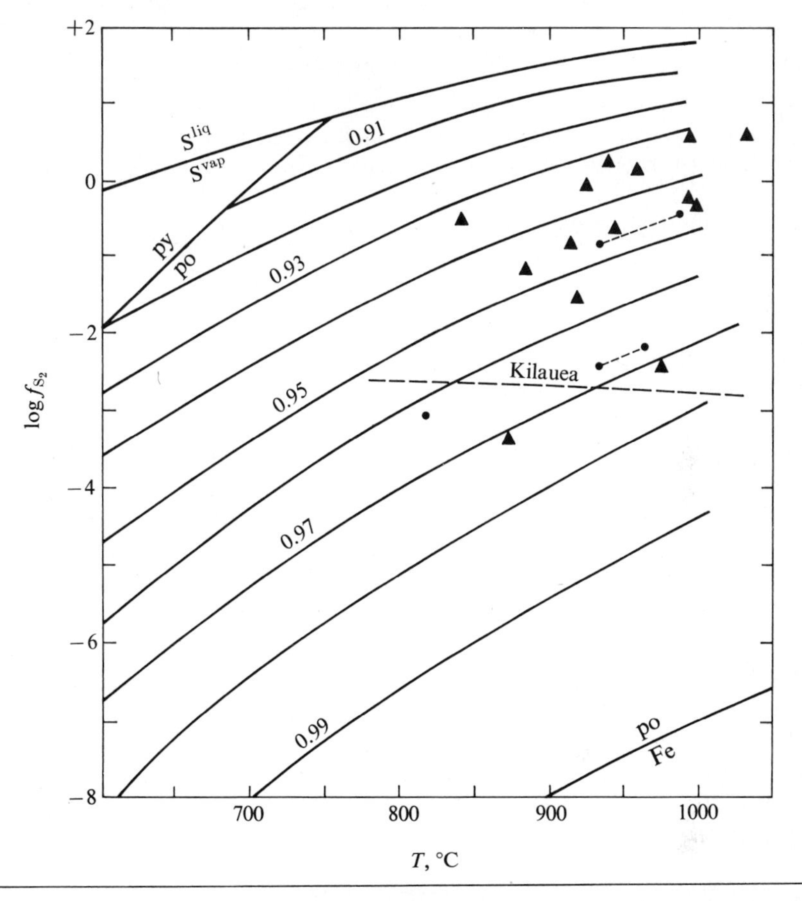


**Fig. 6-2** Composition of pyrrhotite (mole fraction FeS) as a function of $f_{S_2}$ and $T$ (Toulmin and Barton, 1964). Compositions of pyrrhotite in the Rabaul series of andesite-dacite-rhyolite pumice flows are shown as triangles, compositions from the rhyolite extrusions of Mono Craters, California, as filled circles. Values of $p_{S_2}$ for Kilauea gas (Fig. 6-1) are also shown. po is pyrrhotite, py is pyrite, $S_{liq}$ and $S_{vap}$ are sulfur as liquid and as vapor, respectively.

For lavas that contain olivine, magnetite, and pyrrhotite, $\log f_{S_2}$ can be estimated from the reaction

$$4Fe_2SiO_4 + S_2 = 2FeS + 2Fe_3O_4 + 4SiO_2 \tag{6-23}$$

which at equilibrium can be recast to

$$\log f_{S_2} = \frac{\Delta G_r^\circ}{2.303RT} + \log \frac{(a_{FeS}^{pyrr})^2 (a_{Fe_3O_4}^{titanomag})^2 (a_{SiO_2}^{liq})^4}{(a_{Fe_2SiO_4}^{ol})^4} \tag{6-24}$$

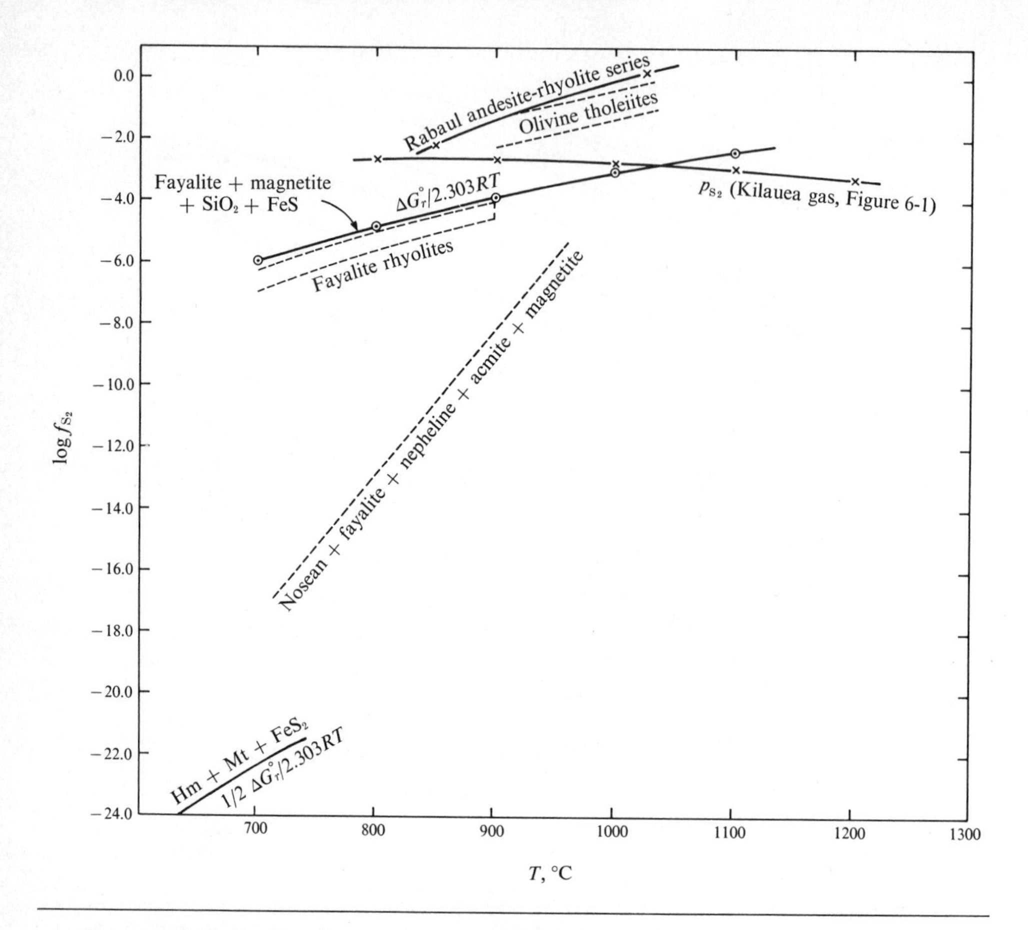


**Fig. 6-3** Variation of $\Delta G_r^\circ/2.303RT(= \log f_{S_2})$ with temperature for the reactions $4Fe_2SiO_2 + S_2 = 2FeS + 2Fe_3O_4 + 4SiO_2$ (FMQ + FeS) and $3Fe_3O_4 + S_2 = FeS_2 + 4Fe_2O_3$ (HM + $FeS_2$) with all components in their standard states; note that the scale is halved for the second reaction. Also shown is the $p_{S_2}$ curve of the Kilauea gas (Fig. 6-1) and $f_{S_2}$ curve for the Rabaul series (Heming and Carmichael, 1973). The fields of $f_{S_2}$ for average olivine tholeiite and fayalite rhyolite obtained by substituting typical mineral data in Eq. (6-24) are plotted together with the calculated variation of $f_{S_2}$ for a nosean-bearing nephelinite obtained by substituting known mineral compositions in Eq. (6-27). Total pressure is 1 bar, except for the Rabaul series, where total pressure is unknown.

The calculated variation of $f_{S_2}$ with temperature (at 1 bar pressure) for all the components in their standard states (unit activity) has been plotted in Fig. 6-3; this calculated curve cuts the Kilauea $p_{S_2}$ gas curve (Fig. 6-1) at about 1025°C.

If mineral compositions typical of both olivine tholeiitic basalts and fayalite rhyolites are substituted for the activity terms ($X_{FeS} = 0.97$ in both cases, and appropriate values for $\log a_{SiO_2}^{liq}$), then the range of $\log f_{S_2}$ for both lava types can be calculated (Fig. 6-3). Acid plutonic rocks which contain the assemblage pyrite + hematite + magnetite in equilibrium have extremely low values of $f_{S_2}$ in comparison with fayalite rhyolites (Fig. 6-3; note half scale).

Some confirmation of the calculated values of $f_{S_2}$ at temperatures close to 1050°C can be obtained from the bulk composition of the immiscible sulfide liquid in the Alae tholeiitic lava lake (Skinner and Peck, 1969), which crystallized to a mixture of pyrrhotite, chalcopyrite, and titanium-poor magnetite. Naldrett (1969) determined the liquidus surface over part of the pyrrhotite field in the system Fe–S–O, which is reproduced in Fig. 6-4. Also plotted in Fig. 6-4 is the estimated bulk composition of the Alae sulfide liquid, the proportions of Cu (4 percent) being included with Fe because Naldrett has shown that small amounts of both Cu and Ni affect the pyrrhotite liquidus to much the same extent as comparable amounts of Fe. An estimated liquidus temperature of 1060°C for the Alae lake sulfide liquid is indicated, as is a value of $f_{S_2}$ of about $10^{-4}$, which is in fair agreement with the data of Fig. 6-3. The estimated range of composition for the sulfide minerals in Hawaiian lavas (Desborough et al., 1968), also shown in Fig. 6-4, are notably richer in Cu and poorer in $O_2$, and would have liquidus temperatures well in excess of 1100°C.

The occurrence of sulfate in igneous rocks is often considered to require relatively high oxygen fugacities, or values higher than those usually associated with sulfide minerals, which in the Kilauea tholeiitic basalts must have been close to those defined by the FMQ buffer (Figs. 6-1 and 6-12). Nosean, the common magmatic sulfate mineral, occurs in phonolites and nepheline syenites, but its more abundant calcium counterpart hauyne has a wider distribution, being found in alkali olivine-basalt bombs, Italian leucite trachytes, and as phenocrysts in the etindite (nephelinite) lavas of Mt. Cameroun. In all these lavas, we can reasonably represent the paragenesis of nosean (or hauyne) by the following reaction (Stormer and Carmichael, 1971*a*):

$$\underset{\text{nosean}}{Na_8Al_6Si_6O_{24}SO_4} + \underset{\text{magnetite}}{Fe_3O_4} + \underset{\text{glass}}{16SiO_2}$$

$$= \underset{\text{troilite}}{FeS} + \underset{\text{acmite}}{2NaFeSi_2O_6} + \underset{\text{albite}}{6NaAlSi_3O_8} + \underset{\text{gas}}{2O_2} \tag{6-25}$$

By LeChatelier's principle, any reduction of silica activity will favor the assemblage nosean-magnetite at the expense of pyrrhotite-acmite if $f_{O_2}$ remains constant. Perhaps nosean is to phonolitic liquids what pyrrhotite is to rhyolitic liquids of higher silica activity. For the purposes of estimating sulfur fugacity for a nosean assemblage, we may write

$$\underset{\text{magnetite}}{4Fe_3O_4} + \underset{\text{acmite}}{4NaFeSi_2O_6} + \underset{\text{nepheline}}{12NaAlSiO_4} + S_2$$

$$= \underset{\text{fayalite}}{8Fe_2SiO_4} + \underset{\text{nosean}}{2Na_8Al_6Si_6O_{24}SO_4} \tag{6-26}$$

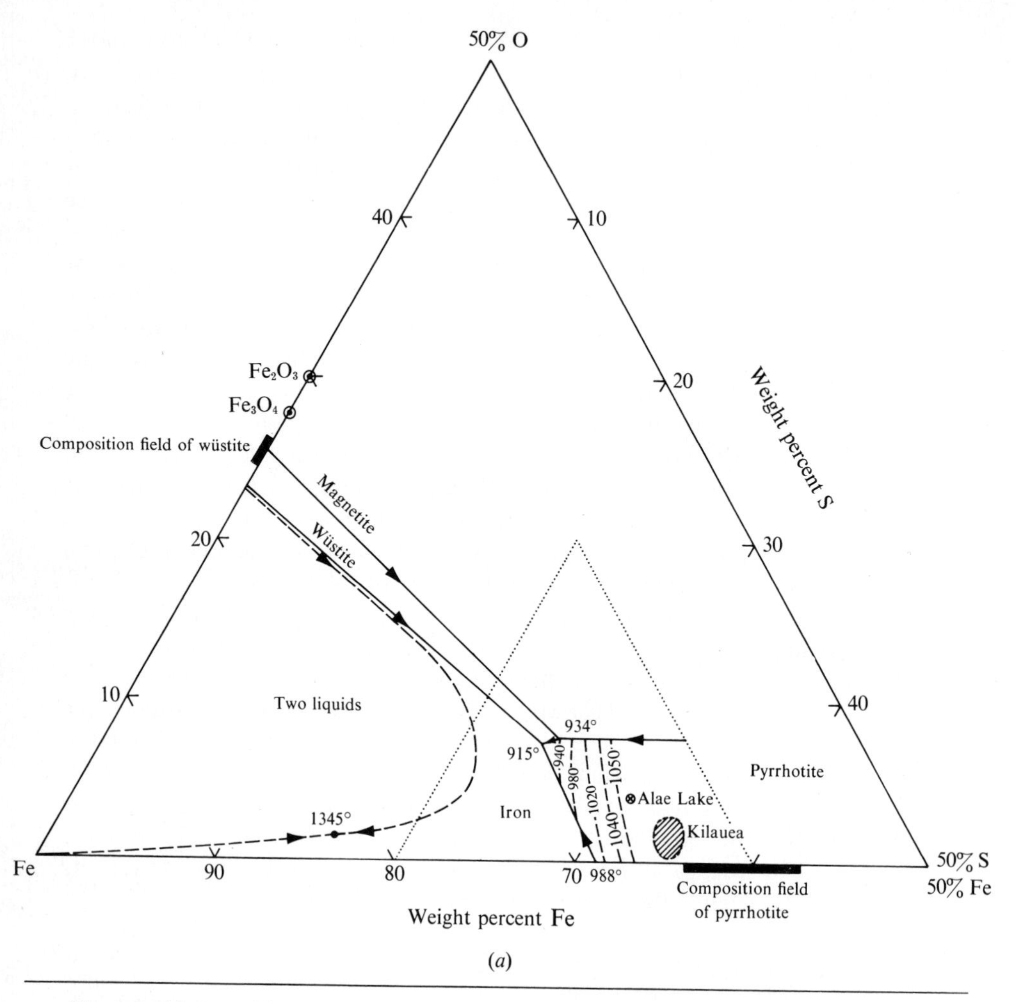


**Fig. 6-4** (*a*) Iron-rich corner of the system Fe–S–O (after Naldrett, 1969) showing the pyrrhotite field bounded by the fields of magnetite, wüstite, and iron, and a region of liquid immiscibility. Isothermals in dotted triangle, °C. (*b*) On opposite p. Triangle shown dotted in (*a*). Isothermals on the liquidus surface (°C) are shown dashed; isobars for minus log $f_{S_2}$ at 1050°C are dotted. The estimated composition of the immiscible sulfide liquid, Alae lava lake, suggests a liquidus temperature 1060°C, $f_{S_2} \sim 10^{-4}$ bar. Estimated compositions of sulfide liquids in other Kilauean tholeiitic lavas are also shown. (Data from Skinner and Peck, 1969; Desborough et al., 1968.)

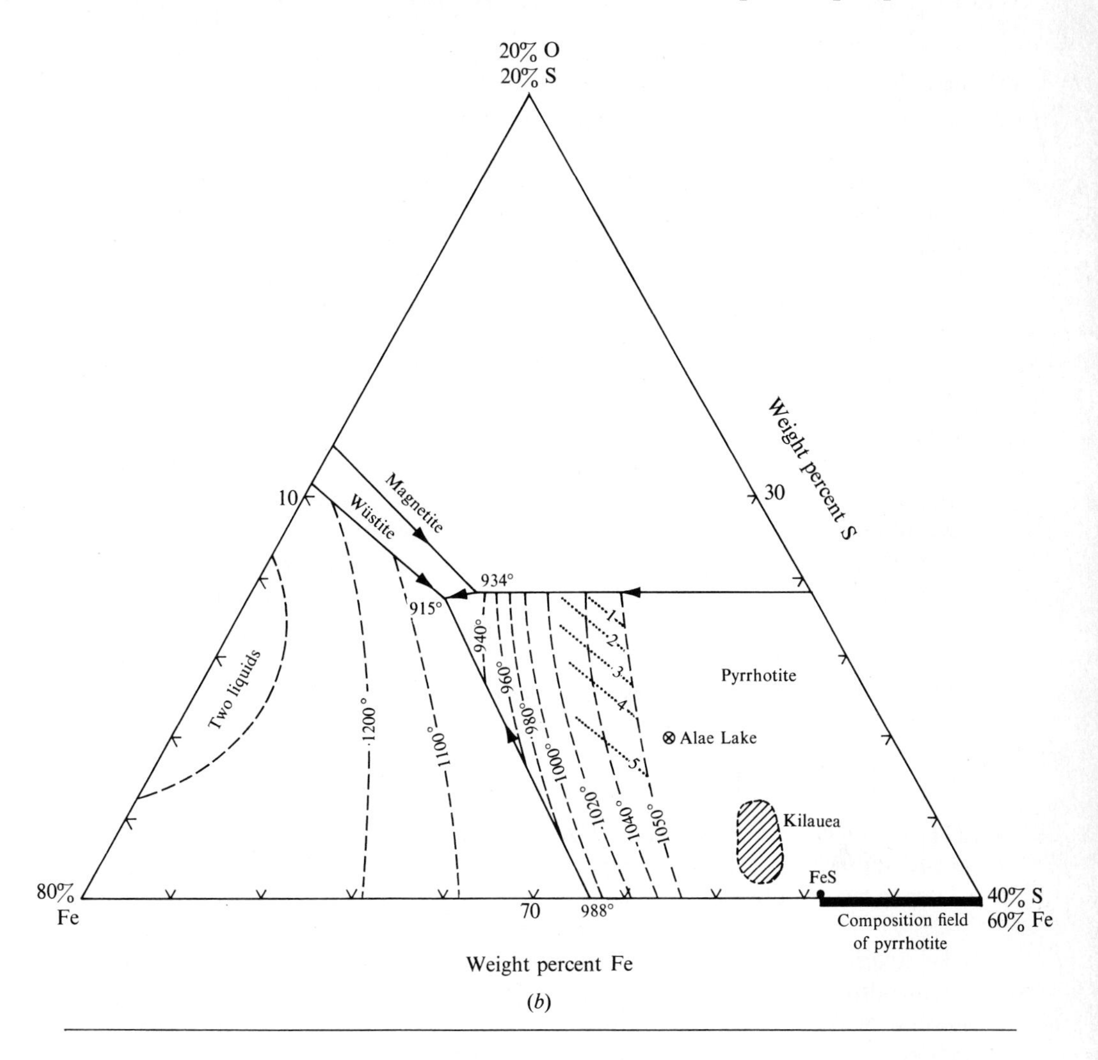


(*b*)

which can be rearranged, assuming equilibrium, to

$$\log f_{S_2} = \frac{\Delta G_r^\circ}{2.303RT} + \log \frac{(a_{Fe_2SiO_4}^{ol})^8 (a_{nosean}^{nosean})^2}{(a_{Fe_3O_4}^{titanomag})^4 (a_{NaFeSi_2O_6}^{pyrox})^4 (a_{NaAlSiO_4}^{neph})^{12}} \tag{6-27}$$

which is a representation of the only olivine-nosean-nepheline nephelinite for which we have data. If values appropriate to the observed mineral compositions are substituted for the activity terms in Eq. (6-27), then $f_{S_2}$ varies with temperature as shown in Fig. 6-3. It is perhaps fortuitous that this calculated $f_{S_2}$ curve cuts the Kilauea $p_{S_2}$ curve at about 1000°C, a temperature which cannot be far wrong for a nephelinite lava.

Schneider (1970) found more sulfate-sulfur in alkali olivine basalts, basanites, and nephelinites than in tholeiitic basalts, which supports the premise that silica activity (or the bulk composition of the liquid), as well as $f_{O_2}$, controls the sulfate-sulfide equilibrium [Eq. (6-25)]. However, for those lavas whose silica activities are not low enough to stabilize nosean or hauyne, it is apatite that, in the absence of a sulfate or sulfide mineral, is the principal repository of sulfur (presumably as sulfate).

To conclude this section on sulfur in magmas: it has been shown by J. G. Moore and Fabbi (1971) that subaerial basaltic lavas contain on average 107 ppm sulfur, whereas the chilled glassy selvage of a submarine lava contains about 800 ppm S. This latter amount is considered to be a reasonable estimate of the juvenile content of sulfur in basaltic magma derived from the mantle.[1] Subaerial basaltic lavas must degas close to or on the surface, and this loss of $SO_2$, the dominant sulfur gas species in basaltic magma, is believed to account for the decrease in $f_{O_2}$ in Kilauea lavas at the time of eruption (A. T. Anderson and Wright, 1972). There is much uncertainty about the role of sulfur in the generation of basaltic magmas. Intuitively it would seem likely that magmas generated at considerable depths in the mantle would become purged of their sulfide components on their ascent to the surface. The high-density sulfide liquid would deplete the magma in Cu and Ni and perhaps in oxygen, which, if the liquidus trends of the system $FeS–Fe_3O_4–FeO–SiO_2$ (Maclean, 1969) are applicable, could in turn increase the solubility of sulfur.

## PHOSPHORUS FUGACITY

The Showa-shinzan gases (Table 6-2) provided the only analyses compiled by White and Waring (1963) that report values for phosphorus; yet apatite comes closer to being a ubiquitous igneous mineral than any other, so that all volcanic gases will contain a phosphorus component. In the tholeiitic lava of Alae lava lake, apatite started to crystallize at approximately 1000°C (Peck et al., 1966); apatite also became one of the accumulating phases in the later stages of crystallization of the Skaergaard intrusion (Wager and Brown, 1967). For lavas or gabbros which contain olivine, or indeed even for rhyolites, the following reaction may be used to estimate phosphorus fugacity:*

$$\underset{\text{magnetite}}{\tfrac{15}{2}Fe_3O_4} + \underset{\text{forsterite}}{\tfrac{1}{2}Mg_2SiO_4} + \underset{\text{diopside}}{2CaMgSi_2O_6} + \underset{\text{anorthite}}{3CaAl_2Si_2O_8} + \underset{\text{glass}}{\tfrac{3}{4}SiO_2}$$

$$+ \underset{\text{steam}}{\tfrac{1}{2}H_2O} + \underset{\text{gas}}{3P} = \underset{\text{apatite}}{Ca_5(PO_4)_3OH} + \underset{\text{spinel}}{3MgAl_2O_4} + \underset{\text{fayalite}}{\tfrac{45}{4}Fe_2SiO_4} \qquad (6\text{-}28)$$

[1] Recent data on submarine basalts from the Reykjanes ridge, Iceland, give an average value of 843 ppm S in basaltic magma (Moore and Schilling, 1973).

* This reaction assumes that the apatite of igneous rocks is a hydroxyl variety, although all the evidence suggests that it is predominantly fluorapatite.

which at equilibrium gives

$$\log f_{\mathrm{P}} = \frac{1}{3}\frac{\Delta G_r}{2.303RT}$$

$$+ \frac{1}{3}\log \frac{a^{\mathrm{ap}}_{\mathrm{Ca_5(PO_4)_3OH}}(a^{\mathrm{mag}}_{\mathrm{MgAl_2O_4}})^3(a^{\mathrm{ol}}_{\mathrm{Fe_2SiO_4}})^{45/4}}{(a^{\mathrm{mag}}_{\mathrm{Fe_3O_4}})^{15/2}(a^{\mathrm{ol}}_{\mathrm{Mg_2SiO_4}})^{1/2}(a^{\mathrm{pyrox}}_{\mathrm{CaMgSi_2O_6}})^2(a^{\mathrm{plag}}_{\mathrm{CaAl_2Si_2O_5}})^3(a^{\mathrm{liq}}_{\mathrm{SiO_2}})^{3/4}(f_{\mathrm{H_2O}})^{1/2}} \tag{6-29}$$

If typical values for the composition of olivine, pyroxene, titanomagnetite, and plagioclase are substituted for the activity terms, and if $f_{\mathrm{H_2O}}$ is assumed to be between 1 and 500 bars, then "zones" of $\log f_{\mathrm{P}}$ can be delineated for the average olivine tholeiite, basanite, and fayalite rhyolite (Fig. 6-5).

Unfortunately there is little correspondence between the calculated values and the phosphorus found at 750°C in the Showa-shinzan fumarolic gas (Table 6-2). This discrepancy could arise because the passage of hot acid gas extracts phosphorus from wall rock, as Murata (1966) showed at Kilauea, and thus the gas no longer represents a composition determined by equilibrium with the

**Fig. 6-5** Variation of $\log f_{\mathrm{P}}$ with temperature for the (standard-state) reaction involving apatite and olivine [Eqs. (6-28) and (6-29)]. The zone of various types of olivine-bearing basalt is restricted to temperatures below 1020°C, the temperature at which apatite is observed to crystallize in Kilauea. The fields of both fayalite rhyolite and basalt are obtained by substituting typical mineral compositions in Eq. (6-29). The estimated $f_{\mathrm{P}}$ in equilibrium with the content of $P_4O_{10}$ in the Showa-shinzan gas at 750°C (Table 6-2) is also shown.

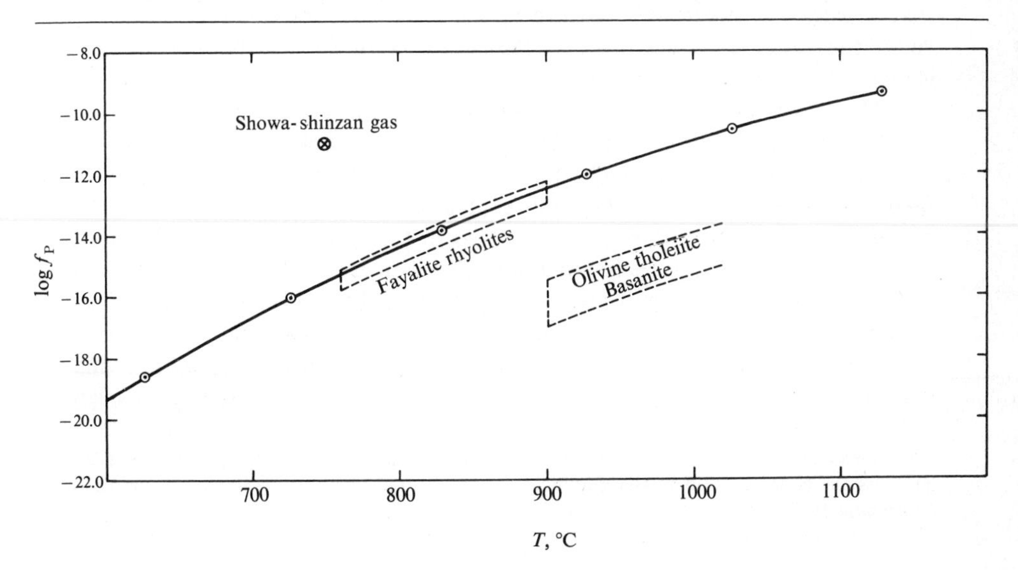

magma. For magmas of low silica activity, carbonatite and leucite syenite (Nash 1972*a*, 1972*b*, 1973), the calculated values of $\log f_P$ are very small.

## CARBON DIOXIDE

Carbon dioxide is usually the predominant component of the "active gas" fraction (D. E. White and Waring, 1963), although in several craters of the Kamchatka volcanoes CO is more abundant. Carbon dioxide has minimal solubility in granitic liquids[1] or synthetic liquids corresponding to a simplified syenite (Wyllie and Tuttle, 1959*a*), but the field association of carbonatite with nepheline syenite, alnoite, and kimberlite suggests that $CO_2$ becomes more soluble as silica activity decreases. The most striking evidence of the insolubility of $CO_2$ in magmas is shown by the olivine phenocrysts of basaltic rocks which contain myriads of inclusions filled with a $CO_2$-rich fluid phase (Roedder, 1965). Estimates of their density indicate that this fluid, once a separate, dispersed phase in the basaltic liquids, was at pressures of between 2500 and 5000 bars when it was incorporated by the growing olivine crystals (Fig. 6-6). These estimates, at temperatures of 1100 to 1200°C, are minimal because the denser inclusions would tend to leak and decrepitate as they are carried into a low-pressure environment. Roedder found no certain indication of the substantial amounts of CO which, for the Kilauea gas samples (Fig. 6-1), would accompany the $CO_2$, but he suggested that the dark margins of many inclusions may be graphite formed during cooling by dissociation of CO to carbon and $CO_2$.

Vesicular submarine lavas dredged from the Kilauea east rift zone (Moore, 1965) at depths down to 4000 m contain too little water (0.45 percent by weight) to exsolve below 800 m at magmatic temperatures (cf. Fig. 6-9). Moore suggests that vesicles found below this depth were caused by the exsolution of a $CO_2$ gas phase—another indication that the solubility of $CO_2$ in basaltic magma is low.

The only igneous suite of silicate composition in which a carbonate mineral is often accounted a primary phase includes kimberlite. Its characteristic brecciated appearance is widely considered to result from the intrusion of a fluidized stream of solid and gas; this explosive stream incorporates fragments of rock through which it passes, and it so disturbs the overlying rocks that fragments of logs and coal founder into the kimberlite (G. C. Kennedy and Nordlie, 1968). Kennedy and Nordlie consider that $CO_2$ is the main propellant of kimberlitic material from its source deep in the mantle, and they suggest that the shape and size of the diamonds could be acquired only by growing from a melt. Kimberlite contains olivine in which $CO_2$ fluid inclusions have been

[1] The effect of small amounts of $CO_2$ in an aqueous fluid in equilibrium with granite is to suppress the concentration of the granitic components, particularly the alkalis, in the fluid (Burnham, 1967).

found; it is also intruded by veins of carbonatite[1] whose isotopic composition ($O^{18}/O^{16}$ and $C^{13}/C^{12}$) indicates magmatic equilibration with diamond (H. P. Taylor et al., 1967). Calcite may therefore have crystallized in equilibrium with a kimberlite assemblage at a pressure where diamond is stable; the inclusions of coesite in diamond (Meyer, 1968) restrict the pressure estimates below the stishovite field.

Olivine, diopside, and magnesian ilmenite are typical of kimberlites, so that this assemblage can be represented by the reaction

$$CO_2 + \underset{\text{diopside}}{CaMgSi_2O_6} + Fe_2O_3 + 3MgTiO_3 = \underset{\text{ilmenite solid solution}}{3FeTiO_3}$$

$$+ \underset{\text{forsterite}}{2Mg_2SiO_4} + \underset{\text{calcite}}{CaCO_3} + \tfrac{3}{4}O_2 \tag{6-30}$$

and, at equilibrium, Eq. (6-30) may be rewritten as

$$\log f_{CO_2} = \frac{\Delta G_r^\circ}{2.303RT} + \log \frac{(a_{FeTiO_3}^{ilm})^3 (a_{Mg_2SiO_4}^{ol})^2 a_{CaCO_3}^{calc} (f_{O_2})^{3/4}}{(a_{MgTiO_3}^{ilm})^3 a_{Fe_2O_3}^{ilm} a_{CaMgSi_2O_6}^{pyrox}} \tag{6-31}$$

Dawson (1962*b*) and Nixon et al. (1963) provide data for the relevant mineral compositions to be substituted in Eq. (6-31); $f_{O_2}$ has been assumed to be that defined by the FMQ buffer, which is rather lower than another estimate based on the occurrence of perovskite (Carmichael and Nicholls, 1967). The calculated variation of $f_{CO_2}$ is plotted in Fig. 6-6, but at the high pressures considered here some account should be taken of the $P\Delta V^{sol}$ term, which would increase the calculated values of $f_{CO_2}$; similarly any increase in $f_{O_2}$, say to that defined by the hematite-magnetite buffer, would also drastically increase $f_{CO_2}$. Obviously the calculated $f_{CO_2}$ curve for a kimberlite is only the crudest type of estimate; if a liquid phase were present, then provided that the mineral assemblage of Eq. (6-30) continued in equilibrium and maintained the same composition, $\log f_{CO_2}$ would remain as plotted in Fig. 6-6. A simple extrapolation of $\log f_{CO_2}$ cuts the graphite-diamond inversion curve at about 1300°C, a temperature also suggested by Boyd (1969) on the basis of the calcium-poor composition of kimberlitic diopside. However, since G. C. Kennedy and Nordlie (1968) maintain that the region of diamond generation growing from a liquid is dictated by the melting curve of garnet lherzolite (Fig. 6-6), further speculation on $f_{CO_2}$ seems profitless.

[1] Carbonatites in alkaline complexes typically have methane, fluorine compounds, and hydrocarbon gases associated with them that are considered to be magmatic (Heinrich and Anderson, 1965).

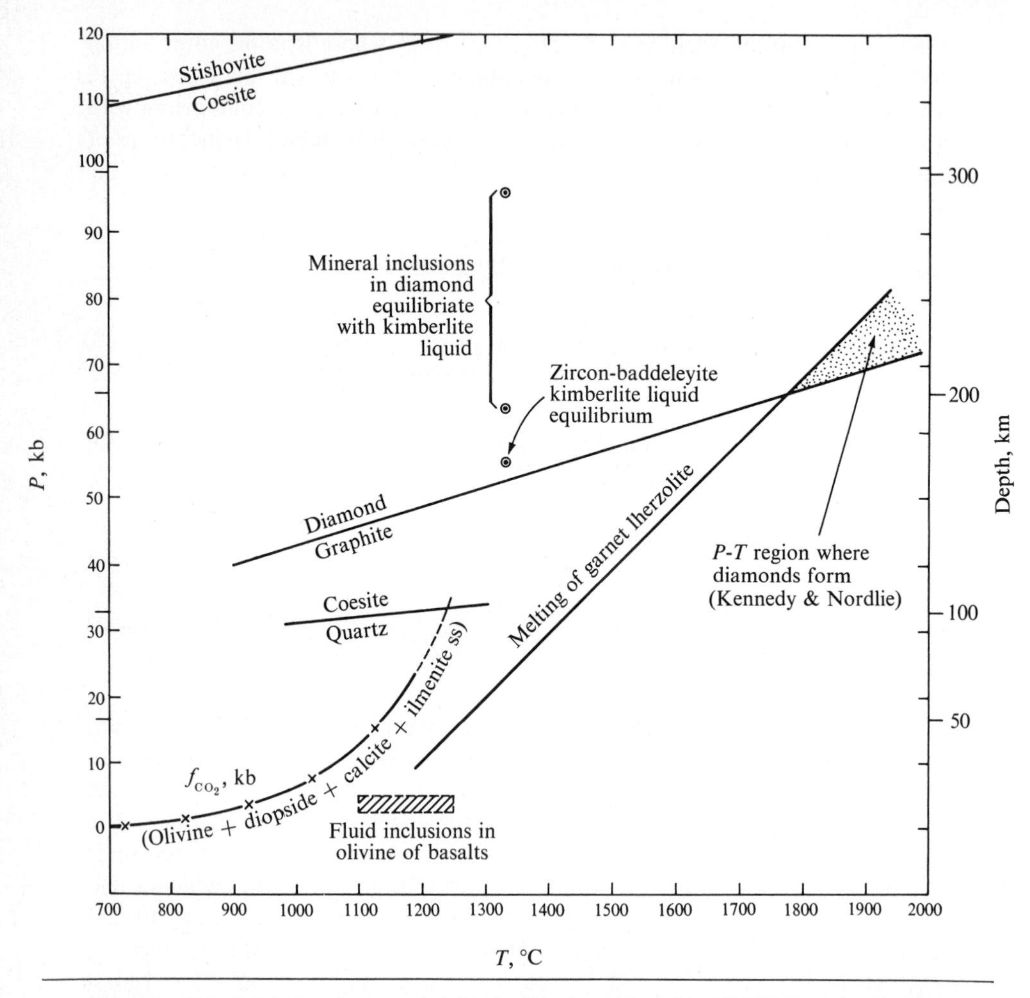


**Fig. 6-6** Calculated $f_{CO_2}$ for a typical kimberlite assemblage [Eq. (6-31)] ilmenite-diopside-calcite (neglecting $P\Delta V^{sol}$) as a function of temperature. Also shown are *PT* curves for coesite-quartz, graphite-diamond, and stishovite-coesite inversions and for melting of garnet lherzolite. Calculated points for equilibration of kimberlite liquid at 1327°C with mineral inclusions in diamond and with the pair zircon-baddeleyite (diamond absent) are from Fig. 3-8.

## THE HALOGENS

HALOGEN COMPONENTS IN THE GAS PHASE Only small amounts of halogen compounds are found in most volcanic gases, although at Katmai, Alaska, HCl was the predominant "active gas." HCl is at least three times more voluminous than HF in volcanic gases, but the proportion varies widely, and in the Showa-shinzan gases (Table 6-2) the ratio increases below 300°C. At magmatic temperatures (>600°C) and low pressures, HCl and HF are very weakly dissociated

(H. L. Barnes et al., 1966), but a cooling fumarolic gas rich in HCl will become strongly acid, and a Kilauea steam condensate of 2N HCl is a good example of this (Murata, 1966). The increasingly corrosive capacity of a cooling fumarolic gas will undoubtedly change its composition as it migrates through rocks on its way to the surface; and so low-temperature gases are likely to be unrepresentative of the original magmatic gas.

We shall use the volume fractions (equal to the partial pressures at STP) of HCl and HF in the Showa-shinzan gases (Table 6-2) as a basis for the subsequent calculation of HF and HCl fugacities. To make comparison easier, we have always included water as a component and taken its fugacity as unity, which is virtually the case of the Showa-shinzan gases, where the partial pressure of $H_2O$ is usually greater than 0.99 bar (Table 6-2). From the example of the Kilauea magmatic gas, much of this water may be a contaminant; however, reducing the fugacity to 0.3 bar changes only slightly the calculated fugacities of the halogen gases.

The contrast in behavior between HF and HCl in igneous rocks and magmas is very striking, and it can be illustrated by two groups of minerals. The first includes micas, amphiboles, and apatite, all of which can interchangeably substitute fluoride, hydroxide, or chloride in their structures. The second comprises such minerals as sodalite ($Na_4Al_3Si_3O_{12} \cdot Cl$), villiaumite (NaF), and fluorite ($CaF_2$); these show no such interchangeable substitution, but their presence in an igneous assemblage will buffer the fugacity of HF ($f_{HF}$) or of HCl ($f_{HCl}$). Some measure of this contrast between the halogens is found in an aqueous phase containing HF and HCl, which, if allowed to equilibrate with granite at liquidus temperatures, will become strongly partitioned; HF will be extracted into the silicate phases (liquid and solid), but HCl will remain exclusively in the aqueous fluid (Burnham, 1967). In lavas which are silica-poor, Cl and F may be completely partitioned between contiguous phases in the groundmass; thus in the leucite basanites of Vesuvius sodalite (with no F) coexists with both fluorphlogopite and fluorapatite (both with no Cl).

The preference of the hydrous minerals mica and apatite for fluoride, hydroxide, and chloride can be modeled by comparing the free energies of formation of the fluoride, hydroxide, and chloride compounds of Li, Mg, Al, and Ca (Stormer and Carmichael, 1971*b*). For example, the exchange reaction

$$Mg_{1/2}OH + HF = Mg_{1/2}F + H_2O \tag{6-32}$$

has $(\Delta G_r^\circ)_{298} = -17.6$ kcal, and similar reactions for Li, Al, and Ca all show that the fluoride ion has a strong preference for these cations in comparison to hydroxide. For reactions of the type

$$Mg_{1/2}OH + HCl = Mg_{1/2}Cl + H_2O \tag{6-33}$$

values of $(\Delta G_r^\circ)_{298}$ (also for Al and Ca) are all positive. This indicates that Cl will be rejected in preference to hydroxyl; however, the values for the comparable reactions for Na and K are negative and in addition suggest that Cl will combine with K in preference to Na. If we translate these inferences to an igneous environment, we should expect minerals which contain Li, Mg, Al, or Ca in octahedral coordination with OH, F, or Cl, namely micas and apatite, to incorporate F preferentially and to reject Cl, which will either enter an alkali compound or be expelled as part of an exsolved fluid phase. Analyses of micas confirm this conclusion, since lepidolite (Li-mica) invariably contains substantial amounts of fluorine while analyses of coexisting magnesian biotite and muscovite show preferential incorporation of F by biotite as suggested by the free-energy data.

Analyses of coexisting phlogopitic mica and apatite in volcanic assemblages show that between 80 and 90 percent of the (OH) group of apatite is filled with F, whereas in unaltered mica there is slightly less F (65 to 75 percent) in the hydroxyl group. Analyses of fresh igneous rocks will therefore closely represent the original F content of the magma, since it is preferentially incorporated by the hydrous minerals; this is exactly the opposite of the case for Cl, which is rejected by hydrous minerals if (OH) or F is present, and unless Cl is incorporated by some alkali compound, it will be expelled from the cooling magma. Chloride will be substantially retained by a crystalline assemblage only if sodalite is stable, as in trachytes or syenites with higher silica activities than nepheline-bearing assemblages. Retention of Cl will not be confined solely to rocks with nepheline (Stormer and Carmichael, 1971*b*).

THE CHLORIDE COMPONENT IN MAGMAS As a guide to the concentration of Cl in various igneous rock types, selected values are given in Table 6-4. Granite

**Table 6-4** Values (in ppm) of Cl in igneous rocks*

| | |
|---|---|
| Average granite | 200 |
| Average basalt | 60 |
| Average phonolite | 2300 |
| Peralkaline acid lavas | |
| Glassy pantellerite | 5600 |
| Crystalline pantellerite | 400 |
| Glassy comendite | 2400 |
| Crystalline comendite | 200 |
| Trachytes and leucite basanites | |
| Mt. Suswa, Kenya | |
| Glassy lava | 1700 |
| Crystalline lava | 2100 |
| Vesuvius | |
| Leucite basanites and tephrites | 4700 |

* Data taken from various sources and Stormer and Carmichael (1971*a*).

and basalt, which constitute the vast majority of crustal igneous rocks, have 200 ppm or less. One type of acid lava, the glassy peralkaline pantellerites and comendites, contains more than ten times as much Cl, but the low Cl of their crystalline counterparts shows that Cl is expelled during crystallization. In contrast, both the glassy and crystalline varieties of sodalite trachytes contain comparable amounts, and high levels of Cl are also typical of the sodalite-bearing leucite basanites and tephrites of Vesuvius.

The solubility of HCl in granitic magmas is probably very small, since their liquidus temperatures are increased by the addition of HCl to the aqueous phase (Wyllie and Tuttle, 1964). Moreover, in liquids of $NaAlSi_3O_8$ composition the addition of HCl to the vapor phase extracts Na, so that NaCl becomes a component of the fluid phase. NaCl is also very insoluble in $NaAlSi_3O_8$ liquids, and this leads to liquid immiscibility (Koster van Groos and Wyllie, 1969). A natural example of NaCl liquid immiscibility in granitic liquids has been found in the fluid inclusions of quartz in the peralkaline granite xenoliths of Ascension Island (Roedder and Coombs, 1967). The generation of a chloride-rich aqueous fluid, together with an immiscible NaCl liquid, would seem to be the normal product of crystallization of Cl-rich magmas of high silica activity; it is the alternative to the crystallization of sodalite in magmas of lower silica activity.

The fugacity of HCl for these contrasted assemblages can be calculated, since $f_{HCl}$ will be defined (at constant temperature and pressure) by either NaCl liquid or sodalite. One representative reaction is

$$\underset{\text{acmite}}{NaFeSi_2O_6} + \underset{\text{gas}}{HCl} = \underset{\text{liquid}}{NaCl} + \underset{\text{glass}}{2SiO_2} + \underset{\text{magnetite}}{\tfrac{1}{3}Fe_3O_4} + \underset{\text{gas}}{\tfrac{1}{12}O_2} + \underset{\text{steam}}{\tfrac{1}{2}H_2O} \tag{6-34}$$

which at equilibrium is recast to

$$\log f_{HCl} = \frac{\Delta G_r^\circ}{2.303RT} + \log \frac{(a_{SiO_2}^{liq})^2 a_{NaCl}^{liq} (a_{Fe_3O_4}^{mag})^{1/3} (f_{O_2})^{1/12} (f_{H_2O})^{1/2}}{a_{NaFeSi_2O_6}^{pyrox}} \tag{6-35}$$

A representative reaction for a sodalite-trachyte assemblage is

$$\underset{\text{acmite}}{NaFeSi_2O_6} + \underset{\text{nepheline}}{3NaAlSiO_4} + \underset{\text{gas}}{HCl}$$

$$= \underset{\text{sodalite}}{Na_4Al_3Si_3O_{12}Cl} + \underset{\text{glass}}{2SiO_2} + \underset{\text{magnetite}}{\tfrac{1}{3}Fe_3O_4} + \underset{\text{gas}}{\tfrac{1}{12}O_2} + \underset{\text{steam}}{\tfrac{1}{2}H_2O} \tag{6-36}$$

and thus

$$\log f_{HCl} = \frac{\Delta G_r^\circ}{2.303RT} + \log \frac{a_{sodal}^{sodal} (a_{SiO_2}^{liq})^2 (a_{Fe_3O_4}^{mag})^{1/3} (f_{O_2})^{1/12} (f_{H_2O})^{1/2}}{a_{NaFeSi_2O_6}^{pyrox} (a_{NaAlSiO_4}^{neph})^3} \tag{6-37}$$

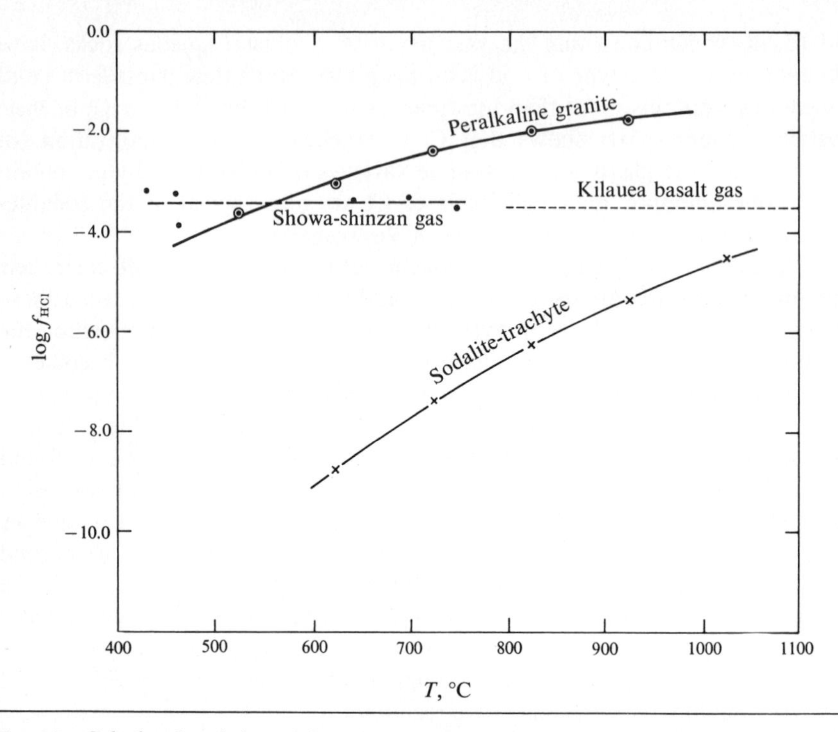


**Fig. 6-7** Calculated variation of $f_{HCl}$ with temperature for a typical peralkaline granite with an immiscible NaCl liquid (preserved in quartz as fluid inclusions) and also for a sodalite trachyte using typical mineral data substituted in Eqs. (6-35) and (6-37). Variation of log $p_{HCl}$ for the Showa-shinzan and Kilauea volcanic gases (Tables 6-2 and 6-3) is also shown. Total pressure 1 bar.

If appropriate values for the respective mineral components are substituted for the activity terms, together with $f_{H_2O}$ equal to unity and $f_{O_2}$ as defined by the FMQ buffer, then the variation of log $f_{HCl}$ with temperature can be plotted for both reactions. This has been done in Fig. 6-7 together with log $p_{HCl}$ of the Showa-shinzan gases (Table 6-2). The intersection of the calculated granite $f_{HCl}$ curve with the volcanic gas curve is appropriate because the gases are derived from a siliceous lava; however, the temperature of intersection is of little significance in view of the approximations used for the mineral compositions of a typical peralkaline granite [Eq. (6-37)]. The much lower values of $f_{HCl}$ for the sodalite assemblage emphasize the role of this mineral, which effectively extracts Cl from the magma and from the gas if one exists.

THE FLUORIDE COMPONENT IN MAGMAS Fluorite, the most widespread and common fluoride mineral, is also found as a typical accessory of hypersolvus granites; many of these rocks are undersaturated with alumina, so that aegirine or an alkali amphibole occur. Fluorite is also found in syenites, nepheline syenites, and in large hydrothermal vein deposits. A reaction analogous to that for $f_{HCl}$ is written below to represent all salic rocks with a sodic pyroxene. It does not contain a $SiO_2$ component, because the paragenesis of fluorite appears to be independent of silica activity.

$$\underset{\text{acmite}}{NaFeSi_2O_6} + \tfrac{1}{2}\underset{\text{anorthite}}{CaAl_2Si_2O_8} + \underset{\text{gas}}{HF}$$

$$= \underset{\text{albite}}{NaAlSi_3O_8} + \tfrac{1}{2}\underset{\text{fluorite}}{CaF_2} + \tfrac{1}{3}\underset{\text{magnetite}}{Fe_3O_4} + \tfrac{1}{12}\underset{\text{gas}}{O_2} + \tfrac{1}{2}\underset{\text{steam}}{H_2O} \tag{6-38}$$

which, assuming equilibrium, can be rearranged to

$$\log f_{HF} = \frac{\Delta G_r^\circ}{2.303RT} + \log \frac{a_{NaAlSi_3O_8}^{feld}(a_{CaF_2}^{fluor})^{1/2}(a_{Fe_3O_4}^{mag})^{1/3}(f_{O_2})^{1/12}(f_{H_2O})^{1/2}}{a_{NaFeSi_2O_6}^{pyrox}(a_{CaAl_2Si_2O_8}^{feld})^{1/2}} \tag{6-39}$$

Very rarely in nepheline syenites poor in calcium, the mineral villiaumite (NaF) takes the place of fluorite.[1] For this type of nepheline syenite, the following reaction fits the paragenesis reasonably well:

$$\underset{\text{acmite}}{NaFeSi_2O_6} + \underset{\text{nepheline}}{NaAlSiO_4} + \underset{\text{gas}}{HF}$$

$$= \underset{\text{villiaumite}}{NaF} + \underset{\text{albite}}{NaAlSi_3O_8} + \tfrac{1}{3}\underset{\text{magnetite}}{Fe_3O_4} + \tfrac{1}{12}\underset{\text{gas}}{O_2} + \tfrac{1}{2}\underset{\text{steam}}{H_2O} \tag{6-40}$$

and at equilibrium

$$\log f_{HF} = \frac{\Delta G_r^\circ}{2.303RT} + \log \frac{a_{NaF}^{vill}a_{NaAlSi_3O_8}^{feld}(a_{Fe_3O_4}^{mag})^{1/3}(f_{O_2})^{1/12}(f_{H_2O})^{1/2}}{a_{NaFeSi_2O_6}^{pyrox}a_{NaAlSiO_4}^{neph}} \tag{6-41}$$

If mineral data for a general fluorite paragenesis and the villiaumite assemblage are substituted in the logarithmic terms, then the calculated variation of $f_{HF}$ for these two assemblages can be plotted as a function of temperature (Fig. 6-8). Again, there is general accordance with the Showa-shinzan gases. If these mineral reactions represent typical igneous fluoride-mineral parageneses, then $f_{HF}$ may vary less than $f_{HCl}$ from one rock type to another.

[1] The only reported locality in the United States is in New Mexico (Stormer and Carmichael, 1970*a*).

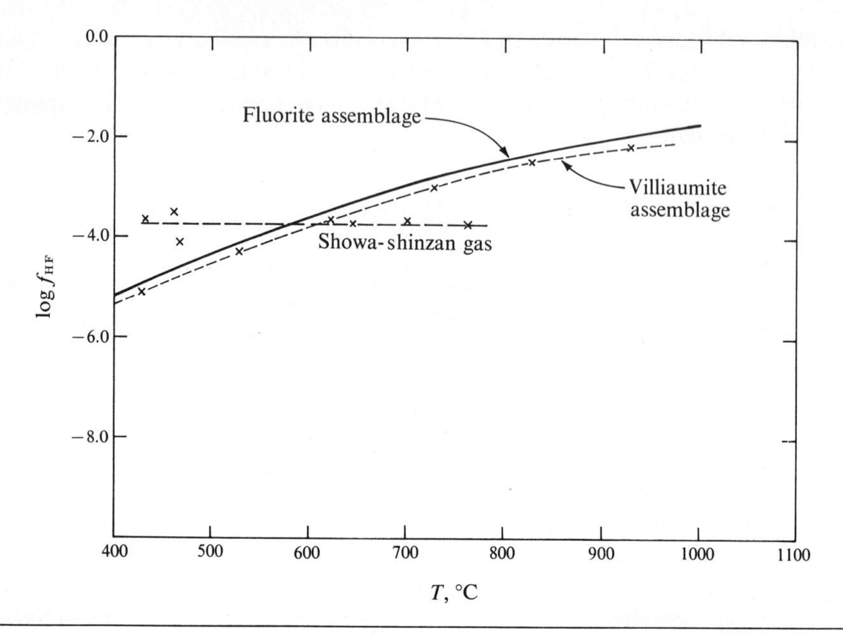


**Fig. 6-8** Calculated variation of $f_{HF}$ with temperature for a typical fluorite-bearing peralkaline granite and for a nepheline syenite with villiaumite (NaF). Typical mineral data for both rock types have been substituted in Eqs. (6-39) and (6-41). Log $p_{HF}$ for the Showa-shinzan volcanic gas (Table 6-2) is also shown. Total pressure 1 bar.

## WATER IN MAGMAS AND IGNEOUS ROCKS

SIGNIFICANCE OF ISOTOPIC COMPOSITION OF HYDROGEN Water is far and away the most abundant constituent of volcanic gases, and although it is not included in the "active gas" component of White and Waring (1963), its origin, whether meteoric or juvenile, is very much open to debate. There are two obvious sources of magmatic contamination: seawater or meteoric water, trapped in sediments or rocks in contact with the magmatic reservoir.

Because magmas or rocks contain very much less hydrogen than oxygen, the hydrogen isotopic ratios of magmas are much more sensitive to the influx of either seawater or groundwater than are the oxygen isotopes; thus, although data showing the combined effects of oxygen and hydrogen exchange between igneous rocks and groundwater are available (Taylor, 1968; Taylor and Epstein, 1968), we shall look here at only deuterium/hydrogen ratios. Isotopic analyses are reported in terms of the difference between the sample and a suitable standard, which for the hydrogen isotopes is a standard mean ocean water (SMOW). Sometimes the results are presented as percentages and sometimes as per mil; all the selected data given here are compared to SMOW* and are given as

* Friedman and Smith's (1958) results for obsidian have been corrected to this standard.

percentages derived from the equation

$$\delta_{D/H} = 100\,\frac{(D/H)^{sample} - (D/H)^{SMOW}}{(D/H)^{SMOW}}$$

Negative values mean that the sample is impoverished in deuterium in comparison to the standard.

Values of $\delta_{D/H}$ given in Table 6-5 show that hydrogen of basic and mafic rocks in general is richer in deuterium than that of acid magmas (represented by obsidians). The pattern for Kilauea lavas is complex and is determined by the relative diffusion rates of hydrogen and deuterium (Friedman, 1967).

All these magmatic $\delta_{D/H}$ ratios (Table 6-5) depart markedly from that of seawater ($\delta_{D/H} \sim 0$). Meteoric D/H ratios can vary between wide limits, because their deuterium content depends on latitude, elevation, evaporation, and temperature; typically rainwater (Friedman and Smith, 1958) is higher in deuterium than those rocks tabulated in Table 6-5. The difficulty of ascribing a juvenile origin to "magmatic" water is exemplified by the data for the Surtsey volcanic gas (Table 6-5). Surface waters in Iceland have $\delta_{D/H}$ ratios of about $-5.2$ percent, whereas subsurface water in a borehole on a neighboring island to Surtsey has $\delta_{D/H}$ equal to $-6.3$ percent; certainly the case cannot be ruled out for meteoric water trapped in the volcanic pile being recycled to the surface.

The isotopic data ($\delta_{D/H}$) for worldwide basic and ultramafic rocks are essentially uniform (Table 6-5); thus H. P. Taylor and Epstein (1968) suggest that all basic rocks with $\delta_{D/H}$ ratios lighter than $-9.0$ percent are contaminated by groundwater. So general do they find this low-temperature isotopic exchange to be in shallow intrusions, that it is imprudent to use analyses of such rocks to estimate the concentration of water in magma near the earth's surface; for this we must return to the volcanic rocks.

WATER CONTENT OF MAGMAS NEAR THE SURFACE Experience shows that the determination of small amounts of water in silicate rocks is both inaccurate and imprecise, so that only analyses obtained by special techniques are likely

**Table 6-5** Selected values of D/H percentages relative to standard mean ocean water*

| | $\delta_{D/H}$, % | $H_2O$, wt % |
|---|---|---|
| 1959 Kilauea summit eruption: basalt pumice (glass) | $-6.6$ to $-7.9$ | 0.064 to 0.099 |
| 1960 Kilauea flank eruption: basalt pumice (glass) | $-5.7$ to $-9.1$ | 0.086 to 0.101 |
| 1964–1967 Surtsey, Iceland, volcanic gas | $-5.53$ | |
| Obsidians from U.S.A., Iceland, and New Zealand | $-8.8$ to $-16.5$ | 0.09 to 0.29 |
| Ultramafic and mafic rocks, worldwide | $-6.0$ to $-8.0$ | |
| Submarine tholeiitic lava, Kilauea eastern rift | $-6.0$ | |

* Data taken from Friedman (1967); Friedman and Smith (1958); Arnason and Sigurgeirsson (1968); H. P. Taylor and Epstein (1968); and J. G. Moore (1970).

to be correct. The analyses reported by Friedman (1967) are of this quality (Table 6-5) and show that tholeiitic magma contains between 0.06 and 0.10 percent $H_2O$ at the time of eruption. J. G. Moore (1965) found rather a constant amount of water (average 0.45 percent $\pm 0.15$ by weight and equivalent to 1.1 mole percent) in submarine lavas dredged from depths down to 4000 m in the Kilauea rift zone. The concentration of water in the olivine-basalt magma of Surtsey, Iceland, is estimated to be about 0.70 percent (weight) using a radioactive tracer technique (Björnsson, 1968). Basaltic magma erupted on the sea floor shows a relationship between water content and lava composition. In round figures, low-potassium tholeiite has 0.25 percent, Hawaiian tholeiite 0.5, and alkali-rich basalt about 0.9 by weight (J. G. Moore, 1970).

Values for rhyolitic obsidians range from 0.09 to 0.29 (Table 6-5), although concentrations as high as 0.4 percent in pumice (equivalent to 1.5 mole percent) are considered by Ross (1964) to represent pristine water. All the hydrated obsidians, perlites or pitchstones, with large amounts of water (2 to 8 percent) have both oxygen and hydrogen isotopic compositions which show extensive low-temperature exchange with groundwater (H. P. Taylor, 1968).

THE SOLUBILITY OF WATER IN MAGMAS It has been over 40 years since the pioneer water solubility experiments of Goranson (1931) were performed on molten Georgian granite. Since then, the majority of experimental measurements, with considerable innovation of technique, have come from Burnham and his coworkers in Pennsylvania and to some extent from Khitarov in Russia (Khitarov et al., 1963). The solubility of water in liquids of granitic, andesitic, and basaltic composition has been measured over a wide range of pressure, with results shown in Fig. 6-9. The curves for basalt and andesite are isothermal (1100°C), whereas the temperature for the granitic pegmatite composition is just above the liquidus and varies from 660 to 820°C. An isothermal curve for pegmatite at 900°C (Fig. 6-9) illustrates the general tendency for the solubility of water to decrease with increasing temperature; for liquids of $NaAlSi_3O_8$ composition at 5000 bars confining pressure, the solubility of water decreases by about 1 percent for a temperature rise of 200°C. The dominant feature of the solubility curves, particularly for the granite pegmatite, is that at pressures equivalent to those of the base of the crust (10 kb) there is an extensive miscibility gap between water-saturated granite melt with 20 percent water and the aqueous fluid which contains about 10 percent (by weight) of silicate material in solution (Burnham, 1967).* Solubility of this silicate material in the aqueous fluid decreases with falling confining pressure; the composition of the fluid also changes in a consistent way with pressure, becoming poorer in silica (relative to the feldspar components) at high pressures.

* Burnham's (1967) essay on the role of water in magmas is an incisive account by the foremost worker in this field.

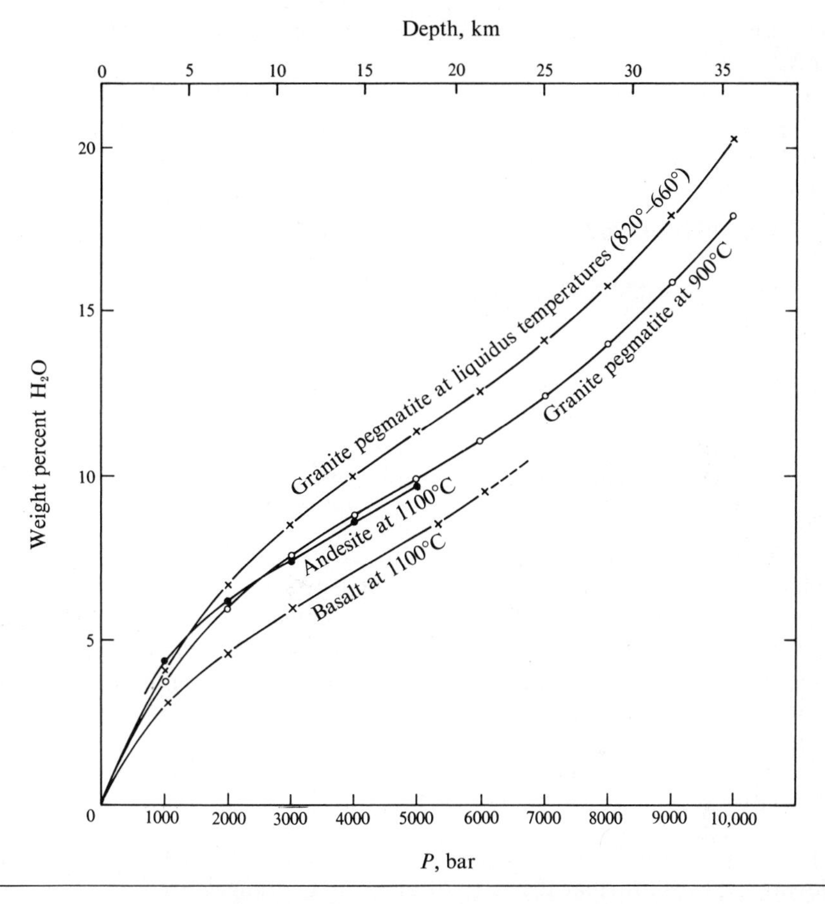


**Fig. 6-9** Isothermal solubility of $H_2O$ in liquids of granite (pegmatite), andesite, and basalt composition. Note that the temperature of the andesite and basalt is 200°C higher than that of the granite; the solubility of $H_2O$ in the granitic pegmatite at liquidus temperatures (820 to 660°C) is also shown. (Data from Burnham and Jahns, 1962; D. L. Hamilton et al., 1964.)

The isothermal solubility of water in a magma is governed by

$$\frac{dX_{H_2O}^{liq}}{dP} = \frac{\hat{V}_{H_2O}^{gas} - \overline{V}_{H_2O}^{liq}}{(\partial\mu_{H_2O}/\partial X_{H_2O}^{liq})_{P,T}} \tag{6-42}$$

where $X_{H_2O}^{liq}$ is the mole fraction and $\overline{V}_{H_2O}^{liq}$ the partial molar volume of $H_2O$ in the magma, and $\hat{V}_{H_2O}^{gas}$ is the molar volume of $H_2O$ in the aqueous gas phase. The presence of silicate material dissolved in the gas phase has been ignored; otherwise $\hat{V}_{H_2O}^{gas}$ becomes $\overline{V}_{H_2O}^{gas}$, the partial molar volume of $H_2O$ in the gas. On the assumption of ideality ($\gamma = 1$), Eq. (6-12) can be differentiated with respect to $X_{H_2O}^{liq}$ to find

$$\left(\frac{\partial \mu_{H_2O}}{\partial X_{H_2O}^{liq}}\right)_{P,T} = \frac{RT}{X_{H_2O}} \tag{6-43}$$

which by substitution in Eq. (6-42) gives, for an ideal solution,

$$\frac{dX_{H_2O}^{liq}}{dP} = \frac{(\hat{V}_{H_2O}^{gas} - \bar{V}_{H_2O}^{liq})X_{H_2O}^{liq}}{RT} \tag{6-44}$$

or

$$2.303RT \log X_{H_2O}^{liq} = \int_1^P (\hat{V}_{H_2O}^{gas} - \bar{V}_{H_2O}^{liq})\, dP$$

As we have noted before, the only experimental data on the partial molar volumes of water in silicate liquids are for $NaAlSi_3O_8$ liquids (Burnham and Davis, 1971). The activity of $H_2O$ ($a_{H_2O}$) in liquids of $NaAlSi_3O_8$, andesitic, and basaltic composition is proportional to $(X_{H_2O})^2$ and not $X_{H_2O}$, which implies that the water in the melt is dissociated. D. L. Hamilton et al. (1964) propose the following reaction for these liquids:

$$\underset{\text{gas}}{H_2O} + \underset{\text{liquid}}{O^{--}} = \underset{\text{liquid}}{2(OH)^-} \tag{6-45}$$

This agrees substantially with Shaw's (1968) data on the state of $H_2O$ in obsidians, which show that the ratio of $H_2O/(OH)^-$ ranges from 0 to 0.05 for a total concentration of $H_2O$ from zero to 6 percent (by weight) at 850°C.

PRESSURE $p_{H_2O}$ AND FUGACITY $f_{H_2O}$ OF WATER IN MAGMAS So many experimental determinations of liquid-solid equilibria in silicate systems are made in the presence of a steam phase that the geologist can become lulled into accepting water saturation as a common state in nature. Although it is obviously a limiting condition to assume that a magma was saturated with water at a particular stage in its cooling history, there is really no good evidence to establish that, in the general case, it was saturated as it started to crystallize at depth. In one intrusion in the Sierra Nevada of California, Putnam and Alfors (1965) estimate that $p_{H_2O}$ became equal to $P_{total}$ only after 92 percent by volume of the magma had crystallized.

Because water pressures in many geological situations could be equal to several thousand bars, it is no longer permissible to equate the partial pressure of water directly with its fugacity, as we have done for many of the volatile components considered so far. It is not appropriate here to look at the ex-

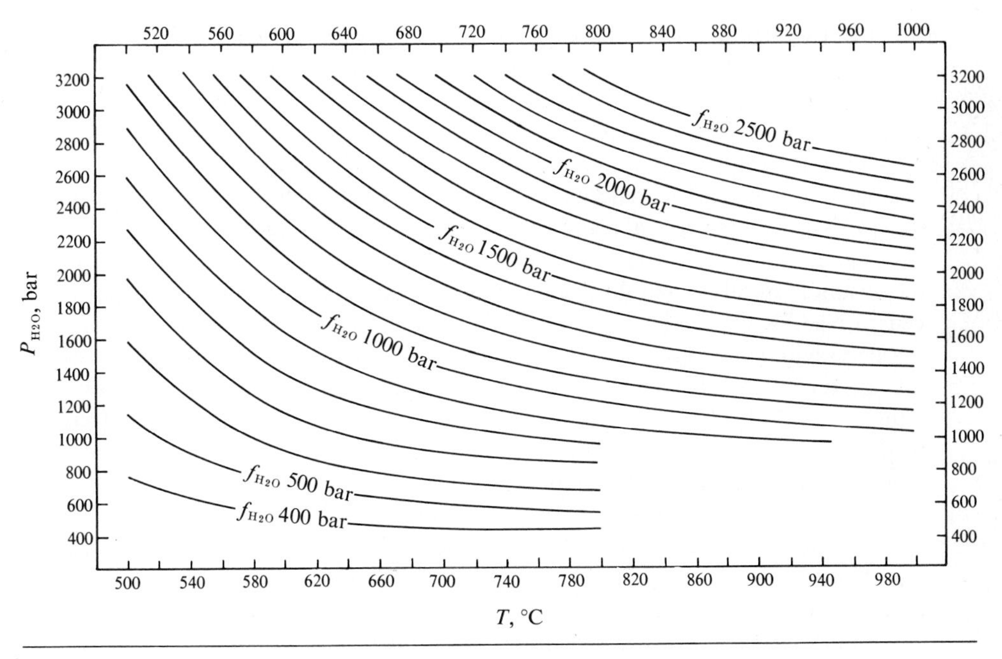


**Fig. 6-10** Variation of $f_{H_2O}$ as a function of $p_{H_2O}$ and temperature. Note that at high temperature $f_{H_2O}$ tends to $p_{H_2O}$. (Data from Burnham et al., 1969.)

perimental techniques or the thermodynamic formulations involved in the determination of water fugacity $f_{H_2O}$ from heat and $PTV$ data. Curves of $f_{H_2O}$ as a function of pressure $p_{H_2O}$ and temperature are shown in Fig. 6-10; note that $f_{H_2O}$ approaches $p_{H_2O}$ at low pressures and at high temperatures where the interaction between the water molecules becomes small. For more detailed data on $f_{H_2O}$ as a function of $P$ and $T$, the reader is referred to the tabulations by Burnham et al. (1969), from which the representative curves in Fig. 6-10 were drawn. As an example of the difference between $f_{H_2O}$ and $p_{H_2O}$, consider a liquid of granitic pegmatite composition which at 900°C contains approximately 6 percent of water (Fig. 6-9) and has an equilibrium vapor pressure of 2000 bars ($p_{H_2O}$), which from the curves of Fig. 6-10 is equivalent to $f_{H_2O}$ of about 1750 bars.

Many attempts have been made to estimate water vapor pressure during part or whole of the crystallization interval of a body of magma. The limiting condition, usually implied but not so stated, in speculations of this type is that the magma was saturated with water, and therefore $p_{H_2O}$ was approximately equal to the confining pressure $P_{total}$, which can be taken to equal the lithostatic load of the overlying column of rock. If an estimate of $P_{total}$ can be made, then by using the water solubility curves shown in Fig. 6-9 it is possible to make a reasonable estimate of the *maximum* water content for a magma of comparable composition. Tilley (1957) has suggested that the water content of pitchstone dikes on the Scottish island of Arran represents saturation at the depth of intrusion, but recent isotopic evidence suggests that much of this water ($\sim$4.3 weight percent) is likely to have been acquired subsequently at low temperatures. A more recent estimate of water pressure in a series of porphyritic ash flows also assumes that the various batches of magma were saturated at the intratelluric stage of phenocryst crystallization. The estimate in this case (Lipman, 1966) depends essentially on matching the normative quartz/alkali-feldspar ratio in the groundmass of an ash flow with the same ratio for synthetic liquids in the system $NaAlSi_3O_8$–$KAlSi_3O_8$–$SiO_2$–$H_2O$ (Tuttle and Bowen, 1958). For these synthetic liquids, increasing water content or $p_{H_2O}$ reduces the silica content of the liquid necessary for quartz and an alkali feldspar to coprecipitate. Lipman's estimates (500 to 1200 bars) for these Nevada ash flows, while possibly not far from reality, not only depend on the assumption of saturation but also ignore the effect of the $CaAl_2Si_2O_8$ component.

If the water contents of basalt and rhyolitic obsidian noted on page 322 are taken to represent the pristine water content of these two types of magma, then from the water solubility curves in Fig. 6-9, magma, even if as much as 1 percent water is assumed, will be unsaturated at depths of more than a kilometer in the crust. It must be emphasized that the occurrence of phenocrysts of a hydrous mineral, such as apatite, mica, or amphibole, does not imply that the magma was saturated with water but only that $H_2O$ was present as a component. By analogy with experimental data on water-unsaturated silicate systems (for example, Luth, 1969), we may expect both the crystallization paths and the coprecipitation boundaries to be rather different in magmas which have $p_{H_2O}$ less than $P_{total}$ in comparison to those more familiar experimental systems where $p_{H_2O}$ equals $P_{total}$. The evidence is accumulating that $p_{H_2O}$ and $P_{total}$ must be treated as independent variables, and except in those magmas close to or on the earth's surface, it is likely that $p_{H_2O}$ will equal $P_{total}$ only toward the last, pegmatitic stages of crystallization, unless water has been extracted from wet country rocks into the cooling magma.

Lavas often contain phenocryst assemblages which indicate that the fugacity of water was defined at the intratelluric (subsurface) stage of crystallization. Phenocrysts of either mica or an amphibole together with their anhydrous breakdown products form a buffer assemblage; however, the presence of substantial amounts of F in the (OH) group of the hydrous minerals will considerably modify and so render unsuitable the calculated values of $f_{H_2O}$. In the following example F accounts for only 4 percent of the (OH) group in the analyzed amphibole. Phenocrysts of cummingtonite together with orthopyroxene, quartz, and iron titanium oxides[1] occur in several rhyolitic pumice

[1] In these examples, the composition of the coexisting iron-titanium-oxide phases is used to give the temperature and $f_{O_2}$ of their equilibration, which is also taken to be coeval with the precipitation of the phenocryst assemblage.

deposits of the Taupo volcanic region of the north island of New Zealand (Ewart et al., 1971). With this mineral assemblage and the temperature estimate of 735 to 745°C for three rhyolite samples, $f_{H_2O}$ can be obtained from

$$\underset{\text{glass}}{SiO_2} + \underset{\text{enstatite}}{7MgSiO_3} + \underset{\text{steam}}{H_2O} = \underset{\text{anthophyllite}}{Mg_7Si_8O_{22}(OH)_2} \tag{6-46}$$

using modified values from Zen (1971) to take account of the presence of silica glass rather than quartz as the standard state of $SiO_2$. From Eq. (6-46)

$$\log f_{H_2O} = \frac{\Delta G_r^\circ}{2.303RT} + \frac{(\Delta V^\circ)^{sol}}{2.303RT}(P-1) + \log \frac{a^{cumm}_{Mg_7Si_8O_{22}(OH)_2}}{(a^{pyrox}_{MgSiO_3})^7 a^{liq}_{SiO_2}} \tag{6-47}$$

Although the known compositions of the phenocrysts can be substituted for the activity terms, no value of $f_{H_2O}$ can be calculated until $P_{total}$ is estimated. This Wood and Carmichael (1973) did by using the silica activity technique discussed in Chap. 3; they obtained an average value of 4.32 kb for the equilibration pressure of the phenocrysts of the three rhyolites. Substitution of this value into Eq. (6-47) gives an average fugacity, $f_{H_2O}$, of 3900 bars, which corresponds to $p_{H_2O}$ about 4.82 kb (Burnham et al., 1969). The slightly higher value of $p_{H_2O}$ in comparison with $P_{total}$ is the result of uncertainties in the temperature, mineral compositions, and thermodynamic data; within these restrictions, however, it has been established that these rhyolites were saturated with water at the stage of phenocryst precipitation, or $P_{total} \approx p_{H_2O}$. This is the only example known to us where this equality has been shown.

In the absence of a value for $P_{total}$, calculations of water fugacity using thermodynamic data on mineral reactions allow only a crude estimate, since the $P\Delta V^{sol}$ term has of necessity to be neglected. However, where $\Delta V^{sol}$ is small, this may be permissible. In many igneous rocks, phenocrysts of biotite coexist with sanidine and iron titanium oxides, especially in lavas of rhyolitic or dacitic composition. If a value of $f_{O_2}$ and $T$ can be derived from the equilibration of the iron titanium oxides, then $f_{H_2O}$ can be estimated from the following reaction:

$$\underset{\text{annite}}{KFe_3AlSiO_{10}(OH)_2} + \tfrac{1}{2}O_2 = \underset{\text{sanidine}}{KAlSi_3O_8} + \underset{\text{magnetite}}{Fe_3O_4} + \underset{\text{steam}}{H_2O} \tag{6-48}$$

Wones (1972) has experimentally determined the value of $\Delta G_r^\circ/2.303RT$ for this reaction, which can be rearranged as

$$\log f_{H_2O} = \frac{7409}{T} + 4.25 + \frac{0.00385}{T}(P-1) + \log \frac{(f_{O_2})^{1/2}(a^{biot}_{annite})^3}{a^{feld}_{KAlSi_3O_8} a^{titanomag}_{Fe_3O_4}} \tag{6-49}$$

and values of $f_{H_2O}$, and hence $p_{H_2O}$, can be calculated for any assemblage containing hydroxyl biotite, alkali feldspar, and titanomagnetite.

If two mineral geothermometers are available, one insensitive and the other sensitive to $p_{H_2O}$, then estimates of $p_{H_2O}$ can be obtained. The plagioclase geothermometer of Kudo and Weill (1970) is affected by $p_{H_2O}$, and they give equations for plagioclase-liquid equilibria without $H_2O$ and with 500, 1000, and 5000 bars $p_{H_2O}$. If the plagioclase-derived temperatures are compared with values totally unaffected by $p_{H_2O}$, then the two sets of temperature estimates will correspond best for a particular $p_{H_2O}$ of the plagioclase geothermometer. Estimates of this type, mainly for the phenocryst assemblages of acid magma, in almost all cases indicate $p_{H_2O}$ of 500 bars or less (Stormer and Carmichael, 1970*b*). Thus these acid magmas would not be saturated with water except high up in the earth's crust.

## OXYGEN FUGACITY IN MAGMAS AND IGNEOUS ROCKS

The only direct measurements of $f_{O_2}$ in cooling lava are those made in drill holes in the upper crust of a cooling Hawaiian lava lake (Sato and Wright, 1966). A specific ion ($O^{--}$) solid electrolyte was lowered down a hole of known temperature profile, and after recording the emf readings, the temperatures were rechecked by thermocouple. Sato and Wright's values for four of the five drill holes are shown in Fig. 6-11. At temperatures above 850°C, only values for hole no. 17 are significantly different, and these higher values of $f_{O_2}$ could possibly result from the short interval of 6 days between drilling the hole (using a water coolant) and the time of measurement; all other holes were measured 4 to 6 weeks after drilling.

The most striking result is that for hole no. 11, which between 750 and 500°C shows a dramatic rise in $f_{O_2}$ to values higher than those defined by the experimental buffer assemblage magnetite-hematite (H-M, Fig. 6-11). Sato and Wright suggest that this increase in $f_{O_2}$ is caused by the cooling of the lava lake to temperatures at which $O_2$ and $H_2O$ can no longer diffuse freely, while $H_2$ continues to escape toward the surface; the crystalline basalt acts as a semi-permeable membrane to $H_2$. The escape of $H_2$ causes further dissociation of $H_2O$ contained in the cooling rock, so that $f_{O_2}$ rises and the basalt becomes oxidized. When the preferential diffusion of $H_2$ ceases at low temperatures, $f_{O_2}$ returns to low values. The diffusion rate of $H_2$ in certain silicate liquids has been found to be 1000 times greater than that of $H_2O$ at the same temperature, which fully supports this hypothesis. In the zone of high $f_{O_2}$, olivine crystals are reddened (oxidized), magnetite is almost pure $Fe_3O_4$ resulting from the oxidation of titanomagnetite, and the assemblage of hematite-pseudobrookite occurs. Similar zones of high oxidation have been found in Icelandic basaltic lavas (Watkins and Haggerty, 1967), where the most intense oxidation is found in the thicker flows.

Considerable data have been acquired in the last few years on the coexisting iron titanium oxides in a wide variety of volcanic rocks, but, regrettably, few alkaline lavas contain an $\alpha$ phase. In Fig. 6-12, the equilibration tem-

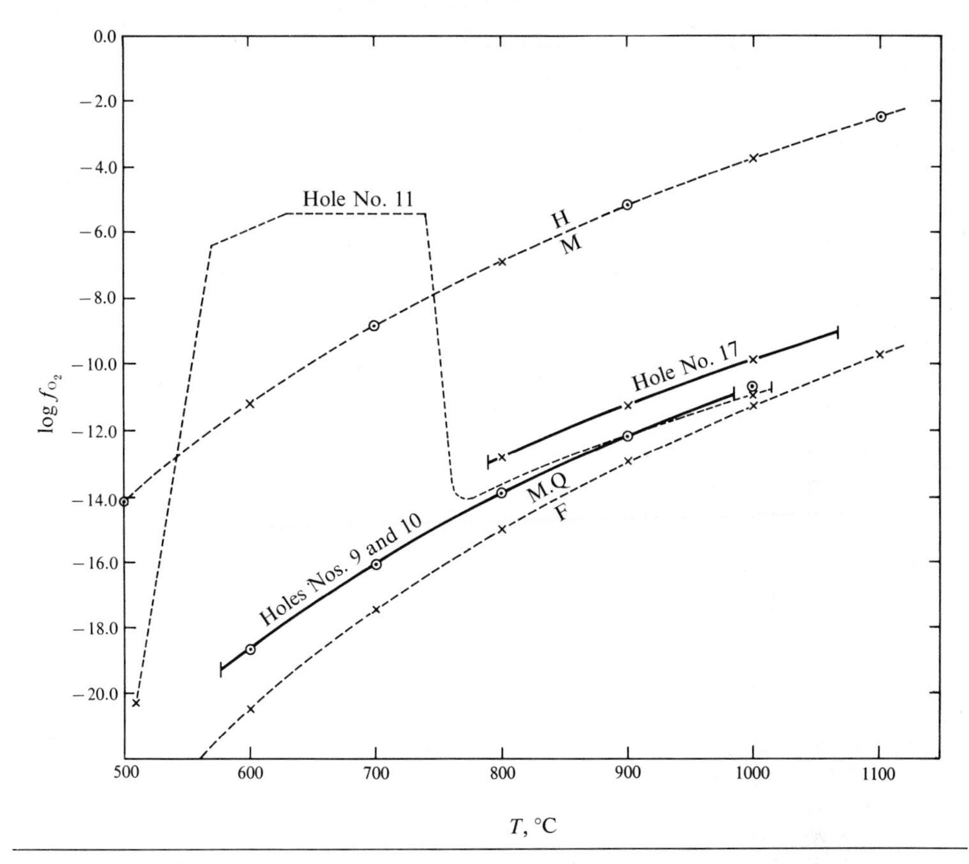


**Fig. 6-11** Variation of $f_{O_2}$ with temperature in four drill holes in the crust of Makaopuhi lava lake, Hawaii (Sato and Wright, 1966). Variation of $f_{O_2}$ with temperature for the two experimental buffers, hematite-magnetite and fayalite-magnetite-quartz, is labeled H–M and F–MQ, respectively.

peratures and $f_{O_2}$'s of the iron-titanium-oxide microphenocrysts from rhyolitic and dacitic lavas or pumices have been plotted. Those lavas with phenocrysts of a fayalitic olivine fall close to or on the FMQ buffer curve, and those with phenocrysts of orthopyroxene fall above the FMQ curve but trend toward it at lower temperatures. Lavas with phenocrysts of biotite and/or amphibole plot above the orthopyroxene-bearing rocks. A group of rhyolitic pumices from New Zealand which contain both orthopyroxene and amphibole appropriately falls between the orthopyroxene curve and the biotite-amphibole curve. Because the total amount of the ferromagnesian phenocrysts is so small (0.1 to 3.0 percent by volume) in many of these acid lavas, the capacity of the condensed phases to buffer the compositional vagaries of a volcanic gas must also be small. However, the pattern of interdependence of the ferromagnesian silicate phenocrysts with a specific set of $f_{O_2}$ conditions (Fig. 6-12) suggests that equilibria

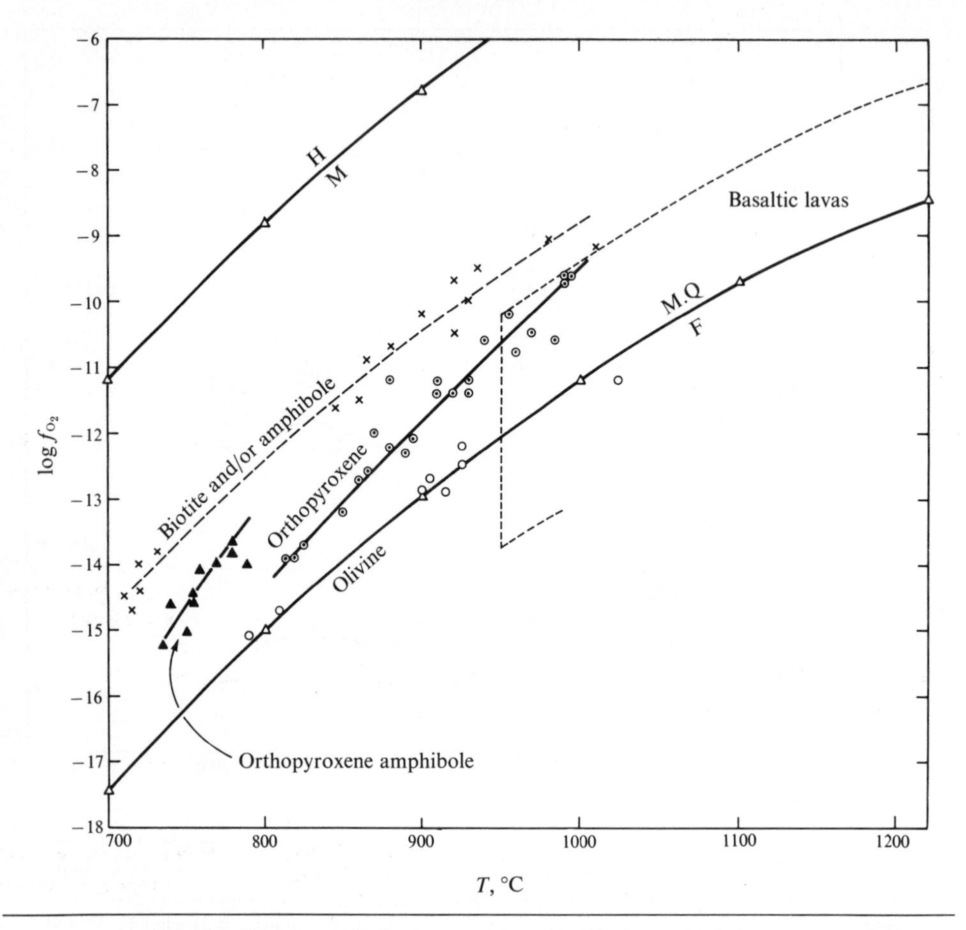


**Fig. 6-12** Equilibration temperature and $f_{O_2}$ of coexisting iron-titanium-oxide microphenocrysts in rhyolites and dacites. Those rhyolites with iron-rich olivine phenocrysts are represented by open circles; those with phenocrysts of orthopyroxene by circled dots; those with biotite and hornblende phenocrysts by crosses; and those with orthopyroxene and amphibole phenocrysts by solid triangles. Variation of $f_{O_2}$ for the two experimental buffers, hematite-magnetite and fayalite-magnetite-quartz, is labeled H–M and F–MQ, respectively. The $T$-$f_{O_2}$ zone of basalt crystallization is taken from Fig. 6-11. (Data mainly from Carmichael, 1967*b*; Ewart et al., 1971; Lipman, 1971; Heming and Carmichael, 1973.)

are determined more by the bulk composition of the magma than by the composition of the volatile components. If these lavas were internally buffered with respect to $f_{O_2}$, then the mass of the volatile component is likely to have been very small. The scatter of points in the New Zealand orthopyroxene-amphibole rhyolites has been attributed to variation in silica activity (Ewart et al., 1971), which is another way of saying that it is the composition of the liquid and solid phases that determines the value of $\log f_{O_2}$.

Data culled from the literature for a wide variety of basic lavas are plotted in Fig. 6-13; the rock types include olivine tholeiite, basanite, trachybasalt, high-alumina basalt, andesite, and ugandite, all of which contain either olivine or orthopyroxene. In most of these rocks the iron titanium oxides are groundmass phases and, in the olivine tholeiites, coexist with olivine and a calcium-poor pyroxene. Only in a rare type of alkali gabbro (shonkinite) does $f_{O_2}$ greatly

**Fig. 6-13** Equilibration temperature and $f_{O_2}$ of coexisting iron-titanium oxides in a wide variety of basic lavas, all of which contain olivine, orthopyroxene, or both. Field of basaltic magma bounded by dashed lines is from Carmichael and Nicholls (1967). One example of $f_{O_2}$ and temperature for an alkali gabbro with primary pseudobrookite ($Fe_2TiO_5$) is shown at $Z$ within an estimated field (stippled) for pseudobrookite occurrence in basic magma. Curves for variation of $f_{O_2}$ with temperature for the two synthetic buffers, hematite-magnetite and fayalite-magnetite-quartz, are labeled H–M and F–MQ, respectively. (Data include material from A. T. Anderson and Wright, 1972.)

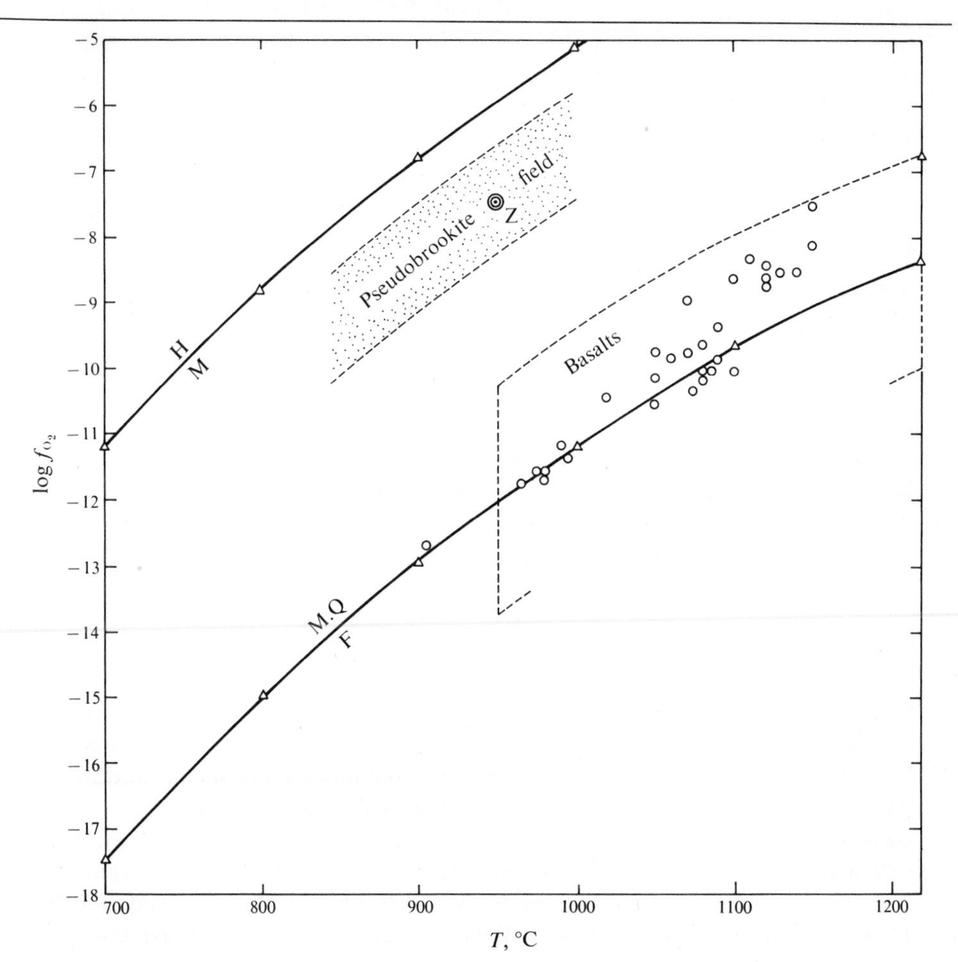

depart from that defined by FMQ, and in these rocks discrete independent grains of pseudobrookite are found with titanohematite in a feldspar-rich groundmass (Carmichael and Nicholls, 1967). The value of $f_{O_2}$ and temperature (using coexisting titanomagnetite phenocryst compositions) for a Hawaiian example of shonkinite is plotted in Fig. 6-13, together with the estimated field in which $Fe_2TiO_5$ crystallizes as a primary phase.

There is no direct evidence that basaltic magma near liquidus temperature has values of $f_{O_2}$ which vary much more than two orders of magnitude on either side of the FMQ buffer. This is not surprising, because basic lavas of diverse types contain ferromagnesian minerals (olivine, pyroxene, and titanomagnetite) whose composition does not vary greatly from one rock type to another, so that variation of $f_{O_2}$ will be determined as much by silica activity [Eq. (6-19)] as by the composition of these minerals.

For magmas other than basaltic and rhyolitic, $\log f_{O_2}$ could be expected to show more variation than that shown in either Fig. 6-12 or 6-13. Unfortunately many of these rarer types, such as leucite phonolite, seldom contain both an $\alpha$ and $\beta$ phase, and $\log f_{O_2}$ must be estimated from free-energy data. Some trachytic and phonolitic lava types contain sphene and melanite garnet, and the following reaction is reasonably representative of their mineral assemblages:

$$\underset{\text{andradite}}{Ca_3Fe_2Si_3O_{12}} + \underset{\text{ulvospinel}}{3Fe_2TiO_4} + \tfrac{5}{6}O_2 = \underset{\text{sphene}}{3CaTiSiO_5} + \underset{\text{magnetite}}{\tfrac{8}{3}Fe_3O_4} \tag{6-51}$$

and at equilibrium

$$\log f_{O_2} = \frac{6}{5}\left[\frac{\Delta G_r^\circ}{2.303RT} + \log \frac{(a_{CaTiSiO_5}^{\text{sphene}})^3 (a_{Fe_3O_4}^{\text{titanomag}})^{8/3}}{(a_{Ca_3Fe_2Si_3O_{12}}^{\text{garnet}})(a_{Fe_2TiO_4}^{\text{titanomag}})^3}\right] \tag{6-52}$$

In general it is the salic magmas—phonolite, trachyte, or rhyolite—which have the greatest diversity in both the species and composition of their ferromagnesian minerals and also of their silica activity, so that their range of $f_{O_2}$ at any given temperature is probably greater than that of basaltic magma, except perhaps rare types of basic lavas for which there are no available data.

## Concluding Statement

Because the concentration of the more abundant components in volcanic gases can be related, through the fugacity function, to the solid phases in a crystallizing magma, we have rather neglected the small amounts of metals and inert gases which also form part of the gas phase. Although the concentration of these inert gases is very small, their presence can bedevil the task of dating a recent lava by the $K^{40}$–$Ar^{40}$ technique. One example of this is the beautiful Rangitoto volcano in the center of Auckland harbor in New Zealand, which from radiocarbon dating is less than 1000 years old with one recent lava erupted about 250 years ago. Yet the K–Ar ages of the same lavas run from 145,000 to 465,000 years old (McDougall et al., 1969). Radiogenic argon has been inherited from either the source region or, in this particular volcano, perhaps from the rare xenoliths of quartz and pyroxene; however, it is present in rather uniform amounts in all the basaltic lavas of this province. If $Ar^{40}$ can be extracted and mixed uniformly in large volumes of magma, then so presumably can $H_2O$, HF, and all the other components, including metals. The task of deciding on the juvenile origin of any volatile component, or a proportion thereof, seems incredibly complex, but it is not difficult to appreciate that magma of one volcanic province can acquire a pattern of trace components distinct from those of other regions.

Of the volcanic gases, water of unknown origin is normally the most abundant component. The estimates discussed before in this chapter suggest that both basaltic and acid magma contain less than 1 percent by weight and will therefore be undersaturated at depths of more than a kilometer or so in the crust. This estimate is supported by G. A. Macdonald (1963*b*), who after years of watching Hawaiian volcanoes considers that, in the early gas-rich phase of basaltic eruptions, the magma contains between 1 and 2.5 percent by weight of volatiles, which falls to between 0.2 and 0.7 percent in the later eruptions. He estimates that basaltic magma at depth has an average gas content closer to 0.5 percent than to 2 percent. Water-saturated gabbro, or even granite, magma intruded at depth in the crust may prove to be the exception rather than the rule.

# Sources of Magma: Geophysical-Chemical Constraints

# 7

## Nature of the Mantle: Geophysical Evidence, Density Distribution, and Composition

### THE MANTLE: A POTENTIAL SOURCE OF MAGMAS

In this chapter we review geophysical data bearing on the composition of the mantle and the generation of magma within it. We also examine the crust as a potential source of magma and particularly those chemical traits that might be passed on in recognizable form to magmas of crustal origin.

The mantle is that part of the earth which lies between the boundary of the core, at a depth of approximately 2900 km, and the Mohorovicic discontinuity at the base of the crust. The latter is a relatively sharp interface at which the velocities $V_P$ and $V_S$ of longitudinal and shear seismic (elastic) waves jump from the relatively low values characteristic of the overlying crust to the higher values (commonly about 8.1 km $s^{-1}$ for $V_P$, 4.5 km $sec^{-1}$ for $V_S$) that are characteristic of the mantle. The mantle is also denser than the crust. The Mohorovicic discontinuity generally lies at a depth of 6 to 8 km beneath the oceanic floor; its depth beneath continents varies from some 30 to about 70 km. The crust is generally thicker where the topographic surface is higher in accordance with the principle of hydrostatic equilibrium or isostasy, which requires the mass of a vertical column of unit cross section to be the same everywhere, regardless of height of the topographic surface.

In oceanic areas where the crust is very thin, basaltic and nephelinitic magmas certainly come from the mantle. The broad similarity in composition between oceanic and continental basalts suggests a similar source for the latter. Arguments based on $Sr^{87}/Sr^{86}$ ratios, as explained later, have been advanced as evidence that some andesitic magmas or some contributions to them come from the mantle, and even that some granitic and granodioritic magmas are not entirely of crustal origin. In fact, in considering the origin of magma of any kind the possibility of some direct contribution from mantle sources cannot be dismissed offhand.

The composition of the mantle and the physical conditions in it, particularly temperature, obviously set limits on the compositional range of magmas derived from it. Conversely, from what is known of the composition of magmas we may infer something about the composition of the mantle. The petrological and geophysical approaches are complementary. As for many other geological problems, considerable uncertainty pervades the interpretation of the data. Geophysical observations from which the composition of the mantle is deduced are generally insensitive to those details of chemical and mineralogical composition that may be petrologically most significant. It is, for example, impossible to deduce from seismic data whether the mantle contains 0.1 or 1.0 percent of $K_2O$ or of $H_2O$. On the other hand, the chemical composition of the liquids that form in the mantle is not exactly known, since there still is some uncertainty regarding the fractionation and differentiation that may occur in magmas between their source and the surface. Progress toward the solution of the problem of magma formation requires a careful and patient procedure of trial and error, or of successive approximations combining geophysical observations, experimental data on the melting behavior of multicomponent systems under pressure, and observation of the petrological nature of materials that have risen from the mantle, such as magmas themselves, nodules in alkali basalts or kimberlites, upthrusted masses of peridotites, etc. The mantle origin of these materials must first be demonstrated, for example by the occurrence in them of minerals or mineral assemblages that could have formed only at pressures greater than those that exist in the crust.

## NATURE OF GEOPHYSICAL EVIDENCE

Observations from which the composition of the mantle is inferred fall into three broad groups: gravity and geodesy, propagation of seismic waves, and free oscillations.

GRAVITY AND GEODESY Measurements of gravity on or near the earth's surface, observations of the motion of the moon and of artificial satellites in the earth's gravitational field, measurements of distances on the earth's surface, and observations of the earth's motions in the gravitational field of the sun and moon (for example, precession of its axis of rotation)—all can be combined to provide the general dimensions of the earth, its total mass, and its moment of inertia about the rotation axis. These observations also establish that the earth is in almost perfect hydrostatic equilibrium under the combined effects of gravity and rotation, except very close to its surface; the difference between the largest and least principal stresses at any point within the earth is very small

(~100 bars) compared to the mean value of these stresses (hydrostatic pressure) which, at the center of the earth, is of the order of $3.6 \times 10^6$ bars. As a consequence of this nearly perfect state of equilibrium, the shape of the geoid[1] is that of a slightly flattened ellipsoid on which are superposed only very small undulations with an amplitude of less than about 100 m. These undulations are generally unrelated to the position of mountain ranges or continents, the visible mass of which appears to be almost perfectly compensated (isostasy) by an equal lack of mass at depth. Geoidal undulations extend over distances of several hundred or several thousand kilometers and are thus distinct from the more localized gravity anomalies that occur, for instance, near island arcs or oceanic trenches. The depth of the density perturbations that are reflected in the undulations of the geoid cannot be precisely located. They may be caused by lateral variations in mantle composition or in temperature, or both.

PROPAGATION OF SEISMIC WAVES Elastic waves, such as may be generated at the focus of an earthquake or by an explosion, travel through an isotropic elastic medium at speeds determined by the density and the two elastic constants $K$ and $\mu$ of the medium. The rigidity $\mu$ is the ratio of a shear stress to the corresponding shear strain; the incompressibility (or bulk modulus) $K$ is the ratio $\rho dP/d\rho$ of the change in pressure $P$ to the relative change in density $\rho$ that it induces. Body waves are of two kinds: compressional (or longitudinal) $P$ waves with velocity $V_P = [(K + \frac{4}{3}\mu)/\rho]^{1/2}$ and shear (or transverse) $S$ waves with velocity $V_S = (\mu/\rho)^{1/2}$. Note that $V_P^2 - \frac{4}{3}V_S^2 = K/\rho$. Surface waves are waves that travel mostly along the surface of a body; their amplitude decreases with increasing depth in the body, and their speed generally depends on frequency or wavelength (dispersion). An elastic wave is generally partly reflected and partly refracted (bent) when it passes from one medium into another with different elastic properties; when going from a slower to a faster medium, the ray path is bent away from the normal to the interface, except for normal incidence for which the direction of propagation remains unchanged.

The variation of $V_P$ and $V_S$ with depth inside the earth is determined from observed travel times, that is, the time taken by a wave to travel from its source at the focus of an earthquake to a recording station. If velocity were uniform throughout the earth, waves would travel along straight lines, and travel times would be proportional to the length of the chord drawn between the focus and the station. Actually, waves that travel further also seem to have traveled faster, implying that velocity generally increases with depth. This implies in turn that waves are refracted in the earth and travel along curved paths that are concave upward; the velocity at the deepest point of the path can be determined from a plot of travel time versus angular distance from the focus to the station. Variation of velocity with depth can also be determined from the variation in speed of

[1] The geoid is the surface of constant gravitational potential that coincides with the surface of the oceans at rest, or mean sea level.

surface waves as a function of their wavelength, since longer waves also penetrate deeper and are propagated at speeds more characteristic of the deeper medium.

General conclusions drawn from the study of thousands of seismograms are that (1) velocities at the top of the mantle are not uniform but vary locally from about 7.8 to about 8.3 km $s^{-1}$ for $V_P$; (2) velocities generally increase downward, the $P$-wave velocity being close to 14 km $s^{-1}$ at the base of the mantle; (3) the rate of increase with depth is not uniform, being particularly rapid in two zones located respectively near 400 and 650 km depth; and (4) there is a zone in the upper mantle, located roughly between 60 and 250 km and called the *low-velocity layer*, in which the velocity $V_S$ first decreases with increasing depth and reaches a minimum before it begins to increase again. Where the velocity decreases with depth, ray paths are refracted downward rather than upward, so that the velocity distribution in the low-velocity layer cannot be precisely determined from travel times; information on the low-velocity layer is obtained mostly from the dispersion of surface waves. Furthermore, the structure of the low-velocity layer varies from place to place, as do the minimum value of $V_S$ and the depth at which this minimum is reached. The velocity $V_P$ in the low-velocity layer is on the whole less affected than $V_S$, although it also shows regional variations. In some regions $V_P$ goes through a minimum, although not necessarily at the same depth as $V_S$, while in others $V_P$ remains nearly constant throughout the zone. One effect of this variability is that velocities cannot be determined with great precision, and it remains difficult to choose among the slightly different velocity distributions proposed by different investigators.

Magmas are commonly considered to be generated in the depth range of the low-velocity layer. The evidence for this is based partly on a few observations of the depth of seismic disturbances preceding eruptions (notably at Kilauea) and partly on the lack of mineralogical evidence (for example, in kimberlites) of pressures greater than 80 kb at the source of the magma. The fact that magmas reach the earth's surface at temperatures generally less than 1200°C probably also points to a relatively shallow source, since liquids generated at great depths, where the melting temperature is higher, are likely to be much hotter than liquids generated nearer to the surface. However, none of these arguments is very convincing, and it may well be that magmas, or some magmas, are generated below the low-velocity layer.

FREE OSCILLATIONS A sufficiently violent disturbance, such as an earthquake of large magnitude, may cause the earth to vibrate as a whole, much as a bell vibrates when struck. These vibrations comprise a large number of superposed "modes" with different periods, the longest of which is about 54 min; these modes are comparable to the several frequencies of a bell that combine to produce its characteristic tone. These "free oscillations" of the earth can be recorded by long-period seismographs or strain meters; vibrations which have a radial component of motion ("spheroidal" oscillations) can also be recorded by means of a gravimeter. The period of a particular mode depends, as does the tone of a bell, on the density and elastic properties of the vibrating material and their distribution with depth; the gravest (longest period) modes depend essentially on average properties of the upper mantle. Since seismic velocities also depend on density and elastic coefficients, there must obviously be a relation between the periods of the several modes and the velocity distribution with depth; thus free oscillations can be used to check velocity distributions obtained from travel times or from the dispersion of surface waves. A difficulty, however, is that periods of free oscillations reflect lateral averages, whereas travel times between the focus of an earthquake and a recording station reflect only local properties which are, as we have seen, variable; a velocity distribution that satisfies free oscillation periods is an average velocity that need not, in fact, exist anywhere.

## DENSITY DISTRIBUTION IN THE MANTLE

A first step in finding what the mantle is made of is to find its density at every depth and discover what materials would have this density at the pressure $P$ and temperature $T$ prevailing at that depth. The determination of $P$ requires knowledge of the density distribution throughout the earth, for, since the earth appears to be very nearly in hydrostatic equilibrium, $P$ increases with depth as $dP = -g\rho dr$, where $r$ is distance from the center, $\rho$ is density, and $g$ is the local value of gravitational acceleration, which depends on the mass inside a sphere of radius $r$. The temperature $T$ is unknown at the start; fortunately, its effect on density and velocities is likely to be small and to a first approximation negligible. The only other constraint on density is that it must be distributed so as to give the correct total mass and moment of inertia of the earth.

The two velocities $V_P$ and $V_S$ depend on the three variables $K$, $\mu$, and $\rho$, all of which depend on $P$ and $T$; thus to determine $\rho$ from the velocity distribution we need one additional relation. Attempts to obtain it can be made along two independent lines.

First, in a layer of the mantle that is chemically and mineralogically homogeneous, the increase in density with depth is purely an effect of increasing pressure, to the extent that the effect of $T$ may be neglected. This increase in density is determined by the incompressibility, since $d\rho/\rho = dP/K$ from the definition of $K$. But we recall that $V_P^2 - \frac{4}{3}V_S^2 = K/\rho$. Let $\phi = V_P^2 - \frac{4}{3}V_S^2$, and note that $\phi$ is known everywhere if $V_P$ and $V_S$ are known; furthermore, $dP = -g\rho dr$. Eliminating $dP$ between these relations, we obtain

$$\frac{d\rho}{\rho} = -\frac{gdr}{\phi}$$

a relation known as the Adams-Williamson equation, within which there is a hidden temperature effect. The incompressibility $K$ that determines the speed of a compressional wave is the "adiabatic" coefficient corresponding to compression without exchange of heat with the surroundings. It differs from the isothermal coefficient $K_T$ corresponding to compression at constant temperature. $K_T$ would control the density variation with depth in a homogeneous layer in which the temperature is uniform; the Adams-Williamson equation applies only to a homogeneous layer in which the temperature increases with depth at the adiabatic rate determined by self-compression. This adiabatic temperature gradient is discussed further later in this chapter.

The Adams-Williamson equation, when integrated through the mantle, leads to an unacceptable value for its moment of inertia, thereby demonstrating that the mantle is not a homogeneous mass in adiabatic equilibrium. The equation can, however, be applied to any layer of the mantle that satisfies these two conditions of adiabacity and homogeneity (both chemical and mineralogical); the difficulty is, however, to prove that a particular layer satisfies these requirements, if indeed it ever does.

A second approach starts with the incompressibility $K$, which measures the repulsive force that develops between adjacent atoms or ions in a crystal as they are pushed closer together by increasing pressure. This force is essentially the sum of the electrostatic (coulomb) repulsions between all the electrons of one ion and those of all adjacent ions. Although these repulsions (and therefore $K$) are theoretically calculable as a function of distance between ions (and therefore of density), calculations turn out to be intractable because of the large number of electrons involved and our ignorance of the wave functions that best describe their spatial distribution. However, some empirical relations between $K$ and $\rho$ may be useful. It appears, for instance, that for many substances the "bulk sound speed" $C = \phi^{1/2}$ is a linear function of density of the form

$$C = a + b\rho$$

where $a$ and $b$ are constants which depend only on the mean atomic weight (m.a.w.) of the substance and not on its chemical composition or crystal structure. A large number of common and less common minerals (quartz, stishovite, corundum, periclase, forsterite, enstatite, etc.) have very nearly the same m.a.w.; if the bulk sound speed for each of these minerals is plotted against its density, all the representative points fall very nearly on a straight line. Conversely, given the value of $C$ at some depth, the local density could be determined if the m.a.w. were known. Wang (1970) has applied this relation to the upper mantle to calculate a density distribution that satisfies seismic and free-oscillation data as well as the general constraints on mass and moment of inertia. Other empirical or partly theoretical equations of state have also been used, and a number of slightly different density distributions have been proposed, among which it is difficult to choose. The density at any depth remains uncertain by perhaps as much as 2 or 3 percent.

## CHEMICAL AND MINERALOGICAL COMPOSITION OF THE MANTLE

Given a density model of the mantle and the corresponding distribution of pressure and elastic constants, the next problem is to determine what materials would have the required density and elastic constants at the given pressure or depth; or conversely, one may attempt to calculate the density and elastic properties of mantle material when brought down to ordinary conditions of pressure and temperature. The main difficulty in doing so is that pressures in the mantle, or at least in its deeper parts, far exceed the greatest pressure that can be attained in static experiments designed to measure density or elastic coefficients; for instance, the pressure at the base of the mantle (1.4 Mb*) is about 10 times greater than the greatest pressures at which these properties can be conveniently measured, particularly at high temperature. Pressures in the megabar range can be created experimentally only by means of explosives (shock waves). In a typical shock-wave experiment, the specimen to be studied is brought irreversibly, and for only a very short time, to a state for which it is possible to determine the pressure, the density, and the internal energy but not the temperature, which may be very high and cannot be estimated without additional data and assumptions. In such experiments, it is also generally difficult to detect changes of state that may be induced by the high pressure or high temperature of the shock wave and to identify the crystallographic nature of resulting phases.

THE UPPER MANTLE The identification of the nature of the uppermost part of the mantle is made easier by the circumstance that pressure and temperatures that prevail there are still within the easily accessible experimental range. At the Mohorovicic discontinuity the pressure is only about 10 to 12 kb under continents, and even less in oceanic regions. The upper-mantle density of 3.3 to 3.4 g $cm^{-3}$ estimated from isostatic considerations, and the "normal" characteristic P-wave velocity of 8.1 km $s^{-1}$, do not differ appreciably from the corresponding values under normal laboratory conditions of pressure and temperature. It turns out that the only common rocks which have this characteristic velocity are peridotites and eclogites. Most eclogites, however, are commonly slightly denser. That both rock types do occur, and with rather wide mineralogical variety, in the upper levels of the mantle is shown by their local appearance as fragments among debris transported rapidly to the surface in kimberlite and nephelinite diatremes (Chap. 10).

Anhydrous peridotites are stable over the whole possible temperature-pressure regime of the upper mantle (Chap. 13). The field of stability of eclogite at pressures up to 15 or 20 kb is more restricted. D. H. Green and Ringwood (1967*a*) find gabbro stable up to 12 or 15 kb, eclogite above 20 kb at temperatures around 1000°C. In the intervening pressure interval the garnet-granulite assemblage pyroxenes-garnet-plagioclase is stable. Crucial in extrapolating

* 1 Megabar (Mb) = $10^6$ bars.

these results to temperatures and pressures at the oceanic Moho is the slope of the transition zone on a *PT* diagram. This slope cannot depart greatly from that of univariant equilibrium curves for solid $\rightleftharpoons$ solid transitions with comparable $\Delta V$ and $\Delta S$. For jadeite + quartz $\rightleftharpoons$ albite and for jadeite $\rightleftharpoons$ albite + nepheline, $dT/dP$ at temperatures below 400°C is about 50° $kb^{-1}$. Green and Ringwood use a slope of 40° $kb^{-1}$, and this would make the anhydrous eclogite assemblage stable at pressures of 2 or 3 kb up to 400°C, that is, under conditions prevailing at the oceanic Moho.[1]

Stability considerations alone do not discriminate between eclogite and peridotite as possible upper-mantle rocks. However, the sharpness of the oceanic Moho discontinuity appears more compatible with a compositional boundary such as gabbro-peridotite than with the complex "phase transition" gabbro-granulite-eclogite that would characterize discontinuity in a system of uniform basaltic composition. This in no way rules out eclogite from a possible major or even the dominant role in the composition of the upper mantle. It could simply reflect a tendency for peridotite, by virtue of its somewhat lower density and experimentally demonstrated capacity for plastic flow in the solid state, to become concentrated upward toward the Moho.

A dozen years of stimulating discussion supported by high-pressure experiment has failed to resolve the problem (cf. Wyllie, 1971*a*, pp. 67–91). Geophysical evidence considered alone is equally consistent with alternative models:

1. A compositionally homogeneous upper mantle composed of peridotite. Aluminum not accommodated by pyroxenes is expressed mineralogically at different depths by anorthite (<30 km), spinel (30 to 70 km), or garnet (>70 km).

2. An eclogite upper mantle, perhaps with minor peridotite immediately below the oceanic Moho.

3. A coarsely heterogeneous eclogite-peridotite upper mantle. Equally permissible are regular lithologic variations in depth or laterally (as from oceanic to continental segments) or again random variation within the very large-scale lateral homogeneity assumed in seismic models. To go further we must turn to the evidence of petrology and experimental geochemistry; the nature of magmas, viewed in the light of experimentally determined high-pressure equilibria, places other and stricter constraints on the chemistry and mineralogy of source rocks in the upper mantle. To this topic, which is the outcome of descriptive igneous petrology, we return in the final chapter.

[1] Yoder and Tilley (1962, p. 498) placed the lower pressure limit of eclogite stability above 10 kb at temperatures of a few hundred degrees. But the slope of their transition zone ($dT/dP$ = 140° $kb^{-1}$) is unrealistic.

THE LOW-VELOCITY LAYER The low-velocity layer, it will be recalled, is that portion of the upper mantle in which the velocity of *S* waves is slightly less than it is immediately above or below; in some regions (for example, western North America), the velocity $V_P$ also seems to reach a minimum in this layer. The density is not precisely known; according to some models density also goes through a minimum in the low-velocity layer, whereas in others it remains constant or increases very slightly with depth. A characteristic feature of the low-velocity layer is lateral variability in its properties. In oceanic regions, for instance, the shear velocity $V_S$ begins to decrease at 60 to 80 km in depth and reaches a minimum near 120 to 150 km. In shield areas, the decrease is best marked and begins somewhat deeper, the broad velocity minimum occurring at a depth of some 200 to 250 km.

A decrease of $V_S$ or $V_P$ with increasing depth could be caused by a particularly steep temperature gradient, since increasing *P* and *T* generally have opposite effects on seismic velocities, which normally increase with increasing *P* or with decreasing *T*; the temperature gradient required is 20° $km^{-1}$ or more for $V_P$, 8 to 18° $km^{-1}$ for $V_S$ (Anderson and Sammis, 1970). The low-velocity layer, however, has two other properties which suggest a different interpretation.

Broadly speaking, the low-velocity layer coincides with the asthenosphere, or "weak sphere," a zone of the mantle in which flow seems to be relatively easy. The existence of the asthenosphere had been postulated, long before the determination of the velocity structure of the low-velocity layer, to account for the mantle flow necessary to establish and maintain isostatic equilibrium in spite of changes in crustal loading, such as may result from erosion or sedimentation, or accumulation and subsequent melting of ice. In particular, the observed gravity anomalies and rate of uplift of Scandinavia since the disappearance of the latest Pleistocene ice sheet that once covered it seem to require the existence at depth of a compensatory flow, first away from the center of loading so as to allow the crust to sink and then toward the center when the load has melted away and the crust rises again. The equivalent viscosity[1] of the asthenosphere calculated from the rate of uplift of Scandinavia is about $10^{21}$ to $10^{22}$ g $cm^{-1}$ $s^{-1}$. A similar viscosity is required beneath the lithosphere[2] to allow lateral displacement of lithospheric plates at the inferred rates of a few centimeters per year.

Another interesting property of the low-velocity zone is that seismic waves and particularly *S* waves are damped more rapidly when traveling through it than when traveling through the lithosphere or the deeper mantle; those modes

[1] That is, the viscosity of a Newtonian fluid that would flow at the observed rate. This equivalent viscosity does not necessarily imply that the asthenosphere indeed behaves as a Newtonian fluid.

[2] The lithosphere comprises the crust and less mobile part of the mantle overlying the low-velocity zone. Its thickness, not precisely known, is probably a function of distance from the oceanic ridge axis where it forms. Plates are fragments of the lithosphere that move as more or less rigid bodies.

of free oscillation for which the amplitude of motion is greatest in the low-velocity layer are also more rapidly damped.

These two properties of relatively low viscosity and rapid damping of elastic oscillations, together with the relatively low velocity of $S$ waves, can be accounted for on the assumption that the low-velocity layer is partly molten, as if the temperature in it reached or slightly exceeded the temperature of incipient melting. The velocity of a shear wave being, of course, zero in a liquid, the presence of a small amount of melt would affect $V_S$ more than $V_P$, as observed. The percentage decrease in shear-wave velocity caused by the presence of a small amount of interstitial fluid depends on the amount of fluid and also on the shape, or "aspect ratio," of the fluid-filled spaces. To account for the observed decrease of $V_S$ in the mantle, it seems that about 10 percent of fluid would be required if the melt-filled cavities were spherical, but only 1 percent if they are about 100 times broader than they are thick. The latter case seems more likely, since incipient melting at a eutectic point occurs only on intergranular surfaces, where two different phases are in physical contact. The presence of an intergranular film of liquid between solid grains understandably also enhances the ability of the material to deform and flow. The important point to bear in mind is the implication that temperatures in the mantle normally reach the melting point (or, better, the melting range) in, and perhaps only in, the low-velocity zone. We also note that there is at present no reason to believe that the low-velocity layer differs petrologically from the mantle above or immediately below it.

THE MANTLE BENEATH THE LOW-VELOCITY ZONE Below the low-velocity layer, both $V_P$ and $V_S$ velocities and density generally increase with depth; as noted above, this increase appears to be particularly rapid in two thin zones located respectively at depths of about 400 and 650 km. It will also be remembered that the Adams-Williamson equation applied to the mantle as a whole leads to the conclusion that it cannot be homogeneous; in particular, density in the mantle increases faster with depth than can be accounted for by self-compression of a homogeneous material with the incompressibility coefficient calculated from the density and the local value of $\phi = V_P^2 - \frac{4}{3}V_S^2 = K/\rho$. Part of the downward density increase in the mantle must be assigned to changes in chemical composition or crystal structure (polymorphic transitions), or both.

A simple polymorphic transition in a one-component system (for example, kyanite $\rightleftharpoons$ sillimanite), as required by the phase rule, occurs at a unique temperature for any given pressure (depth). The slope of the corresponding univariant equilibrium curve on a $PT$ diagram is determined by the Clausius-Clapeyron equation [cf. (Eq. 6-2)]. Such a transition, if it occurred within the mantle, would cause a sharp jump in density and in seismic velocities. Polymorphic transition between solid solutions in a two-component or more complex system [for example, $(Mg, Fe)_2SiO_4$] has a variance greater than 1. It would

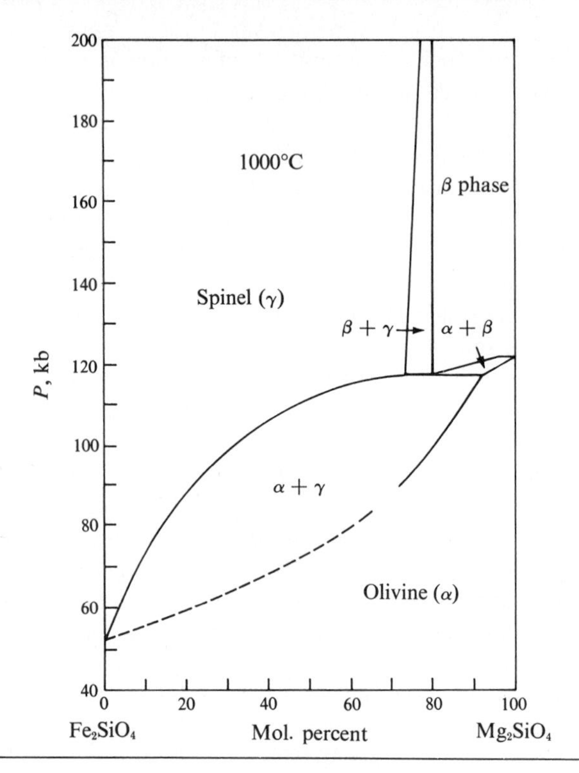


**Fig. 7-1** The $Fe_2SiO_4$–$Mg_2SiO_4$ system at 1000°C. (After Ringwood and Major, 1970.)

be completed over a range of temperature and pressure (depth); and resulting change in density and seismic velocity likewise would be more gradual but confined to some limited range of depth. This is the character of the two transition zones at 400 and 600 km respectively within the upper mantle. They are currently attributed to successive polymorphic transitions in multicomponent mantle rocks.

In particular, the transition near 400 km is generally thought to correspond to the inversion in the $Mg_2SiO_4$–$Fe_2SiO_4$ system from the low-pressure orthorhombic form with the structure of olivine to a high-pressure isometric polymorph with the spinel structure. This system has been extensively studied experimentally. Fayalite ($Fe_2SiO_4$) inverts to the spinel form at $P = 50$ kb, $T = 1000$°C. Pure forsterite ($Mg_2SiO_4$) inverts at a higher pressure of 120 kb at 1000° to a distorted spinel structure called $\beta$–$Mg_2SiO_4$, so that complications occur in the magnesium-rich portion of the phase diagram shown in Fig. 7-1. The temperature gradient of the transition is not well known but is probably close to 30° $kb^{-1}$. If it is indeed the olivine-spinel transition that occurs near 400 km depth, where the pressure is approximately 130 kb, the temperature at that depth must be in the neighborhood of 1600°C; its value depends, of course, on the Fe/Mg ratio in olivine in the mantle at that point, which is not precisely known but which is probably close to 0.1. All this argument depends on the assumption that olivine is a principal constituent of mantle rock at 400 km depth. There are other possible transformations, such as incorporation of $MgSiO_3$ into garnet solid solution, which seems to be possible at pressures around 120 kb (Ringwood and Major, 1966).

There is as yet no clear-cut interpretation of the transition near 650 km; several polymorphic transformations seem possible. A transition induced by increasing pressure must necessarily lead to a form of smaller molar volume, which is most easily effected by increasing the coordination number of the cations and the closeness of packing of the oxygen anions. Thus the distorted spinel structure of $\beta$–$Mg_2SiO_4$, in which Si is still in four-coordination as in olivine, could conceivably break down to a mixture of periclase (MgO) and stishovite ($SiO_2$) in which Si is six-coordinated. Alternatively, $\beta$–$Mg_2SiO_4$ could break down to periclase plus a form of enstatite with the ilmenite ( corundum) structure in which both Mg and Si are in six-coordination. Several other highly coordinated structures (strontium plumbate, perovskite, etc.) have also been suggested. Garnet could transform to a mixture of two phases having the ilmenite and perovskite structure, respectively. Pyroxene *could* invert to a corundum structure (Ringwood and Major, 1966).

Between depths of about 650 and 2700 km or so, both $V_P$ and $V_S$ appear to increase downward at a slow and uniform rate, suggesting that the mantle in this region may be mineralogically homogeneous. The density distribution calculated from the Adams-Williamson equation agrees well with free-oscillation data, indicating that departures from homogeneity and adiabaticity are slight. Recent studies suggest, however, that the velocity distribution may be less uniform and gradual than it was thought to be, perhaps indicating the occurrence in the lower mantle of several more phase transitions with relatively small changes in density. The gross chemical composition, and particularly the Fe/Mg ratio of the lower mantle, remain uncertain. Because iron is heavier than other common elements (Si, Mg, Al, etc.), equilibrium in the earth's gravitational field requires a progressive and marked enrichment in iron with increasing depth. There is, however, no direct evidence that gravitational equilibrium is attained in the mantle any more than it is in thick, intrusive bodies, in which such an enrichment is generally not observed. Thus the lower mantle might well have approximately the same, or nearly the same, gross chemical composition as the upper mantle, although its mineralogy must be quite different. Most calculations of the density that lower-mantle material would have at ordinary pressure yield numbers in the neighborhood of 4.0 ($\pm$0.1) g cm$^{-3}$, as opposed to 3.3 g cm$^{-3}$ for the upper mantle. Thus a large fraction of the actual density difference between top and bottom of the mantle (3.3 to 5.6 or so) is caused by phase transitions, the rest being due to self-compression.

## Temperature and Melting

### TEMPERATURE DISTRIBUTIONS IN THE MANTLE

The ability of the mantle to transmit elastic shear waves, the observed periods of the earth's free oscillations, and the manner in which the earth deforms under the tidal gravitational pull of the sun and moon—all clearly indicate that the mantle as a whole possesses a very high rigidity and is therefore generally at a temperature well below its melting point. The only known exception to this generalization occurs in the low-velocity layer, where the temperature of incipient melting is closely approached or slightly exceeded. To understand how large quantities of melt can form in the mantle, it is thus necessary to examine (1) what the temperature distribution could be at any instant of geologic time, for example, now; (2) how melting temperatures vary with depth or pressure; and (3) what mechanisms could locally cause the temperature of certain portions of the mantle to exceed the local melting temperature. Possible sources for the necessary latent heat of melting must also be examined.

### FACTORS IN HEAT TRANSFER

CONDUCTION If heat is transferred by conduction, the temperature distribution at any time depends on (1) the temperature distribution at prior times; (2) the rate of heat generation, which is determined mainly by the distribution of radioactive elements (U, Th, K) that generate heat; and (3) the values of certain physical parameters, notably thermal conductivity. The temperature distribution at prior times is likely to be important because heat transfer in the earth is notoriously slow. The thermal conductivity of rocks is indeed so small, and the dimensions of the earth are so large, that if heat were transferred only by conduction, temperatures prevailing now at depths greater than a few hundred kilometers would still be controlled mostly by the initial conditions that prevailed when the earth formed. The rate of heat generation in the earth is essentially determined by the abundances of radioactive elements which, in the case of uranium, thorium, and even potassium, occur at such low concentrations that their abundance cannot be determined from density or elastic properties (which they affect only to a negligible degree). Their concentrations must thus be guessed, as from the generally assumed resemblance between the earth and meteorites, or from geochemical observations to the effect that (1) the ratio of uranium to thorium in rocks tends to be generally uniform; and (2) uranium and potassium content generally increase sympathetically, both being higher in rocks with a high silica content (for example, granites) and lower in ferromagnesian rocks of low silica content (for example, dunite).

Most calculations of the temperature distribution in the earth, or at least in the upper mantle, start from the fundamental datum provided by the measurement of surface heat flow, that is, of the rate at which heat escapes through the earth's crust. This heat flow equals, of course, the temperature gradient times the thermal conductivity of the near-surface rocks in which the gradient is measured; it is, on the average, about $1.5 \times 10^{-6}$ cal cm$^{-2}$ s$^{-1}$. A certain distribution of radioactive matter is then assumed, and temperatures are calculated from conduction theory, making allowance for the possibility of heat transfer by radiation (photons) as well as by ordinary lattice conduction (phonons). The thermal conductivity by ordinary lattice conduction is not sensitive to composition or pressure, so that no great error can arise from extrapolation of surface measurements to mantle conditions. Radiative conductivity, by contrast, is very sensitive to temperature and to such structural conditions as grain size, lattice imperfections, impurities, etc., which are not well known. Radiation was once thought to be the dominant mode of heat transfer in the mantle at depth greater than 100 km or so, but recent measurements of optical absorption in minerals at elevated temperatures and pressure suggest that radiation transfer may be much less important than it was thought to be. Because of probable differences in radioactive content of the continental and oceanic crust and mantle, calculations must be carried out separately for continental shield and oceanic areas. The often-reproduced oceanic and shield "geotherms" of S. P. Clark and Ringwood (1964) are products of such calculations, which are no longer thought to be valid.

Thermal data have been used in attempting to determine the composition of the mantle. If the unwarranted assumption is made that the earth at present is neither heating up nor cooling down, the heat now escaping at the surface must be exactly equal to the heat generated inside. Since most of the heat is generated by radioactive decay of U, Th, and $K^{40}$, heat flow measures the total amount of these elements in the earth.

These data by themselves tell little about the total chemistry of the mantle. Further speculation requires the risky assumption that some kind of accidentally accessible planetary matter represents mantle rock. Stony meteorites are of course prime candidates for such extrapolation. Birch (1965) found, for instance, that present total surface heat flow divided by the total mass of the earth is very close to the rate of heat generation in olivine-pigeonite chondrites, but slightly higher than that for average chondrites. He noted, however, that the K/U ratio is much higher in these chondrites than in terrestrial rocks. Thus, because of the common geochemical association of K and U in rocks, we would expect to find much more K in the crust of a chondritic earth than we actually do. By contrast, carbonaceous chondrites such as the meteorite Orgueil have a lower K/U ratio, and the rate of heat generation in them, although higher than in chondrites, is still less than the surface heat flow divided by the mass of the mantle (as opposed to the mass of the earth). Birch therefore suggests that the primitive undifferentiated earth may have resembled the meteorite Orgueil, at least with respect to its radioactive content. This argument for preferring carbonaceous to other chondrites for the primitive earth fails if, as recently suggested by J. F. Lewis (1971), much of the terrestrial K is presently in the earth's core where it could form stable compounds with sulfur (compare to the rare mineral djerfisherite, a potassium-iron sulfide found in some meteorites). Presence of K in the core rather than in the crust would significantly alter previously accepted distributions of heat sources in the earth and corresponding calculated temperatures. The validity of the whole approach is somewhat doubtful, since it seems unlikely that a few random fragments of extraterrestrial rock can be considered in any respect to represent an average sample of mantle rock.

It now appears also that calculations based solely on conduction theory may be invalid, except perhaps for the lithosphere or some parts of it. Elsewhere, and particularly in the low-velocity layer, the dominant mode of heat transfer appears to be by convection.

CONVECTION Convection is mass motion caused by the action of gravity on thermally induced density perturbations, the general idea being that hot fluid, being lighter, tends to rise while cold, dense fluid tends to sink. Convection implies departure from hydrostatic equilibrium, which is equilibrium in a fluid at rest. It can easily be shown that hydrostatic equilibrium in a gravitational field requires that both density and pressure be constant on gravitational equipotential surfaces which are, by definition, normal to the local direction of gravity (gravity is the gradient of the gravitational potential). In a homogeneous fluid in which density depends on both $P$ and $T$, constancy of density on isobaric surfaces also requires that these surfaces be isothermal; thus in a liquid at rest, density, pressure, and temperature must all three be constant on equipotential surfaces. Instability arises and motion occurs when this condition is not satisfied, as when, for instance, temperature varies laterally on an isobaric or equipotential surface.

Equipotential surfaces in the earth are nearly ellipsoidal surfaces on which are superposed very small undulations; they are essentially parallel to the geoid, which is the equipotential surface that coincides with mean sea level. It is unlikely that temperature could be uniform on such surfaces. It is clearly not uniform on the geoid itself, where differences in temperature between equator and poles cause oceanic circulation and currents; nor can it be uniform on the surface that coincides roughly with the ocean floor, since the temperature on the sea floor is about 2°C while in the continental crust it may be as high as 150°C or so at this same level, some 5 km below the surface. Temperature also seems to be variable at the same level beneath continents, as indicated for instance by difference in speeds $V_P$ and $V_S$ in the upper mantle below the western or eastern part of North America. Sharp lateral variations in the electrical conductivity of the upper mantle also seem to indicate lateral variations in temperature. It is noteworthy that calculated geotherms for shield and oceanic areas show differences of a few hundred degrees at depths of the order of 100 km; such differences would be sufficient to induce convection that would invalidate thermal calculations based on conduction theory alone.

How large must lateral temperature variations be to induce convection?

The answer depends of course on local rheological properties. In rigid bodies of great strength, very large horizontal gradients of temperature are needed to induce stress differences greater than the strength; but in a Newtonian fluid of zero strength, horizontal gradients of the order of a fraction of a degree per kilometer are sufficient to maintain flow at a rate of several centimeters per year in a fluid as viscous as the low-velocity layer.

Recently developed concepts of plate theory further tend to invalidate temperature calculations by conduction theory,[1] since these start from the observed surface heat flow and assume that heat flow is essentially vertical, that is, that heat flow at a point on the surface is determined by the temperature distribution and heat sources beneath that point. It now appears much more likely that the heat which escapes from the lithosphere, at least in oceanic areas, is heat liberated by cooling of the lithospheric plate as it slowly moves away from the oceanic ridge where it forms. Lateral mobility of the plate dissociates what happens in it from what happens beneath it; heat flow at a point $A$ is determined not so much by what happens in the mantle beneath $A$ as by what happened, some tens of millions of years ago at some point $B$, a few hundred or a thousand kilometers away, where the lithosphere formed. It is not inconceivable that the temperature at some point in the lithosphere could be higher than it is in the mantle under that same point.

Even when the temperature is uniform on an equipotential surface, instability may arise where there exists a vertical temperature gradient greater than the so-called adiabatic gradient. This gradient measures the rate at which temperature would increase as a result of self-compression in thermally insulated systems in which the heat generated by compression all goes to raise the temperature of the system. For a reversible compression through a pressure range $dP$, adiabaticity means constancy of entropy, and it is easily shown that at constant entropy temperature varies with pressure as

$$\left(\frac{dT}{dP}\right)_S = \frac{\alpha T}{\rho C_p}$$

where $\rho$ is density, $\alpha$ is the coefficient of thermal expansion $-1/\rho(\partial\rho/\partial T)_P$, and $C_p$ is the specific heat at constant pressure. Substituting for $dP$ its hydrostatic value $-g\rho dr$, the adiabatic gradient becomes

$$\left(\frac{dT}{dr}\right)_S = -\frac{\alpha T g}{C_p}$$

which is of the order of a few tenths of a degree per kilometer in the mantle. The actual gradient in excess of the adiabatic value necessary to cause flow

[1] Except in the rigid lithosphere, perhaps 50 to 100 km thick, in which heat transport is by conduction except where magma is injected.

depends on several factors, such as the viscosity (a high value of which tends to inhibit convection) and the linear dimensions of the system. As it turns out for a body of large dimensions such as the mantle, convection becomes possible, in spite of a viscosity as high as $10^{22}$ to $10^{23}$ poises, if the temperature gradient exceeds the adiabatic value by as little as a small fraction of a degree per kilometer.

There is a further reason why the mantle should convect. A powerful theorem due to the astrophysicist H. von Zeipel asserts that hydrostatic equilibrium is generally impossible in a homogeneous fluid containing heat sources (for example, radioactive matter) if it also rotates. McKenzie (1968) has attempted to describe the circulation (or Eddingtonian currents, as they are called) that would ensue and finds it to be negligible. Calculations are notoriously difficult, and McKenzie's results are valid only for a mantle of extremely high viscosity ($10^{27}$ poises) in which convection of any kind is probably impossible, or at least so slow as to be geologically insignificant. He points out, however, that Eddingtonian currents would be symmetrical about the plane of the Equator, with flow confined to meridional planes; because he sees no indication in the earth's surface features of dominantly north-south (or south-north) flow, he dismisses Eddingtonian currents as unimportant. The matter may require reexamination under less restrictive assumptions; but whatever the results may turn out to be, the important conclusion must be kept in mind that not one but several lines of evidence indicate the likelihood of convective motion in the mantle or those portions of it, such as the low-velocity layer, that have an equivalent viscosity of less than $10^{23}$ poises or so.

Under these conditions it becomes very difficult indeed to estimate mantle temperatures, since the temperature at any point in time or space will depend on the local sense and rate of motion; the temperature distribution will be different, for instance, in currents respectively ascending or descending. Nor is the pattern of motion likely to be invariant in time or regular in space. More probably, as Elder (1968) has pointed out, the pattern of motion will be irregular, with blobs of matter rising occasionally here or there in rather random fashion. Much depends of course on the unknown actual rheological properties of the mantle and particularly on the dependence of its viscosity on stress. Although there are theoretical reasons for suggesting that viscosity may be independent of stress, as in a truly Newtonian fluid, these reasons, as pointed out by Weertman, are not compelling.

## TEMPERATURE INDICATORS

If the mineralogical composition of the mantle and the pressure and temperature dependence of $V_P$ and $V_S$ in its constitutive minerals were exactly known, the temperature at any depth could, in theory, be calculated from the difference between observed velocities $V_P$, $V_S$ at any depth and velocities calculated for the corresponding pressure at room temperature. But, as we have seen, very little of the necessary information is presently available.[1] The only indications relevant to the temperature are as follows: (1) temperature distribution in the lower mantle, below approximately 700 km, is probably very close to adiabatic; (2) the temperature near 400 km must be about 1600°C, if the transition that occurs there is indeed the olivine-spinel transition, and if the composition of olivine at that depth is approximately $Mg_{0.89}Fe_{0.11}SiO_4$; and (3) the temperature in the low-velocity layer is everywhere close to that of incipient melting. To this matter of melting temperatures and their dependence on depth we now turn.

## MELTING TEMPERATURES IN THE MANTLE

Neither the temperature of incipient melting (solidus) or of total melting (liquidus) nor their pressure dependence can be ascertained, since the composition of the mantle is not precisely known; in particular, its water content, which may greatly affect its melting behavior, is very uncertain. The water content of inclusions in basalts and kimberlites is not necessarily a good indicator of that in the mantle; for it is not generally agreed that these inclusions do come from the mantle, and if they do, the amounts of water lost or gained by them on their way up may be difficult to estimate.

MELTING OF PERIDOTITE A dry feldspathic peridotite might begin to melt at about 1100°C at ordinary pressure, and its melting point (mp) would increase with pressure at a rate of about 10° $kb^{-1}$. This rate is likely to be increased by any pressure-induced phase transformation, as shown in Fig. 7-2*a*, which refers to a one-component substance with a polymorphic transition, and in Fig. 7-2*b*, which refers to a hypothetical two-component system with mutual solubility of the two components in both the low-pressure and high-pressure solid phases. Thus the initial slope of 10° $kb^{-1}$ of the peridotite solidus is likely to increase with depth as first the proportion of spinel and then that of garnet gradually increases.

In the presence of water vapor, however, the melting behavior becomes more complicated. If there is enough water to saturate the melt, increasing pressure (assumed to be the same on all phases, including the vapor phase) depresses the melting point as it also increases the water concentration in the melt. It is useful to remember that because of the low molecular weight of water as compared to that of silicates, a relatively small amount of water, measured in weight percent, has a relatively large effect on depressing the melting point of silicates. If only a small amount of water is present, it will first depress the melting point, but when all the available water has gone into the melt, the melt's water content cannot further increase with increasing pressure, and the

[1] As noted above, the uncertainty on the density at any depth may be of the order of 2 or 3 percent. Since the coefficient of thermal expansion in the deeper mantle is of the order of $2 \times 10^{-5}$, the uncertainty on the temperature arising from that cause alone is of the order of 1000°.

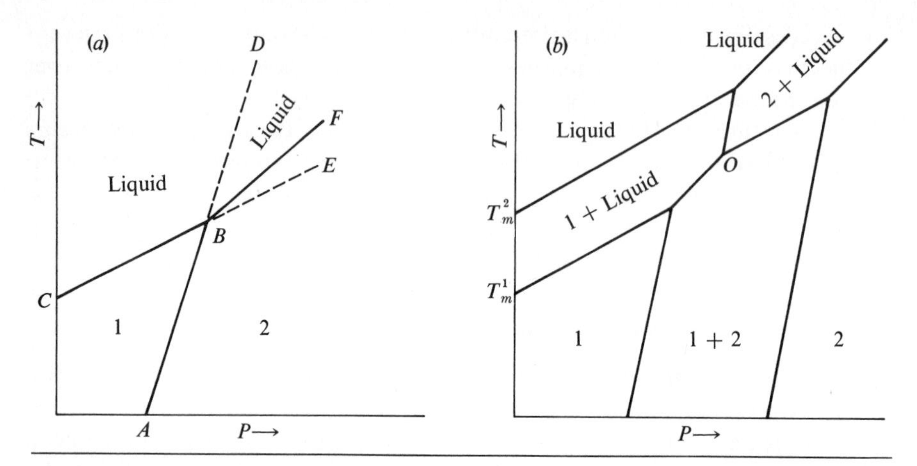


**Fig. 7-2** Phase diagrams for polymorphic transitions. (*a*) A substance that undergoes a polymorphic transition represented by the univariant equilibrium curve *AB*. *CB* is the melting curve of phase 1; *BF* is the melting curve of phase 2. (*b*) A system of two components (for example, $Fe_2SiO_4$–$Mg_2SiO_4$) forming a solid solution that undergoes a phase transition from form 1 (for example, olivine) to form 2 (for example, spinel). The diagram is applicable to a system of fixed composition (for example, 10 percent $Fe_2SiO_4$). $T_m^1$ and $T_m^2$ are, respectively, the temperatures at which melting begins and ends at $P = 1$ bar. Three phases coexist at *O*. In any two-phase region, the phases in equilibrium do not have the same composition.

melting point begins to rise much as it would if the system were anhydrous. A similar change from negative to positive of the slope of the melting point occurs when the wet solidus intersects the dehydration curve of possible hydrous phases (for example, amphibole or phlogopite). This is shown in the hypothetical dehydration curve 1 in Fig. 7-3.

Figure 7-4, taken from Wyllie (1971*a*), approximately represents the melting behavior to be expected for a peridotite with 0.1 percent water. It shows a broad minimum of the solidus at about 1000°C and 35 kb (the equivalent depth is approximately 100 km). The existence of this minimum could perhaps explain why incipient melting is restricted to a relatively narrow zone of the mantle. But remembering that the amount of interstitial melt in the low-velocity layer is nowhere greater than 5 or 10 percent and probably closer to 1 percent, one may wonder what constrains the temperature in that layer to remain so close to the solidus. Presumably melting and the latent heat associated with it serve as a kind of temperature buffer (see below). We note also that the amount of interstitial melt is probably not uniform in the low-velocity layer, as shown by regional differences in the seismic velocity distribution.

MELTING OF ECLOGITE At any given pressure dry eclogite becomes completely melted within a narrow temperature interval—no more than 100°. The liquidus temperature increases from 1300°C at 25 kb to 1500°C at 35 kb. The presence of only 0.1 percent water, held in accessory hornblende, drastically

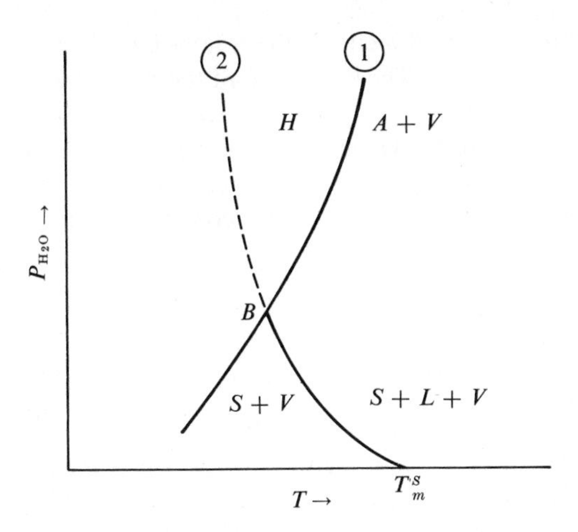


**Fig. 7-3** Melting in a system with a hydrous phase $H$. Curve 1 is the univariant dehydration equilibrium $H = A + V$. Curve 2 is the minimum melting curve for S in the presence of water. $T_m^S$ is the minimum melting point of dry $S$. Beyond point $B$, water pressure is controlled by dehydration of $H$, and partial melting of $S$ follows curve 1. $S$ signifies solids, $L$ liquid, $V$ vapor.

**Fig. 7-4** Melting in a hypothetical peridotite with 0.1 percent water (after Wyllie, 1971, p. 132). Curve 1 (solidus) represents the beginning of melting; curve 3 is the liquidus; curve 2 is close to the solidus for the anhydrous system. Segment of melting curve of hornblende $H$ is shown dashed. In the region between 1 and 2, there is only a very small amount of melt. The stippled area represents the olivine-spinel transition.

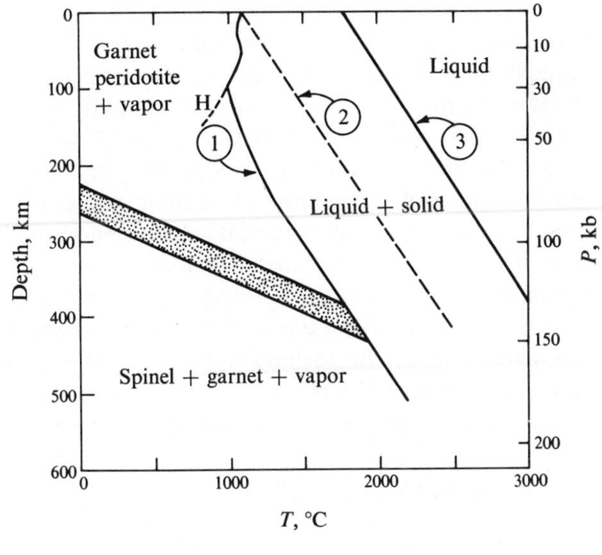

lowers temperatures of initial melting (the solidus curve) in the pressure interval 25 to 35 kb (80 to 120 km). There is a broad temperature span—roughly from 700 to 1400°C—over which a very small fraction of hydrous "andesitic" melt coexists with the normal dry eclogite assemblage (Wyllie, 1970; 1971*a*, p. 122, fig. 6-13). Such behavior meets the requirements of the low-velocity layer without imposing the narrow temperature constraints required by the slightly hydrous peridotite model.

## Formation of Magma: Mantle Influences

### BASALTIC MAGMA

In dry systems, as we have seen, melting temperature increases with increasing pressure. This observation has led to the hypothesis that melting and magma formation are caused by relief of pressure. This relief was commonly assumed to occur by opening of vast and deep fissures or by arching of the lithosphere. Neither mechanism appears likely or even possible. It is clear that, except in the uppermost lithosphere, departures from hydrostatic equilibrium in the mantle are very small even where it flows convectively or in response to loading or unloading of the surface. Nonhydrostatic stresses, that is, differences between maximum and minimum principal stresses, are everywhere small compared to the mean of those principal stresses which is the hydrostatic pressure; they cannot therefore sensibly affect the melting point. But the hydrostatic pressure gradient is a function of the mass and size of the earth and of the density distribution within it, and it cannot be reduced by any mechanism short of reducing the mass of the earth or increasing its radius (with constant mass).

Pressure on a particular element of mass of the mantle can, however, be reduced by moving the element upward, as in a rising convection current. As pointed out long ago by the authors of this book, convection is probably the main cause of melting and magma formation. The principle can be easily understood by referring to Fig. 7-5. Consider, to simplify, a one-component system with a melting curve represented by the solid line $AB$. Suppose an element of mass initially at $M$ at a temperature less than the melting temperature at that depth, and suppose that it is displaced upward without exchanging heat with its surroundings. As it moves to lower pressures, it expands and cools adiabatically along the line $MM'$. Because the adiabatic gradient in the upper mantle is almost certainly much smaller than, and perhaps only about one-tenth of, the melting-point gradient, the adiabatic curve $MM'$ intersects the melting curve $AB$ at $M'$ and melting begins. But melting requires a certain amount of heat (latent heat) that must be supplied by the system itself, which therefore must cool as it rises, its temperature remaining everywhere along the melting curve $AB$.*

* This is the buffer action referred to on page 352.

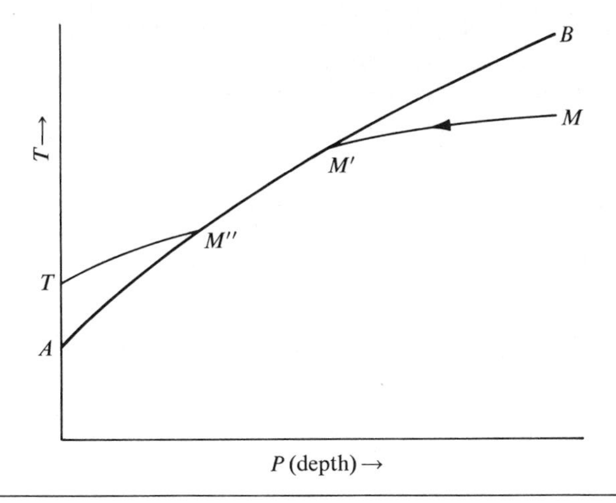


**Fig. 7-5** Melting by upwelling.

At $M''$, cooling from $M'$ to $M''$ has supplied all the necessary latent heat, and the system is now completely liquid. The liquid continues to rise by buoyancy, since it is lighter than solid mantle at the same depth, and reaches the surface at temperature $T$. The vertical distance between $M'$ and $M''$ can be roughly estimated as follows. If the latent heat of melting is about 100 cal $g^{-1}$ and the specific heat is 0.3 cal $g^{-1}$, a cooling of about 330° is required. If the slope of the melting curve is 10° $kb^{-1}$, the pressure difference between points $M'$ and $M''$ is 33 kb. This corresponds to a vertical distance of 100 km if the density is 3.3 g $cm^{-3}$, or slightly less if the density is greater. The vertical distance between $M'$ and $M$ depends on the difference between the melting point and the actual temperature in the mantle; it could easily be of the order of 100 km or more.

Basaltic magma reaches the surface, at the Hawaiian volcanoes for instance, at temperatures approaching 1200°C. The temperature at which it forms, corresponding to the point $M''$ in Fig. 7-5, is not likely to be much less than 1250 or 1300°, although it is difficult to estimate precisely how much adiabatic cooling occurs between $M''$ and the surface or to estimate heating caused by surface oxidation of some of the magmatic components, notably sulfur, which is occasionally seen to burn where magmatic gases come into contact with air. The temperature at the top of the low-velocity layer appears from Fig. 7-5 to be much less than 1300°, from which it may be surmised (1) that the basaltic magma which reaches the earth's surface is not just interstitial melt squeezed out of the low-velocity layer, and (2) that magma formation must begin rather far down, toward the bottom of the low-velocity layer or even below it; the point $M$ where the disturbance starts must lie even deeper. It may be noted in passing that there is very little direct evidence as to the depth at which magmas actually form, except that they must do so below the lithosphere if the linear arrangement of volcanoes of progressively younger age observed, for instance, in the Hawaiian chain and elsewhere, is to be interpreted by the displacement of the lithosphere above a magma source at a fixed point in the mantle.

The depth $M'$ (Fig. 7-5) at which copious melting begins is determined by the temperature at $M$, where the upward motion begins, and the adiabatic gradient through $M$. The deepest possible level for $M'$ would be at the intersection of the melting curve and of the deep-mantle geotherm extrapolated adiabatically upward. The melting curve is not precisely known; in particular, the effect on it of the phase transitions near 400 and 650 km has not been determined. The adiabatic gradient, as we have already shown, is directly proportional to the temperature and is therefore also uncertain. It does not seem unlikely, however, that large-scale melting might be restricted to the upper 200 or 300 km of the mantle, even though the upward motion that causes it may start much deeper.

We thus view the formation of basaltic magma to be a necessary consequence of convective upwelling starting perhaps in the deeper mantle tens or hundreds of kilometers below the low-velocity zone. A great deal has been written on possible causes of convection in the deeper mantle, the main cause presumably being radioactive heat generation (perhaps in the earth's core) leading to a temperature distribution slightly in excess of the adiabatic gradient. Contrariwise, it has also been argued that convection is impossible in the deep mantle for two reasons: first because its viscosity is too great, and in the second place because polymorphic transitions, such as the olivine-spinel transition, act as impenetrable barriers to convection currents. None of these arguments appears to be final, since the viscosity of the mantle below the low-velocity layer is not known and since it has recently been shown by Schubert and Turcotte (1971) that a phase transition may enhance instability rather than hinder it.

But is it always possible to associate basaltic magma production with upwelling in the mantle? Why should there be, for instance, an upwelling of relatively small cross section located below Hawaii or, more precisely, below the southeastern half of that island, which itself stands in the middle of a vast region in which nothing else appears to be happening at present? There is here a striking contrast between the small horizontal dimensions of the source and the much larger dimensions one would expect for any convective motion with such a large vertical scale and a time scale of at least 10 to 20 million years, as demanded by the age of the oldest islands at the northwestern end of the chain (Chap. 8). The contrast becomes even greater if we assume that, as suggested by W. J. Morgan (1971), the upwelling responsible for Hawaiian volcanism is also responsible for the motion of the whole Pacific plate. The localization of basaltic volcanism would seem to require that melting affect only a small proportion of the moving mass, most of which may never cross the melting curve; this would happen if the temperature in the mantle beneath the low-velocity zone were well below the melting range.

Alternatively, Wyllie (1971*b*) has suggested that production of basaltic magma results from diapiric uprise in the low-velocity layer triggered by water rising from the deep mantle into the base of the low-velocity layer where it causes an increasing amount of melt to form and thereby lowers the average density. It has also been suggested that concentration of stress at certain points on the boundaries of grains in a solid mass undergoing plastic flow could locally lower the melting point and cause melting (shear melting).

H. R. Shaw (1969, 1973) has pointed out that melting could result from a thermal feedback mechanism which may occur in thermally insulated systems in which viscosity decreases rapidly with increasing temperatures. In a viscous system with Newtonian viscosity $\eta$, stress $\sigma$ is related to strain rate $\dot{\varepsilon}$ as $\sigma = \eta\dot{\varepsilon}$. Work done on unit volume of the fluid is $\sigma\dot{\varepsilon}$ (force times displacement), and the power $P$ dissipated as heat is

$$P = \sigma\dot{\varepsilon} = \eta\dot{\varepsilon}^2 = \frac{\sigma^2}{\eta}$$

Power dissipation raises the temperature, which lowers the viscosity, which increases $P$ if $\sigma$ is constant; more heat is generated and temperature rises further, until melting of an initially solid system occurs. On the contrary, flow at constant strain rate does not lead to instability, since power dissipation decreases as the temperature rises and the viscosity falls. Shaw believes that melting could occur in the mantle in a shear zone wider than 10 km for a shear stress of about 100 bars and an initial viscosity of $10^{21}$ poises. This mechanism may have some application, particularly to the generation of andesitic magma in a Benioff zone (see below), although it is not clear why flow should occur at constant stress or what causes the localized stress implied by the sporadic occurrence of volcanoes. Why does the process not occur wherever plates are moving, rather than just at very few, very restricted spots? The only point that remains overwhelmingly clear is that the formation of basaltic magma in all its petrological diversity (alkali basalts, tholeiites, etc.) is only very imperfectly understood. As we have seen, much uncertainty remains as to the composition, chemical and mineral, of the mantle, its water content, its melting behavior at relevant pressures, the temperature distribution, and its lateral variations.

## ANDESITIC MAGMA

Physical conditions that lead to the production of copious amounts of andesitic magma are equally or even more mysterious. The common association of andesitic volcanism with island arcs or, more generally, with regions where lithospheric plates are presumed to descend into the mantle (Benioff zones), suggests that melting can also be induced by downward motion. This would clearly be the case if material with a relatively low melting point were to sink from the surface to depths where the temperature exceeds the melting point of that material while still remaining below that of the mantle itself. Thermal and chemical conditions in a descending lithospheric slab are, however, obscure. The slab, like the oceanic lithosphere, presumably consists of (1) a thin (1 km or less) cover of sediments; (2) a thin (5 to 10 km) crust of basaltic, gabbroic, or serpentinitic composition; and (3) a layer of upper mantle rock 50 to 100 km thick. Thus the composition of the descending plate cannot differ very much from that of the upper mantle itself, particularly when it is noted that much of the sedimentary cover appears to be scraped off the plate where it begins to descend and accumulates on the landward side of the oceanic trench which marks the place where descent begins. The main chemical difference between a descending plate and the normal lithosphere lies perhaps in a somewhat higher water content of the former.

Gorshkov reported in 1956 that seismic waves passing under the Kliucheskaya group of volcanoes in Kamchatka were anomalously attenuated at depths between 50 and 70 km; he attributed this attenuation to the presence of a magma reservoir. In a summary of these observations, Gorshkov (1971) also reports the work of Fedotov and Farberov, who observed attenuation of *S* waves at depths between 20 and 80 km under the Avachinskaya volcanic group, also in Kamchatka. Firstov and Shirokov (1971) have confirmed screening of *P* and *S* waves at depths between 35 and 110 km under Kliuchesk-aya, where zones of anomalous attenuation seem to be associated with regions of magma localization along an inclined deep fracture. Kubota and Berg (1967) have observed similar effects near the volcano Katmai, in Alaska; here small magma chambers are thought to occur at depths from 10 to 70 km. In all these instances the depth to the chambers seems to be less than the local depth of the descending slab, suggesting that andesitic magma has its source above the slab or on its uppermost face; however, the interpretation of the seismic observations is difficult, and the exact depth of the magma source is still uncertain.

It is somewhat paradoxical that melting should be associated precisely with the descent of *cold* material, since calculations of the thermal regime of a descending slab essentially show a deep depression of isotherms, even when the heat generated by friction and by self-compression or released by phase transformations is included. The seismic evidence seems indeed clear that the descending plate maintains down to depths of several hundred kilometers a high rigidity and a high $Q$ factor;[1] it seems to cut through the soft, low-$Q$, low-velocity layer which is absent under island arcs (Fig. 7-6). The situation is, however, rather complicated, and the production of magma can presumably not be dissociated from the common occurrence on the inward side of the arc (for example, in the Sea of Japan between the Japanese arc and the mainland of Asia) of anomalously high heat flow. Perhaps, as suggested by McKenzie (1969), the downward motion of the plate induces a compensating upward

[1] The ratio $1/Q$ measures the attenuation or damping of waves and oscillations; high $Q$ (low attenuation) implies absence of viscous or other dissipative processes.

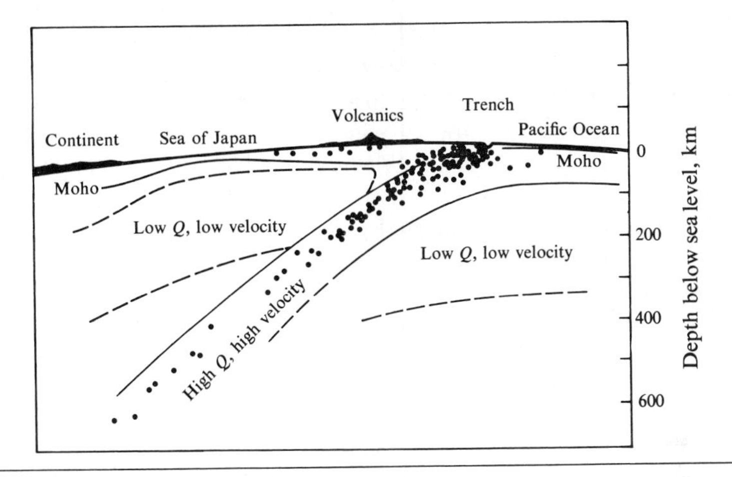


**Fig. 7-6** Deep structure of the northern Japanese arc; dots represent earthquake foci. (After Utsu, 1971.)

motion in the mantle above the inclined plane, which is where the melting might occur. It should be noted also that andesite is now being erupted in regions (for example, the Cascade Range of western North America) where the complete absence of deep-focus earthquakes suggests the absence of a descending slab; it is possible, however, that the andesitic magma that is now coming to the surface may have formed 10 or 20 million years ago when a plate was still descending under this area. The inherent slowness of heat and mass transfer in the earth, and particularly in the lithosphere, is such that present conditions at the surface may reflect conditions as they existed deeper down a long time ago; in any event, the possibility of large time lags does not make the interpretation easier.

## THE MAGMA ENIGMA

In summary, then, the formation of magma remains a geophysical enigma, and much work, particularly experimental work, on the melting behavior of possible mantle materials remains to be done. It would also be interesting to examine more closely possible implications of the fact that magma of peridotitic composition either never forms or, at least, scarcely ever reaches the surface. What is there in the thermal regions of the earth that makes it impossible for magma to reach the surface at temperatures greater than about 1200°C? Geophysical and geochemical implications of this observation have, to our knowledge, never been fully explored, even though they could provide important constraints on temperature distribution in the mantle and on the very mechanism of magma formation.

## The Crust as a Potential Source of Magma: Lithology and Chemistry

Even though most, if not all, basaltic magmas seem to have their source in the mantle, other magmas (granitic, granodioritic, andesitic) may perhaps be wholly or partly of crustal origin. We now examine the evidence for crustal sources of magma.

The continental crust, as inferred from geophysical data, is thicker than oceanic crust (about 30 to 70 km) and more variable. The exposed part of the continental crust is generally less dense than the oceanic crust, but at least locally it appears to be stratified, becoming denser (approaching gabbroic or amphibolitic composition?) downward. Any surface section of the crust, however ancient its component rocks, displays only the upper portion of the crust at that locality. Much of the older material exposed in continental shields is metamorphic. Some of it is volcanic, some sedimentary in origin; and there are great masses of quartzofeldspathic gneiss and granulite whose ultimate origin is hard to decipher. Another important element of the shields is plutonic—great bodies of intrusive granodiorite, gabbro, and anorthosite. All in all, exposed crustal rocks of the continents are more siliceous than the thin "basaltic" crust of the oceans.

In regions of intense heat flow, such as the geothermal province of north-central New Zealand, temperatures in the fusion range of water-saturated crustal rocks may be attained at relatively shallow depths. Here bodies of magma may temporarily exist within the crust and from time to time rise to the surface and burst out as lava and pyroclastic debris. We may envisage similar happenings elsewhere in times past. So it is possible that under favorable conditions crustal rocks, even the superficial covering sometimes called supracrust, may have contributed significantly to magma from the earliest times.

Thus arise questions of paramount importance in igneous petrology. By what criteria can magmas be identified as of crustal versus mantle origin? Or if both sources are involved, is it possible to verify crustal participation and assess its relative contribution? These questions loom large in classic older discussions of the roles of assimilation and contamination in igneous petrogenesis. Answers are based on the reasonable assumption that the crust is composed principally of rocks represented—although the sample admittedly is imperfect—among exposed surface outcrops. Experimental findings reviewed in the previous chapters set limits on the composition range of crust-derived magmas. Most extensively exposed crustal rocks are too siliceous, too rich in alkali, to give rise to magmas more basic than andesite or dacite. Some crustal gneisses are similar chemically to granodiorites, some graywackes to andesites or dacites, some dolomitic limestones to carbonatites. It is conceivable that common crustal rocks could generate or contribute to magmas. That is as far as gross chemical data can take us.

The pattern of minor- and trace-element abundances in a magma may be used to narrow the limits of possible source rocks. Especially significant here are the "incompatible" dispersed elements (page 631) which, because they can be accommodated in common rock minerals only with difficulty, must enter preferentially, even to the point of almost total extraction, into any liquid phase resulting from partial fusion. Attention today focuses on K, Sr, Rb, Pb, Th, and U. The first two can be accommodated easily in feldspars of appropriate composition, but they are incompatible with respect to the various anhydrous phases proposed for mantle rocks—olivine, pyroxenes, garnet, spinel. The sources of indigenous heat in any rock are K, Th, and U. Radioactive isotopes of Rb, Th, and U give birth continuously to daughter isotopes of Sr and Pb; and on this general fact, of course, are based the techniques of radiometric geochronology. But the same data have other potential uses relating to our present problem of magma sources. Before we pursue this topic in the next section, consider briefly some possible patterns of elemental abundance in the upper mantle and especially in the crust.[1]

The key elements K, Th, U, and Rb, like all incompatible elements, tend to enter preferentially not only into silicate melts but also into any fluid phase—magmatic gas, metamorphically generated solution, circulating groundwater—in prolonged contact with crystalline phases in which they reside. If such fluids have been expelled continuously upward for billions of years, vertical concentration gradients conforming to some general pattern might be expected. Elements such as K, U, and Th are far more abundant in the continental crust than they are in peridotites, meteorites, or any likely mantle material. It has been proposed (Harris, 1957)—although this remains an untested model—that potassium for the crust has been drawn preferentially from the uppermost mantle, which is now depleted in K compared with a somewhat more potassic zone a few hundred kilometers deeper. The contrary view that abyssal tholeiitic magmas ascend from a depleted deep source has its proponents. However, if regular vertical composition gradients do exist in the upper mantle, they must terminate at the base of any lithospheric plate currently or recently in lateral motion.

It has been shown more convincingly that within the continental shields large segments of crystalline rocks have been depleted in U, and less so in Th and Rb, by high-grade metamorphism in the granulite facies. This was first demonstrated for parts of the Australian shield (Heier and Adams, 1965; Lambert and Heier, 1968). The Lewisian gneiss of northwestern Scotland has undergone repeated metamorphism, with maximum temperatures in the opening recorded episode—the Scourian, dated at 2600 m.y. minimum, most probably 2900 m.y. The resulting mineral assemblages, where they survive, are in the granulite facies. Moorbath et al. (1969) find that during this episode the Lewisian gneisses became severely depleted in U and have remained so ever since. Abundance values are below 0.3 ppm in most, even below 0.1 ppm in a number of analyzed specimens, as compared with 2.7 ppm commonly cited for average

[1] This topic is taken up in greater detail in Chap. 13.

crustal rock. This severe depletion is reflected in anomalously low abundances of $Pb^{206}$ and $Pb^{207}$ in present-day lead (see below). Thorium and rubidium seem also to have been affected, but not nearly so much so as uranium; lead apparently was not significantly reduced.

### Isotope Chemistry of Magma Sources: Radiogenic "Tracers"[1]

In biology and in medicine, sources and paths of diffusing compounds in living tissue can be traced through artificially introduced radioactive or radiogenic nuclides. Magmatic systems have built-in potential tracers—U, Th, Rb, and their daughter isotopes of Pb and Sr. Present in the source rocks, they are inherited, during genesis and subsequent evolution of the magma and its fractions, in concentration patterns that change predictably in any closed system. Complicating factors peculiar to the geological context of the problem are related to time, possible heterogeneity of source materials, and the roles of contamination, fractionation, and mixing in later magmatic evolution. As the significance of these has become increasingly apparent, we have had to abandon or modify the simple approach that was in vogue a few years ago. Natural tracers nevertheless have great potential. Their use has compelled petrologists to realize that petrogenesis can be—and perhaps in most cases is—a highly complicated process, with unique features in any particular petrogenic province.

The principle of the method is to use the isotopic character of a magma to identify the source rock, since different potential source materials differ markedly in this respect. We look at the present isotopic composition of a rock or group of rocks and calculate what its composition was when it formed $t$ years ago. For a closed system which initially contained none of the daughter element $D$, the atomic abundance $D$ of the daughter nuclide generated from a parent $P$ is given by

$$\log\left(1 + \frac{D}{P}\right) = \frac{t\lambda}{2.3026} \qquad (7\text{-}1)$$

where $\lambda$ is the decay constant of $P$ (cf. Verhoogen et al., 1970, p. 205). Present values of $D$ and $P$ are determined by mass spectroscopy, and $t$ is calculated from them. Complications arise when the system—as almost invariably is the case in geology—initially contains some of the daughter elements or when it has not been closed to both $P$ and $D$; but even in such cases, ages and initial isotopic compositions can be obtained by use of isochrons or concordia curves (Verhoogen et al., 1970, pp. 206–208).

The isotopic composition of strontium is written as the abundance ratio $Sr^{87}/Sr^{86}$; $Sr^{87}$ is the radiogenic isotope from $Rb^{87}$, and $Sr^{86}$ is one of the nonradiogenic isotopes (about 10 percent of average strontium).[2] That of lead

[1] See especially Doe (1970) and Faure and Powell (1972).

[2] Two aspects of analytic data relating to strontium are likely to confuse the nonspecialist. Some early records of $Sr^{87}/Sr^{86}$ values have been adjusted to meet more recently adopted laboratory standards. Increasing analytic precision may require revision of some early statements regarding identity versus small real differences in strontium of associated rocks.

is given as the respective abundance ratios of three radiogenic isotopes $Pb^{206}$ (from $U^{238}$), $Pb^{207}$ (from $U^{235}$), and $Pb^{208}$ (from $Th^{232}$) to the nonradiogenic isotope $Pb^{204}$ (less than 2 percent of young ore lead). Evolution of isotopes in magma source materials, as contrasted to solidified magmas, is commonly discussed in terms of the simplest conceivable model, a hypothetical global reservoir that has remained closed and homogeneous since the earliest times. Present composition of strontium in such a system would depend on time $t$ that has elapsed since its initiation and on initial values of $Sr^{87}/Sr^{86}$ and of $Rb^{87}/Sr^{86}$. The controlling parameters for lead would be initial values of $Pb^{206}/Pb^{204}$ and $U^{238}/Pb^{204}$; of $Pb^{207}/Pb^{204}$ and $U^{235}/Pb^{204}$; and of $Pb^{208}/Pb^{204}$ and $Th^{232}/Pb^{204}$.

To illustrate the principle, consider a reservoir (for example, the mantle or some part of it) formed at the beginning of terrestrial time ($4.55 \times 10^9$ years ago) and containing initially certain amounts of $U^{238}$, $U^{235}$, $Pb^{204}$, $Pb^{206}$, and $Pb^{207}$. These amounts are unknown; so for want of a better alternative $Pb^{206}/Pb^{204}$ and $Pb^{207}/Pb^{204}$ are *assumed* to be respectively 9.56 and 10.42 as found in uranium-free troilite of meteorites (Murthy and Paterson, 1962). These ratios will increase with time at a predictable rate which depends only on the initial amount (relative to $Pb^{204}$) of the uranium isotopes. If we knew the present lead ratios in the reservoir, we could subtract from them the initial ratios (that is, $Pb^{206}/Pb^{204} = 9.56$; $Pb^{207}/Pb^{206} = 10.42$) to find how much $Pb^{206}$ and $Pb^{207}$ has been produced since the formation of the reservoir. From this we could calculate the initial ratios $U^{238}/Pb^{204}$ and $U^{235}/Pb^{204}$ necessary to produce those amounts of $Pb^{207}$ and $Pb^{206}$, and we could also calculate the *present* ratio $U_0^{238}/Pb^{204}$ (commonly called $\mu_0$, the subscript indicating $t = 0$, or the present) in the reservoir. We could also calculate the ratios $Pb^{206}/Pb^{204}$ and $Pb^{207}/Pb^{204}$ at any time in the past, so that if the reservoir had been sampled at known times $t_1, t_2$, the lead isotopic ratios plotted against each other would fall on a growth curve characterized by values of $\mu_0$ and $\varepsilon_0$ ($U_0^{235}/Pb^{204}$).

Consider a postglacial olivine basalt from Iceland (Welke et al., 1968, p. 224, no. 1–4), in which, by virtue of its young age, the $Pb^{206}/Pb^{204}$ and $Pb^{207}/Pb^{204}$ ratios are still essentially those of the reservoir whence it came. The measured values are $Pb^{206}/Pb^{204} = 18.33$ and $Pb^{207}/Pb^{204} = 15.60$, so that during $0.455 \times 10^{10}$ years, the two lead ratios have changed from their initial values (assumed to be 9.56 and 10.42, respectively) by addition of radiogenic $Pb^{206}$ and $Pb^{207}$. The number of added atoms of each, per atom of $Pb^{204}$ is given by

$$Pb^{206}/Pb^{204} = 18.33 - 9.56 = 8.77$$

$$Pb^{207}/Pb^{204} = 15.60 - 10.42 = 5.18$$

and we can estimate the *present* ratio $U_0^{238}/Pb^{204}$ ($\mu_0$) in the reservoir. From

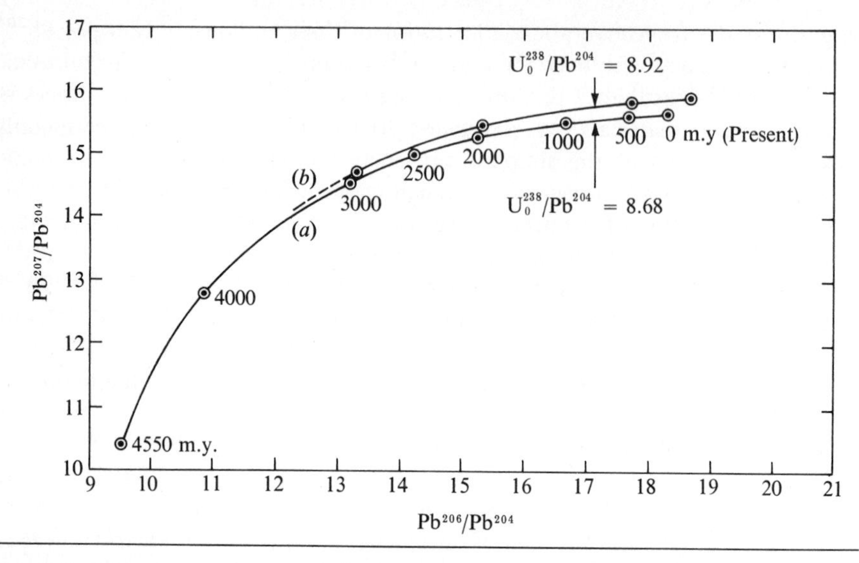


**Fig. 7-7** "Primary growth curves" for lead of two Recent volcanic rocks computed to fit data stated in text. Circled points are dated in millions of years from the present.

the decay equation (7-1) we find

$$\log\left(1 + \frac{Pb^{206}/Pb^{204}}{U_0^{238}/Pb^{204}}\right) = \frac{1.537 \times 10^{-10} \times 0.455 \times 10^{10}}{2.3026}$$

and substituting 8.77 for $Pb^{206}/Pb^{204}$, we obtain a value of $\mu_0$ ($U_0^{238}/Pb^{204}$) of 8.66. (Lead of most analyzed oceanic basalts treated thus gives values of $\mu_0 = 8.6 - 8.8$, while many continental lead ores give slightly higher values—8.9 to 9.0.) By an analogous calculation we obtain a value for $U_0^{235}/Pb^{204}$ of 0.063, which gives a (present-day) $U_0^{238}/U_0^{235}$ ratio of 137.5 for the reservoir from which this Icelandic basalt was drawn.

Lead primary-growth curves are shown in Fig. 7-7, calculated from the following isotopic analyses of two Recent volcanic rocks:

$Pb^{206}/Pb^{204} = 18.35$ $Pb^{207}/Pb^{204} = 15.51$ $U_0^{238}/Pb^{204} = 8.68$

$Pb^{206}/Pb^{204} = 18.60$ $Pb^{207}/Pb^{204} = 15.85$ $U_0^{238}/Pb^{204} = 8.92$

Since the time the earth was formed, about 90 percent of $U^{235}$ has decayed to $Pb^{207}$, compared to about 50 percent of $U^{238}$, so that any single-stage $Pb^{207}/Pb^{204}$ growth curve is now asymptotically approaching a limiting value as seen in Fig. 7-7.

Model single-stage growth curves constructed for lead of many igneous rocks resemble Fig. 7-7 and have been called "normal" or "conformable" lead ratios. This uniformity, when isotopic ratios are corrected for age, has been taken as evidence that all these rocks represent "samples taken at various time intervals from . . . a single, homogeneous, worldwide source of sufficient volume that its composition has not been observably affected by these withdrawals" (Slawson and Russell, 1967, p. 97). The most likely location of such a source, it was once generally believed, is in the mantle.

Nevertheless, models of this kind, in which lead or strontium at any point in time is seen as the product of primary or single-stage growth, obviously are unrealistic. They conflict with accepted fundamental concepts of terrestrial evolution: early differentiation of continental crust from the mantle; its growth by accretion from mantle-derived magmas; convection in the mantle; plate tectonics. Isotopic evolution in any substantial segment of the mantle is likely to have been interrupted from time to time by opening and resetting of the system. Even so, its general course would in the long run involve increase in the relative proportion of radiogenic isotopes with passage of time, provided that Rb/Sr, U/Pb, and Th/Pb were not reduced to vanishingly small values.

Common ore leads represent systems that were radiogenically quenched by swamping with lead (compared with minute traces of uranium and thorium) at the time of deposition. Old ore leads will therefore tend to be notably less radiogenic than young leads. It is not surprising that in a compositional plot of sedimentary ore leads of all ages, many points fall in appropriate chronologic order close to a curve of the kind that would represent single-stage growth in a closed system. Moreover the computed originating composition of lead at the projected 4.55 b.y. point is close to that of meteoritic lead. This strengthens the general geochemical assumption that primordial terrestrial and meteoritic leads (and lead/uranium ratios) were similar. Nevertheless, it would be wrong to conclude that any terrestrial ore lead is the product of single-stage growth from primordial lead. Each could have reached its present state along any one of an infinite number of possible multistage paths that ultimately derive from primordial terrestrial lead. All such paths necessarily trend in the same general pattern.

"Rock leads," found as a few parts per million in common igneous rocks, and many younger ore leads, too, may depart notably from primary growth curves of Fig. 7-7. This is what one would expect in any natural system that has been opened and reset from time to time—by generation of magma, by fractional crystallization, by outward diffusion induced by metamorphism, and so on. In none of these processes is U/Pb or Th/Pb reduced to zero. Leads of this kind, extracted from chemically different cogeneric associated rocks, commonly give a rectilinear plot (Fig. 7-8*a*). This is a necessary consequence of any two-stage growth model: primary growth followed by addition or depletion of U relative to Pb, and then secondary growth in the new system (Slawson and Russell, 1967, pp. 90, 91). Again different leads generated in two

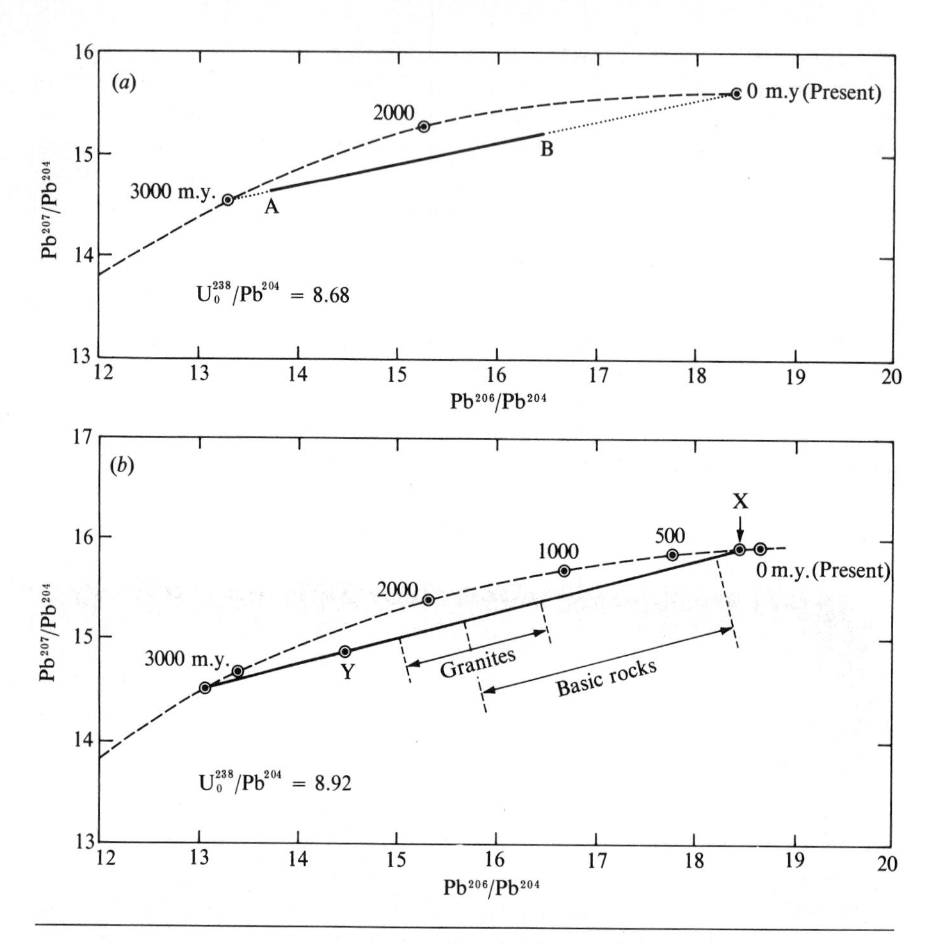


**Fig. 7-8** Isotopic composition of lead (*a*) in a set of cogeneric rocks *A–B* of different composition and (*b*) in a mixed rock system (fused ancient crust, *Y*; basaltic magma, *X*).

systems may become mixed, as when basaltic magma mixes with fused ancient crust (Fig. 7-8*b*).

With strontium there is no counterpart of ore lead. In any rock it is accompanied by significant rubidium, so that its composition is never naturally quenched by isolation from its radioactive progenitor. The rate at which its composition changes is stepped up or slowed down whenever Rb/Sr increases or diminishes as the system is reset. In spite of such possible interruptions, $Sr^{87}/Sr^{86}$ in many oceanic basalts—whose magmas have at least remained free from influence by sialic crust—falls within a narrow range, about 0.7035 for ocean-island basalts. Given an average Rb/Sr ratio of 0.025, round which many island basalts cluster, the composition of such strontium plots back on a linear regression curve (*OA* in Fig. 7-9) to 0.699 at 4.55 b.y. This is the value of primordial meteoritic strontium. So again we find a single-stage growth curve that could explain the composition of strontium in the source rocks of many ocean-island basalts. Again, however, resemblance to the simple model is probably illusory. Starting from 0.699 at 4.55 b.y., the $Sr^{87}/Sr^{86}$ ratio could reach its present value 0.7035 by any of an infinite number of paths (Faure and Powell, 1972, pp. 129–131). One is the single-stage rectilinear growth curve *a*. At the other extreme is curve *b* depicting evolution of strontium in a system where Rb/Sr continuously falls as rubidium is preferentially withdrawn (for example, from mantle to crust). Even curve *a* could represent an open system from which both elements have been withdrawn in proportion to their original respective abundances.

In the early days of isotopic petrochemistry, strontium and lead were viewed as "tracers" whose composition survives the short span of a magmatic cycle unchanged. Lead and strontium of the kind that is characteristic of recent oceanic basalts were thought of in terms of protracted single-stage evolution in a mantle reservoir. Numbers such as 0.703 to 0.704 for $Sr^{87}/Sr^{86}$ and 18.5 to 19.5 for $Pb^{206}/Pb^{204}$, whether determined from oceanic basalts or continental granodiorites, were thought to indicate direct tapping of magma from a primitive mantle source. More radiogenic strontiums or leads were looked on as indices of crustal sources; for indeed there are ancient siliceous crustal rocks with high Rb/Sr and U/Pb ratios and correspondingly high present $Sr^{87}/Sr^{86}$ and $Pb^{206}/Pb^{204}$. As more refined and much more plentiful data have become

**Fig. 7-9** Possible paths of isotopic evolution of strontium from an assumed initial value of $Sr^{87}/Sr^{86}$ = 0.699 at 4.55 b.y. Curve *a*, in a closed system, Rb/Sr = 0.025; curve *b*, in an open system in which Rb/Sr continually decreases from an initial value 0.05 to a present value 0.01.

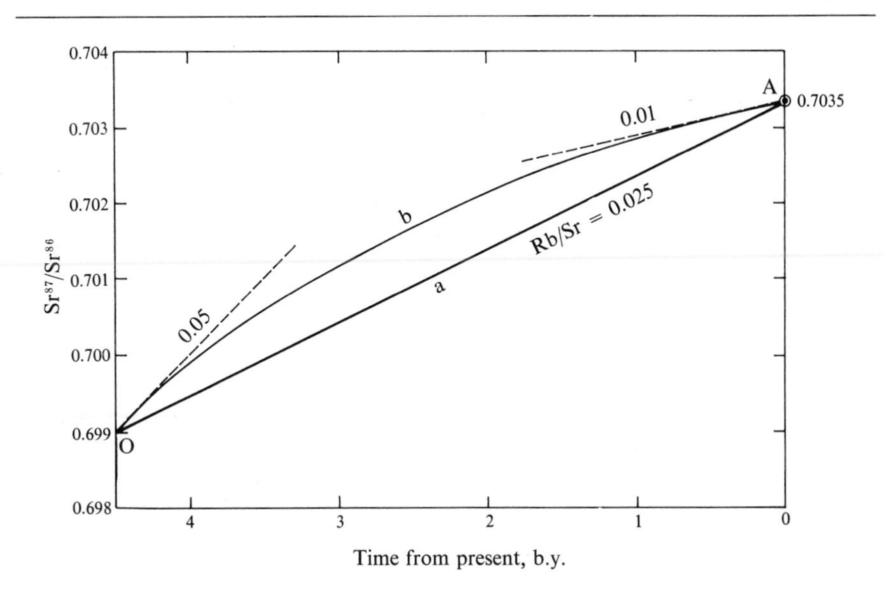

available, anomalies and inconsistencies have multiplied to the point where the older, simpler models have had to be abandoned. The current situation has been reviewed by a veteran in the field:[1]

> Observed Sr and Pb isotope ratios reflect the sum total of *all* processes that have participated in the evolution of a rock, such as: differentiation or melting of mantle-derived sources; partial melting of continental crust; contamination by and assimilation of country rock by magma; various types of wall-rock reaction leading to isotope migration and exchange between magma and wall-rock; interaction with circulating waters; etc.

With these constraints in mind and with due caution, the isotopic compositions of strontium and lead in igneous rocks can still be used most effectively toward solution of many petrogenic problems. Here are some examples:

1. Values determined for oceanic basalts probably reflect compositions in mantle sources from which they were derived. Such data test the degree of isotopic homogeneity of the suboceanic mantle on any desired scale.
2. It is thought that isotopes of any heavy element cannot be significantly fractionated by fractional crystallization of magma. Where the initial[2] isotopic compositions of lead and strontium are uniform throughout a chemically variable igneous rock series, the case for differentiation is thereby strengthened. Isotopic variation indicates the influence of some other factor: lapse of time between differentiation and eruption; contamination by external agencies; or it may even be that the basic assumption is unfounded.
3. In general, parent and daughter nuclides will differ in ionic charge and radius. Thus an ion of $Sr^{87}$ at a lattice site previously occupied by a parent $Rb^{87}$ ion might find itself in an energetically unfavored position; and partial melting might concentrate $Sr^{87}$ with respect to $Sr^{86}$ in the liquid fraction. This effect has not been demonstrated experimentally, but it remains a possible cause of minor isotopic differences between source rocks and their magmatic extracts.
4. Particularly in the continental setting, dual-source or contamination models of magmatic evolution may be tested by isotopic evidence. However convincing such a model may be, the data will always be open to alternative interpretation; but this applies to genetic models of any kind.
5. The magmatic and cooling history of a pluton is more protracted, and more likely to involve external influences, than that of a subaerial volcanic pile. The isotopic record of plutonic rocks is therefore likely to be less legible and more complicated than that of a volcanic series. Lead-strontium data

[1] S. Moorbath: abstract of a review presented to the Geological Society of America, November 1972.

[2] Referred to the date of extrusion or intrusion.

for plutonic rocks are especially suspect where (*a*) there is petrographic evidence of deuteric or hydrothermal alteration; or (*b*) oxygen has anomalous isotopic composition attributable to exchange with circulating waters; or (*c*) where abundances of Sr or Pb in the rock are low compared with those in possible contaminants.

6. Neither the uniform global isotopic reservoir nor the "primary growth curve" of lead have any reality; nor can any isotopic composition of strontium be attributed to single-stage evolution from a primordial value. Yet the conventional lead growth curve and present $Sr^{87}/Sr^{86}$ values commonly given by oceanic basalts (0.702 to 0.704) can serve as standards against which to gauge the radiogenic character of other leads or strontiums.

## ISOTOPIC CHARACTER OF ANCIENT CONTINENTAL CRUST

Sampling of basement rocks gives some idea of the composition of the local crust in continental provinces. With this end in view, Moorbath and his associates investigated the isotopic composition of lead and the uranium content of the ancient Lewisian basement as a potential contributing source of Tertiary magmas in the Hebridean volcanic province (Chap. 9). They found that high-grade Scourian metamorphism (dated at 2600 to 2900 m.y.) has greatly lowered the abundance of U relative to Pb and Th in the Lewisian granulites. Measured $U^{238}/Pb^{204}$ values scatter between 0.10 and 6.3, well below the $U_0^{238}/Pb^{204}$ value 8.3 to 9.3 calculated for mantle sources of most basalts. Rocks at greater depths in the crust, where metamorphic temperatures presumably were higher, must be at least equally depleted in U. As a consequence, even after the lapse of 2900 m.y., lead of Lewisian granulites today is strikingly deficient in the radiogenic isotopes $Pb^{206}$ and $Pb^{207}$: $Pb^{206}/Pb^{204}$ = 13.52 to 16.5; $Pb^{207}/Pb^{204}$ = 14.52 to 15.0. How this isotopic pattern is reflected in the Tertiary granites of Skye we shall see shortly.

Other segments of ancient crust, where metamorphism has been less severe, are composed of granitic rocks and metasediments that have retained initially high U, Th, and Rb. Today these have strongly radiogenic Sr and Pb. Moinian and Dalradian sediments of Scotland, for example, metamorphosed in the amphibolite facies (perhaps at 400 to 600°C) in early Paleozoic times, have high radiogenic Sr compared with ancient Lewisian gneisses further west that lost much of their Rb in the Scourian metamorphism. Among the Precambrian granite gneisses and gneisses tabulated by Doe (1970, pp. 86–92) are many with $Pb^{206}/Pb^{204}$ = 19 to 23 and $Pb^{207}/Pb^{204}$ = 15.7 to 16.2. Clearly numerical values of isotopic ratios alone cannot identify "mantle strontium," "old crustal lead," and so on. All we can say is that while low contents of radiogenic Sr may indicate a mantle source, very low radiogenic Pb suggests the influence of crustal rock severely reworked in the remote past. High abundances of radiogenic isotopes of both Sr and Pb, on the other hand, are consistent with source materials that might well be ancient unmodified crust or possibly mantle rock in which radioactive parent elements have long been concentrated above average level.

SOURCES OF TERTIARY MAGMAS: SKYE, AN EXAMPLE To illustrate sophisticated use of radiogenic tracers, we turn to a single, well-documented case history (Moorbath and Bell, 1965; Moorbath and Welke, 1968). On the island of Skye, as elsewhere in the British Hebridean volcanic province (Chap. 9), profuse outpouring of flood basalts was followed by intrusion of substantial bodies of granitic magma. What, if any, is the genetic relation between the acid and the basic magma series? From what sources were the Skye magmas generated? There is general consensus among geologists that most basaltic magmas form by partial fusion of upper-mantle rocks. Some granitic magmas form by fractional crystallization of basaltic magma. But in the case of the Skye granites, geologic reasoning suggests that a major role in the generation of the acid magmas has been played by large-scale fusion of the Lewisian basement which nearby forms the exposed part of the underlying crust.

Initial $Sr^{87}/Sr^{86}$ values (extrapolated from present values back to 60 m.y., the age of the igneous episode) show a markedly bimodal frequency distribution that is clearly related to rock composition (Fig. 7-10*a*). There are strong peaks at 0.705 to 0.707 (basalts and gabbros) and 0.711 to 0.714 (granites). Values for granites show greater spread than those for the basic rocks. The compound isotopic pattern rules out simple fractionation of granitic from basaltic magma in a closed system. Moorbath and Bell concluded that the granitic magmas formed by melting of source rocks—probably the Lewisian basement—with significantly higher Rb/Sr (and greater spread of Rb abundance) than the sources of associated basaltic magma. The latter would be placed by some geologists, on the basis of rather low $Sr^{87}/Sr^{86}$, in a primitive mantle reservoir of strontium; and this would seem at first to be a satisfactory answer to our simple questions. But isotopic studies on lead in the same rocks (Moorbath and Welke, 1968) reveal a petrogenic picture of previously unsuspected complexity. The respective leads of the two series differ, as might be expected, in isotopic composition. That of the basic rocks tends to be more radiogenic than that of the granites. But there is some overlap. Moreover, a plot of $Pb^{206}/Pb^{204}$ against $Pb^{207}/Pb^{204}$ for all analyzed specimens shows a wide range of composition and is almost rectilinear (Fig. 7-10*b*). It can be interpreted most simply as a curve of mixing between two leads *X* and *Y* of very different ages and origins. *X* is comparable to 60 m.y.-old lead of oceanic basalts (presumably coming direct from the mantle). *Y* is the mean of a group of old two-stage leads in analyzed rocks of the Lewisian—whose nonradiogenic character reflects very early depletion in U (but not Pb) during metamorphism 2600 to 2900 m.y. ago. Moorbath and Welke conclude that leads in both the basic rocks and the granites are mixtures of *X* and *Y*. And for any rock it is possible to calculate for each analyzed lead the ratio of the mantle-derived to the crustal component: between 55:45 and 80:20 for most basic rocks and between 23:77 and 65:35 for granites.

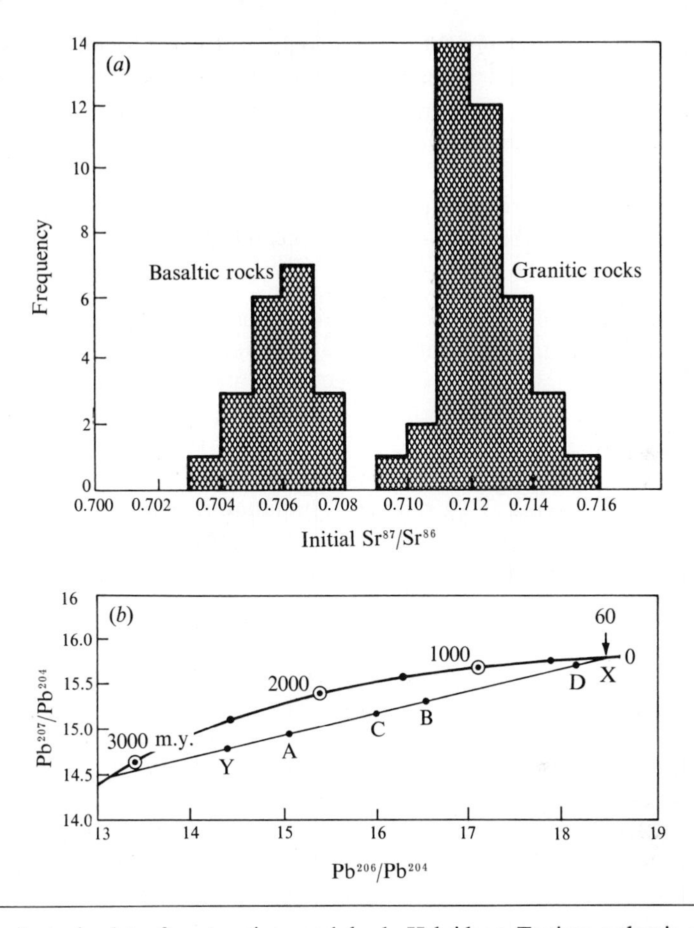


**Fig. 7-10** Isotopic data for strontium and lead, Hebridean Tertiary volcanic province. (*a*) Strontium data for basaltic rocks (basalts, dolerites, mugearites) and granitic rocks (granites, ferrodiorites), Skye (after Moorbath and Bell, 1965, p. 56); (*b*) lead data for the same suite of rocks: acid (granitic) rocks *AB*; basaltic rocks *CD*; ferrodiorites *CB*. Average lead of Lewisian basement is *Y*. Upper curve is the "primary single-stage growth curve" ($\mu_0 = 8.92$) for lead derived from "primordial meteoric lead," intersecting *AD* at *X*, 60 m.y. (age of Hebridean lavas).

In assessing relative roles of mantle melting, crustal melting, crustal contamination (by inward diffusion of Sr and Pb or even differential diffusion of their radiogenic isotopes), and fractionation of magmas, it must be remembered that basaltic magmas are notably richer in total Sr and poorer in total Pb than their granitic differentiates. Most analyzed Lewisian gneisses and granulites are rather close to magmatic granites in both respects. So contamination of basaltic magma by Lewisian rocks will have little effect on the $Sr^{87}/Sr^{86}$ ratio, but it will markedly affect the composition of lead in the product. Admixture of fractionated granitic liquid with primary granodiorite magma formed by melting of Lewisian rocks will cause significant scatter in both strontium and lead compositions. More than one model will satisfy simultaneously the strontium and lead data for Skye magmas. That proposed by Moorbath and Welke involves several processes:

1. Melting of mantle rock to give primary basaltic magma having strontium and lead comparable with those of oceanic basalts—$Sr^{87}/Sr^{86} \sim 0.704$; $Pb^{206}/Pb^{204} \sim 18.6$; $Pb^{207}/Pb^{204} \sim 15.7$; $\mu_0 = U_0^{238}/Pb^{204} \sim 8.92$.
2. Contamination of the primary basaltic magma by extraction of Sr and Pb from the Lewisian basement prior to final ascent. The effect on lead composition is too marked to be an effect of contamination by simple melting of country rock, since the contamination product is still strictly basaltic in composition. So we are forced to fall back on some mechanism by which Sr and Pb, or merely their radiogenic isotopes, diffuse preferentially from wall rock to magma. That such contamination is possible, even though the mechanism is imperfectly understood, is of great importance in interpreting lead-isotope patterns as indicators of magmatic sources in general. The effect on Sr composition will be small compared with that on Pb.
3. Fractionation of the basaltic magma prior to, during, or after contamination, to yield granitic liquid differentiates. These are not necessarily represented in a pure state among Skye granites.
4. Melting of the Lewisian basement, perhaps by heat from underlying basic magma, to yield "primary" granodiorite-granite magmas. The importance of crustal fusion is clinched by the low U/Pb values of Skye granites (cf. Lewisian gneisses) as contrasted to values recorded for granitic magmas that are thought to have formed elsewhere by differentiation—for example, the acid rocks of Iceland.
5. Admixture, to varying degrees, of primary granitic liquids formed by crustal fusion with fractionated granitic liquids.

The preceding model satisfies all the chemical, including isotopic, data. It may later have to be revised or even largely discarded as more information becomes available. The most significant outcome of such studies to date is a growing realization of the complexities of petrogenesis and of the need for caution in attempting to identify magmatic sources by use of single radiogenic tracers.

# Basaltic Associations of Ocean Basins

## The Oceanic Environment

A survey of igneous case histories begins in the ocean basins. Here the earth's crust is uniformly thin. Magmas may be expected to retain compositional traits inherited directly from mantle sources. The oceanic crust itself presumably has solidified from just such effusive magmas. Throughout the vast extent of the oceans there is recognizable uniformity, perhaps partly for insufficiency of data, in many large-scale features of structure, topography, tectonism, and volcanism. Under the thin veneer of young marine sediment, the oceanic abyssal plain (mean depth 5000 m) and the system of ridges and rises (depth about 2500 m) are built of volcanic rocks. So too are the oceanic islands and the innumerable guyots (seamounts) rising above the local level of the ocean floor. There is a good deal of uniformity among the rocks themselves: most by far are some type of basalt.

According to the plate-tectonics concept new crust, essentially basaltic, is generated by volcanic eruption concentrated along the median oceanic ridges. Thence it moves outward—mostly east or west—riding on underlying upper-mantle plates perhaps 100 km in thickness. Piercing the crustal surface of the moving plates, younger volcanoes have added their locally superimposed quota of lavas. Some of these volcanic piles, rising from the abyssal plain to form lofty islands, as in Hawaii and the Canaries, are among the world's most impressive volcanic edifices.

Because observational coverage of the ocean floors is still both erratic and inadequate, delineation of specific petrogenic provinces is necessarily somewhat empirical except in sharply bounded archipelagoes such as the Hawaiian Islands. In the oceans, moreover, there is a special complicating element in the time-versus-place problem that pervades all geology: the life span of a major volcano is perhaps $10^6$ years; within a broader oceanic province, where such can be recognized, the total duration of recorded igneous activity may be $10^7$ years or more. Ocean-floor spreading is thought to be rapid—1 to ~5 cm per year—so that during its life span a volcanic center, riding a lithospheric plate, may move as much as 100 km with respect to the site of deep-seated thermal disturbance that initiated the magmatic cycle. In currently favored plate-tectonics models, heat sources are localized where plumes of mantle rock ascending from great depth impinge on overriding lithospheric plates (Oxburgh and Turcotte, 1968; W. J. Morgan, 1972).

## Ocean-Ridge Basalts (Abyssal Tholeiites)

Lavas dredged from the crests and flanks of oceanic ridges at depths of 2000 to 3500 m below sea level are all basaltic and for the most part fall in the chemical range of olivine tholeiites. Petrographically they rather resemble the alkali basalts in that augite is the sole pyroxene and olivine may be present in two generations.

Many specimens from sites between 22°S and 32°N latitude along the mid-Atlantic ridge (Fig. 8-1) conform to a distinctive chemical pattern and have come to be known as *oceanic* or *abyssal tholeiite* (Engel and Engel, 1964*a*; G. D. Nicholls, 1965; Engel et al., 1965; Miyashiro et al., 1969). Characteristic chemical traits (Tables 8-1 and 8-2, analyses 1–5) are very low $K_2O$ and $TiO_2$, low total iron and $P_2O_5$, low $Fe_2O_3/FeO$, and high Ca; atomic abundances of Ba, Rb, Sr, Pb, Th, U, and Zr are all low to very low; K/Rb is exceptionally high (mostly 700 to 1700); Th/U is very low; and $Sr^{87}/Sr^{86}$ is low (0.7029 to 0.7035*). Basalts (Table 8-3, analyses 1–3) from the crest and slopes of the Pacific rise, across the full width of the tropics, are chemically identical except for somewhat lower normative *ol* (Engel and Engel, 1964*b*).

G. D. Nicholls (1965) first recognized a close analogy between the major-element composition of some rocks of the mid-Atlantic ridge (Table 8-1, analysis 4) and that of high-alumina basalt from Medicine Lake Highlands, California. This, however, does not extend to trace-element and isotope-abundance patterns. Many other Atlantic-ridge basalts on the other hand have a normal content of $Al_2O_3$ (14.5 to 15.5 percent). These show chemical similarities with basalts low in both $K_2O$ and $TiO_2$ from the Hawaiian province—for example, the Pololu series of the Kohala shield volcano, Hawaii (Muir and Tilley, 1966, p. 197).

Basalts from the rift floor of the mid-Atlantic ridge (Fig. 8-1) at latitude 45°45′N (Muir and Tilley, 1964*b*) are more strongly undersaturated olivine basalts (Table 8-1, analysis 8) still chemically within the olivine-tholeiite range. The $TiO_2$ content is uniformly low; but $K_2O$ is several times more abundant

* Values first reported by Tatsumoto et al. (1965) as 0.7021 to 0.7027 and later normalized to 0.7080 for standard E and A $SrCO_3$ (Hedge and Peterman, 1970, p. 119).

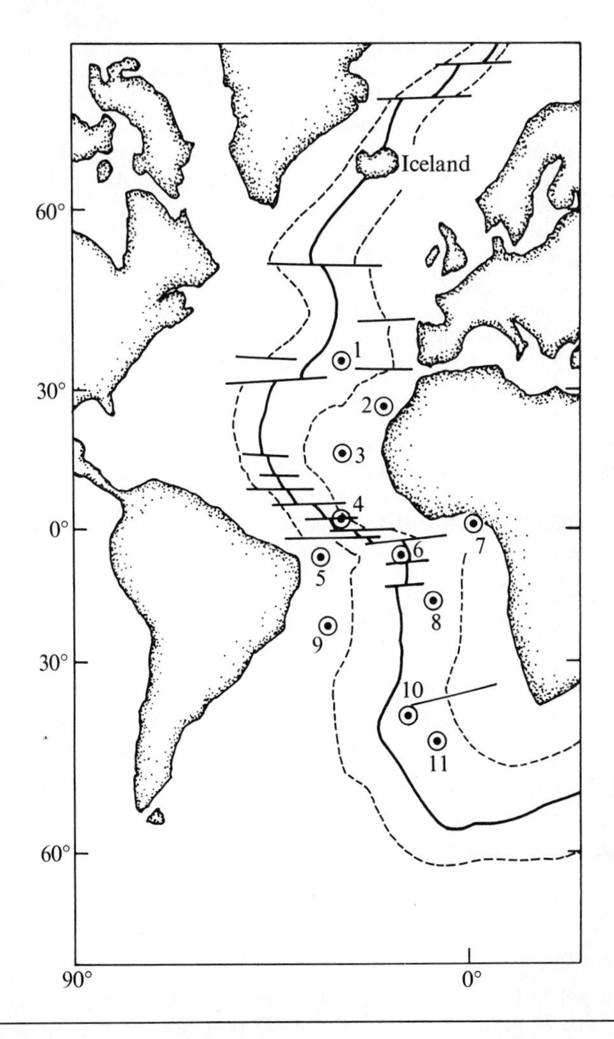


**Fig. 8-1** Mid-Atlantic ridge (outer borders dashed) and Atlantic islands: (1) Azores; (2) Canaries; (3) Cape Verde; (4) St. Paul's Rocks; (5) Fernando de Noronha; (6) Ascension; (7) Gulf of Guinea; (8) St. Helena; (9) Trinidade; (10) Tristan da Cunha; (11) Gough.

**Table 8-1** Chemical compositions (oxides, wt %) and CIPW norms of rocks from mid-Atlantic ridge

| | 1 | 2 | 3 | 4 | 5 | 6 | 7 | 8 | 9 | 10 |
|---|---|---|---|---|---|---|---|---|---|---|
| $SiO_2$ | 49.20 | 49.02 | 49.27 | 47.94 | 49.00 | 47.50 | 49.70 | 48.65 | 43.15 | 48.56 |
| $TiO_2$ | 2.03 | 1.46 | 1.26 | 0.75 | 1.46 | 1.83 | 1.49 | 1.44 | 2.70 | 0.24 |
| $Al_2O_3$ | 16.09 | 18.04 | 15.91 | 17.45 | 15.50 | 16.00 | 14.85 | 15.99 | 13.46 | 18.69 |
| $Fe_2O_3$ | 2.72 | 1.58 | 2.76 | 1.21 | | | 2.16 | 2.18 | 4.52 | 2.27 |
| FeO | 7.77 | 6.22 | 7.60 | 8.47 | 9.77 | 12.20 | 8.27 | 6.19 | 8.22 | 4.30 |
| MnO | 0.18 | 0.13 | 0.13 | 0.13 | — | — | 0.18 | 0.15 | 0.11 | 0.11 |
| MgO | 6.44 | 7.85 | 8.49 | 10.19 | 8.00 | 5.37 | 8.56 | 9.66 | 10.80 | 9.26 |
| CaO | 10.46 | 11.51 | 11.26 | 11.26 | 10.80 | 11.40 | 11.17 | 11.52 | 9.80 | 12.67 |
| $Na_2O$ | 3.01 | 2.92 | 2.58 | 2.37 | 2.90 | 2.57 | 2.69 | 2.71 | 3.47 | 1.88 |
| $K_2O$ | 0.14 | 0.08 | 0.19 | 0.09 | 0.21 | 0.49 | 0.15 | 0.57 | 1.63 | 0.07 |
| $P_2O_5$ | 0.23 | 0.12 | 0.13 | 0.08 | — | — | 0.13 | 0.21 | 0.75 | 0.02 |
| $H_2O^+$ | 0.70 | 0.64 | 0.35 | 0.23 | | | 0.61 | 0.75 | 1.21 | 1.72 |
| $H_2O^-$ | 0.95 | 0.57 | 0.51 | 0.15 | 1.19 | 3.28 | 0.16 | 0.30 | 0.15 | 0.17 |
| Total | 99.92 | 100.14 | 100.44 | 100.32 | 98.83 | 100.64 | 100.12 | 100.32 | 99.97 | 99.96 |
| *Q* | 0.3 | | | | | | | | | |
| *or* | 0.8 | 0.5 | 1.1 | 0.6 | 1.27 | 2.96 | 1.11 | 3.34 | 9.63 | 0.56 |
| *ab* | 25.7 | 24.4 | 21.8 | 20.0 | 25.03 | 22.25 | 23.06 | 23.06 | 9.67 | 15.72 |
| *an* | 29.8 | 36.3 | 31.2 | 36.7 | 29.23 | 31.38 | 27.52 | 29.75 | 16.34 | 42.46 |
| *ne* | | | | | | | | | 10.67 | |
| *di* | 17.4 | 16.6 | 19.2 | 15.2 | 16.19 | 15.42 | 21.81 | 21.10 | 22.23 | 15.96 |
| *hy* | 16.2 | 7.7 | 13.6 | 4.5 | 9.14 | 6.63 | 13.19 | 1.26 | | 14.07 |
| *ol* | | 9.0 | 5.9 | 19.7 | 10.08 | 9.48 | 6.33 | 14.45 | 16.76 | 4.95 |
| *mt* | 4.0 | 2.3 | 4.0 | 1.8 | 2.22 | 2.22 | 3.25 | 3.25 | 6.55 | 3.25 |
| *il* | 3.8 | 2.7 | 2.4 | 1.4 | 2.83 | 3.56 | 2.89 | 2.74 | 5.13 | 0.46 |
| *ap* | 0.5 | 0.3 | 0.3 | 0.2 | 0.33 | 0.34 | 0.34 | 0.48 | 1.64 | 0.05 |
| Total | 98.5 | 99.8 | 99.5 | 100.1 | 96.32 | 94.24 | 99.50 | 99.43 | 98.62 | 97.48 |

*Explanation of column headings*

1 Oceanic tholeiite, depth 2910 m; 20°40′S, 13°16′W (Engel and Engel, 1964*a*, D2–1)
2 Oceanic tholeiite (diabase), depth 2388 m; 9°39′N, 40°27′W (Engel and Engel, 1964*a*, D5–5)
3 Oceanic tholeiite, depth 3566 m, rift floor; 28°53′N, 43°20′W (G. D. Nicholls, 1965, table 1, analysis 1)
4 High-alumina basalt, some locality as 3 (G. D. Nicholls, 1965, table 2, analysis 2)
5 Oceanic tholeiite, depth 4200 m, rift floor; 30°08′N, 43°37′W (Kay et al., 1970, analysis A150–21–1C)
6 Basalt, depth 3700 m; 31°49′N, 42°25′W (Kay et al., 1970, analysis GE160)
7 Basalt, depth 3700 m; 31°49′N, 42°25′W (Muir and Tilley, 1966, p. 195, analysis 3)
8 Basalt, depth 3600 m, rift floor; 45°44′N, 27°44′W (Muir and Tilley, 1964*b*, table 1, analysis 5)
9 Alkali olivine basalt, depth between 2000 and 3000 m; a few kilometers northeast of St. Paul's Rocks; 1°1′N, 29°21′W (Melson et al., 1967)
10 Laminated gabbro, depth 4000 to 5000 m, Romanche trench; 0°14′N, 17°7′W (Melson and Thompson, 1970)

**Table 8-2** Atomic abundances (ppm) and abundance ratios of trace elements and isotopes, lavas of mid-Atlantic ridge*

| | 1 | 1a† | 2 | 2a† | 5 | 6 | 8 | 9 | 10 |
|---|---|---|---|---|---|---|---|---|---|
| Ba | 5 | | 5 | | | 11.7 | 16 | 300 | 10 |
| Co | 32 | | 26 | | | | 38 | | |
| Cr | 220 | | 280 | | | | 700 | 250 | 900 |
| Ni | 87 | | 78 | | 190 | 100 | 220 | 270 | 200 |
| Pb | | 1.29 | | 0.56 | | | | | 2 |
| Rb | 1.14 | 1.42 | < 10 | | 1.90 | 12.9 | 22 | | < 20 |
| Sr | 190 (134) | | 90 | | 90 | 105 | 320 | 500 | 110 |
| Th | 0.15 | | | 0.13 | | | | | |
| U | 0.16 | | | 0.09 | | | | | |
| V | 280 | | 240 | | | | 350 | | 110 |
| Zr | 160 | | 62 | | | | 45 | 200 | 10 |
| K/Rb | 1020 | 950 | | | 700 | 366 | 230 | | |
| Rb/Sr | ~0.01 | | < 0.1 | | 0.02 | 0.12 | 0.07 | | < 0.2 |
| Th/U | 0.9 | | | 1.4 | | | | | |
| $Sr^{87}/Sr^{86}$ | | 0.7032‡ | | | | | | | |
| $Pb^{206}/Pb^{204}$ | | 18.471 | | 18.816 | | | | | |
| $Pb^{207}/Pb^{204}$ | | 15.54 | | 15.68 | | | | | |
| $Pb^{208}/Pb^{204}$ | | 38.01 | | 38.65 | | | | | |
| $U^{238}/Pb^{204}$ | | 7.9 | | 10.5 | | | | | |

* Column headings numbered as in Table 8-1.
† Values reported by Gast (1967) and Tatsumoto (1966).
‡ Normalized to 0.7080 for standard E and A $SrCO_3$ (Hedge and Peterman, 1970, p. 119).

than in typical abyssal tholeiites just described; and there is little resemblance between the respective abundance patterns for trace elements (Table 8-2). A single chemically similar rock (Table 8-1, analysis 6) has been recorded among typical oceanic tholeiites some 1500 km south.

Not all the igneous rocks of the ocean ridges are volcanic. Among material recovered from depths of 3000 to 4500 m at latitude 28°32′N on the mid-Atlantic ridge are serpentinized peridotite and gabbro (Quon and Ehlers, 1963). Some 3000 km to the south, two-pyroxene gabbros, some laminated, and peridotites in various stages of serpentinization are abundantly represented in material dredged from the deep transverse equatorial fracture known as the Romanche trench. It has been inferred (Melson and Thompson, 1970) that the trench cuts a layered basic intrusion of the continental Skaergaard type. An analyzed rock (Table 8-1, analysis 10) is chemically similar to high-alumina oceanic tholeiite; and in its high $Al_2O_3$ content and low $K_2O$, $TiO_2$, and $Fe_2O_3$ it resembles to some degree the chilled marginal gabbros of the Skaergaard intrusion itself (cf. Table 9-16, analysis 2).

Because of their great extent the ocean ridges remain imperfectly sampled. There are a few striking petrographic anomalies: a *ne*-normative alkali basalt

(Table 8-1, analysis 9) from a depth of 3000 km a few kilometers northeast of the equatorial peridotite islet of St. Paul's Rocks (Melson et al., 1967*b*) and an "andesitic" (icelandite?) glass from the East Pacific rise (Table 8-3, analysis 4). But the overwhelming impression of ocean-ridge basalts is chemical and petrographic uniformity, with low-potassium, abyssal, tholeiitic basalt greatly predominant. As coverage is increasing, it seems that some recognizable chemical variations are locally consistent.

Basalts of the mid-Atlantic ridge tend to be highly undersaturated, commonly with normative *ol* above 10 percent. Most basalts so far recorded from the oceanic ridges of the eastern Pacific (Kay et al., 1970) have normative *ol* below 6 percent. Basalts of the Gordo rise, 300 km west of the northern California coast (Fig. 8-2), are consistently high in $Al_2O_3$ (15.5 to 17.5 percent; Table 8-3, analysis 5) and in most cases have $Sr^{87}/Sr^{86}$ values between 0.7025 and 0.7031 (Kay et al., 1970; Hedge and Peterman, 1970). Some 500 km to the northwest, basalts of the Juan de Fuca rise are low-alumina types ($Al_2O_3$ 12 to 15 percent; Table 8-3, analysis 6), with $Sr^{87}/Sr^{86}$ (in three analyzed samples) 0.7022 to 0.7025.

In summary it now seems to be established that volcanic rocks from the deep-sea floor along the oceanic ridges (Figs. 8-1, 8-2) are predominantly olivine tholeiites (abyssal tholeiites) exceptionally low in $K_2O$ and $TiO_2$ and with a trace-element pattern in some respects resembling that of stony meteorites. Low $Sr^{87}/Sr^{86}$ values are universal, and lead seems somewhat less radiogenic than in most island basalts.

Almost nothing is known of the petrology and structure of the innumerable "abyssal hills" that dot the floor of the deep ocean. It is generally assumed that their origin is volcanic, perhaps connected with shallow intrusion of magma into sea-floor sediments. One such hill, at a depth of 4890 m on the floor of the Indian Ocean, was found by D. S. Korzhinsky[1] to consist of basalt within the general composition range of abyssal tholeiite, but with unusually high MgO (10.45). Another hill, depth between 4300 and 4400 m, in the northeastern Pacific (32°25′N, 125°45′W, Luyendyk and Engel, 1969) also consists of basalt whose identified mineral constituents are labradorite, pigeonite, and opaque ores and which chemically resembles Hawaiian tholeiites.

## Volcanic Rock Series of Oceanic Islands

### PETROGRAPHIC-CHEMICAL SYNOPSIS

Petrologic sampling of oceanic islands (Figs. 8-1, 8-2) has been virtually limited to subaerially exposed rocks—no more than 5 or 10 percent of any individual volcanic pile. All the island rocks are young: many have been extruded within the last few million years, and numerous volcanoes are still active.

The most plentiful rocks by far are olivine basalts, although on some islands

[1] Cited by Engel and Engel (1964*b*, p. 484).

**Table 8-3** Chemical compositions (oxides, wt %), CIPW norms,* and atomic abundances and abundance ratios of trace elements and isotopes of ocean-floor lavas, rises of East Pacific Ocean

| | 1 | 2 | 3 | 4 | 5 | 6 |
|---|---|---|---|---|---|---|
| $SiO_2$ | 49.80 | 49.13 | 48.30 | 59.00 | 49.90 | 50.10 |
| $TiO_2$ | 2.02 | 1.23 | 2.19 | 1.75 | 1.08 | 2.18 |
| $Al_2O_3$ | 14.88 | 14.97 | 14.30 | 12.60 | 17.30 | 13.80 |
| $Fe_2O_3$ | 1.55 | 3.28 | 11.70 | 12.00 | 7.60 | 12.30 |
| FeO | 10.24 | 5.72 | | | | |
| MnO | 0.21 | 0.16 | | | | |
| MgO | 6.74 | 7.68 | 6.70 | 1.70 | 7.08 | 6.11 |
| CaO | 10.72 | 12.68 | 10.10 | 5.60 | 12.78 | 10.90 |
| $Na_2O$ | 2.91 | 2.37 | 2.75 | 4.25 | 2.45 | 2.83 |
| $K_2O$ | 0.24 | 0.16 | 0.18 | 0.65 | 0.18 | 0.16 |
| $P_2O_5$ | 0.28 | 0.15 | | | | |
| $H_2O^+$ | 0.54 | 1.06 | 1.29 | 1.78 | 0.80 | |
| $H_2O^-$ | 0.06 | 1.25 | | | | |
| Total | 100.19 | 99.84 | 97.51 | 99.33 | 99.17 | 98.38 |
| *Q* | | 0.79 | | 13.66 | | 0.04 |
| *or* | 1.1 | 0.89 | 1.10 | 3.92 | 1.08 | 0.96 |
| *ab* | 24.6 | 20.01 | 24.08 | 36.70 | 20.99 | 24.25 |
| *an* | 26.7 | 29.75 | 27.06 | 13.66 | 36.12 | 24.79 |
| *di* | 22.0 | 25.73 | 19.59 | 11.84 | 21.90 | 23.96 |
| *hy* | 13.7 | 12.79 | 17.73 | 14.27 | 13.16 | 19.27 |
| *ol* | 5.7 | | 3.54 | | 2.12 | |
| *mt* | 2.3 | 4.76 | 2.25 | 2.22 | 2.20 | 2.20 |
| *il* | 3.7 | 2.36 | 4.30 | 3.39 | 2.08 | 4.19 |
| *ap* | 0.6 | 0.35 | 0.34 | 0.33 | 0.33 | 0.33 |
| Total | 100.4 | 97.43 | 99.99 | 99.99 | 99.98 | 99.99 |

| | 1*a* | 3*a* | 4*a* |
|---|---|---|---|
| Ba | 25 | 19.4 | 54.8 |
| Ce | | 16.5 | 75 |
| Cs | | 0.074 | 0.082 |
| Co | 35 | | |
| Cr | 160 | | |
| Ni | 58 | 58 | 10 |
| Pb | 0.49 | | |
| Rb | 1.06 | 5 | 7.25 |
| Sr | 110 (86) | 107 | 105 |
| Th | 0.21 | | |
| U | 0.09 | | |
| V | 400 | | |
| Zr | 150 | | |
| K/Rb | 1890 | 310 | 770 |
| Cs/Rb | | 0.014 | 0.011 |
| Th/U | 2.3 | | |
| $Sr^{87}/Sr^{86}$ | 0.7025 | | |
| $Pb^{206}/Pb^{204}$ | 18.24 | | |
| $Pb^{207}/Pb^{204}$ | 15.53 | | |
| $Pb^{208}/Pb^{204}$ | 38.03 | | |
| $U^{238}/Pb^{204}$ | 6.4 | | |

* For analyses 3–6, norms are calculated (Kay et al., 1970) assuming a low state of oxidation of Fe ($Fe_2O_3$ = 1.50 percent) and reasonable values of MnO (0.18 percent) and $P_2O_5$ (0.15 percent).

*Explanation of column headings*

1 Glassy basalt, East Pacific rise, depth 2300 m; 12°52′S, 110°57′W (Engel et al., 1965, table 1, analysis PVD-3)

1*a* Trace-element and isotopic data for analysis 1 (Engel et al., 1965; Tatsumoto et al., 1965; Tatsumoto, 1966)

2 Basalt, Mohole drill core, depth 3746 m; off Guadalupe Island (East Pacific rise), 28°59′N, 117°30′W (Engel and Engel, 1961, p. 1799, analysis 1)

3, 3*a* Basalt, East Pacific rise, depth 3120 m; 7°08′N, 103°15′W (Kay et al., 1970, p. 1593, analysis V2023)

4, 4*a* "Andesite" glass, East Pacific rise, depth 3182 m; 5°31′S, 106°46′W (Kay et al., 1970, p. 1593, analysis V2140)

5 Basalt, Gordo rise, depth 2500 m; 41°15′N, 127°28′W (Kay et al., 1970, p. 1592, analysis 13E)

6 Basalt, Juan de Fuca rise, depth 2502 m; 44°36′N, 130°19′W (Kay et al., 1970, p. 1591, analysis 2C)

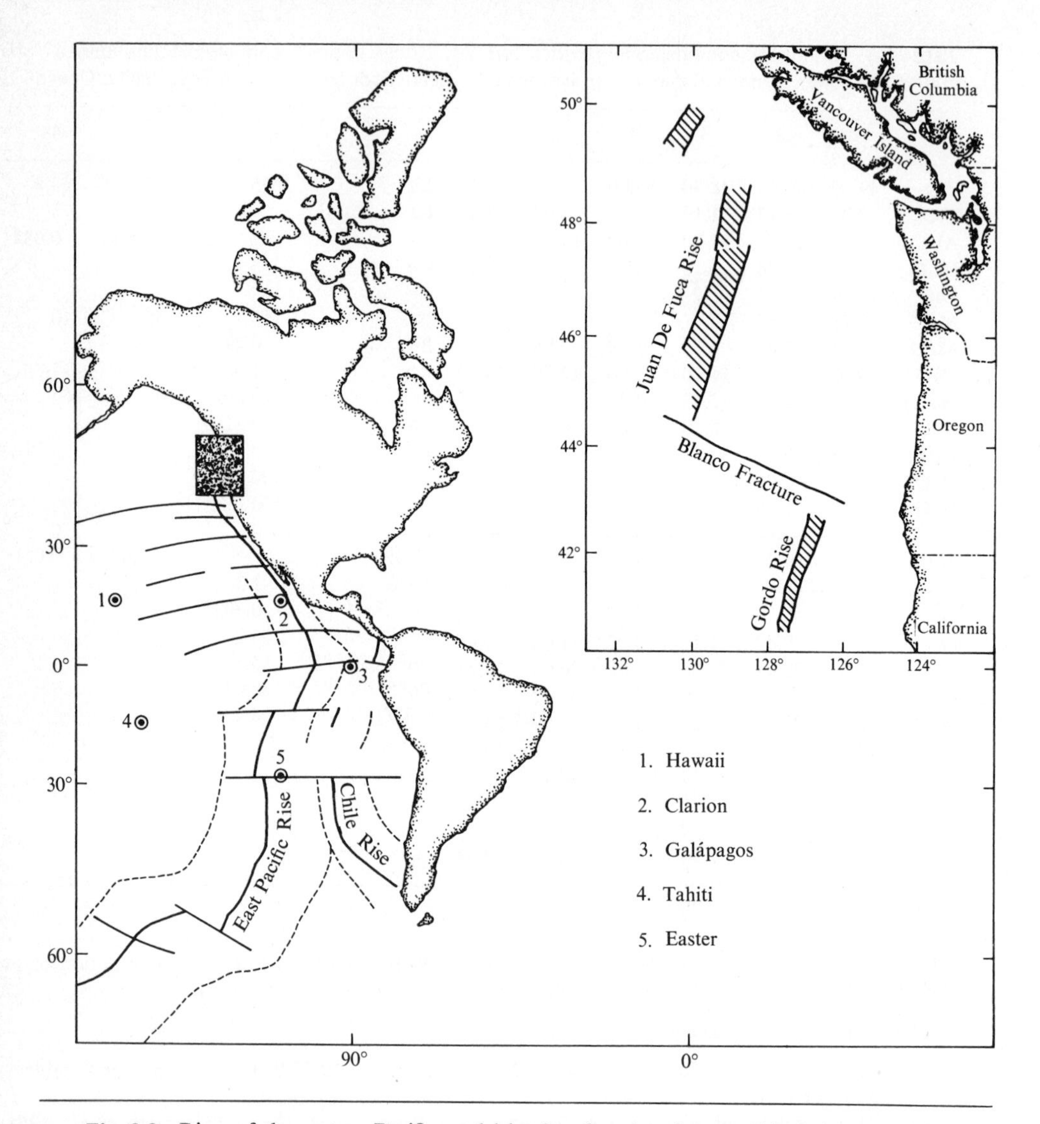


**Fig. 8-2** Rises of the eastern Pacific, and islands referred to in text. Stippled area off the northwestern American coast is enlarged, top right.

or in particular eruptive episodes the predominant basic lavas are more alkaline and have high normative and even modal *ne*. These are nephelinites, basanites, trachybasalts, and related rocks. Oceanic basalts in general, and often locally, cover a wide range of chemical composition. Some are oversaturated but most are undersaturated in silica. As undersaturation increases *ol* rises, *hy*/*di* falls off to zero, and at this point *ne* appears and then increases. There is broad correlation between the degree of silica saturation of oceanic basalts, their mineralogical makeup, and the nature of more felsic associated rocks. This is most apparent at either end of the saturation range and less consistent for moderately undersaturated basalts with high normative *ol* but no *ne*.

Oversaturated or slightly undersaturated basalts with high normative *hy* can be classed unequivocally as tholeiitic (Tilley and Muir, 1967). Olivine is present only as phenocrysts; there are two pyroxenes, one calcic (diopsidic) and the other low in Ca (hypersthene or pigeonite). At the opposite compositional extreme are alkali basalts with normative *ne*. These have more abundant modal olivine, which typically appears in two generations—as phenocrysts and in the groundmass. There is a single pyroxene phase, augite (commonly titaniferous). Strongly undersaturated basalts that lack *ne* but have rather low *hy* in the norm are mineralogically indistinguishable from alkali basalts. As normative *ne* increases, alkali basalts merge into basanites. But it is notoriously difficult to distinguish the two groups petrographically, since it is more or less a matter of accident whether nepheline appears in the modal composition or remains as an occult component of groundmass glass. There is no significant difference between basanites (with modal nepheline) and some rocks variously called basanitoid (for example, in Tahiti) or alkali basalt (for example, in the Canaries).

Terminology of basic oceanic lavas gradational toward trachyte and phonolite [with high ($Na_2O$ + $K_2O$) and $Si_2O$ near 50 percent] also is confused and redundant. We tend to use *mugearite* and *hawaiite* for rocks of basaltic aspect containing alkali feldspar (usually calcium-bearing anorthoclase) and a plagioclase more sodic than the typically basaltic labradorites and bytownites—oligoclase-andesine in mugearites, medium to calcic andesine in hawaiites (Muir and Tilley, 1961). These rocks may also contain amphibole (phenocrystic kaersutite, groundmass barkevikite) and minor late biotite. Roughly synonymous terms are *trachybasalt* and *trachyandesite*. Among oceanic lavas there are no true andesites. Oversaturated rocks in the same silica range as andesite but significantly lower in $Al_2O_3$ and with higher Fe/Mg are termed *icelandites*.

We now turn to selected provinces treated more or less in order of increasing silica saturation of what are generally considered the parent basaltic magmas. References to other well-documented island provinces are grouped somewhat arbitrarily to bring out chemical and petrologic characteristics in common. But the reader who delves more deeply into the literature will find that no two individual provinces are identical. Nowhere in petrology is the unique character of the individual petrogenic province more obvious than in the oceanic islands.

## NEPHELINITE PHONOLITE PROVINCES

IHLA DA TRINIDADE The Atlantic island of Trinidade (Almeida, 1961) rises from the oceanic abyssal plain (depth 5000 m) 1140 km east of the Brazilian coast at 20°30′S, 29°20′W. It is situated beyond the eastern end of a chain of seamounts that dot the crest of a submarine ridge running westward along parallel 20°30′S to the continental edge. Only 6 $km^2$ in area and 600 m in maximum elevation, the island is the eroded tip of a major volcano where activity has continued into late Pleistocene and Holocene times.

Chemically and mineralogically the rocks of Trinidade are unique and extreme among oceanic volcanic suites. All are exceptionally high in alkali and low in silica (Table 8-4, analyses 1–4). Many of the mafic types are augite- and olivine-rich nephelinites with high *ne* (exceeding 25 percent) and even *lc* in the norm. Even the most felsic rocks—members of the phonolite family—contain less than 52 percent $SiO_2$. Throughout the whole rock suite, plagioclase is virtually absent; every rock contains feldspathoids (one or more of nepheline,

**Table 8-4** Chemical compositions (oxides, wt %) of volcanic rocks from Trinidade and Fernando de Noronha

| | 1 | 2 | 3 | 4 | 5 | 6 | 7 | 8 | 9 |
|---|---|---|---|---|---|---|---|---|---|
| $SiO_2$ | 39.00 | 40.08 | 44.80 | 51.16 | 38.42 | 42.68 | 44.23 | 54.82 | 60.81 |
| $TiO_2$ | 3.60 | 2.30 | 1.60 | 0.49 | 4.01 | 2.00 | 4.33 | 0.50 | 0.65 |
| $Al_2O_3$ | 11.86 | 15.67 | 17.76 | 21.53 | 13.55 | 16.65 | 10.12 | 22.46 | 18.88 |
| $Fe_2O_3$ | 6.20 | 6.75 | 5.55 | 2.64 | 3.32 | 5.08 | 3.50 | 1.84 | 2.57 |
| FeO | 9.55 | 5.17 | 3.81 | 1.86 | 9.40 | 8.11 | 6.58 | 0.72 | 0.00 |
| MnO | 0.19 | 0.15 | 0.20 | 0.07 | 0.21 | 0.20 | 0.18 | 0.12 | — |
| MgO | 12.31 | 4.49 | 3.47 | 0.68 | 12.54 | 5.57 | 11.70 | 0.07 | 0.61 |
| CaO | 10.40 | 10.60 | 7.80 | 1.92 | 11.75 | 11.00 | 11.45 | 1.42 | 1.70 |
| $Na_2O$ | 3.68 | 6.49 | 6.87 | 10.53 | 3.72 | 5.04 | 3.20 | 10.22 | 6.20 |
| $K_2O$ | 1.80 | 1.35 | 3.87 | 5.69 | 0.86 | 1.69 | 1.12 | 5.93 | 5.80 |
| $P_2O_5$ | 0.55 | 1.40 | 0.88 | 0.05 | 1.01 | 0.67 | 0.78 | 0.12 | |
| $H_2O^+$ | 0.50 | 3.40 | 1.10 | 1.72 | 1.15 | 0.52 | 2.04 | 0.82 | |
| $H_2O^-$ | 0.30 | 2.20 | 2.30 | 0.45 | n.d. | 0.78 | 0.50 | 0.02 | 2.22 |
| Cl | | | | 0.34 | | | | 0.28 | |
| $SO_3$ | | | | 1.14 | | | 0.31* | 0.98 | |
| Total | 99.94 | 100.05 | 100.01 | 100.27 | 99.94 | 99.99 | 100.04 | 100.32 | 99.44 |

* $CO_2$.

*Explanation of column headings*

1 Nephelinite (ankaratrite), Trinidade (Almeida, 1961, p. 168, table 19, no. 6)
2 Nephelinite, Trinidade (Almeida, 1961, p. 122, table 9, no. 1; p. 168, table 19, no. 10)
3 Sanidine nephelinite (grazinite), Trinidade (Almeida, 1961, p. 137, table 13, no. 1; p. 168, table 19, no. 17)
4 Phonolite (tinguaite), Trinidade (Almeida, 1961, p. 108, table 6, no. 3; p. 168, table 19, no. 24)
5 Nephelinite (ankaratrite), Fernando de Noronha (Almeida, 1955, p. 154, no. 30)
6 Alkali basalt, Fernando de Noronha (Almeida, 1955, p. 150, no. 18)
7 Nepheline basanite, Fernando de Noronha (W. C. Smith and Burri, 1933, p. 430; Almeida, 1955, p. 150, no. 19)
8 Sodalite phonolite, Fernando de Noronha (W. C. Smith and Burri, 1933, p. 412; Almeida, 1955, p. 145, no. 3)
9 Trachyte, Fernando de Noronha (Almeida, 1955, p. 147, no. 10)

sodalite, and analcite). Alkali feldspar appears in basic rocks transitional from nephelinite to phonolite (analysis 3) and is an essential constituent of the phonolites themselves. All the rocks contain pyroxene (titanaugite or aegirine-augite); biotite and hornblende are widely distributed. Olivine is found only in the ultrabasic nephelinites. The great bulk of the exposed rocks are phonolitic tuffs and breccias. There are also numerous domes, necks, and dikes of phonolite. Nephelinites occur principally as dikes, but also as surface flows.

Plutonic rocks, represented by xenoliths in volcanic and by large blocks in pyroclastic rocks, include nepheline syenites and various ultramafic and mafic rocks with such assemblages as biotite, aegirine-augite, nepheline, and sphene; aegirine-augite, titanomagnetite, biotite, sphene, and apatite; pyroxene, magnetite, sphene, and apatite (cf. jacupirangite of Brazil); olivine and aegirine-augite. No such rocks outcrop at the surface. Almeida concludes that they come from deep within the volcanic edifice. From this it may perhaps be inferred that the volcanic pile as a whole is composed of rock types whose chemical range (but not relative volumes) is not very different from that of rocks that crop out in the exposed tip.

FERNANDO DE NORONHA This linear archipelago extends about 15 km from northeast to southwest at 3°50′S, 32°25′W, 345 km off the northeastern corner of Brazil (Almeida, 1955; W. C. Smith and Burri, 1933). The range of rock types is rather similar to that on Trinidade but somewhat less extreme (Table 8-4, analyses 5–9). The relative volumes of mafic and phonolitic rocks, however, are reversed; augite-rich nephelinites (analysis 5) predominate. Basic volcanics include basanites and equivalent alkali basalts. Phonolites are much less extensive than on Trinidade, and in addition there are approximately saturated sodalite trachytes ($SiO_2 > 60$ percent).

Here too a diversified xenolithic suite of plutonic rocks ranges from nepheline syenites to mafic rocks rich in pyroxene and hornblende. Some nepheline basanites contain xenoliths of another type—peridotite composed of magnesian olivine and subordinate enstatite and chrome spinel, such as we shall encounter again and again in rocks of the nephelinite-basanite family the world over.

Almeida tentatively derives the rocks from parent magmas that are but poorly represented among exposed rocks—alkali basalt (analysis 6) or basanite (analysis 7)—but as in other basalt-nephelinite provinces it is difficult to develop any adequate single-lineage model of magmatic evolution.

### BASANITE ALKALI-BASALT PHONOLITE PROVINCES

TAHITI Tahiti (Williams, 1933; McBirney and Aoki, 1968), situated in the mid-Pacific at 17°40′S, 149°20′W, consists of two deeply eroded volcanic centers 35 km apart and rises straight from the Pacific floor (depth 4000 m) to a maximum elevation of 2200 m above sea level. The bulk of the island mass is composed of lavas; but in the eroded cores of each volcano there is a suite of plutonic

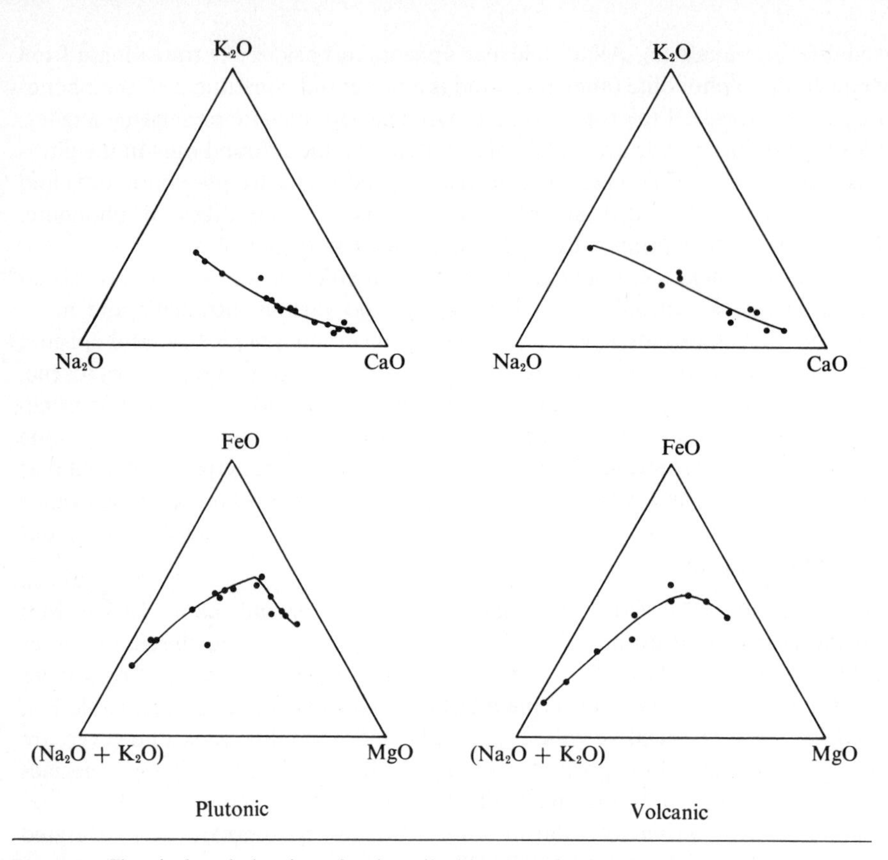


**Fig. 8-3** Chemical variation in volcanic and plutonic (subvolcanic) rocks of Tahiti. (After McBirney and Aoki, 1968, p. 545.) Total iron shown as FeO.

rocks chemically equivalent to the volcanic series (Fig. 8-3) but mineralogically very different. Chemical compositions of typical rocks are given in Table 8-5. The commonest rocks by far are very basic lavas, varying greatly as to texture but all with *ne* in the norm. They include alkali basalts (analysis 3), basanites (analysis 1), and olivine- and augite-rich ankaramites (analysis 2). Greatly subordinate are distinctive, more siliceous but still basic, highly alkaline lavas called *tahitites* (analysis 4). They are less calcic than basanites and more potassic than mugearites, but they have affinities with both types of rock; the presence of hauyne is characteristic. Also quite subordinate, but widespread in their field occurrence, are phonolites and slightly undersaturated trachytes (analyses 5, 6). Plutonic rocks range from alkali gabbros (analysis 7)—with various combinations of labradorite, titanaugite, kaersutite, olivine, alkali feldspar, and nepheline—to syenites and nepheline syenites. Chemically the whole Tahitian suite conforms broadly to the conventional model of differentiation of a series of increasingly alkaline and siliceous melts from some sort of parent alkali-basalt or basanite magma. To be more specific is hardly justified—especially since chemically equivalent volcanic and plutonic rocks differ so strongly mineralogically. There seems to be a real divergence, at the felsic end of the series, between rock types respectively strongly and weakly undersaturated—for example, phonolites versus trachytes.

ISLANDS IN THE GULF OF GUINEA From the head of the Gulf of Guinea a line of volcanic islands extends 700 km southwestward down the continental slope of West Africa. This is the seaward extension of a chain of major volcanoes—the Cameroon line—stretching deep into continental Africa toward Lake Chad. Not much is known about the field relations of different volcanic rocks in the islands. However, collections from the islands of São Tomé and

**Table 8-5** Chemical compositions (oxides, wt %) of igneous rocks of Tahiti

| | 1 | 2 | 3 | 4 | 5 | 6 | 7 | 8 | 9 |
|---|---|---|---|---|---|---|---|---|---|
| $SiO_2$ | 42.53 | 43.26 | 45.53 | 48.50 | 55.81 | 61.18 | 43.17 | 53.23 | 54.22 |
| $TiO_2$ | 4.78 | 3.40 | 3.18 | 2.68 | 0.26 | 0.76 | 4.92 | 1.10 | 1.70 |
| $Al_2O_3$ | 16.61 | 9.69 | 12.83 | 17.53 | 22.00 | 19.16 | 14.71 | 20.54 | 21.00 |
| $Fe_2O_3$ | 3.80 | 3.66 | 3.92 | 4.43 | 1.68 | 1.70 | 4.53 | 2.86 | 2.97 |
| FeO | 7.84 | 8.97 | 8.93 | 3.73 | 0.89 | 1.50 | 7.94 | 1.81 | 2.73 |
| MnO | 0.13 | 0.16 | 0.16 | 0.12 | 0.06 | 0.12 | 0.15 | 0.07 | 0.11 |
| MgO | 5.56 | 12.64 | 8.90 | 4.32 | 0.21 | 0.78 | 5.39 | 0.77 | 1.18 |
| CaO | 10.03 | 12.10 | 12.38 | 6.21 | 0.93 | 1.58 | 10.64 | 3.26 | 3.40 |
| $Na_2O$ | 3.85 | 1.59 | 2.30 | 6.79 | 9.26 | 6.36 | 3.43 | 7.09 | 5.62 |
| $K_2O$ | 1.96 | 1.18 | 1.05 | 3.54 | 5.45 | 5.17 | 1.84 | 5.40 | 4.13 |
| $P_2O_5$ | 0.90 | 0.61 | 0.32 | 0.62 | 0.02 | 0.13 | 1.71 | 0.13 | 0.29 |
| $H_2O^+$ | 1.27 | 1.79 | 0.71 | 0.91 | 2.17 | 0.83 | 1.13 | 3.45 | 1.70 |
| $H_2O^-$ | 0.45 | 0.67 | 0.34 | 0.47 | 0.46 | 0.61 | 0.22 | 0.39 | 0.61 |
| Total | 99.71 | 99.72 | 100.55 | 99.85 | 99.20 | 99.88 | 99.78 | 100.10 | 99.66 |

*Explanation of column headings*

All analyses taken from McBirney and Aoki (1968, tables 1 and 3). Numbers in parentheses refer to specimens described in that work.

*Volcanic rocks*

1 Basanite (no. 71)
2 Ankaramite (with plentiful olivine, no. 38)
3 Olivine-augite basalt with normative *ne* (no. 52)
4 Tahitite (no. 53)
5 Phonolite (no. 45)
6 Trachyte (no. 69)

*Plutonic rocks*

7 Alkali gabbro (biotite-rich theralite, no. 57); cf. basanite, analysis 1
8 Nepheline syenite (no. 61B)
9 Syenite (no. 46)

Principe (Neiva, 1954; Barros, 1960) demonstrate the existence of a diversified alkaline suite ranging from alkali basalts and basanites through intermediate types ("tephrites") to basic phonolites ($SiO_2$ 53 to 58 percent) and phonolitic trachytes. The whole suite (Table 8-6) is characterized by low $SiO_2$, and nepheline is conspicuous in the norm of most as well as in the mode of many rocks. In both islands it is the basic rocks that predominate, except locally (for example, in southern Principe) where phonolites are more abundant.

Although the Guinea suite is highly alkaline and characterized throughout by low $SiO_2$, it is closer in its overall character to the Tahitian series than to the alkaline suites of Trinidade and Fernando de Noronha across the Atlantic. Geographic separation of the Brazilian from the West African continental margin began of course long before eruption of the island lavas.

## ALKALI-BASALT TRACHYTE (PHONOLITE) PROVINCES

ST. HELENA The island of St. Helena (Daly, 1927; I. Baker et al., 1967; I. Baker, 1969), at latitude 16°S, longitude 6°W, lies about 800 km east of the crest of the mid-Atlantic ridge (Fig. 8-1). Some 800 m in maximum elevation and 120 $km^2$ in area, the island comprises the uppermost 20 $km^3$ of a volcanic pile whose total bulk is 400 $km^3$, rising from the deep ocean floor 4000 to 4500 m below sea level. There are two eruptive centers, both active in Pliocene times and now deeply eroded.

At the older center, toward the northeast end of the island, subaerial volcanism—dated by the K/A technique—spans $3 \times 10^6$ years ($>14$ to $11\frac{1}{2}$ m.y.). The earlier lavas are alkali basalts, but in the upper levels more siliceous rocks (mugearites) also are common. At the younger center, 12 km to the southwest, several distinct volcanic episodes collectively cover a further $4 \times 10^6$ years ($11\frac{1}{2}$ to $7\frac{1}{2}$ m.y.). Each episode began with extrusion of alkali basalt, and in each hawaiites and mugearites appeared later in the cycle. The final episode culminated in intrusion and extrusion of dikes and domes of trachyte, phonolite, and hawaiite. Of the total exposed rocks, alkali basalts (with subordinate ankaramites) make up 70 to 80 percent, mugearitic lavas 15 to 25 percent, and trachytes and phonolites together no more than 1 percent. Pyroxene in the lavas of St. Helena is titanaugite except in trachytes and phonolites, where it is aegirine-augite. Except for minor chrome spinel in the ankaramites, the iron-oxide phase in all rocks is titanomagnetite.

Chemical variation diagrams (Fig. 8-4) are consistent with a fractional-crystallization model advocated by I. Baker (1969); the postulated parent liquid is alkali olivine basalt corresponding to the value 35 on the (*or* + *ab* + *ne*) abscissa of Fig. 8-4. Chemical analyses of representative rocks are given in Table 8-7 (analyses 1–4). Trace-element abundances can also be correlated with the fractional-crystallization model (Fig. 8-5). Ni and, to a lesser degree, Co drop sharply as they are withdrawn by olivine crystals; Mn is unaffected until it begins, at a late stage, to enter preferentially into crystals of iron-rich olivine

**Table 8-6** Chemical compositions (oxides, wt %) and CIPW norms of volcanic rocks of Principe

| | 1 | 2 | 3 | 4 | 5 | 6 | 7 |
|---|---|---|---|---|---|---|---|
| $SiO_2$ | 39.50 | 41.84 | 42.11 | 43.01 | 48.88 | 53.60 | 61.36 |
| $TiO_2$ | 0.90 | 2.69 | 0.85 | 2.99 | 3.25 | 0.55 | 0.15 |
| $Al_2O_3$ | 13.88 | 12.59 | 17.25 | 14.55 | 16.43 | 24.25 | 18.08 |
| $Fe_2O_3$ | 5.68 | 5.33 | 5.32 | 5.60 | 4.88 | 0.64 | 3.41 |
| FeO | 9.56 | 7.25 | 7.89 | 8.55 | 5.21 | 1.36 | 0.64 |
| MnO | 0.36 | 0.49 | 0.29 | 0.01 | 0.12 | 0.04 | 0.15 |
| MgO | 11.46 | 11.43 | 7.12 | 8.45 | 2.86 | 0.61 | 0.32 |
| CaO | 12.04 | 12.29 | 10.24 | 10.21 | 8.32 | 2.29 | 0.65 |
| $Na_2O$ | 3.17 | 2.06 | 3.06 | 1.82 | 6.79 | 9.21 | 5.82 |
| $K_2O$ | 0.37 | 1.06 | 2.80 | 1.02 | 2.18 | 6.08 | 7.42 |
| $P_2O_5$ | 0.95 | 0.89 | 1.04 | 0.53 | 0.97 | 0.08 | 0.49 |
| $H_2O^+$ | 1.49 | 1.74 | 1.86 | 2.87 | 0.50 | 0.70 | 1.27 |
| $H_2O^-$ | 0.86 | 0.10 | 0.65 | 0.15 | 0.42 | 0.72 | 0.79 |
| Total | 100.22 | 99.76 | 100.48 | 99.76 | 100.81 | 100.13 | 100.55 |
| *or* | 2.22 | 6.67 | 16.68 | 6.12 | 12.79 | 36.14 | 43.92 |
| *ab* | 1.57 | 6.81 | 1.41 | 15.30 | 29.87 | 16.20 | 45.85 |
| *an* | 22.24 | 21.68 | 25.02 | 28.50 | 7.88 | 6.65 | 0.28 |
| *ne* | 13.92 | 5.96 | 13.35 | | 15.00 | 33.20 | 1.80 |
| *di* | 25.29 | 27.36 | 15.98 | 15.08 | 18.58 | 3.83 | |
| *hy* | | | | 8.47 | | | |
| *ol* | 20.53 | 14.70 | 13.92 | 8.34 | | 0.50 | |
| *mt* | 8.35 | 7.66 | 7.66 | 8.12 | 7.19 | 0.93 | 3.94* |
| *il* | 1.67 | 5.17 | 1.67 | 5.65 | 6.23 | 1.06 | 0.46 |
| *ap* | 2.35 | 2.02 | 2.35 | 1.34 | 2.35 | | 1.34 |
| Total | 98.14 | 98.03 | 98.04 | 96.92 | 99.79 | 98.51 | 97.59 |

* *mt* 1.86 + *hm* 2.08.

*Explanation of column headings*

Analyses are cited from Barros (1960), to whose work page and specimen numbers refer.

1 Limburgite (rare); Roca Esperanca (33-P, p. 22)
2 Alkali basalt ("basanitoid basalt," widely represented on Principe); Roca Sundy (51-P, p. 20)
3 Basanite; Belmonte (12-P, p. 27)
4 Olivine tholeiite (olivine basalt widely present); Roca Paciência (19-P, p. 23)
5 Mugearite (analcite-sanidine tephrite); Roca Sundy (01-P, p. 30)
6 Sodalite phonolite; Roca Infante D. Henrique (24-P, p. 38)
7 Biotite trachyte; near Abade (46-P, p. 34)

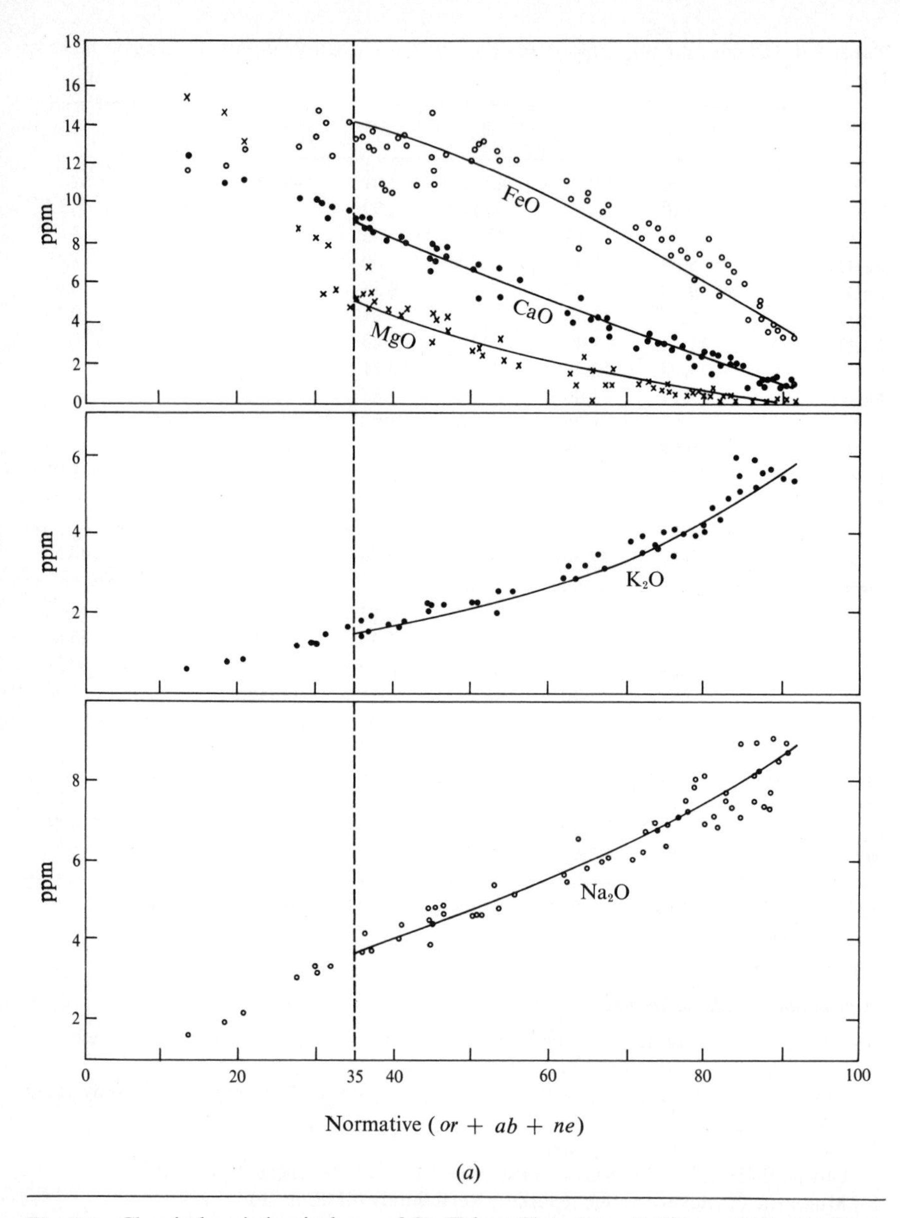


**Fig. 8-4** Chemical variation in lavas of St. Helena (data from I. Baker, 1969). Total iron plotted (page opposite) as FeO. Lavas of Tristan da Cunha (from Le Maitre, 1962, p. 1338) plot in stippled band of $K_2O$–$Na_2O$–CaO triangle.

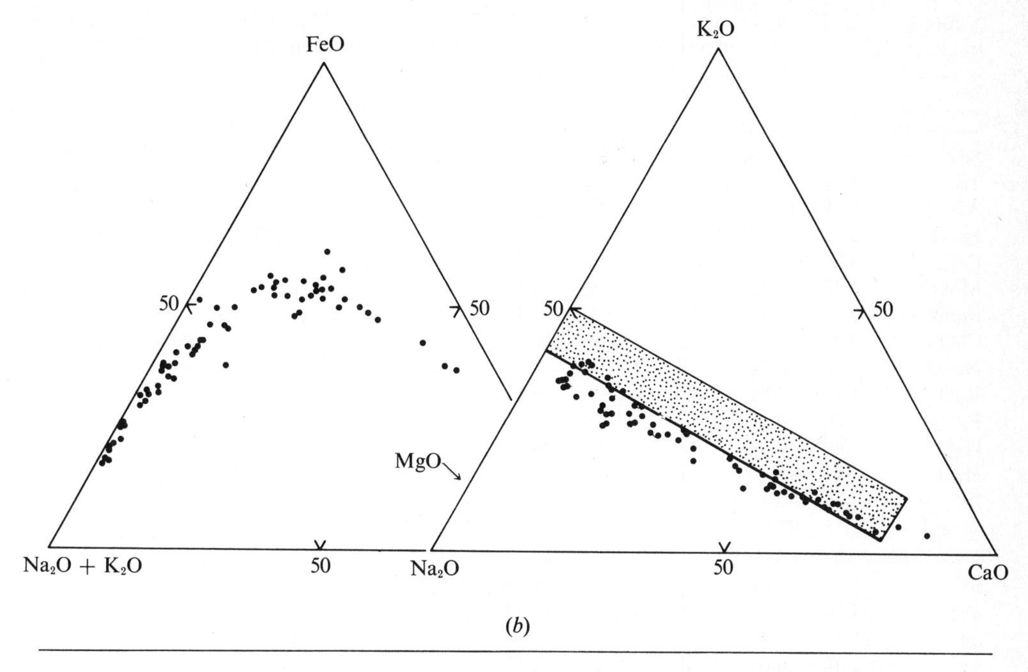


(*b*)

**Fig. 8-4** *Continued.*

and pyroxene. Early crystallization of picotite effectively eliminates Cr; the liquid fractions are steadily depleted in Ti, which is taken up by crystallizing augite. Ba, Sr, and Rb build up in the liquid until Sr begins to be withdrawn by plagioclase and Ba and Rb together enter the final crystalline phase, alkali feldspar. Zr increases steadily to the end. Convincing though this total picture undoubtedly is, anomalies to be discussed in more detail later in this chapter arise from the isotopic composition of Sr in rocks of different composition.

Magma series broadly comparable with that of St. Helena are more widely distributed than any other in the ocean basins.

THE AZORES The Azores (Esenwein, 1929) are a group of ten volcanic islands, some with active centers, that trend eastward from the mid-Atlantic ridge between latitudes 39 and 37°N. They rise from a ridge some 700 km long (between 24 and 33°W longitude) with a mean depth of 2000 m below sea level. Olivine basalts include *ne*-normative types like those of St. Helena—though perhaps somewhat more potassic—as well as slightly less alkaline rocks with minor *hy* in the norm. Lavas of intermediate composition are hawaiites (Girod and Lefevre, 1972), grading into slightly oversaturated trachytes. On the whole, then, the volcanic suite of the Azores has chemical and petrographic affinities with that of St. Helena on the one hand, and on the other with the more siliceous series of Ascension Island.

**Table 8-7** Chemical composition (oxides, wt %), CIPW norms, and atomic abundances (ppm) of trace elements of volcanic rocks of St. Helena (I. Baker, 1969, table 2) and Mauritius (Shand, 1933)

| | 1 | 2 | 3 | 4 | 5 | 6 |
|---|---|---|---|---|---|---|
| $SiO_2$ | 45.50 | 54.88 | 59.64 | 59.92 | 46.90 | 60.69 |
| $TiO_2$ | 3.44 | 1.11 | 0.42 | 0.06 | 3.31 | 0.15 |
| $Al_2O_3$ | 15.71 | 17.41 | 17.67 | 19.86 | 15.05 | 19.74 |
| $Fe_2O_3$ | 3.61 | 2.44 | 2.59 | 1.69 | 1.11 | 1.92 |
| FeO | 8.64 | 6.79 | 3.92 | 1.70 | 10.46 | 2.33 |
| MnO | 0.22 | 0.23 | 0.22 | 0.18 | 0.16 | 0.17 |
| MgO | 5.37 | 1.88 | 0.50 | 0.05 | 8.41 | 0.01 |
| CaO | 9.43 | 3.48 | 2.09 | 1.07 | 10.92 | 1.02 |
| $Na_2O$ | 3.47 | 6.04 | 7.62 | 8.94 | 3.13 | 7.95 |
| $K_2O$ | 1.38 | 2.89 | 4.02 | 4.93 | 0.27 | 5.50 |
| $P_2O_5$ | 0.29 | 0.41 | 0.29 | 0.17 | Tr. | Tr. |
| $H_2O^+$ | 0.60 | 0.77 | 0.08 | 0.07 | 0.27 | 0.58 |
| $H_2O^-$ | 2.49 | 1.53 | 1.04 | 1.29 | 0.23 | 0.28 |
| Total | 100.15 | 99.86 | 100.10 | 99.93 | 100.22 | 100.34 |
| *or* | 8.16 | 17.08 | 23.76 | 29.14 | 1.67 | 32.80 |
| *ab* | 24.79 | 51.07 | 53.68 | 46.68 | 23.35 | 47.16 |
| *an* | 23.22 | 11.86 | 2.14 | | 26.13 | 1.67 |
| *ne* | 2.48 | 0.02 | 5.85 | 15.18 | 2.27 | 10.79 |
| *ac* | | | | 0.83 | | |
| *di* | 15.97 | 0.32 | 6.65 | 4.29 | 21.84 | 2.97 |
| *ol* | 9.60 | 10.15 | 1.96 | 0.03 | 16.74 | 1.81 |
| *mt* | 5.23 | 3.54 | 3.76 | 2.04 | 1.62 | 2.55 |
| *il* | 6.53 | 2.11 | 0.80 | 0.11 | 6.23 | 0.30 |
| *ap* | 1.42 | 1.82 | 0.19 | 0.17 | — | — |
| Total | 97.40 | 97.97 | 98.79 | 98.47 | 99.85 | 100.05 |

| | 1*a* | 2*a* | 3*a* | 4*a* |
|---|---|---|---|---|
| Ba | 290 | 790 | 1100 | 1100 |
| Cr | 36 | | | |
| Rb | 40 | 50 | 90 | 220 |
| Sr | 510 | 545 | 280 | 85 |
| Zr | 195 | 490 | 645 | 1085 |
| Rb/Sr | 0.08 | | | |
| K/Rb | 300 | 400 | 350 | 180 |

*Explanation of column headings*

1, 1*a* Basalt, St. Helena (803)
2, 2*a* Mugearite (trachyandesite), St. Helena (822)
3, 3*a* Trachyte, St. Helena (763)
4, 4*a* Phonolite, St. Helena (11)
5 Olivine basalt, Mauritius (A, p. 5)
6 Phonolitic trachyte, Mauritius (A, p. 8)

A characteristic feature of Holocene volcanism in the Azores is independent eruption of basaltic and trachytic ejecta (Booth et al., 1970). Thus on the small island of Faial—a basaltic cone 20 km in diameter and 1000 m high—large volumes of trachytic pumice have been periodically ejected with explosive violence from the summit caldera over the last 10,000 years. Meanwhile centers along a lateral fissure a few kilometers west and 500 m below the summit have erupted a comparable volume of basaltic lava and tuff. On São Miguel, 300 km

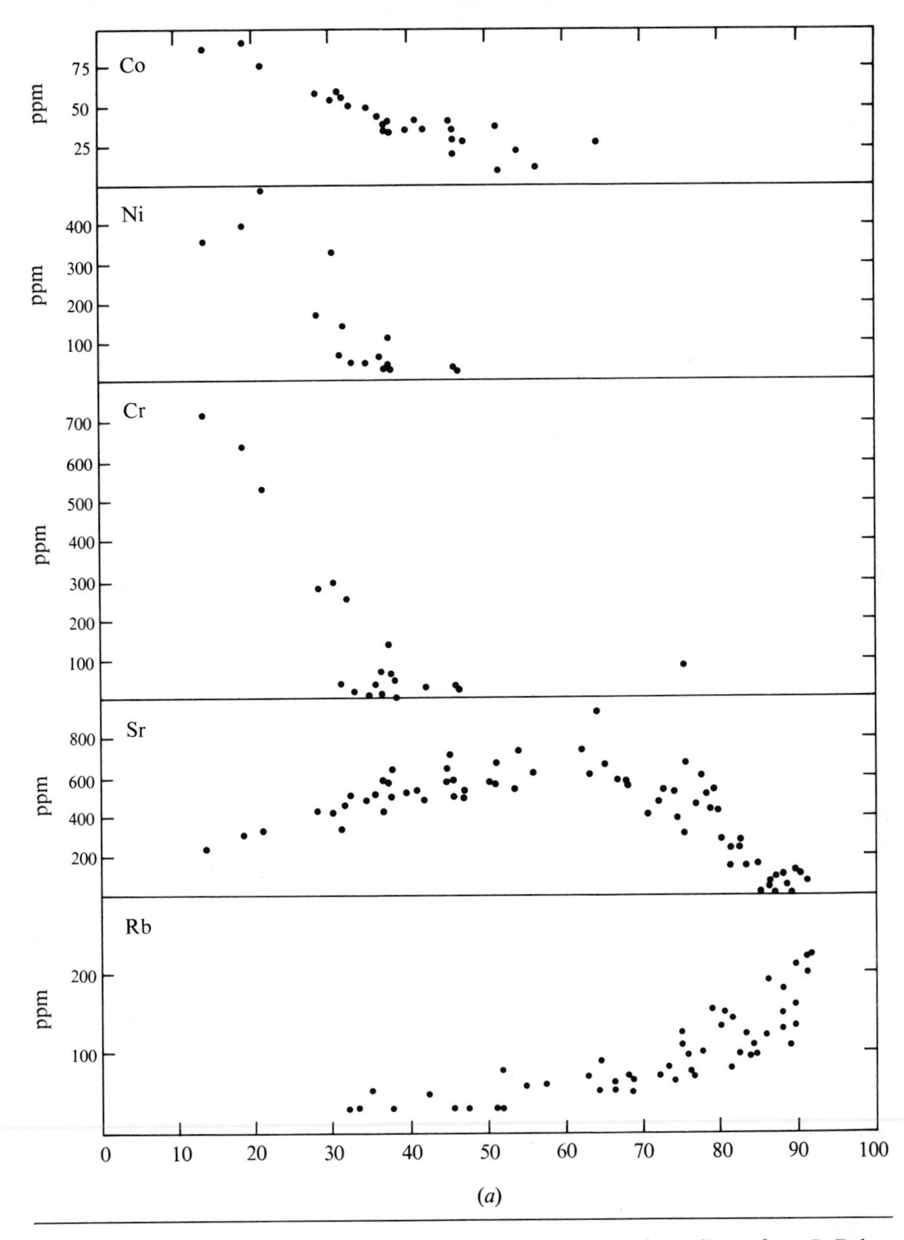


**Fig. 8-5** Trace- and minor-element variation in lavas of St. Helena. (Data from I. Baker, 1969.) Abundances (ppm) plotted against normative (*or* + *ab* + *ne*). (*a*) Co, Ni, Cr, Sr, Rb.

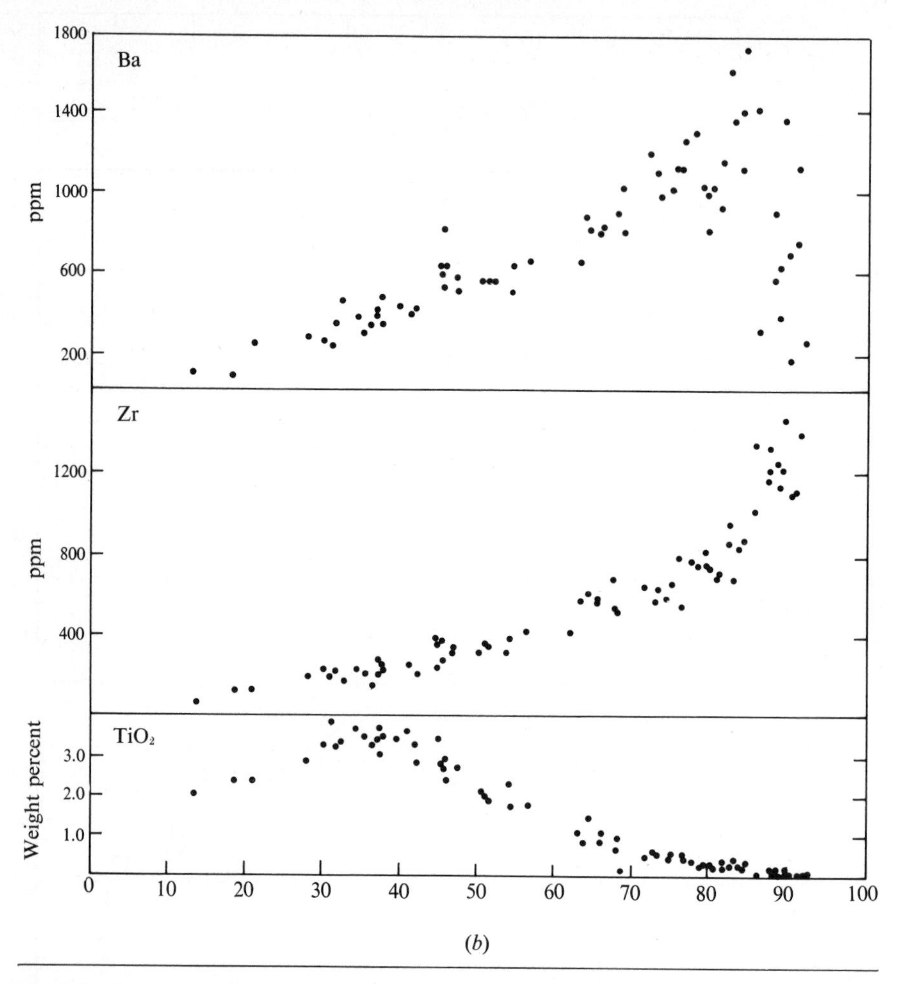


**Fig. 8-5** *Continued* (*b*) Ba, Zr, $TiO_2$.

east-southeast of Faial, we find much the same eruptive pattern over the past 5000 years: half a dozen explosive summit eruptions of trachyte from each of three volcanoes and intermittent eruption of basalt from flanking fissures 500 m lower down.

ISLANDS OF THE INDIAN OCEAN East of Madagascar at 21°S are three volcanic islands a few hundred kilometers apart, aligned more or less east-west in the order Rodriguez (Upton et al., 1965), Mauritius (Shand, 1933), Réunion (McDougall and Compston, 1965; MacDougall, Upton, and Wadsworth, 1965). Olivine basalts are the dominant rocks on all three islands: mostly of olivine-tholeiitic composition on Réunion and alkali basalts on the other two islands. On all three there are plentiful mugearites; phonolitic trachytes have been recorded only on Mauritius, but there is alkali syenite in the eroded core of a volcanic center on Réunion. In Pleistocene lavas (dated $1\frac{1}{2}$ m.y.) on Rodriguez there are numerous xenoliths of gabbro, anorthosite, pyroxenite, and dunite, hinting at the possible existence beneath of a differentiated mafic (tholeiitic) pluton.

In the subantarctic Crozet group, the lavas are olivine and augite-rich picritic alkali basalts with minor normative *ne* (Gunn et al., 1970).

PACIFIC ISLANDS Volcanic rocks reminiscent of the St. Helena series have been reported from many tropical islands of the Pacific Ocean (for example, Lacroix, 1928). East Pacific examples (McBirney and Williams, 1969, table 21, p. 176), from west to east, are Rapa (Smith and Chubb, 1927), Juan Fernandez (McBirney and Williams, 1969, p. 167), and San Felix (Willis and Washington, 1924). From the western Pacific we cite two island clusters in the eastern portion of the Caroline group: Truk (Stark and Hay, 1963) and Ponape (Yagi, 1960; Ishikawa and Yagi, 1970). The basic members in these two provinces include rocks of the melilite-basalt and nephelinite families as well as more plentiful alkali basalts; there are mugearites and trachytes but no phonolites.

## POTASSIC ALKALI-BASALT TRACHYTE (PHONOLITE) PROVINCES

In four volcanic provinces of the southern oceans the range of rock types is much the same as on St. Helena or Tahiti; but the rock series as a whole are unmistakably more potassic. This is no trivial characteristic; the K/Na ratio of a primitive magma is thought to give significant information as to the nature and environment of the source material. Moreover with high potassium we expect to find correspondingly high values for Rb, Ba, Sr, Th, and U—and also for radiogenic Sr.

The island groups are these: Tristan da Cunha (37°S, 12°W) and Gough (40°S, 10°W), both on the eastern flank of a westerly bulge in the mid-Atlantic ridge; the Kerguelen group (49°S, 70°E) and Heard Island (53°S, $73\frac{1}{2}$°E) some thousands of kilometers east-southeast in the southern Indian Ocean.

TRISTAN DA CUNHA This is a group, 50 km in diameter, comprising three islands—the emergent summits of immense volcanoes based on the ocean floor, here at a mean depth of 3000 m. The largest, Tristan da Cunha itself, is close to 2000 m high. It was the scene of a spectacular eruption in 1961, subsequent investigation of which has yielded an unusually comprehensive report on the petrology of the island group (P. E. Baker et al., 1964).

On all three islands the most basic lavas—alkali basalts (Table 8-8, analysis 1) and ankaramites—are greatly exceeded in volume by trachybasalts unusually rich in alkali (analysis 2) and with $K_2O/Na_2O$ mostly in the range 0.7 to 0.8 (cf. shaded band, $K_2O$–$Na_2O$–CaO triangle, Fig. 8-4). Many of these rocks contain leucite, a mineral rare in oceanic lavas; all have high normative *ne*.

Minor but widely distributed rocks include *ne*-normative hawaiites (trachyandesites) and trachytes close to saturation in silica (analyses 3, 4). As might be expected from their potassic composition, the lavas of Tristan have high Rb, Sr, Ba, and Nb (Table 8-8).

Coarse-grained gabbroid rocks are widely distributed as xenoliths and fragmented blocks in the lavas and pyroclastic rocks of Tristan. Their essential constituents, in varying proportions, are plagioclase, augite, kaersutite, and magnetite (less commonly biotite)—minerals that are also the principal constituents of the trachybasaltic lavas. Crude layering and other cumulate textural traits suggest that they are cognate xenoliths torn from a subjacent layered basic pluton.

OTHER PROVINCES The potassic character of the lavas of Gough Island (Le Maitre, 1962) is not quite so pronounced as in the Tristan da Cunha suite. Nevertheless it is clearly expressed by general predominance of trachytes and trachybasalts (hawaiites) with $K_2O/Na_2O$ = 0.7 to 0.8 (Table 8-8, analysis 6) and mineralogically by the presence of alkali feldspar even in mafic rocks. These are less alkaline than the basalts of Tristan and include both alkali basalts and olivine basalts lacking normative *ne* (Table 8-8, analysis 5). At the felsic end of the series, trachyandesites (hawaiites) grade into trachytes, which are unusually abundant. Both types have normative *ne*. The trace-element pattern of the suite as a whole resembles that of the Tristan lavas. A remarkable feature of Gough is the widespread occurrence of xenoliths and fragmented blocks of gabbro whose petrographic and chemical character seems alien to that of the volcanic suite (Le Maitre, 1965). Their mineral components are plagioclase, diopsidic augite, hypersthene, and olivine—an assemblage characteristic, the world over, of layered plutons of tholeiitic parentage (cf. Chap. 9); chemically they are olivine tholeiites: $K_2O/Na_2O$ = 0.15 to 0.25.

The volcanic rocks of the Kerguelen Islands are unusually diverse (Edwards, 1938*a*). At the mafic end are alkali basalts, mugearites, and basanites, mostly (but not all) highly potassic; rarely they contain leucite. The felsic members include not only phonolites but a variety of trachytes, some of which are silica-saturated and pass into soda rhyolites. Xenoliths seem to be mineralogically related to the lavas in which they occur: peridotites and gabbro in the basaltic rocks, hornblende peridotite in trachybasalts and phonolites.

In most respects Heard Island is a petrological replica of St. Helena. However, its trachybasalts tend to be more potassic; in half the analyzed rocks $K_2O/Na_2O$ is between 0.75 and 1.

### OLIVINE-THOLEIITE QUARTZ-TRACHYTE (RHYOLITE) PROVINCES

ASCENSION ISLAND The island of Ascension (Daly, 1925) rises from the crest of the mid-Atlantic ridge at 8°S latitude. It covers an area of 100 km$^2$ and its maximum elevation is 860 m. The most extensive rocks are olivine basalts with augite as the sole pyroxene. Chemically these rate as olivine tholeiites, since there is significant *hy* in the norm (Table 8-9; analyses 1, 2). Analyses of associated mugearites and hawaiites (Table 8-9; analyses 3, 4) likewise show normative *hy* and even *Q*. Felsic lavas, as usual, are greatly subordinate to basalts, and they are represented by numerous domes and by tuffs and a few thick flows. All are strongly oversaturated rocks (Table 8-9; analyses 5, 6) composed mainly of anorthoclase with aegirine the common mafic mineral. They range from trachytes (66 percent $SiO_2$) to aegirine pantellerites (71 percent $SiO_2$). Granitic blocks in trachyte tuffs prove to be cognate sodic rocks chemically and mineralogically equivalent to the Ascension pantellerites (Tilley, 1950, p. 43).

PACIFIC ISLANDS In several Pacific island groups there are volcanic rock series that recall the association on Ascension Island. Olivine basalts everywhere preponderate; mostly they are olivine tholeiites (Table 8-9; analyses 7, 8), but a few recorded analyses show minor *ne*. As examples we cite the Samoan islands of Tutuila (Macdonald, 1944), Upolu, and Savaii (Hedge et al., 1972) in the southwest and the islands of Guadalupe (Engel and Engel, 1964*b*) and Clarion (Bryan, 1967) in the eastern Pacific. The last-named islands are situated on the East Pacific rise, off the coast of Mexico. Their component basalts, like those dredged from the swarm of guyots just northwest of Guadalupe, are notably richer in alkali ($K_2O$ especially) than are oceanic tholeiites recovered in deep water nearby from the crest and flanks of the East Pacific rise. The trace-element pattern, too, is totally different (Engel et al., 1965): Ba, Sr, Rb, and Zr are many times higher, and Cr and Ni notably lower (Table 8-9, analysis 8*a*), than in basalts from the rise.

### THOLEIITE ICELANDITE QUARTZ-TRACHYTE (RHYOLITE) PROVINCES

WESTERN GALÁPAGOS VOLCANOES The Galápagos Islands (C. Richardson, 1933; McBirney and Williams, 1969) are a cluster of Pleistocene and Recent volcanoes 400 km or so in diameter, situated in the East Pacific on the Equator 1000 km off the coast of South America. They are based on a submarine platform 1500 m below sea level that juts eastward at this point from the east Pacific rise. Today this is one of the most active volcanic regions of the world.

The islands are composed almost entirely of basaltic lavas and pyroclastics. Chemically these range from tholeiites to alkali basalts, and composition conforms rather sharply to a recognizable geographic distribution. The central and eastern islands are built mainly of alkali basalts. Tholeiitic lavas on the other hand make up the two large western islands Albemarle and Narborough; and differentiated lavas of tholeiitic parentage are the main constituents of the small islands Duncan and Jervis, immediately adjacent to Albemarle on the east. It is this tholeiitic subprovince of the Galápagos that is the topic of present discussion.

**Table 8-8** Chemical compositions (oxides, wt %), CIPW norms, and atomic abundances (ppm) of trace elements in volcanic rocks of Tristan da Cunha (P. E. Baker et al., 1964) and Gough Island (Le Maitre, 1962, table 10)

| | 1 | 2 | 3 | 4 | 5 | 6 | 7 | | 1*a* | 2*a* | 3*a* | 4*a* | 5*a* | 6*a* | 7*a* |
|---|---|---|---|---|---|---|---|---|---|---|---|---|---|---|---|
| $SiO_2$ | 42.43 | 46.01 | 54.95 | 58.0 | 47.73 | 48.79 | 54.41 | Ba | 750 | 950 | 1200 | 1000 | 800 | 700 | 1400 |
| $TiO_2$ | 4.11 | 2.19 | 1.58 | 1.2 | 3.30 | 3.18 | 1.67 | Co | 40 | 20 | | | 22 | 100 | 5 |
| $Al_2O_3$ | 14.15 | 16.84 | 19.63 | 19.5 | 15.53 | 17.39 | 17.37 | Cr | 65 | 18 | | | 220 | 100 | |
| $Fe_2O_3$ | 5.84 | 7.61 | 1.62 | 1.7 | 2.02 | 2.48 | 4.02 | Li | 4 | 4 | 10 | 15 | 16 | 5 | 7 |
| FeO | 8.48 | 5.37 | 3.31 | 2.2 | 8.95 | 7.39 | 3.29 | Ni | 50 | | | | 100 | 30 | 2 |
| MnO | 0.17 | 0.18 | 0.18 | 0.1 | 0.14 | 0.10 | 0.12 | Pb | 10 | 18 | 14 | 28 | | | |
| MgO | 6.71 | 4.75 | 1.42 | 1.0 | 8.37 | 4.00 | 2.27 | Rb | 110 | 110 | 230 | 350 | 100 | 40 | 100 |
| CaO | 11.91 | 9.36 | 5.73 | 3.3 | 8.71 | 8.97 | 4.36 | Sr | 1000 | 1100 | 1300 | 650 | 650 | 1000 | 850 |
| $Na_2O$ | 2.77 | 3.74 | 5.89 | 6.5 | 2.89 | 3.28 | 4.94 | V | 400 | 200 | 95 | 50 | 140 | 250 | 80 |
| $K_2O$ | 2.04 | 2.72 | 4.95 | 5.3 | 1.70 | 2.28 | 4.69 | Zr | 200 | 300 | 350 | 350 | 125 | 220 | 450 |
| $P_2O_5$ | 0.58 | 1.18 | 0.43 | 0.2 | 0.29 | 0.26 | 0.46 | | | | | | | | |
| $H_2O^+$ | 0.34 | 0.01 | 0.00 | 0.2 | 0.18 | 0.98 | 0.86 | | | | | | | | |
| $H_2O^-$ | 0.44 | 0.08 | 0.01 | 0.1 | 0.06 | 0.76 | 1.50 | | | | | | | | |
| Total | 99.97 | 100.04 | 100.05* | 99.3 | 99.87 | 99.86 | 100.09† | | | | | | | | |
| *or* | 12.06 | 16.08 | 29.27 | 31.33 | 10.01 | 13.34 | 27.80 | | | | | | | | |
| *ab* | 6.95 | 21.63 | 29.99 | 40.12 | 24.10 | 27.77 | 39.30 | | | | | | | | |
| *an* | 20.16 | 21.13 | 13.57 | 8.38 | 24.46 | 26.13 | 11.40 | | | | | | | | |
| *ne* | 8.93 | 5.43 | 9.66 | 8.06 | | | 1.42 | | | | | | | | |
| *di* | 28.01 | 13.73 | 9.60 | 5.36 | 13.69 | 13.57 | 5.19 | | | | | | | | |
| *hy* | | | | | 0.86 | | | | | | | | | | |
| *ol* | 5.48 | 4.03 | 0.94 | 0.55 | 16.57 | 6.99 | 2.40 | | | | | | | | |
| *mt* | 8.47 | 11.03 | 2.34 | 2.46 | 3.02 | 3.71 | 5.80 | | | | | | | | |
| *il* | 7.81 | 4.16 | 3.00 | 2.28 | 6.23 | 6.08 | 3.19 | | | | | | | | |
| *ap* | 1.37 | 2.78 | 1.01 | 0.47 | 0.67 | 0.67 | 1.34 | | | | | | | | |
| Total | 99.24 | 100.00 | 99.38 | 99.01 | 99.61 | 98.26 | 97.84 | | | | | | | | |

* Including Cl = 0.27, F = 0.08.
† Including F = 0.13.

*Explanation of column headings*

Numbers in parentheses refer to specimen numbers in sources cited.

1, 1*a* Alkali basalt (6), Tristan da Cunha
2, 2*a* Trachybasalt (369), Tristan da Cunha
3, 3*a* Trachyandesite (657), Tristan da Cunha
4, 4*a* Trachyte (560), Tristan da Cunha
5, 5*a* Olivine basalt (G111), Gough Island
6, 6*a* Trachybasalt (G22), Gough Island
7, 7*a* Trachyandesite (G86), Gough Island

**Table 8-9** Chemical compositions (oxides, wt %) and CIPW norms of volcanic rocks of Ascension Island and of Guadalupe and Clarion Islands

| | 1 | 2 | 3 | 4 | 5 | 6 | 7 | 8 | | 8*a* |
|---|---|---|---|---|---|---|---|---|---|---|
| $SiO_2$ | 47.69 | 48.64 | 54.04 | 58.00 | 66.12 | 71.88 | 47.52 | 50.48 | Ba | 540 |
| $TiO_2$ | 2.79 | 3.52 | 0.94 | 3.38 | | 0.25 | 2.85 | 2.25 | Co | 21 |
| $Al_2O_3$ | 16.23 | 15.54 | 19.58 | 14.92 | 15.51 | 12.85 | 16.06 | 18.31 | Cr | 7 |
| $Fe_2O_3$ | 2.20 | 5.31 | 5.09 | 1.73 | 3.27 | 3.60 | 3.90 | 3.21 | Li | 10 |
| FeO | 9.93 | 7.73 | 3.75 | 5.78 | 0.93 | 0.05 | 7.71 | 6.03 | Ni | 18 |
| MnO | 0.17 | 0.17 | — | 0.11 | | 0.29 | 0.16 | 0.21 | Rb | 60 |
| MgO | 7.15 | 4.96 | 1.99 | 2.23 | 0.17 | 0.18 | 6.26 | 4.21 | Sr | 680 |
| CaO | 10.02 | 9.03 | 5.54 | 4.50 | 1.05 | 0.60 | 9.00 | 7.21 | V | 180 |
| $Na_2O$ | 2.87 | 3.60 | 4.70 | 5.88 | 6.31 | 5.32 | 2.92 | 4.80 | Zr | 420 |
| $K_2O$ | 0.64 | 1.24 | 3.48 | 2.76 | 5.40 | 4.78 | 1.55 | 1.93 | K/Rb | 267 |
| $P_2O_5$ | 0.59 | 0.64 | 0.31 | 0.71 | | 0.05 | 0.73 | 0.74 | Rb/Sr | 0.09 |
| $H_2O^+$ | 0.19 | 0.18 | 1.16 | 0.31 | 1.98 | 0.17 | 0.89 | 0.46 | | |
| $H_2O^-$ | 0.09 | 0.16 | | 0.09 | | 0.18 | 0.31 | 0.38 | | |
| Total | 100.56 | 100.72 | 100.58 | 100.40 | 100.74 | 100.20 | 99.86 | 100.22 | | |
| *Q* | | | | 2.76 | 8.34 | 23.40 | | | | |
| *or* | 3.33 | 7.23 | 20.57 | 16.68 | 31.69 | 28.36 | 9.45 | 10.56 | | |
| *ab* | 24.63 | 30.39 | 39.82 | 49.78 | 49.78 | 38.78 | 24.63 | 35.11 | | |
| *an* | 29.47 | 22.52 | 21.96 | 5.84 | | | 26.13 | 27.24 | | |
| *ac* | | | | | 3.23 | 5.08 | | | | |
| *wo* | | | | | 1.74 | | | | | |
| *di* | 13.60 | 15.14 | 2.87 | 9.98 | 0.86 | 0.86 | 11.18 | 3.29 | | |
| *hy* | 8.83 | 9.14 | 3.43 | 4.55 | | | 8.94 | 7.93 | | |
| *ol* | 10.48 | 0.15 | 0.90 | | | | 5.76 | 2.36 | | |
| *mt* | 3.25 | 7.66 | 7.42 | 2.55 | 3.02 | 2.12* | 5.57 | 6.03 | | |
| *il* | 5.32 | 6.69 | 1.82 | 6.38 | | 0.46 | 5.47 | 5.47 | | |
| *ap* | 1.24 | 1.24 | 0.67 | 1.55 | | 0.09 | 1.68 | 1.51 | | |
| Total | 100.15 | 100.16 | 99.46 | 100.07 | 98.66 | 99.15 | 98.81 | 99.50 | | |

* Sum of magnetite 0.46, hematite 1.66.

*Explanation of column headings*

1 Olivine basalt, Ascension Island (Daly, 1925, p. 44, table 2, no. 1)
2 Olivine-poor basalt, Ascension Island (Daly, 1925, p. 44, table 2, no. 2)
3 Hawaiite (trachydolerite), Ascension Island (Daly, 1925, p. 46, table 3, no. 1)
4 Mugearite (trachyandesite), Ascension Island (Daly, 1925, p. 49, table 4, no. 1)
5 Trachyte, Ascension Island (Daly, 1925, pp. 51, 52, tables 5, 6, no. 3)
6 Rhyolite, Ascension Island (Daly, 1925, p. 58, table 7, no. 1)
7 Olivine basalt, Clarion Island (Bryan, 1967, pp. 1467, 1468, analysis 2)
8 Olivine basalt, Guadalupe Island (Engel et al., 1965, p. 725, table 5, Gu77)
8*a* Same specimen as 8. Trace-element abundances (ppm)

Albemarle (3000 $km^2$) and Narborough (500 $km^2$) together form a group of seven major shield volcanoes—six of them active—whose maximum elevations are between 1100 and 1700 m. An average composition of their basaltic lavas computed by McBirney and Williams shows minor normative *Q*; but in the norms of other recorded rocks (Table 8-10, analysis 1) there is significant *ol*. Total iron tends to be high—12 to 13 percent in most Albemarle tholeiites. Mineralogically there is not much to indicate a tholeiitic composition: olivine, a magnesian type, is usually but a minor constituent; the common pyroxene diopsidic augite is accompanied in just a few rocks by hypersthene or pigeonite. Olivine basalts from the northernmost, now long extinct volcano of Albemarle have alkali feldspar as well as the usual plagioclase in the groundmass and have also a relatively high normative *ol*. There is one recorded exception to the sequence of tholeiitic eruptions on Albemarle. From the smallest volcano, Alcado, a very recent eruption covered the slopes with an extensive blanket of quartz-trachyte pumice (Table 8-10, analysis 5).

On the two small islands, Jervis and Duncan, basaltic rocks are abundant. Very few match the tholeiitic lavas of Albemarle. Most are slightly more siliceous *Q*-normative tholeiites or ferrobasalts (Table 8-10, analysis 3) with unusually high total iron (13.5 to 14 percent). Icelandites, too, are plentiful on these islands (Table 8-10, analysis 4). Most Galápagos icelandites contain two pyroxenes (diopsidic augite and pigeonite), plagioclase phenocrysts are abundant, and olivine is typically absent. Greatly subordinate to ferrobasalts and icelandites are siliceous trachytes with high normative *Q*.

Pyroclastic accumulations at minor explosion centers on Albemarle and Jervis Islands contain blocks of cognate plutonic rocks: basic olivine-bytownite-hornblende gabbro cumulates, hornblende-augite ferrogabbros with olivine $Fo_{80}$ to $Fo_{30}$, and plagioclase-rich diorites and quartz syenites in which augite, hypersthene, and olivine (in diorites only) are very rich in iron. As in so many basalt-dominated provinces, ultramafic inclusions composed of olivine, pyroxene, and spinel are confined to areas (in this case the southerly island, Charles) where the basaltic lavas are alkaline.

Close and exclusive field association between trachytes, icelandites, and tholeiitic basalts in the western Galápagos is taken by McBirney and Williams (1969, pp. 152–153) to indicate an underlying genetic relationship. Tholeiitic magma is assigned the parent role; more siliceous rocks are thought to have crystallized from derivative (differentiated) magmas. Most petrologists would accept this argument, and there are clues as to some aspects of the differentiation process. The cumulate texture of calcic gabbro xenoliths suggests that there must be complementary liquids depleted in some or all of the mineral phases olivine, magnesian augite, and calcic plagioclase. A marked drop in Ni abundance in ferrobasalts and ferrogabbros probably reflects earlier removal of olivine; but there is no diminution in Sr such as should accompany withdrawal of plagioclase crystals. In lavas transitional from ferrobasalt to icelandite, and again in the dioritic xenoliths, sudden depletion in V suggests removal of

**Table 8-10** Chemical compositions (oxides, wt %), CIPW norms, and atomic abundances of trace elements; volcanic rocks of western Galápagos

| | 1 | 2 | 3 | 4 | 5 | | 2*a* | 4*a* |
|---|---|---|---|---|---|---|---|---|
| $SiO_2$ | 47.01 | 48.45 | 48.54 | 55.34 | 66.87 | Ba | 76 | 285 |
| $TiO_2$ | 3.20 | 3.39 | 4.20 | 1.93 | 0.66 | | (65) | (280) |
| $Al_2O_3$ | 15.57 | 13.75 | 14.49 | 14.02 | 12.55 | Co | 55 | 54 |
| $Fe_2O_3$ | 2.32 | 4.72 | 4.67 | 3.31 | 1.84 | Cr | 47 | 17 |
| FeO | 11.57 | 8.60 | 9.16 | 8.73 | 2.53 | | (142) | |
| MnO | 0.20 | 0.20 | 0.16 | 0.17 | 0.09 | Mn | 1440 | 1600 |
| MgO | 5.25 | 6.05 | 4.58 | 2.71 | 0.60 | Ni | 70 | 14 |
| CaO | 9.77 | 10.71 | 8.29 | 6.54 | 1.10 | Sr | 235 | 250 |
| $Na_2O$ | 3.00 | 2.79 | 3.58 | 4.54 | 5.32 | | (323) | (266) |
| $K_2O$ | 0.31 | 0.50 | 0.84 | 1.33 | 3.08 | V | 310 | 98 |
| $P_2O_5$ | 0.32 | 0.36 | 0.47 | 0.65 | 0.05 | Zr | (255) | (538) |
| $H_2O^+$ | 1.40 | 0.36 | 0.68 | 0.64 | 4.66 | | | |
| $H_2O^-$ | 0.24 | 0.02 | 0.16 | 0.18 | 0.33 | | | |
| Total | 100.16 | 99.90 | 99.82 | 100.09 | 99.68 | | | |
| *Q* | — | 2.21 | 2.04 | 5.95 | 20.82 | | | |
| *or* | 1.83 | 2.95 | 4.96 | 7.86 | 18.20 | | | |
| *ab* | 25.39 | 23.61 | 30.29 | 38.42 | 45.02 | | | |
| *an* | 28.10 | 23.52 | 20.99 | 13.95 | 1.27 | | | |
| *di* | 15.17 | 22.00 | 13.86 | 11.97 | 3.24 | | | |
| *hy* | 11.92 | 11.12 | 11.00 | 11.16 | 2.15 | | | |
| *ol* | 5.93 | — | — | — | — | | | |
| *mt* | 3.36 | 6.84 | 6.77 | 4.80 | 2.67 | | | |
| *il* | 6.08 | 6.44 | 7.98 | 3.67 | 1.23 | | | |
| *ap* | 0.76 | 0.75 | 1.11 | 1.54 | 0.12 | | | |
| Total | 98.54 | 99.44 | 99.00 | 99.32 | 94.72 | | | |

*Explanation of column headings*

1 Basalt, Albemarle Island (McBirney and Williams, 1969, p. 124, table 3, no. 8)

2 Tholeiitic basalt, Albemarle Island (McBirney and Williams, 1969, p. 121, table 2, no. 63)

3 Ferrobasalt, Jervis Island (McBirney and Williams, 1969, p. 147, table 10, no. 65)

4 Icelandite, Duncan Island (McBirney and Williams, 1969, p. 146, table 10, no. 71).

5 Siliceous trachyte pumice, Alcado volcano, Albemarle Island (McBirney and Williams, 1969, p. 146, table 10, no. 130)

2*a*, 4*a* Trace-element abundances, same rocks as analyses 2 and 4. Determinations by emission spectrography. Values in parentheses determined by neutron activation (Cr) or x-ray fluorescence (Ba, Sr, Zr).

magnetite. The most obvious result of differentiation has been to preserve or increase the state of silica saturation with which the parent basaltic magmas were endowed. Herein we see conformity to Lacroix's generalization of 40 years' standing (cf. McBirney and Williams, 1969, pp. 171–172).

There are close resemblances between the volcanic suite of the western Galápagos and lavas of Easter Island (Bandy, 1937) in the southeastern Pacific 2000 km to the south. There, the felsic members of the series include not only trachytes but also peralkaline rhyolites.

ICELAND Iceland covers 50,000 $km^2$ south of the Arctic Circle on the northeastern fork of the mid-Atlantic ridge. The island is entirely volcanic and, since mid-Tertiary times the pattern of volcanism has been simple and uniform. Great volumes of tholeiitic magma have been erupted from fissures and central volcanoes. Most of the basalts are oversaturated in silica, but some have minor normative as well as modal olivine. At local eruptive centers these are accompanied by considerable quantities of rhyolite and smaller amounts of icelandite. Consistently low $Sr^{87}/Sr^{86}$ values for basalts, and even for siliceous lavas, underscore the essentially oceanic character of the whole volcanic pile; there is no evidence to postulate the existence of an underlying continental crust.

Chemically and mineralogically the Icelandic volcanic series, exemplified by the dissected Tertiary volcano Thingmuli, conform nicely to a model of crystal fractionation (Carmichael, 1964*b*). Removal of olivine from parental olivine-tholeiite magma causes minor increase in $SiO_2$ and a pronounced rise in FeO/MgO in the derivative liquids. Titanomagnetite now crystallizes, and the liquid fraction is greatly enriched in silica as it reaches the composition of icelandite. Differentiation toward rhyolite, with increase in alkali and silica, is attributed mainly to feldspar fractionation. The trace-element pattern supports the model; for example, successive liquids are depleted first in Ni, then in V, and finally in Sr as olivine, magnetite, and plagioclase separate in that order.

Yet there are some puzzling features. The volume of rhyolite at individual centers is too large to be explained by local differentiation. If one regards the rhyolites as regional differentiates—complementary to the vastly more extensive basalts of the whole province—by what means has the siliceous magma become concentrated for localized eruption? However this anomaly is to be explained, the mechanism, like that of differentiation itself, must be sought at no great depth, since high-pressure experimental data show that the proposed fractionation scheme could operate only at pressures below about 5 kb (cf. Chap. 13).

## LOCAL AND CHRONOLOGICAL VARIATION IN ISLAND PROVINCES

The western Galápagos volcanoes erupt magma of tholeiitic composition. Elsewhere in the archipelago the products of contemporary eruption consistently are alkali basalts. In the Hawaiian province, variation in magma composition is more regularly related to time than to place (Verhoogen et al., 1970,

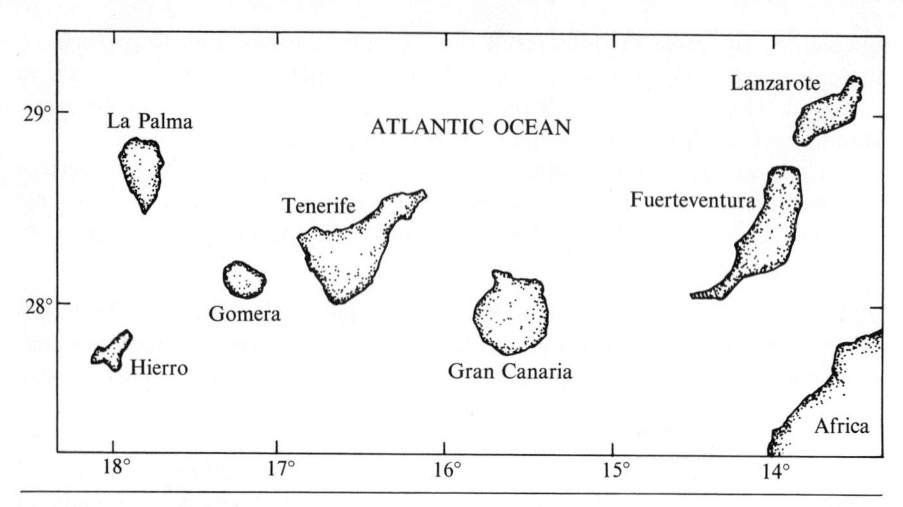


**Fig. 8-6** The Canary Islands.

p. 188): there seems to be a marked tendency for early massive eruption of tholeiitic magma to give way in the later stages of the volcanic life cycle to limited outpouring of basalts of alkaline composition. For a time the Hawaiian pattern of variation was accepted as general, and it was sought in other oceanic provinces. But the generalization has not been sustained as universal, or even as typical of oceanic volcanism. Here we briefly examine the changing petrogenic character of two provinces, the Canaries and the Hawaiian Islands.

### CANARY ISLANDS

*Geologic environment* The Canary Islands, seven in number (Fig. 8-6), are strung out east-west for 400 km off the Atlantic coast of North Africa at latitude 28 to 29°N.* The eastern islands, Lanzarote and Fuerteventura, are within 100 km of the continental coast and seem to be underlain by thick continental sediments blanketing "continental" crust (Dash and Bosshard, 1968). Indeed folded sediments with Mesozoic fossils outcrop from beneath the volcanic complex of Fuerteventura. It has been conjectured therefore that the continental margin extends beneath these two islands. By contrast, from Gran Canaria westward the geophysical picture indicates dense "oceanic" crust devoid of significant sedimentary cover, with the Moho discontinuity no more than 11 to 15 km deep.

*Unity and divergence in volcanism* Throughout the province a common regional datum is recognized—a stage of early shield building whose absolute age decreases from east to west. This is dated at 12 m.y. or more on Lanzarote,

* Other alkaline island provinces close to the West African coast are the Cape Verde group (15 to 17°N) and the islands of the Cameroon line in the Gulf of Guinea.

16 to 12 m.y. on Gran Canaria, 12 to 5 m.y. on Gomera, and <2 m.y. on La Palma. The early shield lavas on every island—called Series 1 in current literature—are "alkali basalts," chemically equivalent to basanite and monotonously uniform in composition (Table 8-11, analyses 1–3).

Lavas of successive younger volcanic episodes, continuing to the present day, include abundant felsic as well as mafic rocks. Most are highly alkaline; phonolites and trachytes are prominent among the later lavas of the central islands. This stage of development has not been reached on La Palma. It is most completely represented on Tenerife and Gran Canaria. On the easternmost islands, though activity continues today, the earlier products of the post-shield-building phases have been partially destroyed by erosion. While there is obvious

**Table 8-11** Chemical analyses (oxides, wt %) and CIPW norms of rocks from the Canary Islands

| | 1 | 2 | 3 | 4 | 5 | 6 | 7 | 8 | 9 |
|---|---|---|---|---|---|---|---|---|---|
| $SiO_2$ | 42.50 | 43.20 | 46.10 | 45.86 | 47.80 | 49.30 | 52.55 | 59.20 | 64.20 |
| $TiO_2$ | 2.20 | 3.93 | 2.38 | 2.95 | 2.22 | 2.06 | 1.23 | 0.73 | 0.35 |
| $Al_2O_3$ | 13.50 | 16.30 | 14.24 | 15.81 | 13.19 | 13.43 | 19.32 | 20.10 | 16.29 |
| $Fe_2O_3$ | 5.16 | 8.09 | 3.95 | 4.65 | 2.21 | 1.02 | 2.76 | 1.26 | 2.77 |
| FeO | 7.14 | 4.69 | 7.90 | 5.24 | 8.88 | 9.23 | 2.33 | 1.12 | 0.62 |
| MnO | 0.17 | 0.16 | 0.15 | 0.18 | 0.14 | 0.15 | 0.18 | 0.15 | 0.05 |
| MgO | 10.44 | 5.16 | 9.25 | 6.31 | 10.50 | 11.39 | 1.74 | 0.91 | 1.05 |
| CaO | 10.99 | 10.74 | 10.45 | 9.77 | 9.90 | 9.39 | 4.97 | 2.40 | 2.36 |
| $Na_2O$ | 3.28 | 3.12 | 3.10 | 5.04 | 3.19 | 2.48 | 8.12 | 7.32 | 7.00 |
| $K_2O$ | 1.20 | 1.64 | 0.63 | 3.11 | 0.93 | 0.74 | 4.88 | 5.00 | 4.40 |
| $P_2O_5$ | 1.04 | 0.88 | 0.53 | 0.70 | 0.58 | 0.36 | 0.34 | 0.16 | 0.05 |
| $H_2O^+$ | 2.67 | 2.03 | 1.12 | | 0.24 | 0.17 | | | |
| $H_2O^-$ | | | | | | | }0.40 | 1.36 | 0.66 |
| Rest | | | 0.10* | 0.37 | | | 1.18† | 0.20* | |
| Total | 100.29 | 99.94 | 99.90 | 99.99 | 99.78 | 99.72 | 100.00 | 99.91 | 99.80 |

* $CO_2$ † Cl = 0.24; $SO_3$ = 0.94.

*Explanation of column headings*

Analyses 1–6 are cited from Ibarrola (1969), analysis 7 from Hernández-Pacheco (1969), and analyses 8 and 9 from Muñoz (1969*a*). Tables and numbers given in parentheses refer to these works.

1 Olivine basalt (first episode), Lanzarote (table 1, no. 7)
2 Olivine basalt (first episode), Gomera (table 4, no. 8)
3 Olivine basalt (first episode), Fuerteventura (table 2, no. 12)
4 Average Quaternary basalt (tephrite), Gran Canaria (table M4, no. 2)
5 Olivine basalt, last Quaternary episode, Lanzarote (table 5, no. 27)
6 Olivine basalt, historic flow of 1730–1736 eruptions, Lanzarote (table 5, no. 30)
7 Tahitite, Gran Canaria (table 1, no. 4)
8 Nepheline syenite, central stock, Pájara, Fuerteventura (table 1, no. 3)
9 Syenite, La Peñitas ring dike, Pájara, Fuerteventura (table 1, no. 16)

unity in volcanic pattern across the whole province, each island has unique elements. So it would be unsafe to generalize further toward a comprehensive pattern of Canary volcanism.[1]

*Central Islands: volcanism after Series 1* Tenerife, 2000 $km^2$ in area, is the largest island. Upon a basalt shield of alkali basalts (Series 1) a major Pliocene volcano has been built by highly alkaline *ne*-normative lavas: most are trachybasanites and phonolites of intermediate content; and today they are exposed in the walls of an immense caldera 15 km wide at 2000 m elevation. Within this caldera are two large Quaternary volcanoes, both still active. They are built almost exclusively of highly alkaline phonolitic lavas (Table 8-12). Ridley (1970) presents a convincing case for differentiation of a parental basanitic magma toward phonolitic liquids by removal first of olivine and titanomagnetite, later of plagioclase and calcic anorthoclase.

A rather similar sequence of events, though somewhat earlier chronologically, occurred 120 km to the east on Gran Canaria. In the southwestern part of this island a late Miocene (14 to 9 m.y.) central edifice (Tejeda volcano) developed upon an eroded remnant of an early basaltic shield (of Series 1). It grew by repeated eruption of trachytic and soda-rhyolitic ash, with interspersed flows of phonolite and in the earlier stages alkali basalt. The structure has been dissected to reveal a collapsed caldera and a stock and cone-sheet complex of syenitic rocks. Episodic activity was renewed in the Quaternary, with extensive flooding by basanitic lavas ("tephrites"; Table 8-11, analysis 4) poured from vents east of the dissected Miocene volcano. The youngest of these lavas include extremely alkaline types with modal hauyne (Table 8-11, analysis 7) reminiscent of the tahitites of Tahiti (Hernández-Pacheco, 1969).

*Eastern Islands* On Lanzarote and Fuerteventura the volcanic (Miocene to Recent) suite, as in the rest of the archipelago, is dominated by alkali basalts. Erosion has bitten deeper into the pre-Miocene understructure. And here we see the record of preliminary igneous events, some of them unrecorded or but obscurely indicated in islands to the west:

1. In the presumably pre-Miocene basement of Fuerteventura there is an extensive basic pluton whose chemical, mineralogical, and textural characteristics duplicate those of stratified tholeiitic intrusions in many continental regions. Intrusions such as these must have supplied gabbro and peridotite xenoliths, many with obvious cumulate texture, that are abundant in younger and chemically unrelated alkali basalts elsewhere, notably on Lanzarote (Fúster et al., 1969).

[1] For subsequent sections we have drawn especially from Abdel-Monen et al. (1967); Schmincke (1967); Rothe and Schminke (1968); Fúster et al. (1969); Ibarrola (1969); Lopez Ruis (1969); Muñoz (1969*a*, 1969*b*); Ridley (1970).

**Table 8-12** Chemical analyses (oxides, wt %), CIPW norms, and atomic abundances (ppm) of trace elements for volcanic rocks of Tenerife

| | 1 | 2 | 3 | 4 | 5 | | 1*a* | 2*a* | 3*a* | 4*a* |
|---|---|---|---|---|---|---|---|---|---|---|
| $SiO_2$ | 41.15 | 48.83 | 54.17 | 61.60 | 57.62 | Ba | 350 | 700 | 750 | 480 |
| $TiO_2$ | 4.40 | 2.43 | 1.88 | 0.84 | 1.30 | Co | 50 | 20 | — | |
| $Al_2O_3$ | 13.27 | 18.70 | 19.14 | 19.97 | 19.71 | Cr | 200 | — | — | |
| $Fe_2O_3$ | 4.69 | 4.19 | 2.87 | 1.76 | 1.63 | Ni | 10 | 10 | — | |
| FeO | 9.01 | 3.12 | 2.22 | 0.69 | 2.03 | Nb | 50 | 90 | 110 | 160 |
| MnO | 0.18 | 0.14 | 0.17 | 0.25 | 0.15 | Rb | — | 80 | 60 | 120 |
| MgO | 10.07 | 2.85 | 2.18 | 0.35 | 1.32 | Sr | 1400 | 2000 | 1600 | 40 |
| CaO | 10.50 | 7.92 | 4.48 | 0.88 | 2.41 | V | 300 | 200 | 70 | |
| $Na_2O$ | 3.69 | 5.34 | 7.21 | 7.08 | 7.95 | Zr | 500 | 550 | 850 | 950 |
| $K_2O$ | 1.55 | 2.52 | 3.69 | 5.84 | 4.75 | | | | | |
| $P_2O_5$ | 0.80 | 0.88 | 0.44 | 0.22 | 0.14 | | | | | |
| $H_2O^+$ | 0.69 | 2.11 | 0.61 | 0.13 | 0.44 | | | | | |
| $H_2O^-$ | 0.21 | 1.22 | 0.22 | 0.35 | 0.14 | | | | | |
| Total | 100.21 | 100.25 | 99.28 | 99.96 | 99.59 | | | | | |
| *or* | 9.16 | 14.89 | 21.81 | 34.52 | 28.02 | | | | | |
| *ab* | 5.32 | 30.91 | 37.16 | 52.35 | 41.33 | | | | | |
| *an* | 15.07 | 20.55 | 8.97 | 2.93 | 4.10 | | | | | |
| *ne* | 14.03 | 7.73 | 12.98 | 4.09 | 14.05 | | | | | |
| *di* | 25.29 | 8.81 | 8.19 | — | 5.48 | | | | | |
| *ol* | 13.43 | 2.11 | 1.65 | 0.61 | 0.86 | | | | | |
| *mt* | 6.80 | 6.08 | 4.16 | 2.55 | 2.36 | | | | | |
| *il* | 8.35 | 4.61 | 3.57 | 1.59 | 2.47 | | | | | |
| *ap* | 1.89 | 2.08 | 1.04 | 1.52 | 0.33 | | | | | |
| Total | 99.34 | 97.77 | 99.53 | 100.16 | 99.00 | | | | | |

*Explanation of column headings*

Analyses cited from Ridley (1970, table 4). Numbers in parentheses refer to this table.

1 Pyroxene-olivine basanite, lower Las Canadas series (1)
2 Trachybasanite, lower Las Canadas series (5)
3 Phonolite, lower Las Canadas series (9)
4 Phonolite, upper Las Canadas series (23)
5 Phonolite, Teida central vent (40)
1*a*–4*a* Same samples as 1–4

2. Swarms of dikes cut the pre-Miocene basement of Fuerteventura. The most extensive swarm, probably predating extrusion of the Miocene basalts, strikes north-northeast by south-southwest parallel to the length of the island and to the nearby African coast. They consist of olivine-augite basalt. Their regular alignment and great abundance (locally 80 percent of the volume of exposed rock) imply intrusion under crustal tension.

3. In Fuerteventura gabbros of the layered intrusion and basaltic dikes of the basement swarm are cut by still younger syenite ring-dike complexes that perhaps may be regarded as a more deeply dissected equivalent of the syenite-trachyte complex of the old Tejeda volcano of Gran Canaria. In Fuerteventura, too, there is complete gradation from *ne*-normative to oversaturated felsic rocks, nepheline syenites, and quartz syenites (Table 8-11; analyses 8, 9).

Alkali basalts (Table 8-11, analysis 5), generally similar to those of the opening late Tertiary episode, are the main product of two Quaternary episodes in both islands. But in the last phases of a later third cycle, including the historic eruptions of 1730 to 1736 on Lanzarote, the magma type has changed markedly: the composition is that of olivine tholeiite with high normative *hy* (Table 8-11, analysis 6). Rocks of this composition appear too, along with alkali basalts, in the penultimate Pleistocene cycle on Fuerteventura. This late change from alkali basalt toward olivine tholeiite is in marked contrast to the synchronous trend toward increasing alkalinity and lower silica saturation in basalts of the central islands. It contravenes two widely accepted tenets of modern petrologic thought: first that with passage of time successive basaltic magmas become increasingly undersaturated in silica (more alkaline); second that within Atlantic island provinces dominant basaltic magmas become increasingly saturated in silica with increasing proximity to the mid-Atlantic ridge.

### HAWAIIAN ISLANDS

*Volcanic and petrologic pattern* The Hawaiian ridge, a narrow topographic high on the Pacific Ocean floor, extends west-northwestward from the island of Hawaii (20°N, 155°W) to beyond Midway (28°N, 178°W)—a distance close to 3000 km. Within 100 km of the crest the sides fall sharply to depths of 4500 m or more. Along its full length the ridge crest is studded with volcanoes—extinct and mostly submerged in the west, still active in the east. Those that project above sea level along the eastern 500-km segment constitute the Hawaiian Islands (Fig. 8-7*a*). We have summarized the general character of the Hawaiian petrogenic province elsewhere (Verhoogen et al., 1970, pp. 187–188, 287–294), and in recent literature there are a number of concise statements of various aspects of Hawaiian volcanism and petrology (Macdonald, 1949*a*, 1949*b*, 1968; Tilley, 1950; Macdonald and Katsura, 1964; McDougall, 1964*b*; Stearns, 1966; R. W. White, 1966; Murata and Richter, 1966*a*, 1966*b*; Jackson, 1968). To these sources the reader is referred also for the early classic accounts of Washington, Daly, Stearns, Powers, and others.

Throughout the literature there is a strong emphasis on search for pattern in Hawaiian volcanism and petrogenesis. Here and there the issue tends to be clouded by confusion in terminology: chemically distinct rocks from different volcanoes lose something of their individuality when classed together as tholeiitic; one man's "pigeonite" turns out to be another's "augite"; "andesites"

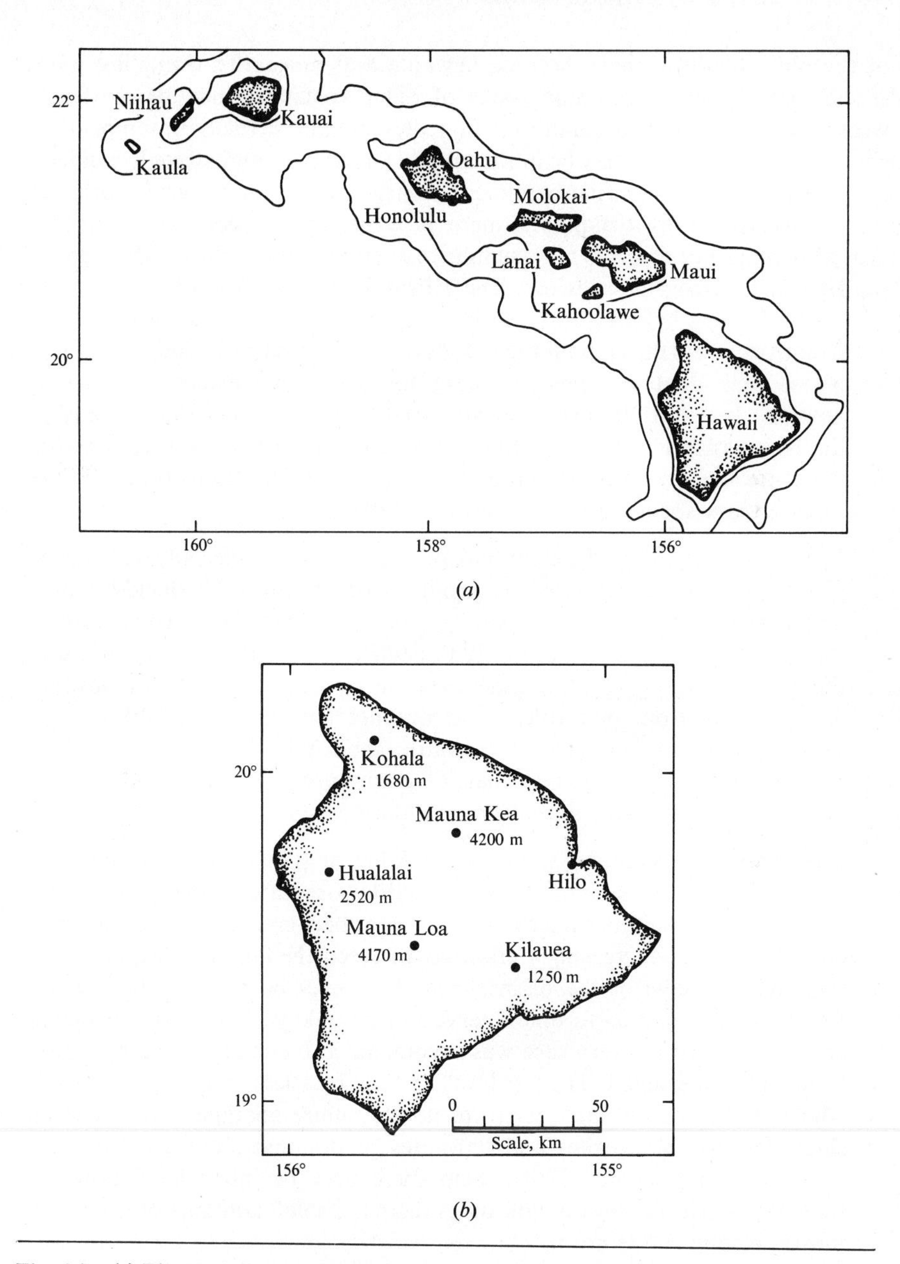


**Fig. 8-7** (*a*) The Hawaiian Islands (depth contours 1000, 4000 m); (*b*) the island of Hawaii.

of the older literature have become hawaiite and mugearite. Sampling poses special problems: the common rocks of older volcanoes may be so deeply weathered under tropical conditions that they remain virtually unsampled for chemical analysis; what may be in the interior of younger undissected volcanoes or in their hidden submarine basements is largely a matter for guesswork. All these difficulties notwithstanding there is a perceptible common pattern—though perhaps not as sharp nor as uniform as some would have it—throughout the Hawaiian petrogenic province. The salient features are these:

1. Eruptive centers throughout the islands are massive basaltic shield volcanoes whose individual volumes—allowing for submarine extension—may be anything between 3000 (Kauai) and 40,000 $km^3$ (Mauna Loa). Many of the larger islands are compound. In the easternmost, Hawaii (Fig. 8-7*b*), there are five partially overlapping shields, all built within the last half-million years. (McDougall and Swanson, 1972).

2. In Hawaiian literature the great bulk of exposed lavas, especially in dissected older volcanoes, are treated as products of "primitive"[1] shield-building magmas. They are invariably basaltic; and it has been asserted repeatedly that their composition is uniformly tholeiitic. Only in the broadest sense of the word, however, can this generalization be sustained. The composition varies widely from one volcano to another (contrast Mauna Loa with Mauna Kea and Hualalai on Hawaii). The total compositional range is comparable to the combined span of predominating basaltic magmas in the western Galápagos, Ascension, and the Azores.

3. The late eruptive pattern following building of the shield is variable. On some volcanoes the summit collapsed and lava repeatedly filled and spilled over from the resulting caldera. As activity waned, erosion between eruptive episodes became increasingly effective. Between the final lava capping and underlying basalts of the main shield there may be recognizable unconformity corresponding to time intervals of 200,000 years or less. Change in eruptive pattern in every case was accompanied by change in composition of the magma erupted. The late lavas, in total bulk less than 1 percent of the whole shield, for the most part are notably more alkaline than the main shield basalts. Mugearites and hawaiites predominate, but alkali basalts are also well presented. With them there may be intercalated flows of tholeiitic basalt. On several volcanoes there are small amounts of trachytes, mostly with *ne* in the norm.[2]

[1] The term *primitive* so used denotes parental as opposed to derivative status, implied by great abundance during the early stages of the volcanic cycle. Here we substitute for *primitive* the adjective *shield-building* (descriptive) or *parental* (genetic).

[2] A unique exception is a rhyolitic lava on Mauna Kuwale, Oahu—a protruding remnant of an early volcano (8 m.y.) subsequently largely engulfed by basaltic lavas of the Waianae shield (McDougall, 1964*b*, p. 115).

4. The exposed parts of most Hawaiian volcanoes were built individually in less than 500,000 years. The longest recorded life history is that of Kauai—1.8 m.y. (McDougall, 1964*b*). Subaerial activity of the Waianae volcano, Oahu, covered 1.2 m.y. (3.6 to 2.4 m.y.) and was overlapped by emergence from the sea of the adjacent Koolau edifice whose exposed lavas have been dated at 2.6 to 1.8 m.y. (Doell and Dalrymple, 1973). The record of subaerial activity at Kohala volcano, Hawaii, covers about 100,000 years (McDougall and Swanson, 1972). Following profound erosion, on four islands—Kauai, Niihau, Oahu, and western Maui—volcanism flared up once more after a long interval (1 to 2 m.y.). The nature of the erupted magma had now changed drastically. The volcanic products without exception include highly alkaline, very basic rocks—nephelinites, melilite basalts, and basanites.

5. Throughout the province peridotite and gabbro xenoliths occur in profusion in many of the lavas and pyroclastic rocks (R. W. White, 1966; E. D. Jackson, 1968). Their composition and distribution correlate closely with the chemical composition of the host rocks.

   a. Where the host is nephelinite or basanite, xenoliths are almost exclusively peridotites. They show wide mineralogical variety, including all combinations of olivine, enstatite, diopsidic augite, and chrome spinel; many consist of all four phases and are called lherzolite. Most bear the imprint of strong deformation and metamorphic recrystallization. Almost certainly their source is in the mantle (cf. Chap. 13). In the nephelinites of Oahu (E. D. Jackson and Wright, 1970) are unique garnet pyroxenites[1] also derived from the mantle.

   b. Xenoliths of the early shield basalts are much less numerous and are completely different in character. They are mainly gabbros (olivine-orthopyroxene-diopside-plagioclase) with layered and other cumulate textures duplicating those that characterize the great stratified basic plutons of the continents. There can be little doubt that these xenoliths are cognate and have come from floored chambers of crystallizing basaltic magma deep beneath the volcanoes.

   c. Both kinds of xenoliths are found in alkali basalts: hawaiites and mugearites. But the ultramafic assemblages here are simple—olivine-augite or almost pure olivine (dunite). Most remarkable was the historic alkali basalt flow of 1801 on Hualalai, which brought up thousands of tons of angular gabbro and dunite xenoliths torn from a layered intrusion below (Richter and Murata, 1961).

[1] Until recently confused with eclogite.

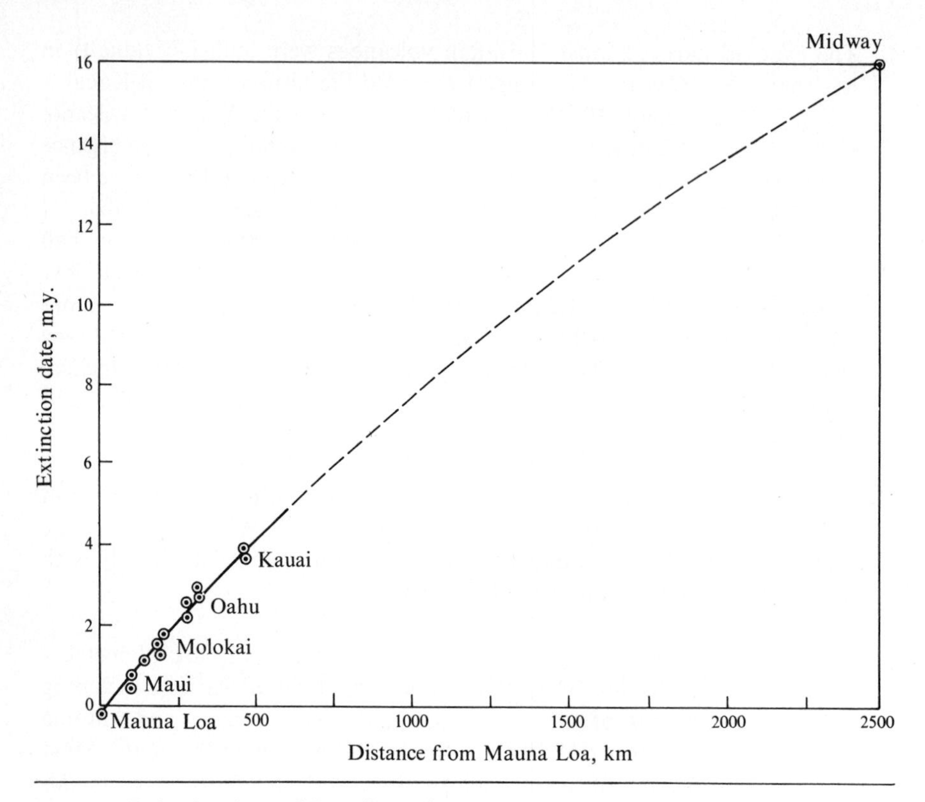


**Fig. 8-8** Extinction dates of Hawaiian volcanoes.

*Chronological pattern* At each Hawaiian volcano, as we have seen, development of the edifice, nature of eruptive activity, and trends of compositional variation in erupted magmas conform in some degree to common chronological pattern. The time scale of the individual volcanic cycle is small—normally 0.5 to 1.5 m.y. (E. D. Jackson et al., 1972, pp. 606–607).

There is also a more comprehensive chronological pattern of Hawaiian volcanism on a scale that is more extensive both in time (15 to 20 m.y.) and in space (2500 km). The complete linear array of volcanoes along the length of the Hawaiian ridge records a chronological sequence of major volcanic events (Fig. 8-8). At the eastern extremity, witness to the most recent of these—events still in progress—stand the highly active volcanoes Mauna Loa and Kilauea. Northwestward come mature and still feebly active centers such as Mauna Kea and, on Maui, Haleakala, and then the extinct and progressively more dissected volcanoes of Molokai, Oahu, and Kauai. The northwestern segment of the ridge is dotted by atolls and banks extending to Midway Island—the submerged coral-capped summits of volcanoes presumably still more ancient. As we pass west-northwestward from Hawaii to Midway we see, as Dana recognized long ago, a vista of volcanic episodes extending progressively further back in time.

Radiometric dates have confirmed the regularity of geographic-chronological correlation and have calibrated the relevant time scale. One convenient common comparative datum is *extinction* (McDougall, 1964b). This can be defined somewhat arbitrarily as the age of the youngest lava formations (commonly hawaiites and mugearites)—excluding products of minor revived activity (nephelinites and basanites) at a few centers. Extinction dates in geographic order from east-southeast to west-northwest are as follows: Mauna Loa and other southeastern volcanoes of Hawaii, 0; Kohala (west Hawaii), 0.06 m.y.; east Maui (Haleakala), 0.8 to 0.4 m.y.; west Maui, 1.16 m.y.; east Molokai, 1.5 to 1.3 m.y.; west Molokai, 1.8 m.y.; east Oahu, 2.5 to 2.2 m.y.; west Oahu,[1] 2.9 to 2.7 m.y.; Kauai, 3.9 to 3.8 m.y.; Midway Island (Ladd et al., pp. 1088–1094, 1967), 16 m.y. Extinction dates plotted against distance from Mauna Kea fall on a curve whose gradient—10 to 15 cm per year—depicts the mean velocity of propagation of volcanism relative to fixed surface coordinates.

More recent radiometric data on the *inception* of volcanism at each center have elaborated the geographic-chronological picture (E. D. Jackson et al., 1972; Shaw, 1973), which turns out to be compound rather than continuously linear. Successive volcanic episodes each covering several million years have been recognized. Each is geographically restricted and represents a succession of shorter eruptive events progressing northwest to southeast down a curved segment of the island chain. Segments are arranged *en echelon*, not in a continuous linear sequence. Most clearly revealed is that which progresses from Kauai (5.5 m.y.)—via Oahu (3.5 and 2.5 m.y.), Molokai (2.6 and 1.5 m.y.), and Maui (1.4 and 1.3 m.y.)—to Hawaii (0.7 m.y. at Kohala to somewhat younger at Kilauea). This episode is marked by accelerating volume rates of eruption culminating today in 0.11 $km^3$ per year at Kilauea. The total picture of volcanic progression down the Hawaiian ridge (and of preceding volcanism southward down the Emperor seamount chain from the Aleutians to its junction with the Hawaiian ridge) has been related to motions of the Pacific lithospheric plate in what is perhaps the most convincing volcanic-tectonic chronological synthesis to date in the field of plate tectonics.

*Parental magmas* The repeatedly asserted dictum that the parental Hawaiian magma is tholeiitic rests on the assumption that there is in fact a uniform pattern of Hawaiian petrogenesis. Its validity depends on criteria of parental status and definition of the term *tholeiitic*—neither of them free from ambiguity.

Hawaiian geologists (for example, H. A. Powers, 1955, p. 78; Macdonald and Katsura, 1964, pp. 82–83, 110; Stearns, 1966, p. 220) consider the basalt flows that build the bulk of each shield to represent parental ("primitive") basaltic magmas. They rely on the abundance criterion, and since the magmas

[1] Not including the trachyte of Mauna Kuwale, dated by McDougall at 8.5 to 8.3 m.y.

so designated also conform to the requirement of maximum liquidus temperature as compared with associated lavas of other composition, the concept is valid. Indeed it is difficult to frame any other objective criterion of parental status. Nevertheless at Kilauea, where observations have been conducted in exceptional detail, the basalts themselves display significant chemical variation that can be correlated with chronology and site (relative elevation) of eruption (for example, Murata and Richter, 1966*a*, 1966*b*). Here we see effects of short-term differentiation in high-level reservoirs, and perhaps of mixing of magmas drawn from independently differentiating reservoirs (T. L. Wright, 1973).

Turning now to composition, shield-building lavas on Kauai, Oahu (Waianae and Koolau), western Maui, and on some volcanoes of Hawaii (Kohala, Mauna Loa, Kilauea) are *Q*-normative basalts (Table 8-13; analyses 2, 3). In this respect and in mineral composition they are tholeiitic: many contain hypersthene, and in some there is pigeonite as well as augite and olivine. Even the olivine-enriched picrite basalts of Kilauea have hypersthene in the groundmass and high *hy* as well as *ol* in the norm. Basalts of other shields—Mauna Kea and Hualalai on Hawaii, Haleakala on Maui—cover the whole chemical range of "olivine tholeiite." Mineralogically they resemble alkali basalts. In the norm *ol* is high and there may even be minor *ne* (Table 8-13; analyses 4, 5). Thus some of the shield basalts (notably on Hualalai) are alkali basalts comparable to the lavas of St. Helena. We conclude that the parental shield-building magmas of Hawaiian volcanoes range from oversaturated tholeiite through olivine tholeiite to alkali basalt. The case for uniform tholeiitic composition rests on the assumption that all volcanoes conform strictly to uniform pattern and that beneath the shield lavas of Mauna Kea and Hualalai there must be more voluminous concealed basalts akin to those of nearby Mauna Loa and Kilauea.

Even among volcanoes whose lavas fall within the tholeiite class *sensu stricto*, the unique individuality of each stands out on plots of simple chemical parameters:

1. Using a criterion devised by Tilley (1950), Macdonald and Katsura (1964, pp. 87, 103) plotted weight percentages ($Na_2O + K_2O$) against $SiO_2$ (Fig. 8-9). There is a broad scatter of points for basaltic rocks in the $SiO_2$ range 42 to 54 percent (Fig. 8-9*b*); but the line *AB* drawn in an area of minimal population was used to divide somewhat arbitrarily a "tholeiitic" from an alkali-basalt field.[1] Macdonald and Katsura (1964, p. 103) showed that lavas of some individual volcanoes plot in distinct but partially overlapping fields of Fig. 8-9*b*. Those of the Hualalai shield fall well within the field of alkali basalts—along with hawaiites, mugearites, and nephelinites, which collectively account for most of the points in this sector of Macdonald and Katsura's diagram.

[1] The slope of *AB* is given by $(Na_2O + K_2O)/(SiO_2 - 39) = 0.37$. Although its significance is strictly local, the same line is becoming increasingly used as a measure of "alkalinity" in basalts of other oceanic provinces.

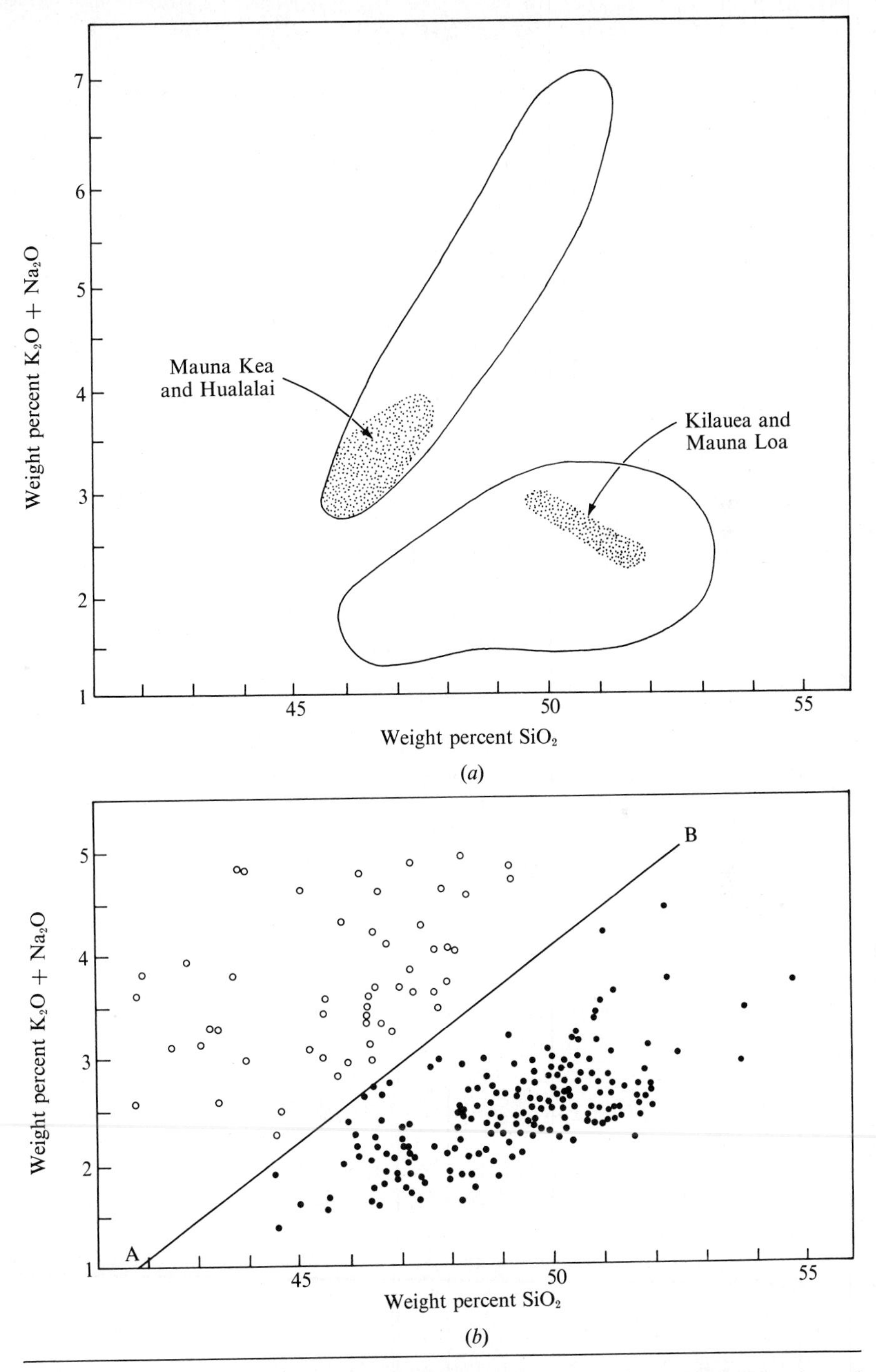


**Fig. 8-9** Alkali-silica plots (wt %) for Hawaiian basalts. (*a*) Fields of alkali basalts and mugearites (above) and tholeiitic basalts (below) (after Tilley, 1950). (*b*) Alkali basalts (above *AB*) and tholeiitic basalts (below *AB*) as defined by G. A. Macdonald and Katsura (1964).

**Table 8-13** Chemical analyses (oxides, wt %), CIPW norms, and atomic abundances (ppm) of trace elements of Hawaiian lavas

| | 1 | 2 | 3 | 4 | 5 | 6 | 7 | 8 | 9 | 10 | | 1*a* | 3*a* | 7*a* | 8*a* |
|---|---|---|---|---|---|---|---|---|---|---|---|---|---|---|---|
| $SiO_2$ | 47.25 | 49.86 | 50.52 | 45.14 | 46.01 | 47.26 | 51.99 | 62.02 | 66.78 | 36.34 | Ba | 150 | 150 | 1000 | 800 |
| $TiO_2$ | 1.61 | 2.43 | 3.63 | 3.04 | 1.80 | 3.58 | 3.02 | 0.31 | 0.59 | 2.87 | Co | 70 | 30 | 7 | 2 |
| $Al_2O_3$ | 9.07 | 15.11 | 13.85 | 13.49 | 15.40 | 17.19 | 16.30 | 18.71 | 15.69 | 10.14 | Cr | 1000 | 400 | — | 20 |
| $Fe_2O_3$ | 1.45 | 3.66 | 0.98 | 3.60 | 1.22 | 2.87 | 2.75 | 4.30 | 1.45 | 6.53 | Li | 1 | 1 | 15 | 30 |
| FeO | 10.41 | 7.82 | 9.77 | 9.27 | 8.15 | 9.36 | 7.44 | 0.10 | 1.40 | 10.66 | Ni | 400 | 80 | 10 | 15 |
| MnO | 0.13 | 0.17 | 0.14 | 0.18 | 0.08 | 0.22 | 0.11 | | 0.05 | 0.20 | Rb | — | — | 70 | 300 |
| MgO | 19.96 | 6.00 | 7.07 | 10.02 | 13.25 | 5.08 | 3.19 | 0.40 | 1.28 | 10.68 | Sr | 400 | 800 | 4000 | 100 |
| CaO | 7.88 | 10.34 | 11.33 | 10.60 | 10.74 | 7.82 | 6.67 | 0.86 | 2.61 | 13.10 | V | 300 | 400 | 30 | — |
| $Na_2O$ | 1.38 | 2.05 | 1.51 | 2.24 | 2.30 | 3.50 | 5.64 | 6.90 | 4.49 | 4.54 | Zr | 100 | 100 | 1000 | 1500 |
| $K_2O$ | 0.35 | 0.26 | 0.47 | 0.80 | 0.67 | 1.40 | 2.13 | 4.93 | 3.60 | 1.78 | | | | | |
| $P_2O_5$ | 0.21 | 0.27 | 0.22 | 0.26 | 0.62 | 0.77 | 1.25 | 0.24 | 0.58 | 1.02 | | | | | |
| $H_2O^+$ | 0.08 | 1.87 | 0.04 | 0.67 | 0.19 | 0.30 | 0.29 | 0.80 | 0.59 | 1.00 | | | | | |
| $H_2O^-$ | 0.04 | 0.65 | — | 1.00 | 0.07 | 0.56 | 0.07 | 0.31 | 0.66 | 1.00 | | | | | |
| Total | 99.82 | 100.49 | 99.53 | 100.31 | 100.50 | 99.91 | 100.85 | 99.88 | 99.77 | 99.86 | | | | | |
| *Q* | | 7.08 | 5.88 | | | | | 1.80 | 20.76 | | | | | | |
| *or* | 2.22 | 1.67 | 2.78 | 5.00 | 3.89 | 8.34 | 12.23 | 28.91 | 21.13 | | | | | | |
| *ab* | 11.53 | 17.29 | 12.58 | 18.86 | 15.72 | 29.34 | 44.01 | 58.16 | 38.25 | | | | | | |
| *an* | 17.51 | 31.14 | 29.47 | 24.19 | 29.75 | 27.24 | 13.07 | 2.22 | 9.17 | 2.78 | | | | | |
| *ne* | | | | | 1.99 | | 1.99 | | | 19.88 | | | | | |
| *lc* | | | | | | | | | | 8.28 | | | | | |
| *di* | 16.65 | 14.84 | 21.29 | 21.19 | 16.03 | 4.80 | 9.58 | | | 22.30 | | | | | |
| *hy* | 18.12 | 15.35 | 18.79 | 1.06 | | 9.36 | | 1.00 | 3.60 | | | | | | |
| *ol* | 28.17 | | | 16.66 | 26.22 | 7.46 | 6.23 | | | 18.32 | | | | | |
| *mt* | 2.09 | 5.34 | 1.39 | 5.34 | 1.86 | 4.18 | 3.94 | | 2.09 | 9.51 | | | | | |
| *il* | 3.04 | 4.56 | 6.84 | 5.78 | 3.50 | 6.84 | 5.78 | 0.15 | 1.22 | 5.47 | | | | | |
| *ap* | 0.34 | 0.67 | 0.34 | 0.67 | 1.34 | 2.02 | 3.02 | 0.67 | 1.34 | 2.35‡ | | | | | |
| Total | 99.67 | 97.94 | 99.36 | 98.75 | 100.30 | 99.58 | 99.85 | 92.91* | 97.56† | 88.89‡ | | | | | |

* Add corundum, 1.00; hematite, 4.30; rutile, 0.24; total, 98.45.
† Add corundum, 1.02; total, 98.58.
‡ Add $Ca_2SiO_4$, 8.0; total, 96.89.

*Explanation of column headings*

1 Picrite basalt, Kilauea, Hawaii (Macdonald, 1949*b*, p. 74, no. 1)**
2 Basalt, lower member of Waianae series, Oahu (Macdonald and Katsura, 1964, table 1, no. 20)
3 Basalt, Kilauea, Hawaii (Macdonald, 1949*b*, p. 74, no. 12)**
4 Basalt, Hamakua series, Mauna Kea, Hawaii (Macdonald and Katsura, 1964, table 4, no. 4)
5 Basalt, Hualalai, Hawaii (Macdonald, 1949*b*, p. 78, no. 1)
6 Hawaiite, Haleakala, Maui (Macdonald and Katsura, 1964, table 7, no. 7)
7 Mugearite, Kohala, Hawaii (Macdonald, 1949*b*, p. 87, no. 11)**
8 Trachyte, Hualalai, Hawaii (Macdonald, 1949*b*, p. 78, no. 8)**
9 Rhyolite (rhyodacite), Mauna Kuwale, Oahu (Macdonald and Katsura, 1964, table 2, no. 12)
10 Melilite nephelinite, Honolulu series, Oahu (Winchell, 1947, pp. 30, 31, no. 19)

** Trace-element data for same rocks (correspondingly numbered 1*a*, 3*a*, 7*a*, 8*a*) are taken from Wager and Mitchell (1953, p. 218).

2. A consistent trait of the lavas of some shields is the content of $K_2O$—mostly <0.2 percent (cf. oceanic tholeiites, Table 8-1) for Kohala, 0.25 to 0.4 percent for Waianae, 0.8 percent for Mauna Kea, and so on. McDougall 1964*b*, p. 122) notes a tendency for sharply restricted ranges of $K_2O$ in lavas of individual formations of a single shield. In Fig. 8-10 we have combined the chemical parameters (1) and (2) in a plot of $SiO_2/(K_2O + Na_2O)$ against $Na_2O/K_2O$.

3. Coombs (1963) devised, as a general criterion of alkalinity and of differentiation potentialities, an index based on molecular normative parameters.[1] Adapting this to the CIPW norm, we use the ratio $(hy + 4Q)/(hy + di + 4Q)$. Values for Hawaiian shield lavas [mainly average compositions from Macdonald and Katsura (1964, table 8) and Coombs (1963, table 1)] are as follows. Provinces elsewhere whose basalts are comparable on this basis are noted in parentheses:

   Kauai: 0.77 (Columbia River, Yakima basalts)
   Koolau, Oahu: 0.69 (Tasmanian diabase)
   Kohala, Hawaii: 0.67 (Palisade diabase)
   Kilauea, Hawaii: 0.59 (Deccan)
   Haleakala, Maui: 0.37 (Ascension)
   Mauna Kea, Hawaii: 0.09 (Réunion)
   Hualalai, Hawaii: $\leq 0$* (St. Helena)

*Role of differentiation* In the Hawaiian province, trends of differentiation in tholeiitic magmas crystallizing near the surface are clear. Here linear compositional variation has been directly traced over short vertical distances or through short intervals of time. Successive lavas of historic eruptive episodes of a few months' duration on Kilauea and Mauna Loa show distinct though limited compositional trends that correlate inversely with chronological order (for example, Muir and Tilley, 1963, Murata and Richter, 1966*b*). The slightly more siliceous (further differentiated) lavas are those erupted first in the short-term cycle of activity. In vertical sections of lava lakes that from time to time have filled pit craters along the southeast rift of Kilauea, early formed crystals of magnesian olivine have become notably concentrated downward under gravity (for example, Moore and Evans, 1967; Evans and Moore, 1968). Pegmatoid offshoots and segregations from tholeiitic rocks (for example, Kuno et al., 1957)

[1] Coombs' "indicator ratio" is

$$\frac{\mathrm{Hy} + 2\mathrm{Qz}}{\mathrm{Hy} + 2(\mathrm{Qz} + \mathrm{Di})}$$

He demonstrated strong correlation between the value of this ratio and the course of crystallization of pyroxenes. Care has to be taken to be certain that the iron titanium oxides have not been oxidized during cooling, since this can greatly affect these normative parameters.
* Indicator ratio 0 shows incoming of normative nepheline.

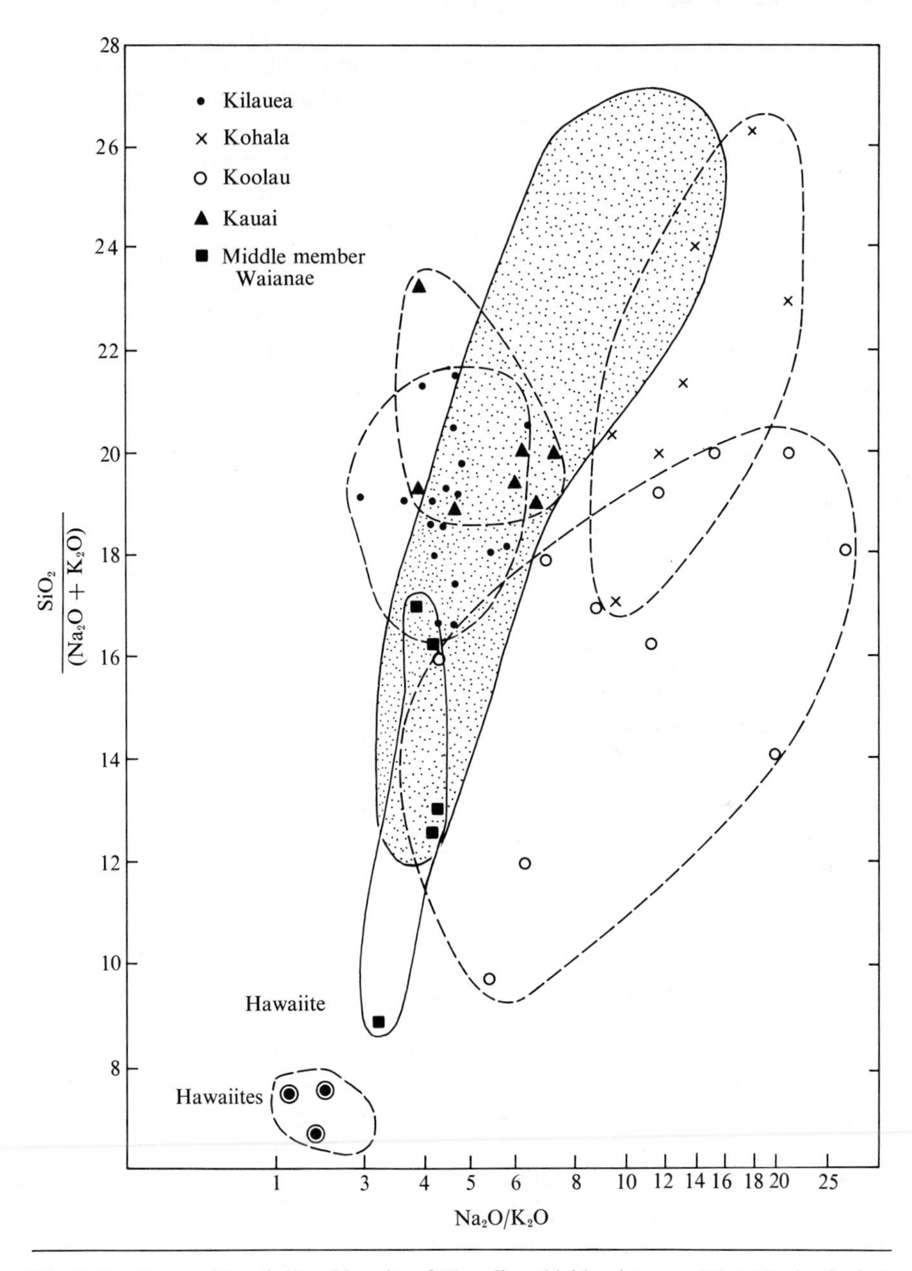


**Fig. 8-10** Composition fields of basalts of Hawaiian shield volcanoes. West Maui, stippled; Mauna Loa is closely similar to Kilauea. Three hawaiites shown as circled dots, lower left, are from Mauna Kea and Kohala. Note that $Na_2O/K_2O$ is plotted on a logarithmic scale.

show the trend of differentiation of water-saturated magmas of this class. All lines of evidence point in the same direction. Differentiation yields liquids enriched in $SiO_2$ and with significantly higher Fe/Mg and lower MgO than the parent magma. The magnitude of variation may not be great—even in a lava lake 75 m deep that took 50 years to solidify. But the trend is clear, and it is consistent with the data of laboratory experiment at low pressures. Following this direction a tholeiitic magma of the Kilauea or Koolau type could ultimately yield magmas similar in composition to the rhyolite of Mauna Kuwale (Table 8-13, analysis 9).

It can hardly be doubted that the olivine-rich picrite basalts that accompany tholeiitic basalts on Kilauea and other volcanoes have formed by accumulation of early olivine crystals separating from liquids saturated or slightly oversaturated in silica. Petrographic and chemical evidence (Muir and Tilley, 1957) is borne out by the abundance pattern of trace elements (Wager and Mitchell, 1953): the picrite basalts show marked enrichment in Ni, Cr, and to a lesser degree Co (Table 8-13; analyses 1, 1*a*).

Particularly well documented is the case for low-pressure differentiation as the cause of limited but clearly discernible variation among tholeiitic magmas of Mauna Loa and Kilauea (for example, Murata and Richter, 1966*b*; T. L. Wright, 1971). Depletion and enrichment of complementary magma fractions in olivine crystals account remarkably well for observed variation. Indeed the MgO content provides an excellent inverse index of differentiation. Short-term variation attributable to sinking of olivine crystals during short residence of magma in high-level reservoirs is illustrated by the succession of magmas erupted at the summit of Kilauea during a 5-week period at the end of 1959 (Murata and Richter, 1966*a*, 1966*b*).[1] In the first phase, a week in duration, $30 \times 10^6$ $m^3$ of lava was erupted at increasing rate of discharge. *Q*-normative lava with 7 to 9 percent MgO gave way to *ol*-normative types with 12 to 20 percent MgO as this episode reached its climax. Less voluminous magmas of subsequent eruptive phases, drawn from the emptying reservoir, were likewise *ol*-normative and magnesia-rich. Short-term variation among lavas of individual eruptive cycles on Mauna Loa is of the same order of magnitude but is rather more complex. It has been explained by a more complex mechanism of low-pressure fractionation in which crystalline olivine, pyroxenes, and plagioclase all participate (T. L. Wright, 1971). Comparison of prehistoric, eighteenth- and nineteenth-century, and twentieth-century lavas on Kilauea reveals a long-term trend in which $K_2O$, $P_2O_5$, and $TiO_2$ (in rocks of similar MgO content) increase in chronological sequence. This trend has been explained in terms of deep-seated fractionation at high pressure (T. L. Wright, 1971). Such a model, however, has yet to be tested by trace-element criteria, and differences in chemistry of source rocks or change in the regime of melting with passage of

[1] Different aspects of this eruption at Kilauea-Iki crater and immediately ensuing flank eruptions in early 1960 are covered in *U.S.G.S. Prof. Paper* 537, 1966–1970.

time could equally well be responsible for long-term variation of Kilauean magma.

Most writers take the view (cf. Muir and Tilley, 1961) that the lithologic series alkali basalt, hawaiite, mugearite, trachyte (Table 8-13, analyses 6–8) represents a liquid line of descent, possibly branching, determined essentially by fractional crystallization of parent alkali-basalt magma.[1] Generally consistent with this thesis is the late appearance of the more felsic lavas in the full-length volcanic cycle, their small aggregate volume, and decreasing liquidus temperatures down the proposed line. There is additional support in the trace-element abundances—high Ba, Li, Rb, Sr, and Zr (Table 8-13; analyses 7*a*, 8*a*)—and in nearly uniform isotopic compositions of Sr and Pb.

The most disputed point regarding Hawaiian petrogenesis is the role of alkali-basalt magmas. Those who seek uniformity of pattern tend to regard the alkali basalts as derivatives of parental tholeiitic magma—an intermediate link in the process leading to the mugearite-trachyte line of descent. Macdonald (1968, p. 518), for example, appeals to fractionation of olivine-tholeiitic magma at 30 to 50 km depth to yield the alkali-basalt liquid (in contrast to silica-saturated liquid at shallow depth). We see no compelling evidence for so simple a relation. Nor is there any hint of it in what we know of how Hawaiian tholeiitic magma crystallizes under near-surface conditions. Rather we invoke Powers' (1955) concept of independent magma batches recurrently generated at each volcanic site from mantle materials whose Pb and Sr have almost uniform isotopic compositions. If, as we think likely, alkali basalts and likewise lavas of the nephelinite family are not derivative, then the conditions of magma generation beneath individual volcanoes must have changed in time according to uniform pattern, since there is no denying the regularity with which eruption of alkaline magmas closes the life cycle of Hawaiian volcanoes. Moreover a broadly similar chronological succession of magmas, though far from universal, has been recognized in other ocean-island provinces. One such is the Comores Archipelago, straddling the northern opening of the Mozambique Channel (Flower, 1973).

## Oceanic-island Petrogenesis

### PARENTAL MAGMAS

Oceanic petrogenesis hinges on a sequence of phenomena involving crystal-melt equilibria:

1. Partial or complete fusion of mantle rocks to yield mafic magmas
2. Modification of such magma by fractional crystallization at various stages during ascent through the upper mantle

[1] This view has been challenged by Chayes (1963), and the basis of the argument has been elaborated (Chayes, 1964) and also criticized (Harris, 1963; Macdonald, 1963*b*).

3. Differentiation by fractional crystallization and possibly other processes in shallow reservoirs at the base of or immediately below the crust

The parental basaltic magma of any oceanic rock series is the ultimate product of the first two sets of processes. These are hidden from direct observation. Consideration of their possible nature must be deferred to a broader context (Chap. 13) covering also more or less comparable continental magmas.

We put forward the proposition, based on the survey just concluded, that the ultimate parental magmas of all ocean-island provinces are mafic and that they cover a wide compositional spectrum. Alkali-basalt magmas (cf. St. Helena) seem to be most widespread; but olivine-tholeiitic magmas dominate some extensive provinces (for example, Hawaii, Iceland). Also important is a wide range of magmas high in normative *ne*—basanitic (for example, Tahiti) or nephelinitic (Fernando de Noronha, Hawaiian Islands). And there are island groups (for example, Tristan da Cunha, Western Samoa) where the parental alkali-basalt magma is distinctively potassic.

There is some variation, too, in the isotopic character of island basalts (Table 8-14). Many give values of $Sr^{87}/Sr^{86}$ 0.703 to 0.704 and of $Pb^{206}/Pb^{204}$ 18.5 to 19.5. But potassic mafic magmas like those of Tristan da Cunha have decidedly more radiogenic strontium ($Sr^{87}/Sr^{86}$, 0.705 to 0.706). To these we shall return when considering the sources of basaltic magmas (Chap. 13).

The total compositional (including isotopic) range of parental mafic magmas of oceanic islands is much broader than and scarcely overlaps that of abyssal basalts of oceanic ridges in the deep ocean. This broadness is reflected in corresponding wide variety among associated derivative rock series.

## DIFFERENTIATED MAGMA SERIES

CHEMICAL VARIATION Variation diagrams show that in any oceanic province chemical variation in any series of arbitrarily selected rock samples is generally consistent with a fractional-crystallization model. The crystalline phases in most such models are those stable at relatively low pressures. Removal (for example, by sinking) of early magnesian olivine and augite from a parental basaltic melt must leave liquid fractions enriched in $SiO_2$, $Na_2O$, $K_2O$, and in Fe relative to Mg, and depleted in Mg. Picrite basalts and ankaramites demonstrate the efficiency of such a mechanism. The pattern of trace-element variation likewise fits the model.

But other differentiation mechanisms cannot be excluded. Gas may have played some part. It is possible that strong composition gradients unrelated to crystal fractionation may have developed at some stage.

MINERALOGICAL CRITERIA Consider now the mineral assemblages in a rock series based on some chemical index (such as $SiO_2$ percentage) arbitrarily selected with a view to the possibility of differentiation. It is found that liquidus temperatures (at 1 bar) fall regularly from the basaltic to the felsic end of the series. Olivines and augites become enriched in iron; plagioclase becomes progressively more sodic. High-temperature phases such as olivine and diopsidic pyroxene cut out; low-temperature minerals (hornblende, alkali feldspar) come in. At the end of the "line of descent" from all but highly undersaturated parental magmas appears quartz—a phase whose low melting entropy postpones crystallization to the last stage of fractionation in silicate-melt systems. The most compelling evidence in support of the fractional-crystallization model, as Bowen realized long ago, is inherent in the mineral assemblages of the rocks themselves.

Reexamined against the mineralogical background, the pattern of trace-element variation takes on added clarity. As olivine (with its spinel inclusions) drops out of the mineral assemblage, rocks further down the line are sharply depleted in Ni, Co, and Cr. Titanomagnetite enters the assemblage, and V thereafter is eliminated. As calcic plagioclase gives way to andesine, there is a sharp drop in Sr but not in Ba. Minor elements that build up continuously to the end—Ba, Rb, Zr—are those that could not be accommodated in the earlier crystalline fractions.

Lately computer methods have become increasingly used to match the observed chemical line of descent—as shown on variation diagrams—with a predicted line of descent based on removal of successive batches of analyzed phenocryst minerals. This method can generate a model based on many components (Bryan et al., 1969). So for a model to be acceptable today it is necessary to establish *numerically* a close fit between observed and predicted trends. Even then, all that is shown is that fractional crystallization is one of what may be several viable mechanisms of petrogenesis. A model of this kind has recently been established for the basalts and trachytes of Gough Island (Zielinski and Frey, 1970). It is based primarily on major-element variation, and the analyzed crystalline phases concerned are olivine, pyroxene, feldspar, and apatite. This model is then used to predict variation in a number of key trace elements—rare-earth elements, Sr, Ba, Cr, and Ni—in successive liquid fractions (based on observed crystal/liquid distribution coefficients). Observed and predicted trace-element trends agree remarkably well.

PERSISTENCE OF SATURATION PATTERN It was Lacroix (1928, pp. 57–61) particularly who stressed correlation between the chemical character (silica saturation) of oceanic basalts and the nature of their more felsic minor associates (cf. also Barth, 1931). His generalizations have survived the test imposed in recent years by growing coverage of oceanic provinces and vastly augmented and diversified chemical data (cf. Coombs, 1963).

The undersaturated peralkaline character of nephelinites and basanites persists in their felsic phonolitic associates. With *ne*-normative alkali basalts we find hawaiites, mugearites, trachytes, and phonolites, also in the main having normative *ne*. Tholeiitic basalts, icelandites, and rhyolites constitute an oversaturated, consistently *Q*-normative magma series. We find a similar pattern of silica saturation, too, in those continental provinces in which basaltic magmas play the dominant role. For 40 years[1] this increase in silica has been attributed to low-pressure fractional crystallization in which pyroxene as well as olivine has been influential. Basaltic augites have minor *ol* in their norm; where the host rock is alkali basalt, the augites have a little *ne* as well. Coombs (1963) shows that removal of such pyroxenes would accentuate both the oversaturated character of tholeiitic residues and the undersaturated character of liquids derived from alkali basalts.

THE CUMULATE ROCKS Fractional crystallization of basaltic magma at pressure below about 20 to 25 kb implies formation of ultramafic and basic cumulates complementary to felsic differentiated liquids. Few would doubt the cumulate origin of phenocrystic olivine and augite in picrite basalts and ankaramites. And there is increasing evidence that beneath the major oceanic volcanoes there are extensive bodies of layered cumulate plutonic rocks. One such is the layered gabbro-peridotite intrusion exposed in the eroded pre-Miocene basement of Fuerteventura. It is not directly related to the overlying alkali basalts of the Miocene-Recent Canary Island province; but it has supplied xenoliths that profusely load some of these later lavas. There is evidence of similar layered tholeiitic bodies at no great depth in the oceanic crust.

Xenoliths of the same kind are now familiar in tholeiitic lavas of oceanic islands. They tend to show the same chemical traits that mark the individuality of the local island province (for example, ferrogabbro xenoliths in iron-rich tholeiitic basalts of the Galápagos). But they are not confined to tholeiitic provinces; we find them in profusion in the alkaline lavas of Gough Island and of the eastern Canaries. In Hawaiian basalts xenoliths of this class are so widely distributed that E. D. Jackson (1968, p. 148) came to conclude that "the lower crust directly under the volcanic centers on the Hawaiian Ridge consists of layered cumulates very similar in texture to rocks of the Stillwater Complex in Montana."

Alkaline plutonic xenoliths are abundant in many island provinces where the lavas too are alkaline—alkali basalts, hawaiites, and nephelinites. Many have the characteristic layered structure of cumulates. The mineral phases are those expected to crystallize at low pressures from alkaline basic magmas with low silica activity—kaersutite and biotite as well as titanaugite, olivine, and alkali feldspars. They have no known counterpart among xenoliths of tholeiitic lavas, but comparable rocks are exposed in deeply eroded volcanoes of some islands, for example, on Tahiti.

CHRONOLOGICAL CONSIDERATIONS In many carefully studied volcanoes the composition of erupted magmas tends to change in an orderly manner with passage of time; trend is of the kind to be expected—though sometimes in

[1] For historic views of Barth, Kennedy, and Lehmann on the role of pyroxene crystallization in shaping the line of liquid descent, the reader is referred to Turner and Verhoogen (1960, p. 199).

inverse order—in a fractionated magma series. Here we are on safer ground in interpreting chemical trends as genetic lineages. Within short, historic, eruptive episodes on Kilauea the chronological order of eruption is the reverse of a limited, observed, compositional trend of fractional crystallization. On a larger time scale derivative magmas (mugearites, trachytes, and complementary ankaramites; or icelandites and rhyolites) become abundant only in the late stages of growth of many volcanoes (Hawaii, St. Helena, Ascension, Iceland). The overall case for fractional crystallization is undeniably strengthened by chronological considerations. Consistent with the same thesis is the small total volume of lavas considered to be differentiates.

A somewhat puzzling feature of oceanic volcanism is a tendency for closely consecutive or virtually synchronous eruption of basalt and its presumed differentiates at the same center. Flows of alkali basalt and mugearite alternate in the late lava caps of some Hawaiian volcanoes. Daly (1925, pp. 75, 76) long ago drew attention to the intimate field association between basalt and trachyte on Ascension and other islands. During late Pleistocene eruptions in the Azores, trachytic debris exploded from summit vents while basalt lavas poured from flank fissures a few hundred meters lower down. Prior to eruption there existed a column of magma basaltic at the base, completely differentiated toward the top, but still largely liquid. Such a picture is difficult to reconcile with any simple cycle of fractional crystallization.

Late appearance and small volume alone, without compelling chemical and mineralogical support, are not adequate grounds for classifying a magma as differentiated. Few petrologists would derive the tholeiitic lavas erupted in historic times on Lanzarote from vastly more plentiful alkali basalts that preceded them. There is no obvious parental relation between the oversaturated basalts of the Oahu shield volcanoes and the greatly undersaturated nephelinites and basanites erupted at the same general site one or two million years later (cf. E. D. Jackson and Wright, 1970). We feel that the same may be said of late alkali basalts where they follow more voluminous tholeiites, as on some Hawaiian volcanoes.

ISOTOPIC EVIDENCE Isotopes of lighter elements, notably oxygen and hydrogen, show discernible partitioning between different minerals in any igneous assemblage. Thus $\delta_{O^{18}/O^{16}}$ is low in magnetite and high in quartz compared with values for associated feldspar. It follows that fractional crystallization could significantly change the whole-rock isotopic composition of oxygen along a differentiated lineage. It is thought (H. P. Taylor, 1968, pp. 31–32) that, compared with marked variation in $\delta_{O^{18}/O^{16}}$ due to differences in source materials, contamination processes, and contact with atmospheric water, changes induced purely by fractional crystallization will be small, particularly in the early stages. Exceptions are to be expected where magnetite separated early.

In a suite of lavas from Easter Island (H. P. Taylor, 1968, p. 10) $\delta_{O^{18}/O^{16}}$ (whole rock) is 5.5 to 6.2 for basalts and 5.9 to 6.0 for rhyolitic obsidians. The range of values is narrow and very close to 5.9, which is currently considered an average value for many oceanic basalts. If the oxygen of Easter Island basalts represents that of the mantle source, so also must that of associated rhyolites—especially in view of the broad range of $\delta_{O^{18}/O^{16}}$ values (5.6 to 10.2) recorded for continental rhyolites. Here is strong support for direct relation between Easter Island basalts and rhyolites along a differentiated line of descent. There are other presumed differentiates for which $\delta_{O^{18}/O^{16}}$ is close to 5.9—a rhyolite from Iceland and two mugearites from Maui in Hawaii.

The most significant generalization regarding isotopic chemistry of oceanic basalts is that the composition of strontium in abyssal ocean-ridge tholeiites, even allowing for slight local fluctuation, is incredibly uniform. Rocks unaffected by seawater contamination give values of $Sr^{87}/Sr^{86}$ between 0.7023 and 0.70245—the lowest recorded for igneous rocks of any kind. Island basalts mostly give higher values (Table 8-14), ranging from 0.703 (Iceland) to 0.706 (Tristan da Cunha).

In many island volcanoes—Iceland (Hekla), Samoa, and Easter—$Sr^{87}/Sr^{86}$ values in basalts and in associated felsic lavas are virtually identical. This fact is consistent with a differentiated relationship. More generally, $Sr^{87}/Sr^{86}$ in any series increases perceptibly with Rb/Sr, that is, from the basaltic to the felsic end. Serial variation of this type does not rule out fractional crystallization, for which, on the contrary, independent evidence outlined above is very strong. But differentiation, if it did occur, must have preceded eruptions by a span of time—perhaps 10 m.y.—sufficient for generation of perceptibly higher $Sr^{87}$ in fractions enriched in Rb and depleted in Sr by the differentiation process. Perhaps a two-stage or even multistage differentiation model is generally applicable in crystallization of basaltic magma: freezing of the differentiates; remelting and eruption of felsic derivatives along with new drafts of parental magma. Close approximation of $Sr^{87}/Sr^{86}$ values for basaltic and felsic members of some rock series would merely imply short duration for the whole multistage process.

## CONCLUSION

The classic studies of Daly (1925, 1927) and Lacroix (1928) firmly established the wide distribution for olivine basalts and associated trachytes in island volcanoes across the ocean basins. The felsic end members and rocks of intermediate composition gradational toward basalts were seen from the first as differentiates of basaltic parentage. Bowen (1928, p. 240) wrote regarding "the basalt-trachyte association of oceanic islands": "If our general thesis is correct trachyte may be regarded as a derivative of the basaltic magma, differentiation having occurred through fractional crystallization of quite small bodies." Other petrologists (for example, Daly and Shand), recognizing the characteristically explosive nature of trachyte eruption, invoked active participation of water in the final stages of preceding differentiation high up in the magma column.

**Table 8-14** Isotopic compositions of Sr* and Pb in oceanic volcanic series

| PROVINCE | ROCKS | $Sr^{87}/Sr^{86}$ | $Pb^{206}/Pb^{204}$ |
|---|---|---|---|
| Réunion (McDougall and Compston, 1965) | Basalts (5) | 0.7040–0.7046 | |
| | Mugearites (3) | 0.7042–0.7044 | |
| | Syenite (1) | 0.7046 | |
| Easter Island (Hedge and Peterman, 1970; Tatsumoto, 1966*a*) | Basalts (2) | 0.7030, 0.7036 | 19.280, 19.301 |
| | Icelandite (1) | 0.7030 | 19.253 |
| | Rhyolite obsidian (1) | | 19.308 |
| Hawaiian Islands (Hedge, 1966; Tatsumoto, 1966*b* Hedge and Peterman, 1970) | Hawaii: Kilauea tholeiites (3) | 0.7039–0.7041 | |
| | Hawaii: Hualalai, alkali basalt | | 17.92 |
| | Hawaii: Hualalai, trachyte | 0.7035 | 18.08 |
| | Hawaii: Mauna Kea, picrite basalt | 0.7034 | 18.48 |
| | Hawaii: Mauna Kea, hawaiite | 0.7033 | 18.47 |
| | Oahu: Koolau basalts | 0.7039 | 18.09 |
| | Oahu: Waianae basalts | 0.7032 | |
| | Oahu: Honolulu series (nephelinites) | 0.7031 | 18.17, 18.24 |
| Iceland (Moorbath and Walker, 1965) | Pleistocene-Recent basalts (3) | 0.7028–0.7033 | |
| | Recent obsidians (1) | 0.7017 | |
| | Tertiary basalts (6) | 0.7021–0.7032 | |
| | Tertiary rhyolites (3) | 0.7014–0.7015 | |
| | Late Tertiary granophyres (6) | 0.7010–0.7033 | |
| Guadalupe Island (Tatsumoto, 1966*a*; Peterman and Hedge, 1971) | Alkali basalts (4) | 0.7033–0.7036 | 20.172–20.436 |
| St. Helena (Hedge, 1966) | Alkali basalts (2) | 0.7031, 0.7032 | |
| | Phonolites (2) | 0.7047, 0.7054 | |
| Tutuila, Samoa (Hedge, 1966) | Basalt | 0.7057 | |
| | Trachytes (2) | 0.7062, 0.7066 | |
| Upolu and Savaii, Samoa (Hedge et al., 1972) | Potassic alkali basalts (10) | 0.7051–0.7066 | |
| Gough Island (Gast et al., 1964) | Basalt | 0.7045 | 18.36 |
| | Trachyandesite, trachybasalt | 0.7050, 0.7043 | 18.37, 18.43 |
| | Trachytes (2) | 0.7094, 0.7050 | 18.63, 18.73 |
| Ascension Island (Gast et al., 1964) | Basalts (2) | 0.7025 | 18.43, 19.55 |
| | Trachyandesite | 0.7025 | |
| | Trachytes (2) | 0.7045, 0.7073 | 19.72 |
| | Obsidian | | 19.50 |

* All values normalized to $Sr^{86}/Sr^{88} = 0.1194$, $Sr^{87}/Sr^{86} = 0.7080$, for standard E. and A. $SrCO_3$.

Today evidence for differentiation in oceanic volcanic suites is collectively stronger, much more voluminous, and more diversified in character than in the day of Lacroix and Daly. Essentially, however, opinion is much the same now as half a century ago. The picture is now more complex in that we recognize other kinds of trend than the single olivine-basalt trachyte line of descent. Chemical and mineralogical details of magmatic evolution have sharpened. High-pressure experimental data show that differentiation of the kind we have described must be limited to rather shallow depths, no greater perhaps than 30 to 50 km. The role of water and other volatile constituents of alkaline magmas, although still somewhat vague, is perhaps better understood. Daly and Shand saw water, in the form of streaming gas, taking a direct part in the mechanics of differentiation, thereby diverting the liquid line of descent toward a trachytic rather than an andesitic composition. Bowen (1928, p. 234) had already speculated that dissolved water and other volatile components might disturb the structure of felsic liquid magmas, thereby bringing about undersaturation of late liquid fractions in $SiO_2$ relative to ($Na_2O + K_2O$). And Ringwood (1957) suggested that the structural effect of high concentrations of water suppresses early crystallization of anorthite in favor of diopsidic pyroxene and hornblende, again turning the trend of residual liquids toward soda-enriched trachytes. Measurements on volume percentages of vesicles in ocean-floor basalts extruded at 1 to 2.5 km depth (J. G. Moore, 1970) show conclusively that the water content of the magma rises with alkali ($K_2O$) content: 0.25 percent by weight in K-poor abyssal tholeiite, 0.5 percent in Hawaiian tholeiite, 0.9 percent in more potassic basalts ($K_2O$ = 1.5 percent). Directly correlated with rising $H_2O$ percentage is increase in $P_2O_5$, F, and Cl.

The petrologist of today is much better equipped than his predecessors to assess quantitatively the chemical trends of fractional crystallization. Quantitative models have been set up for a number of oceanic lava series—St. Helena, Gough, Thingmuli in Iceland, some of the Hawaiian volcanoes, Tenerife in the Canaries. Models, however, they remain—and open to competition with yet other models that fit the growing data equally well or better. Many of the chemical data on which they are based can equally well fit more complex models in which the final stage is partial or complete fusion of previously consolidated (possibly already differentiated) magma. In view of this and of uncertainties regarding the site and pressure-temperature conditions of differentiation, no model can establish with certainty the actual course of differentiation. It serves rather to illustrate the kinds of events and processes that might have participated in some unique line of evolution whose varied products are tangibly represented as rocks of a unique volcanic series.

# Continental Tholeiitic Provinces

## Tholeiitic Flood Basalts and Intrusive Equivalents

The most voluminous continental lavas are basalts, and the most extensive of these by far are tholeiitic. They occur as massive accumulations of fissure-fed subhorizontal flows flooding very large areas on a regional scale. To these rocks the terms *plateau* and *flood* basalts have been applied synonymously (Tyrrell, 1937*c*). The total volume of lavas in a single province may be enormous: nearly $10^6$ $km^3$ in the Deccan plateau of western India (Cretaceous-Eocene);[1] 180,000 $km^3$ in the Miocene Columbia River province, northwestern United States; and a comparable volume of younger lavas nearby in eastern Oregon and adjacent states (Verhoogen et al., 1970, p. 299). Most extensive of all perhaps are basalts of the Paraná basin (early Cretaceous) in southern Brazil and adjoining countries, with a total area of $10^6$ $km^2$. Elsewhere erosion has reduced former basaltic plateaus to isolated but still very extensive remnants. Such are the basalts of Lesotho (Cox and Hornung, 1966), still 25,000 $km^2$ of the much larger Karroo province in southern Africa (early Jurassic), and remnants of Tertiary basalts in Western Australia (Edwards, 1938*b*).

Imperfect sampling—a consequence of the vast extent and overall chemical uniformity of flood basalts—possibly masks local chemical fluctuations in most provinces. Minor occurrences of nepheline-bearing lavas have been reported from some, and they no longer seen anomalous. In the Nuanetsi province on the northeastern border of the Karroo basin, peralkaline lavas open and close the Karroo volcanic cycle; and at the climax of volcanism here tholeiitic basalts

[1] Sukheswala and Poldervaart (1958); West (1958).

are accompanied by almost equal volumes of rhyolitic debris (Cox et al., 1965; Stillman, 1970). There are provinces such as the Hebridean of northwestern Britain where tholeiitic and alkali-basalt magmas are almost equally well represented. Elsewhere, again, where all the volcanic rocks are tholeiitic, subtypes may be distinguished by criteria such as $TiO_2$ content, Fe/Mg ratio, and phenocryst phases. There is no evidence to indicate that associated chemically diverse magmas in any of these provinces are genetically related. More likely they are of independent origin.

There are other regions where immense volumes of tholeiitic magma have deployed laterally at shallow depths within the crust. The exposed products are massive sills and dike swarms composed of diabase (dolerite). Individual mid-Jurassic sills in Antarctica and Tasmania are 1000 to 5000 $km^3$ in volume. Deeper down in the crust, tholeiitic plutons are more massive still and display a striking internal stratification due to differentiation. The largest recorded example is the Precambrian Bushveld complex of South Africa, whose volume has been estimated as 100,000 $km^3$.

## Two Volcanic Provinces: Columbia River and Karroo Basin

### COLUMBIA RIVER BASALTS, NORTHWESTERN UNITED STATES

Much of eastern Washington and Oregon and adjoining parts of western Idaho are covered by Miocene-Pliocene tholeiitic basalts of the Columbia River province. Most of the basaltic terrane lies east of the Cascade Range, between 200 and 600 km from the Pacific coast (Verhoogen et al., 1970, p. 298). The total volume of lava, most of it in flows 20 to 30 m thick, is about 180,000 $km^3$. Some individual flows are several hundred kilometers long. It was erupted via fissures, at least two-thirds of it (Yakima basalt) within a period of perhaps 5 million years.

Here Waters (for example, 1961, 1962) has convincingly demonstrated the existence of three distinct and remarkably uniform tholeiitic magmas erupted in successive stages of activity:

1. Picture Gorge basalts, mid-Miocene, total volume 40,000 $km^3$. Chemically these are slightly undersaturated or just-saturated tholeiitic basalts (Table 9-1; analyses 1, 2) but mineralogically akin to alkali basalts: olivine occurs as phenocrysts ($Fo_{80}$) and in the groundmass ($Fo_{45}$ to $Fo_{65}$), and the sole recorded pyroxene is augite.
2. Yakima basalts, late Miocene to early Pliocene, total volume 120,000 $km^3$. These are *Q*-normative basalts (Table 9-1; analyses 3, 4) with typical tholeiite mineralogy: minor resorbed olivine, augite, and pigeonite.
3. Late Yakima and Ellenburg flows, early Pliocene, total volume 20,000 $km^3$. These are iron-rich *Q*-normative lavas (Table 9-1; analyses 5, 6) with ferrous

olivine ($Fo_{30}$ to $Fo_{40}$) and two pyroxenes—augite and pigeonite. Chemically they recall the ferrobasalts of the Galápagos.

The Picture Gorge basalts were erupted from fissures—now represented by dike swarms—along the southern margin of the province. They were buried under 500 m of continental sediments before the next stage of activity began. Then followed copious and rapid outpouring of Yakima basalt flows and slow sinking of the basin in which they now lie. In some sections the basalt filling exceeds 600 m in thickness. The final stage of activity (late Yakima-Ellenburg) was set in a changing tectonic environment. The Cascade Range was already rising along the western border of the province, and the Yakima basalts of the main (second) eruptive phase already were becoming warped and faulted. Waters (1961, p. 583) concluded that the three magma types were of independent origin—"products from separate magmatic hearths, and not differentiates of a hypothetical uniform magma." The complete chronological succession nevertheless involves chemical progression—increase in Fe/Mg and in K/Ca—of the kind that in single bodies of mafic magma is attributed to fractional crystallization.

In the independent and equally extensive flood-basalt province of the eastern Oregon plateau which adjoins the Columbia River province on the south (Verhoogen et al., 1970, p. 299), basalts are of yet another type. Here "high-alumina basalts" (Table 9-1, analysis 7) were erupted continuously from early Pliocene to Recent times. With them are associated much smaller but still significant quantities of andesite and rhyolite.

Volcanism in these two provinces, in spite of its vast scale, must be seen as two successive, limited episodes in a tectonic-magmatic event of much greater magnitude affecting the western margin of the whole continent from later Mesozoic times onward.

## KARROO BASIN PROVINCE, SOUTH AFRICA

The Karroo basin, covering 500,000 km$^2$ in southern Africa (Fig. 9-1), is filled with Permian and Triassic continental sediments locally 5000 m thick. Sedimentation was followed by regional flooding with tholeiitic lavas (Stormberg basalts) and synchronous intrusion of innumerable sills, sheets, and dikes of chemically similar magma. Erosion has since reduced the lavas to isolated remnants and exposed the underlying system of sills and dikes of diabase. The whole magmatic episode has been dated at 190 to 154 m.y. (early to middle Jurassic) and was part of an even more comprehensive event, the tectonic breakup of Gondwanaland (cf. F. Walker and Poldervaart, 1949; Cox, 1972).

Lavas of the Lesotho (Basutoland) remnant covering 25,000 km$^2$ are flows, mostly about 10 m thick, constituting an exposed vertical sequence 2000 m thick (Cox and Hornung, 1966). All are tholeiitic basalts—mostly $Q$-normative

**Table 9-1** Chemical compositions (oxides, wt %) and CIPW norms of representative and average basalts, Columbia River plateau (with one basalt from Medicine Lake highland, northern California)

| | 1 | 2 | 3 | 4 | 5 | 6 | 7 |
|---|---|---|---|---|---|---|---|
| $SiO_2$ | 49.5 | 49.3 | 54.50 | 53.8 | 50.0 | 50.0 | 48.98 |
| $TiO_2$ | 1.5 | 1.6 | 1.95 | 2.0 | 3.2 | 3.2 | 1.16 |
| $Al_2O_3$ | 14.6 | 15.6 | 13.59 | 13.9 | 14.1 | 13.5 | 18.92 |
| $Fe_2O_3$ | 2.5 | 3.5 | 3.28 | 2.6 | 2.1 | 1.9 | 2.22 |
| FeO | 9.4 | 7.8 | 8.80 | 9.3 | 11.6 | 12.5 | 7.12 |
| MnO | 0.22 | 0.2 | 0.18 | 0.2 | 0.24 | 0.25 | 0.09 |
| MgO | 7.1 | 6.5 | 3.84 | 4.1 | 4.3 | 4.4 | 7.42 |
| CaO | 10.8 | 10.3 | 7.22 | 7.9 | 8.1 | 8.3 | 10.04 |
| $Na_2O$ | 2.9 | 2.7 | 3.05 | 3.0 | 2.8 | 2.9 | 3.04 |
| $K_2O$ | 0.45 | 0.5 | 1.45 | 1.5 | 1.4 | 1.4 | 0.44 |
| $P_2O_5$ | 0.27 | 0.3 | 0.21 | 0.4 | 0.8 | 0.7 | 0.14 |
| $H_2O^+$ | 1.0 | 1.8 | 0.78 | 1.2 | 1.8 | 0.9 | 0.34 |
| $H_2O^-$ | | | 0.92 | | | | 0.05 |
| Total | 100.2 | 100.1 | 99.77 | 99.9 | 100.4 | 100.0 | 99.96 |
| *Q* | | 0.8 | 9.27 | 6.9 | 2.94 | 1.3 | |
| *or* | 2.8 | 2.8 | 8.75 | 8.9 | 8.35 | 8.4 | 2.2 |
| *ab* | 24.6 | 23.1 | 26.32 | 25.2 | 23.60 | 24.7 | 25.2 |
| *an* | 25.3 | 29.8 | 19.49 | 20.0 | 22.26 | 19.5 | 37.0 |
| *di* | 21.3 | 16.7 | 12.64 | 13.9 | 10.90 | 14.7 | 9.7 |
| *hy* | 10.7 | 17.2 | 14.42 | 15.3 | 19.09 | 20.1 | 10.0 |
| *ol* | 6.3 | | | | | | 9.6 |
| *mt* | 3.5 | 5.1 | 4.84 | 3.7 | 3.02 | 2.8 | 3.3 |
| *il* | 3.0 | 3.0 | 3.78 | 3.8 | 6.08 | 6.0 | 2.3 |
| *ap* | 0.7 | 0.6 | 0.49 | 0.9 | 1.90 | 1.5 | 0.4 |
| Total | 98.2 | 99.1 | 100.00 | 98.6 | 98.14 | 99.0 | 99.7 |

*Explanation of column headings*

1–6 Columbia River plateau
1 Basalt, Picture Gorge group (Waters, 1961, p. 592, analysis 3)
2 Average Picture Gorge basalt (Waters, 1961, p. 592, analysis A)
3 Basalt of Yakima group (Swanson, 1967, p. 1093, analysis 3)
4 Average Yakima basalt (Waters, 1961, p. 593, analysis B)
5 Basalt, late Yakima-Ellenburg group (Waters, 1961, p. 594, analysis 2)
6 Average late Yakima-Ellenburg basalt (Waters, 1961, p. 594, analysis A)
7 Modoc basalt, subophitic, Medicine Lake highland, northern California; eastern Oregon plateau province (Anderson, 1941, p. 387, analysis 3)

rocks with augite, pigeonite, and minor olivine (Tables 9-2, 9-3; analyses 2, 3). Some flows are enriched in olivine (Tables 9-2, 9-3; analysis 1).

The largest intrusive bodies are gently undulating sheets 1000 m thick and 50 km in diameter. There are many sills and inclined sheets a few hundred meters thick and thin, vertical dikes (3 to 10 m wide), some of which can be traced for

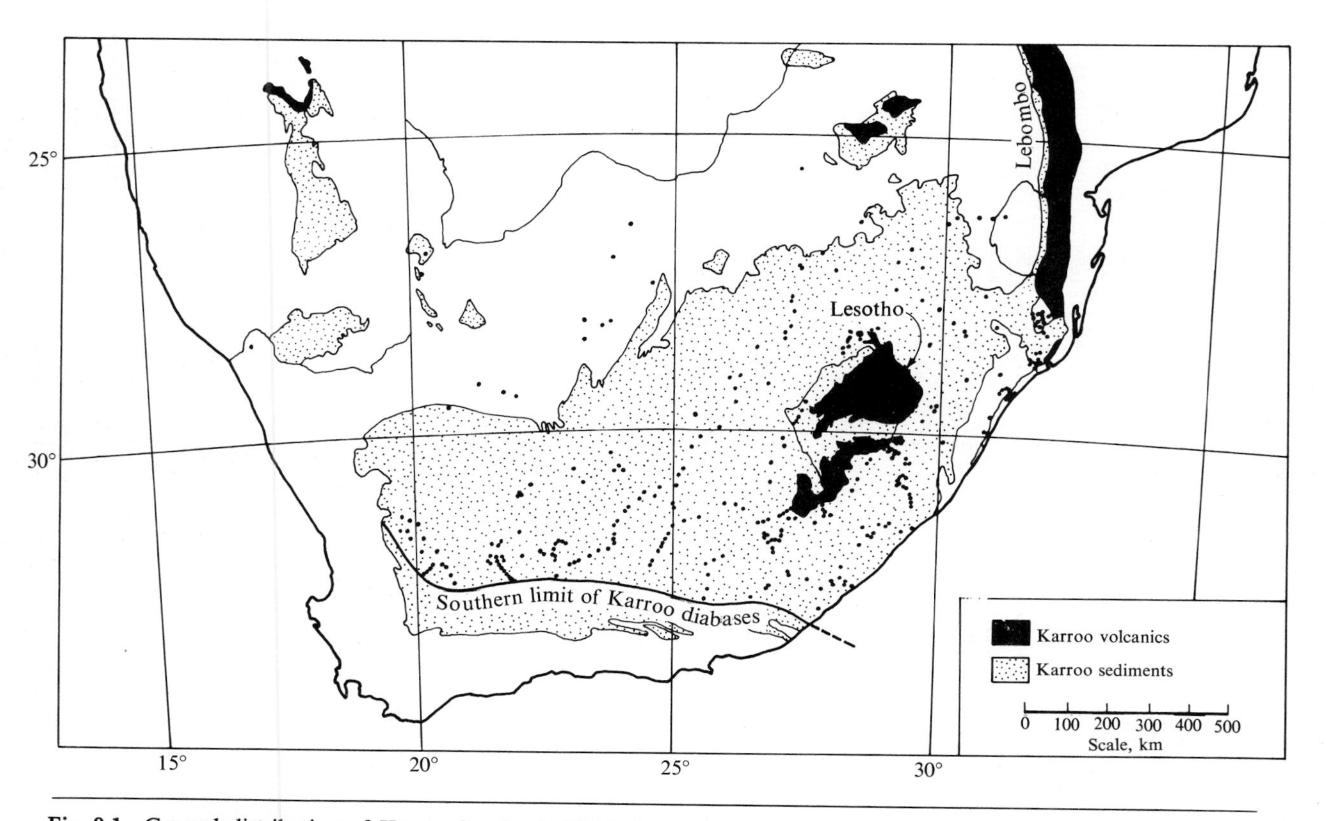


**Fig. 9-1** General distribution of Karroo basalts (solid black areas) and diabase localities (solid circles) within the Karroo system (stippled), southern Africa. (After F. Walker and Poldervaart, 1949.)

**Table 9-2** Chemical compositions (major oxides, wt %) and CIPW norms of flood basalts, diabases, and picrite, Karroo basin, South Africa

| | 1 | 2 | 3 | 4 | 5 | 6 | 7 | 8 |
|---|---|---|---|---|---|---|---|---|
| $SiO_2$ | 49.0 | 51.9 | 49.9 | 51.78 | 50.76 | 53.38 | 52.41 | 43.95 |
| $TiO_2$ | 0.89 | 1.00 | 1.29 | 0.92 | 0.45 | 1.18 | 3.14 | 0.80 |
| $Al_2O_3$ | 11.60 | 14.50 | 13.50 | 15.34 | 14.01 | 15.64 | 12.23 | 6.71 |
| $Fe_2O_3$ | 2.80 | 1.54 | 5.53 | 1.07 | 1.42 | 0.23 | 3.53 | 1.67 |
| FeO | 9.88 | 8.53 | 6.29 | 10.58 | 9.35 | 10.76 | 12.86 | 10.67 |
| MnO | 0.16 | 0.17 | 0.17 | 0.22 | 0.15 | 0.25 | 0.26 | 0.42 |
| MgO | 11.80 | 8.10 | 6.00 | 6.44 | 8.68 | 5.67 | 2.62 | 29.09 |
| CaO | 8.12 | 10.39 | 9.20 | 9.88 | 10.71 | 9.36 | 8.48 | 3.05 |
| $Na_2O$ | 1.85 | 2.18 | 2.37 | 2.54 | 2.62 | 1.72 | 2.01 | 2.32 |
| $K_2O$ | 0.59 | 0.60 | 0.79 | 0.66 | 0.92 | 0.95 | 0.85 | |
| $P_2O_5$ | 0.06 | 0.17 | 0.14 | 0.14 | 0.03 | 0.16 | 0.23 | 0.11 |
| $H_2O^+$ | 3.23 | 1.01 | 4.33 | 0.63 | 1.10 | 0.36 | 0.92 | 0.95 |
| $H_2O^-$ | | | | | 0.61 | 0.12 | 0.14 | 0.40 |
| Total | 99.98 | 100.09 | 99.51 | 100.20 | 100.81 | 99.78 | 99.68 | 100.14 |
| *Q* | | 1.62 | 2.22 | 0.3 | | 6.1 | 12.8 | |
| *or* | 3.34 | 3.34 | 5.00 | 3.9 | 5.6 | 6.1 | 5.3 | 14.9 |
| *ab* | 16.24 | 18.34 | 20.96 | 21.5 | 22.0 | 14.2 | 16.9 | |
| *an* | 22.52 | 28.63 | 25.30 | 28.4 | 23.6 | 32.0 | 21.9 | 9.2 |
| *ne* | | | | | | | | 1.4 |
| *di* | 15.57 | 19.04 | 18.87 | 16.6 | 23.9 | 11.6 | 15.7 | 7.3 |
| *hy* | 30.05 | 25.06 | 22.82 | 25.4 | 8.7 | 26.6 | 14.4 | |
| *ol* | 8.16 | | | | 12.2 | | | 62.0 |
| *mt* | 2.09 | 1.62 | 1.86 | 1.6 | 2.1 | 0.2 | 5.2 | 2.6 |
| *il* | 1.82 | 1.98 | 2.58 | 1.5 | 0.8 | 2.3 | 6.0 | 1.5 |
| *ap* | | | | 0.3 | 0.1 | 0.3 | 0.7 | 0.3 |
| Total | 99.79 | 99.63 | 99.61 | 99.5 | 99.0 | 99.4 | 98.9 | 99.2 |

*Explanation of column headings*

1 Basalt (olivine enriched) near base of sequence, Lesotho (Cox and Hornung, 1966, B67)
2 Basalt near base of sequence, Lesotho (*ibid.*, B69)
3 Basalt near top of sequence, Lesotho (*ibid.*, B36)
4 Diabase, Perdekloof type, Karroo (F. Walker and Poldervaart, 1949, no. 5)
5 Diabase, Blaauwkrans type, Karroo (*ibid.*, no. 8)
6 Chilled marginal facies of sill, Hangnest type, Karroo (*ibid.*, no. 61)
7 Ferrodiabase, upper portion of New Amalfi sheet, Karroo (*ibid.*, no. 54)
8 Picrite, base of Insizwa sheet, Karroo (*ibid.*, no. 43)

50 km or more. Just southeast of the Lesotho volcanic region, vertical dikes are so numerous that they collectively constitute 20 percent of the outcrop area. This region seems to have been a principal focus of igneous activity, with basaltic magma welling up along tensional fractures in a stretched crust and spreading laterally to build massive sheets such as the great Insizwa intrusion.

In this and some other major sheets and sills, vertical compositional

**Table 9-3** Atomic abundances (ppm) and abundance ratios of trace elements in flood basalts and diabases, Karroo province (numbered to correspond with major-element compositions, Table 9-2)

| | 1 | 2 | 3 | 4 | 6 |
|---|---|---|---|---|---|
| Ba | 200 | 160 | 250 | 180 | 360 |
| Co | 55 | 45 | 30 | 30 | 30 |
| Cr | 550 | 500 | 170 | 270 | 200 |
| Ni | 350 | 130 | 45 | 70 | 10 |
| Pb | $<10$ | $<10$ | $<10$ | | |
| Rb | 42 | 38 | 37 | 20 | 20 |
| Sr | 120 | 140 | 180 | 400 | 400 |
| Th | | | | 80 | 160 |
| V | 250 | 300 | 250 | 270 | 400 |
| Zr | 75 | 90 | 85 | | |
| K/Rb | 120 | 130 | 170 | | |

variation (Fig. 9-2) has developed by fractional crystallization in place. Where floors sag downward there are basal picrites containing 25 to 40 percent magnesian olivine, along with labradorite, augite, and orthopyroxene (Table 9-2, analysis 8). The Fe/Mg ratio increases markedly upward as olivine and pyroxenes show progressive enrichment in iron (Table 9-2, analysis 7). The final stage of differ-

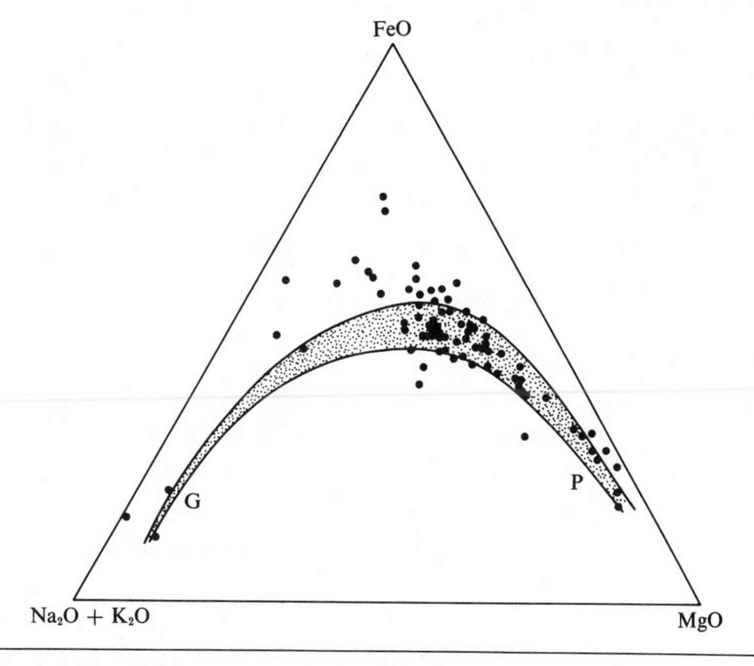


**Fig. 9-2** Plot of analyses of Karroo diabases and associated picrites (P) and granophyres (G); after F. Walker and Poldervaart, 1949. The shaded band encloses compositional range of rocks from the Insizwa sheet (from Scholtz, 1936). FeO includes only $Fe^{++}$.

entiation seems to be represented by coarse pegmatoid streaks in some diabases. These consist of plagioclase, ferropyroxenes, minor iron-rich olivine, and micropegmatitic alkali-feldspar–quartz intergrowths; some contain biotite or hornblende. Granophyres in the roof regions of some sills are interpreted as products of fusion and partial assimilation of Karroo sediments.

Across the whole Karroo basin, tholeiitic magmas consistently have low K, Ti, P, Ba, Sr, and Zr (Table 9-5; analyses 1, 2, 4). This applies as well to high-alumina gabbros (Table 9-4, analysis 3) that form massive ring complexes in the Nuanetsi border zone on the northeast (Fig. 9-3). But here there are also tholeiitic rocks of another chemical type, with uniformly high K, Ti, P, Ba, Sr, and Zr (Table 9-5, analysis 3). This is a region where strong downwarping was more or less synchronous with volcanism. Divergence from the normal chemical pattern and diversification of magma types could reflect differences in source materials or in conditions of melting, or alternatively contamination by crustal rocks.

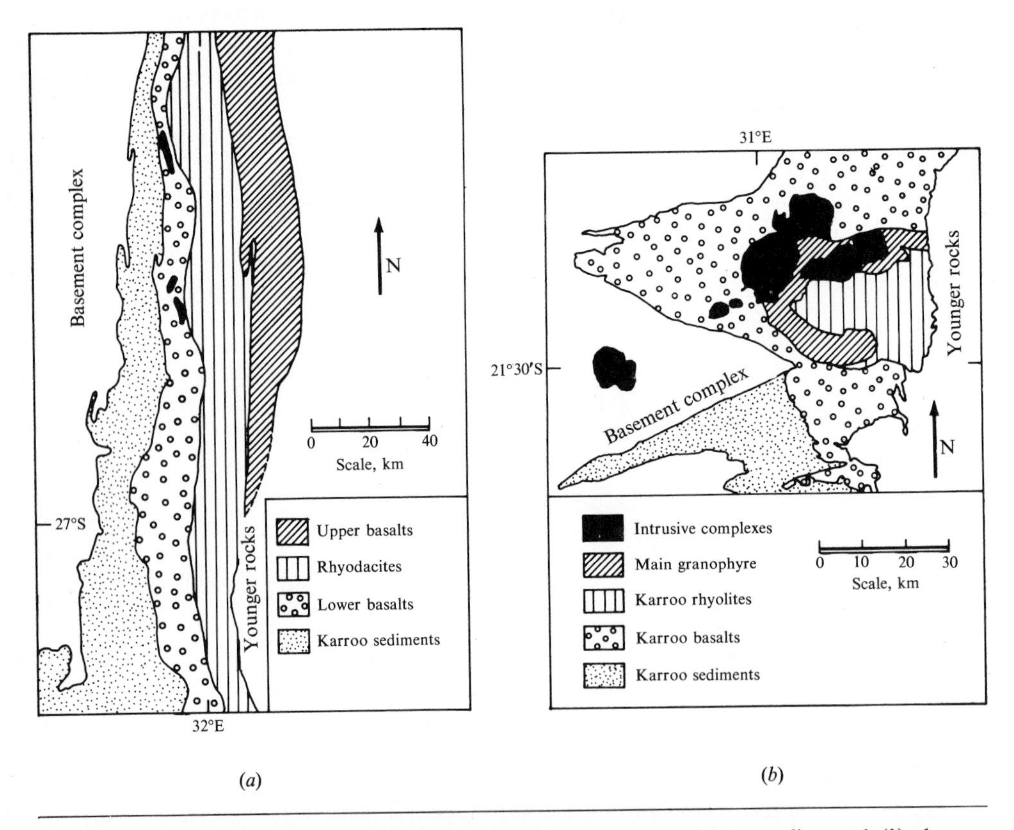


**Fig. 9-3** Karroo volcanics in (*a*) southern part of the Lebombo monocline and (*b*) the Nuanetsi syncline, northeastern extension of Karroo volcanic province. (After Manton, 1968).

**Table 9-4** Chemical compositions (oxides, wt %), CIPW norms, and atomic abundances of trace elements for igneous rocks, Nuanetsi region, Karroo province, southern Africa

| | 1 | 2 | 3 | 4 | 5 | | 1 | 2 | 3 | 4 | 5 |
|---|---|---|---|---|---|---|---|---|---|---|---|
| $SiO_2$ | 49.61 | 52.05 | 51.82 | 71.85 | 72.19 | Ba | 700 | 750 | 180 | 2000 | 1800 |
| $TiO_2$ | 3.54 | 1.70 | 0.66 | 0.47 | 0.45 | Co | 45 | 40 | 30 | | 10 |
| $Al_2O_3$ | 11.28 | 12.43 | 18.38 | 11.15 | 13.07 | Cr | 1100 | <10 | 300 | 30 | ~10 |
| $Fe_2O_3$ | 2.13 | 5.18 | 2.60 | 4.50 | 2.59 | Li | <5 | 30 | ~3 | <5 | 10 |
| FeO | 10.11 | 10.08 | 5.94 | 0.73 | 0.43 | Nb | 50 | 55 | <30 | 120 | 40 |
| MnO | 0.18 | 0.24 | 0.15 | 0.05 | 0.04 | Ni | 300 | 35 | 100 | 10 | 9 |
| MgO | 7.67 | 3.95 | 5.25 | 0.19 | 0.35 | Pb | 25 | 30 | 11 | 25 | 17 |
| CaO | 9.66 | 7.33 | 11.68 | 1.67 | 1.19 | Rb | 72 | 62 | <30 | 140 | 250 |
| $Na_2O$ | 2.37 | 2.76 | 2.58 | 3.03 | 3.23 | Sr | 237 | 183 | 150 | 95 | 45 |
| $K_2O$ | 1.83 | 2.07 | 0.47 | 4.95 | 5.36 | Ta | <300 | <300 | <100 | | |
| $P_2O_5$ | 0.52 | 0.28 | 0.08 | 0.07 | 0.16 | V | 300 | 300 | 150 | 11 | 18 |
| $H_2O^+$ | 0.80 | 1.90 | 0.57 | 0.29 | 0.61 | Zr | 450 | 350 | 70 | 900 | 500 |
| $H_2O^-$ | 0.42 | 0.36 | 0.13 | 0.18 | 0.30 | | | | | | |
| Total | 100.12 | 100.33 | 100.31 | 100.21* | 99.97 | | | | | | |
| *Q* | | 5.88 | 2.8 | 34.14 | | | | | | | |
| *or* | 10.57 | 12.24 | 2.78 | 29.49 | | | | | | | |
| *ab* | 19.92 | 23.59 | 22.01 | 25.69 | | | | | | | |
| *an* | 15.02 | 15.30 | 36.97 | 2.78 | | | | | | | |
| *di* | 23.77 | 15.97 | 17.19 | | | | | | | | |
| *hy* | 17.64 | 13.74 | 12.65 | 0.50 | | | | | | | |
| *ol* | 0.97 | | | | | | | | | | |
| *mt* | 3.01 | 7.41 | 3.71 | 0.93 | | | | | | | |
| *il* | 6.68 | 3.19 | 1.22 | 0.91 | | | | | | | |
| *ap* | 1.34 | 0.67 | | | | | | | | | |
| Total | 98.92 | 97.99 | 99.33 | 100.77* | | | | | | | |

* Including $CO_2$ = 1.08; normative calcite 2.50; normative hematite 3.83.

*Explanation of column headings*

Analyses from Cox et al. (1965), to which references are given in parentheses.

1 Glassy olivine basalt ("limburgite"), lower part of Karroo volcanic sequence (p. 145, LM 434)
2 Basalt, Karroo volcanics (p. 145, LM 126)
3 Noritic gabbro, northern ring complex (p. 156, G 1, 723)
4 Ignimbrite (p. 149, LM 389); Rb and Sr, mean of four analyses (rhyolite) from Manton (1968, p. 30)
5 Granophyre, main granophyre sheet (p. 178, C 437); Rb and Sr, mean of three analyses (granophyre) from Manton (1968, p. 30)

Siliceous rocks in the Karroo province are of three kinds: (1) insignificant streaks of granophyric differentiate in diabase sheets; (2) granophyres thought to have formed by marginal assimilative reaction with Karroo sediments, culminating in partial melting; and (3) extensive granophyric sheets and masses of rhyolitic debris in the northeastern (Nuanetsi) border region (Table 9-4;

**Table 9-5** Some general chemical characteristics of basaltic magmas in the Karroo province

| | 1 | 2 | 3 | 4 |
|---|---|---|---|---|
| $K_2O$ | 0.35–0.90 | 0.5–1.0 | 1.4–3.4 | 0.1–0.6 |
| $TiO_2$ | 0.85–1.40 | 0.5–1.5 | 1.6–3.6 | 0.15–1.0 |
| $P_2O_5$ | <0.17 | <0.15 | 0.3–1.0 | <0.15 |
| Ba | 150–350 | 150–350 | 400–850 | 10–180 |
| Rb | 8–42 | 11–12 | 35–75 | 20–30 |
| Sr | 140–270 | 180–400 | 180–1000 | 100–270 |
| Th | | 1.3–1.5 | | |
| U | | 0.2–0.37 | | |
| Zr | 80–140 | | 350–500 | 10–70 |
| K/Rb | 150–400 | 395–425 | | |
| Th/K | | $2.7–3.0 \times 10^{-4}$ | | |
| U/K | | $0.4–0.77 \times 10^{-4}$ | | |
| $Sr^{87}/Sr^{86}$ (initial) | | 0.7050–0.7065 | 0.706–0.710 | 0.7110–0.7125 |

*Explanation of column headings*

Values cited cover at least 90 percent of recorded analyses in each cited work.

1 Drakensberg basalts, Lesotho: 25 analyses from Cox and Hornung (1966)
2 Karroo diabases: 65 oxide analyses and 4 minor-element abundances from F. Walker and Poldervaart (1949, tables 14, 15); Sr and Rb data (6 analyses) and Th and U data (8 analyses) from Compston et al. (1968, pp. 139, 141)
3 Basalts, including glassy tholeiites, Nuanetsi region: 11 analyses from Cox et al. (1965, pp. 45, 147); Sr and Rb data (7 analyses) from Manton (1968)
4 Gabbros, northern ring complex, Nuanetsi region: 6 analyses from Cox et al. (1965); Sr and Rb data (2 analyses) from Manton (1968)

analyses 4, 5). These last are notably more potassic than granophyres of the first two kinds; and, although isotopic data do not compel such reasoning, it has been suggested that they represent siliceous magmas genetically unconnected with associated tholeiitic rocks (Cox et al., 1965, pp. 206-211; Manton, 1968, p. 36; Cox, 1972).

Strontium of the basic rocks is variable in composition. Most uniform are initial $Sr^{87}/Sr^{86}$ values recorded for diabases (Table 9-5, analysis 2): higher than most oceanic basalts, decidedly lower than in diabases of Tasmania and Antarctica. High values for basalts and gabbros of the tectonically active Nuanetsi zone (Table 9-5; analyses 3, 4) could reflect isotopic exchange with the Archaean basement rocks; but low $K_2O$ in the gabbros rules out bulk assimilation. Nor could melting of the basement play a role in the origin of rhyolites and granites of the same subprovince. Initial $Sr^{87}/Sr^{86}$ values are much too low: 0.7081 on a 206 m.y. isochron for the rhyolites; 0.7085 on a 177 m.y. isochron for ring-complex granites. Strontium data for the Karroo province, while ruling out some simple alternatives, fail to suggest any obvious internally consistent petrogenic model.

Cox (1972) portrays a comprehensive "Karroo cycle" of igneous activity connected with the Mesozoic breakup of the African segment of Gondwanaland. Details of the model presented are open to question along lines discussed in Chap. 13. But it is in keeping with current thinking that a number of independent parental magmas are proposed: potassic tholeiite dominant in the northern sector (Nuanetsi), more sodic "normal" tholeiite in the south and also following the potassic surge in the north; early nephelinitic magmas in the Nuanetsi region; later rhyolitic and allied magmas, attributed to high-level fusion of crustal rocks. Variation in MgO among basaltic magmas is attributed to olivine fractionation at low pressure during ascent. Thus olivine tholeiites with 15 percent MgO in the Nuanetsi region are pictured as representing a magma parental to more copious later basalts with less than 6 percent olivine. Also attributed to fractionation in the 25 to 65 km depth range are minor late feldspathoidal and peralkaline acid lavas.

## Differentiated Tholeiitic Sills

### FORM AND DIMENSIONS

Many individual tholeiitic sills emplaced in continental or epicontinental sediments are very large. The Precambrian (2000 m.y.) Kopinang sill of British Guiana (Hawkes, 1966) is 300 to 400 m thick and has been mapped continuously for 80 km. The late Triassic Palisades sill, New Jersey (F. Walker, 1940; K. R. Walker, 1969*a*) outcrops over an area 80 × 2 km and is about 300 m thick. The Peneplain sill of the mid-Jurassic Ferrar dolerite swarm, Antarctica (Gunn, 1962) is 250 to 400 thick and at least 20,000 km$^2$ in outcrop area; its total volume is no less than 5000 km$^3$ and may be three times greater. In Tasmania (Edwards, 1942; McDougall, 1962, 1964*a*) early Jurassic diabase sheets commonly exceeding 300 m in thickness outcrop over an area of 15,000 km$^2$.

These and most similar sills have been injected almost horizontally in undisturbed sedimentary formations or along subhorizontal surfaces of lithologic discontinuity. Thus the Permian Whin sill of northern England closely follows a Silurian-Carboniferous unconformity for over 100 km. Many large sills of diabase are roofed by laterally extensive slabs of sedimentary rock 2000 to 3000 m thick. The roof rocks are light compared with basaltic magmas: relative densities of 2.3 versus 2.65 are perhaps typical. The mechanics of intrusion evidently involve large-scale flotation of huge blocks of cover rock, buoyed up on denser magma spreading laterally beneath (cf. Bradley, 1965).

### IMPLICATIONS OF INTERNAL VERTICAL VARIATION

Laterally thick sills of diabase commonly maintain a uniform composition for many kilometers. Or where variation is significant, it tends to be abrupt, as in the Great Lake sill of Tasmania, where the upper granophyric layer 20 to 25 m thick is laterally restricted. Vertical variation by contrast is the general rule and may be very marked. Where it is completely gradational through 100 m or more—and this is commonly the case—it must almost certainly be due to some continuous process of magmatic differentiation. Sudden discontinuities or reversals of the vertical compositional trend, on the other hand, suggest multiple intrusion of related magmas or perhaps differentiation during the course of lateral flow.

Two assumptions are generally made in reconstructing the course of differentiation in thick sills:

1. The composition of the parent magma is that of chilled, fine-grained or porphyritic marginal rocks close to the upper and lower contracts. Crystalline phases at liquidus temperatures are those occurring in such rocks as phenocrysts. This assumption has been seriously questioned for massive stratified plutons, but it is probably valid for rapidly injected sills a few hundred meters thick.

2. Chemical variation in the vertical sense—upward from the base or downward from the roof—can be attributed to differentiation where it is broadly consistent with chemical trends of fractional crystallization experimentally established at low pressures: for example, progressively increasing (Na + K)/Ca and Fe/Mg in successively later fractions.

PATTERN OF VERTICAL MINERALOGICAL VARIATION Although every large diabase sheet is to some degree unique, throughout comparable vertical sections mineralogical assemblages in almost all vary with progressive differentiation in a remarkably consistent manner. These bodies in fact supply some of the clearest evidence of how tholeiitic magma differentiates under near-surface conditions. At least in the lower and middle sections of thick sills, the trend of differentiation can be read vertically upward.

The fine-grained texture of chilled border rocks shows that diabase magmas are largely liquid at the time of intrusion. The commonest combination of microphenocrysts is bronzite $En_{80}$ to $En_{85}$, calcic plagioclase, and diopsidic (magnesian) augite. Relatively rarely—for example, the Palisades sill of New Jersey and some sheets of the Karroo province—the place of bronzite is taken by magnesian olivine $Fo_{80}$ to $Fo_{85}$.

Olivine, when not confined to chilled borders and basal zones, becomes increasingly ferrous upward. For silica-saturated olivine diabases of the Karroo province, F. Walker and Poldervaart (1949, table 14) record $Fo_{70}$ to $Fo_{75}$ in central zones of thin sills and in the lower levels of sills 200 m or more thick; $Fo_{40}$ to $Fo_{55}$ in middle and upper zones of thick sills; and $Fo_0$ to $Fo_{50}$ only in granophyric rocks close to the roofs of some sheets (cf. also the uppermost ferrodiabases of the Palisades sill; K. R. Walker, 1969*b*, p. 46).

The crystallization trend of pyroxenes (Fig. 9-4) is almost universal:

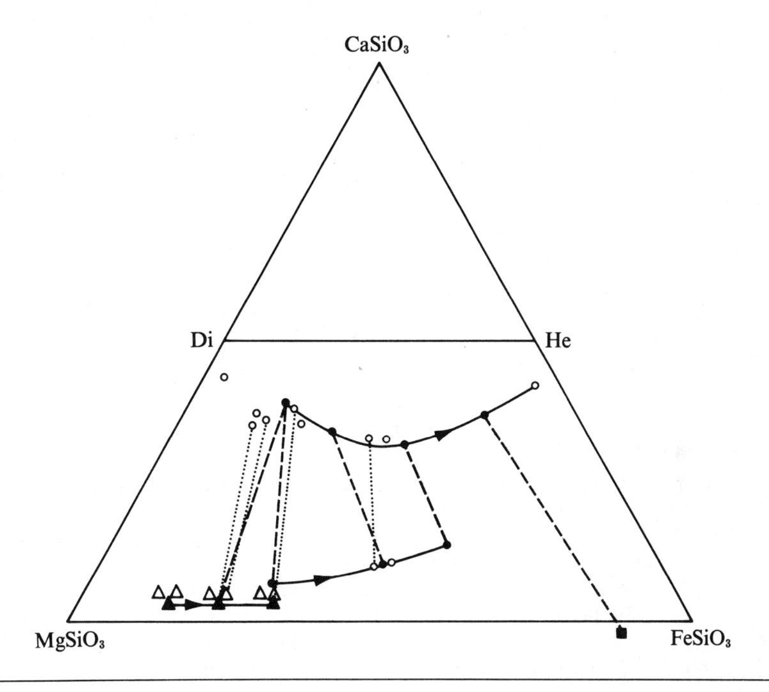


**Fig. 9-4** Molecular plot of composition of pyroxenes (circles, clinopyroxene; triangles, orthopyroxene) of differentiated diabase sheets. Broken (or dotted) ties connect coexisting pyroxenes and olivine (filled square). Filled circles and triangles, Jurassic Tasmanian diabase (differentiation trend shown by arrows); after McDougall (1962, p. 290). Open circles and triangles, Precambrian Kopinang diabase, British Guiana; after Hawkes (1966, p. 1143).

Fe/Mg increases steadily upward in all varieties. At first—in the lower levels—magnesian augite is associated with bronzite, which changes upward from $En_{80}$ to a sharply defined limiting value $En_{65}$ to $En_{70}$. Hypersthenes of slightly more ferrous compositions $En_{60}$ to $En_{70}$ are common at this stage, but only as products of inversion from primary pigeonite since they still retain the distinctive {001} exsolution lamellae and occasional relict {100} twin structure of the parent monoclinic phase. The next stage, typical of the middle levels of most sills, is marked by simultaneous crystallization of augite and pigeonite, both becoming progressively richer in iron. When pigeonite reaches $Wo_{10}En_{35}Fs_{55}$, it cuts out completely. Augite, changing from $Wo_{30}En_{30}Fs_{40}$ through ferroaugites almost to hedenbergite, now crystallizes alone. At the iron-rich extreme there may be two associated phases of nearly identical composition, one (pale green in color) perhaps formed by subsolidus inversion from ferrowollastonite.

Plagioclase occurs mostly in normally zoned crystals in the bytownite-labradorite range. Its bulk composition becomes increasingly sodic upward; but compositions less calcic than $An_{50}$ are reached only in extreme granophyric differentiates (andesine and even oligoclase). More striking is the appearance and steady increase in volume of micrographically intergrown alkali feldspar and

**Table 9-6** Modal compositions of pyroxene- and olivine-enriched diabases

| | 1 | 2 | 3 | 4 |
|---|---|---|---|---|
| Olivine | | | $Fo_{90}$ | $Fo_{77-69}$ |
| % | | | 52 | 23 |
| Orthopyroxene | $En_{82}$ | $En_{81}$–$En_{70}$ | Total pyroxene | $En_{78}$ |
| % | 5 | 29 | 23 | 23 |
| Pigeonite | $En_{58}$ | $En_{73}$–$En_{65}$ | | |
| % | ~23 | 15 | | |
| Augite | $Wo_{42}En_{43}Fs_{15}$ | $Wo_{45}En_{44}Fs_{11}$ | | $Wo_{38}En_{47}Fs_{15}$ |
| % | ~23 | 16 | | 16 |
| Plagioclase (zoned) | $An_{85-61}$ | $An_{81-69}$ | $An_{70}$ | $An_{64}$ |
| % | 40 | 38 | 23 | 30 |

*Explanation of column headings*

1 Diabase of lower 100-m zone, Great Lake sheet (600 m thick), Tasmania (McDougall, 1964*a*, pp. 115, 121; no. 5123–1295). Augite and pigeonite "in about equal amount"
2 Hypersthene diabase, lower zone 50 to 120 m above base, Basement sill (>210 m thick), Antarctica (Gunn, 1962, pp. 832, 835; no. 14910)
3 Picrite, basalt zone, Insizwa sheet (900 m thick), Karroo, South Africa (F. Walker and Poldervaart, 1949, table 15; no. 45)
4 Olivine diabase, 20 m above base, Palisades sill (300 m thick), New Jersey (K. R. Walker, 1969*b*, pp. 32, 92; no. W-824-60)

quartz (micropegmatite) or a chemically equivalent, poorly crystallized mesostasis. Accessory minerals that tend to become more conspicuous at advanced stages of differentiation are magnetite-ilmenite intergrowths, hornblende, and biotite. Rarely (for example, in granophyric diabases high in the Kopinang sheet of British Guiana) hornblende may become an essential phase—20 to 35 percent of the total composition (Hawkes, 1966, pp. 1137, 1144).

### DIFFERENTIATED ROCK SERIES

*Mineral assemblages* Modal compositions of representative rocks at various levels in differentiated sheets are given in Tables 9-6 and 9-7. Low down, though not necessarily at the very base, in most sheets are diabases significantly enriched in pyroxenes (Table 9-6; columns 1, 2) or less commonly in magnesian olivine (Table 9-6; columns 3, 4). Only rarely, as in rocks close to the floor of the Insizwa sheet (nearly 1000 m thick) in the Karroo province, are mafics sufficiently concentrated (> 70 percent) to warrant classification as peridotite or picrite (Table 9-6, column 3).

Most sheets consist mainly of oversaturated diabases with interstitial, micrographically intergrown quartz and alkali feldspars. These rocks contain at least two pyroxenes, the usual combination being augite and pigeonite. The differentiation trend is toward rocks increasingly rich in iron and in micropegmatite or equivalent poorly crystallized mesostasis (Table 9-7).

**Table 9-7** Modal compositions of differentiated Tasmanian diabases

| | 1 | 2 | 3 | 4 | 5 | 6 | 7 |
|---|---|---|---|---|---|---|---|
| Micropegmatite and mesostasis, % | 8 | 17 | 41 | 10 | $14\frac{1}{2}$ | 24 | 31 |
| Plagioclase (zoned) | $An_{85-61}$ | $An_{88-63}$ | $An_{60-51}$ | | $An_{84-49}$ | $An_{80-42}$ | $An_{70-47}$ |
| % | 40 | 47 | 39 | 60 | $59\frac{1}{2}$ | 52 | 49 |
| Clinopyroxene | $Wo_{42}En_{43}Fs_{15}$ $Wo_{8}En_{58}Fs_{34}$ | $Wo_{38}En_{40}Fs_{22}$ $Wo_{7}En_{50}Fs_{43}$ | $Wo_{31}En_{27}Fs_{42}$ $Wo_{14}En_{32}Fs_{54}$ | $Wo_{40}En_{45}Fs_{15}$ $Wo_{8}En_{63}Fs_{29}$ | $Wo_{34}En_{40}Fs_{26}$ $Wo_{11}En_{43}Fs_{46}$ | | $Wo_{31}En_{30}Fs_{39}$ $Wo_{13}En_{32}Fs_{56}$ |
| % (total) | 46 | 34 | 17 | 26 | 21 | 16 | 14 |
| Orthopyroxene | $En_{82}$ | | | $En_{76-67}$ | | | |
| % | 4 | | | 2 | | | |
| Hornblende, % | | | | 0.1 | 0.5 | 0.5 | 0.8 |
| Biotite, % | | | | 0.3 | | | |
| Iron oxides, % | 2 | 2 | 3 | 1 | 2 | $4\frac{1}{2}$ | 3 |
| Mafic index | 41 | 61 | 84 | 53 | 74 | 81 | 87 |
| Felsic index | 13 | 20 | 39 | 16 | 23 | 31 | 35 |

*Explanation of column headings*

1–3 Great Lake diabase sheet (530 m thick), Tasmania (McDougall, 1964*a*)
- 1 Diabase ($SiO_2$ 52.5%), 130 m above base of sheet (5123-1295)
- 2 Diabase ($SiO_2$ 53.2%), 320 m above base of sheet (5123-700)
- 3 Diabase ($SiO_2$ 55.4%), 500 m above base of sheet (5123-100)

4–7 Red Hill dike (1.5 km wide, 400 m thick), Tasmania (McDougall, 1962, p. 286)
- 4 Diabase ($SiO_2$ 53%), 13 m from vertical contact (M-212)
- 5 Diabase ($SiO_2$ 53.9%), center of dike (M-210)
- 6 Diabase ($SiO_2$ 53.9%), center of dike 120 m above 5 (M-222)
- 7 Diabase ($SiO_2$ 55.1%), transitional to granophyre, 135 m above 5 (M-395)

At the end of the differentiated line are granophyric diabases and granophyres with more than 50 percent micropegmatite. The remaining constituents are ferroaugite, sodic plagioclase, and magnetite-ilmenite intergrowth. Some rocks contain small quantities of iron-rich olivine, others hornblende and/or biotite. Granophyres tend to occur close to the roof either as layers 20 to 70 m thick or as small lensoid or veinlike segregations. Some of the larger granophyre bodies that have received closest attention—Dillsburg, Pennsylvania (Hotz, 1953); Red Hill, Tasmania (McDougall, 1962)—must certainly be pure differentiates. They lie within and grade mineralogically into associated predominant diabases (cf. Table 9-7, analysis 7). Fusion of the sedimentary cover can be excluded as a possible factor in their origin; at knife-sharp contacts with chilled diabase, the only contact effect is hornfelsing, limited to a zone a meter or two thick (Hotz, 1953, p. 678; McDougall, 1962, p. 284).

*Chemistry* The most obvious aspects of chemical variation of major elements are shown on conventional diagrams, Figs. 9-5 and 9-6 (cf. also Table 9-8). Mafic and felsic indices increase along the line of differentiation, the first particularly in the diabase range, the second especially as diabases merge into

**Fig. 9-5** Plot of felsic index $100(Na_2O + K_2O)/(Na_2O + K_2O + CaO)$ against mafic index $100(FeO + Fe_2O_3)/(FeO + Fe_2O_3 + MgO)$ for rocks of three differentiated Jurassic diabase sheets, Tasmania (stippled field); after McDougall (1964*a*, p. 119). The arrow shows the trend of differentiation. Solid circles for individual analyses of rocks from Palisades sill, New Jersey; data from K. R. Walker (1969*b*, table 8). Open circles for individual analyses, three Jurassic diabase sills, Antarctica; data from B. M. Gunn (1962).

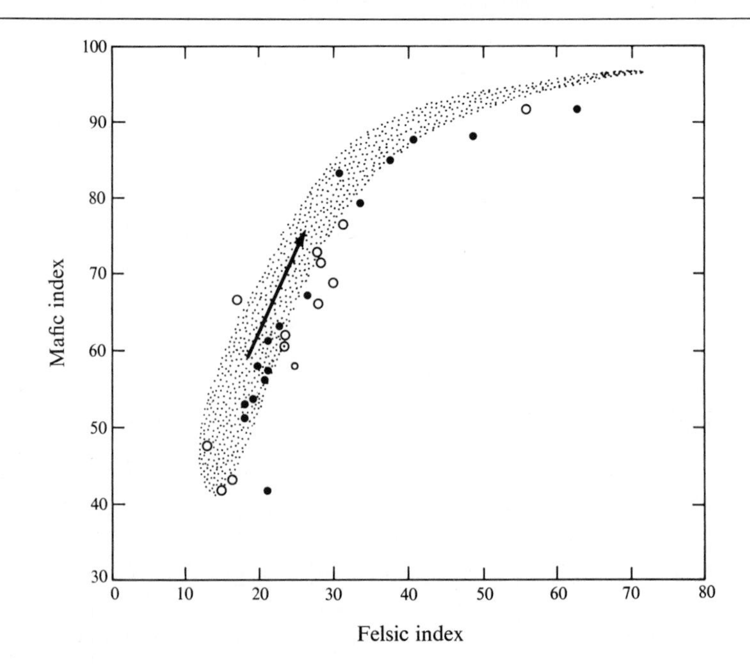

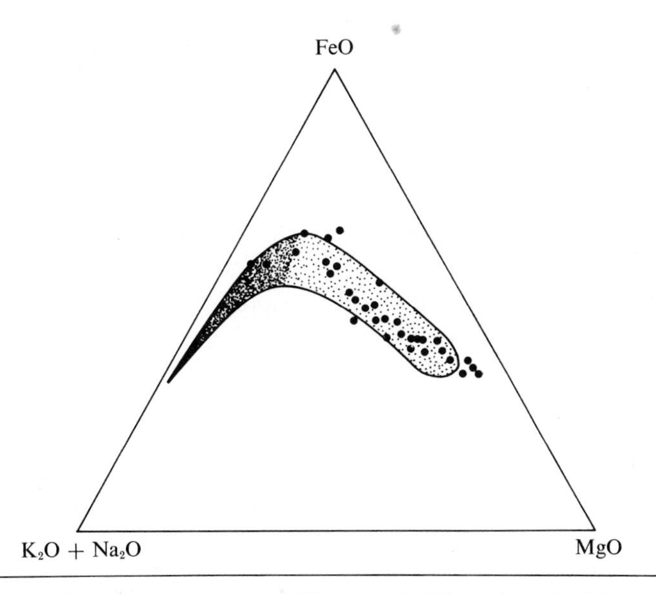


**Fig. 9-6** Compositional range in rocks of differentiated diabase sheets (weight proportions of oxides). Stippled field encloses diabases and granophyres (dense stipple) of three Tasmanian sheets; after McDougall (1964*a*, p. 124). Points are individual analyses, Palisades sill, New Jersey; after K. R. Walker (1969*b*, p. 81). FeO includes only $Fe^{++}$.

granophyres. Silica, which rises to 55 or 56 percent in the more siliceous diabases, typically reaches between 60 and 68 percent in granophyres. In most sills, diabases (Table 9-8, analyses 2–8) maintain normal $Al_2O_3$ values (12 to 15.5 percent) with a slight tendency to increase in the more siliceous varieties. "High-alumina" values are almost unknown: 17 to 19 percent $Al_2O_3$ for a group of quartz diabases in the Red Hill intrusion of Tasmania is a notable exception as well as being locally anomalous. $Na_2O/K_2O$ varies from one province to another, and it falls from initial values of 2.5 to 3 in most cases toward 1 at the granophyre end of the line. Tasmanian diabases (Table 9-8, analyses 5–7) are consistently more potassic than in most provinces: $Na_2O/K_2O$ is 1.2 to 1.6 in the diabases and drops slightly below 1 in associated granophyres (Table 9-8, analysis 10).

Trace-element trends conform to a general pattern now familiar for differentiation series stemming from tholeiitic-basalt parentage. Data for the Palisades sill are given in Table 9-9. Ba builds up steadily and is ultimately concentrated in granophyric diabases and ferrodiabases—mainly in alkali feldspar and biotite. Cr is removed early, mainly in clinopyroxene (no chrome spinel has been recorded). Early removal of Ni, and concentration in the olivine-diabase layer, is effected by magnesian olivine and orthopyroxene. High abundance of Co in the olivine diabase is due solely to preferential accommodation by magnesian olivine. V rises to a peak in ferrodiabases relatively rich in

**Table 9-8** Chemical compositions (oxides, wt %) and CIPW norms of rocks in differentiated diabase sheets

| | 1 | 2 | 3 | 4 | 5 | 6 | 7 | 8 | 9 | 10 |
|---|---|---|---|---|---|---|---|---|---|---|
| $SiO_2$ | 47.41 | 51.98 | 52.67 | 58.59 | 53.32 | 53.18 | 53.20 | 55.08 | 60.37 | 67.62 |
| $TiO_2$ | 0.89 | 1.21 | 2.76 | 1.56 | 0.70 | 0.65 | 0.72 | 1.52 | 1.13 | 0.84 |
| $Al_2O_3$ | 8.66 | 14.48 | 11.94 | 11.26 | 14.18 | 15.37 | 15.59 | 15.36 | 13.77 | 12.31 |
| $Fe_2O_3$ | 2.81 | 1.37 | 3.90 | 3.00 | 0.97 | 0.76 | 0.83 | 3.43 | 2.00 | 5.66 |
| FeO | 11.15 | 8.92 | 11.08 | 10.80 | 8.54 | 8.33 | 8.54 | 9.06 | 8.20 | 3.04 |
| MnO | 0.20 | 0.16 | 0.19 | 0.19 | 0.18 | 0.15 | 0.16 | 0.16 | 0.17 | 0.11 |
| MgO | 19.29 | 7.59 | 3.94 | 1.35 | 7.23 | 6.71 | 6.11 | 1.82 | 0.83 | 0.32 |
| CaO | 6.76 | 10.33 | 8.06 | 3.66 | 11.22 | 11.04 | 11.03 | 7.64 | 6.14 | 2.52 |
| $Na_2O$ | 1.35 | 2.04 | 2.78 | 3.68 | 1.38 | 1.65 | 1.62 | 2.29 | 2.82 | 2.92 |
| $K_2O$ | 0.43 | 0.84 | 1.29 | 2.52 | 0.87 | 1.03 | 1.14 | 1.84 | 2.42 | 3.38 |
| $P_2O_5$ | 0.10 | 0.14 | 0.31 | 0.85 | 0.20 | 0.08 | 0.15 | 0.19 | 0.33 | 0.24 |
| $H_2O^+$ | 1.45 | 0.88 | 1.02 | 1.52 | 1.00 | 0.67 | 1.09 | 1.42 | 1.26 | 0.68 |
| $H_2O^-$ | 0.11 | 0.16 | 0.20 | 0.43 | 0.64 | 0.45 | 0.27 | 0.50 | 0.52 | 0.62 |
| Total | 100.61 | 100.10 | 100.14 | 99.41 | 100.43 | 100.07 | 100.45 | 100.31 | 99.96 | 100.26 |
| *Q* | | 2.80 | 8.09 | 14.33 | 6.58 | 4.80 | 5.39 | 13.20 | 18.18 | 32.46 |
| *or* | 2.39 | 4.73 | 7.62 | 15.12 | 5.14 | 6.09 | 6.74 | 10.56 | 14.46 | 20.02 |
| *ab* | 11.53 | 16.92 | 23.52 | 31.22 | 11.68 | 13.96 | 13.71 | 19.39 | 23.58 | 24.63 |
| *an* | 16.21 | 28.23 | 16.29 | 6.61 | 29.93 | 31.49 | 31.90 | 26.13 | 17.79 | 10.56 |
| *di* | 13.14 | 18.19 | 18.10 | 5.15 | 20.05 | 18.71 | 18.02 | 9.47 | 9.44 | 0.86 |
| *hy* | 25.91 | 23.60 | 13.69 | 15.78 | 22.21 | 21.37 | 20.42 | 11.28 | 8.92 | 0.40 |
| *ol* | 23.77 | | | | | | | | | |
| *mt* | 4.18 | 2.03 | 5.65 | 4.41 | 1.41 | 1.10 | 1.20 | 4.87 | 3.02 | 7.98 |
| *il* | 1.65 | 2.28 | 5.24 | 2.98 | 1.33 | 1.23 | 1.37 | 2.89 | 2.13 | 1.52 |
| *ap* | 0.23 | 0.32 | 0.73 | 2.02 | 0.47 | 0.19 | 0.35 | 0.44 | 0.76 | 0.55 |
| Total | 99.01 | 99.10 | 98.93 | 97.62 | 98.80 | 98.94 | 99.10 | 98.23 | 98.28 | 98.98 |

*Explanation of column headings*

1 Olivine diabase, 20 m above base, Palisades sill (300 m thick), New Jersey; cf. Table 9-6, analysis 4 (K. R. Walker, 1969*a*, table 8, analysis W-824-60)
2 Chilled diabase, base of Palisades sill (300 m thick), New Jersey (K. R. Walker, 1969*a*, table 8, analysis W-899 LC-60)
3 Hypersthene diabase, 210 m above base of Palisades sill, New Jersey (K. R. Walker, 1969*a*, table 8, analysis W-J-60)
4 Granophyric diabase 240 m above base of Palisades sill, New Jersey (K. R. Walker, 1969*a*, table 8, analysis W-F-60)
5 Chilled diabase, upper contact Great Lake sheet, Tasmania (McDougall, 1964*a*, p. 123, analysis 21)
6 Chilled diabase, average of 13 analyses, Tasmanian Jurassic diabases (McDougall, 1962, p. 294)
7 Diabase near middle of Great Lake sheet (530 m thick), Tasmania; cf. Table 9-7, analysis 2 (McDougall, 1964*a*, analysis 5123-700)
8 Diabase transitional to granophyre, Red Hill dike, Tasmania; cf. Table 9-7, analysis 7 (McDougall, 1962, analysis M-395)
9 Fayalite granophyre, Red Hill dike, Tasmania (McDougall, 1962, analysis M-12)
10 Granophyre, Red Hill dike, Tasmania (McDougall, 1962, analysis M-8)

**Table 9-9** Trace-element abundances (ppm) in diabase differentiation series, Palisades sill, New Jersey*

| | 1 | 2 | 3 | 4 | 5 | 6 |
|---|---|---|---|---|---|---|
| Ba | 195 | 150 | 150 | 210 | 310 | 560 |
| Co | 53 | 140 | 56 | 42 | 60 | 27 |
| Cr | 315 | 290 | 715 | 16 | 7 | 7 |
| Ni | 95 | 500 | 135 | 40 | 30 | 7 |
| Pb | 11 | 18 | 13 | 25 | 17 | 16 |
| Sr | 175 | 110 | 160 | 250 | 185 | 185 |
| V | 235 | 155 | 225 | 235 | 490 | 13 |
| Zr | 120 | 90 | 120 | 115 | 180 | 330 |

* Data from K. R. Walker (1969*a*, p. 94, table 13).

*Explanation of column headings*

1 Chilled diabase at base of sheet (W-899 LC-60)
2 Olivine-enriched diabase (W-824-60); height 20 m
3 Bronzite diabase (W-804-60); height 28 m
4 Pigeonite diabase (W-N-60); height 170 m
5 Ferrohypersthene diabase (W-J-60); height 210 m
6 Granophyric diabase (W-F-60); height 240 m

magnetite-ilmenite; it then falls sharply in the final granophyric diabase. There is the usual tendency for Zr to build up with K toward the end of the line.

Heier et al. (1965) found that throughout the complete lithologic range—diabase to granophyre—represented by the Great Lake sheet and the Red Hill dike of Tasmania, Th and U maintain a remarkably constant linear relation to abundance of K. Over a range of K from 0.5 to 2.5 percent, $U/K = 1.25 \times 10^{-4}$ and $Th/K = 5 \times 10^{-4}$. The high initial value of $Sr^{87}/Sr^{86}$ ($0.7115 \pm 0.0007$) likewise is constant within limits of experimental error, and it is the same in granophyres as in the parental magma (chilled diabase). Also consistent throughout is K/Rb ($250 \pm 17$), a parameter which in many other rock series decreases with increasing silica. A similar situation seems to hold for the Jurassic diabases of Antarctica (Compston et al., 1968). Here, however, there are some granophyric pegmatoids with anomalously high $Sr^{87}/Sr^{86}$, perhaps attributable to local contamination by associated Precambrian granitic rocks.

These findings, based as they are on essentially uncontaminated differentiated lineages, lend support to the widely accepted thesis that a rock series within which U/K, Th/K, and initial $Sr^{87}/Sr^{86}$ maintain consistent values is likely to have differentiated from some parental magma having these same chemical traits.

## MECHANICS OF DIFFERENTIATION

*Cooling and solidification model* Clearly one factor that limits the potentiality for differentiation within an intrusive sheet of basaltic magma is the time interval between intrusion and complete solidification. The most important controlling parameter in this connection is the thickness $D$, since the time factor in solidification and cooling by outward conduction is proportional to $D^2$. Jaeger (for example, 1957*a*, 1957*b*; Jaeger and Joplin, 1956) has approached the problem by setting up simplified cooling models based upon available thermal properties of basaltic magma, diabase, and sedimentary rocks such as sandstone and shale. He assumes instantaneous intrusion of a magma that had just reached liquidus temperature. That intrusion of most sills indeed is rapid is suggested by the limited extent of contact metamorphism. And the almost invariable presence of olivine or bronzite microphenocrysts in chilled border rocks is consistent with assumed absence of superheat.

In the simplest possible model, there is no internal motion within the magma body after intrusion. As cooling proceeds, the magma solidifies inward from floor and roof. It is shown that for cover thicknesses of a kilometer or two the rate of cooling is the same at both surfaces. Differentiation effects, due solely to separation of crystalline phases in order of decreasing solidus temperature at the advancing solid-liquid interfaces, must therefore be symmetrical about the median horizontal plane of the sheet.[1] And it is here that the extreme products of differentiation should become segregated.

Order-of-magnitude values calculated for cooling and crystallization of a thick sheet of diabase magma over the interval 1100 to 800°C are as follows:

1. In the early stages before temperatures at the center have fallen perceptibly, the thickness of the completely crystalline border rock, $X$ (in meters), is related to time since intrusion, $t$ (in years), by an expression that simplifies approximately to

$$X \approx 2\sqrt{t}$$

   Thus over the progressive intervals 1, 10, and 100 years after intrusion, the solid border zone thickens to about 2, 6, and 20 m.

2. Time for complete solidification of a sheet, $t_s$ (in years), is related to the thickness $D$ (in meters) by

$$t_s \approx 0.014D^2$$

   Thus a sheet 300 m thick, such as the Palisades sill, would require something like 1300 years to solidify completely. A 100-m sill would take less than 150 years.

*Crystal sinking* It has been almost 70 years since L. V. Lewis suggested that sinking of olivine crystals is responsible for the olivine-enriched diabase layer

[1] Contrast with preferential downward solidification from the roof during asymmetric cooling of Hawaiian lava lakes in the absence of cover (cf. J. G. Moore and Evans, 1967, pp. 210, 211).

20 m above the base of the Palisades sill. Bowen (1928, pp. 71-74) elaborated the idea in the light of experimental data for simple forsterite-silica systems at atmospheric pressure (cf. Verhoogen et al., 1970, pp. 272, 273). He saw in the process a means by which a silica-saturated tholeiitic magma might differentiate into complementary fractions: an undersaturated cumulate enriched in magnesian olivine and pyroxene, and oversaturated residual liquids represented at higher levels by quartz diabase and granophyre. For many years this model found general acceptance; and it has been extended to cover downward concentration of pyroxene in sills completely lacking olivine (cf. Edwards, 1942).

Today there is general agreement that crystal sinking can be an effective differentiation mechanism in sheets more than 300 to 400 m thick. The high MgO content and the mineralogy of low-level rocks in such bodies indeed render this conclusion inevitable (Fig. 9-7; Table 9-6). But this implies one or both of

**Fig. 9-7** Variation of MgO (wt %) with height above base of sill (expressed as ratio height/thickness) in two diabase sheets. Data from McDougall (1964*a*, p. 124) and K. R. Walker (1969*b*, fig. 8). Section in Great Lakes sheet, Tasmania, is from drill hole 5123 extrapolated at either end (dashed lines) from nearby holes.

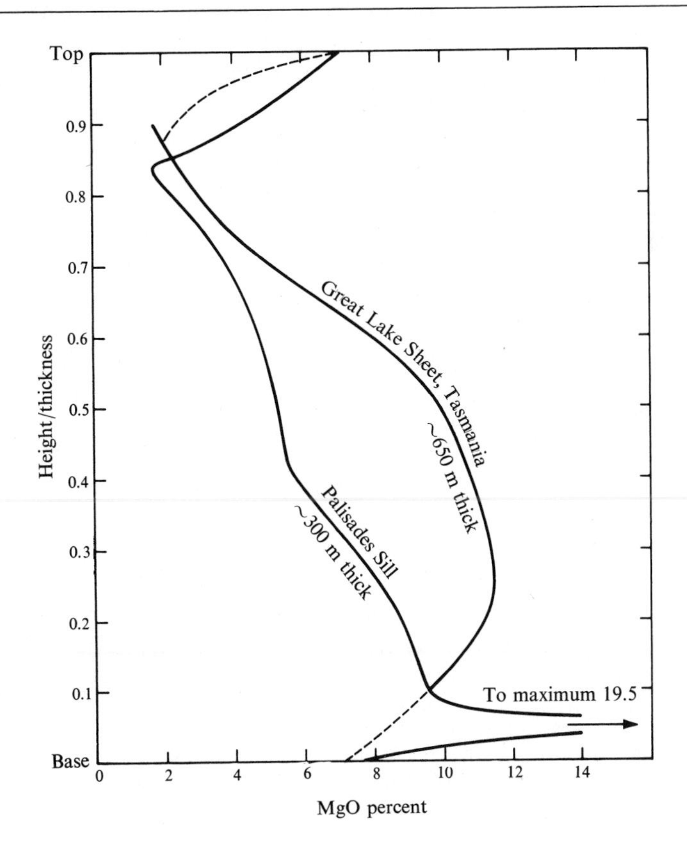

two special conditions if crystals are to survive the slow descent without remelting (cf. Jaeger, 1957*a*; Jaeger and Joplin, 1956): Crystals fall either as relatively large aggregates (clots) or as intratelluric phenocrysts in magma already at liquidus temperature at the time of injection. Jaeger (1957*a*, p. 86) has shown how different injection regimes could greatly affect both the concentration attained by sinking phenocrysts and the thickness of resulting cumulate rocks. For a given sill thickness (say, 400 m) and falling velocity (10 m per year), rapid intrusion into cold rocks favors high concentration of phenocrysts in a thin basal zone; this concentration is enhanced where the main intrusive pulse is along the hot floor provided by a preceding, recently solidified, basal sheet—as seems to have been the case in the Palisades sill (K. R. Walker, 1969*a*).

*Role of convection* H. H. Hess (1960, p. 188) found that in many sills, especially those less than 300 m thick, the trend of differentiation downward from the roof parallels that in the upward sense from the floor. Clearly some process other than crystal sinking has been at work—something perhaps related to Jaeger's simplified model of symmetric cooling from the floor and the roof. But though the two trends are identical, the respective gradients, as shown in a plot of some differentiation index against vertical distance, are very different (Figs. 9-7 and 9-8). Vertical asymmetry in the differentiation profile suggests the influence of gravity. If crystal sinking is minimized, there is still the possibility that some other gravity-controlled factor is effective, for example, the vaguely defined process of "gas streaming" along a pressure gradient or more probably convection.

H. H. Hess (1960) was a strong advocate of convection in cooling sheets of basaltic magma. He favored a model whereby fresh liquid is continuously supplied to solid-liquid interfaces, replacing magma depleted in ions such as Mg and Ca that preferentially enter crystalline phases on the liquidus. In this connection we note that rhythmic alternation of light and dark bands so commonly seen in major basic plutons, although not unknown (Gunn, 1963; Hawkes, 1966, p. 1151), is rare in diabase sheets. So where convection operates, the flow regime[1] of floor currents in such sheets must differ notably from that responsible for crystal sorting on the growing floors of stratified plutons.

TREND REVERSALS AND MULTIPLE INTRUSION In some sills the upward trend may be interrupted or sharply reversed some distance above the floor. This break has been explained by injection of magma in two pulses, separated long enough in time for the earlier (lower) draft to solidify sufficiently to retain much of its identity. K. R. Walker (1969*a*) thus explains the otherwise anomalous

[1] The flow regime is sensitive to the dimensionless Reynolds number $v\rho l/\mu$, where $v$ is velocity, $\rho$ density, $\mu$ the viscosity of the flowing medium, and $l$ a length dimension such as channel depth (Verhoogen et al., 1970, p. 218).

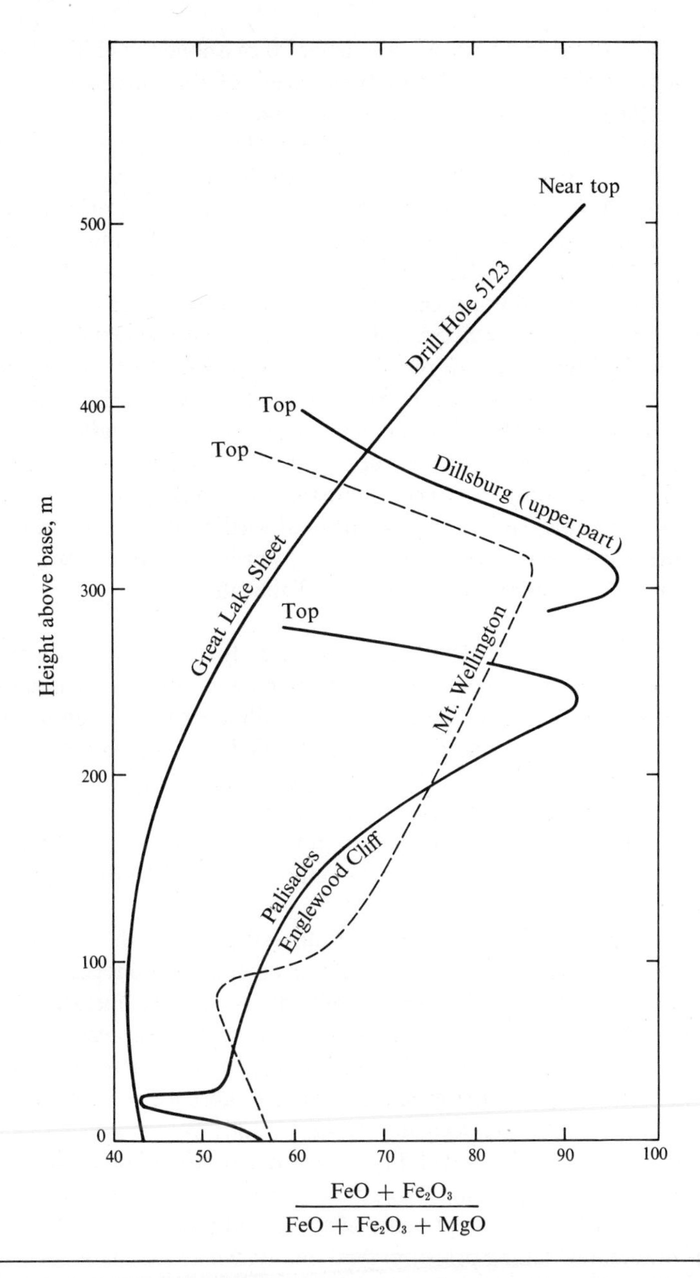


**Fig. 9-8** Plot of mafic index against height above base for four diabase sheets. Data sources: Great Lake sheet, Tasmania, McDougall (1962, pp. 120, 121); Mt. Wellington, Tasmania, Edwards (1942, pp. 467, 470); Dillsburg, Pennsylvania, Hotz (1963, p. 690); Palisades, New Jersey, K. R. Walker (1969*b*, pp. 76, 77).

location of an olivine-enriched diabase layer 20 m above the chilled base of the classic Palisades sill. Among the diabase sheets of the Ferrar dolerite group, Antarctica, there is one—Lake Vanda sill, 340 m thick—that shows a pronounced trend reversal about 100 m above the floor. The lowest layer is pigeonite-augite diabase mostly with less than 50 percent pyroxene and mafic index 50 to 55. It is followed upward by a zone, 70 m thick, of bronzite-augite diabase with a higher pyroxene content (60 percent) and much lower mafic index (38). In this zone K, Ba, and Rb are uniformly low, and Ni and Cr, in sympathy with Mg, maintain maximum abundances. In other differentiated sills all such characteristics are typically displayed in basal layers that crystallized at maximum liquidus temperatures. Gunn (1966) proposes a two-stage intrusive model to account for reversal of the normal trend at 100 m: First came an early pulse of *Q*-normative tholeiitic magma from which crystallized the basal pigeonite-augite diabase. Along its upper, partially solidified surface a more mafic liquid-crystal mush was now injected. This mush was heavily charged with bronzite crystals, which by gravitational settling became concentrated in the bronzite zone; this zone now passes upward into more felsic pigeonite-augite diabase that makes up the top half of the sill.

GENERAL CONCLUSIONS Diabase sills are formed by horizontal injection, under substantial cover, of tholeiitic magma already at liquidus temperature. Relative motion of liquid and crystalline phases, essential to large-scale fractionation, is thenceforward controlled by temperature gradients falling off toward both floor and roof. The system, as density differences develop on all scales with cooling, is inherently unstable. Gravity thus exercises a prime influence on the kinematics of differentiation, and this influence is ultimately expressed in the degree of vertical asymmetry shown by the differentiated rock body (cf. Figs. 9-8 and 9-9).

An obvious element in the mechanism, especially in thick, highly asymmetric sheets, is sinking of heavy crystals or crystalline aggregates (pyroxenes or less commonly olivine). But some other influence must operate, at least in the upper levels, in sills where the differentiation indices rise steadily downward for 100 m or so below the roof. Convection may be one factor here, although some writers exclude it on the grounds of general absence of rhythmic layering (cf. H. H. Hess, 1960; McDougall, 1964*a*, p. 130; Wager and Brown, 1967, p. 531). Then there is the possibility—although it still remains to be argued clearly and quantitatively—that "volatile constituents," whether in the form of gas bubbles or merely as components of the liquid phase, concentrate upward along the gradient of falling pressure. And finally, where the main concentration of mafic minerals is well above the floor, there is a strong possibility that magma, already partially differentiated in some external reservoir, was injected at its present site in more than one pulse.

Even given initial magmas of similar but not identical composition close to liquidus temperatures, such factors as viscosity, pressure, rate of emplacement, and sill dimensions are subject to variation sufficient to make every instance to some degree unique. Rather we might wonder that the differentiation profiles of so many sills conform so closely to a general pattern.

Finally, there is a difficult question regarding inferred liquid lines of descent. A variation diagram such as Fig. 9-5 or 9-6 shows some aspect of chemical variation among rock samples each of which is a differentiate representing some stage of fractional crystallization. But the only compositions that can be referred confidently to some stage in the purely liquid line of descent (as the term was used by Bowen, 1928) are the chilled border rocks and the granophyric segregations. The initial and final liquid fractions are known; but in between are rock compositions that, because of motion of crystals relative to liquid, necessarily depart to some degree from that of any liquid intermediate fraction.

### INTRAPROVINCIAL VARIETY IN MAGMA TYPE

Inadequate though regional sampling may be, it has shown that parental magmas represented by chilled border diabases in extensive provinces display a strong tendency toward chemical uniformity. All such samples from Tasmanian diabase sheets have high $SiO_2$ (and normative $Q$) and $K_2O$; low K/Rb ($\sim$ 200) and U/K ($1.25 \times 10^{-4}$); and exceptionally high initial $Sr^{87}/Sr^{86}$ (0.7115). Similar but less extreme characteristics are shown by the Ferrar dolerites of Antarctica (Tables 9-10 and 9-11, analyses 1–3) throughout a belt 4000 km long.

Yet comparison of individual sills within a province may bring out discernible minor chemical and mineralogical variation from one to another. On the basis of such criteria as $SiO_2$ and $TiO_2$ content, mafic index, and the nature of microphenocryst phases, some writers have even proposed the existence of separate though perhaps related intraprovincial "magma types." Three such types, proposed by Gunn (1966) in the Ferrar dolerite province, are illustrated in Tables 9-10 and 9-11 (analyses 1–3). It is doubtful, however, whether such limited variety within so large a province has any genetic significance other than to demonstrate the independent origin of each major draft of magma. Instead variation more probably expresses the randomness inherent in repeated localized generation and intrusion of magma in the course of a major geologic event that affected much of the Antarctic continent during a short span of mid-Jurassic time.

## Diabase Dike Swarms

There are other extensive tholeiitic provinces within which swarms of regularly aligned, nearly vertical, massive diabasic dikes bear witness to the tensional character of the tectonic environment of magmatism. Such are the northwest-trending Tertiary dikes of northern England and western Scotland, radiating

**Table 9-10** Chemical compositions (oxides, wt %) and CIPW norms of diabases, Antarctica and eastern North America

| | 1 | 2 | 3 | 4 | 5 | 6 | 7 |
|---|---|---|---|---|---|---|---|
| $SiO_2$ | 52.68 | 55.52 | 50.40 | 47.90 | 51.66 | 52.10 | 52.69 |
| $TiO_2$ | 0.51 | 0.87 | 0.44 | 0.59 | 0.76 | 1.12 | 1.14 |
| $Al_2O_3$ | 13.34 | 15.49 | 15.51 | 15.26 | 14.95 | 14.22 | 14.21 |
| $Fe_2O_3$ | 1.81 | 2.20 | 0.99 | 12.10† | 11.77† | 11.65† | 13.87† |
| FeO | 7.36 | 8.36 | 7.83 | | | | |
| MnO | 0.15 | 0.17 | 0.17 | 0.18 | 0.20 | 0.19 | 0.22 |
| MgO | 10.00 | 4.31 | 10.60 | 10.52 | 7.44 | 7.41 | 5.53 |
| CaO | 12.40 | 8.62 | 10.87 | 10.75 | 10.80 | 10.66 | 9.86 |
| $Na_2O$ | 1.17 | 2.24 | 1.42 | 2.00 | 2.23 | 2.12 | 2.51 |
| $K_2O$ | 0.40 | 1.07 | 0.37 | 0.29 | 0.48 | 0.66 | 0.64 |
| $P_2O_5$ | 0.02 | 0.09 | 0.08 | | | | |
| $H_2O^+$ | 0.28 | 0.38 | 1.21 | | | | |
| $H_2O^-$ | 0.33 | 0.95 | 0.34 | | | | |
| Total | 100.45 | 100.27 | 100.23 | 99.59 | 100.29 | 100.13 | 100.67 |
| *Q* | 4.49 | 10.91 | | | 3.38 | 4.73 | 6.23 |
| *or* | 2.34 | 6.35 | 2.19 | 1.71 | 2.84 | 3.90 | 3.78 |
| *ab* | 9.91 | 18.93 | 12.02 | 16.92 | 18.87 | 17.94 | 21.24 |
| *an* | 29.96 | 29.04 | 34.85 | 31.81 | 29.37 | 27.34 | 25.62 |
| *di* | 25.42 | 10.99 | 15.02 | 17.42 | 19.64 | 20.68 | 19.14 |
| *hy* | 24.08 | 17.68 | 31.82 | 15.21 | 18.81 | 17.53 | 15.49 |
| *ol* | | | 0.33 | 9.39 | | | |
| *mt* | 2.62 | 3.20 | 1.44 | 5.25 | 5.12 | 5.06 | 6.03 |
| *il* | 0.97 | 1.65 | 0.84 | 1.12 | 1.44 | 2.13 | 2.17 |
| *ap* | 0.03 | 0.19 | 0.19 | | | | |
| Total | 99.82 | 98.94 | 98.70 | 98.83 | 99.47 | 99.31 | 99.70 |

† Total Fe reported as $Fe_2O_3$*. In all cases $Fe_2O_3$ and FeO for norm were calculated assuming $Fe_2O_3/FeO = 0.3$.

*Explanation of column headings*

1–3 Chilled or marginal diabases, Jurassic Ferrar dolerites, Antarctica (Gunn, 1962, 1966)

1 Hypersthene-tholeiite type, Basement sill, 25 m from contact (Gunn, 1962, p. 837, analysis 14908); microphenocrysts include hypersthene

2 Pigeonite type, lower contact, Peneplain sill (Gunn, 1962, p. 827, analysis 23093); the Ca-poor pyroxene present as microphenocrysts is pigeonite

3 Olivine-tholeiite type, lower contact, Painted Cliff sill (Gunn, 1966, p. 908, analysis 26903); slightly undersaturated in $SiO_2$

4–7 Average compositions of diabase types, dikes of eastern North America (Weigand and Ragland, 1970, p. 200)

4 *Ol*-normative type

5 Low-$TiO_2$ *Q*-normative type

6 High-$TiO_2$ *Q*-normative type

7 High-$Fe_2O_3$* *Q*-normative type

**Table 9-11** Trace-element abundances (ppm) and abundance ratios of diabases, Antarctica and eastern North America

| | 1 | 2 | 3 | 4 | 5 | 6 | 7 |
|---|---|---|---|---|---|---|---|
| Ba | 232 | 430 | 157 | | | | |
| Co | 60 | | 62 | 65 | 53 | 49 | 52 |
| Cr | 142 | 64 | 352 | 766 | 218 | 217 | 94 |
| Ni | 85 | 53 | 249 | 308 | 48 | 81 | 34 |
| Rb | 30 | 56 | 12 | 8 | 15 | 21 | 22 |
| Sr | 126 | 136 | 100 | 115 | 127 | 186 | 178 |
| Th | 4.20* | 5.4 | 1.56 | | | | |
| U | 1.70* | 1.6 | 0.41 | | | | |
| V | 175 | | 126 | | | | |
| Zr | 83 | 147 | 53 | 50 | 60 | 92 | 94 |
| K/Rb | 208 | 196 | 267 | 300 | 250 | 250 | 230 |
| Th/K $\times 10^{-4}$ | 6.8 | 4.9 | 4.9 | | | | |
| U/K $\times 10^{-4}$ | 2.7 | 1.4 | 1.3 | | | | |
| $Sr^{87}/Sr^{86}$ (initial) | 0.7122 | 0.7125 | 0.7113 | | (0.7045)† | | |

* Thorium and uranium values determined on a similar rock from the chilled margin of the Basement sill (4260) are as follows: Th, 3.8; U, 1.6; Th/K, $7.5 \times 10^{-4}$; U/K, $3.2 \times 10^{-4}$.
† Values for Palisades sill (Gast, 1967, p. 343).

*Explanation of column headings*

1–3 Chilled marginal diabases, Antarctica (Gunn, 1966, p. 907; Compston et al., 1968, pp. 133, 141). Specimen numbers as in cited references are given in parentheses.
1 Hypersthene-tholeiite type, Lake Vanda sill (4012)
2 Pigeonitic type, Peneplain sill (4266)
3 Olivine-tholeiite type, Painted Cliff sill (26903)
4–7 Average diabase types, dikes of eastern North America (Weigand and Ragland, 1970, p. 200). Numbered as for identical average analyses (major elements), Table 9-10.

from local centers of more complex contemporary activity in the Hebridean volcanic province (Fig. 9-9*a*). Regionally exposed dike swarms, it is thought, give some idea of the feeder systems that probably underlie flood-basalt accumulations of strictly volcanic provinces.

## APPALACHIAN PIEDMONT PROVINCE, EASTERN NORTH AMERICA

The Appalachian Piedmont belt of Paleozoic and older metamorphic rocks extends for 2000 km northeast to southwest from Nova Scotia to Alabama, about 100 to 200 km inland from the Atlantic margin of the North American continent. Downfaulted within it are a number of elongate basins filled with Triassic continental sediments. Within these are the well-known diabase sills of Pennsylvania, New Jersey, and Connecticut and related surface flows of basalt.

But the main manifestation of late Triassic igneous activity is seen in swarms of tholeiitic dikes, some within but most far beyond the confines of the sediment-filled troughs. Individual dikes may be as much as 100 km long; a few are 100 to 300 m wide, most less than 30 m. Some swarms extend as far as 225 km in a northwest-southeast or north-south direction. Radiometric dates for dikes, sills, and flows cluster around 200 m.y., with an earlier less pronounced cluster at 225 to 230 m.y.

In so extensive a province it is not surprising to find some degree of chemical variation among representative rock types. Weigand and Ragland (1970) propose four types of parental magma (Tables 9-10 and 9-11, analyses 4–7) whose geographic distribution conforms to regular pattern:

1. Throughout the northern third of the province, from Maryland to Nova Scotia, the diabases are *Q*-normative virtually without exception. Here, as throughout the remainder of the province, observed variation in $TiO_2$ and in total iron may possibly be significant (cf. Table 9-10, analyses 5–7): $TiO_2$ values in one group of analyses cluster round 0.76 and in another are close to 1.15; and all analyses with a high mafic index (65 to 75) fall in the second group. Diabase sheets that slightly precede the dikes in time show exactly the same chemical characteristics—with high $TiO_2$ in the Palisades sill, low $TiO_2$ in the Dillsburg sheet. Initial $Sr^{87}/Sr^{86}$ (Gast,* 1967, p. 343) is 0.704 to 0.706 for sill diabases and 0.7075 for Triassic basalts of Connecticut.

2. Over other large regions, for example, in the state of South Carolina, all analyzed diabases prove to be *ol*-normative (cf. Tables 9-10 and 9-11, analysis 4) and low in $TiO_2$. In South Carolina the dikes trend from northwest to southeast, and their close spacing suggests that the magma welled up freely in tensional fractures.

3. Throughout Georgia and Alabama, both *Q*-normative and *ol*-normative types are plentifully represented. Elaborate discussion (Weigand and Ragland, 1970, pp. 204-212) of possible relations between parental magmas and of respective influences of mantle nonhomogeneity and tectonics on magma chemistry seems unwarranted. It is not known whether any of the postulated magmas (each based on analysis averages) ever existed. Nor is it likely that any of them, if they did exist, would be generated again and again at widely separated points according to complex specification prescribed by model calculation. Again we see in chemical variety of rock composition an expression of randomness inherent in regional petrogenesis and stemming from fluctuation to an unknown degree in the many factors that shaped its course.

* Including earlier values corrected where necessary to $Sr^{86}/Sr^{88} = 0.1194$.

## Mixed Tholeiite Alkali-basalt Provinces

### HEBRIDEAN PROVINCE, NORTHWESTERN BRITAIN

SYNOPSIS OF HEBRIDEAN MAGMATISM Along the Atlantic margins north of latitude 54°, basaltic lavas were erupted in profusion in earliest Tertiary times. This whole region from eastern Greenland to northwestern Britain, including intervening subarctic and arctic islands such as Iceland and the Faeroes, was treated in classic literature as a unit—the "Brito-Arctic province." To do justice to increasingly obvious internal diversity, and to comply with current concepts of North Atlantic tectonics, geologists today recognize smaller provinces, each with its own individuality. What they have in common is early Tertiary effusion of one or more kinds of basalt—tholeiitic in the Faeroes (Noe-Nygaard and Rasmussen, 1968), alkaline in the oceanic island Jan Mayen, and several types in western Scotland.

The Eocene Hebridean province (Fig. 9-9) embraces the western islands and adjacent coastal region of Scotland and a large tract in northeastern Ireland. Related basic dike swarms extend across the full width of northern England. This province is classic in igneous petrology.[1] The general Hebridean igneous cycle encompasses three main stages:

1. Regional flooding with basaltic lavas.
2. Locally centralized shallow plutonism and broadly correlated caldera collapse, explosive surface eruption, and development of ring-dike complexes and associated swarms of cone sheets. In these events basic and acid magmas were equally conspicuous.
3. Intrusion of extensive swarms of basic dikes.

What concerns us in the context of this chapter is the presence of tholeiitic and alkaline types of basaltic magma in close mutual association throughout a long and complicated igneous cycle, and the equally important role played by acid magmas throughout the lengthy and varied sequences of events grouped collectively above as centralized activity.

SUCCESSION OF MAGMA TYPES Everywhere in the Hebridean province, volcanism commenced with regional outpouring of olivine basalts. Their most extensive remnant covers 4000 km$^2$ in the plateau of Antrim, northeastern Ireland. In Mull they build the peripheral plateau of the island; and it is lavas of this kind that typify the "plateau magma type" of classic literature. Chemical sampling is still inadequate, and most analyses on which was based so much

[1] Notable among later memoirs of the Geological Survey of Scotland are those that cover Mull (1924), Arran (1928), and Ardnamurchan (1930). Revised ideas on Hebridean petrology will be found in the memoir on northern Skye (1966).

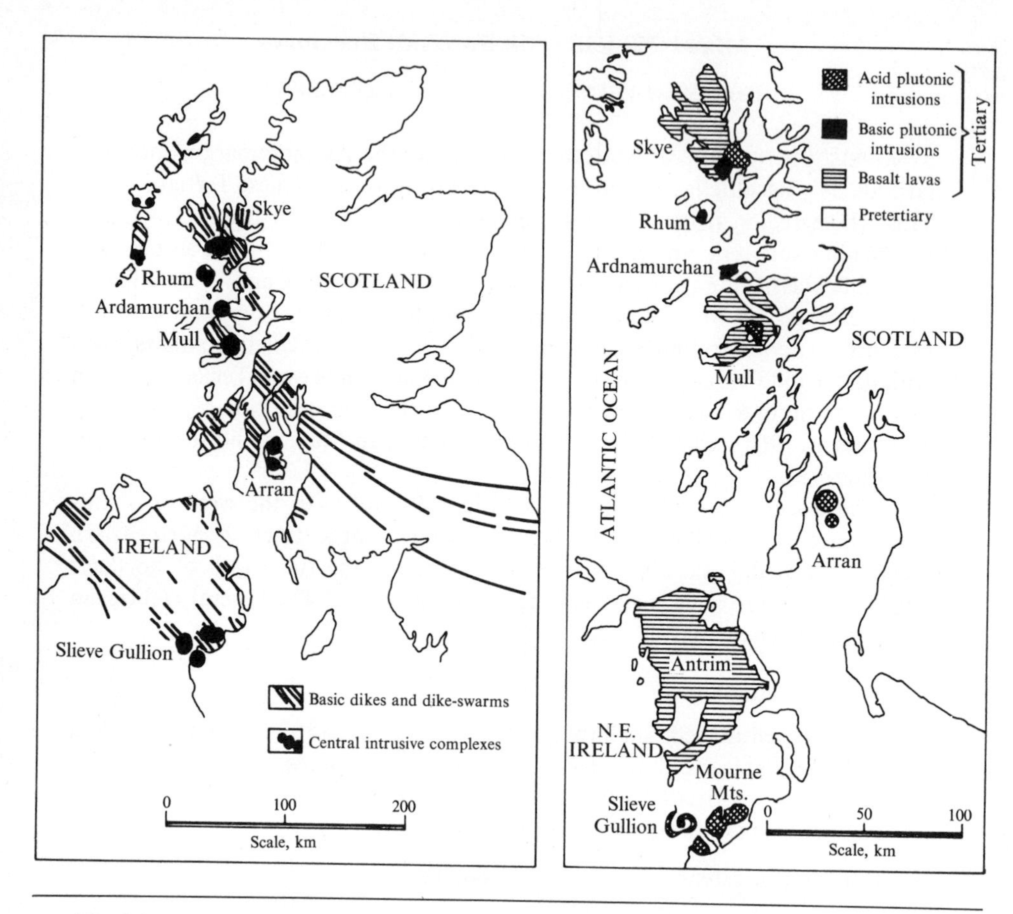


**Fig. 9-9** Hebridean volcanic province of northern Britain (after Richey and Thomas, 1930). (*a*) Dike swarms (simplified) and plutonic centers; (*b*) lavas and plutonic rocks.

petrogenic discussion from 1924 to 1960 are of rocks hydrothermally affected by later granitic plutons. How varied these plateau basalts may be is uncertain (cf. Tilley and Muir, 1962). But they are low in silica (Table 9-12: analyses 1–5), uniformly high in normative *ol*, and have *ne*-normative titaniferous augite as the sole pyroxene. Some have a trace of *ne*, and many have 5 to 15 percent *hy* in the norm. Recent analyses of carefully sampled basalts of this stage in northern Skye (Thompson et al., 1972) show much the same chemistry; half the analyses show 3 percent or less of normative *ne*. The Hebridean plateau basalts recall some of the more alkaline shield lavas of Hawaii (Table 8-13; analyses 4–6). Compared with alkali olivine basalts of most other provinces, they are low in total alkali. Associated with these early olivine basalts, and in general becoming more conspicuous toward the end of the first main volcanic episode, are sub-

**Table 9-12** Chemical compositions (major oxides, wt %) and CIPW norms of basaltic rocks of "plateau magma type" and associated lavas, Hebridean province, northwest Britain

| | 1 | 2 | 3 | 4 | 5 | 6 | 7 |
|---|---|---|---|---|---|---|---|
| $SiO_2$ | 45.34 | 45.52 | 46.12 | 46.97 | 47.90 | 49.68 | 66.13 |
| $TiO_2$ | 1.13 | 1.48 | 1.81 | 1.59 | 1.57 | 2.13 | 0.61 |
| $Al_2O_3$ | 14.67 | 17.58 | 13.94 | 15.00 | 15.28 | 16.99 | 16.03 |
| $Fe_2O_3$ | 2.40 | 4.17 | 1.95 | 1.71 | 1.70 | 3.45 | 3.17 |
| FeO | 9.15 | 8.14 | 10.46 | 8.94 | 9.10 | 8.99 | 0.70 |
| MnO | 0.22 | 0.17 | 0.18 | 0.37 | 0.17 | 0.27 | 0.10 |
| MgO | 13.32 | 8.46 | 11.08 | 10.52 | 7.30 | 2.79 | 0.84 |
| CaO | 9.12 | 9.72 | 9.05 | 10.70 | 12.07 | 5.46 | 1.45 |
| $Na_2O$ | 1.86 | 1.86 | 3.11 | 2.18 | 2.81 | 5.78 | 5.34 |
| $K_2O$ | 0.24 | 0.62 | 0.57 | 0.63 | 0.53 | 1.90 | 4.82 |
| $P_2O_5$ | 0.09 | 0.07 | 0.23 | 0.12 | 0.16 | 0.48 | 0.08 |
| $H_2O^+$ | 1.69 | 1.85 | 1.49 | 0.38 | 1.27 | 1.77 | 0.36 |
| $H_2O^-$ | 1.05 | 0.32 | 0.40 | 0.63 | 0.46 | 0.34 | 0.43 |
| Total | 100.28 | 99.96 | 100.39 | 99.74 | 100.32 | 100.03 | 100.06 |
| *Q* | | | | | | | 12.12 |
| *or* | 1.45 | 3.8 | 3.34 | 3.3 | 2.78 | 11.12 | 28.73 |
| *ab* | 15.72 | 15.6 | 20.96 | 17.8 | 22.27 | 38.90 | 45.06 |
| *an* | 30.86 | 37.7 | 22.38 | 29.8 | 27.80 | 14.46 | 5.28 |
| *ne* | | | 2.84 | | 0.71 | 5.61 | |
| *di* | 11.61 | 7.1 | 17.89 | 18.2 | 25.44 | 6.53 | 0.22 |
| *hy* | 9.24 | 18.6 | | 5.9 | | | 2.26 |
| *ol* | 22.72 | 5.9 | 24.33 | 17.8 | 13.62 | 10.67 | |
| *mt* | 3.48 | 6.0 | 2.90 | 2.6 | 2.55 | 4.98 | 1.37 |
| *il* | 2.13 | 2.9 | 3.50 | 3.0 | 3.04 | 4.10 | 0.67 |
| *ap* | 0.20 | 0.2 | 0.34 | 0.3 | 0.34 | 1.34 | 0.18 |
| Total | 97.41 | 97.8 | 98.48 | 98.7 | 98.55 | 97.71 | 99.09* |

* Including *ac* 3.20.

*Explanation of column headings*

1 Olivine basalt, upper lava series, Antrim (Patterson, 1951, p. 286, analysis 5)
2 Olivine basalt, Ard Bheinn, Arran (King, 1955, p. 328, analysis 1)
3 Olivine basalt, Skye (Tilley and Muir, 1962, p. 212, analysis 1)
4 Diabase ("normal dolerite"), northern Skye (Anderson and Dunham, 1966, p. 147, analysis Q)
5 Olivine basalt, Fingal's Cave, Staffa (Tilley and Muir, 1962, analysis 2)
6 Mugearite, type locality, Skye (Muir and Tilley, 1961, p. 190, tables 4, 5; analysis 1)
7 Trachyte, Skye (Anderson and Dunham, 1966, p. 118, analysis XI)

ordinate flows of hawaiite, mugearite, and locally trachyte (Table 9-12; analyses 6, 7). These have been generally considered—by analogy with similar lavas in oceanic provinces—to be differentiates from undersaturated basaltic magmas of the plateau magma type. Thompson et al. (1972) make a good case for an alkali basalt-hawaiite-mugearite lineage in Skye, the end product of which (benmoreite) is somewhat more siliceous and more sodic than mugearite. They consider that trachyte magmas (with their high normative *Q*) were independent derivatives from a low-alkali tholeiitic parent. There are also basic alkaline sills (teschenite) internally differentiated to picrite low down and to zeolitic diabases and even analcite syenite in the upper levels (Turner and Verhoogen, 1960, pp. 175-184).

Tholeiitic basalts lacking or poor in olivine appear toward the close of the regional eruptive episode in Mull, where they occur mainly in the central part of the island. They exemplify the "nonporphyritic central magma type" of the Mull memoir. Closely similar lavas appear elsewhere in the province at about the same stage—for example, in northern Antrim (Table 9-13, analysis 2) they form an extensive wedge sandwiched between earlier and later olivine basalts of the plateau.[1] Most basalts of this group are slightly oversaturated in silica, but some in the upper part of the main lava series of Skye are olivine tholeiites.

Yet another type of basaltic magma ("porphyritic central type") was recognized on Mull in the form of porphyritic basalt flows extruded just after the opening flood of olivine basalts. These are oversaturated high-alumina basalts with abundant phenocrysts of basic plagioclase. There are also diabase dikes of similar chemistry. In total volume these high-alumina rocks are greatly subordinate to basalts of the other two types. Moreover their porphyritic nature suggests that they are partly cumulate rocks and do not represent any liquid magma of the same composition (Bowen, 1928, p. 139).

In the succeeding phases of locally centralized volcanism and plutonism, acid magmas appear in force—as granitic stocks, granophyric and felsitic ring dikes, and rhyolitic cone sheets, flows, and breccias. Much less abundant are rocks of intermediate silica content, which although generally referred to as andesites are significantly lower in $Al_2O_3$ than are typical orogenic andesites. Basic magmas of all three types are prominent in the stocks, ring dikes, and cone sheets of the central complexes. Thick vertical sections exposed in the lower levels of large central plutons on Rhum and Skye (G. M. Brown, 1956; Wadsworth, 1961; Wager and Brown, 1967) consist of rhythmically layered calcic gabbros and peridotites (to be described shortly). These gravity-sorted accumulations of crystals separated early during solidification of a parental basaltic magma generally believed to have been of the high-alumina porphyritic central type.

In the final stages of activity, great swarms of basic dikes were emplaced in

[1] One of the most carefully studied examples comes from the Giant's Causeway (Holmes, 1932; Table 9–13, analysis 2).

**Table 9-13** Chemical analyses (major oxides, wt %) and CIPW norms of *Q*-normative basic and intermediate rocks, Hebridean province, northern Britain*

| | 1 | 2 | 3 | 4 | 5 | 6 |
|---|---|---|---|---|---|---|
| $SiO_2$ | 50.41 | 50.36 | 53.92 | 57.09 | 54.18 | 58.34 |
| $TiO_2$ | 1.30 | 1.06 | 0.79 | 0.80 | 1.97 | 1.08 |
| $Al_2O_3$ | 15.14 | 14.51 | 15.81 | 13.76 | 13.74 | 14.09 |
| $Fe_2O_3$ | 2.71 | 2.61 | 3.05 | 2.74 | 1.88 | 2.34 |
| FeO | 7.95 | 8.09 | 5.06 | 4.98 | 10.79 | 7.78 |
| MnO | 0.17 | 0.12 | 0.11 | 0.12 | 0.30 | 0.17 |
| MgO | 6.57 | 6.26 | 5.32 | 4.29 | 2.42 | 4.24 |
| CaO | 11.30 | 10.77 | 10.22 | 8.12 | 6.34 | 6.96 |
| $Na_2O$ | 2.29 | 2.48 | 2.09 | 2.51 | 3.46 | 2.66 |
| $K_2O$ | 0.82 | 0.99 | 1.46 | 1.95 | 1.85 | 1.30 |
| $P_2O_5$ | 0.15 | 0.45 | 0.12 | 0.13 | 1.30 | 0.06 |
| $H_2O^+$ | 1.01 | 1.10 | 1.01 | 1.43 | 1.40 | 0.85 |
| $H_2O^-$ | 0.72 | 1.27 | 1.19 | 1.45 | 0.26 | 0.27 |
| Total | 100.54 | 100.07 | 100.15 | 99.37 | 99.89 | 100.14 |
| *Q* | 1.43 | 1.46 | 8.90 | 14.47 | 7.8 | 14.3 |
| *or* | 4.84 | 5.84 | 8.62 | 11.52 | 10.9 | 7.8 |
| *ab* | 19.35 | 20.97 | 17.67 | 20.50 | 29.3 | 22.5 |
| *an* | 28.70 | 25.53 | 29.59 | 21.05 | 16.5 | 22.5 |
| *di* | 21.27 | 20.27 | 15.17 | 11.25 | 5.5 | 9.6 |
| *hy* | 16.37 | 16.63 | 11.46 | 10.91 | 18.7 | 16.5 |
| *mt* | 3.94 | 3.89 | 4.42 | 3.96 | 2.7 | 3.5 |
| *il* | 2.46 | 2.02 | 1.50 | 1.52 | 3.8 | 2.1 |
| *ap* | 0.37 | 0.99 | 0.27 | 0.30 | 3.1 | 0.1 |
| Total | 98.73 | 97.60 | 97.60 | 95.48 | 98.3 | 98.9 |

* Samples 1 to 4 have been equated with the nonporphyrite central magma type.

*Explanation of column headings*

1 Tholeiitic diabase, Kielderhead dike, Northumberland (Holmes and Harwood, 1929, p. 15)
2 Basalt of Giant's Causeway, Nudale lava series, Antrim (Holmes, 1936, p. 91)
3 Tholeiitic diabase, Coley Hill dike, near Newcastle-upon-Tyne (Holmes and Harwood, 1929, p. 32)
4 Tholeiitic diabase, Hebburn dike, Durham (Holmes and Harwood, 1929, p. 35)
5 Ferrodiorite, associated with marscoite, Skye (Wager et al., 1965, p. 283, analysis 6A)
6 "Andesite," Arran (King, 1955, p. 328, analysis 4)

steep fractures fanning out from central intrusive complexes of Skye, Mull, Arran, and Ardnamurchan. Rocks of the Islay swarm, northwest of Arran, are alkaline analcite-olivine diabases (plateau magma type). The much more extensive swarm that radiates from Mull and extends clear across northern England consists of oversaturated two-pyroxene diabases (Table 9-13; analyses 1, 3) assigned to the nonporphyritic central magma type.

Petrogenic discussion in the Mull memoir and in literature of the ensuing two decades centered on attempts to derive the Hebridean magma types and magma series from some common stem (Turner and Verhoogen, 1960, pp. 224-226). No general concensus emerged. Today the same problems are viewed rather differently. The chemical character of the various magma types has been reassessed in the light of new data relating to other provinces—especially the Hawaiian Islands (for example, Tilley and Muir, 1962, 1967). The existence of any direct relationship between associated magma types has been questioned. Indeed it well may be asked whether it is necessary or even probable that over a period of 10 million years three specific kinds of basaltic magma should remain continuously on tap in so large a province. More likely, many batches of basaltic magma were generated locally and from time to time. Seen thus the individual magma type reflects general uniformity in source materials and in conditions of fusion. Magmatic diversity, on the other hand, expressed the scope of fluctuation as to source rocks, conditions and degree of melting, and subsequent fractionations.[1] The origin of Hebridean parental magmas is one aspect of a general problem discussed in Chap. 13.

ROLE OF ACID MAGMA Ideas are changing too on the role of acid magmas in the general scheme of petrogenesis (for example, Carmichael, 1963, pp. 124-126; G. M. Brown, 1963; Wager et al., 1965; Moorbath and Bell, 1965). These magmas appear at many centers in such volume, especially during the centralized phase of activity, that some kind of independent origin seems more probable than descent from basic parentage. This view is supported by general paucity of rocks of intermediate silica content—although there are local exceptions, notably on Arran (King, 1955, p. 352). Seen in the light of naturally established crystallization paths in the quartz-feldspar system, rhyolites with phenocrysts of both sanidine and plagioclase are probably formed from magmas affected by assimilation of acid crustal rocks or even formed by direct fusion of such materials. Experimental melting of rocks available in the Hebridean basement—Lewisian gneiss and Torridonian sandstone—shows that they could indeed yield magmas covering the major-element composition range of the granite-felsite-rhyolite series in the province. The clinching argument comes from studies by Moorbath and coworkers on isotopic composition of strontium and of lead in suites of rocks from two centers on Skye (page 371). These studies confirm the major role played by fusion of the metamorphic Lewisian basement. But this is not the whole story. The granodioritic magmas so formed have been admixed to varying degrees with acid differentiates fractionated from basaltic magmas drawn principally from mantle sources. Only a compound model such as this will explain the compositions of both strontium and lead.

[1] See, for example, the model proposed by Thompson et al. (1972) for the lavas of northern Skye. The Skye magmatic cycle is treated as a complex local event without reference to contemporaneous magmatism at other Hebridean centers; and the notion of magma type no longer dominates discussion.

Although the "normal magma series" no longer stands as a differentiated lineage, differentiation has played some considerable part in magmatic evolution within the province. There are the ultrabasic cumulates of Rhum and other centers; and the hawaiite-mugearite-benmoreite line most probably developed by fractionation of olivine-basaltic magma. Still unresolved is the status of "andesitic" lavas (Table 9-13, analysis 6) and other rocks of intermediate silica content. Are they differentiates from tholeiitic magmas or hybrids of mixed parentage (basic and acid)? Seventy years ago Harker recognized the hybrid nature of some rather basic rocks ("marscoites, with phenocrysts of andesine, potash feldspar, and quartz") from Skye. Recent studies (Wager et al., 1965) indicate that marscoite originated by "mixing of an acid magma containing phenocrysts of quartz and potash feldspar with a basic magma containing phenocrysts of andesine feldspar." The latter magma is represented by closely associated ferrodiorite (Table 9-13, analysis 5). Marscoite yields high initial $Sr^{87}/Sr^{86}$ values ($0.7126 \pm 0.0008$) consistent with hybrid origin. Rather surprisingly, strontium in the ferrodiorite has closely similar composition; so at least this particular rock, chemically reminiscent of the basic icelandites of the Galápagos (Table 8-10, analysis 4), cannot be a pure differentiate of basaltic parentage. Most likely it, too, is a direct product of fusion of basic variants in the Lewisian basement.

## Layered Basic Intrusions

The world's largest basic pluton, with a total volume of the order of 100,000 km$^3$, is the principal component of the Precambrian Bushveld complex of South Africa. The comparable but much smaller Eocene Skaergaard intrusion of Greenland (500 km$^3$) must surely be—through the effort of L. R. Wager—the world's most completely investigated major igneous body. Many plutons of the same kind are now known. In overall composition, basic intrusions of this kind invariably are tholeiitic and for the most part somewhat undersaturated in silica. Internally they exhibit lithological layering on a variety of scales.

Classic examples from which grew early notions regarding layered intrusions are rather small ultramafic central complexes of Skye (Cuillin) and Rhum in the Eocene Hebridean province, the Bushveld "norite" intrusion of South Africa (A. L. Hall, 1932), and the late Precambrian Duluth anorthosite sheet of Minnesota—established by Grout (1918) as the prototype of the lopolith. Current knowledge and ideas stem largely from two sources: studies on the Precambrian (2750 m.y.) Stillwater complex, Montana (H. H. Hess, 1960; E. D. Jackson, 1961), and especially a lifelong investigation by Wager and associates concentrating on the Skaergaard intrusion but also including the ultramafic Hebridean complexes and other layered plutons the world over. Also noteworthy are relatively recent accounts of complex layering in the "critical zone" near the base of the Bushveld intrusion (for example, Cameron, 1963, and papers cited therein), in the ultramafic base of the Great "Dike" of

Rhodesia (Worst, 1958), and in the Muskox intrusion of Canada (C. H. Smith and Kapp, 1963). The whole subject has been treated exhaustively and illustrated beautifully by Wager and Brown (1967), and special facets have been developed in a symposium of the Mineralogical Society of America (Heinrich et al. 1963).

Layered intrusions vary greatly in size and have individual characteristics related partly to their dimensions. They cover much of the span of geologic time: Bushveld, ~2000 m.y.; Stillwater, ~1580 m.y.; Muskox, 1160 m.y.; Skaergaard, ~60 m.y. However, they have sufficient in common to permit generalization as to their nature, structure, and petrogenesis.

## GEOMETRIC FORM

Even the Skaergaard intrusion, although elaborately dissected in a region of glacial topography (relief 1000 to 1300 m) and relatively small for plutons of this class, is still too large for its form and internal structure to be more than partially revealed. Nevertheless, much is known about parts of many plutons: the upper third or half of the Skaergaard, the lower half of the tilted Stillwater complex, detailed sections in the lower part of the Bushveld, segments of the deformed early Paleozoic sheet of northeastern Scotland,[1] the basal ultramafic zones of Tertiary Hebridean complexes (G. M. Brown. 1956; Wadsworth, 1961), the strangely elongate Great "Dike" of Rhodesia and Muskox intrusion of the Canadian Arctic (C. H. Smith and Kapp, 1963). Information from these and other sources—all summarized by Wager and Brown—yields a composite, generalized picture.

In layered basic intrusions we see the most extensive known deployment of basaltic magmas in the plutonic crustal environment. The Bushveld complex in places is 7000 m thick. The exposed remnant of the tilted and partially eroded Stillwater complex is about 400 $km^2$ in area; within it the measured thickness normal to the layering is 6000 m. A similar apparent thickness has been recorded for the Cuillin central complex of Skye, which is less than 10 km in diameter. In fact, thicknesses of the order of 3 to 5 km seem to be general and may be exceeded. Extensive though such plutons are, the largest are only half the size of the Sierra Nevada and southern California granodioritic batholiths.

Following Grout's thinking it was long believed that most layered basic intrusions have the form of a lopolith—"a large, centrally sunken, generally concordant, intrusive mass with its thickness approximately one-tenth to one-twentieth of its width or diameter." Certainly the lateral extent of many bodies (for example, the northeast Scotland sheet) greatly exceeds their proved depth. But the Skaergaard intrusion has the form of a funnel tapering downward (Fig. 9-10*a*). And a triangular cross section with the apex downward proves to be typical even of elongate bodies such as the Muskox intrusion and the Great "Dike"—the latter is certainly no dike and is probably the infaulted basal

[1] *Scottish J. Geol.*, vol. 6, pt. 1, 1970.

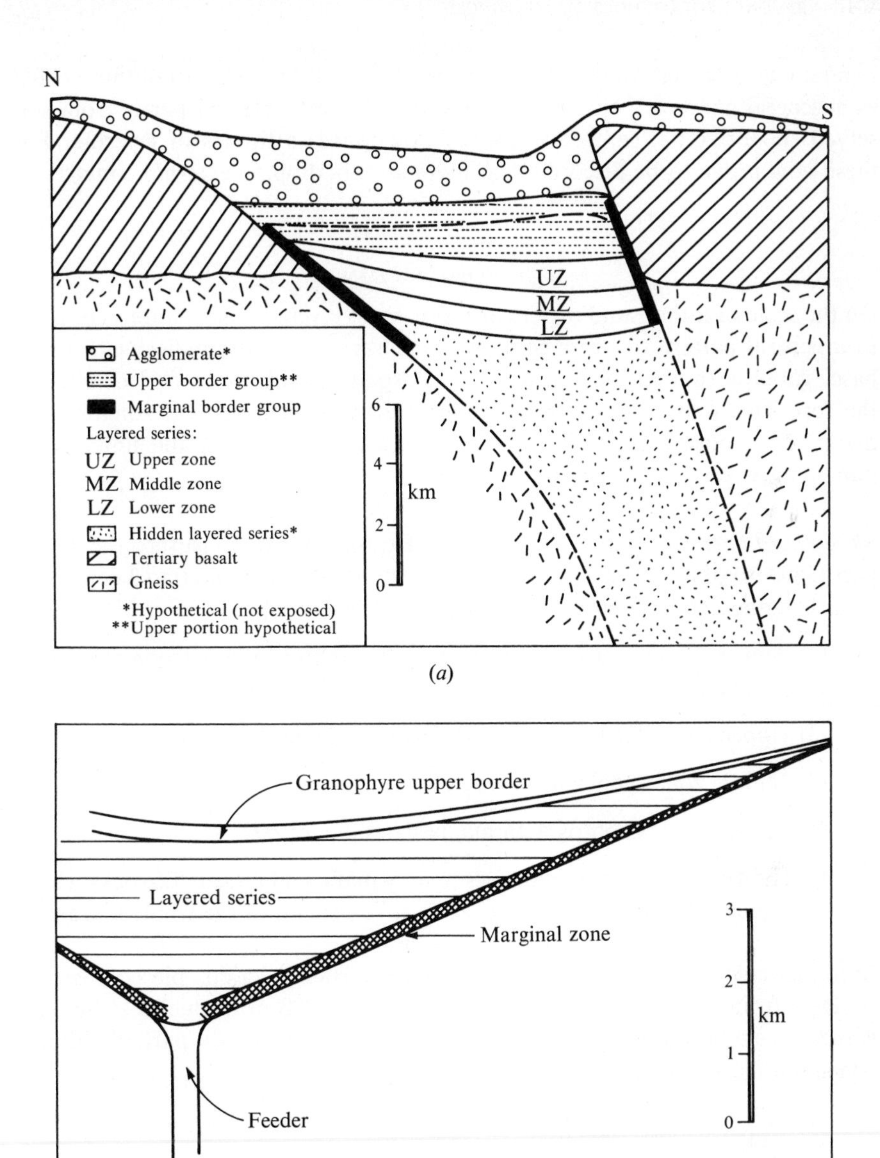


**Fig. 9-10** Simplified vertical sections, extrapolated from field data, through two layered basic plutons. (*a*) Skaergaard, Greenland (after Wager and Brown, 1967, pp. 20, 205); (*b*) Muskox intrusion, Canada (after C. H. Smith and Kapp, 1963, p. 33).

remnant of a larger composite intrusion (Worst, 1958). The lopolithic profile as a general characteristic of basic plutons had been inferred partly from observed subhorizontal (or gently convex downward) attitudes of layering. But these now prove to be discordant to more steeply dipping external contacts.

## MAJOR VERTICAL ZONING

On the scale of a thousand meters or more, most layered basic plutons conform to a general pattern of gross compositional variation: ultramafic rocks at the base, gabbros (norites) in the middle, feldspathic gabbros or ferrodiorites at the top. This pattern permits rather arbitrarily defined vertical zonation for mapping purposes. The following are illustrative examples, the units being numbered 1, 2, . . . , in upward sequence.

*Skaergaard intrustion, eastern Greenland* Exposed rocks constitute the upper part (3500 m) of a vertical section of unknown thickness; zones 0 to 3 constitute what is termed collectively *the layered series* (Fig. 9-10*a*).

(4) Upper border group: contaminated gabbros and ferrodiorites with granophyric streaks and many xenoliths of acid gneiss: 1000 m

(3) Upper zone: ferrogabbros with ferrous olivine: 920 m

(2) Middle zone: gabbros lacking olivine: 780 m

(1) Lower zone: gabbros with magnesian olivine: 800 m

(0) Hidden zone: unexposed, presumably mafic and ultramafic rocks whose total bulk has been estimated as at least 70 percent of the whole pluton

*Stillwater complex, Montana* The following vertical section, pieced together by Hess from several measured sections, exceeds 5000 m in thickness but still is incomplete because the pluton was tilted and the top is still partially hidden beneath a blanket of younger sediments.

(6) Upper gabbro: 600 m

(5) Gabbro and anorthosite (alternating in 200- to 500-m units): 1900 m

(4) Lower gabbro: 650 m

(3) Norite (two-pyroxene gabbro): 850 m

(2) Ultramafic rocks (pyroxenites grading down to peridotites): 1050 m

(1) Border zone, retaining something of the composition of the initial magma: 100 m

*Bushveld pluton* Hall's generalized section through the eastern part of the intrusion exceeds 8000 m. It has been summarized, with some mineralogical and petrographic amplification, by Wager and Brown (1967, pp. 349–382).

(4) Upper zone: ferrodiorites with local minor granophyric differentiates: 1500 m

(3) Main norite zone: two-pyroxene gabbros: 4000 to 5000 m

(2) Critical zone: profusely layered plagioclase-bronzite-chromite rocks: 1000 m

(1) Basal zone: layered bronzite-pyroxenites, harzburgites, dunites: 1300 m

*Cuillin complex, Skye* The generalized section summarized by Wager and Brown (1967, p. 423) exceeds 5500 m.

(4) Gabbro series: 600 m

(3) Eucrite series: 2000 m

(2) Allivalite series: 1700 m

(1) Peridotite series: 1250 m

Distinction between the three basic units—all varieties of olivine gabbro—is based largely on plagioclase composition: $An_{80}$ to $An_{85}$ in allivalite, $An_{70}$ to $An_{80}$ in eucrite, $An_{58}$ to $An_{70}$ in gabbro.

## LAYERING

NATURE OF LAYERING All major basic plutons show subhorizontal compositional layering over a wide range of scale. And as we shall see presently, this is the result of a kind of magmatic "sedimentation." The greater part of each layered pluton has solidified from the base upward by accumulation or precipitation of the component mineral crystals upon the floor, which in the process has slowly grown upward. The rocks so formed are called *cumulates.* Their essential constituents are *cumulus* crystals that have come to rest under gravity. Another initial ingredient of each rock was an *interprecipitate liquid,* trapped much as is pore water in unconsolidated sediment. This liquid ultimately crystallizes, and in so doing may play a number of different roles: it can give rise to recognizable crystalline interprecipitate consisting of phases chemically distinct from the cumulus; or it may modify and enlarge associated cumulus crystals. In some cases the whole solid-liquid system comprising a recently deposited layer may have remained open to exchange with immediately overlying magma. All this is expressed in a wide variety of poikilitic and allied textures in cumulate rocks.

Visually conspicuous lithologic layering is generally termed *rhythmic*, since it is characteristically repeated a dozen, a hundred, or a thousand times in continuous vertical sections. Then there is continuous, gradual change in chemical and mineralogical composition from the base to the top of an intrusion; the resulting contrast between rocks at different levels, because it is not usually obvious, is described as *cryptic* layering. In limited sections it is revealed only by petrographic examination and chemical analysis. The two kinds of layering are controlled by independent influences: rhythmic layering expresses among other influences that of gravity; cryptic layering is largely the response to a vertical gradient of composition in successive residual liquid fractions.

Throughout the long history of crystallization, development of rhythmic and cryptic layering is broadly synchronous. At some levels one may operate to the exclusion of the other. It is not surprising that layering patterns in different plutons and at different levels in a single intrusion display considerable variety.

CRYPTIC LAYERING A universal characteristic of cryptic layering is its pervasive nature. Compositions of mineral solid solutions change gradually upward in the experimentally established order of decreasing liquidus temperature. The only sudden breaks—termed *phase layering* by Hess—are occasioned by exit of high-temperature phases (such as magnesian olivine and chromite) at specific levels and by entry of lower-temperature phases (such as pigeonite and ferrous olivines) at higher levels. Herein is proof beyond doubt that throughout most of the vertical section the plutons have solidified from the floor upward.

In some intrusions, however, the trend of cryptic layering is reversed near the top. In the Skaergaard, an exposed section of 2500 m shows normal cryptic layering, the layered series. Then the trend is sharply reversed, and in the overlying upper border group, 1000 m thick, plagioclase becomes progressively more calcic upward—$An_{30}$ below, $An_{69}$ at the highest exposure. These rocks have grown downward from a roof since removed by erosion.

Cryptic layering has been documented in great detail for the exposed upper portion of the layered series (estimated on compositional grounds as perhaps one-third of the whole) of the Skaergaard intrusion (Wager and Brown, 1967, pp. 20, 26–27, 205). Compositional changes in cumulus phases,[1] as summarized in Table 9-14, are gradual. They show no sign of interruption or reversal through the vertical 2500 m section. The gross rock composition, on a sampling scale that transcends rhythmic layering (see below), ranges from gabbros at the base to ferrodiorites in the upper part of the section. The layered series is

[1] For example, plagioclase in large, unzoned cores of individual crystals.

**Table 9-14** Compositions of coexisting solid-solution series (cumulus) in cryptically layered section, layered series, Skaergaard intrusion*

| ZONE | HEIGHT ABOVE LOWEST EXPOSED LEVEL, m | PLAGIOCLASE An | OLIVINE Fo | AUGITE Ca | AUGITE Mg | AUGITE Fe | INVERTED PIGEONITE Fe |
|---|---|---|---|---|---|---|---|
| | 2500 | 30 | 0 | 43 | 0 | 57 | |
| Upper | 2000 | 38 | 24 | 40 | 21 | 39 | |
| (Ferrodiorites) | 1800 | 40 | 33 | 35 | 33 | 32 | Goes out |
| | 1580 | 45 | 40 comes in | 35 | 37 | 28 | 46 |
| Middle | 1000 | 51 | | 37 | 39 | 24 | n.d. |
| | 800 | 51 | 53 goes out | | n.d. | | n.d. |
| Lower | 600 | 56 | 57 | 38 | 41 | 21 | 35 |
| (Gabbros) | 500 | 58 | 59 | | n.d. | | Comes in |
| | 300 | 62 | 63 | | Comes in | | |
| | 100 | 66 | 67 | | | | |

* Data from Wager and Brown (1967, pp. 26, 27).
n.d. = not determined.

divided into three zones whose boundaries are marked by exit of magnesian olivines at 800 m and reentry of olivines rich in iron at 1580 m.

In the Cuillin section on Skye, as plagioclase changes upward from $An_{88}$ to $An_{58}$, olivine and then augite become increasingly rich in iron (olivine $Fo_{87}$ in the lowermost peridotites, $Fo_{58}$ in the uppermost gabbros). The trend, as always with cryptic layering, passes unchanged across even the sharpest lithologic boundaries of rhythmically contrasted layers on every scale.

In the Bushveld intrusion, the trend of cryptic layering (Wager and Brown, 1967, p. 351) is uninterrupted to the very top. It has its individual character while combining something of both the Skaergaard and the Cuillin patterns. Cumulus plagioclase comes in at the base of the critical zone ($An_{78}$), is increasingly sodic labradorite through the main norite zone, and continues through the andesine range $An_{53}$ to $An_{30}$ at successively higher levels in ferrodiorites of the upper zone. Olivine $Fo_{86}$ to $Fo_{88}$ exits at the top of the basal zone and reappears in the upper zone—$Fo_{49}$ grading upward to $Fo_0$. In both pyroxene series, Fe/Mg increases steadily upward. Cumulus chromite is confined to the basal zone, magnetite to the upper zone. Cryptic layering in the Stillwater is closely comparable with that in the equivalent section (5000 m) of the Bushveld.

In some layered basic plutons, cryptic layering is weak or absent. In a 6000-m rhythmically layered section of the Freetown complex, Sierra Leone (Wells, 1962), the principal mineral phases show only restricted and apparently random variation: plagioclase, $An_{59}$ to $An_{64}$; olivine, $Fo_{58}$ to $Fo_{67}$; Ca-poor pyroxene, $En_{65}$ to $En_{72}$. Yet there is overall upward decrease in the olivine/plagioclase ratio in the general sequence peridotite, troctolite, olivine gabbro, anorthosite; and the same pattern applies to internal rhythms on smaller scales (100 m down to centimeters). Throughout an 800-m section in the Hebridean island of Rhum, 15 units are rhythmically repeated—each peridotite becoming more feldspathic upward. But at every level cumulus olivine is $Fo_{84}$ to $Fo_{86}$, plagioclase $An_{84}$ to $An_{88}$. Gravitational effects are obvious in such instances. Differentiation, nevertheless, is limited to sorting of crystalline phases—mainly olivine and calcic plagioclase—and there is no sign of liquid fractionation within the observed vertical sections.

RHYTHMIC LAYERING Visually conspicuous layering on scales between a few millimeters and a few hundred meters commonly is rhythmically repeated. In the ideal case the repeated unit is compound, and its component layers are mutually gradational across their common boundary surface. But there are degrees of such internal gradation; and at one extreme the internal boundary, even on a microscopic scale, is almost clear-cut. Very commonly there is obvious evidence of gravity stratification, much as in graded sedimentary beds: olivine or pyroxene cumulus crystals, for example, show gradual concentration toward the base while feldspar becomes increasingly abundant toward the top of each compound unit. This condition, widely prevalent on all scales and reinforced by presence of cross-bedding and occasional scour structures, convincingly confirms upward solidification as inferred independently on chemical-mineralogical grounds.

Rhythmic layering of gabbros on the scale of 5 to 40 cm occurs abundantly throughout the main 2500-m section (layered series) of the Skaergaard pluton. Typically the units are graded—feldspathic at the top and increasingly rich in pyroxene or olivine (less commonly iron oxides) toward a sharply defined base. Layered sections may be broken by massive units without layered structure. Toward the top of the middle zone gravity-stratified units become coarser, and boundary surfaces lack definition. At this level macrounits 20 to 60 m thick have been traced laterally for more than 2 km.

Basal segments of other layered plutons display graded layering on an even coarser scale in ultramafic rocks. A section across the Great "Dike" of Rhodesia shows a basal peridotite (dunite) zone (nearly 1000 m) followed by rhythmically repeated macrounits, six in number, each about 200 m thick. Each unit grades from dunite at the base through harzburgite to bronzite pyroxenite at the top. The same sort of pattern is shown in the ultramafic zones of the Bushveld and the Stillwater complex. On Rhum, in the Hebridean province, an 800-m section comprises 15 similar graded macrounits, each between 15 and 150 m thick. Their components are olivine, calcic plagioclase, some augite, and minor chrome spinel. The lower part of each unit is peridotite, which grades upward into increasingly feldspathic rock (allivalite) itself internally layered on the scale of only a few millimeters.

The most famous instance of rhythmic layering with internal grading on the macroscale is unique: the upper 200 m of the critical zone of the Bushveld, in the middle of which lies the Merensky reef, the major source of the world's platinum. There are half a dozen units. In each, bronzite cumulate, with chromite concentrated at the very base, passes gradually upward into anorthosite in which dominant cumulus plagioclase is enclosed poikilitically in subordinate intercumulus bronzite. Platinum and sulfides seem to have accumulated along with other cumulus phases in one particular layer, the Merensky reef, from 1 to 5 m thick. Its lateral extent is phenomenal. It has been traced for 100 km or more through the full eastern and western lengths of outcrop of the Bushveld intrusion. Finally we note remarkably persistent rhythmic layering of another extreme type, where magnetite first appears as a cumulus phase at the base of the upper zone several thousand meters higher in the section. Here there are six internally graded macrounits, each with a lower layer that is 30 cm to 3 m thick and rich in cumulus magnetite and with an upper component that is nearly pure plagioclase.

The fine layering in Rhum allivalites illustrates a common type of rhythmic layering in which gradation between the contrasted components of each unit is minimal. One is almost pure pyroxene, the other plagioclase. "Inch-scale" layering recorded by Hess in parts of the Stillwater is of much the same kind but on a slightly larger scale. In parts of the Bushveld, boundaries between layers alternately rich in opaque oxides and in silicates appear sharp in the field and are gradational only on a microscopic scale. Rhythmic layering of this kind is beautifully displayed throughout the critical zone (Cameron, 1963); the cumulate components involved are plagioclase, bronzite, and chromite, and the scale is mostly 1 to 30 cm. Chromite-cumulate layers a meter or so in thickness can be traced laterally for 50 km or more.

MECHANICS OF LAYERING It was once widely held that major inhomogeneity in large basic plutons implies emplacement by intermittent surges of magmas of different compositions, with complementary gravitational differentiation following each wave of intrusion. This view has now been generally discarded (for example, H. H. Hess, 1960, p. 154; Wager and Brown, 1967, p. 546) because the gradational pattern of cryptic layering so often continues uninterruptedly across zonal boundaries through vertical sections thousands of meters thick. Nevertheless G. M. Brown's (1956, pp. 45–49) hypothesis—that repeated replenishments by draughts of basaltic magma from some deeper source may be responsible for the sequence of identical macrounits in the ultramafic complex of Rhum, where cryptic layering is negligible—has not been completely abandoned (Wager and Brown, 1967, pp. 293, 542).

Most writers today visualize emplacement as a single, short-lived act of intrusion. So the starting material is seen as a very large mass of homogeneous basaltic magma sealed within a frozen skin now represented by chilled border rocks. How so simple an event can be enacted on such a grand scale may be difficult to visualize. There is a hint of oversimplification—as indeed has eventually been proved for so many classic models of other geologic phenomena. But the general similarity of vertical compositional trends in many intrusions and the extraordinary lateral continuity of layering, even with respect to such specialized rocks as chromitites and the unique material of the Merensky reef, point to a seemingly inescapable conclusion. Differentiation begins with an enormous body of homogeneous magma; and homogeneity of the crystallizing magma—above the upwardly encroaching (and perhaps laterally spreading) cumulate floor—was maintained continuously to the end. This does not imply that the magma either initially or at later stages of differentiation must have been entirely liquid. A good case has been made for intrusion of magma strongly charged with suspended olivine crystals as the opening phase in the history of the Muskox intrusion (C. H. Smith and Kapp, 1963). And continous mixing of liquid and crystals in a progressively crystallizing magma seems a likely explanation of the absence of cryptic layering in the Freetown pluton of Sierra Leone.

Whatever the nature of the parental magma, cryptic layering implies gradual upward displacement of the cooling fractionating melt phase. Rhythmic layering, or at least some of its major aspects, are broadly pictured as due to sinking, sorting, and "sedimentation" of mineral crystals of different composition, density, size, and shape in an enclosing silicate-melt medium.

What part in the layering process has been played by convection? A characteristic of basic layered intrusions is widespread structural evidence of current action. Where the dipping walls of a pluton are well exposed, as in the Skaergaard and the Muskox, there are border zones a few hundred meters thick against which the rhythmic layering of the main body pinches out (Fig. 9-10). Except at the chilled contact itself, the rocks of the border zones tend to show fluxional foliation—dimensional preferred orientation of tabular crystals or elongate xenoliths—parallel to the walls and thus sharply discordant with respect to subhorizontal rhythmic layering deeper inside the pluton. Here is the imprint of magmatic flow, either during emplacement or by subsequent convection. Convincing evidence of the flow of horizontal currents across the slowly rising floor is supplied by local current bedding and scour-and-fill structures in finely layered rocks and by fluxional foliation in some massive zones which lack rhythmic layering—notably in the main zone of the Bushveld. All this lends support to—but does not establish completely—a model of crystal sorting by convection currents subject to intermittent bursts of turbulence. Wager and Brown (1967, pp. 546–547) conclude

> that convection currents were probably operative in most magma chambers, and that slow and continuous, or fast and sporadic currents, could best account for the uniform or graded units respectively. In fact, the action of convection currents appears to be the likely mechanism to account for most fine-scale layering.

H. H. Hess (1960, p. 134) goes further than this:

> Megacycles in the mineral variations represented on the floor by thicknesses of crystal accumulates of the order of magnitude of 1,000 feet possibly indicate one complete convective overturn of the magma. The smaller cycles, of which there are hundreds in this thickness, probably represent minor currents in the magma immediately over the floor.

Then there are students who assign the principal role in the layering process to fractional crystallization by gravitational settling (and presumably sorting) of different kinds of crystals. This seems to be the view of Willemse (1959) and of Cameron (1963, p. 106), who goes on to say with respect to the critical zone of the Bushveld:

> To what extent these processes may have been complicated by slow inflow of magma, by successive more or less distinct heaves of magma, or by major convective overturn, it seems impossible at present to say.

Repetition of macrounits without significant internal cryptic layering throughout the series—as on Rhum—presents a special problem whether it is due to spasmodic convectional overturn (cf. H. H. Hess, 1960) or to some other cause such as periodic slight temperature fluctuation or differential nucleation in significantly undercooled melt (Wager and Brown, 1967, pp. 294, 547). The question is more urgent still where a large, layered pluton (for example, Freetown) completely lacks cryptic layering. In such bodies the magma must periodically or ultimately have reached a stage where one or several phases of uniform composition crystallized in profusion for long periods. Inspection of appropriate experimentally studied silicate melt systems shows that this may be possible along cotectic lines where these converge on invariant points. Here equilibrium would necessitate a nearly steady thermal state with outward heat flow nicely counterbalanced by heat released by crystallization.

## PETROGENESIS

GENETIC SIGNIFICANCE OF CUMULATES Layered basic intrusions supply the most detailed and reliable data we have regarding the nature of successive crystalline fractions separating during low-pressure differentiation of large bodies of tholeiitic magma. In any rock specimen the associated cumulus phases can usually be identified and distinguished from interprecipitate phases. The latter have crystallized from a trapped liquid modified to an extent unknown by interdiffusion with cumulus minerals and slightly later superjacent liquid. Given adequate sampling it should be possible to estimate the bulk compositions and volumes of successive large crystalline fractions (5 to 10 percent of the whole body)—parameters used to compute a general liquid line of descent (see below). But the most reliable and readily accessible information concerns the

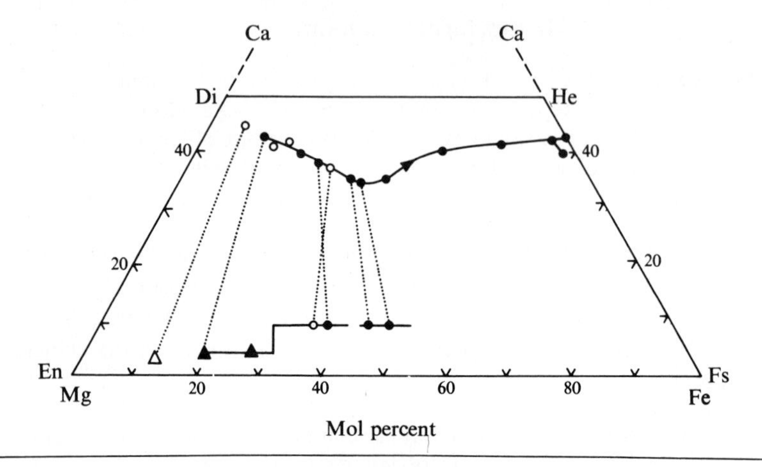


**Fig. 9-11** Pyroxenes (circles, clinopyroxene; triangles, orthopyroxene) of layered basic intrusions. Solid points and compositional trends, Skaergaard intrusion, Greenland (after Wager and Brown, 1967, p. 39). Open points, Bushveld complex, South Africa (after H. H. Hess, 1960, pp. 32, 36). Dotted ties connect coexisting pyroxenes.

order of crystallization of cumulate minerals and their compositional trends in any vertical cryptically layered section.

MINERALOGICAL TRENDS The mineral phases whose crystallization principally determines the trend of differentiation in layered basic intrusions are olivine, pyroxenes, and plagioclase. Separation of chrome spinel in the earlier and of iron titanium oxides in the later stages plays a significant role, especially in shaping the trace-element pattern. Toward the end of the crystallization sequence apatite may appear as a cumulus mineral—as it does in the upper zone of the Skaergaard—and minor quartz, alkali feldspar, zircon, hornblende, or biotite may be formed from ultimate interprecipitate residues high in the pluton.

Olivines and pyroxenes from the lowest cumulates tend to differ significantly from those of ultramafic rocks generally thought to come direct from the mantle —Alpine-type peridotites, and lherzolitic xenoliths in nephelinites and kimberlites (Chap. 13). Both cumulate minerals tend to be slightly less magnesian[1]— commonly $Fo_{85}$ compared with $Fo_{90}$ to $Fo_{92}$ in New Zealand peridotites (Challis, 1965, p. 341) and in Hawaiian lherzolite nodules (R. W. White, 1966, p. 274). The content of $Al_2O_3$ in coexisting early cumulate pyroxenes (1.67 to 1.7 percent in bronzite, 2.7 to 3 percent in diopside of Stillwater ultramafics) overlaps that of some Alpine peridotites but is notably lower than that of others and of spinel and garnet lherzolites. Other differences relate to partitioning of Ca between coexisting pyroxenes.

[1] Exceptional in this respect are optically estimated values $Fo_{94}$ and $En_{93}$ recorded for the lower ultramafic zone of the Great "Dike," Rhodesia.

General trends of compositional change (especially increasing Fe/Mg) in pyroxenes at successively higher levels of the Skaergaard layered series (cf. Table 9-14) are shown in Fig. 9-11. Comparable trends in sections of the Bushveld and Stillwater intrusions are virtually identical and recall those less completely displayed in differentiated tholeiitic diabase sheets (Fig. 9-4).

**Table 9-15** Chemical compositions (oxides, wt %) and CIPW norms of chilled border gabbros (probable liquid compositions) of layered basic intrusions

| | 1 | 2 | 3 | 4 | 5 | | 6 |
|---|---|---|---|---|---|---|---|
| $SiO_2$ | 50.68 | 48.08 | 50.68 | 49.86 | 52.55 | Ba | 25 |
| $TiO_2$ | 1.06 | 1.17 | 0.45 | 2.02 | 0.30 | Co | 55 |
| $Al_2O_3$ | 13.55 | 17.22 | 17.64 | 11.17 | 15.38 | Cr | 170 |
| $Fe_2O_3$ | 1.17 | 1.32 | 0.26 | 2.37 | 0.70 | Ni | 180 |
| FeO | 9.08 | 8.44 | 9.88 | 14.70 | 10.26 | Rb | 4 |
| MnO | 0.18 | 0.16 | 0.15 | 0.28 | 0.21 | Sr | 267 |
| MgO | 9.70 | 8.62 | 7.67 | 3.44 | 6.84 | V | 190 |
| CaO | 11.22 | 11.38 | 10.47 | 10.59 | 10.30 | Zr | 50 |
| $Na_2O$ | 1.79 | 2.37 | 1.87 | 2.46 | 2.70 | | |
| $K_2O$ | 0.63 | 0.25 | 0.24 | 0.91 | 0.22 | | |
| $P_2O_5$ | 0.10 | 0.10 | 0.09 | 1.20 | 0.01 | | |
| $H_2O^+$ | 0.53 | 1.01 | 0.42 | 0.50 | 0.41 | | |
| $H_2O^-$ | 0.06 | 0.05 | 0.06 | 0.08 | 0.06 | | |
| Total | 99.75 | 100.17 | 99.88 | 99.58 | 99.94 | | |
| *Q* | | | | 3.48 | | | |
| *or* | 3.4 | 1.48 | 1.39 | 5.56 | 2.22 | | |
| *ab* | 15.1 | 20.05 | 15.72 | 20.96 | 22.79 | | |
| *an* | 26.9 | 35.62 | 39.06 | 16.40 | 28.77 | | |
| *di* | 23.1 | 16.43 | 9.95 | 24.21 | 18.56 | | |
| *hy* | 23.3 | 7.63 | 31.82 | 18.17 | 26.00 | | |
| *ol* | 2.7 | 13.54 | | | | | |
| *mt* | 1.6 | 1.91 | 0.46 | 3.48 | 0.93 | | |
| *il* | 2.0 | 2.22 | 0.91 | 3.80 | 0.61 | | |
| *ap* | 0.2 | 0.24 | 0.20 | 2.85 | | | |
| Total | 98.3 | 99.12 | 99.51 | 98.91 | 99.88 | | |

*Explanation of column headings*

Analyses 1–3 represent less differentiated, analyses 4 and 5 more differentiated magmatic liquids.

1 Muskox intrusion (average of two analyses; C. H. Smith and Kapp, 1963, p. 33)
2 Skaergaard marginal gabbro (Wager and Brown, 1967, table 7, analysis 4507)
3 Base of Stillwater (H. H. Hess, 1960, p. 162, analysis G70)
4 Roof of Bushveld (H. H. Hess, *ibid.*, BV52)
5 Roof of Great "Dike" (H. H. Hess, *ibid.*, GD29)
6 Atomic abundances of trace elements in chilled border gabbros, Skaergaard (Wager and Brown, 1967, pp. 193–201)

COMPOSITION OF PARENTAL MAGMAS Chilled border rocks (Table 9-15)—and all too few analyses are available—approximate chemically to liquids at some stage of magmatic evolution. But what stage? That generally favored is the earliest, reflecting the composition of the parental magma. This inference merits critical examination.

The northern and southern borders of the Bushveld pluton thin outward from the center; and here the floor, within a thin, chilled border zone, is overlapped outward by cumulates formed at successively later stages of crystallization. Plagioclase of the lowermost cumulate changes outward from $An_{83}$ to $An_{70}$, pyroxene from $En_{80}$ to $En_{60}$. From this, H. H. Hess (1960, pp. 158–161) argued that intrusion was followed, while crystallization was in progress, by lateral spreading of the pluton along a north-south axis. In that case some chilled peripheral border rocks might represent relatively late liquid fractions. It turns out that this most likely is true of at least two roof samples (cf. Table 9-15, analysis 4). In their high mafic indices ($\sim$ 80) and normative $(or + ab)/an$ ratios and their extremely high content of $P_2O_5$, they bear the stamp of liquids late in the fractionation series.

Chilled border rocks from the Muskox, Skaergaard, and Stillwater plutons cited in Table 9-15 (analyses 1–3) in all three respects have a less differentiated chemical character: mafic indices, for example, are 51 to 57. But even the Muskox specimens (mafic index, 51) do not represent the overall bulk composition of the pluton. C. H. Smith and Kapp (1963) picture these border specimens as representing the liquid phase segregated close to the border during intrusion of an olivine-charged liquid-crystal mush. That such segregation is possible has been demonstrated by scale-model experiments (Bhattacharji, 1967).[1]

LIQUID LINES OF DESCENT

*Generalized trend from border samples* Recognizing the close chemical and mineralogical similarity of the Bushveld, Great "Dike", and Stillwater plutons, H. H. Hess (1960, p. 166) deduced a common, generalized "liquid line of descent" from five chilled border rocks. The resulting plot on the $FeO$–$(Na_2O + K_2O)$–$MgO$ triangle (cf. Fig. 9-13, dashed) is a general model rather than an actual differentiation trend. It is based on two assumptions, both questionable: all three parental magmas were identical; and their composition is that of one chilled specimen (from the base of the Stillwater) whose overall chemical character appears to be the most appropriate. Early in this chapter we noted that tholeiitic flood basalts vary significantly from one volcanic province to another and even with passage of a few million years within a single province (for example, contrast the Picture Gorge with the Yakima basalts in the

[1] Inward concentration of phenocrysts is in fact a predictable consequence of mechanical interaction of suspended phenocrysts where the velocity gradient increases with proximity to the confining wall (Komar, 1972). This generates a "grain-dispersive pressure" which is constant for any given flow regime and is proportional to a phenocryst-concentration factor and to the velocity gradient.

Columbia River region). Comparable variation can be expected among the parental magmas of different major tholeiitic plutons. The high $Al_2O_3$ of the "parental" Stillwater magma (Table 9-15, analysis 3) moreover raises the possibility that even this may have been derived from a more primitive magma by removal of magnesian olivine and nonaluminous pyroxene at crustal pressures.

*Computed major-element trends* Petrogenetically, the most significant single item to emerge from the Skaergaard studies is a computed liquid line of descent (Fig. 9-12). This "Skaergaard trend" figures so prominently in ideas concerning

**Fig. 9-12** Oxides (weight percent) in successive computed liquid fractions during solidification of Skaergaard pluton. (After Wager, 1960, fig. 10.)

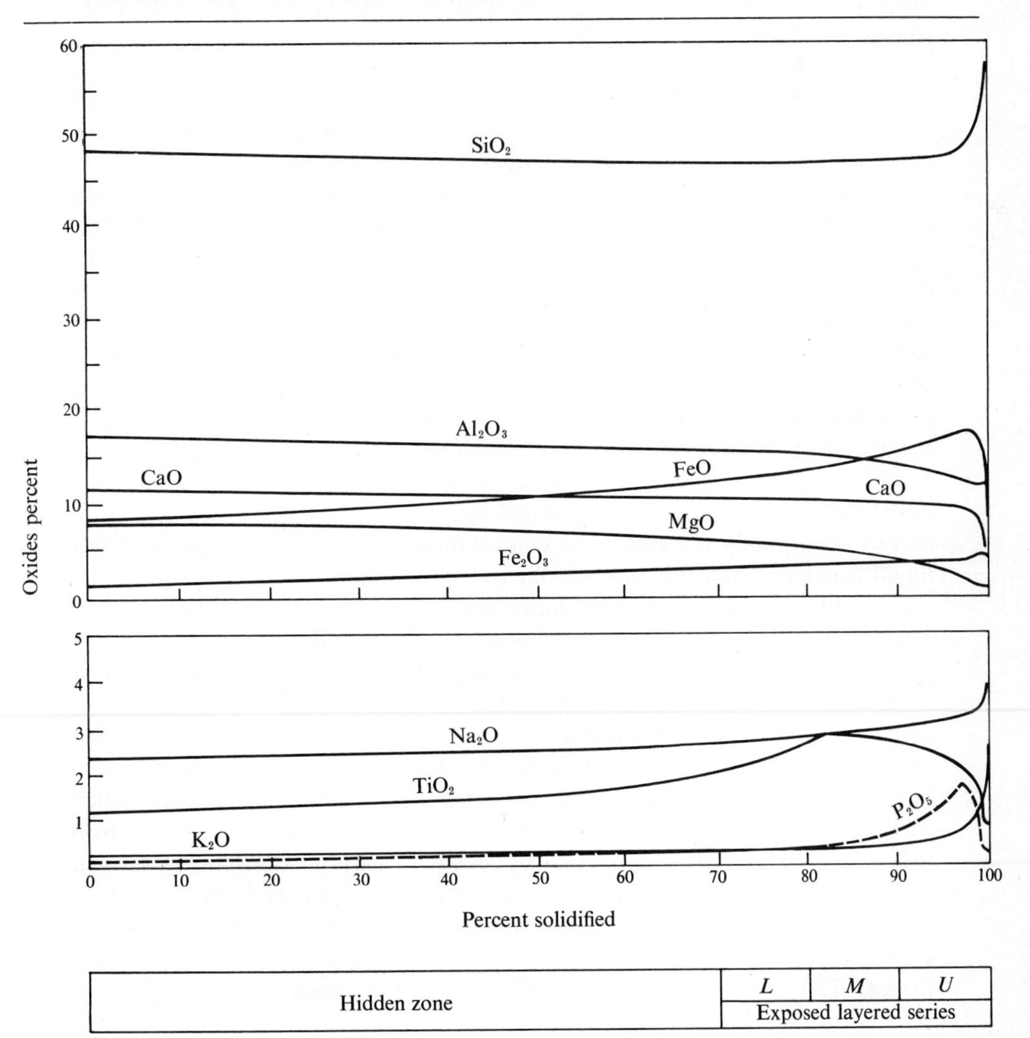

fractionation of tholeiitic magmas at crustal pressures that its validity and limitations deserve critical scrutiny. It is based on data of three kinds: (1) respective volumes and compositions of half a dozen arbitrarily delineated zones collectively constituting the exposed layered series (cf. Table 9-16, analyses 1–3); (2) composition of an assumed parental magma; and (3) compositions of selected samples of granophyre (cf. Table 9-16, analyses 4, 5). From the first two it is possible also to estimate the composition and relative volume of the hidden zone. These parameters place severe additional constraints upon the line of descent computed from the first set of data alone. The procedure first devised by Wager and Deer (1939, pp. 217–224) was later slightly modified by Wager (1960; see also Wager and Brown, 1967, pp. 150–178). Subsequently it has been critically reassessed by Chayes (1970). The Skaergaard trend is most familiar in the form shown in Fig. 9-13.

The composition assigned to the parental magma is that of selected chilled border rock. As Chayes points out, there is hardly any other choice. It is merely a matter of deciding on "best values"; and to this end Wager and Brown finally chose a single specimen (Table 9-16, analysis 2; point O in Fig. 9-13).

Sections *X* to *Y* of curves showing the liquid line of descent represent the path of the changing liquid while the exposed layered series was crystallizing. This path could readily be computed from chemical and geometric (volumetric) data alone provided that these were accurately known and that the system was closed throughout crystallization. Liquid *X* would have the same composition as that of the whole exposed section. The composition of the last residual liquid would be that of the uppermost cumulative zone. Clearly, however, we are not dealing with a closed magmatic system. A substantial fraction of the early magma contributed toward the quartz gabbros of the upper border zone; small quantities of late siliceous residues were segregated from the upper part of the layered series and today are represented by granophyres. Wager was obliged to correct accordingly the estimated compositions of the earliest and the latest liquid fractions—and this can be done only by guessing.

Finally it is necessary to achieve mutual chemical balance between the exposed layered series and estimated volume (exposed plus hidden segments) of parental magma of the composition already stipulated. The obvious geometric procedure for estimating the volume of the hidden zone is to extrapolate downward the external boundaries of the exposed portion of the pluton. But the value so obtained proved much too small to permit the requisite chemical balance. Indeed to achieve this the concentrations of certain components (such as $P_2O_5$ and $TiO_2$) in the lowermost hidden cumulates would have to drop below zero. So the direct approach by geometric extrapolation was discarded. Instead the volume of parental magma was increased, by trial and error, to a point where it could supply the quantity of such components observed in the enriched upper zones and leave a reasonable amount (averaging about 0.8 percent $TiO_2$ and 0.03 percent $P_2O_5$) fixed in the depleted hidden zone. In

**Table 9-16** Chemical compositions (oxides, wt %) and CIPW norms of gabbros and granophyres of three layered intrusions

| | 1 | 2 | 3 | 4 | 5 | 6 | 7 | 8 | 9 |
|---|---|---|---|---|---|---|---|---|---|
| $SiO_2$ | 46.37 | 48.15 | 44.10 | 60.23 | 72.69 | 49.70 | 50.12 | 51.12 | 47.15 |
| $TiO_2$ | 0.79 | 2.64 | 2.43 | 1.18 | 0.46 | 0.16 | 0.19 | 0.19 | 1.21 |
| $Al_2O_3$ | 16.82 | 18.02 | 11.70 | 11.19 | 13.15 | 22.04 | 20.01 | 16.34 | 17.56 |
| $Fe_2O_3$ | 1.52 | 2.52 | 2.05 | 5.52 | 0.92 | 0.66 | 0.80 | 1.09 | 1.46 |
| FeO | 10.44 | 9.50 | 22.68 | 9.11 | 2.80 | 4.02 | 4.29 | 5.84 | 7.60 |
| MnO | 0.09 | 0.12 | 0.21 | 0.24 | 0.04 | 0.09 | 0.09 | 0.15 | 0.15 |
| MgO | 9.61 | 5.25 | 1.71 | 0.51 | 0.16 | 7.03 | 7.91 | 8.49 | 9.76 |
| CaO | 11.29 | 10.17 | 8.71 | 5.11 | 1.80 | 13.59 | 13.97 | 13.63 | 12.41 |
| $Na_2O$ | 2.45 | 3.46 | 2.95 | 3.92 | 4.02 | 1.79 | 1.74 | 2.18 | 2.10 |
| $K_2O$ | 0.20 | 0.14 | 0.35 | 1.94 | 3.26 | 0.07 | 0.05 | 0.06 | 0.18 |
| $P_2O_5$ | 0.06 | 0.05 | 1.85 | 0.27 | 0.03 | 0.02 | Tr. | 0.03 | 0.01 |
| $H_2O^+$ | 0.29 | 0.20 | 0.22 | 0.80 | 0.66 | 0.82 | 0.69 | 0.49 | |
| $H_2O^-$ | 0.09 | 0.02 | 0.20 | 0.10 | 0.09 | 0.09 | 0.04 | 0.07 | 0.76 |
| Total | 100.02 | 100.50* | 99.16 | 100.12 | 100.08 | 100.08 | 99.90 | 99.68 | 100.35 |
| *Q* | | | | 17.74 | 31.25 | | | | |
| *or* | 1.18 | 0.83 | 2.07 | 11.49 | 19.27 | 0.39 | 0.28 | 0.33 | 1.06 |
| *ab* | 17.78 | 29.27 | 24.96 | 33.17 | 34.01 | 15.14 | 14.67 | 18.34 | 17.78 |
| *an* | 34.31 | 33.26 | 17.67 | 7.21 | 8.21 | 51.78 | 46.56 | 34.57 | 37.95 |
| *ne* | 1.60 | | | | | | | | |
| *di* | 17.42 | 13.93 | 11.81 | 14.34 | 0.44 | 12.30 | 18.12 | 26.91 | 19.01 |
| *hy* | | 4.69 | 20.03 | 4.33 | 3.86 | 17.35 | 18.29 | 14.35 | 1.58 |
| *ol* | 23.50 | 9.15 | 10.51 | | | 0.76 | | 2.67 | 17.78 |
| *mt* | 2.20 | 3.65 | 2.97 | 8.00 | 1.33 | 0.93 | 1.16 | 1.62 | 2.11 |
| *il* | 1.50 | 5.01 | 4.62 | 2.24 | 0.87 | 0.30 | 0.38 | 0.38 | 2.29 |
| *ap* | 0.14 | 0.12 | 4.38 | 0.64 | 0.08 | 0.05 | | 0.08 | 0.03 |
| Total | 99.63 | 99.91 | 99.02 | 99.16 | 99.32 | 99.00 | 99.46 | 99.25 | 99.59 |

* Includes $CO_2$, 0.03; S, 0.14; SrO, 0.07; BaO, 0.01; CuO, 0.006.

*Explanation of column headings*

1–5 Skaergaard intrusion (Wager and Brown, 1967, tables 5, 9)
- 1 Average rock (4077, plagioclase-augite-olivine cumulate), middle of lower zone
- 2 Olivine-free gabbro (3662), middle zone
- 3 Ferrodiorite (4145), middle of upper zone
- 4 Melanogranophyre (4332)
- 5 Acid granophyre from transgressive sheet (5259)

6–8 Stillwater complex (H. H. Hess, 1960, pp. 92, 93, 101)
- 6 Analysis of composite sample of 40 weighted rocks, representing bulk composition of exposed portion (4000 m thick) above the ultramafic zone
- 7 Hypersthene gabbro (EB 43) near base of upper norite zone
- 8 Hypersthene gabbro (EB 40) at top of exposed section, upper norite zone

9 Insch layered intrusion, northeastern Scotland (P. D. Clarke and Wadsworth, 1970)
- 9 Olivine-bearing gabbro (1), possibly close to parental magma

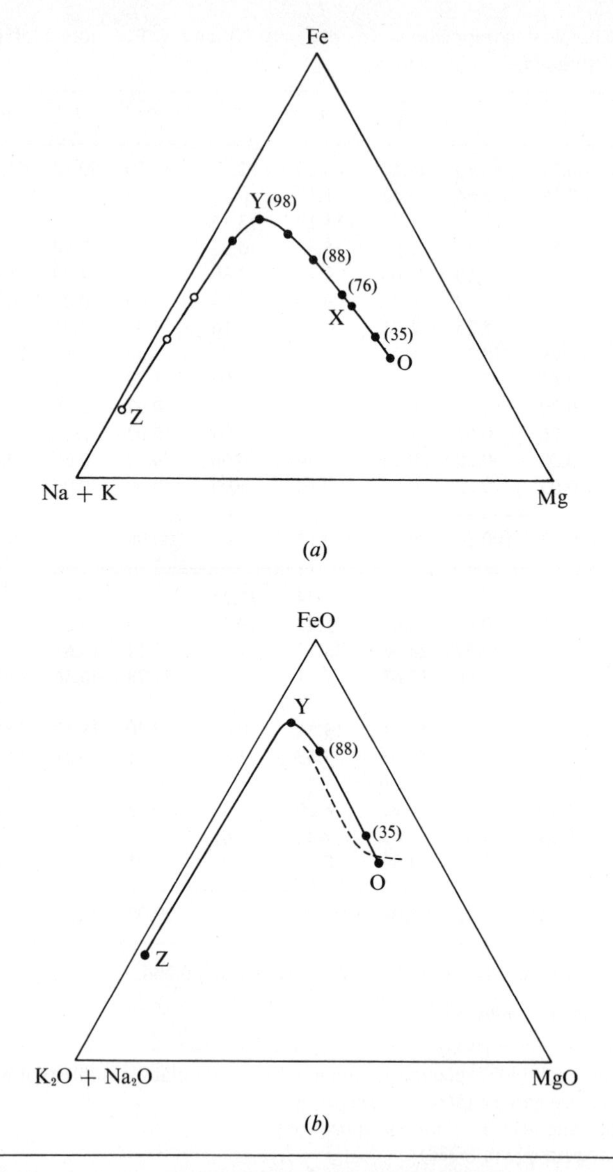


**Fig. 9-13** AFM diagrams for the computed liquid trend of fractionation, Skaergaard pluton. (*a*) Atom percent; (*b*) oxides, weight percent. *O* is parental magma (chilled marginal gabbro); *X* is liquid composition when the lowest exposed rocks began to form (70 percent solidification); *Z* is a late granophyre (other granophyres shown as open circles). Numbers in parentheses show percentage solidified. Dashed curve is the Stillwater "liquid trend" of H. Hess (1960, pl. 11, C). Total iron is shown in (*b*) as FeO.

Wager's model the volume of the hidden zone is 70 percent of the whole pluton.[1] Chayes (1970) prefers a best guess of 81 percent.

The last section of the liquid line, *Y* to *Z* (Fig. 9-13), representing small quantities of the ultimate differentiates, was estimated independently. It is simply an arbitrary plot of selected analyses of late granophyric segregations. That these represent liquid compositions, and that the curve is a fractionation trend, are assumptions—not necessary consequences of the basic data. And the curvature of the *XY* section toward *Y* in Fig. 9-13 reflects these and other assumptions relating to volume that are not universally accepted. Chayes (1970), for example, prefers a model in which enrichment in alkali relative to iron, even at the end of the liquid line of descent, is negligible.

To sum up, no part of the "Skaergaard trend" can be considered as firmly established. Most securely founded is the earlier part of the *XY* section; and the three segments *OX*, *XY*, and *YZ* are open to independent modification as data, especially those relating to volume, become improved. The composite whole, *OXYZ*, is not a reliably established case history in differentiation. Rather it is something perhaps of greater significance: a sophisticated model founded on an impressive body of diversified data, showing how fractionation of tholeiitic magma at low oxygen fugacities[2] is likely to occur at crustal pressures. No doubt the model is in some respects oversimplified—particularly in the assumption that the complete pluton represents an initially homogeneous magma system that remained closed throughout crystallization. Revisions can be expected when more is known about the volume of the hidden zone.

*Computed trace-element trends in the Skaergaard intrusion* The basic chemical data (Wager and Brown, 1967, pp. 179–203) are abundances determined for individual minerals and cumulate rocks and for rocks (chilled gabbros, late granophyres) believed to represent liquid fractions. Cumulate trends for the exposed layered series show a wide scatter reflecting different concentrations in felsic and mafic cumulates at each stage of solidification. A graphic method, essentially that devised to achieve total chemical balance of major elements, was used to construct a possible continuation of the cumulate curve through the hidden zone—*B* to *A* in Fig. 9-14, which illustrates the curve for vanadium. *BCD* is based on abundances observed in rocks at different levels in the layered series (that is, at successive stages of solidification as previously estimated); *L* and *E* are respective mean abundances in chilled gabbro and granophyres. The sector *APB* is drawn arbitrarily so that the stippled area *APBCDE*, expressing total vanadium in the pluton, equals that of the rectangle *OLME*—total vanadium in the initial body of magma. The broken curve *LNE* is a possible liquid trend consistent with the configuration of the cumulate curve.

[1] Even this model will not completely account for high concentrations of $P_2O_5$ in the uppermost zone, where apatite is a cumulus phase. So Wager was obliged to stipulate that these rocks pinch out laterally, thus reducing their estimated volume.

[2] Reflected in low $Fe_2O_3/FeO$ and very strong increase in Fe/Mg along the liquid trend.

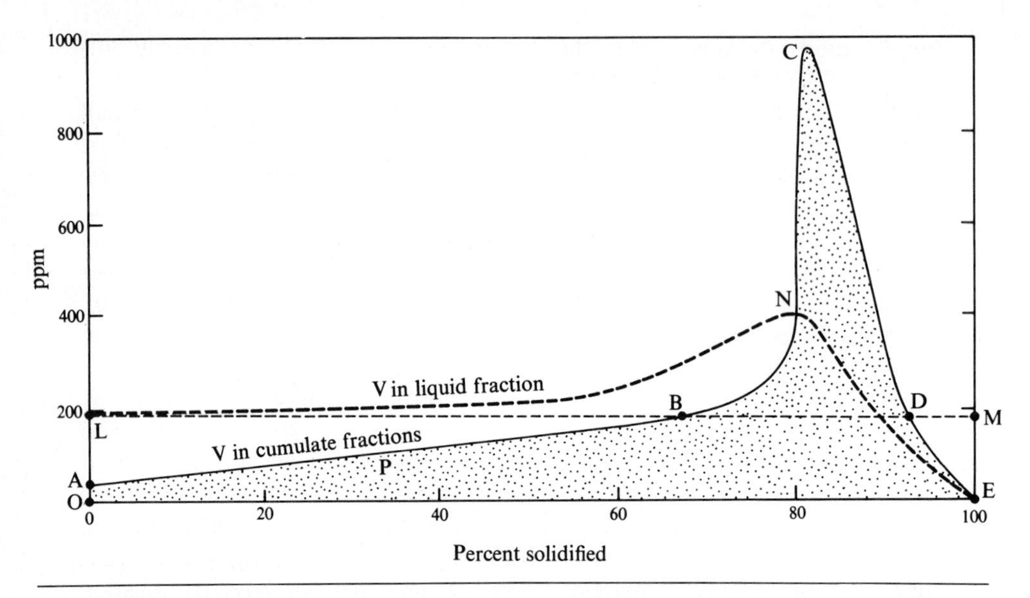


**Fig. 9-14** Graphic estimation of variation in vanadium content of liquid and cumulate fractions with progressive solidification, Skaergaard pluton. After Wager and Brown (1967).

Factors in the fractionation process that have been amply verified by mineral analysis include the following (Fig. 9-15):

1. Early removal of Cr in magnesian pyroxenes and of Ni and Co in olivine, pyroxene, and iron-oxide minerals.
2. Removal of Sr in plagioclase, which becomes most effective at composition $An_{40}$. This implies build-up in the liquid fraction as far as the lower part of the exposed layered series, and subsequent depletion.
3. Relatively late removal of V, Ti, and Mn in titanomagnetite cumulates.
4. Late wholesale concentration of P in apatite cumulates of the upper zone.
5. Steady concentration of Ba, Rb, Li, U, and Zr in the last residual liquids represented by granophyres.

ISOTOPE CHEMISTRY, DIFFERENTIATION, AND CONTAMINATION Differentiation in some layered intrusions is accompanied by changes in isotopic composition of oxygen according to what is generally regarded as a normal pattern in differentiated volcanic series: $\delta_{O^{18}/O^{16}}$ increases along the line of fractionation. This is well exemplified in rocks of the Muskox pluton: $\delta_{O^{18}/O^{16}}$ is 6.8 in chilled border rock, 6 to 8 in olivine gabbros, and 9 to 12 in the final granophyric fractions. The trend in the Bushveld is similar but less pronounced. The Skaergaard intrusion, however, is highly anomalous. Ferrogabbros, although themselves a product of differentiation, have rather low values of $\delta_{O^{18}/O^{16}}$ (between 5 and 6).* Oxygen in the granophyres is completely abnormal: delta values,

* Oxygen-isotope data, and much of the opinion here expressed, come from H. P. Taylor (1963, 1968).

instead of increasing, have fallen to between 1 and 3. Contamination by reaction with invaded Precambrian gneisses cannot be the explanation, since oxygen in these rocks is heavy ($\delta_{O^{18}/O^{16}}$ = 7.8 to 8.4). Considering the extreme nature of Skaergaard differentiation, it is conceivable that the normal trend of oxygen-isotope fractionation was here reversed. More likely, according to H. P. Taylor

**Fig. 9-15** Variation in trace-element abundances (ppm) in analyzed rocks (full lines) and computed liquids (broken lines) during solidification (fractionation) of Skaergaard pluton. Data from Wager and Brown (1967, pl. x).

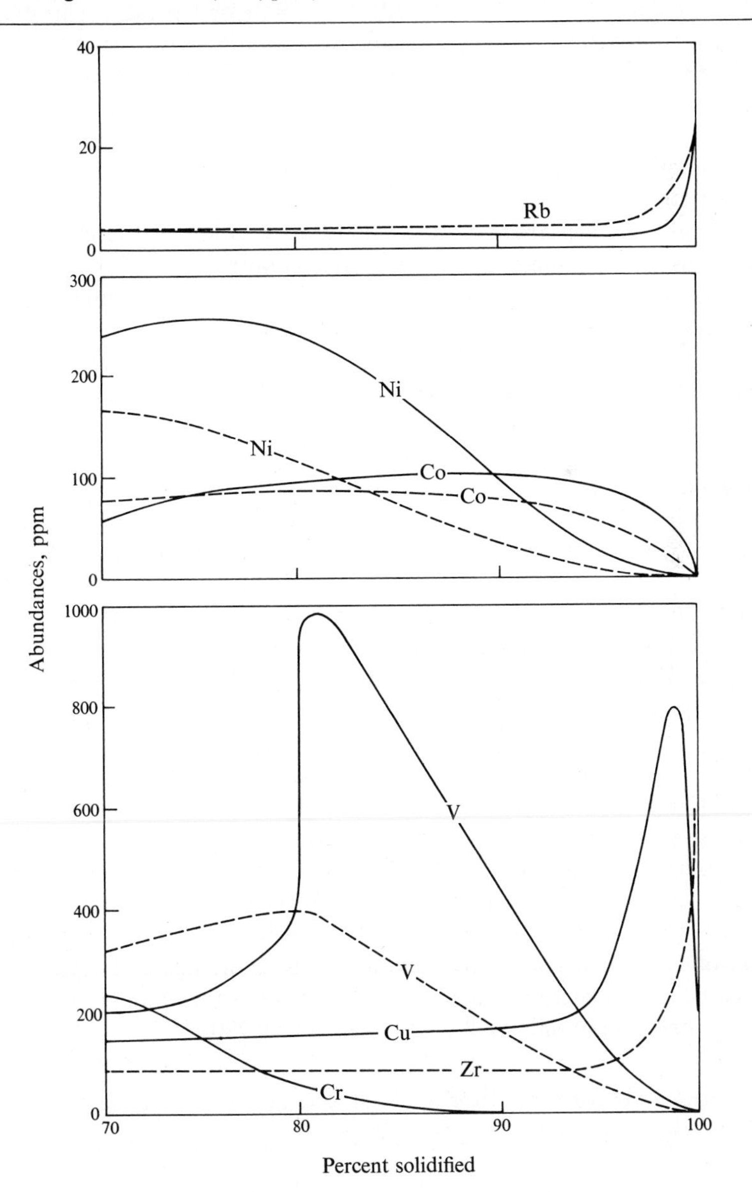

(1968, p. 61) we see the contamination effect of meteoric water sucked into the upper border of the pluton and there exchanging oxygen with a still-liquid fraction that ultimately yielded granophyric end liquids. This interpretation is supported by low values of $\delta_{O^{18}/O^{16}}$ (0.7 to 3.7) in chilled marginal rocks and in some gabbros of the upper border group readily accessible to meteoric water.

Strontium of chilled border rocks and ferrogabbros of the Skaergaard has an initial isotopic composition $Sr^{87}/Sr^{86} = 0.7065 \pm 0.002$—within range of continental basalts (E. I. Hamilton, 1963). Some granophyric differentiates yield anomalously high values (0.7101 to 0.7303) that present a problem still unsolved. Significant contamination by incorporation of or isotopic exchange with gneissic country rock—although apparently contrary to the evidence of oxygen composition—is a possibility.

The influence of contamination and fractional crystallization of basic magma upon the composition of strontium in the ultimate products is illustrated in recent studies of the Insch layered gabbros in northeastern Scotland.[1] This body, whose outcrop area measures about 40 × 8 km, is the largest exposed segment of a basic sheet emplaced in hot country rock near the climax of Dalradian metamorphism dated here at 486 ± 17 m.y. (Pankhurst, 1969, 1970). It is a layered intrusion built of cumulates and showing strong cryptic layering. The end products of differentiation, in the uppermost zone, are cumulates of unusual composition—syenogabbros and syenites in which abundant alkali feldspar is accompanied by andesine and iron-rich pyroxenes and olivine. Initial $Sr^{87}/Sr^{86}$ values in the principal mafic rocks range from 0.703 in chilled basal gabbro to 0.712 in rocks with relatively sodic plagioclase. There is nothing in the major-element composition of these rocks to indicate significant assimilation of country rock. Instead, Pankhurst suggests that there was extensive and continuous exchange of strontium between aureole rocks and the crystallizing magma, progressing toward equilibration between the two systems. The final differentiates—ferrogabbro to syenite—show no further progression in this direction. They fit an isochron for 500 m.y. with initial $Sr^{87}/Sr^{86} = 0.712$. This is one of the few recorded cases in which evidence from the wall-rock system substantiates the model of isotopic exchange. The aureole for a long distance from the gabbro contact itself attained internal isotopic equilibrium following intrusion of the gabbro. Its constituent rocks also fit an isochron—for 500 m.y.—giving an initial $Sr^{87}/Sr^{86}$ value of 0.718. It is thought that the same mechanism that homogenized strontium in the aureole—possibly aqueous diffusion—was responsible too for exchange between wall rocks and magma. Other gabbro bodies in the same area became highly contaminated by partial fusion of Dalradian gneisses containing strongly radiogenic strontium. These are the classic cordierite norites of Haddo House and Arnage; and they are found to have initial $Sr^{87}/Sr^{86}$ ratios as high as 0.730.

[1] For general background and references to extensive classic literature, see *Scottish J. Geol.*, vol. 6, pt. 1, pp. 1–132, 1970.

## Parental Continental Tholeiitic Magmas

Those continental tholeiitic magmas that most clearly qualify—by accepted criteria of volume and chemistry—for parental status range from strongly undersaturated *ol*-normative to oversaturated types. More than one such can be distinguished in some individual provinces. There is a good deal of chemical overlap between continental and oceanic tholeiites. But on the whole, continental magmas seem to be recognizably more potassic; and with this distinction is associated divergence in isotopic composition of strontium and in general abundance pattern of dispersed elements (Table 9-17). In some provinces, such as Tasmania, Antarctica, and the northern Karroo (Cox, 1972), tholeiitic rocks (Table 9-17; analyses 1, 4) are consistently richer in potassium than in any oceanic tholeiitic province. And at the other end of the spectrum low-potassium abyssal tholeiites (Table 9-17, analysis 10), which have such a wide distribution in the oceans, have no continental counterpart.

The normal $Al_2O_3$ content of most continental tholeiitic magmas is between 13 and 15.5 percent by weight. However, in the dominant basalts of some provinces—for example, the Pliocene-Holocene province of eastern Oregon (Verhoogen et al., 1970, pp. 299, 300)—$Al_2O_3$ may exceed 18 percent. Such rocks represent what some have designated a high-alumina basaltic magma type. Chilled border gabbros of many, but not all, stratified basic plutons show the same characteristic. High-pressure experiments have shown that separation of olivine and nonaluminous pyroxene from olivine-tholeiite magma of the usual type at intracrustal pressures could yield *Q*-normative high-alumina liquid fractions. Some high-alumina basalts and gabbros have mafic indices (55 to 60) high enough to suggest some degree of previous differentiation. And massive olivine-pyroxene cumulates in the lower sections of stratified intrusions show that fractionation of the prescribed kind commonly occurs in magma reservoirs within the crust. By contrast there are abyssal oceanic basalts whose low content of potassium and "incompatible" elements implies a parental rather than a derivative condition, but which also qualify as high-alumina basalts. Perhaps concentration of aluminum is achieved in more than one way; and its significance as a criterion of magma type may have been overrated.

## Differentiation of Tholeiitic Magmas

### EVIDENCE FROM BASIC PLUTONS

The large stratified basic plutons, particularly the well-documented Skaergaard intrusion, provide the clearest available picture of the course of fractional crystallization of tholeiitic magma in extensive crustal reservoirs. From such sources, reinforced by rather imperfect information supplied by differentiated diabase sheets, it is possible to draw some general conclusions:

**Table 9-17** Some geochemical data for continental tholeiitic rocks (1–8) compared with oceanic tholeiites (9, 10)

| | 1 | 2 | 3 | 4 | 5 | 6 | 7 | 8 | 9 | 10 |
|---|---|---|---|---|---|---|---|---|---|---|
| K/Na | 0.63 | 0.26–0.48 | 0.23–0.55 | 0.70 | 0.15–0.41 | 0.17 | 0.10 | 0.1–0.3 | 0.10–0.30 | 0.05 |
| K/Rb | 200–220 | 200–270 | 400–425 | 200–270 | | 350 | 500 | 280 | 400–500 | 1300–1500 |
| Rb/Sr | 0.25 | 0.12–0.4 | 0.03–0.07 | 0.32 | | 0.028 | 0.015 | 0.058 | 0.28 | 0.01 |
| $U/K \times 10^4$ | 1.25 | 1.4–3.2 | 0.4–0.77 | | 0.38–0.64 | | | 1.2–1.4 | 0.3–1.0 | 0.75 |
| $Th/K \times 10^4$ | 5 | 5–7.5 | 2.7–3.0 | | 1.8–4.2 | | | 3.5–3.9 | 2.18–2.8 | 1.5 |
| $Sr^{87}/Sr^{86}$ (initial) | 0.7115 | 0.712 | 0.7057 | 0.706–0.710 | 0.704–0.706 | 0.7029† | 0.7065 | 0.7078 | 0.7039–0.7040 | 0.7026 |
| Sr (ppm) | 130 | 100–140 | 180–400 | 180, 230 | 120–180 | 13 | 267 | 260 | 250–800* | 100–200 |
| Ba (ppm) | | 160–430 | 150–350 | 700, 750 | 200 | | 25 | | 150* | 14 |
| U (ppm) | 0.9 | 0.6–1.6 | 0.2–0.4 | | 0.35 | | | 1.9 | 0.2 | 0.1 |
| Th (ppm) | 3.3 | 2.2–5.4 | 1.3–1.5 | | 1.8 | | | 0.53, 0.78 | 0.7 | 0.2 |

* Tholeiitic basalt, Kilauea (Wager and Mitchell, 1953; cf. Table 8-13).
† Present ratio 0.7063, recalculated assuming age 2700 m.y.

*Explanation of column headings*

1. Diabases of Tasmania (143–167 m.y.). Data from Heier et al. (1965); Compston et al. (1968)
2. Diabases (Ferrar dolerites) of Antarctica (147–163 m.y.). Cf. Tables 9-10, 9-11. Data from Gunn (1966) and Compston et al. (1968)
3. Diabases of Karroo, South Africa (154–190 m.y.). Cf. Tables 9-2, 9-3, 9-5. Data from Compston et al. (1968)
4. Two tholeiitic basalts, Nuanetsi region, Karroo province, South Africa. Cf. Tables 9-4, 9-5.
5. Triassic diabases, New Jersey. Cf. Tables 9-7 to 9-10
6. Peridotite, ultramafic zone, Stillwater complex, Montana. Data from Stueber and Murthy (1966, pp. 1247, 1249, 1252)
7. Marginal gabbro, Skaergaard intrusion, Greenland. Data from Wager and Brown (1967); cf. Table 9-16)
8. Tertiary tholeiitic basalts, Brighton, Tasmania. Data from Edwards (1950) and Compston et al. (1968, pp. 140–141)
9. Hawaiian tholeiitic basalts. Data from Faure and Hurley (1963, p. 38); G. A. Macdonald and Katsura (1964); Heier et al. (1965); and Hedge and Peterman (1970).
10. Some abyssal tholeiites of the ocean. Data from Engel et al. (1965) and Hedge and Peterman (1970).

1. The chemical trend in successive crystalline (cumulate) fractions is portrayed by the pattern of cryptic layering. The ratios Fe/Mg and (*or* + *ab*)/*an* increase steadily, and trace-element concentrations (Fig. 9-15) follow trends consistent with early removal of olivine, pyroxene, and calcic plagioclase. Increasing Na/Ca in plagioclase and increasing Fe/Mg in olivines and pyroxenes with progressive fractionation (Table 9-14) are consistent with the data of laboratory experiment.
2. The liquid line of descent is to some extent a matter of inference. At any stage, however, Fe/Mg and (*or* + *ab*)/*an* ratios must both be higher in the residual liquid than in the coexisting crystalline fraction. For all its imperfections, the "Skaergaard liquid trend" portrays the kind of liquid line of descent to be expected from some tholeiitic magmas fractionating by convection in large crustal reservoirs. Extreme iron enrichment in the late Skaergaard liquids probably reflects low oxygen fugacities, which would inhibit early separation of magnetite. Correlated with this are low values for $Fe_2O_3/(Fe_2O_3 + FeO)$—0.13 and 0.11 in chilled gabbros of the Skaergaard and Muskox intrusions compared with 0.25 or higher in tholeiitic basalts of the northwestern United States. Higher oxygen fugacities presumably would depress the liquid line of descent somewhat below *OXY* in Fig. 9-13.
3. Significant increase in silica is seen only in liquids remaining when crystallization is far advanced. In fact, in the computed Skaergaard trend $SiO_2$ actually falls until solidification is more than 80 percent complete. Liquids in this respect comparable to andesite-dacite-rhyolite volcanic series develop only when their total bulk is less than 2 percent of the original volume of magma.

## EVIDENCE FROM VOLCANIC SERIES

There are thick tholeiitic lava sequences in which the chemical character takes a somewhat more "differentiated" aspect at successively higher levels: the mafic index tends to increase, the content of olivine falls off, normative *Q* may become conspicuous. Alternatively, as in eastern North America, the pattern of such variation may be geographic rather than stratigraphic. Elsewhere again, as in the "normal magma series" of the Hebridean province, the scope of chemical variation may be much broader; some lavas have silica contents in the andesite-dacite range. All such variation patterns have been attributed by one writer or another to differentiated liquid lines of descent derived from parental tholeiite magmas. But this is not the only possible interpretation. Equally viable are alternative models involving fluctuations in the chemistry of source rocks or in melting regimes at the source, or multistage fusion and crystallization.

Let us tentatively accept the proposition, nevertheless, that variation in some tholeiitic volcanic provinces indeed reflects the trend of liquid descent.

How does this compare with the more soundly based Skaergaard trend? The answer, as first realized by Wager and Deer in 1939, is that there is no obvious similarity whatever. Rocks of the "normal magma series" of Mull and Arran plot in a dispersed band well below the *OXY* section of the Skaergaard trend on a standard *AFM* diagram (Fig. 9-13). Also they are much more siliceous and voluminous than any possible Skaergaard counterpart. Divergence from the Skaergaard trend in all these respects is even more marked in any orogenic andesite-dacite-rhyolite series. We can only conclude, following Wager and Deer, that continental lavas of andesitic and dacitic composition (in the broad sense), whether associated with tholeiitic basalts or not, have some other parentage or, less likely, reflect a differentiation trend completely different from that responsible for layering of large basic plutons. Fusion of siliceous basement rock has certainly played a part in the genesis of siliceous magmas (granitic rocks) in Skye.

To sum up, the evidence afforded by chemical variation in tholeiite-dominated volcanic series is equivocal—even more so than in most ocean-island provinces because of possible crustal influences. It may be equally compatible with models of differentiation from a single source or others of multiple magmatic parentage. Its overall impact, regardless of any preferred genetic model, is negative but interesting; available volcanic data imply that late residual magmas analogous to those near the end of the Skaergaard trend are never erupted upon the continental surface. True, there are a few volcanic series that show some similarity to the Skaergaard lines—for example, "icelandites" and rhyolites of Iceland and perhaps of the Galápagos. But these in reaching the surface have risen through dense oceanic crust, not lighter sialic continental crust such as encloses the Skaergaard intrusion.

# Continental Mafic Magmas from Deep Sources

## Preliminary Postulation of Deep Origin

The immediate sources of alkali-basalt and nephelinite magmas, whether continental or oceanic, are placed by general consensus at depths—between 40 and 100 km—rather greater than those of tholeiitic magmas. This view is based partly on laboratory behavior of crystal-melt equilibria in basaltic systems at high pressure. It receives strong support from the common occurrence of large xenocrysts of high-pressure minerals—aluminous pyroxenes, spinel, kaersutite (for example, Binns, 1969; Binns et al., 1970)—and of lherzolite xenoliths (for example, Leeman and Rogers, 1970) in alkaline basic magmas. Such xenoliths, whatever their origin, could coexist with *ne*-normative basic magmas at pressures of 10 to 20 kb or even higher (for example, O'Hara, 1968, pp. 89, 92, 110, 126). They have also been recorded, though much less commonly, in trachytic and phonolitic lavas associated with alkali basalts (for example, J. B. Wright, 1970, 1971).

There are three other continental rock kindreds, possibly in some way related, that are also assigned to deep subscrustal sources underlying stabilized or rifted continental plates: kimberlites, carbonatites, and a varied group of highly potassic rocks. Evidence for a deep source is especially compelling for kimberlites. These contain abundant fragments of garnet peridotite and of eclogite, whose distinctive mineral phases and detailed mineral chemistry testify unequivocally to crystallization at high pressure (Verhoogen et al., 1970, p. 313). Carbonatites and highly potassic magmas are placed in a similar high-pressure regime because of recognizable chemical affinities and some tendency for mutual field association. In some volcanic fields, such as the late Miocene chain of volcanoes in eastern Uganda, carbonatite and nephelinite magmas have risen up at the same eruptive centers. The field connection between carbonatites and nephelinites is indeed very close, and there is a growing conviction that the two are genetically connected.

Evidence of an independent character now strengthens the general idea that all these magmas are drawn from deep sources. This evidence is provided by data relating to the effect of pressure and temperature on the activities of silica and alumina in magmas and to the calculated pressure and temperature at which a particular magma could equilibrate with spinel- or garnet-bearing peridotite (Chap. 3). Representative values for various magmas are as follows:

| | SPINEL PERIDOTITE | | GARNET PERIDOTITE | |
|---|---|---|---|---|
| | *T*, °C | *P*, kb | *T*, °C | *P*, kb |
| Trachybasalt (basanitoid) | 1396 | 24.2 | 1321 | 22.7 |
| Ugandite (silica-poor, K-rich lava) | 1340 | 45.2 | 1658 | 51.6 |

Some lavas, or plutonic rocks, have assemblages which allow pressure to be calculated only at some assumed temperature. Extremely potassic lavas (orendites) equilibrate with lherzolite at pressures between 18 and 35 kb (1100 and 1300°C, respectively) while kimberlite at 1327°C traverses 64 kb (diamond present) and 56 kb (diamond absent). The corresponding depth of these magmas is roughly 70 to 200 km—which most students of plate tectonics would place near and below the bases of continental plates.

For the reasons just given we somewhat arbitrarily treat these chemically diverse rock kindreds in a single chapter. Thus considered it is a miscellaneous group. Nothing is thereby implied regarding possible mutual relationships within it. Common to all four groups of magmas is a deep subcrustal environment of origin. Only one, albeit the most extensively developed (alkali-basalt nephelinite), is also represented within the ocean basins.

## Continental Alkali-basalt and Nephelinite Magma Series

### GEOLOGIC SETTINGS IN SIX PROVINCES

Unity of petrochemical pattern tends to be geographically restricted. Within an extensive alkaline province each eruptive center may have its own chemical individuality, and over a span of a few million years this may change abruptly at a given site. There are large tholeiitic or andesitic provinces within which contemporaneous alkali-basalt volcanism may be localized at scattered centers.

To give an idea of the wide variety encompassed by common alkaline lava series, some illustrative examples of mutually gradational general types will be reviewed. Most have been drawn from six typical major provinces, in all of which—in conformity with Harker's classic generalization—the contemporary tectonic environment was one of regional faulting.

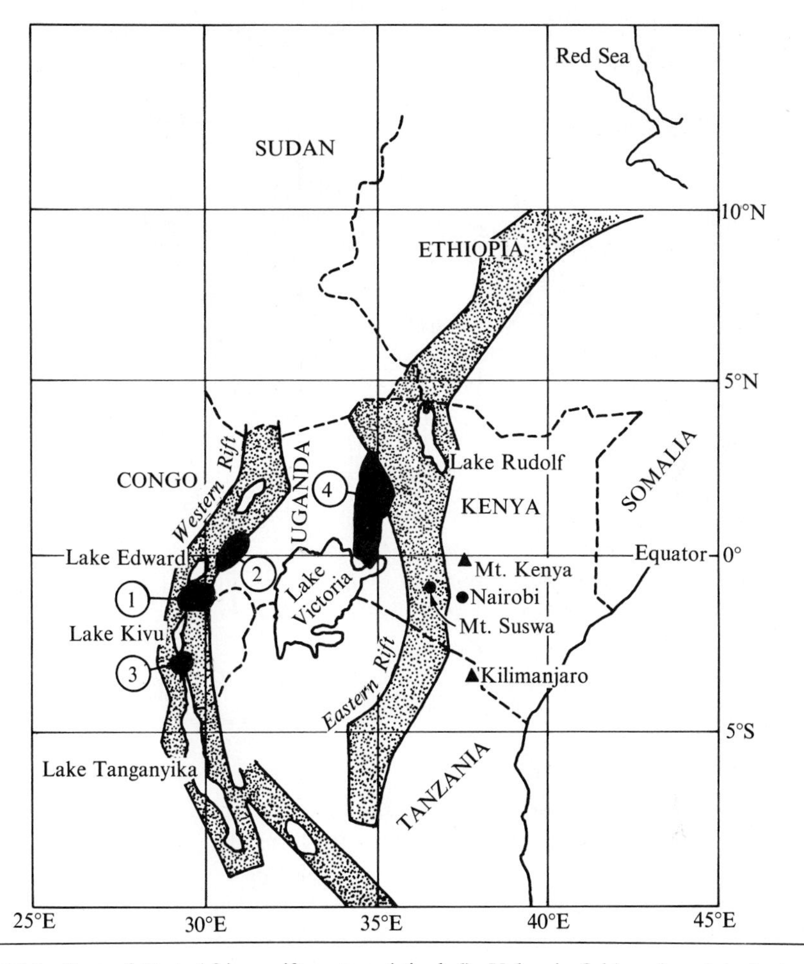


**Fig. 10-1** Part of East African rift system (stippled). Volcanic fields referred to in text: 1, Birunga; 2, Toro-Ankole; 3, Kivu; 4, eastern Uganda.

*East African rifts* Here we consider a part of the East African rift-valley system straddling the Equator between longitudes 25 and 38°E (Fig. 10-1). This was the site of major tectonic dislocation and broadly associated volcanism from Miocene times to the present day (King, 1960, 1970; Wilcockson, 1964). Downwarping along the sites of the main rifts began early in the Miocene. The first major phase of faulting occurred at the opening of the Pliocene. Broad interdependence between volcanism and tectonism is obvious; but there is no persistent relation in detail. Parts of the rifts are devoid of volcanoes. Two of the mightiest Quaternary edifices, Kenya and Kilmanjaro, lie 100 km east of the eastern (Kenya) rift. Equally far to the west of the same structure are the late Miocene volcanoes of eastern Uganda.

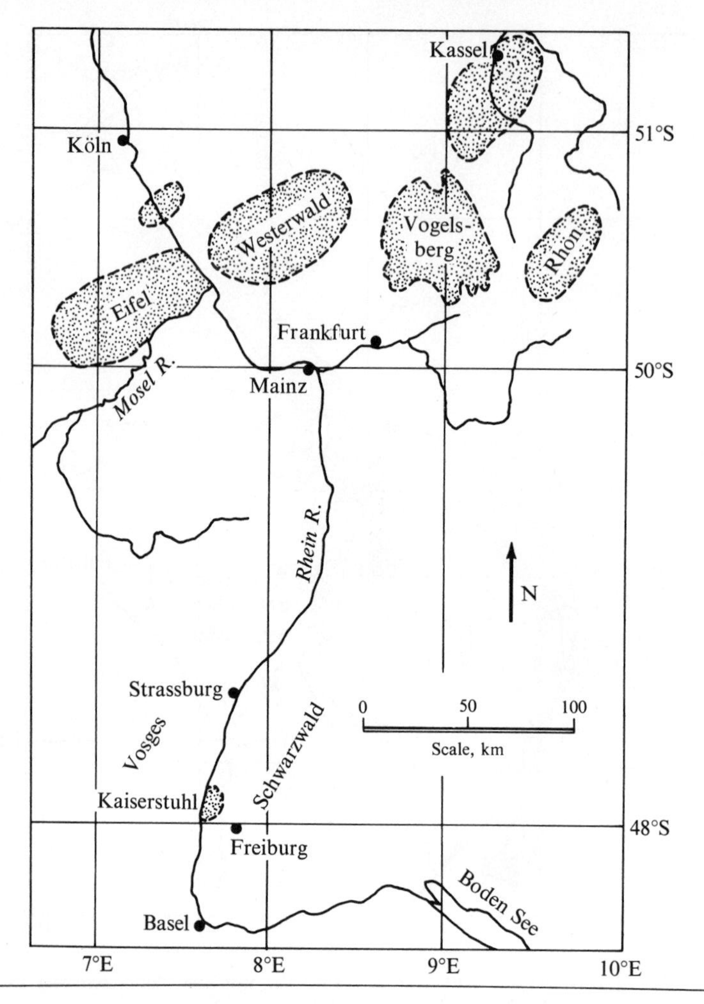


**Fig. 10-2** Rhine volcanic province, western Germany. Volcanic rocks outcrop within stippled areas.

*Late Tertiary Rhine province, western Germany* Miocene and Pliocene volcanism was widespread across much of western Europe beyond (that is, north and west of) the region of Alpine folding. The magmas erupted were largely alkaline. Classic provinces are the Auvergne (central massif of France), the lands bordering the Rhine in southwestern Germany, and the Bohemian Mittelgebirge of Czechoslovakia. The Rhine province (Fig. 10-2) borders the upper Rhine on the east and reaches its greatest development further north, on both sides of the valley where it runs northwestward from Mainz. Here on the east is the Vogelsberg volcanic pile, as big as Etna; to the west is the classic Eifel region, in which much of the volcanic topography still survives. The basement is a folded mid-Paleozoic geosynclinal filling. Except in the northwestern sector—the Eifel and the Westerwald—this basement is thickly mantled with Mesozoic sediments. The tectonic environment of Tertiary volcanism was one of normal faulting—as in the Rhine-graben rift system.

*Miocene volcanics, eastern Australia* In northern New South Wales and adjoining parts of Queensland, Australia (Fig. 10-3), there are half a dozen separate extensive volcanic fields scattered within an area of 100,000 km$^2$. Volcanism has been dated radiometrically as Miocene. The chemical character of the province is essentially alkaline. But there is one large field (Tweed volcano) in which, after local eruption of alkali basalts, the main volcanic edifice was built of tholeiitic lavas associated with strongly oversaturated andesites and rhyolites (J. G. F. Wilkinson, 1968). Basement rocks throughout the province consist partly of folded Paleozoic geosynclinal sediments and intrusive granites and partly of Mesozoic strata which locally include rhyolitic tuffs.

**Fig. 10-3** Distribution of Miocene volcanic rocks, northeastern New South Wales. (After McDougall and Wilkinson, 1967, p. 226.)

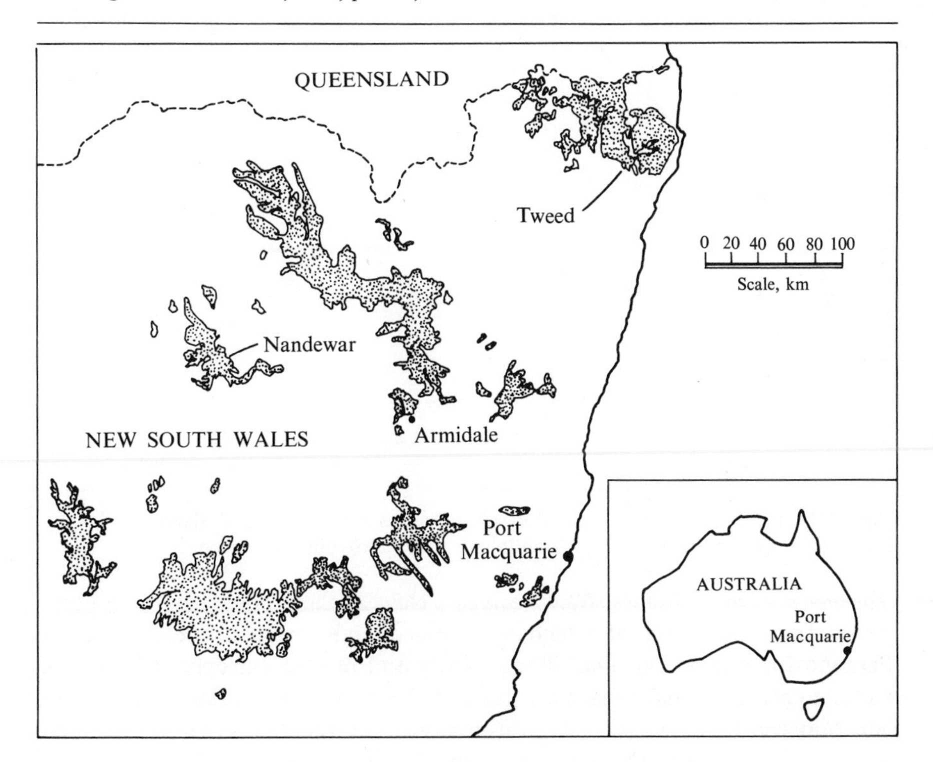

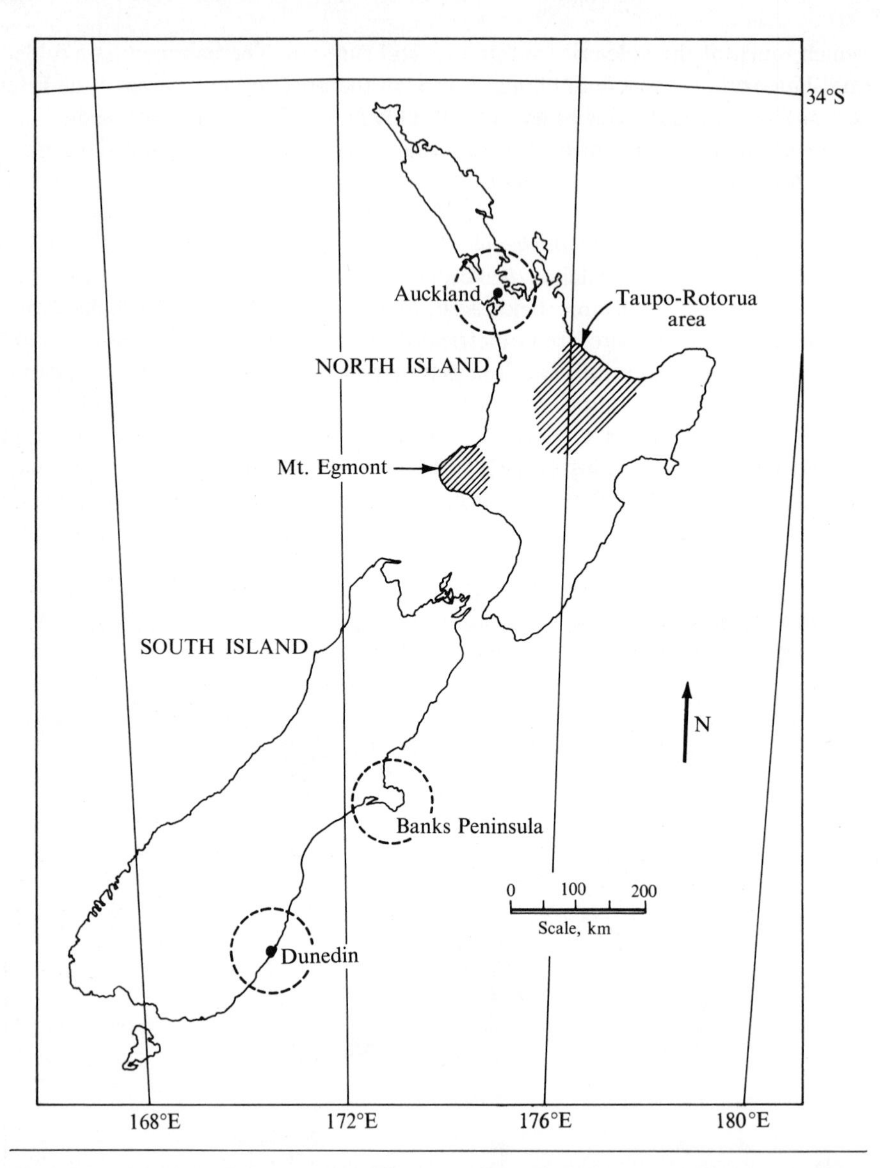


**Fig. 10-4** Three later Cenozoic alkali-basalt provinces (circled), New Zealand. Shaded area is one of contemporaneous and still continuing andesite-rhyolite volcanism.

*Younger alkaline volcanics, New Zealand* New Zealand (Fig. 10-4) is a continental fragment much of which is a folded and variously metamorphosed Permian-Jurassic geosynclinal filling locally and in places deeply covered with Cretaceous and Tertiary marine sediments. There are three principal centers of late Miocene to Pleistocene volcanism where the erupted magmas were consistently alkaline. But this effusion of alkaline magmas was overshadowed by contemporaneous massive eruption of andesitic and rhyolitic magmas in intervening areas, notably in the Taupo-Rotorua field of North Island, where volcanic and geothermal activity is still vigorous.

*Basin-Range Cenozoic province, western United States* Nevada, Utah, and parts of adjoining states constitute an extensive region with strong geologic individuality—physiographic, tectonic, and volcanic. This is the Basin-Range province. From Cambrian times to the opening of the Paleocene this was the site of marine sedimentation in the Cordilleran geosyncline, broken by intermittent episodes of strong orogeny. Throughout the Cenozoic, continental sediments and great effusions of volcanic materials accumulated on the eroded surface of the geosynclinal filling. The tectonic pattern, too, changed drastically to one of shallow faulting in a relatively thin (30 to 35 km) continental crust. Early and middle Cenozoic volcanism gave rise to massive dacitic and rhyolitic ignimbrites. During the Miocene there was a sudden change in the nature of erupted magmas. Everywhere the late Cenozoic lavas are alkaline olivine basalts. Their geologic setting, chemistry, and origin, in relation to chemically different contemporary volcanic series in individually distinct adjoining provinces, has been summarized by Leeman and Rogers (1970).

*Oslo graben, Norway* The city of Oslo in southern Norway lies in the midst of a block of Paleozoic rocks over 200 km long (north-northeast by south-southwest), 30 to 50 km wide, and several kilometers thick, downfaulted into Precambrian gneisses of the Baltic shield. This consists of folded Paleozoic sediments, mostly Cambrian to Devonian, locally capped by lower Permian lavas and cut by numerous large subvolcanic plutons of the same age (radiometrically dated at 276 m.y.). This igneous province was made classic through the early comprehensive work of Brögger, more recently amplified through a series of memoirs by Barth (for example, 1945) and Oftedahl (for example, 1953, 1959) and through a discussion of age considerations by Heier and Compston (1969). It is remarkable for extrusion and intrusion of peralkaline felsic magmas in unusually large volume.

## *Hy*-NORMATIVE ALKALI-BASALT SERIES

As in some oceanic islands, the basic lavas of some continental alkaline provinces are *hy*-normative olivine basalts. Prevalence of titanaugite as the sole pyroxene, a high alkali/silica ratio, and low $hy/(hy + di)$ in the norm ($< 0.45$) distinguish these rocks from undersaturated tholeiites (Coombs, 1963, p. 237). More siliceous associates (Table 10-1; analyses 2–5) are slightly undersaturated hawaiites and mugearites, grading into oversaturated trachytes and peralkaline sodic rhyolites (comendites, pantellerites). In volume these rocks commonly equal or predominate over basalts proper—as they do in the Miocene Nandewar volcano of New South Wales (Abbott, 1969; see also McDougall and Wilkinson,

**Table 10-1** Chemical compositions, CIPW norms, and trace-element abundances (ppm) of *hy*-normative alkaline volcanic rocks, New South Wales and Kenya

| | 1 | 2 | 3 | 4 | 5 | 6 | 7 |
|---|---|---|---|---|---|---|---|
| $SiO_2$ | 47.51 | 50.40 | 53.76 | 62.90 | 73.46 | 63.10 | 62.5 |
| $TiO_2$ | 2.82 | 2.23 | 1.80 | 0.41 | 0.25 | 0.69 | 0.75 |
| $Al_2O_3$ | 16.85 | 16.32 | 16.27 | 16.15 | 13.91 | 12.41 | 15.21 |
| $Fe_2O_3$ | 4.25 | 3.84 | 4.75 | 1.37 | 0.96 | 2.05 | 1.15 |
| FeO | 6.71 | 6.35 | 4.99 | 3.09 | 0.09 | 6.97 | 5.55 |
| MnO | 0.15 | 0.17 | 0.21 | 0.14 | Tr. | 0.35 | 0.28 |
| MgO | 5.62 | 4.96 | 1.89 | 0.61 | 0.13 | 0.16 | 0.33 |
| CaO | 9.28 | 7.43 | 5.16 | 1.83 | 0.04 | 1.12 | 1.45 |
| $Na_2O$ | 3.47 | 4.26 | 5.80 | 5.17 | 5.33 | 8.43 | 7.18 |
| $K_2O$ | 1.08 | 1.84 | 3.19 | 5.57 | 5.27 | 4.57 | 5.20 |
| $P_2O_5$ | 0.93 | 0.92 | 0.96 | 0.17 | 0.08 | 0.09 | 0.14 |
| $H_2O^+$ | 0.59 | 1.01 | 0.19 | 0.13 | 0.12 | 0.14 | 0.30 |
| $H_2O^-$ | | | 0.06 | | | | |
| Total | 99.26 | 99.73 | 99.03 | 98.54 | 99.64 | 100.67* | 100.31† |

| | 1 | 2 | 3 | 4 | 5 | 6 | 7 |
|---|---|---|---|---|---|---|---|
| *Q* | | | | 2.35 | 22.66 | 6.9 | 0.3 |
| *or* | 6.38 | 10.87 | 18.85 | 32.92 | 31.14 | 27.2 | 30.6 |
| *ab* | 29.36 | 36.05 | 49.08 | 52.06 | 42.21 | 38.3 | 49.3 |
| *an* | 27.21 | 19.98 | 8.94 | | | | |
| *ac* | | | | 0.13 | 2.55 | 6.0 | 3.2 |
| *ns* | | | | | | 5.7 | 1.7 |
| *di* | 10.16 | 8.76 | 4.34 | 2.27 | | 1.9 | 4.6 |
| *hy* | 7.48 | 5.57 | 0.72 | 4.52 | 0.32 | 11.7 | 7.9 |
| *ol* | 4.32 | 5.51 | 3.25 | | | | |
| *mt* | 6.16 | 5.57 | 6.89 | 1.92 | 0.08 | | |
| *il* | 5.36 | 4.24 | 3.42 | 0.78 | 0.19 | 1.4 | 1.4 |
| *ap* | 2.23 | 2.18 | 2.27 | 0.40 | 0.19 | 0.3 | 0.3 |
| Total | 98.66 | 98.73 | 97.77 | 97.35 | 99.34 | 99.4 | 99.3 |

| | 1 | 2 | 3 | 4 | 5 |
|---|---|---|---|---|---|
| Ba | 868 | 2735 | 1324 | 478 | 8 |
| Rb | 19 | 33 | 60 | 107 | 185 |
| Sr | 813 | 736 | 645 | 82 | 1 |
| Zr | 127 | 208 | 393 | 936 | 903 |
| K/Na | 0.36 | 0.49 | 0.63 | 1.03 | 1.13 |
| K/Rb | 474 | 464 | 442 | 432 | 236 |
| Rb/Sr | 0.023 | 0.045 | 0.093 | 1.3 | 185 |

* Including F, 0.36; Cl, 0.23.
† Including F, 0.17; Cl, 0.10.

*Explanation of column headings*

Numbers in parentheses refer to analyses in papers cited.

1–5 Nandewar volcano, New South Wales, Australia (Abbott, 1969, pp. 126, 127)
- 1 Olivine basalt (1)
- 2 Hawaiite (4)
- 3 Mugearite ("benmoreite"; 9)
- 4 Trachyte (12)
- 5 Rhyolite (17)

6–7 Eastern rift, Kenya (Macdonald et al., 1970, p. 511)
- 6 Trachyte glass (43/1/S/5), Menegai volcano
- 7 Trachytic glass (43/1/S/6a), Menegai volcano

1967; Binns et al., 1970). In the opening volcanic phase of igneous activity in the Oslo region, mildly alkaline plateau basalts were followed by equally extensive peralkaline trachytes ("rhomb porphyries").

In the central sector of the eastern rift valley, Kenya, *Q*-normative trachytes and pantellerites occur in great volume almost to exclusion of more basic lavas. There is a basal series of flood trachytes surmounted by central volcanoes consisting predominantly of trachyte and pantellerite (Table 10-1; analyses 6, 7). The apparently minor role played here by basaltic magma has led some writers (R. Macdonald et al., 1970; Sutherland, 1970) to question the long accepted argument of Bowen (1937) that the East African petrochemical series basalt-trachyte-pantellerite represents a liquid line of descent controlled by fractional crystallization. Details of trace-element patterns, however, convincingly support derivation of the felsic magmas as extreme liquid residues fractionated from parental basalt (Weaver et al., 1972).

## SHOSHONITE SERIES

In some volcanic fields alkaline lavas erupted immediately after postorogenic uplift of stabilized geosynclines tend to have a persistently potassic character. This is manifested in appearance of sanidine as well as labradorite even in the mafic members and by high $K_2O/Na_2O$ ($\geq 1$). These rocks are relatively high in $SiO_2$ (50 to 54 percent); mostly they are *hy*-normative and close to silica saturation. Their alkaline character resides in a high content of total alkali: $(K_2O + Na_2O) = 6.5$ to 7 percent. Such are the shoshonites and latites of late Tertiary volcanic provinces in the western United States. Commonly associated, and presumably cogeneric, are dikes of mica lamprophyre and felsic lavas of the trachyte and rhyolite families, likewise high in $K_2O$ (cf. Turner and Verhoogen, 1960, pp. 241–243).

These and comparable rock series in eastern Australia have been referred by Joplin (1964, 1965) to a shoshonitic "magma type." There is no known equivalent in oceanic provinces. Even the relatively potassic basalts of Tristan da Cunha are much less siliceous, much lower in total alkali, and have $K_2O/Na_2O$ ratios somewhat lower than 1. Joplin also recognizes a distinct under-saturated division to include rarer feldspathoidal lavas and shallow plutonic rocks (shonkinites) with markedly high $K_2O/Na_2O$, considered with other ultrapotassic rocks in a later section.

## *Ne*-NORMATIVE ALKALI-BASALT SERIES

In many continental provinces the volcanic rocks are predominantly or even exclusively alkaline olivine basalts with normative *ne*. Basalts of this kind in the Basin-Range province (Leeman and Rogers, 1970) have rather high silica ($SiO_2$ 48.5), and *ne* in the norm averages only a few percent. They are accompanied by less abundant, somewhat more siliceous "andesitic" lavas (probably differentiates) with modal andesine rather than labradorite. Alkali basalts (Table 10-3, analysis 1) of a more alkaline type (basanitoid) are the sole products of late Pleistocene volcanism in a small province just south of Auckland, New Zealand (Searle, 1960, 1961; Coombs, 1963, pp. 236, 238; Rodgers and Brothers, 1969). Even more alkaline are basanitic lavas and pyroclastics in a group of Quaternary volcanoes (Korath Range) of the East African rift, north of Lake Rudolf. These have $SiO_2$ 43.5 to 48.5 percent and 15 percent normative *ne* (F. H. Brown and Carmichael, 1969).

Alkaline olivine basalts in yet other fields have minor trachytic associates. This is the case over large segments of the western rift of East Africa, exemplified by the Kivu area (Turner and Verhoogen, 1960, p. 39) and the Rungwe field (Harkin, 1960), where trachytes, however, become locally abundant at Rungwe volcano itself. Elsewhere, undersaturated trachytes and phonolites assume a much more conspicuous role, for example, in the great East African volcano Kenya (Pliocene-Pleistocene). Large-scale outpouring of *ne*-normative trachytic or phonolitic magmas without accompanying basalts is rare but not unknown. The most spectacular recorded instance is regional flooding of great areas along the western margin of the eastern rift of southern Kenya by late Miocene phonolitic lavas and subsequent outpouring of trachytic floods across the floor of the rift during the Pleistocene (Fig. 10-5). Mt. Suswa in the eastern rift north of the Tanzania-Kenya border is a Quaternary volcano built entirely of sodalite trachytes and phonolites (Table 10-2; analyses 9, 10) erupted from independent parental sources in four episodes (Nash et al., 1969). Beneath the floor of trachyte lie Pliocene alkaline basalts of an earlier regional eruptive episode and Miocene phonolites.

Then there are provinces where eruption of felsic lavas substantially preceded outpouring of geographically associated alkali basalts. In northeastern Nigeria (Grant et al., 1972), phonolitic and trachytic lavas were erupted from many centers during successive episodes dated at between 22 and 11 m.y. During the last 7 million years of geologic time, lavas erupted in the same region are alkali basalts and basanites.

## NEPHELINITE SERIES

There are volcanic fields where the mafic lavas are exclusively nephelinites; associated felsic associates, usually in greatly subordinate volume, are phonolites and *ne*-normative trachytes. Much of the erupted material may be pyroclastic; very commonly it contains fragments of chemically similar plutonic rocks such as ijolite and nepheline syenite. In the East African province, nephelinite series (Table 10-2) are exemplified by dissected late Miocene volcanoes in eastern Uganda (King, 1965) and along the east-west Kavirondo rift of western Kenya 100 km or so to the south (Le Bas, 1970). In one of the Uganda volcanoes, Napak, erosion has revealed a central intrusive complex 2 km wide at the site of the original vent. It is composed of coarse-grained rocks of the ijolite family

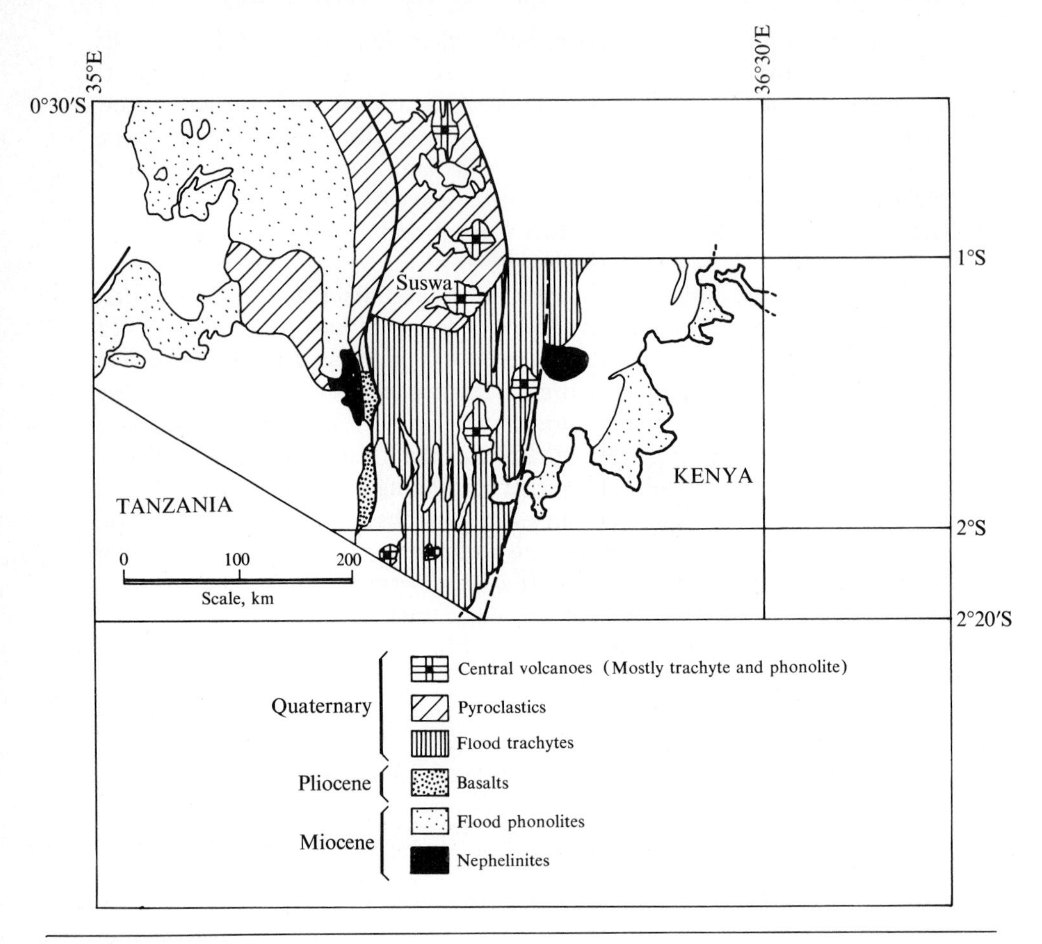


**Fig. 10-5** Trachyte and phonolite flood lavas, southern end of Kenya rift. (After J. B. Wright, 1963, p. 166.)

(pyroxene-nepheline-melanite) with a small central boss of carbonatite. The same sort of rock association appears, too, in much younger volcanoes. One of these, the still active Oldoinyo Lengai (Dawson, 1962*a*) located in the eastern rift just below the Kenya-Tanzania border, is remarkable for eruption of sodic carbonatite lavas in 1960. Carbonatite tuffs are widely developed in this part of Tanzania (Dawson, 1964*a*).

In some provinces leucite accompanies nepheline in nephelinites, basanites, tephrites, and even phonolites. This is a strictly continental chemical pattern, placed by Joplin (1965) in an undersaturated division of her shoshonite series. It is transitional between the nonpotassic nephelinite series described above and much rarer highly potassic kindreds to be considered separately in a later

**Table 10-2** Chemical compositions and CIPW norms of some East African alkaline rock series

| | 1 | 2 | 3 | 4 | 5 | 6 | 7 | 8 | 9 | 10 |
|---|---|---|---|---|---|---|---|---|---|---|
| $SiO_2$ | 42.1 | 48.9 | 38.35 | 39.59 | 54.38 | 45.2 | 50.5 | 56.47 | 61.09 | 56.13 |
| $TiO_2$ | 2.6 | 1.0 | 2.18 | 1.06 | 0.58 | 2.3 | 1.7 | 0.22 | 0.91 | 1.07 |
| $Al_2O_3$ | 12.4 | 17.6 | 13.29 | 13.63 | 17.02 | 16.0 | 18.0 | 19.93 | 16.64 | 16.13 |
| $Fe_2O_3$ | 3.3 | 4.6 | 7.26 | 8.64 | 4.67 | 6.5 | 3.9 | 2.32 | 3.37 | 4.00 |
| FeO | 7.4 | 3.1 | 6.22 | 3.89 | 2.57 | 8.1 | 5.5 | 3.43 | 2.40 | 4.57 |
| MnO | 0.11 | 0.19 | 0.17 | 0.10 | 0.12 | 0.17 | 0.1 | 0.18 | 0.18 | 0.35 |
| MgO | 7.8 | 2.2 | 5.99 | 3.57 | 1.09 | 7.6 | 2.4 | 0.27 | 0.59 | 1.59 |
| CaO | 12.4 | 6.4 | 15.34 | 19.94 | 3.79 | 9.3 | 6.7 | 1.65 | 1.52 | 2.43 |
| $Na_2O$ | 4 | 9.5 | 3.72 | 4.66 | 5.55 | 3.2 | 5.5 | 9.13 | 6.49 | 7.24 |
| $K_2O$ | 2 | 3.4 | 3.21 | 3.80 | 9.54 | 0.81 | 2.8 | 4.97 | 5.89 | 4.99 |
| $P_2O_5$ | 0.80 | 0.24 | 0.82 | 0.21 | 0.35 | 0.39 | 0.89 | 0.13 | 0.13 | 0.20 |
| $H_2O^+$ | 3.5 | 2.8 | 3.31 | — | — | 1.0 | 0.3 | 0.80 | 0.68 | 0.86 |
| $H_2O^-$ | 1.3 | | 0.34 | — | — | | 2.5 | 0.10 | 0.28 | 0.35 |
| Total | 99.7 | 99.9 | 100.20 | 99.09 | 99.66 | 100.6 | 100.8 | 99.85* | 100.25* | 100.11* |
| *or* | 11.7 | 20.0 | | | 39.5 | 5.0 | 16.7 | 28.5 | 34.81 | 29.49 |
| *ab* | | 9.7 | | | | 26.2 | 33.8 | 36.4 | 50.29 | 35.74 |
| *an* | 10.0 | | 10.0 | 5.0 | | 26.7 | 16.1 | | | |
| *lc* | | | 14.8 | | 13.1 | | | | | |
| *kp* | | | 17.0 | 12.6 | | | | | | |
| *ne* | 18.5 | 33.4 | | 21.3 | 18.7 | 0.3 | 6.7 | 24.3 | 1.36 | 10.53 |
| *ac* | | 7.9 | | | 11.1 | | | 0.8 | 1.34 | 4.05 |
| *di* | 37.0 | 16.9 | 26.1 | 19.2 | 12.3 | 13.4 | 9.8 | 5.6 | 5.11 | 8.97 |
| *wo* | | 4.6 | | 22.4 | 1.7 | | | | | |
| *la* | | | 8.4 | 4.0 | | | | | | |
| *ol* | 6.1 | | 3.5 | | | 13.3 | 4.1 | 1.2 | | 3.42 |
| *mt* | 4.6 | 2.8 | 10.4 | 9.5 | 1.2 | 9.5 | 5.6 | 2.0 | 4.21 | 3.77 |
| *il* | 5.0 | 2.0 | 4.1 | 2.1 | 1.1 | 4.4 | 3.2 | 0.3 | 1.73 | 2.03 |
| *ap* | 2.0 | 0.7 | 2.0 | 0.3 | 1.4 | 1.0 | 2.0 | 0.3 | 0.31 | 0.47 |
| Total | 94.9 | 98.0 | 96.3 | 98.5† | 100.1 | 99.8 | 98.0 | 99.4 | 99.16 | 98.47 |

* Including Cl = 0.20 (8), 0.08 (9), 0.20 (10).
† Including *hm* 2.1.

*Explanation of column headings*

Numbers in parentheses refer to analyses in papers cited.

1, 2 Nephelinite series, Moroto, eastern Uganda (Varne, 1968, p. 175)
- 1 Olivine nephelinite (3)
- 2 Felsic nephelinite (6)

3–5 Nephelinite-ijolite series, Napak, eastern Uganda (King, 1965, p. 77)
- 3 Melilite nephelinite (9)
- 4 Ijolite (3)
- 5 Nepheline syenite (7)

6–8 Alkali olivine-basalt series, Moroto, eastern Uganda (Varne, 1968, p. 172)
- 6 Olivine basalt (1)
- 7 Trachyandesite (6)
- 8 Phonolite (11)

9, 10 Trachyte-phonolite series, Mt. Suswa (Nash et al., 1969, pp. 426, 427)
- 9 Primitive underlying flood trachyte (W300)
- 10 Glassy phonolite (W118)

section. Leucite basanites are the main constituents of some major equatorial volcanoes in the western rift close to the Congo-Uganda border (Turner and Verhoogen, 1960, pp. 236–237). There is a broader range of rock types in some of the Tertiary volcanic fields of Europe—notably in the Bohemian Mittelgebirge (Knorr, 1932) and the Eifel region of the Rhine province. In the latter lies the Laacher See volcanic center, celebrated for the mineralogical variety displayed by its trachytic breccias and tuffs (Schürmann, 1960; Jasmund and Seck, 1964). The typical mineral assemblage in uncontaminated trachytes of Laacher See is sanidine, plagioclase, hauyne, sodic augite, hornblende, and titanomagnetite (sphene, apatite, zircon). Associated basic lavas are leucite nephelinites, some containing sphene, others melilite and perovskite; phonolites contain sanidine, leucite, and nepheline, along with hauyne, aegirine-augite, melanite, and hornblende.

In many nephelinite provinces spinel-lherzolite xenoliths are abundantly represented—just as in oceanic nephelinite provinces. The Auvergne and Rhine provinces of Europe are in fact the classic areas for this paragenesis (Ernst, 1935). Curiously enough, in the whole East African province—the most extensive known region of nephelinite volcanism—ultramafic nodules of this type seem to be completely lacking.

## MULTIPLE SERIES

It is by no means unusual to find two or more chemically distinct series closely associated in geographically restricted areas or even in a single volcanic center. Tholeiites and alkali basalts are both abundant in the Vogelsberg pile north of the Rhine-graben rift system (T. Ernst et al., 1969, 1970). Twenty km or so to the east, in the outlying Rhön area (Ficke, 1961), early effusion of tholeiitic lavas was followed by a main eruptive phase whose products are alkaline hornblende basalts, basanites, limburgites, and some nephelinites. Still younger are phonolites, believed to be differentiates from the alkali-basalt stem.

Nephelinites are commonly associated with what seem to be chemically and genetically distinct alkali-basalt series. In the great active volcanic complex Kilimanjaro, east of the eastern rift of Kenya, the predominant series is alkali basalt → trachyandesite → phonolite; but there are also nephelinites (P. Wilkinson, 1966). Both series appear at all levels in the dissected Miocene volcano Moroto, in eastern Uganda (Varne, 1968). The central volcanic complex surrounding Dunedin in the Miocene-Pliocene East Otago volcanic province, southern New Zealand, is only 30 km in diameter. Yet within it there is great diversity of rock types (Turner and Verhoogen, 1960, pp. 165–175); and Coombs and Wilkinson (1969) have demonstrated coexistence of a number of independent lineages, apparently derived from independent parental magmas (Table 10-3; analyses 2–10): alkali basalt → hawaiite → mugearite → trachyte (with independent sodic and potassic variants); basanite → phonolite (likewise potassic or sodic); nephelinites (much less plentiful).

**Table 10-3** Chemical compositions and CIPW norms of *ne*-normative volcanic series, New Zealand

| | 1 | 2 | 3 | 4 | 5 | 6 | 7 | 8 | 9 | 10 |
|---|---|---|---|---|---|---|---|---|---|---|
| $SiO_2$ | 44.80 | 46.29 | 47.49 | 48.00 | 57.53 | 42.49 | 47.56 | 52.97 | 43.03 | 50.41 |
| $TiO_2$ | 1.96 | 1.90 | 2.47 | 1.98 | 0.55 | 2.98 | 2.14 | 0.93 | 2.66 | 2.00 |
| $Al_2O_3$ | 13.86 | 17.17 | 15.82 | 17.11 | 18.37 | 13.57 | 17.93 | 18.54 | 13.82 | 18.57 |
| $Fe_2O_3$ | 2.91 | 2.80 | 3.73 | 3.74 | 2.39 | 4.17 | 4.64 | 3.91 | 3.70 | 2.73 |
| FeO | 9.63 | 10.03 | 8.42 | 9.43 | 3.57 | 9.04 | 6.20 | 3.02 | 8.95 | 6.37 |
| MnO | 0.17 | 0.19 | 0.19 | 0.23 | 0.19 | 0.18 | 0.21 | 0.17 | 0.21 | 0.16 |
| MgO | 11.07 | 6.95 | 4.94 | 3.11 | 0.83 | 9.79 | 3.13 | 1.94 | 8.59 | 2.42 |
| CaO | 10.16 | 7.31 | 9.32 | 6.40 | 2.30 | 11.47 | 7.13 | 3.39 | 9.23 | 5.80 |
| $Na_2O$ | 3.19 | 3.90 | 4.07 | 5.17 | 6.92 | 3.52 | 6.26 | 8.21 | 3.91 | 5.00 |
| $K_2O$ | 1.09 | 1.48 | 1.20 | 2.02 | 4.61 | 0.60 | 2.27 | 3.90 | 2.14 | 3.20 |
| $P_2O_5$ | 0.55 | 0.40 | 0.57 | 0.89 | 0.17 | 0.59 | 0.74 | 0.40 | 0.73 | 0.48 |
| $H_2O^+$ | 0.73 | 1.15 | 0.78 | 1.10 | 1.40 | 1.02 | 0.66 | 1.59 | 2.37 | 2.53 |
| $H_2O^-$ | | | 0.73 | 0.24 | 0.86 | 0.94 | 1.15 | 0.91 | 0.78 | 0.60 |
| Cl | | | | 0.09 | 0.22 | | | 0.35 | | |
| Total | 100.12 | 99.57 | 99.73 | 99.51 | 99.91 | 100.36 | 100.02 | 100.23 | 100.12 | 100.27 |
| *or* | 6.5 | 8.75 | 7.09 | 11.94 | 27.22 | 3.55 | 13.42 | 23.05 | 12.65 | 18.91 |
| *ab* | 13.0 | 24.56 | 31.47 | 34.99 | 46.46 | 11.17 | 26.87 | 34.86 | 8.37 | 32.44 |
| *an* | 20.3 | 24.97 | 21.35 | 17.87 | 6.34 | 19.46 | 14.12 | 3.59 | 13.84 | 18.78 |
| *ne* | 7.6 | 4.58 | 1.61 | 4.38 | 5.65 | 10.08 | 14.14 | 17.35 | 13.39 | 5.35 |
| *hl* | | | | 0.15 | 0.36 | | | 0.58 | | |
| *di* | 21.7 | 7.12 | 14.41 | 4.49 | 3.42 | 26.25 | 12.50 | 8.59 | 22.19 | 5.69 |
| *ol* | 21.3 | 19.86 | 10.15 | 12.55 | 3.30 | 14.65 | 4.45 | 1.32 | 14.50 | 7.11 |
| *mt* | 4.2 | 4.06 | 5.41 | 5.42 | 3.47 | 6.05 | 6.73 | 5.67 | 5.36 | 3.96 |
| *il* | 3.7 | 3.61 | 4.69 | 3.76 | 1.05 | 5.66 | 4.06 | 1.77 | 5.05 | 3.80 |
| *ap* | 1.2 | 0.93 | 1.25 | 2.06 | 0.40 | 1.29 | 1.62 | 0.93 | 1.69 | 1.12 |
| Total | 99.5 | 98.44 | 97.43 | 97.61 | 97.67 | 98.16 | 97.91 | 97.71 | 97.04 | 97.16 |

*Explanation of column headings*

Numbers in parentheses refer to analyses in papers cited.

1 Average alkali basalt (13 analyses), Auckland, New Zealand (Searle, 1960; Coombs, 1963, p. 236, no. 7); $SiO_2$ range 43.0 to 48.2 percent

2–5 Alkali-basalt trachyte series, East Otago province (Coombs and Wilkinson, 1969, pp. 475, 476)
- 2 Alkali olivine basalt (2)
- 3 Hawaiite (3)
- 4 Average mugearite (6)
- 5 Trachyte (8)

6–8 Basanite-phonolite series, East Otago province (Coombs and Wilkinson, 1969, p. 480)
- 6 Basanite (1)
- 7 Nepheline hawaiite (3)
- 8 Phonolite (5)

9, 10 Relatively potassic volcanic series, East Otago province (Coombs and Wilkinson, 1969, p. 487)
- 9 Sanidine basanite (1)
- 10 Trachyandesite (7)

## SHALLOW INTRUSIVE SERIES

DIFFERENTIATED BASIC SILLS Basic alkaline magmas, deploying laterally at shallow depth, commonly give rise to sills 100 m or so thick with strongly differentiated vertical profiles. These first became widely known through the work of British petrologists in two Scottish provinces: the Carboniferous of Ayrshire and the Tertiary Hebridean. The general picture as set out by Turner and Verhoogen (1960, pp. 175–184) remains essentially unchanged in the light of further studies in many parts of the world. More recent accounts (for example, Kushiro, 1964; J. G. F. Wilkinson, 1965; Wilshire, 1967) contribute mineralogical and chemical details and emphasize upward enrichment of residual magmas in alkalis, iron, and water as being to some degree independent of and supplemental to gravitational sinking of olivine and augite. The general petrochemical pattern remains remarkably constant. Alkali-basalt magmas with only minor normative *ne* crystallize as analcite-bearing olivine diabases with titanaugite the sole pyroxene. Their complementary differentiates are olivine-enriched picrites near the base, analcite-syenite segregations toward the top. Basanitic magmas give rise to rocks of the theralite family, mostly with modal nepheline as well as analcite. Their picritic differentiates contain significant amounts of alkali hornblende; felsic segregations include nepheline and analcite syenites (or tinguaites). A computed liquid line of descent is illustrated in Fig. 10-6.

The pyroxenes of some analcite-olivine diabases (Atumi dolerite) from the Miocene of northwestern Japan present an interesting anomaly (Kushiro, 1964, pp. 165–174). The parental magmas are characterized by high alumina and a low content of either *ne* or *hy* in the norm. Analcite appears in the mode, and felsic differentiates are analcite syenites. Yet many of the diabases have two pyroxenes—diopsidic titanaugite and bronzite (or pigeonite)—a feature generally considered diagnostic of tholeiitic magmas.

CENTRAL INTRUSIVE COMPLEXES We saw earlier that the core of the partially dissected Miocene nephelinite volcano Napak in eastern Uganda is an intrusive body of alkaline basic plutonic rocks. These belong to the ijolite family: the main constituents are nepheline and pyroxene, but there is a considerable variety of minor minerals, among them melanite, wollastonite, sphene, apatite, less commonly magnetite-ilmenite, perovskite, biotite, and secondary cancrinite, pectolite, calcite, or zeolites (King, 1965, p. 72). Chemically these rocks tend to resemble nephelinites. A similar situation is revealed in other eroded volcanoes elsewhere. We cite one instance, the Kaiserstuhl complex piercing and overlying the Oligocene Rhine-graben filling near Freiburg at the southern end of the Rhine province (Wimmenauer, 1957, 1959, 1962, 1963, 1966). Here the remnants of the volcanic edifice are composed of nephelinites, tephrites, and phonolites—both lava flows and tuffs. The central core consists of plutonic rocks of the essexite-theralite family, composed principally of augite, plagioclase, alkali feldspar, and feldspathoid (nepheline and/or leucite), with biotite and in some rocks olivine as minor constituents. Both at Napak and in the Kaiserstuhl the center of the plutonic core is occupied by carbonatite (Fig. 10-8*b*).

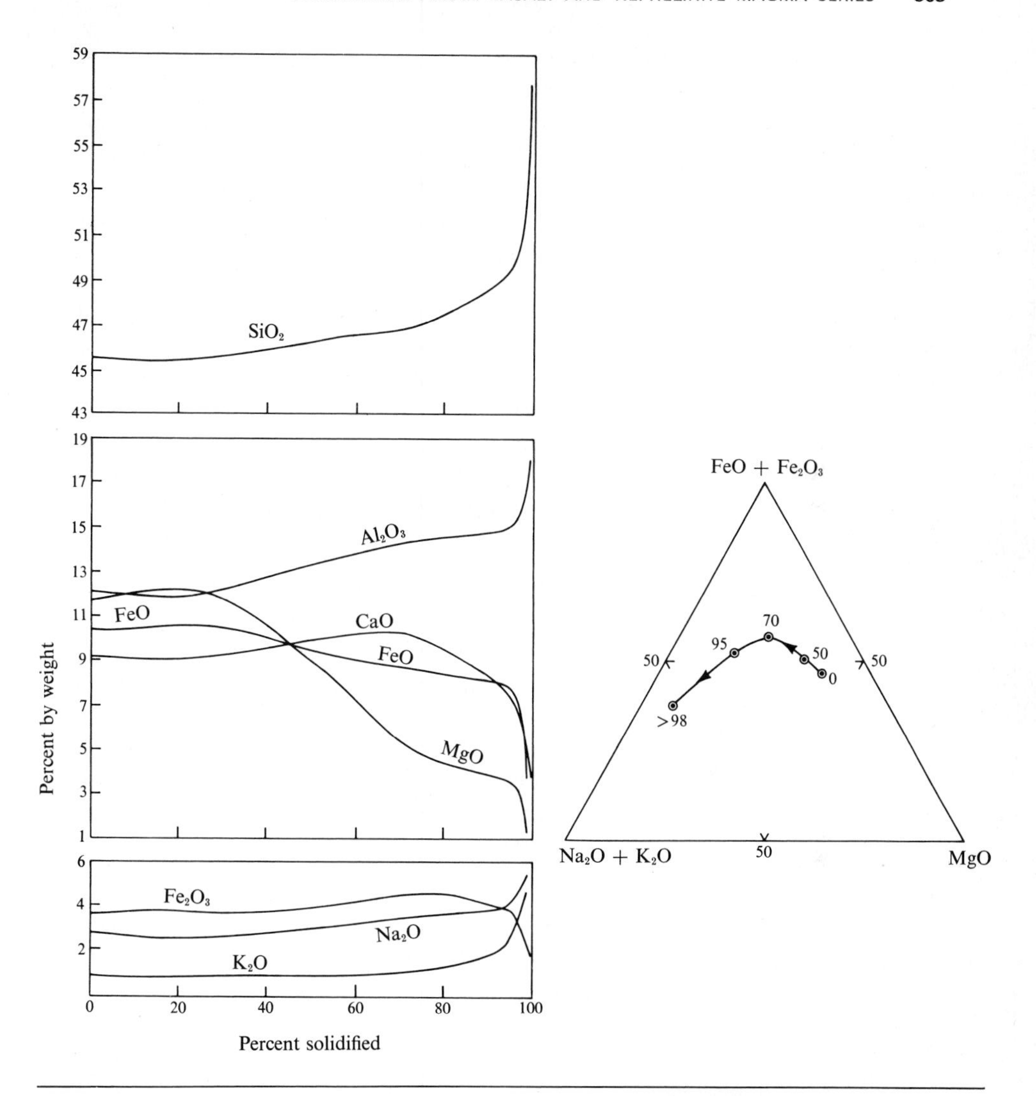


**Fig. 10-6** Computed liquid differentiation trend, Prospect alkaline diabase sheet (100 m thick) near Sydney, Australia. (After Wilshire, 1967.) Assumptions: chilled margins represent parental magma; syenitic and aplitic segregations represent final liquids; fractional crystallization under influence of gravity was entirely postintrusive. Numerals in triangular plot indicate percentage solidified.

The north-south chain of Miocene volcanoes in eastern Uganda is prolonged southward by more deeply eroded centers of early Tertiary volcanism. Here the edifices have been stripped away. What now remain are subvolcanic complexes circular in plan and not more than a few kilometers in diameter. Their component rocks fall into two series (King, 1965, p. 75): pyroxenite, ijolite, and carbonatite; nepheline syenite and and syenite. These are chemically more or less equivalent to the volcanic nephelinite and phonolite-trachyte series of Miocene volcanoes to the north. There are also some essentially volcanic provinces where plutonic rocks in alkaline ring complexes have the same general compositional range as the more voluminous lavas and tuffs with which they are associated in the field—for example, in the Permian province of the Oslo graben.

There can be no doubt that in all these cases, and in other regions where only plutonic complexes remain, the alkaline intrusions are the fillings of subvolcanic reservoirs and conduits. Something of their mineralogical and chemical diversity has been summarized elsewhere (Turner and Verhoogen, 1960, pp. 345–346, 390, 394). More extensive and up-to-date information relating to those alkaline complexes that have carbonatite members is available in recent works by Heinrich (1966) and by Tuttle and Gittins (1966). From all this material emerge some genetically significant generalizations—none, however, without its exceptions:

1. Highly typical is a continental environment stable except for regional rifting. Again and again the alkaline magmas have punched their way to the surface through crystalline Precambrian basement rocks. In some provinces, intrusion of alkaline magmas and reactivation of ancient faults have recurred over long intervals: the whole of Cretaceous time in southern Brazil (Amiral et al., 1967). In the Oslo graben, on the other hand, although faulting recurred at intervals from Precambrian to late Paleozoic times, the whole sequence of eruptive and effusive events covered no more than a few million years in the early Permian.

2. With some exceptions, and with due allowance for differences related to environment of crystallization, there are broad chemical analogies between intrusive and volcanic alkaline rock series:

   {Pyroxenite, ijolite, carbonatite<br>
   {Nephelinite, phonolite

   {Malignite, nepheline syenite<br>
   {Basanite, phonolite

   {Essexite, syenite, nordmarkite<br>
   {Alkali basalt, mugearite, trachyte

3. Rather rarely, intrusion of alkaline magmas is interrupted or closely followed by a surge of calc-alkaline magma of some apparently unrelated type. Again we cite the Oslo province. Here there are extensive late plutons of biotite granite ("Drammen type") whose mineralogical and chemical composition contrasts sharply with that of immediately associated bodies of sodic granite belonging to the prevailing peralkaline series. Curiously enough, the isotopic composition of strontium in these biotite granites is the same as in the alkaline plutonics (initial $Sr^{87}/Sr^{86} = 0.7041 \pm 0.0002$).

4. Effects of processes that in volcanic series can only be inferred may be observed more directly in equivalent intrusive bodies. Contacts between mafic and felsic rocks commonly tend to be sharp rather than gradational; differentiation, if responsible for chemical diversity, must have occurred mainly at levels below the present site of intrusion. Metasomatism of country rocks is widespread and may extend hundreds of meters from contacts (Turner and Verhoogen, 1960, pp. 397–398, 400). It commonly takes the form of fenitization of quartzofeldspathic rocks—especially where carbonatites are components of the intrusion (Heinrich, 1966, pp. 68–92; McKie, 1966). The end products, fenites, are aggregates of alkali feldspars and aegirine (or sodic amphiboles). Marbles may become converted to silicate rocks mineralogically akin to nepheline syenites or even to ijolites (Gittins, 1961; Tilley and Gittins, 1961).

5. There is a close field relation between alkaline intrusions, especially those of mafic composition, and carbonatites. Indeed the concept of an intrusive carbonatite magma developed among Swedish and Norwegian geologists from studies of alkaline intrusive complexes. We shall return later to possible genetic relations between carbonatites and mafic alkaline magmas.

LAYERED NEPHELINE-SYENITE PLUTONS Two remarkable peralkaline nepheline-syenite plutons—Ilimaussaq (150 $km^2$) in southern Greenland and Lovozero (650 $km^2$) in the Kola peninsula of the U.S.S.R.—display spectacular subhorizontal layering (Sörensen, 1969). This involves variation in relative proportions of nepheline, microcline, and mafic minerals (mostly sodic pyroxenes and amphiboles). Zones dominated by a single rock type may be hundreds of meters thick. Internally they may be rhythmically layered on the scale of a few meters or even centimeters; here the layers show graded and cumulus structures reminiscent of the Skaergaard pattern. In both intrusions eudialite and other Zr–Ti–silicates are widely disseminated; and in some zones, where rhythmic layering is especially obvious, these minerals may make up 50 percent or more of individual laminae. Exactly comparable lamination is shown locally by nepheline syenites of the Poços de Caldas ring complex in south-central Brazil. Here, too, the laminated rocks are rich in eudialite and allied zirconium silicates. The layered alkaline plutons are treated in the present context, in spite of general absence of possible mafic parental magmas, because they are chemically similar to phonolites and nepheline syenites of nephelinite- or ijolite-dominated provinces.

## COMPARISON WITH OCEAN-ISLAND VOLCANIC SERIES

In most respects there is a close mineralogical-chemical parallel between the commoner alkaline lava series of the continents and those of oceanic islands. In both geologic settings nephelinites are accompanied by basanites and phonolites, while alkali basalts have more felsic associates ranging from hawaiite (trachyandesite) through trachyte to either phonolite or peralkaline rhyolite (pantellerite). And in both, the mafic lavas and pyroclastics are locally loaded with xenoliths of spinel lherzolite. The trace-element pattern of continental alkali basalts with $K_2O/Na_2O$ in the general range 0.25 to 0.5 duplicates that of similar ocean-island basalts. This pattern is illustrated by the Nandewar basalts of eastern Australia and again by the alkali basalts of the Basin-Range province, for which Leeman and Rogers (1970) have recorded mean abundances (ppm) Rb 25, Sr 800, Th 5, and U 1; and ratios K/Rb 470, K/Na 0.27. Initial $Sr^{87}/Sr^{86}$ values for some continental alkaline lavas overlap the upper range recorded for comparable oceanic volcanoes: 0.703 to 0.706 in Basin-Range basalts (Leeman and Rogers, 1970); 0.703 to 0.706 in nephelinites, 0.7042 to 0.7052 in phonolites of eastern Uganda (K. Bell and Powell, 1970); and 0.7041 in monzonites and syenites, 0.7050 in sodic granites of the Oslo graben. Clearly the petrogenic regimes familiar in oceanic islands have also been duplicated repeatedly in all essentials in nonorogenic continental settings.

Yet the total continental picture shows greater variety, and in some respects it departs from the familiar oceanic pattern. On the whole, potassic alkaline lavas are more widespread and their potassic character more pronounced in continental than in oceanic provinces. Correlated with higher potassium we find in the shoshonite series a higher content of Ba, Sr, Rb, and some other "incompatible" dispersed elements than in even the most potassic of ocean-island rocks. Initial $Sr^{87}/Sr^{86}$ values, notwithstanding the overlap just noted, tend to be higher in continental lavas; and there is a tendency for wider fluctuation among associated rocks of individual provinces (for example, K. Bell and Powell, 1970). It is tempting to view all such features as in some way reflecting the influence of sialic crust.

There is no known oceanic equivalent of the leucite-bearing mafic lavas that appear with nephelinites in the Tertiary fields of the Rhine and Czechoslovakia, nor of even more potassic (ultrapotassic) lavas treated later. Again there is no record of oceanic eruption of peralkaline trachytic or phonolitic magma on a scale remotely comparable with that displayed in the Kenya rift or at the plutonic level by intrusion of sodic granites and syenites in the Oslo graben. Finally carbonatites, which so commonly accompany continental nephelinites, are virtually unknown within the oceans.

## MAGMATIC EVOLUTION

DIVERSITY OF PARENTAGE Much of what has been said regarding petrogenesis of oceanic alkaline volcanic series must apply also to their petrochemical counterparts on the continents. It would seem most unlikely that the source materials and line of descent of the basalt-hawaiite-trachyte-pantellerite series of Nandewar volcano in eastern Australia were radically different from those represented by similar lava series in Samoa or on Ascension Island. The same may be said for basalt-basanite-phonolite-trachyte associations of East Africa and New Zealand compared with those of St. Helena.

Numerous recently proposed petrogenic models for continental series, as for oceanic alkaline series, stress differentiation—with emphasis on fractional crystallization—from mafic parental magmas. It is characteristic of current thinking, as contrasted with the simpler general outlook of Bowen's day, that wide variety among postulated differentiation models stems in large part from diversity among accepted parental magmas. These collectively embrace the compositional range of alkali olivine basalts (with or without normative *ne*), basanites, nephelinites, and potassic (shoshonitic) variants of these. There is no compelling reason to derive these diverse magmas from identical source materials nor to fractionate them at high pressure from parental magma of some other composition, held in mantle reservoirs and ascending thence from the depths. Each of the proposed parental magmas could equally well have been generated in its own fusion regime.

Some writers would go further, viewing voluminous trachytic and phonolitic magmas of some African fields as effusions of magma generated directly from mantle sources (cf. D. K. Bailey and Schairer, 1964; Grant et al., 1972). Underlying experimental evidence that has been cited, however, is scarcely applicable. It derives from silicate-melt equilibria at atmospheric pressure. Nor is the "primitive" composition of strontium that characterizes some of these lavas any certain criterion nor even a generally credible index of mantle sources. Much more likely, somewhere in the course of evolution of all trachytic and phonolitic magmas an essential role has been played by differentiation.

DIFFERENTIATION VERSUS CRUSTAL CONTAMINATION Accepted emphasis on differentiation in alkaline petrogenesis stems largely from general internal consistency in proposed differentiation models based on rock and mineral analyses viewed in the light of experimental data on appropriate silicate-melt equilibria. Supporting evidence of a more direct kind comes from mafic sills of some continental provinces. These show that differentiation—in which gravitationally controlled fractionation plays at least some part—may yield liquid residues of syenitic (trachytic) and nepheline-syenitic (phonolitic) compositions. The late differentiates tend to retain some characteristics inherited from the parental magma—degree of silica saturation (silica activity) and dominance of potassium over sodium or vice versa. They also show a characteristic terminal buildup in certain "incompatible" dispersed elements in patterns similar to those found in trachytes and phonolites elsewhere.

A question still not satisfactorily answered is raised by the greater chemical diversity of continental as contrasted with oceanic lavas and by rather persistent differences in dispersed-element and isotopic chemistry. Could all this perhaps reflect deviation from pure fractionation trends as a result of crustal melting or allied contamination processes? The simplest model—bulk incorporation of sialic crust by melting—seems incompatible with the undersaturated condition of most felsic alkaline lavas. It is difficult, too, to reconcile with the strongly curved configuration of chemical variation diagrams (cf. Le Maitre, 1968, pp. 231–233)—which are much the same for both continental and oceanic rock series. Nevertheless there remains a possiblity (to which we return later) of selective chemical exchange between magma and country rocks, by which "incompatible" components of the latter diffuse preferentially into the liquid (magmatic) phase with which they are in contact.

DIFFERENTIATION MODELS: SOME POSTULATED LINEAGES The numerous differentiation models that have been proposed for continental alkaline volcanic series are depicted graphically, following customary procedure, by chemical variation diagrams based on representative rock analyses. Some have been tested and elaborated by fitting the proposed trend to trace-element data and to phenocryst mineralogy. Combinations of phenocryst phases, in proportions computed with this end in view, are assumed to have been removed from the evolving liquid phase at successive stages along the line of descent. Most such models therefore represent low-pressure crystal-liquid fractionation. Where they fail to account for the full detail of the postulated trend, more remote and less tangible differentiation mechanisms may be invoked—"alkali transfer" or separation of high-pressure crystalline phases in the upper-mantle environment. That felsic alkaline magmas may in fact develop at such depths is suggested by rather common occurrence of kaersutite and occasional presence of spinel-lherzolite xenoliths in phonolites and trachytes (cf. J. B. Wright, 1971). As in ocean-island provinces, $Sr^{87}/Sr^{86}$ values for felsic lavas may be significantly higher than in associated presumably parental mafic magmas (Hoefs and Wederpohl, 1968).

Among proposed differentiated lineages that meet the general test of broad internal coherence are the following (Fig. 10-7):

1. Alkali olivine basalt (*hy*-normative) → hawaiite (trachyandesite) → trachyte → peralkaline rhyolite (for example, Abbott, 1969; Cox et al., 1970; Sutherland, 1970).
2. Alkali olivine basalt (*ne*-normative) → trachybasalt → mugearite → trachyte (*ne*-normative) → phonolite (for example, Varne, 1968, pp. 178–182; Coombs and Wilkinson, 1969; Nash et al., 1969; F. H. Brown and Carmichael, 1969).
3. Nephelinite → felsic nephelinite → phonolite (for example, Wimmenauer, 1963; Varne, 1968, pp. 182–186).

## Lamprophyres

Lamprophyres are mafic dike rocks traditionally grouped together, in spite of considerable chemical and mineralogical variety, on the basis of a distinctive texture. They are strongly porphyritic, with mafic silicates occurring in euhedral

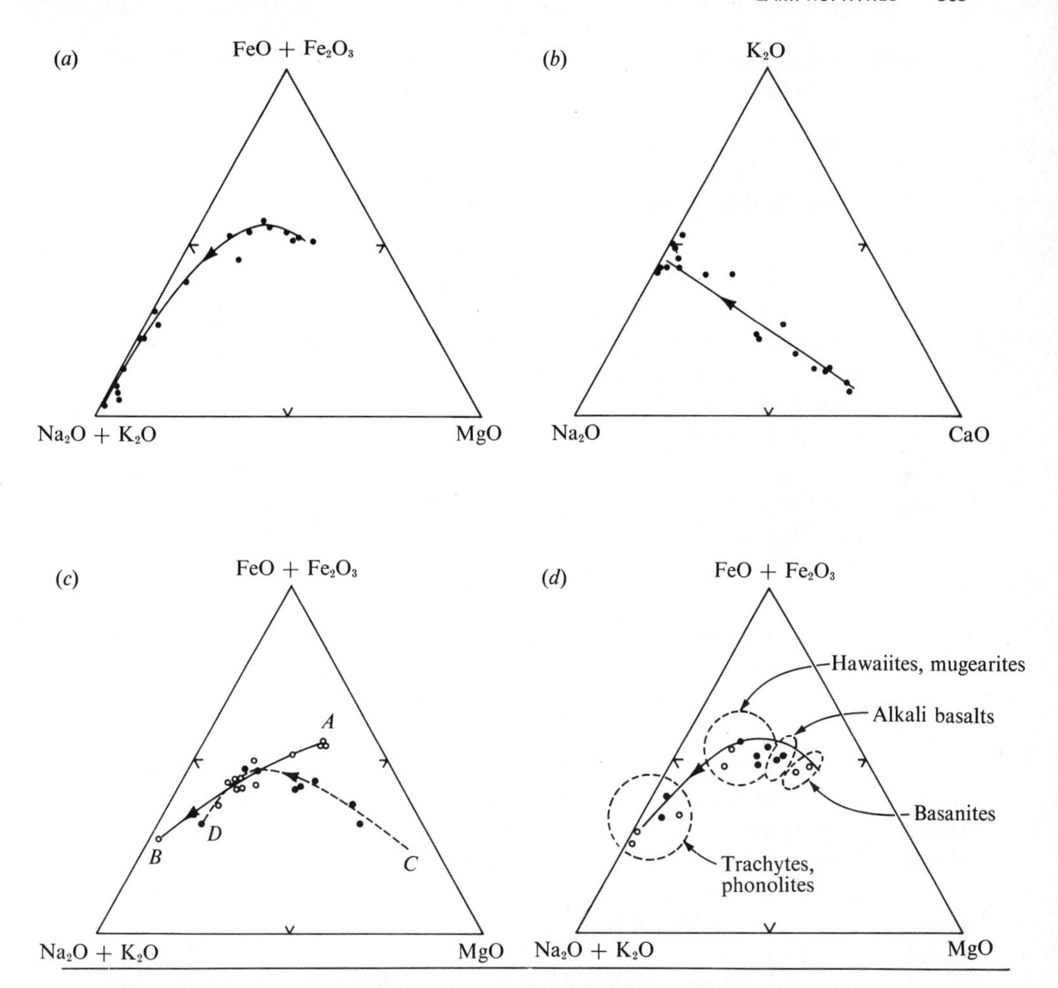


**Fig. 10-7** Chemical variation in continental alkaline lava series. (*a*), (*b*) Nandewar volcano, New South Wales (after Abbott, 1969, p. 130). (*c*) Moroto, eastern Uganda: *AB* and open circles, alkali olivine-basalt series; *CD* and filled circles, olivine-nephelinite series. (*d*) East Otago (Dunedin), New Zealand: filled circles, alkali basalt-hawaiite-mugearite-trachyte series; open circles, basanite-mugearite-phonolite series (after Coombs and Wilkinson, 1969, p. 481).

crystals of two generations; feldspars are confined to a fine-grained groundmass. Their chemical and mineralogical character, mode of occurrence, and classic ideas regarding petrogenesis have been presented elsewhere (Turner and Verhoogen, 1960, pp. 250-256). Here we restate these salient characteristics:

1. Chemically lamprophyres have low $SiO_2$ (mostly 40 to 47 percent), high ($MgO + FeO$), and high ($Na_2O + K_2O$). In these respects they resemble nephelinites, leucitites, and basanites; and indeed analyses of many lampro-

phyres, calculated water-free, are indistinguishable from one or other of these rock-types.

2. There is a high content of water and other "volatile" components. This is reflected mineralogically in abundance of hornblende (commonly barkevikite) and/or biotite; these may be accompanied by augite, less commonly by olivine.
3. Many granitic batholiths and plutons of smaller size are cut by lamprophyric dikes that may occur in radial swarms centering on individual plutons of granite, granodiorite, or syenite. Yet there is no recognizable petrochemical relation between lamprophyres and their felsic plutonic associates nor with other dike rocks (for example, aplite or porphyrite) of the same minor intrusive swarm. It would seem that major granitic plutonism may be followed, usually after a long hiatus, by generation of small quantities of other magmas, lamprophyric among them, beneath the same site. These then rise along tensional fractures whose pattern is controlled by residual stress related to earlier intrusion of granitic plutons.
4. Lamprophyric dikes also appear as integral products of volcanism and subvolcanic intrusion in nephelinite and alkali-basalt provinces. Here they seem to be essentially hydrous variants of nephelinite or ijolite-syenite series.

Considering the rather arbitrary basis on which lamprophyres are grouped together, more than one mode of origin appears likely. Many have in fact been proposed: divergence by some hydration mechanism from nephelinitic and allied magmas; melting of hornblende- or biotite-rich components of the lower crust; contamination of nephelinite or alkali-basalt magmas by reaction with granitic crust. Many lamprophyres are in fact crowded with partially digested granitic xenoliths.

Alnoites, classed petrographically with lamprophyres in spite of peculiar and distinctive mineralogy (presence of melilite and monticellite), have proved affinities with kimberlites. And there are normal biotite-lamprophyre provinces, such as the Navajo Buttes area of Arizona, in which breccia fillings of some diatremes are sufficiently similar to kimberlite to have been classified as such. It looks as if petrogenic processes contributing to formation of kimberlite magmas deep beneath the crust can also give rise to magmas which, rising rapidly toward the surface, crystallize at shallow depth as mica lamprophyres.

## Volcanic Series of Extreme Composition: Ultrapotassic Series, Carbonatites, Kimberlites

Highly potassic mafic lavas ($K_2O/Na_2O > 3$), carbonatites, and kimberlites differ obviously from one another in their gross chemistry and mineralogy. Yet they also seem to have much in common. They are found in the same structural environment, and the first two may be intimately associated in the field. There are striking similarities in the respective trace-element patterns, which, when compared with those of nephelinites, are extreme to the point of exaggeration. This similarity suggests some common factor in petrogenesis, to which we return in Chap. 13.

## ULTRAPOTASSIC SERIES

OCCURRENCE One way in which highly potassic volcanic rocks occur is in localized provinces within much larger regions of nephelinite and alkali olivine basalt. One such is the Eifel province of western Germany (Fig. 10-2); and there are several others, notably the Birunga and Toro-Ankole fields, in the western rift region of East Africa. There are other extensive regions—the Rocky Mountain strip from New Mexico to Wyoming is a well-known example—where the lavas of most individual provinces are to varying degrees potassic. Here the dominant rocks may be shoshonitic lavas (latites), mica lamprophyres, or more locally extreme rocks with $K_2O/Na_2O$ between 3 and 10. Elsewhere again, notably in Western Australia and southern Spain, there are small, isolated provinces where the whole rock suite has this exceptionally potassic character. This is true also of the Roman province of Recent volcanism, comprising 10,000 $km^2$ of west central Italy and including the Somma-Vesuvius volcano.

GENERAL MINERALOGY AND CHEMISTRY Salient features of the mineralogy and chemistry of some highly potassic lavas have been considered in Chap. 5. Here we add further data, mostly geochemical, recorded in recent accounts of several classic late Cenozoic provinces: Birunga and Toro-Ankole fields, East Africa[1] (K. Bell and Powell, 1969; Cundari and Le Maitre, 1970); Shonkin Sag laccolith, Highwood province, Montana (Nash and Wilkinson, 1970, 1971); Leucite Hills, Wyoming (Carmichael, 1967*c;* Powell and Bell, 1970); Jumilla, southern Spain (Borley, 1967; Powell and Bell, 1970); and the Roman province, Italy (Cundari and Le Maitre, 1970).

The composition of the dominant lavas varies greatly from one province to another regarding both silica content and $K_2O/Na_2O$ ratio (cf. Tables 10-4, 10-5). In both East African provinces they are ultramafic alkaline lavas: in the Toro-Ankole field, olivine melilitites (katungites) and leucitites (ugandites and kalsilite-bearing mafurites) with $K_2O/Na_2O$ between 3 and 5 (Table 10-4; analyses 1, 2); in the Birunga province, mostly nephelinites and melilitites very high in total alkali, with $K_2O/Na_2O \approx 1$ (analysis 3), commonly containing modal leucite or even kalsilite. The shonkinites of Montana are *ne*-normative mafic rocks consisting mainly of augite, sanidine, biotite, olivine, and zeolites; $K_2O/Na_2O$ is less than 2 (analysis 6) except in the chilled pseudoleucite-bearing

[1] See also Higazy (1954).

**Table 10-4** Chemical compositions of highly potassic mafic volcanic rocks

| | 1 | 2 | 3 | 4 | 5 | 6 | 7 | 8 | 9 |
|---|---|---|---|---|---|---|---|---|---|
| $SiO_2$ | 35.37 | 39.28 | 39.51 | 45.58 | 47.00 | 48.15 | 50.23 | 55.14 | 43.56 |
| $TiO_2$ | 3.87 | 4.29 | 3.25 | 1.54 | 0.80 | 0.76 | 2.30 | 2.58 | 2.31 |
| $ZrO_2$ | | | | 0.11 | | | 0.25 | 0.27 | 0.27 |
| $Al_2O_3$ | 6.50 | 7.90 | 14.83 | 8.35 | 12.91 | 11.33 | 10.15 | 10.35 | 7.85 |
| $Cr_2O_2$ | 0.01 | 0.09 | | 0.04 | | | 0.06 | 0.04 | 0.04 |
| $Fe_2O_3$ | 7.23 | 4.88 | 3.95 | 3.18 | 1.30 | 3.24 | 3.65 | 3.27 | 5.57 |
| FeO | 5.00 | 5.23 | 8.96 | 4.07 | 7.20 | 5.22 | 1.21 | 0.62 | 0.85 |
| MnO | 0.24 | 0.27 | 0.30 | 0.11 | 0.15 | 0.15 | 0.09 | 0.06 | 0.15 |
| NiO | 0.19 | 0.08 | | | | | | | |
| MgO | 14.08 | 17.58 | 4.29 | 14.75 | 7.55 | 9.00 | 7.48 | 6.41 | 11.03 |
| CaO | 16.79 | 11.03 | 12.15 | 9.0 | 9.70 | 10.82 | 6.12 | 3.45 | 11.89 |
| SrO | 0.04 | 0.22 | | 0.23 | 0.29 | 0.19 | 0.32 | 0.26 | 0.40 |
| BaO | 0.25 | 0.21 | | 0.41 | 0.58 | 0.53 | 0.61 | 0.52 | 0.66 |
| $Na_2O$ | 1.32 | 1.05 | 4.98 | 1.55 | 1.85 | 2.90 | 1.29 | 1.21 | 0.74 |
| $K_2O$ | 4.09 | 4.98 | 5.19 | 3.7 | 6.45 | 4.85 | 10.48 | 11.77 | 7.19 |
| $P_2O_5$ | 0.74 | 0.36 | 1.73 | 1.92 | 0.91 | 1.06 | 1.81 | 1.40 | 1.50 |
| $H_2O^+$ | 2.78 | 2.36 | 0.33 | 3.85 | 1.74 | 1.60 | 2.34 | 1.23 | 2.89 |
| $H_2O^-$ | 1.15 | 0.40 | 0.18 | 0.85 | 0.40 | 0.13 | 1.09 | 0.61 | 2.09 |
| $CO_2$ | 0.09 | 0.14 | | 0.7 | 1.01 | 0.19 | | | |
| $SO_3$ | | | | | | | 0.35 | 0.40 | 0.52 |
| S | 0.35 | 0.12 | 0.17 | | 0.06 | 0.06 | | | |
| Cl | 0.02 | | | | | | | 0.04† | 0.03† |
| F | 0.16 | 0.09 | | | | | | 0.49† | 0.47† |
| Other | 0.03* | | | | | | | | |
| Total | 100.30 | 100.56 | 99.82 | 99.94 | 99.90 | 100.18 | 99.83 | 99.59 | 99.51 |

* $V_2O_3$.
† Recorded in analyses of similar rocks. Not included in present tables.

*Explanation of column headings*

1 Olivine melilitite (katungite), Katunga volcano, Toro-Ankole province, East Africa (Holmes, 1950, p. 780, analysis 9)
2 Kalsilite leucitite (mafurite), Bunyaruguru, Toro-Ankole province, East Africa (Holmes, 1942; cited by Bell and Powell, 1969, p. 552, analysis 30)
3 Leucite nephelinite, Nyiragongo, Birunga province, East Africa (Sahama and Meyer, 1958; cited by Bell and Powell, 1969, p. 549, analysis 2)
4 Jumillite (average of two analyses), Murcia, southern Spain (Borley, 1967, p. 365, analyses 1, 2)
5 Chilled shonkinite border rock, Shonkin Sag sheet, Montana (Nash and Wilkinson, 1970, p. 244, analysis SS2)
6 Shonkinite, Shonkin Sag sheet (Nash and Wilkinson, 1970, p. 244, analysis SS3)
7 Wyomingite, Leucite Hills, Wyoming (Carmichael, 1967*c*, p. 50, analysis LH 1)
8 Orendite, Leucite Hills, Wyoming (Carmichael, 1967*c*, p. 50, analysis LH 12)
9 Madupite, Leucite Hills, Wyoming (Carmichael, 1967*c*, p. 50, analysis LH 16)

**Table 10-5** CIPW norms of highly potassic mafic volcanic rocks*

| | 4 | 5 | 6 | 7 | 8 | 9 |
|---|---|---|---|---|---|---|
| *Q* | | | | | 1.90 | |
| *Z* | | | | 0.37 | 0.40 | 0.40 |
| *or* | 21.7 | 26.14 | 28.66 | 47.76 | 56.50 | 1.67 |
| *ab* | 13.1 | | 0.69 | | | |
| *an* | 4.7 | 7.87 | 3.57 | | | |
| *lc* | | 9.39 | | 6.02 | | 32.00 |
| *ne* | | 8.48 | 12.92 | | | 0.17 |
| *th* | | | | 0.62 | 0.71 | 0.92 |
| *ac* | | | | 7.58 | 6.70 | 2.17 |
| *ks* | | | | 1.80 | 3.62 | |
| *di* | 20.1 | 24.17 | 34.63 | 13.54 | 2.38 | 36.73 |
| *hy* | 0.5 | | | | 14.86 | |
| *ol* | 21.5 | 13.47 | 8.40 | 8.65 | | 7.32 |
| *mt* | 4.4 | 1.88 | 4.70 | | | |
| *cm* | | | | 0.09 | 0.07 | 0.07 |
| *hm* | | | | 1.02 | 0.88 | 4.82 |
| *il* | 3.0 | 1.52 | 1.44 | 2.68 | 1.38 | 2.06 |
| *tn* | | | | | 4.55 | |
| *pf* | | | | 1.51 | | 2.08 |
| *py* | | 0.11 | 0.11 | | | |
| *ap* | 4.4 | 2.16 | 2.51 | 4.30 | 3.33 | 3.56 |
| *cc* | 1.6 | 2.30 | 0.43 | | | |
| Total | 95.0 | 97.49 | 98.06 | 95.94 | 97.28 | 93.97 |

* Column headings correspond to those in Table 10-4.

margin of the Shonkin Sag sheet (analysis 5), which also has significant normative *lc*.

The lavas of Leucite Hills, Wyoming (Table 10-4; analyses 7–9) illustrate a rare and uniquely potassic rock suite otherwise recorded only in the West Kimberley province of Western Australia and in southern Spain.[1] Most are relatively siliceous peralkaline rocks with high normative (*ac* + *ks*). Many have *ol;* a few have minor *lc* in the norm. Others, even some wyomingites with plentiful modal leucite, are slightly oversaturated in silica, which in the mode is concentrated in glass. The principal minerals are leucite, sanidine,[2] diopside, phlogopite, and potassic amphibole (magnophorite); characteristic accessories are perovskite, wadeite $Zr_2K_4Si_8O_{16}$, priderite $(K, Ba)_{1.5}$ $(Ti, Fe^{++}, Mg)_{8.25}$ $O_{16}$, and apatite. Some rocks carry a little olivine. Notable features are the presence of many potassic phases (also containing Ba), absence of iron oxides,

[1] The Spanish jumillites, although chemically less extreme, show marked affinity with lavas of the other two provinces.

[2] Not found in the Western Australian rock suite.

and almost complete accommodation of iron in sanidine and leucite, of sodium in amphibole and perovskite, and of zirconium in wadeite. Much less plentiful are ultrabasic lavas (madupites) devoid of sanidine and leucite and consisting principally of diopside and phlogopite set in a matrix of diopside, chlorite, and glass, with minor magnetite and perovskite. Texturally these recall Arkansas kimberlites except that they completely lack olivine. They rank as the potassic extreme among ultrabasic lavas (Tables 10-4 and 10-5, analysis 9). Their particular interest lies in their possible parental relation to more siliceous associates —orendite and wyomingite—and they supply one of those tenuous links that seem to connect ultrapotassic lavas with kimberlites.

Evidence cited in support of differentiation in ultrapotassic series is of the usual equivocal kind: close field association of more felsic lavas with supposedly parental ultrabasic rocks, and "trends" on standard variation diagrams. Among the more siliceous lavas of the two East African provinces are leucite basanites (kivites, murambites) with $SiO_2$ between 46 and 48 percent, shoshonites and latites ($SiO_2 \approx 50$ percent), and even trachytes ($SiO_2$ 55 to 60 percent). In two areas of the Toro-Ankole field there are carbonatite breccias. It can scarcely be doubted that all these rocks are to some degree congeneric, at least to the extent that they come from some common general source. The same may be said of syenitic bodies within the Shonkin Sag sheet. Cundari and Le Maitre (1970) have shown that, in the lavas of the Birunga and the Roman provinces, 90 percent of total chemical variation (of major elements) approximates a single chemical trend which could well represent a differentiated lineage. Simple models of fractional crystallization, however, encounter difficulties, as we shall see shortly, in the realm of trace-element chemistry.

TRACE-ELEMENT CHEMISTRY The trace-element pattern of ultrapotassic lavas (Table 10-6) is as remarkable as their mineralogy and general chemistry. High Ni and Cr, elsewhere typical only of ultramafic and mafic rocks rich in olivine and spinel, are characteristic even of the more siliceous lavas ($SiO_2$ 52 to 55 percent) of Leucite Hills. Here these elements are accommodated mainly in phlogopite ($Cr_2O_3$ 0.4 to 0.8, NiO 0.15 to 0.20 percent). On the other hand, Rb, Ba, Sr, Nb, Zr, F, and P, all of which elsewhere tend to become concentrated in acid magmas, consistently maintain high abundances even in the ultrabasic members of ultrapotassic series. With them, moreover, is concentrated Sr, which in the absence of calcic plagioclase—its usual host in other rocks—resides in amphibole of wyomingites and in alkali feldspar of shonkinites. All in all the trace-element pattern of ultrapotassic lavas has a good deal in common with that of the commoner sodic kind, but on a greatly exaggerated scale. It recalls in some respects the geochemistry of kimberlites, and it is the very antithesis of what is found in abyssal oceanic tholeiites.

The isotopic composition of strontium is subject to much wider and more erratic variation than has been recorded for *ne*-normative sodic volcanic series (Bell and Powell, 1969; Powell and Bell, 1970). Rocks from both the East

**Table 10-6** Trace-element abundances (ppm) and ratios (by weight) for ultrapotassic volcanic rocks

| | 1 | 2 | 3 | 4 | 5 | 6 | 7 | 8 | 9 | 10 | 11 |
|---|---|---|---|---|---|---|---|---|---|---|---|
| Ba | | | | | 6410 | 4940 | | | | | |
| Nb | 223 | 162 | 188 | | 15 | 15 | | | 110 | 116 | 89 |
| Rb | 118 | 206 | 122 | 162 | 200 | 155<br>(133) | 460 | 205 | 138 | 211 | 173 |
| Sr | 2560 | 1174 | 2265 | 1510 | 2195 | 1325<br>(1620) | 2700 | 3400 | 1243 | 665 | 1216 |
| Zr | 580 | 318 | 628 | | 190 | 160 | | | 460 | 354 | 418 |
| Rb/Sr | 0.046 | 0.176 | 0.054 | 0.017 | 0.09 | (0.082) | 0.170 | 0.060 | 0.111 | 0.318 | 0.142 |
| K/Rb | 270 | 200 | 340 | 230 | 320 | 310 | 190 | 295 | 210 | | 240 |
| $Sr^{87}/Sr^{86}$ (initial) | 0.7051 | 0.7059 | 0.7050 | 0.7151 | | (0.7072) | 0.7070 | 0.7066 | 0.7082 | 0.7098 | 0.7074 |

*Explanation of column headings*

Numbers in parentheses below refer to comparable analyses similarly numbered in Tables 10–4 and 10–5. K/Rb for East African rocks is based on $K_2O$ values from Bell and Powell (1969, table 3).

1 Katungite, Katunga, Toro-Ankole province (1); Bell and Powell (1969, p. 548, no. 101)
2 Mafurite, Bunyaruguru, Toro-Ankole province (2); Bell and Powell (1969, p. 547, no. 48)
3 Leucite nephelinite, Nyiragongo, Birunga province (3); Bell and Powell (1969, p. 546, no. 4)
4 Jumillite, southern Spain (4); Powell and Bell (1970, no. J14)
5 Shonkinite, chilled lower border, Shonkin Sag sheet (5); Nash and Wilkinson (1970, no. SS2)
6 Lower shonkinite, Shonkin Sag sheet (6); Nash and Wilkinson (1970, no. SS3). Values in parentheses from Powell and Bell (1970, p. 5, no. 65–9)
7 Wyomingite, Leucite Hills (7); Carmichael (1967*c*, pp. 50, 52, no. LH1); Powell and Bell (1970, p. 4, no. LH1)
8 Madupite, Leucite Hills (9); Carmichael (1967*c*, pp. 50, 52, no. LH16); Powell and Bell (1970, p. 4, no. LH16)
9 Shoshonite, Sabinyo, Birunga province; Bell and Powell (1969, p. 547, no. 63)
10 Latite, Sabinyo, Birunga province; Bell and Powell (1969, p. 548, no. 107)
11 Banakite (latite), Karisimbi, Birunga province; Bell and Powell (1969, p. 547, no. 69)

African and the Rocky Mountain regions give values of $Sr^{87}/Sr^{86}$ between 0.703 and 0.710, many of these beyond the limit hitherto conventionally conceded for mantle-derived strontium. Extreme values are characteristic of both Spanish jumillites (0.7136 to 0.7158) and wyomingites of Western Australia (0.7125 to 0.7215).

Nowhere in igneous petrology does composition of strontium present such challenging anomalies as in ultrapotassic lavas. And nowhere is there so little consensus regarding answers to the genetic implications of the strontium "tracer." Here are some of the anomalies uncovered to date:

1. In both East African provinces, initial $Sr^{87}/Sr^{86}$ values have a bimodal frequency distribution. For the ultrabasic lavas the total range is 0.7035 to 0.708, with a strong maximum 0.7045 to 0.706 recalling strontium of normal sodic nephelinites and other alkaline lava series of the continents. The range for feldspar-bearing lavas (basanites, shoshonites) is mostly 0.7055 to 0.7095, with a strong maximum at 0.7065 to 0.708. The discrepancy does not rule out differentiation, nor does it preclude fractional crystallization as the main mechanism responsible for lithologic and gross chemical variation in the rock series. But to rescue the differentiation model it is necessary to appeal to the operation of some other process, such as "zone refining" by which "incompatible" elements, possibly from a crustal source, are said to become preferentially introduced into the more siliceous magmas.
2. In both East African provinces, there is strong positive correlation between $Sr^{87}/Sr^{86}$ and Rb/Sr; and this correlation also is consistent with contamination by old sialic rock. There are other provinces where correlation is but dimly perceptible—for example, in the Rocky Mountain region (Powell and Bell, 1970, p. 3)—or even negative. Thus rocks from two eruptive centers in the West Kimberley province respectively give mean values $Sr^{87}/Sr^{86}$ 0.7136, Rb/Sr 0.323 (Mt. North); $Sr^{87}/Sr^{86}$ 0.7203, Rb/Sr 0.213 (Mamilu Hill). And Spanish jumillite with low Rb/Sr (0.017) has $Sr^{87}/Sr^{86} = 0.7151$. Such values are hard to explain by crustal contamination.
3. The extreme ultrapotassic suite of Leucite Hills, Wyoming, is matched in all respects of its bizarre mineralogy and gross chemistry by a volcanic series in Western Australia (West Kimberley province). It can scarcely be doubted that both have originated from chemically identical source rocks under identical conditions very seldom realized in petrogenesis. Yet with respect to isotopic composition of strontium the two suites are utterly different: for Leucite Hills rocks, $Sr^{87}/Sr^{86}$ is 0.7055 to 0.707 and Rb/Sr is 0.07 to 0.170 for rocks from West Kimberley, $Sr^{87}/Sr^{86}$ is 0.7125 to 0.7215 and Rb/Sr is mostly 0.19 to 0.36. Whatever the sources of the unique concentrations of K, Ba, Sr, and Rb in the two series, and whatever may be the mechanism

by which $Sr^{87}$ reached such high levels in the West Kimberley rocks, one must assume greatly different ages for the material supplying radiogenic strontium. This material could be the mantle source of the magmas themselves; or it could be the rocks of the underlying crust—Precambrian in Australia, a recently folded geosynclinal filling in Wyoming.

## CARBONATITES[1]

NATURE AND MAGMATIC AFFINITIES Carbonatites are igneous rocks composed chiefly of carbonates—most commonly calcite, dolomite, or ankerite but on occasion containing also carbonates of Fe, Mn, or even Na (Turner and Verhoogen, 1960, pp. 399, 400). They may also contain silicate minerals (for example, feldspar, pyroxene, olivine) and a wide variety of accessory constituents, among them apatite, perovskite, barite, fluorite, and pyrochlore.

Half a century ago, opinion on the origin of carbonatites—magmatic versus metasomatic—was sharply divided. This divergence is reflected in Bowen's (1926) statement of the problem with special reference to the type Fen area of southern Norway:[2]

> Brögger [in 1920] concluded that calcite and dolomite magmas had been there injected, that to some extent they had become mixed with silicate magmas, and that from these mixed magmas silicates and carbonates had crystallized side by side in the ordinary manner of crystallization of minerals from molten magmas.
>
> A study of my own rather limited collection of these mixed rocks led me [Bowen] to state the conclusion that the carbonates were introduced by aqueous solution, that they bear a replacing relation to the silicates, and that the rock masses consisting solely of carbonates are the result of the same process carried to completion.

So the matter stood (cf. Turner and Verhoogen, 1951, p. 341) for the next 30 or 40 years. To some petrologists, known fusion temperatures of carbonates seemed impossibly high to be maintained in nature. Experiment had shown, moreover, that some carbonates are sufficiently ductile at high confining pressures to permit ready intrusion in the solid state into fractures in more brittle rocks. Proponents of the magmatic model, on the other hand, showed that the diverse forms assumed by mapped carbonatite bodies exactly parallel those of igneous bodies in subvolcanic intrusive complexes—dikes, plugs, even cone-sheet swarms.

Today the dilemma has been successfully resolved on the combined evidence of field and experimental observation (Tuttle and Gittins, 1966):

[1] See comprehensive monographs by Heinrich (1966) and Tuttle and Gittins (1966).

[2] An isolated, late Precambrian, alkaline complex piercing the crystalline basement southwest of the Oslo graben.

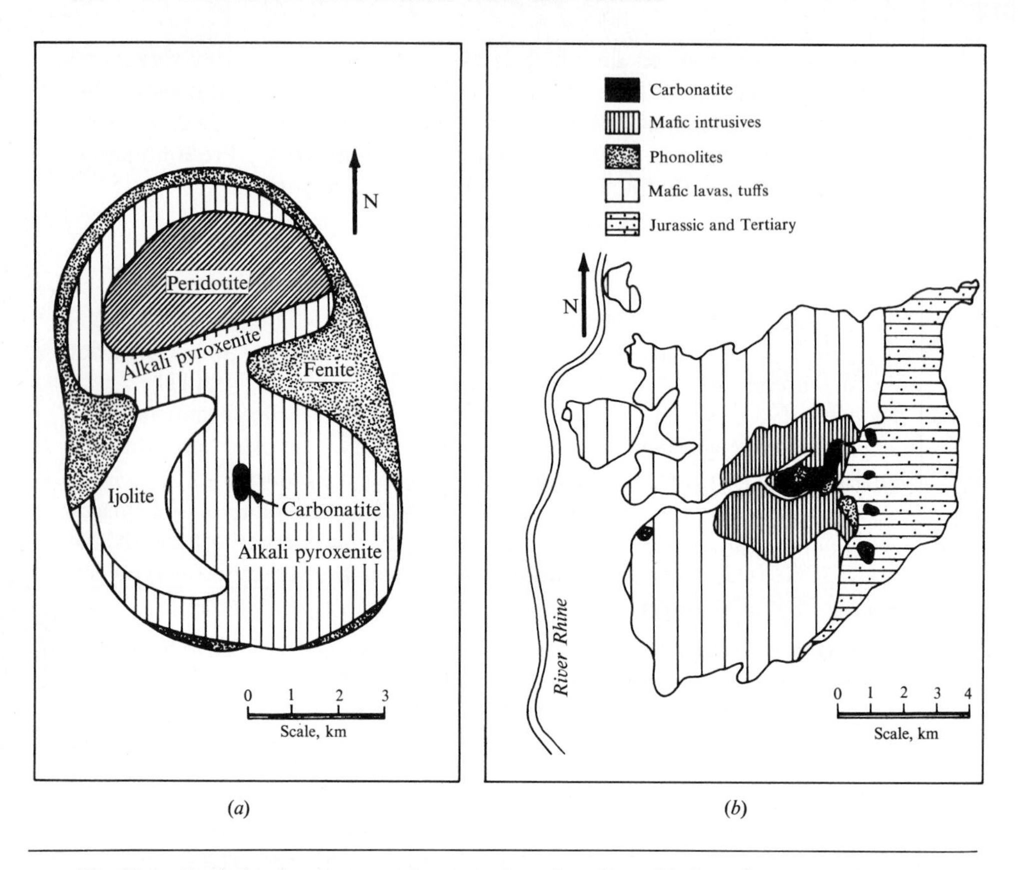


**Fig. 10-8** Geologic sketch maps of two carbonatite plugs. (*a*) Jacupiranga, southeastern Brazil; intrusive into Precambrian granitic gneisses and schists (after Melcher, 1965). (*b*) Kaiserstuhl, south end of Rhine graben, near Freiburg, Germany; intrusive into gravels and sediments of Rhine Valley (after Sutherland, 1967; Wimmenauer, 1957).

1. Mapping of numerous partially eroded volcanic centers, especially in eastern and southwestern Africa and in southern Brazil, has revealed the essential role played by shallow intrusion of central carbonatite plugs late in the eruptive cycle of nephelinite and some ultrapotassic magmas. No one can doubt this who has seen the impressive plug of calcite-apatite carbonatite—an elliptical outcrop (1000 × 400m) stripped and revealed by mining operations—that forms the core of the Jacupiranga pyroxenite-ijolite ring complex in Brazil (Melcher, 1965). This is a compound intrusion (Fig. 10-8), 65 km$^2$ in outcrop area, composed mainly of titanaugite pyroxenite (jacupirangite) enveloping two independent plugs, one of serpentinized peridotite, the other of carbonatite. Tangentially foliated round parts of the perimeter, and with steep outward-dipping contracts proved by drilling to nearly 300 m, this carbonatite body is the very epitome of an igneous plug.

Finally the existence of carbonatite magmas in nature has been clinched by direct observation: recent eruptions of the active Tanzanian volcano Oldoinyo Lengai produced flows of sodium-rich carbonatite lava (Dawson, 1962*a*). Lavas approximating the commoner kind of carbonatite have been recognized in the Toro-Ankole field, Uganda (Von Knorring and Du Bois, 1964).

2. Although carbonatites are habitually associated with alkaline mafic rocks, the two types—the one a carbonate, the other a silicate phase assemblage—tend to retain sharply their respective identities. There is nothing in mutual relationships in the field to indicate derivation at the present site from a common parental magma, although such may have occurred by liquid unmixing prior to intrusion (cf. Nash, 1972*c*).

3. Experiments over the last decade have shown that in appropriate aqueous systems the liquidus temperatures of water-saturated calcium- and sodium-carbonate melts at fluid pressures of about 1 kb may be well below 700°C. The essential factor is a high partial pressure of $H_2O$ in the $CO_2$–$H_2O$ vapor phase (for example, Wyllie and Tuttle, 1959*b*, 1960; Wyllie and Haas, 1965, pp. 890–891; Koster van Groos and Wyllie, 1968; and related contributions in Tuttle and Gittins, 1966). In more complex systems with $NaAlSi_3O_8$ and $CaAl_2Si_2O_8$ as additional components, liquid immiscibility is encountered over a wide compositional range and, other things being equal, is favored by high $p_{CO_2}$ in the fluid phase and by falling temperature. Crystalline phases such as albite, cancrinite, or carbonates can then coexist simultaneously with three fluid phases: undersaturated alkali-silicate melt; carbonate liquid with a limited content of dissolved silicate; and gas, in which $SiO_2$ and $Na_2O$ become increasingly soluble as $p_{H_2O}$ rises at constant $P_{fluids}$.

PETROGENESIS: OUTSTANDING PROBLEMS Granting now their magmatic origin, carbonatites still present some interesting problems of petrogenesis. What is the mechanism of alkali metasomatism that so characteristically accompanies intrusion of the carbonatite magma? Is it possible that some of the resulting rocks may later be remelted to give magmas of unusually potassic composition which are later erupted as trachytic lava? What is the genetic connection between carbonatites and associated alkaline igneous rocks composed of silicates? What is the ultimate source of the $CO_2$ component: mantle materials or recycled calcareous sediments? This last question is postponed for later consideration.

*The fenite problem* Since Brögger's day, petrologists have been intrigued by spectacular alkali metasomatism seen in the vicinity of carbonatite alkali-syenite complexes (cf. McKie, 1966). On one point all are agreed: Granitic gneisses become transformed to fenites whose principal constituents are alkali feldspar (orthoclase or albite) and aegirine or sodic amphibole; and there are numerous cases in which the fenite envelope is clearly related to a carbonatite contact (for example, Von Eckermann, 1961; Dawson, 1964*b*, p. 107; Sutherland, 1965, p. 367; Paarma, 1970). Fenites of this kind, it will be noted, have the chemical composition of peralkaline syenite. Comparable fenites are related to contacts with bodies of ijolite, pyroxenite, or nepheline syenite (cf. Fig. 10-8*a*).

For long the chemistry of metasomatism was obscure. Now, however, it has been established at Oldoinyo Lengai that hydrous multi-cation carbonatite magmas (Na > Ca > K > Sr) can be naturally generated and do on occasion retain their identity to the point of surface eruption. Experimental data lead us to expect, among the varied fractionation products of such magmas, residual alkali-rich fluids presumably capable of producing the kind of metasomatic effects observed in fenites. A variety of metasomatic rocks, some composed of virtually the same mineralogical assemblages as truly igneous counterparts, can form where residual alkaline fluids, expelled from a fractionating carbonatite magma, encounter rocks with which they are in violent disequilibrium. So one finds fenites, mineralogically akin to syenites, forming from granitic gneiss; metasomatic "nepheline-syenite" assemblages replacing marble (Gittins, 1961); and "ijolitic" borders to carbonatite veins cutting syenite.

Bowen questioned the magmatic status of carbonatites in the Fen area. Von Eckermann (1948) reversed the question:[1] Are ijolites, nepheline syenites, and syenites of the Alnö province, he asked, the ultimate products of metasomatism effected by intrusive carbonatite magmas? Such an origin, although far from universal, can scarcely be denied for some "syenitic" and "ijolitic" fenites. There are also well-documented instances of fenite halos in granite gneiss bordering intrusive nepheline-syenite complexes.

Finally, it has been suggested that syenitic fenites may become locally fused and so give rise to palingenic magmas that ultimately appear at the surface as trachytic lavas. This is the origin proposed by Sutherland (1965) for trachytic flow remnants surrounding a central carbonatite-fenite complex at the dissected Tertiary volcanic center of Toror Hills, Uganda. There is supporting textural evidence. And especially intriguing is the exceptionally potassic composition both of the trachytes themselves and of fenites surrounding the subjacent carbonatite pluton (cf. Table 10-7; analyses 3–5).

*Relation to alkaline ultrabasic magmas* Carbonatites are habitually associated, both on the regional scale and at individual eruptive centers, with nephelinites and allied ultrabasic alkaline rocks such as ijolite and alkali pyroxenite. Less commonly the alkaline lavas associated with carbonatites are phonolitic, as in the Eocene complex of Amba Dongas in India (Sukheswala and Avasia, 1971). The logical inference is that the carbonatite and silicate magmas, in spite of their contrasted compositions, must be genetically related. They have much in

[1] See also Dawson (1964*b*).

**Table 10-7** Chemical compositions of some African carbonatites (lavas), trachytes, fenites, and kimberlites

| | 1 | 2 | 3 | 4 | 5 | 6 | 7 |
|---|---|---|---|---|---|---|---|
| $SiO_2$ | 12.99 | Tr. | 58.43 | 62.73 | 55.50 | 36.58 | 27.93 |
| $TiO_2$ | 1.74 | 0.10 | 0.34 | | | 2.67 | 2.73 |
| $Al_2O_3$ | 3.03 | 0.08 | 17.84 | | 13.77 | 7.15 | 4.47 |
| $Fe_2O_3$ | 7.93 | 0.26 | 5.09 | | | 6.69 | 7.04 |
| FeO | 4.44 | | 0.00 | | | 4.99 | 5.12 |
| MnO | 0.40 | 0.04 | 0.42 | | | 0.34 | 0.23 |
| MgO | 8.55 | 0.49 | 0.43 | | | 22.55 | 25.42 |
| CaO | 35.97 | 12.74 | 0.80 | | | 6.05 | 10.01 |
| SrO | 0.63 | 1.24 | | | | | |
| BaO | 0.15 | 0.95 | 0.18 | | | | |
| $Na_2O$ | 0.73 | 29.53 | 0.38 | 0.38 | 3.61 | 0.28 | 0.21 |
| $K_2O$ | 0.20 | 7.58 | 13.90 | 14.97 | 9.20 | 0.47 | 1.18 |
| $P_2O_5$ | 3.32 | 0.83 | 0.35 | | 0.16 | 0.38 | 1.07 |
| $H_2O^+$ | 3.45 | 8.59 | 1.05 | | | 8.51 | 7.89 |
| $H_2O^-$ | 1.65 | | 0.11 | | | 3.31 | 0.68 |
| $CO_2$ | 14.79 | 31.75 | | | | | 5.61 |
| $SO_2$ | 0.35 | 2.00 | | | | | |
| Cl | | 3.86 | | | | | |
| F | 0.30 | 2.69 | | | | | 0.89 |
| Total | 100.62* | 100.73† | 99.32 | | | 99.97 | 100.56‡ |

* S = 0.35.
† Total less O = 2.00 for F and Cl.
‡ Including $Cr_2O_3$ = 0.08.

*Explanation of column headings*

1 Vesicular carbonatitic lava, Kalyango volcano, Fort Portal area, Toro-Ankole province, Uganda (Von Knorring and Du Bois, 1961)
2 Carbonatite lava (pahoehoe), extruded from Oldoinyo Lengai, Tanzania (Dawson, 1964*b*, p. 106, no. BD 114)
3 Trachyte, Toror Hills, Uganda (Sutherland, 1965, p. 370, no. 3)
4 Feldspathic fenite, Toror Hills, Uganda (Sutherland, 1965, p. 371, no. 15); partial analysis
5 Fenitized granitic basement, Toror Hills, Uganda (Sutherland, 1965, p. 371, no. 16); partial analysis
6 Kimberlite, Lesotho (Dawson, 1962*b*, p. 551, analysis 1)
7 Micaceous kimberlite, Lesotho (Dawson, 1962*b*, p. 551, analysis 6)

common, too, with respect to trace- and minor-element chemistry. High abundance of Ba, Sr, Nb, Ce, La, Zr, P, and Cl, which are characteristic of nephelinites and ijolites and are heightened in ultrapotassic lavas, is even further intensified in carbonatites (for example, Sethna, 1971). These indeed are the mother rocks for rare-earth and apatite ores. This fact alone rules out the possibility, once widely entertained, that fusion of carbonate sediments plays a significant role in the genesis of carbonatite magmas. A mean value given for initial $Sr^{87}/Sr^{86}$ is 0.7032 ± 0.001. Individual values range from 0.7025 to

0.7050 in nine East African samples (Powell, 1966, p. 59), all close to or within the range characteristic of associated nephelinites (0.703 to 0.706 in eastern Uganda) and highly potassic ultrabasic lavas (0.7035 to 0.7065 in western Uganda). Strontium of most limestones and marbles is more radiogenic: $Sr^{87}/Sr^{86}$ values between 0.707 and 0.710 are typical (Powell, 1966, p. 60). However, taken alone the composition of strontium is not an infallible criterion for recognizing carbonatites as such; calcite of metamorphic marbles may give $Sr^{87}/Sr^{86}$ values within the carbonatite range (Gittins et al., 1970).

Carbonatites and their alkaline ultrabasic associates obviously differ strongly in bulk chemistry. Rocks of intermediate composition, commonly brecciated, tend to show effects of carbonization of silicate phases—a situation that lent support to older ideas on supposedly secondary origin of carbonatites in general. Now that the magmatic nature of carbonatites is generally recognized, there are still alternative genetic models. In one, carbonatite and nephelinite magmas are independently generated, although perhaps at much the same level in the mantle. Equally feasible in the light of recently established high-pressure phase equilibria is the model proposed by Koster van Groos and Wyllie (1968). A primary water-charged, highly ultrabasic, silicate-carbonate magma splits with falling temperature into two immiscible fractions. One ultimately crystallizes as nephelinite or ijolite, the other as carbonatite. It is still possible that some carbonatite bodies—by analogy with "cold intrusions" of serpentinite—were finally emplaced by a process of solid flow at temperatures below 500°C (cf. Jennings and Mitchell, 1969).

In conclusion we note that the total bulk of carbonatite in any individual complex is usually small compared with that of associated ijolites or nephelinites (cf. Fig. 10-8). The classic Fen complex, 2 km in diameter, is largely composed of carbonatites at the outcrop. However, a strong associated positive gravity anomaly suggests that at no great depth the carbonatites pass down into much denser rocks, probably comparable with ijolites that outcrop locally at the rim (I. B. Ramberg, 1973). The total gravity-lithologic picture is considered by Ramberg to be consistent with upward accumulation of a light carbonatite liquid fraction separating from a vertical pipe or fissure filled with ijolitic magma.

## KIMBERLITES[1]

NATURE AND FIELD OCCURRENCE According to current usage (for example, Dawson and Watson, in Wyllie, 1967, pp. 242, 313–314), kimberlites are serpentinized porphyritic phlogopite peridotites that habitually occur in the form of minor intrusions. Their aggregate bulk in any province is small; yet kimberlite provinces themselves can cover very large areas. In Africa clusters of diatremes and small dikes are scattered across the continent from the northern

[1] Much of the information in this section has been culled from summary accounts by Dawson, Davidson, and Watson in Wyllie (1967).

margin of its folded southern tip (latitude 33°S) to latitudes well north of the Equator. Hundreds of small bodies of kimberlite dot the Siberian platform between latitudes 60 and 70°N from longitude 100 to 136°E. In eastern North America, kimberlites occur at many points along an arcuate belt more than 2000 km long that stretches south from the Precambrian shield of Quebec, skirts the western margins of the Appalachian folds in New York and western Pennsylvania, and ultimately swings southwest and west into Arkansas and Kansas.

Kimberlites are virtually confined to the nonorogenic environment of continental platforms.[1] Yet only restricted segments of these constitute provinces of kimberlite intrusion. Curiously enough, in some provinces kimberlite magmas have burst up from the depths in sharp eruptive episodes separated by long intervals of time whose total span may be 1000 or 2000 m.y. (cf. Dawson, in Wyllie, 1967, p. 251). Most kimberlites that have been dated stratigraphically or radiometrically are of Mesozoic age, the majority of these definitely Cretaceous.

*South African kimberlites* The typical kimberlite body of South African diamond fields is a vertical pipe seldom more than 0.2 $km^2$ in cross section, tapering downward into a filled fissure. Narrow dikes are also known and a sill complex has been reported near Kimberley itself. The pipe filling is breccia consisting of fragments of kimberlite proper and a wide assortment of xenolithic blocks. These have been torn from every level along the path of ascent. Most abundant are rounded fragments of high-grade metamorphic rocks representing the crystalline basement. And there are angular fragments, some very large, derived from the intruded sedimentary cover—even from sedimentary formations hundreds of meters stratigraphically above the sites of present occurrence. Particular interest attaches to a minority of well-rounded blocks whose source, as we shall presently see, must lie in the mantle itself. These are garnet peridotite and eclogite, and rare specimens of both have been found to contain diamond as a primary phase. These mantle rocks must be the source of diamonds—and also of some garnet and chrome-diopside xenocrysts—in the kimberlite matrix.

South Africa kimberlite pipes are deeply weathered: oxidized "yellow ground" is followed downward by "blue ground." Less altered rocks below display a mineralogical and chemical character as consistent as it is striking. This is seen in Dawson's (1962*b*) account of "hardbank" kimberlite in Lesotho: a deuterically altered porphyritic rock with phenocrysts of olivine, chrome diopside, enstatite, phlogopite, pink garnet, and ilmenite. Olivine phenocrysts especially are partially or completely replaced by serpentine minerals. The groundmass consists of serpentine minerals and calcite with small grains of

[1] "Kimberlites" reported from active orogenic zones are atypical: diamond-bearing peridotite breccias in Borneo, alnoites with xenocrysts of pyrope and chrome diopside in the Solomons.

ilmenite, phlogopite, and perovskite. Most rocks from Lesotho have only minor phenocrystic phlogopite and are of a type designated "basaltic." Some, however, have more plentiful mica in two generations, giving the rock a lamprophyric appearance; these are called micaceous kimberlites. There are other South African fields, for example, Orange Free State, where micaceous kimberlites predominate. The two types are mutually gradational, even in a single small body.

Kimberlite intrusions in Lesotho cut the early Jurassic Stormberg flood basalts. Some South African bodies have been dated certainly as Cretaceous, and this may well be the age of most others. Closely associated coeval igneous rocks of other compositions are rare; they include melilitite, nephelinite, and carbonatite (Dawson, in Wyllie, 1967, p. 250).

*Eastern North American kimberlites* Throughout the whole of the eastern province of North America (Watson, in Wyllie, 1967, pp. 312–323), rocks of the kimberlitic suite are of two petrographic types—kimberlite and alnoite. Both occur as dikes, veins, and sills; and intrusion temperatures, estimated from natural coking effects on coals, cannot have exceeded 600°C. Typical unweathered kimberlites from Bachelor Lake, Quebec, are of the micaceous type and contain three minerals in abundance: olivine (much of it serpentinized), phlogopite, and calcite. The latter is largely a primary phase that crystallized directly from the magma. Minor ubiquitous minerals are magnetite, ilmenite, perovskite, augite, apatite, and chlorites. Alnoites are ultrabasic lamprophyres consisting of melilite, olivine (in some rocks both forsteritic olivine and monticellite), serpentine, calcite, pyroxene, ilmenite, magnetite, and perovskite. Kimberlite and alnoite are mutually gradational through rocks with a low content of melilite.

In most North American kimberlite fields there is no record of xenolithic garnet peridotite or eclogite, nor of diamond. However, kimberlite diatremes in Kansas contain blocks of eclogite mineralogically identical with those of African and Siberian kimberlite (Meyer and Brookins, 1971). And diamonds are not unknown. They have been recovered from soil debris at a kimberlite locality in Arkansas and have even been found there in the rock itself.

*Kimberlite variants* Kimberlites as just described are known in almost every continent and are rocks of striking and consistent individuality. Within the class there is, to be sure, an element of local or regional variety. Some American kimberlites contain a little melilite; some, for example, in the Katanga province of equatorial Africa, contain monticellite as well as forsteritic olivine; and although kimberlite is the ultimate source of most of the world's diamonds, there are extensive kimberlite fields where diamond is unknown.

There are other ultramafic rocks whose identification as kimberlite is questionable. Such, as noted earlier, are the diamond-bearing peridotite breccias of Borneo and some alnoitic breccias in the Solomons. Also far from typical are a few serpentinous diatreme fillings, containing low-temperature eclogite xenoliths, in the late Tertiary potash-lamprophyre Navajo province of eastern Arizona (Watson, in Wyllie, 1967, pp. 261–269). Then there are diamond-bearing dikes of unusually potassic character in the Ivory Coast province of West Africa. These dikes contain plentiful leucite as well as phlogopite and are reminiscent of ultrapotassic volcanics (fitzroyites) in the West Kimberley field of Australia. Such variants lend support to the possibility of some genetic connection between kimberlites and alkaline, more especially potassic volcanic rocks. They will not be considered further. Discussion that follows concerns only kimberlites in the more restricted sense as treated in previous sections.

CHEMISTRY OF KIMBERLITES Kimberlites resemble all other peridotites (Dawson, 1962*b*, pp. 553–554; Heinrich, 1966, p. 95) in their content of Si, Mg, Fe, and Na (cf. Table 10-7; analyses 6, 7); but Ti, Al, and Ca are high, and K extremely so. $K_2O/Na_2O$ typically is 2 to 5. Expectably, abundances of Cu, Cr, Ni, and Co are high, as in all ultramafic magnesian rocks. But kimberlites also show abnormally high Ba, La, Li, Pb, Rb, Sr, Th, U, and Zr—elements that normally attain high concentrations in siliceous members of other rock series. This combination of characteristics recalls the trace-element pattern of highly potassic ultrabasic and basic volcanic rocks (Table 10-6). Kimberlites in fact have something in common with potassic lamprophyres, with which they are sometimes classed on purely petrographic grounds. But in kimberlites, $MgO/K_2O$ is far higher (mostly 20 to 70) and $(K_2O + Na_2O)$ significantly lower ($<$ 2 percent by weight).

The isotopic composition of strontium in kimberlite is variable. Powell (1966) gave initial whole-rock values of $Sr^{87}/Sr^{86}$ 0.705 to 0.721, and he concluded that at least some kimberlites contain strontium derived from the sialic crust. However, more recent work by Berg and Allsopp (1972) suggests that the higher ratios are restricted to altered specimens and that initial $Sr^{87}/Sr^{86}$ values for fresh kimberlite are in the range 0.7037 to 0.7046. An alternative explanation of high values, already offered to explain very high $Sr^{87}/Sr^{86}$ in some ultrapotassic lavas, is early concentration of K (and with it Rb) in the source rock from which the kimberlite magma originated much later. Further speculation is unwarranted until we can assess—through oxygen- and hydrogen-isotope analyses—the possible role of meteoric waters in serpentinization of kimberlite.

AFFINITIES, SOURCES, AND EMPLACEMENT OF MAGMA Although there are obvious broad chemical and mineralogical analogies between kimberlites and some potassic basic and ultrabasic rocks, the two classes are not mutually gradational. Affinities with carbonatites are reinforced by textual indications that calcite in some kimberlites (notably those of Quebec) is probably a primary magmatic mineral. Most remarkable in this respect are differentiated sills recently described from near Kimberley itself (Dawson and Hawthorne, 1973). These sills show layered structures due to multiple injection and graded cumulus structures within the layers; and it seems that the mobile intercumulus liquid phase at the time of intrusion was essentially carbonatite. Intimate association of kimberlite, melilitite, and carbonatite in composite bodies, though rare, has been well documented in a few cases (for example, Ukharnov, 1965; cited by Dawson, 1968). Finally there is a marked tendency for mutual overlap (both in space and time) of kimberlite fields and provinces studded with near-surface intrusions of alkaline magmas. This tendency is obvious in the southern half of Africa. And the extensive diamond fields of Brazil, though but sparsely populated with known intrusions of kimberlite, are the site of one of the world's most spectacular provinces of alkaline intrusion.

It would seem, then, that kimberlitic, carbonatite, and ultrapotassic magmas have been generated—especially in Cretaceous and later times—in the mantle beneath specific extensive segments of the continental platforms. We cannot be sure whether this reflects some chemical peculiarity of subcrustal rocks in such regions, or a unique geothermal regime restricted to the environment of kimberlite intrusion over a limited span of time (cf. Harris and Middlemost, 1970). Nor do the three kinds of magma necessarily belong to any single lineage. All three perhaps may be independently generated under rather similar conditions.

In any case, kimberlites provide unique information about the deepest proved levels of the naturally sampled mantle (cf. D. H. Green, 1970*a*, pp. 28-29).

> The diamond pipes [say Kennedy and Nordlie, 1968, p. 499] serve as a window that gives us a look into the earth. There is probably no other group of rocks that originated from even remotely as great a depth as have these. We see, in the pipe, fragmental samples of certainly the upper 200 km of the earth.

This statement, uncompromising to a degree unusual in geological writing, is based upon (and justified by) the universal presence of xenolithic blocks of garnet lherzolite and of eclogite in kimberlite. The pair olivine-garnet, critical of garnet peridotites, is stable with respect to the more familiar peridotite assemblage orthopyroxene-diopside-spinel only at pressures greater than 15 to 20 kb in the temperature range 1000 to 1300°C (for example, O'Hara et al., 1971). Supporting evidence, in the light of high-pressure experimental data, is provided by the $Al_2O_3$ content of both pyroxenes and the $CaSiO_3$ content of clinopyroxene coexisting with pyrope garnet (for example, O'Hara, in Wyllie, 1967, pp. 401-402; D. H. Green, 1970*a*, p. 28). These data show that immediately prior to incorporation in the kimberlite magma, the garnet-peridotite assemblage reached equilibrium at temperatures about 1000 to 1100°C in the pressure range 30 to 50 kb—several hundred degrees below the temperature of dry melting. Over the same range of temperature the eclogite assemblage is stable with respect to the plagioclase-pyroxene assemblage of granulites in rocks of basaltic composition only at pressures exceeding 20 kb. Most compelling of all is the presence of diamond, which becomes more stable than graphite at pressures greater than 40 to 50 kb in the temperature interval 1000 to 1250°C (Fig. 3-8). Moreover, where diamond contains silica inclusions these take the form of the high-pressure phase coesite. The cumulative evidence is impressive. The minimum depth from which the kimberlite magma, loaded with mantle debris, began its explosive ascent is 100 km. The figure 200 km quoted above is not unrealistic.

The form of the typical kimberlite pipe and its chaotically mixed contents derived from every level along the path of ascent speak eloquently for violent turbulence during explosive upward flow. Reaction rates observed in the laboratory underscore the high velocity of emplacement: O'Hara et al. (1971, pp. 62, 65) conclude that well-preserved garnet-lherzolite blocks could not have survived more than a few hours during transport from 40 km depth at temperatures round 850°C. Emplacement of a kimberlite pipe is seen as the upward rush of a highly mobile fluidized system composed of solid xenoliths, a liquid phase (kimberlite magma) internally exploding into fragmented blobs, and a lubricating gas phase continuously emitted from the liquid (cf. Harris and Middlemost, 1970, pp. 84–86). The dike-and-sill systems of northeastern North America represent somewhat less violent intrusion—although even here the injected heterogeneous magma seems to have been highly mobile (Watson in Wyllie, 1967, p. 322) at temperatures below 600°C.

The accepted model of intrusion implies that gas can be generated in quantity—even though only locally—at depths below the Moho discontinuity. Abundance of hydrous silicates and carbonates in the final product indicates that the chief components of the kimberlite gas phase are water and $CO_2$.

From field evidence it is becoming increasingly clear that there are genetic connections of some kind between nephelinite and carbonatite magmas; between each of these and rarer ultrapotassic magmas; and between some of the latter and kimberlites. The collective chemical spectrum is much too broad to permit derivation of all from some common parental magma. More likely is the alternative that all four types are generated to some degree independently from different source materials in the upper mantle deep down beneath continental cratons (see Chap. 13).

# Andesites and Associated Volcanic Rocks of Island Arcs and Continental Margins

# 11

## General Tectonic-Volcanic Pattern

### RECENT ANDESITE VOLCANISM AND BENIOFF ZONES

Volcanic activity today is concentrated along arcuate chains of volcanoes, some of them thousands of kilometers long, located round the borders of the Pacific and parts of other oceans. These chains form strings of islands (island arcs) such as the Aleutians, Kuriles, and the Caribbean arc; or, as in the Andes, they may rise upon the continental margin itself; or again they may crown islands of less than continental size but of continental lithology and structure—for example, Japan, Indonesia, northern New Zealand. Much, and in some instances (for example, in the Caribbean) most, of the erupted magma is andesitic. This environment indeed is the home of andesite. Nevertheless in many arcs andesites are associated with equal or greater quantities of other volcanic rocks—picritic basalts, tholeiites, dacites, or rhyolites. In the Scotia arc they are subordinate to basalts.

The overwhelming majority of active or recent andesitic volcanoes are located near the epicenters of earthquakes with intermediate or deep foci (> 75 km) which define *Benioff zones* dipping toward the continent at angles of between 15 and 85°. A Benioff zone may extend from near the surface to a depth of 700 km; andesitic volcanism is, however, usually restricted to areas where the depth to the seismically defined Benioff zone is between roughly 100 and 300 km. Note that in some areas (for example, in central Asia north of India) deep-focus earthquakes are not associated with andesitic volcanism.

In the framework of sea-floor spreading and plate tectonics, a Benioff zone represents an oceanic lithospheric plate, perhaps 100 km thick, which descends into the mantle.[1] Descent is believed to begin in the neighborhood of the narrow submarine trench, occasionally 10 km deep, which typically is found to be some 200 km away on the ocean (convex) side of the volcanic arc.[2] Commonly the trench is partially filled with sediments, sometimes deformed, sometimes not. Parallel to the trench and arc are associated belts of remarkably large gravity anomaly. A narrow belt of pronounced negative anomaly occurs on the outside near the zone where the projected Benioff zone intersects the surface; it may coincide with the deepest portion of the trench but commonly lies slightly inland from it. A less well marked but much broader positive anomaly occurs further in; the distance from maximum to minimum gravity appears everywhere to be near 115 km. The negative anomaly is believed to be caused by accumulation of oceanic sediments that have been, so to speak, scraped off the plate as it starts to move downward under the plate with which it collides. The positive gravity anomaly, on the other hand, is believed to be caused by the high density of the descending plate, which is colder (and therefore denser) than the surrounding mantle into which it sinks. Relative coldness of the descending plates is confirmed by the generally low ($\sim$ 1 $\mu$cal cm$^{-2}$s$^{-1}$) heat flow observed in the trench. Contrasted high (2 $\mu$cal cm$^{-2}$s$^{-1}$ or more) heat flow observed in deep basins inside and behind the volcanic arc—as in the Sea of Japan—is presumably related to the formation of abundant magma and has not been satisfactorily explained.

Chains of volcanoes located along the margins of some continents duplicate many of the features of island arcs. Thus the line of volcanoes extending along the Andes of Peru and Chile[3] is associated with a well-defined seismic zone and and an offshore oceanic trench; and much the same situation holds in Central America from Mexico southward. In the northwest of the United States, the active andesite volcanoes of the High Cascades in Oregon and Washington (Turner and Verhoogen, 1960) have a less well-defined seismic zone, and the associated trench is filled with sediment. Deep-focus earthquakes ($>$ 50 km) demarcating a dipping seismic zone have yet to be found underlying the andesite volcanoes of Shasta and Lassen in northern California, although the presence of a spreading ridge, the Gordo, in the Pacific Ocean to the west indicates a setting similar to that of the oceanic arcs.

The association of andesites with descending oceanic plates (or subduction zones, as they are also called) seems so well established that there is a tendency

[1] In Chile, for example, the focal mechanisms of the deep-focus earthquakes indicate that the descent of the oceanic plate under this part of South America takes place in discrete and localized episodes, and that the lithospheric slab itself is broken into a series of segments that are being absorbed independently in the mantle (Stauder, 1973).

[2] The distance between the arc and the trench is believed to increase progressively throughout the life of an active arc-trench system (Dickinson, 1973).

[3] The Andean geosyncline, on whose eroded filling many of these recent volcanoes rest, is an older trough itself filled to a depth of 30 km with andesitic lavas and rhyolitic debris during the Jurassic and Cretaceous.

to assume the presence (or recent occurrence) of a subduction zone under andesitic volcanic chains that are not associated with deep seismic activity, such as the Caucasus or the Cascades of western North America. Absence of seismic activity over a period of a few decades does not necessarily imply its absence over a much larger time span. It is also possible that downward motion did indeed stop some $10^6$ years ago and that present eruption of andesite represents merely the last outpouring of magma previously generated and stored at a depth sufficient to prevent it from cooling and crystallizing. A detailed analysis of the geological record may perhaps show that andesites also form, like mountains, at the junction of colliding continental plates and that subduction is not essential to its genesis.

Oceanic lithospheric plates are believed to begin their descent where they meet either a continental or another oceanic plate. The Marianas arc is an example of the latter type; the Andes of South America are of the former type. One would expect these two types also to be petrographically different in view of presumed petrological differences between oceanic and continental plates; this is borne out to some extent, as we shall see. An oceanic plate is typically composed of

1. A thin upper rind a fraction of a kilometer thick; layer 1; $V_P = 2.2$ km s$^{-1}$
2. Basaltic lavas totaling perhaps 1.75 km in thickness: layer 2; $V_P =$ 4.7 km s$^{-1}$
3. The deeper oceanic crust, probably gabbros, diabases, and metabasalts (including spilites), 4 km thick: layer 3; $V_P = 6.6$ km s$^{-1}$
4. Mantle rock beneath the Moho, consisting of peridotite and eclogite (cf. Chap. 7); $V_P = 8.0$ km s$^{-1}$

The rate at which a lithospheric plate descends (is subducted) into the mantle has been deduced from the location and chronology of the magnetic anomalies which symmetrically parallel a mid-ocean spreading ridge. It varies between 1 and 12 cm per year (Le Pichon, 1968), and the lithospheric plates consumed beneath the island arcs at the highest rates are usually associated with seismic zones which extend to 700 km. Plates with the lowest rates often have undeformed sediments in the ocean trenches, which are the junctions on the earth's surface between two plates, one containing the volcanic arc and the other, the oceanic plate, underthrusting it.

## ASSOCIATED BASALTS AND RHYOLITES

Although island arcs are usually dominated by andesite, this is not always the case, as can be seen in the histograms showing the volumes of various rock types in the Caribbean and Scotia arcs (Fig. 11-1). P. E. Baker (1960*a*), from whose paper these data were taken, considers that in young arcs basalt is the principal rock type, but in older, more mature arcs andesite becomes dominant.

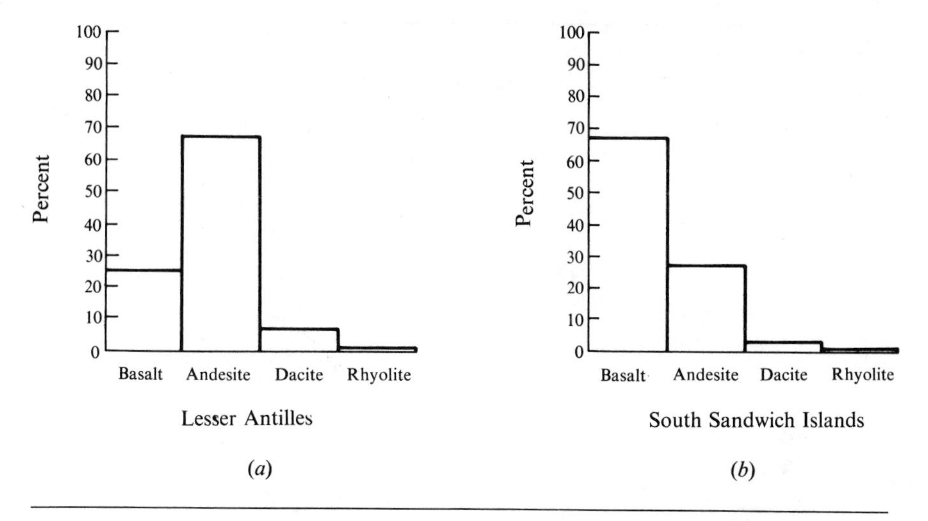


**Fig. 11-1** Estimated relative proportions (percent) of basalt, andesite, dacite, and rhyolite occurring in the Lesser Antilles (Caribbean) and Scotia arcs. (After P. E. Baker, 1968*a*.)

In certain arcs the products of contemporary volcanism display a regular pattern transverse to the arc trend. In the Japanese and Kurile arcs, basalts erupted along the volcanic front are tholeiitic types, often showing iron enrichment in their more siliceous associates, whereas further back from the volcanic front high-alumina basalts are characteristic;[1] these in turn give way to alkali olivine basalts, sometimes carrying large numbers of fragmented ultramafic nodules. This systematic pattern of basaltic composition for the Japanese arc is shown in Fig. 11-2*a* together with the chemical criteria used by Kuno to distinguish the three major basalt types in this region (Fig. 11-2*b*). If the estimates for the Japanese arc (Sugimura et al., 1963) are typical, there also appears to be a progressive decrease in the volume of basaltic lava erupted away from the volcanic front.

Active volcanic arcs, standing on what is presently considered to be oceanic crust, sometimes extend into the continents where they overlap continental crust. Thus the volcanoes of Katmai, Iliamna, and Mt. Spurr represent the eastward extension of the Aleutian arc along the Alaskan peninsula; the Kurile arc continues into Kamchatka, and the Tonga-Kermadec arc into the Taupo volcanic zone of the North Island of New Zealand (Healy, 1962). In each case the continental segment of the arc is associated with ignimbrite (dacite-rhyolite composition) flows, those of New Zealand being particularly voluminous and well studied (Ewart, 1967, 1968; Ewart and Stipp, 1968). In the northern Chilean and Peruvian Andes, the main products of the youngest eruptive cycle

[1] Kuno (1950) has used the names *pigeonite series* and *hypersthenic series* for the tholeiitic and high-alumina series here described. The term *calc-alkali* is often used to denote absence of iron enrichment; but we do not use the term here except in the original sense of Peacock (1931).

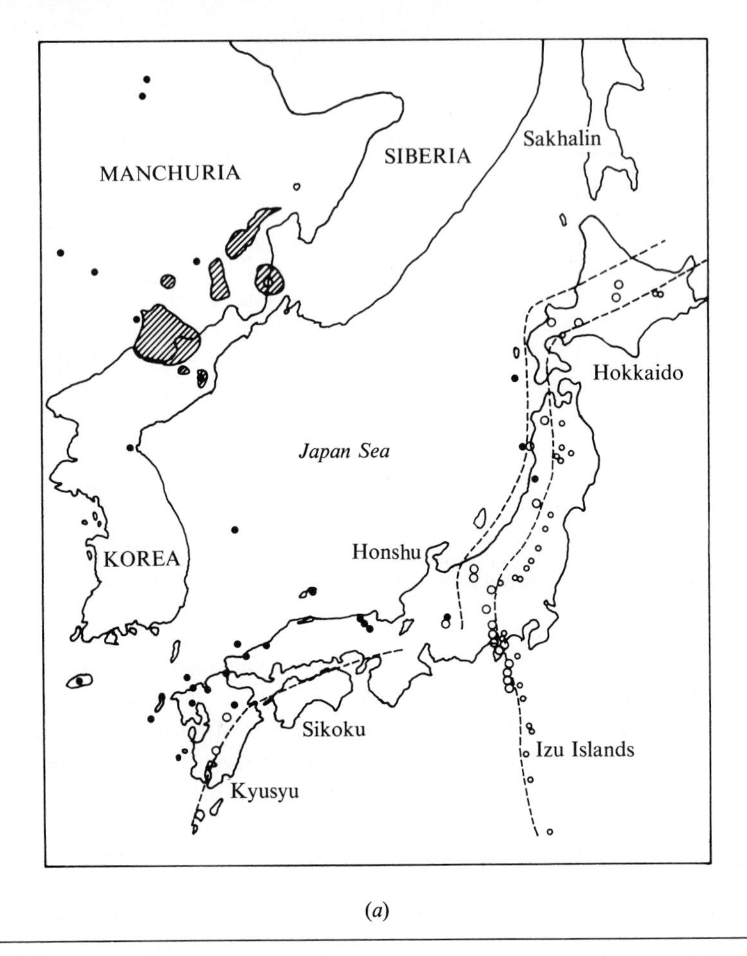


(*a*)

**Fig. 11-2** (*a*) Distribution of tholeiite (small circles), high-alumina basalt (large open circles), and alkali olivine basalt (solid circles) in Quaternary volcanoes of Japan, Korea, and Manchuria. Shaded areas in Manchuria are Tertiary alkali basalts (after Kuno, 1966). (*b*) Total alkali plotted against $SiO_2$ for the late Cenozoic volcanic rocks of the Japanese arc (*a*). The two lines denote the general boundaries between the tholeiite series (solid circles), the high-alumina series (crosses), and the alkali olivine-basalt series (open circles) (after Kuno, 1966).

are sheets of rhyolitic tuff 70,000 $km^2$ in extent. At the southern end of the chain, active volcanoes are erupting mainly basaltic and andesitic lavas.

## GEOGRAPHIC DISTRIBUTION OF ISLAND ARCS

Volcanoes of the Pacific margin include three-quarters of the active volcanoes of the world. Among island arcs that have been comprehensively described we mention the Aleutian arc (Coats, 1952, 1962*), the Kurile-Kamchatka arc

* A remarkably prescient account of what today is widely accepted as the new global tectonics.

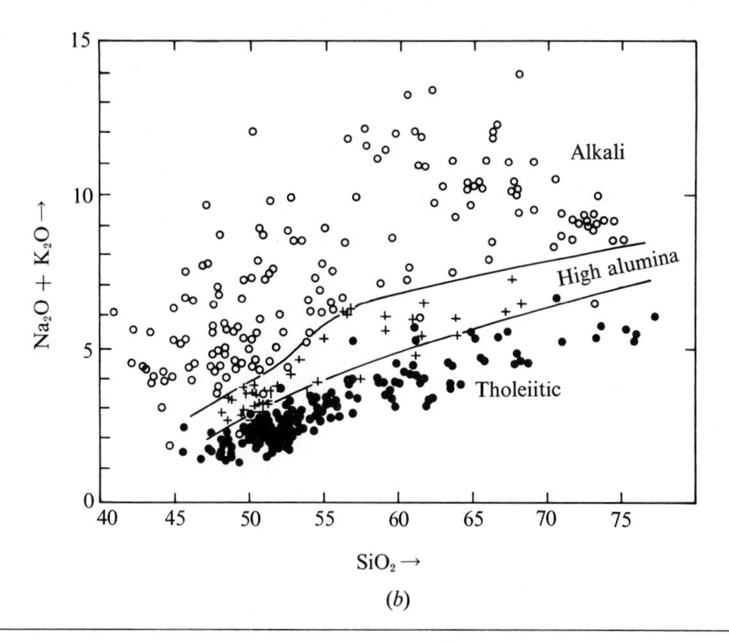


**Fig. 11-2** *Continued.*

(Gorskov, 1972), and the Japanese-Izu arc (Kuno, 1966, 1968).

South of the Japanese-Izu arc system are the two semiparallel arcs of the Marianas (Schmidt, 1957) and the Philippines, separated from one another by the oceanic crust of the Philippine Sea. Again southward lies the great Indonesian-Andaman arc system stretching into the Indian Ocean, and the Southwest Pacific arc system, composed of many segments, which surrounds the Australian lithospheric plate to the north and east. It is the young volcanic arcs of the Southwest Pacific which are considered as an example in this chapter.

Although the Pacific Ocean contains the majority of active oceanic arcs, two occur in the Atlantic: the Caribbean arc (P. E. Baker, 1968*b*, 1969; MacGregor, 1938; Martin-Kaye, 1969; Robson and Tomblin, 1966; Lewis, 1971*a*)[1] and the Scotia arc of the South Atlantic (P. E. Baker, 1968*a*). The western Mediterranean has the Cyclades arc (Lort, 1971), which includes the volcano Santorini (I. A. Nicholls, 1971), one eruption of which is believed to have changed the course of Western civilization in Crete.

## Volcanic Arcs of the Southwest Pacific

The active volcanoes and submarine trenches which border the Australian continent to the north and east are shown in Fig. 11-3; they are the New Guinea–New Britain arc, shown in more detail in Fig. 11-4, the Solomon Islands arc,

[1] See also Sigurdsson et al. (1973).

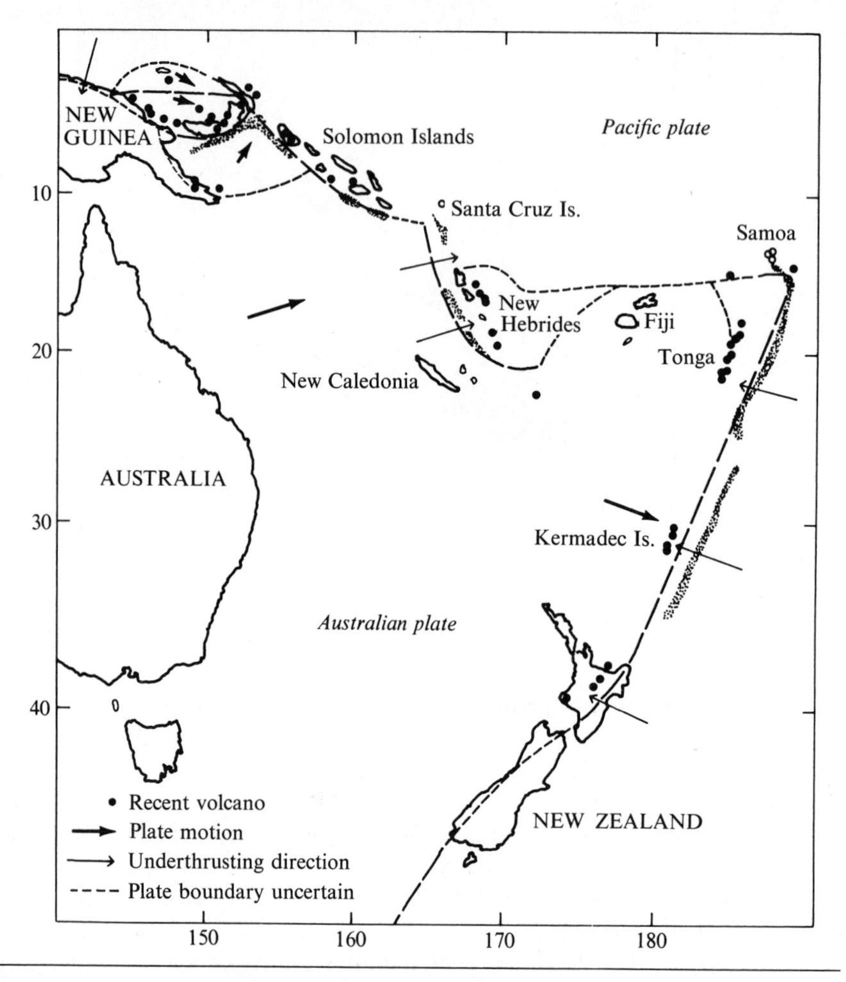


**Fig. 11-3** Active volcanoes (solid circles), plate boundaries (dashed) and submarine trenches (stippled) of the southwest Pacific. Plate boundaries and their motions taken from Le Pichon (1968) and T. Johnson and Molnar (1972).

and the New Hebrides arc, the last being tectonically related, possibly by a transform fault, to the Tonga-Kermadec arc, which continues into the active volcanic region in the North Island of New Zealand.

## NEW GUINEA–NEW BRITAIN ARC

The large-scale geological features of the New Guinea–New Britain arc are shown in Fig. 11-4. To the south of New Britain is a well-defined submarine trench, and the associated Benioff zone, down which the Solomon Sea plate is being consumed, dips to the north. At the eastern end of the island, this seismic

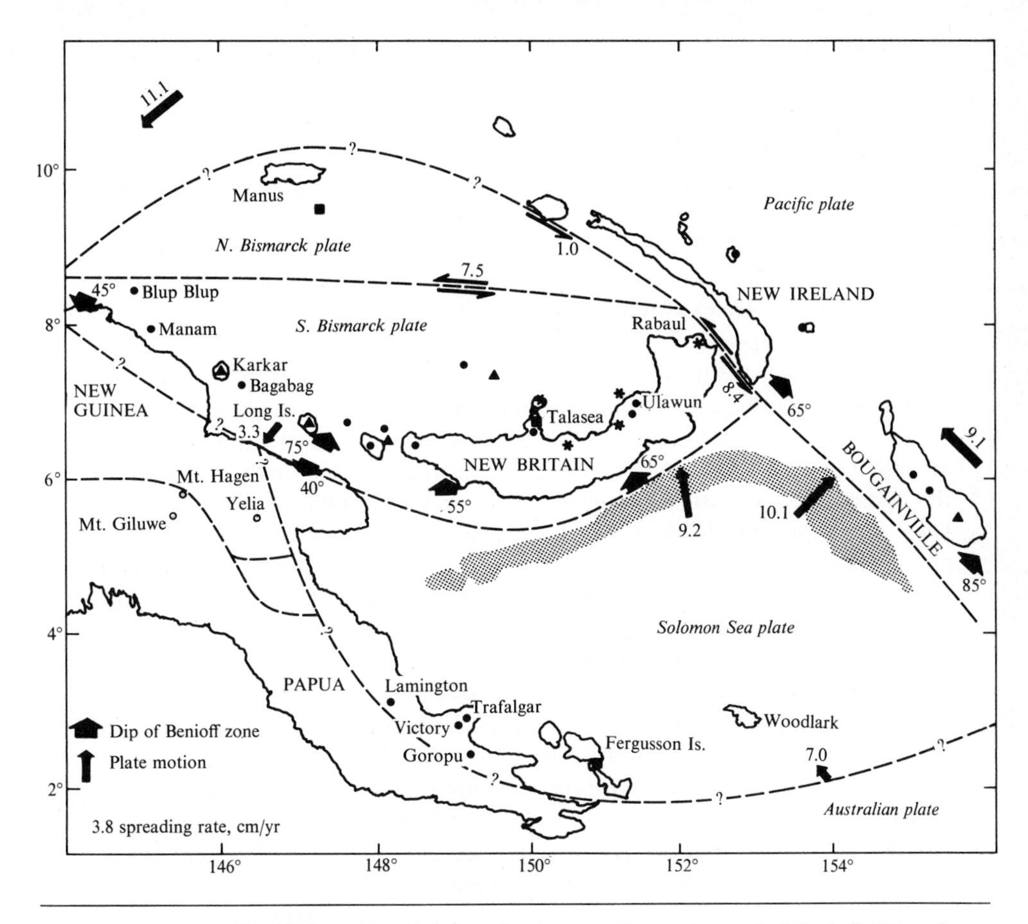


**Fig. 11-4** The New Guinea–New Britain volcanic arc and oceanic trench (stippled). Dips of Benioff zones (broad arrow heads), plate motions (arrows) and plate boundaries taken from Le Pichon (1968) and T. Johnson and Molnar (1972). Active volcanoes: solid circles; solid squares (with acid rocks); triangles (calderas); stars (calderas with acid rocks).

zone turns abruptly south following the ocean trench on the west coast of Bougainville. At the west end of New Britain, the Benioff zone has a dip component toward both the north and south, although further west toward West Irian it dips at a low angle toward Australia (Denham, 1969). Note that there is no trench associated with the volcanoes along the north coast of New Guinea.

Although there is a geographically continuous arc of active or dormant volcanoes along the north coasts of New Guinea and New Britain, there is a conspicuous contrast in the volcanic products of the two islands. Acid lavas and pumice flows are found only from Talasea eastward to Rabaul in New Britain and on the northern edge of the Bismarck Sea near Manus. In western New Britain and along the north coast of North Guinea, acid lavas are absent, and the olivine tholeiites and basaltic andesites which predominate in these volcanoes can be distinguished chemically from lavas of similar type to the east.

Of the 20 or so active volcanoes of this arc, 9 have foundered to produce calderas (Fisher, 1957). The eastern ones such as Talasea have numerous associated extrusions of rhyolite with obsidian carapaces, whereas another, the Rabaul caldera, has been completely draped by extensive pyroclastic flows, the final episode of which occurred between A.D. 445 and 670 (Heming and Carmichael, 1973).

The Recent lavas of the active volcanoes of the north coast of New Guinea are basaltic andesites and andesites with phenocrysts of olivine, orthopyroxene, and plagioclase (W. R. Morgan, 1966); microphenocrysts of titanomagnetite are sporadic accessories, but ilmenite is absent. There are no published data on the composition of either the phenocrysts or the groundmass phases of the lavas of this part of the chain. Of the analyses given by W. R. Morgan (1966), all the latest flows from Manam Island are basaltic andesites (Table 11-1, analysis 1), although andesite has been found. Eastward along the volcanic chain, basaltic andesites and andesites (Table 11-1, analyses 2 and 3) predominate.

More is known about the lavas of New Britain since both the Talasea peninsula (Lowder and Carmichael, 1970; Lowder, 1970; Peterman et al., 1970) and the Rabaul caldera (Heming, in press) have been the subject of detailed studies; the recent eruption (1970) of Ulawun has also been described (R. W. Johnson et al., 1972). The lava types forming the Quaternary volcanoes and caldera of the Talasea peninsula range from basalt to rhyolite (Table 11-1, analyses 4–7). The volumes of the lava types are estimated as: 9 percent basalt, 23 percent basaltic andesite, 55 percent andesite, 9 percent dacite, and 4 percent rhyolite. The basic lavas contain olivine ($Fo_{85}$ to $Fo_{75}$), plagioclase ($An_{92}$ to $An_{50}$), and orthopyroxene ($En_{75}$ to $En_{60}$) phenocrysts, with a few titanomagnetite microphenocrysts set in a groundmass of augite, subcalcic augite, pigeonite, plagioclase, and titanomagnetite, often with a siliceous glassy residuum. Two types of basaltic andesite occur: one fine-grained with few phenocrysts and the other coarser-grained with abundant phenocrysts. Dacite and rhyolite lavas contain amphibole, biotite, and pyroxene phenocrysts together with plagioclase and sometimes quartz. Compared to an oceanic tholeiitic series—for example, Thingmuli, Iceland (Carmichael, 1964*b*)—this tholeiitic series is low in $TiO_2$ ($<$ 1 percent), a little richer in $Al_2O_3$, and shows only mild iron enrichment (Fig. 11-5). Although the Talasea series has contents of $Na_2O + K_2O$ which transgress the tholeiitic–high-alumina-basalt boundary for the Japanese arc (Fig. 11-6), this is only another illustration of the difficulty of extending simple chemical parameters from one province to another; a glance at the analyses of the Talasea lavas in Table 11-1 shows that they are not high-alumina types.

The $Sr^{87}/Sr^{86}$ ratios of the Talasea series range between 0.7034 and 0.7038 (Peterman et al., 1970*a*), a statistically insignificant variation. Since this isotopic

**Table 11-1** Representative analyses and trace-element abundances (ppm) of lavas from the New Guinea–New Britain active volcanic arc

| | NEW GUINEA | | | TALASEA | | | | RABAUL | | |
|---|---|---|---|---|---|---|---|---|---|---|
| | 1 | 2 | 3 | 4 | 5 | 6 | 7 | 8 | 9 | 10 |
| $SiO_2$ | 53.05 | 60.10 | 55.60 | 51.57 | 58.60 | 66.34 | 75.33 | 50.11 | 55.27 | 64.95 |
| $TiO_2$ | 0.35 | 0.56 | 0.81 | 0.80 | 0.89 | 0.71 | 0.27 | 0.95 | 0.86 | 0.84 |
| $Al_2O_3$ | 13.90 | 16.40 | 16.50 | 15.91 | 15.38 | 14.63 | 12.58 | 18.89 | 17.40 | 15.25 |
| $Fe_2O_3$ | 6.07 | 4.25 | 2.90 | 2.74 | 2.22 | 1.63 | 1.58 | 3.67 | 3.80 | 2.04 |
| FeO | 3.99 | 5.30 | 8.20 | 7.04 | 6.71 | 3.77 | 0.88 | 5.80 | 4.17 | 3.00 |
| MnO | 0.16 | 0.12 | 0.21 | 0.17 | 0.18 | 0.14 | 0.07 | 0.18 | 0.15 | 0.24 |
| MgO | 8.60 | 3.10 | 3.10 | 6.73 | 3.22 | 1.48 | 0.24 | 4.64 | 3.80 | 1.42 |
| CaO | 10.68 | 7.30 | 7.30 | 11.74 | 7.02 | 4.04 | 1.25 | 10.86 | 8.22 | 3.64 |
| $Na_2O$ | 2.57 | 1.75 | 3.13 | 2.41 | 3.84 | 4.50 | 4.02 | 2.76 | 3.42 | 4.88 |
| $K_2O$ | 0.90 | 0.84 | 1.93 | 0.44 | 1.46 | 2.14 | 3.82 | 0.94 | 1.53 | 2.89 |
| $P_2O_5$ | 0.25 | 0.15 | 0.36 | 0.11 | 0.25 | 0.21 | 0.02 | 0.19 | 0.23 | 0.30 |
| $H_2O^+$ | Nil | 0.11 | 0.13 | 0.35 | 0.30 | 0.35 | 0.31 | 0.79 | 0.79 | 0.55 |
| $H_2O^-$ | Nil | 0.16 | 0.23 | 0.10 | 0.07 | 0.12 | 0.09 | 0.20 | 0.23 | 0.04 |
| Total | 100.52 | 100.14 | 100.40 | 100.11 | 100.14 | 100.06 | 100.46 | 99.98 | 99.87 | 100.04 |

| | 4 | 5 | 6 | 7 |
|---|---|---|---|---|
| Zr | 30 | 75 | 115 | 150 |
| Sr | 355 | 395 | 270 | 200 |
| Rb | 5 | 15 | 25 | 55 |
| Ni | 100 | 70 | 75 | 40 |
| V | 275 | 235 | 40 | 35 |
| Cr | 125 | 15 | 5 | 5 |
| Ba | 150 | 345 | 350 | 645 |
| K/Rb | 740 | 807 | 712 | 576 |
| Sr/Rb | 0.014 | 0.038 | 0.093 | 0.275 |
| $Sr^{87}/Sr^{86}$* | 0.7035 | 0.7036 | 0.7036 | 0.7035 |

*Strontium isotopic data taken from Peterman, et al. (1970*a*).

*Explanation of column headings*

1 Manam Island, basaltic-andesite flow from main crater, 1957–1958 eruption (W. R. Morgan, 1966)
2 Karkar Island, pyroxene-olivine andesite lava (W. R. Morgan, 1966)
3 Long Island, pyroxene-andesite lava (W. R. Morgan, 1966)
4 Basalt, Lake Dakataua, Talasea (Lowder and Carmichael, 1970, no. 311)
5 Pyroxene andesite, historic flow (1890s) of Mt. Makalia, Talasea (Lowder and Carmichael, 1970, no. 114)
6 Dacite from Lake Dakataua, Talasea (Lowder and Carmichael, 1970, no. 306)
7 Rhyolite obsidian, Volupai road, Talasea (Lowder and Carmichael, 1970, no. 343)
8 Basalt, Rabaul caldera (Heming, in press)
9 Basaltic andesite, Rabaul caldera (Heming, in press)
10 Bomb from 1937 eruption of Vulcan, Rabaul caldera (Heming, in press)

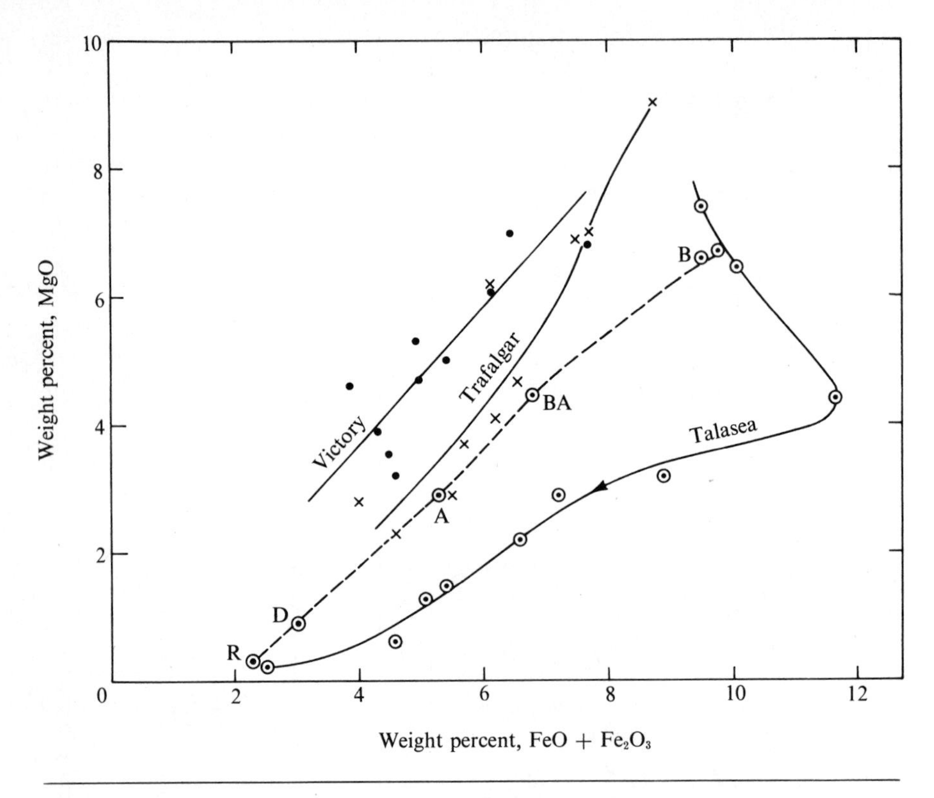


**Fig. 11-5** Plot of MgO against ($FeO + Fe_2O_3$) for lavas of Talasea and of Mt. Victory and Mt. Trafalgar, southeastern Papua. Average compositions of lavas of the Cascades province, North America are shown as B (basalt), BA (basaltic andesite), A (andesite), D (dacite), and R (rhyolite).

evidence suggests that all the lavas come from a common source, it was simple to calculate that the liquid line of descent could be achieved by fractionation of varying amounts of the analyzed phenocryst phases in the different lavas, so that the generation of this suite by crystal fractionation at low pressures is a viable postulate. Only the exposed volumes of the various lava types, particularly the subordinate role of basalt, is at variance with this conclusion, a paradox unresolved in other oceanic arcs (Hedge and Lewis, 1971).

The lavas of the Rabaul caldera, at the eastern tip of New Britain, also range in composition from basalt to rhyolite, again with andesite as the most voluminous lava type. However, this lava series is slightly richer in alkali than that of Talasea, clearly falls into the high-alumina field delineated for the Japanese arc (Fig. 11-6), and consistently has more $Al_2O_3$ than the Talasea equivalents (Table 11-1). A reflection of this more alkaline character, or lower silica activity, is seen in the composition of the groundmass pyroxenes, particularly the calcium-rich pyroxenes, which are more aluminous (Chap. 6) than those of Talasea.

There is a consistent pattern of lateral, or east-west, variation in this volcanic arc. From Manam in the west to Rabaul in the east, the basic lavas become more alkali-rich, eventually becoming high-alumina types at Rabaul; there is also a concomitant small progressive increase in $TiO_2$ from west to east (Fig. 11-7)—a component which, with others (Ba, Zr, Rb), becomes more concentrated with increasing alkali in basic magma.

There is little evidence in the lavas of this arc of conditions at depth; the Moho is 22 km beneath Talasea (Finlayson et al., 1972), but fragments of mantle origin are unknown, a typical feature of andesite volcanoes. Small sporadic xenoliths composed of phases that are found also as phenocrysts occur in many of the basic lavas; sometimes these xenoliths include a hastingsitic amphibole which has yet to be recorded as a phenocryst in the basic lavas. The only phase indicating high pressures, or considerable depths, is quartz, which, as will be

**Fig. 11-6** Plot of total alkali against silica for lavas of Talasea and of the Rabaul caldera, New Britain. Dashed lines show boundaries of Japanese basalt types (named in parentheses) taken from Fig. 11-2*b*). Lettered points refer to average lavas of Cascades province, North America (cf. Fig. 11-5).

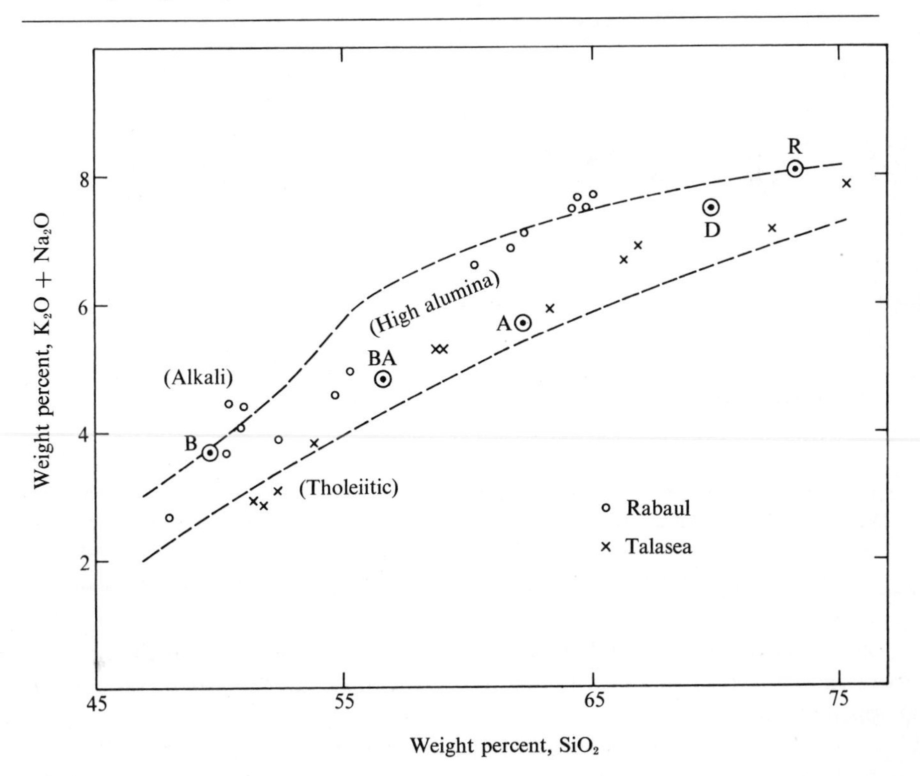

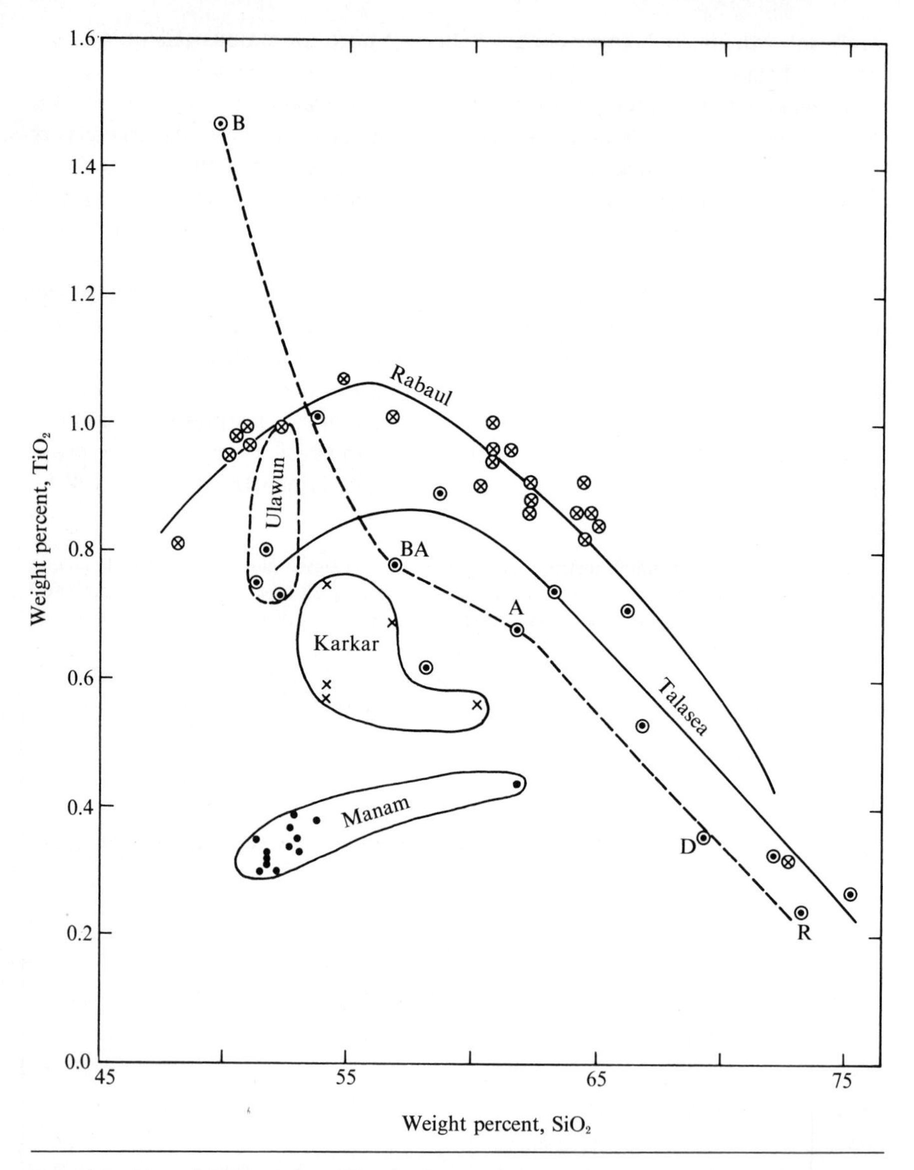


**Fig. 11-7** Plot of $TiO_2$ against $SiO_2$ for lavas of the northern coast of New Guinea and New Britain. *B, BA, A, D,* and *R* refer respectively to the average basalt, basaltic andesite, andesite, dacite, and rhyolite for the North American Cascades.

shown later, can be a stable phase at depth; in many oceanic arcs the presence of quartz "xenocrysts" is in common, but not universal, antipathy to amphibole phenocrysts in basic lavas.

South of the New Guinea arc, there is considerable variation in the composition of lava types. In the New Guinea highlands, the Cenozoic stratovolcanoes Mt. Hagen and Mt. Giluwe (Fig. 11-4) have erupted shoshonitic lavas (Jakěs and White, 1969). Although some of these lavas have been extensively analyzed, there is no published account of their mineralogy; judging from some of the analyses, which show up to 3.3 percent $H_2O$ and comparable amounts of normative $C$ (Table 11-2), the mineral assemblages are unlikely to be pristine. These lavas contain olivine, augite, and amphibole phenocrysts, and their high

**Table 11-2** Analyses and CIPW norms of lavas from the New Guinea highlands and south-eastern Papua

| | 1 | 2 | 3 | 4 | | 3 | 4 |
|---|---|---|---|---|---|---|---|
| $SiO_2$ | 48.94 | 55.45 | 50.59 | 62.98 | Sr | 450 | 950 |
| $TiO_2$ | 0.87 | 0.75 | 1.05 | 0.64 | Ba | 400 | 1230 |
| $Al_2O_3$ | 13.84 | 18.87 | 16.29 | 16.26 | Ni | 150 | 29 |
| $Fe_2O_3$ | 2.53 | 2.60 | 3.66 | 3.13 | V | 216 | 68 |
| FeO | 6.52 | 3.60 | 5.08 | 1.02 | Cr | 360 | 115 |
| MnO | 0.15 | 0.13 | 0.17 | 0.03 | Zr | 130 | 28 |
| MgO | 12.66 | 4.26 | 8.96 | 1.98 | | | |
| CaO | 7.17 | 5.19 | 9.50 | 2.59 | | | |
| $Na_2O$ | 2.29 | 2.55 | 2.89 | 4.45 | | | |
| $K_2O$ | 2.33 | 2.39 | 1.07 | 3.80 | | | |
| $P_2O_5$ | 0.59 | 0.44 | 0.21 | 0.35 | | | |
| Loss on ignition | 2.40 | 3.32 | 0.81 | 2.76 | | | |
| Total | 100.29 | 99.55 | 100.28 | 99.99 | | | |
| *Q* | — | 13.66 | — | 15.04 | | | |
| *C* | — | 3.71 | — | 0.96 | | | |
| *or* | 13.77 | 14.12 | 6.32 | 22.46 | | | |
| *ab* | 19.38 | 21.58 | 24.45 | 37.65 | | | |
| *an* | 20.60 | 22.87 | 28.32 | 10.56 | | | |
| *di* *wo* | 4.64 | — | 7.28 | — | | | |
| *di* *en* | 3.31 | — | 5.39 | — | | | |
| *di* *fs* | 0.92 | — | 1.18 | — | | | |
| *hy* *en* | 7.68 | 10.61 | 11.50 | 4.93 | | | |
| *hy* *fs* | 2.13 | 3.47 | 2.52 | — | | | |
| *fo* | 14.39 | — | 3.80 | — | | | |
| *fa* | 4.39 | — | 0.92 | — | | | |
| *mt* | 3.67 | 3.77 | 5.31 | 1.53 | | | |
| *il* | 1.65 | 1.42 | 1.99 | 1.22 | | | |
| *ap* | 1.40 | 1.04 | 0.50 | 0.83 | | | |
| *hm* | — | — | — | 1.22 | | | |

*Explanation of column headings*

1 Olivine-augite absarokite from Mt. Giluwe, New Guinea (Jakes and White, 1969, no. 99)
2 Amphibole-augite shoshonite, Mt. Hagen (Jakěs and White, 1969, no. 109)
3 Basalt flow, Mt. Trafalgar (Jakěs and Smith, 1970, no. 3544)
4 Andesite flow, Mt. Victory (Jakěs and Smith, 1970, no. 3555)

$K_2O$ content (Table 11-2) is presumably reflected in the presence of sanidine, a characteristic mineral in the shoshonite rock type. Jakěs and White suggest that the New Guinea shoshonites have the same relationship to the volcanic front of the New Guinea arc as the alkali olivine basalts of the Japan Sea do to the Japanese arc (Fig. 11-2*a*). But apart from high $K_2O/Na_2O$ compared with tholeiitic lavas, the chemical character of modern, fresh, shoshonitic lavas in this environment is either unknown or unrecognized; it is an aspect of igneous petrology deserving further work.

In southeastern Papua (Fig. 11-4) there are three active andesite volcanoes,[1] Mt. Goropu, Mt. Victory, and Mt. Lamington, the last well known since the catastrophic eruption of 1951 (G. A. Taylor, 1958). Since these volcanoes are close to the southern boundary of the Solomon Sea plate, which T. Johnson and Molnar (1972) conceive to correspond with the boundary of the Papuan ultramafic belt (Davies, 1968), they may not be part of the New Guinea–New Britain arc, but rather related to the weakly seismic zone where the Australian plate is underthrusting the Solomon Sea plate in southeastern Papua. The lavas of Mt. Victory, and those of the neighboring extinct volcano Mt. Trafalgar, range from olivine basalts to dacites, the most voluminous type being amphibole andesite, which forms half the exposed lavas (Jakěs and Smith, 1970). All are rich in plagioclase phenocrysts, although varieties of basaltic andesites may be less so. Plagioclase and amphibole phenocrysts predominate in the andesites and dacites; the phenocryst assemblage includes augite, iron titanium oxides, and rarely sanidine and orthopyroxene. Two representative analyses are given in Table 11-2, where high potassium (also Ba, Sr, and Ni) is clearly seen in contrast to the Talasea series (Table 11-1). The lavas of each volcano show no iron enrichment relative to magnesium, but each has its own reasonably well-defined trend (contrasted with that of Talasea in Fig. 11-5). Cognate inclusions in the andesitic lavas are interpreted by Jakěs and Smith to represent early crystal accumulates of the lavas at depth; they contain plagioclase, augite, amphibole, iron titanium oxides, and sometimes biotite.

## BOUGAINVILLE AND THE SOLOMON ISLANDS ARC

Bougainville, the northernmost island of the Solomon Islands arc, contains 17 post-Miocene stratovolcanoes (D. H. Blake, 1968*b*). Only one, Bagana, is presently active, its basaltic-andesite and andesite lavas being rich in plagioclase phenocrysts; small amounts of olivine, amphibole, and orthopyroxene phenocrysts are also present. The analyses given by Blake show that the Bagana lavas are more alkali-rich than those of Talasea in the New Britain arc.

The British Solomon Islands (Fig. 11-3) include the New Georgia group of Recent volcanoes, about 250 km long. According to P. J. Coleman (1970)

[1] To the southeast of these andesite volcanoes, in the Ferguson Islands (Fig. 11-4), are lavas of mildly peralkaline rhyolite (W. R. Morgan, 1966) of which little is known mineralogically.

some of the Solomon Islands have a Mesozoic basement of amphibolite and pelagic sediment intruded by gabbro and granite. This succession is overlain by submarine and subaerial flows and pyroclastic rocks of basalt and andesite; reef sediments and ashes cover this sequence. Other islands have a basement of basic lavas, overlain by deep-water foraminiferal oozes, and calcareous clays which include alnoitic and ankaratritic lavas; thrust slices of serpentinite are found in some islands. This marine assemblage, probably old sea floor, is believed to underlie the New Georgia group, whose Recent lavas (Stanton and Bell, 1969) are radically different from any so far described from the New Guinea–New Britain arc.

Picrite basalts and ankaramites are common, and these are associated with olivine basalts and hornblende-bearing basaltic andesites. The picrite basalts contain abundant (up to 50 percent) phenocrysts of magnesian olivine ($Fo_{88}$), up to 20 percent augite phenocrysts, and only rarely plagioclase; the groundmass can be almost entirely glass, but when it is crystalline it contains plagioclase, augite, olivine, and iron titanium oxides. Stanton and Bell believe that there is a continuous gradation between these picritic-ankaramitic types, in which olivine or augite is the dominant phenocryst, and the porphyritic pyroxene-olivine basalts, which are the most voluminous type on the islands. Phenocrysts of chrome diopside, ranging from 40 to 10 percent in volume, are accompanied by about 7 percent olivine and by about 20 percent plagioclase, which together with 1 percent titanomagnetite microphenocrysts are enclosed by a groundmass of plagioclase, augite, minor olivine, and small amounts of glass. Nonporphyritic olivine basalts are also found, many with subophitic intergrowth between augite and plagioclase, and this lava type occasionally has orthopyroxene.

Basaltic andesites occur in two varieties: one, a feldsparphyric type with abundant zoned labradorite phenocrysts (often full of inclusions), less augite, and 1 to 2 percent olivine; the other variety, a hornblende basaltic andesite, contains 30 percent strongly zoned labradorite phenocrysts, 10 percent green and brown hornblende, and small amounts of augite, orthopyroxene, and sometimes olivine. The groundmass of basaltic andesites may be all glassy or, if crystalline, composed of plagioclase, augite, and titanomagnetite.

Representative analyses of the lavas from New Georgia are given in Table 11-3; in common with all lavas of volcanic arcs, they are low in $TiO_2$, and the olivine-rich basalts have substantially more $K_2O$ relative to $Na_2O$ than their Hawaiian counterparts. Although these lavas plot close to the boundary between the Japanese alkali and high-alumina basalts (Fig. 11-8), most of them contain normative $Q$ and substantial amounts of *hy*. Although orthopyroxene has been found only rarely and pigeonite has not been reported, the general chemical characteristics of this suite suggest affinity with the broad tholeiitic type whose common oceanic representatives are nevertheless rather different from the basalts of this volcanic arc. This association of tholeiitic-type picrites and ankaramites with hornblende basaltic andesites is rare in volcanic arcs, although it is also found in the New Hebrides.

**Table 11-3** Representative analyses and CIPW norms of lavas from the New Georgia group, Solomon Islands

| | 1 | 2 | 3 | 4 | 5 |
|---|---|---|---|---|---|
| $SiO_2$ | 44.99 | 48.68 | 49.95 | 53.07 | 55.60 |
| $TiO_2$ | 0.33 | 0.43 | 0.65 | 0.70 | 0.62 |
| $Al_2O_3$ | 6.54 | 12.44 | 14.01 | 18.21 | 16.90 |
| $Cr_2O_3$ | 0.34 | 0.08 | — | 0.03 | — |
| $Fe_2O_3$ | 2.67 | 3.90 | 6.41 | 2.98 | 4.65 |
| FeO | 7.29 | 6.49 | 5.59 | 4.76 | 2.27 |
| MnO | 0.19 | 0.20 | 0.21 | 0.22 | 0.15 |
| MgO | 28.10 | 11.36 | 6.57 | 4.28 | 5.44 |
| CaO | 6.56 | 11.17 | 10.16 | 8.93 | 6.64 |
| $Na_2O$ | 0.92 | 1.93 | 2.65 | 3.10 | 3.63 |
| $K_2O$ | 0.57 | 1.51 | 1.75 | 2.21 | 1.61 |
| $P_2O_5$ | 0.08 | 0.24 | 0.29 | 0.34 | 0.23 |
| $H_2O^+$ | 1.06 | 1.20 | 1.33 | 0.41 | 1.31 |
| $H_2O^-$ | 0.30 | 0.48 | 0.71 | 0.68 | 1.21 |
| Total | 100.09* | 100.21† | 100.28 | 100.08‡ | 100.26 |
| *Q* | — | — | 1.04 | 2.49 | 8.01 |
| *or* | 3.37 | 8.92 | 10.34 | 13.06 | 9.52 |
| *ab* | 7.78 | 16.33 | 22.42 | 26.23 | 30.71 |
| *an* | 12.03 | 20.82 | 21.17 | 29.25 | 25.07 |
| *di* *wo* | 8.35 | 13.53 | 11.42 | 4.94 | 2.66 |
| *di* *en* | 6.45 | 9.56 | 8.23 | 3.06 | 2.30 |
| *di* *fs* | 1.01 | 2.80 | 2.15 | 1.58 | — |
| *hy* *en* | 7.00 | 5.00 | 8.13 | 7.60 | 11.25 |
| *hy* *fs* | 1.10 | 1.46 | 2.13 | 3.92 | — |
| *fo* | 39.62 | 9.62 | — | — | — |
| *fa* | 6.84 | 3.11 | — | — | — |
| *mt* | 3.87 | 5.65 | 9.29 | 4.32 | 6.01 |
| *il* | 0.63 | 0.82 | 1.23 | 1.33 | 1.18 |
| *ap* | 0.19 | 0.57 | 0.68 | 0.80 | 0.54 |

| MODAL ANALYSES, vol. % | 1 | 2 | 3 | 4 | 5 |
|---|---|---|---|---|---|
| Olivine | 52.0 | 12.0 | 7.5 | 1.3 | 3.4 |
| Plagioclase | — | 8.0 | 14.4 | 26.3 | 30.9 |
| Augite | 7.1 | 25.2 | 24.4 | 10.0 | 7.4 |
| Opaques | 1.1 | 0.7 | 1.1 | 0.4 | 3.4 |
| Hornblende | — | — | — | — | 10.5 |
| Groundmass | 39.8 | 54.1 | 52.6 | 62.0 | 44.4 |

* Includes 0.15% NiO.
† Includes 0.10% $CO_2$.
‡ Includes 0.16% $CO_2$

*Explanation of column headings*

1 Picrite basalt; Stanton and Bell (1969, no. NG 143/2)
2, 3 Porphyritic pyroxene-olivine basalt; Stanton and Bell (1969, no. 450 and 111/2)
4 Feldsparphyric basaltic andesite; Stanton and Bell (1969, no. 428)
5 Hornblende basaltic andesite; Stanton and Bell (1969, no. 349)

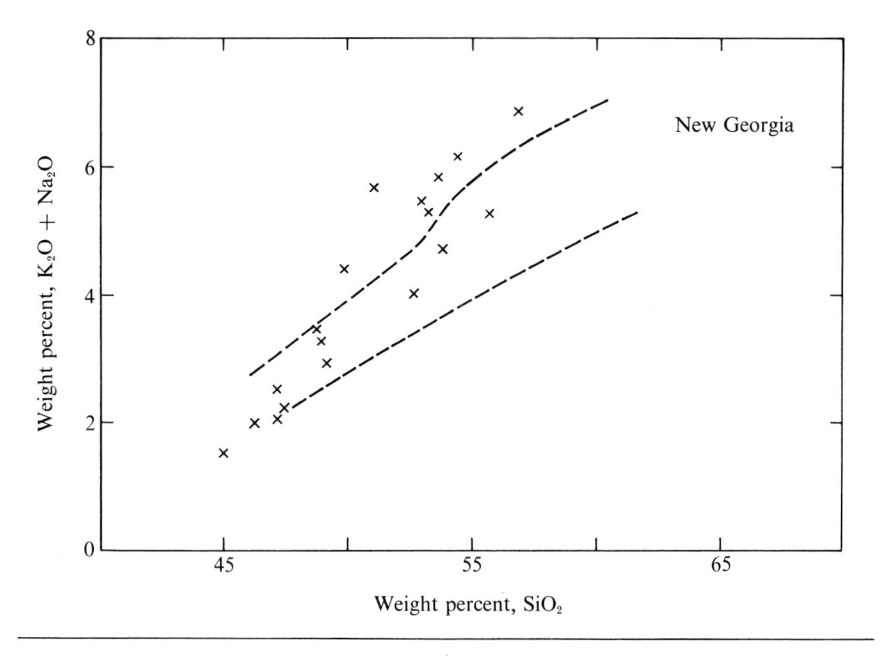


**Fig. 11-8** Plot of total alkali against silica for the lavas of the New Georgia group, British Solomon Islands. Dashed lines represent boundaries of Japanese basalt types (Fig. 11-2*b*).

## NEW HEBRIDES ARC

The active volcanic chain in the New Hebrides is flanked on each side by islands active in the Tertiary (Mitchell and Warden, 1971). The western chain of islands, closest to the submarine trench (Fig. 11-3), have pelagic red mudstones overlain by basaltic pillow lavas, ashes, and marine volcanoclastic rocks. This sequence was intruded by andesites and diorites and underwent zeolite-facies metamorphism. In the Pliocene, reef limestones accumulated. The eastern islands have more extensive pillow basalts interbedded with marine deep-water sediments, and they are intruded by gabbros, serpentinites, and amphibolites. The active island chain consists of volcanoes on top of a thick succession of pillow lavas and volcanoclastic rocks. Of the eight or ten active volcanoes in this arc, the lavas of three—Aoba (Warden, 1970), Ambrym (G. J. H. McCall et al., 1971), and Lopevi (Warden, 1967)—are well known. Basalt is the predominant type in contrast to the large amounts of andesite found on islands where volcanism is now extinct.

Aoba, whose summit has two oval calderas which collapsed without substantial eruptions of pumice, was built up above sea level in the late Pliocene. The lavas of the later shield-forming stage are picrite basalts and ankaramites. The picrite basalts may contain more than 35 percent olivine phenocrysts, with less abundant augite, but no phenocrysts of plagioclase; the groundmass is glassy, or finely crystalline, with plagioclase, olivine, pyroxenes, and dark brown residual glass. Ankaramites have 30 to 40 percent augite phenocrysts and only rarely plagioclase; lavas with less abundant olivine and pyroxene phenocrysts are also found. The latest flows, erupted from fissures on the flanks, are olivine basalts with sporadic plagioclase phenocrysts; they have a groundmass of olivine, augite, plagioclase, and iron titanium oxides.

Chemically the Aoba basalts (Table 11-4) are not unlike those of New Georgia. They have low $TiO_2$, high $K_2O$ relative to $Na_2O$ in comparison to Hawaiian equivalents, and their slightly higher alumina content in comparison to New Georgia basalt (Fig. 11-9) is reflected in small amounts of normative *ne* (Warden, 1970), an unusual feature in lavas of volcanic arcs.

Ambrym, about 100 km to the south of Aoba, also has a large summit caldera, which quietly subsided about 2000 years ago (G. J. H. McCall et al., 1971). Inside it are two active cones whose lavas (Warden, 1970) are porphyritic olivine basalts (Table 11-4, analysis 3).

Lopevi, about 40 km to the south of Ambrym, was active in 1960, from 1963 to 1965, and in 1967. The 1967 lavas contain phenocrysts of bytownite, augite, and olivine with a little orthopyroxene set in a groundmass of plagioclase, augite, iron titanium oxides, and brown glass. Previous eruptions (1963 to 1965)

**Fig. 11-9** Plot of total alkali against silica for the lavas of Aoba, Ambrym, and Lopevi in the New Hebrides. Dashed lines represent boundaries of Japanese basalt types (Fig. 11-2*b*).

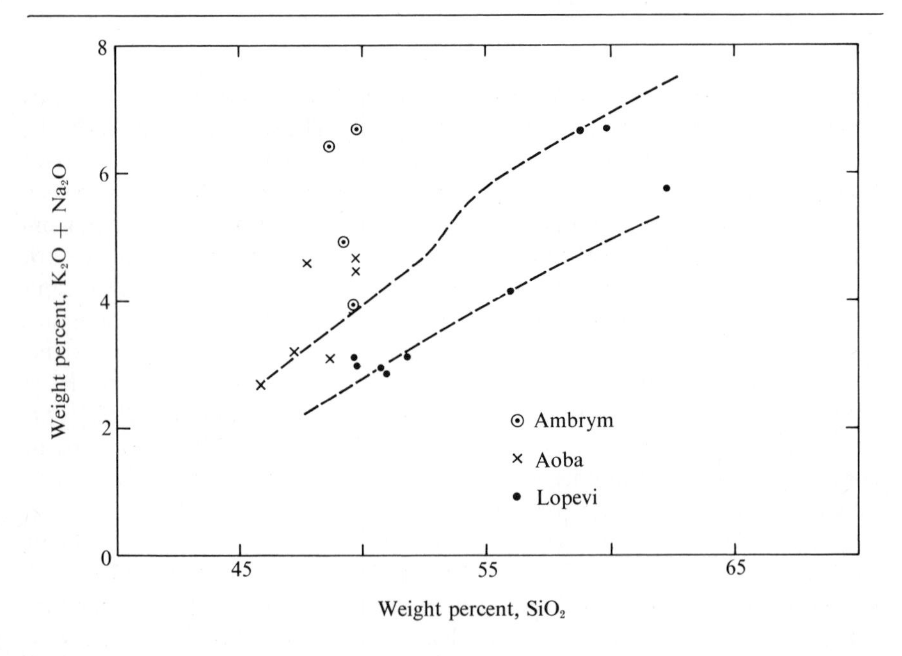

have simultaneously produced both hypersthene andesite and olivine-augite basaltic andesite (Warden, 1967). Representative analyses of the Lopevi basalts (Table 11-4, analyses 4 and 5), basaltic andesites, and andesites have high *hy* (and sometimes *Q*) in marked contrast with the *ne*-normative olivine basalts of Aoba.

**Table 11-4** Representative analyses and CIPW norms of lavas from the active volcanoes of the New Hebrides

| | 1 | 2 | 3 | 4 | 5 |
|---|---|---|---|---|---|
| $SiO_2$ | 46.0 | 48.59 | 49.26 | 50.88 | 49.6 |
| $TiO_2$ | 0.50 | 0.67 | 0.89 | 0.72 | 0.72 |
| $Al_2O_3$ | 8.8 | 11.91 | 17.18 | 18.18 | 16.8 |
| $Fe_2O_3$ | 6.23 | 4.81 | 5.47 | 2.10 | 2.72 |
| FeO | 4.39 | 4.97 | 6.10 | 6.94 | 6.77 |
| MnO | 0.21 | 0.19 | — | 0.16 | 0.18 |
| MgO | 21.29 | 13.18 | 4.28 | 7.20 | 7.92 |
| CaO | 8.85 | 12.31 | 10.78 | 11.22 | 12.04 |
| $Na_2O$ | 1.78 | 2.16 | 3.20 | 2.36 | 2.35 |
| $K_2O$ | 0.88 | 0.95 | 1.76 | 0.54 | 0.72 |
| $P_2O_5$ | 0.13 | 0.14 | 0.37 | 0.01 | 0.11 |
| $H_2O^+$ | 0.43 | 0.19 | 0.90 | 0.03 | 0.29 |
| $H_2O^-$ | 0.08 | 0.26 | 0.12 | 0.07 | 0.03 |
| Total | 99.57 | 100.33 | 100.31 | 100.41 | 100.25 |
| *or* | 5.20 | 5.61 | 10.40 | 3.19 | 4.25 |
| *ab* | 15.06 | 18.10 | 27.08 | 19.97 | 19.89 |
| *an* | 13.42 | 20.00 | 27.32 | 37.42 | 33.17 |
| *ne* | — | 0.10 | — | — | — |
| *di* *wo* | 12.37 | 16.77 | 9.92 | 7.59 | 10.79 |
| *di* *en* | 10.33 | 13.15 | 6.25 | 4.59 | 6.86 |
| *di* *fs* | 0.48 | 1.76 | 3.05 | 2.59 | 3.24 |
| *hy* *en* | 1.45 | — | 0.11 | 12.54 | 6.02 |
| *hy* *fs* | 0.07 | — | 0.05 | 7.08 | 2.85 |
| *fo* | 28.90 | 13.79 | 3.02 | 0.56 | 4.80 |
| *fa* | 1.49 | 2.04 | 1.63 | 0.35 | 2.50 |
| *mt* | 9.03 | 6.97 | 7.93 | 3.04 | 3.94 |
| *il* | 0.95 | 1.27 | 1.69 | 1.37 | 1.37 |
| *ap* | 0.31 | 0.33 | 0.88 | 0.02 | 0.26 |

*Explanation of column headings*

1 Porphyritic picrite basalt, precaldera, Aoba, New Hebrides (Warden, 1970)
2 Olivine-augite glassy basalt, precaldera, Aoba, New Hebrides (Warden, 1970)
3 Basalt lava of 1913, Ambrym, New Hebrides (Warden, 1970)
4 Porphyritic basalt bomb, 1963 eruption, Lopevi, New Hebrides (Warden, 1967)
5 1967 lava of Lopevi, New Hebrides; phenocrysts of plagioclase, olivine, and augite (Warden, 1967)

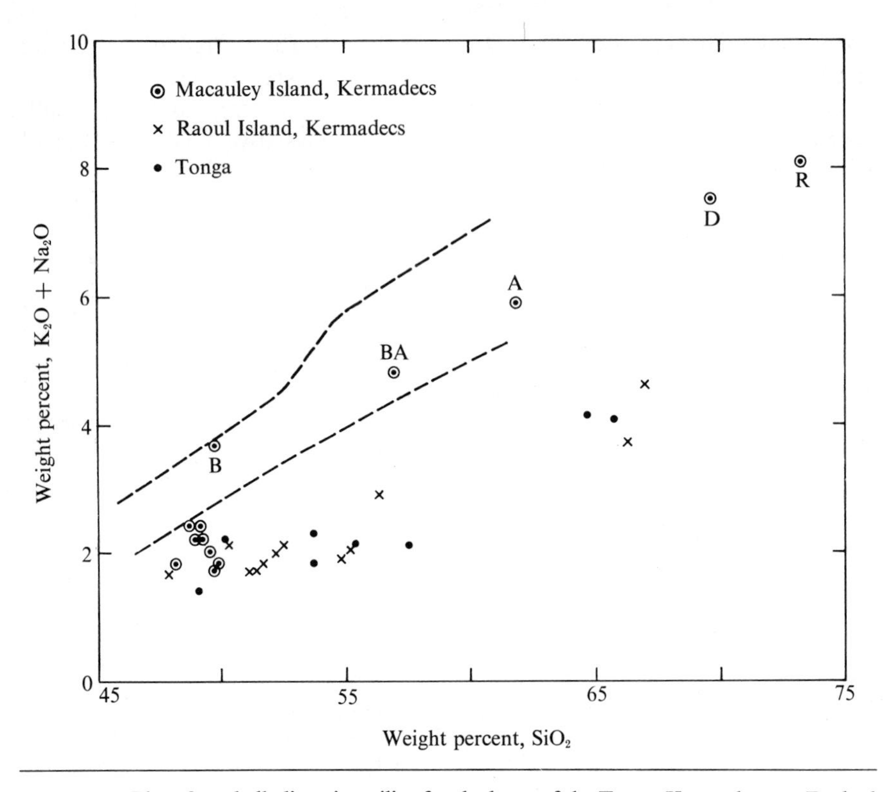


**Fig. 11-10** Plot of total alkali against silica for the lavas of the Tonga-Kermadec arc. Dashed lines represent boundaries of Japanese basalt types (Fig. 11-2*b*). *B*, *BA*, *A*, *D*, and *R* refer to the average basalt, basaltic andesite, andesite, dacite, and rhyolite of the North American Cascades.

## TONGA-KERMADEC ARC

The active volcanoes of the Tonga and Kermadec island groups (Fig. 11-3) are associated with submarine trenches, and although they are apparently part of a single volcanic arc, there is no volcanic activity in the intervening stretch of 750 km that separates the two island groups.

In the Tonga group, there are nine or so active volcanoes, but only six form permanent islands; Bryan et al. (1972) have examined the lavas from four of these. Older (Tertiary) basalts have plagioclase phenocrysts averaging $An_{95}$ which are virtually unzoned except at the rims; augite phenocrysts are less abundant, and the groundmass contains plagioclase, augite, and small amounts of titanomagnetite. Younger basaltic andesites contain phenocrysts of plagioclase and augite, with rare orthopyroxene set in a glass groundmass containing pyroxene, titanomagnetite, and rare olivine. Andesites have fewer phenocrysts than the basaltic andesites, hypersthene is more common, and titanomagnetite microphenocrysts are present. The dacites have a similar assemblage of phenocrysts: bytownite, orthopyroxene, and augite set in a crystalline groundmass in which potassium feldspar has been detected.

As a series the Tongan lavas (Table 11-5, analyses 1–3) are low in alkali, as are the lavas of the whole Tonga-Kermadec arc (Fig. 11-10). The more siliceous members, basaltic andesites, can contain 12 percent or more FeO + $Fe_2O_3$, and coupled with their enrichment in iron is normative *Q* and high normative *hy* relative to *di*. There is a compositional gap between 57 and 65 percent $SiO_2$, which is found also in the lava series of the other islands of this arc.

Further south in the Kermadec group, Raoul Island (Brothers and Searle, 1970) has been built on submarine pillow lavas with intercalated marine sediments of Plio-Pleistocene age. These lavas are overlain by subaerial basaltic and andesitic lavas and ashes; one of the latest is a pumice layer covering the island, which is believed to have been deposited at the time of the caldera collapse in 200 B.C. The lavas range from olivine basalts to dacites (again with no representatives between 57 and 66 percent $SiO_2$) and are petrographically similar to those of Tonga. As a group, the lavas are low in alkali (Fig. 11-10), are sometimes nonporphyritic, and can show extreme enrichment in iron; the basaltic andesite whose analysis is given in Table 11-5 (analysis 5) is one of the most aluminum-poor and iron-rich lavas on earth.

Rounded blocks of plutonic rocks up to 70 cm in diameter are found on the beaches of Raoul Island and are of two types: The accidental xenoliths include gabbro (augite, orthopyroxene, and plagioclase with secondary actinolite), diorites, and quartz diorites, neither of which contain biotite; the cognate xenoliths of cumulative rocks are friable boulders of ferromagnesian minerals and plagioclase with interstitial glass. Olivine is magnesian ($Fo_{80}$ to $Fo_{70}$) and occurs with augite, orthopyroxene ($En_{70}$), and bytownite.

The lavas of Macauley Island (Brothers and Martin, 1970), about 100 km south-southwest of Raoul Island, are distinctly different: high-alumina basalts contrast to tholeiitic basalts of Raoul Island. Many of the basic lavas have few phenocrysts, which are bytownite and olivine, the matrix being labradorite, augite, rare pigeonite, iron titanium oxides, and glass; orthopyroxene has not been found. Dacite pumice with andesine phenocrysts also occurs in a tuff with a variety of other exogenous rocks. The analyses of two lavas from Macauley Island (Table 11-5, analyses 7 and 8) show high alumina and low alkali (Fig. 11-10), and normative *Q* is accompanied by large amounts of *hy*. Brothers and Martin were able to establish that these high-alumina basalts, younger than the lavas of Raoul Island, cap a tholeiitic series on Macauley, since fragments of these are found in the pyroclastic rocks.

## SUMMARY OF THE SOUTHWEST PACIFIC VOLCANIC ARCS

Before considering the extension of the Tonga-Kermadec arc into the continental coast of northern New Zealand, it is appropriate to summarize the salient features of the lavas of the Southwest Pacific volcanic arc system.

*New Guinea arc* Olivine basalts and basaltic andesites of tholeiitic type become richer in $TiO_2$ eastward (Fig. 11-7). Acid lavas are absent. In the New Guinea highlands, shoshonitic lavas may be related to this arc but are poorly known mineralogically.

**Table 11-5** Representative analyses and CIPW norms of lavas from the Tonga-Kermadec volcanic arc

| | 1 | 2 | 3 | 4 | 5 | 6 | 7 | 8 |
|---|---|---|---|---|---|---|---|---|
| $SiO_2$ | 50.37 | 55.35 | 64.79 | 50.42 | 52.51 | 67.00 | 49.8 | 49.7 |
| $TiO_2$ | 1.75 | 0.60 | 0.55 | 1.0 | 1.1 | 1.20 | 0.6 | 0.6 |
| $Al_2O_3$ | 14.65 | 15.57 | 14.48 | 13.7 | 10.5 | 13.60 | 18.4 | 18.8 |
| $Fe_2O_3$ | 2.19 | 3.30 | 1.91 | 4.81 | 5.21 | 1.9 | 5.6 | 4.7 |
| FeO | 9.24 | 7.24 | 6.07 | 8.44 | 11.00 | 4.3 | 4.5 | 5.3 |
| MnO | — | 0.19 | 0.16 | 0.30 | 0.41 | 0.2 | 0.12 | 0.12 |
| MgO | 7.13 | 4.80 | 1.45 | 5.1 | 6.0 | 2.0 | 6.1 | 4.3 |
| CaO | 11.74 | 9.82 | 6.04 | 13.4 | 11.2 | 4.5 | 12.2 | 13.3 |
| $Na_2O$ | 1.88 | 1.82 | 3.03 | 1.7 | 1.8 | 3.8 | 1.4 | 1.3 |
| $K_2O$ | 0.31 | 0.35 | 1.11 | 0.4 | 0.3 | 0.8 | 0.4 | 0.4 |
| $P_2O_5$ | 0.17 | 0.07 | 0.11 | 0.08 | 0.12 | 0.25 | 0.2 | 9.13 |
| $H_2O^+$ | 0.13 | 0.51 | <0.10 | 0.96 (}) | 0.15 (}) | 0.5 | 0.6 | 0.4 |
| $H_2O^-$ | 0.03 | 0.33 | 0.10 | | | | 0.2 | 0.3 |
| Total | 99.59 | 99.95 | 99.80 | 100.31 | 100.30 | 100.05 | 100.12 | 99.35 |
| *Q* | 2.82 | 14.18 | 25.36 | 6.23 | 9.10 | 28.41 | 7.7 | 8.24 |
| *or* | 1.67 | 2.23 | 6.68 | 2.36 | 1.77 | 4.72 | 2.36 | 2.36 |
| *ab* | 16.24 | 15.21 | 25.70 | 14.38 | 15.23 | 32.15 | 11.84 | 11.00 |
| *an* | 30.30 | 33.39 | 22.54 | 28.56 | 19.68 | 17.68 | 42.74 | 44.28 |
| *di* *wo* | 11.37 | 6.16 | 2.90 | 15.60 | 14.65 | 1.25 | 6.87 | 8.70 |
| *di* *en* | 6.40 | 3.31 | 0.90 | 8.30 | 7.21 | 0.62 | 5.19 | 5.52 |
| *di* *fs* | 4.49 | 2.64 | 2.11 | 6.81 | 7.16 | 0.59 | 0.98 | 2.62 |
| *hy* *en* | 11.40 | 8.64 | 2.71 | 4.39 | 7.73 | 4.35 | 9.99 | 5.18 |
| *hy* *fs* | 7.92 | 7.26 | 6.73 | 3.61 | 7.67 | 4.12 | 1.88 | 2.45 |
| *mt* | 3.25 | 4.77 | 2.78 | 6.97 | 7.55 | 2.75 | 8.11 | 6.81 |
| *il* | 3.34 | 1.21 | 1.06 | 1.89 | 2.08 | 2.27 | 1.13 | 1.13 |
| *ap* | 0.34 | 0.20 | 0.27 | 0.18 | 0.27 | 0.58 | 0.46 | 0.30 |

*Explanation of column headings*

1 Olivine-basalt lava, 1929 flow, Tonga (Bryan et al., 1972, no. 2)
2 Bytownite-andesite flow, Tonga (Bryan et al., 1972, no. 7)
3 Bytownite-hypersthene dacite, Tonga (Bryan et al., 1972, no. 9)
4 Basalt lava, Raoul Island (Brothers and Searle, 1970, no. 7101)
5 Basaltic andesite, iron-rich, Raoul Island (Brothers and Searle, 1970, no. 7119)
6 Obsidian, Raoul Island (Brothers and Searle, 1970, no. 7006)
7 Basaltic lava, Macauley Island (Brothers and Martin, 1970, no. 10371)
8 Basaltic lava, Macauley Island (Brothers and Martin, 1970, no. 10373)

*New Britain arc* The only described example of a complete series from olivine basalt to rhyolite in the Southwest Pacific occurs at Talasea. Andesite is the most voluminous rock type, but crystal fractionation producing iron enrichment has been shown to be a viable mechanism for the origin of this tholeiitic series. The $Sr^{87}/Sr^{86}$ ratios show no significant change from basalt to rhyolite. The basic lavas of Rabaul are high-alumina types and have acid pumice flows associated with them.

*Solomon Islands arc* Bougainville has basaltic andesites and andesites. New Georgia has a tholeiitic suite of picrite and ankaramitic basalts, together with olivine basalts and amphibole-bearing basaltic andesites. This series is high in $K_2O$ and low in $TiO_2$ compared to oceanic equivalents. No acid lavas have been reported.

*New Hebrides arc* Picrites, ankaramites, and olivine basalts with normative *ne* characterize one volcano, whereas another has erupted olivine basalt (*Q*-normative) and hypersthene andesite. No contemporary acid lavas have been described.

*Tonga-Kermadec arc* Of the younger lavas of volcanoes along the Tongan segment, the basaltic andesites are iron-rich, and dacite is the only acid representative. One island of the Kermadecs has tholeiitic lavas showing extreme iron enrichment in the basaltic andesites; again dacite is the only acid lava type. The other island has high-alumina basalts capping a tholeiitic group. There are no lavas in the range 56 to 65 percent $SiO_2$ (Brothers, 1970) nor have rhyolites been found.

It is obviously difficult to generalize about the petrographic features of each arc when there is so much diversity between the lavas of adjoining volcanoes within one arc. There does, however, seem to be a general, but by no means universal, pattern. Tholeiitic lavas, often showing varying degrees of iron enrichment, seem to be the earlier products of the presently active volcanic arcs of the Southwest Pacific, as they are in the young Scotia arc (P. E. Baker, 1968*a*); this magma type is eventually displaced by the high-alumina type as volcanic activity continues. Further than that it is difficult to go, for just within the New Hebrides there is much the same pattern of lava types as Kuno found transverse to the whole Japanese arc (Fig. 11-2*a*).

## Taupo Volcanic Zone, New Zealand

Quaternary volcanism is widespread in the northern part of North Island, New Zealand. The city of Auckland encompasses a group of small alkali-basalt (basanitoid) volcanoes of late Pleistocene-Holocene age. A hundred kilometers north are well-preserved Pleistocene basaltic and dacitic volcanoes. But the most extensive and spectacular activity is in the Taupo volcanic zone, south and southeast of Auckland and passing within 100 km of the small alkali-basalt province of the city. Here the continuation of the Tonga-Kermadec arc cuts clear across the continental fragment of New Zealand. The Taupo province as here described extends 250 km between the active volcanoes White Island in the northeast and Ruapehu in the southwest. Still further southwest is the great Pleistocene andesitic volcano Egmont. Volcanism in the Taupo province began in the Pliocene and has produced 15,000 $km^3$ of rhyolitic debris, 800 $km^3$ of andesite, and less than 50 $km^3$ of basalt. During later Tertiary time, however, the whole northern segment of North Island extending from the margin of the Taupo province several hundred kilometers northward was the site of andesitic volcanism, much of it submarine. The surviving products are pillow lavas (non-spilitic) and lahar breccias.

Rhyolitic volcanism in the Taupo province is concentrated largely in four centers (Fig. 11-11), three of which are controlled by ring fractures (Healy, 1962). They are known as the Lake Taupo volcanic center, the Mokai ring structure, the Rotorua caldera (15 km in diameter), and the Okataina volcanic center. Each one has concentrations of rhyolitic domes and is flanked by ignimbrite flows, some of which can amount to 200 $km^3$ or more. Excellent accounts of the volcanic history and petrology of this province are given by Healy (1962), B. N. Thompson et al. (1965), and Ewart (1963, 1967, 1968). Ewart and Stipp (1968) have given the following generalized sequence of eruptions in each volcanic center: (1) ignimbrite eruptions, possibly from circumferential fractures peripheral to the center; (2) rhyolite domes, typically extruded around the peripheries of the centers; (3) glowing avalanche eruptions resulting in partly welded or unwelded pumice deposits; (4) a younger phase of rhyolite domes and flows, typically erupted in the central portions of the volcanic centers; and (5) violent ash-fall eruptions which blanket the surrounding areas in pumice deposits. The most extensive of these was erupted from the Lake Taupo center, and the latest deposit has been dated as A.D. 130.

Almost all the acid volcanic rocks are porphyritic with phenocrysts of plagioclase, orthopyroxene, titanomagnetite, and ilmenite, less frequently augite, quartz, calcic hornblende, and biotite, and rarely sanidine and cummingtonite. The following assemblages of ferromagnesian phenocrysts have been recognized:

1. Orthopyroxene + augite
2. Orthopyroxene + calcic hornblende
3. Orthopyroxene + cummingtonite
4. Biotite + calcic hornblende + orthopyroxene

Although there is some overlap between each group, these assemblages are well defined and show significant correlation between the total amount of phenocrysts and the phenocryst ratio plagioclase/quartz of the rocks in which

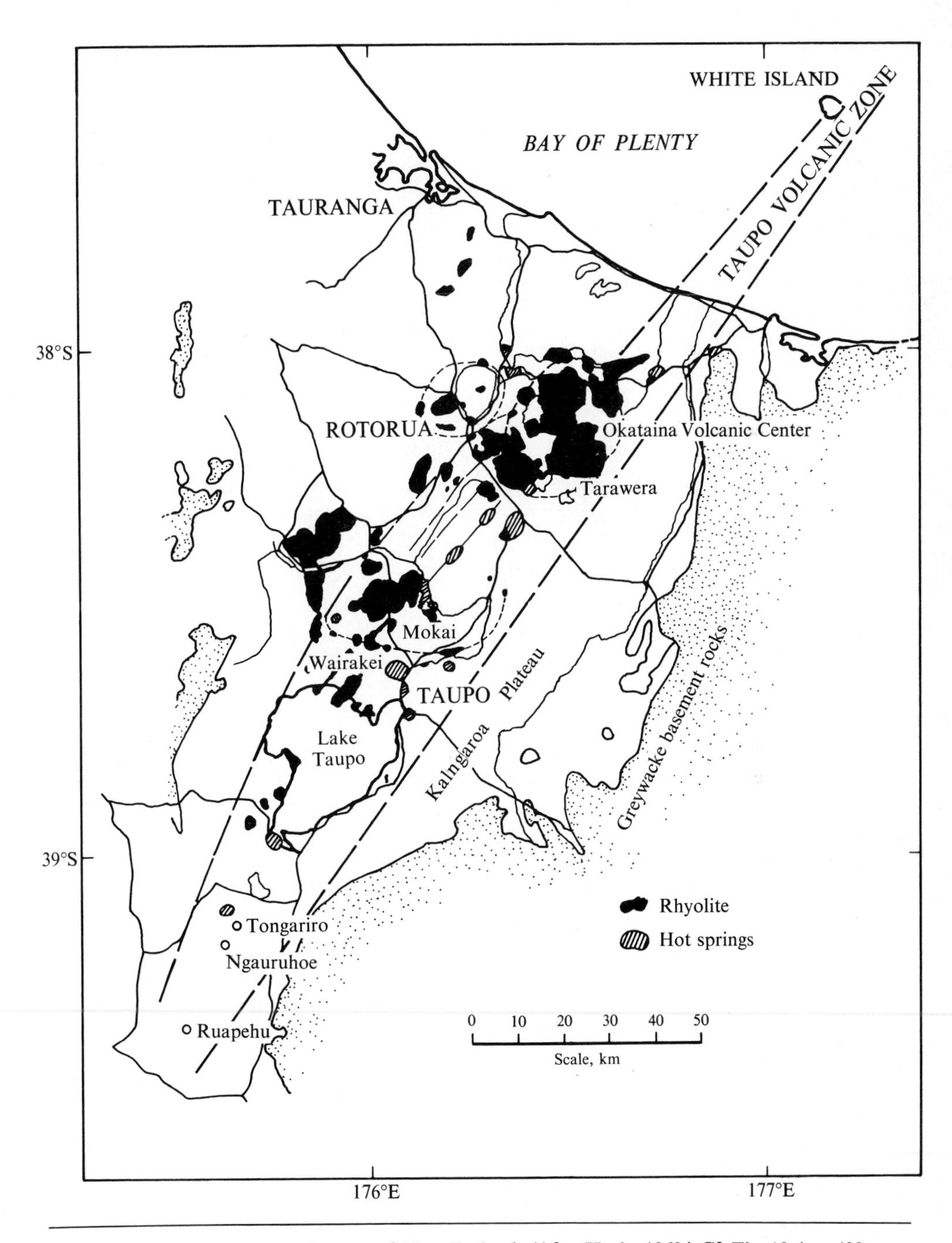


**Fig. 11-11** Taupo volcanic zone of New Zealand. (After Healy, 1962.) Cf. Fig. 10-4, p. 492.

**Table 11-6** Representative analyses and CIPW norms of representative rocks from the Taupo volcanic zone, New Zealand

| | 1 | 2 | 3 | 4 | 5 | 6 | 7 |
|---|---|---|---|---|---|---|---|
| $SiO_2$ | 51.16 | 55.90 | 62.16 | 67.23 | 74.22 | 73.85 | 70.60 |
| $TiO_2$ | 0.80 | 0.76 | 0.71 | 0.34 | 0.28 | 0.23 | 0.20 |
| $Al_2O_3$ | 17.12 | 16.90 | 14.32 | 15.12 | 13.27 | 13.55 | 13.23 |
| $Fe_2O_3$ | 2.40 | 2.10 | 1.64 | 3.33 | 0.88 | 1.25 | 0.75 |
| FeO | 7.25 | 6.30 | 4.33 | 0.50 | 0.92 | 0.60 | 1.39 |
| MnO | 0.18 | 0.15 | 0.15 | 0.09 | 0.05 | 0.05 | 0.11 |
| MgO | 6.12 | 5.20 | 3.97 | 2.02 | 0.28 | 0.30 | 0.30 |
| CaO | 11.41 | 8.40 | 5.89 | 4.25 | 1.59 | 1.53 | 1.56 |
| $Na_2O$ | 2.28 | 2.60 | 2.49 | 3.55 | 4.24 | 3.71 | 4.58 |
| $K_2O$ | 0.54 | 1.00 | 2.37 | 2.20 | 3.18 | 3.60 | 2.86 |
| $P_2O_5$ | 0.13 | 0.10 | 0.12 | 0.11 | 0.05 | 0.05 | 0.04 |
| $H_2O^+$ | 0.27 | Tr. | 1.21 | 0.68 | 0.80 | 0.59 | 3.26 |
| $H_2O^-$ | 0.15 | 0.06 | 0.29 | 0.49 | 0.23 | 0.38 | 0.70 |
| Total | 99.80 | 99.47 | 99.89* | 99.91 | 99.99 | 99.69 | 99.58 |
| *Q* | 2.50 | 9.53 | 19.20 | 26.70 | 33.45 | 34.97 | 28.61 |
| *C* | — | — | — | — | 0.09 | 0.91 | — |
| *or* | 3.19 | 5.91 | 14.46 | 13.00 | 18.81 | 21.26 | 16.92 |
| *ab* | 19.28 | 21.99 | 20.96 | 30.02 | 35.86 | 31.35 | 38.75 |
| *an* | 34.89 | 31.49 | 20.57 | 18.83 | 7.54 | 7.23 | 7.09 |
| *di* *wo* | 8.71 | 3.97 | 3.36 | 0.64 | — | — | 0.15 |
| *di* *en* | 4.98 | 2.26 | 2.00 | 0.55 | — | — | 0.05 |
| *di* *fs* | 3.34 | 1.55 | 1.19 | — | — | — | 0.10 |
| *hy* *en* | 10.25 | 10.69 | 7.90 | 4.48 | 0.69 | 0.74 | 0.69 |
| *hy* *fs* | 6.87 | 7.31 | 4.49 | — | 0.59 | — | 1.71 |
| *mt* | 3.48 | 3.04 | 2.32 | 0.63 | 1.27 | 1.46 | 1.09 |
| *il* | 1.67 | 1.44 | 1.37 | 0.65 | 0.53 | 0.42 | 0.38 |
| *ap* | 0.31 | 0.24 | 0.34 | 0.26 | 0.13 | 0.13 | 0.10 |
| Trace elements (Ewart and Stipp, 1968) | | | | | | | |
| U | 0.37† | — | 1.22† | 1.93 | 2.64 | 2.48 | 2.23 |
| Th | 1.53† | — | 5.23† | 7.88 | 11.5 | 11.4 | 10.2 |
| Sr | 356.4† | 228 | 242.7† | 315 | 106.5 | 140.1 | 164.4 |
| Rb | 12.6† | 35.1 | 49.7† | 63.2 | 106.6 | 112.8 | 90.6 |
| Rb/Sr | | 0.15 | | 0.20 | 1.0 | 0.8 | 0.55 |
| K/Rb | 451† | 236 | 281† | 280 | 251 | 251 | 244 |
| $Sr^{87}/Sr^{86}$ | 0.7042† | 0.7059 | 0.7055† | 0.7048 | 0.7053 | 0.7059 | 0.7060 |

* Includes 0.10% BaO.

† Average values for basalts and andesites, respectively.

*Explanation of column headings*

1 Basaltic bomb, 1886 eruption of Tarawera (Challis, 1971)
2 Andesite lava, 1954 eruption of Ngauruhoe (Steiner, 1958)
3 Hypersthene andesite, White Island (Black, 1970, no. 1)
4 Dacite from Tauhara (Ewart and Stipp, 1968)
5 Average rhyolitic lava (Ewart and Stipp, 1968)
6 Average ignimbrite (Ewart and Stipp, 1968)
7 Average Taupo pumice deposit (Ewart and Stipp, 1968)

they occur (Ewart, 1967). Rhyolitic lavas containing pyroxene as the only ferromagnesian phenocryst are characterized by small amounts of phenocrysts of all types and little or no phenocryst quartz. On the other hand, biotite phenocrysts typically occur in phenocryst-rich rhyolites with low plagioclase/quartz phenocryst ratios. Ewart et al. (1971) have shown that the iron titanium oxide (microphenocryst) equilibration temperatures of the rhyolites with amphibole phenocrysts span the range 735 to 780°C, whereas the very young pumice flows without amphibole (orthopyroxene + augite) have equilibration temperatures in the range 860 to 890°C.

Average analyses of rhyolitic lavas, ignimbrites, and air-fall Taupo pumice are shown in Table 11-6 (analyses 5–7); the first two types have small amounts of normative *C*, which is believed to result from the source rocks of these magmas. Domes of dacite lava (Table 11-6, analysis 4) form only a small proportion of the volume of acid extrusions.

Andesites and basaltic andesites are the predominant rock type of the Tongariro group of volcanoes (Ruapehu, Tongariro, and Ngauruhoe) at the southern end of the province (Fig. 11-11); the volcano forming White Island at the northern end is also andesitic (P. M. Black, 1970). These lavas typically contain abundant phenocrysts of plagioclase ($An_{80}$ to $An_{60}$), together with orthopyroxene, augite, and microphenocrysts of titanomagnetite; hornblende, olivine, and quartz "xenocrysts" are less common, the whole being contained in a groundmass which includes pigeonite. Analyses of two andesite lavas are given in Table 11-6 (analyses 2, 3).

Basalts are rare in this province. A bomb thrown up in the 1886 eruption of Tarawera (Table 11-6, analysis 1) contains far fewer phenocrysts than the andesites, and they are plagioclase and olivine.

Steiner (1958) and R. H. Clark (1960) have both emphasized the existence of two parental magmas in the Taupo province, one a primary basaltic magma, and the other, acid magma, derived by crustal fusion. The extensive $Sr^{87}/Sr^{86}$ data given by Ewart and Stipp (1968), shown in Table 11-7, taken in conjunction with Pb isotopic data (Armstrong and Cooper, 1971) support the conclusion of Ewart and Stipp that the voluminous rhyolitic magma resulted from fusion of the greywacke-argillite sediments which underlie the province. Estimates of the pressure at which the phenocrysts of the acid lavas equilibrated (J. Nicholls et al., 1971) show that the acid magma sources are deeper than 23 km.

The andesites of this province are unlikely to be the products of the fractional crystallization of basaltic magma, since their average $Sr^{87}/Sr^{86}$ ratios (0.7055) are higher than those of basalt (0.7042; Table 11-7); the $Pb^{206}/Pb^{204}$ ratios (Table 11-8) of both andesites and basalts are similar, but the ratio $Pb^{208}/Pb^{206}$ of basalt is slightly higher. Ewart and Stipp consider that the andesitic magma was formed by assimilation of crustal material by basaltic magma, although an independent origin of andesitic magma cannot be discounted. Armstrong and Cooper (1971) showed that the Pb isotopic ratios of Taupo basalts were systematically different from those of the alkali olivine basalts of the nearby contemporaneous Auckland volcanic province.

## Active Volcanic Arcs: Petrographic and Chemical Summary

Taken as a whole, the dominant lava type of the world's active volcanic arcs is monotonously basaltic-andesite or andesite (e.g., the Aleutian arc; Coats, 1952); rhyolite lavas are absent except in those segments underlain by continental crust, and basalt is dominant only in very young arcs. Only in tectonically complex arcs are picritic and ankaramitic lavas erupted.

Basic and intermediate lavas of volcanic arcs typically are poor in $TiO_2$ (< 1.2 percent; Chayes, 1964*a*) compared with oceanic equivalents; mineralogically this is reflected in titanomagnetite (rarely present as microphenocrysts) which is poor in the ulvospinel ($Fe_2TiO_4$) component and in the absence of ilmenite as a primary phase. Plagioclase phenocrysts ($An_{95}$ to $An_{70}$) are usually riddled with glass and ferromagnesian inclusions, often in zonary arrangement, and complex oscillatory zoning is common. Lavas with 50 percent or more phenocrysts are not uncommon, of which plagioclase is usually the predominant phase. The absence of a europium anomaly in the rare-earth pattern (S. R. Taylor et al., 1969) indicates that these lavas have not accumulated plagioclase phenocrysts, and thus their high alumina content is a primary characteristic of andesitic magma. Since plagioclase has been found to be the liquidus phase of Parícutin (Mexico) andesites only when the water content is less than about 2 percent (Eggler, 1972), there is little evidence for the presence of large amounts of water in andesitic magma, provided that these experimental results are applicable to lavas with less than 58 percent $SiO_2$.

Hornblende phenocrysts, when present, are usually titaniferous, contain noticeable fluorine, and according to Jakěs and White (1972) are more magnesian in lavas of island arcs than those of continental margins. Eggler (1972) found that hornblende was not stable with the other phenocryst phases in his hydrous melting experiments on Parícutin andesite. Presumably the small amounts of fluorine can raise the stability range of amphibole (Chap. 3). Hornblende phenocrysts typically show resorption effects around their margins. Resorbed quartz "xenocrysts" occur in some basaltic andesites but are absent in many hornblende-bearing lavas. Few details are published on the pyroxenes, but orthopyroxene phenocrysts are by no means ubiquitous, although augite is; pigeonite or subcalcic augite is more common than orthopyroxene in the groundmass.

Xenoliths of ultrabasic rocks are almost invariably absent though accidental xenoliths in the lavas of Mt. Victory in Papua are an exception. Cognate xenoliths, often interpreted to represent an accumulation of phenocrysts, are common in most lava types. Invariably these have precipitated at pressures less than 9 kb.

The trace-element pattern of lavas from volcanic arcs varies considerably; andesites can contain between 10 and 110 ppm Ni, although they invariably show little depletion in V compared to the associated basalts. This lack of V depletion precludes early fractionation of titanomagnetite from basaltic magma as a means to generate andesite. Strontium can vary between 200 and 1500 ppm in andesites. In the andesites of Mt. Shasta (Cascades of California) Sr is far higher than in the associated basalts, which appears to exclude plagioclase as a residual phase in the derivation of these andesites (Peterman et al., 1970*b*). The rare-earth pattern is generally similar to that of the abyssal tholeiites (that is, low abundances of La, Ce, Pr, Nd) and shows no depletion of Eu.

Isotopic data for strontium and lead with few exceptions show little or no departure from patterns now familiar in tholeiitic basalts of the Hawaiian Islands. Strontium in most provinces maintains a uniform composition through the whole basalt-andesite-rhyolite range (Table 11-7) and duplicates that of Hawaiian tholeiites (Table 8-14). In the Caribbean arc the individuality of each volcanic center is marked by slight departure from the average, and this uniqueness is maintained regardless of rock type, for example, in basalts and associated andesites alike (Hedge and Lewis, 1971). Lead of Japanese lavas (Table 11-8) becomes less radiogenic from east to west transverse to the volcanic front. The alkali olivine basalts give $Pb^{206}/Pb^{204}$ and $Pb^{208}/Pb^{204}$ values significantly lower than those recorded for other basalts and for pyroxene andesites. In this finding Tatsumoto and Knight (1969) see possible evidence of sialic contamination of the tholeiitic and high-alumina basalt-andesite series. In New Zealand, variation is in the opposite sense: lead of basalts and andesites of the Taupo province is less radiogenic than that of nearby alkali olivine basalts of the Auckland city province (Table 11-8).

**Table 11-7** Average initial strontium isotopic values ($Sr^{87}/Sr^{86}$) for selected rock types of active volcanic arcs and continental margins (all data adjusted to a value of 0.7080 for MIT $SrCO_3$)*

| | BASALT (< 52% $SiO_2$) | BASALTIC ANDESITE (52–55% $SiO_2$) | ANDESITE (55–63% $SiO_2$) | DACITE (63–68% $SiO_2$) | RHYOLITE (> 68% $SiO_2$) |
|---|---|---|---|---|---|
| New Britain | 0.7035 | 0.7036 | 0.7036 | 0.7036 | 0.7035 |
| Tonga | — | 0.7037 | 0.7042 | 0.7043 | |
| Marianas | 0.7042 | — | 0.7042 | 0.7038 | — |
| Izu Islands | 0.7036 | — | 0.7040 | — | 0.7034 |
| Caribbean | | | | | |
| St. Kitts | 0.7036 | 0.7040 | 0.7038 | — | — |
| St. Vincent | 0.7042 | 0.7040 | 0.7039 | — | — |
| Carriacou | 0.7052 | — | 0.7054 | — | — |
| North Japan | 0.7043 | — | 0.7041 | — | — |
| California | | | | | |
| Mt. Shasta | — | 0.7039 | 0.7030 | 0.7032 | — |
| Mt. Lassen | 0.7039 | 0.7032 | 0.7040 | — | — |
| Medicine Lake | 0.7034 | 0.7037 | — | — | 0.7040 |
| Central America | 0.7035 | 0.7042 | 0.7036 | 0.7036 | 0.7042 |
| New Zealand | | | | | |
| Taupo | 0.7042 | — | 0.7055 | 0.7051 | 0.7053 |
| Average† | 0.7040 | 0.7038 | 0.7040 | 0.7037 | 0.7038 |

* Data taken from Pushkar (1968), Peterman et al. (1970*a*, 1970*b*), Hedge (1966), Hedge and Knight (1969), Hedge and Lewis (1971), Ewart and Stipp (1968), Oversby and Ewart (1972), Ewart et al. (1973).
† Excludes New Zealand average andesite, dacite, and rhyolite.

## Geosynclinal Volcanic Rocks

### THE EUGEOSYNCLINAL FILLING

The new plate tectonics provides the structural-dynamic setting of Quaternary volcanism along the ocean ridges, in island arcs, and on continental margins. On the other hand the older classic concepts of volcanism in relation to orogeny have developed through a century's cumulative study of eroded geosynclines. An immediate problem, perhaps the most urgent in modern geology, is to correlate concepts and accumulated data in the two areas of study. There have been several stimulating essays in this direction, among them those of Mitchell and Reading (1969) and of Dewey and Bird (1970). The topic has also been treated comprehensively by Wyllie (1971*a*, pp. 366–372).

In this chapter we are concerned only with eugeosynclines, a major component of whose fillings, by definition, is volcanic. Whether the basement rock is metamorphosed oceanic crust or sialic metasediment of continental crust (as in the Mesozoic Andean geosyncline of Chile), we find in geosynclines a cumulative sample of material—much of it reworked by erosion—ejected from vanished volcanoes of the island-arc and continental-margin type. Allowing for

**Table 11-8** Average isotopic composition of lead and average concentrations of Pb, U, and Th (ppm) in selected volcanic rocks of active volcanic arcs and continental margins*

| | Pb | U | Th | $Pb^{206}/Pb^{204}$ | $Pb^{207}/Pb^{204}$ | $Pb^{208}/Pb^{204}$ |
|---|---|---|---|---|---|---|
| New Zealand (Taupo) | | | | | | |
| Basalt | — | — | — | 18.749 | 15.618 | 38.62 |
| Andesite | — | — | — | 18.773 | 15.604 | 38.611 |
| Tonga | | | | | | |
| Andesite | 1.83 | 0.19 | — | 18.528 | 15.549 | 38.118 |
| Dacite | 3.85 | 0.37 | — | 18.525 | 15.544 | 38.131 |
| Kermadec | | | | | | |
| Basalt | 1.88 | 0.18 | — | 18.552 | 15.560 | 38.436 |
| Caribbean (St. Kitts) | | | | | | |
| Basalt | — | — | — | 18.868 | 15.640 | 38.629 |
| Andesite | — | — | — | 18.952 | 15.658 | 38.711 |
| Japan (tholeiitic) | | | | | | |
| Basalt | 3.22 | 0.15 | 0.26 | 18.507 | 15.68 | 38.75 |
| Andesite | 4.04 | 0.38 | 0.38 | 18.385 | 15.66 | 38.65 |
| Japan (high $Al_2O_3$) | | | | | | |
| Basalt | 4.32 | 0.38 | 1.04 | 18.384 | 15.67 | 38.67 |
| Andesite | 5.07 | 0.59 | 1.69 | 18.364 | 15.68 | 38.67 |
| Dacite | 6.02 | 1.46 | 4.11 | 18.346 | 15.67 | 38.65 |
| Japan (alkali) | | | | | | |
| Olivine basalt | 3.35 | 0.68 | 3.34 | 18.08 | 15.56 | 38.60 |
| New Zealand (Auckland) | | | | | | |
| Alkali olivine basalt | — | — | — | 19.130 | 15.580 | 38.764 |

* Data taken from Armstrong and Cooper (1971), Tatsumoto and Knight (1969), Oversby and Ewart (1972).

universal effects of folding and burial metamorphism, volcanic ingredients of geosynclinal fillings show much of the pattern and variety that characterize Recent volcanism in island arcs and along continental margins. Thick sequences of basaltic flows, many of them submarine spilitic pillow lavas, are typical of the lower levels in many geosynclines. Such are the Permian spilites of the New Zealand geosyncline (Permian to Jurassic). In other geosynclines the volcanic filling is largely andesitic. This is so in the late Paleozoic and Mesozoic geosynclinal belt that extends along the western margin of North America from Alaska to California (Dickinson, 1962). Much of the Mesozoic Andean geosyncline of Chile is filled with massive basaltic-andesite lavas. In other segments of the same geosyncline there are great thicknesses of rhyolite (keratophyre) tuff; and the same is true of other geosynclines of other ages.

## SPILITES AND KERATOPHYRES: GENERAL CHARACTERISTICS

Spilites are basic lavas consisting principally of highly sodic plagioclase (albite or oligoclase) and augite or its altered equivalent (actinolite, chlorite epidote, chlorite hematite, etc.). Olivine typically is absent or is represented sparingly by serpentine pseudomorphs. Evidence of hydration and carbonation (for example, alteration of pyroxene, infilling of vesicles with epidote, calcite, and so on) is usually conspicuous, while persistence of relict patches of labradorite or andesine within crystals of albite shows conclusively that in some cases the present condition of the feldspars is a result of albitization of initially more calcic plagioclase. Pure albite ($An_{<5}$) of spilites has anomalous optic and x-ray diffraction parameters (for example, $2V_\gamma$ = 85 to 95°) compared with albite ($2V_\gamma$ = 76 to 78°) of green schists, from which has been inferred a low to intermediate structural state (Vallance, 1960, p. 24; Levi, 1969, p. 34). All spilites are highly sodic rocks ($Na_2O$ = 4.5 to 5.5 percent). Common associates of spilitic lavas are intrusive, chemically equivalent albite diabases, as well as flows, minor intrusions, and tuffs of highly sodic keratophyre and quartz keratophyre. In many of these rocks, too, albite seems to be of secondary origin.

## THE SPILITE-KERATOPHYRE PROBLEM

During the 60 years since Dewey and Flett first recognized the strongly sodic character of many geosynclinal basalts (spilites) and associated rhyolitic debris (keratophyres, quartz keratophyres), its significance has been strongly debated. Do these rocks represent the products of exceptionally sodic magmas (also low in $K_2O$), or is this chemical trait due to posteruptive metasomatism? The nature of the spilitic associations and alternative petrogenic hypotheses have been reviewed at some length in our earlier book (Turner and Verhoogen, 1960, pp. 258–272). During the past few decades, stimulated by Gilluly's (1935) classic examination of the situation in eastern Oregon, opinion has veered steadily away from the concept of the "primary spilite magma" and the role of metasomatism has found increasing favor. The problem has been studied exhaustively by Vallance (1960, 1969), whose final synthesis leaves scarcely any room for further doubt. Spilites and keratophyres of geosynclinal belts can be regarded as basalts, andesites, and rhyolites whose high content of sodium is due to metasomatism (probably through the agency of trapped seawater) during partial low-grade load metamorphism, which now seems ubiquitous in the lower levels of such accumulations (for example, Levi, 1969).

## Magma Sources

It is difficult to generalize on the origin of the magma of island arcs—no universal model will satisfy all the restrictions of composition which detailed investigation is currently exposing. Some island arcs, such as the Scotia, have basalts as the main lava type whereas others, such as the Caribbean, have predominant andesite. Some, such as the Tonga-Kermadec arc, have no rhyolite and only small amounts of dacite, and they lack a continuous series of lava compositions between basalt and dacite; and yet the extension of this arc into New Zealand is associated with an enormous volume of acid magma, much of it ignimbrite. As a generalization, it appears that ignimbrite and extensive amounts of rhyolite lava are associated with continental crust, although not all these acid magmas betray, by their $Sr^{87}/Sr^{86}$ ratios, sialic interference.

It is the origin of the andesites and basaltic andesites which has long been a matter of debate and speculation, although the theories proposed are few and many of the data are permissive. Bowen (1928) suggested that andesites were derived by fractional crystallization of basaltic magma, but an immense volume of basalt would be required to produce the volumes of andesite exposed on the surface. The paucity of associated basalt has always been a matter of concern to geologists mapping andesitic terranes, and in the absence of any convincing theory as to why the necessary large volume of basalt was prevented from reaching the earth's surface, the hypothesis of crystal fractionation has fallen into disfavor. And yet it is the only theory which can be treated with any rigor at this time. If the liquid line of descent, defined by the serial composition of the lavas, major and trace component alike, can be reproduced (by calculation) solely by subtraction of the early crystallizing phases in each lava type, then crystal fractionation of a basaltic parent is a plausible, and viable, mechanism. Nevertheless this conclusion, obtained for the Talasea series of New Britain, apparently founders on the dominance of the andesitic lava type there. Perhaps the eruption of andesite magma, so often accompanied by large amounts of pyroclastic material, requires a substantial amount of water whereas the relatively anhydrous more basic magma beneath it lacks the propellant to achieve eruption. Certainly if it can be shown in the future that the failure to erupt large volumes of basaltic magma in these andesitic arcs is due to some property of basaltic magma, then much of the current hesitation about accepting the crystal fractionation hypothesis (in a low-pressure environment) will evaporate.

An early suggestion (Daly, 1933) to account for the predominance of andesite is that it represents basaltic magma contaminated with sialic crustal material, a suggestion later developed by Tilley (1950), Kuno (1950), and Waters (1955). However, this hypothesis fails as a general explanation because andesites are found in volcanic arcs built on oceanic crust; they also have $Sr^{87}/Sr^{86}$ ratios (Table 11-7) not sensibly different from ratios for tholeiites of the Hawaiian Islands, where sialic crust is generally conceded to be absent. Indeed the uniform $Sr^{87}/Sr^{86}$ isotopic ratios of all lava types in volcanic arcs and continental margins (Table 11-7), with the exception of New Zealand and possibly other ignimbrites, require that all types of magmas were derived from isotopically similar source materials.

The concept of descending lithospheric plates has prompted the idea that magmas may be generated, perhaps by frictional heating, in Benioff zones. Certainly there is good correlation between seismic activity at depths greater than 80 or 100 km and the occurrence of volcanism in island arcs. Dickinson (1968) demonstrated, moreover, a tendency for $K_2O$ in basaltic andesites and andesites at 55 to 60 percent $SiO_2$ to increase with depth of the underlying Benioff zone in individual island arcs (Fig. 11-12, *a*, *b*). More recently it has been shown (Nielson and Stoiber, 1973) that the relation between $K_2O$ and depth is unique in each arc (Fig. 11-12*c*). One of the factors concerned may perhaps be the age of onset of volcanism—Java, 150 to 175 m.y.; Izu Islands, 25–50 m.y.; the trend in the young Scotia and Tongan arcs is similar to that in the Izu Islands. Correlation may possibly be influenced, too, by reaction between ascending magmas and rocks above the Benioff zone. Clearly the composition of acid intrusions is no secure basis for delineating palaeoseismic zones (Dickinson, 1970).

Progressive fusion of sanidine-bearing quartz eclogite (T. H. Green and Ringwood, 1968) duplicates a correlation between $K_2O$ and $SiO_2$† found by Nielson and Stoiber (1973) in andesitic lavas of the Aleutian and some other arcs. (By contrast there is inverse correlation between $K_2O$ and $SiO_2$ in liquids generated from phlogopitic assemblages). Correlation between $K_2O$ and pressure (depth of source) is borne out finally by calculated pressure-temperature conditions for equilibration of basaltic-andesitic magmas with sanidine-quartz eclogites. As matters stand today data from a number of independent sources are consistent with the general model of origin of basaltic-andesitic magmas by fusion of eclogite rocks in Benioff zones. However, lavas of volcanic arcs seldom display in their phenocryst assemblages evidence of origin from such deep sources. A possible exception is provided by quartz. Calculations show that this could be in equilibrium with hydrous magma of basaltic andesite composi-

† Ewart et al. (1973) noted that in the Tongan lavas, Ba, Sr, and Pb increase with $K_2O$ and $SiO_2$.

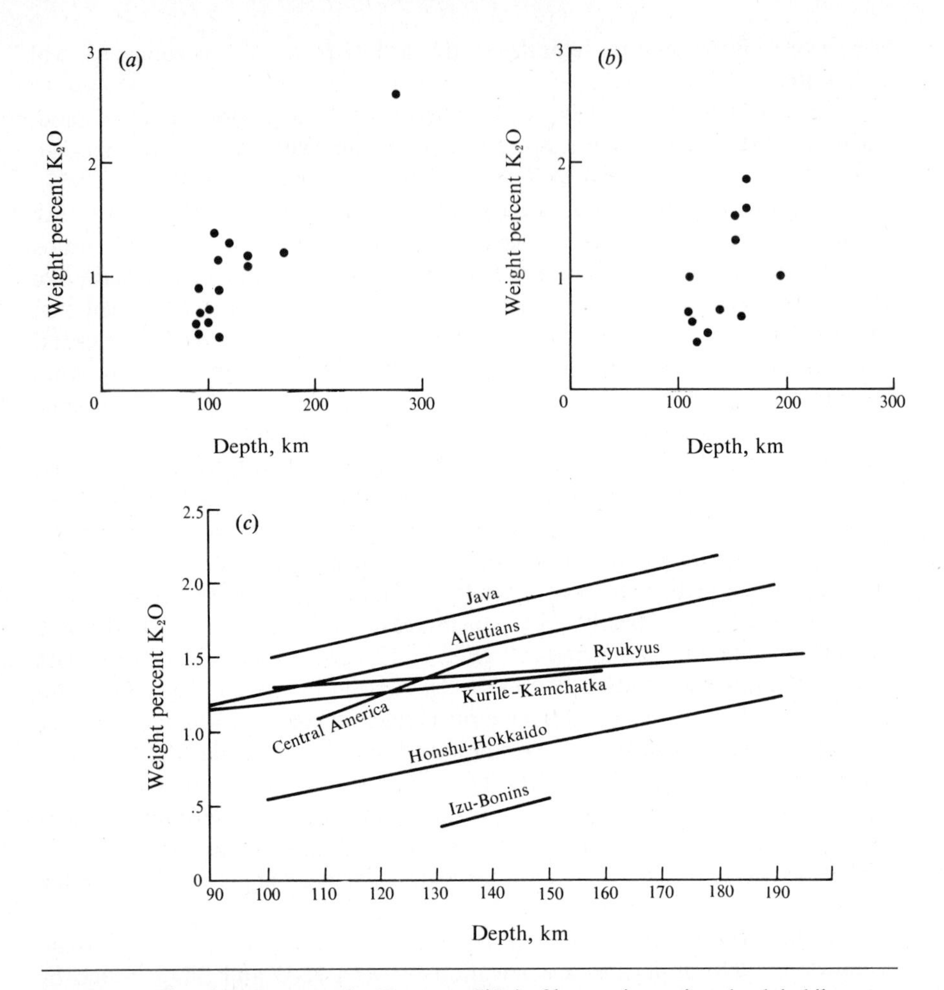


**Fig. 11-12** Plot of $K_2O$ content (at 55 percent $SiO_2$) of lava series against depth in kilometers from the volcano to the inclined seismic zone beneath: (a) Kuriles-Kamchatka; (b) Honshu, Japan; (c) Circum-Pacific volcanic arcs. Length of lines represent range of data for each region, and their position is obtained by statistical regression of published analyses (after Dickinson, 1970 and Nielson and Stoiber, 1973).

tion at 1300 to 1400°C and pressures prevailing in underlying Benioff zones (Marsh and Carmichael, in press).

Another obvious possibility is generation of andesitic magma by fusion of oceanic crust and some of its veneer of pelagic sediment in a subduction zone. If chemically unaltered abyssal tholeiite retaining its typical $Sr^{87}/Sr^{86}$ ratio of 0.7026* is the main component of the crust, the value 0.704 typical of island-arc andesites could be achieved by an increment of strontium with $Sr^{87}/Sr^{86} = 0.709$

* Using average values of Rb and Sr (Kay et al., 1970), this value could be increased to 0.7027 in $200 \times 10^6$ years.

(Armstrong, 1968) from the overlying sediment. There are no unequivocal data to support or deny the possibility that pelagic sediment is indeed involved in the process of magma generation (as suggested by Coats, 1962). Mass balance calculations are hindered by uncertainty and variation in the chemistry of sediments concerned. Church (1973) has evaluated sediment mixing, using the concentrations of Pb, Th, U, and Ba together with the strontium and lead isotopic ratios, and has concluded that it amounts to less than 2 percent in the andesites of the High Cascades; a similar conclusion regarding the small contribution of pelagic sediments to the generation of magma in the Tonga arc was reached by Oversby and Ewart (1972).

It would seem that oxygen isotopes would furnish a sensitive indication of sediment incorporation; deep-sea clays have $\delta_{O^{18}/O^{16}}$ of about + 21 per mil, whereas basalts of oceanic islands have values in the range between +5.9 and 6.8 per mil (H. P. Taylor, 1968). On this basis few lavas from the Cascades province suggest incorporation of deep-sea sediment. Perhaps there is something to be learned from H, C, and N isotopes in the lavas of volcanic arcs. The problem is complicated by the possibility of isotopic contamination of abyssal basalts by exchange with seawater (e.g., Bass et al., 1973; see also Chap. 13).

A solution to the problem of the dominance of andesite in volcanic arcs, and the lack of iron enrichment which these lava suites frequently show, was found by T. H. Green and Ringwood (1968). Their experimental work on a range of liquid compositions (high-alumina basalt, basaltic andesite, andesite, dacite, and rhyodacite) showed that the near-liquidus phases of plagioclase and pyroxene (water absent) were replaced by garnet and pyroxene (rather jadeite-poor) at pressures equivalent to depths of 100 to 150 km. The composition with the lowest liquidus temperature at such pressures was found to be andesite, so that andesitic liquid would be the first to form from the fusion of anhydrous quartz eclogite.

Thus the fractional crystallization, by separation of garnet and pyroxene, of any liquid more basic than andesite will drive the residual liquid toward andesite composition, but not to more acidic liquids; dacites and rhyolites can be derived from an andesitic parent only at low pressures. The liquid paths would show little iron enrichment since the combination of almandine-rich garnet and magnesian augite effectively restricts this. Green and Ringwood have suggested that the presence of almandine-rich garnet phenocrysts in some dacites represents evidence of crystallization at considerable depths. The generation of quartz eclogite itself is considered to result from the recrystallization of subducted oceanic crust, so that this hypothesis of the origin of andesites has come to be known as the two-stage model. The first stage is eruption of basalt, dominantly abyssal tholeiite on the ocean ridges; the second is its recrystallization to quartz-eclogite and later fusion to produce the andesite-dominated suites of the volcanic arcs and continental margins.

This and other models, together with the possible role of water and trace-element patterns, will be considered in Chap. 13, which is concerned principally with the mantle-magma system.

# Rocks of Continental Plutonic Provinces

# 12

## General Distribution of Plutonic Rocks

Exposure of plutonic complexes at the earth's surface requires erosion. So we find them in dissected continental terranes in the form of simple or compound batholiths, immense sheets, swarms of discrete plutons, and diffusely bounded migmatite complexes. Collectively plutonic rocks constitute a major portion of the exposed continental crust. Oversaturated rocks with modal quartz predominate.

No great depth is necessarily required for the condition (slow cooling at liquidus to subsolidus temperatures) implied by the textures and mineralogy of plutonic rocks. Much of the known variety of plutonic rocks is displayed, in fact, in relatively small, shallow subvolcanic complexes. Such rocks, exposed in eroded volcanic centers or present as xenoliths in lavas and tuffs, have already been treated in their pertinent volcanic context.

Bodies of plutonic rock presumably exist in the suboceanic crust. It may be inferred from geophysical data that these must be mainly of basic and ultrabasic composition. Blocks of gabbroid and ultramafic rocks dredged from the ocean floor must have come from such sources; and their probable metamorphosed equivalents may be represented by rocks of basic and ultramafic composition prominent in the crystalline basements of some older eugeosynclines. At present not much more can be said of oceanic plutonism.

Known major plutonic complexes, then, are continental phenomena. They can be grouped in four broad categories:

1. "Granitic" batholiths and migmatite complexes
2. Anorthositic batholiths of Precambrian shields

3. Ultramafic plutons (mainly of peridotite) and serpentinite sheets in orogenic belts

4. Layered basic plutons

In this chapter we treat the first three categories only. Layered basic intrusions have already been considered as plutonic equivalents of tholeiitic volcanic series.

## Granitic Complexes

### GRANITIC PLUTONS IN RELATION TO GEOLOGIC SETTINGS

Here we use the term *granitic* in its time-honored general connotation to cover the whole gamut of plutonic rocks with 10 percent or more of modal quartz. Granitic rocks are the main component of continental shields. They occur too as great compound batholiths in folded geosynclinal belts. So wide is their extent, so varied their relations to crustal depth, thermal environment, metamorphism, tectonism, volcanism, and the flow of time, that generalization inevitably obscures with fictitous clarity one of the most complex of igneous phenomena. No wonder opinion has fluctuated on all aspects of granite genesis and controversy periodically explodes as it did under the stimulus of H. H. Read's writings in the forties (Read, 1957, and literature referred to therein).

Our earlier account of granitic rocks includes essentially factual aspects of the form, size, internal structure, contact relations, and chemical composition of granitic plutons (Turner and Verhoogen, 1960, pp. 329–375). These facts, admittedly presented from a "magmatist" standpoint, will serve as background for reviewing generalizations that have since appeared, relating granite plutonism to geologic environments and tectonic pattern.

GRANITIC PLUTONS AND VOLCANISM In many provinces emplacement of granitic plutons is associated with synchronous eruption of chemically similar, presumably cogeneric volcanic rocks (Buddington, 1959, pp. 680–685; Raguin, 1965, pp. 212–226; Haslam, 1968). Commonly the plutons take the form of ring complexes 10 km or so in diameter with volcanic remnants preserved in central blocks that subsided into the granite cauldrons. This seems to be the principal mode of occurrence of peralkaline granites and quartz syenites as exemplified in the Permian Oslo-graben province (Chap. 10) and in Jurassic ring complexes scattered over an area of 60,000 $km^2$ in northern Nigeria (Jacobson et al., 1958; D. C. Turner, 1963). Normal biotite granites may accompany and even predominate over peralkaline types, as they do in the White Mountain magma series[1] of New Hampshire (cf. Turner and Verhoogen, 1960, pp. 336–338).

[1] Now dated as early Jurassic (185 m.y.) and preceded by intrusion of comparable magmas in the later Paleozoic (Toulmin, 1961).

Where dacites and rhyolites are plentifully represented, granites and granophyres of cogeneric ring complexes lack alkaline affinities, as for example in the late Karroo Nuanetsi province of Rhodesia and in the Caledonian granodiorite complex of Ben Nevis (Haslam, 1968). Then there are andesite-dominated provinces where the close of the volcanic cycle is marked by copious intrusion of chemically similar magma as stocks and somewhat larger plutons of granodiorite. This is the case in the Eocene-Miocene Western Cascades province of Oregon and Washington (for example, Waters, 1962, p. 167; Verhoogen et al., 1970, p. 299, fig. 6-19).

The main interest of minor granitic intrusions is that they demonstrate the purely magmatic origin of some plutonic rocks covering the tonalite-granodiorite-granite spectrum. Conclusions and speculations as to the origin and evolution of associated volcanic magmas must apply equally to magmas represented in these, their plutonic counterparts. But the great bulk of granodiorite-granite bodies are found in completely different settings in which any relation that may exist between plutonism and volcanism is more obscure.

Migmatite complexes typically occur in a high-temperature metamorphic environment far separated in the vertical sense from surface volcanism. Any model that attempts to relate these two kinds of magmatic activity—Read's (1949, p. 149) "granite series," for example (Turner and Verhoogen, 1960, p. 387)—necessarily rests on speculation. All that is certainly known is that the magmatic ("granitic") component of many migmatites includes rock types broadly similar to, and conventionally classified with, rocks of some acid subvolcanic plutons. No direct relation between the two has been established (cf. Buddington, 1959); and Raguin (1965, p. 228) even concludes that "the anatectic granites do not seem likely to generate volcanism."

There is stronger evidence for some kind of relation between emplacement of granitic batholiths and extensive, more or less synchronous volcanism in some orogenic belts. How direct is the relationship? Is volcanism simply the surface manifestation of an upward surge of granitic magmas that elsewhere consolidated in depth as batholiths? Or are the two forms of magmatic activity partially independent, each having its own origin and particular regime in time?

Chemically there is an obvious overall resemblance between granitic rocks of most batholiths and at least some of their volcanic associates—the andesite-dacite-rhyolite series of systematic petrography. This resemblance appears in conventional variation diagrams (Figs. 12-1 and 12-2) and is borne out by distribution patterns of trace elements. So in some provinces it may well be that granites and chemically equivalent felsic volcanics come from a common source. This has been suggested for Devonian (360 ± 30 m.y.) granites, lavas, and tuffs synchronous with the Acadian orogeny in the northern Appalachians (for example, Lyons and Faul, 1968, p. 313). A similar relation is likely for the Boulder batholith, Montana, and scarcely older dacites that make up much of its cover (Klepper et al., 1971), both dated as upper Cretaceous.

Seen in more extensive provinces and in deeper chronological perspective,

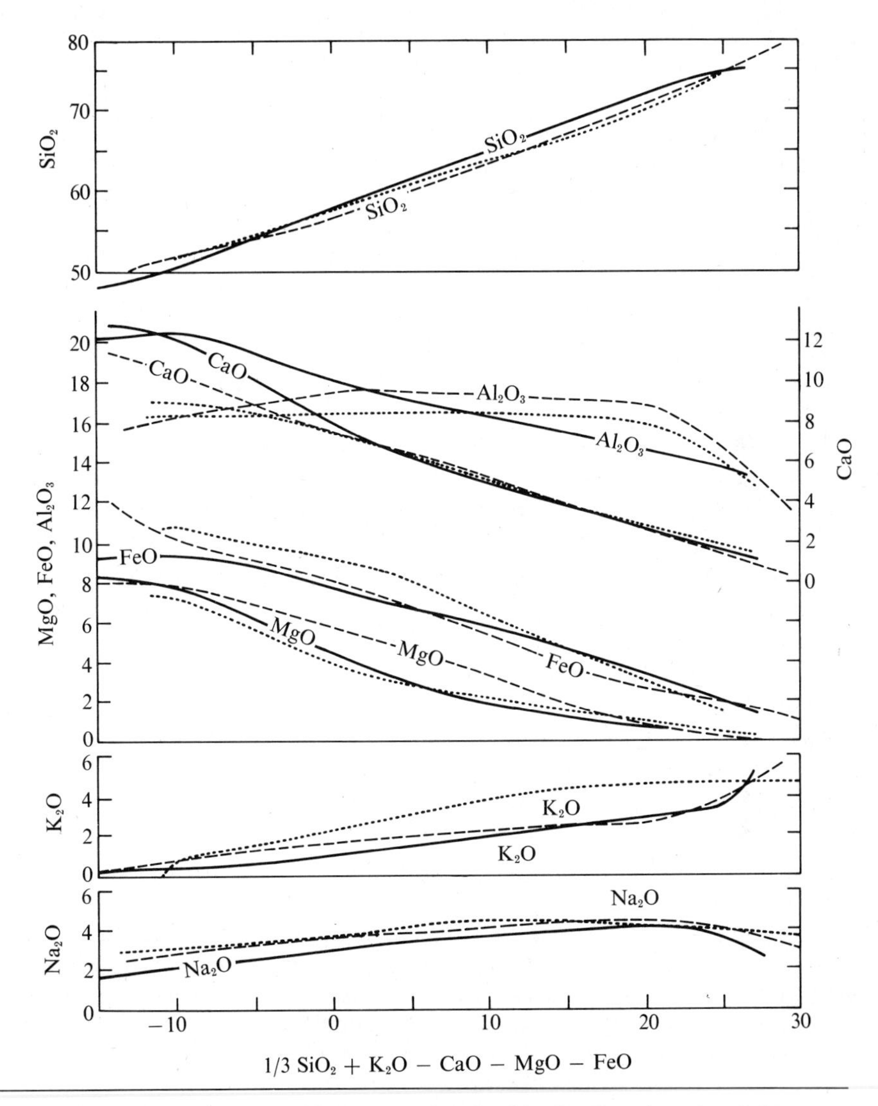


**Fig. 12-1** Variation diagrams weight percentages of oxides for plutonic rocks of the Lower California batholith (full curves) and volcanic rocks of Yellowstone (dashed curves) and San Juan provinces (dotted curves). (Data from Larsen, 1948, etc.)

the respective pictures of granite plutonism and geographically associated volcanism may differ significantly in two respects. One relates to chemical composition of the *dominant* magmas, the other to the detailed regimes of magmatic activity in time. Gilluly (1963) finds that in the western United States during its 600 m.y. of recorded tectonic history, volcanism has been more extensive, more continuous, and more diversified than granite plutonism.

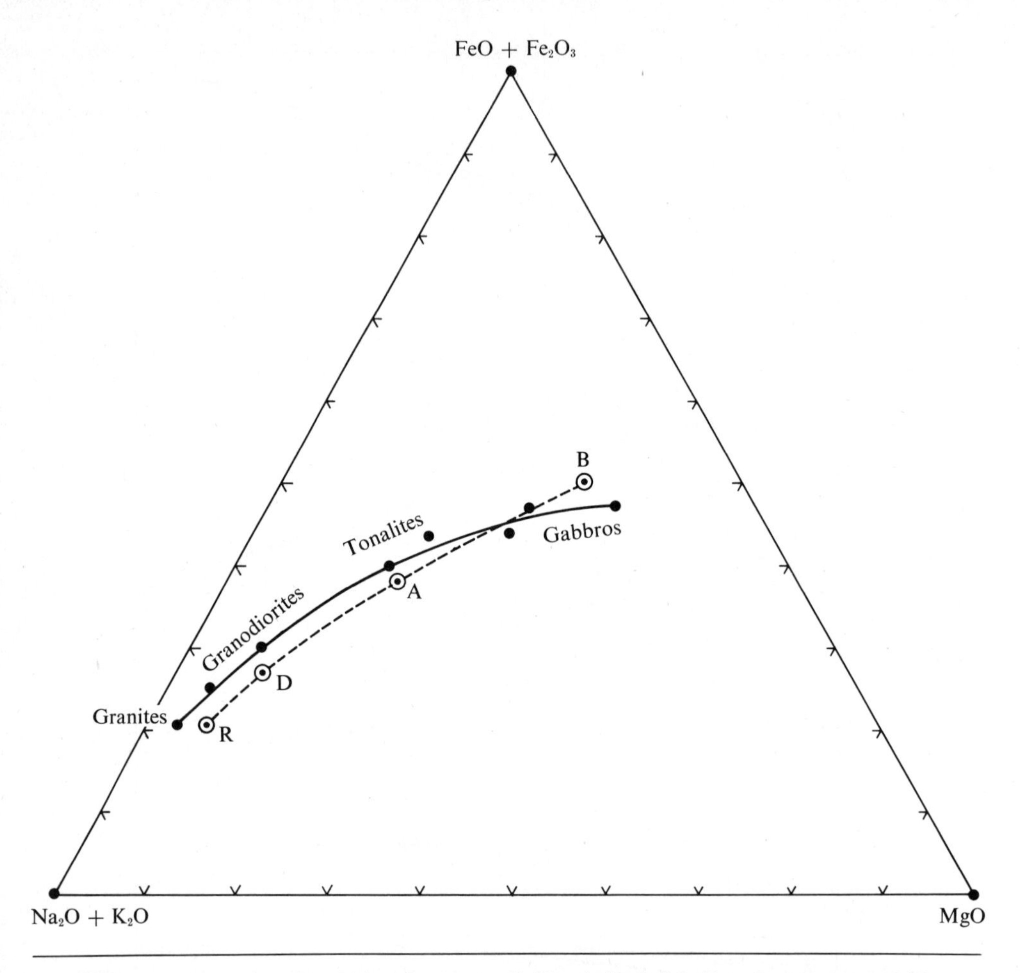


**Fig. 12-2** AFM plot of typical rocks, Lower California batholith (data from Larsen, 1948). Dashed curve shows for comparison a plot of lavas from the Cascades province, northwestern United States (*B*, basalts; *A*, andesites; *D*, dacites; *R*, rhyolites). Oxide proportions by weight.

Volcanic products over large areas and for long periods were mainly basaltic. Elsewhere and at other times they were essentially andesitic, elsewhere again dacite and rhyolite. All these types are represented on a very large scale. In the great batholiths, on the other hand, the dominant rocks by far are quartz diorites, granodiorites, and quartz monzonites. Chemically these collectively resemble the volcanic rhyolites, dacites, and siliceous andesites. Gilluly (1963, p. 168) concludes that "plutonism must depend on processes whose time-scale is of an essentially different order of magnitude from those of volcanism and tectonism." He further suggests (p. 166) that volcanic magmas have here been drawn largely from simatic oceanic crust, and plutonic magmas from more siliceous materials. Something of the same pattern is seen in the Mesozoic Andean geosyncline of Chile (cf. Verhoogen et al., 1970, pp. 303, 306–307). One conclusion seems inescapable. If batholithic granodiorites and volcanic rhyolitic series have a common source, then the basic andesites and basalts so prominent in many geosynclines must have some other origin since rocks of comparable composition are lacking or insignificant in most major granite batholiths.

GRANITIC PLUTONS AND OROGENY Many granitic batholiths are located in orogenic belts of folded and in some instances regionally metamorphosed strata. So common is this situation that one of the most firmly established tenets of tectonic geology is the existence of some kind of fundamental connection between orogeny and large-scale batholithic intrusion of granitic magmas. This relationship, discussed early in the century by such writers as Daly, has been the subject of a number of more recent essays, notable among them those of W. R. Browne (1931) and Gilluly (1963, 1966). Recent discussion centers on two questions. Is granite plutonism on a large scale invariably associated with orogeny and vice versa? Now that the time scale of diastrophism has been refined and qualified by radiometric dating, what can be said about chronological relations between granite plutonism and orogeny? Current differences in opinion are rooted partly in semantics (what is *orogeny*?) and partly in inherent uncertainty as to the identity of each radiometrically dated "event."[1]

*Orogeny* means different things to different geologists (cf. Billings, 1960, pp. 365, 366). Here we use the term (cf. Gilluly, 1966) to cover the whole prolonged event during which geosynclinal rocks become folded and then elevated to form a mountain system. Most major orogenies, such as the Caledonian of northern Britain (McKerrow, 1962) and the Taconic in the Appalachians (J. Rodgers, 1971), are compound events. They involve a succession of stratigraphically and structurally defined episodes of folding, and collectively these may span 50 or 100 m.y. or even more. Although the same sequence of episodes can be recognized for hundreds of kilometers along a single orogenic zone, corresponding episodes at widely separate points within such a belt need not be synchronous. Viewed on this scale, granitic batholiths of fold-mountain systems are synorogenic. But intrusion of granitic magmas, like folding, occurs in separate pulses; and in general these are fewer than episodes of folding in the same orogeny and need not be synchronous with any one of them.

On a more refined scale it may be possible to identify granitic plutons as syntectonic (synchronous, synkinematic) or posttectonic (subsequent, postkinematic) with respect to one particular episode of folding. The criteria are essentially structural, with emphasis on degree of concordance or discordance between internal structure and configuration of plutons and the geometry and

[1] For Rb-Sr dates there is further complicating uncertainty as to the value of the decay constant $\lambda$ of $Rb^{87}$.

style of folding in the enclosing rocks (for example, W. R. Browne, 1931; Buddington, 1959, pp. 671, 677–680, 695–696, 714–715; Turner and Verhoogen, 1960, pp. 350–358). The essential role of very detailed field observation in assessing the structural situation is brought out emphatically in Pitcher's (1970) essay on ghost stratigraphy in granitic plutons. Supplemental evidence comes from radiometric dating[1] of granite emplacement and of regional metamorphism that can be identified with some stage in the folding sequence. A classic instance is the pattern of granite plutonism in northern Scotland and Ireland during the 150 m.y. covered by the Caledonian orogeny (discussed in greater detail later in this chapter). A much more condensed sequence of events has been demonstrated for the Acadian orogeny (Devonian), in Vermont (Naylor, 1971): recumbent folding, regional metamorphism, and finally granite intrusion, all within 30 m.y.

The time scale of Gilluly's (1963, 1966) essays is relatively refined. He refers specifically to the western United States, where granite plutonism is restricted to the last 200 m.y. So he views the subject of discussion close in. From this standpoint, Gilluly maintains that although granite plutonism is synorogenic in the preceding sense, orogeny is by no means invariably accompanied by emplacement of granitic batholiths. Others have found elsewhere a more general one-to-one correlation between the two sets of phenomena (for example, Lyons and Faul, 1967, in Appalachian orogenies of the Paleozoic). Students of Precambrian geology tend to regard mutual association of granite plutonism and orogeny as universal. Some even define orogenies in terms of clustered radiometric dates given by granites and their metamorphic associates (Verhoogen et al., 1970, pp. 211–213). Such divergent opinion partly reflects difference in perspective of time. The Precambrian "Grenville orogeny," radiometrically defined, covered 300 m.y. So did the total span of the several Paleozoic orogenies that have been deciphered in the Appalachian fold belt—each with its attendant granite plutonism. Stratigraphers (for example, McKerrow, 1962) see the Caledonian orogeny of northern Britain as a sequence of localized episodes of folding collectively covering perhaps 150 m.y. (late Cambrian to mid-Devonian). The western American orogenies treated by Gilluly individually covered short intervals between early Mississippian and late Cenozoic times. Comparable phenomena of more ancient date, telescoped in time and blurred by repeated resetting of radiometric clocks, could be taken by some as evidence of a universal reciprocal interdependence of granite plutonism and orogeny.

Finally we return to the relation between plutonism and orogeny *in place*. Most granitic plutons are synorogenic with respect to the fold belts within which they lie. However, Buddington (1959, p. 731) has noted that some American

[1] The date in each case is the point in time at which the radioactive system was finally sealed against loss of radiogenic isotopes. Depending on rates of postintrusive and postmetamorphic cooling, especially in K-A dating, this may be significantly later than the date of the primary event.

batholiths "occur in an old eugeosynclinal structure but may have been emplaced following the development of a miogeosynclinal or other type of structure in the same region."

We find the same thing in southern New Zealand. Here the Permian-Jurassic filling of the New Zealand geosyncline was folded repeatedly and regionally metamorphosed in the mid-Cretaceous. Within the belt so affected, granites are completely lacking (Verhoogen et al., 1970, pp. 570). But there are numerous granitic plutons in a previously folded, immediately adjoining, earlier geosyncline. And these too have been dated radiometrically as mid-Cretaceous.

GRANITIC PLUTONS, METAMORPHIC ENVIRONMENT, AND DEPTH Major folding, regional metamorphism, and syntectonic emplacement of granitic plutons commonly overlap during culmination of orogeny. The metamorphic mineral assemblages in rocks of different composition collectively define a metamorphic facies which can tell us something of the physical conditions prevailing in the metamorphic-plutonic environment (Turner, 1968, pp. 349–374). Local conditions, partly induced by intrusion of magma, likewise are reflected in other facies characterizing metamorphism in aureoles surrounding posttectonic plutons invading folded rocks in which effects of previous regional metamorphism may be negligible.

Further inference as to depth of granite emplacement, based on regional metamorphic facies alone, is at best speculative and in most cases unjustified. Yet this plays an essential role in Read's concept of a unified granite series related to depth in orogenic belts (for example, Turner and Verhoogen, 1960, pp. 387–388). It is the specific basis too for Buddington's (1959) correlation of style of emplacement with depth in terms of Grubenmann's schematic and long-outmoded depth sequence (a downward sequence: epi, meso, and katazones). The character of a metamorphic facies, it is now thought, reflects the influence of several variables, some completely, some partially independent. Most important are pressure, temperature, and activities of volatile components ($H_2O$, $CO_2$, and others) in part determined by pressure. Collectively these variables bear no simple universal relation to depth.

Calibration of aureole temperatures and pressures by various available geothermometers (cf. Turner, 1968, pp. 17–22; Verhoogen et al., 1970, pp. 556–558, 590–591) suggests that many Phanerozoic plutons were emplaced at shallow depth—perhaps between 2 and 8 km (Turner, 1968, p. 258, fig. 6-35). Preintrusive regional metamorphism of enclosing rocks, on the other hand, may have occurred at much greater depth (cf. S. W. Richardson, 1970)—perhaps 15 or even 25 km in the Dalradian terrane of Scotland.

Where folding, regional metamorphism, and intrusion of granite are essentially synchronous, the granitic plutons tend to appear mostly where the metamorphic grade is highest. We visualize the overall thermal pattern as one of abnormally high heat flow with $dT/dP$ (for any given depth) highest in zones of maximum metamorphic grade (cf. S. W. Richardson, 1970). Two propositions follow:

1. Maximum temperatures of regional metamorphism at depths greater than 10 km ($P > 3$ kb) exceed 650 to 700°C, as indicated by prevailing dehydration reactions. So, given a sufficient supply of water, pelitic and semipelitic sediments at maximum metamorphic temperatures could fuse to yield granodioritic melts (Turner, 1968, pp. 376–379).
2. There is no consistent direct relation between metamorphic grade and depth. Tentative pressure-temperature gradients (not depth gradients) constructed for individual facies series (for example, Turner, 1968, pp. 254–255) reach 600°C at pressures corresponding to a wide depth range (5 to > 20 km). Other possible gradients could increase depth, but without affecting the total depth span. Depths would be reduced, on the other hand, if we chose to draw *PT* curves for some facies series with $dP/dT$ zero or even negative (Fig. 12-3).* High grade, then, is no sure criterion of great depth of synchronous plutonism (cf. A. Hall, 1971). More satisfactory in this respect as indices of relatively low pressure (depth less than 10 to 15 km and perhaps as low as 5 km) are such minerals as andalusite and cordierite. Higher pressures (depths perhaps exceeding 25 km) are indicated by mineral assemblages with kyanite and rather magnesian almandine.

GRANITIC MIGMATITES AND ANATEXIS In migmatites two contrasted component rocks are intimately mixed on the scale of an outcrop: one light-colored and granitic (aplite, pegmatite, granite, granodiorite), the other dark and metamorphic (for example, biotite schist, amphibolite). We have summarized elsewhere (Turner and Verhoogen, 1960, pp. 370–374, 386–388) the essential character of migmatites and their possible genetic implications regarding granitic magmas. To appreciate their structural complexity and variety—the key to origin and modes of emplacement of the granitic material—one must turn to classic accounts of Sederholm, Holmquist, Read, and others and to a new and profusely illustrated work by Mehnert (1968). Since the turn of the century, migmatites have loomed large in controversial discussion on the origin of granitic magmas and the mode of emplacement of granitic bodies on all scales in areas affected by synchronous deformation and high-grade regional metamorphism. It is in this environment—Buddington's "katazone"—that many well-studied migmatite complexes formed and that migmatite phenomena are most varied. Controversy has been heightened and issues clouded by semantic obstacles stemming from confused usage of widely used terms such as

* Since nothing is known about the attitude of the present erosion surface (on which isograds are mapped) with respect to the topographic surface at the time of metamorphism, we cannot say for certain in most cases that $dP/dT$ of a geologically mapped facies gradient is necessarily positive.

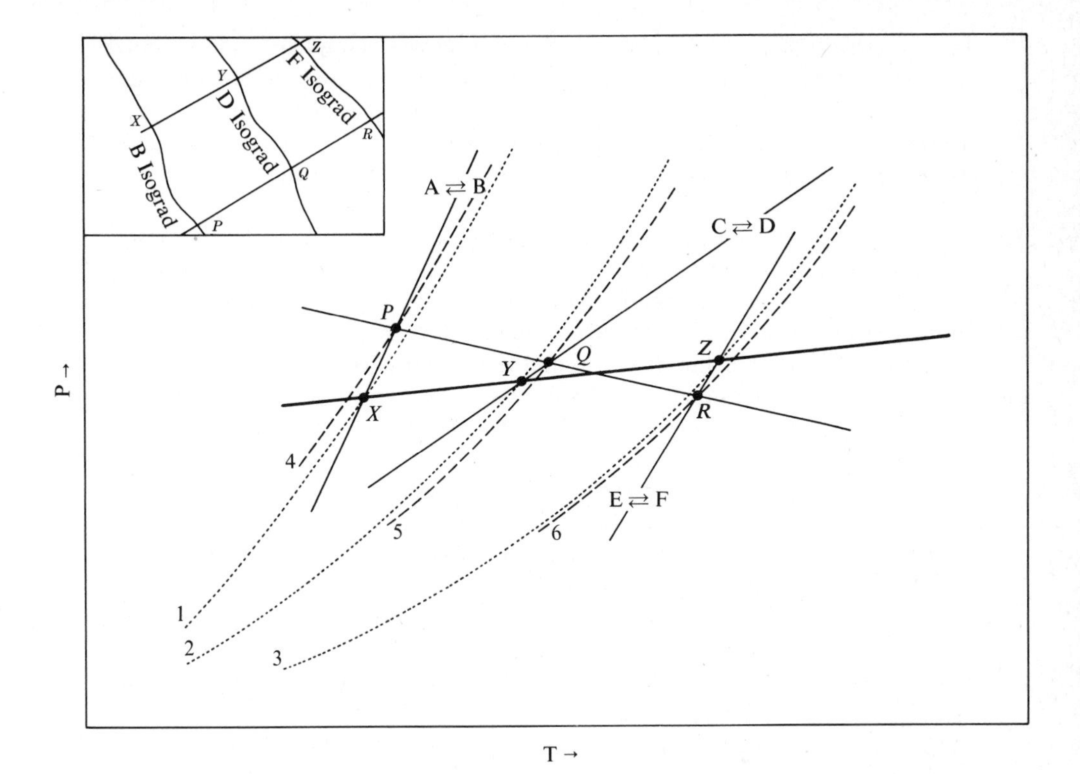


**Fig. 12-3** Possible relations between isograds (marked by incoming of index minerals *B*, *D*, *F* in inset map, upper left) and *P-T* of critical equilibria, $A \rightleftharpoons B$ and so on. *XYZ* and *PQR* are *P-T* gradients recorded in two separate traverses across the isograds. Numbered curves are vertical pressure-temperature gradients below *X*, *Y*, *Z* (dotted) and *P*, *Q*, *R* (dashed) at the time of metamorphism. Steeply inclined lines drawn through *X*, *Y*, *Z*, *P*, *Q*, *R* are segments of curves of univariant equilibrium for critical reactions $A \rightleftharpoons B$, and so on. The diagram is highly schematic, and *P* represents pressure or some pressure-sensitive variable such as $f_{H_2O}$.

*granitization*, *mobilization*, and *rheomorphism*. These have been applied to the evolution and relative movements of the granitic component of migmatites on all scales (see comments by Mehnert, 1959, pp. 155–157; 1968, pp. 119, 228–229, 329, 354–356). The confusion arises not from the obvious end states—an augen gneiss, a layered or veined migmatite—but from the various combinations of inferred processes postulated according to the preference of each observer to account for these states.

There is now fair agreement as to the kinds of physical processes that may be involved in bringing about the heterogeneous state of migmatites: partial or complete fusion of rock to yield anatectic "granitic" melts; relative displacement (flow) of melt with respect to associated solid material; diffusion of ions, especially $K^+$ and $Na^+$, facilitated by a pervasive aqueous intergranular film which itself may be in process of diffusion through heated rocks. Differences in current opinion largely concern relative emphasis placed on particular processes and genetic interpretation of inherited structures such as the ghost stratigraphy that pervades some large bodies of migmatite (cf. Buddington, 1959, p. 733).

There is growing emphasis on the role of fusion in migmatite evolution, a general belief that the granitic component of most migmatites is primarily anatectic. It may have formed by partial melting of rock virtually in place and by segregation of the melt from the unfused residue. Or the melt phase may have migrated far from the site of origin and then been injected into the host rock at the point of present exposure. Both possibilities were recognized by Scandinavian geologists—notably Sederholm and Holmquist—who originated and developed the migmatite concept.

If the granitic component of many syntectonic migmatites is anatectic, do we see here granitic and granodioritic magmas in actual process of formation in situ? Mehnert (1963; 1968, pp. 273–275) thinks that this is so in portions of a pre-Variscan metamorphic terrane of the Black Forest, southwestern Germany. Here he has mapped granodioritic massifs up to 10 km in diameter, almost homogeneous at the core and increasingly heterogeneous and streaked with dark nebulous schlieren toward the borders. The external boundaries of the granodiorite are not sharply defined. It grades outward into a migmatite zone which in turn fades into an enveloping quartz-feldspar-biotite gneiss (metagraywacke) with eyes and megacrysts of oligoclase and potassic-feldspar. Mehnert interprets the central granodiorite as almost completely fused metagraywacke and regards its migmatite envelope as the preliminary product of incipient melting in place. Feldspathic eyes and megacrysts in the outer gneiss are essentially metamorphic; but some degree of alkali metasomatism involving small-scale diffusion in an aqueous pore fluid is postulated. There may also have been some regional upward influx of alkalis, since some granodiorites are distinctly richer in Na and K than the parent metagraywacke gneiss which they otherwise chemically resemble.

Much of the structural variety of migmatites reflects flow of mobile products of anatexis with respect to the more rigid solid surrounding medium. Evidence of flow is seen on every scale: in intrusive contacts between anatectic granitic cores and the foliated metamorphic envelopes of granite-gneiss domes, and within aplitic and granitic vein systems of single outcrops. It is even thought that large bodies of migmatite, complete with internal relics of a ghost stratigraphy, may rise by virtue of their partially fused condition and invade the covering rocks as individual heterogeneous plutons. Finally, since the days of Sederholm and Holmquist it has been recognized that the magmatic component of some migmatites, whether anatectic or otherwise, may be of extraneous origin.

A remaining possible factor in migmatite evolution is alkali metasomatism operating through the medium of aqueous pore solutions. This has been inferred, as the immediate precursor of nearby anatexis, from appearance of megacrysts and eyes of potassium feldspar or oligoclase in rocks otherwise lacking these minerals (for example, Mehnert, 1968, pp. 273, 289). On a much larger scale the same sort of process has been held to play an essential role in development of extensive plutons of gneissic granite and of large bodies of migmatitic augen gneiss. Most writers today advocating this model nevertheless derive the necessary alkali from exposed or postulated granitic plutons (for example, Buddington, 1957, p. 302; 1959, pp. 724, 733).

UNIFIED DEPTH SEQUENCE OF EMPLACEMENT? It is tempting to bring together the varied phenomena of anatexis, migmatite evolution, and granite plutonism in some unified sequence of emplacement pattern in relation to depth. This is no new concept. It has been developed independently by Eskola, Wegmann, and Read in classic models presented elsewhere (Turner and Verhoogen, 1960, pp. 384–388), and the same theme underlies Buddington's (1959) survey of emplacement of North American granites. How valid are generalized models of this kind?

The granitic magma of most plutons—with the exception of some migmatite complexes—was not formed at the site of present exposure. It has become frozen to a standstill in the course of upward migration, aided by gravity, from some deeper crustal level. The exposed vertical depth of any major pluton is but a fraction of its lateral dimensions or of the total depth span of any theoretical emplacement sequence. This means that to complete the general vertical picture it is necessary to extrapolate between plutons that differ not only with respect to pattern of emplacement but in geographic situation and age as well. And in the universal depth model there is the unwarranted built-in assumption of vertical continuity and of uniformity in time. Buddington's (1959) survey of American granite plutons revealed at least one unexplained hiatus in the postulated vertical sequence. He could find no smooth upward transition between the respective emplacement patterns of his "katazone" and "mesozone." No doubt this stems in part from differences between thermal regimes of plutonism in the two environments, the one syntectonic and the other posttectonic. But there may be true depth implications as well.

This possibility is beginning to emerge from kinematic models of magma ascent that take into account simple hydrodynamic parameters and the results of scale-model experiments. Discussion on such lines is an intriguing aspect of a Liverpool symposium on mechanics of igneous intrusion (Newall and Rast, 1970). A contribution by Fyfe (1970*a*, pp. 202–204, 212–214) revived and amplified a scale model proposed originally by Grout (1945). The ascending pluton is pictured as a gigantic bubble of magma—a teardrop with tail inverted—impelled upward by gravity through a cooler, denser, more rigid crustal medium. The pattern of ascent, which leaves its structural imprint on both the pluton and its envelope, will reflect the varying contrast in rheologic behavior between the magma and its solid surroundings (H. Ramberg, 1970, p. 285). Fyfe (1970*a*, p. 214) speculates that a rapidly ascending mass of hot magma might leave in its wake a trail of local thermal perturbation conducive to development of migmatites. He wonders "if some migmatite zones might represent a region through which a drop [evolving pluton] passed rather than a region where melting started." Here would be one explanation of the hiatus in Buddington's unified depth model. All this is still in the realm of speculation which can be tested only in the light of individual case histories compiled for specific plutonic provinces.

## MESOZOIC BATHOLITHS OF THE NORTH AMERICAN CORDILLERA

GEOLOGIC SETTING The western border of North America has been tectonically active for the last 600 m.y. Several longitudinal geosynclines have developed in succession, the history of each terminating in orogeny. From late in the Triassic to Miocene times, great volumes of granitic magmas have intermittently invaded the upper crust. Today the products are exposed in a series of great, compound batholiths strung out along or close to the coast from Alaska to the Mexican peninsula of Baja California (Fig. 12-4). The largest exceed 1500 km in length and are nearly 100 km wide. And there are many lesser scattered bodies, some themselves large enough to rank as batholiths. The major part of this intrusive activity was concentrated in part of mid-Cretaceous time (100 to 80 m.y.). There were earlier plutonic episodes—some in southern British Columbia and the Sierra Nevada dating back to late Triassic. And there are localized minor centers of late Cretaceous and Eocene intrusion (70 to 50 m.y.)

For the most part the Cordilleran batholiths are located along the site of a longitudinal geosyncline filled with late Paleozoic and Mesozoic sediments and volcanics that were folded and lightly metamorphosed late in the Jurassic. This was the culminating deformation phase of the Nevadan orogeny. The main surges of granitic magmas, it will be noted, came later (cf. Gilluly, 1963, pp. 164–165). And the batholiths have imprinted upon their previously metamorphosed folded envelope contact aureoles whose mineralogy (especially wide prevalence of andalusite, cordierite, wollastonite, and forsterite) suggests emplacement at depths no greater than 10 km and perhaps as little as 6 km (cf. Turner, 1968, pp. 200–209, 218, 258).

In the account that follows we refer especially to the central section of the Sierra Nevada batholith of California (Fig. 12-4). Data have been drawn mainly from Compton (1955), Bateman (1961), Bateman et al. (1963), and Bateman and Eaton (1967), supplemented by radiometric dates from G. H. Curtis et al. (1958), Kistler et al. (1965), Hurley et al. (1965), and McKee and Nash (1967). Reference is also made to the southern California and Idaho batholiths (Larsen, 1948; Larsen and Schmidt, 1958), to the Boulder batholith, Montana (Knopf,

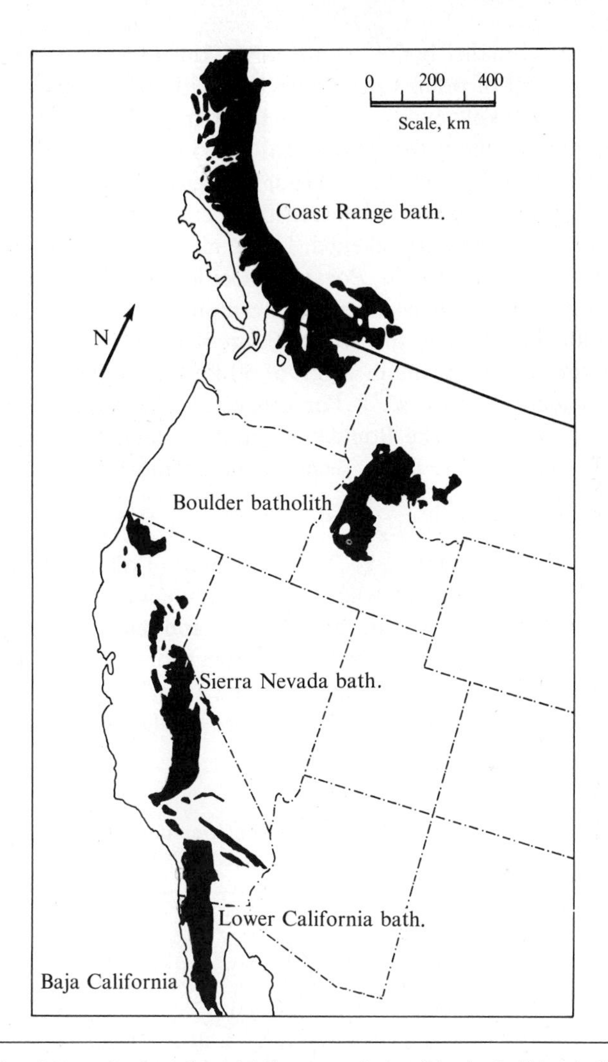


**Fig. 12-4** Major Mesozoic (mostly mid-Cretaceous) granitic batholiths (solid black) of western North America.

1964; Tilling et al., 1968; Tilling and Gottfried, 1969; Klepper et al., 1971), and to dating of granite plutonism in southern British Columbia (Baadsgaard et al., 1961).

GROSS STRUCTURE AND EMPLACEMENT The Sierra Nevada batholith, continuously exposed for 500 km exclusive of outlying masses to the north, trends parallel to the continental margin and to the regional structure of enclosing rocks folded in the Nevadan orogeny. It consists of perhaps 200 individual plutons. Some are several hundred square kilometers in extent; and in between are packed many smaller bodies, some only 1 $km^2$ in area. A similar composite character is typical of other Cordilleran batholiths: there are perhaps a dozen plutons in the 6000 $km^2$ exposure of the Boulder batholith. There is ample and varied evidence to show that individual plutons were emplaced by forceful upward intrusion of light magmas through a denser solid crust.

Seismic profiles show that beneath the Sierra Nevada batholith the continental crust is very greatly thickened. In the model presented by Bateman and Eaton, downward increase in *P*-wave velocities from 6.0 to 6.4 km $s^{-1}$ is correlated with transition between quartz monzonite and granodiorite above and quartz diorite below in the upper 20 km of the vertical section. The lower part of the root (to a maximum depth of 50 km) is pictured as amphibolite or gabbro (*P* velocity 6.9 km $s^{-1}$). For other batholiths there is no convincing evidence of deep roots. The Boulder batholith, which is capped by a roof of only slightly older dacitic lavas, seems to be no more than 15 km in vertical depth and may be a good deal less.

PETROGRAPHY AND CHEMISTRY The great bulk of all the Cordilleran batholiths consists of granitic rocks whose principal mineral constituents are quartz, plagioclase, potassium feldspar, hornblende, and biotite. Current American terminology of rock types is based on relative proportions of plagioclase and potassium feldspar, either modal or normative (Fig. 12-5). Boundaries between rock types have been drawn arbitrarily across what is in fact a completely

**Fig. 12-5** Plot of 40 average modal compositions of granitic rocks, Sierra Nevada batholith, representing 597 analyses. (After Bateman et al., 1963.)

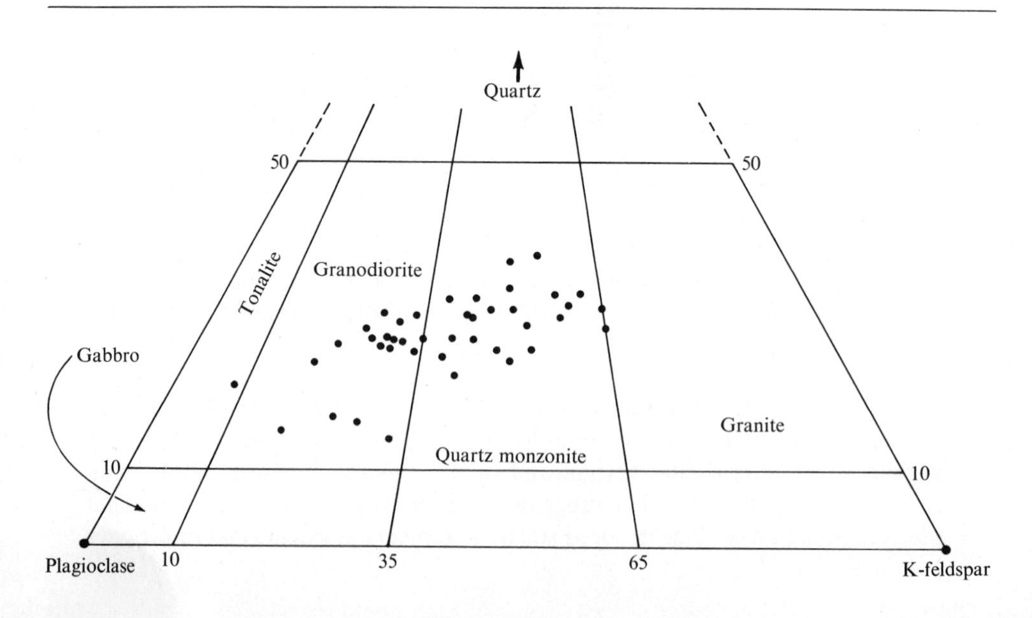

gradational series of rocks. In general, decrease in plagioclase is accompanied by decrease in its content of anorthite ($An_{50}$ in quartz diorite to $An_{20}$ in quartz monzonite), increase in potassium feldspar, decrease in mafic minerals, and increase in the biotite/hornblende ratio.

For the Cordilleran province as a whole, the most abundant rocks by far are granodiorites and quartz monzonites (cf. Bateman et al., 1963, p. 36, fig. 19). Quartz-diorite plutons are important components of some batholiths—southern California and the western portion of the Sierra Nevada. There is one extensive body of hornblende gabbro, 14 percent of the exposed total area, in the southern California batholith. Otherwise, mafic rocks are confined to early small plutons or remnants of these. Granites in the strict sense, as defined above, are insignificant albite-bearing alaskites. Where relative age of contiguous plutons can be established from field observations, the earlier pluton is usually more mafic (basic) than the later of the two

The chemical character of the commonest rocks is illustrated in Table 12-1. Mostly, $SiO_2$ is between 60 and 74 percent by weight. In the norm $ab > or$ except in some of the most siliceous rocks. Considered as a whole, the southern California batholith has an $SiO_2$ content of about 65 percent; that of the Sierra Nevada batholith would be higher (perhaps 68 or 69 percent).

There is a good deal of information about concentrations of radioactive elements and isotopic compositions of Sr and Pb, all of which presumably have some bearing on genesis of granitic magmas and potentialities of granitic plutons as heat sources (Table 12-2). Average contents of uranium (Larsen and Gottfried, 1961) vary between 1.7 ppm (southern California) and 2.7 ppm (Sierra Nevada). Strontium is consistently more radiogenic than that of oceanic basalts, decidedly less so than many old sialic crustal rocks: initial $Sr^{87}/Sr^{86}$, 0.7055 to 0.7092 in the Boulder batholith, 0.7033 to 0.7083 in the Sierra Nevada, mostly 0.7061 to 0.7081 in souther British Columbia. In each composite batholith, the composition of strontium varies from pluton to pluton. So too, but much more markedly, does the composition of lead in the Boulder batholith. Compared with oceanic basalts, lead is notably nonradiogenic (Table 12-2): $Pb^{206}/Pb^{204}$ is as low as $16.9 \pm 0.1$ in some granodiorites and is consistently higher ($18.0 \pm 0.2$) in the Butte quartz-monzonite pluton that comprises 75 percent of the exposed batholith. High radiogenic lead tends to correlate with high radiogenic strontium. Meager available data (Doe, 1967, p. 53) for plutons nearer the coast give significantly higher radiogenic lead ($Pb^{206}/Pb^{204}$, 18.95 to 19.51) comparable with that of oceanic basalts.[1]

EAST-WEST GRADIENTS, SIERRA NEVADA BATHOLITH Throughout the whole Cordilleran province from Alaska to Mexico $SiO_2$, $K_2O/SiO_2$, and $K_2O/Na_2O$ in granitic rocks increase with distance from the coast (for example, Moore

[1] Doe and Delevaux, *Bull. Geol. Soc. Am.*, vol. 84, p. 3513, 1973, give values 18.73 to 19.37 for Sierra Nevada rocks.

et al., 1961). Quartz diorites and mafic granodiorites predominate in the west, granodiorites and quartz monzonites in the east. Along a 70-km west-southwest east-northeast section across the central Sierra Nevada, the mean rock density decreases eastward from 2.72 to 2.64 as the lithology changes (Bateman et al., 1963, p. 32).

**Table 12-1** Chemical compositions (oxides, wt %), CIPW norms, and trace-element abundances (ppm) of typical rocks, Mesozoic batholiths, California

| | 1 | 2 | 3 | 4 | 5 |
|---|---|---|---|---|---|
| $SiO_2$ | 50.78 | 62.2 | 66.92 | 71.42 | 75.4 |
| $TiO_2$ | 0.77 | 0.7 | 0.47 | 0.36 | 0.1 |
| $Al_2O_3$ | 20.40 | 16.6 | 15.19 | 14.03 | 13.3 |
| $Fe_2O_3$ | 1.75 | 1.4 | 1.45 | 0.89 | 0.3 |
| FeO | 6.20 | 4.5 | 2.52 | 1.63 | 0.74 |
| MnO | 0.09 | 0.06 | 0.08 | 0.05 | 0.08 |
| MgO | 6.49 | 2.7 | 1.74 | 0.70 | 0.12 |
| CaO | 10.24 | 5.7 | 3.79 | 1.91 | 0.48 |
| $Na_2O$ | 2.20 | 3.4 | 3.16 | 2.86 | 4.1 |
| $K_2O$ | 0.45 | 1.6 | 3.82 | 5.35 | 4.5 |
| $P_2O_5$ | 0.05 | 0.09 | 0.18 | 0.09 | 0.01 |
| $H_2O^+$ | | | 0.06 | 0.08 | |
| $H_2O^-$ | 0.65 | 0.6 | 0.48 | 0.35 | 0.46 |
| Total | 100.07 | 99.55 | 99.86 | 99.72 | 99.59 |
| *Q* | 3.18 | 12.90 | 22.62 | 28.38 | 32.58 |
| *or* | 2.22 | 8.90 | 22.76 | 31.68 | 26.69 |
| *ab* | 18.34 | 28.82 | 26.72 | 24.11 | 34.58 |
| *an* | 41.98 | 28.91 | 15.84 | 9.46 | 2.50 |
| *di* | 7.20 | 2.07 | 1.83 | | |
| *hy* | 21.42 | 13.57 | 6.18 | 3.55 | 1.36 |
| *mt* | 2.55 | 2.09 | 2.09 | 1.39 | 0.46 |
| *il* | 1.52 | 1.67 | 0.90 | 0.65 | |
| *ap* | 0.33 | 0.34 | 0.35 | | |
| *c* | | | | 0.10 | 0.71 |
| Total | 98.74 | 99.27 | 99.29 | 99.32 | 98.88 |

| | 3 | 4 |
|---|---|---|
| Ba | 1000 | 1000 |
| Co | 10 | 5 |
| Cr | 10 | 6 |
| Ni | 5 | 3 |
| Pb | 20 | 20 |
| Sr | 700 | 300 |
| V | 60 | 40 |
| Zr | 100 | 500 |

*Explanation of column headings*

1 Average composition of San Marcos gabbro, southern California batholith (Larsen, 1948, p. 50, analysis B)

2 Average composition of Bonsall quartz diorite, southern California batholith (Larsen, 1948, p. 67, analysis G)

3 Lamarck granodiorite, east central Sierra Nevada (Bateman et al., 1963, p. 29, analysis 12)

4 Tungsten Hills quartz monzonite, east central Sierra Nevada (Bateman et al., 1963, p. 29, analysis 20)

5 Quartz monzonite, Cathedral Peaks type, east central Sierra Nevada (Bateman et al., 1963, p. 29, analysis 25)

There is also a recognizable west-to-east pattern of emplacement date, and this seems to be independent of rock type. The axial plutons that make up the bulk of the batholith were emplaced in Middle Cretaceous time (90 to 80 m.y.). Quartz diorites and granodiorites of western plutons in the foothills of the northern Sierra Nevada yield dates (140 to 130 m.y.) close to the Jurassic-Cretaceous boundary. In the central sector of the batholith, plutons on both the western and eastern (Inyo Mountains) flanks were emplaced still earlier—in the first half of the Jurassic (180 to 150 m.y.) or even in a few cases mid-Triassic (~210 m.y.).[1]

So in the Sierra Nevada batholith, and probably through the Cordilleran province as a whole, regional structure, gross geometry of plutonic bodies, chemical and lithological composition, and date of emplacement share a common symmetry related to the continental border. It is reassuring, perhaps, to

**Table 12-2** Atomic abundances (ppm), abundance ratios, and isotopic abundances of radioactive and related elements, Mesozoic granitic batholiths, western United States

| | 1 | 2 | 3 | 4 | 5 | 6 | 7 | 8 |
|---|---|---|---|---|---|---|---|---|
| $K \times 10^{-3}$ | 28 | 25 | 34 | 45 | 3.1 | 14.9 | 28.4 | |
| Rb | 83 | 64 | 129 | 218 | | | | |
| Sr | 431 | 610 | 452 | 108 | | | | |
| Th | 11 | 7.3 | 16.2 | 36.3 | | | | |
| U | 3.4 | 1.5 | 4 | 9.2 | 0.22 | 2 | 1.5 | |
| K/Rb | 340 | 390 | 260 | 205 | | | | |
| Rb/Sr | 0.191 | 0.15 | 0.28 | 2.01 | | | | |
| Th/U* | 4 | 4.8 | 4 | 4.9 | | | | |
| $U/K \times 10^4$ | 1.2 | 0.6 | 1.2 | 2.0 | 0.7 | 0.3 | 0.6 | |
| $Th/K \times 10^4$ | 4 | 2.9 | 4.8 | 8 | | | | |
| $Sr^{87}/Sr^{86}$ (initial) | 0.7078 | 0.7063 | 0.7073 | 0.7081 | | | | |
| $Pb^{206}/Pb^{204}$ | 17.98 | 16.94 | 17.96 | 18.14 | | | | 18.95 |
| $Pb^{208}/Pb^{204}$ | 38.23 | 37.68 | 38.23 | 38.41 | | | | 38.52 |

* Computed from individual Th/U ratios.

*Explanation of column headings*

1–4 Boulder batholith (Doe et al., 1968; Tilling and Gottfried, 1969); lead values in determined potassium feldspars
- 1 Mafic granodiorite (Unionville pluton)
- 2 Felsic granodiorite (Rader Creek pluton)
- 3 Butte quartz monzonite (75% of exposed batholith)
- 4 Alaskite

5–7 Southern California batholith (Larsen and Gottfried, 1961, p. 73)
- 5 Hornblende gabbro, San Marcos pluton; $SiO_2$ = 48.16
- 6 Bonsall quartz diorite; $SiO_2$ = 62.28
- 7 Woodson Mountain granodiorite; $SiO_2$ = 72.55

8 Southern California batholith; leucogranite (cited by Doe, 1967, p. 53)

[1] R. W. Kistler and Z. E. Peterman, *Bull. Geol. Soc. Am.*, vol. 84, p. 3492, 1973.

note that geologists had established the essential features of this regional pattern before the present era of plate tectonics.

PETROGENESIS The enormous volume of exposed granitic rocks and contrasted insignificant quantities of diorite and gabbro, reinforced by the isostatic implications of elevated granite mountains and by deep roots of relatively light rocks inferred from seismic data, rule out all possibility that the Cordilleran granites are differentiates of basaltic magmas. This is the view of most geologists who have recently considered the origin of granitic batholiths in this province (for example, Moore, 1959; Gilluly, 1963, pp. 166–167; Bateman and Eaton, 1967). We may further inquire whether it is likely that the chemically varied rocks constituting a great composite batholith should all have been derived from a single parent magma of any kind. The gross chemical and isotopic individuality of each component pluton, and the long interval of time covered by emplacement of the batholith, imply that this cannot have been so. Rather they suggest repeated upsurges of granitic magmas of different compositions generated independently at different loci and probably at different depths. This is the model that we tentatively accept.

What, then, is the significance of chemical variation diagrams that have been constructed for the southern California and Sierra Nevada batholiths (for example, Larsen, 1948, pp. 141–148; Bateman et al., 1963, p. 30)? These follow a common pattern which is also seen in diagrams for chemically comparable andesite-dacite-rhyolite "series." Clearly, if we accept the thesis proposed above, the pattern of conventional variation diagrams is no proof that associated rock types are related through differentiation.

Differentiation, nevertheless, played some part in the scheme of petrogenesis. Many individual plutons are heterogeneous; a relatively mafic border zone passes inward into a more felsic core (for example, Compton, 1955; Bateman et al., 1963, pp. 30–32). This radial variation, which in one instance has been shown to involve inward decrease in Be, Cr, V, and Ni (Putnam and Alfors, 1969), is generally attributed to fractional crystallization—inward concentration of late liquid fractions. In some plutons, there has also been marginal basification by reaction between relatively felsic magma and more mafic earlier plutons or metavolcanic rocks of the envelope.

Purely geological reasoning, based on gross composition and distribution of granitic plutons, strongly suggests that the principal component of the source material of granitic magmas is sialic crust and that this is regionally heterogeneous. Its mean composition could be chemically similar to andesite (Presnall and Bateman, 1973). Isotopic evidence complicates the picture. It could conceivably imply participation of a more basic magma drawn from the deep crust or the mantle (cf. Doe et al., 1968, pp. 904–905). It also sets limits on the isotopic composition of the sialic sources. Compared with many ancient siliceous rocks, this must be relatively low in radiogenic strontium and, at least locally (for example, beneath the Boulder batholith), in radiogenic lead as well. Deep sialic crust, depleted in U, Th, and Rb during granulite-facies metamorphism at some distant date, would seem to fill the bill (Doe et al., 1968, pp. 884, 904). Another suggested source is the graywacke-volcanic filling of eugeosynclines (Hurley et al., 1965, p. 165; Peterman et al., 1967) in much the same way as proposed and documented for rhyolite magmas of New Zealand (Ewart and Stipp, 1968). Such rocks, with initial values of $Sr^{87}/Sr^{86}$ in the range 0.704 to 0.708, could yield strontium of the required composition if melting occurred 100 m.y. or so after sedimentation. But the few available lead measurements cited by Doe (1970, p. 96) yield values of $Pb^{206}/Pb^{204}$ (19.15 to 19.85) much too high to account for the low radiogenic lead of the Boulder batholith.

To conclude on a note of uncertainty, there is still another complicating factor in the field of isotopic geochemistry that could invalidate all the preceding inferences. There is growing evidence that oxygen of many plutons, especially in their border regions, becomes strongly contaminated with heavy oxygen introduced by inward-diffusing meteoric water from enclosing metasediments (for example, Turi and Taylor, 1971). Lead of undergound brines and of natural water in general is radiogenic: $Pb^{206}/Pb^{204}$ is commonly $> 19$ (Doe, 1970, pp. 76, 77). The same is true of strontium in seawater [$Sr^{87}/Sr^{86} \approx 0.709$; Faure and Powell (1972)]. If oxygen, which makes up 90 percent of total rock volume, becomes so strongly affected, is it possible that comparable changes will be wrought also in isotopic compositions of other elements, notably lead and strontium? In the present state of knowledge it would be unsafe to draw positive genetic conclusions from isotopic data available for granitic batholiths of the Cordillera.

## CALEDONIAN GRANITIC PROVINCE, SCOTTISH HIGHLANDS AND NORTHERN IRELAND

NATURE OF THE CALEDONIAN "GRANITES" Numerous and extensive granitic plutons were emplaced across the Scottish Highlands and in northern Ireland (Fig. 12-6) in the course of the Caledonian orogeny during the first half of the Paleozoic era. For nearly 70 years it has been recognized on the basis of structural evidence that there were several episodes of granite plutonism in the Highlands.[1] There are Older Granites, some preceding and some more or less synchronous with the climax of Caledonian deformation and regional metamorphism. Judging from associated metamorphic mineral assemblages these may have been emplaced at depths as great as 30 km (cf. Turner, 1968, p. 363, fig. 8-5). And there are Newer Granite plutons that crosscut the regional structure and have imprinted low-pressure thermal aureoles on enclosing metamorphic

[1] For example, see Read (1949, pp. 132, 133) for a summary of Barrow's views in the first decade of the century.

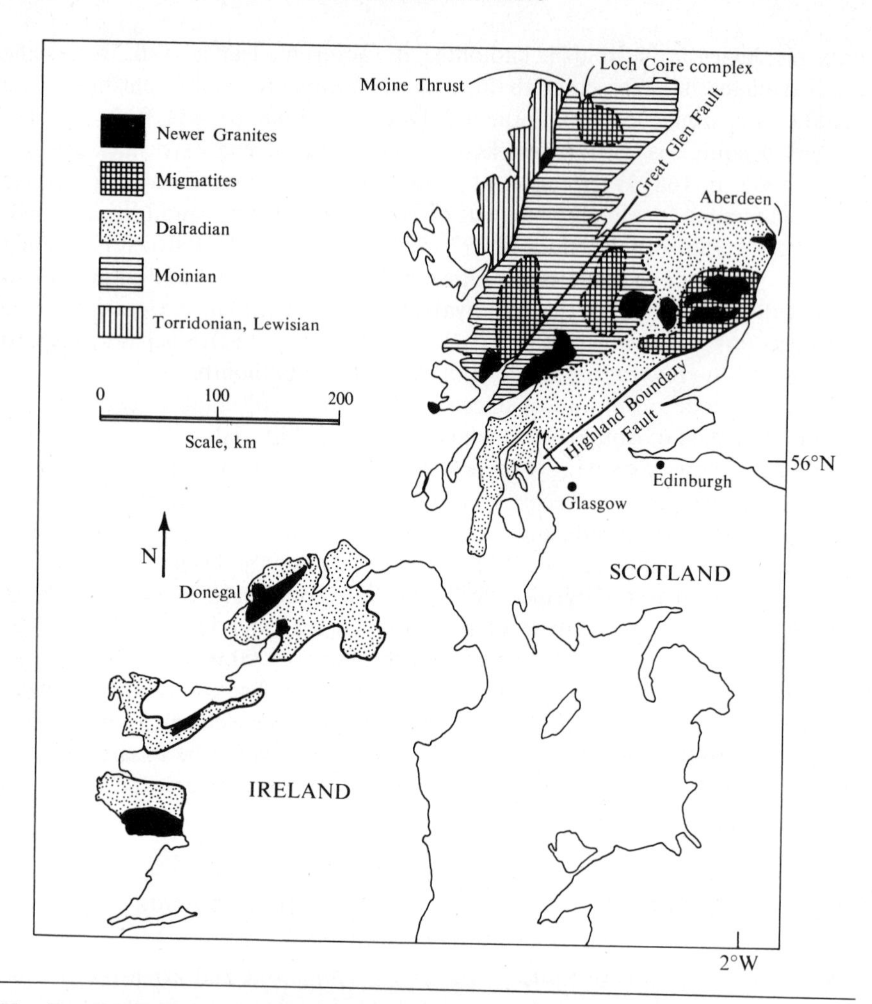


**Fig. 12-6** Main centers of Caledonian granitic plutonism in relation to metamorphic basement, northern Scotland and Ireland. Older Granites are concentrated in migmatite areas.

rocks deformed long previously at much deeper levels. Some are high-level ring complexes with infaulted blocks of cogeneric dacites (for example, Haslam, 1968).

Chemically and mineralogically, most of the "granites," regardless of relative age and depth of emplacement, are granodiorites and quartz monzonites. In some complexes there are quartz diorites and subordinate diorites, and in a few of the Newer Granite plutons there are minor quantities of even more basic rocks. Then there are also closely associated diatremes, insignificant in total bulk, filled with brecciated basic and ultramafic rocks rich in hornblende, augite, or biotite. These are the appinites, whose mode of emplacement is reminiscent of African kimberlites. Their presence seems to suggest that, at all stages of granite plutonism, small quantities of much more basic alkaline magma would burst upward explosively from some deep source.

A characteristic of many Older Granite plutons is extensive peripheral development of regional migmatite. Some of the principal migmatite areas are 20 to 40 km in diameter. One such is the Loch Coire complex of northernmost Scotland, well known through the early accounts of Read and more recently investigated chemically by P. E. Brown (1967). The granitic pluton at the center of the complex is a sodic granodiorite (Na/K $\approx$ 3). Sheets and veins of felsic rock in the surrounding migmatite are even more sodic; one, a pink granitic sheet (plagioclase-quartz-biotite) 30 m thick, has Na/K slightly less than 5. Pelitic and semipelitic components of the migmatite are much more potassic (Na/K mostly 0.3 to 1). Read proposed that the "granitic" element in the migmatite complex is metasomatic, the product of sodic solutions emanating from the central body of granite magma. There is also the possibility that it may be anatectic.

Structural aspects of successive intrusion of magmas at one broad site are beautifully displayed in the Donegal granite complex of northwestern Ireland (Pitcher, 1953, 1970; Pitcher and Read, 1959). This is a composite mass, 50 $\times$ 20 km, consisting of four separate plutons (Fig. 12-7). The oldest (Thorr intrusion) is a tonalite ($SiO_2$ = 58 to 62 percent) crowded with rafts and xenoliths of country rock whose overall pattern of arrangement preserves a ghost stratigraphy that can be traced continuously into the envelope. Here, as shown by Pitcher, a migmatite has developed by magmatic intrusion; it is the roof region of the pluton that is now exposed. Later plutons are two circular ring complexes, each less than 10 km in diameter, one of granodiorite and the other of quartz monzonite ($SiO_2$ = 72 percent). Last to be emplaced, and the most extensive, is the main Donegal granite, also of granodiorite composition. This body, too, preserves an almost perfect ghost stratigraphy due to wedging apart of roof septa of country rock during sheet injection of the rising magma. On purely structural criteria the Donegal granites—at least the younger plutons —would fall in the Newer Granite category.

Extreme lithologic diversification among Newer Granites is displayed by the Garabal Hill complex, some 20 $km^2$ in area, near Loch Lomond (Nockolds, 1940). The great bulk of the pluton, nevertheless, is granodiorite ($SiO_2$ = 62 to 68 percent). There are small masses ($< 1$ $km^2$) of hornblende gabbro and augite peridotite, and even less extensive outcrops of appinite and mica-pyroxene diorite. Emplacement of these basic rocks preceded intrusion of the main granodiorites. There is little to sustain a model of fractional crystallization to connect the rock series as a whole. And once again the relative bulk of granodioritic rocks is consistent with a primary origin for magma of corresponding composition.

PLUTONISM IN RELATION TO OROGENY The main interest of the Caledonian province of northern Britain in the context of this chapter is that it displays in

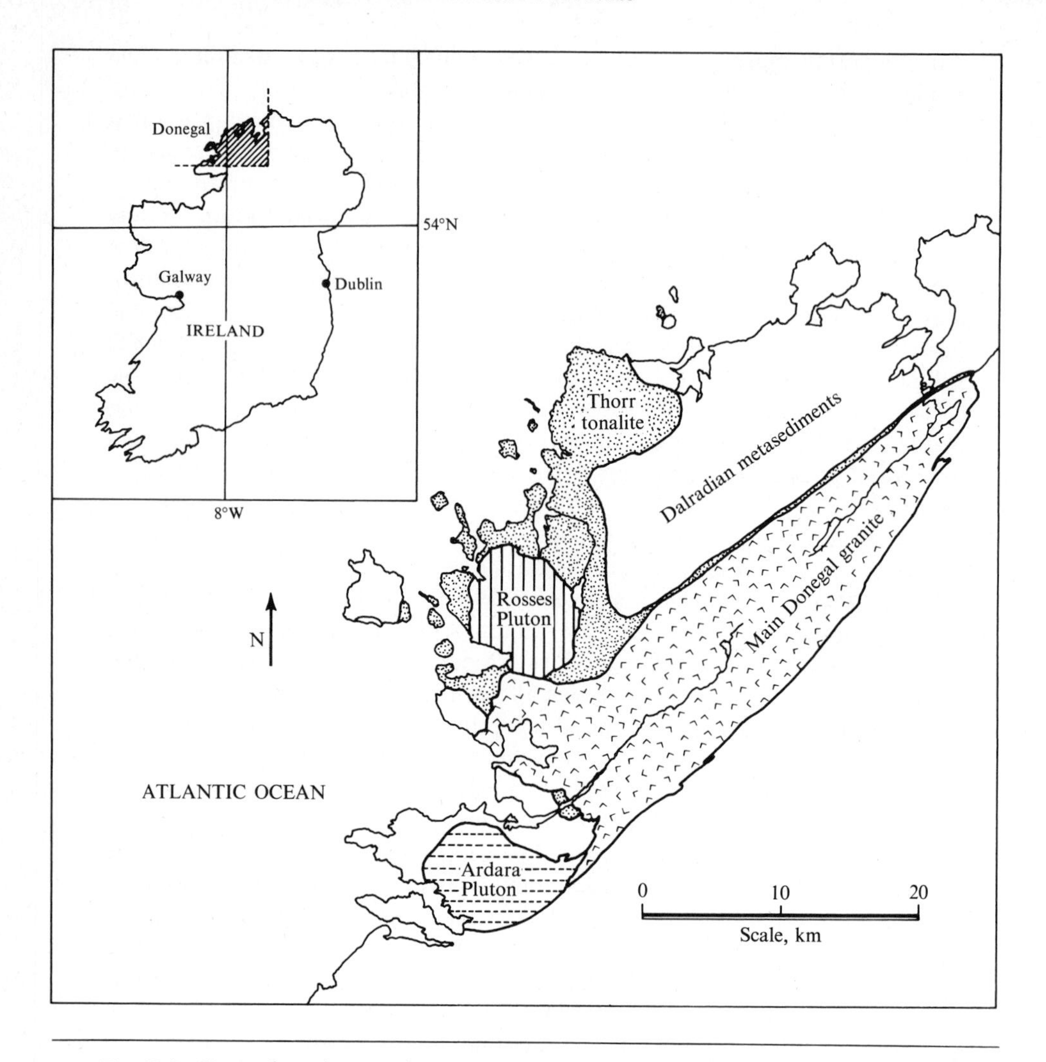


**Fig. 12-7** Donegal granites, northwestern Ireland.

detail how episodic generation and intrusion of granitic magmas at different depths may be related to a complex sequence of tectonic events. These constitute the Caledonian orogeny and span 150 m.y. between later Cambrian and mid-Devonian times (cf. McKerrow, 1962). Their collective imprint is seen in the Moinian and Dalradian series that make up the metamorphic basement of all northern Scotland southeast of the Moine thrust (Fig. 12-6). These metamorphic rocks—excluding perhaps the lower Moinian—were marine sediments, with some associated basic volcanics, that filled a geosynclinal trough during late Precambrian and Cambrian times. This is the Scottish segment of the Caledonian geosyncline. Its southern extension is seen in the Dalradian metamorphic terrane that stretches across the north of Ireland.

We have presented elsewhere (Verhoogen et al., 1970, pp. 559–566) a tentative picture to illustrate the complex interplay between tectonic, metamorphic, and plutonic events in this orogeny. A more elegant and sophisticated synthesis drawing on more recent data (especially Bell, 1968) is that of Dewey and Pankhurst (1970), the essence of which is as follows.

Toward the close of the Cambrian a northeast-southwest geosyncline, floored along its northwestern border by a continental metamorphic (Lewisian) basement and further southeast by oceanic crust, had become filled with a thick mass of sediments (Moinian and Dalradian). Along the southeastern margin of this prism an oceanic trench began to form very early in the Ordovician. Then followed in quick succession four separate episodes of folding ($F_1$ to $F_4$ or $D_1$ to $D_4$), the first of which (gravity-propelled nappe formation) achieved the greatest bulk strain. Regional metamorphism reached its culmination (maximum temperatures) between the $D_3$ and $D_4$ events. The total $D_1$ to $D_4$ sequence, together covering perhaps 30 m.y., constitutes the "climactic episode" of orogeny, dated by Dewey and Pankhurst at 510 to 480 m.y. It may have been somewhat earlier in Ireland than in Scotland. While the climactic episode was still in progress, marginal uplift had started along great border fractures, the Moine thrust on the northwest, the Highland Boundary fault to the southeast. Slow uplift of the whole deformed prism continued from mid-Ordovician to mid-Devonian times—nearly 100 m.y. in all. All this history has been pieced together from stratigraphic and structural observations supplemented by radiometric dating of metamorphic rocks and minerals. Rb-Sr dates give the time of metamorphic or magmatic crystallization. K-A dates show a wider spread and may be significantly later than those given by Rb-Sr because they record the time at which a given rock cooled below the critical level for argon retention—perhaps 200°C for biotite, somewhat higher for muscovite. Cooling at any point is thought to have been controlled by reduction of cover thickness by erosion; and this in turn must have been conditioned by progressive elevation of the landmass following the climactic episode. Contours (thermochrons) drawn through apparent K-A dates of the same value reveal the pattern and deviation of uplift—postorogenic isostatic recovery (Fig. 12-8).

While Dalradian sedimentation was still in progress, basic magma erupted as submarine lavas along the southeastern trench. Later, during the climactic episode, it rose again in the northeast to give the great layered gabbro sheet, later disrupted by $D_3$ folding, of Aberdeenshire. But with this exception magmas generated during the Caledonian orogeny were mainly granitic; and their products are the Older and Newer Granites of classic literature.

CHRONOLOGY AND SOURCE MATERIALS OF GRANITE PLUTONISM Dates of granite emplacement based on Rb-Sr measurements[1] group into at least three episodes:

[1] Data of Bell (1968) revised to an $Rb^{87}$ half-life of $4.85 \times 10^{10}$ m.y. ($\lambda = 1.43 \times 10^{-11}$ years).

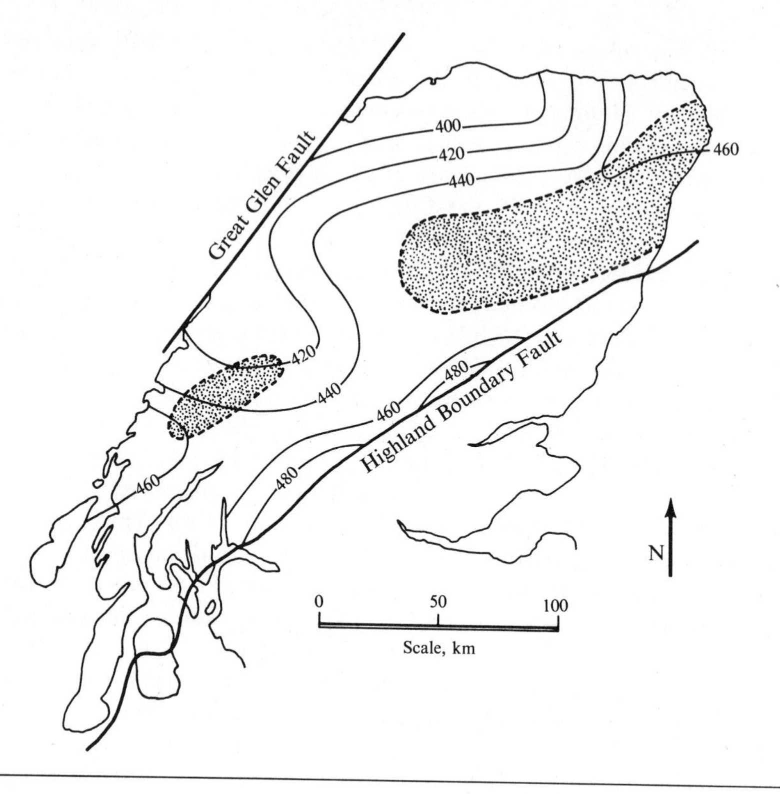


**Fig. 12-8** Thermochrons drawn on K-A dates (in m.y.) for muscovite for Caledonian metamorphism and plutonism, Scottish Highlands, south of Great Glen fault. Newer Granites are concentrated in stippled areas. (After Dewey and Pankhurst, 1970.)

1. Late Cambrian (550 m.y.): premetamorphic granites, some of them typically foliated Older Granites bearing the imprint of later deformation during the climactic episode. High initial $Sr^{87}/Sr^{86}$ (0.715 to 0.719) indicates a subjacent sialic source.

2. Ordovician (500 to 440 m.y.): including some foliated Older Granites of Barrow's sillimanite zone. Their exceptionally high initial $Sr^{87}/Sr^{86}$ values (0.728 to 0.733) match those of associated Dalradian metasediments; this is consistent with origin of the magma by melting of Dalradian rocks virtually in place. On the other hand, the Donegal granites—structurally placed in the Newer Granite clan and dated at about 485 to 500 m.y.† (Leggo et al., 1969; Leggo and Pidgeon, 1970)—have strontium (initial $Sr^{87}/Sr^{86} = 0.708$) only slightly more radiogenic than that of oceanic basalts. Moreover there

† Leggo's alternative dates 470 m.y. (based on $\lambda$ for $Rb^{87} = 1.47 \times 10^{-11}$ years) and 498 m.y. ($\lambda = 1.39 \times 10^{-11}$ years) recalculated to the constant used by Dewey and Pankhurst ($\lambda = 1.43 \times 10^{-11}$ years).

may have been some relative increment in $Sr^{87}$ by assimilation of Dalradian rocks; the main Donegal granite, it will be remembered, is locally crowded with xenoliths of this kind.

3. Silurian to early Devonian (420 to 400 m.y.): the greatest upsurge of granitic magmas came when postmetamorphic elevation and erosion were well advanced.[1] The Ross of Mull granite is at least 420 m.y. old (Beckinsdale and Obradovich, 1973). The plutons are Newer Granites whose mutually concordant K-A and Rb-Sr ages are markedly younger than those of K-A thermochrons based on micas of surrounding metasediments. This could scarcely be the case, nor could the whole chronological-structural picture be so complete and consistent, if strontium in the granites were contaminated by isotopic exchange with circulating waters. Initial $Sr^{87}/Sr^{86}$ values (0.707 to 0.717) of Scottish Newer Granites suggest derivation from, or at least the contaminative influence of, a subjacent sialic source. The Galway granite (about 400 m.y.) of western Ireland, like the much older Donegal granites, has low radiogenic strontium (initial $Sr^{87}/Sr^{86} = 0.704$). Although no such rock is locally exposed, it would seem reasonable to postulate some rubidium-depleted sialic source comparable to the Lewisian basement of northwestern Scotland.

## SOME ASPECTS OF PRECAMBRIAN GRANITE PLUTONISM

Some years ago Knopf (1955, p. 685) wrote that "the oldest bathyliths of early Precambrian (Archaean) age are especially well shown in the Shield areas. . . . Areally they constitute half to three-quarters or more of the exposed Precambrian terranes of these regions. So far they have invariably been found to be intrusive into the Precambrian supracrustal rocks. Consequently no field evidence for a primordial granitic shell has yet been found."

Yet it is in these batholithic rocks—the bulk of them granitic—that we have the best chance of observing materials that may have been drawn directly from a primitive sialic crust, if such exists. To survey the granitic rocks of the shields is beyond our competence or the scope of this book. Instead we draw attention to aspects of some Precambrian granites that may perhaps bear significantly on the problem implicit in Knopf's statement.

CHARNOCKITIC GRANITES Large segments of the shields consist of crystalline rocks whose principal constituents are feldspars and pyroxenes and which lack, or contain but small amounts of, hornblende or micas (Turner and Verhoogen, 1960, p. 346; Turner, 1968, p. 333). Some are clearly metamorphic—the pyroxene granulites. Others form intrusive plutons. Among these latter are

[1] Locally, near Glencoe in the southeast Highlands and at Ben Nevis, the magmas burst out at the surface to give andesite-dacite-rhyolite members of the Old Red Sandstone.

rocks of broadly granitic composition consisting of quartz, microcline perthite, plagioclase, pyroxene, and in some cases garnet. These are charnockites.[1] That intrusive truly plutonic bodies of charnockite exist is generally agreed (cf. Pichamuthu, 1967, pp. 63–64). Associated country rocks invariably show effects of regional metamorphism in the granulite facies. Charnockites and granulites in fact are isofacial; and their mineral assemblages imply regionally prevalent temperatures exceeding 700°C at depths which, in the case of cordierite granulites, may have been only 15 km (cf. Turner, 1968, p. 366, fig. 8-6). Many quartzfeldspathic granulites are pelitic and semipelitic sediments that have become completely dehydrated by breakdown of micas and hornblende at high temperatures. So the mineral assemblages of some charnockite plutons may also possibly be metamorphic derivatives of normal granitic assemblages with hornblende and biotite. On this question—metamorphic versus magmatic origin of the existing paragenesis—there is no general agreement. Nevertheless, the charnockite-granulite association, whatever its history, implies wide prevalence, during Precambrian times, of an almost unique crustal temperature regime recorded in few exposed crystalline rocks of later age.

Plutonic charnockites, using the term in the restricted sense, range from acid granitic types to rocks in the intermediate silica range. In many the $K_2O/Na_2O$ ratio is conspicuously high (Table 12-3, analysis 1), so that acid rocks are truly granitic and those with $SiO_2$ at the 65 to 68 percent level are quartz-pyroxene syenites rather than quartz monzonites or granodiorites.

POTASSIC GRANITES Among Precambrian granites, a potassic character is by no means confined to charnockites. One gets the impression that acid rocks with $K_2O/Na_2O > 2$ are much more extensive in the shields than in younger orogenic belts—where the dominant plutonic rocks are, in fact, granodioritic. This characteristic is illustrated in Table 12-3 by individual analyses, each typical of very extensive Precambrian plutons: voluminous sheets of later Precambrian rapakivi granite in Finland (Turner and Verhoogen, 1960, pp. 368–389); intrusive granites and partially metasomatic derivative elements of fringing migmatites in the regional migmatite complexes of southwestern Finland (Härme, 1965); hornblende granites that make up the bulk of granitic batholiths in the northwestern Adirondacks (Buddington, 1957).

Possibly significant in this connection is the experimentally demonstrated fact that potassic compositions of water-saturated melts in the system $SiO_2$–$KAlSi_3O_8$–$NaAlSi_3O_8$ at minimal liquidus temperatures are favored by relatively low pressures.

ARCHAEAN SODIC GRANITES An essay by Glikson (1972) exemplifies a modern tendency to search for special features peculiar to Archaean petrogenesis. He

[1] Some writers broaden the term *charnockites* to cover pyroxene-bearing rocks ranging from acid to basic and even ultrabasic types. For a discussion of the nature of charnockites from the type area, see Subramaniam (1959).

presents a good case for primordial mafic crust analogous to modern oceanic crust—its components tholeiitic lavas and bodies of peridotite and pyroxenite, themselves perhaps extrusive. Many of the Archaean plutons invading this crust are composed of granitic rocks with high Na/K and unusually high abundances of Cr and Ni. They are thought to represent magmas formed by partial fusion at the base of sagging crustal segments. Their emergence gave rise to sialic islands—embryonic continents with a possible modern analogy in

**Table 12-3** Chemical analyses (oxides, wt %) and CIPW norms of some representative potassic Precambrian granites

| | 1 | 2 | 3 | 4 | 5 |
|---|---|---|---|---|---|
| $SiO_2$ | 71.32 | 73.84 | 69.50 | 77.04 | 70.11 |
| $TiO_2$ | 0.27 | 0.22 | 0.45 | 0.05 | 0.42 |
| $Al_2O_3$ | 14.31 | 13.09 | 12.34 | 12.39 | 14.11 |
| $Fe_2O_3$ | 0.88 | 0.87 | 2.02 | 0.15 | 1.14 |
| FeO | 1.18 | 1.79 | 3.97 | 0.27 | 2.62 |
| MnO | 0.07 | 0.03 | 0.06 | 0.01 | 0.06 |
| MgO | 0.52 | 0.23 | 0.35 | 0.17 | 0.24 |
| CaO | 1.05 | 1.12 | 2.33 | 0.22 | 1.66 |
| $Na_2O$ | 2.62 | 2.32 | 2.50 | 1.96 | 3.03 |
| $K_2O$ | 6.87 | 6.24 | 5.80 | 6.91 | 6.03 |
| $P_2O_5$ | 0.03 | 0.05 | 0.08 | 0.12 | 0.09 |
| $H_2O^+$ | 0.65 | 0.45 | 0.43 | 0.41 | 0.23 |
| $H_2O^-$ | 0.15 | 0.07 | 0.10 | 0.06 | 0.07 |
| F | | 0.17 | 0.11 | | 0.09 |
| Total | 99.92 | 100.49 | 100.04 | 99.76 | 99.90 |
| *Q* | 26.27 | 33.61 | 26.68 | 38.67 | 23.19 |
| *C* | 0.51 | 1.05 | | 1.57 | |
| *or* | 40.49 | 36.74 | 34.52 | 40.83 | 35.58 |
| *ab* | 22.01 | 19.41 | 20.98 | 16.58 | 28.03 |
| *an* | 5.56 | 3.98 | 5.29 | 0.31 | 5.70 |
| *di* | | | 2.87 | | 1.02 |
| *hy* | 2.49 | 2.84 | 4.88 | 0.73 | 3.40 |
| *mt* | 1.39 | 1.16 | 3.01 | 0.22 | 1.79 |
| *il* | 0.46 | 0.46 | 0.91 | 0.10 | 0.76 |
| *ap* | | 0.14 | 0.34 | 0.28 | 0.20 |
| Total | 99.10 | 99.39 | 99.48 | 99.29 | 99.67 |

*Explanation of column headings*

1 Charnockite, Madras, India (Subramaniam, 1959, p. 348, analysis 5)
2 Biotite rapakivi, Ahvenisto massif, Finland (Savolahti, 1956, p. 77, analysis 5)
3 Hornblende rapakivi, Ahvenisto massif, Finland (Savolahti, 1956, p. 77, analysis 10)
4 Potassic granite, Helsinki, Finland (Härme, 1965, p. 12)
5 Hornblende granite, Adirondacks, New York (Buddington, 1957, p. 293, analysis 1)

the Fijian archipelago; and it was by aggregation of these primitive nuclei that the Precambrian shields ultimately developed. Later anatexis of sodic granites and their extrusive equivalents gave rise to potassic granitic magmas of later Precambrian time.

ISOTOPIC COMPOSITION OF STRONTIUM AND LEAD A good deal is known about the present composition of strontium in Precambrian granites, much less about lead. Since Rb/Sr is high in most undoubtedly magmatic granites, points on a plot of $Sr^{87}/Sr^{86}$ against $Rb^{87}/Sr^{86}$ for any granitic complex may be too closely clustered to permit accurate evaluation of initial $Sr^{87}/Sr^{86}$ by drawing a whole-rock isochron. The difficulty may be partially overcome by drawing an isochron that fits the data for granites and is also consistent with the age of enclosing metamorphic rocks of more varied composition. The assumption is that the granites were emplaced during or shortly after metamorphism.[1] In spite of such difficulties it seems that many, though not all, older Precambrian granites have relatively low *initial* radiogenic strontium. A carefully documented example cited in Verhoogen et al. (1970, pp. 213–214) is an Ontario granite giving $Sr^{87}/Sr^{86} = 0.704$ at the time of emplacement, 1725 m.y. ago (Krogh and Davis, 1969). Spooner (1969) finds that low initial $Sr^{87}/Sr^{86}$ (0.701 to 0.708) is characteristic of most charnockites, and he cites Crawford's study of rocks from the type locality near Madras, which gave a value of 0.7039 on a whole-rock 2620-m.y. isochron.

There are several records of low radiogenic lead in ancient Precambrian granitic rocks whose high Pb/(U + Th) values imply that the isotopic composition of lead has been little changed since the time of emplacement. Thus quartz diorite and quartz monzonite from the Bighorn Mountains, Wyoming (Heimlich and Banks, 1968), dated by K-A at 3000 ± 100 m.y., give present lead values $Pb^{206}/Pb^{204} = 13.68$, $Pb^{207}/Pb^{204} = 14.89$, and $Pb^{208}/Pb^{204} = 33.71$. Most interesting of all are gneissic granites, granodiorites, and tonalites from western Greenland (L. P. Black et al., 1971) with extremely low radiogenic strontium and lead. These underwent amphibolite-facies metamorphism at a very early date (3700 ± 140 m.y., on an Rb-Sr isochron).[2] Initial $Sr^{87}/Sr^{86}$ is $0.6992 \pm 0.0010$; minimum whole-rock lead values are $Pb^{206}/Pb^{204} = 11.5$ and $Pb^{207}/Pb^{204} = 13.14$. Black et al. (p. 257) conclude that "it is tempting to regard the parent (pre-metamorphic) igneous rocks as a part of primeval terrestrial crust which still preserves its original age and record."

[1] Commonly the procedure is reversed for dating purposes: by assigning a "reasonable" value to initial $Sr^{87}/Sr^{86}$ (say, 0.705) the age of the granite is determined by drawing the connecting isochron.

[2] Supported by a lead-lead isochron (S. Moorbath, R. K. O'Nions, and R. J. Pankhurst, *Nature*, vol. 245, p. 139, 1973).

## THE GRANITE PROBLEM TODAY

NATURE OF THE PROBLEM Today there is not much support for once-popular genetic models that derive great volumes of granitic magma by fractionation of parental basaltic liquids, nor for others that postulate metasomatic granitization on a large scale without participation of granite melts. The classic "granite controversy" of 20 or 30 years ago has abated. But another major granite problem is with us and is as urgent as ever was its predecessor. Granted that granitic batholiths are the most voluminous solidified products of primary granitic magmas, what are the sources of these magmas and where are they located?

POSSIBILITY OF ULTRAMAFIC SOURCES Granitic magmas cannot form from anhydrous ultramafic source rocks at pressures prevailing in the lower crust or upper mantle. Such liquids, as shown by high-pressure experiment, cannot coexist with the principal component phases generally postulated for mantle rocks at pressures greater than 10 kb. Access of water in sufficient quantity could change the situation (Kushiro, 1972*a*). At depths at least as great as 80 km, partial fusion of garnet lherzolite in the presence of water could generate a hydrous granodioritic liquid fraction. Some such mechanism may have played a significant part, as Kushiro suggests, in early development of the primitive continental crust at a stage of terrestrial evolution when water was being copiously expelled from the mantle. It is much less likely to have contributed to large-scale formation of later granite magmas unless water has been supplied in quantity from external sources. Magmas so formed in any case would be abnormally hydrous.

TESTIMONY OF STRONTIUM AND LEAD ISOTOPES There are granitic rocks of several known ages (determined by K-A dating and stratigraphic-structural evidence) whose highly radiogenic initial strontium matches that of surrounding high-grade regionally metamorphosed sediments. These probably formed by partial fusion of the metamorphic envelope. The Older Granite Glen Clova complex, situated in the sillimanite zone of the southeastern Dalradian of Scotland, seems to be a pluton of this kind. Another example is the Silurian Cooma granite of New South Wales, with initial $Sr^{87}/Sr^{86} = 0.7179$ (Pidgeon and Compston, 1965).

For the most part, however, the source rocks of exposed granitic plutons are hidden. The problem then is to infer something regarding source rocks from granites now situated in another environment—always bearing in mind the possibility that these, or the magmas from which they crystallized, may have been contaminated (with increase in $Sr^{87}/Sr^{86}$) by contact with circulating brines or with more ancient metasediments. It is highly probable that some of these granites, especially if their oxygen has a normal composition, must have come from sources rich in radiogenic strontium (for example, old sialic crust) or alternatively were strongly influenced by such rocks. A fully documented example is the Heemskirk granitic complex of Tasmania (Brooks and Compston, 1965; Faure and Powell, 1972, pp. 48–50). Stratigraphic evidence shows that this cannot be younger than mid-Devonian (400 m.y.). Initial $Sr^{87}/Sr^{86}$ values for different granites in the complex are very high—0.719 to 0.741. The differences between different types of granite are real; and this, taken in conjunction with the nonmetamorphic condition of the envelope and large size of the pluton, seems to rule out the possibility of isotopic exchange between the granitic magmas and immediately surrounding sediments.

At the other end of the spectrum, strontium of many granitic rocks of all ages is poorly radiogenic. Perhaps one-half of analyzed granites give initial $Sr^{87}/Sr^{86}$ between 0.703 and 0.706 (Faure and Powell, 1972, pp. 46, 47, 54). Some of these rocks also have strikingly low radiogenic lead. Many other granitic rocks, including the great collective bulk of all Phanerozoic granites of North America, give initial $Sr^{87}/Sr^{86}$ from 0.706 to 0.708. Of granitic magmas low in radiogenic strontium, Faure and Powell (1972, p. 152) state that they "can contain only relatively minor amounts of older sialic crustal material." And for those with initial values near 0.707, models have been proposed—with emphasis on fusion in subduction zones—which mix strontium drawn from "primitive" oceanic and from sialic crustal sources. All this is speculation, permissible perhaps but in no way demanded by the data. Because of early metamorphic depletion in uranium and rubidium (Heier, 1965; Heier and Thoresen, 1971), some ancient sialic rocks give $Sr^{87}/Sr^{86}$ and $Pb^{206}/Pb^{204}$ ratios covering the lower range found in granitic rocks of all ages. Fusion of these ancient rocks could give rise to granitic magmas with $Sr^{87}/Sr^{86}$ values of 0.707 or even less without involving other and more basic source materials in the melting process (cf. Faure and Powell, 1972, pp. 52–53).

CRUSTAL ANATEXIS AND WATER DEFICIENCY Anatexis, as seen in migmatites, is no simple phenomenon. Certainly it cannot be viewed just in terms of classic experiment designed for the extreme case $p_{H_2O} = P_{total}$. There is now a good deal of experimental evidence regarding the nature and water content of melts that can form from high-grade metamorphic rocks containing micas and/or hornblende (for example, Piwinskii and Wyllie, 1970; G. C. Brown and Fyfe, 1970; Robertson and Wyllie, 1971). Particularly significant is the buffering effect of the hydrous silicates on the water content of the melts with which they are in contact. In the absence of pore fluid it is these minerals that supply water to the melt as they decompose in the order muscovite, biotite, hornblende.[1] In consequence, most granitic melts of anatectic origin will be undersaturated in water, and they can form only at temperatures considerably higher than those

[1] At $P$ = 6 to 8 kb, breakdown temperatures are muscovite ~700 to 750°C, biotite ~800 to 900°C, hornblende ~900 to 950°C.

on the minimum melting curve (water-saturated solidus). Granitic magmas representing an advanced stage of crustal fusion are likely to consist of quartzo-feldspathic melt undersaturated in water and charged with refractory crystalline materials such as calcic plagioclase, hornblende, and biotite or aggregates of these in the form of amphibolite xenoliths that are so common in granitic rocks.

The slope of the *PT* melting curve for a granitic system of fixed water content, like that of a single hydrous silicate, is positive. Thus such a melt, once generated, could ascend to high levels without crystallizing. It is the less hydrous and hotter of these magmas that have the greatest chance of reaching the surface. The more hydrous (cooler) granitic magmas reach the water-saturated liquidus curve at pressures still as high as 2 or 3 kb (Harris et al., 1970). These then emit water and rapidly crystallize to the point of immobilization. Herein may be a partial explanation of the preference of granitic magmas for the plutonic environment.

#### TENTATIVE CONCLUSIONS

1. Some granitic magmas form by fusion of mixed sedimentary rocks at the climax of regional metamorphism.
2. Some are generated by fusion of ancient sialic rocks or partial fusion of more basic rocks (for example, amphibolite) of the continental crust. Such magmas will for the most part be undersaturated in water.
3. Basic magmas derived from the mantle or from fusion of mafic rocks in the deep crust may contribute—through magmatic mixing—to magmas at the tonalite end of the granitic magma series.
4. Large volumes of granitic magma cannot be generated by direct fusion of ultramafic mantle rocks. Some contribution from such sources is possible if a hydrous phase is present in the primary mantle assemblage or if water is supplied in quantity from an external source. This second possibility may have been significant with respect to early differentiation of continental sialic crust from the mantle.
5. The isotopic compositions of lead and strontium in granitic rocks cannot be used alone to identify the source rocks nor to locate these at specific levels within the mantle or the deep crust.

## Precambrian Anorthosite Massifs

### INTRODUCTION

In the Grenville tectonic province of Quebec and Labrador (Verhoogen et al., 1970, p. 212), one-fifth of the total area of several million square kilometers is anorthosite (Fig. 12-9). The largest pluton has an outcrop area of 20,000 $km^2$. A southwestern offshoot of this province is the Adirondacks, and in this lies a massif of anorthosite covering 2000 $km^2$ made classic through the work of

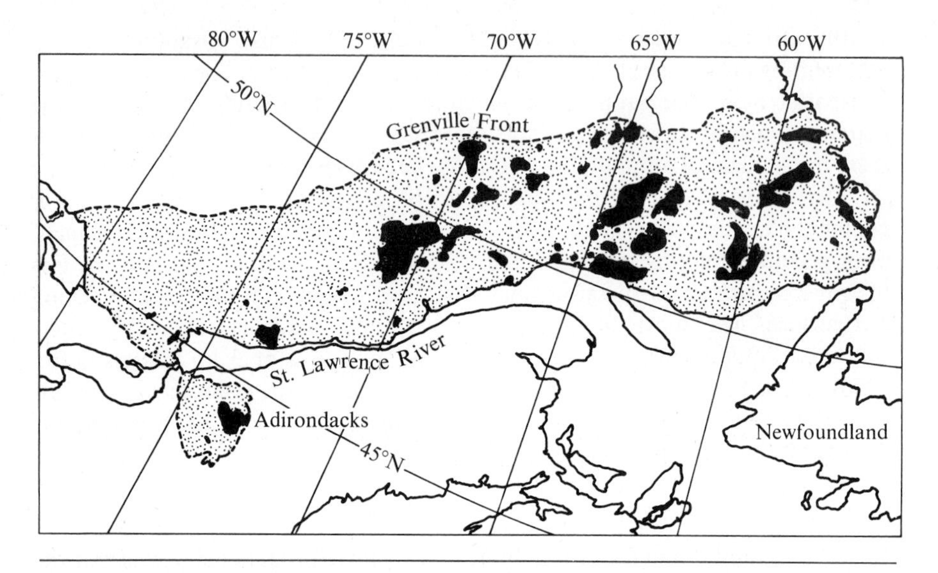


**Fig. 12-9** Main anorthosite plutons (solid black) in the Grenville province (stippled), Canada and New York. (After Martignole and Schrijver, 1970.)

Buddington (1939, pp. 19–52, 201–230). Most of us, in fact, see the Precambrian anorthosite problem in terms of the Adirondack massif and a comparable body at the southwestern tip of Norway (Barth, 1936). Today anorthosite massifs are known to exist in all the continental shields. Many, but not all, duplicate the peculiar mineralogical and structural character of the Adirondack and Norwegian anorthosites; and these we shall call *anorthosites of the Adirondack type.*

Their striking individuality and the genetic problems raised thereby are seen today much as set out in our previous statement (Turner and Verhoogen, 1960, pp. 321–328). But the picture has been elaborated and details modified by work of de Waard, Romey, and others in the Adirondacks and by that of P. and J. Michot in Norway. Buddington (1960; and pp. 215–232 in Isachsen, 1968) has twice reevaluated, in the light of newer and more varied data, the anorthosite problem as it appears in the Adirondacks. Similar massifs have been found in the Canadian and other shields. It is now apparent that not all Precambrian anorthosites, nor indeed all Canadian massifs, conform to the Adirondack pattern. Geochemists have contributed by experiment in the field of high-pressure silicate-melt equilibria (Yoder, pp. 13–22 in Isachsen, 1968; T. H. Green, 1969) and by isotopic analysis of strontium. A large sample of these newer data and a fair spectrum of current ideas on petrogenesis have been brought together in a recent informative symposium under the editorship of Isachsen (1968). The logical starting point for present discussion is to grasp the essential character of anorthosites of the Adirondack type.

## INDIVIDUALITY OF THE ADIRONDACK TYPE

From the first it was realized that the Adirondack and Norwegian massifs and their chief component rocks differ strikingly and consistently from bodies of anorthosite in stratified basic plutons:

1. The typical massif is a thick sheet or lens with a domed roof implying forceful upward intrusion of a pattern familiar in granitic domes and in the component plutons of many granitic batholiths (for example, Martignole and Schrijver, 1970). The core of a typical massif consists of anorthosite with only minor amounts (5 percent) of mafic minerals. Rocks of the border zones, variously called gabbroic anorthosite or norite, may contain 10 to 30 percent pyroxene (or its metamorphic equivalent). Gabbros and peridotites are lacking, and negative gravity anomalies preclude the possibility that they exist in depth (Simmons, 1964; Buddington, p. 216 in Isachsen, 1968).
2. The primary mineral assemblage of anorthosites and associate norites is plagioclase, hypersthene, augite, ilmenite, and titanomagnetite. The plagioclase is relatively sodic: $An_{42}$ to $An_{45}$ is typical (more sodic composition $An_{30}$ to $An_{35}$ is found in metamorphically recrystallized plagioclase). Textures tend to be very coarse, and large crystals of plagioclase are commonly bent and shattered.
3. Closely associated with anorthosite are varied igneous rocks of completely different composition—granitic, granodioritic, and quartz syenitic. These are all pyroxene-bearing varieties: in other words, typical igneous charnockites.
4. Anorthosite massifs occur exclusively in Precambrian terranes. The country rock, at least for those of the Adirondack type, exhibits regional metamorphism in the granulite facies. In fact anorthosites, their charnockitic associates, and regional country rocks are mineralogically isofacial. This implies emplacement of the massifs within a unique geothermal regime the nature of which will not be fully understood until we have a clearer chronological picture of the events in question: emplacement of anorthosite, emplacement of charnockites, deformation, and regional metamorphism. Magnetite-ilmenite compositions indicate a final crystallization temperature of $810 \pm 50°C$ for anorthosite, $650 \pm 50°C$ for regional metamorphism of granulites in the Adirondacks.

## OTHER TYPES OF ANORTHOSITE MASSIFS

One important outcome of recent work is a growing realization that many massifs have a mineralogical and structural character that in some respects stands intermediate between massifs of the Adirondack type and stratified basic plutons of the Stillwater type. In the Canadian shield the best-known examples are found outside the Grenville province. They are gabbroic or troctolitic anorthosites with labradorite ($An_{50}$ to $An_{65}$), not andesine. Mafic minerals commonly make up 10 to 30 percent of the total composition. Gravity-controlled mineral stratification and cumulate structures are present at least locally and may be pervasive through thick sections. One of these bodies is the Duluth "gabbro" complex of Minnesota (Grout and Schwartz, 1939; Phinney, pp. 135–148 in Isachsen, 1968). Another is the Michikaman intrusion, Nairn tectonic province of Labrador (Emslie, pp. 163–174 in Isachsen, 1968). This is an immense sheet, 9 km in thickness, the bulk of which consists of layered troctolitic anorthosite (plagioclase $An_{50}$ to $An_{65}$, 70 to 95 percent; olivine $Fo_{55}$ to $Fo_{70}$, 5 to 30 percent; minor pyroxene). There is an upper layer of true anorthosite and a chilled marginal layer of gabbro; and there are minor differentiates of ferrous diorite, granodiorite, and syenite. Hypersthene-cordierite-quartz and cordierite-sillimanite-spinel assemblages in the contact aureole indicate emplacement at comparatively shallow depth.

## PETROGENESIS

UNITY OF THE ANORTHOSITE PROBLEM Although, as we have just seen, the collective individuality of Precambrian anorthosite massifs is not as sharply defined as was once supposed, all these bodies have much that is striking in common: They are confined to a Precambrian environment; the dominant mineral component by far is plagioclase whose average composition is on the whole less calcic than that of mafic plutons of the Stillwater type; the primary mafic phases are anhydrous; initial isotopic composition of strontium in all cases is $Sr^{87}/Sr^{86} = 0.703$ to 0.706. Present knowledge, we think, scarcely justifies further paraphrasing of Read's once-apt expression to invoke some kind of fundamental distinction between "anorthosites and anorthosites." Broad unity of character among Precambrian massif anorthosites will be presumed in seeking some general petrogenic model.

MAGMATIC CHARACTER OF ANORTHOSITES Like many other crystalline rocks of comparable age, most anorthosites have been affected to some degree by regional metamorphism. Pyroxenes may be partially replaced by hornblende; coarse plagioclase may be converted locally to granular aggregates of more sodic composition ($An_{33}$ to $An_{35}$). In Idaho anorthosites there are local layers with aluminous metamorphic phases—kyanite, sillimanite, andalusite, garnet, cordierite (for example, Hietanen, p. 376 in Isachsen, 1968).

Nevertheless the concept of formation of anorthosite massifs by metamorphism and metasomatism of some other kind of rock is considered untenable for several reasons. Detailed mapping of many anorthosite bodies has established their intrusive nature and plutonic individuality. The layered structure of some, however imperfect it may be, is clearly analogous with that of stratified basic plutons of the Stillwater type. The high positive europium anomaly that characterizes the rare-earth-element pattern of anorthosites (for example,

T. H. Green et al., 1972, p. 253) shows that plagioclase crystals coexisted with a melt phase at some stage in the formation of anorthosite. Consistent with magmatic origin, too, are low $Sr^{87}/Sr^{86}$ ratios[1] and unusually high K/Rb values, both comparable with data from oceanic basalts (Heath and Fairbairn, pp. 99–110 in Isachsen, 1968).

Further discussion will be confined to magmatic models. Admissible source materials are deep crustal or mantle rocks low in radiogenic strontium. A good many such models are currently in vogue. They differ among themselves chiefly according to how one answers these questions:

1. What is the nature of the parent magma of anorthosites?

2. Are anorthosites and charnockitic associates cogeneric, or have they formed from independently generated primary magmas?

3. Can a magma of anorthositic or quasi-anorthositic composition be generated under conditions, and from source materials, that can be reasonably postulated in the deep crust or in the mantle during Precambrian times? Answers to this third question must stand the test of compatibility with data of laboratory experiment. Bowen's reply, based on equilibria at atmospheric pressure, was negative. But the question has since been reopened in the light of newer data on silicate-melt equilibria at high pressures.

HIGH-PRESSURE EXPERIMENTAL DATA Yoder found that at water pressures of 5 to 10 kb the anorthite-diopside eutectic is markedly enriched in anorthite ($\approx An_{75}Di_{25}$) compared with the eutectic at 1 bar. This discovery he followed up with experiments on appropriate silicate systems of several components and on fusion of common basalts at high water pressures (Yoder, pp. 13–22 in Isachsen, 1968). Such mixtures and rocks at pressures prevailing in the uppermost mantle may yield liquids very rich in feldspar components. Subsolidus reactions at high pressures would tend, however, to eliminate plagioclase in favor of pyroxene-rich assemblages in the ultimate crystalline product, so that for anorthositic liquid to yield crystalline anorthosite it would have to rise into the crust and crystallize at depths no greater than 35 km.

T. H. Green (1969) has investigated possibly significant anhydrous systems of two kinds. A gabbroic anorthosite magma had been postulated as parental by Buddington; and Yoder's work on the diopside-anorthite system at high water pressures had shown that the eutectic (with a very high content of water) would indeed approximate a simple gabbroic anorthosite. Green's conclusion, however, is that neither a subsequently dehydrated liquid of this kind nor any high-alumina basalt liquid could yield an anorthosite fraction by differentiation

[1] Present ratios, 0.703 to 0.706, can in this case be equated with initial ratios since Rb/Sr is universally very low.

at any level in the crust ($P < 10$ kb). Green then turned to anhydrous melts of "quartz-diorite" composition, chemically more or less equivalent to andesite (62 percent $SiO_2$, 17 percent $Al_2O_3$) or even to Archaean graywacke. From such liquids relatively sodic plagioclase separates at 1200 to 1250°C in the pressure range 5 to 13 kb; solidus temperatures are 1070 to 1150°C. Thus largely crystalline anorthosite could form either as solid residuum from partial anatexis of "quartz-diorite" or alternatively as an early crystal cumulate separating from quartz-diorite magma at crustal pressures. The complementary liquid fraction in either case could have the composition of quartz syenite or granite. Green's focus on anhydrous systems reflects his concern for a compelling aspect of the anorthosite problem—the anhydrous character of anorthosites and of associated charnockitic rocks and granulites.

ALTERNATIVE GENETIC MODELS IN OUTLINE Students of the anorthosite problem in the thirties were agreed in one essential respect: Anorthosites were seen as accumulations of plagioclase crystals separated during the earlier stages of fractional crystallization from some kind of parental magma. Controversy then centered on the nature of the magma and the status—independent or cogeneric—of associated granitic (charnockitic) magmas (cf. Turner and Verhoogen, 1960, p. 326). Today we are faced with the same situation (cf. Isachsen, 1968, pp. 442–444). But choice of parental magmas today is more varied, and high-pressure experiment and geochemical (especially isotopic) data have supplied new degrees of freedom as well as setting new limitations for manipulation of the evolutionary model. Here are some of the possibilities, each ably argued by its proponents and none yet proved untenable, at least for individual cases (Isachsen, 1968; Savolahti, 1966):

1. A F. Buddington (Adirondacks): Partial melting of mantle or deep crustal rock yields gabbroic anorthosite liquid heavily charged with andesine crystals. This liquid rises to higher levels and in the course of intrusion is differentiated into anorthosite cumulate and a more mafic liquid fraction that crystallizes as anorthositic gabbro (norite). Associated granitic rocks—charnockitic in the Adirondacks, rapakivi in Finland—formed later from independently generated magmas.

2. J. Michot and P. Pasteels (1969): In southern Norway anorthositic ("plagioclasic") magma was generated directly from unspecified mantle rocks from which it inherited strontium with initial $Sr^{87}/Sr^{86} = 0.704$ to 0.705. Associated norites and mangerites with initial $Sr^{87}/Sr^{86} = 0.707$ to 0.713 are anorthositic derivatives contaminated by bordering gneisses (initial $Sr^{87}/Sr^{86} = 0.706$ to 0.719).

3. H. S. Yoder: Hydrous anorthosite magma forms by partial melting of basic mantle rock. The magma rises into the crust and plagioclase crystallizes in the environment of reduced pressure. Gabbroid variants can develop in

response to suitable variation in pressure and water content in the environment of crystallization. Rise and final emplacement of the anorthosite mush is accompanied by crystal brecciation; and residual alkaline liquids are expelled to give associated charnockitic rocks.

4. A. R. Philpotts (1966); T. H. Green (cf. Barth, 1936): The primary magma has the composition of relatively anhydrous quartz diorite. This crystallizes slowly at high pressure in the deep crust to yield a plagioclase cumulate. Rise and intrusion of the mush at higher levels yields crystalline anorthosite and complementary granitic liquid fractions (charnockite).

5. D. de Waard (1967); T. H. Green: Partial melting of quartz diorite in the deep crust yields a granitic liquid fraction that later crystallizes as the charnockitic series.[1] The complementary crystalline residue of partial fusion is anorthosite.

6. N. L. Bowen (Adirondacks); J. F. Olmsted (Mineral Lake, Wisconsin); R. F. Emslie (Michikaman): The presumed parental magma is basaltic. Concentration of plagioclase to yield anorthosite is the result of fractional crystallization by various possible mechanisms—early depletion of liquid in components of olivine or pyroxene, flotation or flow fractionation of crystalline plagioclase, and so on. Late liquid fractions yield various sialic rocks—quartz syenite, ferrogranite.

Clearly any generally acceptable unified genetic model must be framed only in the broadest terms (cf. Michot, 1972): Some form of primary source material rich in the components of plagioclase gives rise to an accumulation of plagioclase crystals at deep crustal levels. The source may be a potentially "feldspathic" magma formed by melting of mantle rocks at even greater depth; or it could be a rock of "quartz-diorite" composition, partial fusion of which, in the deep crust, yields a two-phase mixture—plagioclase crystals suspended in the first-formed liquid fraction. The plagioclase-rich mush, however generated, rises to higher levels under gravity. During its emplacement the liquid fractions are strained away and ultimately crystallize as rocks of other composition. Possibly, but by no means certainly, these are represented by charnockitic granites and syenites.

A PRECAMBRIAN ANORTHOSITE EVENT? Increasing knowledge over the past four decades has eliminated or weakened some models of anorthosite genesis and strengthened others; but the diversity of those that are still in vogue is broader than ever. From the new data there has nevertheless emerged at least one generalization, which if it proves tenable may be uniquely significant

[1] The similar but much earlier suggestion of Winkler and Von Platen (1960) that comparable liquid fractions could form by partial fusion of lime-bearing illitic clay or shale seems untenable in the light of strontium-isotope data.

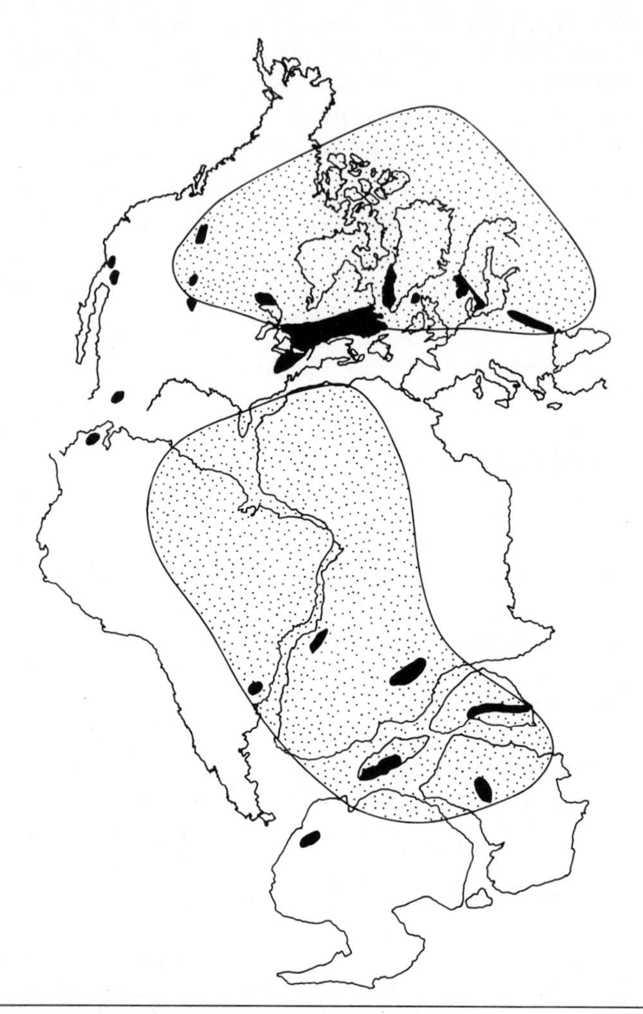

**Fig. 12-10** Pre-Triassic (predrift) configuration of continents, showing main areas within which occur anorthosite massifs (black; after Herz, 1969). Stippled areas enclose known areas of ancient Precambrian rocks (> 1700 m.y.); after Hurley and Rand (1969).

(Herz, 1969). It appears, in spite of peculiar difficulties in radiometric dating,[1] that all the massifs may have been emplaced during a sharply limited span of Precambrian time—1700 to 1100 m.y. (perhaps 1300 ± 200 m.y.). The further suggestion that anorthositic massifs are restricted to two elongate belts, one in each hemisphere, is less convincing. Hurley and Rand (1969) demonstrated a coherent ancient geographic pattern of distribution for older Precambrian rocks. On a pre-Triassic (predrift) reconstruction of the continents, rocks older than 1700 m.y. lie within the confines of two extensive segments (Fig. 12-10). Within

[1] The rubidium content of anorthosites is too low to permit direct dating by the Rb-Sr technique. K-A dates, on the other hand, may simply record the last resetting of the radiometric clock—in the Grenville province at 950 ± 150 m.y.

and bordering these segments are rocks in the range 1700 to 800 m.y. The limited geographic distribution of anorthosites, taking also into account imperfections of exposure and incomplete exploration, may perhaps stem from this general global Precambrian pattern rather than have any significance of its own. Hertz has nevertheless raised a provocative question on which we shall close present discussion of anorthosites: In spite of prevailing uncertainty as to the geochemistry and environment of their genesis, do we see in the Precambrian anorthosite massifs the record of a unique thermal or cataclysmic event in geologic history?

## Ultramafic Rocks of Orogenic Belts

### SCOPE OF INQUIRY

Peridotites are among the igneous rocks that match most closely the physical properties prescribed by geophysicists for mantle materials. If mantle rocks or their magmatic equivalents ever become transported into the accessible crust, they should be sought in dissected geosynclinal fillings of orogenic belts. Peridotites and their hydration products serpentinites are in fact abundantly represented in many such zones of maximum tectonic activity. Do these rocks, or do they not, supply direct information about material existing in the upper mantle? The significance of this question in modern geology is underscored in a number of recently published symposia and collective studies (Burk, 1964; Wyllie, 1967; B. A. Morgan, 1970; *Tectonophysics* vol. 7, no. 5-6, special issue, 1969).[1] There are at least two easily distinguishable kinds of ultramafic plutons in the orogenic setting—the widely distributed "Alpine type" of classic literature and a much less extensive, locally developed type, but recently recognized as such, in "concentrically zoned ultramafic complexes" (Noble and Taylor, 1960; Taylor and Noble, 1960).

### MOBILITY AND FLOW OF PERIDOTITE: EXPERIMENTAL DATA

It is generally agreed—and indeed the field evidence is overwhelming—that ultramafic bodies in orogenic belts have been emplaced by intrusive flow. This could imply, as H. H. Hess (1938, 1955) long maintained, that they were intruded as fluid ultramafic magma (highly magnesian ultramafic melt) or alternatively that the solid rock, under long-sustained stress at crustal temperatures and pressures, becomes capable of plastic or some other type of cohesive flow (cf. Verhoogen et al., 1970, pp. 504–511). It is now possible to assess these alternatives within limits of experimentally determined parameters.

[1] These sources contain carefully documented first-hand accounts of specific cases and varied conclusions reflecting a diversity of approaches—geochemical, geophysical, and strictly geological. Wyllie (1967) has admirably summarized opinion on separate aspects of the general problem and has supplied a lengthy list of modern references.

From his work on olivine-melt equilibria at atmospheric pressure, Bowen consistently denied the existence of magnesian magmas capable of precipitating olivine, except at impossibly high temperatures—perhaps 1500 to 1600°C. He therefore saw all peridotites as cumulates separated from basic magma and postulated the presence of some residual liquid to render the olivine cumulate mobile enough for intrusion. Subsequent work on multicomponent systems with $H_2O$ and $CO_2$ at high pressures lowered the range of possible liquidus temperatures but has not eliminated the problem (cf. Wyllie, 1967, pp. 407–409). Thus R. H. Clark and Fyfe (1961), in testing an early suggestion of Hess that serpentinite magmas might exist at reasonably low temperatures, found instead that a natural serpentinite at 500 to 1000 bars $p_{H_2O}$ became largely molten only at about 1400°C. So if intrusion of ultramafic magma is postulated as an alternative to Bowen's model, we must look for high-temperature contact effects comparable at least with what is seen beneath intrusive sheets of gabbro and diabase. [These, it turns out, may be limited to an unexpectedly narrow zone even where diabase sheets are hundreds of meters thick (Verhoogen et al., 1970, p. 559)].

Certain zoned peridotite bodies (see below) seem to have been emplaced in a liquid state. There are also Precambrian ultramafic rocks whose remarkable texture, marked by aligned, very coarse, bladed, skeletal crystals of olivine, strongly suggests crystallization from rapidly chilled peridotite magma that possibly was even extruded as submarine flows (Pyke et al., 1973; Viljoen and Viljoen, 1969; Nesbitt, 1971).

During the past decade much has been found out regarding strength and flow behavior of crystalline olivine, pyroxenes, and serpentine under long-sustained stress at crustal temperatures and pressures. Plastic flow by translation gliding has been demonstrated in both the anhydrous silicates. And it is likely that yield strengths will be so diminished at low strain rates, and by water weakening in a wet environment (as with quartz), that plastic flow in the solid state now seems a reasonable mechanism for intrusion of peridotite. Serpentinite at some temperature between 300 and 600°C becomes extremely weak with the onset of dehydration reactions and fails by brittle fracture (internal shattering). Herein Raleigh (in Wyllie, 1967) sees a mechanism that conceivably may play a part in intrusion of serpentinite bodies as such.

## CONCENTRICALLY ZONED ULTRAMAFIC COMPLEXES[1]

Concentrically zoned ultramafic plutons are known principally from two orogenic provinces, one in southeastern Alaska and the other along the axis of the Urals. The Alaskan complexes were emplaced in mid-Cretaceous times (110 to 100 m.y.), those of the Urals in the Devonian. But in both regions the sequence of plutonic events was the same and the lithologic pattern within the complexes virtually identical.

[1] See especially the accounts of Wyllie, Irvine, and Taylor in Wyllie (1967, pp. 83–121).

In southeastern Alaska numerous small gabbro plutons are dotted along a coastal belt less than 50 km wide that extends 500 km parallel to the structural trend of enclosing Paleozoic and early Mesozoic rocks. Clearly associated with many of these plutons are younger subcylindrical ultramafic complexes between 1 and 10 km in diameter. Regional emplacement of the Coast Range batholith came later still, about the close of the Cretaceous. The smaller and a few of the larger ultramafic bodies are composed of hornblende pyroxenite. But most of the larger plutons are more varied in lithology, with a crudely concentric distribution of rock types: a central core of dunite enclosed by successive cylindrical shells of peridotite (wehrlite), olivine pyroxenite (Table 12-4, analysis 1), magnetite pyroxenite, and hornblende pyroxenite. The dunite core (Table 12-4, analysis 2) may be nearly 2 km in diameter. Each such complex is enclosed, or almost so, in gabbro, intrusion of which, however, clearly constitutes an independent earlier event.

The mineralogical pattern throughout the province is simple, uniform, and distinctive—contrasting in several respects with that shown in peridotites of the Alpine type. Olivine is $Fo_{93}$ in dunites and becomes increasingly ferrous outward ($Fo_{75}$ in olivine pyroxenites); the sole pyroxene is aluminous diopside (1 percent hedenbergite in dunite, increasing to 30 percent in hornblende pyroxenite); plagioclase ($An_{90}$ to $An_{98}$) is present in small amount only in peripheral hornblendites: abundant (up to 20 percent) magnetite and minor ilmenite are typical of pyroxenites; and minor chromite is the sole oxide phase of dunites.

In all shells except the outermost hornblende pyroxenite, there is horizontal rhythmic layering with "graded bedding" of olivine and pyroxene crystals on a scale of 1 to 20 cm. This persistent structure may be continuously maintained through vertical sections hundreds of meters thick. There can be no doubt that it records crystal sedimentation in ultramafic liquids, and the presence of wide high-temperature aureoles in the surrounding rocks is consistent with this conclusion. Mineralogical evidence shows conclusively that their liquidus temperatures decreased outward through successive zones as in a fractional-crystallization sequence. Hornblende is a late product of deuteric action in the interprecipitate of the outer zone.

The generally accepted petrogenic model postulates independent generation first of tholeiitic and then of ultramafic magmas. The first gave rise to extensive bodies of noritic gabbro which mineralogically and chemically bear no relation to the ultramafic rocks. These latter were emplaced at localized centers by multiple intrusion in order of increasing liquidus temperature: magnetite pyroxenite, wehrlite, and/or dunite. Taylor (in Wyllie, 1967, p. 120) infers that "these [ultramafic] magmas must be related to one another at depth, either through fractional crystallization in a deeper magma chamber or in the conduit itself [cf. the Muskox feeder dike] or by primary fractional fusion in their source region, presumably the upper mantle." The second possibility suggests successively deeper loci of magma generation, starting with production of tholeiitic magma at a relatively high level.

The Urals province duplicates that of Alaska in almost every respect. The early episode of basic intrusion here gave rise to a narrow, almost continuous gabbro belt which was then pierced by numerous small ultramafic complexes. To many of us the chief significance of zoned ultramafic complexes is that they lend credence to the possibility that magnesian ultramafic magmas may indeed form deep down in orogenic belts. Perhaps they can migrate upward only through sufficiently dense country rock (gabbro).

## ULTRAMAFIC BODIES OF THE ALPINE TYPE

GENERAL CHARACTER AND ASSOCIATED ROCKS: OPHIOLITE SUITE At the turn of the century European geologists were fully aware that ultramafic rocks are a chacteristic component of fold-mountain belts, typified by the Alps. Suess termed them *green rocks*, and Steinmann recognized a group of closely associated rocks—serpentinites, gabbros, spilites, and amphibolites[1]—which he called collectively *ophiolites.* From the first, Steinmann insisted that radiolarian cherts and foraminiferal mudstones closely associated with ophiolites demonstrated emplacement of the latter in an abyssal marine environment. Such sediments in fact he termed *abyssites.* And among European geologists the seemingly inseparable association serpentinite-spilite-chert became known as the "Steinmann trinity." It is only rather recently, with growing appreciation of island-arc tectonics and turbidite sedimentation, that American geologists have come fully to accept Steinmann's interpretation of his lithologic trinity. Many students of plate tectonics (cf. Presnall and Bateman, 1973) now regard the ophiolite suite as old oceanic crust and underlying mantle rock.

Benson (1926, p. 6), in his comprehensive survey of basic and ultramafic rocks in relation to tectonics, introduced the term *Alpine type* to cover ultramafic rocks of the Alpine ophiolite suite and similar bodies in other orogenic belts. Today the category has been broadened to include serpentinites and other ultramafic rocks of island arcs and ocean margins; Hess in particular recognized that these and Alpine ultramafics must have much in common as to origin and tectonic environment. Hess was preoccupied especially with serpentinite, which is by far the most widespread type of Alpine ultramafic rock and which he believed also to be the principal component of oceanic crust. But Thayer (for example, in Wyllie, 1967, p. 222), on the basis of field observations in many orogenic belts, has convincingly reasserted the essential though usually subordinate role that gabbros play as integral components of many Alpine ultramafic complexes. (This still leaves open the question of possible genetic relation between the essentially volcanic spilites and ultramafic members of the Steinmann trinity.)

[1] Albitized basalts and diabases here called *spilites* were termed *diabases* in Alpine literature. *Amphibolites* include a variety of actinolitic greenschists and other metabasaltic rocks.

PERIDOTITES AND PYROXENITES: GENERAL MINERALOGY AND INTERNAL STRUCTURE The component ultramafic rocks of Alpine complexes are peridotites, subordinate pyroxenites, and their hydration products, serpentinites. In most provinces the first two are composed of some combination of four phases—in order of abundance, olivine, enstatite, chromian diopside, and a spinel (chromite or picotite). All three silicate phases are highly magnesian: atomic ratio Mg/Fe > 9. Most widely distributed are enstatite peridotites (harzburgites) with little or no diopside and accessory spinel. These grade into dunites in which olivine reaches 95 percent or more. Rarely there are peridotites with accessory highly calcic plagioclase. Pyroxenites, in marked contrast with their abundance in layered basic plutons, are limited to discontinuous lenses and narrow veins. Podlike bodies and streaks of chromite, according to Thayer, are diagnostic of peridotites of the Alpine type.

Also highly characteristic is discontinuous layering, on the scale of 1 to 50 cm, with differential concentration of olivine, pyroxene, or chromite and a tendency to parallel alignment of rather elongate olivine grains. Some writers (for example, Challis, 1965) attribute this structure and accompanying strong preferred orientation of the olivine lattice to gravitational settling of crystals in a liquid (magmatic) medium. Thayer (1963; and pp. 222–223 in Wyllie, 1967) considers much of it to be the result of "extensive flowage of largely crystalline magma during emplacement." Still others see layering and foliation of Alpine peridotites as a postcrystalline phenomenon—the imprint of plastic flow in the solid state (*hot-working* in the parlance of metallurgy). Carefully documented cases for this model have been made for the Twin Sisters dunite slab, Washington (Ragan, pp. 160–167 in Wyllie, 1967), and for banded Norwegian peridotites emplaced in a metamorphic environment of much higher temperature (Lappin, pp. 183–190 in Wyllie, 1967). There is more general agreement that mylonitic structure and crystal-strain effects seen in many dunites are due to deformation in the solid (*cold-working*).

SERPENTINITES AND SERPENTINIZATION Few large bodies of Alpine peridotite are completely anhydrous. In most of them olivine and enstatite, at least locally, have been partially converted to minerals of the serpentine family. In fact the most widely distributed of Alpine ultramafic rocks are serpentinites in which metasomatism is virtually complete. Serpentinites in an environment of low-grade regional metamorphism consist of the assemblage lizardite-chrysotile-brucite with persistent accessory magnetite. Such rocks are typical of the Franciscan terrane, Coast Ranges of California (N. J. Page, 1967; R. G. Coleman, 1971, p. 906), and of much of the great ultramafic belt of southwestern New Zealand (R. G. Coleman, 1966). The surrounding metamorphic rocks have been regionally metamorphosed at high pressure and temperatures of about 250°C. In metamorphic environments of somewhat higher grade, antigorite is a principal serpentine phase, and the ultramafic assemblage is essentially metamorphic in every sense. Schistose antigorite serpentinites are well known in the biotite zone (greenschist facies) of southwestern New Zealand and at grades below the $An_{17}$ isograd as drawn for *Bunderschiefer* in the Lepontine Alps (for example, Evans and Trommsdorff, 1970).

Commonly associated with serpentinites are narrow marginal zones and enclosed small bodies of metasomatites of quite a different kind. But these are recognized as integral by-products of the serpentinization process. Commonest perhaps are rodingites—assemblages of Ca–Al silicates such as hydrogrossular, vesuvianite, zoisite, and prehnite—replacing dikes or enclosed blocks of gabbro or diabase within the serpentinite body (R. G. Coleman, 1966, 1967, 1971, p. 909; Thayer, 1966, p. 701; Leonardos and Fyfe, 1967). The necessary calcium seems to have been released from the ultramafic body itself—mainly from diopside—during serpentinization. Equally characteristic is sodium metasomatism giving rise to narrow marginal zones of albitite, in some cases with aegirine and/or crossite as additional phases (for example, Reed, 1959*b;* R. G. Coleman, 1966, pp. 36–40: Leonardos and Fyfe, 1967*). These sodic rocks replace metasediments and metavolcanics along the serpentinite contact.

Serpentinization undoubtedly is a complex process and cannot be explained by any simple unique reaction. Apart from temperature and pressure there are many possible chemical variables related to composition of participating aqueous solutions. J. Barnes et al. (1972) have analyzed three types of spring water issuing from Californian serpentinite outcrops and have shown by thermodynamic argument how each could be an effective agent of serpentinization. These waters are alike only in rather high pH and unusually low Mg content.

It is easy to write simple equations to give some idea of what may happen during serpentinization of peridotite (for example, Turner and Verhoogen, 1960, pp. 318, 319; R. G. Coleman, 1971, pp. 907–908). Some models involving only addition of water and silica necessarily imply great increase in volume (60 to 70 percent). Equally possible theoretically are open-system reactions in which water is added and substantial MgO and $SiO_2$ are removed from the system; these can be adjusted to require little or no change in volume. Opinion is still divided on the question of volume change in serpentinization (cf. Thayer, 1966 and subsequent critical discussion[1]). Most probably, individual cases differ mutually and lie somewhere between volume-for-volume replacement and metasomatism with significant expansion.

Serpentinization, once considered to be a late magmatic (deuteric) process, is now conceded to result from low-temperature reaction between cooling peridotite and water introduced from external sources (Turner and Verhoogen, 1960, pp. 320–321; R. G. Coleman, 1971; Verhoogen et al., 1970, p. 311). In terrestrial waters the isotopic composition of oxygen and hydrogen varies

* These authors speculate on the complex chemistry of the metasomatic process and raise the question of participation of saline waters in serpentinization.
[1] N. J. Page (1967*b*) and Thayer (1967).

widely with location and geologic environment. Similar variation has been recognized in some serpentinites. Using this criterion, serpentinization of some chrysotile-lizardite serpentinites of western North America has been traced to deeply circulating meteoric water whose hydrogen becomes increasingly heavy with decreasing latitude. Oceanic water, which has a unique combination of heavy hydrogen and light oxygen, seems to be the agent of serpentinization of blocks dredged from the mid-Atlantic ridge. Isotopic patterns of quite another type have been found in antigorite serpentinites. Using the same approach J. Barnes et al. (1972) found that one type of "serpentinizing" spring water issuing from Californian serpentinite is of meteoric origin; others are connate, and one of these could well represent scarcely modified marine water of late Mesozoic seas.

EMPLACEMENT AND CONTACT METAMORPHISM From field observations and petrography much can be inferred about intrusion, crystallization, and metamorphic effects of stratified basic intrusions and postkinematic granitic plutons. With Alpine ultramafic bodies the situation is more complex and much of the plutonic history may be illegible—partly because such rocks are prone to late serpentinization and partly because of their mobility in the solid state, even at low temperatures. We have already noted the difficulty that attends interpretation of layered and foliated structures in peridotites. In serpentinites these tend to be further heightened. Bodies of serpentinite, as demonstrated long ago by Taliaferro in California, may reach their final sites of emplacement only during the last stages of folding; or they may even be squeezed up in a brecciated state along postorogenic faults. It is not surprising that many serpentinized bodies show no trace of a contact aureole or merely show low-temperature mineral assemblages in thin border zones along the path of circulating connate and meteoritic waters. This familiar situation gave rise to the belief—once widely held but now discarded—that ultramafic bodies of all kinds are emplaced at universally low temperature.

Where peridotites have escaped serpentinization, the record of earlier events is more clearly legible. Some of these bodies have high-temperature contact aureoles (Mackenzie, 1960; D. H. Green, 1964). In others only local segments of such aureoles have survived late displacements (Challis, 1965, p. 326; Walcott, 1969, p. 76). Amphibolite and eclogite blocks at the margins of yet other "cold intrusions" of serpentinite in California are perhaps transported fragments of disrupted deep-seated aureoles—relics of an otherwise obliterated early stage of intrusion at high temperature.

TEMPERATURES AND PRESSURES OF INTERNAL EQUILIBRATION High-pressure experimental data for the system $MgO–CaO–Al_2O_3–SiO_2$ have made it possible to place numerical limits on pressures and temperatures at which natural anhydrous ultramafic assemblages reached equilibrium. (This, of course, is not necessarily the temperature of intrusion.) This approach (Fig. 12-11) has been

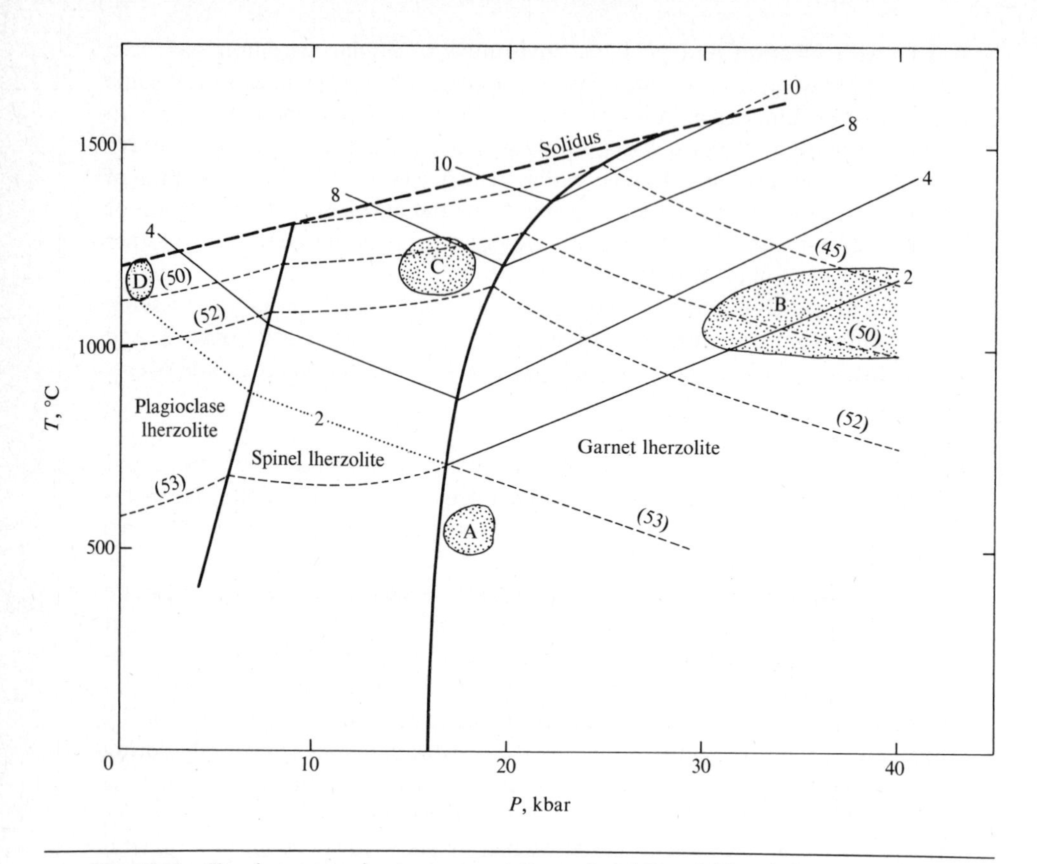


**Fig. 12-11** Showing approximate representation of stability fields of lherzolite assemblages and location of some natural lherzolites on O'Hara's pyroxene-calibrated *PT* grid (modified from O'Hara in Wyllie 1967, pp. 396, 402). Numbers 2–10 give the values of $100Al_2O_3/(CaSiO_3 + MgSiO_3 + Al_2O_3)$ in clinopyroxene; numbers (45)–(53) indicate $100CaSiO_3/(CaSiO_3 + MgSiO_3)$ in clinopyroxene. *A:* Norwegian garnet peridotites (re-equilibriated). *B:* Nodules in kimberlites. *C:* Lizard (primary). *D:* South Island, New Zealand.

used especially by O'Hara (pp. 7–14, 393–403 in Wyllie, 1967), D. H. Green (1963, 1964), and Onuki (Onuki and Tiba, 1965). The commonest Alpine peridotites are harzburgites with 5 to 10 percent normative (*di* + *an*). Lherzolites with abundant normative and modal diopside are much less abundant. At high subsolidus temperatures (say, 800 to 1100°C) and pressures of the deeper crust and upper mantle (5 to 25 kb), the equilibrium assemblages consist of four phases: olivine, enstatite, diopside, and a pressure-dependent aluminous phase. This is anorthite up to 6 to 7 kb; spinel at intermediate pressures up to 15 to 18 kb; garnet at still higher pressures. Lower temperatures favor appearance of garnet at lower pressure. Partitioning of Ca, Mg, and Al between coexisting pyroxenes and spinel provides possible indices of temperature and pressure of final equilibration. O'Hara finds that the most useful indices are given by clinopyroxene: $100CaSiO_3/(CaSiO_3 + MgSiO_3)$, which is especially sensitive to temperature; and $100Al_2O_3/(CaSiO_3 + MgSiO_3 + Al_2O_3)$, which is more sensitive to pressure. His grid (Fig. 12-11), in spite of possible effects of Fe and Cr, has given a consistent pattern that perhaps approximates the spread of actual equlibration conditions.[1]

SPECIFIC INTRUSIVE ENVIRONMENTS

*Ultramafic belt, South Island, New Zealand* This linear array of ultramafic bodies, once nearly continuous for 300 km, is now disrupted by faulting and divided into a northern and a southern segment by mutual right-lateral displace-

**Table 12-4** Chemical compositions (oxides, wt %) of ultramafic rocks from orogenic belts

| | 1 | 2 | 3 | 4 | 5 | 6 |
|---|---|---|---|---|---|---|
| $SiO_2$ | 49.2 | 40.2 | 39.53 | 43.37 | 44.60 | 44.65 |
| $TiO_2$ | 0.2 | | 0.01 | 0.15 | 0.16 | 0.08 |
| $Al_2O_3$ | 2.4 | 1.7 | 0.93 | 1.19 | 4.18 | 3.50 |
| $Fe_2O_3$ | 2.0 | 0.9 | 0.65 | 0.18 | | |
| $Cr_2O_3$ | | | 1.01 | 0.56 | 0.37 | 0.59 |
| FeO | 6.8 | 9.3 | 7.62 | 7.41 | 8.30 | 6.81 |
| MnO | 0.2 | 0.2 | 0.12 | 0.14 | 0.09 | 0.14 |
| NiO | | | 0.32 | 0.14 | | 0.29 |
| MgO | 19.1 | 47.5 | 48.83 | 45.35 | 40.45 | 41.66 |
| CaO | 18.9 | | Tr. | 0.79 | 1.72 | 2.02 |
| $Na_2O$ | 0.2 | | Tr. | 0.15 | 0.11 | 0.23 |
| $K_2O$ | 0.1 | | | Tr. | 0.02 | 0.04 |
| $P_2O_5$ | | | | Tr. | | |
| $H_2O^+$ | 0.6 | 0.3 | 0.89 | 0.64 | | |
| $H_2O^-$ | 0.2 | 0.1 | 0.16 | 0.06 | | |
| Total | 99.9 | 100.2 | 100.15* | 100.13 | 100.00 | 100.01 |

* Including 0.02% CoO, 0.06% $CO_2$.

*Explanation of column headings*

1 Olivine pyroxenite, Union Bay, Alaska (Ruckmick and Noble, 1959, table 1, analysis 9)
2 Dunite, Union Bay, Alaska (Ruckmick and Noble, 1959, table 1, analysis 231)
3 Dunite, Dun Mt., New Zealand (Reed, 1959*a*)
4 Harzburgite, Red Hills, Nelson, New Zealand (Challis, 1965, p. 336, analysis 4)
5 Peridotite (Primary; recalculated water-free), Lizard, Cornwall (D. H. Green, 1964, p. 144, analysis V)
6 Garnet peridotite (mean of 3), Ugelvik, Norway (Carswell, 1968*a*, p. 344)

[1] The value of this model is weakened by the necessity to evaluate effects of Fe and Cr in an empirical manner untested by experiment.

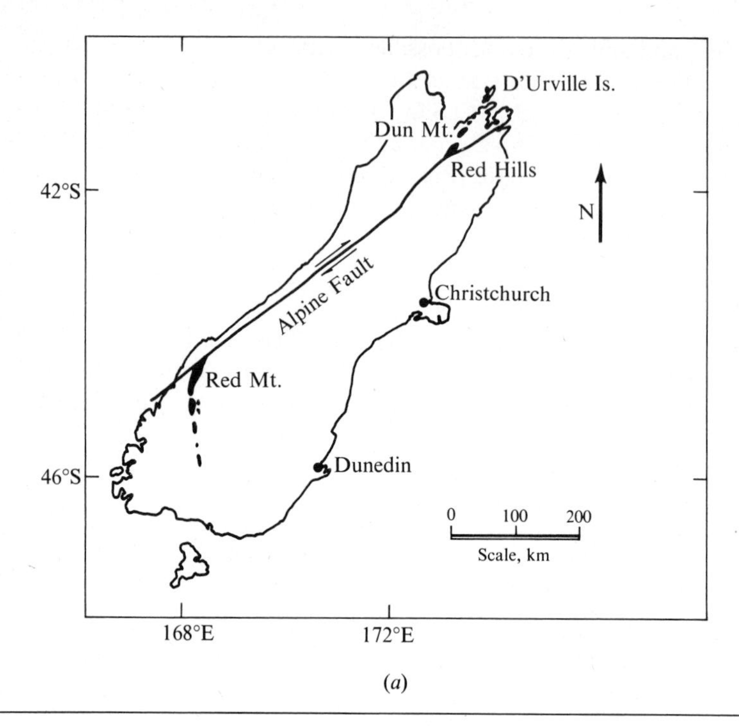


**Fig. 12-12** (*a*) Permian ultramafic rocks (solid black) of South Island, New Zealand; (*b*) ultramafic bodies of Dun Mountain belt, Nelson. Simplified after Walcott (1969, p. 58).

ment along the still-active Alpine fault (Fig. 12-12). The belt trends parallel to the regional structure of Permian and Triassic geosynclinal rocks that were folded during the Cretaceous Rangitatan orogeny (Verhoogen et al., 1970, pp. 309, 310, 574). Accompanying regional metamorphism (pumpellyite-prehnite facies with lawsonite) indicates synorogenic environmental temperatures of perhaps 250°C and pressures of about 4 to 5 kb. Everywhere the ultramafic bodies were emplaced at about the same stratigraphic level, among lower Permian spilitic lavas and breccias. Locally radiolarian cherts complete the Steinmann trinity (Reed, 1950).

The main rock types (Table 12-4; analyses 3, 4) are harzburgite, less plentiful dunite, and serpentinite. The latter is the sole component of minor bodies—some of them probably sills—and in the larger complexes it is concentrated especially along the margins. There are small lenses and crosscutting veins of pyroxenite. Gabbros are confined to thin bands in layered harzburgite-dunite plutons; and in some places there are narrow late dikes of hornblende microgabbro. Broad zones in the larger bodies have strongly cataclastic structure which in places is largely healed by late recrystallization.

One of the largest individual bodies—Red Hill complex, Nelson—is a tilted and faulted sheet 3 to 4 km thick, today exposed over an outcrop area of 120 km$^2$ (Challis, 1965; Wallcott, 1969). The basal zone—1 km thick—is

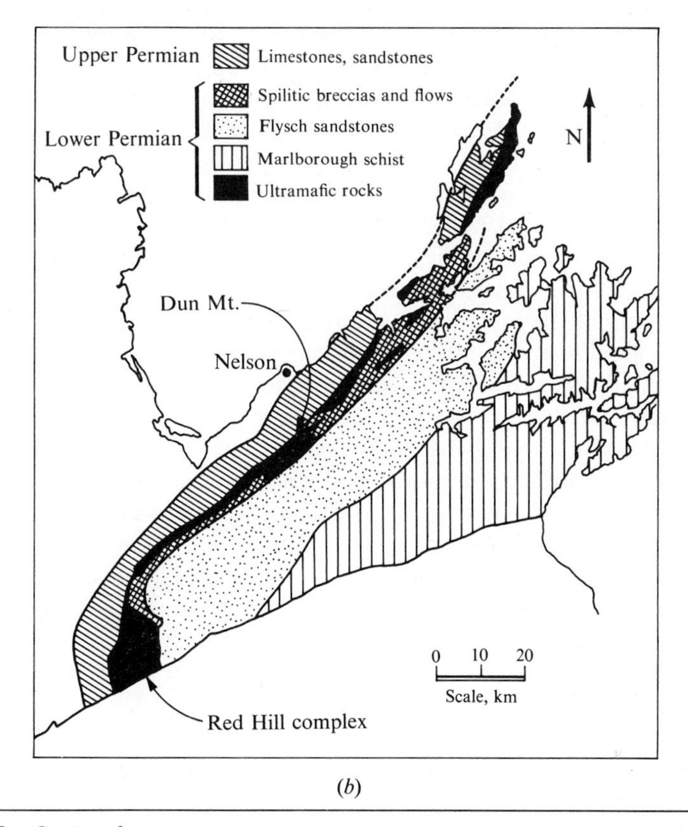


(*b*)

**Fig. 12-12** *Continued.*

partially recrystallized cataclastic harzburgite (olivine 72 percent, enstatite 23 percent, diopside 3 percent, chrome spinel 2 percent). The rest of the pluton is interlayered harzburgite and dunite, with minor bands of anorthite gabbro. Accessory anorthite is commonly present, along with chromian spinel, in both the principal ultramafic assemblages. Where the upper contact with geosynclinal spilites is preserved, there is a wide high-temperature metamorphic aureole.

Diopsides from harzburgite samples collected from the three largest plutons (Challis, 1965) plot on O'Hara's grid (in Wyllie, 1967, p. 402) at equilibration temperatures near the solidus (1100 to 1200°C) and pressures below 4 kb. These conditions are consistent with the nature of the contact aureole and with prevalence of minor anorthite in Red Hill harzburgites. The total picture is one of crystallization at near-fusion temperatures in a low-pressure crustal environment. Layering and local poikilitic texture have been interpreted by some as evidence of crystallization from a magma. But there is nothing to show whether the magmatic stage, if there was one, occurred at the site of present emplacement or earlier at greater depth. The latter alternative would necessarily imply postmagmatic equilibration of pyroxenes at the present site.

*Lizard peridotite intrusion, Cornwall* The Lizard peridotite intrusion of southern Cornwall is a pluglike body, 70 $km^2$ in outcrop area, cutting a hornblende-schist metamorphic complex of uncertain age. It has been examined by D. H. Green (1964), who established it as the type for high-temperature emplacement of ultramafic material (probably crystalline) transported from a much deeper environment, the imprint of which is still legible in the mineral chemistry of the central core. The primary minerals of the core rocks are olivine, enstatite, diopside, and olive-green spinel; the principal rock type is harzburgite, in many places layered by rhythmic concentration of pyroxene on the scale of 1 to 15 cm. The outer shell of the intrusion is strongly sheared and has recrystallized to a secondary peridotite assemblage—olivine-enstatite-diopside, with accessory plagioclase and chromite. Both zones are widely serpentinized.

Pyroxenes of the core zone are notably more aluminous ($Al_2O_3$ 2 to 3.5 percent in enstatite, $\approx$ 4 percent in diopside). On O'Hara's grid (in Wyllie, 1967, p. 402) the primary diopsides plot below the solidus at 1100 to 1200°C and pressure about 17 kb (depth 55 to 60 km). The secondary diopsides plot at 1100°C, 7 to 9 kb (depth 30 km).

Amphibolitic country rocks at the contact have been converted to the high-grade metamorphic assemblage hornblende-hypersthene-diopside-plagioclase diagnostic of the pyroxene-hornfels and granulite facies. This mineral assemblage is consistent with locally elevated aureole temperatures (perhaps 700 to 800°C) at a wide range of crustal depth.

*Coast Ranges province, California and Oregon* Much of the Coast Ranges of California and Oregon is made up of later Jurassic and Cretaceous geosynclinal sediments, volcanics, and associated bodies of ultramafic rocks (Fig. 12-13). There are two coeval marine facies, each perhaps 15 km in stratigraphic thickness (Bailey et al., 1964, 1970). To the west is the eugeosynclinal Franciscan assemblage—graywackes, cherts, and spilitic lavas and tuffs. This is flanked eastward, across a thrust-fault contact, by miogeosynclinal sandstones, shales, and conglomerates along the western margin of the Great Valley of California. This is the Great Valley sequence. Both facies, particularly the Franciscan, were strongly deformed during later Cretaceous times and still later have been affected by major dislocation on spectacular transcurrent faults still active. All this disturbance is currently attributed (for example, M. C. Blake et al., 1969; Medaris and Dott, 1970; B. M. Page, 1972) to locally changing but protracted subduction of a western oceanic plate veneered by Franciscan rocks beneath an eastern essentially continental plate whose outer margin, floored by Jurassic oceanic crust, is mantled by rocks of the Great Valley sequence. Much of the Franciscan terrace has been affected by low-temperature high-pressure metamorphism in the prehnite-pumpellyite and glaucophane-lawsonite-schist facies (perhaps 150 to 300°C, 4 to 10 kb). Local amphibolite and eclogite masses indicate derivation from an environment of somewhat higher temperature and perhaps pressure as well.

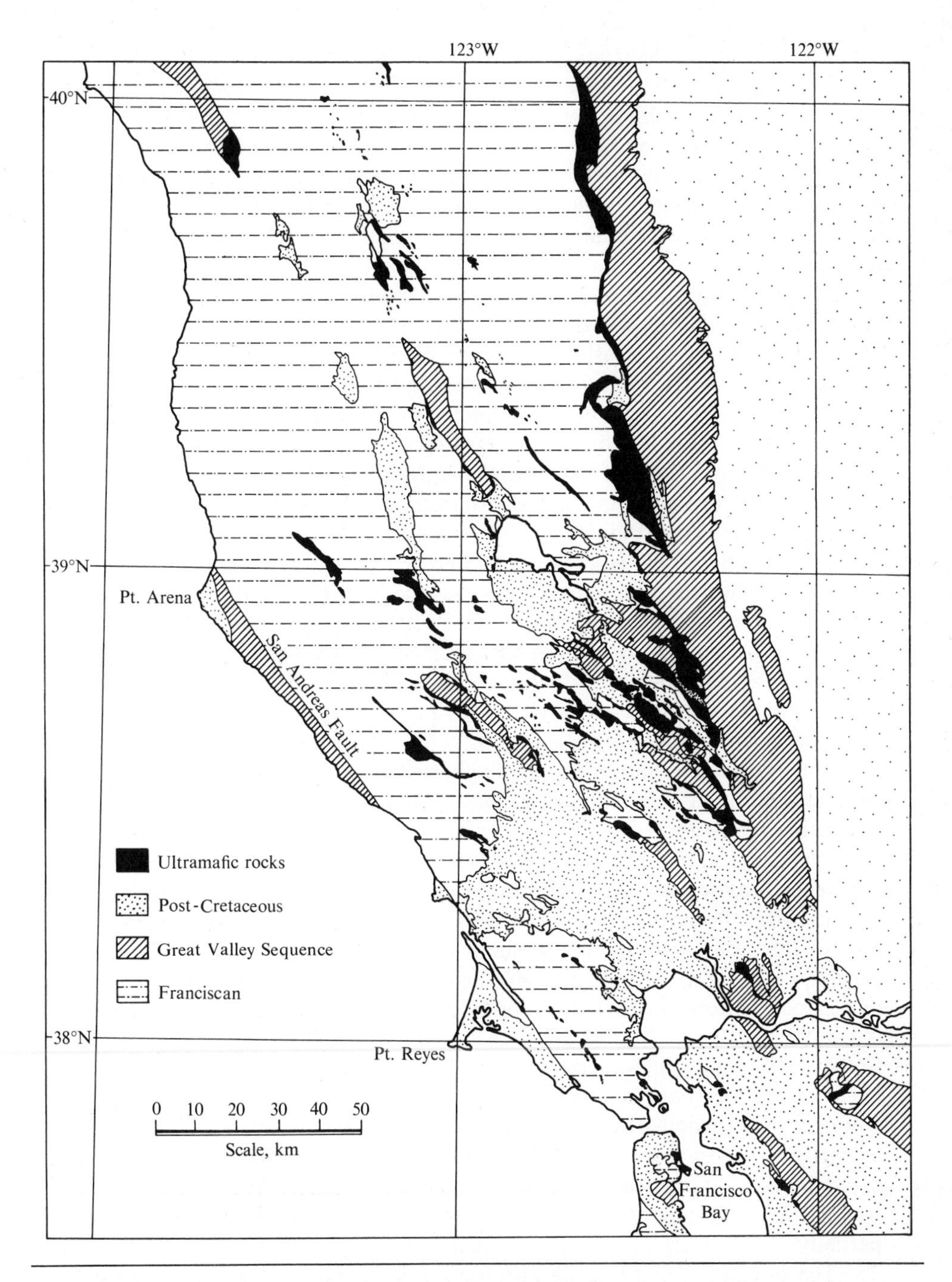


**Fig. 12-13** Distribution of ultramafic bodies (solid black) in coast ranges north of San Francisco, California. Simplified after E. H. Bailey et al. (1964).

The nature of the basement underlying the Franciscan assemblage is unknown. Much of the terrace has been reduced by repeated deformation to a chaotic structural mélange, prominent constituents of which are bodies of ultramafic rock. Most of these have been largely converted to serpentinite, and final emplacement at their sites of present outcrop has been by "cold" (tectonic) intrusion. Some of these bodies in a highly brecciated state seem to be still in motion along major transcurrent faults such as the active Hayward and San Andreas faults close to San Francisco. The great thrust contact between the Franciscan and the Great Valley sequence is marked by an ophiolite association in places 1500 m thick and at the northern end continuous in outcrop for 200 km. This ophiolite series consists of massive basal serpentinized peridotite and pyroxenite overlain by spilitic basalts and diabases, cherts, and even keratophyres. This whole association is interpreted in today's fashion as mid-Jurassic (155 m.y.) oceanic crust upon which were deposited the later Jurassic (140 m.y.) lower members of the Great Valley sequence (Bailey et al., 1970; Lanphere, 1973).

Special interest attaches to a few ultramafic bodies that have largely escaped serpentinization. Trivial low-temperature contact effects show that these, too, have been emplaced at temperatures no greater than a few hundred degrees. But their mineral composition still preserves the imprint of early crystallization in environments of high temperature and in some cases of very high pressure. Two examples, both situated in an environment of mélange, are a block 2 km in diameter at Burro Mountain, latitude 36°S (Loney et al., 1971) and a cluster of somewhat larger bodies at latitude 42 $\frac{1}{2}$°S in southernmost Oregon (Medaris and Dott, 1970; Medaris, 1972). Each is composed mainly of harzburgite with the primary mineral assemblage olivine-enstatite-diopside-spinel. Both are strongly deformed internally, and both have been partially serpentinized. On the basis of Mg partition between enstatite and spinel, Loney and coauthors place the equilibration temperature at Burro Mountain at about 1200°C. Both pyroxenes contain less than 3 percent $Al_2O_3$, implying final crystallization at low pressures. Primary pyroxenes in the Oregon rocks, on the other hand, are highly aluminous, and the composition of diopside (with $Al_2O_3$ 4.9 to 6.4 percent) indicates equilibrium at 1100°C and 16 to 20 kb. Recrystallized pyroxene in sheared matrices has partially adjusted to pressures between 5 and 15 kb.

*Garnet peridotites, southwestern Norway* In southwestern Norway about 250 km north of Bergen a number of small ultramafic bodies occur as tectonic blocks in the Caledonian metamorphic basement. Some of these consist largely or partly of garnet peridotite (Mercy and O'Hara, 1965; O'Hara, pp. 168–171 in Wyllie, 1967; Carswell, 1968*a*, 1968*b*; O'Hara et al., 1971, pp. 62–63). The regional metamorphic environment is one of high temperature (granulite partially retrogressing to amphibolite facies).

The primary garnet-bearing ultramafic assemblage, where it has survived with minimal modification (Carswell, 1968*a*), has the modal composition olivine 76 percent, garnet 12, enstatite 7, and chromian diopside 4 (Table 12-4, analysis 6). All phases are highly magnesian (garnet is 70 percent pyrope), and the diopside has a significant content of jadeite. Both pyroxenes, as to be expected in the garnet-peridotite field, are relatively low in aluminum. Estimated equilibrium conditions are pressure 35 kb and temperature 1000 to 1100°C—well below the solidus. Other peridotites (Fig. 12-11*A*) have nearly reequilibrated at 600°C, 18 to 20 kb (O'Hara, p. 402 in Wyllie, 1967). Above 850°C the garnet-lherzolite assemblage has been found in the laboratory to invert rapidly to spinel herzolite. So survival of garnet peridotites at the surface implies that upward transport through 30 or 40 km in the upper mantle must have been rapid indeed (O'Hara et al., 1971, p. 62).

PETROGENESIS Wyllie (1967, pp. 409–415) has critically reviewed nine separate petrogenic models for Alpine-type peridotites, all with recent or current advocates. Here we present three of these models that, in each particular case, appear to be consistent with available field and geochemical data.

*1. Gravitational differentiation of basic magma in the crust* Layered and poikilitic structure so common among Alpine peridotites, although open to other interpretations, immediately suggest analogy with stratified basic sheets. Such sheets indeed have been emplaced early in the tectonic history of some geosynclines; and in the course of later folding they have been disrupted and deformed to the point where only careful field work can establish the original identity and continuity of severed ultramafic segments. Thus, in northeastern Scotland there are many exposures of layered differentiated gabbro, some partially metamorphosed, enclosed in Dalradian metasediments. At some localities the rocks are essentially ultramafic (feldspathic peridotite), internally layered and now tilted at high angles. These may be parts of one or of several differentiated gabbro sheets; but they were certainly emplaced early in the Caledonian orogeny and were strongly deformed and perhaps disrupted during the culmination of folding (for example, F. H. Stewart, 1970; Weedon, 1970, p. 39). In Connemara, western Ireland, disruption of an early Caledonian differentiated basic sheet went much further. Gabbros were largely amphibolitized. One large layered ultramafic mass, the Dawros peridotite (Rothstein, 1957), was severed from the rest, bodily emplaced as an Alpine-type intrusion, and turned upside down in the process (Leake, 1970). Primary and late-recrystallized pyroxenes in this body respectively record two stages of equilibration: first at temperatures above the solidus and a depth of 25 to 30 km and later, still near the solidus, close to the surface (Rothstein, 1958; Wyllie, 1967, p. 402, fig. 12-6).

It is tempting to go further and propose a similar model for most Alpine peridotites even where connection with associated basic rocks is not obvious. Just such an origin has been argued by Challis (1965; also Challis and Lauder, 1966) for peridotites of New Zealand's ultramafic belt. In associated spilitic lavas she sees the complementary basic liquid fraction decanted from the ultramafic cumulate at no great depth and erupted along a volcanic island arc.

Others have even suggested that the whole ophiolite rock assemblage—to use the words of Wyllie (1967, p. 412)—comprises "massive differentiated lava flows ranging in composition from basic to ultrabasic, together with some intrusive rocks." Here are models that fit many observations and give a nice explanation of the Steinmann trinity. But, to quote Challis and Lauder (1966, p. 2) writing in support of their basalt-differentiation model, "it might thus be expected, if this method of formation is correct, that ultramafic rocks of the Alpine type would not only be genetically related to the ultramafic rocks of the layered intrusions, but also that their chemical and mineralogical constitution would be consistent with derivation from a basaltic magma." It is in this very respect that the hypothesis of basaltic origin runs into severe, but not insuperable, difficulty. Alpine peridotites, including New Zealand dunite, have a peculiar trace-element pattern that has no parallel among common basalts. They are extremely low in alkalis, Rb/Sr is very low, and $Sr^{87}/Sr^{86}$ is unexpectedly high: 0.707 to 0.727 (Stueber and Murthy, 1966).

*2. Serpentinite as oceanic crust* H. H. Hess (1955, 1962, 1964) developed the hypothesis that oceanic crust ("layer 3") is serpentinized mantle peridotite. The water for the hydration process was derived from convecting mantle material and has been expelled continuously from the oceanic ridges along with water now stored in the ocean. Great masses of serpentinite exposed today in Cuba and Puerto Rico were seen as slabs of oceanic crust, naked of cover, thrust above sea level in a tectonically active zone (Greater Antilles arc). It is but a step further to ascribe a similar origin to Alpine serpentinites: They could be blocks of oceanic crust thrust over the edges of moving continental plates.

At least some aspects of Alpine ultramafics cannot be explained thus. Obviously peridotite bodies that are only partially serpentinized could not have come from oceanic crust. Serpentinization in at least some geosynclinal belts has occurred within the present general environment of emplacement and has been effected through the agency of circulating meteoric waters. But records of serpentinite in ocean dredge hauls are increasingly numerous, and here the hydration agent may have been seawater itself. Oxygen-isotope partition between serpentine minerals and magnetite suggests that temperatures of serpentinization were lower in mid-Atlantic than in orogenic serpentinites (Wenner and Taylor, 1971).

*3. Alpine peridotites as mantle fragments* In 1957 De Roever proposed that Alpine-type peridotite bodies might be tectonically transported fragments of a peridotite mantle. The idea has since grown steadily in favor and today is an accepted concept of plate tectonics. Well-documented cases (for example, Loomis, 1972*b*) are supported by consistent field and geophysical data; and some intrusions must have reached their present site at temperatures in excess of 1000°C since the mineralogy of their extensive contact aureoles implies a range of about 700 to 800°C (D. H. Green, 1964; Loomis, 1972*a*).

Estimated pressures of equilibration of mineral assemblages of some spinel peridotites correspond to depths of more than 50 km; corresponding estimated temperatures are high in the subsolidus field. Other spinel peridotites such as those of southern New Zealand have finally equilibrated at high temperature but much lower pressure. These could well have come from much greater depth (in the mantle) and then readjusted while still hot to pressures in the crust. At least one Alpine-type garnet peridotite from Norway retains the imprint of equilibration at about 100 km depth, the temperature being a few hundred degrees below the solidus. The relative rarity of such rocks and their characteristically altered state are consistent with experimentally demonstrated rapid rates of inversion to spinel peridotite as pressure drops below 15 to 20 kb (O'Hara et al., 1971, p. 62).

Alkali-metal abundances, relative abundance patterns of rare-earth elements, and isotopic composition of strontium in Alpine peridotites from widely separated provinces show characteristic and consistent features mainly different from those of ultramafic rocks in layered basic intrusions (for example, Stueber and Murthy, 1966; Frey et al., 1971; Goles, pp. 352–362 in Wyllie, 1967). Abundances of Na, K, Rb, and Sr all are very low (Table 12-5), and K/Rb also is consistently low. A number of rocks show strong depletion in the lighter rare-earth elements compared with the standard pattern in chondrites, whereas in ultramafic rocks of the Stillwater type there is no such depletion. $Sr^{87}/Sr^{86}$ values for many Alpine-type peridotites are surprisingly high, and Rb/Sr is so low that reduction with backward projection in time is insignificant (– 0.001 per $10^9$ years). The composition of such strontium alone—provided

**Table 12-5** Alkali-metal abundances (ppm) and strontium chemistry of ultramafic rocks from orogenic zones (Stueber and Murthy, 1966)

| | 1 | 2 | 3 | 4 | 5 | 6 |
|---|---|---|---|---|---|---|
| Na | 97 | 126 | 81 | 1200 | 119 | 117 |
| K | 24 | 61 | 15 | 26 | 44 | 19 |
| Rb | 0.11 | 0.30 | 0.08 | 0.09 | 0.13 | 0.07 |
| Sr | 4.4 | 6.3 | 3.0 | 3.9 | 9.9 | 2.3 |
| K/Rb | 215 | 203 | 195 | 277 | 337 | 268 |
| Rb/Sr | 0.025 | 0.048 | 0.026 | 0.024 | 0.013 | 0.031 |
| $Sr^{87}/Sr^{86}$ (present) | 0.7091 | 0.7078 | 0.7156 | 0.7084 | 0.7078 | 0.7101 |

*Explanation of column headings*

1 Dunite, Dun Mt., New Zealand
2 Dunite, Papua
3 Dunite, North Carolina
4 Peridotite, Tinaquillo, Venezuela (high-temperature intrusion)
5 Dunite, Almklovdalen, Norway (Caledonian dunite associated with garnet peridotite)
6 Dunite, Cartwell, Alaska (zoned peridotite intrusion)

that it has not been changed by late contamination[1]—rules out the possibility that these Alpine peridotites could have any contemporary genetic connection with most kinds of basalt. Not all Alpine-type peridotites, however, have such highly radiogenic strontium. Initial $Sr^{87}/Sr^{86}$ values between 0.7023 and 0.7047 have been determined for a suite of rocks from a South Carolina locality in the Appalachian peridotite belt (Jones et al., 1973).

Thus the combined evidence of phase equilibria and geochemistry strongly favors derivation of *some* Alpine-type peridotites in the solid state from within the upper mantle. There is at present every reason to suppose that *most* but not all such bodies and their serpentinized equivalents came in the same manner from the same general source. Students of plate tectonics today incline to the view that the whole ophiolite association (the Steinmann trinity) represents old oceanic crust.[2] If this is so, then one ingredient of this crust must be partially serpentinized peridotite brought from deep in the outer mantle. This view is not so far removed from the serpentinite-crust model proposed by H. H. Hess when, as early as 1960, he pioneered thinking along modern global-tectonic lines (H. L. James, 1973, p. 115).

[1] Because total Sr is so low, peridotites are particularly vulnerable to contamination by brines with a high content of Sr ($Sr^{87}/Sr^{86}$ = 0.707 to 0.709 in seawater, 0.714 to 0.720 in some river and lake waters).

[2] For a review of the growth of this concept, see Dewey et al. (1973, pp. 3419–3150).

# Petrologic Assessment of the Mantle-Magma System

# 13

## Scope of Inquiry

The five preceding chapters are largely descriptive. They illustrate the nature and variety of igneous rock associations, with emphasis on tectonic setting and on patterns of internal diversification. The impact of all this information on broader geologic thinking can be gauged in terms of three criteria: consistency and predictability of igneous phenomena; insight given with respect to allied geologic phenomena, orogeny, plate tectonics, and the nature of the mantle; and finally an inventory of major questions of petrology still unanswered. On this basis it should be possible to assess the current status and influence and the future potential of petrologic inquiry within the general framework of geoscience.

One of the ultimate problems of igneous petrology is to identify the source rocks of magmas. At the outset opinions differ as to how a particular magma may be identified as "primary" in the sense that it is a product of direct fusion of mantle or deep crustal rock. It has even been denied that the most likely candidates—oceanic basalts—ever reach the surface without previous significant modification by fractional crystallization en route from the source (O'Hara, 1965). Nevertheless, the possibility—in our opinion the probability—that some common mafic lavas are essentially primary in the above sense must be scrutinized. Even those who consider that all magmas reaching the surface have been modified to some degree along the path of ascent are much concerned with the composition of the original magmas that must have existed prior to such modification (for example, O'Hara, 1965). Also the nature of possible source rocks must be examined and a search made for fragments that may have reached the earth's surface without significant change.

With regard to the nature of rocks in the upper mantle, we start with two propositions that most geologists would accept as axiomatic: The upper mantle is composed of dense silicate rocks ($\rho \geq 3.3$ g cm$^{-3}$); all visible terrestrial rocks are extracts derived directly or indirectly from the mantle. By searching their chemistry and mineralogy it should be possible to isolate and identify some traits that can be extrapolated back to mantle sources. Most promising, because their mode of origin and high density imply minimal departure from mantle composition, are basic and ultrabasic igneous rocks. Attention is focused particularly on rocks of this kind that are found in the oceanic environment; here there is least chance of modification by sialic crust.

From igneous petrology, then, limitations may be set on the gross chemistry and mineralogy of upper-mantle rocks.[1] Essential to this approach is information obtained from high-pressure experiments on appropriate crystal-melt and subsolidus phase equilibria. Models that emerge must be tested against the rapidly accumulating data of trace-element and isotope chemistry of basic and ultrabasic igneous rocks, since relative concentrations of such elements in coexisting melt and residual crystalline fractions can be predicted for different degrees and regimes of partial fusion.

Whatever model may be preferred, a number of fundamental issues will remain to be resolved. There is the problem of original versus secondary nature of common magmas—siliceous as well as basic—and the scope of modification, especially with respect to "incompatible" dispersed elements and isotope chemistry, of magmas ascending through the crust. Most important in their broad geologic implications are the degree and scale of heterogeneity—in the lateral and in the vertical sense—of the upper mantle itself and possible changes in petrogenic processes within it throughout geologic time.

Fundamental to the whole scope of present inquiry is the notion of homogeneity applied to systems under scrutiny. A system is homogeneous when any two samples of equal volume are chemically and physically identical. The size of the sample defines the scale of homogeniety. The seismologist, viewing the mantle on a very large scale, sees a layer of great lateral extent as homogeneous and assigns to it specific properties. Mineralogists and geochemists work on much smaller scales: their samples are hand specimens or even smaller. On these scales, by analogy with the crust, the mantle is likely to be heterogeneous. There need be no conflict between the homogeneous condition of a mantle layer inferred from seismic data and marked small-scale heterogeneity demonstrated by geochemical investigation.

## Chemical Background: High-pressure (Experimental) and Dispersed-element Chemistry

### MAJOR-ELEMENT PHASE CHEMISTRY: DATA OF HIGH-PRESSURE EXPERIMENTS

THE EXPERIMENTAL APPROACH Design of high-pressure experiments has been geared to Bowen's classic proposition: Primary mafic magmas are liquid

[1] A clear and concise essay along these lines is that of Wyllie (1970).

fractions that become separated from more refractory crystalline residues during the fusion of ultramafic mantle rocks. The data that have emerged concern major-element chemistry of coexisting phases at temperatures ranging from liquidus to subsolidus and pressures up to 30 kb in model systems. The liquid phase at any specified stage (mostly 5 to 30 percent) of melting is in equilibrium with coexisting crystalline phases. This is the fusion regime that in mathematical models is termed *equilibrium partial melting* (batch melting).[1] There are three lines of attack:[2]

1. Study of phase equilibria in simple pure systems such as $CaMgSi_2O_6$–$Mg_2SiO_4$–$SiO_2$ or $CaMgSi_2O_6$–$Mg_3Al_2Si_3O_{12}$

2. Experiments on partial fusion of natural peridotites and basalts, with or without water

3. Investigation of crystallization behavior (liquidus-solidus phase equilibria) and subsolidus equilibria in complex artificial systems designed to duplicate possible mantle source rocks and their primary magmatic extracts

IGNEOUS MINERALS AT HIGH PRESSURE The stable existence of feldspar at low pressures in magma is self-evident, but at high pressure the familiar solidus-liquidus loop of the plagioclase feldspar no longer exists. In the absence of water, $CaAl_2Si_2O_8$ melts incongruently above 10 kb to corundum and liquid while $NaAlSi_3O_8$ melts incongruently to jadeite and liquid above 32 kb (D. H. Lindsley, 1968). $KAlSi_3O_8$ is stable to much higher pressures (Seki and Kennedy, 1964), and in the absence of water and fluorine it could probably be stable in many rocks equilibrating at high pressures in the mantle (T. H. Green and Ringwood, 1968). Pyroxenes and garnets will reduce the stability field of plagioclase by accepting the high-pressure breakdown components into solid solution; thus $CaAl_2SiO_6$ and $NaAlSi_2O_6$ enter pyroxene and $Ca_3Al_2Si_3O_{12}$ will go into garnet, and so the low-pressure aluminous minerals of lavas, the plagioclase feldspars, will be displaced at high pressure by solid solutions in garnet, spinel, and pyroxene. Sanidine alone could still persist, although in the presence of olivine, water, and fluorine it would be eliminated in favor of phlogopite; amphibole could also exist at high pressure and incorporate potassium, with, as in phlogopite, various amounts of the trace elements Ba, Ti, Zr, Ni, Rb, and Cr.

[1] To be distinguished from ideal fractional melting.

[2] For experimental detail the reader is referred to Annual Reports of the Geophysical Laboratory in *Carnegie Institution of Washington Year Books* 62, pp. 66–103, 1963; 63, pp. 147–172, 1964; 68, pp. 240–247, 1970. Comprehensive reviews and pertinent data will be found in works by O'Hara and Yoder (1967), D. H. Green and Ringwood (1967*a*, 1967*b*), Ito and Kennedy (1968), O'Hara (1968), and D. H. Green (1970*b*).

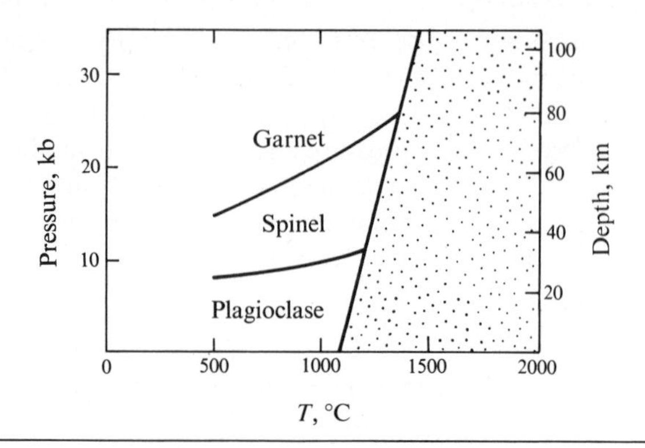


**Fig. 13-1** Generalized diagram to illustrate melting behavior of anhydrous peridotite. Field of coexisting crystals and liquid, stippled (after Wyllie, 1970, p. 22).

CHOICE OF SOURCE ROCKS FOR EXPERIMENTAL MODELS Common highly magnesian peridotites such as harzburgite and dunite turn out to be chemically inadequate to yield melts duplicating the composition of basalt or nephelinite. Even the earliest melt fractions are much too high in Mg. Their normative *an* (an index of potential feldspar) is too low and their alkali content negligible. Such rocks probably exist in the upper mantle, but they cannot be the sources of voluminous basaltic magmas. So experimental models start with more promising material.

Ringwood (D. H. Green and Ringwood, 1967*b*, p. 159; D. H. Green, 1970*a*, p. 31; Wyllie, 1970, p. 18) proposes an upper mantle consisting of "pyrolite," whose chemical composition equals three parts of harzburgite plus one of Hawaiian olivine tholeiite (Table 13-1, analysis 1). Ito and Kennedy (1968, p. 189) have chosen starting materials closer to basalt: one is a "picritic' mixture of peridotite and olivine tholeiite in equal parts (Table 13-1, analysis 2). O'Hara's choice is a natural garnet lherzolite occurring as xenoliths in kimberlite (Table 13-1, analysis 3). Chemically some of these garnet peridotites are not far from Ringwood's pyrolite.[1] Their melting behavior can be compared with that in the artificial diopside-pyrope system. Eclogites, in spite of their chemical similarity to basalts, have received less attention (Yoder and Tilley, 1962, pp. 470–518; O'Hara, 1963*b*; D. H. Green and Ringwood, 1967*b*).

LHERZOLITE AND PYROLITE SYSTEMS

*Solidus and subsolidus phase assemblages* At solidus and subsolidus temperatures and pressures up to 30 kb (100 km depth), stable anhydrous assemblages consist of four phases: olivine, enstatite, diopsidic pyroxene, and a phase high in aluminum. The nature of the aluminous phase depends on pressure more than temperature—with $dP/dT$ positive for transition boundaries.[2] In order of increasing pressure at 800 to 1000°C (Fig. 13-1), plagioclase gives way to

[1] For example, BD 1140 and BD 1150 of Carswell and Dawson (1970, p. 169).
[2] Shown as lines in Fig. 13-1, but, strictly speaking, narrow fields of five-phase assemblages.

**Table 13-1** Chemical compositions (oxides, wt %) and CIPW norms of model materials and rocks relevant to mantle sources of mafic magmas

| | 1 | 2 | 3 | 4 | 5 |
|---|---|---|---|---|---|
| $SiO_2$ | 45.16 | 47.88 | 44.95 | 45.10 | 44.57 |
| $TiO_2$ | 0.71 | 0.88 | 0.11 | 0.13 | 1.76 |
| $Al_2O_3$ | 3.54 | 9.40 | 2.93 | 3.92 | 13.61 |
| $Fe_2O_3$ | 0.46 | — | 2.23 | 1.00 | 4.17 |
| $Cr_2O_3$ | 0.43 | 0.24 | 0.50 | 0.31 | 0.06 |
| FeO | 8.04 | 9.27 | 4.82 | 7.29 | 8.49 |
| MnO | 0.14 | 0.17 | 0.13 | 0.14 | 0.21 |
| NiO | 0.20 | 0.13 | 0.26 | 0.25 | 0.04 |
| MgO | 37.47 | 24.60 | 36.12 | 38.81 | 13.34 |
| CaO | 3.08 | 5.55 | 3.35 | 2.66 | 11.42 |
| $Na_2O$ | 0.57 | 1.54 | 0.38 | 0.27 | 1.69 |
| $K_2O$ | 0.13 | 0.23 | 0.28 | 0.02 | 0.02 |
| $P_2O_5$ | 0.06 | 0.11 | 0.03 | 0.01 | 0.02 |
| $H_2O^+$ | | | 3.56 | 0.07 | 0.29 |
| $H_2O^-$ | | | 0.20* | 0.12 | 0.14 |
| Total | 99.99 | 100.00 | 99.85 | 100.10 | 99.83 |
| *or* | | 1.33 | 1.65 | | 0.12 |
| *ab* | | 13.00 | 3.22 | | 14.30 |
| *an* | | | 5.46 | | 29.49 |
| *ne* | | 18.01 | | | |
| *di* | | 6.96 | 7.62 | | 21.40 |
| *hy* | | 19.56 | 25.97 | | 3.92 |
| *ol* | | 38.77 | 47.86 | | 20.61 |
| *mt* | | | 3.23 | | 6.05 |
| *il* | | 1.67 | 0.21 | | 3.34 |
| *im* | | 0.36 | 0.74 | | 0.09 |
| *ap* | | 0.27 | 0.07 | | 0.05 |
| Total | | 99.93 | 96.03 | | 99.37 |

* $CO_2$.

*Explanation of column headings*

1 Pyrolite: three parts harzburgite (synthetic), one part Hawaiian tholeiite (D. H. Green and Ringwood, 1967*b*, p. 160, table 20)
2 Synthetic picrite: equal parts of peridotite and olivine tholeiite (Ito and Kennedy, 1968, p. 180, B1P1)
3 Garnet lherzolite: xenolith in South African kimberlite (Carswell and Dawson, 1970, p. 169, BD 1150)
4 Spinel lherzolite: xenolith, Honolulu series, Oahu, Hawaii (E. D. Jackson and Wright, 1970, p. 417, D102102)
5 Garnet pyroxenite (clinopyroxene 61 percent, garnet 37 percent): xenolith, Honolulu series, Oahu, Hawaii (Beeson and Jackson, 1970, pp. 99, 101, 68 SAL-11)

spinel at about 10 kb, and spinel to garnet at 18 kb (O'Hara et al., 1971). Corresponding reactions in simplified form are

1. Plagioclase + olivine → aluminous pyroxenes + spinel
2. Orthopyroxene + spinel → pyrope + olivine

Figure 13-1 gives only a generalized picture. Field boundaries will be displaced as chemical parameters are changed. Garnet appears on the solidus only above 27 kb in Ringwood's pyrolite system but comes in at 22 kb in Ito and Kennedy's picrite.

Much of the aluminum of spinel peridotites is accommodated in both pyroxenes. In garnet peridotites, pyroxenes are less aluminous. Above the plagioclase-spinel boundary, sodium is held exclusively in clinopyroxene.

Most peridotites are essentially anhydrous; and except under special conditions melting of peridotite in the upper mantle will also be nearly anhydrous. Possible hydrous phases are pargasite and, if potassium should be present, phlogopite. Both are known in rare occurrences of natural peridotite. Kushiro (1970, pp. 246, 247) finds that at 30 kb (100 km depth) phlogopite remains stable up to about 1000°C; pargasite breaks down at rather lower temperatures.

*Results of equilibrium partial melting* The changing composition of early liquid fractions generated by batch melting of anhydrous peridotites depends on the phases stable at solidus temperatures and on the order in which they become eliminated as the volume of the melt fraction increases with rising temperature (cf. D. H. Green and Ringwood, 1967*b*, pp. 159–167; O'Hara, 1968, pp. 126–128; D. H. Green, 1970*a*, p. 48). Pertinent observations include the following:

1. In the pyrolite model the melt fraction at 25 percent melting must inevitably be olivine tholeiite (what went into the pyrolite system—25 percent olivine tholeiite—must finally come out). We are concerned with the nature of the first-formed melts—at 5 or 10 percent melting—and the path along which each changes to olivine-tholeiite liquid if equilibrium is continuously maintained in a closed system.

2. Incongruent melting of $MgSiO_3$—a key factor in Bowen's scheme of petrogenesis—persists only to 5 kb. So anhydrous peridotites can yield *Q*-normative tholeiite melt fractions only at shallow depths ($< 15$ km). This is an obvious anomaly in view of the vast development of corresponding magmas on the continents, where the Moho discontinuity is much deeper than this.

3. Plagioclase is preferentially eliminated during early melting at pressures up to 10 kb. Between 5 and 10 kb the product, still limited by geologic requirements to the suboceanic mantle, will be high-alumina olivine-tholeiitic melt.

4. In the pressure range 10 to 20 kb—below even the continental Moho discontinuity—preferential melting of aluminous clinopyroxene (which holds all sodium) plays a decisive role. The first melt fraction is *ne*-normative. As melting progresses, enstatite is withdrawn increasingly into the liquid; normative *ne* drops to zero and thenceforward *hy* increases as the liquid moves toward the ultimate olivine-tholeiite composition.
5. From 20 to 30 kb increasing pressure reduces *ne* and increases *ol* in the norm of even the earliest liquid fractions. These have a composition analogous to picritic olivine tholeiite. Ito and Kennedy found much the same result in their picritic system where garnet is stable in the solidus assemblage.
6. In slightly hydrous peridotites, pargasite and/or phlogopite remain stable up to solidus temperatures, at which point their content of water will be withdrawn almost in its entirety into the melt. In the incipient stage of fusion the small quantity of melt phase could thus contain significant water. Kushiro (1970, 1972*a*) finds it to be silica-saturated up to pressures of 20 or even perhaps 30 kb. By later fractionation such a melt could yield "andesitic" derivatives in small quantity. Kushiro's model of hydrous melting could partially dispose of the anomaly indicated under (1) above. Given an abundant influx of water from some external source, quartz-tholeiitic magmas could be generated from mantle pyrolites far below the continental Moho discontinuity, but only in small amounts.

*Results of fractional crystallization* Potentially significant in petrogenesis are consequences of fractional crystallization of the melts formed by partial fusion of mantle rocks (for example, O'Hara, 1965; O'Hara and Yoder, 1967; D. H. Green, 1970*a*, 1970*b*). Fractional crystallization is not simply the reverse of the fusion process; rather it begins in a new chemical system where fusion stopped and the melt, now in a cooling regime, has been separated from contact with its source rock. Crystalline phases of the first potential cumulates are those stable just below the liquidus at this point. The path of liquid evolution in any given system over the same temperature interval will lie somewhere between two extreme differentiation regimes (Neumann et al., 1954). In one the cumulate fraction remains in contact and in equilibrium with the diminishing liquid residue until the latter is decanted as a now-isolated differentiate. In the other—which is approximated perhaps in layered basic intrusions—each small crystalline increment at once becomes isolated from the liquid fraction. Both cases are amenable to treatment by computer models.

The pyrolite model provides many possibilities; and it is not required that the liquid phase (now the parent liquid) should necessarily have been generated in the first place from pyrolite. This is important. A tiny quantity of nephelinitic melt may form from pyrolite (olivine-enstatite-clinopyroxene) at 15 kb. But a large volume of identical magma formed by some other means will embark on the same fractionation course as it cools at the same pressure. Divergent lines of descent in O'Hara's model are controlled partly by compositions of initial melts but are strongly influenced, too, by crystallization of key phases that enter the first cumulate under various pressure regimes. Among these in the pressure range 5 to 30 kb are clinopyroxene and garnet at high pressure, two aluminous pyroxenes at intermediate pressure, and plagioclase at lower pressure.

With so many variables at one's disposal it is possible to generate many liquid lines of descent that duplicate important chemical characteristics of natural magma series (O'Hara, 1965, p. 37, table 1). The value of any such comprehensive model perhaps lies more in the potential diversity it reveals than in demonstration of any actual paths of natural magmatic evolution or any universal composition of upper-mantle rock. Its universal applicability is an obstacle to rigorous testing by attempted disproof and so is a weakness.[1] The same and even an extended range of parental magmas can be generated along simpler paths by postulating greater chemical diversity of the source materials.

## ECLOGITE SYSTEMS

*Solidus and subsolidus phase assemblages* Eclogites cover essentially the same composition range as basalts. Some indeed were once basalts (or gabbros) tectonically dragged down (subducted) to the pressure regime where eclogite is stable and there metamorphosed. Here we are concerned only with eclogites of another kind. These could be primitive components of the mantle from very early times or else peridotite-derived extracts that have crystallized well down in the upper mantle and have remained there. Such rocks will be essentially anhydrous,[2] and we are concerned with them as possible sources of basaltic magmas.

Under physical conditions likely to be encountered to depths of 100 km in the upper mantle—500 to 1200°C, 10 to 30 kb—the sequence of phase assemblages with increasing pressure conforms to the following pattern (D. H. Green and Ringwood, 1967*a;* Ringwood and Green, 1966; Ito and Kennedy, 1970):

1. Low-pressure regime: below 10 to 15 kb
   Gabbro assemblage: plagioclase-pyroxenes ± olivine ± spinel

[1] The greatest living exponent of the logic of scientific discovery, Karl Popper, has said ("Conjectures and Refutations," 4th ed., p. 36, Routledge and Kegan Paul, London, 1972):

> Every good scientific theory is a prohibition: it forbids certain things to happen. The more a theory forbids, the better it is.
>
> A theory which is not refutable by any conceivable event is non-scientific. Irrefutability is not a virtue of a theory (as some people think), but a vice.
>
> Every genuine *test* of a theory is an attempt to falsify or refute it. Some theories are more testable, more exposed to refutation, than others; they take, as it were, greater risks.

[2] Most hornblende eclogites are metamorphic: prograde equivalents of amphibolite or retrograde derivatives of mantle eclogite transported into the crust. Some eclogitic rocks containing kaersutite probably come from the upper mantle.

2. Intermediate-pressure regime: upper limit, 20 ± 3 kb
   Garnet-granulite assemblage: garnet-pyroxenes-plagioclase (sodic)

3. High-pressure regime: > 20 kb
   Eclogite assemblage: garnet-omphacite-quartz ± rutile

The pressures of both critical transitions increase with temperature. They tend to be higher for oversaturated tholeiitic than for *ol*-normative tholeiitic compositions.

In the upper 50 km or so of the mantle, then, gabbro, and at higher pressures garnet granulite, could coexist with spinel lherzolite. At depths beyond about 70 km eclogites can be expected in the same environment as garnet lherzolites.

*Fusion of eclogites* Yoder and Tilley (1962, p. 505) demonstrated the surprisingly narrow temperature interval (85°) between incipient and complete fusion of eclogites. This could have far-reaching petrogenic consequences. If a heterogeneous peridotite-eclogite domain in the mantle were to become heated to the point of fusion, eclogite would be completely melted when associated peridotites were still in the earlier stages of fusion.[1] Subsequent fractional crystallization of the eclogitic (basaltic) liquid would tend to be restricted by the narrowness of the temperature (and so also time) interval before solidification was complete. Moreover compositional differences between clinopyroxene and associated melt are not great insofar as major elements are concerned: as G. C. Kennedy (1959) first found, clinopyroxenes at high pressure can readily accommodate most of the major components of a basaltic system.[2] Yoder and Tilley (1962, p. 518) also suggested that the narrowness of the melting interval of eclogites perhaps implies that the rocks themselves approximate eutectic compositions consistent with origin by partial melting of garnet peridotite.

The melting interval of eclogite is greatly extended in the presence of excess water (Fig. 13-2). For melting of eclogite in the mantle this effect would be insignificant unless water were introduced from some external source. Hornblende can enter the eclogite assemblage only at relatively low pressures and subsolidus temperatures (up to perhaps 20 kb at 1000°C). From Fig. 13-2, it seems that melting of hornblende eclogite or of hornblende-garnet granulite in the uppermost mantle could yield a melt phase beyond the breakdown curve of hornblende. The quantity of melt—more or less "andesitic" in composition—would be restricted by its relatively high content of water compared with the limited supply available from hornblende.

[1] Even though some peridotites will start to melt before some kinds of eclogite.

[2] This is strikingly confirmed by exsolution of plentiful garnet and orthopyroxene in clinopyroxene crystals of garnet-pyroxenite xenoliths in nephelinite (Beeson and Jackson, 1970, p. 97).

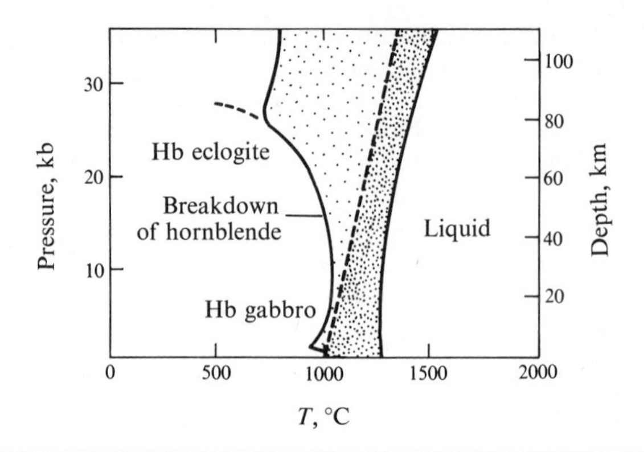


**Fig. 13-2** Generalized diagram to illustrate melting of hornblende eclogite. "Andesitic" liquid in small amount exists in lightly stippled area; basaltic liquid, increasing in quantity toward the right, in heavily stippled area. (After Wyllie, 1970, p. 24.)

CONTROLLING INFLUENCE OF MANTLE ROCKS ON JUXTAPOSED MAGMAS

In the course of ascent mafic magmas are likely to encounter and even remain in prolonged contact with mantle rock of some unrelated kind or in a pressure-temperature regime other than that of its origin. The immediate result will usually be temporary or continuing disequilibrium between the solid phases and juxtaposed melt. Experimental and theoretical considerations outlined earlier in the chapter permit prediction of the kind of chemical adjustments to be expected in specific situations. In general the effect in terms of degree of partial melting will be to offset the composition of the magma toward that of small melt fractions which would otherwise form by incipient fusion within the juxtaposed crystalline system. Thus a basaltic magma of any origin held in prolonged contact with larger volumes of pyrolite or lherzolite on the peridotite solidus at 25 kb will tend to assume the picrite-tholeiitic composition of early melt fractions formed from peridotite alone. Again a *hy*-normative tholeiitic melt would tend to become *ne*-normative if kept in contact with excess quantities of spinel lherzolite at temperatures above the lherzolite solidus at 18 kb. Below 10 kb differential melting of the plagioclase component of mantle peridotite could change the composition of an ascending magma to that of high-alumina basalt.

By analogy with zone refining, as first suggested by Harris, Ringwood particularly has invoked outward diffusion of incompatible dispersed elements from wall rock into magma ascending through narrow fissures in order to explain the high concentrations of such elements observed in many basaltic magmas. This concept has not been tested by experiment nor by observed marginal depletion of wall rock or of xenolithic fragments in elements of this class. The necessity to appeal to this mechanism ad hoc weakens the pyrolitic model.

It seems to exemplify what the philosopher Karl Popper has called a "conventionalist stratagem," of which he says (*op. cit.*, p. 37):

> Some genuinely testable theories, when found to be false, are still upheld by their admirers by introducing ad hoc some auxiliary assumption or by re-interpreting the theory ad hoc in such a way that it escapes refutation. Such a procedure is always possible, but it rescues the theory from refutation only at the price of destroying, or at least lowering, its scientific status.

GENERAL CONCLUSIONS Partial melting of dense ultramafic rocks in the upper mantle could generate a diversity of magmas. The compositions of any melt so formed would depend on that of the source rock, the phase chemistry of the latter (a function also of pressure and temperature), and the degree of melting:

1. At 5 to 10 kb. Plagioclase-bearing peridotites yield high-alumina olivine-tholeiitic melts. Pyroxene-garnet granulites could generate an even more felsic small initial melt fraction.
2. At 10 to 20 kb. Lherzolites (pyrolites) give *ne*-normative initial fractions which as the relative volume of melt increases change to *hy*-normative picritic tholeiite. In the lower part of this pressure range granulites could still generate high-alumina melts.
3. At 20 to 30 kb. Melts derived from lherzolites (pyrolites) mostly have the composition of picritic tholeiite. Eclogites could yield chemically similar basaltic magmas.
4. Basaltic magmas of any origin held in prolonged contact with juxtaposed lherzolite would change toward a *ne*-normative composition while cooling toward the temperature of the lherzolite solidus at high pressures. Olivine-tholeiite magma similarly held in contact with harzburgite at low pressure would change toward a *Q*-normative tholeiite.

## DISPERSED-ELEMENT CHEMISTRY

DISPERSED ELEMENTS: COMPATIBLE AND INCOMPATIBLE The nature of crystalline phases that coexist with a melt in a specific system under given conditions is determined by relative quantities of a few major elements: O, Si, Al, Ca, Mg, Fe, Na, and in particular cases perhaps several others. Much more numerous in magmatic systems are *dispersed* elements—those "that do not exist as a major (stoichiometric) component of any phase in the system" (Gast, 1968, p. 1059). In most igneous systems all the minor or trace elements fall in this category, and in typical ultramafic systems K, Ti, and possibly Na also are dispersed elements. A few are accommodated preferentially in a crystalline rather than in the liquid phase: Ni and Co in olivine, V in magnetite, Cr in spinel, Yb in garnet, and Eu in plagioclase (especially in sodic members). By contrast most dispersed elements are not readily accommodated by any crystalline phase and these preferentially enter the coexisting liquids. These Ringwood has called *incompatible* elements. They include, among others, Kb, Ba, Sr, U, Th, Pb, Zr, the rare-earth elements, P, Cl, and in ultramafic systems K and Ti as well. In mafic lavas their abundance patterns are much more complex and varied than those of major elements (cf. Gast, 1968, pp. 1057-1058); and so perhaps they have more significance in evaluating the petrogenic process.

ABUNDANCE PATTERNS AND ABUNDANCE STANDARDS The chemistry of common igneous rocks is largely determined by a few types of crystal-melt equilibria —olivine-pyroxenes-melt, pyroxenes-plagioclase-melt, olivine-pyroxene-hornblende-melt, and so on. Abundance patterns of dispersed elements vary correspondingly, but they tend to conform to recognizable patterns. It is common practice to refer the various patterns to appropriate standard materials such as average chondritic meteorite, abyssal tholeiitic basalt, Kilauean tholeiitic basalt, and so on. It is true that some standards were selected in the first instance with a view to possibly primordial or primitive chemical attributes. But such implications play no part in setting up standards and they are completely discounted here. The isotopic composition of oxygen in igneous rocks is referred to that of standard seawater with no suggestion of a return to the philosophy of Werner. To take a single illustration, the varied abundance patterns of rare-earth elements in igneous rocks are commonly described with reference to that established for average chondritic meteorites. In Fig. 13-3 each pattern is depicted graphically as a plot of elemental ratios—relative abundance in rock sample compared to that in standard "average chondrite." Because such ratios for the heavier rare-earth elements in many mafic rocks do not deviate greatly from 1, some geochemists have argued that the total pattern of average chondrite may be duplicated in magmatic source rocks. But this is mere speculation that can be rejected without impairing the usefulness of the chondrite standard purely for descriptive purposes. This usage is in fact firmly established in the field of rare-earth geochemistry.

FRACTIONATION DURING PARTIAL MELTING[1] The respective concentrations $c^l$ and $c^\alpha$ of a dispersed element in a coexisting crystalline phase $\alpha$ and melt of

[1] This section is based on Gast's (1968) discussion of dispersed elements in basaltic magmas and later clarification by D. M. Shaw (1970). Symbols and equations are from Shaw and do not conform to standard usage elsewhere in this book.

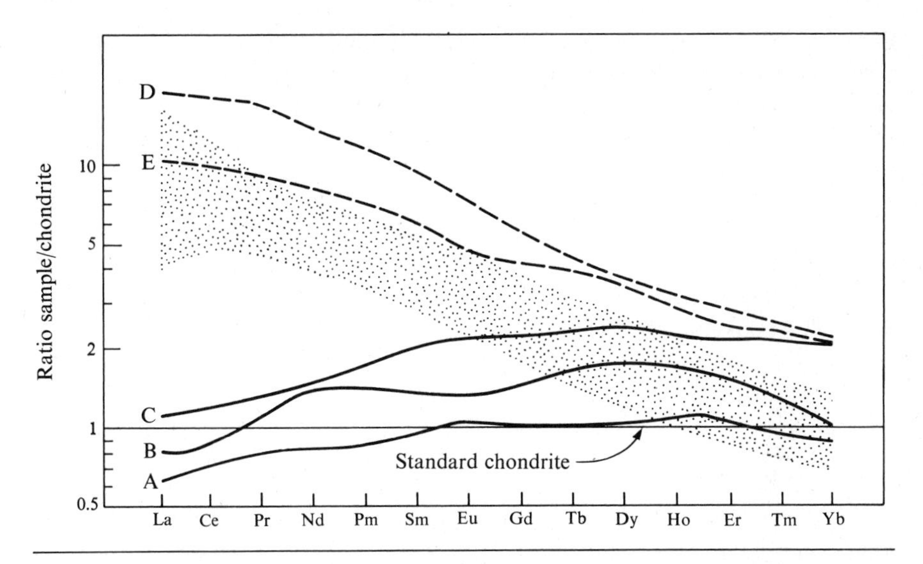


**Fig. 13-3** Relative abundances of rare-earth elements in basalts from mid-Atlantic ridge (*A*, *B*, *C*), Ascension Island (*D*, *E*), and oceanic alkaline basalts in general (stippled), compared with abundances in chondritic meteorites. (After Gast, 1968, p. 1066.)

the same stoichiometric composition—for example, Ni in crystalline and liquid $Mg_2SiO_4$—are fixed, *at constant P and T*, by a distribution coefficient

$$K^{l/\alpha} = \frac{1}{K^{\alpha/l}} = \frac{c^l}{c^\alpha}$$

Petrologists are concerned especially with the concentration of any trace element in the liquid at some stage of melting (denoted by a fraction $F$) relative to the initial concentration $c_\circ$ in the solid before melting. Two extreme cases may be considered: equilibrium partial melting (concentration ratio $c^L/c_\circ$) and ideal fractional melting (concentration ratio $\bar{c}^l/c_\circ$). For equilibrium melting, we have

$$\frac{c^L}{c_\circ} = \frac{1}{K^{\alpha/l} + F(1 - K^{\alpha/l})} \tag{13-1}$$

For ideal fractional melting (sometimes incorrectly called a Rayleigh model), the liquid fraction has formed by accretion of successive small increments, each isolated from contact with the crystalline residue from the moment of formation. In this case

$$\frac{\bar{c}^l}{c_\circ} = \frac{1}{F}\left[1 - (1 - F)^{1/K^{\alpha/l}}\right] \tag{13-2}$$

For melting of a polyphase aggregate (with crystalline phases $\alpha, \beta, \ldots$) we need a bulk distribution coefficient $D_\circ$ to replace $K^{\alpha/l}$. The simplest case is "modal" or eutectic melting where the relative proportions of $\alpha, \beta, \ldots$ in the solid aggregate and in the melt ($X^\alpha, X^\beta, \ldots$) remain constant and $D_\circ$ likewise remains constant throughout melting. Then

$$D_\circ = \frac{X^\alpha}{K^{l/\alpha}} + \frac{X^\beta}{K^{l/\beta}} + \cdots \tag{13-3}$$

Equation (13-1) now becomes

$$\frac{c^L}{c_\circ} = \frac{1}{D_\circ + F(1 - D_\circ)} \tag{13-4}$$

and Eq. (13-2) becomes

$$\frac{\bar{c}^l}{\bar{c}_\circ} = \frac{1}{F}[1 - (1 - F)^{1/D_\circ}] \tag{13-5}$$

For the general case where the solid phases are eliminated in nonmodal proportions, let their relative proportions in the liquid fraction at a stage of melting $F$ be $p^\alpha, p^\beta, \ldots$. Then an additional factor $P$ is introduced, such that

$$P = \frac{p^\alpha}{K^{l/\alpha}} + \frac{p^\beta}{K^{l/\beta}} + \cdots \tag{13-6}$$

Equation (13-4) now becomes

$$\frac{c^L}{c_\circ} = \frac{1}{D_\circ + F(1 - P)} \tag{13-7}$$

and Eq. (13-5) becomes

$$\frac{\bar{c}^l}{c_\circ} = \frac{1}{F}\left[1 - \left(1 - \frac{PF}{D_\circ}\right)^{1/P}\right] \tag{13-8}$$

In both models much of the content of an incompatible element in the solid source is withdrawn into the first small liquid fraction. The concentration ratio $c^L/c_\circ$ or $\bar{c}^l/c_\circ$ falls from a maximum value $(1/D_\circ)$ at the onset to 1 at completion of melting (Fig. 13-4*a*). Similarly, that of a compatible element rises from a minimum value $(1/D_\circ)$ to 1 (Fig. 13-4*b*). There is an inherent defect in all this discussion. For want of information, distribution coefficients—which actually must change exponentially with increasing temperature—are assumed to be constant over considerable temperature intervals of progressive fusion. In D. H. Green's (1970*a*, p. 36) pyrolite model, as the melt yield increases from

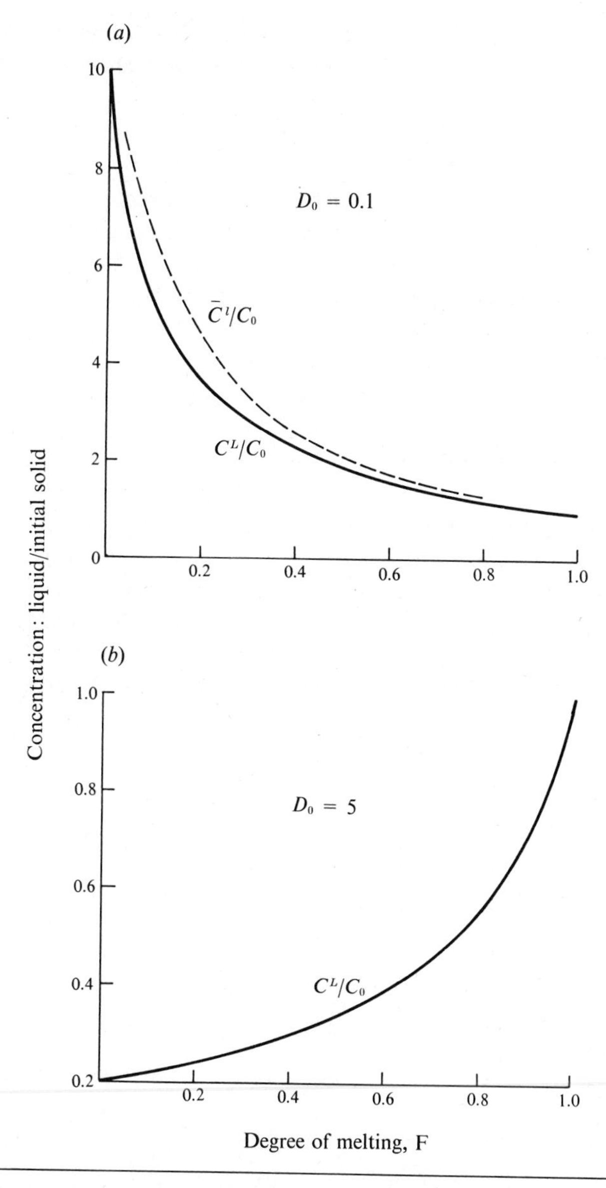


**Fig. 13-4** Change of concentration of dispersed elements, compared with that in an initial polyphase solid, during progressive melting in which the proportions of phases in the solid remain constant (eutectic melting at constant temperature). $c^L/c_o$ values are for equilibrium partial melting (batch melting). $\bar{c}^l/c_o$ values are for ideal fractional melting.

zero to 30 percent, the temperature rises from 1000 to 1400°C. Present application of the distribution law to dispersed-element chemistry of magmatic systems can be only qualitative.

CONSTRAINTS ON DISPERSED-ELEMENT CHEMISTRY OF SOURCE ROCKS **In** the context of this chapter the problem is to reconstruct the dispersed-element pattern and mineralogy of source rocks of common mafic magmas. The main requirement is a set of appropriate distribution coefficients for liquids in equilibrium with such minerals as olivine, pyroxenes, garnet, hornblende, and phlogopite *over a wide range of temperatures.* Tentative values (Table 13-2) have been estimated within rather crude limits from the respective abundances of each element in phenocrysts and in enclosing fine-grained matrix of porphyritic rocks. In spite of obvious imperfections in the data, several valuable qualitative generalizations emerge.

1. For strongly incompatible elements $D_o$ appears to be in the range 0.02 to 0.06. Its influence in equations like (13-4) and (13-5) becomes overshadowed by $F$ when the latter exceeds 0.1 or 0.2. It is only in the early stages of melting, therefore, that liquid/source concentration ratios for such elements are high. Thus $D_o$ for potassium in a lherzolite system (*ol* 60, *di* 20, *en* 20 percent) is about 0.03; and $c^L/c_o$ drops from 20 at 2 percent, to 5 at 20 percent equilibrium melting. For 50 percent melting $c^L/c_o$ for any incompatible element approximates 0.5.

**Table 13-2** Some rough estimates of liquid/solid distribution coefficents, $1/D_o$ (compare $K^{l/\alpha}$), for dispersed elements in mafic systems*

| CRYSTALLINE PHASE ($\alpha$) | K | Rb | Sr | Ba | La | Eu | Yb | Ni |
|---|---|---|---|---|---|---|---|---|
| Olivine | 150 | 220 | 40 | 240 | 100 | | 15 | 0.2 |
| | 180 | 100 | 50–100 | 100 | 100 | | 45 | |
| Orthopyroxene | 10 | 22 | 24 | 12 | 30 | | 3 | 1 |
| | 50 | 50 | 50 | 70 | 30 | | 2 | |
| Clinopyroxene | 10 | 15 | 4 | 12 | 5 | | 1.5 | 1 |
| | 15 | 20 | 6 | 20 | 4 | | 1.2 | |
| Hornblende | 5 | 20 | 12 | 18 | 20 | | 4 | 1.2 |
| | 1–3 | 2–20 | 2 | 1.5–20 | 10 | | 2 | |
| Phlogopite | 0.2–1 | 0.3–1 | 1.5–12 | 0.1–1 | 35 | | 10–30 | |
| Garnet | 40 | 30 | 80 | 24 | 40 | | 0.02 | 15 |
| | 50 | 120 | 70 | 60 | 3 | | 0.025 | |
| Plagioclase | 3–10 | 5–20 | 0.3–0.7 | 2–7 | 3–50 | 0.5–5 | 3–50 | |

* Values in top line of each box from Gast (1968, p. 1669); values in lower line from Philpotts and Schnetzler (1970) and from Schnetzler and Philpotts (1970).

2. For the same reason, abundance ratios between pairs of strongly incompatible elements, for example, Rb/K and Rb/Sr, closely approximate those of the source rock once the melt fraction exceeds 10 or 15 percent (Fig. 13-5). This is a useful generalization indeed, and its validity will not be greatly affected by ignoring the temperature dependence of $D_o$. The wide spread of such values among common types of basalts compels one or some combination of three alternatives: (*a*) there was comparable chemical diversity among source rocks; (*b*) if a single source is postulated, the magmas with extreme abundance ratios (high Rb/K or Rb/Sr) must characterize liquids segregated in the very earliest stages of melting; (*c*) extreme values alternatively could reflect prolonged interchange between magma and large volumes of wall rock during ascent along narrow conduits.

**Fig. 13-5** Fractionation of Rb and K during ideal fractional melting of lherzolite (diopside 10, enstatite 20, olivine 65, spinel 5%). Ratios for 3% equilibrium partial melting are given by circled points *X* (Rb), *Y* (K). (After Gast, 1968, p. 1073.)

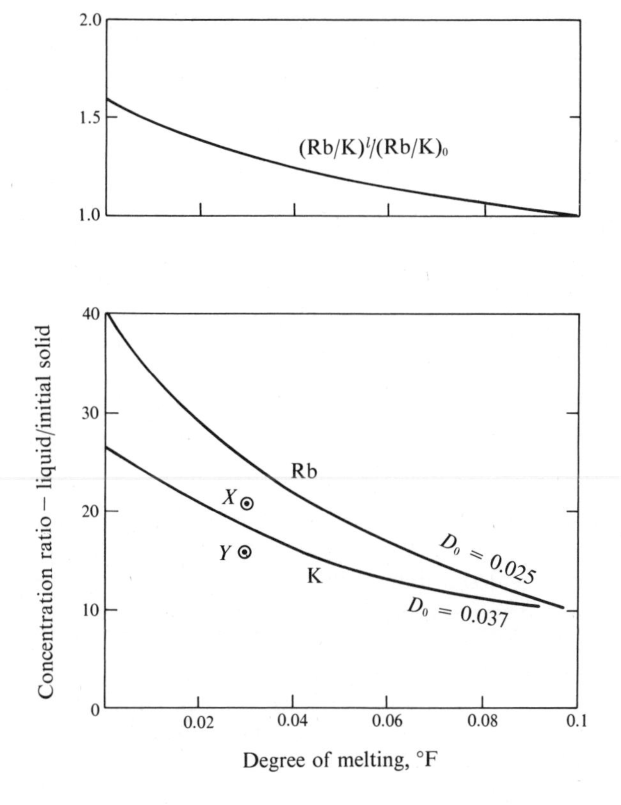

3. Depletion of a magma in compatible elements is largely accomplished during the early stages of fractional crystallization [at high values of $F$ in Eqs. (13-4) and (13-5)]. Crystallization and removal of a 10 percent garnet fraction can reduce the $c^L/c_o$ value for Yb to 0.1 compared with 0.025 in the ultimate minute fraction of residual liquid.

SPECIFIC MINERALOGICAL CONSTRAINTS The data of Table 13-2 show how some incompatible elements are preferentially tolerated by one or more of the mineral phases likely to be found in ultramafic source rocks. Of the group K, Rb, Sr, and Ba, clinopyroxene and to some extent olivine can accommodate Sr in preference to the others. Both minerals, and orthopyroxene as well, tolerate K significantly above Rb. Mica shows extreme preference for K (no longer a dispersed element), Rb and Ba with respect to Sr. Hornblende takes up K rather than Rb, or Ba. Among the rare-earth elements the heavier are tolerated in preference to the lighter in most of the listed minerals. There are important petrologic aspects of this rule. Garnet has a phenomenal preference for Yb and other heavy elements, which it retains in concentrations 100 times greater than those of light elements such as La. Plagioclase accommodates $Eu^{3+}$ alone in strong preference to all the other rare earths.

On this basis abundance ratios of certain elements in magmas may place constraints on the mineralogy of source rocks—especially where they represent liquid fractions isolated during the earlier stages of melting.

1. Garnet- or spinel-lherzolite sources: values of Rb/Sr could be only one-third, and Rb/K rather more than one-half, of the ratios measured in derivative magmas.

2. Hornblende-peridotite sources: K/Rb can be expected to be three or four times higher, and K/Sr somewhat higher, than in the early liquid fractions generated from them.

3. Mica-peridotite sources: K/Sr, Rb/Sr, and Ba/Sr will all be many times higher (perhaps by a factor of 10 or 30) in the source rock than in melt fractions.

4. Garnet-peridotite or eclogite sources: the presence of garnet will be reflected in extreme values of La/Yb in partial melts. O'Nions and Clarke (1972, fig. 2) have presented a model for equilibrium partial melting of lherzolite with 5 percent of garnet which becomes eliminated along with clinopyroxene, when melting is 15 percent complete. They compute La/Yb values that fall from 20 to 6 over the melting interval 5 to 10 percent.

5. Plagioclase-peridotite sources: partial melts should be marked by strong depletion in Sr relative to Rb and Ba and in Eu with respect to other rare-earth elements (the well-known negative europium anomaly).

6. The possible role of fractional crystallization as a magma rises from its source may also be evaluated to some extent in terms of its dispersed-element pattern. Obvious possibilities are depletion in Ni and Co by removal of olivine. If "eclogite fractionation" at high pressure is as effective as some writers claim, it should leave a spectacular imprint in low Yb/La values in magmas from which garnet has separated. This constraint applied to recorded rare-earth patterns in basaltic rocks seems virtually to exclude ideal eclogite fractionation (cf. Gast, 1968, p. 1072: O'Nions and Clarke, 1972, p. 444).

## Rock Samples from the Upper Mantle: Ultramafic Xenoliths and Intrusions

### PRELIMINARY INVENTORY

It is part of today's fashion in petrology to look on certain kinds of ultramafic rocks as fragments torn from the upper mantle and transported to their present sites by volcanic or tectonic means. The most compelling evidence of deep-seated origin lies in the phase chemistry of the various mineral assemblages seen in the light of experimentally investigated high-pressure equilibria. Many of these rocks clearly existed as such for long periods of time in the depths. In their internal fabric they bear the imprint of severe strain—even extensive flow in the solid state—in a manner that can be induced in the laboratory only at high confining pressures. And the isotopic chemistry of some samples suggests residence in the mantle since rather remote times.

Currently recognized mantle samples are of four kinds:

1. Xenoliths of spinel lherzolite, and very rarely of garnet pyroxenite, in nephelinite and other alkaline mafic lavas and pyroclastics
2. Xenoliths of garnet lherzolite and of eclogite in kimberlite
3. Ultramafic plutons, especially of the Alpine type
4. Eclogite bodies in highly deformed rocks of other metamorphic facies

Each category, it is generally agreed, includes samples transported from the upper mantle. But opinion is divided as to whether rocks of some categories are exclusively of mantle origin. And there is no consensus on possible relations between mantle fragments and closely associated magmas, on chronological aspects of their petrogenesis, or on their roles in the general makeup of the upper mantle. A constantly recurring question is this: Is the ultramafic rock cogeneric with associated magma—a cognate deep-seated cumulate—or a refractory residue of partial fusion, or even a sample of the magmatic source rock? Or is the relation purely accidental—the ultramafic sample having been caught up by chance in the path of ascending magma of independent origin (as is certainly the case for many of the varied xenoliths in kimberlite).

Basaltic rocks of many provinces—especially those of oceanic islands—enclose mafic and ultramafic xenoliths whose mineralogy and cumulate structure leave no doubt as to their cognate relation to their basaltic hosts. Presence of plagioclase and low-alumina pyroxenes in some confirms a picture of crystallization at no great depth. In others kaersutite and high-alumina augite indicate origin at greater depth (cf. Aoki, 1970) in the upper levels of the suboceanic mantle. So it seems likely that ultramafic and mafic cumulates of various ages are important components of the uppermost suboceanic mantle and immediately overlying deep crust. This topic will be pursued no further.

In what follows we consider rocks in each of the four categories just mentioned which seem to be samples from an enduring solid mantle rather than new additions from recently generated magmas. For alternative opinions in each case the reader is referred to appropriate literature.

## LHERZOLITE XENOLITHS IN ALKALINE MAFIC LAVAS

Xenoliths of spinel lherzolite (Table 13-1, analysis 4) habitually appear in nephelinites and basanites of oceanic and continental-margin provinces.[1] They may be accompanied by much less plentiful ultramafic xenoliths of other kinds, and their presence in nephelinite provinces is not universal. But association of "lherzolite nodules" with nephelinite host rocks is so general that it cannot be accidental. The chemistry of lherzolite mineral assemblages places equilibration close to the pyrolite solidus at 1100 to 1200°C and 12 to 20 kb (cf. Bacon and Carmichael, 1973). It is most unlikely that lherzolite crystallizes directly from the enclosing nephelinite magma as it ascends through the corresponding depth interval as advocated by O'Hara (1968, p. 127), since the internal fabric characteristically bears the imprint of strong deformation (T. Ernst, 1935, 1936; Talbot et al., 1963; E. D. Jackson and Wright, 1970, p. 416). Such an origin, in specific instances, is unequivocally excluded by isotopic data. Strontium tends to be notably more radiogenic than that of enclosing alkaline lavas (Leggo and Hutchison, 1968; Peterman et al., 1970*b*; D. K. Paul, 1971). The Rb/Sr ratios in such lherzolites, moreover, are so low that the anomalies must be of very long standing. The absolute ages of some nodules are in fact demonstrably much greater than those of their hosts (Kryukov, cited by Kutolin and Frolova, 1970, p. 175). Lead-isotope data, where available, point to the same conclusion. In Quaternary basanites from southern Australia, lead has a composition in the normal range of modern leads: $Pb^{206}/Pb^{204}$, 18.3 to 18.5; $Pb^{207}/Pb^{204}$, 15.6; $Pb^{208}/Pb^{204}$, 38.5. Lead of enclosed lherzolites is much less radiogenic: $Pb^{206}/Pb^{204}$, 16.1 to 16.6; $Pb^{207}/Pb^{204}$, 15.2 to 15.3; $Pb^{208}/Pb^{204}$, 36. Basanite sources and lherzolites may have existed here as independent closed

[1] Ross et al., 1954; Forbes and Kuno, 1965; R. W. White, 1966; O'Hara (pp. 346–349 in Wyllie, 1967); E. D. Jackson and Wright, 1970; Kuno, 1969; Kutolin and Frolova, 1970; Bacon and Carmichael, 1973.

systems for a very long time at temperatures (of lherzolite equilibration) close to 1000°C and a depth of 30 km or more.[1] Finally the detail of rare-earth abundance patterns appears to conflict with a cumulate origin for lherzolite nodules from Hawaii (Nagasawa et al., 1969) and the Eifel (Frey et al., 1971).

Lherzolite xenoliths could be either ancient cumulates or residues of partial fusion, long resident in the upper mantle and unrelated to the magmas which bore them to the surface. But their presence in nephelinite and their isotopic chemistry raise puzzling questions still not satisfactorily answered. Is it possible that the pressure regime most favorable to development of nephelinite-basanite magmas is that at which the lherzolite assemblage is stable (depths around 50 km)? If so, this implies association of nephelinite source rocks and lherzolite—that is, a heterogeneous lithology—in the upper mantle at these depths (cf. E. D. Jackson and Wright, 1970, p. 425, fig. 6). Finally, within the nodules themselves there may be extreme isotopic disequilibrium of very long standing: $Sr^{87}/Sr^{86}$ values recorded for diopside and associated olivine in Quaternary nodules from southern California are 0.7016 (with Rb/Sr 0.014) and 0.7087 (with Rb/Sr 0.14), respectively (Peterman et al., 1970*b*). How can such compositional differences on the single-crystal scale have survived without exchange for 2000 m.y. or so at mantle temperatures?

## GARNET-PYROXENITE XENOLITHS IN NEPHELINITES

The Honolulu series of Oahu, Hawaii (Chap. 8), consists of melilite-nephelinite and alkali-basalt tuffs recently erupted from many separate vents within an area of 400 km$^2$ upon a much larger dissected quartz-tholeiitic shield volcano (Koolau series) that had become extinct some 2 m.y. earlier. The tuffs are locally strewn with dunite and lherzolite xenoliths. But within a limited area there are many xenoliths of garnet pyroxenite ("eclogite" of earlier accounts) of a kind unique among oceanic provinces (Beeson and Jackson, 1970; E. D. Jackson and Wright, 1970). They range from coarse clinopyroxene-garnet aggregates to clinopyroxene-enstatite pyroxenites in which crystals of the monoclinic phase contain exsolved garnet and enstatite. Mineralogical evidence suggests that these have come from depths of perhaps 100 km. Linear compositional trends can be constructed between various kinds of pyroxenite and nephelinite host rocks; and the composition of strontium ($Sr^{87}/Sr^{86} = 0.703$) is the same in xenoliths and nephelinites. This pattern is consistent with some kind of genetic connection between the two—either by partial fusion of pyroxenite or by deep-seated fractional crystallization of nephelinite magma.

There are a few records of hornblende-garnet-pyroxene xenoliths ("eclogites") in isolated diatremes—again with nephelinite or alkali-basalt hosts—in continental-margin settings. One is in eastern Australia (Lovering and Richards, 1964; Lovering and White, 1964), another in southeastern New Zealand

[1] This conclusion holds only if lead of component phases proves to be in isotopic equilibrium.

(Mason, 1968). In each case there is an assortment of mafic and ultramafic xenoliths of other kinds—dunites, lherzolites, pyroxene granulites—some of which came from deep crustal levels rather than the mantle.

Chemically there is little resemblance between xenolithic garnet pyroxenites (for example, Table 13-1, analysis 5) and either the pyrolite or the garnet peridotite of experimentally based mantle models. Though somewhat similar to their nephelinitic hosts, because of higher $SiO_2$ and lower alkali they are *hy*-normative. They qualify better than any natural peridotite as potential sources of mafic magmas. Seen thus, the rarity of their recorded occurrence at the surface is something of an anomaly. Perhaps they can escape fusion during ascent—as liquidus temperatures fall with pressure—only if upward transit is exceptionally rapid (cf. Kutolin and Frolova, 1970, p. 176). Single-phase xenocrysts, on the other hand, would have a better chance of survival. They are, in fact, abundant—large, rounded, polished crystals of garnet, tschermakitic augite, hornblende, and sodic feldspar ("anorthoclase")[1]—in diatremes carrying garnet-pyroxenite xenoliths. And they have a much wider distribution than the latter in alkaline mafic lavas.

## MANTLE-DERIVED XENOLITHS IN KIMBERLITES

Of the assorted xenoliths that crowd kimberlite pipes, two widespread types at least—garnet lherzolites and eclogites—must have come from deep within the upper mantle. They vary erratically in distribution and mutual association. In southern Africa, garnet lherzolite regionally predominates; but there are diamond-rich pipes (for example, at Roberts Victor mine) where the most abundant xenoliths are eclogite. Nearer the Equator in the Katanga field, where diamonds though known are scarce, eclogite xenoliths greatly predominate. In the Yakutia province of Siberia, garnet-peridotite and eclogitic xenoliths are widely represented; but one type of eclogite—"grospydite" consisting of grossular-rich garnet, omphacite, and kyanite—is virtually confined to a single pipe (Sobolev et al., 1968), where it is the dominant kind of mantle-derived material. Cretaceous kimberlite diatremes of Kansas contain eclogite xenoliths, but neither garnet peridotite nor diamond. Yet the high pyrope content of their garnets indicates crystallization in the same temperature-pressure regime as that of South African kimberlitic eclogites. Systematic differences in $Cr_2O_3/Fe_2O_3$ of garnet—for example, in South African compared with Siberian kimberlites—suggest the existence of regional chemical heterogeneity in the mantle (Rickwood, 1969).

Beneath continental cratons, then, both garnet lherzolites and eclogites

[1] It cannot be assumed that these feldspars are components of mantle rocks. They certainly come from great depth (30 km) but may well have crystallized from the magma that encloses them (Bacon and Carmichael, 1973).

collectively covering a wide compositional range are erratically distributed at depths of 50 km or so below the Moho discontinuity. At least in the African region, both kinds of rock have been in separate existence for a long time—perhaps far back into the Precambrian. For in each type of garnet-pyroxene system the present isotopic composition of strontium (preferentially held in pyroxene) departs strongly from equilibrium (Allsopp et al., 1969): $Sr^{87}/Sr^{86}$ values for pyroxene are much lower than for garnet, in which Rb/Sr ratios are many times higher. Almost every conceivable relation between garnet lherzolite, eclogite, and kimberlitic magmas has its recent advocates (cf. Rickwood et al., 1968, pp. 295–299). Isotopic evidence which appears to conflict with some models is meager and moreover is suspect in view of the possibilities of contamination by water at some stage of ascent. The question, therefore, is here left open.[1] But we note that mutual association of ultramafic rocks in kimberlitic diatremes in no way implies genetic relationships or even any regular pattern of distribution in depth (an eclogite above a peridotite layer or vice versa). It does demonstrate small-scale heterogeneity in the subcontinental upper mantle; and xenoliths of kimberlite diatremes extend the lithologic range of subcontinental mantle rocks beyond that beneath the oceans.

MINERAL INCLUSIONS IN DIAMOND A study of mineral inclusions in diamond (Meyer, 1968; Meyer and Boyd, 1969, 1970) has revealed hitherto unknown mineral compositions presumably representing an approach to equilibrium prior to nucleation of the enclosing diamond. The mineral inclusions are usually euhedral crystals which vary little in composition, although kimberlite sources are distributed as far apart as South Africa, Sierra Leone, Ghana, and Venezuela and range in age from the Precambrian to the Mesozoic. Among 200 inclusions in 50 diamonds examined by Meyer and Boyd, the order of abundance is olivine, chrome pyrope, chromite, enstatite, and diopside; coesite, pyrrhotite (Sharp, 1966), and pentlandite have also been reported.

The olivine inclusions are magnesian, showing a narrow compositional range near $Fo_{93}$, and contain small amounts of $Cr_2O_3$ and only traces of CaO. The calcium content of olivine, although not sensitive, is a good indication of pressure; Simkin and Smith (1970) have shown that volcanic olivines rarely contain less than 0.10 percent CaO, whereas plutonic olivines contain less; J. Nicholls et al. (1971) showed that the depletion of CaO in olivine and its incorporation into diopside is a direct function of pressure.

The chrome-pyrope garnet inclusions are not only the most magnesian garnets found in nature, but they also contain up to 30 percent $Mg_3Cr_2Si_3O_{12}$ component; they are generally richer in $Cr_2O_3$ than garnets from peridotite xenoliths in kimberlite and are quite unlike garnets of eclogite in containing

[1] For alternative models see Harris et al. (1967); D. H. Green (1970*a*, p. 29); Allsopp et al. (1969); Carswell and Dawson (1970, pp. 179, 181); Frick (1972).

$Cr_2O_3$ but little $TiO_2$ and MnO. Chromite inclusions, too, are unusually rich in $Cr_2O_3$, and the enstatites ($En_{94}$) also have higher $Cr_2O_3$ and lower $Al_2O_3$ and CaO than enstatites of xenoliths found in kimberlite.

Ferromagnesian inclusions in diamond contrast with comparable phases of associated peridotite xenoliths mainly with respect to their notably higher content of $Cr_2O_3$. Calcium-rich pyroxenes of xenoliths fall into two clearly defined groups: diopsidic and subcalcic augites—the latter indicating equilibration close to 1400°C (Boyd, 1969). Subcalcic augite has yet to be identified among inclusions in diamond.

## ULTRAMAFIC PLUTONS AS MANTLE FRAGMENTS

BODIES OF THE ALPINE TYPE It is now thought that most ultramafic bodies of the Alpine type are large, tectonically transported fragments of upper-mantle rock. Evidence of the deep-seated origin of many has been obliterated by low-pressure serpentinization. In others it has been obscured but not completely destroyed by reequilibration of mineral phases (particularly pyroxenes) and imprint of late deformational fabrics during emplacement in the crust. But there are many records of fresh ultramafic rocks that still retain the clear mineralogical imprint of equilibration at pressures and temperatures of the upper mantle.

Major-element chemistry of most Alpine-type peridotites—especially the common harzburgites and dunites—rules them out as potential sources of basaltic magma, as indeed do consistently high $Sr^{87}/Sr^{86}$ values (0.709 to 0.715). So at least these Alpine-type peridotites cannot be unchanged remnants of a primitive homogeneous mantle—if such ever existed.

They could, on the other hand, be refractory residues left after extraction of mafic magma or alternatively early cumulate fractions derived from such magmas. Rare-earth abundance patterns determined for the primary (high-temperature) zone of the Lizard peridotite (Chap. 12) strongly support the former alternative (Frey et al., 1971). There is a strong overall deficiency in the lighter elements that nicely complements heightened values characteristic of most basalts. But whatever the relation to basaltic magmas may have been, the high values of $Sr^{87}/Sr^{86}$ compared with those of basaltic rocks require explanation. Various multistage petrogenic models have been proposed (Hurley, pp. 374–375 in Wyllie, 1967), postulating enrichment in rubidium or depletion in strontium at some ancient date. There is always the possibility, however, of late contamination perhaps even by connate waters: abundances of both strontium and rubidium in peridotites are very low; and Tertiary peridotites of Skye, whose cumulate origin in relation to basaltic magma is beyond doubt, yield initial $Sr^{87}/Sr^{86}$ values between 0.7065 and 0.7108.* Partial serpentinization of North Carolina peridotites likewise has raised initial $Sr^{87}/Sr^{86}$ values from 0.7023–0.7047 to 0.7058–0.7068 (Jones et al., 1973).

* S. Moorbath in a lecture to the Geological Society of America, November 1972.

Two recent studies attribute internal heterogeneity in peridotite bodies to segregation of a partial melt fraction within a dominantly crystalline ultramafic mass. Evidence rests on the character of major-element trends between all variants present. One case concerns a Norwegian garnet-peridotite body (Carswell, 1968*b*) in which garnet lherzolite is thought to have unmixed to refractory dunites and complementary garnet pyroxenites—representing the melt fraction. The other—a spinel-peridotite pluton in the Spanish Betic Cordillera (Dickey, 1970)—consists mainly of harzburgite, dunite, and lherzolite (the assumed parental material), and it is veined and streaked with a variety of mafic rocks interpreted as partial melts. These include such assemblages as garnet-clinopyroxene-spinel, clinopyroxene-chromite-plagioclase, and others. Major-element data satisfy the models presented: Fe/Mg and Na increase significantly in the "melt fractions," and Ni is concentrated in "parental" rocks and in "refractory residues." But in neither case is there a hint of expectable fractionation of K into the segregated "melt." The melting model could be tested by data on relative abundances of rare-earth and incompatible dispersed elements.

ST. PAUL'S ROCKS, ATLANTIC St. Paul's Rocks, a group of islets on the mid-Atlantic ridge just north of the Equator, consists of mylonitic peridotites generally considered to represent mantle rocks emplaced by solid intrusion through the thin oceanic crust. Most of the rocks are ultramafic, and there are two principal mineral assemblages: olivine-enstatite-diopside-spinel; olivine-pargasite-enstatite-spinel with remarkably varied accessory minerals, among them chlorapatite, scapolite, biotite, allanite, and zircon. These suggest equilibrium crystallization at moderate but not extreme depths, the first at 40 to 100 km, the second perhaps nearer the surface (cf. Melson et al., 1967*a*). The pattern of rare-earth elements (Frey, 1970) resembles that of island basalts but differs from those of Alpine peridotites—there is marked enrichment of La and other light elements relative to the chondritic meteorite standard. Low values of $Sr^{87}/Sr^{86}$ ($\sim$ 0.7040) also overlap the range of oceanic basalts. $K_2O$ is variable, but higher than in most Alpine peridotites—0.01 to 0.17 percent. Here again is a mantle rock whose origin seems to be complex—neither primordial mantle rock nor a refractory residue from melting. It could well be genetically related to a unique highly undersaturated alkaline basalt that occurs in the deep ocean a few kilometers to the northeast.

## ECLOGITE BODIES AS MANTLE FRAGMENTS

Eclogites have been shown to be stable over a wide range of temperature (perhaps 400 to 1200°C) at pressures above about 10 kb. If material of basaltic composition exists in the upper mantle, it will take the form of some kind of eclogite: olivine-bearing if strongly undersaturated, quartz-kyanite eclogite if oversaturated in silica.

We have already noted the presence of eclogite xenoliths in kimberlite diatremes. Here, as in most other types of occurrence, the mineral facies of eclogite (which reflects conditions of crystallization) is out of step with that of its associates. Evidence of retrogressive metamorphic adjustment, moreover, is nearly universal. So it may be impossible in individual cases to tell whether minerals such as hornblende, muscovite, and zoisite are original or metamorphic members of the eclogite assemblage.

In parts of western Norway, eclogites are intimately associated with garnetiferous peridotite bodies in a high-grade metamorphic basement complex. There is every reason to believe that these masses, like their peridotite associates, are tectonically transported mantle fragments.[1] Mineralogically they resemble kimberlitic eclogites, especially in the high pyrope content (> 55 percent) of their garnets. Other eclogites (with 30 to 55 percent pyrope in garnet) occurring as lenses in high-grade metamorphic gneisses, although they may possibly have come from the mantle, alternatively may be deep crustal gabbros that have been depressed to upper-mantle pressures and there metamorphosed (cf. Mysen and Heier, 1972). Yet another group with almandine garnet (pyrope < 30 percent) and acmitic omphacite is habitually associated with glaucophane schists and serpentinized peridotites in orogenic zones of low-temperature, high-pressure metamorphism. Some of these undoubtedly are metamorphosed basalts. In today's fashion they are generally considered to be blocks of cold oceanic crust dragged down and metamorphosed in the thermally disturbed high-pressure regime of tectonic subduction zones. Eclogites of this clan have no bearing on the present problem except that they demonstrate a deficiency of water in some parts at least of the sinking oceanic plate.

## SUMMARY OF DIRECTLY OBSERVED LITHOLOGY AND CHEMISTRY

The most significant collective implication of accessible samples is that the upper mantle is highly heterogeneous. Peridotites are the most abundant rock types that have survived the journey to the surface. Most consist of two to four minerals, in order of general abundance olivine, enstatite, diopside, and minor spinel. Garnet-bearing assemblages, perhaps because of their presumably deeper origin and smaller chance of survival, are much less abundant, and their distribution is more restricted. They include garnet lherzolites, garnet pyroxenites, and eclogites; and there is only one recorded occurrence (garnet pyroxenites) in the oceans. Because of their susceptibility to destruction at high temperatures—in the case of eclogite by melting over a small temperature interval—garnetiferous rocks need not necessarily play a subordinate role to spinel peridotite in the overall makeup of the upper mantle.

Of all these rocks only eclogites, garnet pyroxenites, and perhaps some garnet peridotites have the chemical capacity to generate substantial fractions

[1] Many would term the "garnet-pyroxenite" of Carswell's (1970*b*) study an eclogite.

of basaltic and allied melts. The common peridotites seem to be either refractory residues or cumulates—some, at least, long separated from cogeneric liquid fractions. Many have existed in their present state since very early times. As a potential source of juvenile gas, the suite as a whole is most unpromising. They contain no carbonates. Hydrous components—amphibole or phlogopite—occur sparsely in some peridotites but (except for hornblende of some pyroxenites and eclogites) are rare.

There is greater lithologic and chemical variety among known mantle fragments from continental than from oceanic environments. This discrepancy could well represent deficiencies in sampling of oceanic rocks. Most of the striking mineralogical departures from general lithologic pattern are confined to very limited, even unique occurrences such as grospydites of a single Siberian kimberlite pipe and the hornblende-garnet-pyroxenes of an insignificant nephelinite diatreme at Kakanui, southeastern New Zealand. Comparable materials could well exist undetected within the expanse of the ocean basins. Two by chance are exposed and have been carefully studied—the pargasite peridotite of St. Paul's Rocks and the garnet pyroxenites of Hawaii.

Correlations between rock type and tectonic or magmatic environment are well established: Alpine-type peridotites with old geosynclines and active trenches; garnetiferous ultramafics with kimberlitic magmas; spinel lherzolites with nephelinitic and alkali-basalt magmas. But their significance is equivocal. The surface distribution of Alpine-type peridotites could be explained by correspondingly limited distribution in the mantle or, in our opinion, more probably by the profound tectonic activity necessary to bring large fragments of mantle rock within reach of surface erosion. Limitations of depth or direct mutual genetic connection could equally well explain the xenolithic patterns of kimberlitic diatremes and nephelinite pyroclastics.

Finally the isotope chemistry of the mantle places constraints on that of derivative magmas. It also provides a severe test for independently proposed models of mantle composition, especially those based on supposed analogy with some kind of stony meteorite. There is no compelling reason to suppose that mantle materials are chemically identical to those of any other planetary body—meteorites included. Indeed the precise opposite—that terrestrial rocks are chemically unique—has been argued; and the growing evidence of lunar petrology reinforces this latter view. But in this connection it is appropriate nevertheless to review the testimony of isotope chemistry—mantle samples compared with analyzed stony meteorites. Most convincing and consistent is the composition of oxygen[1]—an element subject to wide isotopic variation in different terrestrial materials. Values of $\delta_{O^{18}/O^{16}}$ for ultramafic mantle rocks are narrowly limited (+ 5.1 to + 6.6); values for component pyroxenes are in the same range, as also are values for Alaskan ultramafics that may have been derived directly from the mantle by melting. Pyroxenes of many stony meteorites

[1] Data and conclusions are cited from H. P. Taylor (pp. 368–370 in Wyllie, 1967).

yield $\delta_{O^{18}/O^{16}}$ values in the same range (+ 5.3 to + 6.3); these include ordinary chondrites, enstatite chondrites, enstatite achondrites, and diopside-olivine achondrites (collectively including by far the greater number of meteorites observed to fall). The oxygen composition of pyroxene in other stony meteorites is totally different: + 3.7 to + 4.4 in most achondrites and some stony irons; − 0.8 to + 8.6 in the highly variable carbonaceous chondrites. If there is any analogy or remote cogeneric relation between the terrestrial upper mantle and the meteoritic samples, it is with common chondrite.[1] Carbonaceous chondrite, favored in some models, seems to be completely excluded by the evidence of oxygen isotopes.

## Mafic Magmas and Mantle Source Rocks

### LIMITING CONSTRAINTS

Mafic magmas give additional information regarding the chemistry and to a lesser degree the mineralogy of mantle source rocks. Backward extrapolation of magma chemistry to source materials is limited by constraints relating to recycling and fractionation of magma, to the problem of random "noise" in the chemical data, and to time.

RECYCLING PROBLEM Although the ultimate sources of mafic magmas are ultramafic mantle rocks, it is neither necessary nor likely that every recently erupted magma has been generated in a single cycle. Oceanic crust, one of the principal products of volcanism today, seems to be constantly recycled—partly at least back into the mantle—by convection and subduction. Much of the basaltic magma that is being erupted today is therefore likely to have come from mantle or deep crustal source rocks (eclogites) of generally similar composition, themselves the product of earlier cycles of partial fusion. Some of the more siliceous mafic lavas, such as hawaiites of some oceanic islands and icelandites of others, likewise could have a multistage origin: The final stage could equally well be partial fusion of eclogite or of gabbro or complete fusion of a previously differentiated rock—in either case at deep crustal rather than at mantle levels. Whether the history of a magma was simple or complex, the ultimate volcanic product gives information relating only to the final genetic episode.

PROBLEM OF DEEP-SEATED DIFFERENTIATION There are no certain chemical criteria by which a magma generated by fusion can be recognized as such in contradistinction to one of derivative origin. In experimentally based fusion models it is possible, by manipulating degrees of melting and later fractional crystallization under different pressure-temperature regimes, to derive almost any kind of mafic magma (expressed in simple normative parameters) from some

[1] Data for silicon and sulfur, although meager, completely agree with this conclusion.

common postulated source rock (cf. Jamieson, 1966; O'Hara and Yoder, 1967; D. H. Green, 1970*a*). Some writers (for example, O'Hara, 1965) arguing along such lines see most erupted basaltic magmas as derivative, and they invoke deep-seated fractionation to explain much of their chemical variation.

In general we hold the contrary view that voluminously erupted mafic magmas of uniform composition although possibly recycled, appear at the surface with their initial chemical character—imparted by fusion of appropriate ultramafic source rocks—unimpaired or at least still recognizable. This view assigns to high-level differentiation a subordinate but in some cases clearly discernible role, seen for example in short-term variation among Kilauean basalts of a single eruptive episode. Even though "practically all erupted lavas of Kilauea are somewhat differentiated" (Murata and Richter, 1966*b*, p. A23) and the same may be said too of lavas of Mauna Loa, the saturated or slightly undersaturated tholeiitic character of parental magmas of both volcanoes stands out clearly. At many large volcanic centers—Mauna Loa, for example—a single type of magma has been generated a thousand times or more and erupted with only minor subsequent modification. The distinctive individuality of such a pile, or even of a more extensive lava series (for example, Yakima basalts of Columbia River province), can scarcely be due to repeated retracing of an identical course through the maze of an ingenious model. We shall find, too, that while a degree of preeruptive fractionation can explain much of the major-element chemistry of a mafic series, it commonly fails to account for more striking characteristics of the dispersed-element pattern (cf. Gast, 1968). Whether a particular erupted magma is essentially unmodified or somewhat fractionated, however, the chemistry of its incompatible dispersed elements —provided that there has been no crustal contamination—will be determined purely by that of the source rocks and by the degree and regime of fusion.

RELEVANCE VERSUS "NOISE" IN CHEMICAL DATA As chemical data have multiplied and diversified, a problem of increasing urgency is to single out what is relevant to a given problem from what is "noise" of random fluctuation. There is no unique answer to such questions. This is brought out by difference of emphasis in current rock classifications and by growth of ideas on the nature and genetic significance of basaltic magma types. In emphasizing the unique character of every basaltic province we have treated as significant what others have dismissed as "noise." This is the implication of an earlier statement (Verhoogen et al., 1970, p. 303) that "strictly speaking the number of basaltic magma types is infinite."

In the classic approach to differentiation, major-element chemistry was considered relevant, trace-element fluctuation largely as "noise." Now, with vastly more information available, variation in trace-element patterns has become increasingly recognized; this, it is now realized, is more relevant to petrogenesis than the "noise" of more obvious parameters such as $Al_2O_3$

content. Only criteria such as trace-element and isotopic chemistry have revealed the sharply separate identities of abyssal tholeiitic basalts, island tholeiites (for example, in Hawaii and the Galápagos), and tholeiitic diabase magmas of Tasmania and Antarctica.

Trivial differences cannot be dismissed completely as "noise" if they consistently distinguish the chemical character of individual volcanoes or plutons within a province. This we have already remarked in tholeiitic lava piles of nearby Hawaiian volcanoes and among the component plutons of the Sierra Nevada batholith. It is the existence rather than the precise nature of such variation that is significant; it betokens individuality in separate source materials or in melting regimes of petrogenesis.

Some readers, comparing this book with more systematic classics or even with its predecessor of 1960, may criticize seeming lack of systematic precision and find the petrogenic picture here presented obscured by apparently over-plentiful detail and difficult to grasp in its entirety. Difference in the individual viewpoint will depend largely on difference in assessment of "noise." A principal task that faces the petrologist today is to probe the complexity and almost infinite variety of the petrogenic process and of the magmatic environment. And this can be done only if much that used to be considered "noise" is now recognized as relevant.

CHRONOLOGICAL LIMITATIONS It is unlikely that the overall pattern of magmatism has remained unchanged throughout the 4.6 b.y. of geologic time. Differentiation of continental crust from the mantle may have been largely accomplished in very early times—some think as long ago as 3900 m.y. The original contribution of felsic magmas to the sialic crust may have been out-weighed by that of subsequent weathering, sedimentation, and metamorphism. Certainly some very old Precambrian rocks have long been strongly depleted, through granulite-facies metamorphism, in incompatible dispersed elements, notably Th, U, Pb, and Rb (cf. Heier and Thoresen, 1971). There are hints in the geologic record of magmatic patterns restricted to Precambrian times or to specific early events. Already noted are highly potassic Precambrian granites and anorthosite batholiths, with no apparent counterpart among younger rocks.

At present, insufficient data are available to bring out any regular global chronological variation in magmatism if such exists. But the reader is reminded that the data of modern igneous petrology, especially in relation to volcanism and volcanic rocks, are heavily biased toward very recent phenomena accompanying mid-Mesozoic fragmentation of continents and ensuing motions of lithospheric plates.

## OCEANIC BASALT-MANTLE SYSTEMS

DIVERSITY OF SUBOCEANIC SOURCE ROCKS Among common mafic lavas, oceanic basalts are those most likely to retain the chemical imprint of their source rocks unobscured by contamination of sialic crust. During the life cycle of some major volcanoes the composition of magma may change markedly and rather abruptly: for example, from earlier tholeiitic to later alkali basalt in Hawaii and in Réunion. This change probably reflects a change in pressure or in degree of melting or in the composition of the source materials.

The unique character of each island province, and even of individual volcanic piles within one large province, suggests that source compositions may vary widely in a lateral sense. So we may ask whether the potassic character of basaltic lavas on Gough and Tristan da Cunha may be inherited from a laterally restricted potassic segment of the mantle. Or does it imply derivation from a potassic stratum in the mantle—unusually deep or unusually shallow according to some favored model of mantle stratification? Or again has the Tristan magma been regenerated from source rocks already differentiated during some earlier magmatic cycle? Differences of this kind cannot be attributed simply to degrees of fractionation during ascent. The erupted products are still essentially basaltic, and only extreme fractionation could account for the markedly different patterns of incompatible dispersed elements that characterize different island provinces.

ABYSSAL OCEANIC THOLEIITE MAGMAS Engel et al. (1965) first demonstrated that abyssal oceanic tholeiites which constitute the great bulk of oceanic lavas (new oceanic crust) have a highly characteristic pattern of incompatible dispersed elements (Table 8-3, analysis 1*a*). Maximum limits of abundances (ppm) were estimated as: K, 1200; Rb $<$ 10; Sr, 130; Ba, 14; U, 0.1; Th, 0.2; Pb, 0.8. Abundance ratios are Ba/Sr $< 0.2$; Rb/K $< 0.008$; Th/U, 2. If, as seems likely from their great volume and regional chemical uniformity over very large areas, these lavas were generated as substantial melt fractions of source rocks, elemental abundances in these latter may be no more than half those measured in basalts; the cited abundance ratios could in fact be much the same in magmas and parental material.

The typical rare-earth pattern of oceanic tholeiites departs but little from that of standard chondrite; La/Yb in general is perhaps a little lower. So if garnet was a component of the source rock, it must have been completely eliminated in generation of abyssal magmas. Values of $Sr^{87}/Sr^{86}$ are consistently low (0.7025 to 0.7030), and Rb/Sr ($\sim$ 0.01) may be much too low to permit extrapolation—to 0.699 at 4.55 b.y. This and the characteristic subdued pattern of incompatible elements have been taken as evidence that the immediate source rocks have become heavily depleted in such elements, and especially in Rb relative to Sr, by early episodes of partial fusion. It is difficult to reconcile this conclusion with the vast volume of abyssal magma that has been generated during the past few million years, welling up from deep sources below the oceanic ridges. How could the refractory residues of ancient fusion yield such quantities of new melt?

ALKALI BASALTS OF ISLAND PROVINCES At the opposite extreme, alkali basalts of island provinces have a very different character (Engel et al., 1965,

p. 725; cf. also Tables 8-7, 8-8, 8-9): K 15 to 20 × $10^3$, Rb 30 to 100, Ba 300 to 800, Sr 500 to 1000 ppm; Rb/K ≈ 0.002, Ba/Sr 0.5 to 0.9; La/Yb ten times that for chondritic meteorites. In melilite nephelinites of Hawaii, fractionation of La relative to Yb is still more pronounced (Schilling and Winchester, 1969, p. 33). These characteristics collectively cannot be reconciled with once-popular models of derivation (by fractional crystallization) from a parental tholeiitic magma; only extreme fractionation could bring about such pronounced divergence in dispersed-element chemistry. Two alternative explanations remain: (1) alkali-basalt magmas represent an early stage of incipient melting compared with more complete melting of the same source rock to yield abyssal tholeiitic magma; or (2) the two magmas have come from radically different sources.

Gast (1968) finds the first alternative adequate to explain the contrasted patterns of dispersed elements, provided that the degree of melting required to generate alkaline liquids were small—only a few percent. But to explain the marked difference between respective rare-earth abundance patterns (cf. Fig. 13-3), Philpotts and Schnetzler (1970) are obliged to restrict the alkali-basalt fraction to about 1 percent—a figure that seems much too low to account for the volume of alkaline lavas in many island provinces. The great chemical diversity among alkaline magmas themselves from one island to another would seem to demand corresponding variety among immediate source rocks. It is difficult to visualize the great piles of potassic lavas on Tristan da Cunha and Gough as products of *incipient* melting of mantle rock whose K abundance is no more than 0.15 percent. It is still possible that some of the implied diversity at the source reflects differentiation during earlier magmatic cycles in the basement of island volcanoes.

This final complication can be examined in the light of isotopic evidence, more particularly with respect to strontium. There are island lavas, such as the tholeiitic basalts of Iceland, whose strontium has much the same composition as that of abyssal tholeiite. More commonly among alkaline basalts and island tholeiites alike (as in Hawaii), present $Sr^{87}/Sr^{86}$ values are slightly but significantly higher—between 0.703 and 0.705. Strongly alkaline lavas of other islands give $Sr^{87}/Sr^{86}$ values that are higher still: $> 0.705$ on Samoa, 0.7045 to 0.7065 on Gough and Tristan da Cunha (cf. Table 8-14). In these rocks there is a strong positive correlation between $Sr^{87}/Sr^{86}$, Rb/Sr, and K abundance. All this adds up as evidence of marked heterogeneity of long standing in the mantle sources on a broad geographic scale. Anomalies such as those that characterize Gough and Tristan da Cunha are just what might be expected if the underlying source were the solidified melt fraction of an earlier cycle of partial fusion—enriched relative to *its* source in K and in Rb with respect to Sr.

The isotopic composition of lead varies among oceanic basalts, but with no obvious relation to magma type. This variation points to heterogeneity in source rocks without systematic relation to geography, depth, or source chemistry. Leads that plot, at $t = 0$, on what was termed the standard "primary growth curve" are sometimes termed "normal." Hawaiian leads are of this kind.

So too are six abyssal tholeiite leads: $Pb^{206}/Pb^{204}$, 17.8 to 18.8; $Pb^{207}/Pb^{204}$, 15.5 to 15.7; $Pb^{208}/Pb^{204}$, 37.5 to 38.7. While rejecting the concept of a single-stage development from primordial lead, we can accept these values as a useful reference standard with which to compare other modern leads.

## ORIGINS OF CONTINENTAL MAFIC MAGMAS

SPECIAL CONDITIONS There is a broad compositional overlap between continental and oceanic basalts and nephelinites. But there also are many classes of rocks which are largely or exclusively continental (Chaps. 10 and 12). Even within the most extensive class of volcanic rocks—tholeiitic basalts—many continental and oceanic members are mutually distinct.

Continental petrogenesis is complicated by a special problem: evaluation of the relative roles of continental crust and upper mantle in determining the distinctive character of continental igneous series. The sialic crust is assigned the major role in the origin and evolution of continental siliceous magmas (Chaps. 11 and 12). Mafic magmas, on the other hand, could originate in a deep simatic crust or in the underlying mantle. If most, by analogy with oceanic basalts, are formed by partial fusion of mantle rocks, then their distinctive continental traits could reflect unique chemical characteristics in the subcontinental mantle; or alternatively these traits may have been imprinted by crustal contamination of the ascending magma. It has been argued that decoupling of lithospheric plates from underlying mantle rock, inherent in plate-tectonics theory, renders improbable any long-standing chemical difference between subcontinental and suboceanic mantle rocks (cf. Jamieson and Clarke, 1970, pp. 200–201). We prefer to leave this question open, at least with respect to the uppermost part of the mantle. In each lithospheric plate there is a lower component of mantle rock whose thickness is unknown (cf. Verhoogen et al., 1970, p. 690).

CONTINENTAL ALKALI BASALTS AND NEPHELINITES The chemistry, including trace-element and isotopic patterns, of many continental alkali basalts and nephelinites so closely parallels that of typical ocean-island provinces that one can only assume that similar source rocks exist, and the same genetic mechanism has operated, in the subcontinental and in the suboceanic upper mantle. But there are also continental variations in trace-element and isotopic chemistry that have no recorded counterpart in alkali basalts of the oceans. For example, late Cenozoic alkali basalts of the southern sector of the Great Basin, western United States (Hedge and Noble, 1971) have the unusual combination of moderately radiogenic strontium ($Sr^{87}/Sr^{86} \sim 0.707$) and unusually high strontium ($\sim$ 1200 ppm; Rb/Sr $\sim$ 0.03). These traits not only rule out isotopic evolution of the source rock by any single-stage model, even of very long duration, but exclude contamination of the magma by normal sialic crust with relatively high Rb/Sr. Hedge and Noble appeal to a mantle source of unusual composition, once with high Rb/Sr and subsequently enriched in Sr during some previous magmatic event. Again we see evidence of present-day mantle heterogeneity.

CONTINENTAL THOLEIITIC MAGMAS Continental tholeiitic flood basalts and related diabases, more than any other class of volcanic rocks, satisfy the two criteria postulated previously for magmas generated directly by fusion—great volume and compositional homogeneity within each province. How else can one explain the uniform character of more than 100,000 $km^3$ of Yakima basalts erupted within 5 m.y. in the Columbia River province, northwestern United States? To derive this magma from picritic basalt of deep-seated origin by low-pressure fractionation (cf. O'Hara, 1965, table 1) requires that again and again each successive draught of magma must rid itself cleanly, while still largely liquid, of the same fraction of crystalline olivine along some identical course of ascent. This seems highly improbable. Fortunately, the issue of derivative magmatic status has little bearing on the source problem or on the difficult question of possible crustal contamination.

At one end of the compositional spectrum there are basalts whose chemistry duplicates that of the much more abundant abyssal tholeiitic basalts of the ocean. The chemistry of one such group from Baffin Island and the immediately facing western coast of Greenland has been found consistent with origin by partial melting of garnet peridotite at 30 kb (D. B. Clarke, 1970). But the rare-earth patterns of the Baffin Island and Greenland rocks differ significantly; and to explain this O'Nions and Clarke (1972) propose a complex model involving different degrees of melting in the two areas, followed by precisely limited "eclogite fractionation" in one. The necessity for this special manipulation would disappear if differences in source materials (for example, as to relative quantities of garnet) were invoked.

In general, it would seem that continental tholeiites as a class tend to be more potassic than oceanic counterparts (Jamieson and Clarke, 1970). With higher potassium go higher abundances of Ba, Sr, Rb, U, and Th; higher Rb/K, Na/K, and Rb/Sr ratios; and a tendency for higher $Sr^{87}/Sr^{86}$. On the other hand Ti, which is commonly grouped with incompatible elements, seems to be twice as high in oceanic as in continental tholeiites (cf. Chayes, 1964*a*, p. 1582; also data of Jamieson and Clarke, 1970, p. 197). Data for lead are most imperfect. In at least one case (Tatsumoto and Snavely, 1969, p. 1090, no. 17–12), lead of Columbia River basalts has been found to be more radiogenic than that of oceanic tholeiites.

Chemical implications of crustal influence, at least in some instances, are too clear to be dismissed. Continental tholeiites differ from their oceanic counterparts in those very characteristics that might be expected from some sort of interaction between a primitive magma and sialic crust. An explanation perhaps may be sought in a multistage genetic process involving recycling of mantle-derived magma via the crust. Consider the chemically identical Jurassic

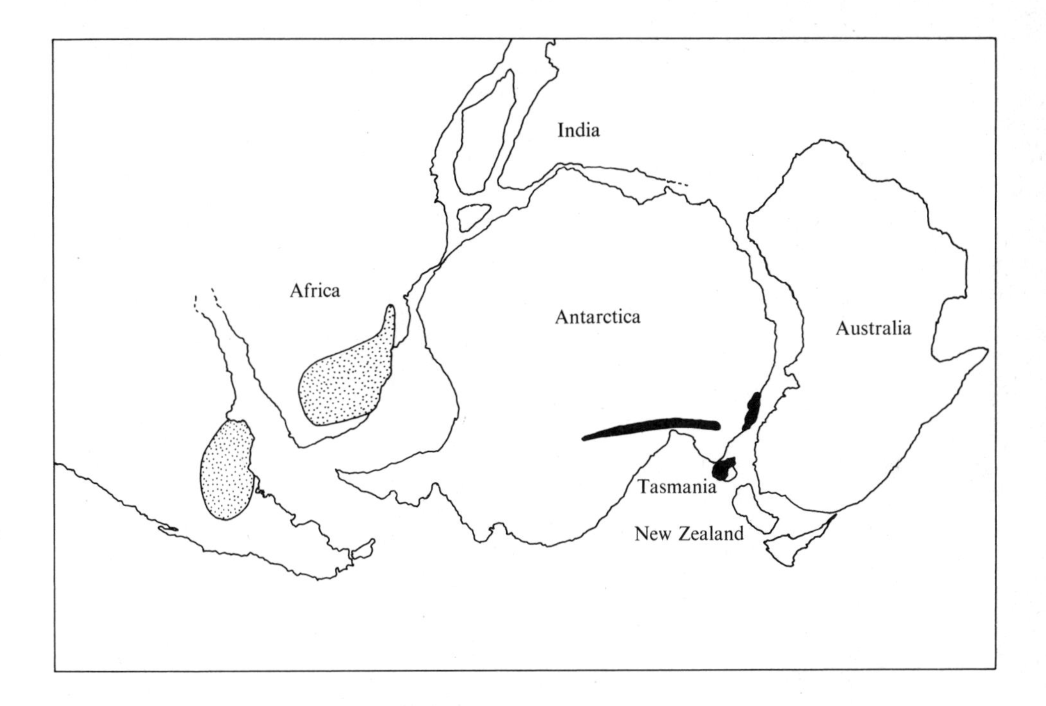


**Fig. 13-6** Configuration of southern continents (Gondwanaland) at close of Paleozoic. Tasmanian and Antarctica diabases (solid black); basaltic provinces of Karroo and Paraná (stippled). (After Engel and Kelm, 1972, p. 2335.)

diabases of Tasmania and Antarctica representing a chemically uniform basaltic magma rapidly deployed within extensive areas at that time geographically contiguous (Fig. 13-6). Some 100 m.y. later, tholeiitic magmas once more rose to the surface in Tasmania. The Tertiary basalts then erupted differ decidedly in dispersed-element and isotopic chemistry from their Jurassic predecessors (cf. Table 9-17; columns 1, 8). This difference might reflect chemical differences related to depth or even to geographic station in time, since in the course of 100 m.y. of drift the Tasmanian plate might conceivably have become uncoupled from the source of its Jurassic magmas or alternatively it might have come to overlie a new source of Tertiary magma in the low-velocity layer.

One model that could account for strontium-isotope and dispersed-element data at present is illustrated in Fig. 13-7.* In a primitive mantle source comparable with that of many oceanic basalts, $Sr^{87}/Sr^{86}$ increases throughout geologic time along *AB*. The source of Tertiary magma, considerably richer in

* From Verhoogen et al. (1970, p. 207, eq. 4–9), given $Rb^{87} = 0.2785$ Rb and $Sr^{86} = 0.0986$ Sr, the rate of increase in $Sr^{87}/Sr^{86}$ in a closed system is 0.0393 Rb/Sr per $10^9$ years.

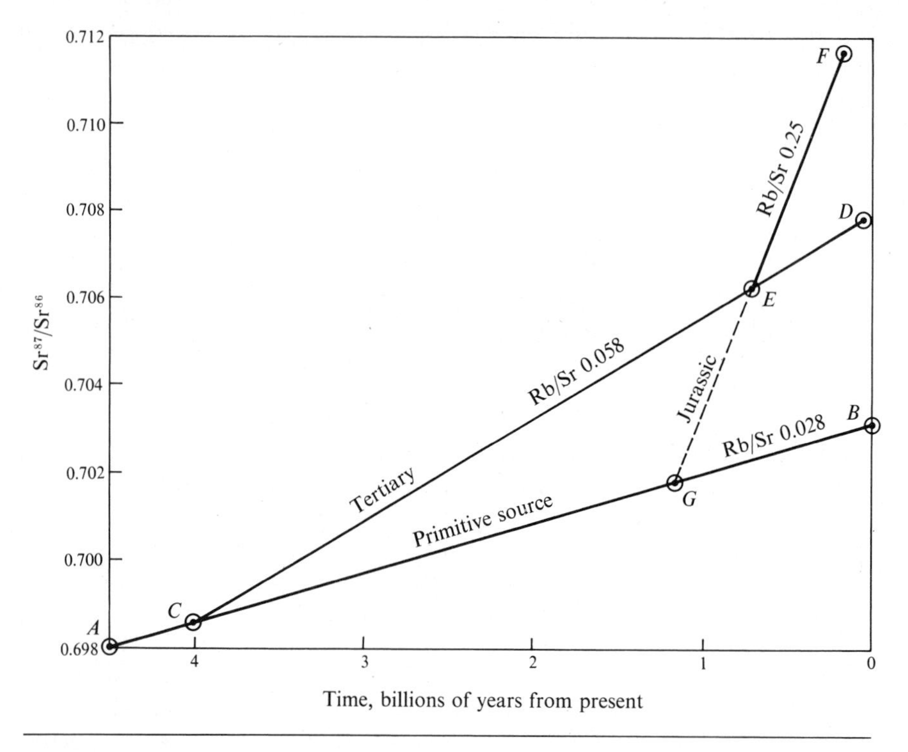


**Fig. 13-7** Possible (but by no means unique) model for evolution of strontium composition, $Sr^{87}/Sr^{86}$, in source rocks of Jurassic and Tertiary tholeiitic magmas of Tasmania. It is assumed that Rb/Sr is the same in erupted magmas as in their sources.

Rb (Rb/Sr = 0.058), is shown as *CD*. This source could represent an Rb-enriched variant of the primitive mantle or a very early offshoot from it (within limits of measurement of Rb/Sr). The Jurassic diabase comes from a source *EF* diverging from the *CD* or possibly the *AB* line late in the Precambrian. Divergence is pictured as upward accretion of a partial melt which by contact with the lower sialic crust became enriched in K, Rb, and related elements by assimilative reaction or some selective diffusive mechanism. Prolonged residence in the deeper crust permitted homogenization in these respects. During the breakup of the southern continents this secondary source became recycled and ascended rapidly as a copious flood into the uppermost sedimentary crust in mid-Jurassic times. In presenting this model we do not imply that we have pictured events exactly as they occurred. Rather we show the kind of thing that may have happened; and we appeal to some kind of multiple recycling of magma and interaction with the crust to explain the "continental" chemistry of Tasmanian diabases and the marked chemical differences that distinguish the Jurassic from the Tertiary magma.

SOURCES OF BASIC MAGMAS OF ISLAND ARCS The origins and possible relationships of basaltic and andesitic magmas of island arcs and continental margins are part of a single unsolved problem. Everything at present points to generation of basaltic-andesitic and tholeiitic magmas at great depth in Benioff zones. But what role, if any, is played by mantle rocks and what by oceanic crust is unknown. Plate-tectonics theory has provided a site for magma generation but has introduced new degrees of freedom relating to time and place which can be manipulated to fit, but not rigorously to test, any preferred model of andesite magmatism.

If oceanic crust is the main starting material, it is presumably in a metamorphic condition when it reaches the site at which basaltic andesite magma is to be generated. If still anhydrous, this direct source material is thus likely to be eclogite; if hydrous, then a hornblende eclogite or a garnet amphibolite. There is no means of knowing whether oceanic abyssal tholeiite on its way to becoming eclogite (or hornblende eclogite) passes through the spilitic stage of low-temperature burial metamorphism. Spilites of old geosynclines—the end products of the continental-margin cycle of magmatism—have high Na/K ratios. Whether this fact partly reflects origin from oceanic crust as the source of magma, or is due purely to the metasomatic activity of seawater after extrusion, it is impossible to tell. Likewise ambiguous is the rather high Na/K ratio shown by some eclogites, for example, in the Franciscan terrane of California.

All this uncertainty currently leaves open a way to explain the marked difference between $Sr^{87}/Sr^{86}$ values of most island-arc tholeiites and basaltic andesites (0.704 to 0.705) and those typical of new oceanic crust (abyssal tholeiite, 0.7025 to 0.703). It would seem likely that in the complete reconstitution of feldspar (replacement of CaAl by NaSi) which accompanies spilitization of basalt, the isotopic composition of strontium should take on something of the character of marine strontium ($Sr^{87}/Sr^{86} = 0.709$). Few data are at present available as to the composition of strontium (or of hydrogen) in young spilites. However, the metamorphic model of spilitization predicts that $Sr^{87}/Sr^{86}$ values for spilites should be significantly higher than those for oceanic tholeiites[1]. If this is verified, and if strontium of metamorphic eclogites of the Franciscan type is found to have a similar composition, it will become possible to explore further the role of oceanic crust in the genesis of basaltic-andesite magmas.

## KIMBERLITE, ULTRAPOTASSIC MAGMAS, AND CARBONATITE

In Chap. 10 we documented close mutual association between nephelinites and carbonatites, between both of these and rarer ultrapotassic lavas, and between these latter and kimberlites. The collective sources of corresponding magmas

[1] Basaltic lavas of the Pacific floor sampled by deep-sea drilling have $Sr^{87}/Sr^{86} = 0.7038$ (Bass et al., 1973). Presumably it has taken about 60 m.y. to raise the likely initial ratio ($\sim 0.7026$) to this value by interaction with seawater.

must lie at the same general level in the upper mantle. Deep-seated origin is most convincingly demonstrated on mineralogical grounds for kimberlites; and the striking chemistry of these rocks, and by inference similar traits appearing in ultrapotassic lavas, is likely to stem largely from mantle sources.

KIMBERLITE PROBLEM The dispersed-element pattern of kimberlites is extreme: K, Rb, Ba, Sr, and some other incompatible elements reach concentrations an order of magnitude above accepted average mantle abundances. Very high relative enrichment in light rare-earth elements in some African kimberlites is what could be expected in magmas coexisting with garnet (Gast, 1968, p. 1074). Values of $Sr^{87}/Sr^{86}$ are variable (Berg and Allsopp, 1972): they prove to be low (0.7037 to 0.704) in fresh specimens, much higher in altered ones. Lead, too, may be strongly radiogenic (Lovering and Tatsumoto, 1968, p. 354). Kennedy's "kimberlite window" through continental cratons reveals, well down in the upper mantle, local operation of extreme chemical processes. This is generally agreed. But as to the precise nature of the process and the mutual relations between kimberlitic magmas, garnet-peridotite and eclogite xenoliths, and postulated magma sources, half a dozen divergent opinions are currently in vogue. Here are four:

1. Kimberlitic magmas are formed by partial melting of mantle rock of unusual composition—possibly containing phlogopite or potassic amphibole as well as olivine, pyroxenes, and garnet. Peridotite and eclogite xenoliths are accidental.
2. The essential feature of the partial-melting model is physical more than chemical. At sites of unusually low heat flow, water and carbon dioxide distilled slowly from the surrounding mantle prolong incipient fusion for long periods of time (Harris and Middlemost, 1970). The characteristic dispersed-element pattern is that of the earliest stages of partial fusion. High $Sr^{87}/Sr^{86}$ values could reflect previous operation of a similar process, but without eruption of magma, in ancient times.
3. Kimberlitic magmas are extreme liquid residues from fractional crystallization of basic magma by removal of garnet and pyroxene in a high-pressure environment (O'Hara, 1968, p. 94; MacGregor, 1970). Xenolithic eclogites, and perhaps garnet peridotites too, represent complementary crystal cumulates.
4. Kimberlitic magmas are products of reaction between aqueous carbonatite (ankeritic) magma and deep crustal rocks (Dawson, pp. 225–227 in Wyllie, 1967).

Our tentative preference is for the first of these models. The chemistry of kimberlites seems to demand local high concentration of potassium within the mantle; and this is consistent too with the chemistry and mode of occurrence of ultrapotassic lavas.

ULTRAPOTASSIC MAGMAS Older ideas regarding the origin of potassium-rich mafic magmas have been reviewed by Turner and Verhoogen (1960, pp. 250–256). The preference there expressed for assimilative reaction between nephelinitic magma and granitic or pelitic crust remains a likely model for genesis of potassic lamprophyres and leucite nephelinites that appear in some continental alkaline provinces. A similar model proposed by Waters (1955) invokes fusion of deep crustal rocks rich in biotite and/or hornblende.

Ultrapotassic lavas of East African provinces show erratic $Sr^{87}/Sr^{86}$ values (0.703 to 0.710) which, however, correlate directly with Rb/Sr. This correlation suggests fusion of deep source rocks perhaps in the crust (K. Bell and Powell, 1969), perhaps a potassic variant of the mantle, or possibly even ancient kimberlite itself (cf. P. G. Harris, cited by K. Bell and Powell, 1969, p. 563). In ultrapotassic rocks further afield (Powell and Bell, 1970), $Sr^{87}/Sr^{86}$ may reach very high values (0.713 to 0.721 in Kimberley, West Australia; 0.706 to 0.707 in Leucite Hills, Wyoming) and there is no direct correlation with Rb/Sr values. These extreme rocks we also tentatively trace to a highly potassic mantle source, very rich in phlogopite; but the characteristically high abundances of Ni and Cr in Leucite Hills lavas imply that olivine if originally present was entirely eliminated before the magma separated from any crystalline residue. Values of $Sr^{87}/Sr^{86}$ are compatible with ages of source materials far back in the Precambrian. Erratic correlation between $Sr^{87}/Sr^{86}$ and Rb/Sr within individual provinces could reflect mutual fractionation of Sr and Rb in the process of magma genesis—which indeed is possible if the source rock contains phlogopite (cf. Table 13-2).

CARBONATITE PROBLEM Accepting the magmatic origin of carbonatites from sources situated at the same level as those of kimberlites and nephelinites, we still face unsolved problems as to how carbonatite magmas form and their possible relation to alkaline silicate magmas (cf. Chap. 10). Two things, however, seem clear. It has been shown experimentally that hydrous alkaline magmas could generate imiscible carbonatite liquids (for example, Wyllie, 1966). In the second place, carbonate liquids containing water and greatly enriched in many "incompatible" dispersed elements are being generated locally within the upper mantle of continental cratons. And strontium, which in such liquids is abundant, maintains a composition consistent with a mantle source. This fact, nevertheless, does not rule out crustal contamination since the abundance of Sr in carbonatites is high enough to mask addition of more radiogenic strontium from crustal rocks.

## SUMMARY OF MAGMATIC CONSTRAINTS ON SOURCE CHEMISTRY

The testimony of mafic magmas ascending from the mantle, and in continents perhaps also from deep crustal sources, in general confirms and in fact amplifies conclusions drawn independently from solid samples of mantle rock:

1. The isotopic composition of oxygen in mafic magmas, especially that of crystallizing pyroxenes and olivine, is narrowly limited ($\delta_{O^{18}/O^{16}} = +5.9 \pm 0.3$ per mil) and matches that of mantle fragments and of common chondritic meteorites (H. P. Taylor, 1967*b*).

2. Assuming that observed Rb/Sr and $Sr^{87}/Sr^{86}$ values for abyssal and island basalts have been inherited unchanged from suboceanic mantle sources,[1] something can be said about the isotopic composition of primordial terrestrial strontium and the questionable concept that reservoirs of primordial strontium survive in the mantle.

   a. Single-stage regression curves for $Sr^{87}/Sr^{86}$ in abyssal tholeiites extrapolate back to values between 0.6998 and 0.7012 at 4.6 b.y. (Fig. 13-8*a*, full lines). Primitive values so estimated for many oceanic basalts—allowing for possible inaccuracy in measured Rb/Sr values—cluster in the range 0.699 to 0.702. All this lends credence to the widely accepted idea that $Sr^{87}/Sr^{86}$ ratio in primordial terrestrial strontium was near or identical to that of meteorites.

   b. To achieve identity with the meteoric ratio, 0.698, it is necessary, however, to invent multistage models of strontium evolution in mantle systems continuously or intermittently disturbed (opened and reset with respect to Rb or Sr). Gradual depletion of the source in Rb relative to Sr by continuous early degassing (broken lines, Fig. 13-8*a*) or by a single degassing event at 3.5 b.y. (broken line, Fig. 13-8*b*) could account for data available for abyssal tholeiitic basalts. To explain relatively high Rb/Sr of more alkaline island basalts, one can postulate single events at *X*, by which the source material becomes enriched in Rb relatively to primordial mantle rock. Generation of the source by partial fusion of primordial rock could give the desired result. Such manipulation, in the absence of critical independent testing, is scarcely scientific. So we come to a third, more general conclusion of greater validity.

   c. If the primordial composition of strontium is indeed identical to that of meteorites, then the sources of present-day basaltic magma cannot represent reservoirs of primordial mantle rock—provided that the initial assumptions set out above are valid and that analytic data are accurate.

3. Mantle source rocks are chemically heterogeneous on a regional scale. Within large individual provinces, on the other hand, source materials tapped over short intervals of geologic time may be remarkably homogeneous. Most striking in this respect is the uniformity of great volumes of abyssal tholeiitic magma rising from depths unknown along the oceanic ridges. Dispersed-element abundances in these rocks are so low that some

[1] Approximately true for Rb/Sr only if the magma represents a substantial melt fraction.

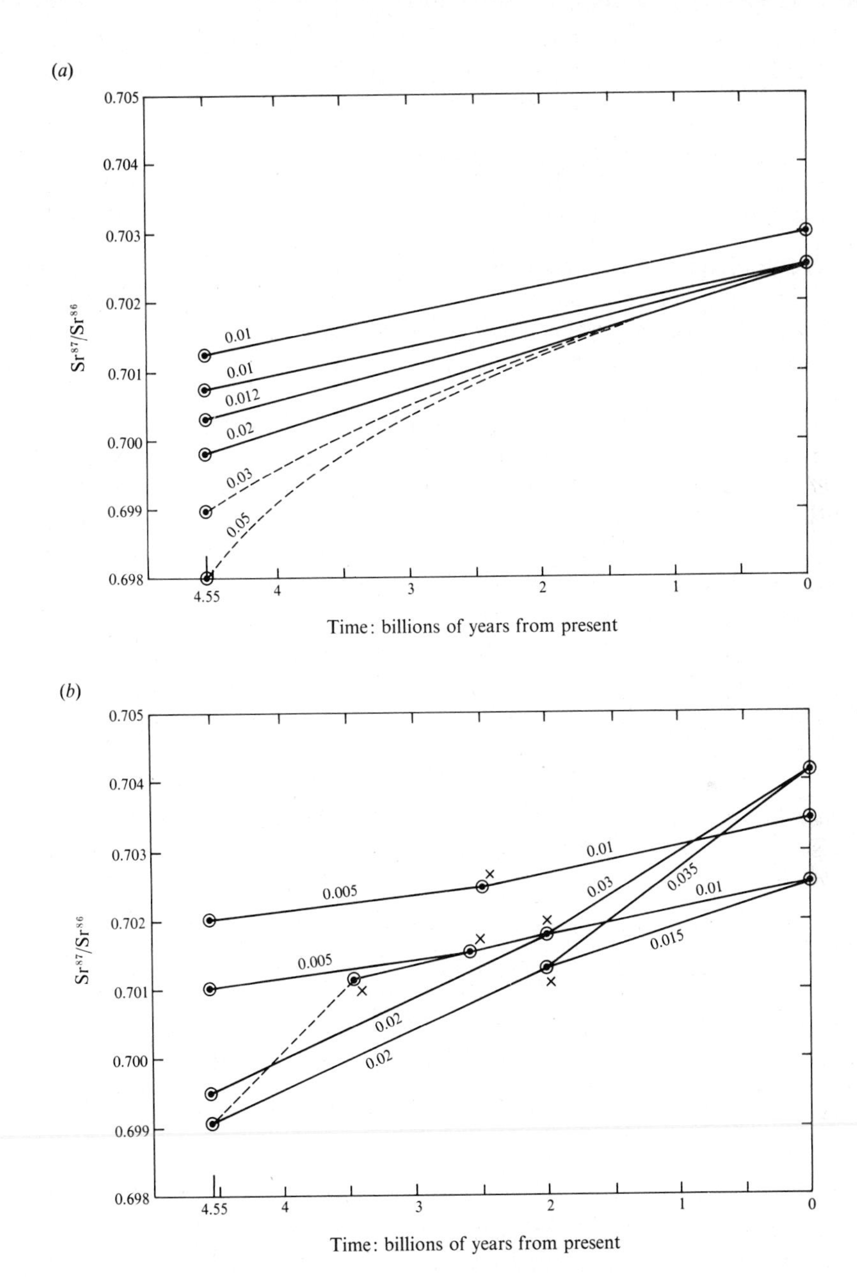


**Fig. 13-8** Some possible alternative models for evolution of strontium composition, $Sr^{87}/Sr^{86}$, in sources of oceanic basalts. Assumed Rb/Sr values indicated on each regression curve. $Sr^{87}/Sr^{86}$ increases by 0.0393 Rb/Sr per $10^9$ years. (*a*) Abyssal tholeiites. Closed systems, full lines; open systems with continuous reduction of Rb/Sr, dashed lines. (*b*) Two-stage models, with systems reset at *X*.

geochemists regard the sources as significantly depleted compared with any postulated average mantle rock.

4. Heterogeneity in a given province may possibly be related to depth. But no global depth pattern can be certainly inferred nor is it to be expected in the light of plate-tectonics theory.

5. Proved diversity of source rocks is considerably greater beneath continents than in the suboceanic upper mantle. Partly this reflects crustal influences, but in part the observed chemical variety (notably as inferred from kimberlites) is inherent in mantle rocks.

Open questions of great significance remain unsolved. The phase chemistry of rocks at depths exceeding 120 km is still experimentally unexplored. In kimberlites we find relics of material from greater depth; and there is no reason to discount the possibility that the direct sources of some magmas are far below the 100 km level. Also undecided is the mineralogical and chemical character of magma sources. Pyrolite is simply a postulated model material unrepresented in nature. Most peridotites could generate only very limited liquid fractions of basaltic composition; and they are incapable of yielding the steady stream of water and carbon dioxide that continues to ascend from the depths, or even the high content of elements such as phosphorus and chlorine observed in alkaline mafic rocks. It is hard to avoid the conclusion that within the upper mantle there must be rocks chemically similar to the basalts themselves—eclogites and at greater depth similar rocks whose mineralogical makeup remains a matter of speculation.

## Petrologic Constraints on Plate-tectonics Models

All roads of geologic inquiry today lead to plate tectonics. Igneous petrology is no exception. The mutual relation between the two fields is all the more direct in that the most compelling and clearly legible data in each concern relatively recent, genetically connected events. Thus it is appropriate to conclude this book by listing, without explaining, some of the basic facts of petrology that must ultimately be accommodated in comprehensive plate-tectonics models:

1. Ocean-ridge lavas and volcanic rocks of the spreading sea floor, although by no means chemically uniform, have a narrow compositional range. They are tholeiitic basalts ("abyssal tholeiites") with a distinctive "subdued" dispersed-element pattern, poorly radiogenic strontium, and lead of a rather consistent isotopic composition.

2. Lavas of oceanic islands cover a much broader spectrum. Parental magmas may be tholeiitic, alkali basaltic, or nephelinitic; and the range of dispersed-element patterns is correspondingly broad. There is a somewhat broader

range, too, in the isotopic composition of strontium and lead. Felsic associates preserving the silica-saturation characteristic of the parental magma are widespread.

3. Chronological trends recognized in ocean-island provinces are of two kinds. Short-term chemical trends are displayed in lavas erupted over periods of the order of $10^3$ to $10^5$ years from single centers. There is a hint in some cases that such trends may partly reflect processes (differentiation, melting) that operated in the crust over periods of longer duration preceding the eruptive cycle.

4. Magmas erupted along island arcs are correlated on seismic evidence with deep sources located in Benioff zones. They are dominantly basaltic (tholeiitic) and andesitic. Where an island arc impinges on or transgresses continental crust, much of the erupted (fragmented) debris may be rhyolitic. The isotopic composition of strontium and lead in island-arc volcanics is variable; in some provinces it overlaps the range found in ocean-island basalts. It in no way demands any role for incorporation of oceanic sediment in the magma-forming process.

5. Similar basaltic, andesitic, and rhyolitic debris, in various stages of low-grade metamorphism, figures prominently in eroded geosynclinal fillings located along ancient or existing continental margins. Eruption of the same volcanic suite may be renewed long after folding, uplift, and partial erosion of earlier geosynclinal prisms. Such eruptions need not be related to any obvious Benioff (subduction) zone. Gilluly (1971) cites in this connection the Eogene volcanics of the San Juan province (Turner and Verhoogen, 1960, p. 273) and the Neogene volcanics of Yellowstone, both more than 1500 km inland from the mobile western edge of the North American continent.

6. The geologic locale of ultramafic bodies of the Alpine type and of the ophiolite association of which they are characteristic members, is in ultramafic belts of active island arcs and their counterparts in zones of older orogeny. The evidence that most such ultramafic rocks are tectonically transported mantle fragments is strong. In today's fashion the whole ophiolite association is widely interpreted as old oceanic crust.

7. Sharp episodes of granitic plutonism are characteristic of the later (orogenic) stages of geosynclinal development. In some instances (for example, Cenozoic batholiths of the Rockies) the site of intrusion may be far inland from active Benioff zones.

8. Rifting of continental plates preliminary to episodes of accelerated drift again and again has been accompanied by eruption and shallow intrusion of immense volumes of mafic magmas. This is predominantly, but by no means exclusively, tholeiitic; and even among tholeiitic provinces there is

a good deal of variation as to dispersed-element and radiogenic-isotope patterns. Well-documented examples (compare episodes of drift delineated by Le Pichon, 1968) include the following:

a. Karroo province, southern Africa: tholeiitic basalts and diabases: age 190 to 154 m.y.

b. Paraná province, Brazil: tholeiitic basalts and diabases; age 135 to to 115 m.y.

c. Tasmanian-Antarctic province: tholeiitic diabases; age 167 to 143 m.y.

d. Hebridean and eastern Greenland provinces: tholeiitic and alkaline basalts; age 60 m.y.

e. Southern half of eastern Australia (cf. Packham and Falvey, 1971): highly alkaline and some tholeiitic basalts; age mostly 25 to 20 m.y.

f. Rift system of eastern Africa and Red Sea: nephelinites, alkali basalts, pantellerite-phonolite floods; age $\sim$ 20 to 0 m.y.

9. Partially overlapping and following Mesozoic continental eruption and shallow intrusion of tholeiitic magmas in southern Africa, eastern North America, and Brazil was widespread localized intrusion of small quantities of kimberlitic, ijolitic, nepheline-syenitic, and carbonatitic magmas derived from deep sources.

10. Although broad conformity to magmatic-tectonic pattern is obvious in the preceding list, every major magmatic episode and province has unique petrochemical characteristics. Strongly implied is heterogeneity and/or change with passage of time in mantle source rocks.

One of the most intriguing aspects of both mantle-magma and global-tectonics models is at the same time a current source of weakness. Degrees of freedom for manipulation are many; and they open equally numerous avenues for the builder of genetic models to invoke Karl Popper's "conventionalist stratagem." More rigorous mutual constraints by which current hypotheses may be tested by attempted falsification are needed before we shall see any enduring general synthesis of a viable mantle-magma system with the phenomena of global tectonics.

# References


ABBOTT, M. J.: Petrology of the Nandewar Volcano, N.S.W., Australia, *Contrib. Mineral. Petrol.*, vol. 20, pp. 115–134, 1969.

ABDEL-MONEN, A., N. D. WATKINS, and P. W. GAST: Volcanic History of the Canary Islands (abstr.), *Am. Geophys. Union Trans.*, vol. 48, pp. 226–227, 1967.

ALLEN, J. B., and T. DEANS: Ultrabasic Eruptions with Alnoitic-Kimberlitic Affinities from Malaita, Solomon Islands, *Mineral. Mag.*, vol. 34, pp. 16–34, 1965.

ALLSOPP, H. L., L. O. NICOLAYSEN, and P. HAHN-WEINHEIMER: Rb/K Ratios and Sr-isotopic Compositions of Minerals in Eclogitic and Peridotitic Rocks, *Earth Plan. Sci. Letters*, vol. 5, pp. 231–244, 1968.

ALMEIDA, F. F. M. DE: Geologia e petrologia do arquipêlago de Fernándo de Noronha, *Div. Geol. Mineral. Dep. Nac. da Product. Mineral*, Monograph 13, Rio de Janeiro, 1955.

———: Geologia e petrologia da Ilha da Trinidade, *Div. Geol. Mineral. Dep. Nac. da Product. Mineral*, Monograph 18, Rio de Janeiro, 1961.

AMARAL, G., U. G. CORDANI, K. KAWASHITA, and J. H. REYNOLDS: Potassium-Argon Dates of Basaltic Rocks from Southern Brazil, *Geochim. Cosmochim. Acta*, vol. 30, pp. 159–190, 1966.

———, J. BUSHEE, U. G. CORDANI, K. KAWASHITA, and J. H. REYNOLDS: Potassium-Argon Ages of Alkaline Rocks from Southern Brazil, *Geochim. Cosmochim. Acta*, vol. 31, pp. 117–142, 1967.

ANDERSON, A. T.: The Oxygen Fugacity of Alkaline Basalt and Related Magmas, Tristan da Cunha, *Am. J. Sci.*, vol. 266, pp. 704–727, 1968.

———, R. N. CLAYTON, and T. K. MAYEDA: Oxygen Isotope Thermometry of Mafic Igneous Rocks, *J. Geol.*, vol. 79, pp. 715–729, 1971.

———, and T. L. WRIGHT: Phenocrysts and Glass Inclusions and Their Bearing on Oxidation and Mixing of Basaltic Magmas, Kilauea Volcano, Hawaii, *Am. Mineral.*, vol. 57, pp. 188–216, 1972.

ANDERSON, C. A.: Volcanoes of the Medicine Lake Highland, California, *Univ. Calif. Bull. Dept. Geol. Sci.*, no. 25, pp. 347–442, 1941.

ANDERSON, D. L. and C. SAMMIS: Partial Melting in the Upper Mantle, *Phys. Earth Planet. Interiors*, vol. 3, pp. 41–45, 1970.

———, ———, and J. JORDAN: Composition and Evolution of the Mantle and Core, *Science*, vol. 171, pp. 1103–1112, 1971.

ANDERSON, F. W., and K. C. DUNHAM: The Geology of Northern Skye, *Mem. Geol. Surv. Gt. Britain*, 1966.

ANDRADE, E. M. DA C.: A Theory of Viscosity of Liquids *1* and *2*, *Phil. Mag.*, vol. 17, pp. 497–511 and 698–732, 1934.

ANGELL, C. A.: Oxide Glasses in Light of the "Ideal Glass" Concept: I. Ideal and Nonideal Transitions, and Departures from Ideality, *J. Am. Ceram. Soc.*, vol. 51, pp. 117–124, 1968.

AOKI, K.: Petrology of Kaersutite-bearing Ultramafic and Mafic Inclusions in Iki Island, Japan, *Contrib. Mineral. Petrol.*, vol. 25, pp. 270–283, 1970.

ARMSTRONG, R. L.: A Model for the Evolution of Strontium and Lead Isotopes in a Dynamic Earth, *Rev. Geophys.*, vol. 6, pp. 175–200, 1968.

——— and J. A. COOPER: Lead Isotopes in Island Arcs, *Bull. Volc.*, vol. 35, pp. 27–63, 1971.

ARNASON, B., and T. SIGURGEIRSSON: Deuterium Content of Water Vapor and Hydrogen in Volcanic Gas at Surtsey, Iceland, *Geochim. Cosmochim. Acta*, vol. 32, pp. 807–814, 1968.

ARNDT, J., and F. HÄBERLE: Thermal Expansion and Glass Transition Temperatures of Synthetic Glasses of Plagioclase-like Compositions, *Contrib. Mineral. Petrol.*, vol. 39, pp. 175–183, 1973.

ATKINS, F. B.: Pyroxenes of the Bushveld Intrusion, South Africa, *J. Petrol.*, vol. 10, pp. 222–249, 1969.

BAADSGAARD, H., R. E. FOLINSBEE, and J. LIPSON: Potassium-Argon Dates of Biotites from Cordilleran Granites, *Bull. Geol. Soc. Am.*, vol. 72, pp. 689–702, 1961.

BACON, C. R., and I. S. E. CARMICHAEL: Stages in the P-T Path of Ascending Basalt Magma: An Example from San Quintin, Baja California, *Contrib. Mineral. Petrol.*, vol. 41, pp. 1–22, 1973.

BAILEY, D. K.: The Stability of Acmite in the Presence of $H_2O$, *Am. J. Sci.*, vol. 267-A (Schairer vol.), pp. 1–16, 1969.

——— and R. MACDONALD: Alkali-Feldspar Fractionation Trends and Derivation of Peralkaline Liquids, *Am. J. Sci.*, vol. 267, pp. 242–248, 1969.

——— and ———: Petrochemical Variations among Mildly Peralkaline (Comendite) Obsidians from the Oceans and Continents, *Contrib. Mineral. Petrol.*, vol. 28, pp. 340–351, 1970.

——— and J. F. SCHAIRER: Feldspar-Liquid Equilibria in Peralkaline Liquids—The Orthoclase Effect, *Am. J. Sci.*, vol. 262, pp. 1198–1206, 1964.

BAILEY, E. B., C. T. CLOUGH, W. B. WRIGHT, J. E. RICHEY, and G. V. WILSON: Tertiary and Post-Tertiary Geology of Mull, Loch Aline and Oban, *Mem. Geol. Surv. Scot.*, 1924.

BAILEY, E. H., M. C. BLAKE, and D. L. JONES: On-land Mesozoic Oceanic Crust in California Coast Ranges, *U.S.G.S. Prof. Paper* 700C, pp. C70–C81, 1970.

———, W. P. IRWIN, and D. L. JONES: Franciscan and Related Rocks and Their Significance in the Geology of Western California, *Bull. Calif. Div. Mines and Geol.*, no. 183, 1964.

BAKER, I.: Petrology of the Volcanic Rocks of Saint Helena Island, South Atlantic, *Bull. Geol. Soc. Am.*, vol. 80, pp. 1283–1310, 1969.

———, N. H. GALE, and J. SIMONS: Geochronology of the Saint Helena Volcanoes, *Nature*, vol. 215, pp. 1451–1456, 1967.

BAKER, P. E.: Comparative Volcanology and Petrology of the Atlantic Island-arcs, *Bull. Volc.*, vol. 32, pp. 189–206, 1968*a*.

———: Petrology of Mt. Misery Volcano, St. Kitts, West Indies, *Lithos*, vol. 1, pp. 124–150, 1968*b*.

———: The Geological History of Mt. Misery Volcano, St. Kitts, West Indies, *Overseas Geol. Min. Res. (G.B.)*, vol. 10, pp. 207–230, 1969.

———, I. G. GASS, P. G. HARRIS, and R. W. LE MAITRE: The Volcanological Report of the Royal Society Expedition to Tristan da Cunha, 1962, *Phil. Trans. Roy. Soc. Lond.*, ser. A, vol. 256, pp. 439–578, 1964.

BAKSI, K., and N. D. WATKINS: Volcanic Production Rates: Comparison of Oceanic Ridges, Islands and the Columbia River Basalts, *Science*, vol. 180, pp. 493–496, 1973.

BANDY, M. C.: Geology and Petrology of Easter Island, *Bull. Geol. Soc. Am.*, vol. 48, pp. 1589–1610, 1937.

BANNO, S., and Y. MATSUI: On the Formulation of Partition Coefficients for Trace Elements Distribution between Minerals and Magma, *Chem. Geol.*, vol. 11, pp. 1–15, 1973.

BARNES, H. L., H. C. HELGESON, and A. J. ELLIS: Ionization Constants in Aqueous Solutions, *Geol. Soc. Am. Mem.*, no. 97, pp. 401–413, 1966.

BARNES, J., J. B. RAPP, J. R. O'NEILL, R. A. SHEPPARD, and A. J. GUDE: Metamorphic Assemblages and Direction of Flow of Metamorphic Fluids in Four Instances of Serpentinization, *Contrib. Mineral. Petrol.*, vol. 35, pp. 263–276, 1972.

BARROS, L. A.: A ilha do Principe e a "Linha dos Camarôes," *Mem. Junta Investig. do Ultramar*, no. 17, Lisbon, 1960.

BARTH, T. F. W.: The Crystallization Process of Basalt, *Am. J. Sci.*, ser. 5, vol. 31, pp. 321–351, 1931.

———: The Large Precambrian Intrusive Bodies in the Southern Part of Norway, *Proc. 16th Int. Geol. Congr.*, pp. 297–309, 1936.

———: The Lavas of Gough Island, *Skr. Norske Vid.-Akad. Oslo*, no. 20, pp. 1–20, 1942.

———: Studies on the Igneous Rock Complex of the Oslo Region: pt. II, Systematic Petrography of the Plutonic Rocks, *Skr. Norske Vid.-Akad. Oslo*, no. 9, 1945.

———: "Theoretical Petrology," Wiley, New York, 1952.

———: The Feldspar Geologic Thermometers, *Norsk Geol. Tidsskr.*, vol. 42, pp. 330–339, 1962.

——— and I. B. RAMBERG: The Fen Circular Complex, pp. 225–260 in O. F. Tuttle and J. Gittins (eds.), "Carbonatites," Interscience (Wiley), New York, 1966.

BASS, M. N., R. MOBERLY, J. M. RHODES, C. SHIH, and S. E. CHURCH: Volcanic rocks cored in the central Pacific, Leg 17, Deep Sea Drilling Project, *Trans. Am. Geophys. Union*, vol. 54, pp. 991–995, 1973.

BATEMAN, P. C.: Granitic Formations in the East-central Sierra Nevada near Bishop, California, *Bull. Geol. Soc. Am.*, vol. 72, pp. 1521–1538, 1961.

———, L. D. CLARK, N. K. HUBER, J. G. MOORE, and C. D. RINEHART: The Sierra Nevada Batholith, *U.S.G.S. Prof. Paper* 414-D, 1963.

——— and J. P. EATON: Sierra Nevada Batholith, *Science*, vol. 158, pp. 1407–1417, 1967.

BECKER, G. F.: Some Queries on Rock Differentiation, *Am. J. Sci.*, ser. 4, vol. 3, pp. 21–40, 1897*a*.

———: Fractional Crystallization of Rocks, *Am. J. Sci.*, ser. 4, vol. 4, pp. 257–261, 1897*b*.

BECKINSDALE, R. D., and J. D. OBRADOVICH: Potassium-Argon Ages for Minerals from the Ross of Mull, Argyllshire, Scotland, *Scott. J. Geol.*, vol. 9, pp. 147–156, 1973.

BEESON, M. H., and E. D. JACKSON: Origin of the Garnet-Pyroxenite Xenoliths at Salt Lake Crater, Oahu, *Mineral. Soc. Am. Spec. Paper*, no. 3, pp. 95–112, 1970.

BELL, K.: Age Relations and Provenance of the Dalradian Series of Scotland, *Bull. Geol. Soc. Am.*, vol. 79, pp. 1167–1194, 1968.

——— and J. L. POWELL: Strontium Isotopic Studies of Alkalic Rocks: The Potassium-rich Lavas of the Birunga and Toro-Ankole Regions, East and Central Equatorial Africa, *J. Petrol.*, vol. 10, pp. 536–672, 1969.

——— and ———: Strontium Isotopic Studies of Alkalic Rocks: The Alkalic Complexes of Eastern Uganda, *Bull. Geol. Soc. Am.*, vol. 81, pp. 3481–3490, 1970.

BELL, T., G. HETHERINGTON, and K. H. JACK: Water in Vitreous Silica: pt. 2, Some

Aspects of Hydrogen-Water-Silica Equilibria, *Physics Chem. Glasses*, vol. 3, pp. 141–146, 1962.
BELOUSOV, V. V.: The Crust and Upper Mantle of Continents, *Int. Geol. Rev.*, vol. 8, pp. 317–330, 1966.
BENSON, W. N.: The Tectonic Conditions Accompanying the Intrusion of Basic and Ultrabasic Plutonic Rocks, *Nat. Acad. Sci. Mem.* (Washington), vol. 19, mem. 1, 1926.
BERG, G. W. and H. L. ALLSOPP: Low $Sr^{87}/Sr^{86}$ Ratios in Fresh South Africa Kimberlites, *Earth Plan. Sci. Letters*, vol. 16, pp. 27–30, 1972.
BERNAL, J. D.: The Structure of Liquids (the Bakerian Lecture, 1962), *Proc. Roy. Soc. Lond.*, ser. A, vol. 280, pp. 299–322, 1964.
BERRANGE, J. P.: Some Critical Differences between Orogenic-Plutonic and Gravity-stratified Anorthosites, *Geol. Rundsch.*, vol. 55, pp. 617–642, 1965.
BHATTACHARJI, S.: Mechanics of Flow Differentiation in Ultramafic and Mafic Sills, *J. Geol.*, vol. 75, pp. 101–112, 1967.
——— and C. H. SMITH: Flowage Differentiation, *Science*, vol. 145, pp. 150–153, 1964.
BILLINGS, M. P.: Diastrophism and Mountain-building, *Bull. Geol. Soc. Am.*, vol. 71, pp. 363–398, 1960.
BINNS, R. A.: High-pressure Megacrysts in Basanitic Lavas near Armidale, New South Wales, *Am. J. Sci.*, vol. 267-A, pp. 33–49, 1969.
———, M. B. DUGGAN, and J. F. G. WILKINSON: High-pressure Megacrysts in Alkaline Lavas from Northeastern New South Wales, *Am. J. Sci.*, vol. 269, pp. 132–168, 1970.
BIRCH, F.: Speculations on the Earth's Thermal History, *Bull. Geol. Soc. Am.*, vol. 76, pp. 133–154, 1965.
———: Compressibility; Elastic Constants, pp. 97–173 in S. P. Clark (ed.), "Handbook of Physical Constants," (*Geol. Soc. Am. Mem.*, no. 97), 1966.
BJÖRNSSON, S.: Radon and Water in Volcanic Gas at Surtsey, Iceland, *Geochim. Cosmochim. Acta*, vol. 32, pp. 815–822, 1968.
BLACK, L. P., N. H. GALE, S. MOORBATH, R. J. PANKHURST, and V. R. MCGREGOR: Isotopic Dating of Very Early Precambrian Amphibolite Facies Gneisses from Godthaab, West Greenland, *Earth Planet. Sci. Letters*, vol. 12, pp. 245–259, 1971.
BLACK, P. M.: Observations on White Island Volcano, New Zealand, *Bull. Volc.*, vol. 34, pp. 158–167, 1970.
BLAKE, D. H.: Gravitational Sorting of Phenocrysts in Some Icelandic Intrusive Sheets, *Geol. Mag.*, vol. 105, pp. 140–148, 1968*a*.
———: Post Miocene Volcanoes in Bougainville Island, Territory of Papua and New Guinea, *Bull. Volc.*, vol. 32, pp. 121–138, 1968*b*.
BLAKE, M. C., W. P. IRWIN, and R. G. COLEMAN: Blueschist-facies Metamorphism Related to Regional Thrust Faulting, *Tectonophysics*, vol. 8, pp. 237–246, 1969.
BOETTCHER, A. L., and P. J. WYLLIE: Phase Relationships in the System $NaAlSiO_4$–$SiO_2$–$H_2O$ to 35 Kilobars Pressure, *Am. J. Sci.*, vol. 267, pp. 875–909, 1969.
BONATTI, E. and D. E. FISHER: Oceanic Basalts: Chemistry versus Distance from Oceanic Ridges, *Earth Planet. Sci. Letters*, vol. 11, pp. 307–311, 1971.
BOOTH, B., R. CROASDALE, and G. P. L. WALKER: Tephrochronology in the Azores, *Geol. Soc. Lond. Proc.*, no. 1663, pp. 156–158, 1970.
BORLEY, G. D.: Potash-rich Potassic Rocks from Southern Spain, *Mineral. Mag.*, vol. 36, pp. 364–379, 1967.

Borom, M. P. and J. A. Pask: Kinetics of Dissolution and Diffusion of the Oxides of Iron in Sodium Disilicate Glass, *J. Am. Ceram. Soc.*, vol. 51, pp. 490–498, 1968.
Bottinga, Y., A. M. Kudo, and D. F. Weill: Some Observations on Oscillatory Zoning and Crystallization of Magmatic Plagioclase, *Am. Mineral.*, vol. 51, pp. 792–806, 1966.
——— and D. F. Weill: Densities of Liquid Silicate Systems Calculated from Partial Molar Volumes of Oxide Components, *Am. J. Sci.*, vol. 269, pp. 169–182, 1970.
——— and ———: The Viscosity of Magmatic Silicate Liquids: A Model for Calculation, *Am. J. Sci.*, vol. 272, pp. 438–475, 1972.
Bowen, N. L.: The Melting Phenomena of the Plagioclase Feldspars, *Am. J. Sci.*, ser. 4, vol. 35, pp. 577–599, 1913.
———: The Crystallization of Haplobasaltic, Haplodioritic and Related Magmas, *Am. J. Sci.*, ser. 4, vol. 40, pp. 161–185, 1915.
———: Crystallization Differentiation in Igneous Magmas, *J. Geol.*, vol. 27, pp. 393–430, 1919.
———: The Carbonate Rocks of the Fen Area in Norway, *Am. J. Sci.*, vol. 12, pp. 499–502, 1926.
———: "The Evolution of the Igneous Rocks," Princeton University Press, Princeton, N.J., 1928.
———: Recent High-temperature Research on Silicates and Its Significance in Igneous Geology, *Am. J. Sci.*, ser. 5, vol. 33, pp. 1–21, 1937.
———: Lavas of the African Rift Valleys and Their Tectonic Setting, *Am. J. Sci.*, ser. 5, vol. 35A, pp. 19–33, 1938.
———: Phase Equilibria Bearing on the Origin and Differentiation of Alkaline Rocks, *Am. J. Sci.*, vol. 243A, (Daly vol.), pp. 75–89, 1945.
———: Magmas, *Bull. Geol. Soc. Am.*, vol. 58, pp. 263–280, 1947.
———: The Granite Problem and the Method of Multiple Prejudices, pp. 79–90 in "Origin of Granite" (*Geol. Soc. Am. Mem.*, no. 28), 1948.
——— and J. F. Schairer: Crystallization Equilibrium in Nepheline-Albite-Silica Mixtures with Fayalite, *J. Geol.*, vol. 46, pp. 397–411, 1938.
———, ———, and E. Posnjak: The System $CaO–FeO–SiO_2$, *Am. J. Sci.*, ser. 5, vol. 26, pp. 193–284, 1933.
——— and O. F. Tuttle: The System $NaAlSi_3O_8–KAlSi_3O_8–H_2O$, *J. Geol.*, vol. 58, pp. 489–511, 1950.
Bowes, D. R., W. R. Skinner, and A. E. Wright: Petrochemical comparison of the Bushveld Igneous Complex with Some Other Mafic Complexes, *Geol. Soc. S. Africa Spec. Publ.* 1, pp. 425–440, 1970.
Boyd, F. R.: Electron-probe Study of Diopside Inclusions from Kimberlite, *Am. J. Sci.*, vol. 267A, pp. 50–69, 1969.
———: Garnet Peridotites and the System $CaSiO_3–MgSiO_3–Al_2O_3$ *Mineral. Soc. Am. Spec. Paper*, no. 3, pp. 63–76, 1970.
Brace, W. F., and J. B. Walsh: Some Direct Measurements of the Surface Energy of Quartz and Orthoclase, *Am. Mineral.*, vol. 47, pp. 1111–1122, 1962.
Bradley, J.: Intrusion of Major Dolerite Sills, *Trans. Roy. Soc. N.Z., Geol.*, vol. 3, pp. 27–55, 1965.
Brooks, C., and W. Compston: The Age and Initial $Sr^{87}/Sr^{86}$ Ratio of the Heemskirk Granite, Western Tasmania, *J. Geophys. Res.*, vol. 71, pp. 5447–5458, 1965.

BROTHERS, R. N.: Petrological Affinities of Volcanic Rocks from the Tonga-Kermadec Island Arc, Southwest Pacific, *Bull. Volc.*, vol. 34, pp. 308–329, 1970.

——— and K. R. MARTIN: The Geology of Macauley Island, Kermadec Group, Southwest Pacific, *Bull. Volc.*, vol. 34, pp. 330–346, 1970.

——— and E. J. SEARLE: The Geology of Raoul Island, Kermadec Group, Southwest Pacific, *Bull. Volc.*, vol. 34, pp. 7–37, 1970.

BROWN, F. H.: Zoning in Some Volcanic Nephelines, *Am. Mineral.*, vol. 55, pp. 1670–1680, 1970.

———: "Volcanic Petrology of the Toro-Ankole Region, Western Uganda," Ph.D. thesis, University of California, Berkeley, 1971.

——— and I. S. E. CARMICHAEL: Quaternary Volcanoes of the Lake Rudolf Region: pt. I, The Basanite-Tephrite Series of the Korath Range, *Lithos*, vol. 2, pp. 239–260, 1969.

——— and ———: Quaternary Volcanoes of the Lake Rudolf Region: pt. II, The Lavas of North Island, South Island, and the Barrier, *Lithos*, vol. 4, pp. 305–323, 1971.

BROWN, G. C., and W. S. FYFE: The Production of Granitic Melts During Ultrametamorphism, *Contrib. Mineral. Petrol.*, vol. 28, pp. 310–318, 1970.

BROWN, G. M.: The Layered Ultrabasic Rocks of Rhum, Inner Hebrides, *Phil. Trans. Roy. Soc. Lond.*, ser. B., vol. 240, pp. 1–53, 1956.

———: Pyroxenes from the Early and Middle Stages of Fractionation of the Skaergaard Intrusion, East Greenland, *Mineral. Mag.*, vol. 31, pp. 511–543, 1957.

———: Melting Relations of Tertiary Granitic Rocks in Skye and Rhum, *Mineral. Mag.*, vol. 33, pp. 533–562, 1963.

———: Experimental Studies on Inversion Relations in Natural Pigeonite Pyroxenes, *Ann. Rept. Geophys. Lab.* (*Carnegie Inst. Yrbk. 66*), pp. 347–353, 1967.

——— and E. A. VINCENT: Pyroxenes from the Late Stages of Fractionation of the Skaergaard Intrusion, East Greenland, *J. Petrol.*, vol. 4, pp. 175–197, 1963.

BROWN, P. E.: Major Element Composition of the Loch Coire Migmatite Complex, Sutherland, Scotland, *Contrib. Mineral. Petrol.*, vol. 14, pp. 1–26, 1967.

BROWNE, W. R.: Notes on Bathyliths and Some of Their Implications, *J. Roy. Soc. New South Wales*, vol. 65, pp. 112–114, 1931.

BRYAN, W. B.: Geology and Petrology of Clarion Island, *Bull. Geol. Soc. Am.*, vol. 78, pp. 1461–1476, 1967.

———, L. W. FINGER, and F. CHAYES: Estimating Proportions in Petrographic Mixing Equations by Least-squares Approximation, *Science*, vol. 163, pp. 926–927, 1969.

———, G. D. STICE, and A. EWART: Geology, Petrography and Geochemistry of Tonga, *J. Geophys. Res.*, vol. 77, pp. 1566–1585, 1972.

BUDDINGTON, A. F.: Adirondack Igneous Rocks and Their Metamorphism, *Geol. Soc. Am. Mem.*, no. 7, 1939.

———: Interrelated Precambrian Granitic Rocks, Northwest Adirondacks, New York, *Bull. Geol. Soc. Am.*, vol. 68, pp. 291–306, 1957.

———: Granite Emplacement with Special Reference to North America, *Bull. Geol. Soc. Am.*, vol. 70, pp. 671–747, 1959.

———: The Origin of Anorthosites Reevaluated, *Rec. Geol. Surv. India*, vol. 86, pp. 421–432, 1960.

——— and D. H. Lindsley: Iron-Titanium Oxide Minerals and Synthetic Equivalents, *J. Petrol.*, vol. 5, pp. 310–357, 1964.

Buie, B. F.: Igneous Rocks of the Highwood Mountains, Montana: Part III, Dikes and Related Intrusives, *Bull. Geol. Soc. Am.*, vol. 52, pp. 1753–1808, 1941.

Burk, C. A. (ed.): "A Study of Serpentinite" (*Nat. Acad. Sci.–Nat. Res. Council Publ.* 1188), 1964.

Burke, J.: "The Kinetics of Phase Transformations in Metals," Pergamon, New York, 1965.

Burn, I., and J. P. Roberts: Influence of Hydroxyl Content on the Diffusion of Water in Silica Glass, *Physics Chem. Glasses*, vol. 11, pp. 106–114, 1970.

Burnham, C. W.: Hydrothermal Fluids at the Magmatic Stage, in H. L. Barnes (ed.), "Geochemistry of Hydrothermal Ore Deposits," Holt, New York, 1967.

——— and N. F. Davis: Thermodynamic Properties of Water-bearing Magmas, *Phys. Earth Planet. Interiors*, vol. 3, p. 332, 1970.

——— and ———: The Role of $H_2O$ in Silicate Melts: pt. 1, P-V-T Relations in the System $NaAlSi_3O_8$–$H_2O$ to 10 Kilobars and 1000°C, *Am. J. Sci.*, vol. 270, pp. 54–79, 1971.

———, J. R. Holloway, and N. F. Davis: Thermodynamic Properties of Water to 1000°C and 10,000 Bars, *Geol. Soc. Am. Spec. Paper*, no. 132, pp. 1–96, 1969.

——— and R. H. Jahns: A Method for Determining the Solubility of Water in Silicate Melts, *Am. J. Sci.*, vol. 260, pp. 721–745, 1962.

Burns, R. G.: "Mineralogical Applications of Crystal Field Theory," Cambridge, London, 1970.

——— and W. S. Fyfe: Site Preference Energy and Selective Uptake of Transition Metal Ions During Magmatic Crystallization, *Science*, vol. 144, pp. 1001–1003, 1964.

——— and ———: Distribution of Elements in Geological Processes, *Chem. Geol.*, vol. 1, pp. 49–56, 1966.

——— and ———: Trace-element Distribution Rules and Their Significance: A Review, *Chem. Geol.*, vol. 2, pp. 89–104, 1967.

Butler, J. R., and A. Z. Smith: Zirconium, Niobium and Certain Other Trace Elements in Some Alkali Igneous Rocks, *Geochim. Cosmochim. Acta*, vol. 26, pp. 945–953, 1962.

——— and A. J. Thompson: Zirconium: Hafnium Ratios in Some Igneous Rocks, *Geochim. Cosmochim. Acta*, vol. 29, pp. 167–175, 1965.

——— and ———: Cadmium and Zinc in Some Alkali Acidic Rocks, *Geochim. Cosmochim. Acta*, vol. 31, pp. 97–105, 1967.

Cameron, E. N.: Structure and Rock Sequences of the Critical Zone of the Eastern Bushveld Complex, *Mineral. Soc. Am. Spec. Paper*, no. 1, pp. 93–107, 1963.

Carman, J. H.: "The Study of the System $NaAlSiO_4$–$Mg_2SiO_4$–$SiO_2$–$H_2O$ from 200 to 5000 Bars and 800°C to 1100°C and Its Petrologic Applications," Ph.D. thesis, Pennsylvania State University, 1969.

——— and O. F. Tuttle: Experimental Verification of Solid Solution of Excess Silica in Sanidine from Rhyolites, *Geol. Soc. Am. Spec. Paper*, no. 115, p. 33, 1967.

Carmichael, I. S. E.: The Pyroxenes and Olivines from Some Tertiary Acid Glasses, *J. Petrol.*, vol. 1, pp. 309–336, 1960.

———: Pantelleritic Liquids and Their Phenocrysts, *Mineral Mag.*, vol. 33, pp. 86–113, 1962.

———: The Crystallization of Feldspar in Volcanic Acid Liquids, *Q.J. Geol. Soc. Lond.*, vol. 119, pp. 95–131, 1963.
———: Natural Liquids and the Phonolitic Minimum, *Geol. J.* (Liverpool), vol. 4, pp. 55–60, 1964*a*.
———: The Petrology of Thingmuli, a Tertiary Volcano in Eastern Iceland, *J. Petrol.*, vol. 5, pp. 435–460, 1964*b*.
———: Trachytes and Their Feldspar Phenocrysts, *Mineral. Mag.*, vol. 34, pp. 107–125, 1965.
———: The Mineralogy of Thingmuli, a Tertiary Volcano in Eastern Iceland, *Am. Mineral.*, vol. 52, pp. 1815–1841, 1967*a*.
———: The Iron-Titanium Oxides of Salic Volcanic Rocks and Their Associated Ferromagnesian Silicates, *Contrib. Mineral. Petrol.*, vol. 14, pp. 36–64, 1967*b*.
———: The Mineralogy and Petrology of the Volcanic Rocks from the Leucite Hills, Wyoming, *Contrib. Mineral. Petrol.*, vol. 15, pp. 24–66, 1967*c*.
——— and W. S. MacKenzie: Feldspar-Liquid Equilibria in Pantellerites: An Experimental Study, *Am. J. Sci.*, vol. 261, pp. 382–396, 1963.
——— and J. Nicholls: Iron-Titanium Oxides and Oxygen Fugacities in Volcanic rocks, *J. Geophys. Res.*, vol. 72, pp. 4665–4687, 1967.
———, ———, and A. L. Smith: Silica Activity in Igneous Rocks, *Am. Mineral.*, vol. 55, pp. 246–263, 1970.
Carswell, D. A.: Possible Primary Upper Mantle Peridotite in Norwegian Basal Gneiss, *Lithos*, vol. 1, pp. 322–355, 1968*a*.
———: Picrite Magma–residual Dunite Relationships in Garnet Peridotite at Kalskaret near Tafjord, South Norway, *Contrib. Mineral. Petrol.*, vol. 19, pp. 97–124, 1968*b*.
——— and J. B. Dawson: Garnet Peridotite Xenoliths in South African Kimberlite Pipes and Their Petrogenesis, *Contrib. Mineral. Petrol.*, vol. 25, pp. 163–184, 1970.
Challis, G. A.: The Origin of New Zealand Ultramafic Intrusions, *J. Petrol.*, vol. 6, pp. 322–364, 1965.
———: Chemical Analyses of New Zealand Rocks and Minerals with C.I.P.W. Norms and Petrographic Descriptions, 1917–1957: pt. 1, Igneous and Pyroclastic Rocks, *N.Z. Geol. Surv. Bull.* 84, 1971.
——— and W. R. Lauder: The Genetic Position of "Alpine" Type Ultramafic Rocks, *Bull. Volc.*, vol. 29, pp. 283–306, 1966.
Chayes, F.: Relative Abundance of Intermediate Members of the Oceanic Basalt-Trachyte Association, *J. Geophys. Res.*, vol. 68, pp. 1519–1534, 1963.
———: A Petrographic Distinction between Cenozoic Volcanics in and around the Open Oceans, *J. Geophys. Res.*, vol. 69, pp. 1573–1588, 1964*a*.
———: Descriptive and Genetic Significance of Normative *ns*, *Ann. Rept. Geophys. Lab.* (*Carnegie Inst. Yrbk. 63*), pp. 190–193, 1964*b*.
———: Variance-covariance Relations in Some Published Harker Diagrams of Volcanic Suites, *J. Petrol.*, vol. 5, pp. 219–237, 1964*c*.
———: On Estimating the Magnitude of the Hidden Zone and Compositions of the Residual Liquids in the Skaergaard Layered Series, *J. Petrol.*, vol. 11, pp. 1–14, 1970.
——— and E. G. Zies: Sanidine Phenocrysts in Some Peralkaline Volcanic Rocks, *Ann. Rept. Geophys. Lab.* (*Carnegie Inst. Yrbk. 61*), pp. 112–118, 1962.

CHRISTIAN, J. W.: "The Theory of Transformations in Metals and Alloys," Pergamon, New York, 1965.
CHURCH, S. E.: Limits of Sediment Involvement in the Genesis of Orogenic Volcanic Rocks, *Contrib. Mineral. Petrol.*, vol. 39, pp. 17–32, 1973.
CLARK, R. H.: The Significance of Flow Structure in Microporphyritic Ophitic Basalts of Arthur's Seat, *Edinburgh Geol. Soc. Trans.*, vol. 15, pp. 69–83, 1952.
———: Andesite Lavas of North Island, New Zealand, *Proc. 21st Int. Geol. Congr.*, vol. 13, pp. 123–131, 1960.
——— and W. S. FYFE: Ultrabasic Liquids, *Nature*, vol. 191, pp. 158–159, 1961.
CLARK, S. P., JR. (ed.): "Handbook of Physical Constants," (*Geol. Soc. Am. Mem.*, no. 97), 1966.
——— and A. E. RINGWOOD: Density Distribution and Constitution of the Mantle, *Rev. Geophys.*, vol. 2, pp. 35–88, 1964.
CLARKE, D. B.: Tertiary Basalts of Baffin Bay: Possible Primary Magma from the Mantle, *Contrib. Mineral Petrol.*, vol. 25, pp. 203–224, 1970.
CLARKE, P. D., and W. J. WADSWORTH: The Insch Layered Intrusion, *Scottish J. Geol.*, vol. 6, pp. 7–25, 1970.
CLIFFORD, T. N., and I. G. GASS, "African Magmatism and Tectonics," Hafner, New York, 1970.
COATS, R. R.: Magmatic Differentiation in Tertiary and Quaternary Volcanic Rocks from Adak and Kanaga Islands, Aleutian Islands, Alaska, *Bull. Geol. Soc. Am.*, vol. 63, pp. 485–514, 1952.
———: Magma Type and Crustal Structure in the Aleutian Arc, pp. 92–109 in "The Crust of the Pacific Basin" (*Am. Geophys. Union Monograph* 6), 1962.
COLEMAN, P. J.: Geology of the Solomon and New Hebrides Islands, as Part of the Melanesian Re-entrant, Southwest Pacific, *Pacific Science*, vol. 24, pp. 289–314, 1970.
COLEMAN, R. G.: New Zealand Serpentinites and Associated Metasomatic Rocks, *N.Z. Geol. Surv. Bull.* 76, 1966.
———: Low-temperature Reaction Zones and Alpine Ultramafic Rocks of California, Oregon, and Washington, *U.S.G.S. Bull.* 1247, 1967.
———: Petrologic and Geophysical Nature of Serpentinites, *Bull. Geol. Soc. Am.*, vol. 82, pp. 897–918, 1971.
COMBE, A. D., and A. HOLMES: The Kalsilite Bearing Lavas of Kabirenge and Lyakauli, Southwest Uganda, *Roy. Soc. Edinburgh Trans.*, vol. 61, pp. 359–379, 1945.
COMPSTON, W., I. MCDOUGALL, and K. S. HEIER: Geochemical Comparison of the Mesozoic Basaltic Rocks of Antarctica, South Africa, South America, and Tasmania, *Geochim. Cosmochim. Acta*, vol. 32, pp. 129–149, 1968.
COMPTON, R. R.: Trondhjemite Batholith near Bidwell Bar, California, *Bull. Geol. Soc. Am.*, vol. 66, pp. 9–44, 1955.
COOMBS, D. S.: Trends and Affinities of Basaltic Magmas and Pyroxenes as Illustrated on the Diopside-Olivine-Silica Diagram, *Mineral Soc. Am. Spec. Paper*, no. 1, pp. 227–250, 1963.
——— and J. F. G. WILKINSON: Lineages and Fractionation Trends in Undersaturated Volcanic Rocks from the East Otago Volcanic Province (New Zealand) and Related Rocks, *J. Petrol.*, vol. 10, pp. 440–501, 1969.
COOPER, J. A., and D. H. GREEN: Lead Isotope Measurements in Lherzolite Inclusions

and Host Basanites from Western Victoria, Australia, *Earth Planet. Sci. Letters*, vol. 6, pp. 67–76, 1969.
Cox, K. G.: The Karroo Volcanic Cycle, *Geol. Soc. J.*, vol. 128, pp. 311–336, 1972.
———, I. G. Gass, and D. I. J. Mallick: The Peralkaline Volcanic Suite of Aden and Little Aden, South Arabia, *J. Petrol.*, vol. 11, pp. 433–462, 1970.
——— and G. Hornung: The Petrology of the Karroo Basalts of Basutoland, *Am. Mineral.*, vol. 51, pp. 1414–1432, 1966.
———, R. L. Johnson, L. J. Monkman, C. J. Stillman, J. R. Vail, and D. N. Wood: The Geology of the Nuantesi Igneous Province, *Phil. Trans. Roy. Soc. Lond.*, ser. A, vol. 257, pp. 71–218, 1965.
Cundari, A., and R. W. Le Maitre: On the Petrogeny of the Leucite-bearing Rocks of the Roman and Birunga Regions, *J. Petrol.*, vol. 11, pp. 33–47, 1970.
Curtis, C. D.: Applications of Crystal Field Theory to the Inclusion of Trace Transition Elements in Minerals During Magmatic Crystallization, *Geochim. Cosmochim. Acta*, vol. 28, pp. 389–402, 1964.
Curtis, G. H., J. F. Evernden, and J. Lipson: Age Determination of Some Granitic Rocks in California, *Calif. Div. Mines, Spec. Rept.* 54, 1958.
Dallwitz, W. B., D. H. Green, and J. E. Thompson: Clinoenstatite in a Volcanic Rock from the Cape Vogel Area, Papua, *J. Petrol.*, vol. 7, pp. 375–403, 1966.
Daly, R. A.: "Igneous Rocks and Their Origin," McGraw-Hill, New York, 1914.
———: Genesis of the Alkaline Rocks, *J. Geol.*, vol. 26, pp. 97–134, 1918.
———: The Geology of Ascension Island, *Proc. Am. Acad. Arts and Sci.*, vol. 60, no. 1, pp. 1–80, 1925.
———: The Geology of Saint Helena Island, *Proc. Am. Acad. Arts and Sci.*, vol. 62, pp. 31–92, 1927.
———: "Igneous Rocks and the Depths of the Earth," McGraw-Hill, New York, 1933.
Darken, L. S., and K. Schwerdtfeger: Activities in Olivine and Pyroxenoid Solid Solutions of the System Fe–Mn–Si–O at 1150°C, *Trans. AIME*, vol. 236, pp. 201–211, 1966.
Dash, B. P., and E. Bosshard: Crustal Studies around the Canary Islands, *Proc. 23rd Int. Geol. Congr.*, vol. 1, pp. 249–260, 1968.
Davies, H. L.: Papuan Ultramafic Belt, *Proc. 23rd Int. Geol. Congr.*, vol. 1, pp. 209–220, 1968.
Davis, B. T. C.: The System Diopside-Forsterite-Pyrope at 40 Kilobars, *Ann. Rept. Geophys. Lab.*, (*Carnegie Inst. Yrbk. 63*), pp. 165–171, 1964.
——— and F. R. Boyd: The Join $Mg_2Si_2O_6$–$CaMgSi_2O_6$ at 30 Kilobars Pressure and Its Application to Pyroxenes from Kimberlites, *J. Geophys. Res.*, vol. 71, pp. 3567–76, 1966.
Dawson, J. B.: The Geology of Oldoinyo Lengai, *Bull. Volc.*, vol. 24, pp. 155–168, 349–387, 1962*a*.
———: Basutoland Kimberlites, *Bull. Geol. Soc. Am.*, vol. 73, pp. 545–560, 1962*b*.
———: Carbonatite Volcanic Ashes in Northern Tanganyika, *Bull. Volc.*, vol. 27, pp. 1–11, 1964*a*.
———: Reactivity of the Cations in Carbonate Magmas, *Proc. Geol. Assn. Canada*, vol. 15, pp. 103–113, 1964*b*.
———: Oldoinyo Lengai—An Active Volcano with Sodium Carbonatite Lava Flows, in O. F. Tuttle and J. Gittins (eds.), "Carbonatites," Interscience (Wiley), 1966.

———: Recent Researches on Diamond and Kimberlite Geology, *Econ. Geol.*, vol. 63, pp. 504–511, 1968.

——— and J. B. HAWTHORNE: Magmatic Sedimentation and Carbonatite Differentiation in Kimberlite Sills at Benfontein, South Africa, *J. Geol. Soc. Lond.*, vol. 129, pp. 61–86, 1973.

DEER, W. A., and D. ABBOTT: Clinopyroxenes of the Gabbro Cumulates of the Kap Edvard Holm Complex, East Greenland, *Mineral. Mag.*, vol. 34, pp. 177–193, 1965.

DENHAM, D.: Distribution of Earthquakes in the New Guinea–Solomon Islands Region, *J. Geophys. Res.*, vol. 74, pp. 4290–4299, 1969.

DENNEN, W. H., W. H. BLACKBURN, and A. QUESADA: Aluminum in Quartz as a Geothermometer, *Contrib. Mineral. Petrol.*, vol. 27, pp. 332–342, 1970.

DEROEVER, W. P.: Sind die Alpinotypen Peridotitmassen vielleicht tectonisch verfrachtete Bruchstücke der Peridotitschale?, *Geol. Rundsch.*, vol. 46, pp. 137–146, 1957.

DESBOROUGH, G. A.: Closed System Differentiation of Sulfides in Olivine Diabase, Missouri, *Econ. Geol.*, vol. 62, pp. 595–613, 1967.

———, A. T. ANDERSON, and T. L. WRIGHT: Mineralogy of Sulfides from Certain Hawaiian Basalts, *Econ. Geol.*, vol. 63, pp. 636–644, 1968.

DE WAARD, D.: On the Origin of Anorthosite by Anatexis, *Konikl. Nederl. Akad. Wetenschap. Proc.*, ser. B, pp. 411–419, 1967.

DEWEY, J. F., and J. M. BIRD: Mountain Belts and the New Global Tectonics, *J. Geophys. Res.*, vol. 75, pp. 2625–2647, 1970.

——— and R. J. PANKHURST: The Evolution of the Scottish Caledonides in Relation to Their Isotopic Age Pattern, *Roy. Soc. Edin. Trans.*, vol. 68, pp. 361–389, 1970.

DICKEY, J. S.: Partial Fusion Products in Alpine-type Peridotites, *Mineral. Soc. Am. Spec. Paper*, no. 3, pp. 33–49, 1970.

DICKINSON, W. R.: Circum-Pacific Andesite Types, *J. Geophys. Res.*, vol. 73, pp. 2261–2269, 1968.

———: Petrogenic Significance of Geosynclinal Andesitic Volcanism along the Pacific Margin of North America, *Bull. Geol. Soc. Am.*, vol. 73, pp. 1241–1256, 1969.

———: Relation of Andesite, Granites and Derivative Sandstones to Arc-trench Tectonics, *Rev. Geophys.*, vol. 8, pp. 813–860, 1970.

———: Width of Modern Arc-trench Gaps Proportional to Past Duration of Igneous Activity in Associated Magmatic Arcs, *J. Geophys. Res.*, vol. 78, pp. 3376–3389, 1973.

DOE, B. R.: The Bearing of Lead Isotopes on the Source of Granitic Magma, *J. Petrol.*, vol. 8, pp. 51–83, 1967.

———: "Lead Isotopes," Springer, New York, 1970.

———, R. I. TILLING, C. E. HEDGE, and M. R. KLEPPER: Lead and Strontium Isotope Studies of the Boulder Batholith, Southwest Montana, *Econ. Geol.*, vol. 63, pp. 884–906, 1968.

DOELL, R. R., and G. B. DALRYMPLE: Potassium-Argon Ages and Paleomagnetism of the Waianae and Koolau Volcanic Series, Oahu, Hawaii, *Bull. Geol. Soc. Am.*, vol. 84, pp. 1217–1242, 1973.

DOUGLAS, R. W., P. NATH, and A. PAUL: Oxygen Ion Activity and Its Influence on the Redox Equilibrium in Glasses, *Physics Chem. Glasses*, vol. 6, pp. 216–223, 1965.

DUNNE, J. C.: Volcanology of the Tristan da Cunha Group, *Norske Vid. Akad. Oslo*, no. 2, pp. 1–145, 1941.

EDWARDS, A. B.: Tertiary Lavas from the Kerguelen Archipelago, *Rept. B.A.N.Z., Antarctic Expedition (D. Mawson) 1929–1931*, ser. A, pt. 5, no. 2, pp. 72–100, 1938*a*.

———: The Tertiary Tholeiitic Magma in Western Australia, *J. Roy. Soc. Western Australia*, vol. 24, pp. 1–12, 1938*b*.

———: The Petrology of the Cainozoic Volcanic Rocks of Tasmania, *Proc. Roy. Soc. Victoria*, vol. 62, pp. 97–120, 1950.

EGGLER, D. H.: Water-saturated and Undersaturated Melting Relations in a Parícutin Andesite and an Estimate of Water Content in the Natural Magma, *Contrib. Mineral. Petrol.*, vol. 34, pp. 261–271, 1972.

ELDER, J. W.: Convection, the Key to Dynamical Geology, *Sci. Prog.*, vol. 56, pp. 1–33, 1968.

ELLIS, A. J.: Chemical Equilibrium in Magmatic Gases, *Am. J. Sci.*, vol. 255, pp. 416–431, 1957.

ELSDON, R.: Clinopyroxenes from the Upper Layered Series, Kap Edvard Holm, East Greenland, *Mineral. Mag.*, vol. 38, pp. 49–57, 1971.

EMELEUS, C. H., A. C. DUNHAM, and R. N. THOMPSON: Iron-rich Pigeonites from Acid Rocks in the Tertiary Igneous Province of Scotland, *Am. Mineral.*, vol. 56, pp. 940–951, 1971.

EMMONS, R. C.: The Contribution of Differential Pressures to Magmatic Differentiation, *Am. J. Sci.*, vol. 238, pp. 1–21, 1940.

ENGEL, A. E. J., and C. G. ENGEL: Composition of Basalt Cored in Mohole Project (Guadalupe Site), *Bull. Am. Assn. Petrol. Geol.*, vol. 45, p. 1799, 1961.

——— and ———: Composition of Basalts from the Mid-Atlantic Ridge, *Science*, vol. 144, pp. 1330–1333, 1964*a*.

——— and ———: Igneous Rocks of the East Pacific Rise, *Science*, vol. 146, pp. 477–485, 1964*b*.

———, ———, and R. G. HAVENS: Chemical Characteristics of Oceanic Basalts and the Upper Mantle, *Bull. Geol. Soc. Am.*, vol. 76, pp. 719–734, 1965.

——— and D. L. KELM: Pre-Permian Global Tectonics: A Tectonic Test, *Bull. Geol. Soc. Am.*, vol. 83, pp. 2325–2340, 1972.

EPSTEIN, S., and H. P. TAYLOR: Variation of $O^{18}/O^{16}$ in Minerals and Rocks, in P. H. Abelson (ed.), "Researches in Geochemistry," vol. 2, Wiley, New York, 1967.

ERLANK, A. J., and P. K. HOFMEYR: K/Rb and K/Cs Ratios in Karroo Dolerites from South Africa, *J. Geophys. Res.*, vol. 71, pp. 5439–5445, 1966.

ERNST, T.: Olivinknollen der Basalte als Bruchstücke alter Olivinfelsen, *Nachr. Ges. Wiss. Göttingen*, Math.-Phys. K1. IV, Geol. Mineral. 1, pp. 147–154, 1935.

———: Die Melilit-Basalt des Westberges bei Hofgeismar, *Chemie der Erde*, vol. 10, pp. 631–666, 1936.

———, H. KOHLER, D. SCHULTZ, and R. SCHWAB: The Volcanism of the Vogelsberg (Hessen) in the North of the Rhinegraben Rift System, pp. 143–146 in "Graben Problems" (*Int. Upper Mantle Proj. Rept.* 27), Nägele and Obermiller, Stuttgart, 1970.

——— and H. MÖRTEL: Die Restausscheidung tholeiitischer Basalte der "Maintrapps," *Neues Jahrb. Mineral. Mh.*, vol. 8, pp. 362–379, 1969.

ERNST, W. G.: "Minerals, Rocks and Inorganic Materials: vol. 1, Amphiboles," Springer, New York, 1968.

ESENWEIN, P.: Zur Petrographie der Azoren, *Zeits. Vulkan.*, vol. 12, pp. 108–227, 1929.

ESSENE, E. J., W. S. FYFE, and F. J. TURNER: Petrogenesis of Franciscan Glaucophane Schists and Associated Metamorphic Rocks, California, *Contrib. Mineral. Petrol.*, vol. 11, pp. 695–704, 1965.

EUGENFELD, R., and B. FICKE: Exkursion am 21 September, 1962: Tertiärer Vulkanismus der Rhön, *Deutsche Mineral. Ges., 40 Jahrest. in Würtzburg*, 1962.

EUGSTER, H. P., A. L. ALBEE, A. E. BENCE, J. B. THOMPSON, JR., and D. R. WALDBAUM: The Two-phase Region and Excess Mixing Properties of Paragonite-Muscovite Crystalline Solutions, *J. Petrol.*, vol. 13, pp. 147–179, 1972.

——— and G. B. SKIPPEN: Igneous and Metamorphic Reactions Involving Gas Equilibria, pp. 492–520 in P. H. Abelson (ed.), "Researches in Geochemistry," vol. 2, Wiley, New York, 1967.

——— and D. R. WONES: Stability Relations of the Ferruginous Biotite, Annite, *J. Petrol.*, vol. 3, pp. 82–125, 1962.

EULER, R., and H. G. F. WINKLER, Über die Viskositäten von Gesteins und Silikatschmelzen, *Glastech. Ber.*, vol. 8, pp. 325–332, 1957.

EVANS, B. W.: Application of a Reaction-rate Method to the Breakdown Equilibria of Muscovite and Muscovite Plus Quartz, *Am. J. Sci.*, vol. 263, pp. 647–667, 1965.

——— and J. G. MOORE: Mineralogy as a Function of Depth in the Prehistoric Makaopuhi Tholeiitic Lava Lake, Hawaii, *Contrib. Mineral. Petrol.*, vol. 17, pp. 85–115, 1968.

——— and V. TROMMSDORFF: Regional Metamorphism of Ultramafic Rocks in the Central Alps, *Schweiz. Mineral. Petrogr. Mitteil.*, vol. 50, pp. 481–492, 1970.

——— and T. L. WRIGHT: Composition of Liquidus Chromite from the 1959 (Kilauea Iki) and 1965 (Makaopuhi) Eruptions of Kilauea Volcano, Hawaii, *Am. Mineral.*, vol. 57, pp. 217–230, 1972.

EWART, A.: Petrology and Petrogenesis of the Quaternary Pumice Ash in the Taupo Area, New Zealand, *J. Petrol.*, vol. 4, pp. 392–431, 1963.

———: Review of Mineralogy and Chemistry of the Acidic Volcanic Rocks of the Taupo Volcanic Zone, New Zealand, *Bull. Volc.*, vol. 29, pp. 147–172, 1966.

———: The Petrography of the Central North Island Rhyolitic Lavas: pt. 1, Correlations between the Phenocryst Assemblages, *N.Z. J. Geol. Phys.*, vol. 10, pp. 182–197, 1967.

———: The Petrography of the Central North Island Rhyolitic Lavas: pt. 2, Regional Petrography Including Notes on Associated Ash-flow Pumice Deposits, *N.Z. J. Geol. Geophys., vol.* 11, pp. 478–545, 1968.

———, W. B. BRYAN, and J. B. GILL: Mineralogy and Geochemistry of the Younger Volcanic Islands of Tonga, S.W. Pacific, *J. Petrol.*, vol. 14, pp. 429–465, 1973.

———, D. C. GREEN, I. S. E. CARMICHAEL, and F. H. BROWN: Voluminous Low Temperature Rhyolitic Magmas in New Zealand, *Contrib. Mineral. Petrol.*, vol. 33, pp. 128–144, 1971.

——— and J. J. STIPP: Petrogenesis of the Volcanic Rocks of the Central North Island, New Zealand, as Indicated by a Study of $Sr^{87}/Sr^{86}$ Ratios, and Sr, Rb, K, U and Th Abundances, *Geochim. Cosmochim. Acta*, vol. 32, pp. 699–735, 1968.

———, S. R. TAYLOR, and A. C. CAPP: Geochemistry of the Pantellerites of Mayor Island, New Zealand, *Contrib. Mineral. Petrol.*, vol. 17, pp. 116–140, 1968*a*.

———, ———, and ———: Trace and Minor Element Geochemistry of the Rhyolitic Volcanic Rocks, Central North Island, New Zealand, *Contrib. Mineral. Petrol.*, vol. 18, pp. 76–104, 1968*b*.

EYRING, H., and T. REE: Significant Liquid Structures: pt. VI, The Vacancy Theory of Liquids, *Proc. Natl. Acad. Sci.*, vol. 47, pp. 526–537, 1961.

FAURE, G., and P. M. HURLEY: The Isotopic Composition of Strontium in Oceanic and Continental Basalts, *J. Petrol.*, vol. 4, pp. 31–50, 1963.

——— and J. L. POWELL: "Strontium Isotope Geology," Springer, New York, 1972.

FERNÁNDEZ SANTÍN, S.: Pegmatoides en la serie basáltica fisural de las islas de Lanzarote y Fuerteventura, *Estud. Geol. Inst. Lucas Mallada*, vol. 25, pp. 53–100, 1969.

FICKE, B.: Petrologische Untersuchungen an tertiären basaltischen und phonolitischen Vulkaniten der Rhön, *Tschermaks Mineral. Petrograph. Mitteil.*, ser. 3, vol. 7, pp. 337–436, 1961.

FINDLAY, D. C., and C. H. SMITH: The Muskox Drilling Project, *Geol. Surv. Canada Paper* 64–44, 1965.

FINLAYSON, D. M., J. P. CULL, W. A. WIEBENGA, A. S. FURUMOTO and J. P. WEBB: New Britain–New Ireland Crustal Seismic Refraction Experiments, 1967–1969, *Geophys. J.*, vol. 29, pp. 245–254, 1972.

FIRSTOV, P. P., and V. A. SHIROKOV: Seismic Investigations of the Roots of the Kliuchevskaya Group Volcanoes, Kamchatka, *Bull. Volc.*, vol. 35, pp. 164–172, 1971.

FISHER, N. H.: "Catalogue of the Active Volcanoes of the World: pt. V, Melanesia." Int. Volc. Assn., 1957.

FLOOD, H., and W. J. KNAPP: Structural Characteristics of Liquid Mixtures of Feldspar and Silica, *J. Am. Ceram. Soc.*, vol. 51, pp. 259–263, 1968.

FLOWER, M. F. J.: Evolution of Basaltic and Differentiated Lavas from Anjouan, Comores Archipelago, *Contrib. Mineral. Petrol.*, vol. 38, pp. 237–260, 1973.

FORBES, R. B., and H. KUNO: The Regional Petrology of Peridotite Inclusions and Basaltic Host Rocks, *Int. Union Geol. Sci. Upper Mantle Symposium, New Delhi, 1964*, Copenhagen, pp. 161–179, 1965.

FOUQUET, F., and A. MICHEL-LEVY: "Synthese des mineraux et des roches," Masson, Paris, 1882.

FRANCO, R. R., and J. F. SCHAIRER: Liquidus Temperatures in Mixtures of the Feldspars of Soda, Potash and Lime, *J. Geol.*, vol. 59, pp. 259–267, 1951.

FREY, F. A.: Rare Earth Abundances in Alpine Ultramafic Rocks, *Phys. Earth Planet. Interiors*, vol. 3, pp. 323–330, 1970.

———, L. A. HASKIN, and M. A. HASKIN: Rare-earth Abundances in Some Ultramafic Rocks, *J. Geophys. Res.*, vol. 76, pp. 2057–2070, 1971.

FRICK, C.: The Garnets in Kimberlite and in the Associated Griquaite and Ultramafic Nodules, *Contrib. Mineral. Petrol.*, vol. 35, pp. 63–76, 1972.

FRIEDMAN, I.: Water and Deuterium in Pumice from the 1959–1960 Eruption of Kilauea Volcano, Hawaii, *U.S.G.S. Prof. Paper* 575-B, pp. 120–127, 1967.

——— and R. L. SMITH: The Deuterium Content of Water in Some Volcanic Glasses, *Geochim. Cosmochim. Acta*, vol. 15, pp. 218–228, 1958.

———, W. LONG, and R. L. SMITH: Viscosity and Water Content of Rhyolite Glass, *J. Geophys. Res.*, vol. 68, pp. 6523–6535, 1963.

FRISCHAT, G. H., and H. J. OEL: Diffusion of Neon in a Glass Melt, *Physics Chem. Glasses*, vol. 8, pp. 92–95, 1967.

FUDALI, R. F.: Experimental Studies Bearing on the Origin of Pseudoleucite and Associated Problems of Alkalic Rock Genesis, *Bull. Geol. Soc. Am.*, vol. 74, pp. 1101–1126, 1963.

———: Oxygen Fugacities of Basaltic and Andesitic Magmas, *Geochim. Cosmochim. Acta*, vol. 29, pp. 1063–1075, 1965.

FULLER, R. E.: Gravitational Accumulation of Olivine During the Advance of Basaltic Flows, *J. Geol.*, vol. 47, pp. 303–313, 1939.

FÚSTER, J. M., A. PÁEZ, and J. SAGREDO: Significance of Basic and Ultramafic Rock Inclusions in the Basalts of Canary Islands, *Bull. Volc.*, vol. 33, pp. 665–693, 1969.

FYFE, W. S.: Some Thoughts on Granitic Magmas, pp. 201–216 in G. Newall and N. Rast (eds.), "Mechanics of Igneous Intrusion," Gallery Press, Liverpool, 1970*a*.

———: Lattice Energies, Phase Transformations and Volatiles in the Mantle, *Phys. Earth Planet. Interiors*, vol. 3, pp. 196–200, 1970*b*.

GASS, I. G.: Magma-types of the Tristan da Cunha Group, *Proc. Geol. Soc. Lond.*, no. 1626, p. 147, 1965.

GAST, P. W.: Isotope Geochemistry of Volcanic Rocks, pp. 325–358 in "Basalts (Poldervaart Treatise)," vol. 1, Interscience (Wiley), New York, 1967.

———: Trace Element Fractionation and the Origin of Tholeiitic and Alkaline Magma Types, *Geochim. Cosmochim. Acta*, vol. 32, pp. 1057–1086, 1968.

———, G. R. TILTON, and C. HEDGE: Isotopic Composition of Lead and Strontium from Ascension and Gough Islands, *Science*, vol. 145, pp. 1181–1185, 1964.

GASTESI BASCUÑANA, P.: El complejo plutónico basico y ultrabasico de Betancuria, Fuerteventura (Islas Canarias), *Estud. Geol. Inst. Lucas Malada*, vol. 25, pp. 1–51, 1969.

GIBB, F. G. F.: Flow Differentiation in the Xenolithic Ultrabasic Dykes of the Cuillins and the Strathaird Peninsula, Isle of Skye, Scotland, *J. Petrol.*, vol. 9, pp. 411–443, 1968.

GIBSON, I. L.: The Chemistry and Petrogenesis of a Suite of Pantellerites from the Ethiopian Rift, *J. Petrol.*, vol. 13, pp. 31–44, 1972.

——— and G. P. L. WALKER: Some Composite Rhyolite/Basalt Lavas and Related Dykes in Eastern Iceland, *Proc. Geol. Assn.*, vol. 74, pp. 301–318, 1963.

GILLULY, J.: Keratophyres of Eastern Oregon and the Spilite Problem, *Am. J. Sci.*, vol. 29, pp. 225–252, 336–352, 1935.

———: The Tectonic Evolution of the United States, *Q.J. Geol. Soc. Lond.*, vol. 119, pp. 133–174, 1963.

———: Orogeny and Geochronology, *Am. J. Sci.*, vol. 264, pp. 97–111 (with discussion by L. B. Platt, pp. 745–750), 1966.

———: Plate Tectonics and Magmatic Evolution, *Bull. Geol. Soc. Am.*, vol. 82, pp. 2383–2396, 1971.

GIROD, M., and C. LEFEVRE: A propos des "andesites" des Açores, *Contrib. Mineral. Petrol.*, vol. 35, pp. 159–167, 1972.

GITTINS, J.: Nephelinization in the Haliburton-Bancroft District, Ontario, Canada, *J. Geol.*, vol. 69, pp. 291–308, 1961.

———, A. HAYATSU, and D. YORK: A Strontium Isotope Study of Metamorphosed Limestones, *Lithos*, vol. 3, pp. 51–58, 1970.

GLIKSON, A. Y.: Early Precambrian Evidence of a Primitive Ocean Crust and Island Nuclei of Sodic Granite, *Bull. Geol. Soc. Am.*, vol. 83, pp. 3323–3344, 1972.

GOLDSCHMIDT, V. M.: The Principles of Distribution of Chemical Elements in Minerals and Rocks, *J. Chem. Soc.*, pp. 655–672, 1937.

———: "Geochemistry," Clarendon Press, Oxford, 1954.

GORANSON, R. W.: The Solubility of Water in Granitic Magmas, *Am. J. Sci.*, vol. 22, pp. 481–502, 1931.

GORSHKOV, G. S.: "Volcanism and the Upper Mantle: Investigation in the Kurile Island Arc," Plenum, New York, 1970.

———: Prediction of Volcanic Eruptions and Seismic Methods of Location of Magma Chambers, *Bull. Volc.*, vol. 35, pp. 198–211, 1971.

GRANT, N. K., D. C. REX, and S. J. FREETH: Potassium-Argon Ages and Strontium Isotope Ratio Measurements from Volcanic Rocks in Northeastern Nigeria, *Contrib. Mineral. Petrol.*, vol. 35, pp. 277–292, 1972.

GREEN, D. C.: Transitional Basalts from the Eastern Australian Tertiary Province, *Bull. Volc.*, vol. 33, pp. 930–941, 1970.

GREEN, D. H.: Alumina Content of Enstatite in a Venezuelan High-temperature Peridotite, *Bull. Geol. Soc. Am.*, vol. 74, pp. 1397–1402, 1963.

———: The Petrogenesis of the High-temperature Peridotite Intrusion in the Lizard Area, Cornwall, *J. Petrol.*, vol. 5, pp. 134–188, 1964.

———: The Origin of Basaltic and Nephelinitic Magmas in the Earth's Mantle, *Tectonophysics*, vol. 7, pp. 409–422, 1969.

———: The Origin of Basaltic and Nephelinitic Magmas, *Trans. Leicester Lit. Phil. Soc.*, vol. 64, pp. 26–54, 1970*a*.

———: A Review of Experimental Evidence on the Origin of Basaltic and Nephelinitic Magmas, *Phys. Earth Planet. Interiors*, vol. 3, pp. 221–235, 1970*b*.

———: Composition of Basaltic Magmas as Indicators of Conditions of Origin: Application to Oceanic Volcanism, *Phil. Trans. Roy. Soc. Lond.*, ser. A, vol. 268, pp. 707–725, 1971.

——— and W. HIBBERSON: Experimental Duplication of Conditions of Precipitation of High-pressure Phenocrysts in a Basaltic Magma, *Phys. Earth Planet. Interiors*, vol. 3, pp. 247–254, 1970.

——— and A. E. RINGWOOD: An Experimental Investigation of the Gabbro to Eclogite Transformation and Its Petrological Applications, *Geochim. Cosmochim. Acta*, vol. 31, pp. 767–833, 1967*a*.

——— and ———: The Genesis of Basaltic Magmas, *Contrib. Mineral. Petrol.*, vol. 15, pp. 103–190, 1967*b*.

GREEN, E. J.: On the Perils of Thermodynamic Modelling, *Geochim. Cosmochim. Acta*, vol. 34, pp. 1029–1032, 1970.

GREEN, T. H.: High-pressure Experimental Studies on the Origin of Anorthosites, *Canad. J. Earth Sci.*, vol. 6, pp. 427–440, 1969.

———: Crystallization of Calc-Alkaline Andesite under Controlled High-pressure Conditions, *Contrib. Mineral. Petrol.*, vol. 34, pp. 150–166, 1972.

———, A. O. BRUNFELT, and K. S. HEIER: Rare-earth Element Distribution and K/Rb Ratios in Granulites, Mangerites, and Anorthosites, Lofoten-Vesteraalen, Norway, *Geochim. Cosmochim. Acta*, vol. 36, pp. 241–257, 1972.

———, D. H. GREEN, and A. E. RINGWOOD: The Origin of High-alumina Basalts, *Earth Planet. Sci. Letters*, vol. 2, pp. 41–51, 1967.

——— and A. E. RINGWOOD: Genesis of the Calc-Alkaline Igneous Rock Suite, *Contrib. Mineral Petrol.*, vol. 18, pp. 105–162, 1968.

GROMME, C. S., T. L. WRIGHT, and D. L. PECK: Magnetic Properties and Oxidation of Iron-Titanium Oxide Minerals in Alae and Makaopuhi Lava Lakes, Hawaii, *J. Geophys. Res.*, vol. 74, pp. 5277–5293, 1969.

GROUT, F. F.: The Lopolith: An Igneous Form Exemplified by the Duluth Gabbro *Am. J. Sci.*, vol. 46, pp. 516–522, 1918.

———: Formation of Igneous-looking Rocks by Metasomatism, *Bull. Geol. Soc. Am.*, vol. 52, pp. 1525–1576, 1941.

———: Scale Models of Structures Related to Batholiths, *Am. J. Sci.*, vol. 243A, pp, 260–284, 1945.

——— and G. M. SCHWARTZ: The Geology of the Anorthosites of the Minnesota Coast of Lake Superior, *Minnesota Geol. Surv. Bull.* 28, 1939.

GUGGENHEIM, E. A.: "Mixtures," Oxford, New York, 1952.

GUNN, B. M.: Differentiation in Ferrar Dolerites, Antarctica, *N.Z. J. Geol. Geophys.*, vol. 5, pp. 820–863, 1962.

———: Layered Intrusions in the Ferrar Dolerites of Antarctica, *Mineral. Soc. Am. Spec. Paper*, no. 1, pp. 124–133, 1963.

———: K/Rb and K/Ba Ratios in Antarctic and New Zealand Tholeiites and Alkali Basalts, *J. Geophys. Res.*, vol. 70, pp. 6241–6247, 1965.

———: Modal and Element Variation in Antarctic Tholeiites, *Geochim. Cosmochim. Acta*, vol. 30, pp. 881–920, 1966.

———, R. COY-YLL, N. D. WATKINS, C. ABRANSON, and J. NOUGIER: Geochemistry of an Oceanite-Ankaramite-Basalt Suite from East Island, Crozet Archipelago, *Contrib. Mineral. Petrol.*, vol. 28, pp. 319–339, 1970.

HAGGERTY, J. S., A. R. COOPER, and J. H. HEASLEY: Heat Capacity of Three Inorganic Glasses and Supercooled Liquids, *Physics Chem. Glasses*, vol. 9, pp. 47–51, 1968.

HAGGERTY, S. E.: High-temperature Oxidation of Ilmenite in Basalts, *Ann. Rept. Geophys. Lab.* (*Carnegie Inst. Yrbk. 70*), pp. 165–176, 1971.

HAKLI, T. A., and T. L. WRIGHT: The Fractionation of Nickel between Olivine and Augite as a Geothermometer, *Geochim. Cosmochim. Acta*, vol. 31, pp. 877–884, 1967.

HALL, A.: The Relationship between Geothermal Gradient and the Composition of Granitic Magmas in Orogenic Belts, *Contrib. Mineral Petrol.*, vol. 32, pp. 186–192, 1971.

HALL, A. L.: The Bushveld Igneous Complex of the Central Transvaal, *Geol. Surv. S. Africa Mem.*, no. 28, 1932.

HAMILTON, D. L.: Nephelines as Crystallization Temperature Indicators, *J. Geol.*, vol. 69, pp. 321–329, 1961.

———, C. W. BURNHAM, and E. F. OSBORN: The Solubility of Water and Effects of Oxygen Fugacity and Water Content on Crystallization in Mafic Magmas, *J. Petrol.*, vol. 5, pp. 21–39, 1964.

——— and W. S. MACKENZIE: Phase Equilibrium Studies in the System $NaAlSiO_4$ (Nepheline)–$KAlSiO_4$ (Kalsilite)–$SiO_2$–$H_2O$, *Mineral Mag.*, vol. 34, pp. 214–231, 1965.

HAMILTON, E. I.: The Isotopic Composition of Strontium in the Skaergaard Intrusion, East Greenland, *J. Petrol.*, vol. 4, pp. 383–391, 1963.

———: The Isotopic Composition of Lead in Igneous Rocks: pt. 1, The Origin of Some Tertiary Granites, *Earth Planet. Sci. Letters*, vol. 1, pp. 30–37, 1966.

HAMILTON, W.: Geology and Petrogenesis of the Island Park Caldera of Rhyolite and Basalt, Eastern Idaho, *U.S.G.S. Prof. Paper* 504-C, 1965.

HARKER, A.: "The Natural History of Igneous Rocks," Macmillan, New York, 1909.

HARKIN, O. A.: The Rungwe Volcanics at the Northern End of Lake Nyasa, *Mem. Geol. Surv. Tanganyika*, no. 11, 1960.

HÄRME, M.: On the Potassium Migmatites of Southern Finland, *Bull. Comm. Geol. Finlande*, no. 219, 1965.

HARRIS, P. G.: Zone Refining and the Origin of Potassic Basalts, *Geochim. Cosmochim. Acta*, vol. 12, pp. 195–208, 1957.

———: Comments on a Paper by F. Chayes, *J. Geophys. Res.*, vol. 68, pp. 5103–5107, 1963.

——— and E. A. K. MIDDLEMOST: The Evolution of Kimberlites, *Lithos*, vol. 3, pp. 77–88, 1970.

———, W. Q. KENNEDY, and C. M. SCARFE: Volcanism versus Plutonism—The Effect of Chemical Composition, pp. 187–200 in G. Newall and N. Rast (eds.), "Mechanics of Igneous Intrusion," Gallery Press, Liverpool, 1970.

———, A. REAY, and I. G. WHITE: Chemical Composition of the Upper Mantle, *J. Geophys. Res.*, vol. 72, pp. 6359–6369, 1967.

HASLAM, H. W.: The Crystallization of Intermediate and Acid Magmas at Ben Nevis, Scotland, *J. Petrol.*, vol. 9, pp. 84–104, 1968.

HAWKES, D. D.: Differentiation of the Tumatumari-Kopinang Dolerite Intrusion, British Guiana, *Bull. Geol. Soc. Am.*, vol. 77, pp. 1131–1158, 1966.

HEALD, E. F.: Graphical Representation of Homogeneous Chemical Equilibria in Volcanic Gas Systems, *Am. J. Sci.*, vol. 266, pp. 389–401, 1968.

———, J. J. NAUGHTON, and I. L. BARNES, JR.: The Chemistry of Volcanic Gases: pt. 2, Use of Equilibrium Calculations in the Interpretation of Volcanic Gas Samples, *J. Geophys. Res.*, vol. 68, pp. 545–557, 1963.

HEALY, J.: Structure and Volcanism in the Taupo Volcanic Zone, New Zealand, pp. 151–157 in "The Crust of the Pacific Basin" (*Am. Geophys. Union Monograph* 6), 1962.

HEDGE, C. E.: Variations in Radiogenic Strontium Found in Volcanic Rocks, *J. Geophys. Res.*, vol. 71, pp. 6119–6126, 1966.

——— and R. J. KNIGHT: Lead and Strontium Isotopes in Volcanic Rocks from Northern Honshu, Japan, *Geochim. J. Japan*, vol. 3, pp. 15–24, 1969.

——— and J. F. LEWIS: Isotopic Composition of Strontium in Three Basalt-Andesite Centers along the Lesser Antilles Arc, *Contrib. Mineral. Petrol.*, vol. 32, pp. 39–47, 1971.

——— and D. C. NOBLE: Upper Cenozoic Basalts with High $Sr^{87}/Sr^{86}$ and Sr/Rb Ratios, Southern Great Basin, Western United States, *Bull. Geol. Soc. Am.*, vol. 82, pp. 3503–3510, 1971.

——— and Z. E. PETERMAN: The Strontium Isotopic Composition of Basalts from the Gordo and Juan de Fuca Rises, Northeastern Pacific Ocean, *Contrib. Mineral. Petrol.*, vol. 27, pp. 114–120, 1970.

———, Z. E. PETERMAN, and W. R. DICKINSON: Petrogenesis of Lavas from Western Samoa, *Bull. Geol. Soc. Am.*, vol. 83, pp. 2709–2714, 1972.

HEIER, K. S.: Geochemistry of the Nepheline Syenite of Stjernoy, North Norway, *Norsk Geol. Tidsskr.*, vol. 44, pp. 205–215, 1964.
———: Metamorphism and the Chemical Differentiation of the Crust, *Geol. Fören. Stockholm Förk*, vol. 87, pp. 249–256, 1965.
——— and J. A. S. ADAMS: Concentration of Radioactive Elements in Deep Crustal Material, *Geochim. Cosmochim. Acta*, vol. 29, pp. 53–61, 1965.
——— and W. COMPSTON: Rb–Sr Isotopic Studies on the Plutonic Rocks of the Oslo Region, *Lithos*, vol. 2, pp. 133–156, 1969.
———, ———, and I. MCDOUGALL: Thorium and Uranium Concentrations, and the Isotopic Composition of Strontium in the Differentiated Tasmanian Dolerites, *Geochim. Cosmochim. Acta*, vol. 29, pp. 643–659, 1965.
——— and S. R. TAYLOR: A Note on the Geochemistry of Alkaline Rocks, *Norsk Geol. Tidsskr.*, vol. 44, pp. 197–204, 1964.
——— and K. THORESEN: Geochemistry of High Grade Metamorphic Rocks, Lofoten-Vesterålen, North Norway, *Geochim. Cosmochim. Acta*, vol. 35, pp. 89–99, 1971.
HEIMLICH, R. A., and P. O. BANKS: Radiometric Age Determinations, Bighorn Mountains, Wyoming, *Am. J. Sci.*, vol. 266, pp. 180–192, 1968.
HEINRICH, E. W.: "The Geology of Carbonatites," Rand McNally, Chicago, 1966.
——— and R. J. ANDERSON: Carbonatites and Alkalic Rocks of the Arkansas River Area, Fremont County, Colorado: pt. 2, Fetid Gas from Carbonatite and Related Rocks, *Am. Mineral.*, vol. 50, pp. 1914–1920, 1965.
——— et al. (eds.): Symposium on Layered Intrusions, *Mineral. Soc. Am. Spec. Paper*, no. 1, 1963.
HEMING, R. F.: Geology and Petrology of the Rabaul Caldera, New Britain, *Bull. Geol. Soc. Am.*, in press.
——— and I. S. E. CARMICHAEL: High-temperature Pumice Flows from the Rabaul Caldera, Papua, New Guinea, *Contrib. Mineral. Petrol.*, vol. 38, pp. 1–20, 1973.
HERNÁNDEZ-PACHECO, A.: The Tahitites of Gran Canaria and Hauynitization of Their Inclusions, *Bull. Volc.*, vol. 33, pp. 701–728, 1969.
HERZ, N.: Anorthosite Belts, Continental Drift, and the Anorthosite Event, *Science*, vol. 164, pp. 944–947, 1969.
HESS, H. H.: A Primary Peridotite Magma, *Am. J. Sci.*, vol. 35, pp. 321–344, 1938.
———: Serpentinites, Orogeny and Epeirogeny, *Geol. Soc. Am. Spec. paper*, no. 62, pp. 391–408, 1955.
———: Stillwater Igneous Complex, Montana, *Geol. Soc. Am. Mem.*, no. 80, 1960.
———: History of Ocean Basins, pp. 599–620 in "Petrologic Studies" (*Buddington Vol., Geol. Soc. Am.*), 1962.
———: The Oceanic Crust, the Upper Mantle, and the Mayaguez Serpentinized Peridotite, pp. 169–175 in C. A. Burk (ed.), "A Study of Serpentinite," (*Nat. Acad. Sci.–Nat. Res. Council Publ.* 1188), 1964.
HESS, P. C.: Polymer Models of Silicate Melts, *Geochim. Cosmochim. Acta*, vol. 35, pp. 289–306, 1971.
HIGAZY, R. A.: Trace Elements of Volcanic Ultrabasic Potassic Rocks of South-western Uganda and Adjoining Part of the Belgian Congo, *Bull. Geol. Soc. Am.*, vol. 65, pp. 39–70, 1954.
HOEFS, J., and K. H. WEDERPOHL: Strontium Isotope Studies on Young Volcanic Rocks from Germany and Italy, *Contrib. Mineral. Petrol.*, vol. 19, pp. 328–338, 1968.

HOLGATE, N.: The Role of Liquid Immiscibility in Petrogenesis, *J. Geol.*, vol. 62, pp. 439–480, 1954.

HOLLOWAY, J. R., and C. W. BURNHAM: Melting Relations of Basalt with Equilibrium Water Pressure Less Than Total Pressure, *J. Petrol.*, vol. 13, pp. 1–29, 1972.

HOLMES, A.: The Origin of Igneous Rocks, *Geol. Mag.*, vol. 69, pp. 543–558, 1932.

———: A Record of New Analyses of Tertiary Igneous Rocks (Antrim and Staffa), *Roy. Irish Acad. Proc.*, vol. 43, sec. B, no. 8, pp. 89–94, 1936.

———: A Suite of Volcanic Rocks from Southwest Uganda Containing Kalsilite (a Polymorph of $KAlSiO_4$), *Mineral. Mag.*, vol. 26, pp. 197–217, 1942.

———: Petrogenesis of Katungite and Its Associates, *Am. Mineral.*, vol. 35, pp. 772–792, 1950.

———: The Potash Ankaratrite-Melaleucitite Lavas of Nabugando and Mbuga Craters, Southwest Uganda, *Geol. Soc. Edin. Trans.*, vol. 15, pp. 187–213, 1952.

——— and H. F. HARWOOD: The Tholeiite Dikes of the North of England, *Mineral. Mag.*, vol. 22, pp. 1–52, 1929.

——— and ———: The Volcanic Area of Bufumbira: pt. II, The Petrology of the Volcanic Field of Bufumbira, South-west Uganda and of Other Parts of the Birunga Field, *Geol. Surv. Uganda Mem.* 3, 1937.

HOTZ, P. E.: Petrology of the Granophyre in Diabase near Dillsburg, Pennsylvania, *Bull. Geol. Soc. Am.*, vol. 64, pp. 675–704, 1953.

HUGHES, D. J., and G. C. BROWN: Basalts from Madeira: A Petrochemical Contribution to the Genesis of Oceanic Alkali Rock Series, *Contrib. Mineral. Petrol.*, vol. 37, pp. 91–110, 1972.

HURLEY, P. M., P. C. BATEMAN, H. W. FAIRBAIRN, and W. H. PINSON: Investigations of $Sr^{87}/Sr^{86}$ Ratios in the Sierra Nevada Plutonic Province, *Bull. Geol. Soc. Am.*, vol. 76, pp. 165–174, 1965.

——— and J. R. RAND: Pre-drift Continental Nuclei, *Science*, vol. 164, pp. 1229–1242, 1969.

IBARROLA, E.: Variation Trends in Basaltic Rocks of the Canary Islands, *Bull. Volc.*, vol. 33, pt. 3, pp. 729–777, 1969.

IDDINGS, J. P.: "Igneous Rocks," vol. I, Wiley, New York, 1909.

ISACHSEN, Y. W. (ed.): Origin of Anorthosite and Related Rocks, *N.Y. State Mus. and Sci. Mem.* 18, 1968.

ISHIKAWA, H., and K. YAGI: Geochemical Studies of the Alkali Rocks of the Morotu District, Sakhalin, and Ponape Island, Western Pacific Ocean, *Jap. J. Geol. Geogr.*, vol. 41, no. 1, pp. 15–32, 1970.

ITO, K., and G. C. KENNEDY: Melting and Phase Relationships in the Plane Tholeiite-Lherzolite-Nepheline Basanite to 40 Kilobars with Geological Implications, *Contrib. Mineral. Petrol.*, vol. 19, pp. 177–211, 1968.

——— and ———: The Fine Structure of the Basalt-Eclogite Transition, *Mineral. Soc. Am. Spec. Paper*, no. 3, pp. 77–84, 1970.

JACKSON, E. D.: Primary Textures and Mineral Associations in the Ultramafic Zone of the Stillwater Complex, Montana, *U.S.G.S. Prof. Paper* 358, 1961.

———: The Character of the Lower Crust and Upper Mantle beneath the Hawaiian Islands, *Proc. 23rd Int. Geol. Congr.*, vol. 1, pp. 131–150, 1968.

——— and T. L. WRIGHT: Xenoliths in the Honolulu Volcanic Series, Hawaii, *J. Petrol.*, vol. 11, pp. 405–430, 1970.

———, E. A. SILVER, and G. B. DALRYMPLE: Hawaiian-Emperor Chain and Its

Relation to Cenozoic Circumpacific Tectonics, *Bull. Geol. Soc. Am.*, vol. 83, pp. 601–618, 1972.

JACKSON, K. A.: Current Concepts in Crystal Growth from the Melt, pp. 53–80 in H. Reiss (ed.), "Progress in Solid State Chemistry," vol. 4, Pergamon, London, 1967.

JACOBSON, R. R. E., W. N. MCLEOD, and P. BLACK: Ring Complexes in the Younger Granite Province of Northern Nigeria, *Geol. Soc. Lond. Monograph* 1, 1958.

JAEGER, J. C.: The Solidification and Cooling of Intrusive Sheets, pp. 77–87 in "Dolerite, a Symposium," University of Tasmania, Dept. of Geology, 1957*a*.

———: The Temperature in the Neighborhood of an Intrusive Sheet, *Am. J. Sci.*, vol. 255, pp. 306–318, 1957*b*.

———: Cooling and Solidification of Igneous Rocks, pp. 503–536 in H. H. Hess (ed.), "Basalts," vol. 2, Interscience (Wiley), New York, 1968.

——— and G. A. JOPLIN: Discussion on Magnetic Properties and Differentiation of Dolerite Sills, *Am. J. Sci.*, vol. 254, pp. 443–446, 1956.

JAHNS, R. H., and C. W. BURNHAM: Experimental Studies of Pegmatite Genesis: pt. I, A Model for the Derivation and Crystallisation of Granitic Pegmatites, *Econ. Geol.*, vol. 64, pp. 843–864, 1969.

JAKEŠ, P., and I. E. SMITH: High Potassium Calc-Alkaline Rocks from Cape Nelson, Eastern Papua, *Contrib. Mineral. Petrol.*, vol. 28, pp. 259–271, 1970.

——— and A. J. R. WHITE: Structure of the Melanesian Arcs and Correlation with Distribution of Magma Types, *Tectonophysics*, vol. 8, pp. 223–236, 1969.

——— and ———: Hornblendes from Calc-Alkaline Volcanic Rocks of Island Arcs and Continental Margins, *Am. Mineral.*, vol. 57, pp. 887–902, 1972.

JAMES, H. L.: Harry Hammond Hess, *U.S. Nat. Acad. Sci. Biogr. Mem.*, vol. 43, pp. 109–128, 1973.

JAMES, R. S., and D. L. HAMILTON: Phase Relations in the System $NaAlSi_3O_8$–$KAlSi_3O_8$–$CaAl_2Si_2O_8$–$SiO_2$ at 1 Kilobar Water Vapour Pressure, *Contrib. Mineral. Petrol.*, vol. 21, pp. 111–141, 1969.

JAMIESON, B. G.: Evidence on the Evolution of Basaltic Magma at Elevated Pressures, *Nature*, vol. 212, pp. 243–246, 1966.

——— and D. B. CLARKE: Potassium and Associated Elements in Tholeiitic Basalts, *J. Petrol.*, vol. 11, pp. 183–204, 1970.

JASMUND, K., and H. A. SECK: Geochemische Untersuchungen an Auswürflingen (Gleesiten) des Laacher-See-Gebietes, *Beitr. Mineral. Petrogr.*, vol. 10, pp. 275–295, 1964.

JENNINGS, D. S., and R. H. MITCHELL: An Estimate of the Temperature of Intrusion of Carbonatite at the Fen Complex, South Norway, *Lithos*, vol. 2, pp. 167–169, 1969.

JOHANNSEN, A.: "A Descriptive Petrography of the Igneous Rocks," vol. 1, University of Chicago Press, Chicago, 1939.

JOHNSON, R. W., R. A. DAVIES, and A. J. R. WHITE: Ulawun Volcano, New Britain, *Bur. Min. Res. (Australia) Bull.* 142, 1972.

JOHNSON, T., and P. MOLNAR: Focal Mechanisms and Plate Tectonics of the Southwest Pacific, *J. Geophys. Res.*, vol. 77, pp. 5000–5032, 1972.

JONES, L. M., M. E. HARTLEY, and R. L. WALKER: Strontium Isotope Composition of Alpine-type Ultramafic Rocks in the Lake Chatuge District, Georgia–North Carolina, *Contrib. Mineral. Petrol.*, vol. 38, pp. 321–328, 1973.

JOPLIN, G. A.: "A Petrography of Australian Rocks," Angus and Robertson, Sydney, 1964.

———: The Problem of the Potash-rich Basaltic Rocks, *Mineral. Mag.*, vol. 34, pp. 266–275, 1965.

———: The Shoshonite Association: A Review, *J. Geol. Soc. Australia*, vol. 15, pp. 275–294, 1968.

KÁDIK, A. A., and N. I. KHITAROV: Thermodynamic Conditions of Melting of Silicates at High Water Vapor Pressures (translation), *Geokhimiya*, no. 5, pp. 523–535, 1969.

KAY, R., N. J. HUBBARD, and P. W. GAST: Chemical Characteristics and Origins of Oceanic Ridge Volcanic Rocks, *J. Geophys. Res.*, vol. 75, pp. 1585–1613, 1970.

KEESON, S., and R. C. PRICE: The Major and Trace Element Chemistry of Kaersutite and Its Bearing on the Petrogenesis of Alkaline Rocks, *Contrib. Mineral. Petrol.*, vol. 35, pp. 119–124, 1972.

KELLEY, K. K.: Contributions to the Data on Theoretical Metallurgy: pt. 13, High Temperature Heat Content, Heat Capacity and Entropy Data for the Elements and Inorganic Compounds. *U.S. Bur. Mines Bull.* 584, 1960.

KENNEDY, G. C.: Equilibrium between Volatiles and Iron Oxides in Igneous Rocks, *Am. J. Sci.*, vol. 246, pp. 529–549, 1948.

———: The Origin of Continents, Mountain Ranges, and Ocean Basins, *Am. J. Sci.*, vol. 47, pp. 491–504, 1959.

——— and B. E. NORDLIE: The Genesis of Diamond Deposits, *Econ. Geol.*, vol. 63, pp. 495–503, 1968.

KENNEDY, W. Q.: Trends of Differentiation in Basaltic Magmas, *Am. J. Sci.*, vol. 25, pp. 239–256, 1933.

——— and E. M. ANDERSON: Crustal Layers and the Origin of Magmas, *Bull. Volc.*, vol. 3, pp. 24–41, 1938.

KHITAROV, N. I., E. B. LEBEDEV, and A. A. KÁDIK: Solubility of Water in Granitic Melt at Pressures up to 7000 Atmospheres, *Geochem.*, pp. 992–994, 1963.

KING, B. C.: The Ard Bheinn Area of the Central Igneous Complex of Arran, *Q. J. Geol. Soc. Lond.*, vol. 110, pp. 323–356, 1955.

———: Petrogenesis of the Alkaline Igneous Rock Suites of the Volcanic and Intrusive Centers of Eastern Uganda, *J. Petrol.*, vol. 6, pp. 67–100, 1965.

———: Volcanism in Eastern Africa and Its Structural Setting, *Proc. Geol. Soc. Lond.*, no. 1629, pp. 16–19, 1966.

———: Vulcanicity and Rift Tectonics in East Africa, pp. 263–283 in T. N. Clifford and I. G. Gass (eds.), "African Magmatism and Tectonics," Hafner, New York, 1970.

———, and D. S. SUTHERLAND: Alkaline Rocks of Eastern and Southern Africa, *Sci. Prog.*, vol. 48, pp. 298–321, 504–524, 709–720, 1960.

KISTLER, R. W., P. C. BATEMAN, and W. W. BRANNOCK: Isotopic Ages of Minerals from Granitic Rocks of the Central Sierra Nevada and Inyo Mountains, California, *Bull. Geol. Soc. Am.*, vol. 76, pp. 155–164, 1965.

KLEIN, C., JR.: Coexisting Amphiboles, *J. Petrol.*, vol. 9, pp. 281–330, 1968.

KLEPPER, M. R., G. D. ROBINSON, and H. W. SMEDES: On the Nature of the Boulder Batholith in Montana, *Bull. Geol. Soc. Am.*, vol. 82, pp. 1563–1580, 1971.

KNOPF, A.: Time Required to Emplace the Boulder Bathylith, Montana: A First Approximation, *Am. J. Sci.*, vol. 262, pp. 1207–1211, 1964.

———: Bathyliths in Time, "Crust of the Earth," *Geol. Soc. Am. Spec. Paper*, no. 62, pp. 685–702, 1955.

KNORR, H.: Differentiations- und Eruptionsfolge im Bömischen Mittelgebirge, *Mineral. Petrogr. Mitteil.*, vol. 42, pp. 318–370, 1932.

KOMAR, P. D.: Flow Differentiation in Igneous Dikes and Sills: Profiles of Velocity and Phenocryst Concentration, *Bull. Geol. Soc. Am.*, vol. 83, pp. 3443–3448, 1972.

KOSTER VAN GROOS, A. F., and P. J. WYLLIE: Liquid Immiscibility in the System $Na_2O$–$Al_2O_3$–$SiO_2$–$CO_2$ at Pressures to 1 Kilobar, *Am. J. Sci.*, vol. 264, pp. 234–255, 1966.

——— and ———: Liquid Immiscibility in the Join $NaAlSi_3O_8$–$Na_2CO_3$–$H_2O$, and Its Bearing on the Origin of Carbonatites, *Am. J. Sci.*, vol. 266, pp. 932–967, 1968.

——— and ———: Melting Relationships in the System $NaAlSi_3O_8$–$NaCl$–$H_2O$ at One Kilobar Pressure with Petrological Applications, *J. Geol.*, vol. 77, pp. 581–605, 1969.

KRACEK, F. C.: *Ann. Rept. Geophys. Lab.* (*Carnegie Inst. Yrbk. 32*), pp. 61–63, 1933.

KRETZ, R.: Chemical Study of Garnet, Biotite and Hornblende from Gneisses of Southwestern Quebec, with Emphasis on Distribution of Elements in Coexisting Minerals, *J. Geol.*, vol. 67, pp. 371–402, 1959.

KROGH, T. E., and G. L. DAVIS: Geochronology of the Grenville Province, *Ann. Rept. Geophys. Lab.* (*Carnegie Inst. Yrbk 67*), pp. 224–230, 1969.

———, ———, L. T. ALDRICH, S. R. HART, and A. STUEBER: Geological History of the Grenville Province, *Ann. Rept. Geophys. Lab.* (*Carnegie Inst. Yrbk. 66*), pp. 528–536, 1968.

KUBOTA, S., and E. BERG: Evidence for Magma in the Katmai Volcanic Range, *Bull. Volc.*, vol. 23, pp. 175–214, 1967.

KUDO, A. M., and D. F. WEILL: An Igneous Plagioclase Thermometer, *Contrib. Mineral. Petrol.*, vol. 25, pp. 52–65, 1970.

KUNO, H.: Petrology of Hakone Volcano and the Adjacent Areas, Japan, *Bull. Geol. Soc. Am.*, vol. 61, pp. 957–1020, 1950.

———: Lateral Variation of Basalt Magma Type across Continental Margins and Island Arcs, *Bull. Volc.*, vol. 29, pp. 195–222, 1966.

———: Origin of Andesite and Its Bearing on the Island Arc Structure, *Bull. Volc.*, vol. 32, pp. 141–176, 1968.

———: Mafic and Ultramafic Nodules in Basaltic Rocks of Hawaii, *Geol. Soc. Am. Mem.*, no. 115, pp. 189–234, 1969.

———, K. YAMASAKI, C. IIDA, and K. NAGASHIMA: Differentiation of Hawaiian Magmas, *Jap. J. Geol. Geol. Geogr.*, vol. 28, pp. 179–218, 1957.

KUSHIRO, I.: Si–Al Relations in Clinopyroxenes from Igneous Rocks, *Am. J. Sci.*, vol. 258, pp. 548–554, 1960.

———: Petrology of the Atumi Dolerite, Japan, *J. Fac. Sci. Univ. Tokyo*, sec. 2, vol. 15, pp. 135–202, 1964.

———: The System Forsterite-Diopside Silica, with and without Water at High Pressures, *Am. J. Sci.*, vol. 267A (Schairer vol.), pp. 269–294, 1969.

———: Systems Bearing on Melting of the Upper Mantle under Hydrous Conditions, *Ann. Rept. Geophys. Lab.*, (*Carnegie Inst. Yrbk. 68*), 1970.

———: Effect of Water on the Composition of Magmas Formed at High Pressures, *J. Petrol.*, vol. 13, pp. 311–334, 1972*a*.

———: Determination of Liquidus Relations in Synthetic Silicate Systems with Electron Probe Analysis: The System Forsterite-Diopside-Silica at 1 Atmosphere, *Am. Mineral.*, vol. 57, pp. 1260–1271, 1972*b*.

——— and H. S. Yoder, Jr.: Anorthite-Forsterite and Anorthite-Enstatite Reactions and Their Bearing on the Basalt-Eclogite Transformation, *J. Petrol.*, vol. 7, pp. 337–362, 1966.

Kutolin, V. A., and V. M. Frolova: Petrology of Ultrabasic Inclusions from Basalts of Minusa and Transbaikalian Regions (Siberia, U.S.S.R.), *Contrib. Mineral. Petrol.*, vol. 29, pp. 163–179, 1970.

Lacroix, A.: La constitution lithologique des iles volcaniques de la Polynesie australe, *Acad. Sci. (Paris) Mem.*, vol. 5, pp. 1–80, 1928.

Lacy, E. D.: Atomic Packing in Silicate Glasses, pp. 23–46 in "The Vitreous State," Glass Delegacy, University of Sheffield, England, 1955.

———: A Statistical Model of Polymerization/depolymerization Relationships in Silicate Melts and Glasses, *Physics Chem. Glasses*, vol. 6, pp. 171–179, 1965.

———: The Newtonian Flow of Simple Silicate Melts at High Temperature, *Physics Chem. Glasses*, vol. 8, pp. 238–246, 1967.

———: Structure Transition in Alkali Silicate Glasses, *J. Am. Ceram. Soc.*, vol. 51, pp. 150–157, 1968.

Ladd, H. S., J. I. Tracey, and M. G. Gross: Drilling on Midway Atoll, Hawaii, *Science*, vol. 156, pp. 1088–1094, 1967.

Lambert, I. B., and K. S. Heier: Geochemical Investigations of Deep-seated Rocks in the Australian Shield, *Lithos*, vol. 1, pp. 30–53, 1968.

———, J. K. Robertson, and P. J. Wyllie: Melting Reactions in the System $KAlSi_3O_8$–$SiO_2$–$H_2O$ to 18.5 Kilobars, *Am. J. Sci.*, vol. 267, pp. 609–626, 1969.

——— and P. J. Wyllie: Melting in the Deep Crust and Upper Mantle and the Nature of the Low Velocity Layer, *Phys. Earth. Planet. Interiors*, vol. 3, pp. 316–322, 1970.

Lanphere, M. A.: Age of the Mesozoic Oceanic Crust in the California Coast Ranges, *Bull. Geol. Soc. Am.*, vol. 82, pp. 3209–3212, 1973.

Larsen, E. S.: Some New Variation Diagrams for Groups of Igneous Rocks, *J. Geol.*, vol. 46, pp. 505–520, 1938.

———: Batholith of Southern California, *Geol. Soc. Am. Mem.*, no. 29, 1948.

——— and W. Cross: Geology and Petrology of the San Juan Region, Southwestern Colorado, *U.S.G.S. Prof. Paper* 258, 1956.

——— and D. Gottfried: Distribution of Uranium in Rocks and Minerals of Mesozoic Batholiths in Western United States, *U.S.G.S. Bull.* 1070-C, 1961.

———, J. Irving, and F. A. Gonyer: Petrologic Results of a Study of the Minerals from the Tertiary Volcanic Rocks of the San Juan Region, Colorado, *Am. Mineral.*, vol. 23, pp. 227–257, 417–429, 1938.

———, ———, ———, and E. S. Larsen, III: Petrologic Results of a Study of the Minerals from the Tertiary Volcanic Rocks of the San Juan Region, Colorado, *Am. Mineral.*, vol. 22, pp. 889–905, 1937.

——— and R. G. Schmidt: A Reconnaissance of the Idaho Batholith, *U.S.G.S. Bull.* 1070-A, 1958.

Leake, B. E.: The Fragmentation of the Connemara Basic and Ultrabasic Intrusions, pp. 103–122 in G. Newall and N. Rast (eds.), "Mechanics of Igneous Intrusion," Gallery Press, Liverpool, 1970.

LE BAS, M. J.: The Role of Aluminum in Igneous Clinopyroxenes with Relation to Their Parentage, *Am. J. Sci.*, vol. 260, pp. 267–288, 1962.

———: A Combined Central- and Fissure-type Phonolitic Volcano in Western Kenya, *Bull. Volc.*, vol. 34, pp. 518–536, 1970.

LEE, R. W.: On the Role of Hydroxyl in the Diffusion of Hydrogen in Fused Silica, *Physics Chem. Glasses*, vol. 5, pp. 35–43, 1964.

LEEMAN, W. P., and J. J. W. ROGERS: Late Cenozoic Alkali-Olivine Basalts of the Basin-Range Province, U.S.A., *Contrib. Mineral. Petrol.*, vol. 25, pp. 1–24, 1970.

LEGGO, P. J., and R. HUTCHISON: A Rb–Sr Study of Ultrabasic Xenoliths and Their Basaltic Host Rocks from the Massif Central, France, *Earth Planet. Sci. Letters*, vol. 5, pp. 71–75, 1968.

——— and R. T. PIDGEON: Geochronological Investigations of Caledonian History in Western Ireland, *Eclogae Geologicae Helvetiae*, vol. 63, pp. 207–212, 1970.

———, P. W. G. TANNER, and B. E. LEAKE: Isochron Study of Donegal Granite and Certain Dalradian Rocks of Britain, *Am. Assn. Petroleum Geol. Mem.* 12, pp. 354–362, 1969.

LE MAITRE, R. W.: Petrology of Volcanic Rocks, Gough Island, South Atlantic, *Bull. Geol. Soc. Am.*, vol. 73, pp. 1309–1340, 1962.

———: The Significance of the Gabbroic Xenoliths from Gough Island, South Atlantic, *Mineral. Mag.*, vol. 34, pp. 303–317, 1965.

———: Chemical Variation within and between Volcanic Rock Series—A Statistical Approach, *J. Petrol.*, vol. 9, pp. 220–252, 1968.

LEONARDOS, O. H., and W. S. FYFE: Serpentinites and Associated Albitites, Moccasin Quadrangle, California, *Am. J. Sci.*, vol. 265, pp. 609–618, 1967.

LE PICHON, X.: Sea-floor Spreading and Continental Drift, *J. Geophys. Res.*, vol. 73, pp. 3661–3697, 1968.

LESSING, P., R. W. DECKER, and R. C. REYNOLDS: Potassium and Rubidium Distribution in Hawaiian Lavas, *J. Geophys. Res.*, vol. 68, pp. 5851–5855, 1963.

LEVI, B.: Burial Metamorphism of a Cretaceous Volcanic Sequence West from Santiago, Chile, *Contrib. Mineral. Petrol.*, vol. 24, pp. 30–49, 1969.

LEVIN, E. M., C. R. ROBBINS, and H. F. MCMURDIE: "Phase Diagrams for Ceramists," *Am. Ceram. Soc.*, 1964.

LEWIS, G. N., and M. RANDALL: "Thermodynamics," 2d ed., McGraw-Hill, New York, 1961.

LEWIS, J. F.: Tauhara Volcano, Taupo Zone: pt. 1, Geology and Structure, *N.Z. J. Geol. Geophys.*, vol. 11, pp. 212–224, 1968.

———: Composition, Origin and Differentiation of Basalt Magma in the Lesser Antilles, *Geol. Soc. Am. Mem.*, no. 130, pp. 159–179, 1971*a*.

LEWIS, J. S.: Consequences of the Presence of Sulfur in the Core of the Earth, *Earth Planet. Sci. Letters*, vol. 11, pp. 130–134, 1971*b*.

LINDSLEY, D. H.: Melting Relations of Plagioclase at High Pressures, *N.Y. State Mus. and Sci. Mem.*, vol. 18, pp. 39–46, 1968.

———, G. M. BROWN, and I. D. MUIR: Conditions of the Ferrowollastonite-Ferrohedenbergite Inversion in the Skaergaard Intrusion, East Greenland, *Mineral. Soc. Am. Spec. Paper*, no. 2, pp. 193–201, 1969.

——— and S. E. HAGGERTY: Phase Relations of Fe–Ti Oxides and Aenigmatite; Oxygen Fugacity of the Pegmatoid Zones, *Ann. Rept. Geophys. Lab.* (*Carnegie Inst. Yrbk. 69*), pp. 278–284, 1971.

———, and J. L. Munoz: Subsolidus Relations along the Join Hedenbergite-Ferrosilite, *Am. J. Sci.*, vol. 267-A, pp. 295–324, 1969.

Lipman, P. W.: Water Pressures During Differentiation and Crystallization of Some Ash-flow Magmas from Southern Nevada, *Am. J. Sci.*, vol. 264, pp. 810–826, 1966.

———: Iron-Titanium Oxide Phenocrysts in Compositionally Zoned Ash-flow Sheets from Southern Nevada, *J. Geol.*, vol. 79, pp. 438–456, 1971.

Loney, R. A., G. R. Himmelberg, and R. G. Coleman: Structure and Petrology of the Alpine-type Peridotite at Burro Mountain, California, U.S.A., *J. Petrol.*, vol. 12, pp. 245–309, 1971.

Loomis, T. P.: Contact Metamorphism of Pelitic Rock by the Ronda Ultramafic Intrusion, Southern Spain, *Bull. Geol. Soc. Am.*, vol. 83, pp. 2449–2474, 1972*a*.

———: Diapiric Emplacement of the Ronda High-temperature Ultramafic Intrusion, Southern Spain, *Bull. Geol. Soc. Am.*, vol. 83, pp. 2475–2496, 1972*b*.

Lopez Ruis, J.: Origen de las inclusiones de dunitas y otras rocas ultramáficas en las rocas volcánicas de Lanzarote y Fuerteventura, *Estud. Geol. Inst. Lucas Mallada*, vol. 25, pp. 189–223, 1969.

———: Estudio petrográfico y geoguímicuo del complejo filoniano de Fuerteventura (Ilas Canarias), *Estud. Geol. Inst. Lucas Mallada*, vol. 26, pp. 173–208, 1970.

Lort, J. M.: The Tectonics of the Eastern Mediterranean: A Geophysical Review, *Rev. Geophys. Space Phys.*, vol. 9, pp. 189–216, 1971.

Lovering, J. F., and J. R. Richards: Potassium-Argon Age Study of Possible Lower-crust and Upper-mantle Inclusions in Deep-seated Intrusions, *J. Geophys. Res.*, vol. 69, pp. 4895–4901, 1964.

——— and M. Tatsumoto: Lead Isotopes and the Origin of Granulite and Eclogite Inclusions in Deep-seated Pipes, *Earth Planet. Sci. Letters*, vol. 4, pp. 350–356, 1968.

——— and A. J. R. White: The Significance of Primary Scapolite in Granulitic Inclusions from Deep-seated Pipes, *J. Petrol.*, vol. 5, pp. 195–218, 1964.

Lowder, G. G.: The Volcanoes and Caldera of Talasea, New Britain: Mineralogy, *Contrib. Mineral. Petrol.*, vol. 26, pp. 324–340, 1970.

——— and I. S. E. Carmichael: The Volcanoes and Caldera of Talasea, New Britain: Geology and Petrology, *Bull. Geol. Soc. Am.*, vol. 81, pp. 17–38, 1970.

Ludington, S. D.: Refinement of the Biotite-Apatite Geothermometer, *Geol. Soc. Am. Abstracts*, vol. 5, pp. 493–494, 1973.

Luth, W. C.: Studies in the System $KAlSiO_4$–$Mg_2SiO_4$–$SiO_2$–$H_2O$: pt. 1, Inferred Phase Relations and Petrologic Applications, *J. Petrol.*, vol. 8, pp. 372–416, 1967.

———: The Systems $NaAlSi_3O_8$–$SiO_2$ and $KAlSi_3O_8$–$SiO_2$ to 20 kb and the Relationship between $H_2O$ Content, $P_{H_2O}$ and $P_{total}$ in Granitic Magmas, *Am. J. Sci.*, vol. 267-A (Schairer vol.), pp. 325–341, 1969.

———, R. H. Jahns, and O. F. Tuttle: The Granite System at Pressures of 4 to 10 Kilobars, *J. Geophys. Res.*, vol. 69, pp. 759–773, 1964.

Luyendyk, B. P., and C. G. Engel: Basalt Dredged from an Abyssal Hill in the North-east Pacific, *Nature*, vol. 223, pp. 1049–1050, 1969.

Lyons, J. B., and H. Faul: Isotope Geochronology of the Northern Appalachians, pp. 305–318 in E-an Zen et al. (eds.), "Studies of Appalachian Geology: Northern and Maritime," Interscience (Wiley), New York, 1968.

MACCOLL, R. S.: Geochemical and Structural Studies in Batholithic Rocks of Southern California: pt. 1, Geology of Rattlesnake Mountain Pluton, *Bull. Geol. Soc. Am.*, vol. 75, pp. 805–822, 1964.

MACDONALD, G. A.: Petrography of Maui, *Territory of Hawaii, Div. Hydrogr. Bull.* 7, pp. 275–333, 1942.

———: Petrography of the Samoan Islands, *Bull. Geol. Soc. Am.*, vol. 51, pp. 1333–1362, 1944.

———: Hawaiian Petrographic Province, *Bull. Geol. Soc. Am.*, vol. 60, pp. 1541–1596, 1949*a*.

———: Petrography of the Island of Hawaii, *U.S.G.S. Prof. Paper* 214-D, pp. 51–96, 1949*b*.

———: Physical Properties of Erupting Hawaiian Magmas, *Bull. Geol. Soc. Am.*, vol. 74, pp. 1071–1078, 1963*a*.

———: Relative Abundance of Intermediate Members of the Oceanic Basalt-Trachyte Association—A Discussion, *J. Geophys. Res.*, vol. 68, pp. 5100–5102, 1963*b*.

———: Composition and Origin of Hawaiian Lavas, *Geol. Soc. Am. Mem.*, no. 116, pp. 477–522, 1968.

——— and T. KATSURA: Chemical Composition of Hawaiian Lavas, *J. Petrol.*, vol. 5, pp. 82–133, 1964.

MACDONALD, R., and I. L. GIBSON: Pantelleritic Obsidians from the Volcano Chabbi (Ethiopia), *Contrib. Mineral. Petrol.*, vol. 24, pp. 239–244, 1969.

———, D. K. BAILEY, and D. S. SUTHERLAND: Oversaturated Peralkaline Glassy Trachytes from Kenya, *J. Petrol.*, vol. 11, pp. 507–517, 1970.

MACEDO, P. B., and T. A. LITOVITZ: On the Relative Roles of Free Volume and Activation Energy in the Viscosity of Liquids, *J. Chem. Phys.*, vol. 42, pp. 245–256, 1965.

MACGREGOR, A. G.: The Volcanic History and Petrology of Montserrat, with Observations on Mt. Pelée in Martinique; Royal Society Expedition to Montserrat, B.W.I., *Phil. Trans. Roy. Soc.*, ser. B, vol. 229, pp. 1–90, 1938.

MACKENZIE, D. B.: High-temperature Alpine-type Peridotite from Venezuela, *Bull. Geol. Soc. Am.*, vol. 71, pp. 303–318, 1960.

MACLEAN, W. H.: Liquidus Phase Relations in the FeS–FeO–$Fe_3O_4$–$SiO_2$ System and Their Application to Geology, *Econ. Geol.*, vol. 64, pp. 865–884, 1969.

MANTON, W. I.: The Origin of Associated Basic and Acid Rocks in the Lebombo-Nuanetsi Igneous Province, Southern Africa, as Implied by Strontium Isotopes, *J. Petrol.*, vol. 9, pp. 23–39, 1968.

MARSH, B. D., and I. S. E. CARMICHAEL: Benioff Zone Magmatism, *J. Geophys. Res.*, in press.

MARTIGNOLE, J., and K. SCHRIJVER: The Level of Anorthosites and Its Tectonic Pattern, *Tectonophysics*, vol. 10, pp. 403–409, 1970.

MARTIN-KAYE, P. H. A.: A Summary of the Geology of the Lesser Antilles, *Overseas Geol. Min. Res. (G.B.)*, vol. 10, pp. 172–206, 1969.

MASON, B.: Eclogitic Xenoliths from Volcanic Breccia at Kakanu, New Zealand, *Contrib. Mineral. Petrol.*, vol. 19, pp. 316–327, 1968.

MATHEZ, E. A.: Refinement of the Kudo-Weill Plagioclase Thermometer and Its Application to Basaltic Rocks, *Contrib. Mineral. Petrol.*, vol. 41, pp. 61–72, 1973.

MATSUO, S.: On the Origin of Volcanic Gases, *J. Earth Sci.*, Nagoya Univ., vol. 8, pp. 222–245, 1960.

MATUSITA, K., and M. TASHIRO: Rate of Homogeneous Nucleation in Alkali Disilicate Glasses, *J. Non-Crystalline Solids*, vol. 11, pp. 471–484, 1973.

MCBIRNEY, A. R., and K. AOKI: Petrology of the Island of Tahiti, *Geol. Soc. Am. Mem.*, no. 116, pp. 523–556, 1968.

——— and T. MURASE: Factors Governing the Formation of Pyroclastic Rocks, *Bull. Volc.*, vol. 34, pp. 372–384, 1971.

——— and H. WILLIAMS: Volcanic History of Nicaragua, *Univ. Calif. Pub. Geol. Sci.*, vol. 55, pp. 1–65, 1965.

——— and ———: Geology and Petrology of the Galápagos Islands, *Geol. Soc. Am. Mem.*, no. 118, 1969.

MCCALL, G. J. H., R. W. LE MAITRE, A. MALAHOFF, G. P. ROBINSON, and P. J. STEPHENSON: The Geology and Geophysics of the Ambrym Caldera, New Hebrides, *Bull. Volc.*, vol. 34, pp. 681–696, 1971.

MCDOUGALL, I.: Differentiation of the Tasmanian Dolerites: Red Hill Dolerite-Granophyre Association, *Bull. Geol. Soc. Am.*, vol. 73, pp. 279–316, 1962.

———: Potassium-Argon Age Measurements on Dolerites from Antarctica and South Africa, *J. Geophys. Res.*, vol. 68, pp. 1535–1545, 1963.

———: Differentiation of the Great Lake Dolerite Sheet, Tasmania, *J. Geol. Soc. Australia,* vol. 11, pp. 107–132, 1964*a*.

———: Potassium-Argon Ages of Lavas from the Hawaiian Islands, *Bull. Geol. Soc. Am.*, vol. 75, pp. 107–128, 1964*b*.

——— and W. COMPSTON: Strontium Isotope Composition and Potassium Rubidium Ratios in Some Rocks from Réunion and Rodriguez, Indian Ocean, *Nature*, vol. 207, pp. 252–253, 1965.

———, H. A. POLACH, and J. J. STIPP: Excess Radiogenic Argon in Young Sub-aerial Basalts from the Auckland Volcanic Field, New Zealand, *Geochim. Cosmochim. Acta*, vol. 33, pp. 1485–1520, 1969.

——— and N. RUEGG: Potassium-Argon Dates on the Serra Geral Formation of South America, *Geochim. Cosmochim. Acta*, vol. 30, pp. 191–195, 1966.

——— and D. A. SWANSON: Potassium-Argon Ages of Lavas from the Hawi and Pololu Volcanic Series, Kohala Volcano, Hawaii, *Bull. Geol. Soc. Am.*, vol. 83, pp. 3731–3738, 1972.

———, B. G. J. UPTON, and W. J. WADSWORTH: A Geological Reconnaissance of Rodriguez Island, Indian Ocean, *Nature*, vol. 206, pp. 26–27, 1965.

——— and J. G. F. WILKINSON: Potassium-Argon Dates on Some Cainozoic Volcanic Rocks from Northeastern New South Wales, *J. Geol. Soc. Australia*, vol. 14, pp. 225–234, 1967.

MCGREGOR, I. D.: An Hypothesis for the Origin of Kimberlite, *Mineral. Soc. Am. Spec. Paper*, no. 3, pp. 51–62, 1970.

MCKEE, E. H., and D. B. NASH: Potassium-Argon Ages of Granitic Rocks in the Inyo Batholith, East-central California, *Bull. Geol. Soc. Am.*, vol. 78, pp. 669–680, 1967.

MCKENZIE, D. P.: The Influence of the Boundary Conditions and Rotation on Convection in the Earth's Mantle, *Geophys. J.*, vol. 15, p. 457, 1968.

———: Speculations on the Consequences and Causes of Plate Motions, *Geophys. J.*, vol. 18, pp. 1–32, 1969.

MCKERROW, W. S.: The Chronology of Caledonian Folding in the British Isles, *Nat. Acad. Sci. Proc.*, vol. 48, pp. 1905–1913, 1962.

McKie, D.: Fenitization, pp. 261–294 in O. F. Tuttle and J. Gittins (eds.), "Carbonatites," Interscience (Wiley), New York, 1966.
Medaris, L. G.: High-pressure Peridotite in Southwestern Oregon, *Bull. Geol. Soc. Am.*, vol. 83, pp. 41–58, 1972.
——— and R. H. Dott: Mantle-derived Peridotites in Southwestern Oregon: Relation to Plate Tectonics, *Science*, vol. 169, pp. 971–974, 1970.
Mehnert, K. R.: Der Gegenwartige Stand des Granitproblems, *Fortschr. Mineral.*, vol. 37, pp. 117–206, 1959.
———: Petrographie und Abfolge der Granitization im Schwarzwald, *Neues Jahrb. Mineral. Abh.*, vol. 85, pp. 55–140, 1953; vol. 90, pp. 39–90, 1957; vol. 98, pp. 208–249, 1962; vol. 99, pp. 161–199, 1963.
———: "Migmatites and the Origin of Granitic Rocks," Elsevier, Amsterdam, 1968.
Meiling, G. S., and D. R. Uhlmann: Crystallization and Melting Kinetics of Sodium Disilicate, *Physics. Chem. Glasses*, vol. 8, pp. 62–68, 1967.
Melcher, G. C.: O carbonatito de Jacupiranga, *Univ. São Paulo, Brasil, Bol.*, (Geologica no. 21), 1965.
Melson, W. G., E. Jarosewich, V. T. Bowen, and G. Thompson: St. Peter and St. Paul Rocks: A High-temperature Mantle-derived Intrusion, *Science*, vol. 155, pp. 1532–1535, 1967*a*.
———, ———, R. Cifelli, and G. Thompson: Alkali Olivine Basalt Dredged near St. Paul's Rocks, Mid-Atlantic Ridge, *Nature*, vol. 215, pp. 381–382, 1967*b*.
——— and G. Thompson: Layered Basic Complex in Oceanic Crust, Romanche Fracture, *Science*, vol. 168, pp. 817–820, 1970.
Menard, H. W.: Growth of Drifting Volcanoes, *J. Geophys. Res.*, vol. 74, pp. 4827–4837, 1969.
Mercy, E. L. P., and M. J. O'Hara: *Norsk. Geol. Tiddskr.*, vol. 45, pp. 323–332, 1965.
Métais, D., and F. Chayes: Varieties of Lamprophyre, *Ann. Rept. Geophys. Lab.* (*Carnegie Inst. Yrbk. 62*), pp. 156–157, 1963.
Meyer, H. O. A.: Mineral Inclusions in Diamonds, *Ann. Rept. Geophys. Lab.* (*Carnegie Inst. Yrbk. 66*), pp. 446–450, 1968.
——— and F. R. Boyd: Mineral Inclusions in Diamonds, *Ann. Rept. Geophys. Lab.* (*Carnegie Inst. Yrbk. 67*), pp. 130–135, 1969.
——— and ———: Inclusions in Diamonds, *Ann. Rept. Geophys. Lab.* (*Carnegie Inst. Yrbk. 68*), pp. 315–320, 1970.
——— and D. G. Brookins: Eclogite Xenoliths from Stockdale Kimberlite, Kansas, *Contrib. Mineral. Petrol.*, vol. 34, pp. 60–72, 1971.
Michot, J.: Anorthosite et recherche pluridisciplinaire, *Annal. Soc. Géol. Belg.*, vol. 95, pp. 5–43, 1972.
——— and P. Pasteels: La variation du rapport $Sr^{87}/Sr^{86}$ dans les roches génétiquement associées au magma plagioclasique, *Annal. Soc. Géol. Belg.*, vol. 92, pp. 255–262, 1969.
Minear, J. W., and M. N. Toksoz: Thermal Regime of a Downgoing Slab and the New Global Tectonics, *J. Geophys. Res.*, vol. 75, pp. 1397–1420, 1970.
Mitchell, A. H., and H. S. Reading: Continental Margins, Geosynclines, and Ocean Floor Spreading, *J. Geol.*, vol. 77, pp. 629–646, 1969.
———, and A. J. Warden: Geological Evolution of the New Hebrides Island Arc, *J. Geol. Soc. London*, vol. 127, pp. 501–530, 1971.
Miyashiro, A., F. Shido, and M. Ewing: Diversity and Origin of Abyssal Tholeiite

from the Mid-Atlantic Ridge near 24° and 30° North Latitude, *Contrib. Mineral. Petrol.*, vol. 23, pp. 38–52, 1969.

MOORBATH, S., and J. D. BELL: Strontium Isotope Abundance Studies and Rubidium-Strontium Age Determinations on Tertiary Igneous Rocks from the Isle of Skye, Northwest Scotland, *J. Petrol.*, vol. 6, pp. 37–66, 1965.

———, ———, and N. H. GALE: The Significance of Lead Isotope Studies in Ancient High-grade Metamorphic Complexes, as Exemplified by the Lewisian Rocks of Northwest Scotland, *Earth Planet. Sci. Letters*, vol. 6, pp. 245–256, 1969.

——— and G. P. L. WALKER: Strontium Isotope Investigations of Igneous Rocks from Iceland, *Nature*, vol. 207, pp. 837–840, 1965.

——— and H. WELKE: Lead Isotope Studies on Igneous Rocks from the Isle of Skye, Northwest Scotland, *Earth Planet. Sci. Letters*, vol. 5, pp. 217–230, 1968.

MOORE, J. G.: The Quartz-Diorite Boundary Line in the Western United States, *J. Geol.*, vol. 67, pp. 198–210, 1959.

———: Petrology of Deep-sea Basalt near Hawaii, *Am. J. Sci.*, vol. 263, pp. 40–52, 1965.

———: Water Content of Basalt Erupted on the Ocean Floor, *Contrib. Mineral. Petrol.*, vol. 28, pp. 272–279, 1970.

——— and B. W. EVANS: The Role of Olivine in the Crystallization of the Prehistoric Makaopuhi Tholeiitic Lava Lake, Hawaii, *Contrib. Mineral. Petrol.*, vol. 15, pp. 202–223, 1967.

——— and B. P. FABBI: An Estimate of the Juvenile Sulfur Content of Basalt, *Contrib. Mineral. Petrol.*, vol. 33, pp. 118–127, 1971.

——— and J. G. SCHILLING: Vesicles, Water and Sulfur in Reykjanes Ridge Basalts, *Contrib. Mineral. Petrol.*, vol. 41, pp. 105–118, 1973.

———, A. GRANTZ, and M. C. BLAKE: The Quartz-Diorite Line in Northwestern North America, *U.S.G.S. Prof. Paper* 424-C, pp. 87–90, 1961.

MOORE, W. J.: "Physical Chemistry," 3d ed., Prentice-Hall, Englewood Cliffs, N.J., 1965.

MORGAN, B. A. (ed.): Mineralogy and Petrology of the Upper Mantle, *Mineral. Soc. Am. Spec. Paper*, no. 3 (H. H. Hess vol.), pp. 3–121, 1970.

MORGAN, W. J.: Convection Plumes in the Lower Mantle, *Nature*, vol. 230, pp. 42–43, 1971.

———: Deep Mantle Convection Plumes and Plate Motions, *Bull. Am. Assn. Petrol. Geol.*, vol. 56, pp. 203–213, 1972.

MORGAN, W. R.: A Note on the Petrology of Some Lava Types from East New Guinea, *J. Geol. Soc. Australia*, vol. 13, pp. 583–591, 1966.

MUAN, A.: Phase Equilibria at High Temperatures in Oxide Systems Involving Changes in Oxidation States, *Am. J. Sci.*, vol. 256, pp. 171–207, 1958.

——— and E. F. OSBORN: Phase Equilibria at Liquidus Temperatures in the System $MgO–FeO–Fe_2O_3–SiO_2$, *J. Am. Ceram. Soc.*, vol. 39, pp. 121–140, 1956.

MUELLER, R. F.: Energetics of HCl and HF in Volcanic Emanations, *Geochim. Cosmochim. Acta*, vol. 34, pp. 737–744, 1970.

MUIR, I. D.: Crystallisation of Pyroxenes in an Iron-rich Diabase from Minnesota, *Mineral. Mag.*, vol. 30, pp. 376–388, 1954.

———: Basalt Types from the Floor of the Atlantic Ocean, *Geol. Soc. Lond. Proc.*, no. 1626, pp. 141–143, 1965.

——— and C. E. Tilley: Contributions to the Petrology of Hawaiian Basalts: pt. 1, The Picrite-basalts of Kilauea, *Am. J. Sci.*, vol. 255, pp. 241–253, 1957.

——— and ———: Mugearites and Their Place in Alkali Igneous Rock Series, *J. Geol.*, vol. 69, pp. 186–203, 1961.

——— and ———: Contributions to the Petrology of Hawaiian Basalts: pt. 2, The Tholeiitic Basalts of Mauna Loa and Kilauea, *Am. J. Sci.*, vol. 261, pp. 111–128, 1963.

——— and ———: Iron Enrichment and Pyroxene Fractionation Trends in Tholeiites, *J. Geol.*, vol. 4, pp. 143–156, 1964*a*.

——— and ———: Basalts from the Northern Part of the Rift Zone of the Mid-Atlantic Ridge, *J. Petrol.*, vol. 5, pp. 409–434, 1964*b*.

——— and ———: Basalts from the Northern Part of the Mid-Atlantic Ridge, pt. II, *J. Petrol.*, vol. 7, pp. 193–201, 1966.

Muñoz, M.: Ring Complexes of Pájara in Fuerteventura Island, *Bull. Volc.*, vol. 33, pt. 3, pp. 840–861, 1969*a*.

———: Estudio petrológico de las formationes alcalinas de Fuerteventura (Islas Canarias), *Estud. Geol. Inst. Lucas Mallada*, vol. 25, pp. 257–310, 1969*b*.

Munoz, J. L., and H. P. Eugster: Experimental Control of Fluorine Reactions in Hydrothermal Systems, *Am. Mineral.*, vol. 54, pp. 943–959, 1969.

Murase, T.: Viscosity and Related Properties of Volcanic Rocks at 800° to 1400°C, *J. Fac. Sci., Hokkaido Univ.*, ser. 7, vol. 1, pp. 489–584, 1962.

Murata, K. J.: An Acid Fumarolic Gas from Kilauea Iki Hawaii, *U.S.G.S. Prof. Paper* 537-C, pp. 1–6, 1966.

——— and D. H. Richter: The Settling of Olivine in Kilauean Magma as Shown by Lavas of the 1959 Eruption, *Am. J. Sci.*, vol. 254, pp. 194–203, 1966*a*.

——— and ———: Chemistry of the Lavas of the 1959–1960 Eruption of Kilauea Volcano, Hawaii, *U.S.G.S. Prof. Paper* 537-A, 1966*b*.

Murthy, V. R., and C. C. Patterson: Primary Isochron of Zero Age for Meteorites and the Earth, *J. Geophys. Res.*, vol. 67, pp. 1161–1167, 1962.

Mysen, B. O., and K. S. Heier: Petrogenesis of Eclogites in High Grade Metamorphic Gneisses, Exemplified by the Hareidland Eclogite, Western Norway, *Contrib. Mineral. Petrol.*, vol. 36, pp. 73–94, 1972.

Nafziger, R. H.: High-temperature Activity-Composition Relations of Equilibrium Spinels, Olivines and Pyroxenes in the System MgO-Fe-O-$SiO_2$, *Am. Mineral.*, vol. 58, pp. 457–465, 1973.

Nagasawa, H. H., H. Wakita, H. Higuchi, and N. Onuma: Rare Earths in Peridotite Nodules, *Earth Planet. Sci. Letters*, vol. 5, pp. 377–381, 1969.

Nakamura, Y., and I. Kushiro: Compositional Relations of Coexisting Orthopyroxene, Pigeonite and Augite in a Tholeiitic Andesite from Hakone Volcano, *Contrib. Mineral. Petrol.*, vol. 26, pp. 265–275, 1970.

Naldrett, A. J.: A Portion of the System Fe–S–O between 900 and 1080°C and Its Application to Sulphide Ore Magmas, *J. Petrol.*, vol. 10, pp. 171–201, 1969.

——— and G. Kullerud: A Study of the Strathcona Mine and Its Bearing on the Origin of the Nickel-Copper Ores of the Sudbury District, Ontario, *J. Petrol.*, vol. 8, pp. 453–531, 1967.

Nash, W. P.: Apatite Chemistry and Phosphorous Fugacity in a Differentiated Igneous Intrusion, *Am. Mineral.*, vol. 57, pp. 877–886, 1972*a*.

———: Apatite-Calcite Equilibria in Carbonatites: Chemistry of Apatite from Iron Hill, Colorado, *Geochim. Cosmochim. Acta*, vol. 36, pp. 1313–1319, 1972*b*.
———: Mineralogy and Petrology of the Iron Hill Carbonatite Complex, Colorado, *Bull. Geol. Soc. Am.*, vol. 83, pp. 1361–1382, 1972*c*.
———: Apatite Chemistry and Phosphorous Fugacity in a Differentiated Igneous Intrusion: Correction, *Am. Mineral.*, vol. 58, p. 345, 1973.
———, I. S. E. CARMICHAEL, and R. W. JOHNSON: The Mineralogy and Petrology of Mount Suswa, Kenya, *J. Petrol.*, vol. 10, pp. 409–439, 1969.
——— and J. F. G. WILKINSON: Shonkin Sag Laccolith, Montana: pt. I, Mafic Minerals and Estimates of Temperature, Pressure, Oxygen Fugacity and Silica Activity, *Contrib. Mineral. Petrol.*, vol. 25, pp. 241–269, 1970.
——— and ———: Shonkin Sag Laccolith, Montana: pt. II, Bulk Rock Geochemistry, *Contrib. Mineral. Petrol.*, vol. 33, pp. 162–170, 1971.
NATH, P., and R. W. DOUGLAS: $Cr^{3+}$–$Cr^{6+}$ Equilibrium in Binary Alkali Silicate Glasses, *Physics Chem. Glasses*, vol. 6, pp. 197–202, 1965.
NAYLOR, R. S.: Acadian Orogeny: An Abrupt and Brief Event, *Science*, vol. 172, pp. 558–560, 1971.
NEIVA, J. M. C.: Chemisme des roches eruptives des Islas de S. Thomé et Prince, *Proc. 19th Geol. Congr.*, vol. 21, pp. 321–333, 1954.
NESBITT, R. W.: Skeletal Crystal Forms in the Ultramafic Rocks of Yilgarn Block, Western Australia, *Geol. Soc. Australia Spec. Pub.* 3, pp. 331–347, 1971.
NEUMANN, H., J. MEAD, and C. J. VITALIANO: Trace Element Variation during Fractional Crystallization as Calculated from the Distribution Law, *Geochim. Cosmochim. Acta*, vol. 6, pp. 90–99, 1954.
NEWALL, G., and RAST, N.: "Mechanics of Igneous Intrusion," Gallery Press, Liverpool, 1970.
NEWHOUSE, W. H.: Opaque Oxides and Sulfides in Common Igneous Rocks, *Bull. Geol. Soc. Am.*, vol. 47, pp. 1–52, 1936.
NICHOLLS, G. D.: Basalts from the Deep Ocean Floor, *Mineral. Mag.*, vol. 34, pp. 373–388, 1965.
NICHOLLS, I. A.: Petrology of Santorini Volcano, Cyclades, Greece, *J. Petrol.*, vol. 12, pp. 67–120, 1971.
——— and A. E. RINGWOOD: Effect of Water on Olivine Stability in Tholeiites and the Production of Silica-saturated Magmas in the Island-arc Environment, *J. Geol.*, vol. 81, pp. 285–300, 1973.
NICHOLLS, J., and I. S. E. CARMICHAEL: A Commentary on the Absarokite-Shoshonite-Banakite Series of Wyoming, U.S.A., *Schweiz. Min. Petr. Mitt.*, vol. 49, pp. 47–64, 1969*a*.
——— and ———: Peralkaline Acid Liquids: A Petrological Study, *Contrib. Mineral. Petrol.*, vol. 20, pp. 268–294, 1969*b*.
——— and ———: The Equilibration Temperature and Pressure of Various Lava Types with Spinel- and Garnet-Peridotite, *Am. Mineral.*, vol. 57, pp. 941–959, 1972.
———, ———, and J. C. STORMER: Silica Activity and $P_{total}$ in Igneous Rocks, *Contrib. Mineral. Petrol.*, vol. 33, pp. 1–20, 1971.
NIELSON, D. R. and R. E. STOIBER: Relationship of Potassium Content in Andesitic Lavas and Depth to the Seismic Zone, *J. Geophys. Res.*, vol. 78, pp. 6887–6892, 1973.
NIXON, P. H., O. VON KNORRING, and J. M. ROOKE: Kimberlites and Associated

Inclusions of Basutoland: A Mineralogical and Geochemical Study, *Am. Mineral.*, vol. 48, pp. 1090–1132, 1963.

NOBLE, J. A., and H. P. TAYLOR: Correlation of the Ultramafic Complexes of Southeastern Alaska with Those of Other Parts of America and the World, *Proc. 21st Int. Geol. Congr.*, vol. 13, pp. 188–197, 1960.

NOCKOLDS, S. R.: The Garabal Hill–Glen Fyne Igneous Complex, *Q.J. Geol. Soc. Lond.*, vol. 96, pp. 451–511, 1940.

———: Average Chemical Compositions of Some Igneous Rocks, *Bull. Geol. Soc. Am.*, vol. 65, pp. 1007–1032, 1954.

——— and R. ALLEN: The Geochemistry of Some Igneous Rock Series, *Geochim. Cosmochim. Acta*, vol. 4, pp. 105–142, 1953; vol. 5, pp. 245–285, 1954; vol. 9, pp. 34–77, 1956.

NOE-NYGAARD, A.: Chemical Composition of Tholeiitic Basalts from the Wyville–Thompson Ridge Belt, *Nature*, vol. 212, pp. 272–273, 1966.

——— and J. RASMUSSEN: Petrology of a 3,000 Meter Sequence of Basaltic Lavas in the Faroe Islands, *Lithos*, vol. 1, pp. 286–304, 1968.

NORDLIE, B. E.: The Composition of the Magmatic Gas of Kilauea and Its Behavior in the Near Surface Environment, *Am. J. Sci.*, vol. 271, pp. 417–463, 1971.

OFTEDAHL, C.: Studies on the Igneous Rock Complex of the Oslo Region: pt. XIII, The Cauldrons, *Skr. Norske Vid.-Akad. Oslo*, no. 3, 1953.

———: Volcanic Sequence and Magma Formation in the Oslo Region, *Geol. Rundsch.*, vol. 48, pp. 18–26, 1959.

O'HARA, M. J.: Melting of Garnet Peridotite at 30 Kilobars, *Ann. Rept. Geophys. Lab.* (*Carnegie Inst. Yrbk. 62*), pp. 71–76, 1963*a*.

———: Melting of Bimineralic Eclogite at 30 Kilobars, *Ann. Rept. Geophys. Lab.* (*Carnegie Inst. Yrbk. 62*), pp. 76–77, 1963*b*.

———: Primary Magmas and the Origin of Basalts, *Scottish J. Geol.*, vol. 1, pp. 19–40, 1965.

———: The Bearing of Phase Equilibria Studies in Synthetic and Natural Systems on the Origin and Evolution of Basic and Ultrabasic Rocks, *Earth-sci. Rev.*, vol. 4, pp. 69–133, 1968.

———, S. W. RICHARDSON, and G. WILSON: Garnet-Peridotite Stability and Occurrence in Crust and Mantle, *Contrib. Mineral. Petrol.*, vol. 32, pp. 48–68, 1971.

——— and H. S. YODER: Formation and Fractionation of Basic Magmas at High Pressures, *Scottish J. Geol.*, vol. 3, pp. 67–117, 1967.

O'NIONS, R. K., and D. B. CLARKE: Comparative Trace-element Geochemistry of Tertiary Basalts from Baffin Bay, *Earth Planet. Sci. Letters*, vol. 15, pp. 436–446, 1972.

ONUKI, H., and T. TIBA: Notes on Petrochemistry of Ultrabasic Intrusives—Specially, Aluminum Distribution in Co-existing Pyroxenes, *J. Jap. Assn. Min. Petr. Econ. Geol.*, vol. 53, pp. 215–227, 1965.

OSBORN, E. F.: Role of Oxygen Pressure in the Crystallization and Differentiation of Basaltic Magma, *Am. J. Sci.*, vol. 259, pp. 609–647, 1957.

———: Reaction Series for Subalkaline Igneous Rocks Based on Different Oxygen Pressure Conditions, *Am. Mineral.*, vol. 47, pp. 211–226, 1962.

——— and J. F. SCHAIRER: The Ternary System Pseudowollastonite-Akermanite-Gehlenite, *Am. J. Sci.*, vol. 239, pp. 715–763, 1941.

OVERSBY, V. M., and A. EWART: Lead Isotopic Compositions of Tonga-Kermadec

Volcanics and Their Petrogenetic Significance, *Contrib. Mineral. Petrol.*, vol. 37, pp. 181–210, 1972.

OWEN, D. C., and J. D. C. MCCONNELL: Spinodal Behavior in an Alkali Feldspar, *Nature, Phys. Sci.*, vol. 230, pp. 118–119, 1971.

OXBURGH, E. R., and D. L. TURCOTTE: Mid-ocean Ridges and Geotherm Distribution During Mantle Convection, *J. Geophys. Res.*, vol. 73, pp. 2643–2661, 1968.

PAARMA, H.: A New Find of Carbonatite in North Finland, the Sokli Plug, *Lithos*, vol. 3, pp. 129–133, 1970.

PACKHAM, G. A., and D. A. FALVEY: An Hypothesis for the Formation of Marginal Seas in the Western Pacific, *Tectonophysics*, vol. 11, pp. 79–109, 1971.

PAGE, B. M.: Oceanic Crust and Mantle Fragment in Subduction Complex near San Luis Obispo, California, *Bull. Geol. Soc. Am.*, vol. 83, pp. 957–972, 1972.

PAGE, N. J.: Serpentinization at Burro Mountain, California, *Contrib. Mineral. Petrol.*, vol. 14, pp. 321–342, 1967*a*.

———: Serpentinization Considered as a Constant-volume Process: A Discussion, *Am. Mineral.*, vol. 52, pp. 545–549, 1967*b*.

PANKHURST, R. J.: Strontium Isotope Studies Related to Petrogenesis in the Caledonian Basic Igneous Province of N.E. Scotland, *J. Petrol.*, vol. 10, pp. 115–143, 1969.

———: The Geochronology of the Basic Complexes (of North-east Scotland). *Scottish J. Geol.*, vol. 6, pp. 83–107, 1970.

PATTERSON, E. M.: A Petrochemical Study of the Tertiary Lavas of North-east Ireland, *Geochim. Cosmochim. Acta*, vol. 2, pp. 283–299, 1951.

——— and D. J. SWAINE: A Petrochemical Study of Tertiary Tholeiitic basalts: The Middle Lavas of the Antrim Plateau, *Geochim. Cosmochim. Acta*, vol. 8, pp. 173–181, 1955.

PAUL, A., and R. W. DOUGLAS: Ferrous-ferric Equilibrium in Binary Alkali Silicate Glasses, *Physics Chem. Glasses*, vol. 6, pp. 207–211, 1965*a*.

——— and ———: Cerous-ceric Equilibrium in Binary Alkali Borate and Alkali Silicate Glasses, *Physics Chem. Glasses*, vol. 6, pp. 212–215, 1965*b*.

——— and D. LAHIRI: Manganous-manganic Equilibrium in Alkali Borate Glasses, *J. Am. Ceram. Soc.*, vol. 49, pp. 565–567, 1966.

PAUL, D. K.: Strontium Isotope Studies on Ultramafic Inclusions from Dreiser Weiher, Eifel, Germany, *Contrib. Mineral. Petrol.*, vol. 34, pp. 22–28, 1971.

PEACOCK, M. A.: Classification of Igneous Rock Series, *J. Geol.*, vol. 39, pp. 54–67, 1931.

PECK, D. L., T. L. WRIGHT, and J. G. MOORE: Crystallization of Tholeiitic Basalt in Alae Lava Lake, Hawaii, *Bull. Volc.*, vol. 29, pp. 629–656, 1966.

PETERMAN, Z. E., I. S. E. CARMICHAEL, and A. L. SMITH: $Sr^{87}/Sr^{86}$ Ratios of Quaternary Lavas of the Cascade Range, Northern California, *Bull. Geol. Soc. Am.*, vol. 81, pp. 311–318, 1970*a*.

———, ———, and ———: Strontium Isotopes in Quaternary Basalts of South-eastern California, *Earth Planet. Sci. Letters*, vol. 7, pp. 381–384, 1970*b*.

——— and C. E. HEDGE: Related Strontium Isotopic and Chemical Variations in Oceanic Basalts, *Bull. Geol. Soc. Am.*, vol. 82, pp. 493–500, 1971.

———, ———, R. G. COLEMAN, and P. D. SNAVELY: $Sr^{87}/Sr^{86}$ Ratios in Some Eugeosynclinal Sedimentary Rocks and Their Bearing on the Origin of Granitic Rocks in Orogenic Belts, *Earth Planet. Sci. Letters*, vol. 2, pp. 433–439, 1967.

———, G. G. Lowder, and I. S. E. Carmichael: $Sr^{87}/Sr^{86}$ Ratios of the Talasea Series, New Britain, Territory of New Guinea, *Bull. Geol. Soc. Am.*, vol. 81, pp. 39–40, 1970.

Philpotts, A. R.: Origin of Anorthosite-Mangerite Rocks in Southern Quebec, *J. Petrol.*, vol. 7, pp. 1–64, 1966.

———: Origin of Certain Iron-Titanium Oxide and Apatite Rocks, *Econ. Geol.*, vol. 62, pp. 303–315, 1967.

———: Density, Surface Tension and Viscosity of the Immiscible Phase in a Basic, Alkaline Magma, *Lithos*, vol. 5, pp. 1–18, 1972.

Philpotts, J. A., and C. C. Schnetzler: Phenocryst-Matrix Partition Coefficients for K, Rb, Sr, and Ba with Applications to Anorthosite and Basalt Genesis, *Geochim. Cosmochim. Acta*, vol. 34, pp. 307–322, 1970.

Pichamuthu, C. S.: The Precambrian of India, in "The Geologic Systems," Interscience (Wiley), New York, 1967.

Pidgeon, R. T., and W. Compston.: The Age and Origin of the Cooma Granite and its Associated Metamorphic Zones, New South Wales, *J. Petrol.*, vol. 6, pp. 193–222, 1965.

Pitcher, W. S.: The Migmatitic Older Granodiorite of Thorr District, Co. Donegal, *Q.J. Geol. Soc. Lond.*, vol. 108, pp. 413–446, 1953.

———: Ghost Stratigraphy in Intrusive Granites, pp. 123–140 in G. Newall and N. Rast (eds.), "Mechanics of Igneous Intrusion," Gallery Press, Liverpool, 1970.

——— and H. H. Read: The Main Donegal Granite, *Q.J. Geol. Soc. Lond.*, vol. 114, pp. 259–305, 1959.

Piwinskii, A. J., and P. J. Wyllie: Experimental Studies of Igneous Rock Series, *J. Geol.*, vol. 78, pp. 52–76, 1970.

Powell, J. L.: Isotopic Composition of Strontium in Carbonatites and Kimberlites, *Mineral. Soc. India*, I.M.A. vol., pp. 58–66, 1966.

——— and K. Bell: Strontium Isotopic Studies of Alkalic Rocks: Localities from Australia, Spain, and Western United States, *Contrib. Mineral. Petrol.*, vol. 27, pp. 1–10, 1970.

——— and S. E. DeLong: Isotopic Composition of Strontium in Volcanic Rocks from Oahu, Hawaii, *Science*, vol. 153, pp. 1239–1242, 1966.

Powers, H. A.: Composition and Origin of Basaltic Magma of the Hawaiian Islands, *Geochim. Cosmochim. Acta*, vol. 7, pp. 77–107, 1955.

Presnall, D. C.: The Join Forsterite-Diopside-Iron Oxide and Its Bearing on the Crystallization of Basaltic and Ultramafic Magmas, *Am. J. Sci.*, vol. 264, pp. 753–809, 1966.

———: The Geometrical Analysis of Partial Fusion, *Am. J. Sci.*, vol. 267, pp. 1178–1194, 1969.

——— and P. C. Bateman: Fusion Relations in the System $NaAlSi_3O_8$–$CaAl_2Si_2O_8$–$KAlSi_3O_8$–$SiO_2$–$H_2O$ and Generation of Granitic Magmas in the Sierra Nevada Batholith, *Bull. Geol. Soc. Am.*, vol. 84, pp. 3181–3202, 1973.

Prigogine, I., and R. Defay: Treatise of Thermodynamics: vol. 1, Chemical Thermodynamics," Longmans, New York, 1954.

Pushkar, P.: Strontium Isotope Ratios in Volcanic Rocks of Three Island Arc Areas, *J. Geophys. Res.*, vol. 73, pp. 2701–2714, 1968.

Putnam, G. W., and J. T. Alfors: Depth of Intrusion and Age of the Rocky Hill

Stock, Tulare County, California, *Bull. Geol. Soc. Am.*, vol. 76, pp. 357–364, 1965.

——— and ———: Geochemistry and Petrology of the Rocky Hill Stock, Tulare County, California, *Geol. Soc. Am. Spec. Paper*, no. 120, 1969.

PYKE, D. R., A. J. NALDRETT, and O. R. ECKSTRAND: Archaean Ultramafic Flows in Munro Township, Ontario, *Bull. Geol. Soc. Am.*, vol. 84, pp. 955–978, 1973.

QUON, S. H., and E. G. EHLERS: Rocks of the Northern Part of Mid-Atlantic Ridge, *Bull. Geol. Soc. Am.*, vol. 74, pp. 1–8, 1963.

RAGUIN, E.: "Geology of Granite" (translated from the French, "Géologie du Granite," 1957, by E. H. Kranck and P. R. Eakins), Interscience (Wiley), New York, 1965.

RAHMAN, S., and W. S. MACKENZIE: The Crystallization of Ternary Feldspars: A Study from Natural Rocks, *Am. J. Sci.*, vol. 267-A (Schairer vol.), pp. 391–406, 1969.

RALEIGH, C. B., and S. H. KIRBY: Creep in the Upper Mantle, *Mineral. Soc. Am. Spec. Paper*, no. 3, pp. 113–121, 1970.

RAMBERG, H.: Model Studies in Relation to Intrusion of Plutonic Bodies, pp. 261–286 in G. Newall and N. Rast (eds.), "Mechanics of Igneous Intrusion," Gallery Press, Liverpool, 1970.

RAMBERG, I. B.: Gravity Studies of the Fen Complex, Norway, and Their Petrological Significance, *Contrib. Mineral. Petrol.*, vol. 38, pp. 115–134, 1973.

RAMSAY, J. G.: A Camptonitic Suite at Monar, Ross-shire and Inverness-shire, *Geol. Mag.*, vol. 92, pp. 297–309, 1955.

READ, H. H.: A Contemplation of Time in Plutonism, *Q.J. Geol. Soc. Lond.*, vol. 105, pp. 101–156, 1949.

———: "The Granite Controversy," Murby, London, 1957.

REED, J. J.: Spilites, Serpentinites, and Associated Rocks of the Mossburn District, Southland, *Roy. Soc. N.Z. Trans.*, vol. 78, pp. 106–126, 1950.

———: Chemical and Modal Composition of Dunite from Dun Mountain, Nelson, *N.Z. J. Geol. Geophys.*, vol. 2, pp. 916–919, 1959*a*.

———: Soda-metasomatized Argillites Associated with the Nelson Ultramafic Belt, *N.Z. J. Geol. Geophys.*, vol. 2, pp. 905–915, 1959*b*.

REISS, H., and S. W. MAYER: Theory of the Surface Tension of Molten Salts, *J. Chem. Phys.*, vol. 34, pp. 2001–2003, 1961.

RICHARDSON, C.: Petrology of the Galápagos Islands, *B. P. Bishop Mus. Bull.* 180, pp. 45–57, 1933.

RICHARDSON, S. W.: The Relation between a Petrogenic Grid, Facies Series, and the Geothermal Gradient in Metamorphism, *Fortschr. Mineral.*, vol. 47, pp. 65–76, 1970.

RICHEY, J. E., and H. H. THOMAS: The Geology of Ardnamurchan, Northwest Mull, and Coll, *Mem. Geol. Surv. Scot.*, 1930.

RICHTER, D. H., and K. J. MURATA: Xenolithic Nodules in the 1800–1801 Kaupulehu Flow of Hualalai Volcano, *U.S.G.S. Prof. Paper* 424-B, pp. 215–217, 1961.

RICKWOOD, P. C.: Possible Evidence for Regional Chemical Heterogeneity in the Upper Mantle, *Contrib. Mineral. Petrol.*, vol. 24, pp. 354–358, 1969.

———, M. MATHIAS, and J. C. SIEBERT: A Study of Garnets from Eclogite and Peridotite Xenoliths Found in Kimberlite, *Contrib. Mineral. Petrol.*, vol. 19, pp. 271–301, 1968.

RIDLEY, W. I.: The Petrology of Las Canadas Volcanoes, Tenerife, Canary Islands, *Contrib. Mineral. Petrol.*, vol. 26, pp. 124–160, 1970.

RINGWOOD, A. E.: The Principles Governing Trace Element Distribution During Magmatic Crystallization: pt. I, The Influence of Electronegativity, *Geochim. Cosmochim. Acta*, vol. 7, pp. 189–202, 1955*a*.

———: The Principles Governing Trace Element Distribution During Magmatic Crystallization: pt. II, The Rule of Complex Formation, *Geochim. Cosmochim. Acta*, vol. 7, pp. 242–254, 1955*b*.

———: Composition and Evolution of the Upper Mantle, in P. J. Hart (ed.), "The Earth's Crust and Upper Mantle" (*Am. Geophys. Union Monograph* 13), 1969.

——— and D. H. GREEN: An Experimental Investigation of Gabbro-Eclogite Transformation and Some Geophysical Implications, *Tectonophysics*, vol. 3, pp. 383–427, 1966.

——— and J. F. LOVERING: Significance of Pyroxene-Ilmenite Intergrowths among Kimberlite Xenoliths, *Earth Planet. Sci. Letters*, vol. 7, pp. 371–375, 1970.

——— and A. MAJOR: High-pressure Transformations in Pyroxenes, *Earth Planet. Sci. Letters*, vol. 1, pp. 351–357, 1966.

——— and ———: The System $Mg_2SiO_4$–$Fe_2SiO_4$ at High Pressures and Temperatures, *Phys. Earth Planet. Interiors*, vol. 3, pp. 89–108, 1970.

ROBERTS, G. J., and J. P. ROBERTS: An Oxygen Tracer Investigation of the Diffusion of "Water" in Silica Glass, *Physics Chem. Glasses*, vol. 7, pp. 82–89, 1966.

ROBERTSON, J. K., and P. J. WYLLIE: Rock-Water Systems, with Special Reference to the Water-deficient Region, *Am. J. Sci.*, vol. 271, pp. 252–277, 1971.

ROBIE, R. A., and D. R. WALDBAUM: Thermodynamic Properties of Minerals and Related Substances at 298.15°K (25.0°C) and One Atmosphere (1.013 Bars) Pressure and at Higher Temperatures, *U.S.G.S. Bull.* 1259, 1968.

ROBSON, G. R., and J. F. TOMBLIN: "Catalogue of the Active Volcanoes of the World: pt. XX, West Indies," Int. Volc. Assn., 1966.

ROCKETT, T. J., and W. R. FOSTER: The Thermal Stability of Purified Tridymite, *Am. Mineral.*, vol. 52, pp. 1233–1240, 1967.

RODGERS, J.: The Taconic Orogeny, *Bull. Geol. Soc. Am.*, vol. 82, pp. 1141–1178, 1971.

RODGERS, K. A., and R. N. BROTHERS: Olivine, Pyroxene, Feldspar, and Spinel in Ultramafic Nodules from Auckland, New Zealand, *Mineral. Mag.*, vol. 37, pp. 375–390, 1969.

ROEDDER, E.: Low-temperature Liquid Immiscibility in the System $K_2O$–FeO–$Al_2O_3$–$SiO_2$, *Am. Mineral.*, vol. 36, pp. 282–286, 1951.

———: Liquid $CO_2$ Inclusions in Olivine-bearing Nodules and Phenocrysts from Basalts, *Am. Mineral.*, vol. 50, pp. 1746–1782, 1965.

———: Fluid Inclusions as Samples of Ore Fluids, in H. L. Barnes (ed.), "Geochemistry of Hydrothermal Ore Deposits," Holt, New York, 1967.

——— and D. S. COOMBS: Immiscibility in Granitic Melts, Indicated by Fluid Inclusions in Ejected Granitic Blocks from Ascension Island, *J. Petrol.*, vol. 8, pp. 417–451, 1967.

ROMEY, W. D.: An Evaluation of Some "Differences" between Anorthosites in Massifs and in Layered Complexes, *Lithos*, vol. 1, pp. 230–241, 1968.

ROSENBUSCH, H.: Über des Wesen der körnigen und porphyrischen Struktur bei Massengestein, *Neues Jahrb.*, vol. 2, pp. 1–7, 1882.

Ross, C. S.: Volatiles in Volcanic Glasses and Their Stability Relations, *Am. Mineral.*, vol. 49, pp. 258–271, 1964.

———, M. D. Foster, and A. T. Meyers: Origin of Dunites and of Olivine-rich Inclusions in Basaltic Rocks, *Am. Mineral.*, vol. 39, pp. 693–737, 1954.

Rothe, P., and H. U. Schmincke: Contrasting Origins of the Eastern and Western Islands of the Canarian Archipelago, *Nature*, vol. 218, pp. 1152–1154, 1968.

Rothstein, A. T. V.: The Dawros Peridotite, Connemara, Eire, *Q.J. Geol. Soc. Lond.*, vol. 113, pp. 1–25, 1957.

———: Pyroxenes from the Dawros Peridotite and Some Comments on Their Nature, *Geol. Mag.*, vol. 95, pp. 456–462, 1958.

Ruckmick, J. C., and J. A. Noble: Origin of the Ultramafic Complex at Union Bay, Southeastern Alaska, *Bull. Geol. Soc. Am.*, vol. 70, pp. 981–1018, 1959.

Sahama, T. G.: Kalsilite in the Lavas of Mt. Nyiragongo (Belgian Congo), *J. Petrol.*, vol. 1, pp. 146–171, 1960.

——— and A. Meyer: A Study of the Volcano Nyirangongo; Progress Report, *Exploration du Parc National Albert: Mission d'études vulcanologiques*, fasc. 2, 1958.

Sakka, S., and J. D. Mackenzie: Relation between Apparent Glass Transition Temperature and Liquidus Temperature for Inorganic Glasses, *J. Non-Crystalline Solids*, vol. 6, pp. 145–162, 1971.

Sato, M.: Half-cell Potentials of Semi-conductive Simple Binary Sulphides in Aqueous Solution, *Electrochim. Acta*, vol. 11, pp. 361–373, 1966.

——— and T. L. Wright: Oxygen Fugacities Directly Measured in Volcanic Gases, *Science*, vol. 153, pp. 1103–1105, 1966.

Savolahti, A.: The Ahvenisto Massif in Finland, *Bull. Comm. Geol. Finland*, no. 174, 1956.

———: The Differentiation of Gabbro-Anorthosite Intrusions and the Formation of Anorthosites, *Comptes. rend. Soc. géol. Finlande*, no. 38, pp. 173–197, 1966.

Scarfe, C. M: Viscosity of Basaltic Magmas at Varying Pressures, *Nature*, vol. 241, pp. 101–102, 1973.

Schairer, J. F.: Melting Relations of the Common Rock-forming Oxides, *J. Am. Ceram. Soc.*, vol. 40, pp. 215–235, 1957.

——— and N. L. Bowen: The System Anorthite-Leucite-Silica, *Bull. Comm. Geol. Finland*, vol. 20, pp. 67–87, 1947.

——— and ———: The System $Na_2O$–$Al_2O_3$–$SiO_2$, *Am. J. Sci.*, vol. 254, pp. 129–195, 1956.

Schilling, J. G., and J. W. Winchester: Rare Earth Contribution to the Origin of Hawaiian Lavas, *Contrib. Mineral. Petrol.*, vol. 23, pp. 27–37, 1969.

Schmidt, R. G.: Geology of Saipan, Mariana Islands: chap. B, Petrology of the Volcanic Rocks, *U.S.G.S. Paper* 280-B, pp. 127–175, 1957.

Schmincke, H. U.: Cone Sheet Swarm, Resurgence of Tejeda Caldera, and Early Geologic History of Gran Canaria, *Bull. Volc.*, vol. 31, pp. 153–162, 1967.

Schneider, A.: The Sulfur Isotope Composition of Basaltic Rocks, *Contrib. Mineral. Petrol.*, vol. 25, pp. 95–124, 1970.

Schnetzler, C. C., and J. A. Philpotts: Partition Coefficients of Rare-earth Elements between Igneous Matrix Material and Rock-forming Mineral Phenocrysts, pt. II, *Geochim. Cosmochim. Acta*, vol. 34, pp. 331–340, 1970.

SCHOLTZ, D. L.: The Magmatic Nickeliferous Ore Deposits of East Griqualand and Pondoland, *Trans. Geol. Soc. S. Africa*, vol. 39, pp. 81–210, 1936.
SCHUBERT, G., and D. L. TURCOTTE: Phase Changes and Mantle Convection, *J. Geophys. Res.*, vol. 76, pp. 1424–1432, 1971.
SCHÜRMANN, H.: Petrographische Untersuchung der Gleesite des Laacher-See-Gebietes, *Beitr. Mineral. Petrogr.*, vol. 7, pp. 104–136, 1960.
SCLAR, C. B.: High Pressure Studies in the System $MgO–SiO_2–H_2O$, *Phys. Earth Planet. Interiors*, vol. 3, p. 333, 1970.
SEARLE, E. J.: Petrochemistry of the Auckland Basalts, *N.Z. J. Geol. Geophys.*, vol. 3, pp. 23–40, 1960.
———: The Petrology of the Auckland Basalts, *N.Z. J. Geol. Geophys.*, vol. 4, pp. 165–204, 1961.
SEKI, Y., and G. C. KENNEDY: The Breakdown of Potassium Feldspar, $KAlSi_3O_8$, at High Temperatures and High Pressures, *Am. Mineral.*, vol. 49, pp. 1688–1706, 1964.
SETHNA, S. F.: A Note on the Trace Element Content of Carbonatites of Amba Dongar, *J. Geol. Soc. India*, vol. 12, pp. 311–317, 1971.
SHAND, S. J.: "The Eruptive Rocks," Wiley, New York, 1927.
———: The Lavas of Mauritius, *Q.J. Geol. Soc. Lond.*, vol. 89, pp. 1–13, 1933.
———: "Eruptive Rocks," 2d ed., Wiley, New York, 1943.
SHARP, W. E.: Pyrrhotite: A Common Inclusion in South African Diamonds, *Nature*, vol. 211, pp. 402–403, 1966.
SHAW, D. M.: The Camouflage Principle of Trace Element Distribution in Magmatic Minerals, *J. Geol.*, vol. 61, pp. 142–151, 1953.
———: Trace Element Fractionation During Anatexis, *Geochim. Cosmochim. Acta*, vol. 34, pp. 237–243, 1970.
SHAW, H. R.: Obsidian–$H_2O$ Viscosities at 1000 and 2000 Bars in the Temperature Range 700° to 900°C, *J. Geophys. Res.*, vol. 68, pp. 6337–6343, 1963.
———: Theoretical Solubility of $H_2O$ in Silicate Melts: Quasicrystalline Models, *J. Geol.*, vol. 72, pp. 601–617, 1964.
———: Comments on Viscosity, Crystal Settling, and Convection in Granitic Magmas, *Am. J. Sci.*, vol. 263, pp. 120–152, 1965.
———: Field Determination of Viscosity in Basaltic Magma, in Geological Survey Research 1966, chap. A, *U.S.G.S. Prof. Paper* 550, p. A158, 1966.
———: Chemical States of $H_2O$ and Reaction in Silicate-$H_2O$ Liquids and Glasses, *1968 Ann. Mtg. Geol. Soc. Am.*, pp. 274–275, 1968.
———: Rheology of Basalt in the Melting Range, *J. Petrol.*, vol. 10, pp. 510–535, 1969.
———: Viscosities of Magmatic Silicate Liquids: An Empirical Method of Prediction, *Am. J. Sci.*, vol. 272, pp. 870–893, 1972.
———: Mantle Convection and Volcanic Periodicity in the Pacific: Evidence from Hawaii, *Bull. Geol. Soc. Am.*, vol. 84, pp. 1505–1526, 1973.
———, D. L. PECK, T. L. WRIGHT, and R. OKAMURA: The Viscosity of Basaltic Magma: An Analysis of Field Measurements in Makaopuhi Lava Lake, Hawaii, *Am. J. Sci.*, vol. 266, pp. 225–264, 1968.
SHEPHERD, E. S.: The Gases in Rocks and Related Problems, *Am. J. Sci.*, vol. 35A, pp. 311–351, 1938.

SHIEH, Y. N., and H. P. TAYLOR, JR.: Oxygen and Hydrogen Isotope Studies of Contact Metamorphism, *Contrib. Mineral. Petrol.*, vol. 20, pp. 306–356, 1969.

SHIRAKI, K.: Metamorphic Basement Rocks of Yap Islands, Western Pacific: Possible Oceanic Crust beneath an Island Arc, *Earth. Planet. Sci. Letters*, vol. 13, pp. 167–174, 1971.

SIGURDSSON, H., J. F. TOMBLIN, G. M. BROWN, J. G. HOLLAND, and R. J. ARCULUS: Strongly Undersaturated Magmas in the Lesser Antilles Island Arc, *Earth Planet. Sci. Letters*, vol. 18, pp. 285–295, 1973.

SIMHA, R., and R. F. BOYER: On a General Relation Involving the Glass Temperature and Coefficients of Expansion of Polymers, *J. Chem. Phys.*, vol. 37, pp. 1003–1007, 1962.

SIMKIN, T.: Flow Differentiation in the Picritic Sills of North Skye, in P. J. Wyllie (ed.), "Ultramafic and Related Rocks," Wiley, New York, 1967.

——— and J. V. SMITH: Minor-element Distribution in Olivine, *J. Geol.*, vol. 78, pp. 304–325, 1970.

SIMMONS, G.: Gravity Survey and Geological Interpretation, Northern New York, *Bull. Geol. Soc. Am.*, vol. 75, pp. 81–98, 1964.

SIMPSON, E. S. W.: On the Graphical Representation of Differentiation Trends in Igneous Rocks, *Geol. Mag.*, vol. 91, pp. 238–244, 1954.

SKINNER, B. J.: Thermal Expansion, in S. P. Clark (ed.), "Handbook of Physical Constants," (*Geol. Soc. Am. Mem.*, no. 97), pp. 75–96, 1966.

——— and D. L. PECK: An Immiscible Sulfide Melt from Hawaii, *Econ. Geol. Monograph*, no. 4, pp. 310–322, 1969.

SLAWSON, W. F., and D. R. RUSSELL: Common Lead Isotope Abundances, in H. L. Barnes (ed.), "Geochemistry of Hydrothermal Ore Deposits," Holt, New York, 1967.

SMITH, A. L., and I. S. E. CARMICHAEL: Quaternary Lavas from the Southern Cascades, Western U.S.A., *Contrib. Mineral. Petrol.*, vol. 19, pp. 212–238, 1968.

——— and ———: Quaternary Trachybasalts from Southeastern California, *Am. Mineral.*, vol. 54, pp. 909–923, 1969.

SMITH, C. H., and H. E. KAPP: The Muskox Intrusion, *Mineral. Soc. Am. Spec. Paper*, no. 1, pp. 30–35, 1963.

SMITH, D., and D. H. LINDSLEY: Stable and Metastable Augite Crystallization Trends in a Single Basalt Flow, *Am. Mineral.*, vol. 56, pp. 225–233, 1971.

SMITH, J. V., and W. S. MACKENZIE: The Alkali Feldspars: pt. IV, The Cooling History of High-temperature Sodium-rich Feldspars, *Am. Mineral.*, vol. 43, pp. 872–889, 1958.

SMITH, W. C., and C. BURRI: The Igneous Rocks of Fernando de Noronha, *Schweiz. Mineral. Petrogr. Mitteil.*, vol. 13, pp. 405–434, 1933.

——— and L. J. CHUBB: The Petrography of the Austral or Tubuai Islands, *Q.J. Geol. Soc. Lond.*, vol. 83, pp. 317–341, 1927.

SOBOLEV, N. V., I. K. KUZNETSOVA, and N. I. ZYUZIN: The Petrology of the Grospydite Xenoliths from the Zagadochnaya Kimberlite Pipe in Yakutia, *J. Petrol.*, vol. 9, pp. 253–280, 1968.

SÖRENSEN, H.: Rhythmic Igneous Layering in Peralkaline Intrusions, *Lithos*, vol. 2, pp. 261–283, 1969.

SPEIDEL, D. H.: Effect of Magnesium on the Iron-Titanium Oxides, *Am. J. Sci.*, vol. 268, pp. 341–353, 1970.

SPERA, F. J.: A Thermodynamic Basis for Predicting Water Solubilities in Silicate Melts and Implications for the Low Velocity Zone, *Contrib. Mineral. Petrol.* (in press).

SPOONER, C. M.: Initial Ratios and Whole Rock Ages of Pyroxene Granulites, *M.I.T.—1381—17th Ann. Prog. Rept. for 1969*, pp. 45–93, 1969.

SPRY, A.: Flow Structure and Laminar Flow in Bostonite Dykes at Armidale, New South Wales, *Geol. Mag.*, vol. 90, pp. 248–256, 1953.

STANTON, R. L., and J. D. BELL: Volcanic and Associated Rocks of the New Georgia Group, British Islands Protectorate, *Overseas Geol. Min. Res. (G.B.)*, vol. 10, pp. 113–145, 1969.

STARK, J. T., and R. L. HAY: Geology and Petrography of the Volcanic Rocks of the Truk Islands, East Caroline Islands, *U.S.G.S. Prof. Paper* 409, 1963.

STAUDER, W.: Mechanism and Spatial Distribution of Chilean Earthquakes with Relation to Subduction of the Oceanic Plate, *J. Geophys. Res.*, vol. 78, pp. 5033–5061, 1973.

STEARNS, H. T.: "Geology of the State of Hawaii," Pacific Books, Palo Alto, Calif., 1966.

STEINER, A.: Petrogenetic Implications of the 1954 Ngauruhoe Lava and Its Xenoliths, *N.Z. J. Geol. Geophys.*, vol. 1, pp. 325–363, 1958.

STEVENS, N. C.: The Volcanic Rocks of the Southern Part of the Main Range, Southeast Queensland, *Proc. Roy. Soc. Queensland*, vol. 77, pp. 37–52, 1965.

STEWART, D. B., and E. H. ROSEBOOM, JR.: Low-temperature Terminations of the Three-phase Region Plagioclase Alkali-Feldspar Liquid, *J. Petrol.*, vol. 3, pp. 280–315, 1962.

STEWART, F. H.: The "Younger" Basic Igneous Complexes of North-east Scotland, and Their Metamorphic Envelope: Introduction, *Scottish. J. Geol.*, vol. 6, pp. 3–6, 1970.

STILLMAN, C. J.: Structure and Evolution of the Northern Ring Complex, Nuanetsi Igneous Province, Rhodesia, pp. 33–48 in G. Newall and N. Rast (eds.), "Mechanics of Igneous Intrusion," Gallery Press, Liverpool, 1970.

STOIBER, R. E., and W. I. ROSE, JR.: The Geochemistry of Central American Volcanic Gas Condensates, *Bull. Geol. Soc. Am.*, vol. 81, pp. 2891–2912, 1970.

STORMER, J. C., and I. S. E. CARMICHAEL: Villiaumite and the Occurrence of Fluoride Minerals in Igneous Rocks, *Am. Mineral.*, vol. 55, pp. 126–134, 1970*a*.

——— and ———: The Kudo-Weill Plagioclase Geothermometer and Porphyritic Acid Glasses, *Contrib. Mineral. Petrol.*, vol. 28, pp. 306–309, 1970*b*.

——— and ———: The Free-energy of Sodalite and the Behavior of Chloride, Fluoride and Sulphate in Silicate Magmas, *Am. Mineral.*, vol. 56, pp. 292–306, 1971*a*.

——— and ———: Fluorine-Hydroxyl Exchange in Apatite and Biotite: A Potential Igneous Geothermometer, *Contrib. Mineral. Petrol.*, vol. 31, pp. 121–131, 1971*b*.

STUEBER, A. M., and V. R. MURTHY: Strontium Isotope and Alkali Element Abundances in Ultramafic Rocks, *Geochim. Cosmochim. Acta*, vol. 30, pp. 1243–1259, 1966.

STULL, D. R., and H. PROPHET: JANAF Thermochemical Tables (2d ed.), *Nat. Stand. Ref. Data Ser., U.S. Nat. Bur. Stand.* 37, 1971.

SUBRAMANIAM, A. P.: Charnockites of the Type Area near Madras, a Reinterpretation, *Am. J. Sci.*, vol. 257, pp. 321–353, 1959.

SUGIMURA, A., T. MATSUDA, K. CHINZEI, and K. NAKAMURA: Quantitative Distribution of Late Cenozoic Volcanic Materials in Japan, *Bull. Volc.*, vol. 26, pp. 125–140, 1963.

SUKHESWALA, R. N., and R. K. AVASIA: Carbonatite-Alkalic Complex of Panwad-Kawant, Gujarat, and Its Bearing on the Structural Characteristics of the Area, *Bull. Volc.*, vol. 35, pp. 564–577, 1971.

——— and A. POLDERVAART: Deccan Basalts of the Bombay Area, India, *Bull. Geol. Soc. Am.*, vol. 69, pp. 1475–1494, 1958.

SUTHERLAND, D. S.: Potash-Trachytes and Ultra-potassic Rocks Associated with the Carbonatite Complex of Toror Hills, Uganda, *Mineral. Mag.*, vol. 35, pp. 363–378, 1965.

———: A Note on the Occurrence of Potassium-rich Trachytes in the Kaiserstuhl Carbonatite Complex, West Germany, *Mineral. Mag.*, vol. 36, pp. 334–341, 1967.

———: Pantelleritic Volcanism in the Naivasha Area of Kenya, *Proc. Geol. Soc. Lond.*, no. 1663, pp. 161–162, 1970.

SWALIN, R. A.: "Thermodynamics of Solids," Wiley, New York, 1962.

SWANSON, D. A.: Yakima Basalt of the Teton River Area, South-central Washington, *Bull. Geol. Soc. Am.*, vol. 78, pp. 1077–1110, 1967.

TALBOT, J. L., B. E. HOBBS, H. G. WILSHIRE and T. R. SWEATMAN: Xenoliths and Xenocrysts from Lavas of the Kerguelen Archipelago, *Am. Mineral.*, vol. 48, pp. 159–179, 1963.

TATSUMOTO, M.: Genetic Relations of Oceanic Basalts as Indicated by Lead Isotopes, *Science*, vol. 153, pp. 1094–1101, 1966*a*.

———: Isotopic Composition of Lead in Volcanic Rocks from Hawaii, Iwo-Jima and Japan, *J. Geophys. Res.*, vol. 71, pp. 1721–1723, 1966*b*.

———, C. E. HEDGE, and A. E. J. ENGEL: Potassium, Rubidium, Strontium, Thorium and Uranium in Oceanic Tholeiitic Basalt, *Science*, vol. 150, pp. 886–888, 1965.

——— and R. J. KNIGHT: Isotopic Composition of Lead in Volcanic Rocks from Central Honshu—With Regard to Basalt Genesis, *Geochim. J. Japan*, vol. 3, pp. 53–86, 1969.

——— and P. D. SNAVELY: Isotopic Composition of Lead in Rocks of the Coast Ranges, Oregon and Washington, *J. Geophys. Res.*, vol. 74, pp. 1087–1110, 1969.

TAYLOR, G. A.: The 1951 Eruption of Mt. Lamington, Papua, *Bur. Min. Res.* (*Australia*) *Bull.* 38, 1958.

TAYLOR, H. P.: $O^{18}/O^{16}$ Ratios in Rocks and Coexisting Minerals of the Skaergaard Intrusion, East Greenland, *J. Petrol.*, vol. 4, pp. 51–74, 1963.

———: Oxygen Isotope Studies of Hydrothermal Mineral Deposits, in H. L. Barnes (ed.), "Geochemistry of Hydrothermal Ore Deposits," Holt, New York, 1967*a*.

———: Stable Isotope Studies of Ultramafic Rocks and Meteorites, pp. 362–372 in P. J. Wyllie (ed.), "Ultramafic and Related Rocks," Wiley, New York, 1967*b*.

———: The Oxygen Isotope Geochemistry of Igneous Rocks, *Contrib. Mineral. Petrol.*, vol. 19, pp. 1–71, 1968.

——— and S. EPSTEIN: Hydrogen-isotope Evidence for the Influx of Meteoric Ground-water into Shallow Igneous Intrusions, *1968 Ann. Mtg. Geol. Soc. Am.*, p. 294, 1968.

———, J. FRECHEN, and E. T. DEGENS: Oxygen and Carbon Isotope Studies of Carbonatites from the Laacher See District, West Germany and the Alnö District, Sweden, *Geochim. Cosmochim. Acta*, vol. 31, pp. 407–430, 1967.

——— and J. A. NOBLE: Origin of the Ultramafic Complexes in Southeastern Alaska, *Proc. 21st Int. Geol. Congr.*, vol. 13, pp. 175–187, 1960.

TAYLOR, R. W.: Phase Equilibrium in the System FeO–$Fe_2O_3$–$TiO_2$ at 1300°C, *Am. Mineral.*, vol. 49, pp. 1016–1030, 1964.

TAYLOR, S. R., A. C. CAPP, A. L. GRAHAM, and D. H. BLAKE: Trace Element Abundances in Andesites: pt. II, Saipan, Bougainville and Fiji, *Contrib. Mineral. Petrol.*, vol. 23, pp. 1–26, 1969*a*.

———, M. KAYE, A. J. R. WHITE, A. R. DUNCAN, and A. EWART: Genetic Significance of Co, Cr, Ni, Sc and V Content of Andesites, *Geochim. Cosmochim. Acta*, vol. 33, pp. 275–286, 1969*b*.

THAYER, T. P.: Flow Layering in Alpine Peridotite-Gabbro Complexes, *Mineral. Soc. Am. Spec. Paper*, no. 1, pp. 55–61, 1963.

———: Serpentinization Considered as a Constant-volume Metasomatic Process, *Am. Mineral.*, vol. 51, pp. 685–710, 1966.

———: Serpentinization Considered as a Constant-volume Process: A Reply, *Am. Mineral.*, vol. 52, pp. 549–553, 1967.

THOMPSON, B. N., L. O. KERMODE, and A. EWART: "New Zealand Volcanology, Central Volcanic Region," *N.Z. Geol. Surv. Handbook, Inf. Ser. 50*, 1965.

THOMPSON, J. B., JR.: Thermodynamic Properties of Simple Solutions, pp. 340–361 in P. H. Abelson (ed.), "Researches in Geochemistry II," Wiley, New York, 1967.

THOMPSON, R. N., and W. S. MACKENZIE: Feldspar-Liquid Equilibria in Peralkaline Acid Liquids: An Experimental Study, *Am. J. Sci.*, vol. 265, pp. 714–734, 1967.

———, J. ESSON, and A. C. DUNHAM: Major Elemental Chemical Variation in Eocene Lavas of the Isle of Skye, Scotland, *J. Petrol.*, vol. 13, pp. 219–253, 1972.

THORNTON, C. P., and O. F. TUTTLE: Chemistry of Igneous Rocks: pt. I, Differentiation Index, *Am. J. Sci.*, vol. 258, pp. 664–684, 1960.

TICKLE, R. E.: The Electrical Conductance of Molten Alkali Silicates: pt. 2, Theoretical Discussion, *Physics Chem. Glasses*, vol. 8, pp. 113–124, 1967.

TILLEY, C. E.: An Alkali Facies of Granite at Granite-Dolomite Contacts in Skye, *Geol. Mag.*, vol. 86, pp. 81–93, 1949.

———: Some Aspects of Magmatic Evolution, *Q.J. Geol. Soc. Lond.*, vol. 106, pp. 37–61, 1950.

———: A Note on the Pitchstones of Arran, *Geol. Mag.*, vol. 94, pp. 329–333, 1957.

———: A Note on the Nosean Phonolite of the Wolf Rock, Cornwall, *Geol. Mag.*, vol. 96, pp. 503–504, 1959.

——— and J. GITTINS: Igneous Nepheline-bearing Rocks of the Haliburton-Bancroft Province of Ontario, *J. Petrol.*, vol. 2, pp. 38–48, 1961.

——— and I. D. MUIR: The Hebridean Plateau Magma Type, *Edinburgh Geol. Soc. Trans.*, vol. 19, pp. 208–215, 1962.

——— and ———: Tholeiite and Tholeiitic Series, *Geol. Mag.*, vol. 104, pp. 337–343, 1967.

——— and H. S. YODER: Pyroxene Fractionation in Mafic Magma at High Pressures and Its Bearing on Basalt Genesis, *Ann. Rept. Geophys. Lab.* (*Carnegie Inst. Yrbk. 63*), pp. 114–121, 1964.

———, ———, and J. F. SCHAIRER: Melting Relations of Volcanic Rock Series, *Ann. Rept. Geophys. Lab.* (*Carnegie Inst. Yrbk. 65*), pp. 260–269, 1967.

———, ———, and ———: Melting Relations of Igneous Rock Series, *Ann. Rept. Geophys. Lab.* (*Carnegie Inst. Yrbk. 66*), pp. 450–457, 1968.

TILLING, R. I., and D. GOTTFRIED: Distribution of Thorium, Uranium and Potassium in Igneous Rocks of the Boulder Batholith Region, Montana, *U.S.G.S. Prof. Paper* 614-E, 1969.

———, M. R. KLEPPER, and J. D. OBRADOVICH: K–Ar Ages and Time Span of Emplacement of the Boulder Batholith, Montana, *Am. J. Sci.*, vol. 266, pp. 671–689, 1968.

TOMKEIEFF, S. I.: The Tertiary Lavas of Rum, *Geol. Mag.*, vol. 79, pp. 1–13, 1942.

TOULMIN, P.: Geological Significance of Lead-alpha and Isotopic Age Determination of "Alkalic" Rocks of New England, *Bull. Geol. Soc. Am.*, vol. 72, pp. 775–780, 1961.

——— and P. B. BARTON, JR.: A Thermodynamic Study of Pyrite and Pyrrhotite, *Geochim. Cosmochim. Acta*, vol. 28, pp. 641–671, 1964.

TURCOTTE, D. L., and E. R. OXBURGH: Convection in a Mantle with Variable Physical Properties, *J. Geophys. Res.*, vol. 74, pp. 1458–1474, 1969.

TURI, B., and H. P. TAYLOR: An Oxygen and Hydrogen Isotopic Study of a Granodiorite Pluton from the Southern California Batholith, *Geochim. Cosmochim. Acta*, vol. 35, pp. 383–406, 1971.

TURNBULL, D., and J. C. FISHER: Rate of Nucleation in Condensed Systems, *J. Chem. Phys.*, vol. 17, pp. 71–73, 1949.

TURNER, D. C.: Ring Structures in the Sara-Fier Younger Granite Complex, Northern Nigeria, *Q.J. Geol. Soc. Lond.*, vol. 119, pp. 345–366, 1963.

TURNER, F. J.: "Metamorphic Petrology," McGraw-Hill, New York, 1968.

———: Uniqueness versus Conformity to Pattern in Petrogenesis, *Am. Mineral.*, vol. 55, pp. 339–348, 1970.

——— and J. VERHOOGEN: "Igneous and Metamorphic Petrology," McGraw-Hill, New York, 1951.

——— and———: "Igneous and Metamorphic Petrology," 2d ed., McGraw-Hill, New York, 1960.

——— and L. E. WEISS: "Structural Analyses of Metamorphic Tectonites," McGraw-Hill, New York, 1963.

TUTTLE, O. F.: The Variable Inversion Temperature of Quartz as a Possible Geologic Thermometer, *Am. Mineral.*, vol. 34, pp. 723–730, 1949.

———: Origin of the Contrasting Mineralogy of Extrusive and Plutonic Salic Rocks, *J. Geol.*, vol. 60, pp. 107–124, 1952.

——— and N. L. BOWEN: Origin of Granite in the Light of Experimental Studies in the System $NaAlSi_3O_8$–$KAlSi_3O_8$–$SiO_2$–$H_2O$, *Geol. Soc. Am. Mem.*, no. 74, 1958.

——— and J. GITTINS: "Carbonatites," Interscience (Wiley), New York, 1966.

——— and J. V. SMITH: The Nepheline-Kalsilite System: pt. II, Phase Relations, *Am. J. Sci.*, vol. 256, pp. 571–589, 1958.

TYLER, R. C., and B. C. KING: The Pyroxenes of the Alkaline Igneous Complexes of Eastern Uganda, *Mineral. Mag.*, vol. 36, pp. 5–21, 1967.

TYRRELL, G. W.: "Principles of Petrology," Methuen, London, 1926*a*.

———: The Petrography of Jan Mayen, *Roy. Soc. Edin. Trans.*, vol. 54, pt. 3, pp. 747–765, 1926*b*.

———: The Petrology of Heard Island, *Rept. B.A.N.Z., Antarctic Expedition (D. Mawson) 1929–1931*, ser. A., vol. 2, pt. 3, 1937*a*.

———: The Petrology of Possession Island, *Rept. B.A.N.Z., Antarctic Expedition (D. Mawson) 1929–1931*, ser. A. vol. 2, pt. 4, 1937*b*.

———: Flood Basalts and Fissure Eruptions, *Bull. Volc.*, vol. 1, pp. 89–111, 1937c.

UPTON, B. G. J., and W. J. WADSWORTH: Geology of Réunion Island, Indian Ocean, *Nature*, vol. 207, pp. 151–154, 1965.

———and T. C. NEWMAN: The Petrology of Rodriguez Island, Indian Ocean, *Bull. Geol. Soc. Am.*, vol. 78, pp. 1495–1506, 1967.

UTSU, T.: Seismological Evidence for Anomalous Structure of Island Arcs with Special Reference to the Japanese Region, *Rev. Geophys. Space Phys.*, vol. 9, pp. 859–890, 1971.

VALLANCE, T. G.: Concerning Spilites, *Linnean Soc. New South Wales Proc.*, vol. 85, pp. 8–52, 1960.

———: Spilites Again: Some Consequences of the Degradation of Basalts, *Linnean Soc. New South Wales Proc.*, vol. 94, pp. 8–50, 1969.

VARNE, R.: The Petrology of Moroto Mountain, Eastern Uganda, and the Origin of Nephelinites, *J. Petrol.*, vol. 9, pp. 169–190, 1968.

VERHOOGEN, J.: Les eruptions 1938–1940 du volcan Nyamuragira, *Exploration du Parc National Albert: Mission J. Verhoogen (1938 and 1940)*, fasc. 1, 1948.

———: Thermodynamics of a Magmatic Gas Phase, *Univ. Calif. Bull. Dept. Geol. Sci.*, vol. 28, pp. 91–136, 1949.

———: Distribution of Titanium between Silicates and Oxides in Igneous Rocks, *Am. J. Sci.*, vol. 260, pp. 211–220, 1962.

———, F. J. TURNER, L. E. WEISS, C. WAHRHAFTIG, and W. S. FYFE: "The Earth: An Introduction to Physical Geology," Holt, Rinehart, and Winston, New York, 1970.

VILJOEN, M. J., and R. P. VILJOEN: Evidence of the Existence of a Mobile Extrusive Peridotite Magma, *Upper Mantle Project* (*Geol. Soc. S. Africa Spec. Pub.* 2), pp. 87–112, 1969.

VINCENT, E. A., and R. PHILLIPS: Iron-Titanium Oxide Minerals in Layered Gabbros of the Skaergaard Intrusion, East Greenland: pt. I, Chemistry and Ore Microscopy, *Geochim. Cosmochim. Acta*, vol. 6, pp. 1–26, 1954.

VOGT, J. H. L.: The Physical Chemistry and Magmatic Differentiation in Igneous Rocks, *J. Geol.*, vol. 29, pp. 319–350, 1921.

———: The Physical Chemistry of the Magmatic Differentiation of Igneous Rocks: pt. II, On the Feldspar Diagram Or:Ab:An, *Skr. Norske Vid.-Akad. Oslo*, no. 4, 1926.

VON ECKERMANN, H.: The Alkaline District of Alnö Island, *Sver. Geol. Undersök.*, no. 36, 1948.

———: The Petrogenesis of the Alnö Alkaline Rocks, *Bull. Geol. Inst. Uppsala*, vol. 40, pp. 25–36, 1961.

VON KNORRING, O., and C. G. B. DU BOIS: Carbonatitic Lava from Fort Portal Area in Uganda, *Nature*, vol. 192, pp. 1064–1065, 1961.

WADSWORTH, W. J.: The Layered Ultrabasic Rocks of Southwest Rhum, Inner Hebrides, *Phil. Trans. Roy. Soc. Lond.*, ser. B, vol. 244, pp. 21–64, 1961.

WAGER, L. R.: A Chemical Definition of Fractionation Stages as a Basis for Comparison of Hawaiian, Hebridean, and Other Basic Lavas, *Geochim. Cosmochim. Acta*, vol. 9, pp. 217–248, 1956.

———: The Major Element Variation of the Layered Series of the Skaergaard Intrusion, *J. Petrol.*, vol. 1, pp. 364–398, 1960.

———: A Note on the Origin of Ophitic Texture in the Chilled Olivine Gabbro of the Skaergaard Intrusion, *Geol. Mag.*, vol. 98, pp. 353–366, 1961.

——— and G. M. BROWN: "Layered Igneous Rocks," Freeman, San Francisco, 1967.

———, ———, and W. J. WADSWORTH: Types of Igneous Cumulates, *J. Petrol.*, vol. 1, pp. 73–85, 1960.

——— and W. A. DEER: Geological Investigations in East Greenland: pt. III, The Petrology of the Skaergaard Intrusion, Kangerdlugssuaq, East Greenland, *Medd. Grönland*, vol. 105, no. 4, 1939.

——— and R. L. MITCHELL: The Distribution of Trace Elements During Strong Fractionation of Basic Magma—A Further Study of the Skaergaard Intrusion, East Greenland, *Geochim. Cosmochim. Acta*, vol. 1, pp. 129–208, 1951.

——— and ———: Trace Elements in a Suite of Hawaiian Lavas, *Geochim. Cosmochim. Acta*, vol. 3, pp. 217–223, 1953.

———, E. A. VINCENT, G. M. BROWN, and J. D. BELL: Marscoite and Related Rocks of the Western Red Hills Complex, Isle of Skye, *Phil. Trans. Roy. Soc. Lond.*, ser. A, vol. 257, pp. 273–307, 1965.

———, ———, and A. A. SMALES: Sulfides in the Skaergaard Intrusion, East Greenland, *Econ. Geol.*, vol. 52, pp. 855–903, 1957.

WAGSTAFF, F. E.: Crystallization Kinetics of Internally Nucleated Vitreous Silica, *J. Am. Ceram. Soc.*, vol. 51, pp. 449–452, 1968.

———: Crystallization and Melting Kinetics of Cristobalite, *J. Am. Ceram. Soc.*, vol. 52, pp. 650–654, 1969.

WALCOTT, R. I.: Geology of the Red Hill Complex, Nelson, New Zealand, *Roy. Soc. N.Z. Trans., Earth Sci.*, vol. 7, pp. 57–88, 1969.

WALDBAUM, D. R.: The Configurational Entropies of $Ca_2MgSi_2O_7$–$Ca_2SiAl_2O_7$ Melilites and Related Minerals, *Contrib. Mineral. Petrol.*, vol. 39, pp. 33–54, 1973.

——— and J. B. THOMPSON, JR.: Mixing Properties of Sanidine Crystalline Solutions: pt. II, Calculations Based on Volume Data, *Am. Mineral.*, vol. 53, pp. 2000–2017, 1968.

——— and ———: Mixing Properties of Sanidine Crystalline Solutions: pt. IV, Phase Diagrams from Equations of State, *Am. Mineral.*, vol. 54, pp. 1274–1298, 1969.

WALKER, F.: The Geology of the Shiant Isles, *Q.J. Geol. Soc. Lond.*, vol. 86, pp. 388–398, 1930.

———: The Differentiation of the Palisade Diabase, New Jersey, *Bull. Geol. Soc. Am.*, vol. 51, pp. 1059–1106, 1940.

———: The Magnetic Properties and Differentiation of Dolerite Sills—A Critical Discussion (with further discussion by J. Jaeger and G. Joplin, and by H. H. Hess), *Am. J. Sci.*, vol. 254, pp. 433–451, 1956.

——— and A. POLDERVAART: Karroo Dolerites of the Union of South Africa, *Bull. Geol. Soc. Am.*, vol. 60, pp. 591–706, 1949.

WALKER, K. R.: A Mineralogical, Petrological, and Geochemical Investigation of the Palisades Sill, New Jersey, *Geol. Soc. Am. Mem.*, no. 115, pp. 175–187, 1969*a*.

———: The Palisades Sill, New Jersey: A Reinvestigation, *Geol. Soc. Am. Spec. Paper*, no. 111, 1969*b*.

WANG, C. Y.: Density and Constitution of the Mantle, *J. Geophys. Res.*, vol. 75, pp. 3264–3284, 1970.

WARDEN, A. J.: The Geology of the Central Islands, *New Hebrides Condominium Geol. Surv. Rept.* 5, 1967.

———: Evolution of Aoba Caldera Volcano, New Hebrides, *Bull. Volc.*, vol. 34, pp. 107–140, 1970.

WASHINGTON, H. S.: The Volcanoes and Rocks of Pantelleria, *J. Geol.*, vol. 21, pp. 653–713, 1913.

———: The Volcanoes and Rocks of Pantelleria, *J. Geol.*, vol. 22, pp. 16–27, 1914.

WATERS, A. C.: Volcanic Rocks and the Tectonic Cycle, pp. 703–722 in A. Poldervaart (ed.), "Crust of the Earth" (*Geol. Soc. Am. Spec. Paper*, no. 62), 1955.

———: Stratigraphic and Lithologic Variations in the Columbia River Basalt, *Am. J. Sci.*, vol. 259, pp. 583–611, 1961.

———: Basalt Magma Types and Their Tectonic Associations: Pacific Northwest of the United States, pp. 158–170 in "The Crust of the Pacific Basin" (*Am. Geophys. Union Monograph* 6), 1962.

WATKINS, N. D., and S. E. HAGGERTY: Primary Oxidation Variation and Petrogenesis in a Single Lava, *Contrib. Mineral. Petrol.*, vol. 15, pp. 251–271, 1967.

WATKINSON, D. H., and P. J. WYLLIE: Phase Equilibrium Studies Bearing on the Limestone-assimilation Hypothesis, *Bull. Geol. Soc. Am.*, vol. 80, pp. 1565–1576, 1969.

WEAVER, S. D., J. S. C. SCEAL, and J. L. GIBSON: Trace-element Data Relevant to the Origin of Trachytic and Pantelleritic Lavas in the East African Rift System, *Contrib. Mineral. Petrol.*, vol. 36, pp. 181–194, 1972.

WEEDON, D. S.: The Layered Ultrabasic Rocks of Sgurr Dubh, Isle of Skye, *Scottish J. Geol.*, vol. 1, pp. 41–68, 1965.

———: The Ultrabasic and Basic Igneous Rocks of the Huntly Region, *Scottish. J Geol.*, vol. 6, pp. 26–40, 1970.

WEERTMAN, J.: The Creep Strength of the Earth's Mantle, *Rev. Geophys. Space Phys.*, vol. 8, pp. 145–169, 1970.

WEIGAND, P. W., and P. C. RAGLAND: Geochemistry of Mesozoic Dolerite Dikes from Eastern North America, *Contrib. Mineral. Petrol.*, vol. 29, pp. 195–214, 1970.

WEILL, D. F., and M. J. DRAKE: Europium Anomaly in Plagioclase Feldspar: Experimental Results and Semi-quantitative Model, *Science*, vol. 180, pp. 1059–1060, 1973.

WELKE, H., S. MOORBATH, G. L. CUMMING, and H. SIGURDSON: Lead Isotopic Studies on Igneous Rocks from Iceland, *Earth Planet. Sci. Letters*, vol. 4, pp. 221–231, 1968.

WELLS, M. K.: Structure and Petrology of the Freetown Layered Basic Complex of Sierra Leone, *Overseas Geol. Mineral. Res. (G.B.) Bull. Suppl.* 4, 1962.

WENNER, D. B., and H. P. TAYLOR: Temperature of Serpentinization of Ultramafic Rocks Based on $O^{18}/O^{16}$ Fractionation between Coexisting Serpentine and Magnetite, *Contrib. Mineral. Petrol.*, vol. 32, pp. 165–185, 1971.

WENTWORTH, C. K., and H. WINCHELL: Koolau Basalt Series, Oahu, Hawaii, *Bull. Geol. Soc. Am.*, vol. 58, pp. 49–78, 1947.

WEST, W. D.: The Petrography and Petrogenesis of Forty-eight Flows of Deccan Trap, *Nat. Inst. Sci. India Trans.*, vol. 4, pp. 1–56, 1958.

WHITE, A. J. R., B. W. CHAPPELL, and P. JAKĚS: Coexisting Clinopyroxene, Garnet, and Amphibole from an "Eclogite," Kakanui, New Zealand, *Contrib. Mineral. Petrol.*, vol. 34, pp. 185–191, 1972.

WHITE, D. E., and G. A. WARING: Data of Geochemistry (6th ed.): chap. K, Volcanic Emanations, *U.S.G.S. Prof. Paper* 440-K, pp. 1–29, 1963.

WHITE, R. W.: Ultramafic Inclusions in Basaltic Rocks from Hawaii, *Contrib. Mineral. Petrol.*, vol. 12, pp. 245–314, 1966.

WILCOCKSON, W. H.: Some Aspects of East African Vulcanology, *Advance. Sci.*, vol. 21, pp. 1–13, 1964.

WILCOX, R. E.: Petrology of Parícutin Volcano, Mexico, *U.S.G.S. Bull.* 965-C, 1954.

WILKINSON, J. G. F.: Some Feldspars, Nephelines, and Analcimes from the Square Top Intrusion, Nundle, N.S.W., *J. Petrol.*, vol. 6, pp. 420–444, 1965.

———: The Magmatic Affinities of Some Volcanic Rocks from the Tweed Shield Volcano, S.E. Queensland–N.E. New South Wales, *Geol. Mag.*, vol. 105, pp. 275–289, 1968.

WILKINSON, P.: The Kilimanjaro-Meru Region, *Proc. Geol. Soc. Lond.*, no. 1629, pp. 28–30, 1966.

WILLEMSE, J.: The "Floor" of the Bushveld Igneous Complex, *Geol. Soc. S. Africa Proc.*, vol. 72, pp. xxi–lxxx, 1959.

WILLIAMS, H.: Geology of Tahiti, Moorea, and Maiao, *B. P. Bishop Mus. Bull.* 105, 1933.

———: The Geology of Crater Lake National Park, Oregon, *Carnegie Inst. Washington Publ.* 540, 1942.

———, F. J. TURNER, and C. M. GILBERT: "Petrography," Freeman, San Francisco, 1955.

WILLIAMS, R. J.: Activity-composition Relations in the Fayalite-Forsterite Solid Solution between 900° and 1300° at Low Pressures, *Earth. Planet. Sci. Letters*, vol. 15, pp. 296–300, 1972.

WILLIS, B., and H. S. WASHINGTON: San Felix and San Amrosio: Their Geology and Petrology, *Bull. Geol. Soc. Am.*, vol. 35, pp. 365–384, 1924.

WILSHIRE, H. G.: The Prospect Alkaline Diabase-Picrite Intrusion, New South Wales, Australia, *J. Petrol.*, vol. 8, pp. 97–163, 1967.

———: Mineral Layering in the Twin Lakes Granodiorite, Colorado, *Geol. Soc. Am. Mem.*, no. 115, pp. 235–261, 1969.

WILSON, J. T.: Evidence from Islands on the Spreading of the Sea Floors, *Nature*, vol. 197, pp. 336–338, 1963.

WIMMENAUER, W.: Beiträge zur Petrographie des Kaiserstuhl, *Neues Jahrb. Mineral Abh.*, pt. 1, vol. 91, pp. 131–150, 1957; pt. 2, 3, vol. 93, pp. 133–173, 1959; pt. 4, 5, vol. 98, pp. 367–415, 1962; pt. 6, 7, vol. 99, pp. 231–276, 1963.

———: The Eruptive Rocks and Carbonatites of the Kaiserstuhl, Germany, pp. 183–204 in O. F. Tuttle and J. Gittins (eds.), "Carbonatites," Interscience (Wiley), New York, 1966.

WINCHELL, H.: Honolulu Series, Oahu, Hawaii, *Bull. Geol. Soc. Am.*, vol. 58, pp. 1–48, 1947.

WINKLER, H. G. F.: Kristallgrösse und Abkühlung, *Heidelberger Beitr. Mineral. Petrogr.*, vol. 1, pp. 86–104, 1947.

———: "Petrogenesis of Metamorphic Rocks," Springer, New York, 1965.

——— and H. VON PLATEN: Experimentelle Gesteinsmetamorphose: pt. III, *Geochim. Cosmochim. Acta*, vol. 18, pp. 294–316, 1960.

WONES, D. R.: A Low Pressure Investigation of the Stability of Phlogopite, *Geochim. Cosmochim. Acta*, vol. 31, pp. 2248–2253, 1967.

———: Stability of Biotite: A Reply, *Am. Mineral.*, vol. 57, pp. 316–317, 1972.

——— and H. P. EUGSTER: Stability of Biotite: Experiment, Theory and Application, *Am. Mineral.*, vol. 50, pp. 1228–1272, 1965.

——— and M. E. GILBERT: The Fayalite-Magnetite-Quartz Assemblage between 600°C and 800°C, *Am. J. Sci.*, vol. 267-A (Schairer vol.), pp. 480–488, 1969.

Wood, B. J., and S. Banno: Garnet-Orthopyroxene and Orthopyroxene-Clinopyroxene Relationships in Simple and Complex Systems, *Contrib. Mineral. Petrol.*, in press.

——— and I. S. E. Carmichael: $P_{total}$, $P_{H_2O}$ and the Occurrence of Cummingtonite in Volcanic Rocks, *Contrib. Mineral. Petrol.*, vol. 40, pp. 149–158, 1973.

Worst, B. G.: The Differentiation and Structure of the Great Dyke of Rhodesia, *Geol. Soc. S. Africa Trans.*, vol. 61, pp. 283–354, 1958.

Wright, J. B.: A Note on Possible Differentiation Trends in Tertiary to Recent Lavas of Kenya, *Geol. Mag.*, vol. 100, pp. 164–180, 1963.

———: High Pressure Phases in Nigerian Cenozoic Lavas, *Bull. Volc.*, vol. 34, pp. 833–847, 1970.

———: The Phonolite-Trachyte Spectrum, *Lithos*, vol. 4, pp. 1–5, 1971.

Wright, T. L.: Chemistry of Kilauea and Mauna Loa Lava in Space and Time, *U.S.G.S. Prof. Paper* 735, 1971.

———: Magma Mixing as Illustrated by the 1959 Eruption, Kilauea Volcano, Hawaii, *Bull. Geol. Soc. Am.*, vol. 84, pp. 849–958, 1973.

——— and R. S. Fiske: Origin of the Differentiated and Hybrid Lavas of Kilauea Volcano, Hawaii, *J. Petrol.*, vol. 12, pp. 1–66, 1971.

———, W. T. Kinoshita, and D. L. Peck: March 1965 Eruption of Kilauea Volcano and the Formation of Makaopuhi Lava Lake, *J. Geophys. Res.*, vol. 73, pp. 3181–3205, 1968.

Wyllie, P. J.: Experimental Data Bearing on the Petrogenic Links between Kimberlites and Carbonatites, *Mineral. Soc. India*, I.M.A. vol., pp. 67–82, 1966.

——— (ed.): "Ultramafic and Related Rocks," Wiley, New York, 1967.

———: Ultramafic Rocks and the Upper Mantle, *Mineral. Soc. Am. Spec. Paper*, no. 3, pp. 3–32, 1970.

———: "The Dynamic Earth," Wiley, New York, 1971*a*.

———: Role of Water in Magma Generation and Initiation of Diapiric Uprise in the Mantle, *J. Geophys. Res.*, vol. 76, pp. 1328–1338, 1971*b*.

———: Petrologic Aspects of Plate Tectonics, *Am. Geophys. Union Trans.*, vol. 52, IUGG 62–66, 1971*c*.

——— and J. L. Haas: The System $CaO$–$SiO_2$–$CO_2$–$H_2O$: pt. I, Melting Relationships with Excess Water at 1 Kilobar Pressure, *Geochim. Cosmochim. Acta*, vol. 29, pp. 871–892, 1965.

——— and O. F. Tuttle: Effect of Carbon Dioxide on the Melting of Granite and Feldspars, *Am. J. Sci.*, vol. 257, pp. 648–655, 1959*a*.

——— and ———: Synthetic Carbonatite Magma, *Nature*, vol. 183, p. 770, 1959*b*.

——— and ———: The System $CaO$–$CO_2$–$H_2O$ and the Origin of Carbonatites, *J. Petrol.*, vol. 1, pp. 1–46, 1960.

——— and ———: Experimental Investigation of Silicate Systems Containing Two Volatile Components: pt. III, The Effects of $SO_3$, $P_2O_5$, HCl and $Li_2O$, in Addition to $H_2O$, on the Melting Temperature of Albite and Granite, *Am. J. Sci.*, vol. 262, pp. 930–939, 1964.

Yagi, K.: Petrochemical Studies on the Alkalic Rocks of the Morotu District, Sakhalin, *Bull. Geol. Soc. Am.*, vol. 64, pp. 769–810, 1953.

———: Petrochemistry of the Alkalic Rocks of the Ponape Island, Western Pacific Ocean, *Proc. 21st Int. Geol. Congr.*, pt. 13, pp. 108–122, 1960.

———: The System Acmite-Diopside and Its Bearing on the Stability Relations of Natural Pyroxenes of the Acmite-Hedenbergite-Diopside Series, *Am. Mineral.*, vol. 51, pp. 976–1000, 1966.

——— and K. ONUMA: The Join $CaMgSi_2O_6$–$CaTiAl_2O_6$ and Its Bearing on the Titanaugites, *J. Fac. Sci., Hokkaido Univ.*, ser. 4, vol. 8, pp. 463–483, 1967.

YODER, H. S., D. B. STEWART, and J. R. SMITH: Ternary Feldspars, *Ann. Rept. Geophys. Lab.* (*Carnegie Inst. Yrbk. 56*), pp. 206–217, 1957.

——— and C. E. TILLEY: Origin of Basaltic Magmas: An Experimental Study of Natural and Synthetic Rock Systems, *J. Petrol.*, vol. 3, pp. 342–532, 1962.

YUND, R. A., and G. KULLERUD: Thermal Stability of Assemblages in the Cu–Fe–S System, *J. Petrol.*, vol. 7, pp. 454–488, 1966.

——— and R. H. MCCALLISTER: Kinetics and Mechanisms of Exsolution, *Chem. Geol.*, vol. 6, pp. 5–30, 1970.

ZEN, E-AN: Comments on the Thermodynamic Constants and Hydrothermal Stability Relations of Anthophyllite, *Am. J. Sci.*, vol. 270, pp. 136–150, 1971.

ZIELINSKI, R. A., and F. A. FREY: Gough Island: Evaluation of a Fractional Crystallization Model, *Contrib. Mineral. Petrol.*, vol. 29, pp. 242–254, 1970.

ZIES, E. G.: Temperature Measurements at Parícutin Volcano, *Am. Geophys. Union Trans.*, vol. 27, pp. 178–180, 1946.

# Name Index

# Name Index

# Subject Index

# Subject Index